Lecture Notes in Artificial Intelligence 9851

Subseries of Lecture Notes in Computer Science

More information about this series at http://www.springer.com/series/1244

Paolo Frasconi · Niels Landwehr
Giuseppe Manco · Jilles Vreeken (Eds.)

Machine Learning and Knowledge Discovery in Databases

European Conference, ECML PKDD 2016
Riva del Garda, Italy, September 19–23, 2016
Proceedings, Part I

 Springer

Editors

Paolo Frasconi
Università degli Studi di Firenze
Florence
Italy

Giuseppe Manco
National Research Council (ICAR-CNR)
Rende
Italy

Niels Landwehr
Computer Science
University of Potsdam
Potsdam
Germany

Jilles Vreeken
MPI for Informatics
Saarland University
Saarbrucken, Saarland
Germany

ISSN 0302-9743   ISSN 1611-3349   (electronic)
Lecture Notes in Artificial Intelligence
ISBN 978-3-319-46127-4   ISBN 978-3-319-46128-1   (eBook)
DOI 10.1007/978-3-319-46128-1

Library of Congress Control Number: 2016950748

LNCS Sublibrary: SL7 – Artificial Intelligence

Printed on acid-free paper

This Springer imprint is published by Springer Nature
The registered company is Springer International Publishing AG
The registered company address is: Gewerbestrasse 11, 6330 Cham, Switzerland

Preface

These are the proceedings of the 15th European Conference on Machine Learning and Principles and Practice of Knowledge Discovery in Databases (ECML PKDD 2016), held in Riva del Garda, Italy, during September 19–23, 2016. This event is the premier European Machine Learning and Data Mining conference and builds upon a very successful series of 26 ECML and 19 PKDD conferences, which have been jointly organized for the past 15 years.

The response to our call for paper was very good. We received 353 papers for the main conference track, of which 100 were accepted, yielding an acceptance rate of about 28 %.

Traditionally, ECML PKDD provides an extensive technical program that consists of several focused tracks:

- the conference track, featuring regular conference papers, published in these proceedings;
- the journal track, featuring papers that satisfy the quality criteria of journal papers and at the same time lend themselves to conference talks (these papers are published separately in the journals *Machine Learning* and *Knowledge Discovery and Data Mining*);
- the industrial track, aiming to bring together participants from academia, industry, government, and NGOs (non-governmental organizations) in a venue that highlights practical and real-world studies of machine learning, knowledge discovery, and data mining.
- the demo track, presenting innovative prototype implementations or mature systems that use machine learning techniques and knowledge discovery processes in a real setting;
- the nectar track, offering conference attendees a compact overview of recent scientific advances at the frontier of machine learning and data mining with other disciplines, as published in related conferences and journals.

Moreover, the conference program included 3 discovery challenges, 13 workshops, and 10 tutorial presentations. The discovery challenges were organized by Elio Masciari and Alessandro Moschitti. Fabrizio Costa, Matthijs van Leeuwen, and Albrecht Zimmermann had the responsibility of selecting workshop and tutorial proposals. The PhD Forum, where junior PhD students exchange ideas, experiences, and get advise from senior researchers, was organized by Leman Akoglu and Tijl De Bie.

The program included six plenary keynotes by invited speakers Susan Athey (Stanford Graduate School of Business), Zoubin Ghahramani (University of Cambridge and Alan Turing Institute), Thore Graepel (Google DeepMind and University College London), Ravi Kumar (Google), Rasmus Pagh (IT University of Copenhagen), and Alex "Sandy" Pentland (MIT).

Putting together the program of this conference would have been impossible without the help of a large and supportive team. Our thirty Area Chairs nominated reviewers, moderated the discussion among them to find a consensus over each paper, and made a final accept/reject decision. A total of 315 reviewers (listed in this book) helped to select papers. Two best student papers were selected by Toon Calders and Hendrik Blockeel. The associated awards were sponsored by Springer and the journals *Machine Learning* and *Data Mining and Knowledge Discovery*.

For the fourth time, the conference used a double submission model: next to the regular conference tracks, papers submitted to the Springer journals Machine Learning (MACH) and Data Mining and Knowledge Discovery (DAMI) were considered for presentation at the conference. These papers were submitted to the ECML PKDD 2016 special issue of the respective journals, and underwent the normal editorial process of these journals. Those papers accepted for one of these journals were assigned a presentation slot at the ECML PKDD 2016 conference. A total of 120 original manuscripts were submitted to the journal track during this year. Some of these papers are still being refereed. Of the fully refereed papers, 8 were accepted in DAMI and 10 in MACH, together with 10 papers from last year's call, which were also scheduled for presentation at this conference.

There were two major innovations at this year's conference. First, we decided to have a full day of plenary presentation on September 21st, while the usual four parallel session tracks were run on September 20th and 22nd. These plenary oral presentations were selected by the Program and Journal Track Co-chairs from the pool of all accepted papers according to criteria such as: (1) novelty and significance of the results and their expected impact; (2) breadth of interest for both machine learners and data miners. It is our belief that this will strengthen the synergy between the ML and the DM sub-communities, allowing papers of general interest for both to be presented to the whole audience.

The second major difference is the adoption of the practices of Reproducible Research (RR). Authors were encouraged to adhere to such practices by making available data and software tools for reproducing the results reported in their papers. In total, 29 papers with accompanying software and/or data are flagged as RR-papers on the conference website http://ecmlpkdd2016.org/, which provides links to such additional material (links are also available within the paper bodies in these proceedings).

Part I and Part II of the proceedings of the ECML PKDD 2016 conference contain the full papers of the contributions presented in the scientific track and the abstracts of the scientific plenary talks. Part III of the proceedings of the ECML PKDD 2016 conference contains the full papers of the contributions presented in the industrial track, short papers describing the demonstrations, the nectar papers, and the abstracts of the industrial plenary talks. First of all, we would like to express our gratitude to the general chairs of the conference, Fosca Giannotti and Andrea Passerini, as well as to all members of the Organizing Committee, for managing this event in a very competent and professional way. In particular, we thank the demo, workshop and tutorial, industrial, and nectar track chairs. Special thanks go to the proceedings chairs, Marco Lippi and Stefano Ferilli, for the hard work of putting these proceedings together. We thank the PhD Forum organizers, the Discovery Challenge organizers, and all the people involved in the conference, who worked hard for its success. We would like to thank Microsoft for allowing us to use their CMT software for conference management. Last but not least, we would like to sincerely thank the authors for submitting their work to the conference and the reviewers and area chairs for their tremendous effort in guaranteeing the quality of the reviewing process, thereby improving the quality of these proceedings.

September 2016

Paolo Frasconi
Niels Landwehr
Giuseppe Manco
Jilles Vreeken

Organization

ECML PKDD 2016 Organization

General Chairs

Andrea Passerini   University of Trento, Italy
Fosca Giannotti   National Research Council (ISTI-CNR), Italy

Program Chairs

Paolo Frasconi   University of Florence, Italy
Niels Landwehr   University of Potsdam, Germany
Giuseppe Manco   National Research Council (ICAR-CNR), Italy
Jilles Vreeken   Cluster of Excellence MMCI, Saarland University & Max
     Planck Institute for Informatics, Germany

Journal Track Chairs

Thomas Gärtner   University of Nottingham, UK
Mirco Nanni   National Research Council (ISTI-CNR), Italy
Andrea Passerini   University of Trento, Italy
Céline Robardet   National Institute of Applied Science in Lyon, France

Industrial Track Chairs

Björn Bringmann   Deloitte GmbH, Germany
Gemma Garriga   Inria, France
Volker Tresp   Siemens AG & Ludwig Maximilian University of Munich,
     Germany

Local Organization Chairs

Simone Marinai   University of Florence, Italy
Gianluca Corrado   University of Trento, Italy
Katya Tentori   University of Trento, Italy

Workshop and Tutorial Chairs

Matthijs van Leeuwen   Leiden University, Netherlands
Fabrizio Costa   University of Freiburg, Germany
Albrecht Zimmermann   University of Caen, France

Awards Committee Chairs

Toon Calders   Free University of Bruxelles, Belgium
Hendrik Blockeel   University of Leuven, Belgium

Nectar Track Chairs

Bettina Berendt   University of Leuven, Belgium
Pauli Miettinen   Max Planck Institute for Informatics, Germany

Demo Chairs

Nikolaj Tatti   Aalto University School of Science, Finland
Élisa Fromont   Jean Monnet University, France

Discovery Challenge Chairs

Elio Masciari   National Research Council (ICAR-CNR), Italy
Alessandro Moschitti   Qatar Computing Research Institute, HBKU,
                              Qatar & University of Trento, Italy

Sponsorship Chairs

Michelangelo Ceci   University of Bari, Italy
Chedy Raïssi   Inria, France

Publicity and Social Media Chairs

Olana Missura   University of Bonn, Germany
Nicola Barbieri   Yahoo!, UK
Gianluca Corrado   University of Trento, Italy

PhD Forum Chairs

Leman Akoglu   Stony Brook University, USA
Tijl De Bie   Ghent University, Belgium

Proceedings Chairs

Marco Lippi   University of Bologna, Italy
Stefano Ferilli   University of Bari, Italy

Web Chair

Daniele Baracchi   University of Florence, Italy

Area Chairs

Hendrik Blockeel   KU Leuven, Belgium
Francesco Bonchi   ISI Foundation Turin, Italy
Karsten Borgwardt   ETH Zurich, Switzerland
Toon Calders   University of Antwerp, Belgium
Aaron Courville   University of Montreal, Canada
Ian Davidson   University of California at Davis, USA
Tijl De Bie   Ghent University, Belgium
Luc De Raedt   KU Leuven, Belgium
Carlotta Domeniconi   George Mason University, USA
Peter Flach   University of Bristol, UK
Claudio Gentile   Università dell'Insubria, Italy
Mohammad   Inria Lille, France
  Ghavamzadeh
Aristides Gionis   Aalto University, Finland
Bart Goethals   University of Antwerp, Belgium
Geoff Holmes   University of Waikato, New Zealand
Andreas Hotho   University of Wurzburg, Germany
Eyke Hüllermeier   University of Paderborn, Germany
Manfred Jaeger   Aalborg University, Denmark
George Karypis   University of Minnesota, USA
Samuel Kaski   Aalto University, Finland
Kristian Kersting   University of Dortmund
Donato Malerba   University of Bari, Italy
Pauli Miettinen   Max-Planck Institute for Informatics, Germany
Dino Pedreschi   University of Pisa, Italy
Bernhard Pfahringer   University of Waikato, New Zealand
Guido Sanguinetti   University of Edinburgh, UK
Arno Siebes   University of Utrecht, Netherlands
Guy Van den Broeck   University of California at Los Angeles, USA
Marco Wiering   University of Groningen, Netherlands
Stefan Wrobel   University of Bonn, Germany

Reviewers

Prashanth A.
Mohammad Al Hasan
Carlos Alzate
Aijun An
Aris Anagnostopoulos
Fabrizio Angiulli
Annalisa Appice
Ira Assent
Martin Atzmueller
Antonio Bahamonde
Jose Balcázar
Nicolas Ballas
Nicola Barbieri
Christian Bauckhage
Roberto Bayardo
Martin Becker
Srikanta Bedathur
Jessa Bekker
Vaishak Belle
András Benczúr
Michael Berthold
Albert Bifet
Konstantinos Blekas
Paul Blomstedt
Dean Bodenham
Mario Boley
Gianluca Bontempi
Henrik Bostrom
Jean-François Boulicaut
Marc Boulle
Pavel Brazdil
Ulf Brefeld
Robert Busa-Fekete
Rui Camacho
Longbing Cao
Francisco Casacuberta
Michelangelo Ceci
Peggy Cellier
Loic Cerf
Tania Cerquitelli
Edward Chang
Thierry Charnois
Duen Horng Chau

Keke Chen
Ling Chen
Silvia Chiusano
Arthur Choi
Frans Coenen
Fabrizio Costa
Vítor Santos Costa
Bruno Crémilleux
Botond Cseke
Boris Cule
Tomaz Curk
James Cussens
Claudia d'Amato
Maria Damiani
Jesse Davis
Martine De Cock
Colin de la Higuera
Juan del Coz
Anne Denton
Christian Desrosiers
Nicola Di Mauro
Tom Diethe
Ying Ding
Stephan Doerfel
Frank Dondelinger
Anton Dries
Madalina Drugan
Wouter Duivesteijn
Robert Durrant
Ines Dutra
Sašo Džeroski
Tapio Elomaa
Dora Erdos
Floriana Esposito
Nicola Fanizzi
Elaine Faria
Fabio Fassetti
Ad Feelders
Stefano Ferilli
Carlos Ferreira
Cesar Ferri
Maurizio Filippone
Asja Fischer

Eibe Frank
Élisa Fromont
Fabio Fumarola
Johannes Fürnkranz
Victor Gabillon
Esther Galbrun
Patrick Gallinari
Joao Gama
Byron Gao
Paolo Garza
Eric Gaussier
Ricard Gavalda
Rainer Gemulla
Konstantinos Georgatzis
Pierre Geurts
Aris Gkoulalas-Divanis
Dorota Glowacka
Mehmet Gonen
Michael Granitzer
Caglar Gulcehre
Francesco Gullo
Stephan Gunnemann
Maria Halkidi
Jiawei Han
Xiao He
Denis Helic
Jose Hernandez-Orallo
Thanh Lam Hoang
Frank Hoeppner
Jaakko Hollmen
Arjen Hommersom
Tamas Horvath
Yuanhua Huang
Van-Anh Huynh-Thu
Dino Ienco
Bhattacharya Indrajit
Frederik Janssen
Nathalie Japkowicz
Szymon Jaroszewicz
Alipio Jorge
Giuseppe Jurman
Robert Jeschke
Hachem Kadri

Bo Kang
U Kang
Andreas Karwath
Hisashi Kashima
Ioannis Katakis
Yoshinobu Kawahara
Mehdi Kaytoue
John Keane
Latifur Khan
Arto Klami
Levente Kocsis
Yun Sing Koh
Alek Kolcz
Irena Koprinska
Frederic Koriche
Walter Kosters
Lars Kotthoff
Meelis Kull
Nicolas Lachiche
Helge Langseth
Thomas Lansdall-Welfare
Pedro Larranaga
Silvio Lattanzi
Niklas Lavesson
Nada Lavrač
Sangkyun Lee
Florian Lemmerich
Jiuyong Li
Juanzi Li
Limin Li
Jefrey Lijffijt
Felipe Llinares López
Daniel Hernandez Lobato
Corrado Loglisci
Peter Lucas
Elio Masciari
Andres Masegosa
Wannes Meert
Ernestina Menasalvas
Rosa Meo
Mehdi Mirza
Karthika Mohan
Anna Monreale
Guido Montufar
Joao Moreira
Katharina Morik

Mohamed Nadif
Ndapa Nakashole
Amedeo Napoli
Sriraam Natarajan
Benjamin Nguyen
Thomas Nielsen
Xia Ning
Kjetil Nørvåg
Eirini Ntoutsi
Andreas Nürnberger
Francesco Orsini
George Paliouras
Apostolos Papadopoulos
Evangelos Papalexakis
Panagiotis Papapetrou
Ioannis Partalas
Nikos Pelekis
Jing Peng
Ruggero Pensa
Francois Petitjean
Nico Piatkowksi
Andrea Pietracaprina
Gianvito Pio
Marc Plantevit
Pascal Poncelet
Philippe Preux
Kai Puolamaki
Buyue Qian
Chedy Raïssi
Jan Ramon
Huzefa Rangwala
Zbigniew Rás
Chotirat Ratanamahatana
Jan Rauch
Steffen Rendle
Chiara Renso
Achim Rettinger
Fabrizio Riguzzi
Matteo Riondato
Pedro Rodrigues
Juan Rodriguez
Simon Rogers
Damian Roqueiro
Fabrice Rossi
Juho Rousu
Celine Rouveirol

Stefan Rueping
Salvatore Ruggieri
Yvan Saeys
Alan Said
Lorenza Saitta
Ansaf Salleb-Aouissi
Scott Sanner
Claudio Sartori
Lars Schmidt-Thieme
Christoph Schommer
Matthias Schubert
Giovanni Semeraro
Sohan Seth
Vinay Setty
Junming Shao
Sameer Singh
Andrzej Skowron
Kevin Small
Marta Soare
Yangqiu Song
Mauro Sozio
Myra Spiliopoulou
Papadimitriou Spiros
Eirini Spyropoulou
Jerzy Stefanowski
Daria Stepanova
Florian Stimberg
Gerd Stumme
Mahito Sugiyama
Einoshin Suzuki
Panagiotis Symeonidis
Sandor Szedmak
Andrea Tagarelli
Domenico Talia
Letizia Tanca
Nikolaj Tatti
Maguelonne Teisseire
Aika Terada
Georgios Theocharous
Kai Ming Ting
Ljupco Todorovski
Hannu Toivonen
Luis Torgo
Roberto Trasarti
Daniel Trejo-Banos
Panagiotis Tsaparas

Vincent Tseng
Grigorios Tsoumakas
Karl Tuyls
Niall Twomey
Theodoros Tzouramanis
Antti Ukkonen
Jan Van Haaren
Matthijs van Leeuwen
Maarten van Someren
Iraklis Varlamis
Michalis Vazirgiannis
Julien Velcin
Shankar Vembu
Celine Vens
Deepak Venugopal

Vassilios Verykios
Herna Viktor
Fabio Vitale
Christel Vrain
Willem Waegeman
Jianyong Wang
Ding Wei
Cheng Weiwei
Zheng Wen
Joerg Wicker
Makoto Yamada
Jeffrey Yu
Philip Yu
Bianca Zadrozny
Marco Zaffalon

Gerson Zaverucha
Demetris Zeinalipour
Filip Železný
Bernard Zenko
Junping Zhang
Min-Ling Zhang
Nan Zhang
Shichao Zhang
Ying Zhao
Mingjun Zhong
Djamel Zighed
Arthur Zimek
Albrecht Zimmermann
Indre Zliobaite
Blaž Zupan

Sponsors

Gold Sponsors

Google   http://research.google.com
IBM   http://www.ibm.com

Silver Sponsors

Deloitte   http://www.deloitte.com
Siemens   http://www.siemens.com
Unicredit   http://www.unicreditgroup.eu
Zalando   http://www.zalando.com

Award Sponsors

Deloitte   http://www.deloitte.com
DMKD   http://link.springer.com/journal/10618
MLJ   http://link.springer.com/journal/10994

Badge Lanyard

Knime   http://www.knime.org

Additional Supporters

Springer   http://www.springer.com

Institutional Supporters

ICAR-CNR	http://www.icar.cnr.it
DISI-UNITN	http://www.disi.unitn.it
ISTI-CNR	http://www.isti.cnr.it
COGNET	http://www.cognet.5g-ppp.eu

Organizing Institutions

UNITN	http://www.unitn.it
UNIFI	http://www.unifi.it
ISTI-CNR	http://www.isti.cnr.it
ICAR-CNR	http://www.icar.cnr.it

Abstracts of Invited Talks

Causal Inference and Machine Learning: Estimating and Evaluating Policies

Susan Athey

Stanford Graduate School of Business

Abstract. In many contexts, a decision-making can choose to assign one of a number of "treatments" to individuals. The treatments may be drugs, offers, advertisements, algorithms, or government programs. One setting for evaluating such treatments involves randomized controlled trials, for example A/B testing platforms or clinical trials. In such settings, we show how to optimize supervised machine learning methods for the problem of estimating heterogeneous treatment effects, while preserving a key desiderata of randomized trials, which is providing valid confidence intervals for estimates. We also discuss approaches for estimating optimal policies and online learning. In environments with observational (non-experimental) data, different methods are required to separate correlation from causality. We show how supervised machine learning methods can be adapted to this problem.

Bio. Susan Athey is The Economics of Technology Professor at Stanford Graduate School of Business. She received her bachelor's degree from Duke University and her Ph.D. from Stanford, and she holds an honorary doctorate from Duke University. She previously taught at the economics departments at MIT, Stanford and Harvard. In 2007, Professor Athey received the John Bates Clark Medal, awarded by the American Economic Association to "that American economist under the age of forty who is adjudged to have made the most significant contribution to economic thought and knowledge." She was elected to the National Academy of Science in 2012 and to the American Academy of Arts and Sciences in 2008. Professor Athey's research focuses on the economics of the internet, online advertising, the news media, marketplace design, virtual currencies and the intersection of computer science, machine learning and economics. She advises governments and businesses on marketplace design and platform economics, notably serving since 2007 as a long-term consultant to Microsoft Corporation in a variety of roles, including consulting chief economist.

Automating Machine Learning

Zoubin Ghahramani

University of Cambridge and Alan Turing Institute

Abstract. I will describe the "Automatic Statistician"[1], a project which aims to automate the exploratory analysis and modelling of data. Our approach starts by defining a large space of related probabilistic models via a grammar over models, and then uses Bayesian marginal likelihood computations to search over this space for one or a few good models of the data. The aim is to find models which have both good predictive performance, and are somewhat interpretable. The Automatic Statistician generates a natural language summary of the analysis, producing a 10–15 page report with plots and tables describing the analysis. I will also link this to recent work we have been doing in the area of Probabilistic Programming (including an new system in Julia) to automate inference, and on the rational allocation of computational resources (and our entry in the AutoML conference).

Bio. Zoubin Ghahramani FRS is Professor of Information Engineering at the University of Cambridge, where he leads the Machine Learning Group, and the Cambridge Liaison Director of the Alan Turing Institute, the UK's national institute for Data Science. He studied computer science and cognitive science at the University of Pennsylvania, obtained his PhD from MIT in 1995, and was a postdoctoral fellow at the University of Toronto. His academic career includes concurrent appointments as one of the founding members of the Gatsby Computational Neuroscience Unit in London, and as a faculty member of CMU's Machine Learning Department for over 10 years. His current research interests include statistical machine learning, Bayesian nonparametrics, scalable inference, probabilistic programming, and building an automatic statistician. He has published over 250 research papers, and has held a number of leadership roles as programme and general chair of the leading international conferences in machine learning including: AISTATS (2005), ICML (2007, 2011), and NIPS (2013, 2014). In 2015 he was elected a Fellow of the Royal Society.

---

[1] http://www.automaticstatistician.com/.

AlphaGo - Mastering the Game of Go
with Deep Neural Networks and Tree Search

Thore Graepel

Google DeepMind and University College London

Abstract. The game of Go has long been viewed as the most challenging of classic games for artificial intelligence owing to its enormous search space and the difficulty of evaluating board positions and moves. Here we introduce a new approach to computer Go that uses 'value networks' to evaluate board positions and 'policy networks' to select moves. These deep neural networks are trained by a novel combination of supervised learning from human expert games, and reinforcement learning from games of self-play. Using this search algorithm, our program AlphaGo achieved a 99.8 % winning rate against other Go programs and beat the human European Go champion Fan Hui by 5 games to 0, a feat thought to be at least a decade away by Go and AI experts alike. Finally, in a dramatic and widely publicised match, AlphaGo defeated Lee Sedol, the top player of the past decade, 4 games to 1. In this talk, I will explain how AlphaGo works, describe our process of evaluation and improvement, and discuss what we can learn about computational intuition and creativity from the way AlphaGo plays.

Bio. Thore Graepel is a research group lead at Google DeepMind and holds a part-time position as Chair of Machine Learning at University College London. He studied physics at the University of Hamburg, Imperial College London, and Technical University of Berlin, where he also obtained his PhD in machine learning in 2001. He spent time as a postdoctoral researcher at ETH Zurich and Royal Holloway College, University of London, before joining Microsoft Research in Cambridge in 2003, where he co-founded the Online Services and Advertising group. Major applications of Thore's work include Xbox Live's TrueSkill system for ranking and matchmaking, the AdPredictor framework for click-through rate prediction in Bing, and the Matchbox recommender system which inspired the recommendation engine of Xbox Live Marketplace. More recently, Thore's work on the predictability of private attributes from digital records of human behaviour has been the subject of intense discussion among privacy experts and the general public. Thore's current research interests include probabilistic graphical models and inference, reinforcement learning, games, and multi-agent systems. He has published over one hundred peer-reviewed papers, is a named co-inventor on dozens of patents, serves on the editorial boards of JMLR and MLJ, and is a founding editor of the book series Machine Learning & Pattern Recognition at Chapman & Hall/CRC. At DeepMind, Thore has returned to his original passion of understanding and creating intelligence, and recently contributed to creating AlphaGo, the first computer program to defeat a human professional player in the full-sized game of Go, a feat previously thought to be at least a decade away.

Sequences, Choices, and Their Dynamics

Ravi Kumar

Google

Abstract. Sequences arise in many online and offline settings: urls to visit, songs to listen to, videos to watch, restaurants to dine at, and so on. User-generated sequences are tightly related to mechanisms of choice, where a user must select one from a finite set of alternatives. In this talk, we will discuss a class of problems arising from studying such sequences and the role discrete choice theory plays in these problems. We will present modeling and algorithmic approaches to some of these problems and illustrate them in the context of large-scale data analysis.

Bio. Ravi Kumar has been a senior staff research scientist at Google since 2012. Prior to this, he was a research staff member at the IBM Almaden Research Center and a principal research scientist at Yahoo! Research. His research interests include Web search and data mining, algorithms for massive data, and the theory of computation.

Dimensionality Reduction with Certainty

Rasmus Pagh

IT University of Copenhagen

Abstract. Tool such as Johnson-Lindenstrauss dimensionality reduction and 1-bit minwise hashing have been successfully used to transform problems involving very high-dimensional real vectors into lower-dimensional equivalents, at the cost of introducing a random distortion of distances/similarities among vectors. While this can alleviate the computational cost associated with high dimensionality, the effect on the outcome of the computation (compared to working on the original vectors) can be hard to analyze and interpret. For example, the behavior of a basic kNN classifier is easy to describe and interpret, but if the algorithm is run on dimension-reduced vectors with distorted distances it is much less transparent what is happening. The talk starts with an introduction to randomized (data-independent) dimensionality reduction methods and gives some example applications in machine learning. Based on recent work in the theoretical computer science community we describe tools for dimension reduction that give stronger guarantees on approximation, replacing probabilistic bounds on distance/similarity with bounds that hold with certainty. For example, we describe a "distance sensitive Bloom filter": a succinct representation of high-dimensional boolean vectors that can identify vectors within distance r with certainty, while far vectors are only thought to be close with a small "false positive" probability. We also discuss work towards a deterministic alternative to random feature maps (i.e., dimension-reduced vectors from a high-dimensional feature space), and settings in which a pair of dimension-reducing mappings outperform single-mapping methods. While there are limits to what performance can be achieved with certainty, such techniques may be part of the toolbox for designing transparent and scalable machine learning and knowledge discovery methods.

Bio. Rasmus Pagh graduated from Aarhus University in 2002, and is now a full professor at the IT University of Copenhagen. His work is centered around efficient algorithms for big data, with an emphasis on randomized techniques. His publications span theoretical computer science, databases, information retrieval, knowledge discovery, and parallel computing. His most well-known work is the cuckoo hashing algorithm (2001), which has led to new developments in several fields. In 2014 he received the best paper award at the WWW Conference for a paper with Pham and Mitzenmacher on similarity estimation, and started a 5-year research project funded by the European Research Council on scalable similarity search.

Social Learning

Alex "Sandy" Pentland

MIT

Abstract. Human decisions are heavily influenced by social interaction, so that predicting or influencing individual behavior requires modeling these interaction effects. In addition the distributed learning strategies exhibited by human communities suggest methods of improving both machine learning and human-machine systems. Several practical examples will be described.

Bio. Professor Alex "Sandy" Pentland directs the MIT Connection Science and Human Dynamics labs and previously helped create and direct the MIT Media Lab and the Media Lab Asia in India. He is one of the most-cited scientists in the world, and Forbes recently declared him one of the "7 most powerful data scientists in the world" along with Google founders and the Chief Technical Officer of the United States. He has received numerous awards and prizes such as the McKinsey Award from Harvard Business Review, the 40th Anniversary of the Internet from DARPA, and the Brandeis Award for work in privacy.

He is a founding member of advisory boards for Google, AT&T, Nissan, and the UN Secretary General, a serial entrepreneur who has co-founded more than a dozen companies including social enterprises such as the Data Transparency Lab, the Harvard-ODI-MIT DataPop Alliance and the Institute for Data Driven Design. He is a member of the U.S. National Academy of Engineering and leader within the World Economic Forum.

Contents – Part I

Contents – Part II

Contents – Part III

Industrial Track Contributions

adaQN: An Adaptive Quasi-Newton Algorithm for Training RNNs

Nitish Shirish Keskar[1]([⊠]) and Albert S. Berahas[2]

[1] Department of Industrial Engineering and Management Sciences,
Northwestern University, Evanston, IL, USA
keskar.nitish@u.northwestern.edu
[2] Department of Engineering Sciences and Applied Mathematics,
Northwestern University, Evanston, IL, USA
albertberahas@u.northwestern.edu

Abstract. Recurrent Neural Networks, or RNNs, are powerful models that achieve exceptional performance on a plethora pattern recognition problems. However, the training of RNNs is a computationally difficult task owing to the well-known "vanishing/exploding" gradient problem. Algorithms proposed for training RNNs either exploit no (or limited) curvature information and have cheap per-iteration complexity, or attempt to gain significant curvature information at the cost of increased per-iteration cost. The former set includes diagonally-scaled first-order methods such as ADAGRAD and ADAM, while the latter consists of second-order algorithms like Hessian-Free Newton and K-FAC. In this paper, we present ADAQN, a stochastic quasi-Newton algorithm for training RNNs. Our approach retains a low per-iteration cost while allowing for non-diagonal scaling through a stochastic L-BFGS updating scheme. The method uses a novel L-BFGS scaling initialization scheme and is judicious in storing and retaining L-BFGS curvature pairs. We present numerical experiments on two language modeling tasks and show that ADAQN is competitive with popular RNN training algorithms.

1 Introduction

Recurrent Neural Networks (RNNs) have emerged as one of the most powerful tools for modeling sequences [6]. They are extensively used in a wide variety of applications including language modeling, speech recognition, machine translation and computer vision [7,8,18,25]. RNNs are similar to the popular Feed-Forward Networks (FFNs), but unlike FFNs, allow for cyclical connectivity in the nodes. This enables them to have exceptional expressive ability, permitting them to model highly complex sequences. This expressiveness, however, comes at the cost of training difficulty, especially in the presence of long-term dependencies [1,24]. This difficulty, commonly termed as the "vanishing/exploding" gradient problem, arises due to the recursive nature of the network. Depending on the eigenvalues of the hidden-to-hidden node connection matrix during the Back Propagation Through Time (BPTT) algorithm, the errors either get recursively

© Springer International Publishing AG 2016
P. Frasconi et al. (Eds.): ECML PKDD 2016, Part I, LNAI 9851, pp. 1–16, 2016.
DOI: 10.1007/978-3-319-46128-1_1

amplified or diminished making the training problem highly ill-conditioned. Consequently, this issue precludes the use of methods which are unaware of the curvature of the problem, such as Stochastic Gradient Descent (SGD), for RNN training tasks.

Many attempts have been made to address the problem of training RNNs. Some propose the use of alternate architectures; for e.g. Gated Recurrent Units (GRUs) [4] and Long Short-Term Memory (LSTM) [9] models. These network architectures do not suffer as severely from gradient-related problems, and hence, it is possible to use simple and well-studied methods like SGD for training, thus obviating the need for more sophisticated methods. Other efforts for alleviating the problem of training RNNs have been centered around designing training algorithms which incorporate curvature information in some form; see for e.g. Hessian-Free Newton [14,17] and Nesterov Accelerated Gradient [21].

First-order methods such as ADAGRAD [5] and ADAM [12], employ diagonal scaling of the gradients and consequently achieve invariance to diagonal re-scaling of the gradients. These methods have low per-iteration cost and have demonstrated excellent performance on a large number of deep learning tasks. Second-order methods like Hessian-Free Newton [14] and K-FAC [16], allow for non-diagonal-scaling of the gradients using highly expressive Hessian information, but tend to either have higher per-iteration costs or require non-trivial information about the structure of the graph. We defer the discussion of these algorithms to the following section.

In this paper, we present ADAQN, a novel (stochastic) quasi-Newton algorithm for training RNNs. The algorithm attempts to reap the merits of both first- and second-order methods by judiciously incorporating curvature information while retaining a low per-iteration cost. Our algorithmic framework is inspired by that of Stochastic Quasi-Newton (SQN) [3], which is designed for stochastic convex problems. The proposed algorithm is designed to ensure practical viability for solving RNN training problems.

The paper is organized as follows. We end the introduction by establishing notation that will be used throughout the paper. In Sect. 2, we discuss popular algorithms for training RNNs and also discuss stochastic quasi-Newton methods. In Sect. 3, we describe our proposed algorithm in detail and emphasize its distinguishing features. We present numerical results on language modeling tasks in Sect. 4. Finally, we discuss possible extensions of this work and present concluding remarks in Sects. 5 and 6 respectively.

1.1 Notation

The problem of training RNNs can be stated as the following optimization problem,

$$\min_{w \in \mathbb{R}^n} f(w) = \frac{1}{m} \sum_{i=1}^{m} f_i(w). \tag{1}$$

Here, f_i is the RNN training error corresponding to a data point denoted by index i. We assume there are m data points. During each iteration, the algorithm samples data points $\mathcal{B}_k \subseteq \{1, 2, \cdots, m\}$. The iterate at the kth iteration is denoted by w_k and the (stochastic) gradient computed on this mini-batch is denoted by $\hat{\nabla}_{\mathcal{B}_k} f(w_k)$. In particular, the notation $\hat{\nabla}_{\mathcal{B}_j} f(w_k)$ can be verbally stated as the gradient computed at w_k using the mini-batch used for gradient computation during iteration j. For ease of notation, we may eliminate the subscript $\mathcal{B}_k$ whenever the batch and the point of gradient evaluation correspond to the iterate index; in other words, we use $\hat{\nabla} f(w_k)$ to mean $\hat{\nabla}_{\mathcal{B}_k} f(w_k)$. Unless otherwise specified, H_k denotes any positive-definite matrix and the step-length is denoted by α_k. Lastly, we denote the ith component of a vector $v \in \mathbb{R}^n$ by $[v]_i$ and use v^2 to represent element-wise square.

2 Related Work

In this section, we discuss several methods that have been proposed for training RNNs. In its most general form, the update equation for these methods can be expressed as

$$w_{k+1} = w_k - \alpha_k H_k(\hat{\nabla} f(w_k) + v_k p_k) \tag{2}$$

where $\hat{\nabla} f(w_k)$ is a stochastic gradient computed using batch $\mathcal{B}_k$; H_k is a positive-definite matrix representing an approximation to the inverse-Hessian matrix; p_k is a search direction (usually $w_k - w_{k-1}$) associated with a momentum term; and $v_k \geq 0$ is the relative scaling of the direction p_k.

2.1 Stochastic First-Order Methods

Inarguably, the simplest stochastic first-order method is SGD whose updates can be represented in the form of (2) by setting $H_k = I$ and $v_k, p_k = 0$. Momentum-based variants of SGD (such as Nesterov Accelerated Gradient [21]) use $p_k = (w_k - w_{k-1})$ with a tuned value of v_k. While SGD has demonstrated superior performance on a multitude of neural network training problems [2], in the specific case of RNN training, SGD has failed to stay competitive owing to the "vanishing/exploding" gradients problem [1,24].

There are diagonally-scaled first-order algorithms that perform well on the RNN training task. These algorithms can be interpreted as attempts to devise second-order methods via inexpensive diagonal Hessian approximations. ADAGRAD [5] allows for the independent scaling of each variable, thus partly addressing the issues arising from ill-conditioning. ADAGRAD can be written in the general updating form by setting $v_k, p_k = 0$ and by updating H_k (which is a diagonal matrix) as

$$[H_k]_{ii} = \frac{1}{\sqrt{\sum_{j=0}^{k} [\hat{\nabla} f(w_j)]_i^2 + \epsilon}},$$

where $\epsilon > 0$ is used to prevent numerical instability arising from dividing by small quantities.

Another first-order stochastic method that is known to perform well in RNN training is ADAM [12]. The update, which is a combination of RMSPROP [28] and momentum, can be represented as follows in the form of (2),

$$p_k = \sum_{j=0}^{k-1} \beta_1^{(k-j-1)}(1-\beta_1)\hat{\nabla}f(w_j) - \hat{\nabla}f(w_k), \qquad v_k = \beta_1$$

$$r_k = \sum_{j=0}^{k} \beta_2^{(k-j)}(1-\beta_2)\hat{\nabla}f(w_j)^2, \qquad [H_k]_{ii} = \frac{1}{\sqrt{[r_k]_i + \epsilon}}.$$

The diagonal scaling of the gradient elements in ADAGRAD and ADAM allows for infrequently occurring features (with low gradient components) to have larger step-sizes in order to be effectively learned, at a rate comparable to that of frequently occurring features. This causes the iterate updates to be more stable by controlling the effect of large (in magnitude) gradient components, to some extent reducing the problem of "vanishing/exploding" gradients. However, these methods are not completely immune to curvature problems. This is especially true when the eigenvectors of $\nabla^2 f(w_k)$ do not align with the co-ordinate axes. In this case, the zig-zagging (or bouncing) behavior commonly observed for SGD may occur even for methods like ADAGRAD and ADAM.

2.2   Stochastic Second-Order Methods

Let us first consider the Hessian-Free Newton methods (HF) proposed in [14,17]. These methods can be represented in the form of (2) by setting H_k to be an approximation to the inverse of the Hessian matrix ($\nabla^2 f(w_k)$), as described below, with the circumstantial use of momentum to improve convergence. HF is a second-order optimization method that has two major ingredients: (i) it implicitly creates and solves quadratic models using matrix-vector products with the Gauss-Newton matrix obtained using the "Pearlmutter trick" and (ii) it uses the Conjugate Gradient method (CG) for solving the sub-problems inexactly. Recently, KFAC [16], a method that computes a second-order step by constructing an invertible approximation of a neural networks' Fisher information matrix in an online fashion was proposed. The authors claim that the increased quality of the step offsets the increase in the per-iteration cost of the algorithm.

Our algorithm ADAQN belongs to the class of stochastic quasi-Newton methods which use a non-diagonal scaling of the gradient, while retaining low per-iteration cost. We begin by briefly surveying past work in this class of methods.

2.3   Stochastic Quasi-Newton Methods

Recently, several stochastic quasi-Newton algorithms have been developed for large-scale machine learning problems: oLBFGS [20,26], RES [19], SDBFGS [29] and SQN [3]. These methods can be represented in the form of (2) by setting $v_k, p_k = 0$ and using a quasi-Newton approximation for the matrix H_k. The methods enumerated above differ in three major aspects: (i) the update rule for the curvature pairs used in the computation of the quasi-Newton matrix, (ii) the frequency of updating, and (iii) the applicability to non-convex problems. With the exception of SDBFGS, all aforementioned methods have been designed to solve convex optimization problems. In all these methods, careful attention must be taken to monitor the quality of the curvature information that is used.

The RES and SDBFGS algorithms control the quality of the steps by modifying the BFGS update rule [22]. Specifically, the update equations take on the following form,

$$s_k = w_{k+1} - w_k, \tag{3}$$

$$y_k = \hat{\nabla}_{\mathcal{B}_k} f(w_{k+1}) - \hat{\nabla}_{\mathcal{B}_k} f(w_k) - \delta s_k, \tag{4}$$

$$H_{k+1}^{-1} = B_{k+1} = B_k + \frac{y_k^T y_k}{y_k^T s_k} - \frac{B_k s_k s_k^T B_k}{s_k^T B_k s_k} + \delta I. \tag{5}$$

This ensures that the Hessian approximations are uniformly bounded away from singularity, thus preventing the steps from becoming arbitrarily large. Further, in these methods, the line-search is replaced by a decaying step-size rule. Note that at the kth iteration, the gradients used during updates (4) are both evaluated on $\mathcal{B}_k$. oLBFGS, is similar to the above methods except no δ-modification is used. In the equations above, B_k and H_k denote approximations to the Hessian and inverse-Hessian matrices respectively.

Finally, in [3], the authors propose a novel quasi-Newton framework, SQN, in which they recommend the decoupling of the stochastic gradient calculation from the curvature estimate. The BFGS matrix is updated once every L iterations as opposed to every iteration, which is in contrast to other methods described above. The authors prescribe the following curvature pair updates,

$$s_t = \bar{w}_t - \bar{w}_{t-1}, \qquad \text{where } \bar{w}_t = \frac{1}{L} \sum_{i=(t-1)L}^{tL} w_i, \tag{6}$$

$$y_t = \hat{\nabla}_{\mathcal{H}_t}^2 f(\bar{w}_t) s_t, \tag{7}$$

where t is the curvature pair update counter, L is the update frequency (also called the aggregation length) and $\mathcal{H}_t$ is a mini-batch used for computing the sub-sampled Hessian matrix. The iterate difference, s, is based on the average of the iterates over the last $2L$ iterations, intuitively allowing for more stable approximations. On the other hand, the gradient differences, y, are not computed using gradients at all, rather they are computed using a Hessian-vector product representing the approximate curvature along the direction s.

The structure of the curvature pair updates proposed in SQN has several appealing features. Firstly, updating curvature information, and thus the Hessian approximation, every L iterations (where L is typically between 2 and 20) considerably reduces the computational cost. Additionally, more computational effort can be expended for the curvature computation since this cost is amortized over L iterations. Further, as explained in [3], the use of the Hessian-vector product in lieu of gradient differences allows for a more robust estimation of the curvature, especially in cases when $\|s\|$ is small and the gradients are noisy.

The SQN algorithm was designed specifically for convex optimization problems arising in machine learning, and its extension to RNN training is not trivial. In the following section, we describe ADAQN, our proposed algorithm, which uses the algorithmic framework of SQN as a foundation. More specifically, it retains the ability to decouple the iterate and update cycles along with the associated benefit of investing more effort in gaining curvature information.

3   adaQN

In this section, we describe the proposed algorithm in detail. Specifically, we address the key ingredients of the algorithm, including (i) the initial L-BFGS scaling, (ii) step quality control, (iii) the choice of Hessian matrix for curvature pair computation, and (iv) the suggested choice of hyper-parameters. The pseudo-code for ADAQN is given in Algorithm 1.

We emphasize that the storage of $(\hat{\nabla} f(w_k)\hat{\nabla} f(w_k)^T)$ in Step 6 is for ease of notation; in practice it is sufficient to store $\hat{\nabla} f(w_k)$ and compute y in Step 18 without explicitly constructing the matrix. Also, the search direction $p_k = -H_k \hat{\nabla} f_k$ is computed via the two-loop recursion using the available curvature pairs (S, Y), and thus the matrix H_k (the approximation to the inverse-Hessian matrix) is never constructed. Further, in Algorithm 1, we specify a fixed monitoring set $\mathcal{M}$, a feature of the algorithm that was set for ease of exposition. In practice, this set can be changed to allow for lower bias in the step acceptance criterion.

3.1   Choice of $H_k^{(0)}$ for L-BFGS

Firstly, we discuss the most important ingredient of the proposed algorithm: the initial scaling of the L-BFGS matrix. For L-BFGS, in both the deterministic and stochastic settings, a matrix $H_k^{(0)}$ must be provided, which is an estimate of the scale of the problem. This choice is crucial since the relative scale of the step (in each direction) is directly related to it. In deterministic optimization,

$$H_k^{(0)} = \frac{s_k^T y_k}{y_k^T y_k} I \tag{8}$$

is found to work well on a wide variety of applications, and is often prescribed [22, Chapter 6]. Stochastic variants of L-BFGS, including oBFGS and SQN, prescribe the use of this initialization (8). However, this is dissatisfying in the context of RNN training for two reasons. Firstly, as mentioned in the previous sections, the issue of "vanishing/exploding" gradients makes the problems highly ill-conditioned; using a scalar initialization of the L-BFGS matrix does not

---

Algorithm 1. ADAQN

Inputs: w_0, L, α, sequence of batches $\mathcal{B}_k$ with $|\mathcal{B}_k| = b$ for all k, $m_L = 10$, $m_F = 100$, $\epsilon = 10^{-4}$, $\gamma = 1.01$

1: Set $t \leftarrow 0$ and $\bar{w}_o, w_s = 0$
2: Initialize accumulated Fisher Information matrix FIFO container $\tilde{F}$ of maximum size m_F and L-BFGS curvature pair containers S, Y of maximum size m_L.
3: Randomly choose a mini-batch as monitoring set $\mathcal{M}$
4: **for** $k = 0, 1, 2, \dots$ **do**
5:      $w_s = w_s + w_k$ $\triangleright$ Running sum of iterates for average computation
6:      $w_{k+1} = w_k - \alpha H_k \hat{\nabla} f(w_k)$ $\triangleright$ Compute adaQN updates using two-loop recursion
7:      Store $\hat{\nabla} f(w_k) \hat{\nabla} f(w_k)^T$ in $\tilde{F}$
8:      **if** $\mod(k, L) = 0$ **then**
9:          $\bar{w}_n = \frac{w_s}{L}$ $\triangleright$ Compute average iterate
10:         $w_s = 0$ $\triangleright$ Clear accumulated sum of iterates
11:         **if** $t > 0$ **then**
12:            **if** $f_{\mathcal{M}}(\bar{w}_n) > \gamma f_{\mathcal{M}}(\bar{w}_o)$ **then** $\triangleright$ Check for step rejection
13:               Clear L-BFGS memory and the accumulated Fisher Information container $\tilde{F}$.
14:               $w_k = \bar{w}_o$ $\triangleright$ Return to previous aggregated point
15:               **continue**
16:            **end if**
17:            $s = \bar{w}_n - \bar{w}_o$ $\triangleright$ Compute curvature pair
18:            $y = \frac{1}{|\tilde{F}|}(\sum_{i=1}^{|\tilde{F}|} \tilde{F}_i \cdot s)$ $\triangleright$ Compute curvature pair
19:            **if** $s^T y > \epsilon \cdot s^T s$ **then** $\triangleright$ Check for sufficient curvature
20:               Store curvature pairs s and y in containers S and Y respectively
21:               $\bar{w}_o = \bar{w}_n$
22:            **end if**
23:         **else**
24:            $\bar{w}_o = \bar{w}_n$
25:         **end if**
26:         $t \leftarrow t + 1$
27:      **end if**
28: **end for**

---

address this issue. Secondly, since s and y are noisy estimates of the true iterate and gradient differences, the scaling suggested in (8) could introduce adversarial scale to the problem, causing performance deterioration.

To counter these problems, we suggest an initialization of the inverse-Hessian matrix based on accumulated gradient information. Specifically, we set

$$[H_k^{(0)}]_{ii} = \frac{1}{\sqrt{\sum_{j=0}^{k}[\hat{\nabla} f(w_j)]_i^2 + \epsilon}}, \forall i = 1, ..., n. \tag{9}$$

We direct the reader to [22, Chapter 7] for details on how the above initialization is used as part of the L-BFGS two-loop recursion. We emphasize that this initialization is: (i) a diagonal matrix with non-constant diagonal entries, (ii) has a cost comparable to (8), and (iii) is identical to the scaling matrix used by

ADAGRAD at each iteration. This choice is motivated by our observation of ADA-GRAD's stable performance on many RNN learning tasks. By initializing L-BFGS with an ADAGRAD-like scaling matrix, we impart a better scale in the L-BFGS matrix, and also allow for implicit safeguarding of the proposed method. Indeed, in iterations where no curvature pairs are stored, the ADAQN and ADAGRAD steps are identical in form.

3.2   Step Acceptance and Control

While curvature information can be used to improve convergence rates, noisy or stale curvature information may in fact deteriorate performance [3]. SQN attempts to prevent this problem by using large-batch Hessian-vector products in (7). Other methods attempt to control the quality of the steps by modifying the L-BFGS update rule to ensure that H_k is positive definite for all k. However, we have found that these do not work well in practice. Instead, we control the quality of the steps by judiciously choosing the curvature pairs used by L-BFGS. We attempt to store curvature pairs during each cycle but skip the updating if the calculated curvature is small; see [22, Chapter 6] for details regarding skipping in quasi-Newton methods. Further, we flush the memory when the step quality deteriorates, allowing for more reliable steps till the memory builds up again.

The proposed criterion (Line 12 of Algorithm 1) is an inexpensive heuristic wherein the functions are evaluated on a monitoring set, and γ approximates the effect of noise on the function evaluations. A step is rejected if the function value of the new aggregated point is significantly worse (measured by γ) than the previous. In this case, we reset the memory of L-BFGS which allows the algorithm to preclude the deteriorating effect of any stored curvature pairs. The algorithm resumes to take ADAGRAD steps and build up the curvature estimate again. We report that, as an alternative to the proposed criterion, a more sophisticated criterion such as *relative improvement*,

$$\frac{f_{\mathcal{M}}(\bar{w}_n) - f_{\mathcal{M}}(\bar{w}_o)}{f_{\mathcal{M}}(\bar{w}_o)} > \tilde{\gamma} \in (0,1)$$

delivered similar performance on our test problems.

In the case when the sufficient curvature condition (Line 19 of Algorithm 1) is not satisfied, the storage of the curvature pair is skipped. In deterministic optimization, this problem is avoided by conducting a Wolfe line-search. If the curvature information for a given step is inadequate, the line-search attempts to look for points further along the search path. We extend this idea to the RNN setting by not updating $\bar{w}_o$ when this happens. This allows us to move further, and possibly glean curvature information in subsequent update attempts. We have experimentally found this safeguarding to be crucial for the robust performance of ADAQN. That being said, such rejection happens infrequently and the average L-BFGS memory per epoch remains high for all of our reported experiments; see Sect. 4.

3.3   Choice of Curvature Information Matrix

As in SQN, the iterate difference s in our algorithm is computed using aggregated iterates and the gradient difference y is computed through a matrix-vector product; refer to Eqs. (6) and (7). The choice of curvature matrix for the computation of y must address the trade-off between obtaining informative curvature information and the computational expense of its acquisition. Recent work suggests that the Fisher Information matrix (FIM) yields a better estimate of the curvature of the problem as compared to the true Hessian matrix (which is a natural choice); see for e.g. [15, 23].

Given a function f parametrized by a random variable $\mathcal{X}$, the (true) FIM at a point w is given by

$$F(w) = \mathbb{E}_{\mathcal{X}}[\nabla f_{\mathcal{X}}(w)\nabla f_{\mathcal{X}}(w)^T].$$

Since the distribution for $\mathcal{X}$ is almost never known, the *empirical* Fisher Information matrix (eFIM) is often used in practice. The eFIM can be expressed as follows

$$\hat{F}(w) = \frac{1}{|\mathcal{H}|} \sum_{i \in \mathcal{H}} \nabla_i f(w)\nabla_i f(w)^T, \tag{10}$$

where $\mathcal{H} \subseteq \{1, 2, \cdots, m\}$.

Notice from Eq. (10) that the eFIM is guaranteed to be positive semi-definite, a property that does not hold for the true Hessian matrix. The use of the FIM (or eFIM) in second-order methods allows for attractive theoretical and practical properties. We exclude these results for brevity and refer the reader to [15] for a detailed survey regarding this topic.

Given these observations and results, the use of the eFIM may seem like a reasonable choice for the Hessian matrix approximation used in the computation of y_t (see Eq. (7)). However, the use of this matrix, even infrequently, increases the amortized per-iteration cost as compared to state-of-the-art first-order stochastic methods. Further, unlike second-order methods which rely on relatively accurate curvature information to generate good steps, quasi-Newton methods are able to generate high-quality steps even with crude curvature information [22]. In this direction, we propose the use of a modified version of the empirical Fisher Information matrix that uses historical values of stochastic gradients, which were already computed as part of the step, thus reducing the computational cost considerably. This reduction, comes at the expense of storage and potentially noisy estimates due to stale gradient approximations. We call this approximation of the eFIM the *accumulated* Fisher Information matrix (aFIM) and denote it by $\bar{F}$. Given a memory budget of m_F, the aFIM at the kth iteration is given by

$$\bar{F}(w_k) = \frac{1}{\sum_{j=k-m_F+1}^{k} |\mathcal{B}_j|} \sum_{j=k-m_F+1}^{k} \nabla_{\mathcal{B}_j} f(w_j)\nabla_{\mathcal{B}_j} f(w_j)^T. \tag{11}$$

For the purpose of our implementation, we maintain a finite-length FIFO container $\tilde{F}$ for storing the stochastic gradients as they are computed. Whenever the algorithm enters lines 12–16, we reject the step, and the contents of $\tilde{F}$ along with the L-BFGS memory are cleared. By clearing $\tilde{F}$, we also allow for additional safeguarding of future iterates against noisy gradients in the $\tilde{F}$ container that may have contributed in the generation of the poor step.

3.4   Choice of Hyper-Parameters

ADAQN has a set of hyper-parameters that require tuning for competitive performance. Other than the step-size and batch-size, which needs to be tuned for all aforementioned methods, the only hyper-parameter exposed to the user is L. We prescribe L to be chosen from $\{2, 5, 10, 20\}$. We experimentally observed that the performance was not highly sensitive to the choice of α and L. Often, $L = 5$ and the same step-length as used for ADAGRAD gave desirable performance. The other hyper-parameters have intuitive default values which we have found to work well for a variety of applications. Additional details about the offline tuning costs of ADAQN as compared to ADAGRAD and ADAM can be found in Sect. 4.

3.5   Cost

Given the nature of the proposed algorithm, a reasonable question is about the per-iteration cost. Let us begin by first considering the per-iteration cost of other popular methods. For simplicity, we assume that the cost of the gradient computation is $\mathcal{O}(n)$, which is a reasonable assumption in the context of deep learning. SGD has one of the cheapest per-iteration costs, with the only significant expense being the computation of the mini-batch stochastic gradient. Thus, SGD has a per-iteration complexity of $\mathcal{O}(n)$. ADAGRAD and ADAM also have the same per-iteration complexity since the auxiliary operations only involve dot-products and elementary vector operations. Further, these algorithms have $\mathcal{O}(1)$ space complexity. On the other hand, second-order methods have higher per-iteration complexity since each iteration requires an inexact solution of a linear system, and possibly, storage of the pre-conditioning matrices.

The per-iteration time complexity of our algorithm consists of three components: (i) the cost of gradient computation, (ii) the cost of the L-BFGS two-loop recursion, and (iii) the amortized cost of computing the curvature pair. Thus, the overall cost can be written as

$$\underbrace{\mathcal{O}(n)}_{\text{gradient computation}} + \underbrace{4m_L n}_{\text{two-loop recursion}} + \underbrace{m_F n L^{-1}}_{\text{cost of computing curvature pair}}. \tag{12}$$

Given the prescription of $L \approx 5$, $m_F = 100$ and $m_L = 10$, the cost per-iteration remains at $\mathcal{O}(n)$. The memory requirement of our algorithm is also $\mathcal{O}(n)$ since we require the storage of up to $m_F + 2m_L$ vectors of size n.

This result is similar to the one presented in [3]. The difference in the complexity arises in the third term of (12) due to our choice of the accumulated

Fisher Information matrix as opposed to using a sub-sampled Hessian approximation. It is not imperative for our algorithm to use aFIM for the computation of y_t (7). We can instead use the eFIM (10), which would allow for a lower memory requirement (from $(m_F + m_L)n$ to $m_L n$) at the expense of added computation during curvature pair estimation. However, the time complexity would remain linear in n for either choice. As we mention in Sect. 3.3, by using the accumulated Fisher Information matrix, we avoid the need for additional computation at the expense of memory; a choice we have found to work well in practice.

4   Empirical Results

In this section, we present numerical evidence demonstrating the viability of the proposed algorithm for training RNNs[1]. We also present meta-data regarding the experiments which suggests that the performance difference between ADAQN and its competitors (ADAGRAD in particular) can be attributed primarily to the incorporation of curvature.

4.1   Language Modeling

For benchmarking, we compared the performance of ADAQN against ADAGRAD and ADAM on two language modeling (LM) tasks: character-level LM and word-level LM. For the character-level LM task [10], we report results on two data sets: The Tale of Two Cities (Dickens) and The Complete Works of Friedrich Nietzsche (Nietzsche). The former has 792 k characters while the latter has 600 k. We used the Penn-Tree data set for the word-level LM task [30]. This data set consists of 929 k training words with 10 k words in its vocabulary.

For all tasks, we used an RNN with 5 recurrent layers. The input and output layer sizes were determined by the vocabulary of the data set. The character-level and word-level LMs were constructed with 100 and 400 nodes per layer respectively. The weights were randomly initialized from $\mathcal{N}(0, 0.01)$. Unless otherwise specified, the activation function used was tanh. The sequence length was chosen to be 50 for both cases. For readability, we exclude other popular methods that did not consistently perform competitively. In particular, SGD (with or without momentum) was not found to be competitive despite significant tuning. For ADAQN, ADAGRAD and ADAM, all hyper-parameters were set using a grid-search. In particular, step-sizes were tuned for all three methods. ADAM needed coarse-tuning for (β_1, β_2) in the vicinity of the suggested values. For ADAQN, the value of L was chosen from $\{2, 5, 10, 20\}$. The rest of the hyper-parameters $(m_F, m_L, \epsilon, \gamma)$ were set at their recommended values for all experiments (refer to Algorithm 1). It can thus be seen that the offline tuning costs of ADAQN are comparable to those of ADAGRAD and ADAM. We ran all experiments for 100 epochs and present the results (testing error) in Fig. 1.

---

[1] A MATLAB implementation of all discussed stochastic quasi-Newton methods, including ADAQN, on an example logistic regression problem can be found in our GitHub repository [11].

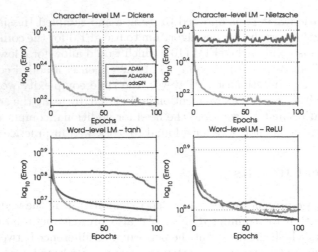

Fig. 1. Numerical results on LM Tasks

It is clear from Fig. 1 that ADAQN presents a non-trivial improvement over both ADAGRAD and ADAM on all tasks with tanh activation function. Specifically, we emphasize the performance gain over ADAGRAD, the method which ADAQN is safeguarded by. On the character-level task with ReLU activation, ADAQN performed better than ADAM but worse than ADAGRAD. We point out that experiments with other (including larger) data sets yielded results of similar nature.

4.2 Average L-BFGS Memory per Epoch

Given the safeguarded nature of our algorithm, a natural question regarding the numerical results presented pertains to the effect of the safeguarding on the performance of the algorithm. To answer this question, we report the average L-BFGS memory per epoch in Fig. 2. This is computed by a running sum initialized at 0 at the start of each new epoch. A value greater than 1 indicates that at least one curvature pair was present in the memory (in expectation) during a given epoch. Higher average values of L-BFGS memory suggest that more directions of curvature were successfully explored; thus, the safeguarding was less necessary. Lower values, on the other hand, suggest that the curvature information was either not informative (leading to skipping) or led to deterioration of performance (leading to step rejection).

The word-level LM task with the ReLU activation function has interesting outcomes. It can be seen from Fig. 1 that the performance of ADAGRAD is similar to that of ADAQN for the first 50 epochs but then ADAGRAD continues to make progress while the performance of ADAQN stagnates. During the same time, the average L-BFGS memory drops significantly suggesting that safeguarding was necessary and that, the curvature information was not informative enough and

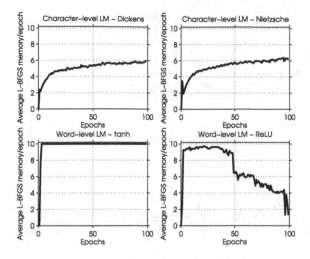

Fig. 2. Average L-BFGS memory per epoch

even caused deterioration in performance (evidenced by occasional increase in the function value).

4.3 MNIST Classification from Pixel Sequence

A challenging toy problem for RNNs is that of image classification given pixel sequences [13]. For this problem, the image pixels are presented sequentially to the network one-at-a-time and the network must predict the corresponding category. This long range dependency makes the RNN difficult to train. We report results for the popular MNIST data set. For this experiment, we used a setup similar to that of [13] with two modifications: we used tanh activation function instead of ReLU and initialized all weights from $\mathcal{N}(0, 0.01)$ instead of using their initialization trick. The results are reported in Fig. 3.

As can be seen from the Fig. 3, ADAM and ADAGRAD struggle to make progress and stagnate at an error value close to that of the initial point. On the other hand, ADAQN is able to significantly improve the error values, and also achieves superior classification accuracy rates. Experiments on other toy problems with long range dependencies, such as the addition problem [9], yielded similar results.

5 Discussion

The results presented in the previous section suggest that ADAQN is competitive with popular algorithms for training RNNs. However, ADAQN is not restricted to this class of problems. Indeed, preliminary results on other architectures (such as Feed-Forward Networks) delivered promising performance. It may be possible to further improve the performance of the algorithm by modifying the update rule

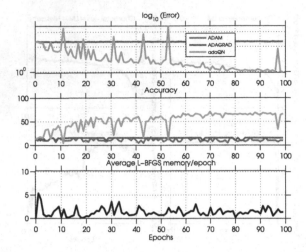

Fig. 3. Numerical results on MNIST with sequence of pixels

and frequency. In this direction, we discuss the practicality of using momentum in such an algorithm and possible heuristics to allow the algorithm to adapt the cycle length L as opposed to tuning it to a constant value.

Recent work by [27] suggests superior performance of momentum methods on a wide variety of learning tasks. These methods, with the right initialization, have been shown to outperform sophisticated methods such as the Hessian-Free Newton method. However, recent efforts suggest the use of second-order methods *in conjunction* with momentum [15,16]. In this case, one interpretation of momentum is that of providing a pre-conditioner to the CG sub-solver. Significant performance gains through the inclusion of momentum have been reported when the gradients are reliable [15]. We hypothesize that performance gains can be obtained through careful inclusion of momentum for methods like ADAQN as well. However, the design of such an algorithm, and the efficacy of using momentum-like ideas is an open question for future research.

Lastly, we discuss the role of the aggregation cycle length L on the performance of the algorithm. If L is chosen to be too large, the aggregation points will be too far-apart possibly leading to incorrect curvature estimation. If L is too small, then the iterates change insufficiently before an update attempt is made leading to skipping of update pairs. Besides the issue of curvature quality, the choice of L also has ramifications on the cost of the algorithm as discussed in Sect. 3.5. Thus, a natural extension of ADAQN is an algorithm where L can be allowed to adapt during the course of the algorithm. L could be increased or decreased depending on the quality of the estimated curvature, while being bounded to ensure that the cost of updating is kept at a reasonable level. The removal of this hyper-parameter will not only obviate the need for tuning, but will also allow for a more robust performance.

6   Conclusions

In this paper, we present a novel quasi-Newton method, ADAQN, for training RNNs. The algorithm judiciously incorporates curvature information while retaining a low per-iteration cost. The algorithm builds upon the framework proposed in [3], which was designed for convex optimization problems. We discuss the key ingredients of our algorithm, such as, the scaling of the L-BFGS matrices using historical gradients, curvature pair updating and step acceptance criterion, and, suggest the use of an accumulated Fisher Information matrix during the computation of a curvature pair. We examine the per-iteration time and space complexity of ADAQN and show that it is of the same order of magnitude as popular first-order methods. Finally, we present numerical results for two language modeling tasks and demonstrate competitive performance of ADAQN as compared to popular algorithms used for training RNNs.

References

1. Bengio, Y., Simard, P., Frasconi, P.: Learning long-term dependencies with gradient descent is difficult. Neural Netw. IEEE Trans. **5**(2), 157–166 (1994)
2. Bengio, Y., Goodfellow, I., Courville, A.: Deep learning (2016). Book in preparation for MIT Press. http://www.deeplearningbook.org
3. Byrd, R.H., Hansen, S.L., Nocedal, J., Singer, Y.: A stochastic quasi-Newton method for large-scale optimization. SIAM J. Optim. **26**(2), 1008–1031 (2016)
4. Cho, K., Van Merriënboer, B., Gülçehre, Ç., Bahdanau, D., Bougares, F., Schwenk, H., Bengio, Y.: Learning phrase representations using RNN encoder-decoder for statistical machine translation. In: Proceedings of the 2014 Conference on Empirical Methods in Natural Language Processing (EMNLP), pp. 1724–1734. Association for Computational Linguistics, Doha, Qatar, October 2014
5. Duchi, J., Hazan, E., Singer, Y.: Adaptive subgradient methods for online learning and stochastic optimization. J. Mach. Learn. Res. **12**, 2121–2159 (2011)
6. Graves, A.: Supervised sequence labelling with Recurrent Neural Networks, vol. 385. Springer, Heidelberg (2012)
7. Graves, A., Mohamed, A., Hinton, G.: Speech recognition with deep recurrent neural networks. In: IEEE International Conference on Acoustics, Speech and Signal Processing (ICASSP 2013), pp. 6645–6649 (2013)
8. Graves, A., Schmidhuber, J.: Offline handwriting recognition with multidimensional recurrent neural networks. In: Advances in Neural Information Processing Systems (NIPS 2009), pp. 545–552 (2009)
9. Hochreiter, S., Schmidhuber, J.: Long short-term memory. Neural Comput. **9**(8), 1735–1780 (1997)
10. Karpathy, A., Johnson, J., Li, F.: Visualizing and understanding recurrent networks. In: International Conference on Learning Representations (ICLR 2016) (2016)
11. Keskar, N., Berahas, A.S.: minSQN: Stochastic Quasi-Newton Optimization in MATLAB (2016). https://github.com/keskarnitish/minSQN/
12. Kingma, D., Ba, J.: Adam: A method for stochastic optimization. In: International Conference on Learning Representations (ICLR 2015) (2015)

13. Le, Q.V., Jaitly, N., Hinton, G.E.: A simple way to initialize recurrent networks of rectified linear units. arXiv preprint (2015). arXiv:1504.00941
14. Martens, J.: Deep learning via hessian-free optimization. In: Proceedings of the 27th International Conference on Machine Learning (ICML 2010) (2010)
15. Martens, J.: New perspectives on the natural gradient method. arXiv preprint (2014). arXiv:1412.1193
16. Martens, J., Grosse, R.: Optimizing neural networks with kronecker-factored approximate curvature. In: Proceedings of the 32th International Conference on Machine Learning (ICML 2015) (2015)
17. Martens, J., Sutskever, I.: Learning recurrent neural networks with hessian-free optimization. In: Proceedings of the 28th International Conference on Machine Learning (ICML 2011), pp. 1033–1040 (2011)
18. Mikolov, T., Kombrink, S., Deoras, A., Burget, L., Cernocky, J.: Rnnlm-recurrent neural network language modeling toolkit. In: Proceedings of the 2011 ASRU Workshop, pp. 196–201 (2011)
19. Mokhtari, A., Ribeiro, A.: RES: Regularized stochastic BFGS algorithm. Sig. Proces. IEEE Trans. **62**(23), 6089–6104 (2014)
20. Mokhtari, A., Ribeiro, A.: Global convergence of online limited memory BFGS. J. Mach. Learn. Res. **16**, 3151–3181 (2015)
21. Nesterov, Y.: A method of solving a convex programming problem with convergence rate o (1/k2). Sov. Math. Dokl. **27**, 372–376 (1983)
22. Nocedal, J., Wright, S.: Numerical optimization. Springer, Heidelberg (2006)
23. Pascanu, R., Bengio, Y.: Revisiting natural gradient for deep networks. In: International Conference on Learning Representations (ICLR 2013) (2013)
24. Pascanu, R., Mikolov, T., Bengio, Y.: On the difficulty of training recurrent neural networks. In: Proceedings of the 31st International Conference on Machine Learning (ICML 2014), pp. 1310–1318 (2013)
25. Robinson, T., Hochberg, M., Renals, S.: The use of recurrent neural networks in continuous speech recognition. Automatic speech and speaker recognition. Springer, Heidelberg (1996)
26. Schraudolph, N.N., Yu, J., Günter, S.: A stochastic quasi-newton method for online convex optimization. In: International Conference on Artificial Intelligence and Statistics, pp. 436–443 (2007)
27. Sutskever, I., Martens, J., Dahl, G., Hinton, G.: On the importance of initialization and momentum in deep learning. In: Proceedings of the 30th International Conference on Machine Learning (ICML 2013), pp. 1139–1147 (2013)
28. Tieleman, T., Hinton, G.: Lecture 6.5-RMSProp: Divide the gradient by a running average of its recent magnitude. COURSERA: Neural Netw. Mach. Learn. **4**, 26–31 (2012)
29. Wang, X., Ma, S., Liu, W.: Stochastic quasi-Newton methods for nonconvex stochastic optimization. arXiv preprint (2014). arXiv:1412.1196
30. Zaremba, W., Sutskever, I., Vinyals, O.: Recurrent Neural Network Regularization. arXiv preprint (2014). arXiv:1409.2329

Semi-supervised Tensor Factorization
for Brain Network Analysis

Bokai Cao[1]([⊠]), Chun-Ta Lu[1], Xiaokai Wei[1], Philip S. Yu[1,2], and Alex D. Leow[3]

[1] Department of Computer Science, University of Illinois, Chicago, IL, USA
{caobokai,clu29,xwei2,psyu}@uic.edu
[2] Institute for Data Science, Tsinghua University, Beijing, China
[3] Department of Psychiatry, University of Illinois, Chicago, IL, USA
alexfeuillet@gmail.com

Abstract. Brain networks characterize the temporal and/or spectral connections between brain regions and are inherently represented by multi-way arrays (tensors). In order to discover the underlying factors driving such connections, we need to derive compact representations from brain network data. Such representations should be discriminative so as to facilitate the identification of subjects performing different cognitive tasks or with different neurological disorders. In this paper, we propose SEMIBAT, a novel semi-supervised Brain network Analysis approach based on constrained Tensor factorization. SEMIBAT (1) leverages unlabeled resting-state brain networks for task recognition, (2) explores the temporal dimension to capture the progress, (3) incorporates classifier learning procedure to introduce supervision from labeled data, and (4) selects discriminative latent factors for different tasks. The Alternating Direction Method of Multipliers (ADMM) framework is utilized to solve the optimization objective. Experimental results on EEG brain networks illustrate the superior performance of the proposed SEMIBAT model on graph classification with a significant improvement 31.60% over plain vanilla tensor factorization. Moreover, the data-driven factors can be readily visualized which should be informative for investigating cognitive mechanisms. The software related to this paper is available at https://www.cs.uic.edu/~bcao1/code/semibat.zip.

Keywords: Brain network · Graph mining · Tensor factorization

1 Introduction

Brain networks (*a.k.a*, connectome [25]) obtained from neuroimaging data have been commonly employed to study neuropsychiatric disorders [3,15,27,30]. Connectivity patterns are usually embedded within the graph structures by a set of vertices and edges where vertices correspond to regions of interest in the brain and edges represent the connectivity strength or correlation between brain regions. Considering the temporal and spectral domain, original brain networks are typically represented in multi-way arrays (*i.e.*, tensors) which make the conventional vector-based classification algorithms inapplicable. Moreover, directly

© Springer International Publishing AG 2016
P. Frasconi et al. (Eds.): ECML PKDD 2016, Part I, LNAI 9851, pp. 17–32, 2016.
DOI: 10.1007/978-3-319-46128-1_2

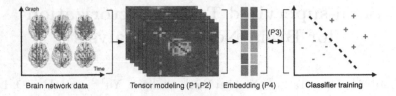

Fig. 1. The framework of tensor-based brain network analysis.

reshaping tensors into vectors would result in the curse of dimensionality, and the number of brain network samples is usually small, thereby making it challenging to train an effective classifier in a high dimensional feature space with a limited number of samples.

In order to apply conventional machine learning algorithms and train pattern classifiers, it is preferable to first derive vector representations from the brain network data. In general, researchers have proposed to extract two types of features: (1) graph-theoretical measures [13,27] and (2) subgraph patterns [6,17]. However, the expressiveness of these features is limited to the predefined formulations. To explore a larger space of potentially informative features to represent brain networks, it motivates us to learn latent representations from the brain network data. It is desirable to let the latent representations be discriminative so that brain networks with different labels can easily be separated. Learning such representations is a non-trivial task due to the following problems:

(P1) Although labeled brain network data for specific tasks or diseases are usually costly to obtain, brain networks under resting-state from healthy subjects are recorded in many neuroimaging experiments. How can we leverage the unlabeled data, *i.e.*, resting-state brain networks, to facilitate classification?

(P2) Existing studies usually compute time-averaged brain networks before further analysis [28] which may result in formidable information loss. How can we directly fully utilize the temporal information in our model?

(P3) In order to obtain discriminative representations, we should incorporate the classifier training procedure into the representation learning process for leveraging the supervision information. How can we effectively fuse these two procedures together?

(P4) Different classes (or tasks) are usually associated with different subsets of the latent factors. How can we achieve feature selection in the latent space?

In this paper, we propose SEMIBAT, a semi-supervised Brain network Analysis approach based on constrained Tensor factorization. The proposed framework is illustrated in Fig. 1. The contributions of this work are fourfold:

- We leverage unlabeled resting-state brain networks together with labeled data to collectively learn a latent space, which alleviates the problem that labeled brain network data for specific tasks or diseases are usually very limited.
- We model brain networks through partially symmetric tensor factorization which is suitable for inherently undirected graphs, *e.g.*, EEG brain networks. The temporal dimension is modeled as one of modes in the fourth-order tensor.

- We blend representation learning and classifier training into a unified optimization problem, which allows classifier parameters to interact with discriminative latent factors by leveraging the supervision information.
- We incorporate the $\ell_{2,1}$ norm to conduct feature selection in the latent space, thereby identifying discriminative latent factors for different tasks.

2   Preliminaries

Table 1 lists some basic symbols that will be used throughout the paper. We introduce the concept of tensors which are higher order arrays that generalize the notions of vectors (the first-order tensors) and matrices (the second-order tensors), whose elements are indexed by more than two indices. Each index expresses a mode of the data and corresponds to a coordinate direction. The number of variables in each mode indicates the dimensionality of a mode. The order of a tensor is determined by the number of its modes. An mth-order tensor can be represented as $\mathcal{X} = (x_{i_1,\cdots,i_m}) \in \mathbb{R}^{I_1 \times \cdots \times I_m}$, where I_i is the dimension of $\mathcal{X}$ along mode i. An overview of tensor notation and operators is given as follows which will be used to formulate the problem.

Table 1. Overview of tensor notation and operators.

Notation	Interpretation
a	scalar
$\mathbf{a}$	vector
$\mathbf{A}$	matrix
$\mathcal{X}$	tensor, set or space
$\mathbf{X}_{(k)}$	matricization of tensor $\mathcal{X}$ along mode k
$*$	Hadamard product (elementwise product)
$\circ$	tensor product (outer product)
$\times_k$	mode-k product
$\otimes$	Kronecker product
$\odot$	Khatri-Rao product
$\|\cdot\|$	norm of a vector, matrix or tensor
$|\cdot|$	cardinality of a set

Definition 1 (Tensor Product). *The tensor product $\mathcal{X} \circ \mathcal{Y}$ of a tensor $\mathcal{X} \in \mathbb{R}^{I_1 \times \cdots \times I_m}$ and another tensor $\mathcal{Y} \in \mathbb{R}^{I_1' \times \cdots \times I_{m'}'}$ is defined by $(\mathcal{X} \circ \mathcal{Y})_{i_1,\dots,i_m,i_1',\dots,i_{m'}'} = x_{i_1,\dots,i_m} y_{i_1',\dots,i_{m'}'}$.*

Tensor product is also referred to as outer product in some literature [9,29]. An mth-order tensor is a rank-one tensor if it can be defined as the tensor product of m vectors.

Definition 2 (Mode-k Product). *The mode-k product $\mathcal{X} \times_k \mathbf{A}$ of a tensor $\mathcal{X} \in \mathbb{R}^{I_1 \times \cdots \times I_m}$ and a matrix $\mathbf{A} \in \mathbb{R}^{J \times I_k}$ is of size $I_1 \times \cdots \times I_{k-1} \times J \times I_{k+1} \times \cdots \times I_m$ and is defined by $(\mathcal{X} \times_k \mathbf{A})_{i_1,\ldots,i_{k-1},j,i_{k+1},\ldots,i_m} = \sum_{i_k=1}^{I_k} x_{i_1,\ldots,i_m} a_{j,i_k}$.*

Definition 3 (Kronecker Product). *The Kronecker product of two matrices $\mathbf{A} \in \mathbb{R}^{I \times J}, \mathbf{B} \in \mathbb{R}^{K \times L}$ is of size $IK \times JL$ and is defined by*

$$\mathbf{A} \otimes \mathbf{B} = \begin{pmatrix} a_{11}\mathbf{B} & \cdots & a_{1J}\mathbf{B} \\ \vdots & \ddots & \vdots \\ a_{I1}\mathbf{B} & \cdots & a_{IJ}\mathbf{B} \end{pmatrix}$$

Definition 4 (Khatri-Rao Product). *The Khatri-Rao product of two matrices $\mathbf{A} \in \mathbb{R}^{I \times K}, \mathbf{B} \in \mathbb{R}^{J \times K}$ is of size $IJ \times K$ and is defined by $\mathbf{A} \odot \mathbf{B} = (a_1 \otimes b_1, \cdots, a_K \otimes b_K)$ where $a_1, \cdots, a_K, b_1, \cdots, b_K$ are the columns of matrices $\mathbf{A}$ and $\mathbf{B}$, respectively.*

Definition 5 (Partially Symmetric Tensor). *A rank-one mth-order tensor $\mathcal{X} \in \mathbb{R}^{I_1 \times \cdots \times I_m}$ is partially symmetric if it is symmetric on modes $i_1, \ldots, i_j \in \{1, \ldots, m\}$ and can be written as the tensor product of m vectors: $\mathcal{X} = \mathbf{x}^{(1)} \circ \cdots \circ \mathbf{x}^{(m)}$ where $\mathbf{x}^{(i_1)} = \cdots = \mathbf{x}^{(i_j)}$.*

Definition 6 (Mode-k Matricization). *The mode-k matricization of a tensor $\mathcal{X} \in \mathbb{R}^{I_1 \times \cdots \times I_m}$ is denoted by $\mathbf{X}_{(k)}$ and arranges the mode-k fibers to be the columns of the resulting matrix. The dimension of $\mathbf{X}_{(k)}$ is $\mathbb{R}^{I_k \times J}$, where $J = I_1 \cdots I_{k-1} I_{k+1} \cdots I_m$. Each tensor element $(i_1, \cdots, i_m)$ maps to the matrix element (i_k, j): $j = 1 + \sum_{p=1, p \neq k}^{m} (i_p - 1) J_p$ with $J_p = \prod_{q=1, q \neq k}^{p-1} I_q$.*

3   SEMIBAT Framework

3.1   Problem Formulation

Let $\mathcal{D} = \{G_1, \cdots, G_n\}$ denote a dynamic graph dataset of brain networks where $|\mathcal{D}| = n$ is the number of graph objects. All graphs in the dataset share a given set of vertices V which corresponds to a brain parcellation scheme. Suppose the brain is parcellated via an atlas into $|V| = m$ regions, and the temporal dimensionality is t. A brain network G_i can be represented by a partially symmetric tensor $\mathcal{Z}_i \in \mathbb{R}^{m \times m \times t}$. We assume that the first l graphs within $\mathcal{D}$ are labeled and $\mathbf{Y} \in \mathbb{R}^{l \times c}$ is the class label matrix where c is the number of class labels. $\mathbf{Y}(i,j) = 1$ if G_i belongs to the j-th class, otherwise $\mathbf{Y}(i,j) = 0$. For convenience, we also denote the labeled graph dataset by $\mathcal{D}_l = \{G_1, \cdots, G_l\}$, and the unlabeled graph dataset as $\mathcal{D}_u = \{G_{l+1}, \cdots, G_n\}$, $\mathcal{D} = \mathcal{D}_l \cup \mathcal{D}_u$. In our experiments, brain networks under emotion regulation tasks compose labeled graphs, while those under resting-state compose unlabeled graphs.

3.2   Tensor Modeling

We first address the problem (P1) discussed in Sect. 1 by stacking the brain network dataset $\mathcal{D}$ of n graphs, $i.e.$, $\{\mathcal{Z}_i\}_{i=1}^{n}$, as a tensor $\mathcal{X} \in \mathbb{R}^{m \times m \times t \times n}$. Through joint tensor factorization, unlabeled graphs in $\mathcal{D}_u$ could facilitate the representation learning of labeled graphs in $\mathcal{D}_l$ by affecting latent factors.

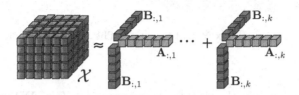

Fig. 2. CP factorization. The fourth-order partially symmetric tensor $\mathcal{X}$ is approximated by k rank-one tensors. The f-th factor tensor is the tensor product of four vectors, $i.e.$, $\mathbf{B}_{:,f} \circ \mathbf{B}_{:,f} \circ \mathbf{T}_{:,f} \circ \mathbf{A}_{:,f}$. The temporal dimension is omitted in the plot.

Note that $\mathcal{X}$ is a fourth-order partially symmetric tensor (symmetric on the first two modes) and it naturally models the temporal dimension discussed as the problem (P2). We assume that $\mathcal{X}$ can be decomposed into k factors in the following manner

$$\mathcal{X} = \mathcal{C} \times_1 \mathbf{B} \times_2 \mathbf{B} \times_3 \mathbf{T} \times_4 \mathbf{A} \tag{1}$$

where $\mathbf{B} \in \mathbb{R}^{m \times k}$ is the factor matrix for vertices, $\mathbf{T} \in \mathbb{R}^{t \times k}$ is the factor matrix for time points, $\mathbf{A} \in \mathbb{R}^{n \times k}$ is the factor matrix for graphs, and $\mathcal{C} \in \mathbb{R}^{k \times \cdots \times k}$ is a fourth-order identity tensor, $i.e.$, $\mathcal{C}(i_1, \cdots, i_4) = \delta(i_1 = \cdots = i_4)$. Basically, Eq. (1) is a CANDECOMP/PARAFAC (CP) factorization [16] as shown in Fig. 2. It is desirable to discover distinct latent factors to obtain more concise and interpretable results, and thus we include orthogonality constraints $\mathbf{A}^\mathrm{T}\mathbf{A} = \mathbf{I}$.[1]

One of the targets is task recognition based on the brain network data. We assume that there is a matrix of regression coefficients $\mathbf{W} \in \mathbb{R}^{k \times c}$ which assigns graphs with labels based on the graph factor matrix $\mathbf{A}$, $i.e.$, $\mathbf{Y} = \mathbf{DAW}$ where $\mathbf{D} = [\mathbf{I}^{l \times l}, \mathbf{0}^{l \times (n-l)}] \in \mathbb{R}^{l \times n}$.

An intuitive idea is to first learn latent representations of brain networks and then train a classifier on them in a serial two-step manner, which however would make these two procedures independent with each other and fail to introduce the supervision information to the representation learning process. Moreover, the advantage is established in [5] of directly searching for classification-relevant structure in the original data, rather than solving the supervised and unsupervised problems independently. To address the problem (P3) discussed in Sect. 1,

---

[1] We considered adding non-negativity constraints to enhance interpretability, but our preliminary results showed that it would degrade performance.

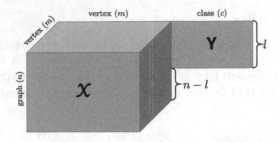

Fig. 3. Partially coupled matrix $\mathbf{Y}$ and tensor $\mathcal{X}$. The temporal dimension is omitted in the plot.

we propose to incorporate the classifier learning process (*i.e.*, $\mathbf{W}$) into the framework of learning latent feature representations of graphs (*i.e.*, $\mathbf{A}$). In this manner, the weight matrix $\mathbf{W}$ and the feature matrix $\mathbf{A}$ can interact with each other in the same learning framework. Note that it is similar to coupled matrix and tensor factorization [2], however $\mathcal{X}$ and $\mathbf{Y}$ are coupled only in part of the graph mode, as shown in Fig. 3.

In summary, the proposed brain network analysis framework can be mathematically formulated as solving the following optimization problem

$$
\min_{\mathbf{B},\mathbf{T},\mathbf{A},\mathbf{W}} \underbrace{\|\mathcal{X} - \mathcal{C}\times_1\mathbf{B}\times_2\mathbf{B}\times_3\mathbf{T}\times_4\mathbf{A}\|_F^2}_{\text{factorization error}} + \alpha\underbrace{\|\mathbf{D}\mathbf{A}\mathbf{W} - \mathbf{Y}\|_F^2}_{\text{classification loss}} + \lambda\underbrace{\|\mathbf{W}^\mathrm{T}\|_{2,1}}_{\text{regularization}}
$$

$$
\text{s.t.} \quad \underbrace{\mathbf{A}^\mathrm{T}\mathbf{A} = \mathbf{I}}_{\text{orthogonality}} \tag{2}
$$

where $\|\mathbf{W}^\mathrm{T}\|_{2,1}$ is the sparsity-promoting regularization term that controls the complexity of $\mathbf{W}$ and has the effects of feature selection thereby addressing the problem (P4), and α, λ are positive parameters which control contributions of classification loss and regularization, respectively.

3.3 Optimization Framework

The model parameters that have to be estimated include $\mathbf{B} \in \mathbb{R}^{m\times k}$, $\mathbf{T} \in \mathbb{R}^{t\times k}$, $\mathbf{A} \in \mathbb{R}^{n\times k}$ and $\mathbf{W} \in \mathbb{R}^{k\times c}$. The optimization problem in Eq. (2) is non-convex with respect to $\mathbf{B}$, $\mathbf{T}$, $\mathbf{A}$ and $\mathbf{W}$ together. There is no closed-form solution for the problem. We now introduce an alternating scheme to solve the optimization problem. The key idea is to optimize the objective with respect to one variable, while fixing others, and decouple constraints using an Alternating Direction Method of Multipliers (ADMM) scheme [4]. The algorithm will keep updating the variables until convergence. First, we define the following notations

$$
\mathbf{E} = \mathbf{P}\odot\mathbf{T}\odot\mathbf{A} \in \mathbb{R}^{(m*t*n)\times k}, \ \mathbf{F} = \mathbf{B}\odot\mathbf{T}\odot\mathbf{A} \in \mathbb{R}^{(m*t*n)\times k}
$$

$$
\mathbf{G} = \mathbf{B}\odot\mathbf{P}\odot\mathbf{A} \in \mathbb{R}^{(m*m*n)\times k}, \ \mathbf{H} = \mathbf{B}\odot\mathbf{P}\odot\mathbf{T} \in \mathbb{R}^{(m*m*t)\times k}
$$

where $\mathbf{P} \in \mathbb{R}^{m \times k}$ is the auxiliary variable.

Update the vertex factor matrix B while fixing T, A and W. Note that $\mathcal{X}$ is a partially symmetric tensor and the objective function in Eq. (2) involves a fourth-order term w.r.t. $\mathbf{B}$ which is difficult to optimize directly. To obviate this problem, we use a variable substitution technique and minimize the following objective function

$$\min_{\mathbf{B},\mathbf{P}} \ \|\mathbf{X}_{(1)} - \mathbf{B}\mathbf{E}^{\mathrm{T}}\|_F^2$$

$$\text{s.t. } \mathbf{P} = \mathbf{B} \tag{3}$$

The augmented Lagrangian function for problem in Eq. (3) is

$$\mathcal{L}(\mathbf{B},\mathbf{P}) = \|\mathbf{X}_{(1)} - \mathbf{B}\mathbf{E}^{\mathrm{T}}\|_F^2 + \frac{\upsilon}{2}\|\mathbf{B} - \mathbf{P} - \frac{1}{\upsilon}\varUpsilon\|_F^2 \tag{4}$$

where $\varUpsilon \in \mathbb{R}^{m \times k}$ are Lagrange multipliers, υ is the penalty parameter which can be adjusted efficiently according to [19].

By setting the derivative of Eq. (4) w.r.t. $\mathbf{B}$ to zero, we obtain the closed-form solution

$$\mathbf{B} = (2\mathbf{X}_{(1)}\mathbf{E} + \upsilon\mathbf{P} + \varUpsilon)(2\mathbf{E}^{\mathrm{T}}\mathbf{E} + \upsilon\mathbf{I})^{-1} \tag{5}$$

To efficiently compute $\mathbf{E}^{\mathrm{T}}\mathbf{E}$, we consider the following property of the Khatri-Rao product of two matrices [16]

$$\mathbf{E}^{\mathrm{T}}\mathbf{E} = (\mathbf{P} \odot \mathbf{T} \odot \mathbf{A})^{\mathrm{T}}(\mathbf{P} \odot \mathbf{T} \odot \mathbf{A}) = \mathbf{P}^{\mathrm{T}}\mathbf{P} * \mathbf{T}^{\mathrm{T}}\mathbf{T} * \mathbf{A}^{\mathrm{T}}\mathbf{A} \tag{6}$$

Similarly, the auxiliary matrix $\mathbf{P}$ can be optimized successively

$$\mathbf{P} = (2\mathbf{X}_{(2)}\mathbf{F} + \upsilon\mathbf{B} - \varUpsilon)(2\mathbf{F}^{\mathrm{T}}\mathbf{F} + \upsilon\mathbf{I})^{-1} \tag{7}$$

The Lagrange multipliers $\varUpsilon$ can be updated using gradient ascent

$$\varUpsilon \leftarrow \varUpsilon + \upsilon(\mathbf{P} - \mathbf{B}) \tag{8}$$

Update the temporal factor matrix T while fixing B, A and W. Since there is no constraint on $\mathbf{T}$, we directly obtain the closed-form solution

$$\mathbf{T} = (\mathbf{X}_{(3)}\mathbf{G})(\mathbf{G}^{\mathrm{T}}\mathbf{G})^{-1} \tag{9}$$

Update the graph factor matrix A while fixing B, T and W. By variable substitution, we need to minimize the following objective function

$$\min_{\mathbf{A},\mathbf{Q}} \ \|\mathbf{X}_{(4)} - \mathbf{A}\mathbf{H}^{\mathrm{T}}\|_F^2 + \alpha\|\mathbf{D}\mathbf{A}\mathbf{W} - \mathbf{Y}\|_F^2$$

$$\text{s.t. } \mathbf{Q}^{\mathrm{T}}\mathbf{A} = \mathbf{I}, \ \mathbf{Q} = \mathbf{A} \tag{10}$$

The augmented Lagrangian function for problem in Eq. (10) is

$$\mathcal{L}(\mathbf{A}, \mathbf{Q}) = \|\mathbf{X}_{(4)} - \mathbf{A}\mathbf{H}^\mathrm{T}\|_F^2 + \alpha\|\mathbf{D}\mathbf{A}\mathbf{W} - \mathbf{Y}\|_F^2$$

$$+ \frac{\phi}{2}\|\mathbf{A} - \mathbf{Q} - \frac{1}{\phi}\Phi\|_F^2 + \frac{\psi}{2}\|\mathbf{I} - \mathbf{Q}^\mathrm{T}\mathbf{A} - \frac{1}{\psi}\Psi\|_F^2 \qquad (11)$$

where $\Phi \in \mathbb{R}^{n \times k}$ and $\Psi \in \mathbb{R}^{k \times k}$ are Lagrange multipliers, ϕ and ψ are penalty parameters. By setting the derivative of Eq. (11) w.r.t. $\mathbf{A}$ to zero, we obtain the Sylvester equation

$$X\mathbf{A} + \mathbf{A}Y = Z$$
$$X = \psi\mathbf{Q}\mathbf{Q}^\mathrm{T}$$
$$Y = 2\mathbf{H}^\mathrm{T}\mathbf{H} + 2\alpha\mathbf{W}\mathbf{W}^\mathrm{T} + \phi\mathbf{I}$$
$$Z = 2\mathbf{X}_{(4)}\mathbf{H} + 2\alpha\mathbf{D}^\mathrm{T}\mathbf{Y}\mathbf{W}^\mathrm{T} + (\phi + \psi)\mathbf{Q} + \Phi - \mathbf{Q}\Psi \qquad (12)$$

which can be solved by several numerical approaches, *e.g.*, the *lyap* function in MATLAB.

The closed-form update for $\mathbf{Q}$ is

$$\mathbf{Q} = (\psi\mathbf{A}\mathbf{A}^\mathrm{T} + \phi\mathbf{I})^{-1}((\phi + \psi)\mathbf{A} - \Phi - \mathbf{A}\Psi^\mathrm{T}) \qquad (13)$$

The Lagrange multipliers Φ and Ψ can be updated by

$$\Phi \leftarrow \Phi + \phi(\mathbf{Q} - \mathbf{A}), \quad \Psi \leftarrow \Psi + \psi(\mathbf{Q}^\mathrm{T}\mathbf{A} - \mathbf{I}) \qquad (14)$$

Update the weight matrix W while fixing B, T and A. According to the analysis of the $\ell_{2,1}$ norm in [22], we need to minimize the following objective function

$$\mathcal{L}(\mathbf{W}) = \|\mathbf{D}\mathbf{A}\mathbf{W} - \mathbf{Y}\|_F^2 + \gamma\|\mathbf{W}^\mathrm{T}\|_{2,1} = \|\mathbf{D}\mathbf{A}\mathbf{W} - \mathbf{Y}\|_F^2 + \gamma\mathrm{tr}(\mathbf{W}\Omega\mathbf{W}^\mathrm{T}) \quad (15)$$

where $\Omega \in \mathbb{R}^{c \times c}$ is an auxiliary diagonal matrix of the $\ell_{2,1}$ norm. The diagonal elements of Ω are computed as $\Omega(i,i) = \frac{1}{2\sqrt{\|\mathbf{W}(:,i)\|_2^2 + \epsilon}}$ where ϵ is a smoothing term which is usually set to a small constant.

By setting the derivative of Eq. (15) w.r.t. $\mathbf{W}$ to zero, we obtain the Sylvester equation

$$X\mathbf{W} + \mathbf{W}Y = Z$$
$$X = 2\mathbf{A}^\mathrm{T}\mathbf{D}^\mathrm{T}\mathbf{D}\mathbf{A}$$
$$Y = \gamma\Omega$$
$$Z = 2\mathbf{A}^\mathrm{T}\mathbf{D}^\mathrm{T}\mathbf{Y} \qquad (16)$$

Based on the above analysis, we develop the optimization framework for brain network analysis based on tensor factorization, as described in Algorithm 1. The code has been made available at the author's homepage[2].

---

[2] https://www.cs.uic.edu/~bcao1/code/semibat.zip.

Algorithm 1. SEMIBAT

Input: $\mathcal{X}, \mathbf{Y}, \alpha, \lambda$
Output: $\mathbf{B}, \mathbf{T}, \mathbf{A}, \mathbf{W}$
1: Set $v_{max} = \phi_{max} = \psi_{max} = 10^6, \rho = 1.15$
2: Initialize $\mathbf{B}, \mathbf{T}, \mathbf{A}, \mathbf{W} \sim \mathcal{U}(0, 1), \varUpsilon = \varPhi = \varPsi = \mathbf{0}, v = \phi = \psi = 10^{-6}$
3: **repeat**
4:      Update $\mathbf{B}$ and $\mathbf{P}$ by Eq. (5) and Eq. (7)
5:      Update $\mathbf{T}$ by Eq. (9)
6:      Update $\mathbf{A}$ and $\mathbf{Q}$ by Eq. (12) and Eq. (13)
7:      Update $\mathbf{W}$ by Eq. (16)
8:      Update $\varUpsilon$, $\varPhi$ and $\varPsi$ by Eq. (8) and Eq. (14)
9:      $v \leftarrow \min(\rho v, v_{max})$, $\phi \leftarrow \min(\rho \phi, \phi_{max})$, $\psi \leftarrow \min(\rho \psi, \psi_{max})$
10: **until** convergence

3.4 Time Complexity

Each ADMM iteration consists of simple matrix operations. Therefore, rough estimates of its computational complexity can be easily derived [18].

- The estimate for the update of $\mathbf{B}$ according to Eq. (5) is: (1) $O(m^2ntk)$ for the computation of the term $2\mathbf{X}_{(1)}\mathbf{E} + v\mathbf{P} + \varUpsilon$, (2) $O((m+n+t)k^2)$ for the computation of the term $2\mathbf{E}^T\mathbf{E} + v\mathbf{I}$ due to Eq. (6), $O(k^3)$ for its Cholesky decomposition, and (3) $O(mk^2)$ for the computation of the system solution that gives the updated value of $\mathbf{B}$. An analogous estimate can be derived for the update of $\mathbf{P}$ and $\mathbf{T}$ which cost $O(k^3 + (m + n + t)k^2 + m^2ntk)$.
- Considering $l < n$ and c is usually a small constant, the estimate for the update of $\mathbf{A}$ according to Eq. (12) is: (1) $O(n^2k)$ for the computation of the term X, (2) $O((m+n+t)k^2)$ for the computation of the term Y, (3) $O(nk^2+m^2ntk+n^2)$ for the computation of the term Z, and (4) $O(nk^2 + n^2k)$ for the computation of the Sylvester equation [14].
- The estimate for the update of $\mathbf{Q}$ according to Eq. (13) is $O(nk^2 + n^2k + n^3)$.
- The estimate for the update of $\mathbf{W}$ according to Eq. (16) is $O(nk^2 + n^2k)$.

Overall, the updates of all model parameters require $O(k^3 + (m+n+t)k^2 + (m^2nt + n^2)k + n^3)$ arithmetic operations in total.

4 Experiments

4.1 Data Collections

Data were collected from 22 healthy participants at the University of Illinois at Chicago (UIC) and from 11 healthy participants at the University of Michigan (UMich), respectively. Each participant underwent an emotion regulation task, while UIC participants further underwent an eight-minute resting-state recording session which served as unlabeled data. During the ERT session, participants

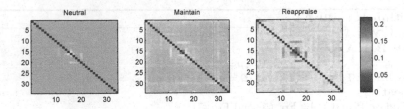

Fig. 4. Average brain networks during NEUTRAL, MAINTAIN and REAPPRAISE. (Color figure online)

were instructed to look at pictures displayed on the screen. Emotionally neutral pictures (*e.g.*, landscape, everyday objects) and negative pictures (*e.g.*, car crash, natural disasters) would appear on the screen for seven seconds in random orders. One second after the picture on display, a corresponding auditory guide would instruct the participant to *look*: viewing the neutral pictures; to *maintain*: viewing the negative pictures as they normally would; or to *reappraise*: viewing the negative pictures while attempting to reduce their emotion response by re-interpreting the meaning of pictures. All EEG data were recorded using the Biosemi system equipped with an elastic cap with 34 scalp channels. A detailed description about data acquisition and preprocessing is available in [28].

Overall, the dataset contains $n = 121$ EEG brain network samples that are based upon $m = 34$ vertices and $t = 130$ time points. The target is to train a classifier on the UIC source (66 training samples and 22 unlabeled samples) to predict which task (NEUTRAL, MAINTAIN, or REAPPRAISE) a subject in the UMich source (33 test samples) is performing. The average brain networks are shown in Fig. 4 where the x and y axes represent the vertex id, and the color of the cell represents the strength of the connection between vertex x and y. Although the group difference appears to be significant, it is non-trivial to identify the tasks for each individual. It will be validated in the experiments that simply using edge values as features to train a classifier could not lead to a good classification performance.

4.2 Compared Methods

The compared methods are summarized as follows:

- SEMIBAT: the proposed semi-supervised brain network analysis approach based on constrained tensor factorization.
- BAT-RIDGE: replacing the $\ell_{2,1}$ norm in SEMIBAT with a regular ridge term.
- BAT-SUPV: a fully supervised variant of SEMIBAT without leveraging the unlabeled data.
- BAT-UNSUPV: an unsupervised variant of SEMIBAT that first learns latent representations of brain networks and then trains a classifier on them in a serial two-step manner.
- BAT-3D: applying SEMIBAT on time-averaged brain networks.

- ALS: plain vanilla tensor factorization using alternating least squares without any constraint [8].
- SUBGRAPH: a discriminative subgraph selection method for uncertain graph classification [7, 17].
- CC: extracting local clustering coefficients as features, one of the most popular graph-theoretical measures that quantify the cliquishness of the vertices [24].
- EDGE: using edge values as features by flatting adjacency matrices of brain networks into vectors.

For a fair comparison, we used a regularized regression in SEMIBAT as the base classifier for all the compared methods. The parameters α and λ were tuned in the range of $2^{-10}, ..., 2^{10}$, the rank k was tuned in the range of $1, ..., 20$. The accuracy with the best parameter configuration was reported, as well as the corresponding precision, recall and F1 score.

4.3   Classification Performance

Experimental results in Table 2 show the classification performance of compared methods on distinguishing the three tasks. EDGE serves as the basis for comparison that treats a brain network as a collection of edges, thereby blinding the connectivity structures of brain networks, which surprisingly outperforms CC. Although clustering coefficients have been widely used to identify Alzheimer's disease [13, 27], they appear to be less useful for distinguishing the emotion regulation tasks. SUBGRAPH achieves a better performance by extracting connectivity patterns within brain networks.

Factorization models demonstrate themselves with significantly better accuracy. According to the low-rank assumption, a low-dimensional latent factor of each graph is obtained by first stacking all the brain network data and then factorizing the constructed tensor. ALS is a direct application of the alternating least

Table 2. Classification performance. N, M and R stand for tasks: NEUTRAL, MAINTAIN and REAPPRAISE, respectively. The best performance on each metric is in bold.

Methods	Evaluation metrics							
	Accuracy	Precision			Recall			F1
		N	M	R	N	M	R	
SEMIBAT	**0.758**	0.833	**0.889**	0.667	**0.833**	0.667	**0.833**	**0.765**
BAT-RIDGE	0.697	**0.909**	0.700	0.600	**0.833**	0.583	0.750	0.706
BAT-SUPV	0.697	0.714	0.750	**0.700**	**0.833**	**0.750**	0.583	0.706
BAT-UNSUPV	0.576	0.818	0.600	0.467	0.750	0.500	0.583	0.588
BAT-3D	0.545	0.857	0.538	0.500	0.500	0.583	0.667	0.559
ALS	0.576	0.750	0.636	0.462	0.750	0.583	0.500	0.588
SUBGRAPH	0.515	0.800	0.500	0.444	0.667	0.333	0.667	0.529
CC	0.364	0.286	0.667	0.391	0.167	0.333	0.750	0.382
EDGE	0.455	0.462	0.700	0.385	0.500	0.583	0.417	0.471

squares technique to the standard tensor factorization problem without incorporating any constraint or supervision. A significant improvement of 31.60 % by SEMIBAT over ALS can be observed, mainly due to the fact that the unsupervised ALS approach fails to interact with the classifier training procedure which shows comparable performance with BAT-UNSUPV. It indicates the importance of addressing the problem (P3) discussed in Sect. 1. Moreover, SEMIBAT outperforms BAT-RIDGE thereby demonstrating that it is critical to apply feature selection in the tensor factorization framework (*i.e.*, the problem (P4)). The advantages of SEMIBAT over BAT-SUPV and BAT-3D are attributed to leveraging unlabeled resting-state brain network data (*i.e.*, the problem (P1)) and modeling the temporal dimension (*i.e.*, the problem (P2)), respectively.

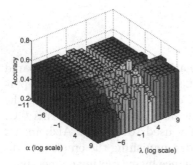

Fig. 5. Sensitivity w.r.t. α and λ.   **Fig. 6.** Sensitivity w.r.t. k.

4.4 Parameter Sensitivity

In all experiments, the regularization parameter λ was tuned for all the baselines, the rank k was tuned for all the factorization models, and α was tuned for SEMIBAT and its variants. We first investigate the influence of α and λ in SEMIBAT and present the results in Fig. 5. It illustrates that neither a small nor a large α or λ would be preferred, and in general, a good choice of α and λ can be found in the range of $2^5, ..., 2^7$ and $2^0, ..., 2^2$, respectively. Moreover, experimental results of factorization models with different k are shown in Fig. 6. In general, a small k would rarely be a wise choice, and the best performance can usually be achieved around $k = 17$.

4.5 Factor Analysis

We first investigate the factor matrices derived from SEMIBAT in a row-wise manner. Note that initially with the best parameter configuration as reported in the last section where $k = 17$, we obtain a 17-dimensional feature vector for each brain network (*i.e.*, $\mathbf{A}(i, :)$) and each time point (*i.e.*, $\mathbf{T}(i, :)$). For visualization we use t-SNE [21] to reduce them into a 2-dimensional space. In Fig. 7, we show the distribution of brain networks, where there are 99 points representing

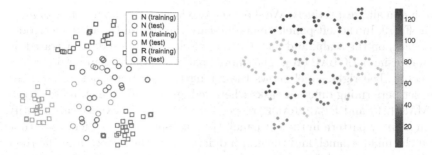

Fig. 7. Embedding of brain networks.   **Fig. 8.** Embedding of time points. (Color figure online)

33 samples from each of the three tasks (22 resting-state samples are omitted). A relatively clear separation between NEUTRAL and REAPPRAISE can be observed, while MAINTAIN usually mix with the other two conditions which make the classification problem challenging. Figure 8 illustrates the distribution of time points, where there are 130 points and each of them represents an exact time point indicated by the color. Basically, adjacent time points are colored similarly. From this figure, we can see that continuous time points form distinct clusters and brain activities change over time, so it is important to capture the temporal dimension explicitly.

Next, we visualize and interpret the factor matrices in a column-wise manner. A k-factor SEMIBAT model extracts the factors $\mathbf{B}(:, i)$, $\mathbf{T}(:, i)$ and $\mathbf{A}(:, i)$, for $i = 1, ..., k$, where these factors indicate the signatures of sources in vertex, time

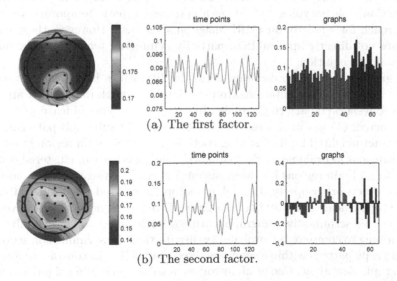

Fig. 9. The two largest factors in terms of magnitude derived from SEMIBAT model for task recognition. (Color figure online)

and graph domain, respectively. We show the two largest factors in terms of magnitude in Fig. 9. In the left panel, points indicate the spatial layout of electrodes (*i.e.*, vertices) on the scalp, and factor values of electrodes are demonstrated on a colormap using EEGLAB [10]. The middle panel shows the temporal changes of the factor. The right panel shows the strength of 66 brain networks in the training set performing different tasks where red, green and blue stand for NEUTRAL, MAINTAIN and REAPPRAISE, respectively. A domain expert can identify the brain activity pattern in the left panel, the corresponding coefficients of time points in the middle panel, and the graph difference on such pattern in the right panel. We can see that different latent factors capture activity of different brain regions. The first factor appears to highlight a quantitative anterior-posterior gradient (maximum values of the first factor appear in the occipital lobe) that is shared across all three conditions, thus may be related to visual processing, while the second factor, which primarily differentiates neutral from maintain and reappraisal, predominantly involves electrodes around the frontal-parietal junction and thus may be related to the late positive potential [11,12].

5   Related Work

Tensor factorization has become an effective technique in many healthcare applications. For example, Acar et al. identify spatial, spectral and temporal signatures of an epileptic seizure as well as an artifact through the application of tensor models [1]. Davidson et al. propose a constrained alternating least squares framework for network discovery of fMRI data [9]. Papalexakis et al. present a scalable solution for the coupled matrix-tensor factorization problem, and find latent variables that jointly explain both the brain activity and the behavioral responses [23]. Wang et al. introduce knowledge guided tensor factorization for computational phenotyping [26]. Ma et al. propose a spatio-temporal tensor kernel approach for whole-brain fMRI image analysis [20]. However, these frameworks are not directly applicable to partially symmetric tensor factorization or further task recognition.

For graph classification on brain networks, literatures have been focused on first deriving vector presentations from the brain network data which are then fed into conventional pattern classifiers. In general, two types of features are usually extracted: (1) graph-theoretical measures and (2) subgraph patterns. Wee et al. extract weighted local clustering coefficients of each brain region in relation to other regions in brain networks to quantify the prevalence of clustered connectivity around brain regions for diagnosis on Alzheimer's disease [27]. In addition to the local network property, Jie et al. use a topology-based graph kernel to measure the topological similarity between paired fMRI brain networks [13]. Kong et al. propose a discriminative subgraph feature selection method based on dynamic programming to compute the probability distribution of discrimination scores for each subgraph pattern within a set of weighted graphs [17]. In contrast to focusing on the graph view alone, Cao et al. introduce a subgraph mining algorithm using side information guidance to find an optimal set of subgraph features for graph classification [6]. However, the expressiveness of these features is limited to the predefined formulations. It is critical to explore a larger space of potentially informative features to represent brain networks through data-driven approaches.

6   Conclusion

This paper presents SEMIBAT, a novel semi-supervised brain network analysis approach based on constrained tensor factorization. It leverages unlabeled resting-state brain networks for task recognition, explores the temporal dimension to capture the progress, incorporates classifier learning procedure to introduce supervision from labeled data, and selects discriminative latent factors for different tasks. ADMM is used to solve the optimization problem. In the experiments on EEG datasets, we demonstrate the superior performance of SEMIBAT on graph classification tasks over the state-of-art methods.

Acknowledgement. This work is supported in part by NSF through grants III-1526499.

References

1. Acar, E., Aykut-Bingol, C., Bingol, H., Bro, R., Yener, B.: Multiway analysis of epilepsy tensors. Bioinformatics **23**(13), i10–i18 (2007)
2. Acar, E., Kolda, T.G., Dunlavy, D.M.: All-at-once optimization for coupled matrix and tensor factorizations (2011). arXiv:1105.3422
3. Ajilore, O., Zhan, L., GadElkarim, J., Zhang, A., Feusner, J.D., Yang, S., Thompson, P.M., Kumar, A., Leow, A.: Constructing the resting state structural connectome. Front. Neuroinformatics **7**, 1427–1437 (2013)
4. Boyd, S., Parikh, N., Chu, E., Peleato, B., Eckstein, J.: Distributedoptimization and statistical learning via the alternating direction method ofmultipliers. Found. Trends. Mach. Learn. **3**(1), 1–122 (2011)
5. Bzdok, D., Eickenberg, M., Grisel, O., Thirion, B., Varoquaux, G.: Semi-supervised factored logistic regression for high-dimensional neuroimaging data. In: NIPS, pp. 3330–3338 (2015)
6. Cao, B., Kong, X., Zhang, J., Yu, P.S., Ragin, A.B.: Mining brain networks using multiple side views for neurological disorder identification. In: ICDM, pp. 709–714. IEEE (2015)
7. Cao, B., Zhan, L., Kong, X., Yu, P.S., Vizueta, N., Altshuler, L.L., Leow, A.D.: Identification of discriminative subgraph patterns in fMRI brain networks in bipolar affective disorder. In: Brain Informatics and Health, pp. 105–114. Springer, Heidelberg (2015)
8. Comon, P., Luciani, X., De Almeida, A.L.: Tensor decompositions, alternating least squares and other tales. J. Chemometr. **23**(7–8), 393–405 (2009)
9. Davidson, I., Gilpin, S., Carmichael, O., Walker, P.: Network discovery via constrained tensor analysis of fMRI data. In: KDD, pp. 194–202. ACM (2013)
10. Delorme, A., Makeig, S.: EEGLAB: an open source toolbox for analysis of single-trial EEG dynamics including independent component analysis. J. Neurosci. Methods **134**(1), 9–21 (2004)

11. Dennis, T.A., Hajcak, G.: The late positive potential: a neurophysiological marker for emotion regulation in children. J. Child Psychol. Psychiatry **50**(11), 1373–1383 (2009)
12. Hajcak, G., MacNamara, A., Olvet, D.M.: Event-related potentials, emotion, and emotion regulation: an integrative review. Dev. Neuropsychol. **35**(2), 129–155 (2010)
13. Jie, B., Zhang, D., Gao, W., Wang, Q., Wee, C., Shen, D.: Integration of network topological and connectivity properties for neuroimaging classification. Biomed. Eng. **61**(2), 576 (2014)
14. Jonsson, I., Kågström, B.: Recursive blocked algorithms for solving triangular systems part i: One-sided and coupled sylvester-type matrix equations. ACM Trans. Math. Softw. **28**(4), 392–415 (2002)
15. Kim, J., Calhoun, V.D., Shim, E., Lee, J.H.: Deep neural network with weight sparsity control and pre-training extracts hierarchical features and enhances classification performance: Evidence from whole-brain resting-state functional connectivity patterns of schizophrenia. NeuroImage **124**, 127–146 (2016)
16. Kolda, T.G., Bader, B.W.: Tensor decompositions and applications. SIAM Rev. **51**(3), 455–500 (2009)
17. Kong, X., Yu, P.S., Wang, X., Ragin, A.B.: Discriminative feature selection for uncertain graph classification. In: SDM (2013)
18. Liavas, A.P., Sidiropoulos, N.D.: Parallel algorithms for constrained tensor factorization via the alternating direction method of multipliers (2014). arXiv:1409.2383
19. Lin, Z., Liu, R., Su, Z.: Linearized alternating direction method with adaptive penalty for low-rank representation. In: NIPS, pp. 612–620 (2011)
20. Ma, G., He, L., Lu, C.T., Yu, P.S., Shen, L., Ragin, A.B.: Spatio-temporal tensor analysis for whole-brain fMRI classification. In: SDM. SIAM (2016)
21. Van der Maaten, L., Hinton, G.: Visualizing data using t-sne. J. Mach. Learn. Res. **9**(2579–2605), 85 (2008)
22. Nie, F., Huang, H., Cai, X., Ding, C.H.: Efficient and robust feature selection via joint 2, 1-norms minimization. In: NIPS, pp. 1813–1821 (2010)
23. Papalexakis, E.E., Mitchell, T.M., Sidiropoulos, N.D., Faloutsos, C., Talukdar, P.P., Murphy, B.: Turbo-smt: Accelerating coupled sparse matrix-tensor factorizations by 200x. In: SDM. SIAM (2014)
24. Rubinov, M., Sporns, O.: Complex network measures of brain connectivity: uses and interpretations. Neuroimage **52**(3), 1059–1069 (2010)
25. Sporns, O., Tononi, G., Kötter, R.: The human connectome: a structural description of the human brain. PLoS Comput. Biol. **1**(4), e42 (2005)
26. Wang, Y., Chen, R., Ghosh, J., Denny, J.C., Kho, A., Chen, Y., Malin, B.A., Sun, J.: Rubik: Knowledge guided tensor factorization and completion for health data analytics. In: KDD, pp. 1265–1274. ACM (2015)
27. Wee, C.Y., Yap, P.T., Zhang, D., Denny, K., Browndyke, J.N., Potter, G.G., Welsh-Bohmer, K.A., Wang, L., Shen, D.: Identification of MCI individuals using structural and functional connectivity networks. Neuroimage **59**(3), 2045–2056 (2012)
28. Xing, M., Tadayonnejad, R., MacNamara, A., Ajilore, O., Phan, K.L., Klumpp, H., Leow, A.: EEG based functional connectivity reflects cognitive load during emotion regulation. In: ISBI. IEEE (2016)
29. Ye, J., Chen, K., Wu, T., Li, J., Zhao, Z., Patel, R., Bae, M., Janardan, R., Liu, H., Alexander, G., Reiman, E.: Heterogeneous data fusion for Alzheimer's disease study. In: KDD. pp. 1025–1033. ACM (2008)
30. Zhang, J., Cao, B., Xie, S., Lu, C.T., Yu, P.S., Ragin, A.B.: Identifying connectivity patterns for brain diseases via multi-side-view guided deep architectures. In: SDM. SIAM (2016)

Scalable Hyperparameter Optimization with Products of Gaussian Process Experts

Nicolas Schilling[✉], Martin Wistuba, and Lars Schmidt-Thieme

Information Systems and Machine Learning Lab,
Universitätsplatz 1, 31141 Hildesheim, Germany
{schilling,wistuba,schmidt-thieme}@ismll.uni-hildesheim.de

Abstract. In machine learning, hyperparameter optimization is a challenging but necessary task that is usually approached in a computationally expensive manner such as grid-search. Out of this reason, surrogate based black-box optimization techniques such as sequential model-based optimization have been proposed which allow for a faster hyperparameter optimization. Recent research proposes to also integrate hyperparameter performances on past data sets to allow for a faster and more efficient hyperparameter optimization.

In this paper, we use products of Gaussian process experts as surrogate models for hyperparameter optimization. Naturally, Gaussian processes are a decent choice as they offer good prediction accuracy as well as estimations about their uncertainty. Additionally, their hyperparameters can be tuned very effectively. However, in the light of large meta data sets, learning a single Gaussian process is not feasible as it involves inversion of a large kernel matrix. This directly limits their usefulness for hyperparameter optimization if large scale hyperparameter performances on past data sets are given.

By using products of Gaussian process experts the scalability issues can be circumvented, however, this usually comes with the price of having less predictive accuracy. In our experiments, we show empirically that products of experts nevertheless perform very well compared to a variety of published surrogate models. Thus, we propose a surrogate model that performs as well as the current state of the art, is scalable to large scale meta knowledge, does not include hyperparameters itself and finally is even very easy to parallelize. The software related to this paper is available at https://github.com/nicoschilling/ECML2016.

Keywords: Hyperparameter optimization · Sequential model-based optimization · Product of experts

1 Introduction

In recent years, machine learning and data mining has been gaining more and more attention by showing very good prediction performance in areas such as recommender systems, pattern, speech and visual object recognition and many

© Springer International Publishing AG 2016
P. Frasconi et al. (Eds.): ECML PKDD 2016, Part I, LNAI 9851, pp. 33–48, 2016.
DOI: 10.1007/978-3-319-46128-1_3

more. The lift in prediction performance is usually due to the development of more complex models as we see for example in the area of deep learning. However, developing more complex models usually has drawbacks, which is the increasing time that is spent for learning the model plus the increasing dimensionality of the hyperparameter space of the associated model. By hyperparameters we denote parameters of a model that can not explicitly be learned from the data by a well-defined optimization criterion such as the minimization of a regularized loss functional. These hyperparameters can be continuous, the reader might consider a positive learning rate of a gradient descent optimization approach, or a regularization constant of a Tikhonov regularization term. However, by hyperparameters we also consider discrete choices, such as the dimensionality of a low-rank factorization or the number of nodes and layers in a deep feedforward neural network. Additionally, hyperparameters can also be categorical, for instance the choice of kernel function in a support vector machine, or even the choice of loss function to optimize within the optimization criterion. Finally, even model choice as well as preprocessing of the data can be understood as hyperparameters of a general learner. What all of these parameters have in common is that they cannot be optimized in a straightforward fashion, but usually their correct setting renders methods from producing weak predictions to state-of-the-art predictions. Due to this impact, practicioners that do not know the underlying techniques very well usually have a hard time optimizing hyperparameters and therefore rely on either choosing standard hyperparameters or on performing a grid-search, which tries many hyperparameters and in the end chooses the one that performs best. In this way, a lot of unnecessary computations are created.

Out of this reason, recent research proposes to use black-box optimization techniques such as sequential model-based optimization (SMBO) to allow for a more directed search in the hyperparameter space. Essentially, SMBO treats the hyperparameter configuration as input for a black box function and uses a surrogate model to learn on a few observed performances to then predict the performance of any arbitrary hyperparameter configuration. The predicted performance as well as the uncertainty of the surrogate model are then used within the context of an acquisition function to finally predict a hyperparameter configuration that likely performs better, while keeping a good balance between exploitation and exploration. On the one hand, exploitation is attained whenever the acquisition function chooses hyperparameter configurations that are very close to already observed well-performing configurations and therefore the surrogate model is quite certain about its estimation. On the other hand, exploration is met if the acquisition function chooses configurations that are very distant to all observed configurations, i.e. explores new areas of the hyperparameter space, where the surrogate model is quite uncertain about its prediction. Given that usually only a few initial observations are present and the amount of overall queries for hyperparameter configurations is limited, a decent tradeoff between both exploration and exploitation is desired.

More recent work is inspired by the area of meta learning, where the goal is to transfer knowledge for parameters of a given model from having learned this model already on other data sets [4]. Thus, these methods propose to also take into account the knowledge of hyperparameter performances on different (past) data sets, where hyperparameter opimization has already been done. This is quite intuitive, as every experienced practitioner, who has already learned a model many times on different data sets probably comes up with better hyperparameter configurations for the target data set to test initially. In many works, the surrogate model is then learned on the hyperparameter performances of past data sets and therefore has a better knowledge of well-performing hyperparameters to choose. In order for the surrogate model to not confuse performances of the same hyperparameter configuration on different data sets, the meta knowledge is usually augmented by additional meta features that describe characteristics of a data set.

Many surrogate models have been proposed, but one of the simplest surrogates is probably a Gaussian process (GP), as it is relatively simple to learn, delivers good predictions and furthermore, due to its probabilistic nature, allows for a direct estimation of uncertainties, which is a key ingredient for SMBO. Another advantage of using Gaussian processes compared to other surrogate models is that they are basically hyperparameter free, as all the parameters that we have to specify for the kernel can be learned by optimizing their marginal log likelihood. However, Gaussian processes have one huge drawback which lies in their scalability. In order to learn a Gaussian process, the kernel matrix computed over all observed instances has to be inverted which is an operation with cubic expense in the number of observations. Thus, if we seek to include meta knowledge of many past data sets into the training data of the Gaussian process, learning the Gaussian process might even take more time and memory than learning the model we seek to optimize the hyperparameters for, which then renders a Gaussian process infeasible, despite its advantages.

In this paper, we propose to use a product of Gaussian process experts as surrogate model, where basically an independent GP is learned for all the observation of one past data set and in the end all the predictions of the individual experts are assembled to predict hyperparameter performances of the target data set. Following this approach, our work has four main contributions:

▶ We learn a product of GP experts, which allows for the inclusion of a large amount of meta information,
▶ by using GPs as base surrogate model, we employ surrogates that are very easy and fast to learn, and do not require much memory
▶ additionally, by using GPs, we do not introduce additional surrogate-hyperparameters in opposition to many state of the art methods,
▶ finally, we show empirically that products of GP experts perform very competitively for hyperparameter optimization against a variety of published competitors, as well as make both the implementation and the meta data publicly available.

2   Related Work

As already mentioned, in the recent years there has been a growing interest in research regarding hyperparameter optimization. Random search has been proposed as an alternative to grid-search and works well in cases of low effective dimensionality, where a subspace of the hyperparameter space does not influence the results as much as the remaining hyperparameter dimensions [3].

In the context of SMBO, many different surrogate models have been proposed in a variety of papers. At first, an independent Gaussian process [17] was used. We denote it as independent as it does not learn across data [20]. Secondly, random forests have been proposed as surrogates and inherit the ability to work well with non numerical as well as hierarchical hyperparameters [13]. Regarding hyperparameter optimization using meta knowledge, a stacking of a GP on top of a ranking SVM was proposed [2], as well as a Gaussian process with a multi task kernel in two closely related works [21,26]. Furthermore, a mixture of a multilayer perceptron and a factorization machine has been employed as surrogate model [18], which automatically learns data set representations and therefore does not necessarily need meta features.

A different aspect of using meta knowledge is conducted through learning an initialization of well-performing hyperparameters. The first work in this context is [8] where the initial hyperparameters are chosen based on data sets that are closest with respect to the Euclidean distance evaluated on the meta features of the respective data sets. This intuition has been extended by [24] which uses a differentiable plug in estimator to compute initial hyperparameters. Finally, [25] employs a static sequence of hyperparameters that is learned using meta knowledge and does not need a surrogate model at all, however, it has the drawback that it needs meta information over different data sets evaluated on the same hyperparameter grids.

There is a plethora of other approaches that are either model specific [1] or use genetic algorithms [7,15], or do both in conjunction [9]. As these approaches are not embedded in the context of SMBO, we will leave them out of further discussions.

Since we are seeking to employ product of experts models in the framework of SMBO-based hyperparameter optimization, we also review the related work in this field as well as various techniques to speed up Gaussian process learning. Initially, product of experts models have been proposed by [11] alongside with a learning algorithm [12] to train the parameters of such a model. The *generalized* product of experts [5] introduces additional weighting factors within the product in order to reduce the overconfidence of the product of experts in unknown areas. Another model that also estimates a joint probability density given by a set of experts is the Bayesian committee machine [22], which includes the prior in its predictions. Finally, the work by [6] combines both the idea of the generalized POE with its weighting factors with the Bayesian committee machine. We do want to highlight that all of this work is not specifically tailored to Gaussian processes, however, [6] argues that using products of experts is an easy way to make Gaussian processes more scalable to larger training data sets.

Additionally, many efforts have been made by the means of sparse GPs, namely Gaussian processes learned on subsets of the original training data such as [19] which employs kd-trees for subsampling. There are many more works in this area such as [10, 23] or [16], however, as we want to make use of the rich meta information of hyperparameter performance on other data sets using only a subset of the meta information seems counterintuitive. Due to this reason, we do not intend to use sparse GPs as surrogates for Bayesian hyperparameter optimization.

3 Background

In this section we first review hyperparameter optimization and sequential model-based optimization in general, secondly, we discuss Gaussian processes shortly and lastly we give a review of product of experts models which we ultimately seek to employ as surrogate models.

3.1 Problem Setting

Let $\mathcal{D}$ denote by the space of all data sets, following the notation by [3], we denote a learning algorithm for a fixed model class $\mathcal{M}$ by a mapping $\mathcal{A} : \Lambda \times \mathcal{D} \longrightarrow \mathcal{M}$. Thus, an algorithm $\mathcal{A}$ is essentially a mapping from a given hyperparameter configuration and training data to a model which is learned by minimizing a loss functional. In many cases, the hyperparameter space Λ is the cartesian product of lower dimensional spaces. Now we can define the problem of *hyperparameter optimization* as choosing the hyperparameter configuration λ^* which minimizes the loss of a learned model on given validation data:

$$\lambda^* := \arg\min_{\lambda \in \Lambda} \mathcal{L}(\mathcal{A}(\lambda, D^{\mathrm{train}}), D^{\mathrm{val}}) =: \arg\min_{\lambda \in \Lambda} b(\lambda, D). \tag{1}$$

Please note that we use the short b as notation for the process of learning a model on training data with given hyperparameters and evaluating it on validation data. Clearly, b is the black box function that we seek to optimize using Bayesian optimization.

3.2 Sequential Model-Based Optimization

The SMBO framework is depicted in Algorithm 1. It starts by learning a surrogate model denoted by Ψ such that $\Psi \approx b$ on a set of given hyperparameter performances which are stored in the observation history $\mathcal{H}$. Secondly, the surrogate model will be used to predict the hyperparameter performance of unknown hyperparameters, these predictions as well as the uncertainties will be forwarded to the acquisition function a, which then picks a hyperparameter configuration to test. The most commonly used acquisition function is Expected Improvement (EI) and can be computed analytically if one assumes the probability of improvement to be Gaussian [14]. Having chosen a candidate configuration, b

---

Algorithm 1. Sequential model-based optimization across data sets

---

Input: Hyperparameter space Λ, observation history $\mathcal{H}$, target data set D, number of iterations T, acquisition function a, surrogate model Ψ, initial best hyperparameter configuration λ^{best}.
Output: Best hyperparameter configuration λ^{best} for D
1: **for** $t = 1$ to T **do**
2:     Fit Ψ to $\mathcal{H}$
3:     $\lambda^{\text{new}} = \arg\max_{\lambda \in \Lambda} a\left(\Psi(\lambda, D), \mathcal{H}\right)$
4:     Evaluate $b\left(\lambda^{\text{new}}, D\right)$
5:     **if** $b(\lambda^{\text{new}}, D) < b(\lambda^{\text{best}}, D)$ **then**
6:         $\lambda^{\text{best}} = \lambda^{\text{new}}$
7:     $\mathcal{H} = \mathcal{H} \cup \left(\lambda^{\text{new}}, b\left(\lambda^{\text{new}}, D\right)\right)$
8: **return** λ^{best}

---

will be evaluated for the proposed hyperparameter configuration, the result will be fed into the observation history and the process is repeated for T many times until finally a best hyperparameter configuration λ^{best} is found. Additionally, the surrogate model's feature vector is usually augmented by meta features, which are descriptive features of a data set, to allow the surrogate model to distinguish between different data sets.

3.3 Gaussian Processes

We introduce Gaussian processes as we use them as base models in a product of experts. Given is a regression problem of the form

$$y(x) = f(x) + \epsilon, \tag{2}$$

where we assume i.i.d. noise $\epsilon \sim \mathcal{N}(0, \sigma^2)$. A Gaussian process assumes that for a given set of input variables $X = (x_1, ..., x_N)$ with associated labels $y = (y_1, ..., y_N)$ the labels are multivariate Gaussian distributed $y \sim \mathcal{N}(0, K)$, where K is a covariance matrix that is defined through a positive semidefinite kernel function $k(x, x')$. A very common choice for k is the squared exponential kernel

$$k(x, x') = \exp\left(\frac{-\|x - x'\|^2}{2\sigma_l^2}\right) + \sigma^2\delta(x = x'), \tag{3}$$

where $\theta = (\sigma_l, \sigma)$ are denoted as the hyperparameters and the δ function returns 1 if its predicate is true and 0 otherwise. Given a set of known observations, the conditional distribution of a label $f_\star$ given its input $x_\star$ is Gaussian distributed with mean and covariance

$$\mu(f_\star) = k_\star^\top K^{-1} y \tag{4}$$

$$\sigma^2(f_\star) = k_{\star\star} - k_\star^\top K^{-1} k_\star, \tag{5}$$

where $k_\star = (k(x_1, x_\star), ..., k(x_n, x_\star))$ is the vector of kernel evaluations of the new input $x_\star$ to all observed inputs and $k_{\star\star} = k(x_\star, x_\star)$ is the prior covariance of $f_\star$.

As we can see, using a GP for predictions requires inverting the kernel matrix of size N, which is an operation of $\mathcal{O}(N^3)$ and thus becomes infeasible for data sets with many instances. Recalling that our primary goal was to include large scale meta knowledge of past hyperparameter performances, this boundary might be reached very soon, which would force us to throw valuable data away or rely on other surrogate models. However, solving the linear system of equations for inversion of K can be reduced to $\mathcal{O}(N^2)$ by using a Cholesky decomposition of the kernel matrix [17].

The kernel hyperparameters can be learned by maximizing their marginal log likelihood using standard optimization techniques such as gradient ascent. Again, for optimizing the kernel hyperparameters, we have to invert the kernel matrix as well as compute its determinant, which also both scale cubically in the dimension of K. As gradient ascent might use several iterations to converge to a useful θ, this inversion becomes even more the bottle neck with respect to both computational speed as well as memory usage.

3.4   (Generalized) Product of Experts (POE)

In order to scale Gaussian processes to a large training data set (i.e. observation history) we will use product of experts models [11], of which several variants have been proposed. Within a product of GP experts, a set of M invididual Gaussian processes are learned on M disjoint subsets of the training data, so let us decompose our training data as

$$X = (X^{(1)}, ..., X^{(M)}) \qquad y = (y^{(1)}, ..., y^{(M)}), \tag{6}$$

such that the individual subsets of instances and labels are disjoint. Then, following the independence assumption, the marginal joint likelihood factorizes into a product of single likelihoods

$$p(y \,|\, X, \theta) = \prod_{i=1}^{M} p_i \left(y^{(i)} \,|\, X^{(i)}, \theta^{(i)} \right). \tag{7}$$

Thus, in order to learn the individual experts, we only need to invert kernel matrices of the size of roughly N/M, thus learning the individual experts can be done in $\mathcal{O}(N^3/M^3)$ which is a reasonable reduction for a sufficiently large enough M. In this way, we also learn M many different sets of kernel hyperparameters.

As the M experts have been learned, we can compute the marginal likelihood by multiplying all individual likelihoods. The *generalized* product of experts [5] introduces additional weighting factors β_i such that:

$$p(y \,|\, X, \theta) = \prod_{i=1}^{M} p_i^{\beta_i} \left(y^{(i)} \,|\, X^{(i)}, \theta^{(i)} \right). \tag{8}$$

Naturally, if all $\beta_i = 1$, we arrive at the initial formulation of Eq. 7. Computing the product of the individual likelihoods yields a density that is proportional to a Gaussian with following mean and precision:

$$\mu^{\text{poe}}(f_\star) = (\sigma^{\text{poe}}(f_\star))^2 \sum_{i=1}^{M} \beta_i \sigma_i^{-2}(f_\star)\mu_i(f_\star) \tag{9}$$

$$(\sigma^{\text{poe}}(f_\star))^{-2} = \sum_{i=1}^{M} \beta_i \sigma_i^{-2}(f_\star) \tag{10}$$

Essentially, by replacing $\sigma_i^{-2}(f_\star) = \tau_i(f_\star)$ with the precision, we see that the mean predicted by the product of experts is a sum of means, weighted by the product of the individual β_i and the precision τ_i, which is then divided by the total sum of weighting factors. Usually, the β_i are set such that $\sum_i \beta_i = 1$, surprisingly, this already works quite well for $\beta_i \equiv 1/M$. This does not change the mean as the multiplication with the precision cancels out the effect, however, the precisions effectively get weighed down and decrease the overconfidence of the initial product of experts without any weights.

3.5   Product of Experts in SMBO

Having introduced the product of experts models, their implementation for hyperparameter optimization in the SMBO framework seems straightforward. However, a few questions still remain unanswered. At first, we split the meta knowledge into all the instances belonging to hyperparameter performances of one data set. In this way, each expert will be learned on the meta information of one distinct data set. If this would still be too large for a GP to learn, we could further subdivide them into smaller subsets.

Secondly, the question of how the information on the target data set will be incorporated into the surrogate model remains. We seek for two alternatives, in the first one we simply add the information of new points on the target data set to all the experts in the ensemble. Doing this, we effectively train all experts to be expert for two data sets, the initial one they have been trained on plus the target data set. In our implementation, we then follow the intution of all weights summing up to one, thus we set all $\beta_i = 1/M$.

As an independent Gaussian process that is learned without any meta knowledge already behaves reasonably well as a surrogate model, we also tried another alternative. We still feed the target data set information into all experts learned in the ensemble but additionally create a new GP that carries only the information of the target data set and is weighed much higher than the individual experts. Specifically, we use $\beta_i = 1/2M$ for the individual experts and $\beta_{M+1} = 1/2$ for the GP learned on only the target data set responses. In this way, we use the meta information as well as the strength of an independent Gaussian process.

Additionally, we seek to scale the hyperparameter performances observed in the meta data as well as the hyperparameter performances of the target data set. This is due to the fact that the range of b naturally depends on the data set. Consider for example a classification problem where b models the misclassification rate of a classifier for some test data. Naturally, for some data sets, very low misclassification rates might be achieved in contrast to other data sets that are simply harder to classify. A POE might then be biased towards choosing hyperparameter configurations that produce good results on simple data sets, which is something we want to prevent. In order to do so, we scale the labels of the meta data to become standard Gaussian distributed, for the target data set, we do this on-the-fly every time we see a new response of b as was also proposed by [26].

4 Experiments

To evaluate the proposed surrogate models for hyperparameter optimization, we conduct hyperparameter optimization within the SMBO framework including a variety of published baselines. The experiments are performed on two meta data sets that we have created ourselves.

4.1 Meta Data Set Creation

We have created two meta data sets for the task of classification using two distinct classifiers, namely being a support vector machine (SVM) and AdaBoost. These meta data sets consists of a complete grid search for both classifiers on 50 classification data sets that we have taken from the UCI repository[1]. If splits were already given, we merged them into one complete data set, shuffled the resulting data set and then took 80 % of the data for training and the remaining 20 % for testing. The AdaBoost meta data set was created by running AdaBoost[2] with hyperparameters $I \in \{2, 5, 10, 20, 50, 100, 200, 500, 1000, 2000, 5000, 10000\}$ and $M \in \{2, 3, 4, 5, 7, 10, 15, 20, 30\}$. This yields 108 meta instances per data set and therefore the overall meta data set contains 5400 instances.

The second meta data set was created by running an SVM[3] on all of the data sets, with four hyperparameters. The first one resembles the choice of kernel and is categorical between a linear, a polynomial and an RBF kernel, thus introduces three binary hyperparameters. The second hyperparameter is the tradeoff parameter, usually denoted as C, the third and fourth hyperparameter are the degree d of the polynomial kernel and the width γ of the RBF kernel. If the kernel hyperparameters are not used, i.e. the polynomial degree for an RBF kernel, we set them to a constant value of zero. As for the AdaBoost meta data set, we computed the misclassification rates using grid-search, where C was chosen from the set $\{2^{-5}, \ldots, 2^6\}$, the polynomial degree d was chosen from $\{2, \ldots, 10\}$ and γ was chosen from $\{0.0001, 0.001, 0.01, 0.05, 0.1, 0.5, 1, 2, 5, 10, 20, 50, 100, 1000\}$. This results in 288 runs per data set, and therefore the overall meta data set contains up to $14, 400$ instances.

Finally, we also added meta features to the meta data set, to allow the surrogate models to distinguish between the same hyperparameter configurations

---

[1] http://archive.ics.uci.edu/ml/index.html.
[2] http://www.multiboost.org.
[3] http://svmlight.joachims.org.

Table 1. List of all meta-features used.

Number of Classes	Log Inverse Data Set Dimensionality	Kurtosis Mean
Number of Instances	Class Cross Entropy	Kurtosis Standard Deviation
Log Number of Instances	Class Probability Min	Skewness Min
Number of Features	Class Probability Max	Skewness Max
Log Number of Features	Class Probability Mean	Skewness Mean
Data Set Dimensionality	Class Probability Standard Deviation	Skewness Standard Deviation
Log Data Set Dimensionality	Kurtosis Min	
Inverse Data Set Dimensionality	Kurtosis Max	

evaluated on different data sets. A list of all employed meta features can be seen in Table 1. For our experiments, all features in the meta data set, namely the computed meta features as well as the hyperparameter configurations have been scaled to values in $[0, 1]$.

4.2   Competing Surrogate Models

Random Search (RANDOM). This is a surrogate that simply picks a random point out of the grid.

Random Forests (RF). Sequential Model-based Algorithm Configuration [13] employs a random forest as surrogate model and computes uncertainties using the learned ensemble by estimating empirical means and standard deviations.

Independent Gaussian Process (IGP). An independent Gaussian process with SE-ARD kernel that is only learned on the observations on the target data set, this was proposed by [20].

Surrogate-based Collaborative Tuning (SCOT). This surrogate model is effectively a stacking of an SVM^{RANK} and a Gaussian process and was proposed by [2]. The ranking SVM learns how to rank hyperparameter configurations across data sets, uncertainties are estimated by stacking a GP on the ranked output.

Full Gaussian Process (FGP). A Gaussian process with SE-ARD kernel that is learned on the whole meta data set. This is basically the model we seek to approximate by learning a product of experts.

Gaussian Process with MKL (MKLGP). This surrogate was proposed by [26] and learns basically a full GP over the whole meta data set using a combination of an SE-ARD kernel and a kernel function that models the distances between data sets based on meta features.

Factorized Multilayer Perceptron (FMLP). The surrogate model that was proposed by [18], which learns a multilayer perceptron and factorizes the weights in the first layer in order to learn latent data set and hyperparameter representations. Uncertainties are estimated by learning an ensemble of FMLPs.

Product of Gaussian Process Experts (POGPE). This surrogate model learns a product of GP experts as described in Sect. 3.4. Each expert employs an SE-ARD kernel. Information of the target data set is distributed to all experts, which are all weighted equally.

Single Gaussian Process Expert (SGPE). This surrogate model also learns a product of GP experts as described in Sect. 3.4, however, also learns an independent GP for the target data set only and weighs the target GP as much as the whole set of experts.

4.3   Experimental Setup

Our experiments are performed in a leave-one-out fashion, meaning that we train the surrogate model on 49 data sets and use the meta knowledge to start SMBO on the remaining test data set. To cancel out random effects, we ran all experiments for a total of 100 times and averaged the results in the end. In total, each SMBO run was allowed to test $T = 70$ different hyperparameter configurations on the test data. As acquisition function we employed the popular expected improvement, which is by now the most widely used acquisition function in hyperparameter optimization using the SMBO framework.

   As evaluation metric, we use the average rank, where, for each target data set, we rank all competing surrogate models based on the best misclassification rate they have found so far. Ties are being solved by granting the average rank, i.e. if one surrogate models find the misclassification rates 0.2, another two find 0.25 and a third one finds only 0.5, we would rank the surrogates with 1, 2.5, 2.5 and 4. As we run the experiments for 50 different target data sets, we report the average of all average ranks.

   The implementations were largely done by ourselves, except for SMAC and SCOT, where we used MLTK[4] for the former and the implementation by Joachims[5] for the ranking SVM used in SCOT. All hyperparameters of the GP based models have been automatically tuned by maximizing their marginal likelihood, for FMLP we used the setting proposed by the authors. For SMAC, SCOT and MKLGP we used leave-one-out cross validation to tune the hyperparameters. For all GP-based models, we implemented the Cholesky decomposition to speed up the inversion of kernel matrices. In order to facilitate reproducibility of our experimental results, we make the program code as well as the employed meta data sets publicly available on Github[6].

---

[4] http://www.cs.cornell.edu/~yinlou/projects/mltk/.
[5] http://svmlight.joachims.org/.
[6] https://github.com/nicoschilling/ECML2016.

4.4 Performance in SMBO

The average rank among all competing methods can be seen in Fig. 1, where
the left plot shows the average rank on AdaBoost versus the number of trials
conducted, and the right one shows the results for the SVM meta data set. First
of all, we see that for both meta data sets the random baseline shows the worst
performance as is expected. Surprisingly, for both meta data sets, POGPE and
SGPE find the best hyperparameter among the competitors in the first trial.
During the SMBO procedure, both the full Gaussian process as well as MKLGP
perform better on the AdaBoost data set, however, we observe that POGPE
performs better on the SVM data set which is quite a surprise. POGPE, despite
its good starting point, is being outperformed by FMLP on the SVM data set
in the first 15 trials, which then degrades in performance and performs worse
than both full GP approaches. Comparing both of these with each other, we
see that they perform almost equally, however, MKLGP tends to have worse
starting points than a simple full GP. For both meta data sets, we see the lift of
including meta knowledge through comparison with the independent GP, which
performs reasonably on AdaBoost but degrades on SVM. This observation leads
us to the conclusion that optimizing the hyperparameters of AdaBoost seems an
easier task than on SVM.

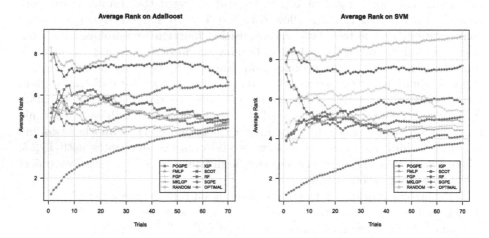

Fig. 1. Average Rank of all competing methods. The left plot shows results for
AdaBoost, the right plot shows results for the SVM meta data set.

In contrast to POGPE, SGPE does not seem to perform that well, maybe the
tradeoff between product of experts and single GP has to be adjusted for each
trial, however, this would introduce another hyperparameter for the surrogate
model, which we do not seek to do.

Overall, we conclude that POGPE, FMLP, FGP and MKLGP are among
the best performing surrogate models, so for these models we also computed the

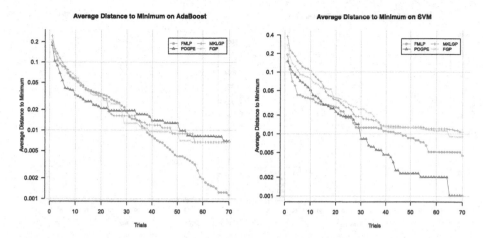

Fig. 2. Average Distance to Minimum of the best performing methods. The left plot shows results for AdaBoost, the right plot shows results for the SVM meta data set.

average distance to the minimum in terms of b. For each data set, we scale all values of b (in this case accuracies) to be in $[0, 1]$. Then, again for each data set, we compute the distance of the best hyperparameter response so far to the best on the overall grid. This value is then averaged to become the average distance to the minimum, which gives an idea of how a surrogate model makes use of the responses it gets on the target data set. The results can be seen in Fig. 2, where the left plot shows the results for AdaBoost and the right for SVM. Overall, we see the same behaviour as we have seen in average rank, however, FMLP seems to be a little bit better here. For AdaBoost, it achieves the lowest average distance, this is due to FMLP winning severely against its competitors on the data set *sonar-scale*, where in the average rank, this win does not count that much. In conclusion, we see that POGPE works really well on both evaluatuion metrics, especially when we consider its simplicity in the light of POGPE actually being an approximation of a full GP.

4.5   Runtime Experiments

In order to demonstrate the scalability of using POGPE, we have also conducted a runtime experiment. We have measured the runtime of the most competitive methods, namely being POGPE, FMLP, and both of the full Gaussian process approaches FGP and MKLGP. Experiments were conducted on a Xeon E5-2670v2 with 2.50 GHz clock speed and 64 GB of RAM, where we again performed a total of 70 trials of an SMBO run on the SVM meta data set. To account for measurement noise, we repeated all experiments 10 times.

The results can be seen in Fig. 3, where the left plot shows the cumulative runtime in seconds without the initial training time of the surrogate in opposition to the right plot which includes it. We plot both results as the training of the surrogate model can be performed in an offline fashion while waiting for new

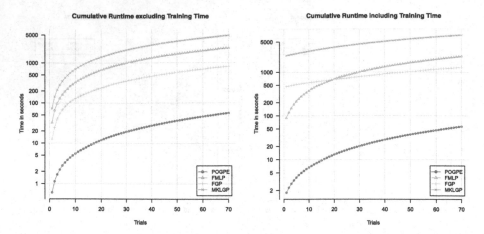

Fig. 3. Runtime comparison among the most competitive surrogate models. The left plot shows the cumulative runtime in seconds.

data. As we do not take into account the learning of the actual model, i.e. the evaluation of b, both plots can be understood as the total overhead time of running SMBO instead of using default hyperparameters. We can observe that POGPE consumes drastically less time than all its competitors, simply due to the fact that we have to invert much smaller kernel matrices. By excluding the potentially offline training time, a full GP is faster than FMLP, however, if the full GP needs to be trained first an FMLP is faster but gets overtaken with respect to computation time if only enough trials are performed. In both plots, MKLGP requires the most computation time. Considering that these differences will be bigger if we use more meta information, we conclude that POGPE is very fast while performing also very well in the SMBO procedure.

5   Conclusions

In this paper, we proposed to choose POE models as surrogate models for hyperparameter tuning, specifically we chose to employ Gaussian processes because of their fairly easy implementation as well as their predictive performance in the field of Bayesian optimizazion. We do acknowledge that POGPE is not the best model in all experiments, but is quite competitive which is a surprise due to its simplicity and its approximative nature. In the very first trial both POGPE and SGPE (as they start out the same) on average pick the best hyperparameter configuration compared to all competitor methods, which shows how efficient usage of the meta data can simply be made by learning a product of experts on each invidual data set and querying the committee. Moreover, the other competitive surrogate models such as FMLP and MKLGP introduce additional hyperparameters for the surrogate model that need to be optimized. For FMLP, tuning of the network architecture such as number of layers and number of nodes per layer as well as setting correct learning rates is demanded. MKLGP requires tuning of the number of neighboring data sets and the tradeoff term between both employed kernels. In comparison, a simple product of GP experts does not require any hyperparameter tuning, as the GP parameters can be learned by maximizing their marginal likelihood quite effectively.

As we have seen in the results, POGPE can also be trained much faster than the other competitive surrogate models. In the light of big data we will probably have access to also growing meta data sets that we can employ for hyperparameter optimization, which makes scalable use of the meta data a necessity. Moreover, POE models are easy to parallelize which allows easy usage in distributed scenarios. Out of all these reasons we see them as a very reasonable choice to pick as surrogate models for hyperparameter optimization including large scale meta data.

Acknowledgments. The authors gratefully acknowledge the co-funding of their work by the German Research Foundation (DFG) under grant SCHM 2583/6-1.

References

1. Adankon, M.M., Cheriet, M.: Model selection for the LS-SVM. Appl. Handwriting Recogn. Pattern Recognit. **42**(12), 3264–3270 (2009)
2. Bardenet, R., Brendel, M., Kegl, B., Sebag, M.: Collaborative hyperparameter tuning. In: Dasgupta, S., Mcallester, D. (eds.) Proceedings of the 30th International Conference on Machine Learning (ICML-13). vol. 28, pp. 199–207. JMLR Workshop and Conference Proceedings, May 2013
3. Bergstra, J., Bengio, Y.: Random search for hyper-parameter optimization. J. Mach. Learn. Res. **13**, 281–305 (2012)
4. Brazdil, P., Carrier, C.G., Soares, C., Vilalta, R.: Metalearning: Applications to data mining. Springer, Heidelberg (2008)
5. Cao, Y., Fleet, D.J.: Generalized product of experts for automatic and principled fusion of gaussian process predictions. arXiv preprint (2014). arXiv:1410.7827
6. Deisenroth, M.P., Ng, J.W.: Distributed gaussian processes. Int. Conf. Mach. Learn. (ICML) **2**, 5 (2015)
7. Escalante, H.J., Montes, M., Sucar, L.E.: Particle swarm model selection. J. Mach. Learn. Res. **10**, 405–440 (2009)
8. Feurer, M., Springenberg, J.T., Hutter, F.: Initializing bayesian hyperparameter optimization via meta-learning. In: Proceedings of the Twenty-Ninth AAAI Conference on Artificial Intelligence (2015)
9. Guo, X.C., Yang, J.H., Wu, C.G., Wang, C.Y., Liang, Y.C.: A novel ls-svms hyperparameter selection based on particle swarm optimization. Neurocomput **71**(16–18), 3211–3215 (2008)
10. Hensman, J., Fusi, N., Lawrence, N.D.: Gaussian processes for big data. arXiv preprint (2013). arXiv:1309.6835
11. Hinton, G.E.: Products of experts. In: Ninth International Conference on (Conf. Publ. No. 470) Artificial Neural Networks, ICANN 99. vol. 1, pp. 1–6. IET (1999)
12. Hinton, G.E.: Training products of experts by minimizing contrastive divergence. Neural Comput. **14**(8), 1771–1800 (2002)

13. Hutter, F., Hoos, H.H., Leyton-Brown, K.: Sequential model-based optimization for general algorithm configuration. In: Dhaenens, C., Jourdan, L., Marmion, M.-E. (eds.) LION 2015. LNCS, vol. 8994, pp. 507–523. Springer, Heidelberg (2011). doi:10.1007/978-3-642-25566-3_40

14. Jones, D.R., Schonlau, M., Welch, W.J.: Efficient global optimization of expensive black-box functions. J. Global Optim. **13**(4), 455–492 (1998)

15. Koch, P., Bischl, B., Flasch, O., Bartz-Beielstein, T., Weihs, C., Konen, W.: Tuning and evolution of support vector kernels. Evol. Intell. **5**(3), 153–170 (2012)

16. Quinonero-Candela, J., Rasmussen, C.E.: A unifying view of sparse approximate gaussian process regression. J. Mach. Learn. Res. **6**, 1939–1959 (2005)

17. Rasmussen, C.E., Williams, C.K.I.: Gaussian Processes for Machine Learning (Adaptive Computation and Machine Learning). The MIT Press (2005)

18. Schilling, N., Wistuba, M., Drumond, L., Schmidt-Thieme, L.: Hyperparameter optimization with factorized multilayer perceptrons. In: Machine Learning and Knowledge Discovery in Databases, pp. 87–103. Springer, Heidelberg (2015)

19. Shen, Y., Ng, A., Seeger, M.: Fast gaussian process regression using kd-trees. In: Proceedings of the 19th Annual Conference on Neural Information Processing Systems. No. EPFL-CONF-161316 (2006)

20. Snoek, J., Larochelle, H., Adams, R.P.: Practical bayesian optimization of machine learning algorithms. In: Pereira, F., Burges, C., Bottou, L., Weinberger, K. (eds.) Advances in Neural Information Processing Systems 25, pp. 2951–2959. Curran Associates, Inc. (2012)

21. Swersky, K., Snoek, J., Adams, R.P.: Multi-task bayesian optimization. In: Burges, C., Bottou, L., Welling, M., Ghahramani, Z., Weinberger, K. (eds.) Advances in Neural Information Processing Systems 26, pp. 2004–2012. Curran Associates, Inc. (2013)

22. Tresp, V.: A bayesian committee machine. Neural Comput. **12**(11), 2719–2741 (2000)

23. Williams, C., Seeger, M.: Using the nyström method to speed up kernel machines. In: Proceedings of the 14th Annual Conference on Neural Information Processing Systems, pp. 682–688. No. EPFL-CONF-161322 (2001)

24. Wistuba, M., Schilling, N., Schmidt-Thieme, L.: Learning hyperparameter optimization initializations. In: IEEE International Conference on Data Science and Advanced Analytics (DSAA), 36678 2015, pp. 1–10. IEEE (2015)

25. Wistuba, M., Schilling, N., Schmidt-Thieme, L.: Sequential model-free hyperparameter tuning. In: 2015 IEEE International Conference on Data Mining (ICDM), pp. 1033–1038. IEEE (2015)

26. Yogatama, D., Mann, G.: Efficient transfer learning method for automatic hyperparameter tuning. In: International Conference on Artificial Intelligence and Statistics (AISTATS 2014) (2014)

Incremental Commute Time Using Random Walks and Online Anomaly Detection

Nguyen Lu Dang Khoa[1(✉)] and Sanjay Chawla[2,3]

[1] Data61, CSIRO, Sydney, Australia
khoa.nguyen@data61.csiro.au
[2] Qatar Computing Research Institute, HBKU, Doha, Qatar
[3] University of Sydney, Sydney, Australia
sanjay.chawla@sydney.edu.au

Abstract. Commute time is a random walk based metric on graphs and has found widespread successful applications in many application domains. However, the computation of the commute time is expensive, involving the eigen decomposition of the graph Laplacian matrix. There has been effort to approximate the commute time in offline mode. Our interest is inspired by the use of commute time in online mode. We propose an accurate and efficient approximation for computing the commute time in an incremental fashion in order to facilitate real-time applications. An online anomaly detection technique is designed where the commute time of each new arriving data point to any data point in the current graph can be estimated in constant time ensuring a real-time response. The proposed approach shows its high accuracy and efficiency in many synthetic and real datasets and takes only 8 milliseconds on average to detect anomalies online on the DBLP graph which has more than 600,000 nodes and 2 millions edges.

Keywords: Commute time · Random walk · Incremental learning · Online anomaly detection

1 Introduction

Commute time is a well-known measure derived from random walks on graphs [10]. The commute time between two nodes i and j in a graph is the expected number of steps that a random walk, starting from i will take to visit j and then come back to i for the first time. Commute time has been used as a robust metric for different learning tasks such as clustering [14] and anomaly detection [7]. It has also found widespread applications in personalized search [16], collaborative filtering [3] and image segmentation [14]. The fact that the commute time is averaged over all paths (and not just the shortest path) makes it more robust to data perturbations.

More advanced measures generally require more expensive computation. Estimating commute time involves the eigen decomposition of the graph Laplacian matrix and resulting in an $O(n^3)$ time complexity which is impractical for large

© Springer International Publishing AG 2016
P. Frasconi et al. (Eds.): ECML PKDD 2016, Part I, LNAI 9851, pp. 49–64, 2016.
DOI: 10.1007/978-3-319-46128-1_4

graphs. Saerens, Pirotte and Fouss [15] used subspace approximation to approximate the commute time. Sarkar and Moore [13] introduced a notion of truncated commute time and a pruning algorithm to find nearest neighbors in the truncated commute time. Recently, Spielman and Srivastava [17] proposed an approximation algorithm to create a structure in nearly linear time so that the pairwise commute time can be approximated in $O(\log n)$ time.

However, all the above-mentioned approximation techniques all work in a **batch** fashion and therefore have a high computation cost for **online** applications. We are interested in the following scenarios: a dataset or a graph D is given from an underlying domain of interest such as data from a network traffic log or a social network graph. A new data point p arrives and we want to determine if p is an anomaly with respect to D in commute time. A data point is an anomaly if it is *far away from its nearest neighbors in commute time measure* (as described in [7]). This particular application requires the computation of commute time in an online fashion. In this paper, we propose a method called iECT to incrementally estimate the commute time and use it to design an online anomaly detection application. The method makes use of the recursive definition of commute time in terms of random walk measures. The commute time from a new data point to any data point in the existing data D is computed based on the current commute times among points in D. The method is novel and reveals insights about commute time which is independent of the applications.

The contributions of this paper are as follows:

- We use characteristics of random walk measures to propose a method to estimate the commute time incrementally in constant time. Then we design an online anomaly detection technique using the incremental commute time. To the best of our knowledge, this is the first method to estimate the commute time in an online fashion.
- The proposed technique is verified by experiments in different applications using several synthetic and real datasets. The experiments show the effectiveness of the proposed methods in terms of accuracy and performance.
- The methods can be applied directly to graph data and can be used in any application that utilizes the commute time (e.g. classification and graph ranking using commute time).

The remainder of the paper is organized as follows. Section 2 reviews notations and concepts related to random walks and commute time and a method to approximate the commute time offline in large graphs. Section 3 presents a simple motivation example to tie up all the definitions and ideas, and proposes a method to incrementally estimate the commute time. In Sect. 4, we propose an online anomaly detection algorithm which uses the incremental commute time. We evaluate our approaches using experiments on synthetic and real datasets in Sect. 5. Sections 6 and 7 cover the related work and a summary of our work.

2 Background

2.1 Random Walks on Graphs and Commute Time

We provide a self-contained introduction to random walks with an emphasis on commute time. Assume we are given a connected undirected and weighted graph $G = (V, E, W)$.

Definition 1. *Let i be a node in G and $N(i)$ be its neighbors. The degree d_i of a node i is $\sum_{j \in N(i)} w_{ij}$. The volume V_G of the graph is defined as $\sum_{i \in V} d_i$.*

Definition 2. *The transition matrix $M = (p_{ij})_{i,j \in V}$ of a random walk on G is given by*

$$p_{ij} = \begin{cases} \frac{w_{ij}}{d_i}, & \text{if } (i,j) \in E \\ 0, & \text{otherwise} \end{cases}$$

Definition 3. *The Hitting Time h_{ij} is the expected number of steps that a random walk starting at i will take before reaching j for the first time.*

Definition 4. *The Hitting Time can be defined in terms of the recursion*

$$h_{ij} = \begin{cases} 1 + \sum_{l \in N(i)} p_{il} h_{lj} & \text{if } i \neq j \\ 0 & \text{otherwise} \end{cases}$$

Definition 5. *The Commute Time c_{ij} between two nodes i and j is given by $c_{ij} = h_{ij} + h_{ji}$.*

Fact 1. *The commute time can be expressed in terms of the Laplacian of G.*

$$c_{ij} = V_G(l_{ii}^+ + l_{jj}^+ - 2l_{ij}^+) = V_G(e_i - e_j)^T L^+(e_i - e_j) \tag{1}$$

where l_{ij}^+ is the (i,j) element of L^+ (the pseudo-inverse of the Laplacian L) and e_i is the $|V|$ dimensional column vector with 1 at location i and zero elsewhere [3]. L^+ can be computed from the eigensystem of L: $L^+ = \sum_{i=2}^{|V|} \frac{1}{\lambda_i} v_i v_i^T$.

2.2 Approximation of Commute Time Embedding (Batch Mode)

Computing commute time involves the eigen decomposition of the graph Laplacian matrix which is impractical for large graphs. Recently, Spielman and Srivastava [17] proposed an approximation algorithm utilizing random projection and a SDD solver to create a structure in nearly linear time so that the pairwise commute time can be approximated in $k_{RP} = O(\log n)$ time (k_{RP} is the reduced dimension in random projection). The fast SDD solver [18] for linear systems is a new class of near-linear time methods for solving a system of equations $Ax = b$ when A is a symmetric diagonally dominant (SDD) matrix.

The idea is based on the fact that $\theta = \sqrt{V_G} L^+ B^T W^{1/2}$ is a commute time embedding where the commute time c_{ij} is a squared Euclidean distance between points i and j in θ. Here m be the number of edges in G, B is a signed edge-vertex incidence matrix and W is a diagonal matrix whose entries are the edge weights. For the details of the embedding creation, refer to [17].

3   Incremental Commute Time

3.1   Problem and Scope

Problem: Given a dataset or a graph D from an underlying domain of interest. When a new data instance p comes in, we want to compute the commute time from p to any data instance in D.

In an Euclidean space, an insertion of a new point does not change the features of existing points. However, an insertion of a new node in an original feature space or a graph will change the features of existing points in the commute time embedding space, which is spanned by eigenvectors of the graph Laplacian matrix. Updating an eigensystem of a graph Laplacian is costly and not suitable for online applications. In this work, we use the characteristics of random walk measures to estimate the commute time incrementally in constant time and use it to design online applications.

There are some notes regarding the scope of this work. Firstly, the proposed method is only suitable for applications which do not need to update the training model overtime (i.e. a representative training data are available). That means we treat the new data one by one, estimate its corresponding commute time and leave the trained model intact. Secondly, in case of graph data, we only deal with the case of node insertion, not node deletion or weight update.

3.2   Motivation Examples

Consider a graph G shown in Fig. 1a where all the edge weights equal 1. The sum of the degree of nodes, $V_G = 8$. We will calculate the commute time c_{12} in two different ways:

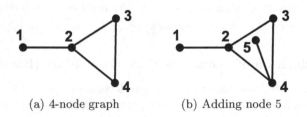

(a) 4-node graph   (b) Adding node 5

Fig. 1. c_{12} increases after an addition of node 5 even though the shortest path distance remains unchanged.

1. Using random walk approach: note that the expected number of steps for a random walk starting at node 1 and returning back to it is $\frac{V_G}{d_1} = \frac{8}{1} = 8$ [10]. But the walk from node 1 can only go to node 2 and then return from node 2 to 1. Thus $c_{12} = 8$.

2. Using algebraic approach: the Laplacian matrix and its pseudo-inverse are

$$
L = \begin{pmatrix} 1 & -1 & 0 & 0 \\ -1 & 3 & -1 & -1 \\ 0 & -1 & 2 & -1 \\ 0 & -1 & -1 & 2 \end{pmatrix} \quad \text{and} \quad L^+ = \begin{pmatrix} 0.69 & -0.06 & -0.31 & -0.31 \\ -0.06 & 0.19 & -0.06 & -0.06 \\ -0.31 & -0.06 & 0.35 & 0.02 \\ -0.31 & -0.06 & 0.02 & 0.35 \end{pmatrix}
$$

Since $c_{12} = V_G(e_1 - e_2)^T L^+(e_1 - e_2)$ and $(e_1 - e_2)^T L^+(e_1 - e_2) =$

$$
\begin{pmatrix} 1 \\ -1 \\ 0 \\ 0 \end{pmatrix}^T \begin{pmatrix} 0.69 & -0.06 & -0.31 & -0.31 \\ -0.06 & 0.19 & -0.06 & -0.06 \\ -0.31 & -0.06 & 0.35 & 0.02 \\ -0.31 & -0.06 & 0.02 & 0.35 \end{pmatrix} \begin{pmatrix} 1 \\ -1 \\ 0 \\ 0 \end{pmatrix} = 1,
$$

$c_{12} = V_G \times 1 = 8$.

Suppose we add a new node (labeled 5) to node 4 with a unit weight as in Fig. 1b. Then $c_{12}^{new} = V_G^{new}/d_1 = 10/1 = 10$. The example in Fig. 1b shows that by adding an edge, i.e. making the 'cluster' which contains node 2 denser, c_{12} increases. This shows that commute time between two nodes captures not only the distance between them (as measured by the edge weights) but also the data densities. For the proof of this claim, see [7]. This property of commute time has been used to simultaneously discover global and local anomalies in data - an important problem in the anomaly detection literature.

In the above example, we exploited the specific topology (degree one node) of the graph to calculate the commute time efficiently. This can only work for very specific instances. The general, more widely used but slower approach for computing the commute time is to use the Laplacian formula as in Eq. 1. One key contribution of this paper is that for an incremental computation of commute time we can use insights from this example to efficiently approximate the commute time using random walk in much more general situations.

3.3 Incremental Estimation of Commute Time

In this section, we derive a new method for computing the commute time in an incremental fashion. This method uses the definition of commute time based on the hitting time. The basic intuition is to expand the hitting time recursion until the random walk has moved a few steps away from the new node and then use the *old* values. In Sect. 5 we will show that this method results in remarkable agreement between the batch and online modes.

We deal with two cases shown in Fig. 2.

1. Rank one perturbation corresponds to the situation when a new node connects with one other node in the existing graph.
2. Rank k perturbation deals with the situation when the new node has k neighbors in the existing graph.

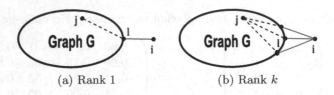

(a) Rank 1   (b) Rank k

Fig. 2. Rank 1 and rank k perturbation when a new data point arrives.

Rank One Perturbation

Proposition 1. *Let i be a new node connected by one edge to an existing node l in the graph G. Let w_{il} be the weight of the new edge. Let j be an arbitrary node in the graph G. Then*

$$c_{ij} = c_{lj}^{old} + \frac{V_G}{w_{il}} + O(\frac{1}{k}) \tag{2}$$

where 'old' represents the commute time in graph G (k nearest neighbor graph) before adding i.

Proof. (Sketch) Since the random walk needs to pass l before reaching j, the commute distance from i to j is:

$$c_{ij} = c_{il} + c_{lj}. \tag{3}$$

It is known that:

$$c_{il} = \frac{(V_G + 2w_{il})}{w_{il}} \tag{4}$$

where V_G is volume of graph G [7]. We also know $c_{lj} = h_{jl} + h_{lj}$ and $h_{jl} = h_{jl}^{old}$. The only unknown factor is h_{lj}. By definition:

$$h_{lj} = 1 + \sum_{q \in N(l)} p_{lq} h_{qj} = 1 + \sum_{q \in N(l), q \neq i} p_{lq} h_{qj} + p_{li} h_{ij}.$$

Since commute time is robust against small changes or perturbation in data, we have $h_{qj} \approx h_{qj}^{old}$. Moreover, $p_{lq} = (1 - p_{li})p_{lq}^{old}$, and $h_{ij} = 1 + h_{lj}$. Therefore,

$$h_{lj} \approx 1 + \sum_{q \in N(l), q \neq i} (1 - p_{li})p_{lq}^{old} h_{qj}^{old} + p_{li}(1 + h_{lj})$$

$$= 1 + (1 - p_{li}) \sum_{q \in N(l), q \neq i} p_{lq}^{old} h_{qj}^{old} + p_{li}(1 + h_{lj})$$

$$= 1 + (1 - p_{li})(h_{lj}^{old} - 1) + p_{li}(1 + h_{lj}).$$

After simplification, $h_{lj} = h_{lj}^{old} + \frac{2p_{li}}{1-p_{li}}$. Then $c_{lj} \approx h_{jl}^{old} + h_{lj}^{old} + \frac{2p_{li}}{1-p_{li}}$. Since there is only one edge connecting from i to G, i is likely an isolated point and thus $p_{li} = O(\frac{1}{k})$ (G is the k nearest neighbor graph). Then

$$c_{lj} = h_{jl}^{old} + h_{lj}^{old} + O(\frac{1}{k}) = c_{lj}^{old} + O(\frac{1}{k}). \tag{5}$$

As a result from Eqs. 3, 4 and 5:

$$c_{ij} = \frac{(V_G + 2w_{il})}{w_{il}} + c_{lj}^{old} + O(\frac{1}{k}) = c_{lj}^{old} + \frac{V_G}{w_{il}} + O(\frac{1}{k})$$

Rank k perturbation The rank k perturbation analysis is more involved but the final formulation is an extension of the rank one case.

Proposition 2. *Denote $l \in G$ be one of k neighbors of i, and j be a node in G. The approximate commute time between nodes i and j is:*

$$c_{ij} \approx \sum_{l \in N(i)} p_{il} c_{lj}^{old} + \frac{V_G}{d_i} + O(\frac{1}{k}) \tag{6}$$

For the proof, see Appendix in the supplement document. When $k = 1$ (rank one case), the Eq. 6 becomes Eq. 2.

4   Online Applications Using Incremental Commute Time

We return to our original motivation for computing incremental commute time. We are given a dataset D which is a *representative* of the underlying domain of interest. We need to find nearest neighbors of a new data point p in commute time metric incrementally. We want to check if p is an anomaly in D.

We train the dataset D using Algorithm 1. First, a mutual k_1-nearest neighbor graph is constructed from the dataset. This graph connects nodes u and v if u belongs to k_1-nearest neighbors of v *and* v belongs to k_1-nearest neighbors of u [11]. Then the approximate commute time embedding θ is computed as in Sect. 2.2. Finally, a distance-based anomaly detection with a pruning rule proposed by Bay and Schwabacher [2] is used in θ to find the top N anomalies. That means the distance-based method uses commute time, instead of Euclidean distance. The anomaly score used is the average commute time of a data instance to its k_2 nearest neighbors.

Algorithm 1. Approximate Commute Time Distance Based Anomaly Detection (for training).

Input: Data matrix X, the numbers of nearest neighbors k_1 (for building the k-nearest neighbor graph) and k_2 (for estimating the anomaly score), the number of random vectors k_{RP}, the numbers of anomalies to return N
Output: Top N anomalies, anomaly threshold τ

1: Construct a mutual k-nearest neighbor graph G from the dataset (using k_1)
2: Compute the approximate commute time embedding θ from G
3: Find top N anomalies using a distance-based technique with pruning rule described in [2] on θ (using k_2)
4: Return top N anomalies and the anomaly threshold τ

Pruning Rule [2]: A data point is not an anomaly if its score (e.g. the average distance to its k nearest neighbors) is less than an anomaly threshold. The threshold can be fixed or be adjusted as the score of the weakest anomaly found so far. Using the pruning rule, many non-anomalies can be pruned without carrying out a full nearest neighbors search.

After training, the corresponding graph G, the commute time embedding θ, and the anomaly threshold τ are obtained (τ is the score of the weakest anomaly found among top N anomalies). We propose a method shown in Algorithm 2 (denote as iECT) to detect anomalies online given the trained model.

Algorithm 2. Online Anomaly Detection using the incremental Estimation of Commute Time (iECT)

Input: Graph G, the approximate commute time embedding θ and the anomaly threshold τ computed in the training phase, and a new arriving data point p
Output: Determine if p is an anomaly or not

1: Add p to G satisfying the property of the mutual nearest neighbor graph
2: Determine if p is an anomaly or not by estimating its anomaly score incrementally using the method described in Sect. 3.3. Use pruning rule with threshold τ to reduce the computation
3: Return whether p is an anomaly or not

When a new data point p arrives, it is connected to graph G created in the training phase so that the property of the mutual nearest neighbor graph is held. The commute times are incrementally updated to estimate the anomaly score of p using the approach in Sect. 3.3. The embedding θ is used to compute the commute time c_{lj}^{old}. The pruning is used as follows: p is not anomaly if its average distance to k nearest neighbors is smaller than the anomaly threshold τ. Generally commute time is robust against small changes or perturbation in data. Therefore, only the anomaly score of a new data point needs to be estimated and be compared with the anomaly threshold computed in the training phase. This claim will be verified by experiments in Sect. 5.

4.1 Analysis

The incremental estimation of commute time in Sect. 3.3 requires $O(k_{RP})$ for each query of c_{lj}^{old} in θ. So if there are k edges added to the graph due to the addition of a new node, it takes $O(kk_{RP})$ for each query of c_{ij}.

As explained earlier, we only need to compute the anomaly score of the new data point. Using pruning rule with the known anomaly threshold, it takes only $O(k_2)$ nearest neighbor search to determine if the test point is an anomaly or not where k_2 is the number of nearest neighbors for estimating the anomaly score. For each commute time query, it takes $O(kk_{RP})$ as described above. Therefore, iECT takes $O(k_2kk_{RP})$ to determine if a new arriving point is an anomaly or not. [19] has suggested that $k_{RP} = 2\ln n/0.25^2$ which is just 442 for a dataset of a million data points. Therefore $k_{RP} \ll n$. Since $k, k_2 \ll n$, $O(k_2 k k_{RP}) = O(1)$ resulting in a near constant time complexity for iECT.

Note that this constant time complexity of iECT does not depend on the complexity of $O(k_{RP})$ for each query of c_{lj}^{old} using the method in [17]. If we query c_{lj}^{old} using Eq. 1 with just $O(k_{EV})$ eigenvectors of Laplacian matrix L (as described in [7]), each query only takes $O(k_{EV} \ll n)$ also resulting in a constant time complexity for iECT.

5   Experiments and Results

In this section, we determined and compared the effectiveness of online anomaly detection application using incremental commute time. The experiments were carried out on synthetic as well as real datasets. In all experiments, the numbers of nearest neighbors were $k_1 = 10$ (for building the nearest neighbor graph), $k_2 = 20$ (for estimating a nearest neighbor score or an anomaly score in anomaly detection applications), and the number of random vectors was $k_{RP} = 200$ (for creating the commute time embedding) unless otherwise stated. We used Koutis's CMG solver [9] as an implementation of the SDD-Solver for creating the embedding. The solver is used for SDD matrices which is available online at http://www.cs.cmu.edu/~jkoutis/cmg.html. The choice of parameters was determined from the experiments and it was also analyzed in Sect. 5.5.

5.1   Approach

We split a dataset into two parts: a training set and a test set. We trained the training set to find top N anomalies and the threshold value τ using Algorithm 1. Then an anomaly score of each instance p in the test set was calculated based on its k_2 neighbors in the training set. If this score was greater than τ then the test instance was reported as an anomaly.

Baseline: in all experiments, the batch method (Algorithm 1) was used as the benchmark since there is no other method to estimate commute time incrementally. Note that for both the batch and incremental methods, we need to compute only the anomaly score of the new arriving data instance and pruning was also applied using τ. The difference is in the batch method, the new approximate commute time embedding was recomputed and the anomaly score was estimated using the new embedding space. The incremental method, on the other hand, estimated the score incrementally using the method described in Sect. 3.3.

5.2   Synthetic Datasets

We created six synthetic datasets with 1000, 10000, 20000, 30000, 40000 and 50000 data points. Each dataset contained several clusters generated from Normal distributions and 100 random points generated from uniform distribution which were likely anomalies. The number of clusters, the sizes, and the locations of the clusters were also chosen randomly. Each dataset was divided into a training set and a test set. There were 100 data points in every test set and half of them were random anomalies mentioned above.

Experiments on Robustness: We first tested the robustness of commute time between nodes in an existing graph when a new node is introduced. As the commute time c_{ij} is a measure of expected path distance, the hypothesis is that the addition of a new point will have minimal influence on c_{ij} and thus the anomaly scores of data points in the existing set are relatively unchanged.

Table 1 shows the average, standard deviation, minimum, and maximum of anomaly scores of points in graph G before and after a new data point was added to G. Graph G was created from the training set of a 1000 point dataset described above. The result was averaged over 100 test points in the test set. The result shows that the anomaly scores of data instances in G do not change much when a new point is added to G (the change of the average score was only about 0.7%).

Table 1. Robustness of commute time. The anomaly scores of data instances in existing graph G are relatively unchanged when a new point is added to G.

	Average	Std	Min	Max
Without test point	15,362.57	50,779.71	916.27	538,563.38
With test point	15,257.38	50,286.20	904.49	534,317.52

In the following experiments, the change in eigensystem of the graph Laplacian L of the training data due to an addition of a new node was analyzed. Figures 3a shows average changes in the top 50 eigenvalues before and after an addition of each test point in the test set in the 1000 point dataset. The changes are small for most of them (most of them were less than 1% and all of them were less than 6%). Figures 3b shows dot products of eigenvectors with the second smallest eigenvalues (the smallest is zero) before and after an addition of each test point. The eigenvectors did not change much due to node addition. As shown in Eq. 1, since the change in eigensystem of the Laplacian is small, the commute times between existing training nodes do not change much.

All these results show commute time is a robust measure: a small change or perturbation in the data will not result in large changes in commute times. Therefore, only the anomaly score of the new point needs to be estimated.

Experiments on Effectiveness: We applied iECT to all six datasets mentioned earlier. The effectiveness of iECT and the commute time approximation were reported and discussed.

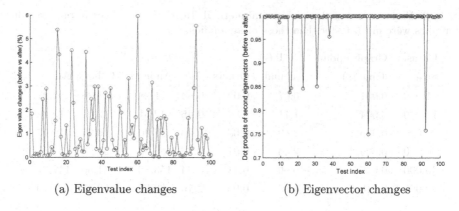

(a) Eigenvalue changes   (b) Eigenvector changes

Fig. 3. Change in eigensystem when new nodes were added to the graph.

Table 2 presents the results in accuracy and performance of iECT in six synthetic datasets. Average score was the average anomaly score with pruning rule over 100 test points. The precision and recall were for the anomalous class. The time was the average time to process each of 100 test points. iECT captured all the anomalies, had a few false alarms, and was much more efficient than the batch method. Note that the scores shown here were the anomaly scores with pruning rule and the scores for anomalies are always much higher than scores for normal points. Therefore the average scores shown in the table were dominated by the scores of anomalies.

Table 2. Effectiveness of the incremental method. iECT captured all the anomalies, had a few false alarms and was much more efficient than the batch method.

Dataset	iECT				Batch	
Size	Precison (%)	Recall (%)	Avg score	Time (s)	Avg score	Time (s)
1,000	82.4	100	2.30×10^5	0.04	1.88×10^5	1.27
10,000	100	100	7.70×10^6	0.45	6.70×10^6	12.32
20,000	96.0	100	2.36×10^7	0.95	1.93×10^7	16.97
30,000	98.0	100	1.39×10^7	1.22	1.14×10^7	38.68
40,000	95.8	100	2.29×10^7	5.24	1.67×10^7	147.27
50,000	100	100	3.11×10^7	2.16	2.41×10^7	61.90

There is an interesting dynamic at play between the pruning rule and the number of anomalies in the data. The reason is there was a high proportion of anomalies in the test set (about 50 %). We know that the pruning rule only works for non-anomalies and therefore, the time to process anomalies should be much longer than the others. Table 3 shows the details of time to process data points in the test set. For batch and iECT methods, the average time to process

Table 3. Performance of the incremental method. In iECT, the times to process non-anomalies were much faster than those of anomalies.

Dataset	Graph update	iECT (s)			Batch (s)		
Size	Time (s)	Anomaly	Others	All	Anomaly	Others	All
1,000	0.001	0.07	0.02	0.04	1.28	1.26	1.27
10,000	0.004	1.11	0.02	0.45	12.81	12.01	12.32
20,000	0.006	1.89	0.08	0.95	17.40	16.57	16.97
30,000	0.009	2.46	0.07	1.22	39.46	37.96	38.68
40,000	0.047	10.90	0.41	5.24	153.86	141.66	147.27
50,000	0.018	4.35	0.06	2.16	63.38	60.47	61.90

only anomalies, only non-anomalies, and all data instances are listed in the table. There was not much difference in batch method between time to process anomalies and non-anomalies since for each new data point the time to create the new commute time embedding was much higher than that of the nearest neighbor search. On the other hand, this gap was very high for iECT so that the times to process non-anomalies were much faster than those of anomalies. In practice, since most of the data points are not anomalies, iECT is very efficient.

Another cost we have not mentioned is the time to update the graph. That is the time to add a new data point to an existing graph satisfying the property of the mutual nearest neighbor graph. Since we stored the kd tree corresponding to the training data, the update cost was very low as shown in Table 3.

5.3   Graph Dataset

In this section, we evaluated the iECT method on a large DBLP co-authorship network to show the scalability of the method. In this graph, nodes are authors and edge weights are the number of collaborated papers between the authors. Since the graph is not fully connected, we extracted its biggest component. It has 612,949 nodes and 2,345,178 edges in a snapshot in December 5th, 2011 which is available at http://dblp.uni-trier.de/xml/. We randomly chose a test set of 50 nodes and removed them from the graph. We ensured that the graph remained connected. After training, each node was added back into the graph along with their associated edges.

We trained the graph using Algorithm 1, stored the approximate embedding in order to query the c_{lj}^{old} in iECT algorithm. The batch method use the approximate embedding created from a new graph after adding each test point.

The result shows that it took 0.008 seconds on average over 50 test data points to detect whether each test point was an anomaly or not. The batch method, which is the fastest approximation of commute time to date, required 1,454 seconds on average to process each test data point. This dramatically highlights the constant time complexity of iECT algorithm and suggests that iECT is highly suitable for the computation of commute time in an incremental fashion. Since there was no anomaly information in the random test set, we cannot report the detection accuracy here. The average anomaly score over all the test points of iECT was 8.6 % higher than the batch method. This shows the high accuracy of iECT approximation even in a very large graph.

5.4   Real Datasets

In this experiment, we report the results for online anomaly detection using real datasets in different application domains. They are applications in spam detection, network intrusion detection and bridge damage detection.

Spambase dataset: The Spambase dataset provided by Machine Learning Repository [4] was investigated. There are 4,601 emails in the data with 57 features each. The task is check whether a email is spam or not. Since the dataset has duplicated data instances, and the numbers of spams and non-spams are not imbalanced, we removed duplicated data, kept the non-spams, and sampled 100 spams from the dataset. Finally we had 2631 data instances.

Computer network anomaly detection: The dataset is from a wireless mesh network at the University of Sydney which was deployed by NICTA [20]. It used a traffic generator to simulate traffic on the network. Packets were aggregated into one-minute time bins and the data was collected in 24 hours. There were 391 origin-destination flows and 1270 time bins. Anomalies were introduced to the network including DOS attacks and ping floods. After removing duplications in the data, we had 1193 time-bin instances.

Damage detection on bridge: The Sydney Harbour Bridge is one of major bridges in Australia, which was opened in 1932. As the bridge is aging, it is critical to ensure it stays structurally healthy. Vibration data caused by passing vehicles have been recorded by accelerometers installed on the joints under the deck of bus lane. For this case study, only six instrumented joints were considered (named 1 to 6). The data were obtained in the period from early August until late October in 2012. A known crack existed in joint 4 while the other joints were in good conditions. The feature extraction was used as described in [8]. A dataset was created to include vibration events from all healthy joints and 100 events from the damaged joint (totally 2523 events).

Each dataset was divided into a training set and a test set with 100 data points. The results of using iECT and batch methods are shown in Table 4. It shows that iECT has a high detection accuracy and is much more efficient than the batch method. Also the commute time scores between iECT and batch method were quite similar.

5.5   Impact of Parameters

In this section, we investigate how the parameters k_1, k_2, and k_{RP} affect the effectiveness of the proposed method. Parameters k_1 and k_2 only affect the accuracy

Table 4. The effectiveness of iECT in real datasets. It shows that iECT has a high detection accuracy and is much more efficient than the batch method.

Dataset	Precision	Recall	iECT		Batch	
	(%)	(%)	Avg Score	Time (s)	Avg Score	Time (s)
Spambase	100	100	1.91×10^6	0.004	1.80×10^6	1.98
Network	100	100	6.40×10^5	0.005	6.39×10^5	0.83
Bridge	97.9	100	2.53×10^{12}	0.10	2.54×10^{12}	1.30

of computing commute time in batch mode and were analyzed in [7]. Therefore, this section analyses impact of k_{RP} to the incremental commute time.

We conducted an experiment with different k_{RP} for the three real datasets mentioned in the previous section. The results in Fig. 4 show that the method can achieve high accuracy with small k_{RP} and is not sensitive to k_{RP}.

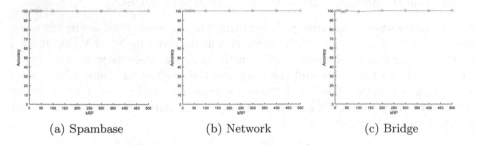

(a) Spambase   (b) Network   (c) Bridge

Fig. 4. Anomaly detection accuracy with different k_{RP}.

5.6   Summary and Discussion

The experimental results show that iECT can accurately approximate the commute time in constant time. It is much more efficient than the batch method using Algorithm 1. The results on real datasets collected from different domains and applications also have similar tendency showing the reliability and effectiveness of the proposed method. One weakness of iECT is that it can only be used in online applications where the update of the graph is given by the addition of a new node, not by updating the edge weights. However, in the case of updating edge weights, the method by Ning et al. in [12] can be used. This method incrementally updates the eigenvalues and eigenvectors of the graph Laplacian matrix based on a change of an edge weight on the graph. Then we can use the new eigen pairs of the Laplacian to update the commute time.

6   Related Work

Khoa and Chawla [7] proposed a new method to find anomalies using commute time. They showed that unlike Euclidean distance, commute time between two nodes can capture both the distance between them and their densities so that it can capture both global and local anomalies using distance based methods.

Incremental learning using an update on eigen decomposition has been studied for a long time. Early work studied the rank one modification of the symmetric eigen decomposition [5,6]. The authors reduced the original problem to the eigen decomposition of a diagonal matrix. Though they can have a good approximation of the new eigenpair, they are not suitable for online applications nowadays since they have at least $O(n^2)$ computation for the update.

More recent approach was based on the matrix perturbation theory [1]. It used the first order perturbation analysis of the rank-one update for a data covariance matrix to compute the new eigenpair. These algorithms have a linear time computation. The advantage of using the covariance matrix is if the perturbation involving an insertion of a new point, the size of the covariance matrix is unchanged. This approach cannot be applied directly to increasing matrix size due to an insertion of a new point. For example, in spectral clustering or commute time based anomaly detection, the size of the graph Laplacian matrix increases when a new point is added to the graph.

Ning et al. [12] proposed an incremental approach for spectral clustering to monitor evolving blog communities. It incrementally updates the eigen system of the graph Laplacian matrix based on a change of an edge weight on the graph using the first order error of the generalized eigen system. This algorithm is only suitable for cases of weight update, not for an addition of a new node.

7 Conclusion

In this paper, we proposed a method to approximate commute time incrementally and used it to design an online anomaly detection application. The method incrementally estimates the commute time in constant time using properties of random walk and hitting time. The main idea is to expand the hitting time recursion until the random walk has moved a few steps away from the new node and then use the old values. The experimental results in synthetic and real datasets show the effectiveness of the proposed approach in terms of performance and accuracy. iECT can incrementally estimate the commute time accurately, resulting in high accuracy in several datasets from different applications. It only took 8 milliseconds on average to process a new arriving node in a graph of more than 600,000 nodes and two millions edges. Moreover, the idea of this work can be extended in other applications which utilize the commute time.

References

1. Agrawal, R.K.: Karmeshu: Perturbation scheme for online learning of features: Incremental principal component analysis. Pattern Recogn. **41**, 1452–1460 (2008)
2. Bay, S.D., Schwabacher, M.: Mining distance-based outliers in near linear time-with randomization and a simple pruning rule. In: KDD 2003: Proceedings of the Ninth ACM SIGKDD International Conference on Knowledge Discovery and Data Mining, pp. 29–38. ACM, New York, NY, USA (2003)

3. Fouss, F., Renders, J.M.: Random-walk computation of similarities between nodes of a graph with application to collaborative recommendation. IEEE Trans. Knowl. Data Eng. **19**(3), 355–369 (2007)
4. Frank, A., Asuncion, A.: Uci machine learning repository (2010)
5. Golub, G.H.: Some modified matrix eigenvalue problems. SIAM Rev. **15**(2), 318–334 (1973)
6. Gu, M., Eisenstat, S.C.: A stable and efficient algorithm for the rank-one modification of the symmetric eigenproblem. SIAM J. Matrix Anal. Appl. **15**, 1266–1276 (1994)
7. Khoa, N.L.D., Chawla, S.: Robust outlier detection using commute time andeigenspace embedding. In: PAKDD 2010: Proceedings of the The 14th Pacific-AsiaConference on Knowledge Discovery and Data Mining, pp. 422–434. Springer, Berlin/Heidelberg (2010)
8. Khoa, N.L.D., Zhang, B., Wang, Y., Chen, F., Mustapha, S.: Robust dimensionality reduction and damage detection approaches in structural health monitoring. Struct. Health Monit. **13**(4), 406–417 (2014)
9. Koutis, I., Miller, G.L., Tolliver, D.: Combinatorial preconditioners andmultilevel solvers for problems in computer vision and image processing. In: Proceedings of the 5th International Symposium on Advances in VisualComputing: Part I, pp. 1067–1078. ISVC 2009, Springer-Verlag, Berlin, Heidelberg (2009)
10. Lovász, L.: Random walks on graphs: a survey. Comb. Paul Erdös is Eighty **2**, 1–46 (1993)
11. von Luxburg, U.: A tutorial on spectral clustering. Stat. Comput. **17**(4), 395–416 (2007)
12. Ning, H., Xu, W., Chi, Y., Gong, Y., Huang, T.: Incremental spectral clustering with application to monitoring of evolving blog communities. In: SIAM International Conference on Data Mining (2007)
13. Purnamrita Sarkar, A.W.M.: A tractable approach to finding closest truncated-commute-time neighbors in large graphs. In: The 23rd Conference on Uncertainty in Artificial Intelligence (UAI) (2007)
14. Qiu, H., Hancock, E.: Clustering and embedding using commute times. IEEE TPAMI **29**(11), 1873–1890 (2007)
15. Saerens, M., Fouss, F., Yen, L., Dupont, P.: The principal components analysisof a graph, and its relationships to spectral clustering. In: Proceedings of the15th European Conference on Machine Learning (ECML 2004), pp. 371–383. Springer-Verlag, Heidelberg (2004)
16. Sarkar, P., Moore, A.W., Prakash, A.: Fast incremental proximity search in large graphs. In: Proceedings of the 25th International Conference on Machine Learning, pp. 896–903. ICML 2008, NY, USA. ACM, New York (2008)
17. Spielman, D.A., Srivastava, N.: Graph sparsification by effective resistances. In: Proceedings of the 40th Annual ACM Symposium on Theory of Computing, pp. 563–568. STOC 2008, NY, USA. ACM, New York (2008)
18. Spielman, D.A., Teng, S.H.: Nearly-linear time algorithms for preconditioning and solving symmetric, diagonally dominant linear systems. CoRR abs/cs/0607105 (2006)
19. Venkatasubramanian, S., Wang, Q.: The johnson-lindenstrauss transform: An empirical study. In: Mller-Hannemann, M., Werneck, R.F.F. (ed.) ALENEX, pp. 164–173. SIAM (2011)
20. Zaidi, Z.R., Hakami, S., Landfeldt, B., Moors, T.: Real-time detection of traffic anomalies in wireless mesh networks. Wirel. Netw. **16**, 1675–1689 (2009)

Online Density Estimation of Heterogeneous Data Streams in Higher Dimensions

Michael Geilke$^{(\boxtimes)}$, Andreas Karwath, and Stefan Kramer

Johannes Gutenberg University Mainz, Mainz, Germany
{geilke,karwath,kramer}@informatik.uni-mainz.de

Abstract. The joint density of a data stream is suitable for performing data mining tasks without having access to the original data. However, the methods proposed so far only target a small to medium number of variables, since their estimates rely on representing all the interdependencies between the variables of the data. High-dimensional data streams, which are becoming more and more frequent due to increasing numbers of interconnected devices, are, therefore, pushing these methods to their limits. To mitigate these limitations, we present an approach that projects the original data stream into a vector space and uses a set of representatives to provide an estimate. Due to the structure of the estimates, it enables the density estimation of higher-dimensional data and approaches the true density with increasing dimensionality of the vector space. Moreover, it is not only designed to estimate homogeneous data, i.e., where all variables are nominal or all variables are numeric, but it can also estimate heterogeneous data. The evaluation is conducted on synthetic and real-world data. The software related to this paper is available at https://github.com/geilke/mideo.

1 Introduction

In the context of discrete densities, Geilke *et al.* [6,7] presented online density estimators that not only capture the distribution of data streams but also support data mining tasks. The presented density estimates were described using (ensembles of (weighted)) classifier chains, where each classifier predicts one variable of the stream and is built using the variables of the previous classifiers. This relationship is inspired by the chain rule of densities, according to which the dependencies between the variables are modeled. As long as the density has only a few variables, this method provides an accurate description of the data [6]. But as soon as the dimensionality increases, the number of classifiers and their size grows quickly – making this approach unsuitable for data of high dimensionality. High-dimensional data streams, however, are becoming more and more frequent with the constantly increasing number of interconnected devices that try to measure aspects of their environment to make intelligent decisions. For example, future smart homes may have many sensors measuring various parameters such as temperature or humidity. By learning from past measurements, machine learning algorithms have the possibility of distinguishing between typical and abnormal behavior and can suggest appropriate actions to the user.

© Springer International Publishing AG 2016
P. Frasconi et al. (Eds.): ECML PKDD 2016, Part I, LNAI 9851, pp. 65–80, 2016.
DOI: 10.1007/978-3-319-46128-1_5

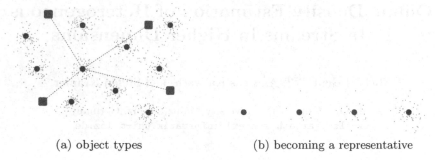

|(a) object types | (b) becoming a representative|

Fig. 1. On the left are the main object types: *landmarks* (dark gray squares), *representatives* (red), and *instances* (gray dots). On the right is the process of becoming a representative: If a distance vector cannot be assigned to a representative or candidate, it is first considered a candidate (big dark gray circle). Over time more instances appear in its neighborhood. If a predefined number is reached, the candidate is turned into a representative (big red circle). (Color figure online)

Whereas an increase of the humidity in the basement can be explained by a tumble dryer that has been started some time ago, such an increase in the bed room could be due to water entering the room through an open window. The former situation is probably quite normal for a household and requires no action, the latter situation needs attention by the user. Providing an estimate that captures the density of the sensor measurements and provides facilities to perform data mining tasks can be useful to develop such applications.

In this paper, we address the problem of estimating the joint density of heterogeneous data streams with many variables. In data mining, one would usually consider variables in the hundreds or thousands as many variables. For density estimation, however, even 50 binary variables is already considered high-dimensional, as there are 2^{50} value combinations. For each of these combinations, the estimator has to assign a density value, which makes the estimation a challenging task. To perform density estimation on data with many variables, we designed an algorithm, called RED (Representative-based online Estimation of Densities). The main idea is to project the data stream into a vector space of lower dimensionality by computing distances to well-defined reference points. In particular, we distinguish between three types of objects (see Fig. 1 for an illustration): *landmarks*, *representatives*, and *instances*. Landmarks are reference points spanning a vector space for representatives and instances, so that the position of each object can be defined in terms of distances to the landmarks (e.g., the position of the red circle in Fig. 1 that is connected to the four landmarks (red circles) can be specified as a four tuple (d_1, d_2, d_3, d_4), where d_i is the distance to landmark i, $i \in [1; 4]$). Representatives stand for clusters of instances and will be the main components for estimating probabilities. The landmarks will be used to compute the relative distance of instances, and the representatives will maintain statistical information about instances that have been observed in their neighborhood. To maintain this information, we employ an extended version of the density estimators proposed by Geilke *et al.* [6][1], called *EDO* (Estimation of Densities Online), and estimate the distances to nearby instances for each representative. Compared to EDO, which directly estimates the density of the instances, RED reduces the dimensionality of the dataset to the number of landmarks. If this number is substantially smaller than the number of variables, the model size of the estimators can be substantially reduced, thereby making the approach suitable for data of higher dimensionality.

The main contributions of the paper are:

1. an online density estimator for heterogeneous data with many variables, i.e., data with many nominal and/or numerical variables,
2. a theoretical analysis for the choice of the landmarks,
3. a theoretical analysis for the consistency of the estimates.

2 Related Work

Whereas many data mining tasks have received considerable attention in the context of stream mining recently, only little is known about the estimation of joint densities in an online setting. In this setting, the algorithm has to learn a joint density f solely from the instances of a data stream. Interdependencies between instances are usually not taken into account, which distinguishes it from online learning protocols where the outcome of a given variable is predicted based on past outcomes [3].

Offline density estimation includes recent work based on decision trees [13], where the leaves contain piecewise constant estimators. A similar approach was pursued by Davies and Moore [4] as part of a conditional density estimator. Work towards the estimation of conditional densities has been pursued among others by Frank and Bouckaert [5] and Holmes *et al.* Multivariate densities are frequently estimated using kernel density estimators [8,14], which is also the predominant direction of the few online variants of density estimation so far. For example, Kristan *et al.* [9–11] proposed a method yielding results that are comparable to the corresponding batch approaches. Xu *et al.* [16] introduced sequences of kernel density estimators to address density estimation from a data stream with only a few variables. The approach presented in this work differs from kernel density estimators in two aspects: (1) the data is projected into a vector space of lower dimensionality by computing distances to reference points, and (2) the basic density estimators are online density estimators that are able to represent complex non-parametric densities.

Datasets with many instances were considered by Peherstorfer *et al.* [12]. They proposed to use a sparse grid where basis functions are not centered around the instances but at grid points. Partitioning the space of data instances was also the strategy pursed by RS-Forest by Wu *et al.* [15], who used a forest of trees

---

[1] The density estimators are designed for discrete variables, but they can be extended to mixed types of variables by using the conditional density estimators proposed by Eibe and Frank [5]. Details will be provided in a forthcoming journal publication.

to partition the data space. Density estimation on data streams with a greater number of variables has – to the best of our knowledge – not been considered so far, but the methods by Kristan *et al.* [11] and RS-Forest come closest to the requirements and are, therefore, the methods we considered for a comparison.[2]

Although distances and representative instances have been used before to project data into a space of lower dimensionality (e.g., multidimensional scaling), the approach presented in the paper is different. Whereas other techniques try to preserve the relevant characteristic properties of the data when embedding it into a space of lower dimensionality, RED characterizes the data using landmarks and provides a back translation to the original data. This back translation is a crucial and necessary part to enable density estimation.

The approach pursued by RED is also different from micro-clustering [1]. Whereas micro-clusters maintain simple statistical properties of the data such as the linear sum, RED uses landmarks and Gaussian mixtures to partition the data space and then estimates the full joint density of each partition.

3   Density Estimation Using Representatives

Let $X_1, \ldots, X_n$ be a set of variables and let x be an instance defined over these variables. Given a possibly infinite stream of instances with many variables, we address the problem of estimating its density, $f : X_1 \times \ldots \times X_n \to [0;1]$, in an online fashion, i.e., only the current instance and its current estimate is provided. In order to determine a density estimate $\hat{f}$, we propose a method that reduces their dimensionality by using a small set of reference points $L := \{L_1, \ldots, L_m\}$, so-called landmarks. With these reference points, it projects the data into a vector space of dimensionality $m < n$ by applying a mapping, $h_L : X_1 \times \ldots \times X_n \to \mathbb{R}^m$, to each instance x. Here, the i-th component of $h_L(x)$ is defined as the distance between x and landmark L_i. The estimate $\hat{f}$ of f is then expressed as the product of two independent components: an online estimate $\hat{g}$ that captures the density in the vector space and a *correctionFactor*, which is the expected number of instances that are mapped to the same distance vector – without the correction factor, we would only estimate g. Hence, $\hat{f}(x) = \hat{g}(h_L(x)) \cdot correctionFactor$. In the remainder of this section, we give a detailed description of these components and provide a theoretical analysis.

3.1   The Density of the Vector Space

In order to estimate the density g, RED distinguishes three kinds of objects in the vector space: *distance vectors*, *representatives*, and *candidates*. A distance vector is a projected data stream instance, which is determined by computing the distances to the landmarks. A representative is a distance vector together with a density estimator and a covariance matrix, where the density estimator

---

[2] Unfortunately, even after several emails, the authors of RS-Forest did not respond to our request to share their program.

Algorithm 1. updateDensityEstimate

Input: landmarks L, instance x, mapping h_L, number of neighbors $k \in \mathbb{N}$,
candidate threshold $\theta_{C \to R} \in \mathbb{N}$, $C := \{(c, \Sigma, b) \mid$ candidate c covariance,
matrix Σ, recent instances $b\}$, $R := \{(r, \Sigma, e, b) \mid$ representative r,
covariance matrix Σ, estimator e, recent instances $b\}$

    // check representatives
1 Let $(r, \Sigma, e, b) \in R$ with r being closest to $h_L(x)$
2 e.update($h_L(x)$) if r.isMember(x, Σ)
    // no matching representative?
3 if $\nexists(r, \Sigma, e, b) \in R : r.isMember(inst, \Sigma)$ then
       // no matching candidate?
4   if $\nexists(c, \Sigma, b) \in C : c.isMember(inst, \Sigma)$ then
          // find k closest neighbors
5       $pq \leftarrow priorityQueue()$
6   for $(r, \Sigma, e, b) \in R$ do
7          $nb \leftarrow (r, e, \|r - h_L(x)\|, b)$
8   pq.insert($nb, \|r - h_L(x)\|$)
9       $neighbors \leftarrow \{pq.peekMin() \mid k \text{ times}\}$
         // compute covariance matrix
10      $sample \leftarrow \emptyset$
11   for $(r, e, \|r - h_L(x)\|, b) \in neighbors$ do
12         $sample \leftarrow sample \cup \{x' \mid x' \in b\}$
         // initialize candidate with empty buffer
13   C.append($(x, \text{covariance}(sample), [x])$)
       // matching candidates
14   else
15   for $(c, \Sigma, b) \in C$ do
16   if c.isMember($h_L(x), \Sigma$) then $b \leftarrow b \cup \{x\}$; break
17   for $(c, \Sigma, b) \in C$ with $|b| \geq \theta_{C \to R}$ do
18         $e \leftarrow$ initialize EDO estimator
19   e.update($h_L(x)$)
20   R.append((c, Σ, e, b))

---

is supposed to provide a density estimate of nearby distance vectors. Whether
a distance is nearby is decided based on its Mahalanobis distance to the representative. A candidate is a precursor of a representative. It will be turned into
a representative, if it gathers enough distance vectors around it. The intuition
behind these objects is that the landmarks provide a space with certain properties and guarantees, and the representatives and candidates are responsible for
modeling the density.

Given the current $\hat{g}$, the estimate is updated as follows (illustrated by Fig. 1):
The next instance x is first projected into the vector space by applying h_L.
Then the resulting distance vector v is tested against all representatives. If a
representative is found, v is forwarded to the corresponding density estimator.
Otherwise, v is tested against all candidates. If a candidate is found, v is assigned
to the candidate. Otherwise, v becomes a candidate itself. Algorithm 1 describes

the update procedure of RED in more detail. In lines 1–2, the EDO estimates are updated if the new instance is considered as a member of that representative. If it does not belong to any representative (line 5), we distinguish two cases: the instance does not belong to any existing candidate (lines 3–16) or it does (line 17–26). In the first case, the given instance becomes a candidate itself, as no matching representative nor candidate has been found. For this purpose, we determine the k closest representatives and compute a covariance matrix from their most recent samples. The result is a new covariance matrix, which will be used in the future to decide whether or not an instance belongs to the new candidate. In the second case (there are candidates to which the instance belongs), there is no need to create a new candidate. Instead, the buffers of matching candidates are simply updated. However, since the recent updates could result in buffer sizes of $\theta_{C \to R}$ instances, the update procedure finishes with checking the buffer size of all candidates (lines 21–25) and turning candidates to representatives that have more than $\theta_{C \to R}$ instances.

Membership Tests. Whether an instance belongs to a candidate or representative is decided by employing a multivariate normal density $\mathcal{N}(v; \Sigma)$, where v is the distance vector of the representative instance and Σ is a covariance matrix computed from instances in the neighborhood – when a new candidate is created, it is computed from recent samples of neighboring representatives. The membership decision is based on the Mahalanobis distance, which is computed as follows: $\sqrt{(x - r)^T \Sigma^{-1}(x - r)}$. Any vector with a Mahalanobis distance less than a user-defined threshold is then considered a member.

Handling of Noise. Almost all real-world applications suffer from certain degrees of noise. Hence, handling noise is of paramount importance for density estimators but in many cases difficult. In an online setting, the problem is even more severe, since future instances cannot be included into the decision making process. The EDO estimators employed by RED are able to handle noise, but in order to keep them as clean as possible, it is important that a noisy instance does not become a representative in the first place. Otherwise, this instance and every instance in its neighborhood becomes inevitably a part of the estimate. Therefore, RED distinguishes between candidates and representatives. If an instance cannot be assigned to an existing representative, it is first considered a candidate for becoming a representative. Only if enough instances are gathered around the candidate, it becomes an actual representative.

Concept Drift. In real-world applications, the distribution of data streams is changing constantly and a density estimator has to adapt to these changes to provide reliable estimates. In order to address this problem, RED pursues a timestamp-based solution. However, old instances are not simply discarded when they become too old, but candidates and representatives are discarded if no instance has been assigned to them within a certain period of time. This time period is specified as a parameter and can be adjusted according to the smallest probability values that should be covered by the estimate – using Chernoff bounds, the parameter can be computed with high confidence.

When setting this parameter, one should also consider the parameter $\theta_{C \to R}$, as a high value for $\theta_{C \to R}$ prevents rare instances from becoming a part of the density estimate. In our experiments, we usually set $\theta_{C \to R}$ to 100, which is large enough for a statistical test but not too large to exclude less frequent instances.

3.2   Distance Measure

With landmarks, high-dimensional data can be mapped to a lower dimensional vector space. But dependent on the number and the choice of the landmarks, the resulting vector space could still be relatively large, so that the distance measure has to be chosen with care. For high-dimensional spaces, the Manhattan distance (1-norm) or a fractional distance measure is usually the best choice [2], so that we prefer p-norms with small p. Employing a p-norm, the mapping $h_L : X_1 \times \ldots \times X_n \to V_1 \times \ldots \times V_m$ with $V_j \subseteq \mathbb{R}$, $1 \le j \le m$, is defined as

$$
h_L(\boldsymbol{x})[V_i] := \left(\sum_{X_j \in X} \| l_i[X_j] - \boldsymbol{x}[X_j] \|^p \right)^{\frac{1}{p}},
$$

where $\| \cdot \|$ computes the distance for the given variable values and is defined as the difference $\frac{l_i[X_j] - \boldsymbol{x}[X_j]}{\max (X_j) - \min (X_j)}$ for numeric values and $\frac{l_i[X_j] - \boldsymbol{x}[X_j]}{\#values} \in [0; 1]$ for nominal values. For p, we select values from the range $(0; 2]$, which corresponds to the Euclidean distance for $p = 2$, the Manhattan distance for $p = 1$, and to fractional norms for $0 < p < 1$. The denominator $\max (X_j) - \min (X_j)$ can only be estimated, as the currently observed minimum and maximum values cannot be determined with certainty in a streaming setting, making a correct normalization impossible. For typical applications, however, an estimate is probably more than sufficient, since extreme deviations are most likely due to a concept drift.

3.3   Choice of the Landmarks

If the data stream is projected into a vector space with lower dimensionality, information about the original instances will possibly be lost. In particular, some of the variable interdependencies are no longer visible, as the mapping h_L only adds up the distances of individual variables. As a consequence, instances may be projected into the same point of the vector space, i.e., there are $\boldsymbol{x}$ and $\boldsymbol{x}'$, such that $h_L(\boldsymbol{x}) = h_L(\boldsymbol{x}')$ but $\boldsymbol{x} \ne \boldsymbol{x}'$. If the original instances has only nominal variables ($\{X_1, \ldots, X_k\}$) and each variable has $|X_i|$, $1 \le i \le k$, many values, then there are already $\prod_{i=1}^{k} 2 \cdot (|X_i| - 1)$ many possible instances that are be mapped to the same distance value. RED would treat all of these instances equally when computing their density value, which poses no problem to the density estimate, as long as similar distances to the landmarks correspond to similar density values of the instances. But it implies that the information encoded by the distances to the landmarks needs to be sufficiently good. Therefore, we propose to choose the landmarks in such a way that the estimate approaches the accuracy of the underlying density estimator as $|L|$ approaches n:

Definition 1. *Let* $X := \{X_1, \ldots, X_n\}$ *be the set of variables, and let* m *be the requested number of landmarks. Then the first landmark* L_1 *is defined as* $(0)_{1 \le j \le n}$ *and landmark* L_{i+1}, $1 \le i < m$, *is defined as*

$$\left((i - j + 1) \cdot \frac{\max(X_j) - \min(X_j)}{m} \right)_{1 \le j < i} \circ (1) \circ (0)_{j > i}$$

where $\max(X_j)$ *and* $\min(X_j)$ *are the currently observed maximum and minimum, respectively. The set of landmarks is denoted by* $L := \{L_i \mid 1 \le i \le m\}$.

By construction of h_L and by the choice of the landmarks, the mapping h_L projects any two instances $\boldsymbol{x}$ and $\boldsymbol{x}'$ to different points in the vector space as long as $\boldsymbol{x} \neq \boldsymbol{x}'$, $|L| = n + 1$, and certain assumptions hold. (An example is given below.) Hence, the mapping becomes injective:

Lemma 1. *If* $|L| = n + 1$, *landmark* L_i, $i \in [1; n + 1]$, *is defined as in Definition 1, and* $\max$ *is the actual maximum for all* $X_j \in X$, *then the projection mapping* $h_L : X_1 \times \ldots \times X_n \to V_1 \times \ldots \times V_m$ *with* $V_j = \mathbb{R}$, $1 \le j \le m$, *is injective.*

Proof. Under the assumption that $\max(X_j)$ is the actual maximum for all $X_j \in X$, $L_j[X_j]$ is always larger than $h_L(\boldsymbol{x})[V_j]$. Hence, for $x_{j_1} \neq x_{j_2} \in X_j$, $h_L(\boldsymbol{x}_{j_1})[V_j]$ is not equal to $h_L(\boldsymbol{x}_{j_2})[V_j]$. (Please notice that this would not have been the case, if we had chosen $L_j[X_j]$ to be $\frac{1}{2} \cdot \max(X_j)$, since h_L does not consider the sign of differences, e.g., $(\min(X_j) + 0.2) - \frac{1}{2} \cdot \max(X_j)$ and $(max(X_j) - 0.2) + \frac{1}{2} \cdot \max(X_j)$ result in the same distance.)

So individual variable values do not cause two different instances to have the same distance vector. It remains to show that this property is preserved when computing the summation over all variable differences in h_L. For this purpose, we project $X_1 \times \ldots \times X_n$ into $\mathbb{R}^n$ and first show that, if $|L| = n + 1$, all vectors $\boldsymbol{x}$ that start in the origin of $\mathbb{R}^n$ and are mapped to the same distance vector have the same length. Let $h_L(\boldsymbol{x}) = (v_1, v_2, \ldots, v_m)$ be the distance vector of $\boldsymbol{x}$. We can determine the possible lengths of all instance vectors $\boldsymbol{x}$ that are mapped to $h_L(\boldsymbol{x})$ by finding the solution for the following system of equations:

$$\left(\begin{array}{ccccc|c} 1 & 0 & 0 & \ldots & 0 & v_1 \\ 1 \cdot s_1 & 1 & 0 & \ldots & 0 & v_2 \\ 2 \cdot s_1 & 1 \cdot s_2 & 1 & \ldots & 0 & v_3 \\ & & & \vdots & & \\ n \cdot s_1 & (n-1) \cdot s_2 & (n-2) \cdot s_3 & \ldots & 1 & v_n \end{array} \right),$$

where s_j equals $\frac{\max(X_j) - \min(X_j)}{m}$. Due to the choice of the landmarks, the left-hand side is a squared matrix with rank n. Hence, the system of equations has only one unique solution: $\boldsymbol{x} = A^{-1}\boldsymbol{b}$.

Dependent on the norm employed by h_L, there are fewer or more instances having the same length in $\mathbb{R}^n$. In case of the Euclidean norm, for example, the corresponding vectors having the same distances to the landmarks $L_2, \ldots, L_{n+1}$ lie on the border of a $(n-1)$-dimensional norm sphere (notice that L_1 is excluded here). In order to ensure that the mapping h_L is injective, we simply have to include the landmark L_1, which introduces another dimension and reduces the border of the $(n-1)$-dimensional norm sphere to a single point. The same approach is also valid for arbitrary p-norms, which we prove by constructing a contradiction: Let L be defined as in Definition 1. Assume that there are $\boldsymbol{x} \neq \boldsymbol{y}$ in the projected vector space, such that $\forall L_i \in L : \left(\sum_{X_j \in X} \|L_i[X_j] - x[X_j]\|^p \right)^{\frac{1}{p}} = \left(\sum_{X_j \in X} \|L_i[X_j] - y[X_j]\|^p \right)^{\frac{1}{p}}$. Due to the projection of $\boldsymbol{x}$ and $\boldsymbol{y}$ into $\mathbb{R}^n$ and due to the definition of $\| \cdot \|$, $\| \cdot \|$ becomes $|\cdot|$ in $\mathbb{R}^n$. From the first part of the proof, we can conclude that for all $L_i \in L$ and for all $X_j \in X$:

$$|L_i[X_j] - x[X_j]|^p = |L_i[X_j] - y[X_j]|^p$$
$$\Leftrightarrow |L_i[X_j] - x[X_j]| = |L_i[X_j] - y[X_j]|,$$

As this equation also has to hold for L_1 ($L_1 = (0)_{1 \leq i \leq n} \in L$) and as for all $X_j \in X : x[X_j], y[X_j] \geq 0$, this implies that for all $X_j \in X : x[X_j] = y[X_j]$, which contradicts the assumption that $\boldsymbol{x} \neq \boldsymbol{y}$.   □

This theorem makes a valuable statement about the validity of the method: Although $n + 1$ landmarks is probably infeasible for extremely high-dimensional data streams, we know that additional landmarks are beneficial for the accuracy of the method. Hence, it is up to the user whether the accuracy or memory consumption is more critical.

3.4   Correction Factor

If $|L| < n+1$, the projection mapping h_L maps several instances from the original data stream into the same point of the vector space. When RED estimates the density value of that point, it has actually estimated the density value of all instances that are mapped to it. Since we have no additional information on how to divide the density value among those points, we will divide it equally.

To obtain the density value from the original data stream, we have to multiply it by the integral $\int_{-\infty}^{+\infty} \int_{-\infty}^{+\infty} \ldots \int_{-\infty}^{+\infty} (x_i, x_{i+1}, \ldots, x_p) dx_{i+1} dx_{i+2} dx_n$, which is the expected density value of the variables $X_i, X_{i+1}, \ldots, X_n$. In case where all variables are discrete, this would simply be $\frac{1}{\prod_{i=1}^{k} |X_i|}$. For a continuous variable X_j, we will approximate the integral using a sampling technique. In particular, we employ a conditional density estimator for $f(X_j \mid V_1, V_2, \ldots V_m)$ and sum $f(x_j \mid v_1, v_2, \ldots, v_m)$ over $\min(X_j) \leq x_j \leq \max(X_j)$ with step size $\frac{\max(X_j) - \min(X_j)}{|S|}$ and sample size $|S|$. Hence, the correction factor is 1, if $|L| \geq n + 1$.

3.5   Illustrative Example

To illustrate the density estimation RED, we give a small example: We generated
a synthetic data stream of dimensionality 3, for which we selected the landmarks
$L_1 = (1, 0, 0)$, $L_2 = (0.2, 1, 0)$, and $L_3 = (0.3, 0.2, 1)$ and projected it into a vector
space of dimensionality 3. Subsequently, we also projected the data into a vector
space of dimensionality 2 – this time with the landmarks $L_1 = (1, 0, 0)$ and
$L_2 = (0.2, 1, 0)$. As Fig. 2 illustrates, the instance clusters are still visible after
applying h_L to the data (see (a) and (d)). The relative positions remain roughly
the same, but due to the landmarks, they capture a different area in the vector
space. For the vector space that is induced by two landmarks, the instance
clusters are arranged similarly. When the instances have been mapped to the
vector space, it is up to the representatives to model the density. The density
values of the original data stream can then be computed by retranslating the
density value using the *correctionFactor*, where the *correctionFactor* accounts
for the instances that have been mapped to the same point in the vector space.

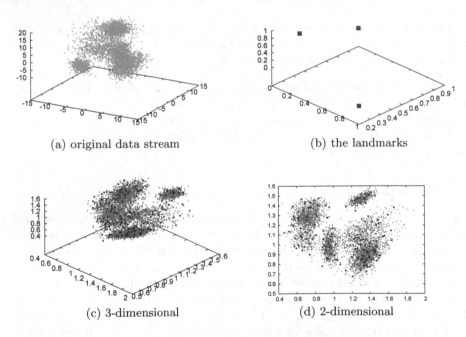

(a) original data stream   (b) the landmarks

(c) 3-dimensional   (d) 2-dimensional

Fig. 2. Illustrates a synthetic data stream (see Plot (a)) that is projected with three
landmarks (see Plot (b)) into a vector space of dimensionality 3 (see Plot (c)) and with
two landmarks to dimensionality 2 (see Plot (d)). Please notice the dark gray dots
in Plot (c) and (d) are the representatives. Due to the Mahalanobis distance, dense
regions have more representatives than sparse regions, which helps to model the density
more accurately. (Color figure online)

3.6 Consistency

The consistency of a density estimator is a desirable property, as it ensures that the estimator approaches the true density. We have already shown that the number of instances that are mapped to the same point in the vector space can be controlled by the number of landmarks. In the following, we will prove that RED estimates are consistent, if some further conditions hold:

Theorem 1. *If $|L| = n + 1$, L is defined as in Definition 1, min is the actual minimum for all $X_j \in X$, max is the actual maximum for all $X_j \in X$, and $\hat{g}$ is consistent, then $\hat{f}$ is consistent.*

Proof. If $|L|$ equals $n+1$, then the correction factor is 1, so that it remains to show that $\hat{f}(\boldsymbol{x}) = \hat{g}(h_L(\boldsymbol{x}))$. Furthermore, as Algorithm 1 partitions the vector space into independent subregions where each subregion has its own online density estimator, it suffices to show that the estimator of each subregion is consistent.

Since $|L|$ equals $n + 1$ and the assumptions for min and max hold, h_L is injective according to Lemma 1. Hence, for any two instances $\boldsymbol{x}_1$ and $\boldsymbol{x}_2$ that only differ in variable X_j by a small amount, it follows that

$$h_L(\boldsymbol{x}_1[V_i])^p - h_L(\boldsymbol{x}_2[V_i])^p = \sum_{k=1}^{p} c_k \cdot \|x_1 - x_2\|,$$

by definition of h_L, where $1 \leq i \leq n + 1$ and c_k are constants that have two factors: l_i and a constant that results from the binomial theorem. In other words, the density in the vector space is only shifted and compressed, but the original information of f is completely contained in g. Therefore, we can conclude that $\hat{f}(\boldsymbol{x}) = \hat{g}(h_L(\boldsymbol{x}))$. □

4 Evaluation

In this section, we analyze the behavior of RED[3] with respect to its parameters on synthetic datasets and evaluate its capability to estimate joint densities on several real-world datasets. RED has several parameters that have to be chosen: the number of landmarks $|L|$, the Mahalanobis distance, the threshold of becoming a representative $\theta_{C \to R}$, and the distance measure p. As $\theta_{C \to R}$ is mostly relevant for drift detection or for application tasks such as outlier detection, we do not discuss this parameter here. Also not discussed is the parameter p, as a detailed analysis would be in conflict with the given space constraints. Hence, we focus our experimental analysis of the parameters on the number of landmarks $|L|$ and the Mahalanobis distance. For $|L|$, we consider the values 2, 3, 5, 10, and 20. For the Mahalanobis distance, we consider the values 0.1, 0.5, 1.0, 2.0, 5.0, and 10.0.

---

[3] An implementation of RED is available as part of the MiDEO framework: https://github.com/geilke/mideo.

As datasets, we generated synthetic data consisting of 1, 2, 5, or 10 multi-variate Gaussians in a d-dimensional vector space with $d \in \{2, 3, 5, 10, 20\}$. The mean variables and covariance matrices were drawn independently and uniformly at random, where the values for the mean have been drawn from the interval $[-10; 10)$ and the values for the covariance matrix from the interval $[0.5; 3)$.

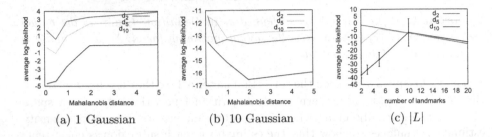

| (a) 1 Gaussian | (b) 10 Gaussian | (c) $|L|$ |

Fig. 3. For data of different dimensionality, the figures give some details about the behavior of RED when the Mahalanobis or the number of landmarks is increased. All plots are aggregated over all synthetic datasets. (Color figure online)

The influence of the number of landmarks is illustrated by Fig. 3. The shapes of the curves show that the performance is increasing with the number of landmarks until it reached the dimensionality of the dataset. So, for d_2, the peak performance is reached at 2, for d_5, the peak performance is reached at 5, and, for d_{10}, the peak performance is reached at 10. The increase for $|L| \leq d$ is completely in line with Theorem 1. The decrease for $|L| > d$ can be explained by the increase of the vector space: Due to higher dimensionality of the vector space, fewer and fewer instances share the same space and, hence, fewer instances are available to provide an estimate for this region. This effect is also responsible for the increase of the variance for increasing numbers of landmarks (as visible for d_{10}). Due to the small number of instances per region, the density estimators are more sensitive to smaller changes, which results in an increased variance. So generally, up to d landmarks are beneficial for the performance, but the closer $|L|$ gets to d, the more instances are required to compensate for the higher variance.

The effect of the Mahalanobis distance is summarized by Figs. 3 and 4. For lower dimensional datasets (e.g., d_2 and d_3), the performance is slightly degrading for minor increases of the Mahalanobis distance. For $M_{5.0}$ and $M_{10.0}$, however, we already see substantial improvements, which can be explained by fewer numbers of representatives, which have more instances at their disposal to provide a good estimate. For higher dimensional datasets (e.g., d_{10} and d_{20}), this trend is reverted, and we consistently observe that a low Mahalanobis distance is the better choice among the given selection. This can be explained by the possibility for each representative to specialize on regions with many instances. Otherwise, one representative would be responsible for a diverse set of instances. This observation is further supported by Fig. 3. If there is only one Gaussian, a

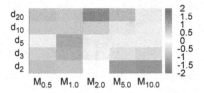

Fig. 4. The heatmap summarizes the effects of the Mahalanobis distance for data of different dimensionality d_i ($i \in \{2, 3, 5, 10, 20\}$). It shows the improvement (red) and degradation (blue) if RED uses a specific Mahalanobis distance ($M_{0.5}, M_{1.0}, M_{2.0}, M_{5.0}, M_{10.0}$) compared with a Mahalanobis distance of 0.1. (Color figure online)

higher Mahalanobis distance is beneficial. But if we have several Gaussians (e.g., 10), a larger Mahalanobis distance leads to an degradation of the performance. So generally, one can say the higher the dimensionality of the data, the lower should be the Mahalanobis distance.

If data mining and machine learning should be performed on RED estimates, the estimates have to describe the density of the data as accurately as possible. Since several compromises have been made to enable the estimation of data with many variables, we do not expect to outperform other density estimators. However, the performance should be in the same order of magnitude.

In order to evaluate RED on real-world data, we selected the state-of-the art online density estimation method, which is the online kernel density estimator oKDE by Kristan *et al.* [11], and compared the performance on four publicly available datasets: covertype (581, 012 instances, 54 attributes), electricity (45, 313 instances, 9 attributes), letter (19, 999 instances, 17 attributes), and shuttle (58, 000 instances, 10 attributes). For every parameter setting and for every dataset, the average log-likelihood is computed 15 times. In order to take possible concept drifts into account, the log-likelihood was computed in a prequential way, i.e., the log-likelihood of a given instance has been computed before using it for training. The instances used to compute the initial estimator (the first 100) were excluded from this computation.

The results are summarized in Fig. 5. The most apparent observation is the performance increase with increasing numbers of landmarks. As already observed on synthetic data, this performance increase is accompanied with an increase of the variance. Dependent on the dataset the effect is visible to different degrees and does not even depend on the number of variables (electricity vs. shuttle). But most surprising is probably the abrupt increase on the covertype dataset. A more detailed analysis of this matter revealed that this is due to the nature of its instances. Covertype has 10 numerical variables, 43 binary variables, and one further nominal variable. As most of the binary variables are 0 for almost all instances, they do not provide sufficient additional information to justify more landmarks. Hence, with every new landmark, the estimators have fewer instances and provide estimates that are more sensitive to individual instances.

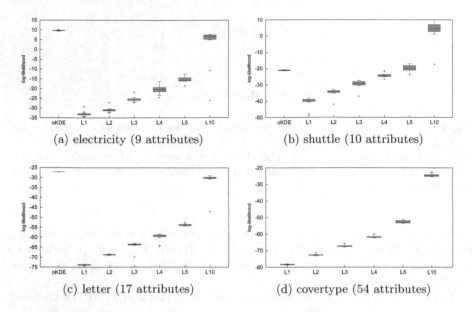

(a) electricity (9 attributes)   (b) shuttle (10 attributes)

(c) letter (17 attributes)   (d) covertype (54 attributes)

Fig. 5. The figure shows a comparison of oKDE with RED for varying numbers of landmarks ($|L| \in \{1, 2, 3, 4, 5, 10\}$) in the case of RED. On the y-axis is the average log-likelihood computed in a prequential way. (Color figure online)

When we compare RED to oKDE, we observe that RED performs surprisingly well. For electricity and letter, RED is approaching the performance of oKDE. For shuttle, the performance is even better than that of oKDE, if RED uses 5 or 10 landmarks. For covertype, oKDE was not even able to process the dataset within 15 hours, whereas RED was able to produce results for all landmark sizes. Hence, RED is able to compete with oKDE on low dimensional data, when a sufficient number of landmarks is chosen, while it can also handle high-dimensional data (e.g., more than 50 attributes). How many landmarks are sufficient for a specific datasets depends on two aspects: its intrinsic dimension and the selected landmarks (because the landmarks determine which dimensions are considered for evaluating the distance function).

Considering that we made several compromises to enable the density estimation of data streams with many variables, RED performed very good on low- and medium-sized data streams. The insights we gained about the parameters should be a useful guide for applications to other data streams. Alternatively, one could also follow a multi-layer approach where three or four RED estimators are initialized with different parameter settings. When enough instances are available, one could then choose the estimator with highest log-likelihood. This is enabled by excellent runtime behavior.

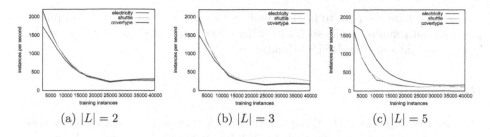

(a) $|L| = 2$ (b) $|L| = 3$ (c) $|L| = 5$

Fig. 6. Number of instances processed per second. (Color figure online)

4.1 Runtime

RED offers many opportunities for parallelism. For our evaluation, however, we wanted to keep the implementation simple and avoided any advanced optimizations. The current implementation is still fast enough for data streams applications, as Fig. 6 summarizes. The general behavior is the same for all tested datasets (electricity, shuttle, and covertype). In the beginning, when almost no representatives are discovered and the density estimators are still very simple, the RED estimate is able to process 1400 or more instances per second. Then, with increasing numbers of training instances, it drops to several hundred instances per second before it stabilizes (at 100 to 300 instances per second, depending on the number of landmarks). This is in line with the expected behavior of the method. First, the instances are required to find a partitioning of the vector space. When this partitioning is converging and the corresponding density estimators have received a larger number of instances, the processing speed of the estimator becomes more and more constant.

5   Conclusions

We proposed a new approach for estimating densities of heterogeneous data streams with many variables, which reduces the dimensionality of the data by projecting it into a vector space. In particular, the algorithm chooses a small number of instances, called landmarks, and creates a vector space in which the position of each instance is computed in terms of distances to the landmarks. Subsequently, the density is described by partitioning the vector space with representatives and estimating the density of each partition by employing online density estimators. In the theoretical analysis, we showed the validity and consistency of the presented density estimator. With experiments on synthetic and real-world data, we showed that – despite the compromises that had to be made to enable the estimation of densities with many variables – RED produces estimates having a comparable performance to that of state-of-the art density estimates. Keeping in mind that other approaches are possibly not able to handle large numbers of variables, RED could be, at this point, the only available option for some data streams.

In the future, we plan to further analyze the choice of landmarks and plan to perform data mining tasks such as outlier detection, the detection of emerging trends and inference on RED estimates.

References

1. Aggarwal, C.C., Han, J., Wang, J., Yu, P.S.: A framework for clustering evolving data streams. In: Proceedings of the 29th International Conference on Very Large Data Bases (VLDB), pp. 81–92 (2003)
2. Aggarwal, C.C., Hinneburg, A., Keim, D.A.: On the surprising behavior of distance metrics in high dimensional spaces. In: Proceedings of the 8th International Conference on Database Theory, pp. 420–434 (2001)
3. Cesa-Bianchi, N., Lugosi, G.: Prediction, Learning, and Games. Cambridge University Press (2006)
4. Davies, S., Moore, A.W.: Interpolating conditional density trees. In: Proceedings of the 18th Conference in Uncertainty in Artificial Intelligence, pp. 119–127 (2002)
5. Frank, E., Bouckaert, R.R.: Conditional density estimation with class probability estimators. In: Proceedings of the First Asian Conference on Machine Learning, pp. 65–81 (2009)
6. Geilke, M., Karwath, A., Frank, E., Kramer, S.: Online estimation of discrete densities. In: Proceedings of the 13th IEEE International Conference on Data Mining, pp. 191–200 (2013)
7. Geilke, M., Karwath, A., Kramer, S.: A probabilistic condensed representation of data for stream mining. In: Proceedings of the 1st International Conference on Data Science and Advanced Analytics, pp. 297–303 (2014)
8. Hwang, J.N., Lay, S.R., Lippman, A.: Nonparametric multivariate density estimation: a comparative study. IEEE Trans. Signal Process. **42**(10), 2795–2810 (1994)
9. Kim, J., Scott, C.D.: Robust kernel density estimation. J. Mach. Learn. Res. **13**, 2529–2565 (2012)
10. Kristan, M., Leonardis, A.: Online discriminative kernel density estimation. In: 20th International Conference on Pattern Recognition, pp. 581–584 (2010)
11. Kristan, M., Leonardis, A., Skocaj, D.: Multivariate online kernel density estimation with Gaussian kernels. Pattern Recogn. **44**(10–11), 2630–2642 (2011)
12. Peherstorfer, B., Pflüger, D., Bungartz, H.: Density estimation with adaptive sparse grids for large data sets. In: Proceedings of the 2014 SIAM International Conference on Data Mining, pp. 443–451 (2014)
13. Ram, P., Gray, A.G.: Density estimation trees. In: Proceedings of the 17th ACM SIGKDD International Conference on Knowledge Discovery and Data Mining, pp. 627–635 (2011)
14. Scott, D.W., Sain, S.R.: Multi-Dimensional Density Estimation, pp. 229–263. Elsevier, Amsterdam (2004)
15. Wu, K., Zhang, K., Fan, W., Edwards, A., Yu, P.S.: Rs-forest: a rapid density estimator for streaming anomaly detection. In: Proceedings of the 14th International Conference on Data Mining, pp. 600–609 (2014)
16. Xu, M., Ishibuchi, H., Gu, X., Wang, S.: Dm-KDE: dynamical kernel density estimation by sequences of KDE estimators with fixed number of components over data streams. Front. Comput. Sci. **8**(4), 563–580 (2014)

Graphical Model Sketch

Branislav Kveton[1]([✉]), Hung Bui[1], Mohammad Ghavamzadeh[1],
Georgios Theocharous[1], S. Muthukrishnan[2], and Siqi Sun[3]

[1] Adobe Research, San Jose, CA, USA
{kveton,hubui,ghavamza,theochar}@adobe.com
[2] Department of Computer Science, Rutgers, New Brunswick, NJ, USA
muthu@cs.rutgers.edu
[3] TTI, Chicago, IL, USA
siqi.sun@ttic.edu

Abstract. Structured high-cardinality data arises in many domains, and poses a major challenge for both modeling and inference. Graphical models are a popular approach to modeling structured data but they are unsuitable for high-cardinality variables. The count-min (CM) sketch is a popular approach to estimating probabilities in high-cardinality data but it does not scale well beyond a few variables. In this work, we bring together the ideas of graphical models and count sketches; and propose and analyze several approaches to estimating probabilities in structured high-cardinality streams of data. The key idea of our approximations is to use the structure of a graphical model and approximately estimate its factors by "sketches", which hash high-cardinality variables using random projections. Our approximations are computationally efficient and their space complexity is independent of the cardinality of variables. Our error bounds are multiplicative and significantly improve upon those of the CM sketch, a state-of-the-art approach to estimating probabilities in streams. We evaluate our approximations on synthetic and real-world problems, and report an order of magnitude improvements over the CM sketch.

1 Introduction

Structured high-cardinality data arises in numerous domains, and poses a major challenge for modeling and inference. A common goal in online advertising is to estimate the probability of events, such as page views, over multiple high-cardinality variables, such as the location of the user, the referring page, and the purchased product. A common goal in natural language processing is to estimate the probability of n-grams over a dictionary of 100k words. Graphical models [9] are a popular approach to modeling multivariate data. However, when the cardinality of random variables is high, they are expensive to store and reason with. For instance, a graphical model over two variables with $M = 10^5$ values each may consume $M^2 = 10^{10}$ space.

A *sketch* [17] is a data structure that summarizes streams of data such that any two sketches of individual streams can be combined space efficiently into

© Springer International Publishing AG 2016
P. Frasconi et al. (Eds.): ECML PKDD 2016, Part I, LNAI 9851, pp. 81–97, 2016.
DOI: 10.1007/978-3-319-46128-1_6

the sketch of the combined stream. Numerous problems can be solved efficiently by surprisingly simple sketches, such as estimating the frequency of values in streams [3,4,15], finding heavy hitters [5], estimating the number of unique values [7,8], or even approximating low-rank matrices [12,18]. In this work, we sketch a graphical model in a small space. Let $(x^{(t)})_{t=1}^n$ be a stream of n observations from some distribution P, where $x^{(t)} \in [M]^K$ is a K-dimensional vector and P factors according to a known graphical model $\mathcal{G}$. Let $\bar{P}$ be the maximum-likelihood estimate (MLE) of P from $(x^{(t)})_{t=1}^n$ conditioned on $\mathcal{G}$. Then our goal is to approximate $\bar{P}$ with $\hat{P}$ such that $\hat{P}(x) \approx \bar{P}(x)$ for any $x \in [M]^K$ with at least $1 - \delta$ probability; in the space that does not depend on the cardinality M of the variables in $\mathcal{G}$. In our motivating examples, x is an n-gram or the feature vector associated with page views.

This paper makes three contributions. First, we propose and carefully analyze three natural approximations to the MLE in graphical models with high-cardinality variables. The key idea of our approximations is to leverage the structure of the graphical model $\mathcal{G}$ and approximately estimate its factors by "sketches". Therefore, we refer to our approximations as *graphical model sketches*. Our best approximation, GMFactorSketch, guarantees that $\hat{P}(x)$ is a constant-factor multiplicative approximation to $\bar{P}(x)$ for any x with probability of at least $1 - \delta$ in $O(K^2 \log(K/\delta)\Delta^{-1}(x))$ space, where K is the number of variables and $\Delta(x)$ measures the hardness of query x. The dependence on $\Delta(x)$ is generally unavoidable and we show this in Sect. 5.4. Second, we prove that GMFactorSketch yields better approximations than the count-min (CM) sketch [4], a state-of-the-art approach to estimating the frequency of values in streams (Sect. 6). Third, we evaluate our approximations on both synthetic and real-world problems. Our results show that GMFactorSketch outperforms the CM sketch and our other approximations, as measured by the error in estimating $\bar{P}$ at the same space.

Our work is related to Matusevych *et al.* [13], who proposed several extensions of the CM sketch, one of which is GMFactorSketch. This approximation is not analyzed and it is evaluated only on a graphical model with three variables. We present the first analysis of GMFactorSketch, and prove that it is superior to other natural approximations and the CM sketch. We also evaluate GMFactorSketch on an order of magnitude larger problems than Matusevych *et al.* [13]. McGregor and Vu [14] proposed and analyzed a space-efficient streaming algorithm that tests if the stream of data is consistent with a graphical model. Several recent papers applied hashing to speeding up inference in graphical models [1,6]. These papers do not focus on high-cardinality variables and are only loosely related to our work, because of using hashing in graphical models. We also note that the problem of representing conditional probabilities in graphical models efficiently has been studied extensively, as early as in Boutilier *et al.* [2]. Our paper is different from this line of work because we do not assume any sparsity or symmetry in data; and our approximations are suitable for the streaming setting.

We denote $\{1, \ldots, K\}$ by $[K]$. The cardinality of set A is $|A|$. We denote random variables by capital letters, such as X, and their values by small letters, such as x. We assume that $X = (X_1, \ldots, X_K)$ is a K-dimensional variable; and we refer to its k-th component by X_k and its value by x_k.

2   Background

This section reviews the two main components of our solutions.

2.1   Count-Min Sketch

Let $(x^{(t)})_{t=1}^n$ be a stream of n observations from distribution P, where $x^{(t)} \in [M]^K$ is a K-dimensional vector. Suppose that we want to estimate:

$$\tilde{P}(x) = \frac{1}{n} \sum_{t=1}^n \mathbb{1}\Big\{x = x^{(t)}\Big\}, \tag{1}$$

the frequency of observing any x in $(x^{(t)})_{t=1}^n$. This problem can be solved in $O(M^K)$ space, by counting all unique values in $(x^{(t)})_{t=1}^n$. This solution is impractical when K and M are large. Cormode and Muthukrishnan [4] proposed an approximate solution to this problem, the *count-min (CM) sketch*, which estimates $\tilde{P}(x)$ in the space independent of M^K. The sketch consists of d hash tables with m bins, $c \in \mathbb{N}^{d \times m}$. The hash tables are initialized with zeros. At time t, they are updated with observation $x^{(t)}$ as:

$$c(i, y) \leftarrow c(i, y) + \mathbb{1}\Big\{y = h^i(x^{(t)})\Big\}$$

for all $i \in [d]$ and $y \in [m]$, where $h^i : [M]^K \to [m]$ is the i-th *hash function*. The hash functions are *random* and *pairwise-independent*. The frequency $\tilde{P}(x)$ is estimated as:

$$P_{\mathrm{CM}}(x) = \frac{1}{n} \min_{i \in [d]} c(i, h^i(x)). \tag{2}$$

Cormode and Muthukrishnan [4] showed that $P_{\mathrm{CM}}(x)$ approximates $\tilde{P}(x)$ for any $x \in [M]^K$, with at most ε error and at least $1 - \delta$ probability, in $O((1/\varepsilon) \log(1/\delta))$ space. Note that the space is independent of M^K. We state this result more formally below.

Theorem 1. *Let $\tilde{P}$ be the distribution in (1) and P_{CM} be its CM sketch in (2). Let $d = \log(1/\delta)$ and $m = e/\varepsilon$. Then for any $x \in [M]^K$, $\tilde{P}(x) \leq P_{\mathrm{CM}}(x) \leq \tilde{P}(x) + \varepsilon$ with at least $1 - \delta$ probability. The space complexity of P_{CM} is $(e/\varepsilon) \log(1/\delta)$.*

The CM sketch is popular because high-quality approximations, with at most ε error, can be computed in $O(1/\varepsilon)$ space.[1] Other similar sketches, such as Charikar *et al.* [3], require $O(1/\varepsilon^2)$ space.

---

[1] https://sites.google.com/site/countminsketch/.

2.2 Bayesian Networks

Graphical models are a popular tool for modeling and reasoning with random variables [10], and have many applications in computer vision [16] and natural language processing [11]. In this work, we focus on Bayesian networks [9], which are directed graphical models.

A *Bayesian network* is a probabilistic graphical model that represents conditional independencies of random variables by a directed graph. In this work, we define it as a pair $(\mathcal{G}, \theta)$, where $\mathcal{G}$ is a directed graph and θ are its parameters. The graph $\mathcal{G} = (V, E)$ is defined by its nodes $V = \{X_1, \dots, X_K\}$, one for each random variable, and edges E. For simplicity of exposition, we assume that $\mathcal{G}$ is a *tree* and X_1 is its root. We relax this assumption in Sect. 3. Under this assumption, each node X_k for $k \geq 2$ has one parent and the probability of $x = (x_1, \dots, x_K)$ factors as:

$$P(x) = P_1(x_1) \prod_{k=2}^{K} P_k(x_k \mid x_{\mathsf{pa}(k)}),$$

where $\mathsf{pa}(k)$ is the *index of the parent variable* of X_k, and we use shorthands:

$$P_k(i) = P(X_k = i), \quad P_k(i, j) = P(X_k = i, X_{\mathsf{pa}(k)} = j), \quad P_k(i \mid j) = \frac{P_k(i, j)}{P_{\mathsf{pa}(k)}(j)}.$$

Let $\mathrm{dom}(X_k) = M$ for all $k \in [K]$. Then our graphical model is parameterized by M *prior probabilities* $P_1(i)$, for any $i \in [M]$; and $(K - 1)M^2$ *conditional probabilities* $P_k(i \mid j)$, for any $k \in [K] - \{1\}$ and $i, j \in [M]$.

Let $(x^{(t)})_{t=1}^{n}$ be n observations of X. Then the *maximum-likelihood estimate (MLE)* of P conditioned on $\mathcal{G}$, $\bar{\theta} = \arg\max_{\theta} P((x^{(t)})_{t=1}^{n} \mid \theta, \mathcal{G})$, has a closed-form solution:

$$\bar{P}(x) = \bar{P}_1(x_1) \prod_{k=2}^{K} \bar{P}_k(x_k \mid x_{\mathsf{pa}(k)}), \tag{3}$$

where we abbreviate $P(X = x \mid \bar{\theta}, \mathcal{G})$ as $\bar{P}(x)$, and define:

$$\forall i \in [M] : \bar{P}_k(i) = \frac{1}{n} \sum_{t=1}^{n} \mathbb{1}\left\{x_k^{(t)} = i\right\},$$

$$\forall i, j \in [M] : \bar{P}_k(i, j) = \frac{1}{n} \sum_{t=1}^{n} \mathbb{1}\left\{x_k^{(t)} = i, x_{\mathsf{pa}(k)}^{(t)} = j\right\},$$

$$\forall i, j \in [M] : \bar{P}_k(i \mid j) = \bar{P}_k(i, j) / \bar{P}_{\mathsf{pa}(k)}(j).$$

3   Model

Let $(x^{(t)})_{t=1}^{n}$ be a stream of n observations from distribution P, where $x^{(t)} \in [M]^K$ is a K-dimensional vector. Our objective is to approximate $\bar{P}(x)$ in (3), the frequency of observing x as given by the MLE of P from $(x^{(t)})_{t=1}^n$ conditioned on graphical model $\mathcal{G}$. This objective naturally generalizes that of the CM sketch in (1), which is the MLE of P from $(x^{(t)})_{t=1}^n$ without any assumptions on the structure of P. For simplicity of exposition, we assume that $\mathcal{G}$ is a tree (Sect. 2.2). Under this assumption, $\bar{P}$ can be represented exactly in $O(KM^2)$ space. This is not feasible in our problems of interest, where typically $M \geq 10^4$.

The key idea in our solutions is to estimate a surrogate parameter $\hat{\theta}$. We estimate $\hat{\theta}$ on the same graphical model as $\bar{\theta}$. The difference is that $\hat{\theta}$ parameterizes a graphical model where each factor is represented by $O(m)$ hashing bins, where $m \ll M^2$. Our proposed models consume $O(Km)$ space, a significant reduction from $O(KM^2)$; and guarantee that $\hat{P}(x) \approx \bar{P}(x)$ for any $x \in [M]^K$ and observations $(x^{(t)})_{t=1}^n$ up to time n, where we abbreviate $P(X = x \mid \hat{\theta}, \mathcal{G})$ as $\hat{P}(x)$. More precisely:

$$\bar{P}(x) \prod_{k=1}^K [1 - \varepsilon_k] \leq \hat{P}(x) \leq \bar{P}(x) \prod_{k=1}^K [1 + \varepsilon_k] \tag{4}$$

for any $x \in [M]^K$ with at least $1 - \delta$ probability, where $\hat{P}$ is factored in the same way as $\bar{P}$. Each term ε_k is $O(1/m)$, where m is the number of hashing bins. Therefore, the quality of our approximations improves as m increases. More precisely, if m is chosen such that $\varepsilon_k \leq 1/K$ for all $k \in [K]$, we get:

$$[2/(3e)]\bar{P}(x) \leq \hat{P}(x) \leq e\bar{P}(x) \tag{5}$$

for $K \geq 2$ since $\prod_{k=1}^K (1 + \varepsilon_k) \leq (1 + 1/K)^K \leq e$ for $K \geq 1$ and $\prod_{k=1}^K (1 - \varepsilon_k) \geq (1 - 1/K)^K \geq 2/(3e)$ for $K \geq 2$. Therefore, $\hat{P}(x)$ is a constant-factor multiplicative approximation to $\bar{P}(x)$. As in the CM sketch, we do not require that $\hat{P}(x)$ sum up to 1.

4   Summary of Main Results

The main contribution of our work is that we propose and analyze three approaches to the MLE in graphical models with high-cardinality variables. Our first proposed algorithm, GMHash (Sect. 5.1), approximates $\bar{P}(x)$ as the product of $K - 1$ conditionals and a prior, one for each variable in $\mathcal{G}$. Each conditional is estimated as a ratio of two hashing bins. GMHash guarantees (5) for any $x \in [M]^K$ with at least $1 - \delta$ probability in $O(K^3 \delta^{-1} \Delta^{-1}(x))$ space, where $\Delta(x)$ is a query-specific constant and the number of hashing bins is set as $m = \Omega(K^2 \delta^{-1})$. We discuss $\Delta(x)$ at the end of this section. Since δ is typically small, the dependence on $1/\delta$ is undesirable.

Our second algorithm, GMSketch (Sect. 5.2), approximates $\bar{P}(x)$ as the median of d probabilities, each of which is estimated by GMHash. GMSketch guarantees (5) for any $x \in [M]^K$ with at least $1 - \delta$ probability in $O(K^3 \log(1/\delta)\Delta^{-1}(x))$ space, when we set $m = \Omega(K^2 \Delta^{-1}(x))$

and $d = \Omega(\log(1/\delta))$. The main advantage over GMHash is that the space is $O(\log(1/\delta))$ instead of $O(1/\delta)$.

Our last algorithm, GMFactorSketch (Sect. 5.3), approximates $\bar{P}(x)$ as the product of $K - 1$ conditionals and a prior, one for each variable. Each conditional is estimated as a ratio of two count-min sketches. GMFactorSketch guarantees (5) for any $x \in [M]^K$ with at least $1 - \delta$ probability in $O(K^2 \log(K/\delta)\Delta^{-1}(x))$ space, when we set $m = \Omega(K\Delta^{-1}(x))$ and $d = \Omega(\log(K/\delta))$. The key improvement over GMSketch is that the space is $O(K^2)$ instead of being $O(K^3)$. In summary, GMFactorSketch is the best of our proposed solutions. We demonstrate this empirically in Sect. 7.

The query-specific constant $\Delta(x) = \min_{k \in [K]-\{1\}} \bar{P}_k(x_k, x_{\mathsf{pa}(k)})$ is the minimum probability that the values of any variable-parent pair in x co-occur in $(x^{(t)})_{t=1}^n$. This probability can be small and our algorithms are unsuitable for estimating $\bar{P}(x)$ in such cases. Note that this does not imply that $\bar{P}(x)$ cannot be small. Unfortunately, the dependence on $\Delta(x)$ is generally unavoidable and we show this in Sect. 5.4.

The assumption that $\mathcal{G}$ is a tree is only for simplicity of exposition. Our algorithms and their analysis generalize to the setting where $X_{\mathsf{pa}(k)}$ is a vector of parent variables and $x_{\mathsf{pa}(k)}$ are their values. The only change is in how the pair $(x_k, x_{\mathsf{pa}(k)})$ is hashed.

5 Algorithms and Analysis

All of our algorithms hash the values of each variable in graphical model $\mathcal{G}$, and each variable-parent pair, to m bins up to d times. We denote the i-th hash function of variable X_k by h_k^i and the associated hash table by $c_k(i, \cdot)$. This hash table approximates $n\bar{P}_k(\cdot)$. The i-th hash function of the variable-parent pair $(X_k, X_{\mathsf{pa}(k)})$ is also h_k^i, and the associated hash table is $\bar{c}_k(i, \cdot)$. This hash table approximates $n\bar{P}_k(\cdot, \cdot)$. Our algorithms differ in how the hash tables are aggregated.

We define the notion of a *hash*, which is a tuple $h = (h_1, \ldots, h_K)$ of K randomly drawn hash functions $h_k : \mathbb{N} \to [m]$, one for each variable in $\mathcal{G}$. We make the assumption that hashes are pairwise-independent. We say that hashes h^i and h^j are *pairwise-independent* when h_k^i and h_k^j are pairwise-independent for all $k \in [K]$. These kinds of hash functions can be computed fast and stored in a very small space [4].

5.1 Algorithm GMHash

The pseudocode of our first algorithm, GMHash, is in Algorithm 1. It approximates $\bar{P}(x)$ as the product of $K - 1$ conditionals and a prior, one for each variable X_k. Each conditional is estimated as a ratio of two hashing bins:

$$\hat{P}_k(x_k \mid x_{\mathsf{pa}(k)}) = \frac{\bar{c}_k(h_k(x_k + M(x_{\mathsf{pa}(k)} - 1)))}{c_{\mathsf{pa}(k)}(h_{\mathsf{pa}(k)}(x_{\mathsf{pa}(k)}))},$$

Algorithm 1. GMHash: Hashed conditionals and priors.

Input: Point query $x = (x_1, \ldots, x_K)$

$$\hat{P}_1(x_1) \leftarrow \frac{c_1(h_1(x_1))}{n}$$

for all $k = 2, \ldots, K$ **do**

$$\hat{P}_k(x_k \mid x_{\mathsf{pa}(k)}) \leftarrow \frac{\bar{c}_k(h_k(x_k + M(x_{\mathsf{pa}(k)} - 1)))}{c_{\mathsf{pa}(k)}(h_{\mathsf{pa}(k)}(x_{\mathsf{pa}(k)}))}$$

$$\hat{P}(x) \leftarrow \hat{P}_1(x_1) \prod_{k=2}^{K} \hat{P}_k(x_k \mid x_{\mathsf{pa}(k)})$$

Output: Point answer $\hat{P}(x)$

where $\bar{c}_k(h_k(x_k + M(x_{\mathsf{pa}(k)} - 1)))$ is the number of times that hash function h_k maps $(x_k^{(t)}, x_{\mathsf{pa}(k)}^{(t)})$ to the same bin as $(x_k, x_{\mathsf{pa}(k)})$ in n steps, and $c_k(h_k(x_k))$ is the number of times that h_k maps $x_k^{(t)}$ to the same bin as x_k in n steps. Note that $(x_k, x_{\mathsf{pa}(k)})$ can be represented equivalently as $x_k + M(x_{\mathsf{pa}(k)} - 1)$. The prior $\bar{P}_1(x_1)$ is estimated as:

$$\hat{P}_1(x_1) = \frac{1}{n} c_1(h_1(x_1)).$$

At time t, the hash tables are updated as follows. Let $x^{(t)}$ be the observation. Then for all $k \in [K], y \in [m]$:

$$c_k(y) \leftarrow c_k(y) + \mathbb{1}\left\{ y = h_k(x_k^{(t)}) \right\},$$

$$\bar{c}_k(y) \leftarrow \bar{c}_k(y) + \mathbb{1}\left\{ y = h_k(x_k^{(t)} + M(x_{\mathsf{pa}(k)}^{(t)} - 1)) \right\}.$$

This update takes $O(K)$ time.

GMHash maintains $2K - 1$ hash tables with m bins each, one for each variable and one for each variable-parent pair in $\mathcal{G}$. Therefore, it consumes $O(Km)$ space. Now we show that $\hat{P}$ is a good approximation of $\bar{P}$.

Theorem 2. *Let $\hat{P}$ be the estimator from Algorithm 1. Let h be a random hash and m be the number of bins in each hash function. Then for any x:*

$$\bar{P}(x) \prod_{k=1}^{K} (1 - \varepsilon_k) \leq \hat{P}(x) \leq \bar{P}(x) \prod_{k=1}^{K} (1 + \varepsilon_k)$$

holds with at least $1 - \delta$ probability, where:

$$\varepsilon_1 = 2K[\bar{P}_1(x_1)\delta m]^{-1}, \quad \forall k \in [K] - \{1\} : \varepsilon_k = 2K[\bar{P}_k(x_k, x_{\mathsf{pa}(k)})\delta m]^{-1}.$$

Proof. The proof is in Appendix. The key idea is to show that the number of bins m can be chosen such that:

$$|\hat{P}_k(x_k \mid x_{\mathsf{pa}(k)}) - \bar{P}_k(x_k \mid x_{\mathsf{pa}(k)})| > \varepsilon_k \tag{6}$$

Algorithm 2. GMSketch: Median of d GMHash estimates.

Input: Point query $x = (x_1, \ldots, x_K)$

for all $i = 1, \ldots, d$ **do**

$$\hat{P}_1^i(x_1) \leftarrow \frac{c_1(i, h_1^i(x_1))}{n}$$

    for all $k = 2, \ldots, K$ **do**

$$\hat{P}_k^i(x_k \mid x_{\mathsf{pa}(k)}) \leftarrow \frac{\bar{c}_k(i, h_k^i(x_k + M(x_{\mathsf{pa}(k)} - 1)))}{c_{\mathsf{pa}(k)}(i, h_{\mathsf{pa}(k)}^i(x_{\mathsf{pa}(k)}))}$$

$$\hat{P}^i(x) \leftarrow \hat{P}_1^i(x_1) \prod_{k=2}^{K} \hat{P}_k^i(x_k \mid x_{\mathsf{pa}(k)})$$

$$\hat{P}(x) \leftarrow \mathrm{median}_{\,i \in [d]} \, \hat{P}^i(x)$$

Output: Point answer $\hat{P}(x)$

---

is not likely for any $k \in [K] - \{1\}$ and $\varepsilon_1, \ldots, \varepsilon_K > 0$. In other words, we argue that our estimate of each conditional $\bar{P}_k(x_k \mid x_{\mathsf{pa}(k)})$ can be arbitrary precise. By Lemma 1 in Appendix, the necessary conditions for event (6) are:

$$\frac{1}{n} c_{\mathsf{pa}(k)}(h_{\mathsf{pa}(k)}(x_{\mathsf{pa}(k)})) - \bar{P}_{\mathsf{pa}(k)}(x_{\mathsf{pa}(k)}) > \varepsilon_k \alpha_k,$$

$$\frac{1}{n} \bar{c}_k(h_k(x_k + M(x_{\mathsf{pa}(k)} - 1))) - \bar{P}_k(x_k, x_{\mathsf{pa}(k)}) > \varepsilon_k \alpha_k,$$

where $\alpha_k = \bar{P}_{\mathsf{pa}(k)}(x_{\mathsf{pa}(k)})$ is the frequency that $X_{\mathsf{pa}(k)} = x_{\mathsf{pa}(k)}$ in $(x^{(t)})_{t=1}^n$. In short, event (6) can happen only if GMHash significantly overestimates either $\bar{P}_{\mathsf{pa}(k)}(x_{\mathsf{pa}(k)})$ or $\bar{P}_k(x_k, x_{\mathsf{pa}(k)})$. We bound the probability of these events using Markov's inequality (Lemma 2 in Appendix) and then get that none of the events in (6) happen with at least $1 - \delta$ probability when the number of hashing bins $m \geq \sum_{k=1}^{K}(2/(\varepsilon_k \alpha_k \delta))$. Finally, we choose appropriate $\varepsilon_1, \ldots, \varepsilon_K$.

Theorem 2 shows that $\hat{P}(x)$ is a multiplicative approximation to $\bar{P}(x)$. The approximation improves with the number of bins m because all error terms ε_k are $O(1/m)$. The accuracy of the approximation depends on the frequency of interaction between the values in x. In particular, if $\bar{P}_k(x_k, x_{\mathsf{pa}(k)})$ is sufficiently large for all $k \in [K] - \{1\}$, the approximation is good even for small m. More precisely, under the assumptions that:

$$m \geq 2K^2[\bar{P}_1(x_1)\delta]^{-1}, \quad \forall k \in [K] - \{1\} : m \geq 2K^2[\bar{P}_k(x_k, x_{\mathsf{pa}(k)})\delta]^{-1},$$

all $\varepsilon_k \leq 1/K$ and the bound in Theorem 2 reduces to (5) for $K \geq 2$.

5.2 Algorithm GMSketch

The pseudocode of our second algorithm, GMSketch, is in Algorithm 2. The algorithm approximates $\bar{P}(x)$ as the median of d probability estimates:

$$\hat{P}(x) = \mathrm{median}_{\,i \in [d]} \, \hat{P}^i(x).$$

Each $\hat{P}^i(x)$ is computed by one instance of GMHash, which is associated with the hash $h^i = (h_1^i, \ldots, h_K^i)$. At time t, the hash tables are updated as follows. Let $x^{(t)}$ be the observation. Then for all $k \in [K], i \in [d], y \in [m]$:

$$c_k(i, y) \leftarrow c_k(i, y) + \mathbb{1}\left\{y = h_k^i(x_k^{(t)})\right\}, \tag{7}$$

$$\bar{c}_k(i, y) \leftarrow \bar{c}_k(i, y) + \mathbb{1}\left\{y = h_k^i(x_k^{(t)} + M(x_{\mathsf{pa}(k)}^{(t)} - 1))\right\}.$$

This update takes $O(Kd)$ time. GMSketch maintains d instances of GMHash. Therefore, it consumes $O(Kmd)$ space. Now we show that $\hat{P}$ is a good approximation of $\bar{P}$.

Theorem 3. *Let $\hat{P}$ be the estimator from Algorithm 2. Let $h^1, \ldots, h^d$ be d random and pairwise-independent hashes, and m be the number of bins in each hash function. Then for any $d \geq 8\log(1/\delta)$ and x:*

$$\bar{P}(x) \prod_{k=1}^{K} (1 - \varepsilon_k) \leq \hat{P}(x) \leq \bar{P}(x) \prod_{k=1}^{K} (1 + \varepsilon_k)$$

holds with at least $1 - \delta$ probability, where ε_k are defined in Theorem 2 for $\delta = 1/4$.

Proof. The proof is in Appendix. The key idea is the so-called median trick on d estimates of GMHash in Theorem 2 for $\delta = 1/4$.

Similarly to Sect. 5.1, Theorem 3 shows that $\hat{P}(x)$ is a multiplicative approximation to $\bar{P}(x)$. The approximation improves with the number of bins m and depends on the frequency of interaction between the values in x.

5.3   Algorithm GMFactorSketch

Our final algorithm, GMFactorSketch, is in Algorithm 3. The algorithm approximates $\bar{P}(x)$ as the product of $K - 1$ conditionals and a prior, one for each variable X_k. Each conditional is estimated as a ratio of two CM sketches:

$$\hat{P}_k(x_k \mid x_{\mathsf{pa}(k)}) = \frac{\hat{P}_k(x_k, x_{\mathsf{pa}(k)})}{\hat{P}_{\mathsf{pa}(k)}(x_{\mathsf{pa}(k)})},$$

where $\hat{P}_k(x_k, x_{\mathsf{pa}(k)})$ is the CM sketch of $\bar{P}_k(x_k, x_{\mathsf{pa}(k)})$ and $\hat{P}_k(x_k)$ is the CM sketch of $\bar{P}_k(x_k)$. The prior $\bar{P}_1(x_1)$ is approximated by its CM sketch $\hat{P}_1(x_1)$.

At time t, the hash tables are updated in the same way as in (7). This update takes $O(Kd)$ time and GMFactorSketch consumes $O(Kmd)$ space. Now we show that $\hat{P}$ is a good approximation of $\bar{P}$.

Algorithm 3. `GMFactorSketch`: Count-min sketches of conditionals and priors.

Input: Point query $x = (x_1, \ldots, x_K)$

// Count-min sketches for variables in $\mathcal{G}$
for all $k = 1, \ldots, K$ **do**
   for all $i = 1, \ldots, d$ **do**
$$\hat{P}_k^i(x_k) \leftarrow \frac{c_k(i, h_k^i(x_k))}{n}$$
$$\hat{P}_k(x_k) \leftarrow \min_{i \in [d]} \hat{P}_k^i(x_k)$$

// Count-min sketches for variable-parent pairs in $\mathcal{G}$
for all $k = 2, \ldots, K$ **do**
   for all $i = 1, \ldots, d$ **do**
$$\hat{P}_k^i(x_k, x_{\mathsf{pa}(k)}) \leftarrow \frac{\bar{c}_k(i, h_k^i(x_k + M(x_{\mathsf{pa}(k)} - 1)))}{n}$$
$$\hat{P}_k(x_k, x_{\mathsf{pa}(k)}) \leftarrow \min_{i \in [d]} \hat{P}_k^i(x_k, x_{\mathsf{pa}(k)})$$

for all $k = 2, \ldots, K$ **do**
$$\hat{P}_k(x_k \mid x_{\mathsf{pa}(k)}) \leftarrow \frac{\hat{P}_k(x_k, x_{\mathsf{pa}(k)})}{\hat{P}_{\mathsf{pa}(k)}(x_{\mathsf{pa}(k)})}$$
$$\hat{P}(x) \leftarrow \hat{P}_1(x_1) \prod_{k=2}^{K} \hat{P}_k(x_k \mid x_{\mathsf{pa}(k)})$$

Output: Point answer $\hat{P}(x)$

Theorem 4. *Let $\hat{P}$ be the estimator from Algorithm 3. Let $h^1, \ldots, h^d$ be d random and pairwise-independent hashes, and m be the number of bins in each hash function. Then for any $d \geq \log(2K/\delta)$ and x:*

$$\bar{P}(x) \prod_{k=1}^{K}(1 - \varepsilon_k) \leq \hat{P}(x) \leq \bar{P}(x) \prod_{k=1}^{K}(1 + \varepsilon_k)$$

holds with at least $1 - \delta$ probability, where:

$$\varepsilon_1 = e[\bar{P}_1(x_1)m]^{-1}, \quad \forall k \in [K] - \{1\} : \varepsilon_k = e[\bar{P}_k(x_k, x_{\mathsf{pa}(k)})m]^{-1}.$$

Proof. The proof is in Appendix. The main idea of the proof is similar to that of Theorem 2. The key difference is that we prove that event (6) is unlikely for any $k \in [K] - \{1\}$ by bounding the probabilities of events:

$$\hat{P}_{\mathsf{pa}(k)}(x_{\mathsf{pa}(k)}) - \bar{P}_{\mathsf{pa}(k)}(x_{\mathsf{pa}(k)}) > \varepsilon_k \alpha_k,$$
$$\hat{P}_k(x_k, x_{\mathsf{pa}(k)}) - \bar{P}_k(x_k, x_{\mathsf{pa}(k)}) > \varepsilon_k \alpha_k,$$

where $\hat{P}_k(x_k, x_{\mathsf{pa}(k)})$ is the CM sketch of $\bar{P}_k(x_k, x_{\mathsf{pa}(k)})$ and $\hat{P}_{\mathsf{pa}(k)}(x_{\mathsf{pa}(k)})$ is the CM sketch of $\bar{P}_{\mathsf{pa}(k)}(x_{\mathsf{pa}(k)})$.

As in Sects. 5.1 and 5.2, Theorem 4 shows that $\hat{P}(x)$ is a multiplicative approximation to $\bar{P}(x)$. The approximation improves with the number of bins m and depends on the frequency of interaction between the values in x.

5.4   Lower Bound

Our bounds depend on query-specific constants $\bar{P}_k(x_k, x_{\mathsf{pa}(k)})$, which can be small. We argue that this dependence is intrinsic. In particular, we show that there exists a family of distributions $\mathcal{C}$ such that any data structure that can summarize any $\bar{P} \in \mathcal{C}$ well must consume $\Omega(\Delta^{-1}(\mathcal{C}))$ space, where:

$$\Delta(\mathcal{C}) = \min_{\bar{P} \in \mathcal{C}, x \in [M]^K, k \in [K] - \{1\} : \bar{P}(x) > 0} \bar{P}_k(x_k, x_{\mathsf{pa}(k)}).$$

Our family of distributions $\mathcal{C}$ is defined on two dependent random variables, where X_1 is the parent and X_2 is its child. Let m be an integer such that $m = 1/\epsilon$ for some fixed $\epsilon \in [0, 1]$. Each model in $\mathcal{C}$ is defined as follows. The probability of any m values of X_1 is ϵ. The conditional of X_2 is defined as follows. When $\bar{P}_1(i) > 0$, the probability of any m values of X_2 is ϵ. When $\bar{P}_1(i) = 0$, the probability of all values of X_2 is $1/M$. Note that each model induces a different distribution and that the number of the distributions is $\binom{M}{m}^{m+1}$, because there are $\binom{M}{m}$ different priors $\bar{P}_1$ and $\binom{M}{m}$ different conditionals $\bar{P}_2(\cdot \mid i)$, one for each $\bar{P}_1(i) > 0$. We also note that $\Delta(\mathcal{C}) = \epsilon^2$. The main result of this section is proved below.

Theorem 5. *Any data structure that can summarize any $\bar{P} \in \mathcal{C}$ as $\hat{P}$ such that $|\hat{P}(x) - \bar{P}(x)| < \epsilon^2/2$ for any $x \in [M]^K$ must consume $\Omega(\Delta^{-1}(\mathcal{C}))$ space.*

Proof. Suppose that a data structure can summarize any $\bar{P} \in \mathcal{C}$ as $\hat{P}$ such that $|\hat{P}(x) - \bar{P}(x)| < \epsilon^2/2$ for any $x \in [M]^K$. Then the data structure must be able to distinguish between any two $\bar{P} \in \mathcal{C}$, since $\bar{P}(x) \in \{0, \epsilon^2\}$. At the minimum, such a data structure must be able to represent the index of any $\bar{P} \in \mathcal{C}$, which cannot be done in less than:

$$\log_2 \left(\binom{M}{m}^{m+1} \right) \geq \log_2 \left((M/m)^{m^2+m} \right) \geq m^2 \log_2(M/m)$$

bits because the number of distributions in $\mathcal{C}$ is $\binom{M}{m}^{m+1}$. Now note that $m^2 = 1/\epsilon^2 = \Delta^{-1}(\mathcal{C})$.

It is easy to verify that `GMFactorSketch` is such a data structure for $m = 5e\Delta^{-1}(\mathcal{C})$ in Theorem 4. In this setting, `GMFactorSketch` consumes $O(\log(1/\delta)\Delta^{-1}(\mathcal{C}))$ space. The only major difference from Theorem 5 is that `GMFactorSketch` makes a mistake with at most δ probability. Up to this factor, our analysis is order-optimal and we conclude that the dependence on the reciprocal of $\min_{k \in [K] - \{1\}} \bar{P}_k(x_k, x_{\mathsf{pa}(k)})$ cannot be avoided in general.

6   Comparison with the Count-Min Sketch

In general, the error bounds in Theorems 1 and 4 are not comparable, because $\tilde{P}$ in (1) is a different estimator from $\bar{P}$ in (3). To compare the bounds, we make the assumption that $(x^{(t)})_{t=1}^n$ is a stream of n observations such that $\bar{P} = \tilde{P}$. This holds, for instance, when $n \to \infty$, because both $\bar{P}$ and $\tilde{P}$ are consistent estimators of P. In the rest of this section, and without loss of generality, we assume that $\bar{P} = \tilde{P} = P$.

In this section, we construct a class of graphical models where GMFactorSketch has a tighter error bound than the CM sketch. This class contains naive Bayes models with $K + 1$ variables:

$$P(x) = P_1(x_1) \prod_{k=2}^{K+1} P_k(x_k \mid x_1). \tag{8}$$

Variable X_1 is binary. For any $k \in [K + 1] - \{1\}$, variable X_k takes values from $[M]$. For simplicity of exposition, we assume that the prior is $P_1(1) = P_1(2) = 0.5$. We fix x and define $C_k = P_k(x_k \mid x_1)$ for any $k \in [K + 1] - \{1\}$.

Suppose that GMFactorSketch represents P_1 exactly, and therefore $\hat{P}_1 = P_1$. Then by Theorem 4, for any x with at least $1 - \delta$ probability:

$$\hat{P}(x) \leq \frac{1}{2} \left[\prod_{k=2}^{K+1} C_k \right] \left[\prod_{k=2}^{K+1} \left(1 + \frac{2e}{C_k m} \right) \right], \tag{9}$$

where m is the number of hashing bins in GMFactorSketch. Since $\hat{P}_1 = P_1$, we can omit $1 + \varepsilon_1$ from Theorem 4. This approximation consumes, up to logarithmic factors in K, $2Km \log(1/\delta)$ space. The CM sketch (Sect. 2.1) guarantees that:

$$P_{\mathrm{CM}}(x) \leq \frac{1}{2} \left[\prod_{k=2}^{K+1} C_k \right] + \frac{e}{m'} = \frac{1}{2} \left[\prod_{k=2}^{K+1} C_k \right] \left(1 + \frac{2e}{m'} \left[\prod_{k=2}^{K+1} \frac{1}{C_k} \right] \right) \tag{10}$$

for any x with at least $1 - \delta$ probability, where m' is the number of hashing bins in the CM sketch. This approximation consumes $m' \log(1/\delta)$ space.

We want to show that the upper bound in (9) is tighter than that in (10) for any reasonable m. Since GMFactorSketch maintains $2K$ times more hash tables than the CM sketch, we increase the number of bins in the CM sketch to $m' = 2Km$, and get the following upper bound:

$$P_{\mathrm{CM}}(x) \leq \frac{1}{2} \left[\prod_{k=2}^{K+1} C_k \right] \left(1 + \frac{e}{Km} \left[\prod_{k=2}^{K+1} \frac{1}{C_k} \right] \right). \tag{11}$$

Now both GMFactorSketch and the CM sketch consume the same space, and their error bounds are functions of m.

Roughly speaking, the bound in (9) seems to be tighter than that in (11) because it contains K potentially large values $1/C_k$, each of which can be offset by a potentially small $1/m$. On the other hand, all values $1/C_k$ in (11) are offset only by a single $1/m$. Now we prove this claim formally. Before we start, note that both upper bounds in (9) and (11) contain $\frac{1}{2}\left[\prod_{k=2}^{K+1} C_k\right]$. Therefore, we can divide both bounds by this constant and get that the upper bound in (9) is tighter than that in (11) when:

$$1 + \frac{e}{Km}\left[\prod_{k=2}^{K+1}\frac{1}{C_k}\right] > \prod_{k=2}^{K+1}\left(1 + \frac{2e}{C_k m}\right). \tag{12}$$

Now we rewrite each $(1+2e/(C_k m))$ on the right-hand side as $(1/C_k)(C_k+2e/m)$ and multiply both sides by $\prod_{k=2}^{K+1} C_k$. Then we omit $\prod_{k=2}^{K+1} C_k$ from the left-hand side and get that event (12) happens when:

$$\frac{e}{Km} > \prod_{k=2}^{K+1}\left(C_k + \frac{2e}{m}\right). \tag{13}$$

If C_k is close to one for all $k \in [K+1] - \{1\}$, the right-hand side of (13) is at least one and we get that m should be smaller than e/K. This result is impractical since K is usually much larger than e and we require that $m \geq 1$. To make progress, we restrict our analysis to a class of x. In particular, let $C_k \leq 1/2$ for all $k \in [K+1] - \{1\}$. Then we can bound the right-hand side of (13) from above as:

$$\prod_{k=2}^{K+1}\left(C_k + \frac{2e}{m}\right) \leq \left(\frac{1}{2}\right)^K\left(1 + \frac{4e}{m}\right)^K \leq e\left(\frac{1}{2}\right)^K$$

for $m \geq 4eK$. This assumption on m is not particularly strong, since Theorem 4 says that we get good multiplicative approximations to $\bar{P}(x)$ only if $m = \Omega(K)$. Now we apply the above upper bound to inequality (13) and rearrange it as $2^K/K > m$. Since $2^K/K$ is exponential in K, we get that the bound in (9) is tighter than that in (11) for a wide range of m and any x where $C_k \leq 1/2$ for all $k \in [K+1] - \{1\}$. Our result is summarized below.

Theorem 6. *Let P be the distribution in (8) and x be such that $P_k(x_k \mid x_1) \leq 1/2$ for all $k \in [K+1] - \{1\}$. Let $m \geq 4eK$ and $m' = 2Km$. Then for any $m < 2^K/K$, the error bound of* GMFactorSketch *is tighter than that of the CM sketch at the same space. More precisely:*

$$P(x)\prod_{k=2}^{K+1}(1 + \varepsilon_k) \leq P(x) + \frac{e}{m'},$$

where ε_k are defined in Theorem 4.

The above result is quite practical. Suppose that $K = 32$. Then our upper bound is tighter for any m such that:

$$4eK < 348 \leq m \leq 2^{27} = 2^{32}/32 = 2^K/K.$$

By the pidgeonhole principle, Theorem 6 guarantees improvements in at least $2(M-1)^K$ points x in any distribution in (8). We can bound the fraction of these points from below as:

$$\frac{2(M-1)^K}{2M^K} = \exp[K\log(M-1) - K\log M] \geq \exp\left[-\frac{K}{M-1}\right] \geq 1 - \frac{K}{M-1}.$$

In our motivating examples, $M \approx 10^5$ and $K \approx 100$. In this setting, the error bound of GMFactorSketch is tighter than that of the CM sketch in at least 99.9 % of x, for any naive Bayes model in (8).

7   Experiments

In this section, we compare our algorithms (Sect. 5) and the CM sketch on the synthetic problem in Sect. 6, and also on a real-world problem in online advertising.

7.1   Synthetic Problem

We experiment with the naive Bayes model in (8), where $P_1(1) = P_1(2) = 0.5$; and:

$$\forall i \in [N] : P_k(i \mid 1) = 1/N, \qquad \forall i \in [M] - [N] : P_k(i \mid 1) = 0,$$
$$\forall i \in [N] : P_k(i \mid 2) = 0, \qquad \forall i \in [M] - [N] : P_k(i \mid 2) = 1/(M-N)$$

for any $k \in [K+1] - \{1\}$ and $N \ll M$. The model defines the following distribution over $x = (x_1, \ldots, x_K)$: when $x_1 = 1$, $P(x) = 0.5N^{-K}$ and we refer to the example x as *heavy*; and when $x_1 = 2$, $P(x) = 0.5(M-N)^{-K}$ and we refer to the example x as *light*. The heavy examples are much more probable when $N \ll M$. We set $M = 2^{16}$.

All compared algorithms are trained on 1M i.i.d. examples from distribution P and tested on 500k i.i.d. heavy examples from P. We report the fraction of imprecise estimates of P as a function of space. The estimate of $P(x)$ is *precise* when $(1/e)P(x) \leq \hat{P}(x) \leq eP(x)$. When the sample size n is large, both $\bar{P} \to P$ and $\tilde{P} \to P$, and this is a fair way of comparing our methods to the CM sketch. We choose $d = 5$. We observe similar trends for other values of d. All results are averaged over 20 runs.

7.2   Easy Synthetic Problem

We choose $K = 4$ and $N = 8$, and then $P(x) = 2^{-13}$ for all heavy x. In this problem, the CM sketch can approximate $P(x)$ within a multiplicative factor of e for any heavy x in about 2^{13} space. This space is small, and therefore this problem is *easy for the CM sketch*.

Our results are reported in Fig. 1a. We observe that all of our algorithms outperform the CM sketch. In particular, note that P_{CM} approximates P well for any heavy x in about 2^{15} space. Our algorithms achieve the same quality of the approximation in at most 2^{13} space. GMFactorSketch consumes 2^{10} space, which is almost two orders of magnitude less than the CM sketch.

7.3   Hard Synthetic Problem

We set $K = 32$ and $N = 64$, and then $P(x) = 2^{-193}$ for all heavy x. In this problem, the CM sketch can approximate $P(x)$ within a multiplicative factor of e for any heavy x in about 2^{193} space. This space is unrealistically large, and therefore this problem is *hard for the CM sketch.*

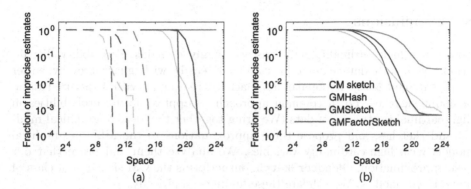

Fig. 1. a. Evaluation of the CM sketch, `GMHash`, `GMSketch`, and `GMFactorSketch` on the easy problem in Sect. 7.2 (dashed lines) and the hard problem in Sect. 7.3 (solid lines). **b.** Evaluation on the real-world problem in Sect. 7.4.

Our results are reported in Fig. 1a and we observe three major trends. First, the CM sketch performs poorly. Second, as in Sect. 7.2, our algorithms outperform the CM sketch. Finally, when the fraction of imprecise estimates is small, our algorithms perform as suggested by our theory. `GMHash` is inferior to `GMSketch`, which is further inferior to `GMFactorSketch`.

7.4   Real-World Problem

We also evaluate our algorithms on a real-world problem where the goal is to estimate the probability of a page view. We experiment with two months of data of a medium-sized customer of *Adobe Marketing Cloud*[2]. This is 65M page views, each of which is described by six variables: COUNTRY, CITY, PAGE NAME, STARTING PAGE NAME, CAMPAIGN, and BROWSER. Variable PAGE NAME takes on more than 42k values and has the highest cardinality. We approximate the distribution P over our variables by a naive Bayes model, where the class variable is $X_1 =$ COUNTRY. Since the behavior of users is often driven by their locations, this approximation is quite reasonable.

All compared algorithms are trained on 1M i.i.d. examples from distribution P and tested on all heavy examples in this sample. We say that the example x is *heavy* when $P(x) > 10^{-6}$. The rest of the setup is identical to that in Sect. 7.1.

---

[2] http://www.adobe.com/marketing-cloud.html.

Our results are reported in Fig. 1b. We observe the same trends as in Sect. 7.3. The CM sketch performs poorly, and our methods outperform it at the same space for any space from 2^{13} to 2^{24}. Also note that none of the compared methods achieve zero mistakes. This is because our sample size n is not large enough to approximate P well in all heavy x. Even if $\hat{P} = \bar{P}$, our methods would still make mistakes.

8   Conclusions

Structured high-cardinality data arises in many domains. Probability distributions over such data cannot be estimated easily with guarantees by either graphical models [9], a popular approach to reasoning with structured data; or count sketches [17], a common approach to approximating probabilities in high-cardinality streams of data. We bring together the ideas of graphical models and sketches, and propose three approximations to the MLE in graphical models with high-cardinality variables. We analyze them and prove that our best approximation, GMFactorSketch, outperforms the CM sketch on a class of naive Bayes models. We validate these findings empirically.

The MLE is a common approach to estimating the parameters of graphical models [9]. We propose, analyze, and empirically evaluate multiple space-efficient approximations to this procedure with high-cardinality variables. In this work, we focus solely on the problem of estimating $\bar{P}(x)$, the probability at a single point x. However, note that our models are constructed from Bayesian networks, which can answer $P(Y = y)$ for any subset of variables Y with values y. We do not analyze such inference queries and leave this for future work.

Our work is the first formal investigation of approximations on the intersection of graphical models and sketches. One of our key results is that GMFactorSketch yields a constant-factor multiplicative approximation to $\bar{P}(x)$ for any x with probability of at least $1-\delta$ in $O(K^2 \log(K/\delta)\Delta^{-1}(x))$ space, where K is the number of variables and $\Delta(x)$ reflects the hardness of query x. This result is encouraging because the space is only quadratic in K and logarithmic in $1/\delta$. The space also depends on constant $\Delta(x)$, which can be small. This constant is intrinsic (Sect. 5.4); and this indicates that the problem of approximating $\bar{P}(x)$ well, for any $\bar{P}$ and x, is intrinsically hard.

References

1. Belle, V., Van den Broeck, G., Passerini, A.: Hashing-based approximate probabilistic inference in hybrid domains. In: Proceedings of the 31st Conference on Uncertainty in Artificial Intelligence (2015)
2. Boutilier, C., Friedman, N., Goldszmidt, M., Koller, D.: Context-specific independence in Bayesian networks. In: Proceedings of the 12th Conference on Uncertainty in Artificial Intelligence, pp. 115–123 (1996)
3. Charikar, M., Chen, K., Farach-Colton, M.: Finding frequent items in data streams. Theor. Comput. Sci. 312(1), 3–15 (2004)

4. Cormode, G., Muthukrishnan, S.: An improved data stream summary: The count-min sketch and its applications. J. Algorithms **55**(1), 58–75 (2005)
5. Cormode, G., Muthukrishnan, S.: What's hot and what's not: Tracking most frequent items dynamically. ACM Trans. Database Syst. **30**(1), 249–278 (2005)
6. Ermon, S., Gomes, C., Sabharwal, A., Selman, B.: Taming the curse of dimensionality: Discrete integration by hashing and optimization. In: Proceedings of the 30th International Conference on Machine Learning, pp. 334–342 (2013)
7. Flajolet, P., Fusy, E., Gandouet, O., Meunier, F.: Hyperloglog: the analysis of a near-optimal cardinality estimation algorithm. In: Proceedings of the 2007 Conference on Analysis of Algorithms, pp. 127–146 (2007)
8. Flajolet, P., Martin, G.N.: Probabilistic counting algorithms for data base applications. J. Comput. Syst. Sci. **31**(2), 182–209 (1985)
9. Jensen, F.: Introduction to Bayesian Networks. Springer, Heidelberg (1996)
10. Koller, D., Friedman, N.: Probabilistic Graphical Models: Principles and Techniques. MIT Press, Cambridge (2009)
11. Lafferty, J., McCallum, A., Pereira, F.: Conditional random fields: Probabilistic models for segmenting and labeling sequence data (2001)
12. Liberty, E.: Simple and deterministic matrix sketching. In: Proceedings of the 19th ACM SIGKDD International Conference on Knowledge Discovery and Data Mining, pp. 581–588 (2013)
13. Matusevych, S., Smola, A., Ahmed, A.: Hokusai - Sketching streams in real time. In: Proceedings of the 28th Conference on Uncertainty in Artificial Intelligence (2012)
14. McGregor, A., Vu, H.: Evaluating Bayesian networks via data streams. In: Proceedings of the 21st International Conference on Computing and Combinatorics, pp. 731–743 (2015)
15. Misra, J., Gries, D.: Finding repeated elements. Sci. Comput. Program. **2**(2), 143–152 (1982)
16. Murphy, K., Torralba, A., Freeman, W.: Using the forest to see the trees: a graphical model relating features, objects, and scenes. Adv. Neural Inf. Process. Syst. **16**, 1499–1506 (2004)
17. Muthukrishnan, S.: Data streams: algorithms and applications. Found. Trend. Theor. Comput. Sci. **1**(2), 117–236 (2005)
18. Woodruff, D.: Low rank approximation lower bounds in row-update streams. Adv. Neural Inf. Process. Syst. **27**, 1781–1789 (2014)

Laplacian Hamiltonian Monte Carlo

Yizhe Zhang$^{(\boxtimes)}$, Changyou Chen, Ricardo Henao, and Lawrence Carin

Department of Electrical and Computer Engineering,
Duke University, Durham, NC, USA
{yz196,cc448,rhenao,lcarin}@duke.edu

Abstract. We proposed a Hamiltonian Monte Carlo (HMC) method with Laplace kinetic energy, and demonstrate the connection between slice sampling and proposed HMC method in one-dimensional cases. Based on this connection, one can perform slice sampling using a numerical integrator in an HMC fashion. We provide theoretical analysis on the performance of such sampler in several univariate cases. Furthermore, the proposed approach extends the standard HMC by enabling sampling from discrete distributions. We compared our method with standard HMC on both synthetic and real data, and discuss its limitations and potential improvements.

1 Introduction

One pivotal question in modern statistical computation is to efficiently sample from an unnormalized probability density function, where the normalization constant (partition function) is intractable. Towards this end, many Markov Chain Monte Carlo (MCMC) [22] methods have been developed. One of the most influential algorithms is Metropolis-Hastings (MH) [15]. Despite its great success, the *random walk* nature often delivers inefficient mixing of the Markov chain [22]. An inappropriate setting of transition kernel would result in either low acceptance ratio or slow moves. Such situation is exaggerated in high dimensional cases, where the samples from the chain can be highly correlated. As a consequence, the effective sample size is usually relatively small. A number of adaptations have been proposed to mitigate these issues [12,20], however, achievable improvements are limited if attempting maintaining the Markov property and reversibility of the chain [1,10,18].

To mitigate the random walk behavior in MH, several approaches have been proposed, such as Hamiltonian Monte Carlo (HMC) [9,20]. HMC augments a target distribution with auxiliary momentum variables, and uses gradient information to propose distant samples, while maintaining ergodic property and detailed balance. The ability of long-range movement with a high acceptance ratio significantly improves mixing performance. However, HMC is sensitive to parameter settings and can only sample continuous distributions. Towards solving these issues, methods were proposed to use adaptive leap-frog steps [13], or automatic stepsize [16], and to relax the discrete distributions sampling tasks to continuous distributions [21,26]. The improvement can be further boosted by leveraging

© Springer International Publishing AG 2016
P. Frasconi et al. (Eds.): ECML PKDD 2016, Part I, LNAI 9851, pp. 98–114, 2016.
DOI: 10.1007/978-3-319-46128-1_7

geometric manifold information [10], by considering better numerical integrators [6], or by relaxing the detailed balance constraint [24].

A different direction towards improving sampling performance is the slice sampler [19]. The slice sampler is related to HMC in the sense that both use auxiliary variables for efficient moves. These moves can be automatically adapted to match the relative scale of the local region being sampled [19]. The sampling procedure alternates between uniformly drawing samples from the target distribution and uniformly drawing the slice variables. Unlike HMC, slice sampling does not require local gradient information. Instead, the primary effort is to locate slice intervals, where the unnormalized density values are greater than the slicing variable. This is typically hard to compute directly, thus requires local search [19]. Further, it is generally less feasible in high-dimensional parameter spaces, because the slice interval is difficult to approximate. For example, using hyper-rectangle estimation may result in high rejection rates [19]. Elliptical slice sampling [17] alleviate this issue by slicing on a high dimensional elliptical curve parameterized by a single scalar. However it assumes the latent variable to be Gaussian distributed.

In this paper, we leverage the Hamiltonian-Jacobi equation from classical physics [11] to unveil a deeper connection between HMC with modified kinetics and standard slice sampling in one-dimensional cases. We propose an equivalent slice sampler, which exploits gradient information without evaluating the slice interval. We formally show that, in several univariate scenarios where theoretical analysis is tractable, the proposed sampler yields lower autocorrelation compared with standard HMC, thus potentially yielding higher effective sample sizes. Finally, we discuss the scenario where our method is most desirable and validate it with synthetic and real-world experiments.

2   Preliminaries

Hamiltonian Monte Carlo. Consider sampling from a probability density function $p(x) \propto \exp[-E(x)]$, where $x \in \mathbb{R}^d$ and $E(x)$ is the potential energy. One can augment the density with an auxiliary momentum random variable $p \in \mathbb{R}^d$. By Assumption, p is independent of x, and has a marginal Gaussian distribution with zero-mean and covariance matrix M. The joint distribution $p(x, p)$ is defined as $p(x, p) \propto \exp[-H(x, p)] = \exp[-E(x) - K(p)]$, where $H(x, p)$ is the total energy or *Hamiltonian*, and $K(p) = \frac{1}{2} p^T M^{-1} p$ is the kinetic energy. Hamiltonian Monte Carlo leverages Hamiltonian dynamics to propose new samples for x, driven by the following ordinary differential equations (ODE):

$$\frac{dx}{dt} = \nabla_p K(p), \qquad \frac{dp}{dt} = -\nabla_x E(x). \qquad (1)$$

The Hamiltonian is preserved under perfect simulation, *i.e*, it is constant over t. However, closed-form dynamic updates are typically infeasible. As a result, one typically employs numerical integrators, *e.g.*, the leap-frog [20], to approximate the Hamiltonian flow. If the integrator is symplectic, by Liouville's theorem, the corresponding sampler is invariant to the target distribution [20].

Slice sampling. Slice sampling [19] was originally proposed as an approach to overcome the need of manually selecting the proposal scale (or stepsize) in the Metropolis-Hastings algorithm. Slice sampling leverages the fact that sampling the unnormalized target distribution $f(x)$ can be perceived as sampling a joint distribution. Therefore, sampling from the points under the unnormalized density curve is the same as sampling from the target distribution. The iterative procedure consists the following *slicing* and *sampling* steps:

$$\text{Slicing}: \qquad p(y_t|x_t) = \frac{1}{f(x_t)}, \qquad\qquad s.t.\ 0 < y_t < f(x_t)$$

$$\text{Sampling}: \qquad q(x_{t+1}|y_t) = \frac{1}{Z_2(y_t)}, \qquad\qquad s.t.\ f(x_t) > y_t, \qquad (2)$$

where y is the augmented slicing variable. $f(x) \triangleq e^{-E(x)}$ is the unnormalized density and $Z_2(y) = \int_{f(x)>y} 1 dx$ is the measure of regions that have functional values greater than the slice variable y. The density function is given by

$$p(x, y) = \begin{cases} \frac{1}{Z_1}, & 0 < y < f(x) \\ 0, & \text{otherwise} \end{cases},$$

where $Z_1 = \int f(x)dx$ is the normalizing constant. The marginal distribution for x exactly recovers the target distribution $f(x)/Z_1$. The evaluation of slice interval $x : f(x) > y$ is typically non-trivial, where iterative procedures to adaptively capture the boundaries of such slice interval are used [19].

3   Canonical Transformation

In this section we use the *canonical transformation* and the *Hamilton-Jacobi equation* (HJE) [11] to reveal a connection between HMC with Laplace kinetics and slice sampling. Without loss of generality to the multivariate cases, for simplicity, here we detail our derivations for the univariate case.

Suppose the kinetic energy function $K(p)$ in HMC can be defined as an arbitrary function of p, as long as the $K(p)$ is convex and symmetric *w.r.t.* p. We consider two particular kinetics forms. The standard HMC uses quadratic kinetics $K(p) = p^2/m$, where m is the *mass parameter* and the marginal distribution of p is proportional to $e^{-K(p)}$, thus is Gaussian distributed with variance m.

We employ the canonical transformation from classical physics to transform the original HMC system (H, x, p, t) in (1), into a new system space, termed as canonical space [11]: (H', x', p', t). The transformation $(H, x, p, t) \to (H', x', p', t)$ satisfies the *Hamilton's principle* [11]:

$$\lambda(p \cdot \dot{x} - H) = p' \cdot \dot{x}' - H' + \frac{\delta G}{\delta t}, \qquad (3)$$

where $\lambda \in \mathbb{R}$ is a constant, $\dot{x} \triangleq dx/dt$, δ denotes functional derivative and G is a user-defined generating function [25]. Such a generating function can be of several types; here we use a type-2 generating function defined as

$$G \triangleq -x' \cdot p' + S(x, p', t).$$

The explicit form of $S(x, p', t)$ is defined below. By substituting G into (3), one can establish the following equations:

$$p = \frac{\partial S}{\partial x}, \quad x' = \frac{\partial S}{\partial p'}, \quad H'(x', p') = H(x, p) + \frac{\partial S}{\partial t}. \tag{4}$$

In the HJE, we let the new Hamiltonian H' to be zero, i.e.,

$$H(x, p) + \frac{\partial S}{\partial t} = H'(x', p') = 0. \tag{5}$$

The Hamilton-Jacobi equation states that after this transformation, the motion of particles collapse into a point in the new space, i.e., (x', p') are constant over time [25].

Consider setting the *Hamilton's principal function* as $S(x, p', t) = W(x) - p't$, where $W(x)$ is an unknown function of x that needs to be solved. Thereby, (5) becomes

$$H(x, p) + \frac{\partial S}{\partial t} = H(x, p) - p' = 0. \tag{6}$$

The implication from (6) is that $p' = H$; i.e., the *generalized momentum* in the new phase space, (x', p'), represents the total Hamiltonian in the original space. We consider the standard Gaussian kinetic function $K(p) = |p|^2/m$. From (4) and (5), we can solve the functional equation in (6) to obtain,

$$W(x) = \int_{x_{min}}^{x(t)} f(z)dz + C, \tag{7}$$

where $f(z) = H - E(z)$ if $z \in \mathbb{X} \triangleq \{x : H - E(x) \geq 0\}$, and 0 otherwise, and $x_{min} = \min\{x : x \in \mathbb{X}\}$. From (4), (6) and (7),

$$x' = \frac{\partial S}{\partial p'} = \frac{\partial W}{\partial H} - t = \frac{1}{2} \int_{x_{min}}^{x(t)} f(z)^{-1/2}dz - t. \tag{8}$$

Note that x' is a constant. In (8), $\int_{x_{min}}^{x(t)} f(z)^{-1/2}dz \in [0, \int_{\mathbb{X}}[H - E(z)]^{-1/2}dz]$. Our objective is to mimic the Hamiltonian dynamics evolving with a random evolution time, t. If we assume a closed contour, the Hamiltonian dynamics has period $T \triangleq \int_{\mathbb{X}}[H - E(z)]^{-1/2}dz$. To sample a new point $x(t)$ on the contour, one can first sample the time, t, constrained to a single period of movement, i.e,

$$t \sim \text{uniform}\left(-x', -x' + \int_{\mathbb{X}}[H - E(z)]^{-1/2}dz\right). \tag{9}$$

where x' can be understood as the "initial" timestamp of x. With a sampled time t from (9), one could solve the Eq. (8) for $x^* \triangleq x(t)$, i.e., the value of x at time t.

However, the integral in (8) is not always tractable. Note that the integral in (8) can be interpreted as (up to normalization) a cumulative density function (CDF) of x. As a result, one can circumvent uniformly sampling t from (9), by directly sampling x^* from the following density function

$$p(x^*|H) \propto [H - E(x^*)]^{-1/2}, \quad s.t., \quad H - E(x^*) \geq 0. \tag{10}$$

Note that p^* is not of interest because it is discard after each dynamic update.

This transformation provides the basic setup to reveal the equivalence between the slice sampler and HMC, which is discussed in Sect. 4.

4   Laplacian HMC

Let $\mathcal{L}(\cdot; m)$ denote the Laplace distribution with scale parameter m, the probability density function is given by

$$\mathcal{L}(p; m) \propto \exp(-|p|/m)$$

We denote the non-standard HMC with Laplace distribution for the momentum variable as Laplacian HMC (L-HMC). Suppose we assume the momentum variable have Laplace kinetics, i.e. employing an ℓ_1 norm for the kinetic function, similar to the derivation in (10), we have

$$p(x^*|H) \propto 1, \quad s.t., \quad H - E(x^*) \geq 0. \tag{11}$$

In light of the above observation, we propose to perform standard HMC and L-HMC with the procedure described in Algorithm 1.

Algorithm 1. HMC/L-HMC in canonical space.

Input: Sample size T, energies $E(x)$ and $K(p; m)$.
Output: Sample results, $\{x_0, \ldots, x_T\}$.
Initialization: Choose initial sample point, x_0.
for $t \in \{1, \ldots, T\}$ **do**
  Sample $p_t \sim \mathcal{N}(p; m)$ (standard HMC) or $\mathcal{L}(p; m)$ (L-HMC).
  Compute Hamiltonian: $H_t = E(x_t) + K(p_t)$.
  Compute $\mathbb{X} \triangleq \{x : x \in \mathbb{R}; E(x) \leq H_t\}$.
  Sample $q(x_{t+1}|H_t) \propto [H_t - E(x_{t+1})]^{-1/2}$(standard HMC) or $q(x_{t+1}|H_t) \propto 1$ (L-HMC), with $x_{t+1} \in \mathbb{X}$.
end for

Denote $y_t = e^{-H_t}$, the conditional updates for the L-HMC sampling procedure in Algorithm 1 share the same formulas as standard slice sampling in (2)

Note that the mass parameter m (scale parameter of the Laplace distribution) is cancelled out.

Accordingly, the equivalent non-standard slice sampling that corresponds to standard HMC can be written as

$$p(y_t|x_t) = \frac{1}{f(x_t)}[\log f(x_t) - \log y_t]^{-\frac{1}{2}}, s.t. \ 0 < y_t < f(x_t) \qquad (12)$$

$$q(x_{t+1}|y_t) = \frac{1}{Z_2(y_t)}[\log f(x_{t+1}) - \log y_t]^{-\frac{1}{2}} .s.t. \ f(x_{t+1}) > y_t \qquad (13)$$

We denote this slice sampler as HMC-SS (the slice sampler corresponding to standard HMC). This iterative procedure yields an invariant joint distribution

$$p(x,y) = \begin{cases} \frac{1}{\sqrt{\pi}Z_1}[\log f(x) - \log y]^{\frac{1}{2}}, \ 0 < y < f(x) \\ 0, \ \text{otherwise} \end{cases},$$

leaving the marginal distribution for x as the desired target distribution, while the marginal distribution of y is given by

$$p(y) = Z_2(y)/(\sqrt{\pi}Z_1). \qquad (14)$$

The equivalent slice sampler for standard HMC and L-HMC is illustrated in Fig. 1. For HMC-SS, the conditional distribution of $q(x_{t+1}|y_t)$ is skewed, so that points that are close to the boundary of the slice interval are more likely to be drawn. In addition, from (12) the conditional draw of slice variable y_t given x_t tends to take values close to $f(x_t)$.

Intuitively, in contrast with the standard slice sampling, the auxiliary variable y_t in HMC-SS tend to stay close with $f(x_t)$, rendering x_{t+1} to be close to x_t. Thus the standard slice sampler with a larger a is expected to be more efficient. Based on the connection between HMC-SS and HMC, as well as standard SS with L-HMC, this seems suggest L-HMC is more efficient that standard HMC. We elaborate more about the mixing performance in Sect. 6.

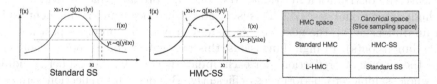

Fig. 1. Standard slice sampling (Left). The equivalent slice sampler of standard HMC, HMC-SS (Middle). Mapping between HMC space and canonical space (Right). $y_t|x_t$ is sampled from (12) (red line) and $x_{t+1}|y_t$ from (13) (blue line). L-HMC is essentially the same but with $y_t|x_t$ and $x_{t+1}|y_t$ sampled from uniform distributions. (Color figure online)

5   Performing L-HMC with Numerical Integrators

Section 4 shows that performing L-HMC in the canonical space can be viewed as performing standard slice sampling. In practice, however, analytically solving the slice interval, $\mathbb{X}$, is typically infeasible. By leveraging the connection between L-HMC and standard slice sampling, one can perform the standard slice sampling in the original space using a numerical integrator, as done in standard HMC. Here we consider the second order Störmer-Verlet integration [20]. The updates for L-HMC[1] are thus given as the following leap-frog steps

$$\mathbf{p}_{t+1/2} = \mathbf{p}_t - \tfrac{1}{2}\varepsilon\nabla E(\mathbf{x}_t)\,, \tag{15}$$

$$\mathbf{x}_{t+1} = \mathbf{x}_t + \varepsilon\,\mathrm{sign}(\mathbf{p})/m\,, \tag{16}$$

$$\mathbf{p}_{t+1} = \mathbf{p}_{t+1/2} - \tfrac{1}{2}\varepsilon\nabla E(\mathbf{x}_{t+1})\,, \tag{17}$$

Note that the mass matrix in our specification is $\mathbf{M} = m\mathbf{I}$. Here we use a random step size, ε, drawn from a uniform distribution with user-defined width, as suggested in [20]. Note that this specification is necessary for L-HMC to avoid moving on a fixed grid determined by ε.

Reflection. Another practical issue that comes with the fact that each contour in the phase space (x, p) has at least 2^D stiff (non-differentiable) points due to the non-differentiable kinetic function $K(p)$. The stiff points occur whenever the contour intersect with hyperplanes $p_d = 0$, for $d \in \{1 \cdots D\}$; D denotes the total dimension.

The naive leap-frog approach of L-HMC in (15)–(17) would lead to high integration errors, comparing with standard HMC, especially when the dimensionality is high. To alleviate this issue, we take a "reflection" action when encountering these stiff points, which shares some similarities with the "bouncing ball" strategy mentioned by [20]. Specifically, in (15) and (17), whenever the d-th component of momentum $p^{(d)}$ changes sign, we set $x_{t+1}^{(d)}$ and $p_{t+1}^{(d)}$ back to $x_t^{(d)}$ and $p_t^{(d)}$, and flip $p_t^{(d)} = -p_t^{(d)}$. A caveat of such a simple remedy lies in the fact that it may not guarantee the conservation of volume in phase space, thus may not leave the distribution invariant. Also, one will probably face "stickiness" in the high dimensional case [20]. This is because when negating the momentum in certain dimension(s), the next sample x_t may stay at the previous position, for those dimension(s). In high dimensions, this problem becomes more prominent since the chance of "reflection" for each update is considerably higher, yielding the whole sampler to perform less efficiently. Besides, this reflection strategy may render the sampler to be less sensitive on tail region of the target distribution. We note that this strategy may violate the invariance property. We hope to remark that the reflection is a first-remedy to ameliorate numerical difficulties. Nevertheless, this approach preserves the total Hamiltonian, and performs well in practice for low-dimensional cases.

---

[1] In the following, we denote L-HMC as the one in the original space, except otherwise explicitly stated.

Sample with constrained domain. As mentioned by [20], one could split the total Hamiltonian, to approach sampling from a bounded domain. An imaginary infinite potential energy can be imposed on regions that violate the constraints, which will give such points zero probability. Whenever the new proposed sample exceeds the constraint, we bounce the sample back. For example, when sampling from a truncated distribution with constraint $x^{(d)} > m$, if at time t the proposed $x_t^{(d)} < m$, the value $2m - x_t^{(d)}$ would be used instead, while the corresponding moment, $p_t^{(d)}$, changes sign.

Partial momentum refreshment. Using fewer number of leap-frog steps would reduce the computational cost of L-HMC, however rendering the algorithm less likely to adequately explore the contour and move to a distant point. [20] described a strategy to partially update the momentum variable, as an approach to further suppress the random-walk behavior when only a small number of leap-frog steps are taken, in which the distribution of the momentum would still be invariant. For the double-exponential kinetic energy form, one could consider a similar strategy to partially refresh the momentum. For the univariate case, without loss of generality to high dimensions, the update for momentum is given by

$$\tilde{p} = \min(p/\alpha, \eta/(1-\alpha))\mathrm{sign}(p), \tag{18}$$

where $\alpha \in (0,1)$ is a tuning parameter and η is an exponential random variable with mean $1/m$. It can be shown that $\tilde{p}$ has the same distribution as p. When α is close to 1, the generated new momentum $\tilde{p}$ would be similar to p. When α is close to zero, the absolute value of the new momentum becomes independent of its previous value. Similar to partial refreshment in standard HMC [20], one iteration applying the modification in (18) consists of three steps: (1) Updating momentum using (18), (2) performing a leap-frog discretization and Metropolis step, and (3) negating the momentum. In practice, the value of α has to be manually selected to achieve good performance.

Sampling discrete distributions. Sampling from discrete distributions such as Poisson, multinomial, Bernoulli, *etc.*, is generally infeasible for standard HMC, primarily due to the lack of gradient information. Recently proposed techniques tackle the discrete case by transforming it into sampling from a continuous distribution [21,26]. We show here that one can directly sample from a discrete distribution with L-HMC.

Notice from Eq. (16) that the update of $\boldsymbol{x}$ for each leap-frog discretization step depend only on the sign of the momentum variable $\boldsymbol{p}$. Based on this observation, one can sample a discrete distribution exactly, in an HMC manner. Consider a scenario, where a multivariate distribution with D dimensions is defined on an infinite grid with equidistant step m. Equation (16) allows the Hamiltonian dynamics to move in such a way, that each update in $\boldsymbol{x}$ moves with multiples of m, so as to stay on the grid. Meanwhile, the gradients in (15) and (17) are substituted with the difference vector $\triangle E(\boldsymbol{x})$, where its d-th component is $\triangle^{(d)} E(\boldsymbol{x}_{t-1/2}) \triangleq E(\boldsymbol{x}_t) - E(x_{t-1}^{(d)}, \boldsymbol{x}_t^{(-d)})$, $x^{(d)}$ denotes the d-th component of

$\boldsymbol{x}$ and $\boldsymbol{x}_t^{(-d)}$ denotes the remaining $D - 1$ components. The iterative updates become

$$\boldsymbol{x}_t = \boldsymbol{x}_{t-1} + \boldsymbol{\varepsilon} \circ \operatorname{sign}(\boldsymbol{p})/m \,, \; \boldsymbol{p}_t = \boldsymbol{p}_{t-1} - \boldsymbol{\varepsilon} \circ \triangle E(\boldsymbol{x}_{t-1/2}) \,,$$

where the stepsize $\boldsymbol{\varepsilon}$ is constrained to $\mathbb{Z}^D$ and $\circ$ is the element-wise product. The reason that this strategy can not be applied to standard HMC is because in L-HMC, each increment $x_{t+1} - x_t$ is a constant that does not depend on the absolute value of momentum p, while in standard HMC, different value of p will yield different increment $x_{t+1} - x_t$. As a result, the sampler may not move on a uniform grid. In practice, one could sample $\boldsymbol{\varepsilon} \in \mathbb{R}^D$ and round it to the closest integer vector. It can be shown that the Hamiltonian is preserved under such procedure in univariate cases. For multivariate cases, the difference vector can be normalized to enforce the conservation of Hamiltonian, i.e. $\sum \triangle E(\boldsymbol{x}_{t-1/2}) = E(\boldsymbol{x}_t) - E(\boldsymbol{x}_{t-1})$. Note that when the Hamiltonian is preserved, the Metropolis-Hasting step can be omitted. As in the continuous scenario, the momentum is negated whenever it would change sign in the next iteration. This specification works well in practice for our tested cases when the dimensionality is low ($D < 5$), however, we remark that this specification would violate the volume preservation and is not the principled way to perform high-dimensional discrete sampling (when the dimensionality increase, the error between $E(\boldsymbol{x}_{t+1}) - E(\boldsymbol{x}_t)$ and $\triangle E(\boldsymbol{x}_t$ would inevitably become larger). How to perform a high dimensional discrete sampling remain as a interesting topic for future investigation. If $E(\boldsymbol{x})$ has well-defined gradient information over the real domain that covers the grid, one can relax the calculation to the continuous space, where the gradient $\nabla E(\boldsymbol{x})$ is computed, instead of D evaluations of the potential energy, $E(\boldsymbol{x})$.

Adaptive search. The fact that updating $\boldsymbol{x}$ does not explicitly involve $\boldsymbol{p}$ may have additional implications. Following [23], this observation enables applying adaptive search for appropriate scale of stepsize, $\boldsymbol{\varepsilon}$, based on the sufficient statistics from previous samples. For example, one could set the relative scale of the stepsize for each coordinate to match the diagonal elements from the empirical covariance matrix. Note that this strategy is particularly suitable to be applied to L-HMC, due to the fact that the update of the dynamics in L-HMC is moving exactly in the direction of the stepsize, $\boldsymbol{\varepsilon}$. This strategy would be expected to perform better than choosing a common stepsize for each dimension, when the landscape has different scales for each dimension. The convergence of adaptive parameters requires establishing regularity conditions [12]. Though it works well in many cases, it is known that this strategy results in a chain that is no longer Markovian, thus it will not always leave the target distribution invariant [23]. Besides, when the distribution has more than one mode, applying this method may render the sampler prone to get trapped into one of the modes.

6   Efficiency Analysis

We note that most of previous work of analyzing the mixing performance of HMC is based on empirical studies. Little work has been done on theoretical analysis [10,20]. Interestingly, we can leverage the implicit connection between HMC and slice sampling, to briefly touch on the analysis of the mixing performance for HMC and L-HMC from examining their corresponding slice samplers. We use the autocorrelation function and effective sample size to monitor mixing performance. We consider sampling from a univariate distribution $p(x) \propto e^{-E(x)}$ for the analysis. The one-time-lag autocorrelation for HMC and L-HMC, $\rho(1)$, is given by

$$\rho(1) = (\mathbb{E}[x_t x_{t+1}] - \mathbb{E}[x]^2)/\mathrm{Var}(x). \qquad (19)$$

$$= (\ \mathbb{E}_{p(y_t)}[\mathbb{E}_{q(x_{t+1}|y_t)}[x_{t+1}]]^2 - \mathbb{E}[x]^2)/\mathrm{Var}(x) \qquad (20)$$

From (12) and stationary assumption, for standard HMC

$$q(x_t|y_t) \propto p(y_t|x_t)p(x_t) \propto [\log f(x_t) - \log y_t]^{-1/2}, s.t.\ \ f(x_t) > y_t$$

For L-HMC, $q(x_t|y_t) \propto 1, s.t.\ \ f(x_t) > y_t$

Given the potential energy form $E(x)$, $\rho(1)$ can be computed from (14), (20) and (2). The h-time-lag autocorrelation function can be obtained as

$$\rho(h) = (\mathbb{E}_{p(x)}[\mathbb{E}_{\kappa_h(x'|x)}[x'x]] - \mathbb{E}[x]^2)/\mathrm{Var}(x),$$

where, $\kappa_h(x_{t+h}|x_t)$ represents the h-order transition kernel, and can be calculated recursively as

$$\kappa_1(x_{t+1}|x_t) = \int q(x_{t+1}|y_t)p(y_t|x_t)dy_t,$$

$$\kappa_h(x_{t+h}|x_t) = \int \kappa_{h-1}(x'|x_t)\kappa_1(x_{t+h}|x')dx'.$$

Finally, the resulting Effective Sample Size (ESS) [5] is given by ESS $= N/(1+2\times \sum_{h=1}^{\infty} \rho(h))$. Analyzing the efficiency of L-HMC for the general case is difficult, however, we can specify a special case where the ESS can be explicitly calculated.

We consider a simple case to assess the efficiency of standard HMC and L-HMC. We aim to sample from a univariate exponential distribution, $\mathrm{Exp}(x; \theta)$, with energy function, $E(x) = \theta x$, for $x > 0$. From the above analysis, for standard HMC

$$\rho(1) = \frac{2}{3}, \quad \rho(h) = (\frac{2}{3})^h, \quad \mathrm{ESS} = \frac{N}{5},$$

For L-HMC, we have

$$\rho(1) = \frac{1}{2}, \quad \rho(h) = (\frac{1}{2})^h, \quad \mathrm{ESS} = \frac{N}{3},$$

We observe that the ESS becomes larger with L-HMC. As a result, under these conditions, and many other univariate cases discussed in the experiments, L-HMC has a theoretical advantage of the mixing rate in stationary period over standard HMC. This observation is consistent with the intuition discussed in Sect. 4.

7   Experiments

7.1   Synthetic Toy Examples

We conduct several experiments to validate the theoretical results, as well as the performance of standard HMC and L-HMC.

Synthetic 1D problems. We first perform our experiments on several univariate distributions, where evaluation of theoretical mixing performance is possible. Our primary objective for this simulation study is to validate that the theoretical results are consistent with the empirical results. Each density is given by $p(x) = \frac{1}{Z_1} \exp(-E(x))$, $s.t$ $x \geq 0$ and

- Exponential distribution: $\text{Exp}(x; \theta)$, where $E(x) = \theta x$.
- Truncated Gaussian: $\mathcal{N}_+(x; 0, \theta)$, where $E(x) = \theta x^2$.

We truncate the Gaussian distribution to the positive side, because for a symmetric distribution the theoretical autocorrelation is always 0, thus rendering the comparison less interesting. Note that for each case, as long as the parameter $\theta > 0$, the performance of the sampler does not depend on θ.

We perform standard HMC and L-HMC, as well as "analytic" slice sampling[2] when available. We collected 30,000 Monte Carlo samples, with 10,000 burn-in samples. The leap-frog steps are set to 100 for each experiment. The mass parameter m and stepsize ε are selected manually to achieve around 0.9 acceptance ratio. We observed that applying the partial momentum refreshment can provided additional help, especially when taking fewer leap-frog steps. However, the improvements are not significant when the number of leap-frog steps is adequate for the tested cases.

As shown in Table 1, in the tested cases, theoretical autocorrelations and ESS match well with empirical performance of standard HMC, L-HMC and analytic slice samplers. In every case, L-HMC obtained better empirical results, which is consistent with our theoretical analysis.

Sampling from a discrete distribution. To demonstrate that the L-HMC can perform sampling of distributions with discrete support, we consider sampling from a univariate Poisson distribution, $\mathcal{P}(\lambda)$, with fixed rate parameter λ (we use $\lambda = 10$ in our experiment). The potential energy is given by

---

[2] Analytic slice sampling is achieved by analytically solving the slice interval and computing the expectation in (20), and is only available for exponential and positive-truncated Gaussian cases.

Table 1. 1D theoretical (Th.) and empirical $\rho(1)$ and ESS. SS denotes the analytical slice sampler corresponding to standard HMC or L-HMC.

	Th. $\rho(1)$	Th. ESS	SS $\rho(1)$	SS ESS	(L-)HMC $\rho(1)$	(L-)HMC ESS
standard HMC (Exp)	0.6667	6000	0.6620	6204	0.6711	6069
L-HMC (Exp)	0.5	10000	0.4868	10227	0.5218	9773
standard HMC ($\mathcal{N}_+$)	0.4787	10576	0.4736	10705	0.4802	10510
L-HMC ($\mathcal{N}_+$)	0.3120	15732	0.3040	15457	0.3061	15595

$E(x) = -x\log\lambda + \log x!$. We apply the update scheme described in Sect. 5, and run 10,000 iterations with 3,000 burn-in samples. The number of iterative dynamic updates, stepsize, and mass parameter m were set to 15, 2, and 1, respectively. Results are shown in Fig. 2. The empirical results match well with the probability mass function of $\mathcal{P}(\lambda)$ with $\lambda = 10$. The acceptance ratio is always one, as during the iterative process, the Hamiltonian is exactly conserved. As a consequence, the Metropolis step can be omitted. The empirical $\rho(1)$ and ESS are 0.024 and 9,984, respectively.

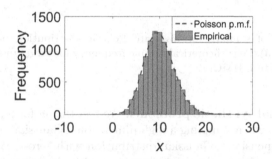

Fig. 2. Histogram of samples for a Poisson distribution, $x \sim \mathcal{P}(\lambda)$ with $\lambda = 10$.

We also apply our methods to sample from a bivariate Poisson distribution [14]. The bivariate Poisson with random covariates (z_1, z_2) can be constructed as $z_1 = y_1 + y_3, z_2 = y_1 + y_2$, where (y_1, y_2, y_3) are three independent Poisson variables with mean parameters $(\lambda_1, \lambda_2, \lambda_3)$. The probability function can be written as

$$\Pr(z_1 = k_1, z_2 = k_2) = \exp(-\lambda_1 - \lambda_2 - \lambda_3)\frac{\lambda_1^{k_1}}{k_1!}\frac{\lambda_2^{k_2}}{k_2!}\sum_{k=0}^{k_1 \wedge k_2}\binom{k_1}{k}\binom{k_2}{k}k!(\frac{\lambda_3}{\lambda_1\lambda_2})^k,$$

We set the ground truth model parameters to $(\lambda_1, \lambda_2, \lambda_3) = (1, 2, 3)$. The dynamic update step, stepsize and mass parameter m are set to be 10, 1 and 1, respectively. When performing the discrete sampling, we normalized the difference vector to enforce the total Hamiltonian to be conserved. We collect 10,000 Monte Carlo samples after 3,000 burn-in samples. The sampled distribution is shown in Fig. 3. The theoretical Pearson correlation for the target bivariate Poisson distribution is given by $\frac{\lambda_3}{\sqrt{\lambda_1+\lambda_3}\sqrt{\lambda_2+\lambda_3}} = 0.6708$. We observed that the empirical Pearson correlation is 0.6983, which matches well with the theoretical value. We also observed that when the dimensionality increases, the discrepancy between target distribution and empirical estimated distribution becomes larger. For this reason, we suggest to consider our method only for low dimensional sampling tasks. How to use HMC to sample from high-dimensional distributions is left for interesting future work.

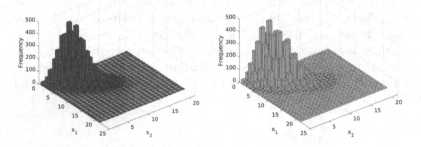

Fig. 3. Histogram of samples for bivariate Poisson distribution parameterized by $(\lambda_1, \lambda_2, \lambda_3) = (1, 2, 3)$. Left: theoretical sample frequency for target distribution. Right: samples from discrete L-HMC.

High-dimensional synthetic problems. We test the performance of standard HMC and L-HMC when sampling a high-dimensional Gaussian distribution. We consider a 100-dimensional Gaussian distribution with zero-mean and diagonal covariance matrix, with its diagonal elements uniformly drawn from $(0, 10]$. We ran 5,000 MC iterations, after 2,500 burn-in samples. For both standard HMC and L-HMC, we use 5 different leap-frog stepsizes, ε_t, $t = \{1, \ldots, 5\}$, where $\varepsilon_{t+1} = 0.8\varepsilon_t$. This scheme allows us to find the elbow points where performance is optimal. The ε_1 and m for standard HMC and L-HMC are set to $(0.025, 2)$ and $(0.015, 1)$, respectively. The sampler was initialized at MLE (estimated by gradient descent) to accelerate burn-in period.

We also compared with the adaptive scheme described in Sect. 5, where the stepsize is automatically tuned at each 500 interactions during the burn-in rounds using an empirically estimated covariance. The adaptation is stopped after burn-in, as suggested by [22]. Both L-HMC and adaptive L-HMC achieved median effective sample size near to the full sample size, and obtained a lower discrepancy between the empirically estimated covariance and the ground truth than standard HMC, see Fig. 4 (left). Employing the adaptive scheme improved the median ESS, probably due to the fact that the stepsize learned from the samples can automatically match the scale of each dimension, Fig. 4 (right).

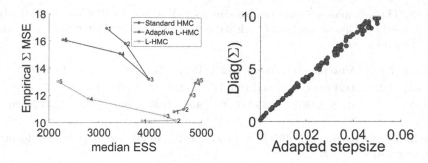

Fig. 4. Standard HMC and L-HMC performance on a 100-dimensional simulated Gaussian distribution. Left: Mean Squared Error (MSE) of estimated Σ vs. median ESS. Labels denote the stepsize index. Right: Elements of diag(Σ) vs. the adapted stepsize after 2,500 burn-in rounds.

7.2 Real Data Analysis

We perform an empirical comparison on two real-world probabilistic modeling tasks: Bayesian Logistic Regression (BLR) and Latent Dirichlet Allocation (LDA).

Bayesian logistic regression. We evaluated the mixing performance of standard HMC and L-HMC on 5 Bayesian logistic regression datasets from the UCI repository [2]. For data $X \in \mathbb{R}^{d \times N}$, response variable $t \in \{0,1\}^N$ and target parameters $\beta \in \mathbb{R}^d$, suppose a Gaussian prior is imposed $\mathcal{N}(0, \alpha I)$ (where $\alpha > 0$) on β, the log posterior is given by [10],

$$\mathcal{L}(\beta) = \beta^T X t - \sum_{n=1}^{N} \log(1 + \exp(\beta^T X_{n,\cdot}^T)) - \frac{\beta^T \beta}{2\alpha}$$

Feature dimensions range from 7 to 15 and the number of data instances are between 250 and $1,000$. All datasets are normalized to have zero mean and unit variance. The sampler was initialized at gradient estimated MLE as in above experiments.

The mass matrix for kinetic function is defined as $M = m \times I$, where m is mass parameter. Gaussian priors $\mathcal{N}(0, 100 I)$ were imposed on the regression coefficients. The leap-frog steps were set to be uniformly drawn from $[1, 100]$, as suggested by [20]. We manually select the stepsize and mass parameter m, so that the acceptance ratios fall in $[0.6, 0.9]$ [3]. On each dataset, the running time for each method is roughly identical, due to the fact that each method took approximately the same number of leap-frog steps. All experiments are based on 5,000 samples, with 1,000 burn-in samples.

Since the MCMC methods that we compared are asymptotically exact to the true posterior, the sample-based estimator is guaranteed to converge to the true expectation over the posterior. ESS indicates the variance of sample based estimator, thus is a good metric for comparison. For this reason, following [6, 10, 21],

Table 2. The minimum effective sample size, as well as the AUROC (in parenthesis) for each method. Dimensionality of each dataset is indicated in parenthesis after the name of each dataset.

Dataset (D)	Australian (15)	German (25)	Heart (14)	Pima (8)	Ripley (7)
Standard HMC	3124 (0.92)	3447 (0.78)	3524 (0.92)	3434 (0.90)	3317 (0.99)
L-HMC	**4308** (0.93)	**4353** (0.79)	**4591** (0.93)	**4664** (0.88)	**4226** (0.99)

Table 3. MNIST results. $D = 101, N = 12,214$. Total sample size is 4,000. AR denotes acceptance ratio.

	ESS min	Median	Max	Time (s)	AR
Standard HMC	2812	3441	3807	287.8	0.978
L-HMC	**3198**	**3808**	**4000**	291.0	0.968

we primarily compare on each method in terms of minimum ESS. We also evaluate the average predictive AUROC based on 10 fold cross-validation, the results showed no significant differences between standard HMC and L-HMC. The results are summarized in Table 2. L-HMC outperforms standard HMC in all datasets.

To further assess the scalability to high-dimensional problems, we also conduct an experiment on the MNIST dataset restricted to digits 7 and 9. We use 12,214 training instances, where the first 100 components from PCA were employed as regression features [6]. We ran $4,000$ MC iterations with $1,000$ burn-in samples, the results are shown in Table 3. L-HMC scales well, and achieved better mixing performance than standard HMC, while taking roughly the same running time. The acceptance ratio of L-HMC decreased by 0.01 *w.r.t.* standard HMC, presumably because the contours for L-HMC are slightly stiffer than those for standard HMC.

Topic modeling. We also evaluate our methods with LDA [4]. LDA models a document as a mixture of multinomial distributions over a vocabulary of size V. The multinomial distributions are parametrized by $\phi_k \in \Delta^V$ for $k = 1, \ldots, K$, where Δ^V denotes the V-dimensional simplex. Each ϕ_k is associated with a symmetric Dirichlet prior with parameter β. Specifically, the generative process for a document is as follows:

- For each topic k, sample a topic-word distribution: $\phi_k | \beta \sim \text{Dirichlet}(\beta)$.
- For each document d, sample a topic distribution: $\theta_d | \alpha \sim \text{Dirichlet}(\alpha)$.
  - For each word i, sample a topic indicator: $z_{di} | \theta_d \sim \text{Discrete}(\theta_d)$.
  - Sample an observed word: $w_{di} | \phi_{z_{di}} \sim \text{Discrete}(\phi_{z_{di}i})$.

To apply the L-HMC and standard HMC, following [8], we re-parametrize ϕ_k with $\tilde{\phi}_k$ as $\phi_{ki} = e^{\tilde{\phi}_{ki}} / (\sum_j e^{\tilde{\phi}_{kj}})$. Similar to [8], a semi-collapsed LDA formulation is used for sampling, where the distribution over topics for each document is integrated out. We use the ICML dataset [7] for the experiment, which

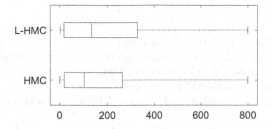

Fig. 5. Empirical distribution of coordinate-wise effective sample size of standard HMC and L-HMC, over 57,540 dimensions.

contains 765 documents corresponding to abstracts of ICML proceedings from 2007 to 2011. After stopword removal, we obtain a vocabulary size of 1,918 and total words of about 44K. We used 80 % of the documents for training and the remaining 20 % for testing. The number of topics is set to 30, resulting in 57,540 parameters. We use a symmetric Dirichlet prior (*i.e.*, all of the elements of parameter vector β have the same value) with parameter $\beta = 0.1$. All experiments are based on 800 MCMC samples with 200 burn-in rounds. We set the stepsizes to be 2.0 for both L-HMC and standard HMC, to obtain acceptance ratios around 0.68. For each iteration we set 20 leap-frog steps. L-HMC has best mixing performance as seen in Fig. 5, and the perplexity is comparable with standard HMC. The perplexities for L-HMC and standard HMC is 958 and 963, respectively.

8   Conclusion

We demonstrated the equivalency between the slice sampler and HMC with a Laplace kinetic energy. This enables us to perform the leap-frog numerical integrator for standard slice sampling in high-dimensional space. We further demonstrated that the resulting sampler can be applied to sampling from discrete distributions, *e.g.*, Poisson. Our method can be seen as a drop-in replacement for scenarios where standard HMC applies, and thus it has many potential extensions. However, our method has its limitations. For high dimensional problems, the numerical issues associated with the sampler are less negligible, and requires carefully selecting the sampler parameters. Future directions include (1) employing more sophisticated numerical methods to reduce the numerical error of our L-HMC approach (2) formal study of the ESS of the proposed L-HMC compared to standard HMC, and (3) exploiting geometric information [10] in the leap-frog updates to further improve the sampling efficiency.

Acknowledgments. The research reported here was supported in part by ARO, DARPA, DOE, NGA and ONR.

References

1. Andrieu, C., Thoms, J.: A tutorial on adaptive mcmc. Stat. Comput. **18**, 4 (2008)
2. Bache, K., Lichman, M.: UCI machine learning repository (2013)
3. Betancourt, M., Byrne, S., Girolami, M.: Optimizing the integrator step size for Hamiltonian Monte Carlo. ArXiv (2014)
4. Blei, D.M., Ng, A.Y., Jordan, M.I.: Latent Dirichlet allocation. J. Mach. Learn. Res. **3** (2003)
5. Brooks, S., Gelman, A., Jones, G., Meng, X.-L.: Handbook of Markov Chain Monte Carlo. CRC Press, Boca Raton (2011)
6. Chao, W.-L., Solomon, J., Michels, D., Sha, F.: Exponential integration for Hamiltonian Monte Carlo. In: ICML (2015)
7. Chen, C., Rao, V., Buntine, W., Whye Teh, Y.: Dependent normalized random measures. In: ICML (2013)
8. Ding, N., Fang, Y., Babbush, R., Chen, C., Skeel, R.D., Neven, H.: Bayesian sampling using stochastic gradient thermostats. In: NIPS (2014)
9. Duane, S., Kennedy, A.D., Pendleton, B.J., Roweth, D.: Hybrid Monte Carlo. Phys. Lett. B **195**, 2 (1987)
10. Girolami, M., Calderhead, B.: Riemann manifold Langevin and Hamiltonian Monte Carlo methods. J. Roy. Stat. Soc. Ser. B (Stat. Method.) **73**, 2 (2011)
11. Goldstein, H.: Classical Mechanics. Pearson Education India, New Delhi (1965)
12. Haario, H., Saksman, E., Tamminen, J.: An adaptive Metropolis algorithm. Bernoulli (2001)
13. Homan, M.D., Gelman, A.: The no-u-turn sampler: adaptively setting path lengths in hamiltonian monte carlo. J. Mach. Learn. Res. **15**, 1 (2014)
14. Karlis, D., Meligkotsidou, L.: Multivariate poisson regression with covariance structure. Stat. Comput. **15**, 4 (2005)
15. Metropolis, N., Rosenbluth, A.W., Rosenbluth, M.N., Teller, A.H., Teller, E.: Equation of state calculations by fast computing machines. J. Chem. Phys. **21**, 6 (1953)
16. Mohamed, S., De Freitas, N., et al.: Adaptive hamiltonian and riemann manifold monte carlo samplers. Arxiv (2013)
17. Murray, I., Adams, R.P., MacKay, D.J.: Elliptical slice sampling. ArXiv (2009)
18. Neal, R.M.: Probabilistic inference using markov chain monte carlo methods. Technical report CRG-TR-93-1 (1993)
19. Neal, R.M.: Slice sampling. Ann. Stat. **31**, 705–767 (2003)
20. Neal, R.M.: MCMC using Hamiltonian dynamics. Handbook of Markov Chain Monte Carlo 2 (2011)
21. Pakman, A., Paninski, L.: Auxiliary-variable exact Hamiltonian Monte Carlo samplers for binary distributions. In: NIPS (2013)
22. Robert, C., Casella, G.: Monte Carlo statistical methods. Springer Science & Business Media, New York (2004)
23. Roberts, G.O., Rosenthal, J.S.: Examples of adaptive MCMC. J. Comput. Graph. Stat. **18**, 2 (2009)
24. Sohl-Dickstein, J., Mudigonda, M., DeWeese, M.R.: Hamiltonian monte carlo without detailed balance. ArXiv (2014)
25. Taylor, J.R.: Classical Mechanics. University Science Books, Colorado (2005)
26. Zhang, Y., Ghahramani, Z., Storkey, A.J., Sutton, C.A.: Continuous relaxations for discrete Hamiltonian Monte Carlo. In: NIPS (2012)

Functional Bid Landscape Forecasting
for Display Advertising

Yuchen Wang[1], Kan Ren[1], Weinan Zhang[1(✉)], Jun Wang[2], and Yong Yu[1(✉)]

[1] Shanghai Jiao Tong University, Shanghai, China
{zeromike,kren,wnzhang,yyu}@apex.sjtu.edu.cn
[2] University College London, London, UK
j.wang@cs.ucl.ac.uk

Abstract. Real-time auction has become an important online advertising trading mechanism. A crucial issue for advertisers is to model the market competition, i.e., *bid landscape forecasting*. It is formulated as predicting the market price distribution for each ad auction provided by its side information. Existing solutions mainly focus on parameterized heuristic forms of the market price distribution and learn the parameters to fit the data. In this paper, we present a functional bid landscape forecasting method to automatically learn the function mapping from each ad auction features to the market price distribution without any assumption about the functional form. Specifically, to deal with the categorical feature input, we propose a novel decision tree model with a node splitting scheme by attribute value clustering. Furthermore, to deal with the problem of right-censored market price observations, we propose to incorporate a survival model into tree learning and prediction, which largely reduces the model bias. The experiments on real-world data demonstrate that our models achieve substantial performance gains over previous work in various metrics. The software related to this paper is available at https://github.com/zeromike/bid-lands.

1 Introduction

Popularized from 2011, real-time bidding (RTB) has become one of the most important media buying mechanism in display advertising [7]. In RTB, each ad display opportunity, i.e., an ad impression, is traded through a real-time auction, where each advertiser submits a bid price based on the impression features and the one with the highest bid wins the auction and display her ad to the user [20]. Apparently, the bidding strategy that determines how much to bid for each specific ad impression is a core component in RTB display advertising [16].

As pointed out in [22], the two key factors determining the optimal bid price in a specific ad auction are *utility* and *cost*. The utility factor measures the value of ad impression, normally quantified as user's response rate of the displayed ad, such as click-through rate (CTR) or conversion rate (CVR) [12]. The cost factor, on the other hand, estimates how much the advertiser would need to pay to win the ad auction [3]. From an advertiser's perspective, the *market price* is defined

© Springer International Publishing AG 2016
P. Frasconi et al. (Eds.): ECML PKDD 2016, Part I, LNAI 9851, pp. 115–131, 2016.
DOI: 10.1007/978-3-319-46128-1_8

as the highest bid price from her competitors[1]. In the widely used second-price auctions, the winner needs to pay the second highest bid price in the auction, i.e., the market price [4]. Market price estimation is a difficult problem because it is the highest bid from hundreds or even thousands of advertisers for a specific ad impression, which is highly dynamic and it is almost impossible to predict it by modeling each advertiser's strategy [2]. Thus, the practical solution is to model the market price as a stochastic variable and to predict its distribution given each ad impression, named as *bid landscape.*

Previous work on bid landscape modeling is normally based on predefining a parameterized distribution form, such as Gaussian distribution [18] or log-normal distribution [3]. However, as pointed out in [19], such assumptions are too strong and often rejected by statistical tests. Another practical problem is the observed market price is right-censored, i.e., only when the advertiser wins the auction, she can observe the market price (by checking the auction cost), and when she loses, she only knows the underlying market price is higher than her bid. Such censored observations directly lead to biased landscape models.

In this paper, we present a novel *functional bid landscape forecasting* model to address the two problems. Decision tree is a method commonly used in data mining [5,17]. By building a decision tree, the function mapping from the auctioned ad impression features to the corresponding market price distribution is automatically learned, without any functional assumption or restriction. More specifically, to deal with the categorical features which are quite common in online advertising tasks, we propose a novel node splitting scheme by performing clustering based on the attribute values, e.g., clustering and splitting the cities. The learning criterion of the tree model is based on KL-Divergence [10] between the market price distributions of children nodes. Furthermore, to model the censored market price distribution of each leaf node, we adopt non-parametric survival models [9] to significantly reduce the modeling bias by leveraging the lost bid information.

The experiments on a 9-advertiser dataset demonstrate that our proposed solution with automatic tree learning and survival modeling leads to a 30.7% improvement on data log-likelihood and a 77.6% drop on KL-Divergence compared to the state-of-the-art model [18].

In sum, the technical contributions of this paper are three-fold.

Automatic function learning: a decision tree model is proposed to automatically learn the function mapping from the input ad impression features to the market price distribution, without any functional assumption.

Node splitting via clustering: the node splitting scheme of the proposed tree model is based on the KL-Divergence maximization between the split data via a K-means clustering of attribute values, which naturally bypasses the scalability problem of tree models working on categorical data.

---

[1] The terms 'market price' and 'winning (bid) price' are used interchangeably in related literature [2,3,18]. In this paper, we use 'market price'.

Efficient censorship handling: with a non-parametric survival model, both the data of observed market prices and lost bid prices are fed into the decision tree learning to reduce the model bias caused by the censored market price observations.

The rest of this paper is organized as follows. We discuss some related work and compare with ours in Sect. 2. Then we propose our solution in Sect. 3. The experimental results and detailed discussions are provided in Sect. 4. We finally conclude this paper and discuss the future work in Sect. 5.

2   Related Work

Bid Landscape Forecasting. As is discussed above, bid landscape forecasting is a crucial component in online advertising framework, however, lacking enough attention. On one hand, researchers proposed several heuristic forms of functions to model the market price distribution. In [22], the authors provided two forms of winning probability w.r.t. the bid price, which is based on the observation of an offline dataset. However, this derivation has many drawbacks since the appropriate distribution of the market price in real world data may deviate much from the simple functional form. On the other hand, some fine-studied distributions are also used in market price modeling. [3] proposed a log-normal distribution to fit the market price distribution. The main drawback is that these distributional methods may lose the effectiveness of handling various dynamic data and they ignore the real data divergency as we will show later in Figs. 1 and 3.

In view of forecasting, [3] presented a template-based method to fetch the corresponding market price distribution w.r.t. the given auction request. However, this paper studied the problem as on the seller side which is quite different from the buyer side as we stand. [18] proposed a regression method to model the market price w.r.t. auction features. However, those methods do not care much about the real data properties, i.e. similarity and distinction among data segments, which may result in poor forecasting performance on different campaigns. Moreover, none of the above methods deal with the data censorship problem in modeling training.

Bid Optimization. Bid optimization is a well studied problem and has drawn many concerns both in RTB environment [1,21,22]. This task aims to optimize the strategies to allocate budget to gain ad display opportunities [1], so it is crucial to model the market competition and make accurate bid landscape prediction [3]. In RTB display advertising with cost-per-impression scheme, the bid decision is made on the level of impression so that the price is also charged on impression level [11]. It again emphasizes key importance of the forecasting task in online bidding. In [21,22], the authors proposed a functional optimization method with the consideration of budget constraints and market price distribution, which led to an optimal bidding strategy. However, the market price distribution adopted in these two papers is under heuristic assumptions, which may not perform well in real-world forecasting tasks.

Learning over Censored Data. In machine learning fields, dealing with censored data is sometimes regarded as handling missing data, which is a well-studied problem [6]. The item recommendation task with implicit feedback is a classic problem of dealing with missing data. [15] proposed a uniform sampling of negative feedback items for user's positive ones [14]. In the online advertising field, the authors in [18] proposed a regression model with censored regression module using the lost auction data to fix the biased data problem. However, the Gaussian conditional distribution assumption turns out to be too strong, which results in weak performance in our experiment. The authors in [2] implemented a product-limit estimator [9] in handling the data censorship in sponsored search, but the bid landscape is built on search keyword level, which is not fine-grained to work on RTB display advertising. We transfer the survival analysis method from [2] to RTB environment and compare with [18] in our experiment.

3 Methodology

3.1 Problem Definition

The goal of bid landscape forecasting is to predict the probabilistic distribution density (p.d.f.) $p_x(z)$ w.r.t. the market price z given an ad auction information represented by a high-dimensional feature vector x.

Table 1. The statistics of attributes.

Attribute	Adex-change	Weekday	Slot-visibility	Slotheight	Slotwidth	Hour	Region	User-agent	Creative	City
Num of values	5	7	11	14	21	24	35	40	131	370

Each auction x contains multiple side information, e.g. user agent, region, city, user tags, ad slot information, etc. In Table 1, we present the attributes contained in the dataset with corresponding numbers of value. We can easily find that different attributes vary in both diversity and quantity. Moreover, the bid price distribution of a given request may be diverse in different attributes. Take the field Region as an example, the bid distribution of the samples with region in Beijing is quite different from that of Xizang, which is illustrated in Fig. 1. Previous work focuses only on the heuristic forms (e.g. log-normal [3] or a unary function [22]) of distribution and cannot effectively capture the divergency within data.

Moreover, in RTB marketplace, the advertiser proposes a bid price b and wins if $b > z$ paying z for the ad impression, loses if $b \leq z$ without knowing the exact value of z, where z represents the market price which is the highest bid price

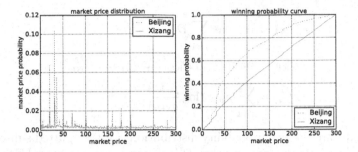

Fig. 1. Market price distribution over different regions.

from the competitors. Apparently, the true market price is only observable for who is winning the corresponding auction. As for the lost auctions, the advertiser only knows the lower bound of the market price, which results in the problem of right censored data [2]. The censorship from the lost auctions may heavily influence the forecasting performance in the online prediction [18].

In this paper, we mainly settle down these two problems. First, we propose to automatically build the function mapping from the given ad auction x to the market price distribution $p_x(z)$ without any functional form assumption, generally represented as

$$p_x(z) = T_p(x). \tag{1}$$

Second, we leverage both the observed market price data of winning auctions and censored one of losing auctions to train a less biased function $T_p(x)$.

We use a binary decision tree to represent $T_p(x)$. More precisely, every node represents a set of auction samples. For each node O_i, we split the contained samples into two sets $\{S_{ij}^t\}$ according to attribute A_j (e.g. Region) value sets (e.g. {Xizang, Beijing, ...}), where $t \in \{1, 2\}, A_j \in \Theta$ and Θ is the attribute space. For each subset S_{ij}^t, the corresponding market price distribution $p_x^t(z)$ can be statistically built. Intuitively, different subsets have diverse distributions and the samples within the same subset are similar to each other, which requires effective clustering and node splitting scheme. Furthermore, KL-Divergence [10] is a reasonable metric to measure the splitted data divergency. So that we choose the best splitting π_i with the highest KL-Divergence value D_{KL}^i calculated between the resulted two subsets $S_{i.}^1$ and $S_{i.}^2$ in node O_i. Essentially, our goal is to seek the splitting strategy $\pi = \cup_{i \in I} \pi_i$, where each splitting action π_i maximizes the KL-Divergence D_{KL} between two child sets in node O_i. Mathematically, our functional bid landscape forecasting system is built as

$$T_p^{\pi}(\boldsymbol{x}) = \arg\max_{\pi} \sum_{i=1}^{l} D_{\mathrm{KL}}^i, \tag{2}$$

$$D_{\mathrm{KL}}^i = \max\{D_{\mathrm{KL}}^{i1}, D_{\mathrm{KL}}^{i2}, ..., D_{\mathrm{KL}}^{ij}, ..., D_{\mathrm{KL}}^{iN}\} \tag{3}$$

$$D_{\mathrm{KL}}^{ij} = \sum_{z=1}^{z_{\max}} p_{\boldsymbol{x}}(z) \log \frac{p_{\boldsymbol{x}}(z)}{q_{\boldsymbol{x}}(z)}, \tag{4}$$

where p and q are the two probability distributions for the splitted subsets, $z_{\max}$ represents the maximum market price, D_{KL}^{ij} means the maximum KL-Divergence of splitting over the sample set of attribute A_j in node i, $N = |\Theta|$ is the number of attributes and l is the number of splitting nodes.

When forecasting, every auction instance will follow a path from the root to the leaf, classified by the dividing strategy according to the attribute value it contains. The bid landscape $p_{\boldsymbol{x}}(z)$ is finally predicted at the leaf node.

3.2 Decision Trees with K-means Clustering

In this section, we propose the K-means clustering method based on KL-Divergence. Then, we will present our iterative optimization algorithm for the decision tree learning model with K-means clustering.

K-Means Clustering with KL-Divergence. For each attribute, the values vary over different samples, as we can see in Table 1. The goal of the binary decision tree spanning is to group similar values w.r.t. one attribute and split the data samples into two divergent subsets. We use KL-Divergence to model the statistics of datasets, which is shown in Eq. (2). KL-Divergence is a measurement assessing the difference between two probability distributions. The problem need to solve is that it requires an effective clustering method to group samples w.r.t. the given metric.

In this paper, the bid samples with the same attribute value are considered as a single point. The goal of K-means clustering is to partition these points into two clusters according to the calculated KL-Divergence. The process of K-means clustering summarize in Algorithm 1.

The input of Algorithm 1 is the attribute A_j and a set of training samples $S = \{s_1, s_2, ..., s_k, ..., s_n\}$, where s_k is the set of training samples with the same value for attribute A_j and n is the number of different values of attribute A_j.

We adopted an iterative algorithm to achieve the clustering goal. First, we randomly split the data into two parts S^1 and S^2. Then we propose an EM-fashion algorithm to iterate two steps below until the whole process converges to the optimal objective, i.e., maximize the KL-Divergence.

E-step: Compute the market price probabilistic distribution Q_1 for S^1 and Q_2 for S^2, which will be discussed in Sect. 3.3.

M-step: Consider the sample data with the same value of attribute A_j as a whole, we will get $\{s_1, s_2, ..., s_k, ..., s_n\}$ if the attribute A_j has n different values. And we will have n corresponding market price probability distributions

---

Algorithm 1. K-Means clustering with KL-Divergence

Input: Training sample $S = \{s_1, s_2, ..., s_n\}$; Attribute A_j;
Output: KL-Divergence D_{KL}^j over attribute A_j, Data clusters S^1 and S^2;
 1: Randomly split the data into two parts S^1 and S^2;
 2: **while** not converged **do**
 3:    **E-step**:
 4:    Get price probability distribution Q_1 for S^1 and Q_2 for S^2;
 5:    **M-step**:
 6:    **for all** $M_k, k \in \{1, 2, 3, ..., n\}$ **do**
 7:       Calculate the K_1 between M_k and Q_1 by Eq. (4);
 8:       Calculate the K_2 between M_k and Q_2 by Eq. (4);
 9:       Update S^1 or S^2 by comparing with K_1 and K_2;
10:    **end for**
11:    Calculate the D_{KL}^j between Q_1 and Q_2 by Eq. (4);
12: **end while**
13: Return D_{KL}^j, S^1 and S^2;

$\{M_1, M_2, ..., M_n\}$. For each market price probability distribution M_k, we will calculate the KL-Divergence K_1 between M_k and Q_1, K_2 between M_k and Q_2, respectively. If $K_1 > K_2$, it means that the probability distribution M_k is more similar with Q_2, thus assign data set s_k to the relatively more similar data set S^2, vice versa.

After each M-step, we will calculate the KL-Divergence D_{KL}^j between Q_1 and Q_2. The EM iteration stops when D_{KL}^j does not change.

In this paper, we only split each node into two subsets, i.e., $k = 2$. To avoid bringing another control variable into the model, we do not discuss about cases of $k >= 3$. As a result, we choose $k = 2$ to make it consistent with the bi-splitting scheme on numeric features.

Building the Decision Tree. The combined scheme of building decision tree based on K-means clustering node splitting is described in Algorithm 2. In Algorithm 2, we first find the splitting attribute with highest KL-Divergence. Then, we perform the binary splitting of the data by maximizing KL-Divergence between two leaf nodes with K-means clustering. The sub-tree keeps growing until the length of sample data in leaf node is less than a predefined value. Finally, we prune the tree by using reduced error pruning method. Compared with the decision tree algorithm, the main difference of our proposed scheme is that the binary node splitting scheme with K-means clustering and the usage of KL-Divergence as the attribute selection criteria.

In the test process, one problem is that there could be new attribute values of some test data instances which do not match any nodes of our decision tree learned from the training data. To handle this, we deploy a randomly choosing method which decides the attribute value of the given test data to randomly goes to one of the two children, which is equivalent to non-splitting on such

---

Algorithm 2. Building Decision Tree with K-Means clustering

Input: Training sample S which contain N attributes;
1: **for** all attribute $A_j, j \in \{1, 2, 3, ..., N\}$ **do**
2:     Calculate the KL-Divergence D_{KL}^j for attribute A_j by Algorithm 1;
3: **end for**
4: $D_{KL}^{best} = \max \{D_{KL}^1, D_{KL}^2, ..., D_{KL}^j, D_{KL}^N\}$;
5: Find A_{best} with D_{KL}^{best};
6: Create a decision node that splits on A_{best};
7: Split the decision node into two nodes S^1 and S^2;
8: Return new nodes as children of the parent node

---

attribute. The experiment results show that such random method works well on the real-world dataset.

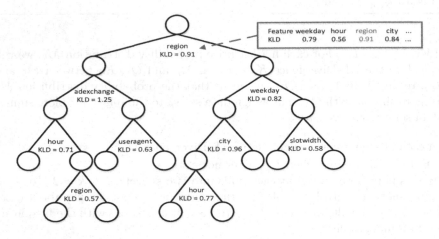

Fig. 2. Illustration of the tree.

Figure 2 is an example of the decision tree. As we can see, for each node, we illustrate its best splitting attribute and the corresponding KL-Divergence. The red box shows the KL-Divergence value for each attribute, and the best splitting attribute with the highest KL-Divergence is chosen.

3.3   Bid Landscape Forecasting with Censored Data

In real-time bidding, an advertiser only observes the market prices of the auctions that she wins. For those lost auctions, she only knows the lower bound of the market price, i.e., her bid price. Such data is named as the right censored data [18]. However, the partial information in lost auctions is still of high value. To better estimate the bid distribution, we introduce survival models [8] to model the censored data. We implement a non-parametric method to model the real market price distribution and transfer survival analysis from keywork search advertising [2] to RTB environment. That is, given the observed impressions and the lost bid requests, the winning probability can be estimated with the non-parametric Kaplan-Meier Product-Limit method [9].

Suppose we have sequential bidding logs in form of $\{b_i, w_i, m_i\}_{i=1,2,...,M}$, where b_i is the bidding price in the auction, w_i is the boolean value of whether we have won the auction or not, and m_i is the market price (unknown if $w_i = 0$). Then we transform our data into the form of $\{b_j, d_j, n_j\}_{j=1,2,...,N}$, where the bidding price $b_j < b_{j+1}$, and d_j represents the number of the winning auctions with bidding price $b_j - 1$, n_j is the number of auctions that cannot not be won with bidding price $b_j - 1$. Then the probability of losing an auction with bidding price b_x is

$$l(b_x) = \prod_{b_j < b_x} \frac{n_j - d_j}{n_j}. \tag{5}$$

Thus the winning probability $w(b_x)$ and the integer[2] market price p.d.f. $p(z)$ are

$$w(b_x) = 1 - \prod_{b_j < b_x} \frac{n_j - d_j}{n_j}, \quad p(z) = w(z+1) - w(z). \tag{6}$$

4 Experiments

In this section, we introduce the experimental setup and analyze the results[3]. We compare the overall performance over 5 different bid landscape forecasting models, and further analyze the performance of our proposed against different hyperparameters (e.g. tree depth, leaf size).

4.1 Dataset

For the following experiments, we use the real-world bidding log from iPinYou RTB dataset[4]. It contains 64.7 M bidding records, 19.5 M impressions, 14.79 K clicks and 16.0 K CNY expense on 9 campaigns from different advertisers during 10 days in 2013. Each bidding log has 26 attributes, including weekday, hour, user agent, region, slot ID etc. More details of the data is provided in [13].

4.2 Experiment Flow

In order to simulate the real bidding market and show the advantages of our survival model, we take the original data of impression log as full-volume auction data, and perform a truthful bidding strategy [12] to simulate the bidding process, which produces the winning bid dataset W and lost bid dataset L respectively. For each data sample $x_{\text{win}} \in W$, the simulated real market price z_{win} is

---

[2] In practice, the bid prices in various RTB ad auctions are required to be integer.

[3] The experiment code is available at http://goo.gl/h130Z0.

[4] Dataset link: http://data.computational-advertising.org.

known for the advertisers, while the corresponding market price z_{lose} remaining unknown for $x_{\text{lose}} \in L$. It guarantees the similar situation as that faced by all the advertisers in the real world marketplace.

In the test phase, the corresponding market price distribution $p_x(z)$ of each sample x in the test data is estimated by all of the compared models respectively. We assess the performance of different settings in several measurements, as listed in the next subsection. Finally we study the performance of our proposed model with different hyperparameters, e.g., the tree depth and the maximum size of each leaf.

4.3 Evaluation Measures

The goal of this paper is to improve the performance of market price distribution forecasting. We use two evaluation methods to measure the forecasting error. The first one is Average Negative Log Probability (ANLP). After we classifying each sample data into different leaves with the tree model, the sum of log probability for all sample data P_{nl} is given by the Eq. (7), and the average negative log probability $\bar{P}_{\text{nl}}$ given by the $\bar{P}_{\text{nl}}$:

$$
P_{\text{nl}} = \sum_{i=1}^{k} \sum_{j=1}^{z_{\max}} (-\log P_{ij}) N_{ij}, \tag{7}
$$

$$
N = \sum_{i=1}^{k} \sum_{j=1}^{z_{\max}} N_{ij}, \quad \bar{P}_{\text{nl}} = P_{\text{nl}}/N, \tag{8}
$$

where k denotes the number of sub bid landscapes, $z_{\max}$ represents the maximum market price, P_{ij} means the probability of training sample in the ith leaf node given price j, N_{ij} is the number of test sample in the ith leaf node given price j. N is the total number of test samples.

We also calculate the overall KL-Divergence to measure the objective forecasting error. D_{KL} is given by the Eq. (9):

$$
D_{\text{KL}} = \frac{1}{N} \sum_{i=1}^{k} N_i \sum_{j=1}^{z_{\max}} P_{ij} \log \frac{P_{ij}}{Q_{ij}}, \tag{9}
$$

where N_i means the number of test sample in the ith leaf node. Q_{ij} means the probability of test sample in the ith leaf node given price j.

4.4 Compared Settings

We compare five different bid landscape forecasting models in our experiment.

NM - The Normal Model predicts the bid landscape based on the observed market prices from simulated impression log W, without using the lost bid request data in L. This model uses a non-parametric method to directly draw the probability function w.r.t. the market price from the winning dataset.

SM - The Survival Model forecasts the bid landscape with survival analysis, which learns from both observed market prices from impression log and the lost bid request data using Kaplan-Meier estimation [2]. The detail has been discussed in Sect. 3.3.

MM - The Mix Model uses linear regression and censored regression to predict the bid landscape respectively, and combines two models considering winning probability into Mixture Model [18] to predict the final bid landscape.

NTM - The Normal Tree Model predicts the bid landscape using only our proposed tree model, without survival analysis. The detailed modeling method has been declared in Sect. 3.2.

STM - The Survival Tree Model predicts the bid landscape with the proposed survival analysis embedded in our tree model, which is our final mixed model.

4.5   Experiment Results

Data Analysis. Table 2 shows the overall statistics of the dataset, where each row presents the statistical information of the corresponding advertiser in the first column. In Table 2, *Num of bids* is the number of total bids, and *Num of win bids* is the number of winning bids in the full simulated dataset $W \cup L$. *WR* is the winning rate calculated by $\frac{|W|}{|W \cup L|}$. *AMP* is the average market price on all bids. *AMP on W* and *AMP on L* are the average market price for the winning bid set W and the lost bid set L, respectively.

We can easily find that the winning rates of all campaigns are low, which is practically reasonable since a real-world advertiser can only win a little proportion of the whole-world volume. The market prices of most impressions are unavailable to the advertiser. We also observe that the average market price on winning bids (*AMP on W*) are much lower than average market price on lost bids (*AMP on L*). This verifies the bias between the observed market price distribution and the true market price distribution.

Table 2. The statistics of the dataset iPinYou.

Advertiser	Num of bids	Num of win bids	WR	AMP	AMP on W	AMP on L
1458	2,055,371	257,077	0.1251	70.2829	29.1130	76.1684
2259	557,038	135,487	0.2432	96.0685	28.3345	117.8383
2261	458,412	176,325	0.3846	92.1654	32.3218	129.5721
2821	881,708	305,134	0.3461	90.6573	35.0455	120.0881
2997	208,292	60,556	0.2907	64.9918	16.6269	84.8163
3358	1,161,403	336,769	0.2900	92.6624	55.6009	107.7978
3386	1,898,535	332,223	0.1750	80.4224	37.4947	89.5276
3427	1,729,177	563,592	0.3259	82.6685	53.1614	96.9359
3476	1,313,574	303,341	0.2309	79.4990	38.7343	91.7393
Overall	10,263,510	2,470,504	0.2407	81.9769	41.1313	94.9256

Bid Landscapes of Leaf Nodes. There are 4 examples of bid landscape between training and testing samples shown on Fig. 3. From the figures, we can find that the bid landscape of each leaf node is quite different from that of other leaf nodes. Especially, some sub bid landscape tends to have a large probability of some price, and the training distribution fit the test distribution very well. This result suggests we can predict the bid landscape more accurately with tree models.

Survival Model. As is mentioned in Sect. 3.3, the observed market price distribution is biased due to the data censorship. Figure 4 shows the comparison of the curves for market price distribution and winning probability. TRUTH represents for the real market price distribution for the test data, which is regarded as the ground truth. FULL curve is built from full-volume data, i.e., assume the advertiser has observed all market prices of $W \cup L$, which is regarded as the upperbound performance of any bid landscape model based on censored data. We observe that (i) FULL curve is the most close to TRUTH since FULL makes use of full-volume training data and is naturally unbiased. However, in the practice, advertisers only have a small number of winning logs W [13]. (ii) Compared to NM, STM curve is much more close to TRUTH, which verifies its advantage of making use of the censored data with survival analysis to improve the performance of market price distribution forecasting.

Performance Comparison. We evaluate on five models described in Sect. 4.4 with evaluation measure given in Sect. 4.3. Table 3 presents the Average negative log probability (ANLP) and KL-Divergence (KLD) of these settings.

For ANLP, we observe that (i) for all campaigns investigated, STM shows the best performance, which verifies the effectiveness of the survival tree model. (ii) SM is better than NM because SM learns from both winning bids and lost bids to handle the censored data problem. (iii) We shall notice that NTM is the tree version of NM, and STM is the tree version of SM. We find that NTM outperforms NM, and STM outperforms SM, which means the tree model effectively improves the performance of bid landscape forecasting. (iv) STM is the combination of SM and NTM, both of which contribute to a better performance as is mentioned in (ii) and (iii). Thus it is reasonable that STM has the best performance. It has both advantages of SM and NTM, i.e., dealing with the bid

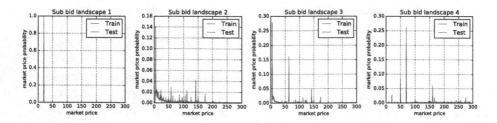

Fig. 3. Examples of different sub bid landscapes.

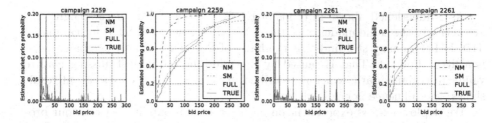

Fig. 4. Comparison of the curves of market price distribution and winning probability.

Table 3. Performance illustration. Average negative probability of five compared settings. ANLP: the smaller, the better. KLD: the smaller, the better.

Campaign	ANLP					KLD				
	MM	NM	SM	NTM	STM	MM	NM	SM	NTM	STM
1458	5.7887	5.3662	4.7885	4.7160	4.3308	0.7323	0.7463	0.2367	0.6591	0.2095
2259	7.3285	6.7686	5.8204	5.4943	5.4021	0.8264	0.9633	0.3709	0.8757	0.1668
2261	7.0205	5.5310	5.1053	4.4444	4.3137	1.0181	0.4029	0.2943	0.3165	0.1222
2821	7.2628	6.5508	5.6710	5.4196	5.3721	0.7816	0.9671	0.3562	0.6170	0.2880
2997	6.7024	5.3642	5.1411	5.1626	5.0944	0.7450	0.4526	0.1399	0.3312	0.1214
3358	7.1779	5.8345	5.2771	4.8377	4.6168	1.4968	0.8367	0.5148	0.8367	0.3900
3386	6.1418	5.2791	4.8721	4.6698	4.2577	0.8761	0.6811	0.3474	0.6064	0.2236
3427	6.1852	4.8838	4.6453	4.1047	4.0580	1.0564	0.3247	0.1478	0.3247	0.1478
3476	6.0220	5.2884	4.7535	4.3516	4.2951	0.9821	0.6134	0.2239	0.5650	0.2238
Overall	6.5520	5.6635	5.0997	4.7792	**4.6065**	0.9239	0.6898	0.2927	0.5834	**0.2160**

distribution difference between different attribute value and learning from the censored data.

For KLD, we can also find that STM achieves the best performance. The results of other models are also similar to those of ANLP, but there are some interesting differences. Note that for campaign 3427, the KL-Divergence values of NM and NTM are equal to each other, so do SM and STM. The KL-Divergence values of SM and STM for Campaign 3476 are also nearly the same. That is because the optimal depth of tree in these cases is 1. We shall notice that actually NM and SM are the special cases of NTM and STM respectively when the tree depthes of the latter two models are equal to 1. The fact arouses the question, how to decides the optimal tree depth? We here take the tree depth as a hyperparameter, and we leave the detailed discussion in the next subsection.

Table 4. p-values in t-tests of ANLP comparison.

Model	MM	NM	SM	NTM
STM	$< 10^{-6}$	$< 10^{-6}$	$< 10^{-6}$	$< 10^{-6}$

As is mentioned above, in terms of KLD, SM and STM for campaign 3476 are actually the same model since the optimal tree depth of STM for campaign 3476 is 1. One may still find that the KLD of SM and STM for campaign 3476 is a little different. That is caused by the handling method of missing feature values in training data, which is described in Sect. 3.2. As the experiment result shows, the influence is negligible.

We deploy a *t-test* experiment on negative log probability between our proposed model STM and each of other compared settings to check the statistical significance of the improvement. Table 4 shows that the p-value of each test is lower than 10^{-6}, which means the improvement is statistically significant. The significant test on KL-Divergence is not performed because KLD is not a metric calculated based on each data instance.

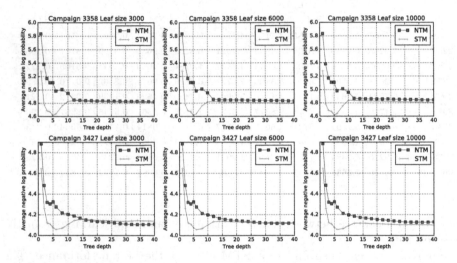

Fig. 5. Average negative log probability with different tree depth.

Hyperparameter Tuning. There are two problems in decision tree algorithms. If we do not limit the size of tree, it will split into many quite small sets, and overfit the training and fail to generalize on new data. However, if the limitation for the size of tree is too much, some nodes that have useful information cannot be split in succession, which is known as horizon effect. In order to avoid both problems, we need to find out the optimal size limitation of the tree. There are two hyperparameters that influence the size of tree, i.e., (i) tree depth (the upperbound of tree depth) and (ii) leaf size (the upperbound of number of training samples in a leaf). In the experiment, we kept changing these two hyperparameters, and compare the average negative log probability with different values. The results are illustrated in Fig. 5. Different pictures represent different limitation on leaf size.

We observe that, for most campaigns, (i) the performance of NTM improves finally converges as the tree depth grows. (ii) The performance of STM improves

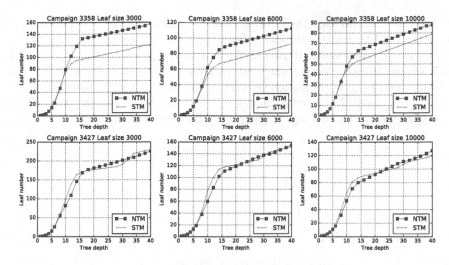

Fig. 6. Relationship between the leaf number and the tree depth.

at first as the tree depth grows, but after tree depth exceeds a certain value, the performance is getting worse, which can be explained as overfitting. (iii) When the tree depth is large enough, the effect of survival model is weakened. The performance of STM and NTM in this case is almost the same. (iv) The leaf size will affect the performance of the tree model, but the overall influence mainly occurs when the tree depth is large. Since the optimal depth for NTM is usually large, the leaf size tends to have a larger influence on performance of NTM. While the optimal depth for STM is usually small, the leaf size will have a smaller influence on STM's performance.

Figure 6 shows the relationship between the leaf number and the tree depth. We can find that the number of leaf increases rapidly at first. When the depth of tree grows up, the growth of leaf number begins to slow down, which corresponds to the convergence of ANLP shown in Fig. 5.

Table 5. The average optimal tree depth and leaf numbers for different models.

	Tree depth		Leaf number	
Model	ANLP	KLD	ANLP	KLD
NTM	20.33	11.33	632.67	398.67
STM	5.89	4.89	25.33	52.11

In the experiment, we use a validation set to find out the optimal tree depth. Table 5 shows the average optimal tree depths and the corresponding leaf numbers for NTM and STM. We can find that the average optimal tree depth and leaf number of STM is lower than that of NTM, because STM learns from both winning bids and lost bids. As the tree grows, it will reach the best performance earlier than NTM, which only learns from the winning bids.

We also experimentally illustrate the EM convergence of the tree model in Fig. 7, which shows the value changes of KL-Divergence over EM training rounds. We observe that our optimization converges within about 6 EM rounds, and the fluctuation is small. In our experiments, the EM algorithm is quite efficient and converges quickly. The average training rounds of our EM algorithm is about 4.

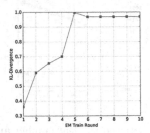

Fig. 7. KL-Divergence convergence w.r.t. EM rounds.

5   Conclusion and Future Work

In this paper, we have proposed a functional bid landscape forecasting methodology in RTB display advertising, which automatically builds a function mapping from the impression features to the market price distribution. The iterative learning framework trains a decision tree by clustering-based node splitting with the KL-Divergence objective. We also incorporate the survival model to handle the model bias problem caused by the censored data observations. The overall model significantly improves the forecasting performance over the baselines and the state-of-the-art models in various metrics.

In the future work, we plan to combine the functional bid landscape forecasting with utility (e.g. click-through rate, conversion rate) estimation model, aiming to make more reasonable and informative decisions in bidding strategy.

References

1. Agarwal, D., Ghosh, S., Wei, K., You, S.: Budget pacing for targeted online advertisements at linkedin. In: KDD (2014)
2. Amin, K., Kearns, M., Key, P., Schwaighofer, A.: Budget optimization for sponsored search: Censored learning in mdps. arXiv preprint arXiv:1210.4847 (2012)
3. Cui, Y., Zhang, R., Li, W., Mao, J.: Bid landscape forecasting in online ad exchange marketplace. In: KDD (2011)
4. Edelman, B., Ostrovsky, M., Schwarz, M.: Internet advertising and the generalized second price auction: Selling billions of dollars worth of keywords. Technical report, National Bureau of Economic Research (2005)

5. Faddoul, J.B., Chidlovskii, B., Gilleron, R., Torre, F.: Learning multiple tasks with boosted decision trees. In: Flach, P.A., Bie, T., Cristianini, N. (eds.) ECML PKDD 2012. LNCS (LNAI), pp. 681–696. Springer, Heidelberg (2012). doi:10.1007/978-3-642-33460-3_49
6. García-Laencina, P.J., Sancho-Gómez, J.L., Figueiras-Vidal, A.R.: Pattern classification with missing data: a review. Neural Comput. Appl. **19**(2), 263–282 (2010)
7. Google: The arrival of real-time bidding (2011)
8. Johnson, N.L.: Survival Models and Data Analysis. Wiley, New York (1999)
9. Kaplan, E.L., Meier, P.: Nonparametric estimation from incomplete observations. J. Am. Stat. Assoc. **53**, 457–481 (1958)
10. Kullback, S.: Letter to the editor: the kullback-leibler distance (1987)
11. Lee, K.C., Jalali, A., Dasdan, A.: Real time bid optimization with smooth budget delivery in online advertising. In: ADKDD (2013)
12. Lee, K.c., Orten, B.B., Dasdan, A., Li, W.: Estimating conversion rate in display advertising from past performance data (2012)
13. Liao, H., Peng, L., Liu, Z., Shen, X.: ipinyou global rtb bidding algorithm competition dataset. In: ADKDD (2014)
14. Marlin, B.M., Zemel, R.S.: Collaborative prediction and ranking with non-random missing data. In: RecSys (2009)
15. Pan, R., Zhou, Y., Cao, B., Liu, N.N., Lukose, R., Scholz, M., Yang, Q.: One-class collaborative filtering. In: ICDM (2008)
16. Perlich, C., Dalessandro, B., Hook, R., Stitelman, O., Raeder, T., Provost, F.: Bid optimizing and inventory scoring in targeted online advertising. In: KDD (2012)
17. Stojanova, D., Ceci, M., Appice, A., Džeroski, S.: Network regression with predictive clustering trees. In: Gunopulos, D., Hofmann, T., Malerba, D., Vazirgiannis, M. (eds.) ECML PKDD 2011. LNCS (LNAI), pp. 333–348. Springer, Heidelberg (2011). doi:10.1007/978-3-642-23808-6_22
18. Wu, W.C.H., Yeh, M.Y., Chen, M.S.: Predicting winning price in real time bidding with censored data. In: KDD (2015)
19. Yuan, S., Wang, J., Chen, B., Mason, P., Seljan, S.: An empirical study of reserve price optimisation in real-time bidding. In: KDD (2014)
20. Yuan, S., Wang, J., Zhao, X.: Real-time bidding for online advertising: measurement and analysis. In: ADKDD (2013)
21. Zhang, W., Wang, J.: Statistical arbitrage mining for display advertising. In: KDD (2015)
22. Zhang, W., Yuan, S., Wang, J.: Optimal real-time bidding for display advertising. In: KDD (2014)

Maximizing Time-Decaying Influence
in Social Networks

Naoto Ohsaka[1,4(✉)], Yutaro Yamaguchi[2,4], Naonori Kakimura[1,4],
and Ken-ichi Kawarabayashi[3,4]

[1] The University of Tokyo, Tokyo, Japan
ohsaka@is.s.u-tokyo.ac.jp, kakimura@global.c.u-tokyo.ac.jp
[2] Osaka University, Osaka, Japan
yutaro_yamaguchi@ist.osaka-u.ac.jp
[3] National Institute of Informatics, Tokyo, Japan
k_keniti@nii.ac.jp
[4] JST, ERATO, Kawarabayashi Large Graph Project, Tokyo, Japan

Abstract. Influence maximization is a well-studied problem of finding a small set of highly influential individuals in a social network such that the spread of influence under a certain diffusion model is maximized. We propose new diffusion models that incorporate the time-decaying phenomenon by which the power of influence decreases with elapsed time. In standard diffusion models such as the independent cascade and linear threshold models, each edge in a network has a fixed power of influence over time. However, in practical settings, such as rumor spreading, it is natural for the power of influence to depend on the time influenced. We generalize the independent cascade and linear threshold models with time-decaying effects. Moreover, we show that by using an analysis framework based on submodular functions, a natural greedy strategy obtains a solution that is provably within $(1 - 1/e)$ of optimal. In addition, we propose theoretically and practically fast algorithms for the proposed models. Experimental results show that the proposed algorithms are scalable to graphs with millions of edges and outperform baseline algorithms based on a state-of-the-art algorithm.

1 Introduction

Recently, the rapidly increasing popularity of online social networks has created opportunities to study information diffusion that models the spread of news, ideas, and product adoption throughout the population. Motivated by applications to marketing, Domingos and Richardson [8] introduced *viral marketing*, which is a cost-effective marketing strategy that promotes products through word-of-mouth. Formally, the *influence maximization problem* [17] asks, for a parameter k, to find a set of k vertices in a social network such that the expected number of activated vertices is maximized.

The *independent cascade (IC)* model and *linear threshold (LT)* model are two of the most basic and widely studied diffusion models in influence maximization.

© Springer International Publishing AG 2016
P. Frasconi et al. (Eds.): ECML PKDD 2016, Part I, LNAI 9851, pp. 132–147, 2016.
DOI: 10.1007/978-3-319-46128-1_9

The IC model proposed by Goldenberg et al. [10] focuses on individual (and independent) interaction among friends in a social network. The LT model [14] focuses on threshold behavior in influence propagation, where a user is influenced when a sufficient number of friends are influenced. The two models are tractable as they are shown to have submodularity [17], which has motivated substantial theoretical and practical follow-up research [2,4,5,20,25,28,29].

Recent research into empirical social networks [11,15,16] reports that time plays an important role in the spread of influence in a network. However, in the IC and LT models, each edge in a network has a fixed power of influence over time. This does not reflect reality. In a practical setting, such as rumor spreading, the power of influence may decay over time. Moreover, the influence may be propagated with delay. In this paper, we incorporate two types of temporal phenomena, i.e., *time-decaying phenomenon* and *time-delay propagation*, into both the IC and LT models.

Time-decaying Phenomenon. First, we consider the "freshness" of information. Intuitively, a rumor has a lifetime, and a new idea is often affected by trends. One may also observe that information becomes less attractive over time. To observe such phenomenon in a real-world social network, we estimate edge probabilities that represent the power of influence *at each time* by applying the method of [13] to the Digg dataset.[1] Figure 1 shows the transition of the average edge probabilities among all edges. As can be seen, the average edge probability clearly decreases over time, especially halving in a day. Thus, the power of word-of-mouth effects strongly depends on the elapsed time. This motivates us to introduce a time-decaying phenomenon to information diffusion models.

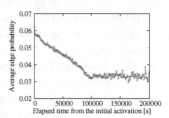

Fig. 1. Time transition of the average edge probability.

Time-delay Propagation. In addition to the time-decaying effect, we incorporate a time-delay effect, which has been studied extensively [11,13]. In many real-world examples, the propagation of influence from one person to another may have a certain time delay due to heterogeneity in human activities. Thus, the speed of influence spread varies. We capture such time-delaying propagation in our model by extending the model [11,22,26] with the time-decaying phenomenon.

1.1 Contributions

In this paper, we address the above temporal issues and extend well-studied diffusion models. We first propose an IC model that incorporates time-decaying probabilities and time-delay propagation. The salient feature of this model is that the power of influence on edges decays over time, as shown in Fig. 1. This model is

---

[1] http://www.isi.edu/~lerman/downloads/digg2009.html.

simple and sufficiently general to deal with various time-decaying probabilities. It should also be noted that our model includes most previous models with temporal effects, such as [12] (see Sect. 3.2). In addition, we propose an LT model with time-decaying probabilities and time-delay propagation, which is another interpretation of temporal phenomena with threshold behavior.

Our main contributions are summarized as follows.

- **Time-varying IC model (Sect. 3):** We extend the IC model with time-decaying probabilities and time-delay propagation. We show that the expected number of activated vertices under the extended model is monotone and submodular with respect to an initial vertex set. Therefore, we can efficiently find a solution that approximates an optimal solution within the ratio $(1 - 1/e - \epsilon)$ using a greedy algorithm [24].
- **Time-varying LT model (Sect. 4):** We introduce the LT model with time-decaying probabilities and time-delay propagation, in which the influences from neighbors decay over time. As with the extended IC model, the expected number of activated vertices is monotone and submodular, and we can efficiently find an approximate solution within the ratio $(1 - 1/e - \epsilon)$ using a greedy algorithm.
- **Scalable algorithms (Sect. 5):** We propose scalable and accurate algorithms for influence maximization under the proposed models by generalizing sketching methods [2,28]. To this end, we design novel dynamic programming that can deal with time-decaying probabilities efficiently.
- **Experimental evaluations (Sect. 6):** We conduct experiments on real-world social networks and demonstrate that the proposed algorithms outperform baseline methods in terms of both efficiency and accuracy.

Due to space limitations, we omit some of the proofs and the experimental results, which will be found in the full version of this paper.

2 Related Work

Inspired by the work of Domingos and Richardson [8], Kempe et al. [17] formulated the influence maximization problem as a discrete optimization problem. They showed that the influence maximization problem is NP-hard for the IC and LT models and that the expected number of activated vertices is a monotone and submodular function with respect to an initial vertex set. This implies that an optimal solution for the influence maximization problem can be approximated efficiently within the ratio $(1 - 1/e - \epsilon)$ with a greedy-type algorithm [24].

Since Kempe et al.'s greedy algorithm suffers from poor scalability, a plethora of scalable algorithms have been proposed. Existing approaches (for the IC and LT models) can be roughly classified into three types. *Simulation-based* methods [5,17,20,25] conduct Monte-Carlo simulations of the diffusion process to estimate the influence spread accurately; however, they suffer from inefficiency. *Heuristic-based* methods [4,5,18] avoid using Monte-Carlo simulations by restricting the spread of influence in a particular group, which often results in poor-quality solutions due to an absence of accuracy guarantees. *Sketch-based* methods [2] resolved the inefficiency of Monte-Carlo simulations while preserving accuracy guarantees. Rather than directly simulating the diffusion process, sketch-based methods build sketches in advance based on an outcome of *reverse simulations*, and efficiently estimate the influence spread. Subsequently, several strategies for bounding the sketch size have been developed [28,29]. In this paper, we generalize sketch-based methods to our proposed models without significant deterioration of efficiency.

Various information diffusion models with time-delay propagation have been proposed in different contexts [3,11,13,22,26,27] to resemble actual cascade distribution. The influence maximization problem in such models has also been studied [3,9,12,28]. We show in Sect. 3.2 that most previous models are included in the proposed model. Note that, the existing models only consider the time difference between two vertices. In contrast, our model considers the time reached from the seeds, which allows us to introduce the time-decaying probabilities, as well as the time difference.

3   Time-Varying IC Model

3.1   Model Definition

Here we define the time-varying IC model formally. Let $G = (V, E)$ be a directed graph, where V is a vertex set of size n and E is an edge set of size m. For a vertex v in V, $N^+(v)$ denotes the set of out-neighbors of v. Each individual vertex can be either *active* (an adopter of the innovation) or *inactive*. In the *time-varying (TV) IC* model, we begin with a seed set A of active vertices. Then, the process unfolds according to the following randomized rule. When a vertex u becomes active at time t_u for the first time, it is given a single chance to activate each current inactive neighbor v of u through the edge $e = (u, v)$. Here unlike the standard IC model, both the distance to v and the probability to activate v depend on time. That is, the conditional likelihood that the influence reaches v at time t is defined by $f_e(t \mid t_u)$. We assume that the likelihood is shift invariant, i.e., $f_e(t \mid t_u) = f_e(t - t_u)$, and nonnegative, i.e., $f_e(s) = 0$ for $s < 0$. Moreover, when v receives the influence at time t, the probability to be activated is given by a nonincreasing function $p_e : \mathbb{R}_+ \to [0, 1]$ of the arrival time, i.e., $p_e(t)$. Thus, the probability that v becomes active at time t is

$$\Pr\left[v \text{ becomes active at time } t \mid u \text{ is active at time } t_u\right] = p_e(t)f_e(t - t_u). \quad (1)$$

When v receives influence from more than one newly activated neighbors simultaneously, their attempts to activate v are sequenced independently in arbitrary order. The process runs until no further activations are possible.

Intuitively, the term $p_e(t)$ represents the decrease in power of influence as time passes, because p_e is a nonincreasing function on elapsed time. On the other hand, $f_e(t - t_u)$ represents the time-delay effect on the edge e. Note that if $p_e(t)$ is a constant c_e for any t, then this model is identical to the IC model.

3.2 Examples of TV-IC Models

Here we present examples of the TV-IC model. The first example is the influence
maximization problem with deadline.

Example 1 (Influence maximization with deadline). Let $c_e \in [0,1]$ be a
constant for each edge e and let T be a positive number. Consider the TV-IC
model where a function p_e is given by $p_e(t) = c_e$ if $t \leq T$ and $p_e(t) = 0$ if $t > T$.
This case means that the influence will expire at time T. Therefore, the influence
maximization problem over such a model is to maximize the expected number
of vertices activated before deadline T.

Moreover, the TV-IC model extends previous models properly.

**Example 2 (Continuous-time independent cascade (CTIC) model [11,
22,26]).** The TV-IC model where a function p_e is a constant includes various
previously proposed models. For example, Saito et al. [26,27] considered the case
where f_e is an exponential function. In their model, the time-delay parameters
$r_e > 0$ and diffusion parameters $c_e \in (0,1)$ for each edge $e = (u,v)$ are given.
When u is activated at time t_u, u will activate an inactive neighbor v with
probability c_e. If it succeeds, a delay time δ is sampled from the exponential distribution $r_e \exp(-r_e\delta)$, and v will become active at time $t_u+\delta$. Gomez-Rodriguez
et al. [11] dealt with more general functions of f_e. However, in their model, $p_e(t)$
is set to 1 for all e and t, which means that a vertex u always activates its neighbor v at some time. A similar model was also proposed in [13,22]. However, in
all these models, p_e is a constant independent of time t.

Example 3 (Independent cascade model with meeting events [3]). In
this model, we are given meeting probabilities m_e and propagation probabilities c_e. When a vertex u is activated at time t_u, u will attempt to meet an
inactive neighbor v with probability m_e, where $e = (u,v)$. Thus a time-delay
$\delta \in \{1,2,\ldots\}$ occurs with probability $m_e(1 - m_e)^{\delta-1}$. When they meet, u will
activate v with probability c_e at that time. This means that the probability that
u will activate v at time $t_u + \delta$ is $c_e m_e(1 - m_e)^{\delta-1}$. Thus, it is included in our
model, where p_e is a constant.

The following example is one in which the probability decays at arrival time t.
However, the time delay effect is not considered.

Example 4. Assume that p_e is given by $p_e(t) = r_e\alpha(t)$ for some constant r_e and
a nonincreasing function $\alpha : \mathbb{R}_+ \to [0,1]$, which represents the influence decay
factor. This is the IC model wherein the probability is decreased by a factor of
$\alpha(t)$ when the influence is reached at time t. This case includes the temporal
factor proposed by Cui et al. [7], where $p_e(t) = r_e \exp(-ct)$ for some constant c.
Note that the model proposed by Cui et al. [7] is more general in order to resemble
actual cascade distribution, which does not clearly possess submodularity. Thus,
their model has no theoretical guarantee for influence maximization.

3.3   Submodularity of the Influence Spread Function

We say that a set function $f : 2^V \to \mathbb{R}$ is *monotone* if $f(S) \leq f(T)$ for all $S \subseteq T \subseteq V$, and *submodular* if $f(S \cup \{v\}) - f(S) \geq f(T \cup \{v\}) - f(T)$ for all $S \subseteq T \subseteq V$ and $v \in V \backslash T$.

Let $\sigma(A)$ be the expected number of vertices activated after running the process of the TV-IC model with an initial seed set A. The following theorem is the main technical result in this section, which generalizes Kempe et al. [17].

Theorem 1. *For the TV-IC model, σ is a monotone submodular function.*

We here present the proof idea. The detailed proof is deferred to the next section. First, we remove the time-delay factor f_e, similar to the proof of [12]. Consider the probability distribution obtained by f_e of all possible time differences between each pair of nodes in the network and sample a length d_e of each edge e from the probability space. Let σ_d be the expected number of vertices activated, assuming that the length of an edge e is d_e. Since σ is the expected value of σ_d, it is sufficient to show that σ_d is monotone and submodular.

Here we focus on the time-decay factor p_e. Note that a standard "coin flipping" technique for the IC model [17] would not work to show the submodularity when p_e depends on time. The key observation was that a set of activated vertices corresponds to the reachability of a random graph generated by "coin flipping" on each edge. Then, the expected size of the reachable vertices is shown to be monotone and submodular. This technique tells us that we do not have to consider time in the IC model. However, due to the time dependency of probability, we cannot directly apply this observation to our problem setting.

To overcome this difficulty, we prepare a random variable x_e in the range $[0, 1]$ on each edge e before the process. Based on these values, we construct a graph in a deterministic manner such that the reachability of the graph is equal to the activated vertices. Note that the obtained graph depends on a seed set, in contrast to Kempe et al. [17]. This requires more careful analysis in the proof.

It should also be noted that the time decay of probabilities is essential to satisfy the submodularity of σ_d. To demonstrate this, consider a graph consisting of a directed triangle $(u, v), (v, w), (u, w)$ with a directed path $(w, w_1), (w_1, w_2), \dots,$ $(w_{\ell-1}, w_\ell)$ of length $\ell > 1$. The length d of the edges is defined as $d_{uw} = 3$ and $d_e = 1$ for any edge $e \neq (u, w)$. The probabilities are set to $p_{vw}(t) = 0$ for $t < 2$, $p_{vw}(t) = 1$ for $t \geq 2$, $p_{ww_1}(t) = 0$ for $t < 4$, $p_{ww_1}(t) = 1$ for $t \geq 4$, and $p_e(t) = 1$ for any other edge e and any time t. This is illustrated in Fig. 2. For this graph, if we take $\{u\}$ as the seed set, then u activates v in time $t = 1$, v activates w in time $t = 2$, and w fails to activate w_1 in time $t = 3$, which stops diffusion. If we take $\{v\}$, then v fails to activate w in time $t = 1$ and diffusion terminates. However, if we take $\{u, v\}$, then v fails to activate w in time $t = 1$, but u succeeds in time $t = 3$, and w activates w_1 in time $t = 4$, which eventually results in influence spreading to all vertices. Thus, we have $\sigma_d(\{u, v\}) - \sigma_d(\{u\}) = (\ell + 3) - 3 > 1 = \sigma_d(\{v\}) - \sigma_d(\emptyset)$, which violates submodularity.

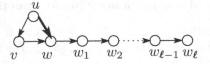

Fig. 2. Example violating submodularity when probability increases.

It follows from Theorem 1 that, using Monte-Carlo simulations to estimate $\sigma(A)$, we can maximize σ within $(1 - 1/e - \epsilon)$ approximation factor by a greedy algorithm [24]. However, naive Monte-Carlo simulations require significant time to estimate $\sigma(A)$. Because the proof of Theorem 1 has a nice combinatorial structure, we provide a theoretically efficient algorithm in Sect. 5.

3.4   Proof of Theorem 1

Here we prove Theorem 1. Let $G = (V, E)$ be a directed graph and $p_e : \mathbb{R}_+ \to [0, 1]$ be a nonincreasing function for each edge e. As described in Sect. 3.3, we may assume that if u becomes active at time t, then the influence reaches a neighbor v at time $t + d_e$, and the probability that v becomes active is $p_e(t + d_e)$.

For each edge e, we choose a number x_e in the range $[0, 1]$ uniformly at random. We assume that we can use the edge $e = (u, v)$ to activate v in the TV-IC model if the arrival time t satisfies $p_e(t) > x_e$. Then, the probability that e can activate v in time t is equal to $p_e(t)$, which is the case when x_e is the range $[0, p_e(t)]$.

Let $X = (x_e)$ be a choice of random numbers x_e for all edges $e \in E$. Then, the number of activated vertices is determined uniquely by such X. $\sigma_X(A)$ is defined as the total number of vertices activated by a seed set A by running the process with X. Since each edge is used at most once in the process, $\sigma(A)$ can be described by the functions $\sigma_X(A)$:

$$\sigma(A) = \int \Pr[X] \sigma_X(A) dX.$$

The function σ_X can be characterized by the reachability of a graph. For each edge e, we say that e is *live in time* t if $p_e(t) > x_e$. For each vertex v, we denote $N_t^+(v) = \{w \in N^+(v) \mid (v, w) \text{ is live in } t + d_{vw}\}$. For a seed set A, we construct a graph $G_X(A)$ as follows.

Procedure to obtain $G_X(A)$ from X and A.
Step 0. Set $r_v = 0$ for each $v \in A$ and $r_v = +\infty$ for each $v \in V \setminus A$. Set $t = 0$.
Step 1. While $t < +\infty$ do the following:
1-1. Define $V_t = \{v \in V \mid r_v = t\}$.
1-2. For each $v \in V_t$ and each $w \in N_t^+(v)$, replace r_w with $\min\{r_v + d_{vw}, r_w\}$.
1-3. Increase t to $\min\{r_v \mid r_v > t\}$.
Step 2. Return $G_X(A) = (R, F)$, where $R = \bigcup_{t < +\infty} V_t$ and $F = \{(v, w) \mid r_w = r_v + d_{vw}\}$.

Note that this procedure simulates the TV-IC model when we fix a choice X. The obtained vertex set R is the set of vertices activated by A.

By the above procedure, we show in Lemma 1 that σ_X is monotone and submodular, which implies Theorem 1. Note that the construction of $G_X(A)$ depends on the given seed set A; thus, we cannot extend the proof in [17] directly, and we must consider the dynamics of reachability.

Lemma 1. *The function σ_X is monotone and submodular.*

4 Time-Varying LT Model

Let $G = (V, E)$ be a directed graph. Each vertex v chooses a *threshold* $\theta_v \in [0, 1]$ uniformly at random. Each edge e has a nonincreasing function $q_e : \mathbb{R}_+ \rightarrow [0, 1]$ and has a function $f_e : \mathbb{R} \rightarrow [0, 1]$ that represents the shift invariant conditional likelihood as in the TV-IC model. We suppose that $\sum_{e:e=(u,v)} q_e(0) \leq 1$ for each $v \in V$.

Given a seed set A, the diffusion process in the *time-varying (TV) LT* model unfolds, similar to the LT model. The difference is that the distance to a neighbor and the amount of influences from neighbors depend on arrival times. Consider the case wherein a vertex u becomes active in time t_u. Then, each edge (u, v) delivers an influence to v, where the likelihood that the influence reaches v at time t is $f_e(t - t_u)$. When the influence reaches v at time t, the amount of influence that v receives is $q_e(t)$. The vertex v becomes activated once the total influence exceeds the threshold θ_v.

Similar to the TV-IC model, $q_e(t)$ represents the time decay of influence, and $f_e(t - t_u)$ represents the time-delay of propagation. Note that if $q_e(t)$ is a constant c_e for any t, this model coincides with the LT model. Moreover, we can consider the same situations as given in Sect. 3.2.

Example 5 (Influence maximization with deadline). Let T be a positive number. For each edge e, define a function q_e to be $q_e(t) = c_e$ if $t \leq T$ and $q_e(t) = 0$ if $t > T$, where $c_e \in [0, 1]$. The TV-LT model with such a function q_e represents that the influence will expire at time T.

Example 6 (Continuous-time diffusion model). For each edge $e = (u, v)$, we are given the time-delay parameters $r_e > 0$ and the power of influence $c_e \in (0, 1)$. Consider the TV-LT model where $q_e(t) = c_e$ and $f_e(t - t_u) = r_e \exp(-r_e(t - t_u))$ for each edge e. The model is a continuous-time variant of the LT model, in which the time-delay on edges occurs based on exponential distribution.

Let $\sigma(A)$ be the expected number of vertices activated after running the process of the TV-LT model with an initial seed set A. Using a technique similar to Theorem 1, σ is shown to be monotone and submodular. As a corollary, an optimal solution for the influence maximization problem under the TV-LT model can be approximated efficiently within the ratio $(1 - 1/e - \epsilon)$.

Theorem 2. *For the TV-LT model, σ is a monotone submodular function.*

5 Scalable Greedy Algorithms for the Proposed Models

In this section, we propose scalable greedy algorithms for influence maximization under the proposed diffusion models by extending a sketching method [2]. We first describe the sketching method and its generalization, and then discuss how to extend it to the proposed models.

5.1 Sketching Method and Generalization

The pseudocode of the sketching method is presented in Algorithm 1. Given a directed graph $G = (V, E)$, a diffusion model $\mathcal{M}$, and a seed size k, the sketching method performs the following two stages. In the first stage, beginning with an empty family $\mathcal{R} = \emptyset$, it repeats the following procedure: sample a *target* vertex z from V uniformly at random, compute the vertex set R that would influence z in an outcome of the diffusion process of $\mathcal{M}$, and add R to $\mathcal{R}$. The above repetition terminates when $\mathcal{R}$ includes a sufficient number of vertex sets for accurate influence estimation. In the second stage, it computes an approximate solution A of the *maximum coverage problem*, which seeks to select a set of k vertices from V that intersects the maximum number of vertex sets in $\mathcal{R}$, by the greedy algorithm. Finally, it returns a solution A.

Here we discuss why A is influential. Let $F_{\mathcal{R}}(A)$ be the fraction of sets in $\mathcal{R}$ intersecting A, i.e., $F_{\mathcal{R}}(A) = \frac{|\{R \in \mathcal{R} \mid R \cap A \neq \emptyset\}|}{|\mathcal{R}|}$. Then, for any vertex set A, $n \cdot F_{\mathcal{R}}(A)$ is an unbiased estimator of $\sigma(A)$, i.e., $\mathbb{E}[n \cdot F_{\mathcal{R}}(A)] = \sigma(A)$ [2], where $\sigma(A)$ is the influence spread of A under $\mathcal{M}$. Therefore, as long as this estimator gives accurate influence estimations, A is likely to have a large influence spread.

Now we consider applying the sketching method to the diffusion models proposed in this paper. There are two main challenges. The first one is to devise a procedure for generating a (random) vertex set that would influence a certain target vertex (line 4 in Algorithm 1) under the proposed models. The second is guaranteeing the accuracy and time complexity of the sketching method with the devised procedure. For the purpose, we adopt *reverse influence (RI) sets*, a model-independent notion introduced by Tang et al. [28], defined as follows.

Algorithm 1. Sketching method for influence maximization.

Require: a directed graph $G = (V, E)$, a diffusion model $\mathcal{M}$, a seed size k.
1: $\mathcal{R} \leftarrow \emptyset$.   ▷ Building sketches.
2: **repeat**
3:      $z \leftarrow$ a vertex chosen from V uniformly at random.
4:      $R \leftarrow$ a vertex set that would influence z in an outcome of the process of $\mathcal{M}$.
5:      $\mathcal{R} \leftarrow \mathcal{R} \cup \{R\}$.
6: **until** $\mathcal{R}$ includes a sufficient number of vertex sets for accurate influence estimation.
7: $A \leftarrow \emptyset$.   ▷ Selecting a seed set.
8: **while** $|A| < k$ **do**
9:      $s \leftarrow \operatorname{argmax}_{v \in V} F_{\mathcal{R}}(v)$.   ▷ $F_{\mathcal{R}}(v)$ is defined as $\frac{|\{R \in \mathcal{R} \mid R \cap \{v\} \neq \emptyset\}|}{|\mathcal{R}|}$.
10:     $A \leftarrow A \cup \{s\}$ **and** remove vertex sets including s from $\mathcal{R}$.
11: **return** A.

Definition 1 (Reverse influence set from Definition 3 in [28]). *For a graph $G = (V, E)$ and a diffusion model $\mathcal{M}$, a reverse influence (RI) set for a vertex z in V is a random vertex set $R \subseteq V$ such that for any vertex set $S \subseteq V$, the probability that $R \cap S \neq \emptyset$ is equal to the probability that the initial activation of vertices in S results in the activation of z under the diffusion process of $\mathcal{M}$.* A random RI set *is defined as an RI set for a vertex randomly sampled from V.*

Thus, if we are given a family $\mathcal{R}$ of random RI sets for $\mathcal{M}$, then we have $n \cdot F_{\mathcal{R}}(S) = \sigma(S)$ for every set S [28]. Furthermore, given a procedure for generating random RI sets under $\mathcal{M}$, Tang et al. [28] proved the time complexity and approximation ratio of a sketching algorithm *IMM* [28] shown as follows.

Theorem 3 (Theorem 5 in [28]). *Under a diffusion model for which a random RI set takes $O(\text{EPT})$ expected time to generate, IMM returns a $(1 - 1/e - \epsilon)$-approximation with probability at least $1 - \frac{1}{n^\ell}$, and runs in $O(\frac{\text{EPT}}{\text{OPT}}(k + \ell)(n + m)\frac{\log n}{\epsilon^2})$ expected time, where $\text{OPT} = \max_{S \subseteq V : |S| = k} \sigma(S)$.*

In summary, it suffices to design efficient and correct computation of RI sets. Remark that such a procedure may not exist depending on $\mathcal{M}$. In the following, we describe an algorithm that produces RI sets under each proposed model and analyze its correctness and computation time.

5.2 Efficient RI Set Generation Under TV-IC Model

Algorithm Description. Here we describe an efficient algorithm for generating RI sets under the TV-IC model. Note that existing approaches for RI set generation, such as a BFS-like algorithm for the IC and LT models [2, 28, 29] and a Dijkstra-like algorithm for the CTIC model [28], cannot be applied to the TV-IC model due to the time dependency of probability.

For this purpose, we exploit the graph introduced in the proof of Theorem 1. Given the choice of d_e's and x_e's, a target vertex z will be activated in the diffusion process with an initial seed vertex v if v can reach z in $G_X(\{v\})$, which is obtained by the procedure discussed in Sect. 3.4. However, a naive implementation of the procedure requires at least quadratic time.

We now present a more efficient algorithm. The key idea is to introduce the *latest activation time* $\tau[v]$ of v, which is defined as the maximum number $\tau[v]$ such that the activation of v within time $\tau[v]$ results in the activation of z given the choice of x_e's and d_e's. Obviously, $\tau[z] = +\infty$. For each vertex u ($\neq z$), u's influence must pass through one of its out-going edges in order to influence z. Specifically, u influences z by passing through (u, v) if u was activated within time $\tau[v] - d_{uv}$ and $p_{uv}(\tau[u] + d_{uv}) > x_{uv}$. Thus, the latest activation time $\tau[u]$ of u is determined by

$$\tau[u] = \max_{v \in N^+(u)} \min\{\tau[v] - d_{uv}, p_{uv}^{-1}(x_{uv})\},$$

where $p_{uv}^{-1}(x_{uv})$ is the maximum number t such that $p_{uv}(t + d_{uv}) > x_{uv}$ (note that $p_e^{-1}(x)$ can be $\pm\infty$). From the equation, the values $\tau[v]$ for all vertices v can be obtained efficiently by performing *dynamic programming*.

Algorithm 2. Efficient RI set generation under the TV-IC model.

Require: a directed graph $G = (V, E)$, edge probability functions $p_e : \mathbb{R}_+ \to [0, 1]$, edge
length likelihoods $f_e : \mathbb{R} \to [0, 1]$, a target vertex z.
1: $\tau[z] \leftarrow +\infty$ and $\tau[v] \leftarrow -\infty$ for all $v \in V \setminus \{z\}$.
2: $Q \leftarrow$ a queue with only one element z.
3: **while** $Q \neq \emptyset$ **do**
4:     Dequeue v with the maximum $\tau[v]$ from Q.
5:     **for all** u with $e = (u, v) \in E$ **do**
6:         $d_e \leftarrow$ an edge length sampled according to f_e.
7:         $x_e \leftarrow$ a uniform random number in $[0, 1]$.
8:         $\rho \leftarrow \min\{\tau[v] - d_e, p_e^{-1}(x_e)\}$.
9:         **if** $\rho > \tau[u]$ **and** $\rho \geq 0$ **then**
10:             $\tau[u] \leftarrow \rho$ **and** enqueue u onto Q.   ▷ $O(\log n)$ time
11: **return** the set of visited vertices, i.e., $\{v \in V \mid \tau[v] \geq 0\}$.

The pseudocode of the RI set generation under the TV-IC model is given in Algorithm 2. Beginning with a queue with a target vertex z with $\tau[z] = +\infty$, we determine the latest activation time of each vertex iteratively. For each iteration, we extract a vertex v with the maximum $\tau[v]$ (≥ 0) from the queue (thereafter, v's latest activation time will not be updated), sample a random number x_{uv} and an edge length d_{uv} of each vertex u in the in-neighbors of v, and update its latest activation time $\tau[u]$ if $\min\{\tau[v] - d_{uv}, p_{uv}^{-1}(x_{uv})\} > \tau[u]$. When $\tau[u] \geq 0$ at that time, we insert u into the queue. When the queue is empty, we return the set of vertices v with $\tau[v] \geq 0$ as an RI set for z. Note that by using a binary heap, both selecting a vertex from the queue (line 4) and inserting a vertex into the queue (line 10) can be performed in $O(\log n)$ time.

Theoretical Analysis. We first give the correctness and time complexity.

Lemma 2. *Algorithm 2 produces an RI set for z for the TV-IC model.*

Proof. We show that for any vertex z and any vertex set $S \subseteq V$, the probability p_1 that the algorithm's output intersects S is equal to the probability p_2 that the initial activation of vertices in S leads to the activation of z.

From the construction of the algorithm, p_1 is the probability of the following event over the choice of x_e's and d_e's: For some vertex s in S, there is a path $v_1 = s, v_2, \ldots, v_{\ell-1}, v_\ell = z$ of length ℓ such that $\tau_1 \geq 0$ where $\tau_\ell = +\infty$ and $\tau_i = \min\{\tau_{i+1} - d_{v_i v_{i+1}}, p_{v_i v_{i+1}}^{-1}(x_{v_i v_{i+1}})\}$ ($1 \leq i \leq \ell - 1$).

From the procedure to obtain $G_X(A)$ in Sect. 3.4, p_2 is the probability of the following event over the choice of x_e's and d_e's: For some vertex s in S, there is a path $v_1 = s, v_2, \ldots, v_{\ell-1}, v_\ell = z$ of length ℓ such that $p_{v_i v_{i+1}}(\tau_i' + d_{v_i v_{i+1}}) > x_{v_i v_{i+1}}$ ($1 \leq i \leq \ell - 1$) where $\tau_1' = 0$ and $\tau_{i+1}' = \tau_i' + d_{v_i v_{i+1}}$ ($1 \leq i \leq \ell - 1$).

It is easy to see that the two events given the choice of x_e's and d_e's are equivalent. Therefore, $p_1 = p_2$ and thus the lemma holds. □

Lemma 3. *Algorithm 2 runs in $O(\frac{m \cdot \text{OPT}}{n} \log n)$ expected time for a randomly selected vertex z.*

Then, by Theorem 3 and Lemmas 2 and 3, we obtain the following.

Theorem 4. *Under the TV-IC model, IMM with Algorithm 2 returns a $(1 - 1/e - \epsilon)$-approximation with probability at least $1 - \frac{1}{n^\ell}$ and runs in $O((k+\ell)(m + \frac{m^2}{n})\frac{\log^2 n}{\epsilon^2})$ expected time.*

Although a factor m^2/n in the time complexity can be $O(m\sqrt{m})$ for dense graphs, real-world social networks are sparse, i.e., m/n is small, and thus the proposed algorithm scales approximately linearly to real-world social networks.

5.3 Efficient RI Set Generation Under TV-LT Model

Similar to the TV-IC model, we develop an efficient algorithm for generating random RI sets under the TV-LT model and obtain the following theorem.

Theorem 5. *Under the TV-LT model, IMM with the above procedure for RI set generation returns a $(1 - 1/e - \epsilon)$-approximation with probability at least $1 - \frac{1}{n^\ell}$ and runs in $O((k+\ell)(m + \frac{m^2}{n})\frac{\log^2 n}{\epsilon^2})$ expected time.*

6 Experimental Evaluations

In this section, we demonstrate the efficiency and accuracy of our algorithms through experiments on real-world networks. We conducted the experiments on a Linux server with an Intel Xeon E5540 2.53 GHz CPU and 48 GB memory. All algorithms were implemented in C++ and compiled using g++ 4.8.2 with the -O2 option. We used five real-world social networks (Table 1).

6.1 Experiments with TV-IC Model

Settings of Edge Probability Functions and Edge Length Likelihoods. Motivated by the empirical evidence shown in Fig. 1, we adopt two nonincreasing functions for edge probabilities. One is the *weighted exponential (WE)* IC model, which assigns $p_{uv}(t) = \frac{1}{d^-(v)}\exp(-ct)$ to each edge (u,v), where c is sampled randomly in the range $[1, 10]$. Here $d^-(v)$ is the in-degree of a vertex v. The other is the *weighted reciprocal (WR)* IC model, which assigns $p_{uv}(t) = \frac{1}{d^-(v)ct}$, where c is sampled randomly in the range $[1, 10]$. Note that these models represent fast and

Table 1. Datasets.

Dataset	n	m	Type
Physicians [1]	241	1,098	Social
ca-GrQc [21]	5,241	28,968	Collaboration
wiki-Vote [21]	7,115	103,689	Social
soc-Epinions1 [21]	75,879	508,837	Social
ego-Twitter [21]	81,306	2,420,744	Social

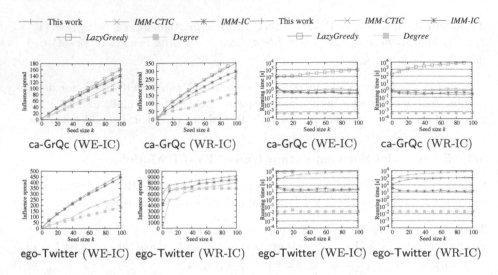

Fig. 3. Influence spreads for TV-IC.   **Fig. 4.** Running times for TV-IC.

slow decay of the power of influence, respectively. We show that such differences in the speed of time-decaying are crucial to the expected size of the cascades.

For each edge e, we set the edge length likelihood to the Weibull distribution [19], whose probability distribution function is defined as:

$$f_e(\delta) = \frac{\alpha_e}{\beta_e} \cdot \left(\frac{\delta}{\beta_e}\right)^{\alpha_e - 1} \cdot \exp\left(-\left(\frac{\delta}{\beta_e}\right)^{\alpha_e}\right), \tag{2}$$

where α_e and β_e are randomly sampled in the range $[0, 10]$. Note that this distribution has been adopted in continuous-time diffusion model literature [9,28].

Comparative Algorithms. For the proposed algorithm for the TV-IC model, i.e., IMM with Algorithm 2, we set $\epsilon = 0.5$ and $\ell = 1$, as described in [28]. Here we compare the proposed algorithm with the following baseline algorithms.

- *LazyGreedy* [23]: An accelerated simulation-based greedy algorithm for monotone submodular function maximization. We conducted Monte-Carlo simulations 10,000 times to estimate the influence spread.
- *IMM-CTIC* [28]: A sketching method for the CTIC model. Since this method takes care of "deadlines" rather than time-decaying edge probabilities, we set its deadline to 1.
- *IMM-IC* [28]: A sketching method for the IC model. We set the probability of each edge e to $p_e(\bar{d}_e)$, where $\bar{d}_e$ is the average edge length.
- *Degree*: Select k vertices in decreasing degree order.

Results. Figure 3 shows the influence spreads for seed sets of sizes $1, 10, 20, \ldots,$ 100 computed by each algorithm.[2] We omitted the results for Physicians, wiki-Vote, and soc-Epinions1, which exhibit similar behaviors, due to space limitations.

---

[2] We take the average after conducting simulations of the TV-IC 10,000 times.

LazyGreedy did not finish in 10,000 seconds with ca-GrQc (WR-IC, $k = 100$), ego-Twitter (WE-IC, $k \geq 30$), and ego-Twitter (WR-IC). Consequently, we were unable to obtain seed sets with these settings. Our method and *LazyGreedy* returned nearly the best results for most settings. Although *IMM-IC* is close to the best results, its influence spread (=4,336) with ego-Twitter (WR-IC, $k = 1$) is 30% worse than the best (=6,279). *IMM-CTIC* provided ineffective seed sets, e.g., with ego-Twitter (WR-IC, $k = 1$). As expected, *Degree* gave poor seed sets. We can also see that the WR-IC setting gives larger influence spreads compared to the WE-IC setting, which demonstrates the critical importance of the time-decaying phenomenon.

Figure 4 shows the running times required to select seed sets of sizes $1, 10, 20, \ldots, 100$ for each algorithm. Note that the running times do not include the time required to read the input graph from secondary storage. *LazyGreedy* even did finish in 10,000 seconds with ca-GrQc ($k = 100$), which is a small network, due to the computation cost of the Monte-Carlo simulations. Our method and *IMM-IC* required only several thousands of seconds for each graph, which is several orders of magnitude faster than *LazyGreedy*.

6.2 Experiments for TV-LT Model

Settings of Edge Weight Functions and Edge Length Likelihoods. Similar to the TV-IC model, we adopt two nonincreasing functions for edge weights, i.e., the *weighted exponential (WE)* LT model, which assigns $q_{uv}(t) = \frac{1}{d^-(v)} \exp(-ct)$, and the *weighted reciprocal (WR)* LT model, which assigns $q_{uv}(t) = \frac{1}{d^-(v)c(t+1)}$, where c is randomly sampled in the range $[1, 10]$.

We set the edge length likelihood to the Weibull distribution (2).

Comparative Algorithms. For the proposed algorithm for the TV-LT model, i.e., *IMM* with the algorithm in Sect. 5.3, we set $\epsilon = 0.5$ and $\ell = 1$ [28]. Since there are no algorithms for continuous-time LT models, we compare our method to *LazyGreedy* [23] and *Degree*, the same as for the TV-IC model, and *IMM-LT* [28], which is a sketching method for the LT model with edge weights $q_e(\bar{d}_e)$.

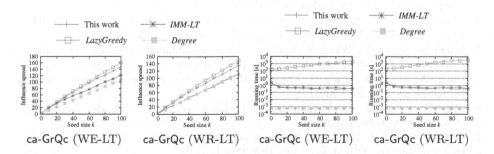

Fig. 5. Influence spreads for TV-LT.   **Fig. 6.** Running times for TV-LT.

Results. Figure 5 shows the influence spreads for seed sets computed by each algorithm. We observe similar behaviors as TV-IC model, *LazyGreedy* gave the best solutions, and the proposed method significantly outperformed *IMM-LT* and *Degree* for most settings.

Figure 6 shows the running times required to select seed sets for each algorithm. As in the case of the TV-IC model, the proposed method has much better scalability than *LazyGreedy*.

7   Conclusions

In this paper, we proposed diffusion models that incorporate time-decaying phenomenon and time-delay propagation by generalizing two standard diffusion models, i.e., independent cascade and linear threshold. We demonstrated that our models include most previous models with temporal effects, and the influence functions are monotone and submodular. Moreover, we devised scalable algorithms for influence maximization under the proposed models and experimentally verified their efficiency and accuracy compared to baseline algorithms.

A possible future direction is to learn edge probability functions from cascade logs. It might also be interesting to consider the influence maximization over diffusion models where cascades may recur [6], i.e., the power of influence is not necessarily nonincreasing. Note, however, that the case does not fall into submodular maximization as described in Sect. 3.3.

Acknowledgments. N. O. was supported by JSPS Grant-in-Aid for JSPS Research Fellow Grant Number 16J09440. Y. Y. was supported by JSPS Grant-in-Aid for JSPS Fellows Grant Number 13J02522. N. K. was partially supported by JSPS KAKENHI Grant Numbers 25730001, 24106002.

References

1. Physicians network dataset - KONECT, March 2016. http://konect.uni-koblenz. de/networks/moreno_innovation
2. Borgs, C., Brautbar, M., Chayes, J., Lucier, B.: Maximizing social influence in nearly optimal time. In: SODA, pp. 946–957 (2014)
3. Chen, W., Lu, W., Zhang, N.: Time-critical influence maximization in social networks with time-delayed diffusion process. In: AAAI, pp. 592–598 (2012)
4. Chen, W., Wang, C., Wang, Y.: Scalable influence maximization for prevalent viral marketing in large-scale social networks. In: KDD, pp. 1029–1038 (2010)
5. Chen, W., Wang, Y., Yang, S.: Efficient influence maximization in social networks. In: KDD, pp. 199–208 (2009)
6. Cheng, J., Adamic, L.A., Kleinberg, J.M., Leskovec, J.: Do cascades recur? In: WWW, pp. 671–681 (2016)
7. Cui, B., Yang, S.J., Homan, C.: Non-independent cascade formation: temporal and spatial effects. In: CIKM, pp. 1923–1926 (2014)
8. Domingos, P., Richardson, M.: Mining the network value of customers. In: KDD, pp. 57–66 (2001)

9. Du, N., Song, L., Gomez-Rodriguez, M., Zha, H.: Scalable influence estimation in continuous-time diffusion networks. In: NIPS, pp. 3147–3155 (2013)
10. Goldenberg, J., Libai, B., Muller, E.: Talk of the network: a complex systems look at the underlying process of word-of-mouth. Market. Lett. **12**(3), 211–223 (2001)
11. Gomez-Rodriguez, M., Balduzzi, D., Schölkopf, B.: Uncovering the temporal dynamics of diffusion networks. In: ICML, pp. 561–568 (2011)
12. Gomez-Rodriguez, M., Schölkopf, B.: Influence maximization in continuous time diffusion networks. In: ICML, pp. 313–320 (2012)
13. Goyal, A., Bonchi, F., Lakshmanan, L.V.: Learning influence probabilities in social networks. In: WSDM, pp. 241–250 (2010)
14. Granovetter, M.: Threshold models of collective behavior. Am. J. Sociol. **83**(6), 1420–1443 (1978)
15. Iribarren, J.L., Moro, E.: Impact of human activity patterns on the dynamics of information diffusion. Phys. Rev. Lett. **103**, 038702 (2009)
16. Karsai, M., Kivelä, M., Pan, R.K., Kaski, K., Kertész, J., Barabási, A.L., Saramäki, J.: Small but slow world: how network topology and burstiness slow down spreading. Phys. Rev. E **83**, 025102 (2011)
17. Kempe, D., Kleinberg, J., Tardos, É.: Maximizing the spread of influence through a social network. In: KDD, pp. 137–146 (2003)
18. Kimura, M., Saito, K.: Tractable models for information diffusion in social networks. In: Fürnkranz, J., Scheffer, T., Spiliopoulou, M. (eds.) PKDD 2006. LNCS, vol. 4213, pp. 259–271. Springer, Heidelberg (2006). doi:10.1007/11871637_27
19. Lawless, J.F.: Statistical Models and Methods for Lifetime Data. Wiley-Interscience, New York (2002)
20. Leskovec, J., Krause, A., Guestrin, C., Faloutsos, C., VanBriesen, J., Glance, N.: Cost-effective outbreak detection in networks. In: KDD, pp. 420–429 (2007)
21. Leskovec, J., Krevl, A.: SNAP datasets: stanford large network dataset collection, June 2014. http://snap.stanford.edu/data
22. Liu, B., Cong, G., Xu, D., Zeng, Y.: Time constrained influence maximization in social networks. In: ICDM, pp. 439–448 (2012)
23. Minoux, M.: Accelerated greedy algorithms for maximizing submodular set functions. In: Stoer, J. (ed.) Optimization Techniques. Lecture Notes in Control and Information Sciences, vol. 7, pp. 234–243. Springer, Heidelberg (1978)
24. Nemhauser, G., Wolsey, L., Fisher, M.: An analysis of the approximations for maximizing submodular set functions. Math. Program. **14**, 265–294 (1978)
25. Ohsaka, N., Akiba, T., Yoshida, Y., Kawarabayashi, K.: Fast and accurate influence maximization on large networks with pruned monte-carlo simulations. In: AAAI, pp. 138–144 (2014)
26. Saito, K., Kimura, M., Ohara, K., Motoda, H.: Learning continuous-time information diffusion model for social behavioral data analysis. In: Zhou, Z.-H., Washio, T. (eds.) ACML 2009. LNCS, vol. 5828, pp. 322–337. Springer, Heidelberg (2009). doi:10.1007/978-3-642-05224-8_25
27. Saito, K., Kimura, M., Ohara, K., Motoda, H.: Selecting information diffusion models over social networks for behavioral analysis. In: Balcázar, J.L., Bonchi, F., Gionis, A., Sebag, M. (eds.) ECML PKDD 2010. LNCS, vol. 6323, pp. 180–195. Springer, Heidelberg (2010). doi:10.1007/978-3-642-15939-8_12
28. Tang, Y., Shi, Y., Xiao, X.: Influence maximization in near-linear time: A martingale approach. In: SIGMOD, pp. 1539–1554 (2015)
29. Tang, Y., Xiao, X., Shi, Y.: Influence maximization: near-optimal time complexity meets practical efficiency. In: SIGMOD, pp. 75–86 (2014)

Learning Beyond Predefined Label Space via Bayesian Nonparametric Topic Modelling

Changying Du[1,2,3](✉), Fuzhen Zhuang[1], Jia He[1,3], Qing He[1],
and Guoping Long[2]

[1] Key Lab of Intelligent Information Processing of Chinese Academy
of Sciences (CAS), Institute of Computing Technology, CAS, Beijing 100190, China
[2] Laboratory of Parallel Software and Computational Science, Institute of Software,
Chinese Academy of Sciences, Beijing 100190, China
changying@iscas.ac.cn
[3] University of Chinese Academy of Sciences, Beijing 100049, China

Abstract. In real world machine learning applications, testing data may contain some meaningful new categories that have not been seen in labeled training data. To simultaneously recognize new data categories and assign most appropriate category labels to the data actually from known categories, existing models assume the number of unknown new categories is pre-specified, though it is difficult to determine in advance. In this paper, we propose a Bayesian nonparametric topic model to automatically infer this number, based on the hierarchical Dirichlet process and the notion of latent Dirichlet allocation. Exact inference in our model is intractable, so we provide an efficient collapsed Gibbs sampling algorithm for approximate posterior inference. Extensive experiments on various text data sets show that: (a) compared with parametric approaches that use pre-specified true number of new categories, the proposed nonparametric approach can yield comparable performance; and (b) when the exact number of new categories is unavailable, i.e. the parametric approaches only have a rough idea about the new categories, our approach has evident performance advantages.

1 Introduction

Human exploration of the world is never-ending, and we never know there still exist how many unknown things beyond our scope. For real-world machine learning applications, we often can only collect limited training instances before we do prediction on a large amount of unlabeled testing instances. Given the temporal and spatial constrictions at the beginning, it is likely that unlabeled new instances observed after a long time involve some meaningful new categories of objects, e.g., the news classification problem studied in [11,24,27], and the bacterial detecting problem in [1,8].

Basically, traditional classification models are unable to recognize new data categories, while clustering models cannot make full use of the supervised information from known categories. An ideal model should simultaneously recognize the new data categories and assign most appropriate category labels to the data

© Springer International Publishing AG 2016
P. Frasconi et al. (Eds.): ECML PKDD 2016, Part I, LNAI 9851, pp. 148–164, 2016.
DOI: 10.1007/978-3-319-46128-1_10

actually from known categories, since these two processes can benefit from each other. Existing models for such a learning scenario typically assume the number of unknown new categories is pre-specified. In [27], Zhuang et al. proposed a double-latent-layered Latent Dirichlet Allocation (DLDA) model, which can utilize supervised information from known categories in a generative manner. While classifying test data into categories acquired from the training data, their model can simultaneously group the remaining data into some pre-specified number of new clusters. In [24], the so-called Serendipitous Learning (SL) model established a maximum margin learning framework that combines the classification model built upon known classes with the parametric clustering model on unknown classes. Though these methods are effective when the true number of unknown new categories is available, their performances can be significantly degraded by a vague or wrong specification of the unknown category information.

Given that the accessibility assumption of the true number of unknown categories often is impractical, in this paper, we propose a Bayesian nonparametric topic model based on the hierarchical Dirichlet process [20] and the notion of latent Dirichlet allocation [4], for semi-supervised text modelling beyond the predefined label space. Unlike existing methods [24, 27] which assume that the number of unknown new categories in test data is known, our model can automatically infer this number via nonparametric Bayesian inference while classifying the data from known categories into their most appropriate categories. Exact inference in our model is intractable, so we provide an efficient collapsed Gibbs sampling algorithm for approximate posterior inference. Extensive experiments on various text data sets show that: (a) compared with parametric approaches that use pre-specified true number of new categories, the proposed nonparametric approach can yield comparable performance; and (b) when the exact number of new categories is unavailable, i.e. the parametric approaches only have a rough idea about the new categories, our approach has evident performance advantages.

In the following, we first review related works, and then present the generative process of our model and its approximate inference; experimental results are discussed in detail, before we conclude the paper and point out future work.

2 Related Work

A special case of the problem studied in this paper is the Positive and Unlabeled (PU) learning [11,15,23], where the goal is to identify usually valuable positive instances from a huge collection of unlabeled ones. Our model generalizes PU learning in that, it not only identifies (multiple) known category of instances but also conducts nonparametric clustering for the remaining instances. It should be noted that the identification of known categories may benefit from a proper grouping of the unknown instance categories.

Another topic closely related to ours is semi-supervised clustering [3], which exploits available knowledge to help partition unlabeled data into groups. Generally, its knowledge is represented in the form of pairwise constraints [3, 10, 17], i.e., cannot-link and must-link, which tends to be inefficient when the number of constraints is very large. Noting that our assumption is plenty of training instances are available from the known categories, these algorithms may suffer from efficiency problems. Moreover, violation of the constraints usually is allowed in these models, so it is not easy to map the resultant data clusters to the known classes. Instead of using constraints as supervision, we directly leverage label information in our model.

Under the nonparametric Bayesian framework, a semi-supervised determinantal clustering process was proposed in [19]. However, in each round of its sampling based inference procedure, its kernelized formulation leads to cubic computational complexity w.r.t. the number of instances to be clustered, which makes it infeasible for large data sets.

In nonparametric Bayesian statistics, the Dirichlet Process (DP) is a popular stochastic process that is widely used for adaptive modelling of the data [21]. Intuitively, it is a distribution over distributions, i.e. each draw from a DP is itself a distribution. Sethuraman [18] explicitly showed that distributions drawn from a DP are discrete with probability one, that is, the random distribution G distributed according to a DP with concentration parameter γ and base distribution H, can be written as

$$G = \sum_{i=1}^{\infty} \pi_i \delta_{\theta_i}, \ \pi_i = v_i \prod_{j=1}^{i-1} (1 - v_j),$$

where $\theta_i \sim H$, $v_i \sim \text{Beta}(1, \gamma)$, and δ_{θ} is an atom at θ. It is clear from this formulation that G is discrete almost surely, that is, the support of G consists of a countably infinite set of atoms, which are drawn independently from H.

Antoniak [2] first introduced the idea of using a DP as the prior for the mixing proportions of simple distributions, which is called the DP Mixture (DPM) model. Due to the fact that the distributions sampled from a DP are discrete almost surely, data generated from a DPM can be partitioned according to their distinct values of latent parameters θ_i's. Therefore, DPM is a flexible mixture model, in which the number of mixture components is random and grows as new data are observed. Teh et al. [20] proposed the Hierarchical DP (HDP), which is a nonparametric Bayesian approach to the modeling of grouped data, where each group is associated with a DPM model, and where we wish to link these mixture models.

3   Learning Beyond Predefined Labels via Generative Modelling

3.1   Problem Specification

Assume we have a labeled training data set $\mathcal{D}_l$ from the known categories $\mathcal{K}$, and an unlabeled test data set $\mathcal{D}_u$ which includes instances from both the known categories $\mathcal{K}$ and some unknown new categories $\mathcal{U}$. The goal is to learn a function $f : \mathcal{D}_u \to \mathcal{K} \cup \mathcal{U}$ that maps any instance in $\mathcal{D}_u$ to its category label in $\mathcal{K} \cup \mathcal{U}$.

Specifically, if an instance comes from the known categories $\mathcal{K}$, we aim to identify its true category label; meanwhile, we aim to group the instances not belonging to the known categories $\mathcal{K}$ into clusters $\mathcal{U}$.

3.2   The Proposed Bayesian Nonparametric Topic Model

For the problem specified above, an ideal model should simultaneously recognize the unknown new data categories and assign most appropriate category labels to the data actually from known categories, since these two processes can benefit from each other. However, it is usually difficult to determine the number of unknown categories in advance, which makes parametric approaches that assume this number is pre-specified impractical. To avoid performance degrading caused by a vague or wrong specification of the category information, in this paper, we propose a Bayesian nonparametric topic model, which can automatically infer the number of unknown new categories underlying test data $\mathcal{D}_u$ while classifying the data from known categories $\mathcal{K}$ into their most appropriate categories. Specifically, focusing on text data, we assume the following generative process for a document corpus:

1. Draw concentration parameters $\gamma \sim \Gamma(\gamma|a_\gamma, b_\gamma)$ and $\alpha \sim \Gamma(\alpha|a_\alpha, b_\alpha)$, where $a.$ and $b.$ are the shape and scale parameter of a Gamma distribution respectively;
2. Draw a discrete distribution $G_0 \sim \mathrm{DP}(\gamma, H)$, where the base distribution H is a L-dimensional Dirichlet with parameter ζ, and G_0 has countable but infinite number of atoms;
3. Draw a discrete category distribution $G_d \sim \mathrm{DP}(\alpha, G_0)$ for the d-th document;
4. Choose a document category $\varphi_{dn} \sim G_d$ for the n-th word in the d-th document[1];
5. Choose a word topic index $y_{dn} \sim \mathrm{Categorical}(\varphi_{dn})$ for the n-th word in the d-th document;
6. Draw word topics $\phi_l \sim \mathrm{Dir}(\beta)$, $l = 1, ..., L$ from a P-dimensional Dirichlet prior with parameter β, where L is the number of topics and P is the vocabulary size;
7. Choose a word $w_{dn} \sim \mathrm{Categorical}(\phi_{y_{dn}})$.

As described above, this generative model integrates the hierarchical Dirichlet process (HDP) [20] with the notion of Latent Dirichlet Allocation (LDA) [4]. However, the difference from standard LDA is that, here the distribution over word topics is conditioned on document categories rather than documents. Placing a DP prior on the document category distribution G_d, a document is allowed to involve an infinite number of categories. Meanwhile, assuming multiple G_d's have the same discrete base distribution G_0 (which also has a DP prior), multiple documents not only can have their distinct categories but also have the

---

[1] Note that, the support of the discrete distribution G_d consists of atoms drawn from G_0, the atoms of which are eventually from H. Thus, φ_{dn} is a vector rather than an index.

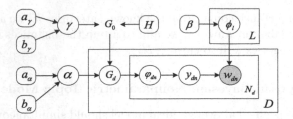

Fig. 1. Graphical representation of the proposed LBPL-NTM model.

chance to share some common ones. The actual number of categories used to model a corpus is determined by nonparametric Bayesian posterior inference. Note that, if the category label of a document is known, we can fix the corresponding category of all words in this document to the category determined by the label during posterior inference. In this way, the supervision from known categories can be injected. For any document without known label, we can infer the most appropriate category for each of its words, and assign this document to the category that generates most of its words.

Since the proposed model for Learning Beyond Predefined Labels (LBPL) is based on Nonparametric Topic Modelling (NTM), it will be denoted by LBPL-NTM in the sequel. The probabilistic generative process of LBPL-NTM is illustrated as a graphical model in Fig. 1.

Note that, LBPL-NTM is conceptually different from the infinite extension of LDA presented in [20], which learns topics in a purely unsupervised manner and cannot make use of the labeled information. From pure modeling perspective, our model introduces an additional topic index layer (y_{dn}) along with L hidden topics to infinite LDA. What's worth mentioning is that, it is not a trivial thing to extend the single-layered infinite LDA to a new two-layered model. With the introduced topic index layer and the hidden topics serving as low level topic modeling module, we can interpret G_d as the distribution over categories (rather than over topics as in infinite LDA) for each document, and then inject labeled information through ϕ_{dn} and infer the number of unknown categories (rather than topics as in infinite LDA) automatically from the data.

Besides, LBPL-NTM also differs from supervised topic models [13,25,26] basically, which train discriminative classification models in the semantic space with pre-specified category labels and cannot identify new categories underlying the test data.

The labeled LDA model proposed in [16], adopted a similar word-label correspondence idea by defining a one-to-one correspondence between LDA's latent topics and labels. However, it was designed to solve the multi-label problem in social bookmarking rather than discover new data categories underlying unlabeled data, thus is different from our model as well.

The double-latent-layered LDA (DLDA) [27] is a more closely related work to ours, where the authors conditioned the distribution over word topics on the document categories as in our model. By utilizing supervised information from known categories in a generative manner, their parametric model can classify unlabeled data into categories acquired from the labeled data, while grouping data into some pre-specified number of new clusters simultaneously. Though DLDA is effective when the true number of new categories is available, its performance can be significantly degraded by a wrong specification of this number. Our key difference with theirs is that our nonparametric model can naturally deal with the scenario where the number of new categories underlying test data is not clear, via allowing an infinite number of categories to model the corpus.

For the model inference of LBPL-NTM, we need to compute the posterior distribution of hidden variables given the data and model hyper-parameters:

$$p(\alpha, \gamma, \phi_l, \varphi_d, \mathbf{Y}_d | a_\alpha, b_\alpha, a_\gamma, b_\gamma, \beta, H, \mathbf{W}_d)$$
$$= \frac{p(\alpha, \gamma, \phi_l, \varphi_d, \mathbf{Y}_d, \mathbf{W}_d | a_\alpha, b_\alpha, a_\gamma, b_\gamma, \beta, H)}{p(\mathbf{W}_d | a_\alpha, b_\alpha, a_\gamma, b_\gamma, \beta, H)}.$$

However, the marginal probability in the denominator is intractable to compute. A popular way to conduct approximate posterior inference is the Markov Chain Monte Carlo (MCMC) method [14]. In the following, we will appeal to the Chinese restaurant franchise representation [20] of HDP for approximate posterior sampling. Note that, the high-dimensional latent topics ϕ_l's and the latent category variables φ_{dn}'s are integrated out to attain efficient collapsed sampling.

3.3   Inference by Collapsed Gibbs Sampling

First we give a brief description of the Chinese restaurant franchise representation of HDP. In the Chinese restaurant franchise, the metaphor of the Chinese restaurant process is extended to allow multiple restaurants which share a set of dishes. A customer entering some restaurant sits at one of the occupied tables with a certain probability, and sits at a new table with the remaining probability. If the customer sits at an occupied table, he eats the dish that has already been ordered. If he sits at a new table, he needs to pick the dish for the table. The dish is picked according to its popularity among the whole franchise, while a new dish can also be tried.

To employ this representation of HDP for posterior sampling, we introduce necessary index variables. Recall that φ_{dn}'s are random variables with distribution G_d. Let $\theta_1, \cdots, \theta_K$ denote K i.i.d. random variables (dishes) distributed according to H, and, for each d, let $\psi_{d1}, \cdots, \psi_{dT_d}$ denote T_d i.i.d. variables (tables) distributed according to G_0. Then each φ_{dn} is associated with one ψ_{dt}, while each ψ_{dt} is associated with one θ_k. Let t_{dn} be the index of the ψ_{dt} associated with φ_{dn}, and let k_{dt} be the index of θ_k associated with ψ_{dt}. Let s_{dt} be the number of φ_{dn}'s associated with ψ_{dt}, m_{dk} is the number of ψ_{dt}'s associated with θ_k, and $m_k = \sum_d m_{dk}$ as the number of ψ_{dt}'s associated with θ_k over all d.

For each d, by integrating out G_d and G_0, we have the following conditional distributions:

$$\varphi_{dn} | \varphi_{d1}, \cdots, \varphi_{dn-1}, \alpha, G_0 \sim \sum_{t=1}^{T_d} \frac{s_{dt}}{n-1+\alpha} \delta_{\psi_{dt}} + \frac{\alpha}{n-1+\alpha} G_0, \qquad (1)$$

$$\psi_{dt}|\psi_{11},\psi_{12},\cdots,\psi_{21},\cdots,\psi_{dt-1},\gamma,H \sim \sum_{k=1}^{K} \frac{m_k}{\sum_k m_k+\gamma}\delta_{\theta_k} + \frac{\gamma}{\sum_k m_k+\gamma}H. \qquad (2)$$

Note that, t_{dn}'s and k_{dt}'s inherit the exchangeability properties of φ_{dn}'s and ψ_{dt}'s, so the conditional distributions in (1) and (2) can be easily adapted to be expressed in terms of t_{dn} and k_{dt}. In the following, we will alternately execute four steps: first sample t_{dn} conditioned on all other variables, then sample k_{dt} for each table of data, thirdly sample y_{dn} for each word, and finally sample hyperparameters γ and α. Note that, if the category label of a document is known, we fix the category index k of all words in this document to the label during the sampling process.

Sampling t. To compute the conditional distribution of t_{dn} given the remaining variables, we make use of exchangeability and treat t_{dn} as the last variable being sampled in the last group. Using (1), the prior probability that t_{dn} takes on a particular previously seen value t is proportional to s_{dt}^{-dn}, whereas the probability that it takes on a new value (say $t^{new} = T_j + 1$) is proportional to α. The likelihood of the data given $t_{dn} = t$ for some previously seen t is simply $f(y_{dn}|\theta_{k_{dt}})$. To determine the likelihood when t_{dn} takes on value t^{new}, the simplest approach would be to generate a sample for $k_{dt^{new}}$ from its conditional prior (2) [14]. If this value of $k_{dt^{new}}$ is itself a new value, say $k^{new} = K+1$, we may generate a sample for $\theta_{k^{new}}$ as well.

Combining all this information, the conditional posterior distribution of t_{dn} is then

$$p(t_{dn} = t|\mathbf{t}^{-dn},\mathbf{k},\mathbf{Y},\Theta) \propto \begin{cases} \alpha f(y_{dn}|\theta_{k_{dt}}), & t = t^{new}, \\ s_{dt}^{-dn}f(y_{dn}|\theta_{k_{dt}}), & t \text{ appeared.} \end{cases} \qquad (3)$$

However, here we show that we don't need to store and update the θ's, i.e., we can get a collapsed sampler. To compute the likelihood that y_{dn} comes from the k-th class θ_k, $1 \le k \le K$, we can first compute the posterior distribution of θ_k given $\mathbf{Y}_{(k)}^{-dn}$ (elements assigned to class k in $\mathbf{Y}^{-dn}$), then integrate over this posterior. Specifically, by conjugacy the posterior of θ_k is also Dirichlet distributed, whose parameter is updated from the prior base distribution H according to $\mathbf{Y}_{(k)}^{-dn}$. If we assume O_{kl} is the number of elements in $\mathbf{Y}_{(k)}^{-dn}$ that equal to l, $1 \le l \le L$, then

$$\theta_k|H,\mathbf{t}^{-dn},\mathbf{k}^{-dt_{dn}},\mathbf{Y}^{-dn} \sim \text{Dir}(\zeta + O_{k\cdot}),$$

where ζ and $O_{k\cdot}$ both are L dimensional vectors. Integrate over this posterior we can get the likelihood for y_{dn},

$$f(y_{dn}|\theta_k : 1 \le k \le K) = \int \theta_{y_{dn}} \cdot \text{Dir}(\theta;\zeta + O_{k\cdot})d\theta = \frac{\zeta_{y_{dn}} + O_{ky_{dn}}}{\sum_l(\zeta_l + O_{kl})}. \qquad (4)$$

To compute the likelihood that y_{dn} comes from a new $k = (K + 1)$-th class θ_{K+1}, we can directly integrate over the prior H:

$$f(y_{dn}|\theta_k : k = K + 1) = \int \theta_{y_{dn}} \cdot \text{Dir}(\theta; \zeta)d\theta = \frac{\zeta_{y_{dn}}}{\sum_l \zeta_l}. \tag{5}$$

Sampling k. Sampling the variables k_{dt} is similar to sampling t_{dn}. Since changing k_{dt} actually changes the component membership of all data items in table t, the likelihood of setting $k_{dt} = k$ is given by $\prod_{n:t_{dn}=t} f(y_{dn}|\theta_k)$, so that the conditional probability of k_{dt} is

$$p(k_{dt} = k|\mathbf{t}, \mathbf{k}^{-dt}, \mathbf{Y}, \Theta) \propto \begin{cases} \gamma \prod_{n:t_{dn}=t} f(y_{dn}|\theta_k), & k = k^{\text{new}}, \\ m_k^{-dt} \prod_{n:t_{dn}=t} f(y_{dn}|\theta_k), & k \text{ appeared}, \end{cases} \tag{6}$$

where $f(y_{dn}|\theta_k)$ can be computed same as above.

Sampling Y. Conditioned on $\mathbf{t}$, $\mathbf{k}$ and $\mathbf{Y}^{-dn}$, the prior of $y_{dn} = l$, $1 \leq l \leq L$ is:

$$p(y_{dn} = l|\mathbf{t}, \mathbf{k}, \mathbf{Y}^{-dn}) = \int \theta_l \cdot \text{Dir}(\theta; \zeta + O_{k_{dt_{dn}} \cdot})d\theta = \frac{\zeta_l + O_{k_{dt_{dn}} l}}{\sum_l (\zeta_l + O_{k_{dt_{dn}} l})}.$$

Assume $\mathbf{W}_{(l)}^{-dn}$ denotes the elements in $\mathbf{W}^{-dn}$ that are generated from topic l, and O_{lw} is the number of elements in $\mathbf{W}_{(l)}^{-dn}$ that equal to w, $1 \leq w \leq P$, $1 \leq l \leq L$, then

$$\phi_l|\beta, \mathbf{Y}^{-dn}, \mathbf{W}_{(l)}^{-dn} \sim \text{Dir}(\beta + O_{l \cdot}),$$

where β and $O_{l \cdot}$ both are P dimensional vectors. Integrating over this posterior, we can get the likelihood that w_{dn} is generated from topic ϕ_l:

$$f(w_{dn}|\mathbf{t}, \mathbf{k}, \mathbf{Y}^{-dn}, \mathbf{W}^{-dn}) = \int \phi_{w_{dn}} \cdot \text{Dir}(\phi; \beta + O_{l \cdot})d\phi = \frac{\beta_{w_{dn}} + O_{lw_{dn}}}{\sum_w (\beta_w + O_{lw})}.$$

The conditional posterior probability of $y_{dn} = l$, $1 \leq l \leq L$ is proportional to the prior times the likelihood:

$$p(y_{dn} = l|\mathbf{t}, \mathbf{k}, \mathbf{Y}^{-dn}, \mathbf{W}) \propto \frac{\zeta_l + O_{k_{dt_{dn}} l}}{\sum_l (\zeta_l + O_{k_{dt_{dn}} l})} \cdot \frac{\beta_{w_{dn}} + O_{lw_{dn}}}{\sum_w (\beta_w + O_{lw})}. \tag{7}$$

Sampling γ and α. In each iteration of our Gibbs sampling, we use the auxiliary variable method described in [20] to sample γ and α.

We summarize the above approximate posterior sampling process in Algorithm 1. After this sampling process converges, we take a sample from the Markov chain and count the words assigned to each category $k = 1, 2, ...$ for each document, and finally a document is assigned to the category that has generated most of its words.

Algorithm 1. Collapsed Gibbs Sampling for LBPL-NTM

Input: the words **W**, the number of topics L, parameter ζ of the base Dirichlet distribution H, the hyper-parameters β, a_γ, b_γ, a_α, b_α, and the maximal number of iterations *maxIter*.
Output: **t**, **k** and **Y**.

1. Initialize the latent variables **t**, **k**, **Y**, γ and α;
2. **for** *iter* = 1 to *maxIter* **do**
3.   Update **t** according to (3), (4), and (5);
4.   Update **k** according to (6), (4), and (5);
5.   Update **Y** according to (7);
6.   Update γ and α using the auxiliary variable method in [20];
7. **end for**
8. Output **t**, **k** and **Y**.

3.4 Computational Complexity

In each round of our collapsed Gibbs sampling, the dominant computation is $O(|\mathbf{W}_u| \cdot (|\bar{\mathbf{t}}| + |\mathbf{k}|) + |\mathbf{W}_a| \cdot L)$, where $|\mathbf{W}_u|$ is the total number of words in the unlabeled documents, $|\mathbf{W}_a|$ is the total number of words in the entire corpus, $|\bar{\mathbf{t}}|$ is the average number of inferred word groups in each document, $|\mathbf{k}|$ is the inferred number of categories, and L is the specified number of topics. Generally, $|\mathbf{k}|$ and $|\bar{\mathbf{t}}|$ are very small, and $L = 128$ throughout the paper[2], thus our model can be seen as scale linearly with the number of words in the corpus.

4   Experiments

In this section, we evaluate the proposed LBPL-NTM model on various text corpora, including the benchmark 20 Newsgroups data set, the imbalanced TDT2 data set and the sparse ODP data set.

4.1 Baselines and Evaluation Metrics

We compare LBPL-NTM with the following algorithms:

- Serendipitous Learning (SL) [24]: a maximum margin learning framework that combines the classification model built upon known classes and the parametric clustering model on unknown classes;

---

[2] For fair comparison with the DLDA model [27], the number of topics L is fixed to the constant 128. We empirically find that L has little performance influence (compared to the number of categories) on the learning problem studied here, as long as it is not too small or too large. This is probably due to the two-layered nature of our model.

- DLDA [27]: a double-latent-layered LDA model, which can utilize supervised information similar as LBPL-NTM when clustering data with pre-specified number of clusters;
- Constrained 1-Spectral Clustering (COSC) [17]: a state-of-the-art graph-based constrained clustering algorithm, which can guarantee that all given constraints are fulfilled;
- Semi-supervised K-means (SSKM) [10]: clustering data with pairwise constraints in original space;
- Unsupervised clustering package CLUTO[3];
- Nonparametric Bayesian unsupervised clustering model Dirichlet Process Gaussian Mixture (DPGM).

Two popular clustering metrics are adopted to compare the clustering quality of these algorithms: normalized mutual information (NMI) [12] and adjusted rand index (ARI) [9]. NMI measures how closely the clustering algorithm could reconstruct the label distribution underlying the data. If A and B represent the cluster assignments and the ground truth class assignments of the data respectively, then NMI is defined as

$$NMI = 2 \cdot I(A; B)/(H(A) + H(B)),$$

where $I(A; B) = H(A) - H(A|B)$ is the mutual information between A and B, $H(\cdot)$ is the Shannon entropy, and $H(A|B)$ is the conditional entropy of A given B.

If a denotes the number of pairs of data points that are in the same cluster in A and in the same class in B, and b denotes the number of pairs of points that are in different clusters in A and in different classes in B, then the Rand Index (RI) is given by $RI = (a + b)/C_2^D$, where C_2^D is the total number of possible pairs in the dataset. Since the expected RI value of two random assignments does not take a constant value, Hubert and Arabie [9] proposed to discount the expected RI of random assignments by defining the ARI as

$$ARI = (RI - Expected\_RI)/(\max(RI) - Expected\_RI).$$

As in [27], we also evaluate the classification accuracy on the data from the known classes with average $F1$ measure. For each known class, the $F1$ score can be computed as follows,

$$F1_i = 2 \cdot Precison_i \cdot Recall_i/(Precison_i + Recall_i), \quad i = 1, ..., k,$$

where $Precison_i$ and $Recall_i$ are the precision and recall on the i-th known class. Then, we use the average $F1$ score over these k known classes as the final measure.

---

[3] http://glaros.dtc.umn.edu/gkhome/cluto/cluto/download.

4.2 Parameter Settings

In all our experiments, we set the parameters and hyper-parameters of LBPL-NTM as follows: $L = 128$, $a_\gamma = 1$, $b_\gamma = 0.001$, $a_\alpha = 5$, $b_\alpha = 0.1$, $\zeta_l = 1$, $l = 1, ..., L$, $\beta_w = 0.01$, $w = 1, ..., P$. We run 3000 Gibbs sampling iterations to sample from the posteriors of LBPL-NTM and DLDA, and use the last sample for classification and clustering performance evaluation[4].

The parameter settings of all compared algorithms follow the instructions in their original papers and are carefully tuned on our data sets. The similarity matrix for COSC is constructed using the cosine value of the angle between each pair of documents[5]. For CLUTO, we use its *direct* implementation for clustering with default parameter settings. PCA is used to reduce the original high dimensionality to 500 for SL and DPGM, due to efficiency problems. Without statement, all algorithms except for DPGM and LBPL-NTM, use the true number of data categories.

4.3 Evaluation Results

Benchmark data—20 Newsgroups: This data set is widely used in text categorization and clustering. It has approximately 20,000 newsgroup documents that are evenly partitioned into twenty different newsgroups. Since some of the newsgroups are very closely related, a part of these twenty newsgroups are further grouped into four top categories, e.g., the top category *sci* contains four subcategories *sci.crypt*, *sci.electronics*, *sci.med* and *sci.space*. We only retain the terms that have document frequency (DF) above 15 and are not in the stop words list. As in Table II of [27], we consider two kinds of 4-way learning problems—the data for each *difficult* problem consist of all 4 subcategories of a top category, and the data for each *easy* problem consist of 4 subcategories from different top categories. Here these problems are denoted as E1-E4 and D1-D4 for short. For each problem, assume we have supervision from the subcategories in bold face in Table II of [27], from which 40 % instances are sampled as training data, and the rest 60 % and all instances from the subcategories without supervision are used as testing data. We independently repeat the experiments 10 times, and the averaged results over these trials are reported in Fig. 2, from which we can see LBPL-NTM and DLDA can significantly outperform other competitors, while these two methods perform similarly. However, it should be noted that DLDA used the actual number of categories, while LBPL-NTM can automatically infer the most appropriate number from data owing to the merits of Bayesian nonparametrics. The posterior frequencies of the inferred numbers of categories by LBPL-NTM are shown in Fig. 3, from which we can see higher frequencies around the true number 4.

---

[4] Such a choice is consistent with the evaluation strategy in [27]. Alternatively, we can also average the classification and clustering scores over multiple posterior samples.

[5] COSC works not well with the k-NN similarity graph [5] on our data sets.

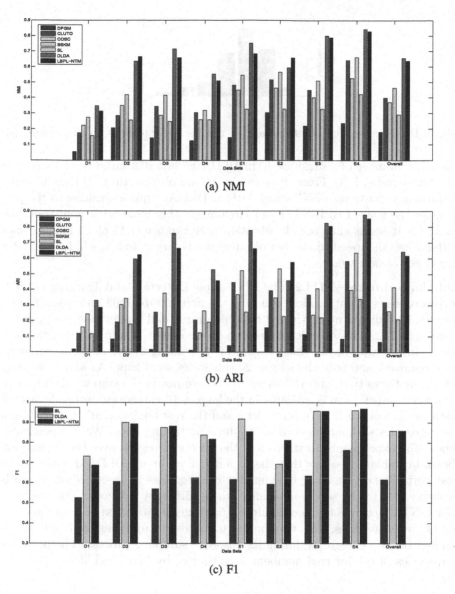

(a) NMI

(b) ARI

(c) F1

Fig. 2. Comparison on the 4-way learning problems (D1-D4, E1-E4) constructed from 20 Newsgroups data. All results are averaged over 10 independent trials in terms of NMI, ARI and F1.

One may naturally question the learning performance of DLDA when actual number of categories is not available. To this end, we further compare LBPL-NTM with DLDA, assuming that we only have a rough idea about the number of unknown categories underlying data. Under the same settings as above, Fig. 4 gives the average results over 10 independent trials on 20 Newsgroups data set

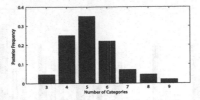

Fig. 3. Posterior frequencies of the inferred numbers of categories on 20 Newsgroups.

when the number of categories K in DLDA is varied from $K = 3$ to $K = 7$ (the true number is 4). From these results we can observe that (1) the clustering performance (in terms of NMI and ARI) of DLDA is quite sensitive to the pre-specified K while LBPL-NTM can circumvent this issue with nonparametric prior; (2) it seems that the classification performance (F1) of DLDA becomes better when the specified number of categories is larger, but as will be seen later this is not always true.

Imbalanced data—TDT2: The NIST Topic Detection and Tracking (TDT2) corpus consists of data collected during the first half of 1998 and taken from 6 sources, including 2 news wires, 2 radio programs and 2 television programs. It consists of 11201 on-topic documents which are classified into 96 semantic categories. In the experiment, those documents appearing in two or more categories were removed, and only the largest 20 categories were kept. As above, we only retain the terms that have DF above 15 and are not in the stop words list. Here we assume supervision is available in the largest 10 categories, from which 40 % instances are sampled as training data, and the rest 60 % and all instances from the categories without supervision are used as testing data. We independently repeat the experiments 10 times, and the averaged results over these trials are shown in Table 1. It seems that the parametric approach DLDA doesn't get its best performance when the true number of categories is pre-specified, which is probably due to the severe imbalance among different categories. Surprisingly, LBPL-NTM achieves the best results without any information of the total number of data categories. This may be due to its ability to dynamically adjust the number of data categories during its posterior sampling process. The posterior frequencies of the inferred numbers of categories by LBPL-NTM are shown in Fig. 5(a).

Sparse data—ODP: This data set is collected by Yin et al. [22], originally for web object classification by exploiting social tags. It contains 5536 web pages from 8 categories, which are detailed in Table 1 in [22]. Since the features on each web page are the social tags on it, these data are extremely sparse. Specifically, the average number of tag words on each web page is 25.76, which is much smaller than that (more than 160) of 20 Newsgroups. Assume that there is supervised information in the categories of Books, Electronic, Health and Garden. As above, we randomly sample 40 % instances as training data from these known categories, and the rest 60 % and all instances from the categories without supervision are used as testing data. We independently repeat the experiments 10 times, and

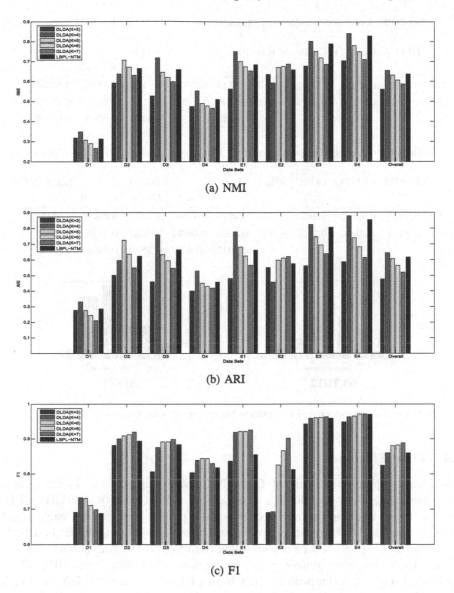

(a) NMI

(b) ARI

(c) F1

Fig. 4. Further comparison with DLDA on the 4-way learning problems constructed from 20 Newsgroups. All results are averaged over 10 independent trials in terms of NMI, ARI and F1.

report the averaged NMI, ARI and F1 values in Table 2, from which we can see LBPL-NTM also has competitive performance on sparse data. The posterior frequencies of the inferred numbers of categories are shown in Fig. 5(b). Note that there is a very small category—Office in ODP, and it is not easy to discover it due to data sparseness.

Table 1. Averaged results over 10 independent trials on the TDT2 data.

	DPGM	CLUTO	COSC	SSKM	SL	DLDA			LBPL-NTM
						$K = 15$	$K = 20$	$K = 25$	
NMI	0.4878	0.8217	0.6042	0.8057	0.7743	0.8157	0.8173	0.8135	**0.8358**
ARI	0.2608	0.6591	0.4375	0.6665	0.7159	0.7804	0.7167	0.6788	**0.7873**
F1	-	-	-	-	0.8443	0.8473	0.8490	0.8068	**0.9075**

Table 2. Averaged results over 10 independent trials on the ODP data.

	DPGM	CLUTO	COSC	SSKM	SL	DLDA			LBPL-NTM
						$K = 5$	$K = 8$	$K = 10$	
NMI	0.2825	0.5302	0.3866	0.5155	0.4523	0.6039	0.6054	**0.6084**	0.5877
ARI	0.0966	0.3983	0.2451	0.4145	0.3684	**0.6034**	0.5480	0.5119	0.5781
F1	-	-	-	-	0.7045	0.7400	**0.7868**	0.7461	0.7715

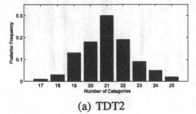

(a) TDT2

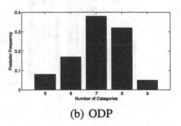

(b) ODP

Fig. 5. Posterior frequencies of the inferred numbers of categories on TDT2 and ODP.

4.4 Time Efficiency

The core sampling procedure of LBPL-NTM was implemented in C++, and all experiments were conducted in Matlab on a desktop with 3.60 GHz CPU. On the 4-way learning problems constructed from 20 Newsgroups, each round of our collapsed Gibbs sampling procedure takes about 0.9 s, which is a little slower than the speed of 0.7 s per sampling round of DLDA (implemented in C). We attribute this speed difference to the nonparametric nature of LBPL-NTM.

It is also observed empirically that both LBPL-NTM and DLDA run much faster than constraints based semi-supervised clustering methods. Besides, as mentioned above, SL and DPGM are quite inefficient for high dimensional text data, and PCA has to be used for them.

5 Conclusion and Future Work

We proposed a nonparametric Bayesian method for learning beyond the prede-fined label space. Unlike existing methods which assume the number of unknown new categories in test data is known, our model can automatically infer this number via nonparametric Bayesian inference. Empirical results show that: (a) compared with parametric approaches that use pre-specified true number of new categories, the proposed nonparametric approach yields comparable performance; and (b) when the exact number of new categories is unavailable, our approach has evident performance advantages. Our model can be extended in several aspects, e.g., (1) adapt it to the online learning scenario with sequential Monte Carlo [6]; and (2) explore multi-source text corpora with cross-domain learning [7,28].

Acknowledgments. This work was supported by the National Natural Science Foundation of China (No. 61473273, 61573335, 91546122, 61303059), Guangdong provincial science and technology plan projects (No. 2015B010109005), and the Science and Technology Funds of Guiyang (No. 201410012).

References

1. Akova, F., Dundar, M., Davisson, V.J., Hirleman, E.D., Bhunia, A.K., Robinson, J.P., Rajwa, B.: A machine-learning approach to detecting unknown bacterial serovars. Stat. Anal. Data Min. ASA Data Sci. J. **3**(5), 289–301 (2010)
2. Antoniak, C.E.: Mixtures of dirichlet processes with applications to bayesian nonparametric problems. Ann. Stat. **2**(6), 1152–1174 (1974)
3. Bilenko, M., Basu, S., Mooney, R.J.: Integrating constraints and metric learning in semi-supervised clustering. In: ICML, pp. 11–19 (2004)
4. Blei, D.M., Ng, A.Y., Jordan, M.I.: Latent dirichlet allocation. J. Mach. Learn. Res. **3**, 993–1022 (2003)
5. Bühler, T., Hein, M.: Spectral clustering based on the graph p-laplacian. In: ICML, pp. 81–88 (2009)
6. Doucet, A., Godsill, S., Andrieu, C.: On sequential monte carlo sampling methods for bayesian filtering. Stat. Comput. **10**(3), 197–208 (2000)
7. Du, C., Zhuang, F., He, Q., Shi, Z.: Multi-task semi-supervised semantic feature learning for classification. In: ICDM, pp. 191–200 (2012)
8. Dundar, M., Akova, F., Qi, A., Rajwa, B.: Bayesian nonexhaustive learning for online discovery and modeling of emerging classes. In: ICML, pp. 113–120 (2012)
9. Hubert, L., Arabie, P.: Comparing partitions. J. Classif. **2**(1), 193–218 (1985)
10. Kulis, B., Basu, S., Dhillon, I., Mooney, R.: Semi-supervised graph clustering: a kernel approach. Mach. Learn. **74**(1), 1–22 (2009)
11. Li, X., Liu, B., Ng, S.K.: Learning to identify unexpected instances in the test set. In: IJCAI, vol. 7, pp. 2802–2807 (2007)
12. Manning, C.D., Raghavan, P., Schütze, H.: Introduction to information retrieval. Cambridge University Press, Cambridge (2008)
13. Mcauliffe, J.D., Blei, D.M.: Supervised topic models. In: NIPS, pp. 121–128 (2008)
14. Neal, R.M.: Markov chain sampling methods for dirichlet process mixture models. J. Comput. Graph. Stat. **9**, 249–265 (2000)
15. du Plessis, M.C., Niu, G., Sugiyama, M.: Analysis of learning from positive and unlabeled data. In: NIPS, pp. 703–711 (2014)
16. Ramage, D., Hall, D., Nallapati, R., Manning, C.D.: Labeled lda: a supervised topic model for credit attribution in multi-labeled corpora. In: EMNLP, pp. 248–256 (2009)
17. Rangapuram, S.S., Hein, M.: Constrained 1-spectral clustering. In: AISTATS, pp. 1143–1151 (2012)

18. Sethuraman, J.: A constructive definition of dirichlet priors. Technical report, DTIC Document (1991)
19. Shah, A., Ghahramani, Z.: Determinantal clustering process-a nonparametric bayesian approach to kernel based semisupervised clustering. In: UAI, pp. 566–576 (2013)
20. Teh, Y.W., Jordan, M.I., Beal, M.J., Blei, D.M.: Hierarchical dirichlet processes. J. Am. Stat. Assoc. **101**(476), 1566–1581 (2006)
21. Teh, Y.W.: Dirichlet process. In: Encyclopedia of Machine Learning, pp. 280–287 (2011)
22. Yin, Z., Li, R., Mei, Q., Han, J.: Exploring social tagging graph for web object classification. In: SIGKDD, pp. 957–966 (2009)
23. Yu, H., Zhai, C., Han, J.: Text classification from positive and unlabeled documents. In: CIKM, pp. 232–239 (2003)
24. Zhang, D., Liu, Y., Si, L.: Serendipitous learning: learning beyond the predefined label space. In: SIGKDD, pp. 1343–1351 (2011)
25. Zhu, J., Ahmed, A., Xing, E.P.: Medlda: maximum margin supervised topic models. J. Mach. Learn. Res. **13**(1), 2237–2278 (2012)
26. Zhu, J., Chen, N., Perkins, H., Zhang, B.: Gibbs max-margin topic models with data augmentation. J. Mach. Learn. Res. **15**(1), 1073–1110 (2014)
27. Zhuang, F., Luo, P., Shen, Z., He, Q., Xiong, Y., Shi, Z.: D-lda: a topic modeling approach without constraint generation for semi-defined classification. In: ICDM, pp. 709–718 (2010)
28. Zhuang, F., Luo, P., Shen, Z., He, Q., Xiong, Y., Shi, Z., Xiong, H.: Mining distinction and commonality across multiple domains using generative model for text classification. IEEE Trans. Knowl. Data Eng. **24**(11), 2025–2039 (2012)

Efficient Bayesian Maximum Margin Multiple Kernel Learning

Changying Du[1,2(✉)], Changde Du[3], Guoping Long[1], Xin Jin[4], and Yucheng Li[1]

[1] Laboratory of Parallel Software and Computational Science, Institute of Software,
Chinese Academy of Sciences, Beijing 100190, China
changying@iscas.ac.cn
[2] Key Lab of Intelligent Information Processing
of Chinese Academy of Sciences (CAS),
Institute of Computing Technology, CAS, Beijing 100190, China
[3] Research Center for Brain-inspired Intelligence, Institute of Automation,
Chinese Academy of Sciences, Beijing 100190, China
[4] Central Software Institute, Huawei Technologies Co. Ltd., Beijing 100085, China

Abstract. Multiple Kernel Learning (MKL) suffers from slow learning speed and poor generalization ability. Existing methods seldom address these problems well simultaneously. In this paper, by defining a multiclass (pseudo-) likelihood function that accounts for the margin loss for kernelized classification, we develop a robust Bayesian maximum margin MKL framework with Dirichlet and the three parameter Beta normal priors imposed on the kernel and sample combination weights respectively. For inference, we exploit the data augmentation idea and devise an efficient MCMC algorithm in the augmented variable space, employing the Riemann manifold Hamiltonian Monte Carlo technique to sample from the conditional posterior of kernel weights, and making use of local conjugacy for all other variables. Such geometry and conjugacy based posterior sampling leads to very fast mixing rate and scales linearly with the number of kernels used. Extensive experiments on classification tasks validate the superiority of the proposed method in both efficacy and efficiency.

1 Introduction

Kernel-based machine learning is a popular technique for dealing with nonlinearities in real prediction tasks. The performance of this kind of learning methods generally is determined by two orthogonal aspects, i.e., the selected kernel function and the learning principle. On one hand, a kernel function implicitly maps the input data points to an infinite-dimensional feature space and actually provides a similarity measure on it. Since the learning process is conducted in the feature space, the appropriateness of the chosen kernel usually is crucial for the final modelling quality. On the other hand, the learning principle, e.g., the maximum margin principle in Support Vector Machine (SVM), defines the searching strategy in the hypothesis space, and thus is responsible for model generalization ability.

© Springer International Publishing AG 2016
P. Frasconi et al. (Eds.): ECML PKDD 2016, Part I, LNAI 9851, pp. 165–181, 2016.
DOI: 10.1007/978-3-319-46128-1_11

Though (single) kernel selection can be done via cross-validation on the training data, there are at least two reasons for studying Multiple Kernel Learning (MKL) [21], an active research topic that aims at learning a linear (or convex) combination of a set of predefined kernels in order to identify a good target kernel for the applications (see [17] for a survey). First, from the perspective of users without sufficient domain knowledge, it is desirable to design algorithms that can learn effective kernels automatically from data. Second and more important, to achieve superior performance it is necessary for many real applications to fully exploit the rich features underlying each sample. A promising way to achieve this is to define a large set of kernel mappings on all features and each individual feature, and then learn the optimal combination of them.

Through the past decade, there have been lots of MKL studies, most of which were focused on seeking (appropriately regularized) max-margin point model estimates [9,20,33,34]. Adopting the max-margin principle, these models essentially have advantage in yielding good generalization performance. However, their deterministic point estimate formulations make them less robust to noisy and small training data. Under the Bayesian framework, there already exist some MKL methods [10,14,15] that estimate the entire posterior distribution of model weights. Unfortunately, these methods either require matrix inversions to compute the posterior covariance of kernel weights or have to perform time-consuming importance sampling, and thus scale poorly with the number of kernels used. Moreover, since the max-margin hinge loss does not lend itself to a convenient description of a likelihood function, the combination of Bayesian MKL and max-margin principle has been deemed as intractable for a long time.

In this paper, by defining a multiclass (pseudo-) likelihood function that accounts for the margin loss for kernelized classification, we develop an efficient Bayesian maximum margin MKL framework. The Bayesian model averaging mechanism along with the max-margin principle allow us to make robust predictions with the guarantee of arguably good generalization performance. Moreover, imposing the sparsity-inducing Three Parameter Beta Normal (TPBN) prior [2] and the Dirichlet prior on the sample and kernel combination weights respectively, the resultant model has good interpretability and adaptivity. For inference, we exploit the data augmentation idea and devise an efficient Markov Chain Monte Carlo (MCMC) algorithm in the augmented variable space, employing the Riemann manifold Hamiltonian Monte Carlo (HMC) technique to sample from the conditional posterior of kernel weights, and making use of local conjugacy for all other variables. Such geometry and conjugacy based posterior sampling leads to very fast mixing rate and scales linearly with the number of kernels used.

Extensive experiments on both binary and multiclass classification data sets show that the proposed Bayesian max-margin MKL model not only outperforms a number of competitors consistently in terms of prediction performance but also requires substantially fewer training time when the number of kernels is large.

2   Related Work

Compared with traditional kernel methods using a single fixed kernel, MKL provides a natural way for the automated kernel parameter tuning, the integration of diverse nonlinear mappings, and the concatenation of heterogeneous data. It was originally formulated as a semi-definite programming (SDP) problem in [21], and then improved with the quadratically constrained quadratic programming (QCQP) [4], and the semi-infinite linear programming (SILP) [34].

Over the past decade, MKL has been actively studied, and a variety of algorithms have been proposed to address the efficiency of MKL, e.g., the adaptive 2-norm regularization formulation [32], the extended level method [36], the group lasso based methods [3,37], the proximal minimization method [35], the online-batch strongly convex two-stage method [29], the spectral projected gradient descent method [19], and the mean-field variational inference method [15]. Besides, a lot of extended MKL techniques have been proposed to improve the regular MKL method, e.g., the localized MKL [7,16,38] that achieve local assignments of kernel weights at the group level, the sample-adaptive MKL [24,28] that switches off kernels at the data sample level, the absent MKL [23] that handles the channel missing problem of individual samples, and the Bayesian MKL [14,15,22] that estimate the entire posterior distribution of model weights. Our method differs from existing efficient MKL algorithms in that, it employs the Riemann manifold HMC technique to sample from the conditional posterior of kernel weights, and makes use of local conjugacy for all other variables. Such geometry and conjugacy based posterior sampling leads to very fast mixing rate and scales linearly with the number of kernels used. Our method also differs from existing Bayesian MKL model since it is based on the max-margin (pseudo-) likelihood and data augmentation idea, and tends to has better generalization performance.

As sparsity-inducing approaches to kernel weight learning rarely outperform trivial baselines in practical applications [8,20], we choose the Dirichlet prior for kernel combination weights in our Bayesian MKL framework. Though it has been considered in [14], our expanded-mean parameterization of the Dirichlet is particularly suitable for HMC based methods, which is more efficient than importance sampling based variational approximation. Besides, our choice of the sparsity-inducing TPBN prior for sample combination weights differs from the Gaussian-inverse-Gamma prior used in [14,15]. As in kernelized SVM, the sparse sample weights is responsible for selecting support vectors actually needed in decision function, and thus good interpretability can be obtained.

We note that the GP-based Bayesian nonlinear SVM model proposed in [18] adopts a similar data augmentation idea for max-margin learning and infers its GP kernel parameters automatically with slice sampling. Its difference from ours is that it was designed for single kernel binary classification, and our model can be seen as an efficient multiclass multi-kernel extension of it. As detailed in the following and observed in the experiments, such extension is not trivial.

3 Bayesian Max-Margin MKL

Suppose we have a set of labeled data $\mathcal{D} = \{\mathbf{x}_i, y_i\}_{i=1}^N$, where $y_i \in \{1, 2, ..., C\}$ and $\mathbf{x}_i \in \mathcal{X}$, a d-dimensional Euclidean space. MKL algorithms typically use a weighted sum of P kernels $\{\mathfrak{K}_m : \mathcal{X} \times \mathcal{X} \to \mathbb{R}\}_{m=1}^P$ to measure data similarity. For multiclass leaning, we adopt the one-versus-all strategy, and learn a shared kernel combination weights vector $\mathbf{w}$ for all binary sub-problems, which corresponds to sharing the similarity measure when jointly learn all sub-problems [7,15,33]. Note that, this sharing not only is essential for efficient computation when the number of classes is large, but also is important to alleviate overfitting when we only have very few training data. Specifically, we define the following decision function for the c-th binary sub-problem:

$$f(\mathbf{x}_i; \Theta_c) = \mathbf{a}_c^\top \left(\sum_{m=1}^P w_m \mathbf{K}_{m,\cdot i} \right) + b_c \tag{1}$$

where $\mathbf{K}_{m,\cdot i} = [\mathfrak{K}_m(\mathbf{x}_i, \mathbf{x}_1), ..., \mathfrak{K}_m(\mathbf{x}_i, \mathbf{x}_N)]^\top$ contains the similarities between $\mathbf{x}_i$ and each training example under the kernel feature mapping $\mathfrak{K}_m$; $\mathbf{a}_c$ and b_c are the sample weights and bias for the c-th binary sub-problem, respectively; $\mathbf{w} = [w_1, ..., w_P]^\top$ is the kernel weights shared by all sub-problems; $\Theta_c = \{\mathbf{a}_c, b_c, \mathbf{w}\}$.

3.1 Max-Margin Pseudo-likelihood

To account for the training error on $(\mathbf{x}_i, y_i)$, we further define the following multiclass max-margin pseudo-likelihood function:

$$L(y_i|\Theta) = \exp\left\{ -2 \sum_{c=1}^C \max(\zeta_{ci}, 0) \right\}, \tag{2}$$

where $\zeta_{ci} = 1 - \delta_{y_i, c} f(\mathbf{x}_i; \Theta_c)$, $\delta_{y_i, c} = 1$ if $y_i = c$ and -1 otherwise. This pseudo-likelihood plays the similar role as that in the Bayesian SVM model [31], which addressed linear binary classification only. Intuitively, the negative margin losses of each binary multiple kernel classifier are summed up and passed through an exponential transformation. The larger the loss, the smaller the likelihood is. Despite its importance in our Bayesian MKL modeling, (2) makes direct posterior inference intractable due to the max function in $L(\Theta)$. Fortunately, the following identity holds [1]:

$$\exp\{-|\zeta|\} = \int_0^\infty \frac{\exp\{\frac{-\zeta^2}{2\varpi} - \frac{\varpi}{2}\}}{\sqrt{2\pi\varpi}} d\varpi. \tag{3}$$

Multiplying through (3) by $\exp\{-\zeta\}$ and noting $\max(\zeta, 0) = \frac{1}{2}(|\zeta|+\zeta)$, we have:

$$L(y_i|\Theta) = \prod_{c=1}^C \int_0^\infty \frac{\exp\{\frac{-1}{2\lambda_{ci}}(\lambda_{ci} + \zeta_{ci})^2\}}{\sqrt{2\pi\lambda_{ci}}} d\lambda_{ci}, \tag{4}$$

which allows us to introduce auxiliary variables to the original inference problem. Thus, by regarding the original posterior as the marginal of a higher dimensional distribution that involves the augmented variables $\boldsymbol{\lambda}$, we can bypass the calculation of the max function. Consequently, efficient algorithms can be designed.

3.2   Priors on Model Parameters

Symmetric Dirichlet Prior on Kernel Weights. Under the Bayesian framework, we can impose either Dirichlet [14] or Gaussian [15] prior on $\mathbf{w}$. While Gaussian prior is convenient for inference, it cannot ensure the positivity of each kernel weight, which sometimes is difficult for interpretation and even degrades performance. Thus, we consider the symmetric Dirichlet prior $\mathbf{w} \sim \text{Dir}(\eta)$ for kernel weights, where $\eta > 0$. An asymmetric prior can be used if the user has a rough idea about kernel importance.

Note that, the Dirichlet prior on kernel weights leads to non-conjugacy even if we can re-express our pseudo-likelihood as the product of C location-scale mixtures of normals. Such a non-conjugacy issue generally complicates posterior inference. To address it, many strategies have been used in literature, ranging from variational approximations to Metropolis-Hastings methods. Unlike the inefficient importance sampling method in [14], we will explore the recently developed Riemann manifold HMC [13] approach. Exploiting the Riemannian geometry of the parameter space, RHMC can efficiently samples from a continuous distribution with its unnormalized probability density.

Sparsity-inducing Prior on Sample Weights. Similar as in kernelized SVM, the sample weights $\mathbf{a}$ is often expected to be sparse for selecting support vectors actually needed in decision function. In this paper, we choose the Three Parameter Beta Normal (TPBN) [2] as the sparsity-inducing prior due to its better mixing properties than priors such as the spike-and-slab, the Student's-t prior, and the double exponential prior. The TPBN prior can be expressed as scale mixtures of normals and favors strong shrinkage of small signals while having heavy tails to avoid over-shrinkage of the larger signals. If $a_{ci} \sim \text{TPBN}(\alpha_a, \beta_a, \kappa)$, $c = 1, ..., C$, $i = 1, ..., N$, then:

$$a_{ci} \sim \mathcal{N}(0, \nu_{ci}), \quad \nu_{ci} \sim \Gamma(\alpha_a, \varsigma_{ci}), \quad \varsigma_{ci} \sim \Gamma(\beta_a, \kappa),$$

where $\mathcal{N}(\cdot)$ and $\Gamma(\cdot)$ denote the Gaussian and Gamma (shape-rate parameterization) distribution respectively. One advantage of this hierarchical shrinkage prior is the full local conjugacy that allows posterior inference easily implemented. For fixed values of α_a and β_a, decreasing the parameter κ encourages stronger shrinkage.

Normal Prior on Biases. Finally, an isotropic normal prior is imposed on the bias vector $\mathbf{b}$, i.e. $\mathbf{b} \sim \mathcal{N}(0, \tau \mathbf{I}_C)$, where $\mathbf{I}_C$ is a C-dimensional identity matrix.

4   Inference via Posterior Sampling

4.1   Augmenting the Posterior

As stated above, (2) makes direct posterior inference intractable due to the max function in $L(\Theta)$. However, it is easy to verify that the posterior of our model is the marginal of

$$q(\Theta, \boldsymbol{\lambda} | \mathcal{D}) = p_0(\Theta) \prod_{i=1}^{N} L(y_i, \boldsymbol{\lambda}_{\cdot i} | \Theta) / Z(\mathcal{D}), \tag{5}$$

where $p_0(\Theta)$ is the model prior, $\lambda_{\cdot i}$ denotes a vector of C augmented variables (each for one class) for $\mathbf{x}_i$, and

$$L(y_i, \lambda_{\cdot i}|\Theta) = \prod_{c=1}^{C} \frac{\exp\{\frac{-1}{2\lambda_{ci}}(\lambda_{ci} + \zeta_{ci})^2\}}{\sqrt{2\pi\lambda_{ci}}}. \tag{6}$$

The above property indicates that we can bypass the calculation of the max function through sampling from the augmented posterior (5), and the interested information about the original posterior can be recovered by discarding λ. Though (5) still is intractable to compute analytically due to the normalization constant, it is not difficult to develop MCMC algorithms by making use of local conjugacy and the Riemann HMC.

4.2   Efficient Geometry-Based MCMC

In the following, we devise a Gibbs sampling algorithm that generates a sample from the posterior distribution of each variable in turn, conditional on the current values of the other variables. It can be shown that the sequence of samples constitutes a Markov chain, and the stationary distribution of that Markov chain is just the joint posterior.

Given λ, $\mathbf{a}$ and $\mathbf{b}$, the conditional (augmented) posterior distribution of $\mathbf{w}$ is

$$q(\mathbf{w}|\lambda, \mathbf{a}, \mathbf{b}, \eta, \mathcal{D}) \propto \mathrm{Dir}(\mathbf{w}; \eta) \cdot \prod_{i=1}^{N} L(y_i, \lambda_{\cdot i}|\Theta),$$

where $L(y_i, \lambda_{\cdot i}|\Theta)$ can be transformed into a Gaussian density of $\mathbf{w}$. Since the Dirichlet prior is not conjugate to the Gaussian distribution, it is hard to get the analytical form of the above distribution. To generate a sample from $q(\mathbf{w}|\lambda, \mathbf{a}, \mathbf{b}, \eta, \mathcal{D})$ with its unnormalized density, we appeal to the Riemann Hamiltonian Monte Carlo (RHMC) [13] approach. As in HMC [27], which simulates the Hamiltonian dynamics, RHMC proposes samples with auxiliary momentum variables $\mathbf{r}$ in a Metropolis-Hastings (MH) framework. The difference from ordinary HMC is that, RHMC explores the underlying geometry of the target distribution to accelerate mixing.

In doing so, the problem is that the Dirichlet prior represents our belief that $\mathbf{w}$ should lie on the probability simplex $\{(w_1, ..., w_P) : w_m \geq 0, \sum_m w_m = 1\} \subset \mathbb{R}^P$, which is compact and has boundaries that has to be accounted for when an update proposes a step that brings the vector outside the simplex. There are several possible ways to simplify boundary considerations via parameterizing the probability simplex, and the performance of RHMC depends strongly on the choice of parameterization. As studied in [30], the expanded-mean parameterization yields higher effective sample size and more efficient computation, so it is adopted here. Specifically, we introduce a P-dimensional unnormalized parameter $\mathbf{e}$ with a product of P independent Gamma distributions, i.e., $p(\mathbf{e}) \propto \prod_{m=1}^{P} e_m^{\eta-1} \exp(-e_m)$. Setting $w_m = e_m / \sum_{m=1}^{P} e_m$ for $m = 1, ..., P$, the prior on $\mathbf{w}$ is still $\mathrm{Dir}(\eta)$, while the conditional posterior of $\mathbf{e}$ is

$$q(\mathbf{e}|\boldsymbol{\lambda}, \mathbf{a}, \mathbf{b}, \eta, \mathcal{D}) \propto \prod_{m=1}^{P} e_m^{\eta-1} \exp\left(-e_m\right) \cdot \prod_{c=1}^{C} \prod_{i=1}^{N} \frac{\exp\{\frac{-1}{2\lambda_{ci}}(\lambda_{ci} + \zeta_{ci})^2\}}{\sqrt{2\pi\lambda_{ci}}},$$

where $\zeta_{ci} = 1 - \delta_{y_i,c}(\mathbf{a}_c^\top \mathbf{h}^i + b_c)$, $\mathbf{h}^i = \sum_{m=1}^{P}\left(\frac{e_m}{\sum_m e_m}\right) \cdot \mathbf{K}_{m,\cdot i}$, $\delta_{y_i,c} = 1$ if $y_i = c$ and -1 otherwise.

Then we consider the Hamiltonian $H(\mathbf{e}, \mathbf{r}) = -\log q(\mathbf{e}|\boldsymbol{\lambda}, \mathbf{a}, \mathbf{b}, \eta, \mathcal{D}) + \frac{1}{2}\mathbf{r}^\top\mathbf{r}$, and use the following transition rule to generate proposals:

$$\mathbf{r}^* = \mathbf{r} + \epsilon\mathbf{G}(\mathbf{e})^{-\frac{1}{2}}\nabla\log q(\mathbf{e}|\boldsymbol{\lambda}, \mathbf{a}, \mathbf{b}, \eta, \mathcal{D}) + \epsilon\nabla\mathbf{G}(\mathbf{e})^{-\frac{1}{2}} - \epsilon\mathbf{G}(\mathbf{e})^{-1}\mathbf{r} + \xi, \quad (7)$$

$$\mathbf{e}^* = |\mathbf{e} + \epsilon\mathbf{G}(\mathbf{e})^{-\frac{1}{2}}\mathbf{r}^*|, \quad (8)$$

where ϵ is the step size, $\xi \sim \mathcal{N}(\mathbf{0}, 2\epsilon\mathbf{G}(\mathbf{e})^{-1})$ is the added Gaussian noise, and $\mathbf{G}(\mathbf{e}) = \mathrm{diag}(\mathbf{e})^{-1}$ is the Riemann manifold used to precondition the dynamics in a locally adaptive manner. Note that, the boundary reflection idea (by taking the absolute value of the proposed new $\mathbf{e}$) is used as in [30] to ensure the positivity.

The other posterior conditional distributions can be derived analytically as follows using the local conjugacy properties.

For $\boldsymbol{\lambda}$: The conditional distribution of λ_{ci} is a generalized inverse Gaussian (GIG) distribution:

$$q(\lambda_{ci}|\Theta, \mathcal{D}) = \mathcal{GIG}\left(\frac{1}{2}, 1, \zeta_{ci}^2\right). \quad (9)$$

See [11] for generating random variates from the GIG distribution.

For $\mathbf{a}$, $\mathbf{b}$: The conditional distribution of $\mathbf{a}_c$ and b_c is

$$q(\mathbf{a}_c, b_c|\boldsymbol{\lambda}, \mathbf{e}, \boldsymbol{\nu}, \tau, \mathcal{D}) \propto \exp\left(-\mathbf{a}_c^\top \Lambda_{\boldsymbol{\nu}_c\cdot}\mathbf{a}_c - \tau b_c^2 - \sum_i \frac{(\lambda_{ci} + \zeta_{ci})^2}{2\lambda_{ci}}\right),$$

a multivariate Gaussian with covariance and mean:

$$\Sigma_{(\mathbf{a},b)} = \left(\begin{bmatrix} \Lambda_{\boldsymbol{\nu}_c\cdot}^{-1}, & 0 \\ 0, & \tau^{-1} \end{bmatrix} + \sum_i \frac{1}{\lambda_{ci}}\begin{bmatrix} \mathbf{h}^i \\ 1 \end{bmatrix}\begin{bmatrix} \mathbf{h}^i \\ 1 \end{bmatrix}^\top\right)^{-1}, \quad (10)$$

$$\mu_{(\mathbf{a},b)} = \Sigma_{(\mathbf{a},b)}\sum_i\left(\delta_{y_i,c} + \frac{\delta_{y_i,c}}{\lambda_{ci}}\right)\begin{bmatrix} \mathbf{h}^i \\ 1 \end{bmatrix}, \quad (11)$$

where $\Lambda_{\boldsymbol{\nu}_c\cdot} = \mathrm{diag}(\boldsymbol{\nu}_c\cdot)$.

For TPBN shrinkage: The conditional distribution of $\boldsymbol{\nu}$ and ς are

$$\nu_{ci}|\mathbf{a}, \varsigma \sim \mathcal{GIG}(\alpha_a - \frac{1}{2}, 2\varsigma_{ci}, a_{ci}^2), \quad (12)$$

$$\varsigma_{ci}|\boldsymbol{\nu}, \kappa \sim \Gamma(\alpha_a + \beta_a, \nu_{ci} + \kappa). \quad (13)$$

4.3   Prediction

For an instance $\mathbf{x}^{new}$ that is unseen during the above sampling process, the posterior predictive probability of its label y^{new} can be estimated as follows:

$$P(y^{new} = c|\mathbf{x}^{new}) = \frac{1}{T}\sum_{t=1}^{T} P(y^{new} = c|\mathbf{x}^{new}, \Theta^{(t)}),$$

$$P(y^{new} = c|\mathbf{x}^{new}, \Theta^{(t)}) = \frac{\exp\{f(\mathbf{x}^{new}; \Theta_c^{(t)})\}}{\sum_{c=1}^{C} \exp\{f(\mathbf{x}^{new}; \Theta_c^{(t)})\}},$$

where $c \in \{1, 2, ..., C\}$ is the class label, T is the number of post-convergence samples $\Theta^{(t)}$ obtained from MCMC.

We finally predict the class label of $\mathbf{x}^{new}$ as

$$y^{new} = \arg\max_c P(y^{new} = c|\mathbf{x}^{new}).$$

5   Further Analysis and Efficient Implementation

5.1   More Informative Prior for λ

With (2) and (4), an improper flat prior distribution on $[0, \infty)$ is implicitly imposed on λ. Alternatively, we can impose an exponential prior $\lambda \sim \text{Exp}(\gamma_0)$ to restrict λ from taking too large values, which is beneficial to discourage $\zeta \ll 0$ for correct classifications [18]. Though not so desirable in linear case [31], such a property is important for robust kernel weights learning in our MKL model, and generally improves mixing. The corresponding new likelihood is

$$L'(y_i|\Theta) = \prod_{c=1}^{C} \int_0^\infty \frac{\gamma_0 \exp\{-\gamma_0 \lambda_{ci}\}}{\sqrt{2\pi\lambda_{ci}}} \exp\left\{\frac{(\lambda_{ci} + \zeta_{ci})^2}{-2\lambda_{ci}}\right\} d\lambda_{ci}$$

$$= \prod_{c=1}^{C} \frac{\gamma_0}{s} \begin{cases} \exp\{-(s+1)\zeta_{ci}\}, & \text{if } \zeta_{ci} \geq 0 \\ \exp\{(s-1)\zeta_{ci}\}, & \text{if } \zeta_{ci} < 0, \end{cases}$$

where $s = \sqrt{1 + 2\gamma_0} > 1$. Note that, the new likelihood always decays faster when the training samples don't satisfy the max-margin criterion $\zeta_{ci} \geq 0$ for each binary classification sub-problem, while it also discourages $\zeta \ll 0$ for correctly classified samples. When $\gamma_0 \to 0$ (hence $s \to 1$), the hinge loss based pseudo-likelihood (2) can be recovered. When $\gamma_0 \gg 0$ (hence $s \gg 1$), $L'(y_i|\Theta)$ will behave more like the mechanism behind proximal SVM [26]. For moderate γ_0, this general likelihood is expected to benefit from both sides, thus will be adopted in our implementation.

For posterior inference, all conditionals remain the same as above, except that the conditional posterior of λ should be modified as

$$q(\lambda_{ci}|\Theta) = \mathcal{GIG}\left(\frac{1}{2}, 1 + 2\gamma_0, \zeta_{ci}^2\right). \tag{14}$$

5.2   Correctness of the Sampler

Here we show the correctness of (7) and (8) using the recent theoretical framework in [25], where a general recipe for constructing MCMC samplers based on continuous Markov processes was developed. The recipe involves defining a (stochastic) system parameterized by two matrices: a positive semidefinite diffusion matrix, $\mathbf{D}(\mathbf{z})$, and a skew-symmetric curl matrix, $\mathbf{Q}(\mathbf{z})$, where $\mathbf{z} = (\mathbf{e}, \mathbf{r})$ with $\mathbf{e}$ model parameters of interest and $\mathbf{r}$ a set of auxiliary variables. The dynamics are then written explicitly in terms of the target stationary distribution and these two matrices. It can be verified that (7) and (8) fall within their framework when

$$
\mathbf{D}(\mathbf{e}, \mathbf{r}) = \begin{bmatrix} 0, & 0 \\ 0, & \mathbf{G}(\mathbf{e})^{-1} \end{bmatrix}, \quad
\mathbf{Q}(\mathbf{e}, \mathbf{r}) = \begin{bmatrix} 0, & -\mathbf{G}(\mathbf{e})^{-1/2} \\ \mathbf{G}(\mathbf{e})^{-1/2}, & 0 \end{bmatrix}.
$$

5.3   Efficient Implementation

Omitting MH correction in Riemann HMC. Inspired by the stochastic gradient HMC method [5] and the general stochastic gradient MCMC framework in [25], we can simplify the leapfrog procedure in RHMC and omit the Metropolis-Hastings correction step by using decreasing step sizes, which guarantees the sampler to yield the correct invariant distribution. However, having to decrease ϵ to zero comes at the cost of increasingly small updates. We can also use a finite, small step size in practice, resulting in a biased (but faster) sampler [25].

Efficiently Sampling. $\boldsymbol{\lambda}$ and $\boldsymbol{\nu}$: Though directly generating random variates from the GIG distribution is straightforward according to [11], it can be inefficient when the data set is very large. In fact, we can make use of the relationship[1] between GIG and inverse Gaussian (IG). For $\boldsymbol{\lambda}$, it is easy to see λ_{ci}^{-1} follows the following inverse Gaussian distribution:

$$
q(\lambda_{ci}^{-1}|\Theta) = \mathcal{IG}\left(\frac{\sqrt{1+2\gamma_0}}{|\varsigma_{ci}|}, \ 1 + 2\gamma_0 \right).
\tag{15}
$$

For $\boldsymbol{\nu}$, consider the case $\alpha_a = 1$, which leads to

$$
\nu_{ci}|\mathbf{a}, \varsigma \sim \mathcal{GIG}\left(\frac{1}{2}, \ 2\varsigma_{ci}, \ a_{ci}^2 \right).
\tag{16}
$$

Consequently, we have

$$
\nu_{ci}^{-1}|\mathbf{a}, \varsigma \sim \mathcal{IG}\left(\frac{\sqrt{2\varsigma_{ci}}}{|a_{ci}|}, \ 2\varsigma_{ci} \right).
\tag{17}
$$

Thus, we can sample from the corresponding IG distribution instead, which is more efficient and adopted in our implementation.

---

[1] Generally, we have [6]: if $\lambda \sim \mathcal{GIG}(1/2, \varrho, \chi)$, then $\lambda^{-1} \sim \mathcal{IG}(\vartheta, \varrho)$, where $\chi = \varrho/\vartheta^2$.

6   Experiments

We conduct extensive experiments on both binary and multiclass classification data sets. For each data set, we run 200 MCMC iterations of the proposed BM^{3}KL model and use the samples collected in the last 20 iterations for prediction. The code for BM^{3}KL was written purely in Matlab and all experiments were performed on a desktop with 3.2 GHz CPU and 12 GB memory.

6.1   Compared Algorithms and Parameter Settings

We compare BM^{3}KL with the following state-of-the-art kernel-based algorithms:

- A data augmentation based Bayesian nonlinear SVM model (BSVM) [18];
- Bayesian efficient MKL [15] with sparse and non-sparse kernel weights (sBE-MKL and BEMKL, respectively);
- Maximum Margin Multiple Kernel (M^{3}K) learning [9] for multiclass classification;
- Efficient and accurate ℓ_p-norm MKL [20], a general max-margin MKL framework with ℓ_p-norm constraint on kernel weights. We consider $p = 1, 2, 4, \infty$ and use its Shogun C++ implementation;
- Online-batch strongly convex MKL (OBSCURE) [29], an efficient two-stage method which learns an online model for batch initialization;
- SimpleMKL [33], a well-known MKL baseline with max-margin principle.

We perform 5-fold cross-validation on training data sets to select the regularization parameter $C \in \{10^{-1}, 10^0, \ldots, 10^4\}$ for SimpleMKL and ℓ_p-norm MKL, and $C \in \{1, 10, 100, 1000\}$ and $p \in \{1.01, 1.05, 1.10, 1.25, 1.50, 1.75, 2\}$ for OBSCURE, following the instructions in their original papers. The recommended setting in the publicly available code for BEMKL and sBEMKL is used. The hyper-parameters of BM^{3}KL are fixed to $\eta = 1$, $\alpha_a = 1$, $\kappa = 10^{-10}$, $\tau = 10^{-4}$ in all experiments.

6.2   Binary Classification on Benchmark Data

In this subsection, we evaluate the performance of the proposed BM^{3}KL on a number of binary classification tasks as shown in Table 1, where five binary classification tasks were constructed from the TRECVID 2003 data set, which has five classes of 165-dimensional manually labeled video shots.

Comparison with single kernel machines. Before we systematically compare BM^{3}KL with various MKL algorithms, we first briefly demonstrate the performance improvements of our multi-kernel method over BSVM [18], which adopts a similar data augmentation idea and infers its GP kernel parameters with slice sampling. Following [18], the data sets were normalized to have zero mean and unit variance, and then randomly split into 10 folds of which one at a time was used as test set to evaluate models trained on the remaining nine folds. For MKL, we predefine a pool of 16 kernel functions on each data set, including

Table 1. The number of instances and the dimensionality for each binary classification data set.

Data set	Ionosphere	Sonar	Wisconsin	Crabs	Bupaliver	Trecvid03				
						1vs3	1vs4	3vs4	3vs5	4vs5
#Insance	351	208	683	200	345	393	340	317	303	250
#Dimension	34	60	9	7	6	165	165	165	165	165

13 Gaussian kernels with widths in $\{2^{-6}, 2^{-5}, \ldots, 2^6\}$ and 3 polynomial kernels with degrees in $\{1, 2, 3\}$. All kernel matrices were normalized to have unit diagonal entries (i.e., spherical normalization). Table 2 shows mean accuracy for the methods under consideration, where the results of BSVM, SVM and Gaussian Process Classification (GPC) are directly cited from [18], and the results of a latest Bayesian MKL model (BEMKL) are also listed for reference. For the proposed BM^{3}KL, we fix $\gamma_0 = 100$ and $\beta_a = 0.1$ for all data sets. The RHMC step size ϵ is set to 0.1 for Sonar, and 0.01 for the others. It is clear that BM^{3}KL can consistently outperform all other competitors, even without bothersome tuning of (hyper-) parameters.

Table 2. Comparison with single kernel learning. Listed results are mean test accuracies (%) from 10-fold cross validation.

Data set	N	BSVM	SVM	GPC	sBEMKL	BEMKL	BM^{3}KL
Ionosphere	315	94.02	94.29	92.59	93.06	92.64	**96.11**
Sonar	187	88.94	88.46	87.50	85.95	86.53	**90.68**
Wisconsin	614	97.07	96.93	97.36	97.78	97.79	**97.91**
Crabs	180	98.50	98.00	97.50	99.00	98.75	**99.50**

Comparison with various MKL algorithms. We then compare the proposed BM^{3}KL with state-of-the-art MKL algorithms. Following the experimental settings in [15], we construct Gaussian kernels with 10 different widths ($\{2^{-3}, 2^{-2}, \ldots, 2^6\}$) and polynomial kernels with 3 different degrees ($\{1, 2, 3\}$) on all features and on each single feature. For the Bupaliver and Sonar data, we randomly select 70 % of each data set as the training set and use the remaining as the test set as in [15]. For the binary classification tasks constructed from Trecvid 2003 (see Sect. 6.3 for its details), the ratio of training/testing split is 20 % vs. 80 % because we have thousands of kernels on them and large training ratios lead to out of memory. All data sets were normalized to have zero mean and unit variance, and all base kernel matrices were normalized to have unit diagonal entries and precomputed before running the algorithm. For the proposed BM^{3}KL, we set[2] $\gamma_0 = 500$, $\epsilon = 0.1/t$ and select $\beta_a \in \{1, 2, 3\}$ via

---

[2] To get decreasing step sizes, we use t to denote the t-th MCMC iteration.

Table 3. Comparison of various MKL methods on binary classification tasks. Each element in the table shows the mean and standard deviation of testing accuracies/training times obtained from 20 independent trials. All experiments were conducted in Matlab, but ℓ_p-MKL and OBSCURE called C++ routines. Thus the results of training time for them may be over optimistic.

Algorithm	Metric	Bupaliver N=241, P=91	Sonar N=144, P=793	Trecvid03 1vs3 N=78, P=2158	Trecvid03 1vs4 N=68, P=2158	Trecvid03 3vs4 N=63, P=2158	Trecvid03 3vs5 N=60, P=2158	Trecvid03 4vs5 N=50, P=2158
SimpleMKL	Test-Acc (%)	70.3 ± 3.3	83.4 ± 4.7	78.8 ± 2.4	77.2 ± 2.8	70.4 ± 2.8	71.8 ± 4.0	71.6 ± 2.7
	Train-Time (s)	18.4 ± 20.1	35.3 ± 15.4	25.2 ± 17.6	22.6 ± 11.7	16.3 ± 10.5	20.8 ± 11.2	14.4 ± 8.9
sBEMKL	Test-Acc (%)	68.4 ± 4.3	76.9 ± 4.0	77.7 ± 2.3	75.6 ± 3.5	69.6 ± 4.7	69.0 ± 3.7	71.3 ± 3.9
	Train-Time (s)	2.8 ± 0.1	17.0 ± 0.3	131.1 ± 0.8	129.9 ± 0.4	129.2 ± 0.4	128.7 ± 0.3	127.1 ± 0.3
BEMKL	Test-Acc (%)	70.7 ± 3.8	82.8 ± 4.1	79.4 ± 2.3	78.5 ± 2.9	71.8 ± 2.6	73.8 ± 3.1	73.6 ± 4.5
	Train-Time (s)	2.9 ± 0.1	16.9 ± 0.2	130.8 ± 0.7	129.7 ± 0.5	129.2 ± 0.4	128.8 ± 0.4	126.9 ± 0.2
ℓ_1-MKL	Test-Acc (%)	68.8 ± 4.2	82.7 ± 3.4	78.8 ± 3.0	77.0 ± 1.6	71.1 ± 4.1	71.8 ± 3.1	70.4 ± 2.8
	Train-Time (s)	2.0 ± 0.7	22.1 ± 3.4	34.7 ± 14.1	30.4 ± 5.0	25.4 ± 6.0	30.3 ± 14.9	14.5 ± 1.8
ℓ_2-MKL	Test-Acc (%)	70.6 ± 4.3	83.3 ± 3.9	79.5 ± 3.4	79.6 ± 1.6	72.1 ± 3.8	74.5 ± 1.5	71.0 ± 3.5
	Train-Time (s)	0.6 ± 0.0	6.2 ± 0.7	7.5 ± 2.5	6.2 ± 0.4	6.6 ± 0.4	6.2 ± 0.2	3.8 ± 0.1
ℓ_4-MKL	Test-Acc (%)	71.0 ± 4.0	83.6 ± 3.5	81.1 ± 3.3	80.1 ± 2.1	72.9 ± 3.7	74.1 ± 1.5	71.3 ± 3.0
	Train-Time (s)	1.2 ± 0.2	8.0 ± 0.7	5.6 ± 0.7	4.4 ± 0.6	5.2 ± 0.7	4.5 ± 0.2	2.6 ± 0.1
ℓ_∞-MKL	Test-Acc (%)	70.5 ± 3.7	83.5 ± 3.1	80.8 ± 3.7	80.4 ± 1.9	73.7 ± 3.4	74.9 ± 1.8	70.5 ± 2.9
	Train-Time (s)	1.1 ± 0.1	6.4 ± 0.5	5.6 ± 1.0	4.2 ± 0.9	4.9 ± 0.6	4.3 ± 0.2	2.8 ± 0.1
OBSCURE	Test-Acc (%)	69.5 ± 4.7	83.4 ± 3.3	81.0 ± 2.2	77.8 ± 2.4	71.0 ± 3.8	72.4 ± 3.0	69.6 ± 2.8
	Train-Time (s)	3.3 ± 0.2	13.5 ± 0.6	4.9 ± 0.3	4.6 ± 0.2	4.4 ± 0.3	3.9 ± 0.1	2.9 ± 0.0
BM^{3}KL	Test-Acc (%)	**71.6 ± 3.5**	**84.3 ± 3.8**	**81.5 ± 1.7**	**81.1 ± 2.8**	**74.0 ± 2.4**	**75.7 ± 2.5**	**75.2 ± 2.9**
	Train-Time (s)	3.1 ± 0.1	5.6 ± 0.1	3.4 ± 0.0	2.6 ± 0.0	2.4 ± 0.0	2.2 ± 0.0	1.5 ± 0.0

5-fold cross-validation for each data set. To get stable results, we independently repeat the random split of each data set, and then run each algorithm on it, for 20 times. The mean and standard deviation are reported in Table 3 in terms of testing accuracy and training time.

The superiority of BM^{3}KL over the competitors is evident. When comparing BM^{3}KL with BEMKL, it is easy to see significant improvements of the former in terms of prediction performance. We attribute this to the max-margin principle underlying our model and its more accurate MCMC-based inference rather than variational approximation with mean-field assumption. As to the computational complexity, BM^{3}KL and BEMKL scale linearly and cubically with the number of kernels P respectively, while they both scale cubically with the number of training samples N. Consequently, as we observed, BEMKL needs considerable more training time when P is large (e.g., on Trecvid03).

Another thing worth to mention is that, with a simple and fixed Dirichlet prior on kernel weights, BM^{3}KL can consistently outperform ℓ_p-MKL and OBSCURE. This indicates that the superiority of our method is not brought in by solely regularizing the weights. We believe this is owing to the inherent advantages of our Bayesian max-margin modeling, and the Riemann manifold based HMC method.

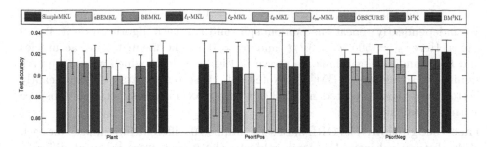

Fig. 1. Multiclass classification performance comparison on protein subcellular localization data sets. All results are averaged over 10 independent trials.

6.3 Multiclass Classification

Protein Subcellular Localization. We first consider three biological data sets (Plant, PsortPos, and PsortNeg) that have been widely used to compare MKL algorithms. The numbers of data instances in Plant, PsortPos, and PsortNeg are $940, 541$, and 1444, respectively, while the numbers of data classes are 4, 4, and 5, respectively. For each of these three data sets, 69 biologically motivated sequence kernels [40] are used and the corresponding base kernel matrices are provided online[3]. To replicate the experiments of a latest multiclass max-margin multiple kernel classification model (M^3K) [9], all kernel matrices were first centered and then normalized to have unit diagonal entries, and the training split fractions for Plant, PsortPos, and PsortNeg were set as 0.5, 0.8, and 0.65, respectively. As above, we set $\gamma_0 = 500$, $\epsilon = 0.1/t$, and select $\beta_a \in \{1, 2, 3\}$ via 5-fold cross-validation. Then we independently repeat the random split of each data set and run each algorithm on it, for 10 times. The mean testing accuracies and the standard deviations are shown in Fig. 1, where the results of M^3K are directly cited from [9]. From the results we can see, (1) BM^3KL achieves highest mean accuracy on each data set; (2) sometimes the performance differences are not so significant since the training sample ratios are large enough; (3) BM^3KL has advantage in terms of robustness.

Video shots classification. The TRECVID 2003 data set has 1078 manually labeled video shots which are categorized into 5 classes. Each of the video shots is represented by a 165-dimensional vector of HSV color histogram. To fully exploit the rich features underlying each sample, we define a large set of kernel mappings on all features and each individual feature, and then learn the optimal combination of them. Specifically, we construct Gaussian kernels with 13 different widths ($\{2^{-6}, 2^{-5}, \ldots, 2^6\}$) and polynomial kernels with 3 different degrees ($\{1, 2, 3\}$) on the 165-dimensional vectors, and construct Gaussian kernels with 7 different widths ($\{2^{-3}, 2^{-2}, \ldots, 2^3\}$) on each single feature. This gives us 1171 kernels in total (to avoid out of memory, we didn't consider even more kernels).

---

[3] http://raetschlab.org//suppl/protsubloc.

As in [12,39], the entire data set was evenly split into training and testing sets. We independently repeat the random split and run each algorithm on it, for 10 times. Here $\gamma_0 = 500$, $\epsilon = 0.1/t$, and $\beta_a = 0.5$. The mean testing accuracies and the standard deviations are shown in Fig. 2(a), from which we can observe significant advantage of BM^{3}KL. Note that, these results are also significantly better than those reported in [12,39], where max-margin supervised subspace learning was studied.

Face recognition. The ORL data set contains 10 different face images for each of 40 distinct subjects. For some subjects, the images were taken at different times with varying lighting and facial details (open/closed eyes, smiling/not smiling, glasses/no glasses). All the images were taken against a dark homogeneous background with the subjects in an upright, frontal position, and were manually aligned, cropped and resized to 32×32 pixels. We construct Gaussian kernels with 13 different widths ($\{2^{-6}, 2^{-5}, \ldots, 2^6\}$) and polynomial kernels with 3 different degrees ($\{1, 2, 3\}$) on the 1024-dimensional pixel vectors, and construct Gaussian kernels with 2 different widths ($\{2^1, 2^2\}$) and polynomial kernel with degree 2 on each single pixel. This gives us 3088 kernels in total (to avoid out of memory, we didn't consider even more kernels). For the training sample ratio, we consider two cases, i.e., 0.2 and 0.7 (accordingly, we have 80 and 280 training samples respectively). We independently repeat the random splits and run each algorithm on them, for 10 times. Here $\gamma_0 = 10000$, $\epsilon = 0.01$, and $\beta_a = 0.1$. The mean testing accuracies and the standard deviations are shown in Fig. 2(b) and (c), from which we see obvious advantages of BM^{3}KL again.

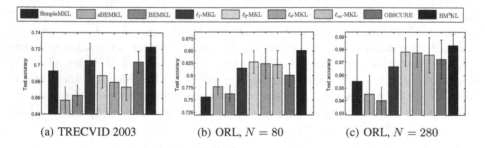

(a) TRECVID 2003   (b) ORL, $N = 80$ (c) ORL, $N = 280$

Fig. 2. Multiclass classification performance comparison on vision data sets. All results are averaged over 10 independent trials.

6.4   Time Efficiency and Convergence Rate

As shown in Table 3 and Fig. 3, the training time efficiency of the proposed BM^{3}KL is either significantly better than or comparable with the state-of-the-art MKL algorithms on all considered data sets. Note that here we couldn't directly compare with M^{3}K since its code is not publicly available. But as shown in [9] it generally needs several times more training time than OBSCURE.

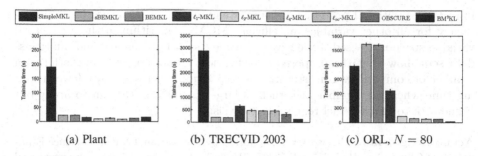

(a) Plant   (b) TRECVID 2003   (c) ORL, $N = 80$

Fig. 3. Training time efficiency comparison on multiclass classification data sets. All results are averaged over 10 independent trials conducted in Matlab, but ℓ_p-MKL and OBSCURE called C++ routines. Thus the results for them may be over optimistic.

Specifically, it is observed that BM³KL generally is more efficient than BEMKL, a latest Bayesian MKL model focusing on efficient inference [15]. The reason is that BEMKL has to perform matrix inversion to compute the posterior covariance of kernel weights in each round of iteration while it is avoided in BM³KL via employing Riemann HMC. Besides, enjoying the Bayesian modelling advantages, BM³KL achieves even faster learning speed than the optimization-based point estimate methods. We attribute this to the fast convergence rate of our geometry and local conjugacy based approximate posterior sampling, which is depicted in Fig. 4.

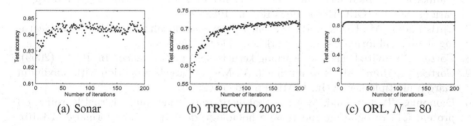

(a) Sonar   (b) TRECVID 2003   (c) ORL, $N = 80$

Fig. 4. Convergence rate of the approximate posterior sampling of BM³KL on Sonar, TRECVID 2003 and ORL ($N = 80$). All settings are the same as above except that the single sample of model weights obtained in each MCMC iteration is used for prediction on each data set.

7   Conclusion and Future Work

By defining a multiclass (pseudo-) likelihood function that accounts for the margin loss for kernelized classification, we have developed a robust Bayesian max-margin MKL framework with Dirichlet and TPBN priors imposed on the kernel and sample weights respectively. Employing Riemann manifold HMC to sample from the conditional posterior of kernel weights, and making use of local conjugacy for all other variables, an efficient MCMC algorithm in the augmented variable space is devised. Extensive experiments on both binary and multiclass data sets show that the proposed classification model not only outperforms a number of competitors consistently but also requires substantially fewer training time when the number of kernels is large. In future, we plan to apply our framework to multi-kernel regression analysis.

Acknowledgments. This work was supported by the National Natural Science Foundation of China (No. 61473273, 61573335, 91546122, 61303059), Guangdong provincial science and technology plan projects (No. 2015B010109005), and the Science and Technology Funds of Guiyang (No. 201410012).

References

1. Andrews, D.F., Mallows, C.L.: Scale mixtures of normal distributions. J. Roy. Stat. Soc. B (Methodol.) **36**(1), 99–102 (1974)
2. Armagan, A., Clyde, M., Dunson, D.B.: Generalized beta mixtures of gaussians. In: NIPS (2011)
3. Bach, F.R.: Consistency of the group lasso and multiple kernel learning. JMLR **9**, 1179–1225 (2008)
4. Bach, F.R., Lanckriet, G.R., Jordan, M.I.: Multiple kernel learning, conic duality, and the smo algorithm. In: ICML (2004)
5. Chen, T., Fox, E., Guestrin, C.: Stochastic gradient Hamiltonian Monte Carlo. In: ICML (2014)
6. Chhikara, R.: The Inverse Gaussian Distribution: Theory, Methodology, and Applications, vol. 95. CRC Press, Boca Raton (1988)
7. Christoudias, M., Urtasun, R., Darrell, T.: Bayesian localized multiple kernel learning. Univ. California Berkeley, Berkeley (2009)
8. Cortes, C.: Invited talk: can learning kernels help performance? In: ICML (2009)
9. Cortes, C., Mohri, M., Rostamizadeh, A.: Multi-class classification with maximum margin multiple kernel. In: ICML, pp. 46–54 (2013)
10. Damoulas, T., Girolami, M.A.: Probabilistic multi-class multi-kernel learning: on protein fold recognition and remote homology detection. Bioinformatics **24**(10), 1264–1270 (2008)
11. Devroye, L.: Random variate generation for the generalized inverse gaussian distribution. Stat. Comput. **24**(2), 239–246 (2014)
12. Du, C., Zhe, S., Zhuang, F., Qi, Y., He, Q., Shi, Z.: Bayesian maximum margin principal component analysis. In: AAAI (2015)
13. Girolami, M., Calderhead, B.: Riemann manifold langevin and hamiltonian monte carlo methods. J. Roy. Stat. Soc. B (Stat. Methodol.) **73**(2), 123–214 (2011)
14. Girolami, M., Rogers, S.: Hierarchic bayesian models for kernel learning. In: ICML (2005)
15. Gönen, M.: Bayesian efficient multiple kernel learning. In: ICML, pp. 1–8 (2012)
16. Gönen, M., Alpaydin, E.: Localized multiple kernel learning. In: ICML, pp. 352–359 (2008)
17. Gönen, M., Alpaydın, E.: Multiple kernel learning algorithms. JMLR **12**, 2211–2268 (2011)

18. Henao, R., Yuan, X., Carin, L.: Bayesian nonlinear support vector machines and discriminative factor modeling. In: NIPS, pp. 1754–1762 (2014)
19. Jain, A., Vishwanathan, S.V., Varma, M.: Spg-gmkl: generalized multiple kernel learning with a million kernels. In: SIGKDD (2012)
20. Kloft, M., Brefeld, U., Sonnenburg, S., Zien, A.: Lp-norm multiple kernel learning. JMLR **12**, 953–997 (2011)
21. Lanckriet, G.R., Cristianini, N., Bartlett, P., Ghaoui, L.E., Jordan, M.I.: Learning the kernel matrix with semidefinite programming. JMLR **5**, 27–72 (2004)
22. Lázaro-gredilla, M., Titsias, M.K.: Spike and slab variational inference for multitask and multiple kernel learning. In: NIPS, pp. 2339–2347 (2011)
23. Liu, X., Wang, L., Yin, J., Dou, Y., Zhang, J.: Absent multiple kernel learning. In: AAAI (2015)
24. Liu, X., Wang, L., Zhang, J., Yin, J.: Sample-adaptive multiple kernel learning. In: AAAI (2014)
25. Ma, Y.A., Chen, T., Fox, E.: A complete recipe for stochastic gradient mcmc. In: NIPS (2015)
26. Mangasarian, O.L., Wild, E.W.: Proximal support vector machine classifiers. In: SIGKDD (2001)
27. Neal, R.M.: Mcmc using hamiltonian dynamics. Handbook of Markov Chain Monte Carlo, vol. 2 (2011)
28. Ni, B., Li, T., Moulin, P.: Beta process multiple kernel learning. In: CVPR (2014)
29. Orabona, F., Jie, L., Caputo, B.: Multi kernel learning with online-batch optimization. JMLR **13**(1), 227–253 (2012)
30. Patterson, S., Teh, Y.W.: Stochastic gradient riemannian langevin dynamics on the probability simplex. In: NIPS, pp. 3102–3110 (2013)
31. Polson, N.G., Scott, S.L.: Data augmentation for support vector machines. Bayesian Analysis **6**(1), 1–23 (2011)
32. Rakotomamonjy, A., Bach, F., Canu, S., Grandvalet, Y.: More efficiency in multiple kernel learning. In: ICML, pp. 775–782 (2007)
33. Rakotomamonjy, A., Bach, F., Canu, S., Grandvalet, Y.: Simplemkl. JMLR **9**, 2491–2521 (2008)
34. Sonnenburg, S., Rätsch, G., Schäfer, C., Schölkopf, B.: Large scale multiple kernel learning. JMLR **7**, 1531–1565 (2006)
35. Suzuki, T., Tomioka, R.: Spicymkl: a fast algorithm for multiple kernel learning with thousands of kernels. Machine learning **85**(1–2), 77–108 (2011)
36. Xu, Z., Jin, R., King, I., Lyu, M.: An extended level method for efficient multiple kernel learning. In: NIPS, pp. 1825–1832 (2009)
37. Xu, Z., Jin, R., Yang, H., King, I., Lyu, M.R.: Simple and efficient multiple kernel learning by group lasso. In: ICML (2010)
38. Yang, J., Li, Y., Tian, Y., Duan, L., Gao, W.: Group-sensitive multiple kernel learning for object categorization. In: CVPR (2009)
39. Zhu, J., Chen, N., Xing, E.P.: Bayesian inference with posterior regularization and applications to infinite latent svms. JMLR **15**, 1799–1847 (2014)
40. Zien, A., Ong, C.S.: Multiclass multiple kernel learning. In: ICML, pp. 1191–1198 (2007)

Transactional Tree Mining

Mostafa Haghir Chehreghani[1]([✉]) and Morteza Haghir Chehreghani[2]

[1] Department of Computer Science, KU Leuven, Leuven, Belgium
mostafa.chehreghani@gmail.com
[2] Xerox Research Centre Europe (XRCE), Meylan, France
morteza.chehreghani@xrce.xerox.com

Abstract. In the transactional setting of finding frequent embedded patterns from a large collection of tree-structured data, the crucial step is to decide whether a tree pattern is subtree homeomorphic to a database tree. Our extensive study on the properties of real-world tree-structured datasets reveals that while many vertices in a database tree may have the same label, no two vertices on the same path are identically labeled. In this paper, we exploit this property and propose a novel and efficient method for deciding whether a tree pattern is subtree homeomorphic to a database tree. Our algorithm is based on a compact data-structure, called **EMET**, that stores all information required for subtree homeomorphism. We propose an efficient algorithm to generate **EMET**s of larger patterns using **EMET**s of the smaller ones. Based on the proposed subtree homeomorphism method, we introduce TTM, an effective algorithm for finding frequent tree patterns from rooted ordered trees. We evaluate the efficiency of TTM on several real-world and synthetic datasets and show that it outperforms well-known existing algorithms by an order of magnitude.

Keywords: Transactional tree mining · Rooted ordered trees · Frequent patterns · Subgraph homeomorphism test

1 Introduction

Mining frequent tree patterns is useful in different domains such as XML document mining, user web log data analysis and classification. For example, Zaki et al. [19] presented XRule, a structural classifier based on frequent tree patterns, and showed its high performance compared to the classifiers such as *SVM* and *random forest*.

The most challenging phase in frequent pattern mining is the frequency counting step. It has two settings: (i) the *transactional* setting where it is decided whether a tree pattern is subtree, under a matching operator, to a database tree, and (ii) the *per-occurrence* setting where all occurrences of a tree pattern in a database tree are counted. The concern of the current paper is the *transactional setting*. A widely used matching operator between a tree pattern and a database tree is *subtree homeomorphism*, where a parent-child relationship in the tree pattern is mapped onto an ancestor-descendant relationship in the database tree.

© Springer International Publishing AG 2016
P. Frasconi et al. (Eds.): ECML PKDD 2016, Part I, LNAI 9851, pp. 182–198, 2016.
DOI: 10.1007/978-3-319-46128-1_12

Most of real-world trees have properties that distinguish them from arbitrary trees. Such properties may be exploited in designing a more efficient algorithm for finding frequent tree patterns. A property that we observed in widely used real-world tree datasets is that while many vertices of a database tree have the same label, no two vertices on the same path starting from the root have the same label. This property can be seen e.g., in CSLOG1 [19], CSLOG2 [19], CSLOG3 [19], and the NASA MBONE multicast data [3,4] and Prions [12]. In this paper, we exploit this property to develop a more effective algorithm for subtree homeomorphism.

Existing tree pattern mining algorithms usually employ specific data-structures for storing information that are used to decide whether a tree pattern is subtree homeomorphic to database trees. When the pattern is extended to a larger one, its stored information is also extended, to generate the information required for the subtree homeomorphism tests of the extended pattern. This technique is sometimes called *vertical frequency counting*. In our recent work [6], we introduced an efficient vertical approach for counting all occurrences of tree patterns in database trees under subtree homeomorphism. With P a pattern, d the size of the rightmost path of P and T a database tree, an $O(d)$ space data structure was defined that represents all occurrences of P in T that share the rightmost path. Based on that data structure, we defined efficient join operations that derives a larger pattern from a smaller one and computes its frequency. However, in the transactional setting it is not necessary to count all occurrences, and it suffices to decide whether the tree pattern is subtree homeomorphic to the database tree. So, a more compact data structure can be used to support frequency counting.

In the current paper, we present a novel and efficient algorithm, called TTM, for discovering frequent tree patterns from rooted ordered trees in the transactional setting, where vertices on the same path have distinct labels. We propose a novel subtree homeomorphism algorithm, based on a more compact data-structure, called **EMET** (which is an abbreviation for **EM**bedding **E**ncoder for **T**ransactional tree mining). The **EMET** data-structure encodes and stores all information required for the subtree homeomorphism test. We then present efficient join operations, defined on **EMET**s, to compute **EMET**s of larger patterns based on **EMET**s of smaller ones. We show that our method has a time complexity better than the most efficient existing methods. To empirically evaluate the efficiency of TTM, we perform extensive experiments, on both real-world and synthetic datasets, and compare TTM against the most efficient existing algorithms for transactional tree mining, that are TreeMinerD [18] and MB3Miner-T [13]. Our experiments reveal that TTM outperforms the other algorithms by an order of magnitude. In particular, there are several cases where TreeMinerD and MB3Miner-T fail, but they are handled effectively by TTM.

The rest of this paper is organized as follows. In Sect. 2, preliminaries and definitions related to the tree pattern mining problem are introduced. In Sect. 3 a brief overview on related works is given. In Sect. 4, we present the **EMET** data-structure, our subtree homeomorphism algorithm, its complexity analysis and the TTM algorithm. We empirically evaluate the effectiveness of TTM in Sect. 5. Finally, the paper is concluded in Sect. 6.

2 Preliminaries

We assume the reader is familiar with the basic concepts in graph theory. The interested reader can refer to e.g., [10]. We use the notations $V(G)$, $E(G)$ and λ_G to refer to the set of vertices, the set of edges and the labeling function of a graph G, respectively. The *size* of G is defined as the number of vertices of G. A *rooted tree* is a directed acyclic graph (DAG) in which: (1) there is a distinguished vertex, called *root*, that has no incoming edges, (2) every other vertex has exactly one incoming edge, and (3) there is an unique path from the root to any other vertex. In a rooted tree T, u is the *parent* of v (v is the *child* of u) if $(u, v) \in E(T)$. The transitive closure of the parent-child relation is called the *ancestor-descendant* relation. A *rooted ordered tree* is a rooted tree such that there is an order over the children of every vertex. Throughout this paper, we refer to rooted ordered trees where no two vertices on the same path have the same label, simply as trees. The position of a vertex in the list of visited vertices during a preorder traversal is called its *preorder number*. We use $p(v)$ to refer to the preorder number of vertex v. *Rightmost vertex* of T is the vertex with the largest preorder number, and *second rightmost vertex* of T is the vertex with the second largest preorder number. *Rightmost path* of T is the path from $root(T)$ to the rightmost vertex of T, and *second rightmost path* of T is the path from $root(T)$ to the second rightmost vertex of T. Two vertices u and v are *relatives* if u is neither an ancestor nor a descendant of v. With $p(u) < p(v)$, u is a *left relative* of v, otherwise, it is a *right relative*. The *depth* of a vertex v in a tree T, denoted by $dep(T, v)$, is defined as the number of edges of the path between the root of T and v. A tree T is a *rightmost path extension* of a tree T' iff there exist vertices u and v such that: (i) $\{v\} = V(T) \setminus V(T')$, (ii) $\{(u, v)\} = E(T) \setminus E(T')$, (iii) u is on the rightmost path of T' and (iv) in T, v is a right relative of all children of u.

A tree C is *subtree homeomorphic* to a tree T (denoted by $C \preceq_h T$) iff there is a mapping $\varphi : V(C) \to V(T)$ such that: (i) $\forall v \in V(C) : \lambda_C(v) = \lambda_T(\varphi(v))$, (ii) $\forall u, v \in V(C)$: u is the parent of v in C iff $\varphi(u)$ is an ancestor of $\varphi(v)$ in T, and (iii) $\forall u, v \in V(C)$: $p(u) < p(v) \Leftrightarrow p(\varphi(u)) < p(\varphi(v))$. Under subtree isomorphism the ancestor-descendant relationship between $\varphi(u)$ and $\varphi(v)$ is strengthened into the parent-child relationship. C is *isomorphic* to T iff C is subtree isomorphic to T and $|V(C)| = |V(T)|$. Note that a pattern C under a subtree morphism can have several mappings to the same database tree T. When the matching operator is subtree homeomorphism, every mapping is called an *occurrence* (or *embedding*) of C in T. An occurrence (embedding) of a vertex v is an occurrence (embedding) of the pattern consisting of the single vertex v. Given an occurrence φ of C in T, the *occurrence tree* $\mathcal{OT}(\varphi)$ is defined as follows: (i) $V(\mathcal{OT}(\varphi)) = \{\varphi(v) : v \in V(C)\}$, (ii) $root(\mathcal{OT}(\varphi)) = \varphi(root(C))$, (iii) for every $v \in V(\mathcal{OT}(\varphi))$, $\lambda_{\mathcal{OT}(\varphi)}(v) = \lambda_T(v)$, and (iv) $E(\mathcal{OT}(\varphi)) = \{(\varphi(v_1), \varphi(v_2)) | (v_1, v_2) \in E(C)\}$ [6].

Given a database D consisting of trees and a tree pattern C, *support* (or *frequency*) of C in D, denoted by $sup(C, D)$, is defined as: $|T \in D : C \preceq_h T|$. C is *frequent* (C is a *frequent embedded pattern*), iff its *support* is greater than or equal to an user defined threshold $minsup > 0$. The problem studied in this paper is as follows: given a database D consisting of trees (where vertices on the same path starting from the root have distinct labels) and an integer $minsup$, find every frequent pattern C such that $sup(C, D) \geq minsup$.

3 Related Work

Mining frequent patterns under homeomorphism. Zaki presented TreeMinerD [18] to find embedded patterns from rooted ordered trees in the transactional setting. For frequency counting, he used an efficient data structure, called *SV-list*. Later, by proposing the SLEUTH algorithm, he extended the work to mine embedded patterns from rooted unordered trees [17]. Tan et al. [13] proposed the MB3Miner-T algorithm, where a unique occurrence list representation of the tree structure is used for efficient implementation of their Tree Model Guided (TMG) candidate generation. Recently, Chehreghani and Bruynooghe [6] presented an efficient algorithm for frequent pattern mining when *per-occurrence support* is used. They proposed the **occ-list** data-structure and effective join operators that are used to count all occurrences of tree patterns. In the current paper, we propose more effective data-structure and algorithms for mining rooted ordered trees in the *transactional setting*. As we discuss in Sect. 4, our algorithm has a better time complexity than the most efficient existing algorithms.

Mining frequent patterns under isomorphism. Asai et al. [1] developed FreqT for mining frequent tree patterns. Chi et al. [8] proposed FreeTreeMiner for mining induced unordered and induced free tree patterns in the transactional setting. Later, Chi et al. proposed CMTreeMiner [9] for mining closed and maximal frequent tree patterns. Their algorithm uses an enumeration DAG to prune the branches of the enumeration tree that do not correspond to closed or maximal patterns. Tatikonda et al. [15] proposed a generic approach for mining embedded or induced patterns. They developed TRIPS and TIDES algorithms using two widely used sequential encodings of trees. In [7], Chehreghani et al. presented the OInduced algorithm for finding frequent tree patterns. They introduced three novel encodings for rooted ordered trees and showed that subtree isomorphism can be done effectively, using these tree encodings. Their method works with both *support* and *per-occurrence support.* Later, in [5], Chehreghani proposed other encodings for rooted unordered trees.

4 Effective Subtree Homeomorphism for the Transactional Setting

In this section, we propose a new approach for efficiently deciding whether a tree pattern is subtree homeomorphic to a database tree. First in Sect. 4.1, we introduce the notions of *AD-join* and *relative-join*. Then in Sect. 4.2, we present the **EMET-List** data-structure and, in Sect. 4.3, the operators for this data structure. We present the TTM algorithm in Sect. 4.4 and an optimization technique in Sect. 4.5.

4.1 AD-join and Relative-join

Two trees with the same size belong to the same *equivalence class* iff their subtrees induced by all but the rightmost vertices are isomorphic. Zaki [18] proposed every two trees in an equivalence class join to generate larger trees. Let C_1 and C_2 be two trees of size at least 2 and v_1 and v_2 their rightmost vertices, respectively. Then:

- If $dep(C_1, v_1) = dep(C_2, v_2)$, by the join of C_1 with C_2, two trees C and C' are generated, where: C is built by adding v_2, as the rightmost child, to the vertex in the depth $dep(C_1, v_1) - 1$ of the rightmost path of C_1, and C' is built by adding v_2, as the rightmost child, to v_1,
- If $dep(C_1, v_1) > dep(C_2, v_2)$, by the join of C_1 with C_2, a tree C is built by adding v_2 to the vertex in the depth $dep(C_1, v_1) - 1$ of the rightmost path of C_1, as its rightmost child.
- If $dep(C_1, v_1) < dep(C_2, v_2)$, no tree is generated by the join of C_1 with C_2.

This provides a complete procedure to generate all frequent tree patterns [18]. Since in the current paper we suppose that in all database trees, vertices on the same path (starting from the root) have distinct labels, the above mentioned procedure can be simplified as follows: when C_1 and C_2 are the same tree, by the join of C_1 and C_2, only one tree C is built by adding v_2, as the rightmost child, to the vertex in the depth $dep(C_1, v_1) - 1$ of the rightmost path of C_1.

In all cases, when v_2 is added to the vertex in the depth $dep(C_1, v_1) - 1$ of C_1, v_1 and v_2 become relatives, and when v_2 is added to the vertex in the depth $dep(C_1, v_1)$ of C_1, v_1 becomes an ancestor of v_2. Therefore, we can consider two types of joins: *ancestor-descendant join*, or *AD-join* in short, where in the generated tree, v_2 becomes a descendant of v_1, and *relative-join*, where in the generated tree, v_2 becomes a relative of v_1. In Propositions 1 and 2, we introduce two properties of AD-join and relative-join that are useful for our subtree homeomorphism algorithm.

Proposition 1. *Let T be a database tree and C a tree pattern generated by the AD-join of two trees C_1 and C_2. Let also v_1 and v_2 be the rightmost vertices of C_1 and C_2, respectively. Suppose that $C_1 \preceq_h T$ and $C_2 \preceq_h T$. $C \preceq_h T$ iff there exists occurrences φ_1 of C_1 and φ_2 of C_2 in T such that $\varphi_1(v_1)$ is an ancestor of $\varphi_2(v_2)$ in T.*

Proof. First, we assume there exists an occurrence φ of C in T and prove there exist occurrences φ_1 of C_1 and φ_2 of C_2 both in T such that $\varphi_1(v_1)$ is an ancestor of $\varphi_2(v_2)$ in T. On the one hand, removing $\varphi(v_2)$ from φ, gives an occurrence φ_1 of C_1 in T. On the other hand, removing $\varphi(v_1)$ from φ and adding $\varphi(v_2)$ to the parent of $\varphi(v_1)$ in φ, yields an occurrence φ_2 of C_2 in T. Furthermore, clearly, $\varphi(v_1)$ is an ancestor of $\varphi(v_2)$. Therefore, the proof is done. Second, we assume there exist occurrences φ_1 of C_1 and φ_2 of C_2 both in T such that $\varphi_1(v_1)$ is an ancestor of $\varphi_2(v_2)$ in T, and prove $C \preceq_h T$. An occurrence tree $\mathcal{OT}(\varphi)$ of C in T is generated by adding $\varphi_2(v_2)$ to $\mathcal{OT}(\varphi_1)$ as the child of $\varphi_1(v_1)$. Note that since in T, $\varphi_2(v_2)$ is a descendant of $\varphi_1(v_1)$, $\varphi_2(v_2)$ does not already belong to φ_1. ∎

Proposition 2. *Let T be a database tree and C a tree pattern generated by the relative-join of two trees C_1 and C_2. Let also v_1 and v_2 be the rightmost vertices of C_1 and C_2, respectively, and w_1 and w_2 the second rightmost vertices of C_1 and C_2, respectively. Suppose C_1 and C_2 are subtree homeomorphic to T where the mappings are φ_1 and φ_2, respectively. $C \preceq_h T$ iff (i) $\varphi_1(w_1) = \varphi_2(w_2)$, and (ii) $\varphi_2(v_2)$ is a right relative of $\varphi_1(v_1)$ in T.*

Proof. First, we suppose there exists an occurrence φ of C in T and prove there exist occurrences φ_1 of C_1 and φ_2 of C_2 both in T such that $\varphi_1(w_1) = \varphi_2(w_2)$ and $\varphi_2(v_2)$ is a right relative of $\varphi_1(v_1)$ in T. On the one hand, removing $\varphi(v_2)$ from φ, gives an occurrence φ_1 of C_1 in T and on the other hand, removing $\varphi(v_1)$ from φ yields an occurrence φ_2 of C_2 in T. Furthermore, since v_2 is a right relative of v_1 in C, $\varphi_2(v_2)$ is a right relative of $\varphi_2(v_1)$ in T; and since φ_1 and φ_2 differ only at $\varphi_1(v_1)$ and $\varphi_2(v_2)$, $\varphi_1(w_1)$ is the same as $\varphi_2(w_2)$. Therefore, the proof is done.

Second, we assume there exist occurrences φ_1 of C_1 and φ_2 of C_2 both in T such that $\varphi_1(w_1) = \varphi_2(w_2)$ and $\varphi_2(v_2)$ is a right relative of $\varphi_1(v_1)$ in T, and prove $C \preceq_h T$. Consider the subtree $\mathcal{OT}(\varphi)$ of T built as follows: $\mathcal{OT}(\varphi)$ is initialized by $\mathcal{OT}(\varphi_1)$ and then, $\varphi_2(v_2)$ is added to the vertex in depth $dep(C, v_2) - 1$ of $\mathcal{OT}(\varphi)$, as the rightmost child. $\mathcal{OT}(\varphi)$ is an occurrence tree of C in T. The reason is that on the one hand, since vertices in the second rightmost path of $\mathcal{OT}(\varphi)$ have distinct labels and φ_1 and φ_2 share the second rightmost vertex and C_1 and C_2 share the second rightmost path, $\mathcal{OT}(\varphi_1)$ and $\mathcal{OT}(\varphi_2)$ share the second rightmost path. This yields that $\varphi(v_2)$ is a descendant of all vertices in depths $0, \ldots, dep(C, v_2) - 1$ of the second rightmost path of $\mathcal{OT}(\varphi)$, and it is a right relative of all vertices in depths $dep(C, v_2), \ldots, dep(C, v_1) - 1$ of the second rightmost path of $\mathcal{OT}(\varphi)$. On the other hand, $\varphi(v_2)$ is a right relative of $\varphi(v_1)$ in T. Therefore, φ satisfies all properties of subtree homeomorphism and $\mathcal{OT}(\varphi)$ satisfies all properties of occurrence tree (as mentioned in Sect. 2). ∎

Note that while Proposition 1 holds for general trees, Proposition 2 holds only for the trees whose vertices on the same path have different labels. To exploit Propositions 1 and 2 for the subtree homeomorphism test, we use Dietz numbering scheme [11]. This scheme associates each vertex v with a pair of numbers $\langle p(v), p(v) + size(v) \rangle$, where $size(v)$ is e.g., the number of descendants of v. For two vertices u and v in a database tree, u is an ancestor of v iff $p(u) < p(v)$ and $p(v) + size(v) \le p(u) + size(u)$, and v is a right relative of u iff $p(u) + size(u) < p(v)$. Several algorithms, such as VTreeMiner and TreeMinerD [16,18] and TPMiner [6] have used this scheme in different forms. In this paper, we introduce the **EMET** data structure based on this scheme. We continue with introducing some more notations. The scope of a vertex x in a database tree T is a pair (l, u), where l is the preorder number of x in T and u is the preorder number of the rightmost descendant of x in T. We use the notations $x.scope.l$ and $x.scope.u$ to refer to l and u of x, respectively.

4.2   EMET-List: An Efficient Data Structure for Transactional Tree Mining

In this section, we present the **EMET-List** data structure for transactional tree mining under subtree homeomorphism. All occurrences represented by an **EMET** share the second rightmost vertex and the rightmost vertex. **EMET** has three components: (i) TId: the identifier of the database tree that contains the occurrences represented by the **EMET**, (ii) *scope*: the scope (in the database tree TId) of the rightmost vertex of the occurrences represented by the **EMET**, (iii) *srv*: the preorder number of the second rightmost vertex of the occurrences represented by the **EMET**. For all occurrences of a pattern C that have the same TId, *scope* and *srv*, one **EMET** in the **EMET-List** of C is generated. Every occurrence is represented by exactly one **EMET**. We refer to the **EMET-List** of C by **EMET-List**(C). It is easy to see that support of C is equal to the number of elements in **EMET-List**(C) that have different TIds.

As an example of **EMET-List**, consider Fig. 1 where $T0$ and $T1$ are two database trees and minimum-support is equal to 2. In the trees presented in all figures of this paper, the numbers assigned to the vertices show their preorder numbers and the assigned letters present their labels. Frequent patterns of size 1 are "a", "b", "c", "d" and "e", and frequent patterns of size 2 are "a b", "a c", "a d", "a e", "b c", "b d" and "c e". Consider the first occurrence of "a b"; it occurs at $T0$ and the preorder number of its second rightmost vertex is 0. The preorder number of its rightmost vertex is 1, and the preorder number of the rightmost descendant of its rightmost vertex is 2. Therefore, the **EMET** formed for this occurrence is $\{0, (1, 2), 0\}$. In a similar way, the **EMET** data structure for other occurrences of "a b" and the **EMET-List**s of other tree patterns are generated.

4.3   Operations on the EMET-Lists

As mentioned above, a tree C is generated by either *AD-join* or *relative-join* of two trees C_1 and C_2. For each case, in Propositions 3 and 4, we present a different

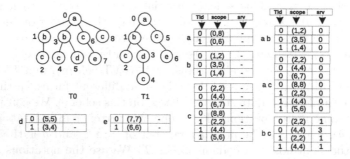

Fig. 1. An example of **EMET-List**. $T0$ and $T1$ are two database trees and minimum-support is 2. The figure presents **EMET-List**s of some of frequent patterns of size 1 or 2.

method to generate the **EMET-List** of C based on the **EMET-Lists** of C_1 and C_2. The proofs of the propositions are straight forward by Propositions 1 and 2.

Proposition 3. *Let C_1, C_2 and C be three trees such that C is generated by AD-join of C_1 and C_2. Let also em_1 be an **EMET** in the **EMET-List** of C_1 and em_2 an **EMET** in the **EMET-List** of C_2. If: (i) $em_1.TId = em_2.TId$, (ii) $em_1.scope.l < em_2.scope.l$ and (iii) $em_2.scope.u \leq em_1.scope.u$, then em with $em.TId = em_1.TId$, $em.scope = em_2.scope$ and $em.srv = em_1.scope.l$ is an element of **EMET-List(C)**.*

Every **EMET** em_1 in the **EMET-List** of C_1 and every **EMET** em_2 in the **EMET-List** of C_2 such that em_1 and em_2 have the same TId are checked to see whether or not they can join by AD-join.

Figure 2 illustrates how AD-join works on **EMET-Lists**. In this figure we want to generate the **EMET-List** of C by the AD-join of the **EMET-Lists** of C_1 and C_2. Suppose that minimum-support is 2 and the database consists of two trees $T0$ and $T1$ presented in Fig. 1. Since the rightmost vertex of C_2 is added as a child, to the rightmost vertex of C_1, two **EMET-Lists** are joined via AD-join. Consider joining the first **EMET** in **EMET-List**(C_1) and the first **EMET** in **EMET-List**(C_2), denoted by em_1 and em_2, respectively. Both em_1 and em_2 occur in $T0$, furthermore, the lower bound of the scope of em_2 (i.e., 2) is greater then the lower bound of the scope of em_1 (i.e., 1) and the upper bound of the scope of em_2 (i.e., 2) is less than or equal to the upper bound of the scope of em_1 (i.e., 2). Therefore, em_1 and em_2 can join together and in the resulted **EMET**, denoted by em, TId is set to 0, $scope$ is set to the $scope$ of em_2 (i.e., $(2,2)$), and srv is set to the lower bound of $scope$ of em_1 (i.e., 1).

Proposition 4. *Let C_1, C_2 and C be three trees such that C is generated by the relative-join of C_1 and C_2. Let also em_1 be an **EMET** in **EMET-List**(C_1) and em_2 an **EMET** in **EMET-List**(C_2). If: (i) $em_1.TId = em_2.TId$,*

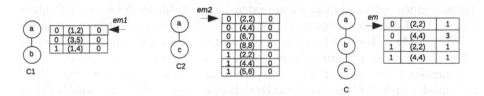

Fig. 2. AD-join of **EMET-Lists**.

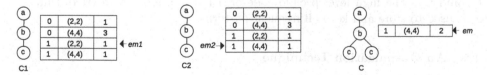

Fig. 3. Relative-join of **EMET-Lists**.

(ii) $em_1.srv = em_2.srv$ and *(iii)* $em_1.scope.u < em_2.scope.l$, then em with $em.TId = em_1.TId$ (or $em_2.TId$), $em.scope = em_2.scope$ and $em.srv = em_1.scope.l$ is an element of **EMET-List(C)**.

Every **EMET** em_1 in the **EMET-List** of C_1 and every **EMET** em_2 in the **EMET-List** of C_2 such that em_1 and em_2 have the same TId are checked to see if they can join by relative-join.

Figure 3 shows how relative-join works on **EMET-Lists**. In this figure, we want to compute the **EMET-List** of C by the relative-join of the **EMET-Lists** of C_1 and C_2. Suppose that minimum-support is 2 and the database contains two trees $T0$ and $T1$ presented in Fig. 1. Since the rightmost vertex of C_2 is not added as a child of the rightmost vertex of C_1, **EMET-Lists** of C_1 and C_2 join via *relative-join*. Consider the relative-join of the third **EMET** in **EMET-List(C_1)**, denoted by em_1, and the fourth **EMET** in **EMET-List(C_2)**, denoted by em_2. Both em_1 and em_2 occur in $T1$, they have the same srv (i.e., 1), and the lower bound of the scope of em_2 (i.e., 4) is greater than the upper bound of the scope of em_1 (i.e., 2). Thus, em_1 and em_2 can join together and in the resulted **EMET**, denoted by em, TId is set to 1, $scope$ is set to the $scope$ of em_2 (i.e., $(4, 4)$), and srv is set to the lower bound of $scope$ of em_1 (i.e., 2).

4.4   TTM: An Efficient Algorithm for Transactional Tree Mining

In this section, we introduce TTM, an efficient algorithm for finding frequent embedded tree patterns in the transactional setting. TTM (an abbreviation for Transactional Tree Miner) takes as input a minimum-support value and a forest of trees. It uses *equivalence class extension* [18] for non-redundant and complete candidate generation.

First, TTM extracts frequent patterns of size 1 (frequent vertex labels) and constructs their **EMET-Lists**. This step can be done by one scan of the database. Every **EMET** of a pattern of size 1 represents one occurrence of the pattern, where its $scope$ is the scope of the occurrence (and its srv is empty). Then, frequent patterns of size 2 are extracted. For this purpose, every frequent pattern C of size 1 is joined, via AD-join, with all other frequent patterns of size 1. Then, **EMET-Lists** of frequent patterns of size 2 are constructed by AD-join. Then, every larger pattern C is generated by either AD-join or relative-join of two smaller patterns C_1 and C_2 that are in the same equivalence class. If the rightmost vertex of C_2 is added to the rightmost vertex of C_1, the **EMET-List** of C is computed using AD-join of the **EMET-Lists** of C_1 and C_2. Otherwise, the **EMET-List** of C is computed using relative-join of the **EMET-Lists** of C_1 and C_2. The high level pseudo code of TTM is presented in Algorithm 1. $\mathcal{F}$ is used to store and keep all frequent patterns.

4.5   An Optimization Technique

For a pattern C, the size of **EMET**(C) is $O(|D| \times e)$, where e is the maximum number of occurrences, with distinct pairs *rightmost vertex* and *second rightmost vertex*,

Algorithm 1. High level pseudo code of the TTM algorithm.

1: TTM
2: **Input.** a database D consisting of trees, a user-defined integer $minsup$.
3: **Output.** $\mathcal{F}$ {the set of frequent embedded tree patterns.}
4: Compute the set $\mathcal{F}_1$ of frequent patterns of size 1 and their **EMET-Lists**
5: Compute the set $\mathcal{F}_2$ of frequent patterns of size 2 and their **EMET-Lists**
6: $\mathcal{F} \leftarrow \mathcal{F}_1 \cup \mathcal{F}_2$
7: **for each** $F \in \mathcal{F}_2$ **do**
8:     Extend($F, minsup, \mathcal{F}$)
9: **end for**

1: Extend
2: **Input.** a tree pattern C_1, a user-defined integer $minsup$.
3: **Input and Output.** $\mathcal{F}$: the set of frequent embedded tree patterns
4: **Side effect.** $\mathcal{F}$ is updated with the frequent patterns generated by the extensions of C_1
5: $\mathcal{Q} \leftarrow \emptyset$
6: **for each** $C_2 \in \mathcal{F}$ so that C_1 and C_2 are in the same equivalence class **do**
7:     {Let v_1 and v_2 be the rightmost vertices of C_1 and C_2, respectively}
8:     **if** $dep(C_1, v_1) = dep(C_2, v_2)$ **then**
9:         $C \leftarrow$ AD-join of C_1 and C_2
10:        Compute **EMET-List**(C) by AD-$join$ of **EMET-List**(C_1) and **EMET-List**(C_2)
11:        **if** $sup(C) \geq minsup$ **then**
12:            $\mathcal{F} \leftarrow \mathcal{F} \cup \{C\}, \mathcal{Q} \leftarrow \mathcal{Q} \cup \{C\}$
13:        **end if**
14:        $C \leftarrow$ relative-join of C_1 and C_2
15:        Compute **EMET-List**(C) by $relative$-$join$ of **EMET-List**(C_1) and **EMET-List**(C_2)
16:        **if** $sup(C) \geq minsup$ **then**
17:            $\mathcal{F} \leftarrow \mathcal{F} \cup \{C\}, \mathcal{Q} \leftarrow \mathcal{Q} \cup \{C\}$
18:        **end if**
19:    **end if**
20:    **if** $dep(C_1, v_1) > dep(C_2, v_2)$ **then**
21:        $C \leftarrow$ relative-join of C_1 and C_2
22:        Compute **EMET-List**(C) by $relative$-$join$ of **EMET-List**(C_1) and **EMET-List**(C_2)
23:        **if** $sup(C) \geq minsup$ **then**
24:            $\mathcal{F} \leftarrow \mathcal{F} \cup \{C\}, \mathcal{Q} \leftarrow \mathcal{Q} \cup \{C\}$
25:        **end if**
26:    **end if**
27: **end for**
28: **for each** $F \in \mathcal{Q}$ **do**
29:     Extend($F, minsup, \mathcal{F}$)
30: **end for**

that C has in a database tree. If C is generated via AD-join, e is $O(n)$, therefore, the size of **EMET**(C) is $O(|D| \times n)$. However, if C is generated via relative-join, e is in the worst case $O(n^2)$, which means the size of **EMET-List**(C) will be $O(|D| \times n^2)$.

In this section, we present an optimization technique that helps us to keep the worst case size of **EMET-List**s linear. The technique is based on storing only *uncovered* *EMET s*.

Definition 1 (Covered EMET). *Let C be a tree pattern and T a database tree. An **EMET** em in **EMET-List**(C) is covered, iff:*

1. *there exists an **EMET** $em' \neq em$ in **EMET-List**(C) such that $em.TId = em'.TId$ and $em.scope.l = em'.scope.l$, and*
2. *there exists another **EMET** $em'' \neq em$ in **EMET-List**(C) such that $em.TId = em''.TId$ and $em.srv = em''.srv$.*

*If an **EMET** is not covered, it is called uncovered.*

Clearly, a pattern C has at least one **EMET** in a database tree T if and only if it has at least one uncovered **EMET** in T. This means storing only uncovered **EMET**s suffices to count frequency of C. In the following, we show that it is also sufficient to generate uncovered **EMET**s of C using only uncovered **EMET**s of C_1 and C_2. For this purpose, we slightly revise the **EMET** data-structure by adding a new component rv_par, which contains the parent of the rightmost vertex of all occurrences represented by the **EMET**.

Algorithm 2 shows how uncovered **EMET**s of C_1 and C_2 join to generate uncovered **EMET**s of C. If C is generated via AD-join, using Proposition 1, every pair of uncovered **EMET**s of C_1 and C_2 that have the same $TIds$ are checked to see if they can form an **EMET** of C. In the case of relative-join, the procedure becomes slightly more complicated as in this case, if every pair of uncovered **EMET**s of C_1 and C_2 that have the same TId are checked to see whether they can form an **EMET** of C, some uncovered **EMET**s of C might still not be generated. An example is depicted in Fig. 4. In this figure, suppose since em_2 is covered by the other **EMET**s of C_2, it is not generated. Then, when we generate the **EMET**s of C by the relative-join of the uncovered **EMET**s of C_1 and C_2, the **EMET** em of C is not generated, even though it is uncovered. If C is further extended to the tree C' depicted in Fig. 4, the **EMET** em' of C' will not be generated, and this makes C' non-frequent, while it is a frequent pattern.

In Algorithm 2, we use a different method to generate all uncovered **EMET**s of tree patterns. Suppose trees C_1 and C_2 join to generate a tree C. First, the uncovered **EMET**s of C_2 are partitioned, so that all **EMET**s with the same TId and rv_par are put on the same partition. Then, every uncovered

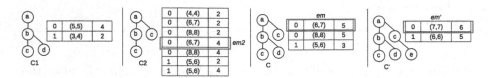

Fig. 4. An example of uncovered **EMET-List**s and their join.

Algorithm 2. High level pseudo code of the optimized subtree homeomorphism algorithm.

1: AD-join
2: **Input.** Two trees C_1 and C_2 that belong to the same equivalence class and a tree C which is generated by AD-join of C_1 and C_2.
3: **Output.** The uncovered **EMET**s of C.
4: **for each** $em_1 \in$ **EMET-List**(C_1) and $em_2 \in$ **EMET-List**(C_2) such that $em_1.TId = em_2.TId$ **do**
5:    **if** $em_1.scope.l < em_2.scope.l$ and $em_2.scope.u \le em_1.scope.u$ **then**
6:       Generate an **EMET** em with: $em.TId = em_1.TId$, $em.srv = em_1.scope.l$, $em.scope = em_2.scope$, and $em.rv\_par = em_1.scope.l$
7:       **if** em is not covered by the existing members of **EMET-List**(C) **then**
8:          Add em to **EMET-List**(C)
9:       **end if**
10:    **end if**
11: **end for**

1: Relative-join
2: **Input.** Two trees C_1 and C_2 that belong to the same equivalence class and a tree C which is generated by relative-join of C_1 and C_2.
3: **Output.** The uncovered **EMET**s of C.
4: Partition **EMET-List**(C_2) based on TId and rv_par, i.e., **EMET**s with the same TId and rv_par are put on the same partition.
5: **for each EMET** em_1 in **EMET-List**(C_1) and partition P of **EMET-List**(C_2) that has a member p such that $em_1.TId = p.TId$ and $em_1.srv = p.srv$ **do**
6:    **for each** member em_2 of P **do**
7:       **if** $em_2.scope.l > em_1.scope.u$ **then**
8:          Generate an **EMET** em with: $em.TId = em_1.TId$, $em.srv = em_1.scope.l$, $em.scope = em_2.scope$, and $em.rv\_par = em_2.rv\_par$
9:          **if** em is not covered by the existing members of **EMET-List**(C) **then**
10:            Add em to **EMET-List**(C)
11:          **end if**
12:       **end if**
13:    **end for**
14: **end for**

EMET em_1 of C_1 and every partition P of the **EMET**s of C_2 can join iff P has at least one member p such that $em_1.TId = p.TId$ and $em_1.srv = p.srv$. The join of **EMET** em_1 and partition P is done as follows: for every member em_2 of P, if the rightmost vertex of the occurrences represented by em_2 is a right relative of the rightmost vertex of the occurrences represented by em_1 (i.e., $em_2.scope.l > em_1.scope.u$), an **EMET** em of C is generated with: $em.TId = em_1.TId$, $em.srv = em_1.scope.l$, $em.scope = em_2.scope$, and $em.rv\_par = em_2.rv\_par$; and is added to the **EMET-List** of C if it is not already generated and it is not covered by the already generated **EMET**s of C. Proposition 5 shows that Algorithm 2 generates a complete and valid list of uncovered **EMET**s of C. Its proof is omitted due to lack of space.

Proposition 5. *Let C_1 and C_2 be two tree patterns in the same equivalence class and suppose that tree C is generated by the join of C_1 and C_2. Algorithm 2 generates all uncovered **EMET**s of C by joining only uncovered **EMET**s of C_1 and C_2.*

Complexity analysis. Compared to most efficient existing vertical algorithms such as TreeMinerD [18] and TPMiner [6], TTM has a better time complexity. With n the size of a database tree and C a pattern, while for the database trees studied in this paper worst case time complexity of both TreeMinerD and TPMiner is $O(n^2 \times |V(C)|)$, it is $O(n^2)$ for TTM. Note that this time complexity holds for TPMiner [6] even if it is implemented efficiently (using linked lists for the RP components of **occ-list**s to store only once, those elements of RP that also belong to the parent pattern) so that its frequency counting data structure becomes linear for the trees studied in this paper.

Algorithms that perform a homeomorphism test from the scratch between two trees have a better worst case time complexity. For example, the algorithm of Bille and Gortz [2] runs in linear space and subquadratic time. However, in practice *vertical* approaches that store compact information for each pattern C and when C is extended to a larger pattern C', the stored information is also extended for C', are more efficient. The reason is that in such approaches frequency of many patterns can be easily computed by extending a little information stored for the parent pattern. An empirical comparison of these two types of methods can be found in e.g., [18].

5   Experimental Results

We performed extensive experiments to evaluate the effectiveness of TTM, using data from real applications as well as synthetic datasets. The experiments were done on one core of a single AMD Processor 270 clocked at 2.0 GHz with 16 GB main memory. All programs were compiled by the GNU C++ compiler 4.8.4. TreeMinerD [18] is the state of the art algorithm for finding frequent embedded patterns. Therefore, we choose it for our comparisons. Based on our best knowledge and to the time of writing this paper, MB3Miner-T [13] is the most efficient recent algorithm for finding frequent embedded tree patterns. Tatikonda et al. [14] proposed algorithms for parallel mining of trees on multicore systems. Since we do not aim at parallel tree mining, their algorithms are not proper for our comparisons. TPMiner [6] is developed to count all occurrences, hence it is not proper for our comparisons.

In our experiments, we state minimum support as a percentage of the number of trees in the database. We test the algorithms at very low values of minimum support: in all datasets, except NASA, the values of minimum support are less than 1 %. When choosing minimum support values, we consider the following: on the one hand, at least one of the algorithms produce the output and on the other hand, the mining setting does not become trivial such that all algorithms produce the output very quickly. Hence, we may use different minimum support values for different datasets.

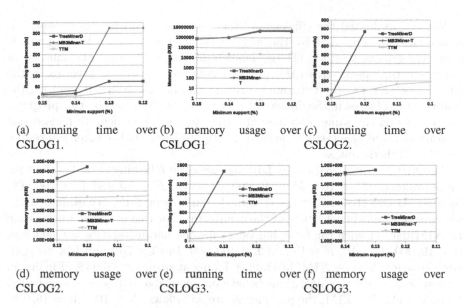

(a) running time over CSLOG1.

(b) memory usage over CSLOG1

(c) running time over CSLOG2.

(d) memory usage over CSLOG2.

(e) running time over CSLOG3.

(f) memory usage over CSLOG3.

Fig. 5. Comparison over the CSLOG datasets.

The first real-world dataset is CSLOGS that contains the web access trees of the CS department of the Rensselaer Polytechnic Institute [18]. It has 59,691 trees, 716,263 vertices and 13,209 unique vertex labels. Each distinct label corresponds to the URLs of a web page. In [19], the log file of every week is separated into a different dataset and 3 different datasets are generated: CSLOG1 for the first week, CSLOG2 for the second week and CSLOG3 for the third week. CSLOG1 contains 8,074 trees, CSLOG2 contains 7,404 trees and CSLOG3 contains 7,628 trees. Figure 5 compares TTM against TreeMinerD and MB3Miner-T over the CSLOG datasets.

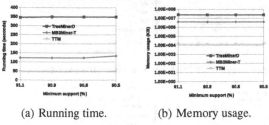

(a) Running time.

(b) Memory usage.

Fig. 6. Comparison over NASA.

The next real-world dataset used in this paper consists of MBONE multicast data that was measured during the NASA shuttle launch between the 14th and 21st of February, 1999 [3,4]. It has 333 distinct vertex labels where every vertex label is an IP address. The data was sampled from this NASA dataset with 10 min sampling interval and has 1,000 trees. It is dense in the sense that there exist strong correlations among database trees. Hence, a lot of large frequent patterns are found at high values of minimum support. Figure 6 compares the algorithms over this dataset.

We also evaluated the efficiency of the proposed algorithm on several synthetic datasets generated by the method described in [16]. The program is
adjusted by 5 parameters: (1) the number of labels (N), (2) the number of
vertices in the master tree (M), (3) the maximum fan-out of a vertex in the
master tree (F), (4) the maximum depth of the master tree (D), and (5) the
total number of trees in the dataset (T). In the trees generated by this program,
two vertices on the same path may find the same label. Hence, we revised this
program to avoid generating of such trees. Figure 7 reports the empirical results
over the synthetic datasets.

The first synthetic dataset is D10 and uses the following default values for
the parameters: $N = 100$, $M = 10,000$, $D = 10$, $F = 10$ and $T = 100,000$.
Figures 7(a) and (b) present the empirical results over D10. We evaluated the
influence of N on the efficiency of the algorithms. For this purpose, we generate
the N1M dataset with the following values for the parameters: $N = 1,000,000$,
$M = 10,000$, $D = 10$, $F = 10$ and $T = 100,000$. In Figs. 7(c) and (d), we compare the algorithms on N1M. MB3Miner-T fails at $minsup = 0.03\,\%$. Finally, in
order to study how the algorithms behave on very large datasets, we evaluated
them on T1M. For T1M, the parameters are as follows: $N = 100$, $M = 10,000$,
$D = 10$, $F = 10$, and $T = 1,000,000$. Figures 7(e) and (f) compare the algorithms over this dataset.

Discussion. Our extensive experiments show that TTM significantly outperforms
well-known existing algorithms. It is due to the more effective subtree homeomorphism algorithm it uses. Moreover, TTM does not generate the tree patterns
that have two vertices on the same path with the same label. One can observe
the followings in our experiments. On the one hand, over CSLOG2, CSLOG3,

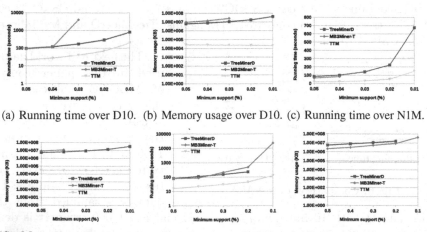

(a) Running time over D10. (b) Memory usage over D10. (c) Running time over N1M.

(d) Memory usage over N1M. (e) Running time over TM. (f) Memory usage over T1M.

Fig. 7. Comparison over synthetic datasets.

D10 and N1M, MB3Miner-T behaves poorly in terms of both running time and memory usage. The reason is that the Tree Model Guided technique used by MB3Miner-T does not significantly reduce the number of candidates that are generated but are infrequent. On the other hand, over CSLOG2, CSLOG3 and T1M, TreeMinerD shows a poor performance. This is due to efficiency of the subtree homeomorphism algorithm used by TTM.

6   Conclusion

In this paper, we presented a new algorithm for deciding whether a tree pattern is subtree homeomorphic to a database tree. Our method is based on a compact data-structure, called **EMET**, that encodes and stores all information required for the subtree homeomorphism test. **EMET**s of larger patterns are generated by efficient join operations, defined on the **EMET**s of their proper subtrees. Based on the proposed subtree homeomorphism method, we introduced TTM, an effective algorithm for finding frequent tree patterns from tree databases. We evaluated the efficiency of TTM over several real-world and synthetic datasets and showed its high efficiency compared to well-known existing algorithms.

Acknowledgements. We are thankful to Prof. Mohammed Javeed Zaki for providing the TreeMinerD code, the CSLOGS datasets and the TreeGenerator program, to Dr Henry Tan for providing the MB3Miner-T code, to Professor Jun-Hong Cui for providing the NASA dataset.

References

1. Asai, T., Abe, K., Kawasoe, S., Arimura, H., Satamoto, H., Arikawa, S.: Efficient substructure discovery from large semi-structured data. In: Proceedings of the Second SIAM International Conference on Data Mining (SDM), pp. 158–174. SIAM (2002)
2. Bille, P., Gortz, I.: The tree inclusion problem: in linear space and faster. ACM Trans. Algorithms **7**(3), 1–47 (2011)
3. Chalmers, R., Almeroth, K.: Modeling the branching characteristics and efficiency gains of global multicast trees. In: Proceedings of the 20th IEEE International Conference on Computer Communications (INFOCOM), pp. 449–458 (2001)
4. Chalmers, R.C., Member, S., Almeroth, K.C.: On the topology of multicast trees. IEEE/ACM Trans. Networking **11**, 153–165 (2003)
5. Chehreghani, M.H.: Efficiently mining unordered trees. In: Proceedings of the 11th IEEE International Conference on Data Mining (ICDM), pp. 111–120 (2011)
6. Chehreghani, M.H., Bruynooghe, M.: Mining rooted ordered trees under subtree homeomorphism. Data Mining and Knowledge Discovery, to appear. http://arxiv.org/abs/1412.1470, doi:10.1007/s10618-015-0439-5
7. Chehreghani, M.H., Chehreghani, M.H., Lucas, C., Rahgozar, M.: OInduced: an efficient algorithm for mining induced patterns from rooted ordered trees. IEEE Trans. Syst. Man Cybern., Part A **41**(5), 1013–1025 (2011)
8. Chi, Y., Yang, Y., Muntz, R.R.: Indexing and mining free trees. In: Proceedings of the Third IEEE International Conference on Data Mining (ICDM), pp. 509–512 (2003)

9. Chi, Y., Yang, Y., Xia, Y., Muntz, R.R.: CMTreeMiner: mining both closed and maximal frequent subtrees. In: Bailey, J., Khan, L., Washio, T., Dobbie, G., Huang, J.Z., Wang, R. (eds.) PAKDD 2016. LNCS (LNAI), vol. 9651, pp. 63–73. Springer, Heidelberg (2004). doi:10.1007/978-3-540-24775-3_9

10. Diestel, R.: Graph Theory, 4th ed. Springer, Heidelberg (2010)

11. Dietz, P.F.: Maintaining order in a linked list. In: Proceedings of the 14th ACM Symposium on Theory of Computing (STOC), pp. 122–127 (1982)

12. Sidhu, A.S., Dillon, T.S., Chang, E.: Protein ontology. In: Ma, Z., Chen, J.Y. (eds.) Database Modeling in Biology: Practices and Challenges, pp. 39–60. Springer, New York (2006)

13. Tan, H., Hadzic, F., Dillon, T.S., Chang, E., Feng, L.: Tree model guided candidate generation for mining frequent subtrees from XML documents. ACM Trans. Knowl. Discov. Data (TKDD) 2(2), 43 (2008)

14. Tatikonda, S., Parthasarathy, S.: Mining tree-structured data on multicore systems. Proc. VLDB Endowment (PVLDB) 2(1), 694–705 (2009)

15. Tatikonda, S., Parthasarathy, S., Kurc, T.M.: TRIPS and TIDES: new algorithms for tree mining. In: Proceedings of the 15th ACM International Conference on Information and Knowledge Management (CIKM), pp. 455–464 (2006)

16. Zaki, M.J.: Efficiently mining frequent trees in a forest. In: Proceedings of the Eighth ACM SIGKDD International Conference on Knowledge Discovery and Data Mining (KDD), pp. 71–80 (2002)

17. Zaki, M.J.: Efficiently mining frequent embedded unordered trees. Fundamenta Informaticae 66(1–2), 33–52 (2005)

18. Zaki, M.J.: Efficiently mining frequent trees in a forest: algorithms and applications. IEEE Trans. Knowl. Data Eng. 17(8), 1021–1035 (2005)

19. Zaki, M.J., Aggarwal, C.C.: XRules: an effective algorithm for structural classification of XML data. Mach. Learn. 62(1–2), 137–170 (2006)

Two-Stage Transfer Surrogate Model
for Automatic Hyperparameter Optimization

Martin Wistuba$^{(\boxtimes)}$, Nicolas Schilling, and Lars Schmidt-Thieme

Information Systems and Machine Learning Lab, University of Hildesheim,
Universitätsplatz 1, 31141 Hildesheim, Germany
{wistuba,schilling,schmidt-thieme}@ismll.uni-hildesheim.de

Abstract. The choice of hyperparameters and the selection of algorithms is a crucial part in machine learning. Bayesian optimization methods have been used very successfully to tune hyperparameters automatically, in many cases even being able to outperform the human expert. Recently, these techniques have been massively improved by using meta-knowledge. The idea is to use knowledge of the performance of an algorithm on given other data sets to automatically accelerate the hyperparameter optimization for a new data set.

In this work we present a model that transfers this knowledge in two stages. At the first stage, the function that maps hyperparameter configurations to hold-out validation performances is approximated for previously seen data sets. At the second stage, these approximations are combined to rank the hyperparameter configurations for a new data set. In extensive experiments on the problem of hyperparameter optimization as well as the problem of combined algorithm selection and hyperparameter optimization, we are outperforming the state of the art methods. The software related to this paper is available at https://github.com/wistuba/TST.

Keywords: Hyperparameter optimization · Meta-learning · Transfer learning

1 Introduction

The tuning of hyperparameters is an omnipresent problem in the machine learning community. In comparison to model parameters, which are estimated by a learning algorithm, hyperparameters are parameters that have to be specified before the execution of the algorithm. Typical examples for hyperparameters are the trade-off parameter C of a support vector machine or the number of layers and nodes in a neural network. Unfortunately, the choice of the hyperparameters is crucial and decides whether the performance of an algorithm is state of the art or just moderate. Hence, the task of hyperparameter optimization is as important as developing new models [2,5,18,23,27].

The traditional way of finding good hyperparameter configurations is by using a combination of manual and grid search. This procedure are a brute force approach of searching the hyperparameter space. They are very time-consuming or even infeasible for high-dimensional hyperparameter spaces. Therefore, methods

© Springer International Publishing AG 2016
P. Frasconi et al. (Eds.): ECML PKDD 2016, Part I, LNAI 9851, pp. 199–214, 2016.
DOI: 10.1007/978-3-319-46128-1_13

to steer the search for good hyperparameter configurations are currently an interesting topic for researchers [3,23,27].

Sequential model-based optimization (SMBO) [13] is a black-box optimization framework and is currently the state of the art for automatic hyperparameter optimization. Within this framework, the trials of already tested hyperparameter configurations are used to approximate the true hyperparameter response function using a surrogate model. Based on this approximation, a promising new hyperparameter configuration is chosen and tested in the next step. The result of this next trial is then used to update the surrogate model for further hyperparameter configuration acquisitions.

Human experts utilize their experience with a machine learning model and try hyperparameter configurations that have been good on *other* data sets. This transfer of knowledge is one important research direction in the domain of automatic hyperparameter optimization. Currently, two different approaches to integrate this idea into the SMBO framework exist. Either by training the surrogate model on past experiments [1,21,25,33], or by using the information on past experiments to initialize the new search [7,32].

We propose a two-stage approach to consider the experiences with different hyperparameter configurations on other data sets. At the first stage, we approximate the hyperparameter response function of the new data set as well as of previous data sets. This approximation is then combined to rank the hyperparameter configurations for the new data set, considering the similarity between the new data set and the previous ones. In two extensive experiments for the problem of hyperparameter optimization and the problem of combined algorithm selection and hyperparameter optimization, we show that our two-stage approach is able to outperform current state of the art competitor methods, which have been recently published on established machine learning conferences.

2   Related Work

The aim of automatic hyperparameter optimization is to enable non-experts to successfully use machine learning models but also to accelerate the process of finding good hyperparameter configurations. Sequential model-based optimization (SMBO) is the current state of the art for automatic hyperparameter optimization. Various approaches exist to accelerate the search for good hyperparameter configurations. One important approach is the use of meta-knowledge. This approach has already proven its benefit for other hyperparameter optimization approaches [9,16,20,29]. One easy way to make use of meta-knowledge is through initialization. This approach is universal and can be applied for every hyperparameter optimization method. Reif et al. [20] suggest to choose those hyperparameter configurations for a new data set as initial trials that performed best on a similar data set in the context of evolutionary parameter optimization. Here, the similarity was defined through the distance among meta-features, which describe properties of a data set. This idea was applied to SMBO by Feurer et al. [7] and later improved [8]. Recently, it was proposed to learn initial hyperparameter configurations in such a way that it is no longer necessary to be limited to choose initial hyperparameter configurations from the set of hyperparameter configurations, which have been chosen in previous experiments [32].

While the initialization can be used for any hyperparameter optimization method, the idea to use transfer surrogate models is specific for the SMBO framework. Bardenet et al. [1] were the first who proposed to learn the surrogate model not only on the current data set but also over previous experiments in order to make use of the meta-knowledge. Soon, this idea was further investigated: specific Gaussian processes [25,33] and neural networks [21] were proposed as surrogate models.

The aforementioned ideas make use of meta-knowledge to accelerate the search for good hyperparameter configurations. Another way of saving time is to stop an hyperparameter configuration evaluation early if it appears to be not promising after few training iterations. Obviously, this is only possible for iterative learning algorithms, which are using gradient-based optimization. Even though this approach is orthogonal to the meta-learning approach, the aim is the same, i.e. accelerating the search for good hyperparameter configurations. Domhan et al. [6] propose to predict the development of the learning curve based on few iterations. If the predicted development is less promising than the currently best configuration, the currently investigated configuration is discarded. A similar approach is proposed by Swersky et al. [26]. Instead of trying different configurations sequentially and eventually discarding them, they learn the models for various hyperparameter configurations at the same time and switch from one learning process to the other if it looks more promising.

3 Background

In this section the hyperparameter optimization problem is formally defined and, for the sake of completeness, the sequential model-based optimization framework is presented.

3.1 Hyperparameter Optimization Problem Setup

A machine learning algorithm $\mathcal{A}_\lambda$ is a mapping $\mathcal{A}_\lambda : \mathcal{D} \to \mathcal{M}$ where $\mathcal{D}$ is the set of all data sets, $\mathcal{M}$ is the space of all models and $\lambda \in \Lambda$ is the chosen hyperparameter configuration with $\Lambda = \Lambda_1 \times \ldots \times \Lambda_P$ being the P-dimensional hyperparameter space. The learning algorithm estimates a model $M_\lambda \in \mathcal{M}$, which minimizes a loss function $\mathcal{L}$ (e.g. residual sum of squares), that is penalized with a regularization term $\mathcal{R}$ (e.g. Tikhonov regularization) with respect to the training set D^{train} of the data set D:

$$\mathcal{A}_\lambda \left(D^{\text{train}} \right) = \arg \min_{M_\lambda \in \mathcal{M}} \mathcal{L} \left(M_\lambda, D^{\text{train}} \right) + \mathcal{R} \left(M_\lambda \right). \tag{1}$$

Then, the task of *hyperparameter optimization* is to find the hyperparameter configuration λ^* that leads to a model M_{λ^*}, which minimizes the loss on the validation data set D^{valid}, i.e.

$$\lambda^* = \arg \min_{\lambda \in \Lambda} \mathcal{L} \left(\mathcal{A}_\lambda \left(D^{\text{train}} \right), D^{\text{valid}} \right) = \arg \min_{\lambda \in \Lambda} f_D \left(\lambda \right). \tag{2}$$

The function f_D is the *hyperparameter response function* of data set D.

$$f_D(\lambda) = \mathcal{L}\left(\mathcal{A}_\lambda\left(D^{\text{train}}\right), D^{\text{valid}}\right) \tag{3}$$

For the sake of demonstration, in the remaining sections, we consider the problem of tuning the hyperparameters of a classifier. Thus, f returns the misclassification rate. This is obviously no limitation, but shall help the reader to understand the concepts given a concrete example.

3.2   Sequential Model-Based Optimization

Sequential model-based optimization (SMBO) [13], originally proposed for black-box optimization, can be used for optimizing hyperparameters automatically by using the SMBO framework to minimize the hyperparameter response function (Eq. 3) [2]. SMBO consists of two components, (i) a surrogate model Ψ, that is used to approximate the function f, which we want to minimize, and (ii) an acquisition function a, that decides which hyperparameter to try next.

Algorithm 1 outlines the SMBO framework for minimizing the function f. For T many iterations different hyperparameters are tried. In iteration t, we approximate f using our surrogate model Ψ_{t+1} based on the observation history $\mathcal{H}_t$, the set of all hyperparameter configurations and performances, which have been evaluated evaluated so far. The surrogate model is an approximation of f with the property that it can be evaluated fast. Based on the predictions of Ψ and the corresponding uncertainties about these predictions, the acquisition function finds a trade-off between exploitation and exploration and determines the hyperparameter configuration to try next. This configuration is then evaluated, and the new observation is added to the observation history. After T trials, the best performing hyperparameter configuration is returned.

Algorithm 1. Sequential Model-based Optimization

Input: Hyperparameter space Λ, observation history $\mathcal{H}$, number of trials T, acquisition
function a, surrogate model Ψ.
Output: Best hyperparameter configuration found.
1: **for** $t = 1$ to T **do**
2:     Fit Ψ_{t+1} to $\mathcal{H}_t$
3:     $\lambda \leftarrow \arg\max_{\lambda \in \Lambda} a\left(\mu\left(\Psi_{t+1}\left(\lambda\right)\right), \sigma\left(\Psi_{t+1}\left(\lambda\right)\right), f^{\min}\right)$
4:     Evaluate $f(\lambda)$
5:     $\mathcal{H}_{t+1} \leftarrow \mathcal{H}_t \cup \{(\lambda, f(\lambda))\}$
6:     **if** $f(\lambda) < f^{\min}$ **then**
7:         $\lambda^{\min}, f^{\min} \leftarrow \lambda, f(\lambda)$
8: **return** $\lambda^{\min}$

Since the acquisition function a needs some certainty about the prediction, common choices are Gaussian processes [1,23,25,33] or ensembles, such as random forests [12]. Typical acquisition functions are the expected improvement

[13], the probability of improvement [13], the conditional entropy of the minimizer [28] or a multi-armed bandit based criterion [24]. The expected improvement is the most prominent choice for hyperparameter optimization and is also the acquisition function, which we choose. Formally, the improvement for a hyperparameter configuration λ is defined as

$$I(\lambda) = \max \{ f^{\min} - Y, 0 \} \tag{4}$$

where $f^{\min}$ is currently the best function value and Y is a random variable modeling our knowledge about the value of the function f for the hyperparameter configuration λ, which depends on $\mathcal{H}_t$. The hyperparameter configuration with highest expected improvement, i.e.

$$\mathrm{E}\left[I(\lambda)\right] = \mathrm{E}\left[\max \{ f^{\min} - Y, 0 \} \mid \mathcal{H}_t\right], \tag{5}$$

is chosen for the next evaluation. Assuming $Y \sim \mathcal{N}\left(\mu\left(\Psi_{t+1}(\lambda)\right), \sigma^2\left(\Psi_{t+1}(\lambda)\right)\right)$, the expected improvement can be formulated in closed-form as

$$\mathrm{E}\left[I(\lambda)\right] = \begin{cases} \sigma\left(\Psi_{t+1}(\lambda)\right)\left(Z \cdot \Phi(Z) + \phi(Z)\right) & \text{if } \sigma\left(\Psi_{t+1}(\lambda)\right) > 0 \\ 0 & \text{otherwise} \end{cases} \tag{6}$$

where

$$Z = \frac{f^{\min} - \mu\left(\Psi_{t+1}(\lambda)\right)}{\sigma\left(\Psi_{t+1}(\lambda)\right)} \tag{7}$$

where $\phi(\cdot)$ and $\Phi(\cdot)$ denote the standard normal density and distribution function, and $\mu\left(\Psi_{t+1}(\lambda)\right)$ and $\sigma\left(\Psi_{t+1}(\lambda)\right)$ are the expected value and the standard deviation of the prediction $\Psi_{t+1}(\lambda)$.

4   Two-Stage Surrogate Model

Our proposed two-stage surrogate model is explained in this section. The first stage of the surrogate model approximates the hyperparameter response functions of a new data set and each data set from the meta-data individually with Gaussian processes. The second stage combines the first-stage models by taking the similarity between the new data set and the data set from previous experiments into consideration. We construct a ranking of hyperparameter configurations as well as a prediction about the uncertainty of this ranking. The proposed two-stage architecture is visualized in Fig. 1.

4.1   Notation

In the following, the prefix *meta* is used to distinguish between the different learning problems. The traditional problem is to learn some parameters θ on a given data set containing instances with predictors. For the hyperparameter optimization problem you can create *meta-data sets* consisting of *meta-instances* with *meta-predictors*. A meta-data set contains meta-instances $\left(\lambda_i, f_D(\lambda_i)\right)$ where

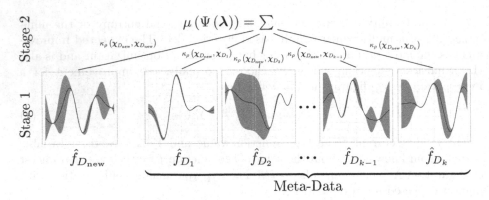

Fig. 1. At the first stage the hyperparameter response functions of the new data set D_{new} as well as data sets $\mathcal{D} = \{D_1, \ldots, D_k\}$ used for previous experiments are approximated using known evaluations. At the second stage the predictions of each individual model $\hat{f}_D$ are taken into account weighted by the similarity between D and D_{new} to determine the final predicted score.

$f_D(\boldsymbol{\lambda}_i)$ is the target and $\boldsymbol{\lambda}_i$ are the predictors. The hyperparameter response function $f_D : \Lambda \to \mathbb{R}$ is a function for a specific classifier and a specific data set D. For a given hyperparameter configuration, it returns the misclassification rate after training the classifier with the respective hyperparameter configuration on the training data set D. The task is to find a good hyperparameter configurations on a new data set D_{new} within T trials. To achieve this, a meta-data set, i.e. meta-instances for other data sets $D \in \mathcal{D}$, is given and this knowledge is transferred to the new problem.

4.2 First Stage - Hyperparameter Response Function Approximation

The first stage of our two-stage surrogate model approximates the hyperparameter response function for each data set. The meta-data set can be used to approximate the hyperparameter response function f_D for all $D \in \mathcal{D}$ by learning a machine learning model $\hat{f}_D$, using the meta-instances of each data set D. Similarly, we can learn an approximation $\hat{f}_{D_{\text{new}}}$ for the new data set, for which we have only few, but a growing number of meta-instances. Before learning $\hat{f}_D$ for all $D \in \mathcal{D}$, the labels of the meta-instances are scaled to $[0, 1]$ per data set. This is done such that each data set has equal influence on the second stage. The labels of the new data set D_{new} remain untouched.

For approximating the hyperparameter response function f_D, any machine learning model can be used, which is able to capture high non-linearity. We decide to use Gaussian processes [19], which are a very prominent surrogate model for SMBO [1, 23, 25, 33].

4.3   Second Stage - Final Hyperparameter Configuration Ranking

The second stage combines all models of the first stage within one surrogate model Ψ to rank the different hyperparameter configurations and predict the uncertainty about the ranking. The predicted score of a hyperparameter configuration is determined using kernel regression [11]. We use the Nadaraya-Watson kernel-weighted average to predict the mean value of the surrogate model

$$\mu\left(\Psi\left(\boldsymbol{\lambda}\right)\right) = \frac{\sum_{D\in\mathcal{D}\cup\{D_{\text{new}}\}} \kappa_\rho\left(\chi_{D_{\text{new}}}, \chi_D\right) \hat{f}_D\left(\boldsymbol{\lambda}\right)}{\sum_{D\in\mathcal{D}\cup\{D_{\text{new}}\}} \kappa_\rho\left(\chi_{D_{\text{new}}}, \chi_D\right)} \tag{8}$$

with the Epanechnikov quadratic kernel

$$\kappa_\rho\left(\chi_D, \chi_{D'}\right) = \delta\left(\frac{\|\chi_D - \chi_{D'}\|_2}{\rho}\right) \tag{9}$$

with

$$\delta\left(t\right) = \begin{cases} \frac{3}{4}\left(1 - t^2\right) & \text{if } t \leq 1 \\ 0 & \text{otherwise} \end{cases} \tag{10}$$

where $\rho > 0$ is the bandwidth and χ_D is a vector describing the data set D. We discuss the description of data sets in-depth in the next section.

The predicted uncertainty for a hyperparameter configuration $\boldsymbol{\lambda}$ is defined as

$$\sigma\left(\Psi\left(\boldsymbol{\lambda}\right)\right) = \sigma\left(\hat{f}_{D_{\text{new}}}\left(\boldsymbol{\lambda}\right)\right) \tag{11}$$

Using Eqs. 8 and 11, the expected improvement for arbitrary hyperparameter configurations can be estimated. Thus, our Two-Stage Transfer surrogate model Ψ can be used within the SMBO framework described in Algorithm 1.

4.4   Data Set Description

In this section we introduce three different ways to describe data sets in vector form.

Description Using Meta-features. The most popular way to describe data sets is by utilizing meta-features [1, 20, 22]. These are simple, statistical or information theoretic properties extracted from the data set. The similarity between two data sets, as defined in Eq. 9, is then dependent on the Euclidean distance between the meta-features of the corresponding data sets. In this work we are using the meta-features listed in Table 1. For a more detailed explanation, we refer the reader to Michie et al. [17]. A well-known problem with meta-features is that it is a difficult problem to find and choose meta-features that are able to adequately describe a data set [15].

Table 1. The list of all meta-features used by us.

Meta-features	
Number of classes	Class probability max
Number of instances	Class probability mean
Log number of instances	Class probability standard deviation
Number of features	Kurtosis min
Log number of features	Kurtosis max
Data set dimensionality	Kurtosis mean
Log data set dimensionality	Kurtosis standard deviation
Inverse data set dimensionality	Skewness min
Log inverse data set dimensionality	Skewness max
Class cross entropy	Skewness mean
Class probability min	Skewness standard deviation

Description Using Pairwise Hyperparameter Performance Rankings.
Describing data sets based on pairwise hyperparameter performance rankings
has been used in few approaches [16,29]. The idea is to select all paired combinations of hyperparameter configurations (λ_i, λ_j) evaluated on the new data
set D_{new} and estimate how often two data sets D and D' agree on the ranking.
Usually, it is assumed that the hyperparameter configurations evaluated on D_{new}
have also been evaluated on all data sets from the meta-data set $D \in \mathcal{D}$. In the
context of general hyperparameter tuning, this is likely not the case. Therefore,
we propose to use the first stage predictors to approximate the performances to
overcome this problem.

Formally, given t many observations on D_{new} for the hyperparameter config-
urations $\lambda_1, \ldots, \lambda_t$, D_{new} can be described as $\chi_{D_{\text{new}}} = ((\chi_{D_{\text{new}}})_i)_{i=1,\ldots,t^2} \in \mathbb{R}^{t^2}$,

$$\left(\chi_{D_{\text{new}}}\right)_{j+(i-1)t} = \begin{cases} 1 & \text{if } f_{D_{\text{new}}}(\lambda_i) > f_{D_{\text{new}}}(\lambda_j) \\ 0 & \text{otherwise} \end{cases}. \tag{12}$$

Similarly, using the same t hyperparameter configurations, we can define for all
$D \in \mathcal{D}$

$$\left(\chi_D\right)_{j+(i-1)t} = \begin{cases} 1 & \text{if } \hat{f}_D(\lambda_i) > \hat{f}_D(\lambda_j) \\ 0 & \text{otherwise} \end{cases}. \tag{13}$$

Please note that we use $\hat{f}$ instead of f. As explained before, a hyperparameter
configuration that is evaluated on D_{new} was likely not evaluated on all data
sets $D \in \mathcal{D}$. For this reason we predict the performance using the first stage
predictors. Using this description, the Euclidean distance between two data sets
is the number of discordant pairs [14].

5   Experimental Evaluation

5.1   Tuning Strategies

We introduce all tuning strategies considered in the experiments. We consider strategies that do not use knowledge from previous experiments as well as those that use it.

Random Search. As the name suggests, this strategy chooses hyperparameter configurations at random. Bergstra and Bengio [3] have shown that this outperforms grid search in scenarios with hyperparameters with low effective dimensionality.

Independent Gaussian Process (I-GP). This tuning strategy uses a Gaussian process with squared-exponential kernel with automatic relevance determination (SE-ARD) as a surrogate model [23]. It only uses knowledge from the current data set and is not using any knowledge from previous experiments.

Spearmint. While I-GP is our own implementation of SMBO with a Gaussian process as a surrogate model we also compare to the implementation by Snoek et al. [23]. The main difference to I-GP is the use of the Matérn 5/2 kernel. We added this as a baseline because it is considered to be a very strong baseline.

Independent Random Forest (I-RF). Besides Gaussian processes, random forests are another popular surrogate model [12], which we compare against in the experiments. We compared our own implementation against the original implementation of SMAC. Since our implementation provided stronger results, we will report these results. No knowledge from previous experiments is employed.

Surrogate Collaborative Tuning (SCoT). SCoT [1] uses meta-knowledge in a two step approach. In the first step, an SVMRank is learned over the whole meta-data. Its prediction for the meta-instances are used to replace the labels of the meta-instances. Bardenet et al. [1] argue that this overcomes the problem of having data sets with different scales of labels. On this transformed meta-data set, a Gaussian process with SE-ARD kernel is trained. In the original work it was proposed to use an RBF kernel for SVMRank. For reasons of computational complexity, we follow the lead of Yogatama and Mann [33] and use a linear kernel instead.

Gaussian Process with Multi-Kernel Learning (MKL-GP). Yogatama and Mann [33] propose to use a Gaussian process as a surrogate model for the SMBO framework. To tackle the problem of different scales on different data sets they are normalizing the data. Furthermore, they are using a kernel which is a linear combination of an SE-ARD kernel and a kernel modeling the distance between data sets.

Factorized Multilayer Perceptron (FMLP). FMLP [21] is the most recent surrogate model we are aware of. Published on last year's ECML PKDD, it is using a specific neural network to learn the similarity between the new data set and those from previous ones implicitly in a latent representation.

Two-Stage Transfer Surrogate (TST). This is the surrogate model proposed by us in this work. We consider two variations with two different data set representations. TST-M is using the meta-feature representation for the data sets, TST-R is using the pairwise ranking representation. We are using SE-ARD kernels for the Gaussian processes.

The kernel parameters are learned by maximizing the marginal likelihood on the meta-training set [19]. All hyperparameters of the tuning strategies are optimized in a leave-one-data-set-out cross-validation on the meta-training set.

The results reported estimated using a leave-one-data-set-out cross-validation and are the average of ten repetitions. For strategies with random initialization (Random, I-GP, Spearmint, I-RF), we report the average over thousand repetitions due to the higher variance. Hyperparameter configurations are limited to the precomputed grid which makes the experiment computational feasible for our infrastructure. We do not believe that limiting the black-box search to a grid has any impact on the results. In the end, this can be considered as additional constraints on the search space. In practice, our surrogate model allows finding arbitrary hyperparameter configurations like all other competitor methods. The evaluation was committed in the same way for transferring and non-transferring methods. Meta-hyperparameters for the surrogate models were individually tuned. For those strategies that use meta-features (SCoT, MKL-GP, TST-M), we use those meta-features that are described in Table 1.

5.2 Meta-data Sets

We use two meta-data set introduced in [31] but increase the number of meta-features from three to the 22 listed in Table 1. The support vector machine (SVM) meta-data set was created using 50 classification data sets chosen at random from the UCI repository. Existing train/test splits were merged, shuffled and split into 80 % train and 20 % test.

The SVM [4] was trained for three different kernels (linear, polynomial and Gaussian) such that the hyperparameter dimension is six. Three dimensions are used for kernel indicator variables, one for the trade-off parameter C, one for the degree of the polynomial kernel d and one for the width γ of the Gaussian kernel. If a hyperparameter was not involved, its value was set to 0. The misclassification error was precomputed on the grid $C \in \{2^{-5}, \ldots, 2^6\}$, $d \in \{2, \ldots, 10\}$ and $\gamma \in \{10^{-4}, 10^{-3}, 10^{-2}, 0.05, 0.1, 0.5, 1, 2, 5, 10, 20, 50, 10^2, 10^3\}$ resulting into 288 meta-instances per data set. Creating this meta-data set took about 160 CPU hours.

Furthermore, we use a Weka meta-data set to evaluate on the combined algorithm and hyperparameter configuration problem as tackled in [27]. 59 classification data sets are preprocessed as done for the SVM meta-data set. Using

19 different Weka classifiers [10], we precomputed the misclassification error on a grid which resulted into 21,871 hyperparameter configurations per data set such that the overall meta-data contains 1,290,389 meta-instances. It took us more than 891 CPU hours to create this meta-data set.

To show that the tuning strategies can also deal with hyperparameter configurations they have never seen on other data sets, the tuning strategies only have access on meta-instances on a subset of the meta-instances. The evaluation on meta-test was done using all meta-instances.

To enable reproducibility, we provide a detailed description of the meta-data sets, the meta-data sets itself and our source code on GitHub [30].

5.3 Evaluation Metrics

We compare all tuning methods with respect to two common evaluation metrics: average rank and average distance to the global minimum. The average rank ranks the tuning strategies per data set according to the best found hyperparameter configuration. These ranks are then averaged over all data sets. The average distance to the global minimum after t trials is defined as

$$\text{ADTM}\left(\Lambda_t, \mathcal{D}\right) = \sum_{D \in \mathcal{D}} \min_{\lambda \in \Lambda_t} \frac{f_D\left(\lambda\right) - f_D^{\min}}{f_D^{\max} - f_D^{\min}} \tag{14}$$

where $f_D^{\max}$ and $f_D^{\min}$ are the worst and best value on the precomputed grid, respectively. Λ_t is the set of hyperparameter configurations, that have been evaluated in the first t trials. The performance per data set is scaled between 0 and 1 to get rid of the influence of different misclassification offsets and scales. Finally, the distances between the performance of the best performing hyperparameter configuration found to the best possible performance on the grid is averaged over all data sets.

5.4 Experiments

We compare the different hyperparameter optimization methods in two different scenarios: (i) hyperparameter tuning and (ii) combined algorithm selection and hyperparameter tuning. For the task of hyperparameter tuning, we optimize the hyperparameters of a support vector machine. The results are summarized in Fig. 2. What we can see is that TST-R is outperforming the competitor methods with respect to both evaluation metrics by a large margin. TST-M has a similar good start as TST-R but its performance degenerates after few trials. Because the only difference between TST-R and TST-M is the way the data sets are described, one might argue that meta-features are less descriptive in describing a data set than the approach of pairwise rankings. We do not think that one can infer this from these results. The true reason for this behavior is that the distances for TST-R are updated after each trials and the distance to the data sets from previous experiments is increasing over time. Thus, the influence of the meta-data set vanishes and TST-R is focusing only on the knowledge about the new data set at some point of time. Contrariwise, TST-M is using a constant distance between data set based on the meta-features. While the meta-knowledge is useful especially in the beginning, TST-M keeps relying on this such that the information of the new data set is not optimally taken into account. One simple way of fixing this problem is to decay the influence of the meta-knowledge which would introduce at least one meta-hyperparameter. Because TST-R is performing well without an additional meta-hyperparameter for the decay, we do not follow this idea here.

Spearmint provides stronger results than I-GP due to the choice of a different kernel. This might be an indication that we can further improve TST-R, if we use the Matérn 5/2 kernel instead of the SE-ARD.

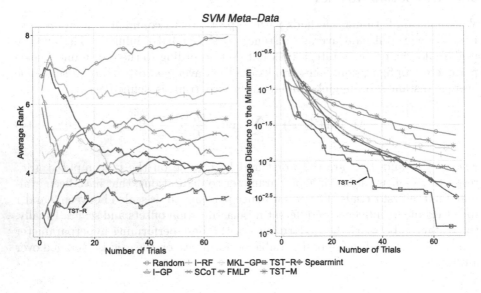

Fig. 2. Our proposed transfer surrogate model TST-R provides the best performance with respect to both evaluation measures for the task of hyperparameter tuning. For both metrics, the smaller the better.

We investigate the performance of the optimization methods also for the problem of combined algorithm selection and hyperparameter tuning on our Weka meta-data set. For this experiment, we remove some methods for different reasons. We remove some weaker methods (Random and TST-M) to improve the readability. Furthermore, we do not compare to methods, which are using one Gaussian process, that is trained on the complete meta-data (SCoT and MKL-GP). The reason for this is that Gaussian processes do not scale to these large meta-data sets (time and memory-wise) [31]. Our approach is learning one Gaussian process for each data set such that each model only needs to be learned on a fraction of the data and thus remains feasible. Nevertheless, we compare to FMLP, the strongest competitor from the previous experiment as well as I-GP

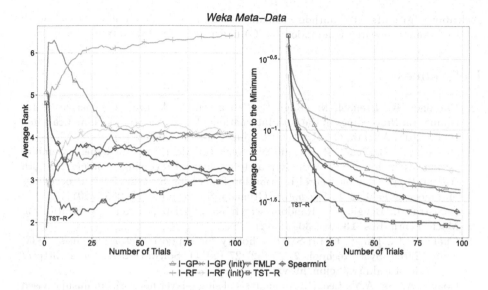

Fig. 3. Our approach TST-R also outperforms the competitor methods for the task of combined algorithm selection and hyperparameter tuning. Surrogate models that use Gaussian processes that train over the whole meta-data are not feasible for this data set [31]. Therefore, we consider I-GP and I-RF with meta-learning initialization.

and I-RF. Furthermore, we also compare to I-GP and I-RF with five initialization steps using a strong meta-initialization technique [32]. The results summarized in Fig. 3 are very similar to our previous experiment. TST-R again is best for both evaluation metrics but FMLP shows to be a strong competitor.

6 Conclusion

In this work, we propose a two-stage transfer surrogate for using meta-knowledge to accelerate the search with the SMBO framework. We propose to approximate the hyperparameter response surface of each data set with an individual model. These individual models are finally combined at the second stage to estimate the score of a hyperparameter configuration. In extensive experiments on two meta-data sets, we compare our method to numerous competitor methods published recently on established machine learning conferences. We show empirically that our two-stage transfer surrogate model is able to outperform all considered competitor methods for the task of hyperparameter tuning as well as the task of combined algorithm selection and hyperparameter tuning.

For future work we are planning to have a deeper look into different ways of describing data sets. Furthermore, we want to investigate whether it is possible to add a decay meta-hyperparameter that enables our approach to also work with typical data set descriptions such as meta-features. Most importantly, we want to investigate the impact of different kernels for TST on the performance. Currently, the Matérn 5/2 seems to be a promising candidate.

Acknowledgments. The authors gratefully acknowledge the co-funding of their work by the German Research Foundation (DFG) under grant SCHM 2583/6-1.

References

1. Bardenet, R., Brendel, M., Kégl, B., Sebag, M.: Collaborative hyperparameter tuning. In: Proceedings of the 30th International Conference on Machine Learning, ICML 2013, Atlanta, GA, USA, 16–21 June, pp. 199–207 (2013)
2. Bergstra, J., Bardenet, R., Bengio, Y., Kégl, B.: Algorithms for hyper-parameter optimization. In: Advances in Neural Information Processing Systems 24: 25th Annual Conference on Neural Information Processing Systems 2011. Proceedings of a meeting held 12–14 December, Granada, Spain, pp. 2546–2554 (2011)
3. Bergstra, J., Bengio, Y.: Random search for hyper-parameter optimization. J. Mach. Learn. Res. **13**, 281–305 (2012)
4. Chang, C.C., Lin, C.J.: LIBSVM: a library for support vector machines. ACM Trans. Intell. Syst. Technol. **2**, 27:1–27:27 (2011). Software available at http://www.csie.ntu.edu.tw/~cjlin/libsvm
5. Coates, A., Ng, A.Y., Lee, H.: An analysis of single-layer networks in unsupervised feature learning. In: Proceedings of the Fourteenth International Conference on Artificial Intelligence and Statistics, AISTATS 2011, Fort Lauderdale, USA, 11–13 April, pp. 215–223 (2011)
6. Domhan, T., Springenberg, J.T., Hutter, F.: Speeding up automatic hyperparameter optimization of deep neural networks by extrapolation of learning curves. In: Proceedings of the Twenty-Fourth International Joint Conference on Artificial Intelligence, IJCAI 2015, Buenos Aires, Argentina, 25–31 July, pp. 3460–3468 (2015)
7. Feurer, M., Springenberg, J.T., Hutter, F.: Using meta-learning to initialize bayesian optimization of hyperparameters. In: ECAI Workshop on Metalearning and Algorithm Selection (MetaSel), pp. 3–10 (2014)
8. Feurer, M., Springenberg, J.T., Hutter, F.: Initializing bayesian hyperparameter optimization via meta-learning. In: Proceedings of the Twenty-Ninth AAAI Conference on Artificial Intelligence, Austin, Texas, USA, 25–30 January, pp. 1128–1135 (2015)
9. Gomes, T.A.F., Prudêncio, R.B.C., Soares, C., Rossi, A.L.D., Carvalho, A.C.P.L.F.: Combining meta-learning and search techniques to select parameters for support vector machines. Neurocomputing **75**(1), 3–13 (2012)
10. Hall, M., Frank, E., Holmes, G., Pfahringer, B., Reutemann, P., Witten, I.H.: The weka data mining software: an update. SIGKDD Explor. Newsl. **11**(1), 10–18 (2009)
11. Hastie, T., Tibshirani, R., Friedman, J.: The Elements of Statistical Learning: Data Mining, Inference and Prediction, 2 edn. Springer, New York (2009)
12. Hutter, F., Hoos, H.H., Leyton-Brown, K.: Sequential model-based optimization for general algorithm configuration. In: Dhaenens, C., Jourdan, L., Marmion, M.-E. (eds.) LION 2015. LNCS, vol. 8994, pp. 507–523. Springer, Heidelberg (2011). doi:10.1007/978-3-642-25566-3_40
13. Jones, D.R., Schonlau, M., Welch, W.J.: Efficient global optimization of expensive black-box functions. J. Global Optim. **13**(4), 455–492 (1998)

14. Kendall, M.G.: A new measure of rank correlation. Biometrika **30**(1/2), 81–93 (1938)
15. Leite, R., Brazdil, P.: Predicting relative performance of classifiers from samples. In: Proceedings of the 22nd International Conference on Machine Learning, ICML 2005, pp. 497–503. ACM, New York (2005)
16. Leite, R., Brazdil, P., Vanschoren, J.: Selecting classification algorithms with active testing. In: Perner, P. (ed.) MLDM 2015. LNCS(LNAI), vol. 9166, pp. 117–131. Springer, Heidelberg (2012). doi:10.1007/978-3-642-31537-4_10
17. Michie, D., Spiegelhalter, D.J., Taylor, C.C., Campbell, J. (eds.): Machine Learning, Neural and Statistical Classification. Ellis Horwood, Upper Saddle River (1994)
18. Pinto, N., Doukhan, D., DiCarlo, J.J., Cox, D.D.: A high-throughput screening approach to discovering good forms of biologically inspired visual representation. PLoS Comput. Biol. **5**(11), e1000579 (2009)
19. Rasmussen, C.E., Williams, C.K.I.: Gaussian Processes for Machine Learning (Adaptive Computation and Machine Learning). The MIT Press, Cambridge (2005)
20. Reif, M., Shafait, F., Dengel, A.: Meta-learning for evolutionary parameter optimization of classifiers. Mach. Learn. **87**(3), 357–380 (2012)
21. Schilling, N., Wistuba, M., Drumond, L., Schmidt-Thieme, L.: Hyperparameter optimization with factorized multilayer perceptrons. In: Appice, A., Rodrigues, P.P., Santos Costa, V., Gama, J., Jorge, A., Soares, C. (eds.) ECML PKDD 2015. LNCS(LNAI), vol. 9285, pp. 87–103. Springer, Heidelberg (2015). doi:10.1007/978-3-319-23525-7_6
22. Smith-Miles, K.A.: Cross-disciplinary perspectives on meta-learning for algorithm selection. ACM Comput. Surv. **41**(1), 6:1–6:25 (2009)
23. Snoek, J., Larochelle, H., Adams, R.P.: Practical bayesian optimization of machine learning algorithms. In: Advances in Neural Information Processing Systems 25: 26th Annual Conference on Neural Information Processing Systems 2012. Proceedings of a meeting held December 3–6, Lake Tahoe, Nevada, United States, pp. 2960–2968 (2012)
24. Srinivas, N., Krause, A., Kakade, S., Seeger, M.W.: Gaussian process optimization in the bandit setting: no regret and experimental design. In: Proceedings of the 27th International Conference on Machine Learning (ICML 2010), Haifa, Israel, 21–24 June, pp. 1015–1022 (2010)
25. Swersky, K., Snoek, J., Adams, R.P.: Multi-task bayesian optimization. In: Advances in Neural Information Processing Systems 26: 27th Annual Conference on Neural Information Processing Systems 2013. Proceedings of a meeting held December 5–8, Lake Tahoe, Nevada, United States, pp. 2004–2012 (2013)
26. Swersky, K., Snoek, J., Adams, R.P.: Freeze-thaw bayesian optimization (2014)
27. Thornton, C., Hutter, F., Hoos, H.H., Leyton-Brown, K.: Auto-weka: combined selection and hyperparameter optimization of classification algorithms. In: Proceedings of the 19th ACM SIGKDD International Conference on Knowledge Discovery and Data Mining, KDD 2013, pp. 847–855. ACM, New York (2013)
28. Villemonteix, J., Vazquez, E., Walter, E.: An informational approach to the global optimization of expensive-to-evaluate functions. J. Global Optim. **44**(4), 509–534 (2009)
29. Wistuba, M., Schilling, N., Schmidt-Thieme, L.: Sequential model-free hyperparameter tuning. In: 2015 IEEE International Conference on Data Mining (ICDM), pp. 1033–1038, November 2015

30. Wistuba, M.: Supplementary website: https://github.com/wistuba/TST, Mar 2016
31. Wistuba, M., Schilling, N., Schmidt-Thieme, L.: Hyperparameter search space pruning – a new component for sequential model-based hyperparameter optimization. In: Appice, A., Rodrigues, P.P., Santos Costa, V., Gama, J., Jorge, A., Soares, C. (eds.) ECML PKDD 2015. LNCS(LNAI), vol. 9285, pp. 104–119. Springer, Heidelberg (2015). doi:10.1007/978-3-319-23525-7_7
32. Wistuba, M., Schilling, N., Schmidt-Thieme, L.: Learning hyperparameter optimization initializations. In: International Conference on Data Science and Advanced Analytics, DSAA 2015, Paris, France, 19–21 October 2015
33. Yogatama, D., Mann, G.: Efficient transfer learning method for automatic hyperparameter tuning. In: International Conference on Artificial Intelligence and Statistics, AISTATS 2014 (2014)

Gaussian Process Pseudo-Likelihood Models for Sequence Labeling

P.K. Srijith[1]([⊠]), P. Balamurugan[2], and Shirish Shevade[3]

[1] Department of Computer Science, University of Sheffield, Sheffield, UK
pk.srijith@dcs.shef.ac.uk
[2] SIERRA Project Team, INRIA-ENS, Paris, France
balamurugan.palaniappan@inria.fr
[3] Computer Science and Automation, Indian Institute of Science, Bangalore, India
shirish@csa.iisc.ernet.in

Abstract. Several machine learning problems arising in natural language processing can be modeled as a sequence labeling problem. Gaussian processes (GPs) provide a Bayesian approach to learning such problems in a kernel based framework. We develop Gaussian process models based on pseudo-likelihood to solve sequence labeling problems. The pseudo-likelihood model enables one to capture multiple dependencies among the output components of the sequence without becoming computationally intractable. We use an efficient variational Gaussian approximation method to perform inference in the proposed model. We also provide an iterative algorithm which can effectively make use of the information from the neighboring labels to perform prediction. The ability to capture multiple dependencies makes the proposed approach useful for a wide range of sequence labeling problems. Numerical experiments on some sequence labeling problems in natural language processing demonstrate the usefulness of the proposed approach.

Keywords: Gaussian processes · Sequence labeling · Variational inference

1 Introduction

Sequence labeling is the task of classifying a sequence of inputs into a sequence of outputs. It arises commonly in natural language processing (NLP) tasks such as part-of-speech tagging, chunking, named entity recognition etc. For instance, in part-of-speech (POS) tagging, the input is a sentence and the output is a sequence of POS tags. The output consists of components whose labels depend on the labels of other components in the output. Sequence labeling takes into account these inter-dependencies among various components of the output [17].

In recent years, sequence labeling has received considerable attention from the machine learning community and is often studied under the general framework of structured prediction. Many algorithms have been proposed to tackle sequence labeling problems. Hidden Markov model (HMM) [20], conditional random field (CRF) [13] and structural support vector machine (SSVM) [25] are the popular

© Springer International Publishing AG 2016
P. Frasconi et al. (Eds.): ECML PKDD 2016, Part I, LNAI 9851, pp. 215–231, 2016.
DOI: 10.1007/978-3-319-46128-1_14

algorithms for sequence labeling. SSVM allows learning a SVM for predicting a structured output including sequences. It is based on a large margin framework and is not probabilistic in nature. HMM is a probabilistic directed graphical model based on Markov assumption and has been widely used for problems in speech and language processing. CRF is also a probabilistic model based on Markov random field assumption. These parametric approaches can provide an estimate of uncertainty in predictions due to their probabilistic nature. However, they do not follow a Bayesian approach as they make a pointwise estimate of their parameters. This makes them less robust and heavily dependent on cross-validation for model selection. Bayesian CRF [19] overcomes this problem by providing a Bayesian treatment to CRF. Approaches like SSVM and maximum margin Markov network (M3N) make use of kernel functions which overcome the limitations arising due to the parametric nature of models such as CRF. Kernel CRF [14] is proposed to overcome this limitation of the CRF, but it is also not a Bayesian approach.

Gaussian processes (GPs) [21] have emerged as a better alternative to offer a non-parametric fully Bayesian approach to solve the sequence labeling problem. An initial work which studied Gaussian process for sequence labeling is [1], where GPs were proposed as an alternative to overcome the limitations of CRF; however they used a maximum a posteriori (MAP) approach instead of a fully Bayesian approach. This caused problems of model selection and robustness issues. A more recent work GPstruct [7] provides a Bayesian approach to general structured prediction problem with GPs. It uses Markov Chain Monte Carlo (MCMC) method to obtain the posterior distribution which slows down the inference. Their approach is based on Markov random field assumption which could not capture long range dependencies among the labels. This difficulty is overcome in [8] which uses an approximate likelihood to reduce the computational complexity arising due to the consideration of larger dependencies. In [8], the proposed model was used to solve grid structured problems in computer vision and was found to be effective in these problems.

In this work, we develop a Gaussian process approach based on pseudo-likelihood to solve sequence labeling problems (which we call GPSL). The GPSL model helps to capture multiple dependencies among the output components in a sequence without becoming computationally intractable. We develop a variational inference method to obtain the posterior which is faster than MCMC based approaches and does not suffer from convergence problems. We also provide an efficient algorithm to perform prediction in the GPSL model which effectively takes into account the dependence on multiple output components. We consider various GPSL models which consider different number of dependencies. We study the usefulness of these models on various sequence labeling problems arising in natural language processing (NLP). The GPSL models which capture more dependencies are found to be useful for these sequence labeling problems. They are also useful in sequence labeling data sets where the labels might be missing for some output components, for example, when the labels are obtained using crowd-sourcing. The main contributions of the paper are as follows:

1. A faster training algorithm based on variational inference.
2. An efficient prediction algorithm which considers multiple dependencies.
3. Application to sequence labeling problems in NLP.

The rest of the paper is organized as follows. Gaussian processes are introduced in Sect. 2. Section 3 discusses the proposed approach, Gaussian process sequence labeling (GPSL), in detail. We provide details of the variational inference and prediction algorithm for the GPSL model in Sects. 4 and 5 respectively. In Sect. 6, we study the performance of various GPSL models on sequence labeling problems and draw several conclusions in Sect. 7.

Notations: We consider a sequence labeling problem over sequences of input-output space pair $(\mathcal{X}, \mathcal{Y})$. The input sequence space $\mathcal{X}$ is assumed to be made up of L components $\mathcal{X} = \mathcal{X}_1 \times \mathcal{X}_2 \times \ldots \mathcal{X}_L$ and the associated output sequence space has L components $\mathcal{Y} = \mathcal{Y}_1 \times \mathcal{Y}_2 \times \ldots \mathcal{Y}_L$. We assume a one-to-one mapping between the input and output components. Each component of the output space is assumed to take a discrete value from the set $\{1, 2, \ldots, J\}$. Each component in the input space is assumed to belong to a P dimensional space $\mathcal{R}^P$ representing features for that input component. Consider a collection of N training input-output examples $\mathbf{D} = \{(\mathbf{x}_n, \mathbf{y}_n)\}_{n=1}^N$, where each example $(\mathbf{x}_n, \mathbf{y}_n)$ is such that $\mathbf{x}_n \in \mathcal{X}$ and $\mathbf{y}_n \in \mathcal{Y}$. Thus, $\mathbf{x}_n$ consists of L components $(\mathbf{x}_{n1}, \mathbf{x}_{n2}, \ldots, \mathbf{x}_{nL})$ and $\mathbf{y}_n$ consists of L components $(y_{n1}, y_{n2}, \ldots, y_{nL})$. The training data $\mathbf{D}$ contains NL input-output components.

2   Background

A Gaussian process (GP) is a collection of random variables with the property that the joint distribution of any finite subset of which is a Gaussian [21]. It generalizes Gaussian distribution to infinitely many random variables and is used as a prior over a latent function. The GP is completely specified by a mean function and a covariance function. The covariance function is defined over latent function values of a pair of inputs and is evaluated using the Mercer kernel function over the pair of inputs. The covariance function expresses some general properties of functions such as their smoothness, and length-scale. A commonly used covariance function is the squared exponential (SE) or the Gaussian kernel

$$cov\big(f(\mathbf{x}_{mi}), f(\mathbf{x}_{nl})\big) = K(\mathbf{x}_{mi}, \mathbf{x}_{nl}) = \sigma_f^2 \exp(-\frac{\kappa}{2}\|\mathbf{x}_{mi} - \mathbf{x}_{nl}\|^2). \qquad (1)$$

Here $f(\mathbf{x}_{mi})$ and $f(\mathbf{x}_{nl})$ are latent function values associated with the input components $\mathbf{x}_{mi}$ and $\mathbf{x}_{nl}$ respectively. $\boldsymbol{\theta} = (\sigma_f^2, \kappa)$ denotes the hyper parameters associated with the covariance function K.

Multi-class classification approaches are useful when the output consists of a single component taking values from a finite discrete set $\{1, 2, \ldots, J\}$. Gaussian process multi-class classification approaches [9,10,26] associate a latent function f^j with every label $j \in \{1, 2, \ldots, J\}$. Let the vector of latent function values associated with a particular label j over all the training examples be $\mathbf{f^j}$. The latent function f^j is assigned an independent GP prior with zero mean and covariance function K^j with hyper parameters $\boldsymbol{\theta}_j$. Thus, $\mathbf{f^j} \sim N(0, \mathbf{K^j})$, where $\mathbf{K^j}$ is a matrix obtained by evaluating the covariance function K^j over all the pairs of training data input components.

In multi-class classification, the likelihood over a multi-class output y_{nl} for an input $\mathbf{x}_{nl}$ given the latent functions is defined as [21]

$$p(y_{nl}|f^1(\mathbf{x}_{nl}), f^2(\mathbf{x}_{nl}), \ldots, f^J(\mathbf{x}_{nl})) = \frac{\exp(f^{y_{nl}}(\mathbf{x}_{nl}))}{\sum_{j=1}^{J} f^j(x_{nl})}. \tag{2}$$

The likelihood (2) is known as multinomial logistic or softmax function and is used widely for the GP multi-class classification problems [9,26]. It is important to note that the likelihood function (2) used for the multi-class classification problems is not Gaussian. Hence, the posterior over the latent functions cannot be obtained in a closed form. GP multi-class classification approaches work by approximating the posterior as a Gaussian using approximate inference techniques such as Laplace approximation [26] and variational inference [9,10]. The Gaussian approximated posterior is then used to make predictions on the test data points. These approximations also yield an approximate marginal likelihood or a lower bound on marginal likelihood which can be used to perform model selection [21].

A sequence labeling problem can be treated as a multi-class classification problem. One can use multi-class classification to obtain a label for each component of the output independently. But this fails to take into account the interdependence among components. If one considers the entire output as a distinct class, then there would be an exponential number of classes and the learning problem becomes intractable. Hence, the sequence labeling problem has to be studied separately from the multi-class classification problems.

3   Gaussian Process Sequence Labeling

Most of the previous approaches [7,13] to sequence labeling use likelihood based on Markov random field assumption which captures only the interaction between neighboring output components. Non-neighboring components also play a significant role in problems such as sequence labeling. In these models, capturing such interactions are computationally expensive due to large clique size. The proposed approach, Gaussian process sequence labeling (GPSL), can take into account interactions among various output components without becoming computationally intractable by using a pseudo-likelihood (PL) model [4].

The PL model defines the likelihood of an output $\mathbf{y_n}$ given the input $\mathbf{x_n}$ as $p(\mathbf{y_n}|\mathbf{x_n}) \propto \prod_{l=1}^{L} p(y_{nl}|\mathbf{x}_{nl}, \mathbf{y_n} \backslash y_{nl})$. where, $\mathbf{y_n} \backslash y_{nl}$ represents all labels in $\mathbf{y_n}$ except y_{nl}. PL models have been successfully used to address many sequence labeling problems in natural language processing [23,24]. They can capture long range dependencies without becoming computationally intractable as the normalization is done for each output component separately. In models such as

CRF, normalization is done over the entire output. This renders them incapable of capturing long range dependencies as the number of summations in the normalization grows exponentially. The PL model is different from a locally normalized model like maximum entropy Markov model (MEMM) as each output component depends on several other output components. Therefore, they do not suffer from the label bias problem [17] unlike MEMM. However, PL models create cyclic dependencies among the output components [11] and this makes prediction hard. We discuss an efficient approach to perform prediction in this case in Sect. 5.

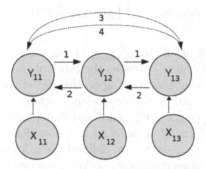

(a) Dependence among input and output components. Dependence on various output components are modelled separately.

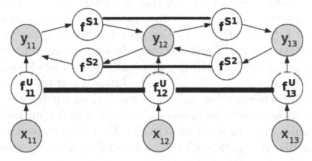

(b) Dependence of local and dependent latent functions. The local latent functions are defined over input-output pairs and dependent latent functions are defined between output components.

Fig. 1. Dependence of latent functions and input-output components in Gaussian process sequence labeling model.

The label of an output component need not depend on the labels of all the other output components. The dependencies among these output components are captured through the set S. Consider the directed graph in Fig. 1a for a sequence labeling problem, where each output component is assumed to depend only on the neighboring output components. Here, the dependency set $S = \{1, 2\}$, where 1 denotes the dependence of an output component on the previous output component and 2 denotes its dependence on the next output component. One can also consider a model where an output component depends on the previous two output components and the next two output components. Let R denote the number of dependency relations in a set S (that is, R is the cardinality of S) and we assume it to be the same for all the output components for the sake of clarity in presentation. Taking into account those dependencies, we can redefine the likelihood as

$$p(\mathbf{y_n}|\mathbf{x_n}) \propto \prod_{l=1}^{L} p(y_{nl}|\mathbf{x}_{nl}, \mathbf{y}_{\mathbf{nl}}^{\mathbf{S}}). \tag{3}$$

Here, $\mathbf{y}_{\mathbf{nl}}^{\mathbf{S}}$ denotes the set of labels $\{y_{nl}^d\}_{d=1}^{R}$ of the output components referred by the dependency set S and y_{nl}^d denotes the label of the dth dependent output component. In (3), instead of conditioning on the rest of the labels, we condition y_{nl} only on the labels defined by the dependency set S.

Now, the likelihood $p(y_{nl}|\mathbf{x}_{nl}, \mathbf{y}_{\mathbf{nl}}^{\mathbf{S}})$ can be defined using a set of latent functions. We use different latent functions to model different dependencies. The dependency of the label y_{nl} on $\mathbf{x}_{nl}$ is defined as a local dependency and is modeled as in GP multi-class classification. We associate a latent function with each label in the set $\{1, 2, \ldots J\}$. The latent function associated with a label j, denoted as f^{Uj}, is called a local latent function. It is defined over all the training input components $\mathbf{x}_{nl}$ for every n and l and the latent function values associated with a particular label j over NL training examples are denoted by $\mathbf{f^{Uj}}$. The local latent functions associated with a particular input component $\mathbf{x}_{nl}$ are denoted as $\mathbf{f}_{\mathbf{nl}}^{\mathbf{U}} = \{f_{nl}^{U1}, \ldots, f_{nl}^{UJ}\}$. We also associate a latent function f^{Sd} with each dependency relation $d \in S$ and call them dependent latent functions. These latent functions are defined over all the values of a pair of labels $(\hat{y}_{nl}, y_{nl})$ where $\hat{y}_{nl} \in \{1, 2, \ldots J\}$ and $y_{nl} \in \{1, 2, \ldots J\}$. The latent function values associated with a particular dependency d over J^2 label pair values are denoted by $\mathbf{f^{Sd}}$. The dependence of various latent functions on the input and output components for the directed graph in Fig. 1a is depicted in Fig. 1b. Given these latent functions we define the likelihood $p(y_{nl}|\mathbf{x}_{nl}, \mathbf{y}_{\mathbf{nl}}^{\mathbf{S}})$ to be a member of an exponential family:

$$p(y_{nl}|\mathbf{x}_{nl}, \mathbf{y}_{\mathbf{nl}}^{\mathbf{S}}, \{\mathbf{f^{Uj}}\}_{j=1}^{J}, \{\mathbf{f^{Sd}}\}_{d=1}^{R}) =$$
$$\frac{\exp(f^{Uy_{nl}}(\mathbf{x}_{nl}) + \sum_{d=1}^{R} f^{Sd}(y_{nl}^d, y_{nl}))}{\sum_{y_{nl}=1}^{J} \exp(f^{Uy_{nl}}(\mathbf{x}_{nl}) + \sum_{d=1}^{R} f^{Sd}(y_{nl}^d, y_{nl}))}. \tag{4}$$

This differs from the softmax likelihood (2) used in multi-class classification in that it captures the dependencies among output components. Given the latent functions and the input $\mathbf{X} = \{\mathbf{x_n}\}_{n=1}^{N}$, the likelihood of the output $\mathbf{Y} = \{\mathbf{y_n}\}_{n=1}^{N}$ is

$$p(\mathbf{Y}|\mathbf{X}, \{\mathbf{f^{Uj}}\}_{j=1}^{J}, \{\mathbf{f^{Sd}}\}_{d=1}^{R}) = \prod_{n=1}^{N} \prod_{l=1}^{L} p(y_{nl}|\mathbf{x}_{nl}, \mathbf{y}_{\mathbf{n}\{D_{nl}\}}, \{\mathbf{f^{Uj}}\}_{j=1}^{J}, \{\mathbf{f^{Sd}}\}_{d=1}^{R}) \tag{5}$$

We impose independent GP priors over the latent functions $\{f^{Uj}\}_{j=1}^{J}$, $\{f^{Sd}\}_{d=1}^{R}$. The latent function f^{Uj} is given a zero mean GP prior with covariance function K^{Uj} parameterized by $\boldsymbol{\theta}_j$. Thus, $\mathbf{f}^{Uj}$ is a Gaussian with mean 0 and covariance $\mathbf{K}^{Uj}$ of size $NL \times NL$, that is $p(\mathbf{f}^{Uj}) = \mathcal{N}(\mathbf{f}^{Uj}; 0, \mathbf{K}^{Uj})$. $\mathbf{K}^{Uj}$ consists of covariance function evaluations over all the pairs of training data input components $\{\{\mathbf{x}_{nl}\}_{l=1}^{L}\}_{n=1}^{N}$. The latent function f^{Sd} is given zero mean GP prior with an identity covariance which is defined to be 1 when inputs are the same and 0 otherwise. Thus $\mathbf{f}^{Sd}$ is a Gaussian with mean 0 and covariance $\mathbf{I}$ of size J^2, that is $p(\mathbf{f}^{Sd}) = \mathcal{N}(\mathbf{f}^{Sd}; 0, \mathbf{I}_{J^2})$. Let $\mathbf{f}^{U} = (\mathbf{f}^{U1}, \mathbf{f}^{U2}, \dots, \mathbf{f}^{UJ})$ be the collection of all local latent functions and $\mathbf{f}^{S} = (\mathbf{f}^{S1}, \mathbf{f}^{S2}, \dots, \mathbf{f}^{SR})$ be the collection of all dependent latent functions. Then the prior over $\mathbf{f}^{U}$ and $\mathbf{f}^{S}$ is defined as

$$p(\mathbf{f}^{U}, \mathbf{f}^{S}|\mathbf{X}) = \mathcal{N}\left(\begin{bmatrix}\mathbf{f}^{U}\\\mathbf{f}^{S}\end{bmatrix}; 0, \begin{bmatrix}\mathbf{K}^{U} & 0\\0 & \mathbf{K}^{S}\end{bmatrix}\right), \tag{6}$$

where $\mathbf{K}^{U} = diag(\mathbf{K}^{U1}, \mathbf{K}^{U2}, \dots, \mathbf{K}^{UJ})$ is a block diagonal matrix and $\mathbf{K}^{S} = \mathbf{I}_{J^2} \otimes \mathbf{I}_R$.

The posterior over the latent functions $p(\mathbf{f}^{U}, \mathbf{f}^{S}|\mathbf{D})$ is

$$p(\mathbf{f}^{U}, \mathbf{f}^{S}|\mathbf{X}, \mathbf{Y}) = \frac{1}{p(\mathbf{Y}|\mathbf{X})}p(\mathbf{Y}|\mathbf{X}, \mathbf{f}^{U}, \mathbf{f}^{S})p(\mathbf{f}^{U}, \mathbf{f}^{S}|\mathbf{X})$$

where $p(\mathbf{Y}|\mathbf{X}) = \int p(\mathbf{Y}|\mathbf{X}, \mathbf{f}^{U}, \mathbf{f}^{S})p(\mathbf{f}^{U}, \mathbf{f}^{S}|\mathbf{X})d\mathbf{f}^{U}d\mathbf{f}^{S}$ is called evidence. Evidence is a function of hyper-parameters $\boldsymbol{\theta} = (\theta_1, \theta_2, \dots, \theta_J)$ and is maximized to estimate them. For notational simplicity, we suppress the dependence of evidence, posterior and prior on the hyper-parameter $\boldsymbol{\theta}$. Due to the non-Gaussian nature of the likelihood, evidence is intractable and the posterior cannot be determined exactly. We use a variational inference technique to obtain an approximate posterior. Variational inference is faster than sampling based techniques used in [7] and does not suffer from convergence problems [16]. It can easily handle multi-class problems and is scalable to models with a large number of parameters. Further, it provides an approximation to the evidence which is useful in estimating the hyper-parameters of the model.

4   Variational Inference

A variational Inference technique [16] approximates the intractable posterior by an approximate variational distribution. It approximates the posterior $p(\mathbf{f}|\mathbf{X}, \mathbf{Y})$ by a variational distribution $q(\mathbf{f}|\boldsymbol{\gamma})$, where $\mathbf{f} = (\mathbf{f}^{U}, \mathbf{f}^{S})$ and $\boldsymbol{\gamma}$ represents the variational parameters. In variational inference, this is done by minimizing the Kullback-Leibler (KL) divergence between $q(\mathbf{f}|\boldsymbol{\gamma})$ and $p(\mathbf{f}|\mathbf{X}, \mathbf{Y})$. This is often intractable and the variational parameters are obtained by maximizing a variational lower bound $L(\boldsymbol{\theta}, \boldsymbol{\gamma})$.

$$KL(q(\mathbf{f}|\boldsymbol{\gamma})\|p(\mathbf{f}|\mathbf{X}, \mathbf{Y})) = -L(\boldsymbol{\theta}, \boldsymbol{\gamma}) + \log p(\mathbf{Y}|\mathbf{X}) \tag{7}$$

where $L(\boldsymbol{\theta}, \boldsymbol{\gamma}) = -KL(q(\mathbf{f}|\boldsymbol{\gamma})\|p(\mathbf{f}|\mathbf{X})) + \int q(\mathbf{f}|\boldsymbol{\gamma})\log p(\mathbf{Y}|\mathbf{X}, \mathbf{f})d\mathbf{f}.$

Maximizing the variational lower bound $L(\boldsymbol{\theta}, \boldsymbol{\gamma})$ results in minimizing the KL divergence $KL(q(\mathbf{f}|\boldsymbol{\gamma})\|p(\mathbf{f}|\mathbf{X},\mathbf{Y}))$, since the evidence $p(\mathbf{Y}|\mathbf{X})$ does not depend on the variational parameters.

We use a variational Gaussian (VG) approximate inference approach [18] where the variational distribution is assumed to be a Gaussian. Variational Gaussian approaches can be slow because of the requirement to estimate the covariance matrix. Fortunately, recent advances in VG inference approaches [18] enable one to compute the covariance matrix using $\mathcal{O}(NL)$ variational parameters. In fact, we use the VG approach for GPs [12] which requires computation of only $\mathcal{O}(NL)$ variational parameters, but at the same time uses a concave variational lower bound. We assume the variational distribution $q(\mathbf{f}|\boldsymbol{\gamma})$ takes the form of a Gaussian distribution and factorizes as $q(\mathbf{f}^{\mathbf{U}}|\boldsymbol{\gamma}^U)q(\mathbf{f}^{\mathbf{S}}|\boldsymbol{\gamma}^U)$ where $\boldsymbol{\gamma} = \{\boldsymbol{\gamma}^U, \boldsymbol{\gamma}^S\}$. Let $q(\mathbf{f}^{\mathbf{U}}|\boldsymbol{\gamma}^U) = \mathcal{N}(\mathbf{f}^{\mathbf{U}}; \mathbf{m}^{\mathbf{U}}, \mathbf{V}^{\mathbf{U}})$ where $\boldsymbol{\gamma}^U = \{\mathbf{m}^{\mathbf{U}}, \mathbf{V}^{\mathbf{U}}\}$ and $q(\mathbf{f}^{\mathbf{S}}) = \mathcal{N}(\mathbf{f}^{\mathbf{S}}; \mathbf{m}^{\mathbf{S}}, \mathbf{V}^{\mathbf{S}})$ where $\boldsymbol{\gamma}^S = \{\mathbf{m}^{\mathbf{S}}, \mathbf{V}^{\mathbf{S}}\}$. Then, the variational lower bound $L(\boldsymbol{\theta}, \boldsymbol{\gamma})$ can be written as

$$L(\boldsymbol{\theta}, \boldsymbol{\gamma}) = \frac{1}{2}(\log |\mathbf{V}^{\mathbf{U}}\boldsymbol{\Omega}^{\mathbf{U}}| + \log |\mathbf{V}^{\mathbf{S}}\boldsymbol{\Omega}^{\mathbf{S}}| - tr(\mathbf{V}^{\mathbf{U}}\boldsymbol{\Omega}^{\mathbf{U}}) - tr(\mathbf{V}^{\mathbf{S}}\boldsymbol{\Omega}^{\mathbf{S}}) \tag{8}$$

$$-\mathbf{m}^{\mathbf{U}^\top}\boldsymbol{\Omega}^{\mathbf{U}}\mathbf{m}^{\mathbf{U}} - \mathbf{m}^{\mathbf{S}^\top}\boldsymbol{\Omega}^{\mathbf{S}}\mathbf{m}^{\mathbf{S}}) + \sum_{n=1}^{N}\sum_{l=1}^{L} \mathbb{E}_{q(\mathbf{f}^{\mathbf{U}}|\boldsymbol{\gamma}^U)q(\mathbf{f}^{\mathbf{S}}|\boldsymbol{\gamma}^S)}[\log p(y_{nl}|\mathbf{x}_{nl}, \mathbf{y}_{nl}^{\mathbf{S}}, \mathbf{f})]$$

where $\boldsymbol{\Omega}^{\mathbf{U}} = \mathbf{K}^{\mathbf{U}^{-1}}$, $\boldsymbol{\Omega}^{\mathbf{S}} = \mathbf{K}^{\mathbf{S}^{-1}}$ and $\mathbb{E}_{q(x)}[f(x)] = \int f(x)q(x)dx$ represents the expectation of $f(x)$ with respect to the density $q(x)$. Since $\mathbf{K}^{\mathbf{U}}$ is block diagonal, its inverse is block diagonal, and hence $\boldsymbol{\Omega}^{\mathbf{U}}$ is block diagonal that is $\boldsymbol{\Omega}^{\mathbf{U}} = diag(\boldsymbol{\Omega}^{\mathbf{U}1}, \boldsymbol{\Omega}^{\mathbf{U}2}, \dots, \boldsymbol{\Omega}^{\mathbf{U}J})$, where $\boldsymbol{\Omega}^{\mathbf{U}j} = \mathbf{K}^{\mathbf{U}j^{-1}}$. Similarly, $\boldsymbol{\Omega}^{\mathbf{S}}$ is also a block diagonal with each block being a diagonal matrix $\mathbf{I}_{J^2}$. The marginal variational distribution of local latent function values $\mathbf{f}^{\mathbf{U}j}$ is a Gaussian with mean $\mathbf{m}^{\mathbf{U}j}$ and covariance $\mathbf{V}^{\mathbf{U}j}$, and that of dependent latent function values $\mathbf{f}^{\mathbf{S}d}$ is a Gaussian with mean $\mathbf{m}^{\mathbf{S}d}$ and covariance $\mathbf{V}^{\mathbf{S}d}$. The variational lower bound $L(\boldsymbol{\theta}, \boldsymbol{\gamma})$ requires computing an expectation of the log likelihood with respect to the variational distribution. However, the integral is intractable since the likelihood is a softmax function. So, we use Jensen's inequality to obtain a tractable lower bound to the expectation of log likelihood. The variational lower bound $L(\boldsymbol{\theta}, \boldsymbol{\gamma})$ can be written as

$$\frac{1}{2}\Big(\sum_{j=1}^{J}(\log |\mathbf{V}^{\mathbf{U}j}\boldsymbol{\Omega}^{\mathbf{U}j}| - tr(\mathbf{V}^{\mathbf{U}j}\boldsymbol{\Omega}^{\mathbf{U}j}) - \mathbf{m}^{\mathbf{U}j^\top}\boldsymbol{\Omega}^{\mathbf{U}j}\mathbf{m}^{\mathbf{U}j})$$

$$+ \sum_{d=1}^{R}(\log |\mathbf{V}^{\mathbf{S}d}\boldsymbol{\Omega}^{\mathbf{S}d}| - tr(\mathbf{V}^{\mathbf{S}d}\boldsymbol{\Omega}^{\mathbf{S}d}) - \mathbf{m}^{\mathbf{S}d^\top}\boldsymbol{\Omega}^{\mathbf{S}d}\mathbf{m}^{\mathbf{S}d}))$$

$$+ \sum_{n=1}^{N}\sum_{l=1}^{L}\Big(m_{nl}^{U y_{nl}} + \sum_{d=1}^{R}m_{(y_{nl}^d, y_{nl})}^{Sd} - \log(\sum_{q=1}^{J}\exp(m_{nl}^{Uj} + \frac{1}{2}V_{(nl,nl)}^{Uj}$$

$$+ \sum_{d=1}^{R}m_{(y_{nl}^d, q)}^{Sd} + \frac{1}{2}V_{((y_{nl}^d, q),(y_{nl}^d, q))}^{Sd}))\Big)\Big). \tag{9}$$

Algorithm 1. Model selection and learning in Gaussian process sequence labeling model

---

1: **Input:** Training data $(\mathbf{X}, \mathbf{Y})$, dependency set S
2: Initialize hyper-parameters $\boldsymbol{\theta}$, variational parameters $\boldsymbol{\gamma}$
3: **repeat**
4:    **repeat**
5:       **for** $j = 1$ to J **do**
6:          Update $\mathbf{m}^{\mathbf{Uj}}$ by maximizing (9) $w.r.t$ $\mathbf{m}^{\mathbf{Uj}}$
7:          Update $\mathbf{V}^{\mathbf{Uj}}$ by maximizing (9) $w.r.t$ $\mathbf{V}^{\mathbf{Uj}}$
8:       **end for**
9:       **for** $d = 1$ to R **do**
10:         Update $\mathbf{m}^{\mathbf{Sd}}$ by maximizing (9) $w.r.t$ $\mathbf{m}^{\mathbf{Sd}}$
11:         Update $\mathbf{V}^{\mathbf{Sd}}$ by maximizing (9) $w.r.t$ $\mathbf{V}^{\mathbf{Sd}}$
12:       **end for**
13:    **until** relative increase in lower bound (9) is small
14:    Update $\boldsymbol{\theta}$ by maximizing (9) $w.r.t$ $\boldsymbol{\theta}$
15: **until** relative increase in lower bound (9) is small
16: **Return:** $\boldsymbol{\theta}, \boldsymbol{\gamma}$

The varia tional parameters $\boldsymbol{\gamma} = \{\{\mathbf{m}^{\mathbf{Uj}}\}_{j=1}^{J}, \{\mathbf{V}^{\mathbf{Uj}}\}_{j=1}^{J}, \{\mathbf{m}^{\mathbf{Sd}}\}_{d=1}^{R}, \{\mathbf{V}^{\mathbf{Sd}}\}_{d=1}^{R}\}$ are estimated by maximizing the variational lower bound (9). The lower bound is jointly concave with respect to all the variational parameters [6] and the optimum can be easily found using gradient based optimization techniques.

The variational parameters are estimated using a co-ordinate ascent approach. We repeatedly estimate each variational parameter while keeping the others fixed. The variational mean parameters $\mathbf{m}^{\mathbf{Uj}}$ and $\mathbf{m}^{\mathbf{Sd}}$ are estimated using gradient based approaches. The variational covariance matrices $\mathbf{V}^{\mathbf{Uj}}$ and $\mathbf{V}^{\mathbf{Sd}}$ are estimated under the positive semi-definite (p.s.d.) constraint. This can be done efficiently using the fixed point approach mentioned in [12]. It is reported to converge faster than other VG approaches for GPs and is based on a concave objective function similar to (9). The approach maintains the p.s.d. constraint on the covariance matrix and computes $\mathbf{V}^{\mathbf{Uj}}$ by estimating only $\mathcal{O}(NL)$ variational parameters. Estimation of $\mathbf{V}^{\mathbf{Uj}}$ using the fixed point approach converges since (9) is strictly concave with respect to $\mathbf{V}^{\mathbf{Uj}}$. The variational covariance matrix $\mathbf{V}^{\mathbf{Sd}}$ is diagonal since $\boldsymbol{\Omega}^{\mathbf{Sd}}$ is diagonal. Hence, for computing a p.s.d. $\mathbf{V}^{\mathbf{Sd}}$ we need to estimate only the diagonal elements of $\mathbf{V}^{\mathbf{Sd}}$ under the element-wise non-negativity constraint. This can be done easily using gradient based methods. The variational parameters $\boldsymbol{\gamma}$ are estimated for a particular set of hyper-parameters $\boldsymbol{\theta}$. The hyper-parameters $\boldsymbol{\theta}$ are also estimated by maximizing the lower bound (9). The variational parameters $\boldsymbol{\gamma}$ and the model parameters $\boldsymbol{\theta}$ are estimated alternately following a variational expectation maximization (EM) approach [16]. Algorithm 1 summarizes various steps involved in our approach.

The variational lower bound (9) is strictly concave with respect to each of the variational parameters. Hence, the estimation of variational parameters using co-ordinate ascent algorithm (inner loop) converges [3]. Convergence of EM for exponential family guarantees the convergence of Algorithm 1. The overall computational complexity of Algorithm 1 is dominated by the computation of $\mathbf{V^{Uj}}$. It takes $\mathcal{O}(JN^3L^3)$ time as it requires inversion of J covariance matrices of size $NL \times NL$. The computational complexity for estimating $\mathbf{V^{Sd}}$ is $\mathcal{O}(RNLJ)$ and is negligible compared to the estimation of $\mathbf{V^{Uj}}$. Note that the computational complexity of the algorithm increases linearly with respect to the number of dependencies R.

5   Prediction

We propose an iterative prediction algorithm which can effectively take into account the presence of multiple dependencies. The variational posterior distributions estimated using VG approximation $q(\mathbf{f^U}) = \prod_{j=1}^{J} q(\mathbf{f^{Uj}})$ $= \prod_{j=1}^{J} \mathcal{N}(\mathbf{f^{Uj}}; \mathbf{m^{Uj}}, \mathbf{V^{Uj}})$ and $q(\mathbf{f^S}) = \prod_{d=1}^{R} q(\mathbf{f^{Sd}}) = \prod_{d=1}^{R} \mathcal{N}(\mathbf{f^{Sd}}; \mathbf{m^{Sd}}, \mathbf{V^{Sd}})$ can be used to predict a test output sequence $\mathbf{y}_*$ given a test input sequence $\mathbf{x}_*$. The predictive probability of assigning a label y_{*l} to a component of the output $\mathbf{y}_*$, given $\mathbf{x}_{*l}$ and rest of the labels $\mathbf{y}_* \backslash y_{*l}$ is

$$p(y_{*l}|\mathbf{x}_{*l}, \mathbf{y}_* \backslash y_{*l}) = \int p(y_{*l}|\mathbf{x}_{*l}, \mathbf{y}_* \backslash y_{*l}, \mathbf{f}_*) p(\mathbf{f}_*) d\mathbf{f}_*$$

$$= \int \frac{\exp(f_{*l}^{U y_{*l}} + \sum_{d=1}^{R} f_*^{Sd}(y_{*l}^d, y_{*l}))}{\sum_{y_{*l}=1}^{J} \exp(f_{*l}^{U y_{*l}} + \sum_{d=1}^{R} f_*^{Sd}(y_{nl}^d, y_{nl}))}$$
$$\{p(f_{*l}^{Uj})\}_{j=1}^{J} \{p(f_*^{Sd})\}_{d=1}^{R} \{df_{*l}^{Uj}\}_{j=1}^{J} \{df_*^{Sd}\}_{d=1}^{R} \quad (10)$$

where $p(\mathbf{f}_*)$ denotes the predictive distribution of all the latent function values for the test input $\mathbf{x}_*$. In (10), $p(f_{*l}^{Uj})$ represents the predictive distribution of the local latent function j for a test input component $\mathbf{x}_{*l}$. This is Gaussian with mean m_{*l}^{Uj} and variance v_{*l}^{Uj} where,

$$m_{*l}^{Uj} = \mathbf{K}_{*l}^{\mathbf{Uj}^\top} \mathbf{\Omega^{Uj} m^{Uj}} \quad \text{and}$$
$$v_{*l}^{Uj} = K_{*l,*l}^{Uj} - \mathbf{K}_{*l}^{\mathbf{Uj}^\top} (\mathbf{\Omega^{Uj}} - \mathbf{\Omega^{Uj} V^{Uj} \Omega^{Uj}}) \mathbf{K}_{*l}^{\mathbf{Uj}}.$$

Here, $\mathbf{K}_{*l}^{\mathbf{Uj}}$ is an NL dimensional vector obtained from the kernel evaluations for the label j between the test input data component $\mathbf{x}_{*l}$ and the training data $\mathbf{X}$ and $K_{*l,*l}^{Uj}$ represents the kernel evaluation of the test data input component $\mathbf{x}_{*l}$ with itself. $\mathbf{f^{Sd}}$ is independent of the test data input and the predictive distribution $p(\mathbf{f_*^{Sd}})$ is the same as $p(\mathbf{f^{Sd}})$. This is a Gaussian with mean $\mathbf{m^{Sd}}$ and covariance $\mathbf{V^{Sd}}$. The computation of the expected value of softmax with respect to the latent functions (10) is intractable. Instead we compute softmax of the expected value of the latent functions and compute a normalized probabilistic score. We refine the normalized score to take into account the uncertainty in true labels associated with the dependencies and compute the refined normalized score (RNS) as

$$RNS(y_{*l}, \mathbf{x}_{*l}) = \frac{\exp(m_{*l}^{U y_{*l}} + \frac{1}{2} v_{*l}^{U y_{*l}} + \sum_{d=1}^{R} \mathbb{E}_{y_{*l}^d} [g^d(y_{*l}^d, y_{*l})])}{\sum_{q=1}^{J} \exp(m_{*l}^{Uj} + \frac{1}{2} v_{*l}^{Uj} + \sum_{d=1}^{R} \mathbb{E}_{y_{*l}^d} [g^d(y_{*l}^d, q)])}$$

Algorithm 2. Prediction in Gaussian process sequence labeling model

---

1: **Input:** Test data $\mathbf{x}_* = (\mathbf{x}_{*1}, \ldots, \mathbf{x}_{*L})$, posterior mean $\{\mathbf{m}^{\mathbf{Uj}}\}_{j=1}^{J}$ and $\{\mathbf{m}^{\mathbf{Sd}}\}_{d=1}^{R}$
   and posterior covariance $\{\mathbf{V}^{\mathbf{Uj}}\}_{j=1}^{J}$ and $\{\mathbf{V}^{\mathbf{Sd}}\}_{d=1}^{R}$

2: Obtain predictive means $\{\{m_{*l}^{Uj}\}_{j=1}^{J}\}_{l=1}^{L}$, and variances $\{\{v_{*l}^{Uj}\}_{j=1}^{J}\}_{l=1}^{L}$

3: **Initialize :** $RNS^0(y_{*l}, \mathbf{x}_{*l}) = \dfrac{\exp(m_{*l}^{U y_{*l}} + \frac{1}{2} v_{*l}^{U y_{*l}})}{\sum_{j=1}^{J} \exp(m_{*l}^{Uj} + \frac{1}{2} v_{*l}^{Uj})}$ $\forall y_{*l} = 1, \ldots, J, \forall l = 1 \ldots, L$

4: **Initialize :** $t = 0$

5: **repeat**

6:    $t = t + 1$

7:    **for** $l = 1$ **to** L **do**

8:       **for** $y_{*l} = 1$ **to** J **do**

9:          $RNS^t(y_{*l}, \mathbf{x}_{*l}) = \dfrac{\exp(m_{*l}^{U y_{*l}} + \frac{1}{2} v_{*l}^{U y_{*l}} + \sum_{d=1}^{R} \mathbb{E}_{y_{*l}^d}[g^d(y_{*l}^d, y_{*l})])}{\sum_{j=1}^{J} \exp(m_{*l}^{Uj} + \frac{1}{2} v_{*l}^{Uj} + \sum_{d=1}^{R} \mathbb{E}_{y_{*l}^d}[g^d(y_{*l}^d, q)])}$

10:          where $\mathbb{E}_{y_{*l}^d}[\cdot] = \sum_{y_{*l}^d=1}^{J} RNS^{t-1}(y_{*l}^d, x_{*l}^d)[\cdot]$

11:       **end for**

12:    **end for**

13: **until** change in RNS^t *w.r.t* RNS^{t-1} is small

14: $(\hat{y}_{*1}, \ldots, \hat{y}_{*L}) = (\text{argmax}_{y_{*1}} RNS^t(y_{*1}, \mathbf{x}_{*1}), \ldots,$
       $\text{argmax}_{y_{*L}} RNS^t(y_{*L}, \mathbf{x}_{*L}))$

15: **Return:** $(\hat{y}_{*1}, \ldots, \hat{y}_{*L})$

Here, $g^d(y^d, y) = \mathbf{m}^{\mathbf{Sd}}_{(y^d, y)} + \frac{1}{2} \mathbf{V}^{\mathbf{Sd}}_{((y^d,y),(y^d,y))}$ determines the contribution of the label y^d of dependency d in predicting the output label y. RNS considers an expected value over all the possible labelings associated with a dependency d. The expectation is computed using the RNS value associated with the labels y_{*l}^d for the input x_{*l}^d, that is, $\mathbb{E}_{y_{*l}^d}[\cdot] = \sum_{y_{*l}^d=1}^{J} RNS(y_{*l}^d, x_{*l}^d)[\cdot]$.

We provide an iterative approach to estimate the labels of a test output in Algorithm 2. An initial RNS value is computed without considering the dependencies. We iteratively refine the RNS value using the previously computed RNS value by taking into account the dependencies. The process is continued until convergence. The final RNS value is used to make prediction separately for each output component by assigning labels with the maximum RNS value. The computational complexity of Algorithm 2 is $\mathcal{O}(J^2 RL)$ and is same as that of Viterbi algorithm [20] for a single dependency case. The convergence of Algorithm 2 follows from the analysis presented in [15] for a similar fixed point algorithm. The algorithm is found to converge in a few iterations in our experiments.

6   Experimental Results

We conduct experiments to study the generalization performance of the proposed Gaussian Process Sequence labeling (GPSL) model. We use the sequence labeling problems in natural language processing to study the behavior of the proposed approach. Although the proposed approach is general and can handle dependencies of any length, we consider three different models of the proposed approach in our experiments. The first model, GPSL1, assumes that the current label depends only on the previous label. The second model, GPSL2, assumes that the current label depends both on the previous and the next label in the sequence. The third model, GPSL4, assumes that the current label depends on the previous two labels and the next two labels.

We consider four sequence labeling problems in natural language processing to study the performance of the proposed approach. The datasets for all these problems are obtained from the CRF++[1] toolbox. We provide a brief description of the tasks in each of these data sets.

Base NP: We need to identify noun phrases in a sentence. The starting word in the noun phrase is given a label B, while the words inside the noun phrase are given a label I. All the other words are given a label O. The task here is to assign each word with a label from the set $\{B, I, O\}$.

Chunking: Shallow parsing or chunking identifies constituents in a sentence such as noun phrase, verb phrase etc. Here, each word in a sentence is labeled as belonging to verb phrase, noun phrase etc. In the *Chunking* dataset, words are assigned a label from a set of size 14.

Segmentation: Segmentation is the process of finding meaningful segments in a text such as words, sentences etc. We consider a word segmentation problem where the words are identified from a Chinese sentence. The *Segmentation* data set assigns each unit in the sentence a label denoting whether it is beginning of a word (B) or inside a word (I). The task is to assign either of these two labels to each unit in a sentence.

Japanese NE: We need to perform Named Entity Recognition (NER) where the task is to identify whether the words in a sentence denote a named entity such as person, place, time etc. We use the *JapaneseNE* dataset where the Japanese words are assigned one of 17 different named entities.

In all these data sets except *Segmentation*, a sentence is considered as an input and words in the sentence as input components. In *Segmentation*, every alphabet is considered as an input component. The features for each input component are extracted using the template files provided in the CRF++ package. The properties of all the data sets are summarized in Table 1. It mentions the number of sentences (N) used for training and testing. The effective sample size (NL) for the GPSL models is obtained by multiplying this quantity by average sentence length which increases the data size by an order of magnitude.

We compare the performance of the proposed approach with popular sequence labeling approaches, structural SVM (SSVM) [2][2], conditional random

---

[1] Available at http://crfpp.googlecode.com/svn/trunk/doc/index.html.
[2] Code available at http://drona.csa.iisc.ernet.in/~shirish/structsvm_sdm.html.

field (CRF) [5][3], and GPstruct [7][4]. All the models used a linear kernel. GPstruct experiments are run for 100000 elliptical slice sampling steps. The performance is measured in terms of average Hamming loss over all the test data points. The Hamming loss between the actual test output $\mathbf{y}_*$ and the predicted test output $\hat{\mathbf{y}}_*$ is given by $Loss(\mathbf{y}_*, \hat{\mathbf{y}}_*) = \sum_{l=1}^{L} \mathbb{I}(y_{*l} \neq \hat{y}_{*l})$, where $\mathbb{I}(\cdot)$ is the indicator function. Table 1 compares the performance (percentage of the average Hamming loss) of various approaches on the four sequence labeling problems. The GPSL models, SSVM, CRF and GPstruct are run over 10 independent partitions of the data set[5] and a mean of the Hamming loss over all the partitions along with the standard deviation are reported in Table 1.

Table 1. Properties of the sequence labeling data sets and a comparison of the performance of various models on these data sets. The approaches GPSL1, GPSL2, GPSL4, SSVM, CRF and GPstruct are compared using average Hamming loss (in percentage). The numbers in bold face style indicate the best results among these approaches. '⋆' and '†' denote if the performance of a method is significantly different from the best performing method and GPstruct repectively, according to paired t-test with 5 % significance level.

	Base NP	Chunking	Segmentation	Japanese NE
#labels	3	14	2	17
#features	6438	29764	1386	102,799
training/ test sentences	150/150	50/50	20/16	50/50
GPSL1	5.73±0.98⋆	13.02±1.87⋆	**23.45 ± 2.96**	8.26±2.63⋆
GPSL2	5.55±0.92⋆	12.69±1.69⋆	23.51±2.93	7.86±2.45 ⋆
GPSL4	5.54±0.94⋆	12.70±1.79⋆	23.53±2.85	7.82±2.56 ⋆
CRF	5.21±0.84†	11.76±1.73⋆†	24.10±3.49⋆†	7.76±2.80 ⋆
SSVM	**5.19 ± 0.91** †	**10.71 ± 1.49** †	23.46±3.45	**6.17 ± 2.60** †
GPstruct	5.66±0.93⋆	12.56±1.82⋆	23.55±2.90	7.79±2.92 ⋆

The reported results show that the GPSL models with multiple dependencies performed better than GPstruct on *BaseNP* and *Segmentation*. In the other two data sets, GPSL models came close to GPstruct. We find that increasing the number of dependencies helped to improve the performance in general except for the *Segmentation* data set. This is due to the difference in nature of the sequence labeling task involved in segmentation. For other data sets, the GPSL model which considered both the previous and next label (GPSL2) gave a better performance. The performance of the GPSL model which considered the previous

---

[3] Code available at http://leon.bottou.org/projects/sgd#stochastic_gradient_descent_ version_2.

[4] Code available at https://github.com/sebastien-bratieres/pygpstruct.

[5] The train and test set partitions are different from those used by [7].

and the next 2 labels (GPSL4) improved only marginally or worsened compared to GPSL2 on these data sets. We note that increasing the number of dependencies beyond four did not bring any improvement in performance for the sequence labeling data sets that we have considered. Overall, the performance of the SSVM is found to be better than other approaches in these sequence labeling data sets. However, GPSL models have the advantage of being Bayesian and can provide a confidence over label predictions which is useful for many NLP tasks.

6.1 Runtime performance of the GPSL models

The proposed GPSL models are implemented in Matlab. The GPSL Matlab programs are run on a 3.2 GHz Intel processor with 4 GB of shared main memory under Linux. The SSVM approach is implemented in C, the CRF approach is coded in C++ and the GPStruct approach is in Python. Since the implementation languages differ, it is unfair to make a runtime comparison of various approaches. Table 2 compares the average runtime (in seconds) for training various GPSL models and GPstruct on the sequence labeling data sets. We find that the GPSL models are an order of magnitude faster than GPStruct. We also find that increasing the dependencies resulted in only a slight increase in runtime.

Table 2. Comparison of average running time (seconds) of various GPSL models and GPstruct

Data	GPSL1	GPSL2	GPSL4	GPstruct
Segmentation	17.13	19.64	22.83	3.82e+03
Chunking	1.09e+03	1.35e+03	1.71e+03	4.56e+04
Base NP	6.01e+03	6.69e+03	7.25e+03	7.54e+04
Japanese NE	1.24e+03	1.56e+03	1.93e+03	4.92e+04

6.2 Experiments with the Prediction algorithm

We conducted experiments to study the performance of Algorithm 2 used to make prediction. The algorithm is compared with the commonly used Viterbi algorithm [20] for the sequence labeling task. Viterbi algorithm consists of a forward phase which calculates the best value attained at the end of the sequence and a backward phase which finds the sequence of labels that lead to it. It is useful only for the setting where one considers a dependency with the previous label. Therefore, we study how the performance of the GPSL1 model differs when Viterbi algorithm is used for prediction instead of the proposed algorithm. We consider an implementation of the Viterbi algorithm provided by the UGM toolkit [22]. Table 3 compares the predictive and runtime performance of the two algorithms. We observe that Algorithm 2 gave a better predictive and runtime performance than the Viterbi algorithm. The predictive performance of Algorithm 2 is significantly better than Viterbi on *Segmentation, Chunking* and

Table 3. Comparison of the prediction algorithms using GPSL1 model

| Data | average Hamming loss | | paired t-test | average runtime (seconds) | | average iterations |
	Algorithm 2	Viterbi	t-value	Algorithm 2	Viterbi	Algorithm 2
Segmentation	23.45	24.26	3.8183	0.1227	0.0856	5
Chunking	13.02	13.69	3.6421	0.2491	0.2628	5
Base NP	5.73	5.75	0.3162	0.5207	0.5338	4
Japanese NE	8.26	8.84	2.475	0.3661	0.5653	3

JapaneseNE. The t-values calculated using paired t-test on these data sets are found to be greater than the critical value of 2.262 for a level of significance 0.05 and 9 degrees of freedom. We also observed that Algorithm 2 converged in 3–5 iterations on an average.

6.3   Experiments with Missing Labels

In many sequence labeling tasks in NLP, the labels of some of the output components might be missing in the training data set. This is common when

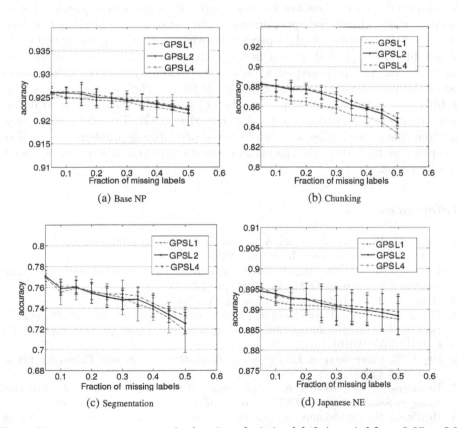

(a) Base NP

(b) Chunking

(c) Segmentation

(d) Japanese NE

Fig. 2. Variation in accuracy as the fraction of missing labels is varied from 0.05 to 0.5

crowd sourcing techniques are employed to obtain the labels. Sequence labeling approaches such as SSVM and CRF are not readily applicable to data sets with missing labels. GPSL models are useful to learn from the data sets with missing labels due to their ability to capture larger dependencies. We learn the GPSL models from the sequence labeling data sets with some fraction of the labels missing. We vary the fraction of missing labels and study how the performance of our model varies with respect to missing labels. Figure 2 provides the variation in performance of various GPSL models as we vary the fraction of missing labels. The performance is measured in terms of accuracy which is obtained by subtracting the average Hamming loss from 1. We find that the performance of the GPSL models does not significantly degrade as the fraction of the missing labels increases. Figure 2 shows that GPSL4 which uses the previous and the next 2 labels provides a better performance than the other GPSL models. GPSL4 learns a better model by considering a larger neighborhood information and is useful to handle data sets with missing labels.

7 Conclusion

We proposed a novel Gaussian Process approach to perform sequence labeling based on pseudo-likelihood approximation. The use of pseudo-likelihood enabled the model to capture multiple dependencies without becoming computationally intractable. The approach used a faster inference scheme based on variational inference. We also proposed an approach to perform prediction which makes use of the information from the neighboring labels. The proposed approach is useful for a wide range of sequence labeling problems arising in natural language processing. Experimental results showed that GPSL models, which capture multiple dependencies, are useful in sequence labeling problems. The ability to capture multiple dependencies makes them effective in handling data sets with missing labels.

References

1. Altun, Y., Hofmann, T., Smola, A.J.: Gaussian process classification for segmenting and annotating sequences. In: ICML (2004)
2. Balamurugan, P., Shevade, S., Sundararajan, S., Keerthi, S.: A Sequential dual method for structural SVMs. In: SDM, pp. 223–234 (2011)
3. Bertsekas, D.P.: Nonlinear programming. Athena Sci. (1999)
4. Besag, J.: Statistical analysis of non-lattice data. Statistician **24**, 179–195 (1975)
5. Bottou, L.: Large-scale machine learning with stochastic gradient descent. In: COMPSTAT (2010)
6. Boyd, S., Vandenberghe, L.: Convex Optimization. Cambridge University Press, New York (2004)
7. Bratieres, S., Quadrianto, N., Ghahramani, Z.: Bayesian structured prediction using gaussian processes. IEEE Trans. Pattern Anal. Mach. Intell. (2014)
8. Bratieres, S., Quadrianto, N., Nowozin, S., Ghahramani, Z.: Scalable gaussian process structured prediction for grid factor graph applications. In: ICML (2014)

9. Chai, K.M.A.: Variational multinomial logit gaussian process. J. Mach. Learn. Res. **13**, 1745–1808 (2012)
10. Girolami, M., Rogers, S.: Variational bayesian multinomial probit regression with gaussian process priors. Neural Comput. **18**(8), 1790–1817 (2006)
11. Heckerman, D., Chickering, D.M., Meek, C., Rounthwaite, R., Kadie, C.: Dependency networks for inference, collaborative filtering, and data visualization. J. Mach. Learn. Res. **1**, 49–75 (2001)
12. Khan, M.E., Mohamed, S., Murphy, K.P.: Fast bayesian inference for non-conjugate gaussian process regression. In: NIPS, pp. 3149–3157 (2012)
13. Lafferty, J.D., McCallum, A., Pereira, F.C.N.: Conditional random fields: probabilistic models for segmenting and labeling sequence data. In: ICML, pp. 282–289 (2001)
14. Lafferty, J.D., Zhu, X., Liu, Y.: Kernel conditional random fields: representation and clique selection. In: ICML (2004)
15. Li, Q., Wang, J., Wipf, D.P., Tu, Z.: Fixed-point model for structured labeling. In: ICML, pp. 214–221 (2013)
16. Murphy, K.P.: Machine learning: A Probabilistic Perspective. The MIT Press, Cambridge (2012)
17. Noah, A.S.: Linguistic Structure Prediction. Morgan and Claypool (2011)
18. Opper, M., Archambeau, C.: The variational gaussian approximation revisited. Neural Comput. **21**, 786–792 (2009)
19. Qi, Y., Szummer, M., Minka, T.P.: Bayesian conditional random fields. In: Proceedings of the AISTATS (2005)
20. Rabiner, L.R.: A tutorial on hidden markov models and selected applications in speech recognition. Proc. IEEE **77**(2), 257–286 (1989)
21. Rasmussen, C.E., Williams, C.K.I.: Gaussian Processes for Machine Learning (Adaptive Computation and Machine Learning). MIT Press, Cambridge (2005)
22. Schmidt., M.: UGM: A Matlab toolbox for probabilistic undirected graphical models (2007). http://www.cs.ubc.ca/schmidtm/Software/UGM.html
23. Sutton, C., McCallum, A.: Piecewise pseudolikelihood for efficient training of conditional random fields. In: ICML, pp. 863–870 (2007)
24. Toutanova, K., Klein, D., Manning, C.D., Singer, Y.: Feature-rich part-of-speech tagging with a cyclic dependency network. In: HLT-NAACL, pp. 252–259 (2003)
25. Tsochantaridis, I., Joachims, T., Hofmann, T., Altun, Y.: Large margin methods for structured and interdependent output variables. J. Mach. Learn. Res. **6**, 1453–1484 (2005)
26. Williams, C.K.I., Barber, D.: Bayesian classification with gaussian processes. IEEE Trans. Pattern Anal. Mach. Intell. **20**(12), 1342–1351 (1998)

OSLα: Online Structure Learning Using Background Knowledge Axiomatization

Evangelos Michelioudakis[1,2][✉], Anastasios Skarlatidis[1], Georgios Paliouras[1], and Alexander Artikis[1,3]

[1] Institute of Informatics and Telecommunications,
NCSR "Demokritos", Athens, Greece
{vagmcs,anskarl,paliourg,a.artikis}@iit.demokritos.gr
[2] School of Electronic and Computer Engineering,
Technical University of Crete, Chania, Greece
[3] Department of Maritime Studies, University of Piraeus, Piraeus, Greece

Abstract. We present OSLα—an online structure learner for Markov Logic Networks (MLNs) that exploits background knowledge axiomatization in order to constrain the space of possible structures. Many domains of interest are characterized by uncertainty and complex relational structure. MLNs is a state-of-the-art Statistical Relational Learning framework that can naturally be applied to domains governed by these characteristics. Learning MLNs from data is challenging, as their relational structure increases the complexity of the learning process. In addition, due to the dynamic nature of many real-world applications, it is desirable to incrementally learn or revise the model's structure and parameters. Experimental results are presented in activity recognition using a probabilistic variant of the Event Calculus (MLN–EC) as background knowledge and a benchmark dataset for video surveillance.

Keywords: Markov Logic Networks · Event Calculus · Uncertainty

1 Introduction

Many real-world application domains are characterized by both uncertainty and complex relational structure. Regularities in these domains are very hard to identify manually, and thus automatically learning them from data is desirable. The field of Statistical Relational Learning (SRL) [7] concerns the induction of probabilistic knowledge by combining the powers of logic and probability. One of the logic-based frameworks that handles uncertainty, proposed in the area of SRL, is Markov Logic Networks (MLNs) [24] which combines first-order logic and probabilistic graphical models.

Structure learning approaches that focus on MLNs have been successfully applied to a variety of applications where uncertainty holds [6]. However, most of these methods are batch algorithms that cannot handle large training sets or large data streams as they are bound to repeatedly perform inference over the entire training set in each learning iteration. This is computationally expensive,

© Springer International Publishing AG 2016
P. Frasconi et al. (Eds.): ECML PKDD 2016, Part I, LNAI 9851, pp. 232–247, 2016.
DOI: 10.1007/978-3-319-46128-1_15

rendering these algorithms inapplicable to real-world applications. Huynh and Mooney [12] proposed an online strategy, called OSL, for updating both the structure and the parameters of the model, in order to effectively handle large training datasets. Nevertheless, OSL does not exploit background knowledge during the search procedure and explores structures that are very common and therefore largely useless for the purposes of learning, yielding models that are not adequate generalizations of the data.

We propose the OSLα online structure learner for MLNs, which extends OSL by exploiting a given background knowledge, in order to effectively constrain the search space of possible structures during learning. The space is constrained subject to characteristics imposed by the rules governing a specific task, herein stated as axioms. To demonstrate the benefits of OSLα we focus on the domain of activity recognition. As a background knowledge we are employing MLN–EC [27], a probabilistic variant of the Event Calculus [20] for event recognition applications.

Running Example. In activity recognition the goal is to recognize *composite events* (CE) of interest given an input stream of *simple derived events* (SDEs). CEs can be defined as relational structures over sub-events, either CEs or SDEs, and capture the knowledge of a target application. Due to the dynamic nature of real-world applications, the CE definitions may need to be refined over time or the current knowledge base may need to be enhanced with new definitions. Manual curation of event definitions is a tedious and cumbersome process and thus machine learning techniques to automatically derive the definitions are essential. The proposed OSLα method is tested on the task of activity recognition from surveillance video footage. The goal is to recognize activities that take place between multiple persons, e.g. people meeting and moving together, by exploiting information about observed activities of individuals. The input stream of SDEs represents people walking, running, staying active, or inactive, and spatial relations, e.g. persons being relatively close to each other.

The remainder of the paper is organized as follows. Section 2 provides background on MLNs and MLN–EC. Section 3 discusses related work on structure learning. Section 4 describes our proposed method for online structure learning. Section 5 reports the experimental results and Sect. 6 proposes directions for future work and concludes.

2 Background

2.1 Markov Logic Networks

Markov Logic Networks (MLNs) [24] consist of weighted first-order formulas. They provide a way of softening the constraints that are imposed by the formulas and facilitate probabilistic inference. Hence, unlike classical logic, all worlds in MLNs are possible and they are quantified by a certain probability. In event recognition the focus is on discriminative MLNs [26]. Let X be a set of evidence atoms, and Y a set of query atoms. The former correspond to the input SDEs while the latter correspond to the CEs of interest in event recognition. Then the conditional probability of **y** given **x** is defined as follows:

$$P(Y = \mathbf{y} \mid X = \mathbf{x}) = \frac{1}{Z(\mathbf{x})} \exp \left(\sum_{i=1}^{|F_c|} w_i n_i(\mathbf{x}, \mathbf{y}) \right)$$

Vectors $\mathbf{x} \in \mathcal{X}$ and $\mathbf{y} \in \mathcal{Y}$ represent a possible assignment of evidence X and query/hidden variables Y, respectively. $\mathcal{X}$ and $\mathcal{Y}$ are the sets of possible assignments that the evidence X and query/hidden variables Y can take. F_c is the set of clauses produced by a knowledge base L and a domain of constants C. The scalar value w_i is the weight of the i-th clause and feature $n_i(\mathbf{x}, \mathbf{y})$ represents the number of satisfied groundings of the i-th clause in $\mathbf{x}$ and $\mathbf{y}$. $Z(\mathbf{x})$ is the partition function that normalizes the probability over all possible assignments $\mathbf{y}' \in \mathcal{Y}$ of query/hidden variables given the assignment $\mathbf{x}$.

2.2 MLN–EC: Probabilistic Event Calculus Based on MLNs

MLN–EC [27] is a probabilistic variant of the discrete Event Calculus [20] in MLNs for event recognition applications. The ontology of MLN–EC consists of *time-points*, *events* and *fluents*, represented by the finite sets $\mathcal{T}$, $\mathcal{E}$ and $\mathcal{F}$, respectively. The underlying time model is linear and represented by integers. A *fluent* is a property whose value may change over time by the occurrence of a particular *event*. MLN–EC comprises the core domain-independent axioms of Event Calculus defining whether a fluent holds or not at a specific time-point. In addition, the domain-independent axiomatization incorporates the common sense *law of inertia*, according to which fluents persist over time, unless they are affected by an event occurrence. MLN–EC axioms (1a) and (2a), shown below, determine when a fluent holds and axioms (1b) and (2b) when a fluent does not hold. Variables and functions start with a lower-case letter and are assumed to be universally quantified. Predicates start with an upper-case letter and predicate Next expresses successive time-points to avoid numerical calculations.

HoldsAt(f, t+1) $\Leftarrow$ ¬HoldsAt(f, t+1) $\Leftarrow$
    InitiatedAt(f, t) $\wedge$ (a)   TerminatedAt(f, t) $\wedge$ (b)   (1)
    Next(t, t+1)   Next(t, t+1)

HoldsAt(f, t+1) $\Leftarrow$ ¬HoldsAt(f, t+1) $\Leftarrow$
    HoldsAt(f, t) $\wedge$ ¬HoldsAt(f, t) $\wedge$
    ¬TerminatedAt(f, t) $\wedge$ (a)   ¬InitiatedAt(f, t) $\wedge$ (b)   (2)
    Next(t, t+1)   Next(t, t+1)

MLN–EC combines composite event definitions with the domain-independent axioms of MLN–EC (1)–(2), generating a compact knowledge base that serves as a pattern for the production of Markov Networks, and enables probabilistic inference and machine learning. The compact knowledge base is generated by performing predicate completion [20] – a syntactic transformation that translates formulas into logically stronger ones. The aim of predicate completion is to rule out all conditions which are not explicitly entailed by the given formulas and thus to introduce closed-world assumption to first-order logic.

3   Related Work

Learning the MLN structure is a task that has received much attention lately. The main approaches to this task stem either from graphical models [8,18,22] or Inductive Logic Programming (ILP) [4,23]. Since MLNs represent probability distributions, better results are obtained by evaluation functions based on likelihood, rather than typical ILP ones like accuracy and coverage [14].

Several approaches have been proposed to date [2,9,13,15–17,19], using various strategies to search the space of possible structures. Most of these approaches are batch learning algorithms that cannot handle very large training sets, due to their requirement to load all data in memory and carry out inference in each iteration. Moreover, most of these algorithms are strictly data-driven and thus they only seek to improve the likelihood of known true worlds.

Huynh and Mooney [12] proposed OSL that updates both the structure and the parameters of the model using an incremental approach whereby training data are consumed in (non-overlapping) micro-batches. Using incorrect predictions of the current model, OSL searches for clauses, using relational pathfinding over a hypergraph [25] constrained by mode declarations [21], and estimates or updates their parameters using the AdaGrad learner [5]. The hypergraph may be seen as a representation of the search space that contains true ground predicates, while the paths found during the mode-guided search may be seen as conjunctions of ground predicates, that are eventually generalized to clauses.

OSL does not exploit background knowledge that may be provided to constrain the search space and typically explores many structures (paths) that are not useful. Specifically, even by performing mode-guided search over the hypergraph, the space of possible paths can become exponentially large. For instance, the Event Calculus is a temporal formalism and therefore data used for training will inevitably contain a large domain of time points (possibly) having multiple complex temporal relations between events. Mode declarations alone cannot handle this large domain. It will be then fundamental to prune a portion of the search space and use only meaningful subspaces that may be found by exploiting the background knowledge axiomatization.

Finally, all aforementioned approaches assume that domains do not contain functions, which are useful in several applications, such as activity recognition.

4   Online Structure Learning Using Background Knowledge Axiomatization

Figure 1 presents the components of OSLα. The background knowledge consists of the MLN–EC axioms (i.e., domain-independent rules) and an already known (possibly empty) hypothesis (i.e., set of clauses). At any step t of the online procedure a training example (micro-batch) $\mathcal{D}_t$ arrives containing simple derived events (SDEs), e.g. two persons walking individually, their distance being less than 34 pixel positions and having the same orientation. Then, $\mathcal{D}_t$ is used together

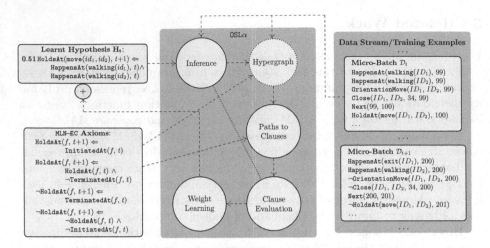

Fig. 1. The procedure of OSLα.

with the already learnt hypothesis to predict the truth values y_t^P of the composite events (CEs) of interest. This is achieved by (maximum a posteriori) MAP inference based on LP-relaxed Integer Linear Programming [10]. Given $\mathcal{D}_t$ OSLα constructs a hypergraph that represents the space of possible structures as graph paths. Then for all incorrectly predicted CEs the hypergraph is searched (guided by MLN–EC axioms) for definite clauses explaining these CEs. The paths discovered during the search are translated into clauses and evaluated. The resulting set of retained clauses is used for weight learning. Finally, the set of weighted clauses is appended to the hypothesis $\mathcal{H}_t$ and the whole procedure is repeated for the next training example $\mathcal{D}_{t+1}$.

4.1   Extracting Templates from Axioms

OSLα begins by partitioning the background knowledge into a set of axioms $\mathcal{A}$ and a set of domain-dependent definitions $\mathcal{B}$, that is the already known hypothesis $\mathcal{H}$ (herein CE definitions). Each axiom $\alpha \in \mathcal{A}$ must not contain any free variables, meaning variables only appearing in a single predicate. It should contain exactly one so-called *template predicate* and at least one *query predicate*. In the case of MLN–EC, $\mathcal{A}$ contains the four axioms (1)–(2), HoldsAt $\in \mathcal{Q}$ are the *query predicates* and InitiatedAt, TerminatedAt $\in \mathcal{P}$ are the *template predicates*. Those latter predicates specify the conditions under which a CE starts and stops being recognized. They form the target CE patterns that we want to learn.

MLN–EC axioms can be used as a template **T** over all possible structures in order to search only for explanations of the template predicates. Upon doing so, OSLα does not need to search over time sequences, instead only needs to find appropriate bodies over the current time-point for the following definite clauses:

$$\text{InitiatedAt}(f, t) \Leftarrow \text{body}$$
$$\text{TerminatedAt}(f, t) \Leftarrow \text{body}$$

The body of these definitions is a conjunction of n literals $\ell_1 \wedge \cdots \wedge \ell_n$, which can be seen as a hypergraph path, as we shall explain in the following sections.

Given the set of axioms $\mathcal{A}$, OSLα partitions it into templates. Each template $\mathbf{T}_i$ contains axioms with identical Cartesian product of domain types over their template predicate variables. MLN–EC axioms (1)–(2) should all belong to one template $\mathbf{T}_1$ because InitiatedAt and TerminatedAt both have joint domain $\mathcal{F} \times \mathcal{T}$. The resulting template $\mathbf{T}_1$ is used during relational pathfinding (see Sect. 4.3) to find an initial search set $\mathcal{I}$ of ground template predicates and search the space of possible structures for specific bodies of the definite clauses. A template $\mathbf{T}_i$ essentially provides mappings of its axioms to the template predicates that appear in the bodies of these axioms. For instance, axiom (1a) of $\mathbf{T}_1$ will be mapped to the predicate InitiatedAt(f, t) since the aim is to construct a rule for this template predicate.

4.2 Hypergraph and Relational Pathfinding

Similar to OSL, at each step t OSLα receives an example $\mathbf{x}_t$, representing the evidence part of $\mathcal{D}_t$ and produces the predicted label $\mathbf{y}_t^P = \operatorname{argmax}_{\mathbf{y}} \in \mathcal{Y}\langle \mathbf{w}, \mathbf{n}(\mathbf{x}_t, \mathbf{y})\rangle$ using MAP inference. It then receives the true label $\mathbf{y}_t$ and finds all ground atoms that are in $\mathbf{y}_t$ but not in $\mathbf{y}_t^P$ denoted as $\Delta y_t = \mathbf{y}_t \backslash \mathbf{y}_t^P$. Hence, Δy_t contains the false positives/negatives of the inference step. In contrast to OSL, OSLα considers all misclassified (false positives/negatives) ground atoms instead of just the true ones (false negatives) in order to find InitiatedAt definitions that correct the false negatives and respectively TerminatedAt for the false positives. OSLα searches the ground-truth world ($\mathbf{x}_t, \mathbf{y}_t$) for clauses specific to the axioms defined in the background knowledge using the constructed templates $\mathbf{T}_i$.

In order to discover useful clauses specific to the set of incorrectly predicted atoms Δy_t, OSLα uses relational pathfinding [25]. It considers $\mathcal{D}_t$ as a hypergraph having constants as nodes and true ground atoms as hyperedges that connect the nodes appearing as its arguments. Hyperedges are a generalization of edges connecting any number of nodes. OSLα searches the hypergraph for paths that connect the arguments of an input incorrectly predicted atom. Functions present in $\mathcal{D}_t$ are transformed into auxiliary predicates (with the prefix AUX) that model the behavior of a function and are required to indirectly include functions in the hypergraph. For example the predicate AUXwalking matches the return values of the function walking and has arity increased by 1 in order to incorporate the return type of the function as an argument of the auxiliary predicate.

A hypergraph representing the training example $\mathcal{D}_t$ of Fig. 1 is presented on the left of Fig. 2. For each incorrectly predicted ground atom in Δy_t (herein incorrectly predicted CEs), relational pathfinding searches for all paths up to a predefined length l. A path of hyperedges corresponds to a conjunction of true ground atoms connected by their arguments and can be generalized into a

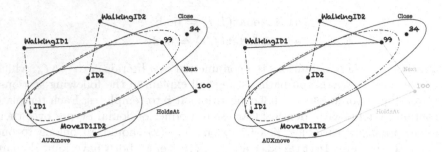

Fig. 2. Initial hypergraph (left) and reduced hypergraph (right). Unlabelled continuous lines represent `HappensAt` predicates, while unlabelled dashed lines and dashed ellipses represent `AUXwalking` and `OrientationMove` respectively.

conjunction of variabilized literals. For example consider that the predicted label y_t^P says that `HoldsAt`$(MoveID_1ID_2, 100)$ is false, while supervision in $\mathcal{D}_t$, that is y_t, says that it is true. Therefore it is an incorrectly predicted atom and the hypergraph should be searched for paths explaining the misclassified CE. Below, we present two of the paths that can be found by searching the left hypergraph of Fig. 2 for paths up to length $l = 7$.

$$\{\texttt{HoldsAt}(MoveID_1ID_2, 100), \texttt{Next}(99, 100), \texttt{HappensAt}(WalkingID_1, 99),$$
$$\texttt{HappensAt}(WalkingID_2, 99), \texttt{AUXwalking}(WalkingID_1, ID_1), \qquad (3)$$
$$\texttt{AUXwalking}(WalkingID_2, ID_2), \texttt{AUXmove}(MoveID_1ID_2, ID_1, ID_2)\}$$

$$\{\texttt{HoldsAt}(MoveID_1ID_2, 100), \texttt{Next}(99, 100), \texttt{Close}(ID_1, ID_2, 34, 99), \quad (4)$$
$$\texttt{AUXmove}(MoveID_1ID_2, ID_1, ID_2)\}$$

Similar to [12], in order to speed up relational pathfinding, OSLα uses path mode declarations as language bias to constrain the space of paths. A $\texttt{modep}(r, p)$ has two components: a recall number $r \in \mathbb{N}_0$, and an atom p whose arguments are place-markers optionally preceded by symbol '#'. A place-marker is '+' (input), '−' (output), or '.' (ignore). The symbol '#' preceding place-markers specifies that this particular predicate argument will remain constant after the generalization of the path. The recall number r limits the number of appearances of the predicate p in a path to r. These place-markers restrict the search of relational pathfinding. A ground atom is only added to a path if one of its arguments has previously appeared as 'input' or 'output' arguments in the path and all of its 'input' arguments are 'output' arguments of previous atoms. We also introduce mode declarations for functions, defined as $\texttt{modef}(r, p)$, that are used to constrain auxiliary predicates in the hypergraph.

The hypergraph is constructed from a training example $\mathcal{D}_t$, by only adding true ground atoms in $\mathcal{D}_t$ that are input or output nodes. There is no point in constructing the entire search space, because only the portion of it defined by the mode declarations will be eventually searched. Template predicates are not added in the hypergraph because they are not allowed to appear in the body of the definite clause. Hence, OSLα does not support recursive definitions.

4.3   Template Guided Search

Starting from each incorrectly predicted ground atom in Δy_t, we use the templates $\mathbf{T}_i$ constructed at the initial steps of the algorithm in order to find the corresponding ground template predicates for which the axioms belonging in $\mathbf{T}_i$ are satisfied by the current training example. As stated in Sect. 4.1 there is only one template $\mathbf{T}_1$ containing all the axioms of MLN–EC. OSLα considers each axiom $\alpha \in \mathbf{T}_1$ in turn. Assume, for example, that one of these is axiom (1a) and that we have predicted that the ground atom HoldsAt(CE, T_4) is false (false negative). We substitute the constants of HoldsAt(CE, T_4) into axiom (1a). The result of the substitution will be the following partially ground axiom:

$$\text{HoldsAt}(CE, T_4) \Leftarrow \text{Next}(t, T_4) \land \text{InitiatedAt}(CE, t) \tag{5}$$

If after the substitution there are no variables left in the template predicate of the axiom, OSLα adds the ground template predicate to the initial search set $\mathcal{I}$, containing all ground template predicates, and moves to the next axiom in the template $\mathbf{T}_1$. In case there are variables left, such as in axiom (5) were InitiatedAt has one remaining variable t, OSLα searches for all literals in the axiom sharing variables with the template predicate. Here the only literal sharing the remaining variable t is Next. For those literals, it searches the training data for all jointly ground instantiations among those satisfying the axiom. Because t represents time-points and Next describes successive time-points, there will be only one true grounding of Next in the training data having as argument the constant T_3. OSLα substitutes the constant T_3 into axiom (5) and adds InitiatedAt(CE, T_3) to the initial search set $\mathcal{I}$. The same applies for axioms (1b) and (2b) determining the termination conditions in the case of a false positive.

For each ground template predicate in the resulting initiation set $\mathcal{I}$, the mode-guided relational pathfinding is used to search the hypergraph for an appropriate body. It recursively adds to the path hyperedges (i.e., ground atoms) that satisfy the mode declarations. The search terminates when the path reaches a specified maximum length or when no new hyperedges can be added.

By employing this procedure, the hypergraph is essentially reduced to contain only ground atoms explaining the template predicates. Consider the hypergraph presented on the left of Fig. 2. By exploiting the Event Calculus axioms, the hypergraph is reduced to contain only predicates that explain the InitiatedAt and TerminatedAt predicates as presented in the right of Fig. 2. The paths (3) and (4) are pruned by removing the Next and HoldsAt predicates, resulting into the paths (6) and (7) shown below. The pruning resulting from the template guided search is essential to learn Event Calculus definitions, because the size of the search space becomes independent of time.

$\{$InitiatedAt($MoveID_1ID_2$, 99), HappensAt($WalkingID_1$, 99),

HappensAt($WalkingID_2$, 99), AUXwalking($WalkingID_1, ID_1$),   (6)

AUXwalking($WalkingID_2, ID_2$), AUXmove($MoveID_1ID_2, ID_1, ID_2$)$\}$

$\{$InitiatedAt($MoveID_1ID_2$, 99), Close(ID_1, ID_2, 34, 99),   (7)

AUXmove($MoveID_1ID_2, ID_1, ID_2$)$\}$

4.4 Clause Creation and Evaluation

In order to generalize paths into first-order clauses, we replace each constant k_i in a conjunction with a variable v_i, except for those declared constant in the mode declarations. Then, these conjunctions are used as a body to form definite clauses using as head the template predicate present in each path. The auxiliary predicates are converted back into functions. Therefore, from the paths (6) and (7), the following definite clauses will be created:

$$\text{InitiatedAt}(\text{move}(id_1, id_2), t) \Leftarrow$$

$$\text{HappensAt}(\text{walking}(id_1), t) \wedge \text{HappensAt}(\text{walking}(id_2), t) \qquad (8)$$

$$\text{InitiatedAt}(\text{move}(id_1, id_2), t) \Leftarrow \text{Close}(id_1, id_2, 34, t) \qquad (9)$$

According to the definitions (8) and (9), the move CE is initiated either when both entities are walking or the distance between them is less than 34 pixel positions. These definite clauses can be used together with the axioms of the background knowledge in order to eliminate all template predicates by exploiting equivalences resulting from predicate completion.

After the elimination process all resulting formulas are converted into clausal normal form (CNF). Therefore the resulting set of clauses is independent of the template predicates. Evaluation takes place for each clause c individually. The difference between the number of true groundings of c in the ground-truth world $(\mathbf{x}_t, \mathbf{y}_t)$ and those in predicted world $(\mathbf{x}_t, \mathbf{y}_t^P)$ is then computed (note that $\mathbf{y}_t^P$ was predicted without c). Only clauses whose difference in the number of groundings is greater than or equal to a predefined threshold μ will be added to the MLN:

$$\Delta n_c = n_c(\mathbf{x}_t, \mathbf{y}_t) - n_c(\mathbf{x}_t, \mathbf{y}_t^P) \geq \mu \qquad (10)$$

The intuition behind this measure is to add to the hypothesis $\mathcal{H}$ clauses whose coverage of the ground-truth world is significantly (according to μ) greater than that of the clauses already learnt.

Subsequently, it may be necessary to perform again predicate completion and template predicate elimination because the resulting set of formulas returned by this transformation may change entirely if any one definite clause is removed during evaluation. To illustrate these changes in the resulting hypothesis, consider the domain-dependent definitions of move – i.e., rules (8)–(9). After predicate completion, these rules will be replaced by the following formula:

$$\text{InitiatedAt}(\text{move}(id_1, id_2), t) \Leftrightarrow$$
$$(\text{HappensAt}(\text{walking}(id_1), t) \wedge \text{HappensAt}(\text{walking}(id_2), t)) \vee \quad (11)$$
$$\text{Close}(id_1, id_2, 34, t)$$

The resulting rule (11) defines all conditions under which the move CE is initiated. Based on the equivalence in formula (11), the domain-independent axiom (1a) of MLN–EC automatically produces the following free of template predicates (i.e., InitiatedAt, TerminatedAt) rules:

$$\text{HoldsAt}(\text{move}(id_1, id_2), t{+}1) \Leftarrow$$
$$\quad \text{HappensAt}(\text{walking}(id_1), t) \wedge \text{HappensAt}(\text{walking}(id_2), t) \quad (12)$$
$$\text{HoldsAt}(\text{move}(id_1, id_2), t{+}1) \Leftarrow \text{Close}(id_1, id_2, 34, t) \quad (13)$$

Similarly, the inertia axiom (2) produces:

$$\neg\text{HoldsAt}(\text{move}(id_1, id_2), t{+}1) \Leftarrow$$
$$\quad \neg\text{HoldsAt}(\text{move}(id_1, id_2), t) \wedge$$
$$\quad \neg((\text{HappensAt}(\text{walking}(id_1), t) \wedge \text{HappensAt}(\text{walking}(id_2), t)) \vee \quad (14)$$
$$\quad \text{Close}(id_1, id_2, 34, t))$$

Consider now, that during the evaluation process the definite clause (13) yields a score less than μ and therefore must be discarded. Then, the resulting hypothesis is reduced to rule (12) produced by axiom (1a), as well as rule (15) produced by axiom (2b) presented below:

$$\neg\text{HoldsAt}(\text{move}(id_1, id_2), t{+}1) \Leftarrow$$
$$\quad \neg\text{HoldsAt}(\text{move}(id_1, id_2), t) \wedge \quad (15)$$
$$\quad \neg(\text{HappensAt}(\text{walking}(id_1), t) \wedge \text{HappensAt}(\text{walking}(id_2), t))$$

4.5 Weight Learning

The weights of all retained clauses are optimized by the AdaGrad online learner [5]. At each step t of OSLα the learnt hypothesis may be updated by adding new clauses found during the hypergraph search and therefore the resulting set of clauses $\mathcal{C}_t$ may be different from the set $\mathcal{C}_{t-1}$. In order for AdaGrad to be able to apply weight updates to a constantly changing theory, OSLα searches for clauses in the current theory $\mathcal{C}_t$ that are θ-subsumed [3] by a clause in the previous theory, in order to inherit its weight. This way the already learnt weight values are transferred to the next step of the procedure. All other clauses are considered new and their weights are set to an initial value close to zero. To illustrate the procedure consider a set of definite clauses $\mathcal{C}_{t-1}$ learnt at step $t-1$, including rules (8) as well as rule (16) presented below:

$$\text{TerminatedAt}(\text{move}(id_1, id_2), t) \Leftarrow$$
$$\quad \text{HappensAt}(\text{inactive}(id_1), t) \wedge \quad (16)$$
$$\quad \text{HappensAt}(\text{active}(id_2), t)$$

By performing predicate completion upon the set C_{t-1} and using the MLN–EC axioms to eliminate the template predicates, the following hypothesis arises:

$$\Sigma_{t-1} = \begin{cases} \texttt{HoldsAt(move}(id_1, id_2), t{+}1) \Leftarrow \\ \quad \texttt{HappensAt(walking}(id_1), t) \wedge \texttt{HappensAt(walking}(id_2), t) \\ \neg\texttt{HoldsAt(move}(id_1, id_2), t{+}1) \Leftarrow \\ \quad \texttt{HappensAt(inactive}(id_1), t) \wedge \texttt{HappensAt(active}(id_2), t) \end{cases}$$

$$\Sigma'_{t-1} = \begin{cases} \texttt{HoldsAt(move}(id_1, id_2), t{+}1) \Leftarrow \\ \quad \texttt{HoldsAt(move}(id_1, id_2), t) \wedge \\ \quad \neg(\texttt{HappensAt(inactive}(id_1), t) \wedge \texttt{HappensAt(active}(id_2), t)) \\ \neg\texttt{HoldsAt(move}(id_1, id_2), t{+}1) \Leftarrow \\ \quad \neg\texttt{HoldsAt(move}(id_1, id_2), t) \wedge \\ \quad \neg(\texttt{HappensAt(walking}(id_1), t) \wedge \texttt{HappensAt(walking}(id_2), t)) \end{cases}$$

The set Σ_{t-1} contains specialized definitions of axioms (1a) and (1b), specifying that a fluent holds (or does not hold) when its initiation (or termination) conditions are met. The set Σ'_{t-1} contains specialized definitions of the inertia axioms (2a) and (2b), determining whether a specific fluent continues to hold or not at any instance of time. Weights for both sets are estimated. In the next learning step t of OSLα the set of definite clauses C_t may be expanded by the following learnt definite clause:

$$\texttt{TerminatedAt(move}(id_1, id_2), t) \Leftarrow \texttt{HappensAt(exit}(id_1), t) \qquad (17)$$

Similarly to C_{t-1}, by applying predicate completion to C_t and eliminating the template predicates using the MLN–EC axioms, a different hypothesis arises. Σ_t includes the rules of Σ_{t-1}, as well as the following, resulting from rule (17):

$$\neg\texttt{HoldsAt(move}(id_1, id_2), t{+}1) \Leftarrow \texttt{HappensAt(exit}(id_1), t)$$

Σ'_t includes the first rule appearing in Σ'_{t-1}, as well as the following rule:

$$\neg\texttt{HoldsAt(move}(id_1, id_2), t{+}1) \Leftarrow \\ \neg\texttt{HoldsAt(move}(id_1, id_2), t) \wedge \\ \neg((\texttt{HappensAt(walking}(id_1), t) \wedge \texttt{HappensAt(walking}(id_2), t)) \\ \vee \texttt{HappensAt(exit}(id_1), t))$$

Note that in the set Σ_t a new rule has appeared and in the set Σ'_t the second rule changed by incorporating a new literal. Therefore in order to refine the weights of the current theory at step t a mapping of the previous learned weights onto the current theory is required so that the already learned values are retained. Using θ-subsumption, OSLα searches for clauses in C_t that are subsumed by clauses in C_{t-1} to inherit their weights. In the example above, the first rule of Σ_t and Σ'_t, as well as the second rule of Σ_t are identical to the previous ones. Moreover, the second rule of Σ'_t is θ-subsumed by the second rule of Σ'_{t-1}.

Hence the weights of the old rules will be used for the new ones. The last rule of Σ_t is completely new and its weight is set to a default initial value.

At the end of the OSLα learning we can choose to remove clauses whose weights are smaller than a predefined threshold ξ. Hence, the hypothesis may be pruned significantly, with negligible penalty in accuracy.

All algorithms composing OSLα (e.g., hypergraph construction), in pseudo-code, are available from iit.demokritos.gr/~vagmcs/pub/osla/appendix.pdf.

5   Empirical Evaluation

We evaluate OSLα in activity recognition, using the publicly available benchmark dataset of the CAVIAR project[1]. The dataset comprises 28 surveillance videos, where each frame is annotated by human experts from the CAVIAR team on two levels. The first level contains SDEs that concern activities of individual persons or the state of objects. The second level contains CE annotations, describing the activities between multiple persons and/or objects, i.e., people meeting and moving together, leaving an object and fighting.

5.1   Experimental Setup

The input to the learning methods being compared is a stream of SDEs along with the CE annotations. The SDEs represent people walking, running, staying active, or inactive. The first and last time that a person is tracked is represented by the enter and exit SDEs. Additionally, the coordinates of tracked persons are also used to express qualitative spatial relations, e.g. two persons being relatively close to each other. The CE supervision indicates when each of the CEs holds. The structure of the training sequences is presented Fig. 1. Each sequence is composed of input SDEs (HappensAt), precomputed spatial constraints (Close), and the corresponding CE annotations (HoldsAt). Negated predicates in the sequence state that the truth value of the corresponding predicate is False.

From the 28 videos, we have extracted 19 sequences that are annotated with the meet and/or move CEs. The rest of the sequences in the dataset are ignored, as they do not contain positive examples of the target CEs. Out of the 19 sequences, 8 are annotated with both meet and move activities, 9 are annotated only with move and 2 only with meet. The total length of the extracted sequences is 12869 frames. Each frame is annotated with the (non-)occurrence of a CE and is considered an example instance. The whole dataset contains a total of 63147 SDEs and 25738 annotated CE instances. There are 6272 example instances in which move occurs and 3722 in which meet occurs. Consequently, for both CEs the number of negative examples is significantly larger than the number of positive examples, specifically 19466 for move and 22016 for meet.

Throughout the experimental analysis, the evaluation results were obtained using MAP inference, as per [10] and are presented in terms of True Positives (TP),

---

[1] http://homepages.inf.ed.ac.uk/rbf/CAVIARDATA1.

False Positives (FP), False Negatives (FN), Precision, Recall and F_1 score. All reported statistics are micro-averaged over the instances of recognized CEs using 10-fold cross validation over the 19 sequences. The average SDEs per fold are 56832 and the average positive CEs are 3350 and 5600 for meet and move respectively. The experiments were performed in a computer with an Intel i7 4790@3.6 GHz processor (4 cores and 8 threads) and 16 GiB of RAM, running Apple OSX version 10.11. All weight and structure learning methods are implemented in LoMRF[2], an opensource implementation of MLNs.

We ran experiments using the AdaGrad [5] and CDA [11] online weight learners as well as a batch max-margin learner [10], using manual definitions developed in [1][3]. These definitions take the form of common sense rules and describe the conditions under which a CE starts or ends (InitiatedAt, TerminatedAt). For example, when two persons are walking together with the same orientation, then move starts being recognized. Similarly, when two persons walk away from each other, then move stops being recognized. We also include in the experiments the results of the logic-based activity recognition method of [1], hereafter EC_{crisp}, that employs a different variant of the Event Calculus, uses the same manual definitions of CEs and cannot perform probabilistic reasoning.

5.2 Experimental Results

We ran structure learning using 10-fold cross validation over 5 distinct values of the evaluation threshold μ—see formula (10). (All other numerical thresholds were manually set.) The highest accuracy is achieved by using $\mu = 4$ and $\mu = 1$ for the meet and move CEs respectively. See Table 1a and b. The batch max-margin weight learning yields the best overall accuracy due the fact that it uses all the data at once to estimate the weights. AdaGrad is the second best choice among the weight learners as it yields more accurate results as opposed to CDA. It also outperforms the unweighted manual knowledge base EC_{crisp}. OSLα achieves very good results, outperforming AdaGrad in the meet CE and achieving a similar F_1 score with it in the move CE. This is very encouraging given that OSLα does not use manually curated rules.

Table 2 presents the averaged training times for the two CEs. The training time for move is much higher than that for meet. This is because move includes the predicate OrientationMove in its predicate mode declarations, leading to a larger search space. We also attempted to perform probabilistic structure learning on this dataset using OSL. Specifically, we began running experiments for the meet CE and we terminated the experimentation after 25 h. During this time OSL had processed only 4 training examples (micro-batches) out of the 17 of the first fold. OSLα on the other hand performed 10 fold cross validation for the meet CE in about 4 h.

In order to secure efficient CE recognition, we prune a portion of the learned weighted structures having absolute weights below a certain threshold ξ, for various values of ξ, and present the results in terms of both accuracy and testing

---

[2] https://github.com/anskarl/LoMRF.
[3] The MLN–EC definitions and CAVIAR dataset can be found in www.iit.demokritos.gr/~anskarl/pub/mlnec/MLN-EC_CAVIAR-20130319-00_07_20.tar.bz2.

Table 1. Recognition accuracy for the two CEs.

Method	Precision	Recall	F₁ score	Method	Precision	Recall	F₁ score
EC_crisp	0.6868	0.8556	0.7620	EC_crisp	0.9093	0.6390	0.7506
MaxMargin	**0.9189**	0.8133	**0.8629**	MaxMargin	0.8443	**0.9410**	**0.8901**
CDA	0.9061	0.4878	0.6342	CDA	0.9032	0.6706	0.7697
AdaGrad	0.7228	**0.8547**	0.7833	AdaGrad	**0.9172**	0.6674	0.7726
OSLα	0.8192	0.8509	0.8347	OSLα	0.8056	0.7522	0.7780

(a) Results for the meet CE ($\mu = 4$)   (b) Results for the move CE ($\mu = 1$))

Table 2. Average training times for meet and move CE.

Method	meet	move
OSLα	00 h 23 m 04 s	1 h 59 m 06 s
OSL	> 25 h 00 m 00 s	-

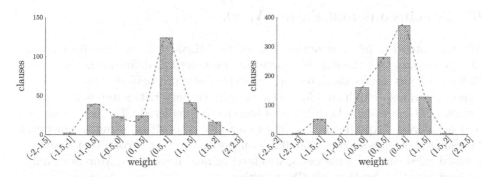

Fig. 3. Weight distribution learned for meet (left) and move (right).

time. We begin by running OSLα on all 19 sequences of the dataset and present a histogram for each CE representing the distribution of weights learned (Fig. 3). The histograms inform us about the portion of the theory that will be pruned for each ξ value. Note that there is a larger number of clauses with weight values in the range $(-1, 1)$. Some of these clauses may be pruned in order to simplify the model without significantly hurting the accuracy, but yielding better inference times. We pruned the resulting structure for 3 distinct values of ξ and present the results obtained over 10 folds.

Figure 4 presents the reduction in the number of clauses in the resulting theory and the effect in accuracy and testing time as ξ increases. It is worth noting that $\xi = 0.5$ results in a slight reduction in accuracy for move and no reduction for meet, but test time is improved a lot. Therefore, we can safely prune a subset of the resulting theory in order to improve inference performance.

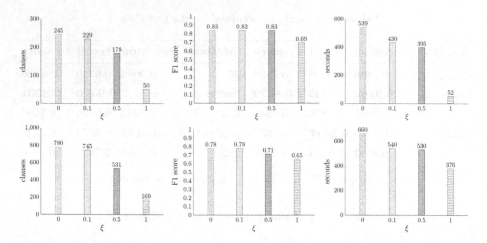

Fig. 4. Effect in the number of clauses learned (left), accuracy (center), and test time (right) as ξ increases for the meet CE (top) and the move CE (bottom).

6   Conclusions and Future Work

We presented the OSLα structure learner for MLNs that exploits background knowledge and uses the MLN–EC axioms to construct CE definitions. The use of MLN–EC axioms allows OSLα to constrain the space of possible structures (i.e., hypergraph) and search only for clauses having characteristics imposed by these axioms. OSLα considers both types of incorrectly predicted CEs (false positives and negatives). Experimental results in activity recognition using a real-world benchmark dataset showed that OSLα outperforms event recognition based on manual rules, and, in some cases, weighted manual definitions. Moreover, OSLα outperforms OSL by learning CE definitions orders of magnitude faster.

We are exploring several directions for future work, such as improving the hypergraph search further using a heuristic or randomized (parallel) graph search procedure, and learning definitions that include negated predicates. We are also studying the problem of structure learning in the presence of unobserved data.

Acknowledgments. This work has been funded by the EU FP7 project SPEEDD (619435).

References

1. Artikis, A., Skarlatidis, A., Paliouras, G.: Behaviour Recognition from Video Content: a Logic Programming Approach. In: IJAIT, pp. 193–209 (2010)
2. Biba, M., Ferilli, S., Esposito, F.: Discriminative structure learning of markov logic networks. In: Železný, F., Lavrač, N. (eds.) ILP 2008. LNCS (LNAI), vol. 5194, pp. 59–76. Springer, Heidelberg (2008). doi:10.1007/978-3-540-85928-4_9
3. De Raedt, L.: Logical and Relational Learning. Springer, Heidelberg (2008)

4. De Raedt, L., Dehaspe, L.: Clausal discovery. Mach. Learn. **26**, 99–146 (1997)
5. Duchi, J., Hazan, E., Singer, Y.: Adaptive subgradient methods for online learning and stochastic optimization. J. Mach. Learn. Res. **12**, 2121–2159 (2011)
6. Domingos, P., Lowd, D.: Markov Logic: An Interface Layer for Artificial Intelligence. Morgan & Claypool Publishers, San Rafael (2009)
7. Getoor, L., Taskar, B.: Introduction to Statistical Relational Learning (2007)
8. Heckerman, D.: Learning in Graphical Models. A Tutorial on Learning with Bayesian Networks, pp. 301–354 (1999)
9. Huynh, T.N., Mooney, R.J.: Discriminative structure and parameter learning for Markov logic networks. In: Proceeding of the 25th ICML, pp. 416–423 (2008)
10. Huynh, T.N., Mooney, R.J.: Max-margin weight learning for Markov logic networks. In: Proceeding of the ECML PKDD, pp. 564–579 (2009)
11. Huynh, T.N., Mooney, R.J.: Online max-margin weight learning for Markov logic networks. In: Proceeding of the 11th SDM, pp. 642–651 (2011)
12. Huynh, T.N., Mooney, R.J.: Online structure learning for Markov logic networks. In: Proceeding of the ECML PKDD, pp. 81–96 (2011)
13. Khot, T., Natarajan, S., Kersting, K., Shavlik, J.: Gradient-based boosting for statistical relational learning: the markov logic network and missing data cases. Mach. Learn. **100**, 75–100 (2015)
14. Kok, S., Domingos, P.: Learning the structure of Markov logic networks. In: Proceeding of 22nd ICML, pp. 441–448 (2005)
15. Kok, S., Domingos, P.: Learning markov logic network structure via hypergraph lifting. In: Proceeding of the 26th ICML, pp. 505–512 (2009)
16. Kok, S., Domingos, P.: Learning Markov logic networks using structural motifs. In: Proceeding of the 27th International Conference on Machine Learning (ICML), pp. 551–558 (2010)
17. Khosravi, H., Schulte, O., Man, T., Xu, X., Bina, B.: Structure learning for Markov logic networks with many descriptive attributes. In: Proceeding of the 24th AAAI (2010)
18. McCallum, A.: Efficiently inducing features of conditional random fields. In: Proceeding of the 19th UAI, pp. 403–410 (2003)
19. Mihalkova, L., Mooney, R.J.: Bottom-up learning of Markov logic network structure. In: Proceeding of the 24th ICML, pp. 625–632 (2007)
20. Mueller, E.T.: Event calculus. In: Handbook of Knowledge Representation, vol. 3. of Foundations of Artificial Intelligence, pp. 671–708. Elsevier (2008)
21. Muggleton, S.: Inverse entailment and progol. New Gen. Comput. **13**(3), 245–286 (1995)
22. Della Pietra, S., Della Pietra, V., Lafferty, J.: Inducing features of random fields. IEEE Trans. Pattern Anal. Mach. Intell. **19**(4), 380–393 (1997)
23. Quinlan, J.R.: Learning logical definitions from relations. Mach. Learn. **5**(3), 239–266 (1990)
24. Richardson, M., Domingos, P.: Markov logic networks. Mach. Learn. **62**(1), 107–136 (2006)
25. Richards, B.L., Mooney, R.J.: Learning relations by pathfinding. In: Proceeding of the 10th AAAI, pp. 50–55 (1992)
26. Singla, P., Domingos, P.: Discriminative training of Markov logic networks. In: Proceeding of the 20th AAAI, pp. 868–873 (2005)
27. Skarlatidis, A., Paliouras, G., Artikis, A., Vouros, G.A.: Probabilistic event calculus for event recognition. ACM Trans. Comput. Logic **16**, 1–37 (2015)

Beyond the Boundaries of SMOTE

A Framework for Manifold-Based Synthetically Oversampling

Colin Bellinger[1,2]([✉]), Christopher Drummond[1,2], and Nathalie Japkowicz[1,2]

[1] School of Electrical Engineering and Computer Science,
University of Ottawa, Ottawa, Canada
cbell052@uottawa.ca, nat@site.uottawa.ca
[2] National Research Council of Canada, Ottawa, Canada
Christopher.Drummond@nrc-cnrc.gc.ca
http://www.uottawa.ca
http://www.nrc-cnrc.gc.ca

Abstract. Problems of class imbalance appear in diverse domains, ranging from gene function annotation to spectra and medical classification. On such problems, the classifier becomes biased in favour of the majority class. This leads to inaccuracy on the important minority classes, such as specific diseases and gene functions. Synthetic oversampling mitigates this by balancing the training set, whilst avoiding the pitfalls of random under and oversampling. The existing methods are primarily based on the SMOTE algorithm, which employs a bias of randomly generating points between nearest neighbours. The relationship between the generative bias and the latent distribution has a significant impact on the performance of the induced classifier. Our research into gamma-ray spectra classification has shown that the generative bias applied by SMOTE is inappropriate for domains that conform to the manifold property, such as spectra, text, image and climate change classification. To this end, we propose a framework for manifold-based synthetic oversampling, and demonstrate its superiority in terms of robustness to the manifold with respect to the AUC on three spectra classification tasks and 16 UCI datasets.

Keywords: Machine learning · Class imbalance · Synthetic oversampling · Manifold and embeddings

1 Introduction

In problems such as radioactive threat classification, oil spill classification, gene function annotation, medical and text classification, the class distribution is imbalanced and the minority class is rare [5,6,18]. Rarity, in this sense, breaks the general assumption of machine learning that demands a representative set of instances from each class. Failure to satisfy this leads to the induction of a decision boundary that is biased in favour of the majority class, thereby causing

© Springer International Publishing AG 2016
P. Frasconi et al. (Eds.): ECML PKDD 2016, Part I, LNAI 9851, pp. 248–263, 2016.
DOI: 10.1007/978-3-319-46128-1_16

weak classification accuracy [15, 25]. Given the practical importance, and the significant challenge posed by domains of this nature, class imbalance has been identified as one of the essential problems in machine learning [26] and has spawned workshops, conferences and special issues [8, 9].

The obvious solution to this problem is more training samples. This is not possible in cases of imbalance arising due to domain properties, such as acquisition cost and class probability. Thus, we turn to the generation of synthetic instances based on the available training instances. Within class imbalance, this is known as synthetic oversampling, and was originally devised to compensate for the weakness of random oversampling [10].

Synthetic oversampling offers a means of balancing the training classes without discarding useful instances from the majority class via random undersampling and without risking overfitting by replicating examples with random oversampling. Instead, the training instances that belong to the minority class are used as the foundation from which to synthesize additional training instances. This avoids overfitting and effectively expands the minority space. How the space is expanded depends on the bias of the synthetic oversampling method, which dictates the way in which the probability mass of the training instances is spread through the feature space.

The state-of-the-art methods in synthetic oversampling are based on the SMOTE algorithm. The two major criticisms of SMOTE are that in some cases it synthesizes instances inside the majority class, thus causing the induced classifier to overcompensate by pushing the decision boundary into the majority space, and in other cases it does not synthesize instances close enough to the majority class. This results from the fact that the instances are synthesized in the convex-hull formed by the minority training points [3]. These negative effects grow quickly with absolute imbalance and dimensionality. In a well-sampled low-dimensional dataset, SMOTE can be expected to interpolate synthetic points between training instances that are in the same local neighbourhood of the feature space. Therefore, the likelihood that the synthetic instances are representative of the latent distribution is high. When there are very few samples of the class, however, the samples are more likely to be dispersed around the feature space. Thus, interpolating synthetic instances between them is likely to be error prone.

In an attempt to manage this, a set of ad-hoc modifications have been proposed to remove minority instances generated in the majority space, whilst others have been proposed to promote the generation of instances close to the majority space [2, 3, 14, 20]. We see these alternatives as addressing symptoms resulting from a generative bias that is inappropriate for the data rather than treating the root cause of the weaknesses. Specifically, these methods have been designed and applied without giving consideration to properties of the data to which that are applied.

In order to maximize the likelihood of generating effective instances from a small training set, we argue that it is essential to design synthetic oversampling methods with biases that match the properties of the target data. The benefit of the correct bias is effectively demonstrated with the analogous problem of inducing a representative function from the training data in Fig. 1. To induce a function, like a generative model, we start with a bias, such as a linear or

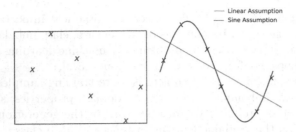

Fig. 1. Left: Training instance. Right: Approximating a sine function with and without prior knowledge.

non-linear function, and a set of free parameters that accompany the bias. The induction process quantifies the free parameters so that they best fit the training data. Selecting the correct bias, in the case of our example, a sine function, increases our likelihood of inducing a good representation, whereas selecting an incorrect bias, such as a linear function with Gaussian noise, will produce a very weak approximation. Similarly, utilizing an incorrect bias in the context of synthetic oversampling can lead to inaccurate synthetic instances that negatively impact classifier induction.

Based on our practical experience in applying synthetic oversampling methods to gamma-ray spectral classification problems, we were able to identify the manifold property as one that has a negative impact on the existing methods. A dataset conforms to the manifold property when its probability density resides in a lower-dimensional space that is embedded in the feature space [7]. The embedded space is thus constructed by combining a subset of features from the feature space. For data that conforms to the manifold property, the embedded representation offers a more concise form than the feature space, much like the grammar and syntax of a programming language provide a much more concise representation of the program to the computer than the pseudo code intended for human consumption. Whilst the embedded space resulting from manifold learning is a form of dimension reduction, it is much more than simple feature selection. Feature selection can, at best find, a subset of the existing features in the feature space. Alternatively, manifold learning discovers a completely new set of features to better represent the data.

Data with the manifold property is common within a diverse set of machine learning domains, ranging from global climate change to medicine. The bias applied by SMOTE uses the straight line distance between training points in the minority class. This is generalized as the Minkowski distance, which is an inaccurate measure for manifolds. Therefore, choosing SMOTE to synthetically oversample data that conforms to the manifold property is similar to choosing a linear model to represent the sine function in Fig. 1; the best we can hope for is synthetic data that is very weakly related to the target distribution. To address this, we propose a framework for synthetically oversampling data that conforms to the manifold property.

The contributions of this paper include: (1) identifying a general weakness in synthetic oversampling methods on data that conforms to the manifold property, (2) illustrating the cause of this weakness for SMOTE, (3) articulating the benefit of synthetic oversampling with a manifold bias, (4) proposing a framework for manifold-based synthetic oversampling, and (5) demonstrating the superiority of the framework using two distinct formalization on artificial data, gamma-ray spectra data and UCI data that conforms to the manifold property.

2   Problem Overview

Our research was originally inspired by our collaboration with the Radiation Protection Bureau at Health Canada where we applied machine learning for safety in regards to radiation. The primary challenges were the high-dimensionality of the domain and the degree of imbalance. These are features that are common to a large number of classification domains, such as global climate change, image recognition, human identification, text classification and spectral classification.

We recognized that domains with this property can often be better represented in a lower-dimensional embedded space. This concept takes advantage of the reality that instances are not spread throughout the feature space but are concentrated around a lower-dimensional manifold. A simple example of a manifold in machine learning comes from handwritten digit recognition, where the digits are recorded in a high-dimensional feature space, but can be effectively represented in a lower-dimensional embedded space that encodes the various orientations and rotations of the digit [12]. Thus, manifold learning provides a gateway to the embedded space in which all possible handwritten digits can be encoded.

A significant amount of research has been dedicated to the development of manifold learning methods [17]. The resulting algorithms utilize a diverse set of assumptions and biases, such as the complexity of the curvature of the manifold and the nature of the noise. Classic methods such as PCA and MDS are simple and efficient. These are guaranteed to determine the structure of the data on or near the embedded manifold. These traditional methods assume a linear manifold [21]. Other, more algorithmically complex methods, such as kernel PCA and autoencoding, enable the induction of non-linear manifolds. Manifold learning has demonstrated great potential in clustering, classification and dimension reduction [4,23,27]. However, in spite of their potential, manifold learning methods have gone unconsidered in problems of class imbalance. We address this gap in the literature with a framework for manifold-based synthetic oversampling.

We illustrate the weaknesses of SMOTE using a one-dimensional manifold embedded in a two-dimensional space. This is visualized in Fig. 2. Because the more recent methods that have been proposed to improve SMOTE all apply the same bias, they suffer from the same weaknesses on data that conforms to the manifold property. For this reason, when we refer to SMOTE, we intend for it to include its derivatives.

The top left graphic in Fig. 2 shows the manifold in red with samples from the manifold appearing as black circles. Each instance can be represented by its one-dimensional coordinate m in the manifold space. In machine learning, we often have data in the feature space, not the embedded space. Manifold learning induces a model of the embedded space, and from this we can focus the generation of instances in high probability regions. This is visualized in the top right graphic where the blue shading illustrates the probability mass being spread along the manifold. In the subsequent section, we demonstrate how this is achieved with our proposed framework.

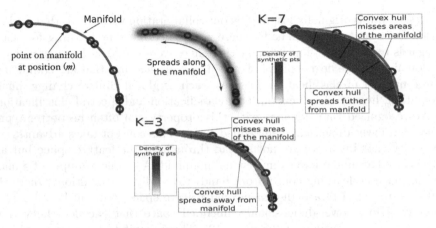

Fig. 2. Erroneous spread of instances away from the manifold with SMOTE.

The bottom graphics demonstrate the result of synthetic oversampling with SMOTE with $k = 7$ and $k = 3$. It balances the training set by interpolating points between k nearest neighbours in the minority training set [10]. As a result, the k value indirectly affects the area covered by the convex hull. The convex hull is represented by the blue area. A larger k value will uniformly spread points over a larger area, whereas a smaller k value creates dense, small clusters of synthetic points. This is emphasized with the shading of the convex hulls.

SMOTE uses the straight line distance to calculate the kNN set for each instance in the minority class, and generates new instances at random points on the edges connecting these neighbours. Due to the topological structure of a manifold, this will only produce an accurate kNN set if the query instances are close together [13]. In problems of class imbalance there are few minority training instances and as a result, this is unlikely to occur. When SMOTE is applied in this context, the convex-hull can extend away from the manifold. In our example, we see that it extends well below the red line representing the latent distribution that we hope to synthetically oversample.

3 Framework

Figure 3 presents the three components of our framework for manifold-based synthetic oversampling. Our objective is to provide a standalone synthetic over-sampler. Therefore, although the data is generated in a hidden embedded space, it is provided to the user in the original feature space. Subsequently, the user can apply a pre-processing method that is appropriate for the classifier.

The first element of the framework induces a manifold representation of the minority class via a well-suited method, such as PCA, kernel PCA, autoencoding, local linear embedding, *etc.* Data is synthesized along the induced manifold during the second phase of the framework, and the final phase maps the synthesized data to the original feature space and returns it to the user.

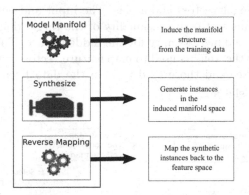

Fig. 3. General framework for synthetic oversampling.

The number of training examples and the complexity of the latent manifold are two factors to consider when selecting a manifold learning method to employ in the framework. If the learning objective involves a linear manifold, or the training data is extremely rare, a linear method is appropriate. Alternatively, non-linear problems with more training data are well-suited for methods that can represent the complexity. Our experiments focus on PCA and denoising autoencoder because together they can model linear and non-linear manifolds that are simple or complex. Moreover, they offer effective and easy-to-implement means of sampling from the induced manifold.

Formalization with PCA: PCA is a linear mapping from the d-dimensional input space to a k-dimensional embedded space where $k \ll d$. The standard process is a result of calculating the leading eigenvectors E corresponding to the k largest eigenvalues λ from the sample covariance matrix Σ of the target data.

In the PCA realization of the framework, a model $pca = \{\mu, \Sigma, E, \lambda\}$ of the d-dimensional target class T with m instances is produced. We produce a synthetic set S of n instances in the manifold-space by randomly sampling n

instances from $T' = T \times E$ (T in the PCA-space) with replacement. In order
to produce unique samples on the manifold, we apply *i.i.d.* additive Gaussian
noise $\mathcal{N}(0, \mathcal{I})$ to each sampled instance prior to adding it to the synthetic set
S. The covariance matrix for the Gaussian noise is a diagonal matrix with each
$\sigma_{i,i}$ specified by $\beta \lambda_i$, where β is the scaling factor applied to the eigenvalues.
This controls the spread of the synthetic instances relative to the manifold, and
can be thought of as a geometric transformation of points along the manifold,
thereby producing new synthetic samples on the manifold. Finally, we map the
synthetic instances S into the feature space as $S' = S \times E^{-1}$ and return them
to the user for use in classifier induction.

Formalization with Autoencoders: Autoencoders are a form of artificial
neural networks commonly used in one-class classification [16]. They have an
input layer, hidden layer and output layer, with each layer connected to the
next via a set of weight vectors and a bias. The input and output layers have a
number of units equal to the dimensionality of the target domain, and the user
specifies an alternate dimensionality for the hidden space. The learning process
involves optimizing the weights used to map feature vectors from the target class
into the hidden space and those used to map the data from the hidden space
back to the output space.

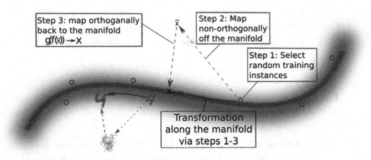

Fig. 4. Three steps of synthesization for the autoencoder formalization with generic
points and handwritten 4s. (Color figure online)

A manifold bias is incorporated in the autoencoding process through its map-
ping from the feature space to the hidden-space and back via $f_\theta(\cdot)$ and $g_{\theta'}(\cdot)$,
where:

$$f_\theta(x) = s(\mathbf{W}x + b)$$
$$g_{\theta'}(y) = s'(\mathbf{W}'y + b'). \tag{1}$$

Here, θ and θ' represent the induced encoding and decoding parameter set,
respectively. Specifically, $\mathbf{W}$ is a $d \times d'$ weight matrix and b is a d-dimensional
bias vector. The function s, is a non-linear squashing function, such as the sig-
moidal. In the decoding parameter set, $\mathbf{W}'$ and b' represent the weight matrix
and the bias vector that cast the encoded vector back to the original space. The s'

function is typically linear in autoencoders. As is standard with artificial neural networks, the weights are learnt using backpropagation and gradient descent. In addition, we utilize denoising during the training process as a form of regularization to promote the learning of key aspects of the input distribution [24]. We add Gaussian noise to the input and the network learns to reconstruct the clean instances.

The learning processes prioritizes the dual objective of a reconstruction function $g(f(\cdot))$ that is as simple as possible, but capable of accurately representing neighbouring instances from the high-density manifold [1]. This promotes accurate reconstruction of points on the manifold, whilst the reconstruction error $|x - g(f(x))|^2$ rises quickly for examples orthogonal to the manifold. Given a point, p, on the manifold, the output $g(f(p))$ remains on the manifold in essentially the same location. Conversely, when an arbitrary point, q, is sampled from off the manifold, the output $g(f(q))$ is mapped orthogonally to the manifold. This is demonstrated in Fig. 4 as $g(f(\tilde{x})) \rightarrow x$, where $\tilde{x}$ is a point off the manifold, with the manifold depicted in red.

The mapping $g(f(\tilde{x})) \rightarrow x$ is key to the formalization of the autoencoder version of our framework. The basic objective is to induce the manifold representation of the minority class and use its ability to perform orthogonal mappings to the manifold to generate samples. Generally speaking, we take an arbitrary minority class instances x, apply a non-orthogonal mapping off the manifold $x \rightarrow \tilde{x}$ and map it orthogonally back to the manifold via $g(f(\tilde{x})) \rightarrow y$. The result is a transformation along the manifold from a training instances x to synthetic instances y. This is illustrated graphically in Fig. 4. The non-orthogonal mapping is produced by adding noise to the training instance x. A greater amount of noise leads to a larger transformation along the manifold. By sampling n instances from the minority class with replacement and performing the transformation, we produce the synthetic set. We note that $g(\cdot)$ maps the synthetic set returned to the user into the target feature space. Algorithm 1 formalizes the method.

Prior to calling Algorithm 1, we perform model selection with the reconstruction error by randomly searching the parameter-space using the minority training data $\mathcal{X}$. This facilitates a simple and effective form of model selection and is the standard means of model selection for autoencoders. Nonetheless, we are exploring alternate forms of model selection for this novel application of the autoencoder. The model selection process of the autoencoder provides the ability to set the free parameters according to the target class, whereas this is not possible with the SMOTE-based methods. As a result, the user cannot know if they have specified a good value for k until they apply the classifiers after synthetic oversampling.

Given the few training instances in problems of class imbalance, we prefer a simple model rather than an overly complex model of the manifold. To encourage this, we conduct the parameter search over a relatively small number of hidden units and training epochs. For the spectra data, we searched 5–30 hidden units with fewer than a thousand epochs of training.

Algorithm 1. dae-SyntheticOversampling($\mathcal{X}$, $DAE_{\{\mathbf{W},b\}}$, n, σ)

Input:

  i) $\mathcal{X}$, an m by d dimensional data matrix.
  ii) $DEA_{\{\mathbf{W},b\}}$, a trained denoising autoencoder with weight matrix $\mathbf{W}$ and bias b.
  iii) n, the number of instances to synthesize.
  iv) σ, variance of the Gaussian sample initiation noise.

Output:

  i) $\mathcal{Y}$, the synthetic samples.

Method:

  1: $\mathcal{X}'$: column normalization of $\mathcal{X}$ between $[-1, 1]$.
  2: $normParams$: column normalization parameters of $\mathcal{X}$.
  3: $\mathcal{Z}$: normalized $\mathcal{X}$ plus sample initiation noise $\mathcal{N}(0, \sigma)$.
  4: $\mathcal{Y}' = DAE_{\{\mathbf{W},b\}}(Z)$: samples $\mathcal{Y}'$ from the induced manifold.
  5: $\mathcal{Y}$: denormalization of $\mathcal{Y}$ based on $normParams$.
  6: $Return(\mathcal{Y})$

End Algorithm

Computational Overhead: The degree to which the framework will add computational overhead depends on the manifold learning method selected. Building a PCA model, for example, requires eigenvalue decomposition of the covariance matrix of the feature vectors. Using Jacobis method for diagonalization requires $\mathcal{O}(d^3 + d^2 m)$ computations; however, the efficiency can be improved [19]. More sophisticated manifold learning methods, such as autoencoders, that involve iterative learning can take longer. The key point to remember here is that learning is being performed on a small training set. This significantly limits the training time because there are few examples to look at, and we want to avoid overfitting. This applies to SMOTE as well. Although SMOTE has the potential to be very slow due to its nearest neighbour search, the small training set means that, in practice, it is reasonable fast. Unlike our proposed framework, however, the adaptions of SMOTE take significant performance hits because they search for nearest neighbours in the entire training set.

4 Gamma-Ray Spectral Classification

Spectra classification for the Radiation Protection Bureau at Health Canada sparked our initial interest in the relationship between manifolds and synthetic oversampling.

4.1 Data

Two gamma-ray spectra datasets from the Canadian national environmental monitoring system, and one collected as part of event security at the Vancouver Olympics are utilized in our experiments. The environmental monitoring datasets were recorded at Thunder Bay and Saanich. These cities were selected for testing because they are geographically, geologically and atmospherically very distinct. This provides for very distinct data distributions.

During a four month period, 19, 112 spectra were recorded at Saanich, 44 of which were from the minority class. At Thunder Bay, 11, 602 spectra instances were recorded, with 29 belonging to the minority class. The Vancouver Winter Games data was recorded and monitored to ensure that no radioactive material entered the venue. There are 39, 000 background instances in the dataset and 39 isotopes of interest in the minority class. The two environmental datasets are 250-dimensional and the Vancouver data is 500-dimensional.

Through discussions with our colleagues at the Radiation Protection Bureau at Health Canada, we inferred the conformance of this data to the manifold property. In particular, we know that the radioactive occurrences that form the minority class will affect subset specific energy levels in the spectra, forming an embedded space.

4.2 Evaluation

We utilize the SVM, MLP, kNN, naïve Bayes and decision tree classifiers in the following experiments. Synthetic oversampling is performed by the autoencoder and PCA formalizations of the framework. These are compared to SMOTE and SMOTE with the removal of Tomek links [22]. The latter is performed in order to remove synthetic instances generated in, or too close to, the majority class. This will potentially assist SMOTE by removing erroneously synthesized instances.

We perform 5 × 2-fold cross validation and report the mean and standard deviations of the AUC performance. This form of cross validation method is ideal for large datasets such as these, and has been shown to have lower probability of issuing a Type I error as compared to k-fold cross validation [11].

4.3 Experimental Results

The mean and standard deviation of the AUC after the application of manifold-based synthetic oversampling and SMOTE-based synthetic oversampling is reported for each classifier on each dataset in Table 1. We specifically show the results of the best manifold-based (PCA or autoencoder) and SMOTE-based (SMOTE or SMOTE with the removal of Tomek links) synthetic oversampling implementation in these tables. This is done to emphasize the relative performance of the two approaches, and shows that the manifold-based framework is superior on the gamma-ray datasets. The combination of the manifold-based method with each classifier produces higher mean AUCs on the Vancouver and Thunder Bay datasets. This is also the case on the Saanich dataset for all except with the SVM classifier. In addition, we report the mean AUC across all classifiers. This shows that our framework is generally superior regardless of the classifier.

Table 1. The mean AUC results for each method on the gamma-ray dataset.

	Vancouver				Thunder Bay			
	Manifold-based		Smote-based		Manifold-based		Smote-based	
	Mean	SD	Mean	SD	Mean	SD	Mean	SD
MLP	**0.829**	0.056	0.721	0.059	**0.948**	0.011	0.771	0.059
NB	**0.820**	0.023	0.762	0.073	**0.945**	0.010	0.733	0.073
DT	**0.778**	0.031	0.761	0.077	**0.942**	0.013	0.723	0.077
SVM	**0.802**	0.042	0.710	0.060	**0.945**	0.010	0.728	0.060
KNN	**0.854**	0.062	0.500	0.056	**0.934**	0.010	0.784	0.056
Mean	**0.817**		0.691		**0.943**		0.739	

	Saanich			
	Manifold-based		Smote-based	
	Mean	sd	Mean	sd
MLP	**0.739**	0.067	0.727	0.059
NB	**0.829**	0.041	0.699	0.073
DT	**0.791**	0.031	0.714	0.077
SVM	0.627	0.042	**0.714**	0.060
KNN	**0.677**	0.062	0.625	0.056
Mean	**0.733**		0.696	

With respect to the specific methods, the autoencoder formalization is better than PCA on the Vancouver and Saanich datasets, whereas the PCA implementation is superior on the Thunder Bay dataset. Interestingly, SMOTE is always the better than its counterpart using the removal of Tomek links.

5   UCI Classification

In order to generalize our findings, we now shift to examine the impact of the manifold on synthetic oversampling over benchmark datasets from the UCI repository. To paint a clearer picture of the impact of the manifold, we artificially control the degree of conformance of the datasets to the manifold property. This is done using a process that we refer to as *manifold augmentation*, which we detail later in this section. Performing manifold augmentation on the UCI datasets enables us to run experiments where we gradually increase the conformance in order to witness the impact of the manifold on each synthetic oversampling method, whilst holding the other aspects of complexity, such as modality and overlap, constant. This enables us to demonstrate the causal link between the increase in conformance and the change in performance.

5.1   UCI Data

The sixteen UCI datasets specified in the first column of Table 3 were selected to ensure a diverse range of dimensionalities and complexities. When required, the datasets are converted to a binary task by selecting a single class to form the minority class, and the remaining classes are merged into one.

For each experiment, we train on 25 minority training instances and 250 majority training instances; thus, we render each domain as an imbalanced classification task involving the concept of absolute imbalance. We have selected constant values for the training distribution, rather then specifying a percentage for the minority class, in order to ensure that the performance differences between datasets are not the result of having access to different numbers of minority instances. If we set the minority portion to 10 %, for example, then a dataset with 1,000 instances would have many more examples in the training set than a dataset with 200 instances. This can have a great impact on performance. Finally, we perform a series of augmentations to each dataset to increasingly strengthen the conformance to the manifold property.

5.2   Manifold Augmentation of UCI Data

Our manifold augmentation process is contingent on the notion that the probability mass resides in a lower-dimensional space. We introduce this by adding columns of uniformly distributed random variables that span both classes to the data matrix. In this case, the augmentation is suggestive of a feature selection problem; however, feature selection is not an effective means of solving manifold problems. This is because they will only find a subset of the features. A manifold space is a more general subspace that is formed from combinations of the original features. These combinations may be simple linear combinations:

$$f'_i = a_1 f_1 + a_2 f_2 + ... + a_d f_d, \tag{2}$$

where f'_i $i \in \{1,..,k\}$ is one of k components of the manifold-space embedded in the d-dimensional feature space; other manifolds are formed of much more complex combinations. In these cases, no subset of the original feature-space will represent the manifold.

5.3   Evaluation

In this set of experiments, we apply the same synthetic oversampling methods that were used in the previous section to balance the training sets prior to the application of the five classifiers. Our primary interest in this set of experiments is to elicit the affect of the manifold. In order to achieve this, we apply the augmentation method described above, in which each UCI dataset is augmented to increase conformance to the manifold property with:

$$p = \{0\,\%, 15\,\%, 30\,\%, 45\,\%, 60\,\%, 75\,\%, 90\,\%\}, \tag{3}$$

where $p = 0\,\%$ is the unchanged UCI data and $p = 90\,\%$ returns a modified dataset with the dimensionality increased by 90 %. Therefore, for each of the 16 UCI datasets, we create 7 augmented versions, where the increasing p values indicates increasing conformance to the manifold assumption.

Thirty repeated trials are run for each augmented dataset. Because we have limited space and we interested in studying the impact of the manifold, we record the mean performance for each synthetic oversampling method on each dataset and calculate the average over all of the classifiers (similar to what we reported in the last row for Table 1). We provide these aggregated results to demonstrate the relative strength of our proposed method. Our analysis of the individual classifiers is postponed for a longer paper.

The first set of results reports the ranking of each synthetic oversampling method based on the AUC. The rankings are tabulated for the performance on the original UCI datasets ($p = 0$) and for the mean of the AUC produced on the augmented datasets ($p = \{15, .., 90\}$). This demonstrates how the relative performance of the methods changes when conformance to the manifold assumption is increased up to $p = 90$.

In the second set of experiments, we include only the best manifold-based method and SMOTE-based method for each dataset in our results. We compare the change in the performance resulting from the increased conformance to the manifold property from p_0 to p_{90}. We refer to this as the loss score for each dataset D, where:

$$loss(D_{p_0}, D_{p_k}) = \overline{AUC(D_{p_0})} - \overline{AUC(D_{p_k})}. \tag{4}$$

This shows the degradation caused by the manifold. If the manifold has no impact, then the loss score is zero. The loss score increases with the relative impact of the manifold

5.4 Experimental Results

AUC Results: Table 2 presents the number of times each synthetic oversampling system produced the highest mean AUC on the UCI datasets. In the case of a tie between two methods, 0.5 is attributed to each. The first column ($p = 0$) refers to the original UCI datasets, and the last column shows the results after augmentation with $p = \{15, .., 90\}$. In both cases, the manifold-based methods are superior. For $p = 0$ the manifold-based methods are better $7 + 4 = 11$ times out of 16 and tied once with a SMOTE-based method. The real strength of the manifold-based method, however, is shown when the conformance to the manifold property is increased. The manifold-based methods are always better when the conformance to the manifold property is increased. Specifically, the autoencoder is the best on 13 of the 16 datasets and PCA is superior on the others.

Loss Results: Table 3 displays the mean loss values for the manifold-based system and the SMOTE-based system on the 16 UCI datasets. Specifically, we report the loss for each dataset with respect to $loss(D_{p_0}, D_{p_{90}})$ as described in Eq. 4. Fourteen of the sixteen datasets have lower loss scores when the manifold-based system is applied; these are highlighted in grey. This shows that in addition to its superiority in terms of the AUC, the proposed framework is more robust with respect to loss caused by the manifold. Specifically, the manifold causes less of a decrease in performance for the manifold-based approach than it causes for SMOTE.

Table 2. Total number of AUC wins for each synthetic oversampling method on the augmented UCI data.

Dataset	Wins	
	$p = 0$	mean ($p = \{15, .., 90\}$)
SMOTE	1	0
Tomek	3.5	0
PCA	7	3
AE	4.5	13

Table 3. The degradation between $p = 0$ and $p = 90$ of the classifiers after the application of synthetic oversampling.

Dataset	Manifold-Based	SMOTE-Based
Letter	0.078	0.116
Musk2	0.018	0.133
Opt Digits	0.001	0.014
Ozone 1hr	0.051	0.055
Pima	0.018	0.028
Sonar	0.061	0.074
Vehicle	0.056	0.062
Wave Form	0.038	0.074
Yeast	0.018	0.063
Satlog	0.067	0.085
Breast	0.002	0.001
Ecoli	0.016	0.022
Heart-Statlog	0.001	0.008
Ionosphere	0.055	0.033
Pen Digits	0.016	0.031
Segment	0.035	0.040

6 Conclusion

We demonstrate that the existing methods of synthetic oversampling based on SMOTE do not achieve their full potential on data that conforms to the manifold property, and argue that a manifold-based approach to synthetic oversampling is required. We address this by proposing a framework for manifold-based synthetic oversampling, which enables users to incorporate the wide variety of methods from manifold learning into the framework. We demonstrate the framework with a PCA and autoencoder formalization. These are selected for their simplicity in use and their abilities to represent a wide variety of manifolds.

We show that the implementations outperform the SMOTE-based methods in terms of the AUC on three gamma-ray spectra datasets that conform to the manifold property. In order to generalize our findings, we use 16 UCI datasets and show that the framework outperforms SMOTE in terms of the AUC and that it is more robust to the manifold property in terms of the loss score. In addition to its strength on data that conforms to the manifold property, these experiments suggest that the framework is generally a good choice for synthetic oversampling.

References

1. Alain, G., Bengio, Y.: What regularized auto-encoders learn from the data generating distribution. J. Mach. Learn. Res. **15**(1), 3563–3593 (2014)
2. Batista, G.E.A.P.A., Bazzan, A.L.C., Monard, M.C.: Balancing training data for automated annotation of keywords: a case study. In: Brazilian Workshop on Bioinformatics, pp. 10–18 (2003)
3. Batista, G.E.A.P.A., Prati, R.C., Monard, M.C.: A study of the behavior of several methods for balancing machine learning training data. ACM SIGKDD Explor. Newsl. **6**(1), 20 (2004). http://portal.acm.org/citation.cfm?doid=1007730.1007735
4. Belkin, M., Niyogi, P.: Laplacian eigenmaps for dimensionality reduction and data representation. Neural Comput. **15**(6), 1373–1396 (2003)
5. Bellinger, C., Japkowicz, N., Drummond, C.: Synthetic oversampling for advanced radioactive threat detection. In: International Conference on Machine Learning and Applications (2015)
6. Blondel, M., Seki, K., Uehara, K.: Tackling class imbalance and data scarcity in literature-based gene function annotation. In: Proceedings of the 34th International ACM SIGIR Conference on Research and Development in Information - SIGIR 2011, pp. 1123–1124. ACM Press (2011). http://portal.acm.org/citation.cfm?doid=2009916.2010080
7. Chapelle, O., Scholkopf, B., Zien, A.: Semi-supervised Learning. MIT Press (2006)
8. Chawla, N.V., Japkowicz, N., Drive, P.: Editorial: special issue on learning from imbalanced data sets. ACM SIGKD Explor. Newsl. **6**(1), 2000–2004 (2004). Special Issue on Learning from Imbalanced Datasets
9. Chawla, N.V., Japkowicz, N., Kolcz, A. (eds.) In: ICML 2003 Workshop on Learning from Imbalanced Data Sets (2003)
10. Chawla, N., Bowyer, K., Hall, L., Kegelmeyer, W.P.: SMOTE: synthetic minority over-sampling technique. J. Artif. Intell. Res. **16**, 321–357 (2002)
11. Dietterich, T.G.: Approximate statistical tests for comparing supervised classification learning algorithms. Neural Comput. **10**(7), 1895–1923 (1998)
12. Domingos, P.: A few useful things to know about machine learning. Commun. ACM **55**(10), 78 (2012)
13. Gauld, D.B.: Topological properties of manifolds. Am. Math. Monthly **81**(6), 633–636 (2008)
14. Han, H., Wang, W., Mao, B.: Borderline-SMOTE: a new over-sampling method in imbalanced data sets learning. In: Huang, D., Zhang, X., Huang, G. (eds.) ICIC 2005. LNCS, vol. 3644, pp. 878–887. Springer, Heidelberg (2005)
15. He, H., Garcia, E.A.: Learning from imbalanced data. IEEE Trans. Knowl. Data Eng. **21**(9), 1263–1284 (2009)
16. Japkowicz, N.: Supervised versus unsupervised binary-learning by feedforward neural networks. Mach. Learn. **42**(1), 97–122 (2001)
17. Ma, Y., Fu, Y.: Manifold Learning Theory and Applications (2011)
18. Nguwi, Y.Y., Cho, S.Y.: Support vector self-organizing learning for imbalanced medical data. In: 2009 International Joint Conference on Neural Networks, pp. 2250–2255. IEEE, June 2009

19. Sharma, A., Paliwal, K.K.: Fast principal component analysis using fixed-point algorithm. Pattern Recogn. Lett. **28**(10), 1151–1155 (2007)
20. Stefanowski, J., Wilk, S.: Improving rule-based classifiers induced by MODLEM by selective pre-processing of imbalanced data. In: ECML/PKDD International Workshop on Rough Sets in Knowledge Discovery (RSKD 2007), pp. 54–65 (2007)
21. Tenenbaum, J.B., de Silva, V., Langford, J.C.: A global geometric framework for nonlinear dimensionality reduction. Science **290**(5500), 2319–23 (2000). (New York, N.Y.), http://www.ncbi.nlm.nih.gov/pubmed/11125149
22. Tomek, I.: Modifications of CNN. IEEE Trans. Syst. Man Cybern. **6**(11), 769–772 (1976)
23. Tuzel, O., Porikli, F., Mee, P.: Human detection via classification on Riemannian manifolds. In: Computer Vision and Pattern Recognition, pp. 1–8 (2007)
24. Vincent, P.: Stacked denoising autoencoders: learning useful representations in a deep network with a local denoising criterion. J. Mach. Learn. Res. **11**, 3371–3408 (2010)
25. Weiss, G.M.: Mining with rarity: a unifying framework. SIGKDD Explor. Newsl. **6**(1), 7–19 (2004)
26. Yang, Q., Wu, X., Elkan, C., Gehrke, J., Han, J., Heckerman, D., Keim, D., Liu, J., Madigan, D., Piatetsky-shapiro, G., Raghavan, V.V., Rastogi, R., Stolfo, S.J., Tuzhilin, A., Wah, B.W.: 10 challenging problems in data mining research. Int. J. Inf. Technol. Decis. Making **5**(4), 597–604 (2006)
27. Zhang, D., Chen, X.: Text classification with kernels on the multinomial manifold. In: Proceedings of the 28th Annual International ACM SIGIR Conference on Research and Development in Information Retrieval, pp. 266–273 (2005)

M-Zoom: Fast Dense-Block Detection in Tensors with Quality Guarantees

Kijung Shin[✉], Bryan Hooi, and Christos Faloutsos

School of Computer Science, Carnegie Mellon University, Pittsburgh, PA, USA
{kijungs,christos}@cs.cmu.edu, bhooi@andrew.cmu.edu

Abstract. Given a large-scale and high-order tensor, how can we find dense blocks in it? Can we find them in near-linear time but with a quality guarantee? Extensive previous work has shown that dense blocks in tensors as well as graphs indicate anomalous or fraudulent behavior (e.g., lockstep behavior in social networks). However, available methods for detecting such dense blocks are not satisfactory in terms of speed, accuracy, or flexibility. In this work, we propose M-Zoom, a flexible framework for finding dense blocks in tensors, which works with a broad class of density measures. M-Zoom has the following properties: (1) **Scalable**: M-Zoom scales linearly with all aspects of tensors and is up to **114×** **faster** than state-of-the-art methods with similar accuracy. (2) **Provably accurate**: M-Zoom provides a guarantee on the lowest density of the blocks it finds. (3) **Flexible**: M-Zoom supports multi-block detection and size bounds as well as diverse density measures. (4) **Effective**: M-Zoom successfully detected edit wars and bot activities in Wikipedia, and spotted network attacks from a TCP dump with near-perfect accuracy ($\mathbf{AUC = 0.98}$). The data and software related to this paper are available at http://www.cs.cmu.edu/~kijungs/codes/mzoom/.

Keywords: Dense-block detection · Anomaly/Fraud detection · Tensor

1 Introduction

Imagine that you manage a social review site (e.g., Yelp) and have the records of which accounts wrote reviews for which restaurants. How do you detect suspicious lockstep behavior: for example, a set of accounts which give fake reviews to the same set of restaurants? What about the case where additional information is present, such as the timestamp of each review, or the keywords in each review?

Such problems of detecting suspicious lockstep behavior have been extensively studied from the perspective of dense subgraph detection. Intuitively, in the above example, highly synchronized behavior induces dense subgraphs in the bipartite review graph of accounts and restaurants. Indeed, methods which detect dense subgraphs have been successfully used to spot fraud in settings ranging from social networks [5,10,13,14], auctions [20], and search engines [8].

Additional information helps identify suspicious lockstep behavior. In the above example, the fact that reviews forming a dense subgraph were also written

© Springer International Publishing AG 2016
P. Frasconi et al. (Eds.): ECML PKDD 2016, Part I, LNAI 9851, pp. 264–280, 2016.
DOI: 10.1007/978-3-319-46128-1_17

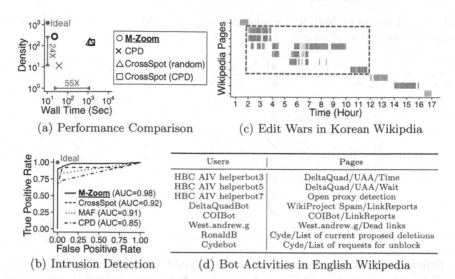

(a) Performance Comparison   (c) Edit Wars in Korean Wikipdia

(b) Intrusion Detection   (d) Bot Activities in English Wikipedia

Fig. 1. M-Zoom is fast, accurate, and effective. Fast: (a) M-ZOOM was 55×
faster with denser blocks than CROSSSPOT in Korean Wikipedia Dataset. **Accurate:**
(a) M-ZOOM found 24× denser blocks than CPD. (b) M-ZOOM identified network
attacks with near-perfect accuracy (AUC = 0.98). **Effective:** (c) M-ZOOM spotted edit
wars, during which many users (distinguished by colors) edited the same set of pages
hundreds of times within several hours. (d) M-ZOOM spotted bots, and pages edited
hundreds of thousands of times by the bots. (Color figure online)

at about the same time, with the same keywords and number of stars, makes the
reviews even more suspicious. A natural and effective way to incorporate such
extra information is to model data as a tensor and find dense blocks in it [12, 19].

However, neither existing methods for detecting dense blocks in tensors nor
simple extensions of graph-based methods are satisfactory in terms of speed,
accuracy, or flexibility. Especially, the types of fraud detectable by each of the
methods are limited since, explicitly or implicitly, each method is based on only
one density metric, which decides how dense and thus suspicious each block is.

Hence, in this work, we propose M-ZOOM (Multidimensional Zoom), a general and flexible framework for detecting dense blocks in tensors. M-ZOOM allows
for a broad class of density metrics, in addition to having the following strengths:

- **Scalable:** M-ZOOM is up to **114× faster** than state-of-the-art methods with
  similar accuracy (Fig. 2) thanks to its linear scalability with all aspects of
  tensors (Fig. 4).
- **Provably accurate:** M-ZOOM provides a guarantee on the lowest density
  of blocks it finds (Theorem 4), as well as shows high accuracy similar with
  state-of-the-art methods in real-world datasets (Fig. 1a).
- **Flexible:** M-ZOOM works successfully with high-order tensors and supports
  various density measures, multi-block detection, and size bounds (Table 1).

Table 1. M-Zoom is flexible. Comparison between M-ZOOM and other methods for dense-block detection. ✓ represents 'supported'.

		M-Zoom	CROSSSPOT [12]	CPD [17]	Subgraph [16]
Data	Matrix	✓	✓	✓	✓
	Tensor	✓	✓	✓	
Density measure	Average mass (ρ_{ari})	✓			✓
	Average mass (ρ_{geo})	✓			✓
	Suspiciousness	✓	✓		
Features	Accuracy guarantee	✓			✓
	Multiple blocks	✓	✓	✓	✓
	Size bounds	✓			✓

- **Effective:** M-ZOOM successfully detected edit wars and bot activities in Wikipedia (Figs. 1c and d), and also detected network attacks with near-perfect accuracy (**AUC = 0.98**) based on TCP dump data (Fig. 1b).

Reproducibility: Our open-sourced code and the data we used are available at http://www.cs.cmu.edu/~kijungs/codes/mzoom.

Section 2 presents preliminaries and problem definitions. Our proposed M-ZOOM is described in Sect. 3 followed by experimental results in Sect. 4. After discussing related work in Sect. 5, we draw conclusions in Sect. 6.

2 Preliminaries and Problem Definition

In this section, we introduce definitions and notations used in the paper. We also discuss density measures and give a formal definition of our problems.

2.1 Definitions and Notations

Let $\mathcal{R}(A_1, A_2, ..., A_N, X)$ be a relation with N dimension attributes A_1, A_2, ..., A_N, and a nonnegative measure attribute X (see the supplementary document [1] for a running example and its pictorial description). We use $\mathcal{R}_n$ to denote the set of distinct values of A_n in $\mathcal{R}$, and use $a_n \in \mathcal{R}_n$ for a value of A_n. The value of A_n in tuple t is denoted by $t[A_n]$, and the value of X is denoted by $t[X]$. The relation $\mathcal{R}$ can be represented as an N-way tensor. In the tensor, each n-th mode has length $|\mathcal{R}_n|$, and each cell has the value of attribute X, if the corresponding tuple exists, and 0 otherwise. Let $\mathcal{B}_n$ be a subset of $\mathcal{R}_n$. Then, we define a *block* $\mathcal{B}(A_1, A_2, ..., A_N, X) = \{t \in \mathcal{R} : 1 \leq \forall n \leq N, t[A_n] \in \mathcal{B}_n\}$, the set of tuples where each dimension attribute A_n has a value in $\mathcal{B}_n$. $\mathcal{B}$ is called 'block' because it forms a subtensor where each n-th mode has length $|\mathcal{B}_n|$ in the tensor representation of $\mathcal{R}$. The set of tuples of $\mathcal{R}$ with attribute $A_n = a_n$ is denoted by $\mathcal{R}(a_n) = \{t \in \mathcal{R} : t[A_n] = a_n\}$. We define the mass of $\mathcal{R}$ as $M_{\mathcal{R}} = Mass(\mathcal{R}) = \sum_{t \in \mathcal{R}} t[X]$, the sum of the values of attribute X in $\mathcal{R}$. We

Table 2. Table of symbols.

Symbol	Definition
$\mathcal{R}(A_1, A_2, ..., A_N, X)$	A relation with N dimension attributes and a measure attribute
N	The number of dimension attributes in a relation
a_n	A value of attribute A_n
$\mathcal{R}_n$	The set of distinct values of attribute A_n in $\mathcal{R}$
$t[A_n]$ (or $t[X]$)	The value of attribute A_n (or X) in tuple t
$\mathcal{R}(a_n)$	The set of tuples with attribute $A_n = a_n$ in $\mathcal{R}$
$M_\mathcal{R}$ (or $Mass(\mathcal{R})$)	The mass of $\mathcal{R}$
$S_\mathcal{R}$ (or $Size(\mathcal{R})$)	The size of $\mathcal{R}$
$V_\mathcal{R}$ (or $Volume(\mathcal{R})$)	The volume of $\mathcal{R}$
$\rho(\mathcal{B}, \mathcal{R})$	Density of block $\mathcal{B}$ in $\mathcal{R}$
k	The number of blocks we aim to find
$[x]$	$\{1, 2..., x\}$

also define the size of $\mathcal{R}$ as $S_\mathcal{R} = Size(\mathcal{R}) = \sum_{n=1}^{N} |\mathcal{R}_n|$ and the volume of $\mathcal{R}$ as $V_\mathcal{R} = Volume(\mathcal{R}) = \prod_{n=1}^{N} |\mathcal{R}_n|$. Lastly, we use $[x] = \{1, 2..., x\}$ for convenience. Table 2 lists frequently used symbols.

2.2 Density Measures

In this paper, we consider three specific density measures although our method is not restricted to them. Two of the density measures (Definitions 1 and 2) are natural multi-dimensional extensions of classic density measures which have been widely used for subgraphs. The merits of the original measures are discussed in [7,15], and extensive research based on them is discussed in Sect. 5.

Definition 1 (Arithmetic Average Mass [7]). *The arithmetic average mass of a block $\mathcal{B}$ of a relation $\mathcal{R}$ is defined as $\rho_{ari}(\mathcal{B}, \mathcal{R}) = M_\mathcal{B}/(S_\mathcal{B}/N)$.*

Definition 2 (Geometric Average Mass [7]). *The geometric average mass of a block $\mathcal{B}$ of a relation $\mathcal{R}$ is defined as $\rho_{geo}(\mathcal{B}, \mathcal{R}) = M_\mathcal{B}/V_\mathcal{B}^{(1/N)}$.*

The other density measure (Definition 3) is the negative log likelihood of $M_\mathcal{B}$ on the assumption that the value on each cell (in the tensor representation) of $\mathcal{R}$ follows a Poisson distribution. This proved useful in fraud detection [12].

Definition 3 (Suspiciousness [12]). *The suspiciousness of a block $\mathcal{B}$ of a relation $\mathcal{R}$ is defined as $\rho_{susp}(\mathcal{B}, \mathcal{R}) = M_\mathcal{B}(\log(M_\mathcal{B}/M_\mathcal{R}) - 1) + M_\mathcal{R}V_\mathcal{B}/V_\mathcal{R} - M_\mathcal{B}\log(V_\mathcal{B}/V_\mathcal{R})$.*

Our method, however, is not restricted to the three measures mentioned above. Our method, which searches for dense blocks in a tensor, allows for any density measure ρ that satisfies Axiom 1.

Axiom 1 (Density Axiom). *If two blocks of a relation have the same cardinality for every dimension attribute, the block with higher or equal mass is at least as dense as the other. Formally,*

$$M_{\mathcal{B}} \geq M_{\mathcal{B}'} \text{ and } |\mathcal{B}_n| = |\mathcal{B}'_n|, \forall n \in [N] \Rightarrow \rho(\mathcal{B}, \mathcal{R}) \geq \rho(\mathcal{B}', \mathcal{R}).$$

2.3 Problem Definition

We formally define the problem of detecting the k densest blocks in a tensor.

Problem 1 (k-Densest Blocks). **(1) Given:** a relation $\mathcal{R}$, the number of blocks k, and a density measure ρ, **(2) Find:** k distinct blocks of $\mathcal{R}$ with the highest densities in terms of ρ.

We also consider a variant of Problem 1 which incorporates lower and upper bounds on the size of the detected blocks. This is particularly useful if the unrestricted densest block is not meaningful due to being too small (e.g. a single tuple) or too large (e.g. the entire tensor).

Problem 2 (k-Densest Blocks with Size Bounds). **(1) Given:** a relation $\mathcal{R}$, the number of blocks k, a density measure ρ, lower size bound S_{min}, and upper size bound S_{max}, **(2) Find:** k distinct blocks of $\mathcal{R}$ with the highest densities in terms of ρ **(3) Among:** blocks whose sizes are at least S_{min} and at most S_{max}.

Even when we restrict our attention to a special case ($N=2$, $k=1$, $\rho=\rho_{ari}$, $S_{min}=S_{max}$), exactly solving Problems 1 and 2 takes $O(S_{\mathcal{R}}^6)$ time [9] and is NP-hard [3], resp., infeasible for large datasets. Thus, we focus on an approximation algorithm which (1) has linear scalability with all aspects of $\mathcal{R}$, (2) provides accuracy guarantees at least for some density measures, and (3) produces meaningful results in real-world datasets, as explained in detail in Sects. 3 and 4.

3 Proposed Method

In this section, we propose M-ZOOM (Multidimensional Zoom), a scalable, accurate, and flexible method for finding dense blocks in a tensor. We present the details of M-ZOOM in Sect. 3.1 and discuss its efficient implementation in Sect. 3.2. After analyzing the time and space complexity in Sect. 3.3, we prove the quality guarantees provided by M-ZOOM in Sect. 3.4.

3.1 Algorithm

Algorithm 1 describes the outline of M-ZOOM. M-ZOOM first copies the given relation $\mathcal{R}$ and assigns it to $\mathcal{R}^{ori}$ (line 1). Then, M-ZOOM finds k dense blocks one by one from $\mathcal{R}$ (line 4). After finding each block from $\mathcal{R}$, M-ZOOM removes the tuples in the block from $\mathcal{R}$ to prevent the same block from being found again (line 5). Due to these changes in $\mathcal{R}$, a block found in $\mathcal{R}$ is not necessarily

---

Algorithm 1. M-ZOOM

Input : relation: $\mathcal{R}$, number of blocks: k, density measure: ρ,
   lower size bound: S_{min}, upper size bound: S_{max}
Output: k dense blocks
1 $\mathcal{R}^{ori} \leftarrow copy(\mathcal{R})$
2 $results \leftarrow \emptyset$
3 **for** $i \leftarrow 1..k$ **do**
4     $\mathcal{B} \leftarrow find\_single\_block(\mathcal{R}, \rho, S_{min}, S_{max})$ $\triangleright$ see Algorithm 2
5     $\mathcal{R} \leftarrow \mathcal{R} - \mathcal{B}$
6     $\mathcal{B}^{ori} \leftarrow \{t \in \mathcal{R}^{ori} : \forall n \in [N], t[A_n] \in \mathcal{B}_n\}$
7     $results \leftarrow results \cup \{\mathcal{B}^{ori}\}$

8 **return** $results$

Algorithm 2. $find\_single\_block$ in M-ZOOM

Input : relation: $\mathcal{R}$, density measure: ρ,
   lower size bound: S_{min}, upper size bound: S_{max}
Output: a dense block
1 $\mathcal{B} \leftarrow copy(\mathcal{R})$
2 $\mathcal{B}_n \leftarrow copy(\mathcal{R}_n), \forall n \in [N]$
3 $snapshots \leftarrow \emptyset$
4 **while** $\exists n \in [N]$ $s.t.$ $\mathcal{B}_n \neq \emptyset$ **do**
5   **if** $\mathcal{B}$ is in $size$ $bounds$ (i.e., $S_{min} \leq S_{\mathcal{B}} \leq S_{max}$) **then**
6        $snapshots \leftarrow snapshots \cup \{\mathcal{B}\}$
7     $a_i^* \leftarrow a_i \in \bigcup_{n=1}^{N} \mathcal{B}_n$ with maximum $\rho(\mathcal{B} - \mathcal{B}(a_i), \mathcal{R})$ $\triangleright$ see Algorithm 3
8     $\mathcal{B} \leftarrow \mathcal{B} - \mathcal{B}(a_i^*)$
9     $\mathcal{B}_i \leftarrow \mathcal{B}_i - \{a_i^*\}$

10 **return** $\mathcal{B} \in snapshots$ with maximum $\rho(\mathcal{B}, \mathcal{R})$

---

a block of the original relation $\mathcal{R}^{ori}$. Thus, instead of returning the blocks found in $\mathcal{R}$, M-ZOOM returns the blocks of $\mathcal{R}^{ori}$ consisting of the same attribute values with the found blocks (lines 6–7). This also enables M-ZOOM to find overlapped blocks, i.e., a tuple can be included in two or more blocks.

Algorithm 2 describes how M-ZOOM finds a single dense block from the given relation $\mathcal{R}$. The block $\mathcal{B}$ is initialized to $\mathcal{R}$ (lines 1–2). From $\mathcal{B}$, M-ZOOM removes attribute values one by one in a greedy way until no attribute value is left (line 4). Specifically, M-ZOOM finds the attribute value a_i that maximizes $\rho(\mathcal{B} - \mathcal{B}(a_i), \mathcal{R})$, which corresponds to the density when tuples with $A_i = a_i$ are removed from $\mathcal{B}$ (line 7). Then, the attribute value, denoted by a_i^*, and the tuples with $A_i = a_i^*$ are removed from $\mathcal{B}_i$ and $\mathcal{B}$, respectively (lines 8–9). Before removing each attribute value, M-ZOOM adds the current $\mathcal{B}$ to the snapshot list if $\mathcal{B}$ satisfies the size bound (i.e., $S_{min} \leq S_{\mathcal{B}} \leq S_{max}$) (lines 5–6). As the final step of finding a block, M-ZOOM returns the block with the maximum density among those in the snapshot list (line 10).

Algorithm 3. Greedy Selection Using Min-Heap in M-Zoom

Input : current block: $\mathcal{B}$, density measure: ρ, min-heaps: $\{H_n\}_{n=1}^N$
Output: attribute value to remove

1 **for** each dimension $n \in [N]$ **do**
2 $\quad\lfloor\ a_n' \leftarrow a_n$ with minimum key in H_n $\qquad\qquad\qquad\qquad\qquad\qquad \triangleright$ key$= M_{\mathcal{B}(a_n)}$
3 $a_i^* \leftarrow a_i' \in \{a_n'\}_{n=1}^N$ with maximum $\rho(\mathcal{B} - \mathcal{B}(a_i'), \mathcal{R})$
4 delete a_i^* from H_i
5 **for** each tuple $t \in \mathcal{B}(a_i^*)$ **do**
6 $\quad$ **for** each dimension $n \in [N]\backslash\{i\}$ **do**
7 $\quad\quad\lfloor\ $ decrease the key of $t[A_n]$ in H_n by $t[X]$ $\qquad\qquad\quad \triangleright$ key$= M_{\mathcal{B}(t[A_n])}$

8 **return** a_i^*

3.2 Efficient Implementation of M-Zoom

In this section, we discuss an efficient implementation of M-Zoom focusing on the greedy attribute value selection and the densest block selection.

Attribute Value Selection Using Min-Heaps. Finding the attribute value $a_i \in \bigcup_{n=1}^N \mathcal{B}_n$ that maximizes $\rho(\mathcal{B} - \mathcal{B}(a_i), \mathcal{R})$ (line 7 of Algorithm 2) can be computationally very expensive if all possible attribute values (i.e., $\bigcup_{n=1}^N \mathcal{B}_n$) should be considered. However, due to Axiom 1, which is assumed to be satisfied by considered density measures, the number of candidates is reduced to N if $M_{\mathcal{B}(a_i)}$ is known for each attribute value a_i. Lemma 1 states this.

Lemma 1. *If we remove a value of attribute A_n from $\mathcal{B}_n$, removing $a_n \in \mathcal{B}_n$ with minimum $M_{\mathcal{B}(a_n)}$ results in the highest density. Formally,*

$$M_{\mathcal{B}(a_n')} \leq M_{\mathcal{B}(a_n)}, \forall a_n \in \mathcal{B}_n \Rightarrow \rho(\mathcal{B} - \mathcal{B}(a_n'), \mathcal{R}) \geq \rho(\mathcal{B} - \mathcal{B}(a_n), \mathcal{R}), \forall a_n \in \mathcal{B}_n.$$

Proof. Let $\mathcal{B}' = \mathcal{B} - \mathcal{B}(a_n')$ and $\mathcal{B}'' = \mathcal{B} - \mathcal{B}(a_n)$. Then, $|\mathcal{B}_n'| = |\mathcal{B}_n''|, \forall n \in [N]$. In addition, $M_{\mathcal{B}'} \geq M_{\mathcal{B}''}$ since $M_{\mathcal{B}'} = M_{\mathcal{B}} - M_{\mathcal{B}(a_n')} \geq M_{\mathcal{B}} - M_{\mathcal{B}(a_n)} = M_{\mathcal{B}''}$. Hence, by Axiom 1, $\rho(\mathcal{B} - \mathcal{B}(a_n'), \mathcal{R}) \geq \rho(\mathcal{B} - \mathcal{B}(a_n), \mathcal{R})$. $\qquad\qquad\square$

By Lemma 1, if we let a_n' be $a_n \in \mathcal{B}_n$ with minimum $M_{\mathcal{B}(a_n)}$, we only have to consider values in $\{a_n'\}_{n=1}^N$ instead of $\bigcup_{n=1}^N \mathcal{B}_n$ to find the attribute value maximizing density when it is removed. To exploit this, our implementation of M-Zoom maintains a min-heap for each attribute A_n where the key of each value a_n is $M_{\mathcal{B}(a_n)}$. This key is updated, which takes $O(1)$ if Fibonacci Heaps are used as min-heaps, whenever the tuples with the corresponding attribute value are removed. Algorithm 3 describes in detail how to find the attribute value to be removed based on these min-heaps, and update keys in them. Since Algorithm 3 considers all promising attribute values (i.e., $\{a_n'\}_{n=1}^N$), it is guaranteed to find the value that maximizes density when it is removed, as Theorem 1 states.

Theorem 1. *Algorithm 3 returns $a_i \in \bigcup_{n=1}^N \mathcal{B}_n$ with maximum $\rho(\mathcal{B} - \mathcal{B}(a_i), \mathcal{R})$.*

Proof. Let a_i^* be $a_i \in \bigcup_{n=1}^N \mathcal{B}_n$ with maximum $\rho(\mathcal{B} - \mathcal{B}(a_i), \mathcal{R})$. By Lemma 1, a_i^* exists among $\{a_n'\}_{n=1}^N$, all of which are considered in Algorithm 3.   □

Densest Block Selection Using Attribute Value Ordering. As explained in Sect. 3.1, M-ZOOM returns the densest block among snapshots of $\mathcal{B}$ (line 10 of Algorithm 2). Explicitly maintaining the list of snapshots, whose length is at most $S_{\mathcal{R}}$, requires $O(N|\mathcal{R}|S_{\mathcal{R}})$ computation and space for copying them. Even maintaining only the current best (i.e., the one with the highest density so far) cannot avoid high computational cost if the current best keeps changing. Instead, our implementation maintains the order by which attribute values are removed as well as the iteration where the density was maximized, which requires only $O(S_{\mathcal{R}})$ space. From these and the original relation $\mathcal{R}$, our implementation restores the snapshot with maximum density in $O(N|\mathcal{R}| + S_{\mathcal{R}})$ time and returns it.

3.3   Complexity Analysis

The time and space complexity of M-ZOOM depend on the density measure used. In this section, we assume that one of the density measures in Sect. 2.2, which satisfy Axiom 1, is used.

Theorem 2. *The time complexity of Algorithm 1 is $O(kN|\mathcal{R}|\log L)$ if $|\mathcal{R}_n| = L$, $\forall n \in [N]$, and $N = O(\log L)$.*

Proof. See Appendix B.

As stated in Theorem 2, M-ZOOM scales linearly or sub-linearly with all aspects of relation $\mathcal{R}$ as well as k, the number of blocks we aim to find. This result is also experimentally supported in Sect. 4.4. In our experiments, the actual running time scaled sub-linearly with k as well as L since the number of tuples in $\mathcal{R}$ decreases as M-ZOOM finds blocks (line 5 in Algorithm 1).

Theorem 3. *The space complexity of Algorithm 1 is $O(kN|\mathcal{R}|)$.*

Proof. See the supplementary document [1].   □

M-ZOOM requires up to $kN|\mathcal{R}|$ space for storing k found blocks, as stated in Theorem 3. However, since the blocks are usually far smaller than $\mathcal{R}$, as seen in Tables 4 and 5 in Sect. 4, actual space usage is much less than $kN|\mathcal{R}|$.

3.4   Accuracy Guarantee

In this section, we show lower bounds on the densities of the blocks found by M-ZOOM on the assumption that ρ_{ari} (Definition 1) is used as the density measure. Specifically, we show that Algorithm 2 without size bounds is guaranteed to find a block with density at least $1/N$ of maximum density in the given relation (Theorem 4). This means that each n-th block returned by Algorithm 1 has density at least $1/N$ of maximum density in $\mathcal{R} - \bigcup_{i=1}^{n-1}(i\text{-th block})$. Let $\mathcal{B}^{(r)}$

be the relation $\mathcal{B}$ at the beginning of the r-th iteration of Algorithm 2, and $a_i^{(r)} \in \mathcal{B}_i^{(r)}$ be the attribute value removed in the same iteration.

Lemma 2. *If a block $\mathcal{B}'$ satisfying $\forall a_i \in \bigcup_{n=1}^N \mathcal{B}_n'$, $M_{\mathcal{B}'(a_i)} \geq c$ exists, there exists $\mathcal{B}^{(r)}$ satisfying $\forall a_i \in \bigcup_{n=1}^N \mathcal{B}_n^{(r)}$, $M_{\mathcal{B}^{(r)}(a_i)} \geq c$.*

Proof. See Appendix C.   □

Theorem 4 ($1/N$-Approximation Guarantee for Problem 1). *Given a relation $\mathcal{R}$, let $\mathcal{B}^*$ be the block $\mathcal{B} \subset \mathcal{R}$ with maximum $\rho_{ari}(\mathcal{B}, \mathcal{R})$. Let $\mathcal{B}'$ be the block obtained by Algorithm 2 without size bounds (i.e., $S_{min} = 0$ and $S_{max} = \infty$). Then, $\rho_{ari}(\mathcal{B}', \mathcal{R}) \geq \rho_{ari}(\mathcal{B}^*, \mathcal{R})/N$.*

Proof. $\forall a_i \in \bigcup_{n=1}^N \mathcal{B}_n^*$, $M_{\mathcal{B}^*(a_i)} \geq M_{\mathcal{B}^*}/S_{\mathcal{B}^*}$. Otherwise, a contradiction would result since for a_i with $M_{\mathcal{B}^*(a_i)} < M_{\mathcal{B}^*}/S_{\mathcal{B}^*}$,

$$\rho_{ari}(\mathcal{B}^* - \mathcal{B}^*(a_i), \mathcal{R}) = \frac{M_{\mathcal{B}^*} - M_{\mathcal{B}^*(a_i)}}{(S_{\mathcal{B}^*} - 1)/N} > \frac{M_{\mathcal{B}^*} - M_{\mathcal{B}^*}/S_{\mathcal{B}^*}}{(S_{\mathcal{B}^*} - 1)/N} = \rho_{ari}(\mathcal{B}^*, \mathcal{R}).$$

Consider $\mathcal{B}^{(r)}$ where $\forall a_i \in \bigcup_{n=1}^N \mathcal{B}_n^{(r)}$, $M_{\mathcal{B}^{(r)}(a_i)} \geq M_{\mathcal{B}^*}/S_{\mathcal{B}^*}$. Such $\mathcal{B}^{(r)}$ exists by Lemma 2. $M_{\mathcal{B}^{(r)}} \geq (S_{\mathcal{B}^{(r)}}/N)(M_{\mathcal{B}^*}/S_{\mathcal{B}^*}) = (S_{\mathcal{B}^{(r)}}/N)(\rho_{ari}(\mathcal{B}^*, \mathcal{R})/N)$. Hence, $\rho_{ari}(\mathcal{B}', \mathcal{R}) \geq \rho_{ari}(\mathcal{B}^{(r)}, \mathcal{R}) = M_{\mathcal{B}^{(r)}}/(S_{\mathcal{B}^{(r)}}/N) \geq \rho_{ari}(\mathcal{B}^*, \mathcal{R})/N$.   □

Theorem 4 can be extended to cases where a lower bound exists. In these cases, the approximate factor is $1/(N+1)$, as stated in Theorem 5.

Theorem 5 ($1/(N+1)$-Approximation Guarantee for Problem 2). *Given a relation $\mathcal{R}$, let $\mathcal{B}^*$ be the block $\mathcal{B} \subset \mathcal{R}$ with maximum $\rho_{ari}(\mathcal{B}, \mathcal{R})$ among blocks with size at least S_{min}. Let $\mathcal{B}'$ be the block obtained by Algorithm 2 with lower size bound (i.e., $1 \leq S_{min} \leq S_{\mathcal{R}}$ and $S_{max} = \infty$). Then, $\rho_{ari}(\mathcal{B}', \mathcal{R}) \geq \rho_{ari}(\mathcal{B}^*, \mathcal{R})/(N+1)$.*

Proof. See the supplementary document [1].   □

4   Experiments

We designed and performed experiments to answer the following questions:

- **Q1.** How fast and accurately does M-ZOOM detect dense blocks in real data?
- **Q2.** Does M-ZOOM find many different dense blocks in real data?
- **Q3.** Does M-ZOOM scale linearly with all aspects of data?
- **Q4.** Which anomalies or fraud does M-ZOOM spot in real data?

Table 3. Summary of real-world datasets.

	StackO.	Youtube	KoWiki	EnWiki	Yelp	Netflix	YahooM.	AirForce
N	3	3	3	3	4	4	4	7
$|\mathcal{R}_1|$	545K	3.22M	470K	44.1M	552K	480K	1.00M	3
$|\mathcal{R}_2|$	96.7K	3.22M	1.18M	38.5M	77.1K	17.8K	625K	70
$|\mathcal{R}_3|$	1.15K	203	101K	129K	3.80K	2.18K	84.4K	11
$|\mathcal{R}_4|$	-	-	-	-	5	5	101	7.20K[a]
$|\mathcal{R}|$	1.30M	18.7M	11.0M	483M	2.23M	99.1M	253M	648K

[a] $|\mathcal{R}_5|$=21.5K, $|\mathcal{R}_6|$=512, $|\mathcal{R}_7|$=512

4.1 Experimental Settings

All experiments were conducted on a machine with 2.67 GHz Intel Xeon E7-8837 CPUs and 1TB RAM. We compared M-Zoom with CrossSpot [12], CP Decomposition (CPD) [17] (see Appendix A for details), and MultiAspectForensics (MAF) [19]. M-Zoom and CrossSpot[1] were implemented in Java, and Tensor Toolbox [4] was used for CPD and MAF. Although CrossSpot was originally designed to maximize ρ_{susp}, it can be extended to other density measures. These variants were used depending on the density measure compared in each experiment. In addition, we used CPD as a seed selection method of CrossSpot, which outperformed HOSVD used in [12] in terms of both speed and accuracy. We used diverse real-world datasets, grouped as follows:

- **User behavior logs:** *StackO.(user,post,timestamp,1)* represents who marked which post as a favorite when on Stack Overflow. *Youtube(user,user,date,1)* represents who became a friend of whom when on Youtube. *KoWiki(user,page, timestamp,#revisions)* and *EnWiki(user,page,timestamp,#revisions)* represent who revised which page when how many times on Korean Wikipedia and English Wikipedia, respectively.
- **User reviews:** *Yelp(user,business,date,score,1)*, *Netflix(user,movie,date,score,1)*, and *YahooM.(user,item,timestamp,score,1)* represent who gave which score when to which business, movie, and item on Yelp, Netflix, and Yahoo Music, respectively.
- **TCP dumps:** From TCP dump data for a typical U.S. Air Force LAN, we created a relation *AirForce(protocol,service,src_bytes,dst_bytes,flag,host_count ,src_count,#connections)*. See the supplementary document [1] for the description of each attribute.

Timestamps are in hours in all the datasets. Table 3 summarizes all the datasets.

---

[1] We referred the open-sourced implementation at http://github.com/mjiang89/CrossSpot.

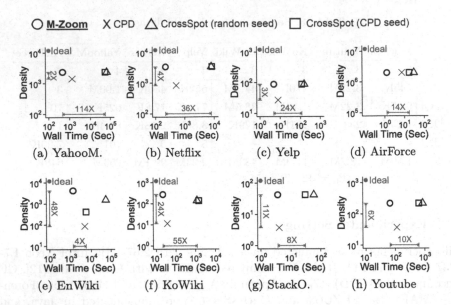

Fig. 2. Only M-Zoom achieves both speed and accuracy. In each plot, points represent the speed of different methods and the highest density (ρ_{ari}) of three blocks found by the methods. Upper-left region indicates better performance. M-ZOOM gives the best trade-off between speed and density. Specifically, M-ZOOM is up to **114 ×** faster than CROSSSPOT with similarly dense blocks.

4.2   Q1. Running Time and Accuracy of M-Zoom

We compare the speed of different methods and the densities of the blocks found by the methods in real-world datasets. Specifically, we measured time taken to find three blocks and the maximum density among the three blocks. Figure 2 shows the result when ρ_{ari} was used as the density measure. M-ZOOM clearly provided the best trade-off between speed and accuracy in all datasets. For example, in YahooM. Dataset, M-ZOOM was 114 times faster than CROSSSPOT, while detecting blocks with similar densities. Compared with CPD, M-ZOOM detected two times denser blocks 2.8 times faster. Although the results are not included in Fig. 2, MAF found several orders of magnitude sparser blocks than the other methods, with speed similar to that of CPD. M-ZOOM also gave the best trade-off between speed and accuracy when ρ_{geo} or ρ_{susp} was used instead of ρ_{ari} (see the supplementary document [1]).

4.3   Q2. Diversity of Blocks Found by M-Zoom

We compare the diversity of dense blocks found by each method. Ability to detect many different dense blocks is useful since distinct blocks may indicate different anomalies or fraud. We define the diversity as the average dissimilarity between the pairs of blocks, and the dissimilarity of two blocks is defined

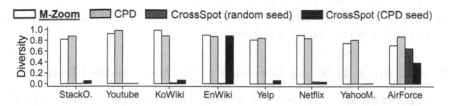

Fig. 3. M-Zoom detects many different dense blocks. The dense blocks found by M-ZOOM and CPD have high diversity, while the dense blocks found by CROSSSPOT tend to be almost same.

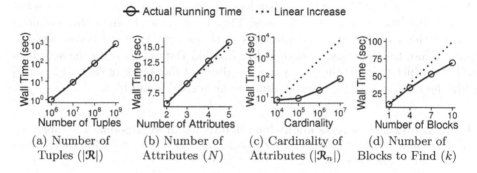

Fig. 4. M-Zoom is scalable. (a) (b) M-ZOOM scales linearly with the number of tuples and the number of attributes. (c) (d) M-ZOOM scales sub-linearly with the cardinalities of attributes and the number of blocks we aim to find.

as $dissimilarity(\mathcal{B}, \mathcal{B}') = 1 - \frac{|(\bigcup_{n=1}^{N} \mathcal{B}_n) \cap (\bigcup_{n=1}^{N} \mathcal{B}'_n)|}{|(\bigcup_{n=1}^{N} \mathcal{B}_n) \cup (\bigcup_{n=1}^{N} \mathcal{B}'_n)|}$. Diversities were measured among three blocks found by each method using ρ_{ari} as the density metric.

As seen in Fig. 3, in all datasets, M-ZOOM and CPD successfully detected distinct dense blocks. CROSSSPOT, however, found the same block repeatedly or blocks with slight difference, even when it started from different seed blocks. Although using CPD for seed-block selection in CROSSSPOT improved the diversity, the effect was limited in most datasets. Similar results were obtained when ρ_{geo} or ρ_{susp} was used instead of ρ_{ari} (see the supplementary document [1]).

4.4 Q3. Scalability of M-Zoom

We empirically demonstrate the scalability of M-ZOOM, mathematically analyzed in Theorem 2. Specifically, we measured the scalability of M-ZOOM with regard to the number of tuples, the number of attributes, the cardinalities of attributes, and the number of blocks we aim to find. We started with finding one block in a randomly generated 10 millions tuples with three attributes each of whose cardinality is 100 K. Then, we measured the running time by changing one factor at a time while fixing the others. As seen in Fig. 4, M-ZOOM scaled linearly with the number of tuples and the number of attributes. Moreover, M-ZOOM scaled sub-linearly with the number of blocks we aim to find as well as the cardinalities of attributes due to the reason explained in Sect. 3.3. These results held regardless of the density measure used.

4.5 Q4. Anomaly/Fraud Detection by M-Zoom in Real Data

We demonstrate the effectiveness of M-Zoom for anomaly and fraud detection by analyzing dense blocks detected by M-Zoom in real-world datasets.

M-Zoom spots edit wars and bot activities in Wikipedia. Table 4 lists the first three dense blocks found by M-Zoom in EnWiki and KoWiki Datasets. As seen in the third dense block visualized in Fig. 1c, the dense blocks detected in KoWiki Dataset indicate edit wars. That is, users with conflicting opinions revised the same set of pages hundreds of times within several hours. On the other hand, the dense blocks detected in EnWiki Dataset indicate the activities of bots, which changed the same pages hundreds of thousands of times. Figure 1d lists the bots and pages corresponding to the second found block.

Table 4. M-Zoom detects anomalous behaviors in Wikipedia. The tables list the first three blocks detected by M-Zoom in KoWiki and EnWiki Datasets, which correspond to edit wars and bot activities, respectively.

Korean Wikipedia (KoWiki)				English Wikipedia (EnWiki)			
#	Volume	Mass	Density (ρ_{ari})	#	Volume	Mass	Density (ρ_{geo})
1	2×2×2	546	273	1	1×1,585×6,733	1.93M	8,772
2	2×2×3	574	246	2	8×12×67.9K	2.43M	13.0K
3	11×10×16	2,305	187	3	1×1×90	17.6K	3,933

M-Zoom spots network intrusions. Table 5 lists the first three blocks found by M-Zoom in AirForce Dataset. Based on the provided ground truth labels, all of the about 3 millions connections composing the blocks were attacks except only one normal connection. This indicates that malicious connections form dense blocks due to the similarity in their behaviors. Based on this observation, we could accurately separate normal connections and attacks based on the densities of blocks they belong (i.e., the denser block a connection belongs, the

Table 5. M-Zoom identifies network attacks with near-perfect accuracy. The first three blocks found by M-Zoom in AirForce Dataset consist of attacks.

#	Volume	Density (ρ_{geo})	# Connections	# Attacks (Ratio)
1	2	2,050,505	2,263,941	2,263,941 (100%)
2	1	263,295	263,295	263,295 (100%)
3	8,100	263,072	952,383	952,382 (99.9%)

more suspicious it is). Especially, we got the highest AUC (Area under the curve) 0.98 with M-ZOOM, as shown in Fig. 1b, because M-ZOOM detects many different dense blocks accurately, as shown in previous experiments. For each method, we used the best density measure that leads to the highest AUC.

5   Related Work

Dense Subgraph/Submatrix/Subtensor Detection. The *densest subgraph problem*, the problem of finding the subgraph which maximizes ρ_{ari} or ρ_{geo} (see Definitions 1 and 2), has been extensively studied in theory (see [18] for surveys). The two major directions are max-flow based exact algorithms [9,16] and greedy algorithms [7,16] giving a 1/2-approximation to the densest subgraph. Variants allow for size restrictions [3], providing a 1/3-approximation to the densest subgraph for the lower bound case. Another related line of research deals with dense blocks in binary matrices or tensors where the definition of density is designed for the purpose of frequent itemset mining [22] or formal concept mining [6,11].

Anomaly/Fraud Detection based on Dense Subgraphs. Spectral approaches make use of eigendecomposition or SVD of the adjacency matrix for dense-block detection. Such approaches have been used to spot anomalous pattens in a patent graph [21], lockstep followers in a social network [14], and stealthy or small-scale attacks in social networks [23]. Other approaches include NETPROBE [20], which used belief propagation to detect fraud-accomplice bipartite cores in an auction network, and COPYCATCH [5], which used one-class clustering and sub-space clustering to identify "Like" boosting in Facebook. In addition, ODDBALL [2] spotted near-cliques in links among posts in blogs based on egonet features. Recently, FRAUDAR [10], which generalizes densest subgraph-detection methods so that the suspiciousness of nodes and edges can be incorporated, spotted follower-buying services in Twitter.

Anomaly/Fraud Detection based on Dense Subtensors. Spectral methods for dense subgraphs can be extended to tensors where tensor decomposition, such as CP Decomposition and HOSVD [17], is used to spot dense subtensors. MAF [19], which is based on CP Decomposition, detected dense blocks corresponding to port-scanning activities based on network traffic logs. Another approach is CROSSSPOT [12], which finds dense blocks by starting from seed blocks and growing them in a greedy way until ρ_{susp} (see Definition 3) converges. CROSSSPOT spotted retweet boosting in Weibo, outperforming HOSVD.

Our M-ZOOM non-trivially generalizes theoretical results regarding the densest subgraph problem, especially [3], for supporting tensors, various density measures, and multi-block detection. As seen in Table 1, M-ZOOM provides more flexibility than other methods for dense-block detection.

6   Conclusion

In this work, we propose M-ZOOM, a flexible framework for finding dense blocks in tensors, which has the following advantages over state-of-the-art methods:

- **Scalable:** M-Zoom is up to **114× faster** than competitors with similar accuracy due to its linear scalability with all input factors (Figs. 2 and 4).
- **Provably accurate:** M-Zoom provides lower bounds on the densities of the blocks it finds (Theorem 4) as well as high accuracy in real data (Fig. 2).
- **Flexible:** M-Zoom supports high-order tensors, various density measures, multi-block detection, and size bounds (Table 1).
- **Effective:** M-Zoom successfully detected fraud based on a TCP dump with near-perfect accuracy (**AUC = 0.98**), and anomalies in Wikipedia (Fig. 1).

Reproducibility: Our open-sourced code and the data we used are at http://www.cs.cmu.edu/~kijungs/codes/mzoom.

Acknowledgments. This material is based upon work supported by the National Science Foundation under Grant No. CNS-1314632 and IIS-1408924. Research was sponsored by the Army Research Laboratory and was accomplished under Cooperative Agreement Number W911NF-09-2-0053. Any opinions, findings, and conclusions or recommendations expressed in this material are those of the author(s) and do not necessarily reflect the views of the National Science Foundation, or other funding parties. The U.S. Government is authorized to reproduce and distribute reprints for Government purposes notwithstanding any copyright notation here on.

A   CP Decomposition (CPD)

In a graph, dense subgraphs lead to high singular values of the adjacency matrix [23]. The singular vectors corresponding to the high singular values roughly indicate which nodes form dense blocks. This idea can be extended to tensors, where dense blocks are captured by components in CP Decomposition [17]. Let $\mathbf{A}^{(1)} \in \mathbb{R}^{|\mathcal{R}_1| \times k}$, $\mathbf{A}^{(2)} \in \mathbb{R}^{|\mathcal{R}_2| \times k}$, ..., $\mathbf{A}^{(N)} \in \mathbb{R}^{|\mathcal{R}_N| \times k}$ be the factor matrices obtained by the rank-k CP Decomposition of $\mathcal{R}$. For each $i \in [k]$, we form a block with every attribute value a_n whose corresponding element in the i-th column of $\mathbf{A}^{(n)}$ is at least $1/\sqrt{|\mathcal{R}_n|}$.

B   Proof of Theorem 2

Proof. In Algorithm 3, lines 1–3 take $O(N)$ for all the density measures considered (i.e., ρ_{ari}, ρ_{geo}, and ρ_{susp}) if we maintain and update aggregated values (e.g., M_B, S_B, and V_B) instead of computing $\rho(\mathcal{B} - \mathcal{B}(a_i'), \mathcal{R})$ from scratch every time. In addition, line 4 takes $O(\log |\mathcal{R}_n|)$ and lines 5–7 take $O(N|\mathcal{B}(a_i^*)|)$ if we use Fibonacci heaps. Algorithm 2, whose computational bottleneck is line 7, has time complexity $O(N|\mathcal{R}| + N \sum_{n=1}^{N} |\mathcal{R}_n| + \sum_{n=1}^{N} |\mathcal{R}_n| \log |\mathcal{R}_n|)$ since lines 1–4 of Algorithm 3 are executed $S_{\mathcal{R}} = \sum_{n=1}^{N} |\mathcal{R}_n|$ times, and line 7 is executed $N|\mathcal{R}|$ times. Algorithm 1, whose computational bottleneck is line 4, has time complexity $O(kN|\mathcal{R}| + kN \sum_{n=1}^{N} |\mathcal{R}_n| + k \sum_{n=1}^{N} |\mathcal{R}_n| \log |\mathcal{R}_n|)$ since Algorithm 2 is executed k times.

Assume $|\mathcal{R}_n| = L$, $\forall n \in [N]$, and $N = O(\log L)$. The time complexity of Algorithm 1 becomes $O(kN(|\mathcal{R}| + NL + L \log L))$. Since $N = O(\log L)$, by assumption, and $L \leq |\mathcal{R}|$, there exists a constant c such that $|\mathcal{R}|+NL+L\log L \leq c|\mathcal{R}|\log L = O(|\mathcal{R}|\log L)$. Thus, the time complexity of Algorithm 1 is $O(kN|\mathcal{R}|\log L)$. □

C   Proof of Lemma 2

Lemma 3. $a_i^{(r)}$ minimizes $Mass(\mathcal{B}^{(r)}(a_j))$ among $a_j \in \bigcup_{n=1}^{N} \mathcal{B}_n^{(r)}$.

Proof. From Theorem 1, $\rho_{ari}(\mathcal{B}^{(r)} - \mathcal{B}^{(r)}(a_i^{(r)}), \mathcal{R}) \geq \rho_{ari}(\mathcal{B}^{(r)} - \mathcal{B}^{(r)}(a_j), \mathcal{R})$, $\forall a_j \in \bigcup_{n=1}^{N} \mathcal{B}_n^{(r)}$. Thus, $Mass(\mathcal{B}^{(r)} - \mathcal{B}^{(r)}(a_i^{(r)})) = \rho_{ari}(\mathcal{B}^{(r)} - \mathcal{B}^{(r)}(a_i^{(r)}), \mathcal{R})$ $(Size(\mathcal{B}^{(r)})-1)/N \geq \rho_{ari}(\mathcal{B}^{(r)}-\mathcal{B}^{(r)}(a_j), \mathcal{R})(Size(\mathcal{B}^{(r)})-1)/N = Mass(\mathcal{B}^{(r)} - \mathcal{B}^{(r)}(a_j))$. Then, $Mass(\mathcal{B}^{(r)}(a_i^{(r)})) = Mass(\mathcal{B}^{(r)}) - Mass(\mathcal{B}^{(r)} - \mathcal{B}^{(r)}(a_i^{(r)})) \leq Mass(\mathcal{B}^{(r)}) - Mass(\mathcal{B}^{(r)} - \mathcal{B}^{(r)}(a_j)) = Mass(\mathcal{B}^{(r)}(a_j))$, $\forall a_j \in \bigcup_{n=1}^{N} \mathcal{B}_n^{(r)}$. □

Proof of Lemma 2.

Proof. Let r be the first iteration in Algorithm 2 where $a_i^{(r)} \in \bigcup_{n=1}^{N} \mathcal{B}_n'$. Since $\mathcal{B}^{(r)} \supset \mathcal{B}'$, $Mass(\mathcal{B}^{(r)}(a_i^{(r)})) \geq Mass(\mathcal{B}'(a_i^{(r)})) \geq c$. By Lemma 3, $\forall a_j \in \bigcup_{n=1}^{N} \mathcal{B}_n^{(r)}$, $Mass(\mathcal{B}^{(r)}(a_j)) \geq Mass(\mathcal{B}^{(r)}(a_i^{(r)})) \geq c$. □

References

1. Supplementary document (examples, proofs, and additional experiments). http://www.cs.cmu.edu/~kijungs/codes/mzoom/supple.pdf
2. Akoglu, L., McGlohon, M., Faloutsos, C.: Oddball: spotting anomalies in weighted graphs. In: Zaki, M.J., Yu, J.X., Ravindran, B., Pudi, V. (eds.) PAKDD 2010. LNCS (LNAI), vol. 6119, pp. 410–421. Springer, Heidelberg (2010). doi:10.1007/978-3-642-13672-6_40
3. Andersen, R., Chellapilla, K.: Finding dense subgraphs with size bounds. In: Kumar, R., Sivakumar, D. (eds.) WAW 2010. LNCS, vol. 6516, pp. 25–37. Springer, Heidelberg (2009). doi:10.1007/978-3-540-95995-3_3
4. Bader, B.W., Kolda, T.G., et al.: Matlab tensor toolbox version 2.6. http://www.sandia.gov/~tgkolda/TensorToolbox/
5. Beutel, A., Xu, W., Guruswami, V., Palow, C., Faloutsos, C.: Copycatch: stopping group attacks by spotting lockstep behavior in social networks. In: WWW (2013)
6. Cerf, L., Besson, J., Robardet, C., Boulicaut, J.F.: Data peeler: constraint-based closed pattern mining in n-ary relations. In: SDM (2008)
7. Charikar, M.: Greedy approximation algorithms for finding dense components in a graph. In: Jansen, K., Leonardi, S., Vazirani, V. (eds.) APPROX 2002. LNCS, vol. 2462, pp. 84–95. Springer, Heidelberg (2000). doi:10.1007/3-540-44436-X_10
8. Gibson, D., Kumar, R., Tomkins, A.: Discovering large dense subgraphs in massive graphs. In: VLDB (2005)
9. Goldberg, A.V.: Finding a maximum density subgraph. Technical report (1984)
10. Hooi, B., Song, H.A., Beutel, A., Shah, N., Shin, K., Faloutsos, C.: Fraudar: bounding graph fraud in the face of camouflage. In: KDD (2016)
11. Ignatov, D.I., Kuznetsov, S.O., Poelmans, J., Zhukov, L.E.: Can triconcepts become triclusters? Int. J. Gen. Syst. **42**(6), 572–593 (2013)

12. Jiang, M., Beutel, A., Cui, P., Hooi, B., Yang, S., Faloutsos, C.: A general suspiciousness metric for dense blocks in multimodal data. In: ICDM (2015)
13. Jiang, M., Cui, P., Beutel, A., Faloutsos, C., Yang, S.: Catchsync: catching synchronized behavior in large directed graphs. In: KDD (2014)
14. Jiang, M., Cui, P., Beutel, A., Faloutsos, C., Yang, S.: Inferring strange behavior from connectivity pattern in social networks. In: Tseng, V.S., Ho, T.B., Zhou, Z.-H., Chen, A.L.P., Kao, H.-Y. (eds.) PAKDD 2014. LNCS (LNAI), vol. 8443, pp. 126–138. Springer, Heidelberg (2014). doi:10.1007/978-3-319-06608-0_11
15. Kannan, R., Vinay, V.: Analyzing the structure of large graphs. Technical report (1999)
16. Khuller, S., Saha, B.: On finding dense subgraphs. In: Loeckx, J. (ed.) ICALP 1974. LNCS, vol. 14, pp. 597–608. Springer, Heidelberg (2009). doi:10.1007/978-3-642-02927-1_50
17. Kolda, T.G., Bader, B.W.: Tensor decompositions and applications. SIAM Rev. 51(3), 455–500 (2009)
18. Lee, V.E., Ruan, N., Jin, R., Aggarwal, C.: A survey of algorithms for dense subgraph discovery. In: Aggarwal, C.C., Wang, H. (eds.) Managing and Mining Graph Data, vol. 40, pp. 303–336. Springer, Heidelberg (2010)
19. Maruhashi, K., Guo, F., Faloutsos, C.: Multiaspectforensics: pattern mining on large-scale heterogeneous networks with tensor analysis. In: ASONAM (2011)
20. Pandit, S., Chau, D.H., Wang, S., Faloutsos, C.: Netprobe: a fast and scalable system for fraud detection in online auction networks. In: WWW (2007)
21. Prakash, B.A., Sridharan, A., Seshadri, M., Machiraju, S., Faloutsos, C.: EigenSpokes: surprising patterns and scalable community chipping in large graphs. In: Zaki, M.J., Yu, J.X., Ravindran, B., Pudi, V. (eds.) PAKDD 2010. LNCS (LNAI), vol. 6119, pp. 435–448. Springer, Heidelberg (2010). doi:10.1007/978-3-642-13672-6_42
22. Seppänen, J.K., Mannila, H.: Dense itemsets. In: KDD (2004)
23. Shah, N., Beutel, A., Gallagher, B., Faloutsos, C.: Spotting suspicious link behavior with fbox: an adversarial perspective. In: ICDM (2014)

Uncovering Locally Discriminative Structure for Feature Analysis

Sen Wang[1]([⊠]), Feiping Nie[2], Xiaojun Chang[3], Xue Li[1], Quan Z. Sheng[5], and Lina Yao[4]

[1] School of Information Technology and Electrical Engineering,
The University of Queensland, Brisbane, Australia
sen.wang@uq.edu.au, xueli@itee.uq.edu.au
[2] Centre for OPTIMAL, Northwestern Polytechnical University, Xi'an, China
feipingnie@gmail.com
[3] Centre for Quantum Computation and Intelligent Systems,
University of Technology Sydney, Ultimo, Australia
cxj273@gmail.com
[4] School of Computer Science, The University of Adelaide, Adelaide, Australia
lina.yao@unsw.edu.au
[5] School of Computer Science and Engineering, The University of New South Wales,
Kensington, Australia
michael.sheng@adelaide.edu.au

Abstract. Manifold structure learning is often used to exploit geometric information among data in semi-supervised feature learning algorithms. In this paper, we find that local discriminative information is also of importance for semi-supervised feature learning. We propose a method that utilizes both the manifold structure of data and local discriminant information. Specifically, we define a local clique for each data point. The k-Nearest Neighbors (kNN) is used to determine the structural information within each clique. We then employ a variant of Fisher criterion model to each clique for local discriminant evaluation and sum all cliques as global integration into the framework. In this way, local discriminant information is embedded. Labels are also utilized to minimize distances between data from the same class. In addition, we use the kernel method to extend our proposed model and facilitate feature learning in a high-dimensional space after feature mapping. Experimental results show that our method is superior to all other compared methods over a number of datasets.

1 Introduction

The performance of machine learning tasks, e.g. classification or clustering, is mainly affected by the input features that are extracted from raw data. Learning distinctive features or effective data representations can without doubt benefit the consequent learning tasks. Over the past decade, feature analysis has

Electronic supplementary material The online version of this chapter (doi:10.1007/978-3-319-46128-1_18) contains supplementary material, which is available to authorized users.

© Springer International Publishing AG 2016
P. Frasconi et al. (Eds.): ECML PKDD 2016, Part I, LNAI 9851, pp. 281–295, 2016.
DOI: 10.1007/978-3-319-46128-1_18

attracted much research attention in different fields, such as machine learning [32,33,41], multimedia analysis [33], biomedical applications [40,42,43], etc. In literature, a number of unsupervised and supervised methods have been developed to learn new features. Clustering algorithms have been widely used as an unsupervised feature learning procedure to obtain statistical data representation. A typical example in multimedia analysis is k-means, which learns a dictionary from a number of image or video training samples without class information [21,23,34]. Even though increasing the size of the dictionary can squeeze out a bit of extra performance, it is still difficult to identify the best choice of the number of centers.

As one of the representative supervised algorithms, Linear Discriminant Analysis (LDA) [8] finds the best data projection maximizing distances between different class centers while making data samples from the same class closer to each other. This is achieved by maximizing the ratio of the between-class covariance to the within-class covariance. When there are sufficient labeled training data, LDA-based features effectively support machine learning algorithms in a variety of applications. For example, LDA has been used to strengthen the class information of face images in many face recognition systems [2,5,11,29,31,37]. Also, LDA has been revisited in [1] and been evaluated in three pipelines over a few face image datasets for the purpose of gender recognition. In [17], authors evaluate three LDA-based variants to obtain a discriminant movement representation for multi-view action recognition. Unfortunately, when the number of training samples is small, LDA suffers from the *small sample size* (SSS) problem. This is because the small number of training samples will make the within-class scatter matrix singular, which would result in computational difficulty. Meanwhile, learning new features from a small number of labeled training samples in a fully supervised manner may lead to the over-fitting problem. To solve these problems, much research attention has been paid over the last few years. For example, subspace learning methods, such as Principal Component Analysis (PCA), are applied to reduce feature dimensionality prior to LDA, with the goal of removing null space of the within-class scatter matrix. However, this preprocessing step may lose discriminant information which means the consequent projection in the subspace by LDA may not be the best. A number of methods [7,15,18,19,35,38,39] have been proposed to tackle the SSS problem without losing discriminant information. Though the SSS problem is dealt with, the over-fitting problem persists. Increasing the number of labeled training samples would be an ideal solution. However, data labeling in the real world is usually time-consuming and expensive. Considering the huge amount of data without labels, it is extremely difficult to obtain massive and comparable label information. For this reason, semi-supervised feature analysis methods [3,6,14,20,22], which make use of both labeled and unlabeled data, have been extensively studied in the past. Unfortunately, most of the existing semi-supervised feature learning algorithms ignore the utilization of both manifold structure and local discriminant information.

In this paper, we propose a new kernel-based feature learning method that can learn new features when the labeled information is limited. Some researchers [25,36] have pointed out that exploiting the local structures is more effective and efficient than learning the global structures. Besides, the manifold structure of data is another crucial property to be considered in feature learning. In this work, the contribution can be briefly summarized as the utilization of both the manifold structure and the local discriminant information to deal with the shortage of labeled data points. Compared with those representative semi-supervised feature learning methods, our proposed method not only makes labeled data within the same class closer to each other, but also incorporates the local discriminant information into a joint framework. We find that the local discriminative information of the manifold structure is very important for feature learning, especially when the label information is scarce. This is omitted in most of the previous works on semi-supervised feature learning. In order to exploit the manifold structure and the local discriminant information, we learn a new graph Laplacian. Specifically, for each data point, we define a local clique in which the data point and its $k - 1$ geometric neighbors are included. To achieve this point, kNN is used to exploit the intrinsic manifold structure of data. Moreover, we employ a variant of the Fisher criterion to each clique to evaluate the local discriminant information. The sum of all cliques will be integrated into a joint framework as a global integration. In this way, a new graph Laplacian that holds manifold information with local discriminant information can be learned. This method has two-fold advantages: Firstly, since there are only k data points in each clique (normally quite smaller than the dimensionality of data points), the overall computational burden is greatly relieved. This is because calculating an inverse of a $k \times k$ matrix n times is faster than a direct inverse calculation on a $n \times n$ matrix when n is big. Secondly, it is easy to extend the local discriminant model to a kernel-based version. In this work, we also extend our proposed model using kernel method to learn both labeled and unlabeled data in a high-dimensional space in which data are linearly separable. Additionally, we have proposed an algorithm to solve the optimization problem.

The rest of paper is organized as follows: Notations and definitions will be presented in Sect. 2. Our proposed method and the optimization algorithm will be elaborated upon in Sect. 3. In Sect. 4, we will report the experimental settings, results, and related analysis. The conclusion will be given in the last section.

2   Notations and Definitions

To give a better understanding of the proposed algorithm, notations and definitions used in this paper are summarized in this section. Matrices and vectors are written as boldface uppercase letters and boldface lowercase letters, respectively. In this paper, we follow the conventional definition in semi-supervised learning. In the training dataset, there are n data samples including m labeled data and $n - m$ unlabeled data. Thus, the training data matrix is defined as:

$$\boldsymbol{X} = \begin{bmatrix} \boldsymbol{X}_l & \boldsymbol{X}_u \end{bmatrix} \tag{1}$$

$X \in \mathbb{R}^{d \times n}$. The labeled data are denoted by $X_l \in \mathbb{R}^{d \times m}$, and the unlabeled data are denoted by $X_u \in \mathbb{R}^{d \times (n-m)}$. d is feature dimensionality. Correspondingly, label assignment matrix $Y \in \mathbb{R}^{n \times c}$ is defined as:

$$Y = \begin{bmatrix} Y_l \\ Y_u \end{bmatrix} \quad (2)$$

$Y_l \in \mathbb{R}^{m \times c}$ is for the labeled data while $Y_u \in \mathbb{R}^{(n-m) \times c}$ is for the unlabeled data. c is the number of classes. Y_u is initialized with all zeros and its entry, $y_{it} \in [0, 1]$, denotes how likely the i-th unlabeled data belongs to the t-th class, where $m + 1 \leq i \leq n$ and $1 \leq t \leq c$.

3 Locally Discriminative Structure Uncovering

3.1 Proposed Method

Inspired by kernel methods, we assume that, after a non-linear mapping function $f = \phi(x)$, the mapped data within the same class are still geometrically close to each other in a high-dimensional space that $\phi(x)$ maps to. The problem can be formulated as:

$$\min_{\phi} \sum_{t=1}^{c} \sum_{x_p, x_q \in \pi_t} ||\phi(x_p) - \phi(x_q)||^2 \quad (3)$$

π_t contains all the data points from the t-th class. Without loss of generality, we can have:

$$\phi(x) = \sum_{i=1}^{m} \alpha_i k(x, x_i), \quad (4)$$

k is a kernel defined on $x_1, \ldots, x_m \in \mathcal{X}$. Note that Eq. (3) needs class information, thus only m labeled data points are considered in Eq. (3). After substituting (4) into (3) by a matrix representation, the objective becomes:

$$\min_{\alpha} \sum_{t=1}^{c} \sum_{x_p, x_q \in \pi_t} ||\alpha^T k(X_l^T, x_p) - \alpha^T k(X_l^T, x_q)||^2$$
$$= \min_{\alpha} \alpha^T K_l L_w K_l \alpha, \quad (5)$$

where

$$L_w = D - W \quad (6)$$

$\alpha = [\alpha_1, \ldots, \alpha_m]^T \in \mathbb{R}^m$ is a vector. The weight matrix $W_l \in \mathbb{R}^{m \times m}$ for labeled data is defined as:

$$W_{ij} = \begin{cases} 1 & x_i \text{ and } x_j \text{ are in the same class;} \\ 0 & \text{otherwise,} \end{cases} \quad (7)$$

D is a diagonal matrix with $D_{ii} = \sum_{j=1}^{m} W_{ij}$. $K_l \in \mathbb{R}^{m \times m}$ is the kernel matrix of labeled samples. Note that the representation of α in (5) is an m-dimensional vector that transforms the original feature into one dimension. Now, we extend the learned feature into r dimensions by using a transformation matrix $a = [\alpha_1, \dots, \alpha_r] \in \mathbb{R}^{m \times r}$. Consequently, the objective function in (5) is equivalent to

$$\min_{a^T K_l a = I} Tr(a^T K_l L_w K_l a), \qquad (8)$$

$Tr(\cdot)$ is trace operator. The constraint $a^T K_l a = I$ is to make the projection orthogonal.

The objective in (8) is based on label information. We aim to make use of both labeled and unlabeled data to improve the performance when the class information is in scarcity. We now extend it into a semi-supervised method by extending the weighted matrix W_l in (7) into a new one:

$$W = \begin{bmatrix} W_l & 0_{m \times (n-m)} \\ 0_{(n-m) \times m} & 0_{(n-m) \times (n-m)} \end{bmatrix} \qquad (9)$$

W_l is the weighted matrix for labeled data. $0_{m \times (n-m)}$ is a zero matrix with m rows and $(n-m)$ columns. In this way, the adjacency graph L_w is enlarged from $m \times m$ to $n \times n$.

As mentioned before, we find that local discriminant information among data, labeled and unlabeled, is quite useful for learning new features especially when the label information is limited. In this work, we aim to learn such local structures and to embed the structures into a joint framework. The objective function in (8) can be re-written as:

$$\min_{a^T Ka = I} Tr(a^T K L_w K a) + \lambda \Omega(\cdot), \qquad (10)$$

$\Omega(\cdot)$ is the regularization term that exploits the local structures among data samples. λ is the regularization parameter. We assume that all data within the same class are put together and define a scaled class assignment matrix $G \in \mathbb{R}^{n \times c}$ as follows:

$$G = Y(Y^T Y)^{-\frac{1}{2}} \qquad (11)$$

$Y = [y_1, \dots, y_n]^T \in \mathbb{R}^{n \times c}$. To exploit the local structure, we define a clique of x_i, denoted as $\mathcal{N}_k(x_i)$, which has k data samples containing x_i itself and its $k-1$ neighbors. Given the data matrix $X \in \mathbb{R}^{d \times n}$, the local data matrix for the i-th data is defined as $X_i = [x_i, x_{i_1}, \dots, x_{i_{k-1}}] \in \mathbb{R}^{d \times k}$. Correspondingly, the local scaled classification matrix for x_i can be defined as $G_i = [g_i, \dots, g_{i_{k-1}}]^T \in \mathbb{R}^{k \times c}$. Note that both X_i and G_i are actually selected from X and G respectively. We define a selection matrix $S_i \in \mathbb{R}^{n \times k}$ for x_i where $S_i^{pq} = 1$, if x_p is the q-th element of $\mathcal{N}_k(x_i)$, $S_i^{pq} = 0$ otherwise. $1 \le q \le k$. Thus, the local scaled classification matrix for each data can be re-written as $G_i = S_i^T G = S_i^T Y(Y^T Y)^{-\frac{1}{2}}$.

In linear discriminant analysis, the objective is to maximize the separability of all data points and to minimize the distances among data that are from the same class. According to the definitions of scatter matrix in linear discriminant analysis, the corresponding total scatter and between class scatter matrices for the local data clique can be defined as S_{t_i} and S_{b_i}. To simplify the formulation, we centralize each local data matrix $\bar{X}_i = X_i H$, where $H = I_{k \times k} - \frac{1}{k} 1_k 1_k^T$. 1_k is a k-dimensional column vector with all ones. Thus, we have:

$$
\begin{aligned}
S_{t_i} &= \sum_{i=1}^{n} (x_i - \mu)(x_i - \mu)^T = \bar{X}_i \bar{X}_i^T \\
S_{b_i} &= \sum_{t=1}^{c} n_t (\mu_t - \mu)(\mu_t - \mu)^T = \bar{X}_i G_i G_i^T \bar{X}_i^T
\end{aligned}
\tag{12}
$$

where μ_t is the mean of samples in the t-th class and μ is the global mean that is zero after the centralization. n_t is the number of data points of the t-th class. Inspired by the Fisher criterion [8], the optimal local scaled class assignment matrix G_i^* can be obtained by optimizing the follow objective:

$$
\begin{aligned}
G_i^* &= \arg \max_{G_i} Tr \left((S_{t_i} + \theta I)^{-1} S_{b_i} \right) \\
&= \arg \max_{G_i} Tr \left(G_i^T \bar{X}_i^T (\bar{X}_i \bar{X}_i^T + \theta I)^{-1} \bar{X}_i G_i \right),
\end{aligned}
\tag{13}
$$

where $\theta I (\theta > 0)$ is added to avoid $(S_{t_i} + \theta I)$ be singular. In Eq. (13), G_i^* is a score that evaluates the local discriminant information of each data points. A larger value indicates that the samples in the local clique from different classes are better separated. To control the capacity of local discriminant model, we add a regularization term $Tr(G_i^T H G_i)$. Then the optimization problem in (13) is equivalent to

$$
\arg \min_{G_i} Tr \left(G_i^T H G_i - G_i^T \bar{X}_i^T (\bar{X}_i \bar{X}_i^T + \theta I)^{-1} \bar{X}_i G_i \right)
\tag{14}
$$

It is proved in the supplemental document, the problem of (14) is equivalent to

$$
\arg \min_{G_i} Tr \left(G_i^T L_i G_i \right),
\tag{15}
$$

$$
\text{where } L_i = H (\bar{X}_i^T \bar{X}_i + \theta I)^{-1} H
$$

Because of $G_i = S_i^T G$, we take all local manifold structures into account together by summing (15) over all local cliques. Then, the global local discriminant score can be written as:

$$
\arg \min_{G} \sum_{i=1}^{n} Tr(G_i^T L_i G_i) = \arg \min_{G} Tr(G^T L G),
\tag{16}
$$

where

$$
L = \sum_{i=1}^{n} S_i L_i S_i^T
\tag{17}
$$

By using the graph Laplacian in (17), local discriminant manifold structures among data points can be therefore embedded into a joint framework. The objective function in (10) arrives at:

$$
\begin{aligned}
&\min_{a^T K a = I} Tr(a^T K L_w K a) + \lambda Tr(a^T K L K a) \\
&= \min_{a^T K a = I} Tr(a^T K (L_w + \lambda L) K a)
\end{aligned}
\tag{18}
$$

where λ is a parameter that leverages the proportion of utilization of both manifold structure and local discriminant information in the joint framework.

3.2   Optimization

To solve the objective function in (18), we assume

$$K = V \Lambda V^T, \tag{19}$$

and $\tilde{V}$ is the null space of V. For any a, we can represent

$$a = V\beta + \tilde{V}\gamma, \tag{20}$$

Thus,

$$\begin{aligned} a^T K a &= (V\beta + \tilde{V}\gamma)^T K (V\beta + \tilde{V}\gamma) \\ &= \beta^T V^T K V \beta = \beta^T \Lambda \beta, \end{aligned} \tag{21}$$

Substituting (20) and (21) into (18), the objective function is re-formulated as

$$\begin{aligned} &\min_{\beta^T \Lambda \beta = I} Tr(a^T K (L_w + \lambda L) K a) \\ &= \min_{\beta^T \Lambda \beta = I} Tr(\beta^T \Lambda V^T (L_w + \lambda L) V \Lambda \beta), \end{aligned} \tag{22}$$

Note that Λ is invertible, so the solution β is the eigenvectors corresponding to $\Lambda^T V^T (L_w + \lambda L) V \Lambda$. To enable the solution in the real domain, we can make

$$\beta = \Lambda^{-\frac{1}{2}} \omega \tag{23}$$

and the problem becomes

$$\min_{\omega^T \omega = I} Tr(\omega^T \Lambda^{\frac{1}{2}} V^T (L_w + \lambda L) V \Lambda^{\frac{1}{2}} \omega) \tag{24}$$

The optimal solution ω is the eigenvectors of $\Lambda^{\frac{1}{2}} V^T (L_w + \lambda L) V \Lambda^{\frac{1}{2}}$ with respect to its eigenvalues in an ascending order. We summarize the entire procedure in Algorithm 1 to learn the new features. For any test data x', its l-th new feature is obtained by $\sum_{i=1}^{n} \alpha_{il} k(x', x_i)$, $1 \leq l \leq r$.

4   Experiments

In this section, we will briefly introduce the datasets and the compared methods which are used in the experiments. Afterwards, experimental results are evaluated and analyzed.

Algorithm 1. Local Discriminant Structure Uncovering.

Input:
    The training data matrix $X = [X_l, X_u] \in \mathbb{R}^{d \times n}$
    The classification assignment matrix $Y = [Y_l, Y_u] \in \mathbb{R}^{n \times c}$
Output:
    Feature transformation matrix $a \in \mathbb{R}^{n \times r}$
1: Compute kernel matrix K;
2: Compute graph L_w and L according to (6) and (17);
3: Compute V and Λ by performing eigen-decomposition on K according to (19);
4: Compute ω by performing eigen-decomposition according to (24);
5: Compute β according to (23);
6: Obtain a according to (20);

4.1 Datasets and Compared Methods

To evaluate our algorithm, we have conducted extensive experiments and compared with a number of approaches on five datasets:

- **COIL-20** [24]: It contains 1,440 gray-scale images of 20 objects (72 images per object) under various poses. The objects are rotated through 360 degrees and taken at the interval of 5 degrees.
- **UMIST** [12]: The UMIST, which is also known as the Sheffield Face Database, consists of 564 images of 20 individuals. Each individual is shown in a variety of poses from profile to frontal views.
- **USPS** [16]: This dataset collects 9,298 images of handwritten digits (0–9) from envelops by the U.S. Postal Service. All images have been normalized to the same size of 16 × 16 pixels in grayscale.
- **Yale** [9]: It consists of 2,414 frontal face images of 38 subjects. Different lighting conditions have been considered in this dataset. All images are reshaped into 24 × 24 pixels.
- **MIMIC II** [26]: It consists of 32,536 patient records in Intensive Care Unit (ICU) at the Beth Israel Deaconess Medical Center (BIDMC) collected from 2001 to 2008. We only extract medical notes from the database together with mortality information (if the patient has been expired at ICU for the first admission). A similar feature extraction pipeline used in [10] has been applied. Differently, Bag-of-Words model is used to encode multiple notes at different times for each patient. Empirically, we set the size of the dictionary as 500. Afterwards, we randomly select 1,000 adult patients of which are positive and negative evenly.

For the first four datasets (COIL20, UMIST, USPS and YaleB), we just simply use their pixels as the input features and the typical RBF kernel. For MIMIC II dataset, we use a 500 dimensional Bag-of-Words representation and a χ^2 kernel. There are a number of kernel functions that have been invented so far. In this paper, our focus is to demonstration the effectiveness of our proposed algorithm rather than making comparisons between different kernels regarding classification performance.

To evaluate our proposed algorithm, we choose several methods as the baseline:

- **Kernel Discriminant Analysis (KDA)** [27]: As one of the representative feature dimensionality reduction methods, KDA aims to project data into a direction on which class centers are far from each other while data samples of the same class are close to each other after feature mapping. We use a speed-up version implemented in [4].
- **Kernel Principal Component Analysis (KPCA)** [28]: Compared to KDA, KPCA reduces feature dimensionality by transforming data into a new coordinate system where the top n greatest variances of data correspondingly lie on the first n coordinates in the new subspace. We test different dimensionality reduction scenarios by KPCA across all the datasets. The best classification performance results are reported.
- **Kernel Semi-supervised Discriminant Analysis (KSDA)** [3]: SDA aims to solve the problem of scarcity of label information when performing discriminant analysis. To utilize unlabeled samples, a graph Laplacian is built to approximate the local geometry of the data manifold where both the labeled and unlabeled data reside. Kernel SDA (KSDA) is used in the experiments.
- **Kernel Semi-supervised Local Fisher discriminant analysis (KSELF)** [30]: SELF which leverages supervised Local Fisher Discriminant Analysis and unsupervised Principal Component Analysis, is a linear semi-supervised dimensionality reduction method which makes feature analysis effective when only a small number of labeled samples are available. In the experiments, we use its non-linear extension termed as KSELF.
- **Kernel Locality Preserving Projections (KLPP)** [14]: KLPP is an unsupervised manifold learning method which preserves the local structure of samples, i.e. neighborhood relationship, in the original feature space as well as in the new projected space.
- **Co-regularized Ensemble for Feature Selection (EnFS)** [13]: This method employs a co-regularized framework in which a joint $\ell_{2,1}$-norm of multiple feature selection matrices can alleviate the over-fitting problem when the number of labeled data is small. Furthermore, a subset of feature that is more distinctive can be uncovered by removing irrelevant or noisy features.

For all the methods, corresponding parameters are tuned in the same range of $\{10^{-4}, 10^{-3}, 10^{-2}, 1, 10^2, 10^3, 10^4\}$. Support Vector Machine (SVM) with a linear kernel has been applied as a classifier evaluating the classification performance of each method. The SVM parameter that controls the trade-off between the margin and the size of the slack variables is also tuned in the aforementioned range. The detailed dataset partition is followed by the convention of the semi-supervised learning approaches. Specifically, the training set contains both labeled and unlabeled data, and the testing set is not available during the training phrase. c is denoted as the number of classes for each dataset. In the training dataset, we randomly sample a number of labeled data per class (1, 3, 5, and 10) as different class settings. Therefore the numbers of labeled training data are $1 \times c$, $3 \times c$, $5 \times c$, and $10 \times c$ in the different class settings, while the remaining training data are treated as unlabeled. Particularly, we also investigate conditions in which more labeled information is available by using $25 \times c$, $50 \times c$, $75 \times c$, and $100 \times c$ on MIMIC II dataset. We repeat the experiments five times using the data partitions mentioned above, and report the average results. Because we focus on classification performance, we use Mean Average Precision (MAP) as the effectiveness measure in the comparisons.

4.2 Evaluation

We have made comparisons among all the methods mentioned above over five datasets under different label settings. Generally speaking, our algorithm consistently achieves better classification performance under different settings than all the counterparts. Specifically, from Table 1, we can observe that our approach outperforms all supervised and unsupervised methods, including KDA, KPCA and EnFS, when labeled data are quite few, i.e. $1 \times c$. For instance, a number of relatively large margins can be observed on UMIST and USPS. Compared to the semi-supervised methods (KSDA, KSELF) and the unsupervised manifold learning method (KLPP), our method still has superior performance over all the datasets under different settings. In the conditions where more labels are available, it is observed that our method still performs better than all the compared approaches. However, the difference margins between our method and the other approaches are quite limited. For example, our method, on the MIMIC II dataset, performs quite similarly to EnFS and KSELF.

Apart from the overall classification performance comparisons in Table 1, we also studied the effects of parameters used in our method. Due to the page limit,

Table 1. Performance comparison (Mean Average Precision ± STD) across all datasets with a linear SVM classifier.

Dataset	Settings	KDA	KPCA	KSDA	KSELF	KLPP	EnFS	Ours
COIL-20	$1 \times c$	0.728±0.026	0.699±0.026	0.777±0.016	0.702±0.032	0.780±0.020	0.702±0.012	**0.818±0.001**
	$3 \times c$	0.832±0.013	0.833±0.018	0.840±0.020	0.819±0.014	0.822±0.007	0.825±0.025	**0.850±0.005**
	$5 \times c$	0.888±0.015	0.881±0.020	0.878±0.022	0.877±0.014	0.876±0.015	0.880±0.013	**0.897±0.002**
	$10 \times c$	0.948±0.009	0.949±0.012	0.926±0.015	0.931±0.012	0.932±0.007	0.935±0.012	**0.957±0.001**
UMIST	$1 \times c$	0.574±0.019	0.558±0.021	0.642±0.026	0.573±0.018	0.580±0.020	0.558±0.017	**0.654±0.010**
	$3 \times c$	0.882±0.033	0.851±0.035	0.860±0.034	0.832±0.044	0.812±0.043	0.810±0.037	**0.889±0.001**
	$5 \times c$	0.960±0.013	0.942±0.019	0.945±0.020	0.937±0.019	0.913±0.021	0.909±0.031	**0.963±0.001**
	$10 \times c$	0.995±0.004	0.989±0.003	0.987±0.008	0.990±0.004	0.967±0.006	0.9779±0.01	**0.996±0.002**
USPS	$1 \times c$	0.536±0.079	0.464±0.056	0.653±0.088	0.351±0.023	0.789±0.056	0.523±0.053	**0.801±0.009**
	$3 \times c$	0.727±0.017	0.720±0.019	0.811±0.015	0.625±0.073	0.905±0.012	0.715±0.027	**0.936±0.008**
	$5 \times c$	0.795±0.024	0.788±0.032	0.865±0.031	0.732±0.028	0.939±0.017	0.788±0.024	**0.948±0.007**
	$10 \times c$	0.868±0.018	0.860±0.007	0.911±0.007	0.837±0.011	0.957±0.005	0.864±0.016	**0.964±0.003**
YaleB	$1 \times c$	0.359±0.013	0.358±0.015	0.245±0.007	0.220±0.011	0.354±0.028	0.216±0.014	**0.463±0.004**
	$3 \times c$	0.872±0.023	0.761±0.033	0.555±0.034	0.652±0.033	0.572±0.033	0.567±0.083	**0.891±0.001**
	$5 \times c$	0.951±0.003	0.907±0.006	0.769±0.021	0.856±0.008	0.717±0.020	0.723±0.039	**0.964±0.002**
	$10 \times c$	0.981±0.003	0.978±0.005	0.948±0.005	0.963±0.011	0.826±0.018	0.863±0.032	**0.992±0.007**
MIMIC II	$25 \times c$	0.683±0.029	0.675±0.028	0.666±0.051	0.699±0.017	0.683±0.037	0.696±0.028	**0.700±0.027**
	$50 \times c$	0.727±0.017	0.722±0.026	0.705±0.035	0.733±0.014	0.712±0.004	0.740±0.008	**0.741±0.018**
	$75 \times c$	0.741±0.012	0.749±0.027	0.727±0.024	0.751±0.010	0.737±0.020	0.754±0.028	**0.755±0.015**
	$100 \times c$	0.754±0.025	0.757±0.021	0.759±0.019	0.757±0.028	0.751±0.020	0.762±0.023	**0.766±0.017**

we only studied the sensitivity of parameters on four image datasets. Note that some of the parameters are fixed for demonstration in the following experiments. Thus, the performance results are not as good as the ones in Table 1 in which all parameters are tuned to achieve the best performance. In the first round, we test parameter sensitivity for θ and k which are both required when constructing the new graph Laplacian. We test θ in the range of $\{10^{-4}, 10^{-3}, 10^{-2}, 1, 10^2, 10^3, 10^4\}$ and k in the range of $[1, 3, 5, 10, 15, 20]$. We find that the system is not sensitive to θ and the optimal k should be 3 or 5.

In the second round, we firstly fix the aforementioned two parameters of graph Laplacian, $\theta = 1$ and $k = 3$, respectively. Moreover, the regularization parameter, λ, in (24) is set to 1. We plot classification performance changes over four datasets when dimensionality reduction parameter, r, varies. Our aim is to understand how the new features with reduced dimensionality impact performance by fixing all the parameters except the dimension of the inferred feature. The results show that we do not have to use full dimensional features. For example, in Fig. 1(a), the performance scores 86.04 % when 100 % learned features are preserved on COIL20. It then peaks at 86.67 % when 70 % are preserved. For each dataset, a further improvement has been observed after reducing dimensionality of the new features. This improvement might be because irrelevant

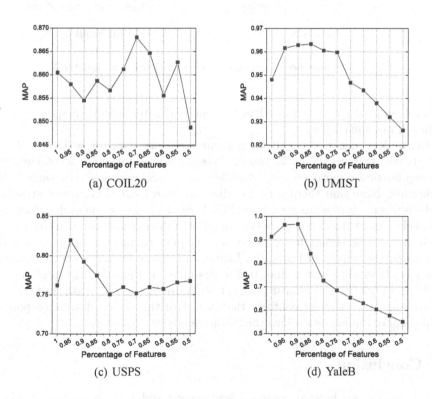

(a) COIL20   (b) UMIST

(c) USPS   (d) YaleB

Fig. 1. Performance variations w.r.t dimensionality reduction parameter r.

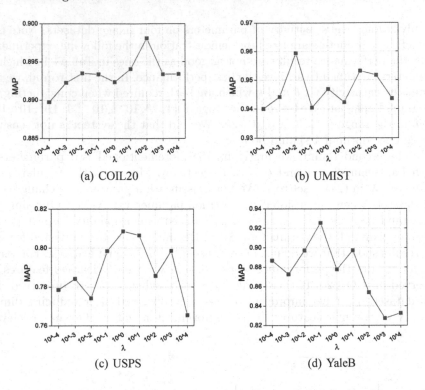

Fig. 2. Performance variations w.r.t regularization parameter λ.

features and noises are removed by our method after feature mapping in the higher dimensional space.

In the last experiment, all three parameters are fixed ($\theta = 1$, $k = 3$ and $r = 1$). We compare the variations of classification performance when changing the regularization parameter, λ, which leverages local manifold structures in the framework. Note that there is no contribution from local discriminant structure analysis when λ is close to zero. From Fig. 2, the performance on each dataset is a relatively lower value when little local manifold information has been considered. With the variations of λ, for each dataset, the performance varies and scores the best when the weight of the graph Laplacian is increased to a certain amount which is obviously greater than 0. For example, in Fig. 1(b), the performance starts around 94 % and almost peaks at 96 % when $\lambda = 0.01$, with a nearly 2 % improvement. This result confirms that our algorithm successfully incorporates local manifold information into the feature analysis procedure.

5   Conclusion

In this paper, we have proposed a semi-supervised feature analysis method. Specifically, our method enforces data from the same class to become closer to each other in a high-dimensional space after feature mapping. In order to take both local discriminant information and manifold structure into account, a local discriminant model has been applied to the local clique of each data point. Our method successfully learns both labeled and unlabeled data via leveraging the new graph Laplacian that holds local discriminant information. It has proven that our method effectively learns features when the number of labeled data points is quite small.

Acknowledgement. This work was supported by Australian Research Council Discovery Project. The project number is DP140100104. Any opinions, findings, and conclusions or recommendations expressed in this material are those of the author(s) and do not necessarily reflect the views of the Australian Research Council.

References

1. Bekios-Calfa, J., Buenaposada, J.M., Baumela, L.: Revisiting linear discriminant techniques in gender recognition. IEEE Trans. Pattern Anal. Mach. Intell. **33**(4), 858–864 (2011)
2. Belhumeur, P.N., Hespanha, J.P., Kriegman, D.: Eigenfaces vs. fisherfaces: recognition using class specific linear projection. IEEE Trans. Pattern Anal. Mach. Intell. **19**(7), 711–720 (1997)
3. Cai, D., He, X., Han, J.: Semi-supervised discriminant analysis. In: ICCV (2007)
4. Cai, D., He, X., Han, J.: Speed up kernel discriminant analysis. VLDBJ **20**(1), 21–33 (2011)
5. Chang, X., Nie, F., Wang, S., Yang, Y., Zhou, X., Zhang, C.: Compound rank-k projections for bilinear analysis. IEEE Trans. Neural Netw. Learn. Syst. **27**(7), 1502–1513 (2016)
6. Chang, X., Nie, F., Yang, Y., Huang, H.: A convex formulation for semi-supervised multi-label feature selection. In: AAAI (2014)
7. Chen, L.F., Liao, H.Y.M., Ko, M.T., Lin, J.C., Yu, G.J.: A new lda-based face recognition system which can solve the small sample size problem. Pattern Recogn. **33**(10), 1713–1726 (2000)
8. Fukunaga, K.: Introduction to statistical pattern recognition. Academic press, San Diego (1990)
9. Georghiades, A.S., Belhumeur, P.N., Kriegman, D.: From few to many: Illumination cone models for face recognition under variable lighting and pose. IEEE Trans. Pattern Anal. Mach. Intell. **23**(6), 643–660 (2001)
10. Ghassemi, M., Naumann, T., Doshi-Velez, F., Brimmer, N., Joshi, R., Rumshisky, A., Szolovits, P.: Unfolding physiological state: Mortality modelling in intensive care units. In: SIGKDD, pp. 75–84 (2014)
11. Goudail, F., Lange, E., Iwamoto, T., Kyuma, K., Otsu, N.: Face recognition system using local autocorrelations and multiscale integration. IEEE Trans. Pattern Anal. Mach. Intell. **18**(10), 1024–1028 (1996)
12. Graham, D., Allinson, N.: Characterizing virtual eigensignatures for general purpose face recognition. Face Recog. Theor. Appl. NATO ASI Seri. F Comput. Syst. Sci. **163**, 446–456 (1998)
13. Han, Y., Yang, Y., Zhou, X.: Co-regularized ensemble for feature selection. In: AAAI, pp. 1380–1386 (2013)

14. He, X., Niyogi, P.: Locality preserving projections. In: NIPS, p. 153 (2004)
15. Huang, R., Liu, Q., Lu, H., Ma, S.: Solving the small sample size problem of lda. ICPR **3**, 29–32 (2002)
16. Hull, J.J.: A database for handwritten text recognition research. IEEE Trans. Pattern Anal. Mach. Intell. **16**(5), 550–554 (1994)
17. Iosifidis, A., Tefas, A., Nikolaidis, N., Pitas, I.: Multi-view human movement recognition based on fuzzy distances and linear discriminant analysis. CVIU **116**(3), 347–360 (2012)
18. Jiang, X., Mandal, B., Kot, A.: Eigenfeature regularization and extraction in face recognition. IEEE Trans. Pattern Anal. Mach. Intell. **30**(3), 383–394 (2008)
19. Kyperountas, M., Tefas, A., Pitas, I.: Weighted piecewise lda for solving the small sample size problem in face verification. TNN **18**(2), 506–519 (2007)
20. Li, S., Fu, Y.: Low-rank coding with b-matching constraint for semi-supervised classification. In: AAAI, pp. 1472–1478 (2013)
21. Lu, Z., Peng, Y.: Learning descriptive visual representation by semantic regularized matrix factorization. In: AAAI, pp. 1523–1529 (2013)
22. Ma, Z., Yang, Y., Nie, F., Sebe, N.: Thinking of images as what they are: Compound matrix regression for image classification. In: AAAI, pp. 1530–1536 (2013)
23. Ma, Z., Yang, Y., Xu, Z., Yan, S., Sebe, N., Hauptmann, A.G.: Complex event detection via multi-source video attributes. In: CVPR, pp. 2627–2633 (2013)
24. Murase, H., Nayar, S.K.: Visual learning and recognition of 3-d objects from appearance. IJCV **14**(1), 5–24 (1995)
25. Roweis, S.T., Saul, L.K.: Nonlinear dimensionality reduction by locally linear embedding. Science **290**(5500), 2323–2326 (2000)
26. Saeed, M., Villarroel, M., Reisner, A., Clifford, G., Lehman, L., Moody, G., Heldt, T., Kyaw, T., Moody, B., Mark, R.: Multiparameter intelligent monitoring in intensive care ii (mimic-ii): a public-access intensive care unit database. Crit. Care Med. **39**(5), 952 (2011)
27. Schölkopf, B., Müller, K.R.: Fisher discriminant analysis with kernels. In: NNSP (1999)
28. Schölkopf, B., Smola, A., Müller, K.R.: Nonlinear component analysis as a kernel eigenvalue problem. Neural Comput. **10**(5), 1299–1319 (1998)
29. Shu, X., Gao, Y., Lu, H.: Efficient linear discriminant analysis with locality preserving for face recognition. Pattern Recogn. **45**(5), 1892–1898 (2012)
30. Sugiyama, M., Idé, T., Nakajima, S., Sese, J.: Semi-supervised local fisher discriminant analysis for dimensionality reduction. Mach. Learn. **78**(1–2), 35–61 (2010)
31. Swets, D.L., Weng, J.J.: Using discriminant eigenfeatures for image retrieval. IEEE Trans. Pattern Anal. Mach. Intell. **18**(8), 831–836 (1996)
32. Wang, S., Nie, F., Chang, X., Yao, L., Li, X., Sheng, Q.Z.: Unsupervised feature analysis with class margin optimization. In: Appice, A., Rodrigues, P.P., Santos Costa, V., Soares, C., Gama, J., Jorge, A. (eds.) ECML PKDD 2015. LNCS (LNAI), vol. 9284, pp. 383–398. Springer, Heidelberg (2015). doi:10.1007/978-3-319-23528-8_24
33. Wang, S., Pan, P., Long, G., Chen, W., Li, X., Sheng, Q.Z.: Compact representation for large-scale unconstrained video analysis. World Wide Web **19**(2), 231–246 (2016)
34. Yang, J., Yu, K., Gong, Y., Huang, T.: Linear spatial pyramid matching using sparse coding for image classification. In: CVPR, pp. 1794–1801 (2009)
35. Yang, Y., Ma, Z., Nie, F., Chang, X., Hauptmann, A.G.: Multi-class active learning by uncertainty sampling with diversity maximization. Int. J. Comput. Vis. **113**(2), 113–127 (2015)

36. Yang, Y., Nie, F., Xu, D., Luo, J., Zhuang, Y., Pan, Y.: A multimedia retrieval framework based on semi-supervised ranking and relevance feedback. IEEE Trans. Pattern Anal. Mach. Intell. **34**(4), 723–742 (2012)
37. Ye, J., Li, Q.: Lda/qr: an efficient and effective dimension reduction algorithm and its theoretical foundation. Pattern Recogn. **37**(4), 851–854 (2004)
38. Yu, H., Yang, J.: A direct lda algorithm for high-dimensional data with application to face recognition. Pattern Recogn. **34**, 2067–2070 (2001)
39. Zhang, T., Fang, B., Tang, Y.Y., Shang, Z., Xu, B.: Generalized discriminant analysis: A matrix exponential approach. IEEE Trans. Cybern. **40**(1), 186–197 (2010)
40. Zhu, X., Li, X., Zhang, S.: Block-row sparse multiview multilabel learning for image classification. IEEE Trans. Cybern. **46**(2), 450–461 (2016)
41. Zhu, X., Li, X., Zhang, S., Ju, C., Wu, X.: Robust joint graph sparse coding for unsupervised spectral feature selection. IEEE Trans. Neural Netw. Learn. Syst. (2016)
42. Zhu, X., Suk, H., Lee, S., Shen, D.: Subspace regularized sparse multitask learning for multiclass neurodegenerative disease identification. IEEE Trans. Biomed. Eng. **63**(3), 607–618 (2016)
43. Zhu, X., Suk, H.I., Wang, L., Lee, S.W., Shen, D., Initiative, Alzheimers Disease Neuroimaging et al.: A novel relational regularization feature selection method for joint regression and classification in ad diagnosis. Med. Image Anal. (2015)

Multi-Objective Group Discovery
on the Social Web

Behrooz Omidvar-Tehrani[1](✉), Sihem Amer-Yahia[2],
Pierre-Francois Dutot[2], and Denis Trystram[2]

[1] The Ohio State University, Columbus, USA
omidvar-tehrani.1@osu.edu
[2] Univ. Grenoble Alps, CNRS, Grenoble, France
{sihem.amer-yahia,pierre-francois.dutot,trystram}@imag.fr

Abstract. We are interested in discovering user groups from collaborative rating datasets of the form $\langle i, u, s \rangle$, where $i \in \mathcal{I}$, $u \in \mathcal{U}$, and s is the integer rating that user u has assigned to item i. Each user has a set of attributes that help find *labeled groups* such as *young computer scientists in France* and *American female designers*. We formalize the problem of finding user groups whose quality is optimized in multiple dimensions and show that it is NP-Complete. We develop α-MOMRI, an α-approximation algorithm, and h-MOMRI, a heuristic-based algorithm, for multi-objective optimization to find high quality groups. Our extensive experiments on real datasets from the social Web examine the performance of our algorithms and report cases where α-MOMRI and h-MOMRI are useful.

1 Introduction

Today's data scientists are faced with large volumes of data to explore. In particular, collaborative rating sites have become essential data resources to make decisions about mundane tasks such as purchasing a book, renting a movie or going to a restaurant. The availability of a number of datasets on the social Web, such as MOVIELENS, a movie rating site, LastFM, a music rating site and BOOKCROSSING, a book rating site, appeals to scientists today who design algorithms that help analysts make better decisions on complex tasks such as crowd data sourcing (which users to ask ratings from), advertisers in determining which items to recommend to which users, and social scientists in validating hypotheses such as *young professionals are more inclined to buying self-help books*, on large datasets.

In practice, however, there does not exist analytics tools that enable the scalable, on-demand discovery of user groups. In this paper, we are given a dataset of rating records in the form $\langle i, u, s \rangle$, where $i \in \mathcal{I}$ (set of items), $u \in \mathcal{U}$ (set of users), and s is the integer rating that user u has assigned to item i. We define the notion of *user group* as a conjunction of demographic attributes over rating records, such as *rich young professionals* or *teachers who live in the countryside*. Given a dataset, e.g., ratings of Woody Allen movies, we formalize

© Springer International Publishing AG 2016
P. Frasconi et al. (Eds.): ECML PKDD 2016, Part I, LNAI 9851, pp. 296–312, 2016.
DOI: 10.1007/978-3-319-46128-1_19

the problem of discovering *high quality* user groups. Quality is formulated as the optimization of two dimensions: *coverage* and *diversity*. Optimizing coverage ensures that most input records $\langle i, u, s \rangle$ will belong to at least one group in the output. Optimizing diversity ensures that found groups are as different as possible from each other, e.g., *males and females* or *young and old*, and unveils ratings by different users. User groups with high coverage and high diversity, can help analysts make a variety of decisions such as audience targeting in advertising or hypothesis validation in social science. Example 1 illustrates a common case in practice.[1]

Example 1. It is generally believed that romantic movies (e.g., *American Beauty*, 1999) are mostly watched by females. This observation is based on *demographic breakdown* reports on IMDb.[2] Anna, who is a social scientist, wants to validate this hypothesis by exploring diverse user groups that cover most ratings for *romance* genre movies. Such a group-centric examination would provide the following 3 user groups: *i. female reviewers from DC* (District of Columbia), *ii. young female reviewers*, and *iii. male teenager reviewers* with average ratings of 4.6, 3.7 and 3.1 (out of 5), respectively. By observing those groups, Anna finds that the hypothesis holds only for a sub-population of female reviewers, *middle-age* or *residents of DC*. Also the results show another group of *romance* genre lovers, *male teenagers*, which contradicts the hypothesis. Anna is confident in her observation (as the results has high coverage) and she can notice different aspects of her hypothesis (as results are diversified).

Beyond coverage and diversity, another interesting dimension of group quality is its *rating distribution*. As it has been argued in previous work [4], groups with *homogeneous* ratings may be more appealing to some applications, while groups with *polarized* ratings are preferred by others. Indeed the rating distribution in a group provides analysts with the ability to tune the quality of found groups according to specific needs. Example 1 is a good case for *homogeneity*. By reporting the average rating of 4.6 for young female reviewers, we know that most individuals in that group have high ratings. The following example shows how tuning the *rating distribution* of discovered groups leads to new discoveries when used alongside coverage and diversity.

Example 2. Following Example 1, Anna then looks at the *variance* of ratings in those groups and finds that *male teenager reviewers* has a higher variance comparing to two other groups. This potentially shows that not all male teenagers like romantic movies. Anna is more interested in a homogenous group, so she can either choose the second or third group or ask the system to find other groups specifically for males or teenagers.

Given an input set of rating records (e.g., Sci-Fi movies from the 90's, David Lynch movies, movies starring Scarlett Johansson), our problem is that of discovering a set of user groups. Even when the number of records is not very high,

---

[1] *We use this example as our running example throughout the paper.*
[2] http://www.imdb.com.

the number of possible groups that could be built may be very large. Indeed, the number of groups is exponential in the number of user attribute values and many groups are very small or empty. Therefore, given the ad-hoc and online nature of group discovery, our challenge is to *quickly* identify high quality user groups. We hence define desiderata that user groups should satisfy (local desiderata) and those that must be satisfied by the set of returned groups (global desiderata).

Local desiderata: *i. (Describability)* Each group should be easily understandable by the analyst. While this is difficult to satisfy through unsupervised clustering of ratings, it is easily enforced in our approach since each group must be formed by rating records of users that share at least one attribute value, which is used to describe that group. *ii. (Size)* Returning groups that contain too few rating records is not meaningful to the analyst. We hence need to impose a minimum size constraint on groups.

Global desiderata: *i. (Coverage)* Together, returned groups should cover most input rating records. While ideally we would like each input record to belong to at least one group, that is not always feasible due to other local and global desiderata associated with the set of returned groups. *ii. (Diversity)* Returned groups need to be different from each other in order to provide complementary information on users. *iii. (Rating Distribution)* Ratings in selected groups should follow a requested distribution (e.g., homogeneity). *iv. (Number of groups)* The number of returned groups should not be too high in order to provide the analyst with an at-a-glance understanding of the data.

A candidate solution is a group-set that verifies all above desiderata. Finding such a group-set is a hard problem because of two reasons. First the pool of candidate group-sets is very large as any possible combination of attribute value pairs can form a group, and any number of groups can form a group-set. By having only 20 attribute value pairs, we end up with $1,048,575$ groups (i.e., $(2^{20}) - 1$) and over 10^{12} group-sets of size 5 (i.e., $1,048,575$ choose 5). The second reason of hardness is that diversity, coverage and rating distribution are conflicting objectives (Sect. 5.1), i.e., optimizing one does not necessarily lead the best values for others. Thus the need for a Multi-Objective optimization approach that will not compromise one objective over another. Such an approach would return *the set of all candidate group-sets* that are not dominated by any other along all objectives.

In this paper, we propose α-MOMRI, an α-approximation algorithm for user group discovery that considers local and global desiderata and guarantees to find group-sets that are α-far from optimal ones. Since α-MOMRI relies on an exhaustive search in the space of all groups, we propose h-MOMRI, a heuristic that exploits the lattice formed by user groups and prunes exploration in order to speed up group-set discovery. Both our algorithms admit a set of rating records of the form $\langle i, u, s \rangle$ and a constrained Multi-Objective optimization formulation [5] and return group-sets that satisfy the formulation and are not dominated by any other group-set. The contributions of this paper are as follows.

1. We formalize specific quality dimensions (coverage, diversity and rating distribution) which we find to be the most natural for discovering user groups on the Social Web. We exploit the semantics of these objectives to go beyond a generic approach.
2. We formalize the problem of discovering user groups as a constrained Multi-Objective optimization problem with quality dimensions as objectives.
3. We develop α-MOMRI, an α-approximation algorithm for user group discovery. Returned group-sets are instances of Pareto plans and are guaranteed to be α-far from optimal ones.
4. We develop h-MOMRI, a heuristic-based algorithm that exploits the lattice formed by user groups to speed up group discovery.
5. In an extensive set of experiments on MOVIELENS and BOOKCROSSING datasets, we analyze different solutions of α-MOMRI and h-MOMRI and show that high quality group-sets are returned by our approximation and very good response time is achieved by our heuristic.

2   Data Model and Preliminaries

We model our database $\mathcal{D}$ as a triple $\langle \mathcal{I}, \mathcal{U}, \mathcal{R} \rangle$, representing the sets of items, reviewers and rating records respectively. Each rating record $r \in \mathcal{R}$ is itself a triple $\langle i, u, s \rangle$, where $i \in \mathcal{I}$, $u \in \mathcal{U}$, and s is the integer rating that reviewer u has assigned to item i. The values of s are application-dependent and do not affect our model.

$\mathcal{I}$ is associated with a set of attributes, denoted as $\mathcal{I}_A = \{ia_1, ia_2, \dots\}$, and each item $i \in \mathcal{I}$ is a tuple with $\mathcal{I}_A$ as its schema. In other words, $i = \langle iv_1, iv_2, \dots \rangle$, where each iv_j is a set of values for attribute ia_j. For example, for the movie *Kazaam* (1996) in MOVIELENS dataset, the set of attribute values are $\langle$Paul M. Glaser, $\{$Comedy, Fantasy$\}\rangle$ for the attribute schema $\langle$director, genre$\rangle$. Note that the attribute genre is multi-valued. We also have the schema $\mathcal{U}_A = \{ua_1, ua_2, \dots\}$ for reviewers, i.e., $u = \langle uv_1, uv_2, \dots \rangle \in \mathcal{U}$, where each uv_j is a value for attribute ua_j. As a result, each rating record, $r = \langle i, u, s \rangle$, is a tuple, $\langle iv_1, iv_2, \dots, uv_1, uv_2, \dots, s \rangle$, that concatenates the tuple for i, the tuple for u, and the numerical rating score s. The set of all attributes is denoted as $A = \{a_1, a_2, \dots\}$. We now define the notion of user group.

Definition 1 (User Group). *A group g is a set of rating records $\langle u, i, s \rangle$ described by a set of attribute value pairs shared among the reviewers and the items of those rating records. The description of a group g is defined as $\{\langle a_1, v_1 \rangle, \langle a_2, v_2 \rangle, \dots\}$ where each $a_i \in A$ (set of all attributes) and each v_i is a set of values for a_i. By $|g|$, we denote the number of rating records contained in g.*

For instance, the first group in Example 1, $g = \{\langle$gender, female$\rangle$, $\langle$location, DC$\rangle$, $\langle$genre, romance$\rangle\}$ contains rating records in MOVIELENS for romance movies whose reviewers are all females in DC. Note that is it au-naturel to combine item attributes (genre) and user attributes (location and gender) together. Figure 1 illustrates an example dataset with 7 rating records. The user

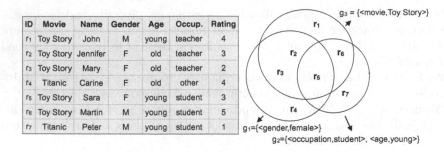

Fig. 1. Example dataset and group-set

group g_1 is for female reviewers with 4 rating records, and g_2 is for young students with 3 rating records. There exists one record in common between two mentioned user groups (r_5). Note that a user group differs from a *where-clause* SQL query, since our objectives and constraints are not expressible as SQL predicates.

Given a rating record $r = \langle v_1, v_2 \ldots, v_k, s \rangle$, where each v_i is a set of values for its corresponding attribute in the schema A, and a group $g = \{\langle a_1, v_1 \rangle, \langle a_2, v_2 \rangle, \ldots, \langle a_n, v_n \rangle\}, n \leq k$, we say that g covers r, denoted as $r \lessdot g$, iff $\forall i \in [1, n], \exists r.v_j$ such that v_j is a set of values for attribute $g.a_i$ and $g.v_j \subseteq r.v_i$. For example, the rating $\langle \texttt{female}, \texttt{DC}, \texttt{student}, 4 \rangle$ is covered by the group $\{\langle \texttt{gender}, \texttt{female} \rangle, \langle \texttt{location}, \texttt{DC} \rangle\}$.

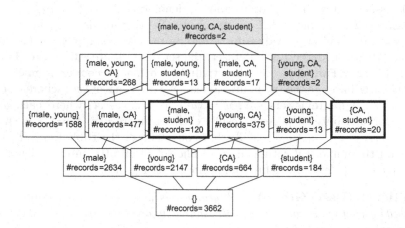

Fig. 2. Partial lattice for the movie *Toy Story*

Similarly to data cubes, the set of all possible groups form a lattice where nodes correspond to groups and edges correspond to parent/child and ancestor/descendant relationships. A partial lattice for rating records of the movie *Toy Story* (1995) is illustrated in Fig. 2 where we have four reviewer attributes to analyze: `gender`, `age`, `location` (CA stands for California) and `occupation`.

For simplicity, exactly one distinct value per attribute is shown in the Figure. The complete lattice contains 15,582 attribute-value combinations.

2.1 Group Quality Dimensions

We now define three quality dimensions for groups, i.e., coverage, diversity and rating distribution. We are given a set of rating records $R \subseteq \mathcal{R}$ and a group-set G.

Coverage is a value between 0 and 1 and measures the percentage of rating records in R contained in groups in G.

$$coverage(G, R) = |\cup_{g \in G} (r \in R, r \lessdot g)|/|R| \tag{1}$$

For instance, in Fig. 1, $coverage(G, R) = 0.8$ where $G = \{g_1, g_2\}$ and R contains rating records for the movie *Toy Story*.

Diversity is a value between 0 and 1 that measures how distinct groups in group-set G are from each other. Diversity penalizes group-sets containing overlapping groups. To prioritize groups with few overlaps, the overlapping penalty is considered as an exponentiation with a negative exponent.

$$diversity(G, R) = 1/(1 + \Sigma_{g_1, g_2 \in G} |r \in R, r \lessdot g_1 \wedge r \lessdot g_2|) \tag{2}$$

For instance, in Fig. 1, $diversity(G, R) = 0.5$.

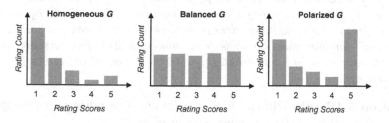

Fig. 3. Different rating distributions for a group-set

Rating Distribution. A group-set G may be characterized by its rating distribution. Figure 3 illustrates some examples of distributions. A rating distribution is a function over the set of ratings in the rating records of groups in G. Equation 3 shows an example of such a function which computes the average *diameter* of ratings. Other aggregation functions could be defined.

$$diameter(G) = avg_{g \in G}(max_{r \in g}(r.s) - min_{r' \in g}(r'.s)) \tag{3}$$

The two most common rating distributions are groups whose members have a consensus (homogeneous distribution, Fig. 3 left), and groups whose members have very different points of view (polarized distribution, Fig. 3 right). A small value of $diameter(G)$ leads a homogeneous group-set G and a high value leads a polarized group-set G. In Fig. 1, $diameter(G) = 3$.

The *diameter* function can capture homogeneity and polarization, but not some other distributions such as "balanced". A detailed discussion on different functions for capturing rating distributions is provided in our technical report [12].

2.2   Multi-Objective Optimization Principles

We propose to use the quality dimensions (coverage, diversity and rating distribution) defined as optimization objectives. When dealing with more than one dimension to optimize, there may be many incomparable group-sets. For instance, for a set of ratings R, we can form two group-sets, G_1 with $coverage(G_1, R) = 0.8$ and $diversity(G_1, R) = 0.4$ and G_2 with $coverage(G_2, R) = 0.5$ and $diversity(G_2, R) = 0.7$. Each group-set has its own advantage: the former has higher coverage and the latter has higher diversity. Another group-set G_3 with $coverage(G_3, R) = 0.5$ and $diversity(G_3, R) = 0.2$ has no advantage compared to G_1, hence it can be ignored. In other words, G_3 is dominated by G_1. In this section, we borrow the terminology of Multi-Objective optimization [5] and define these concepts.

Definition 2 (Plan). *Plan p_i, associated to a group-set G_i for a set of rating records $R \subseteq \mathcal{R}$, is a tuple*
$\langle |G_i|, coverage(G_i, R), diversity(G_i, R), diameter(G_i) \rangle$.

Definition 3 (Sub-plan). *Plan p_i is the sub-plan of another plan p_j if their associated group-sets satisfy $G_i \subseteq G_j$.*

Definition 4 (Dominance). *Plan p_1 dominates p_2 if p_1 has better or equivalent values than p_2 in every objective. The term "better" is equivalent to "greater" for maximization objectives (e.g., diversity, coverage and polarization), and "lower" for minimization ones (e.g., homogeneity). Furthermore, plan p_1 strictly dominates p_2 if p_1 dominates p_2 and the values of objectives for p_1 and p_2 are not equal.*

Definition 5 (Pareto Plan). *Plan p is Pareto if no other plan strictly dominates p. The set of all Pareto plans is denoted as $\mathcal{P}$.*

3   Problem Definition

We define our constrained Multi-Objective optimization problem as follows: for a given set of rating records R and integer constants σ and k (number of groups), the problem is to identify all group-sets, such that each group-set G satisfies:

- $coverage(G, R)$ is maximized;
- $diversity(G, R)$ is maximized;
- $rDistb(G)$ is optimized;
- $|G| \leq k$;
- $\forall g \in G : |g| \geq \sigma$.

Note that our problem focuses on *group-sets* in opposition to *individual groups*, which is a clear distinction from the literature. The last constraint in our problem states that a group g should contain at least σ rating records, an application-defined threshold. For example, if we fix σ to 10 rating records, the groups highlighted in gray in Fig. 2 will not be returned. Note that while we always maximize coverage and diversity, we may either minimize (e.g., in case of homogeneity) or maximize (e.g., in case of polarization) the diameter based on the analyst's needs. We state the complexity of our problem as follows.

Theorem 1. *The decision version of our problem is NP-Complete.*

Proof (sketch). It is shown in [4] that a single-objective optimization problem for user group discovery is NP-Complete by a reduction from the Exact 3-Set Cover problem (EC3). There, homogeneity is maximized and a threshold on coverage is satisfied. In our case, two new conflicting dimensions (diversity and coverage) are added. This means that the problem in [4] is a *special case* of ours, hence our problem is obviously harder.   □

4   Algorithm

The main challenge in designing an algorithm for user group discovery, is the Multi-Objective nature of the problem. A Multi-Objective problem can be easily solved if it is possible to combine all objective dimensions into a single dimension (scalarization), or if optimizing one dimension leads an optimized value for other dimensions.

Both following transformations are infeasible for our problem because our objectives are *conflicting*, i.e., optimizing one does not necessarily lead to an optimized value for others (Sect. 5.1). For instance, a group-set may cover almost all input rating records but contains highly overlapping groups thereby hurting its diversity.

In this paper, we discuss 3 different algorithms for our problem: exhaustive, approximation and heuristic.

4.1   Exhaustive and Approximation Algorithms

The exhaustive algorithm starts by calculating Pareto plans for single groups. Then it iteratively calculates plans for group-sets containing more than one group by combining single groups. At each iteration, dominated plans are discarded. The algorithm combines sub-plans to obtain new plans and exploits the *optimality principle* (POO) for pruning [15]. This approach makes an exhaustive search over all combinations of user groups to find Pareto plans, i.e., both time and space consuming [6].

We propose to improve the complexity of the exhaustive algorithm with our *approximation-based* algorithm which makes less enumerations and guarantees the quality of results. Another way of improvement is *heuristic-based* which will be discussed in Sect. 4.2. For our approximation algorithm, we exploit the near-optimality principle (PONO) [15].

Definition 6 (PONO). *Given a maximization objective f (e.g., diversity, coverage, polarization) and $\alpha \geq 1$, let p_1 be a plan with sub-plans p_{11} and p_{12}. Derive*

Algorithm 1. α-approximation MOMRI (α-MOMRI)

Input: $\sigma, k, \alpha > 1, R$
Output: Pareto result set $\mathcal{P}_\alpha$
1 $\alpha \leftarrow \emptyset$
2 **for** *all user groups g whose size is at least σ* **do**
3 $\quad\mid\quad p_g \leftarrow construct\_plan(g)$
4 $\quad\mid\quad$ **if** p_g *is not α-dominated by any other plan in $\mathcal{P}_\alpha$* **then** $\mathcal{P}_\alpha.add(p_g)$
5 **end**
6 **for** $n \in [2, k]$ **do**
7 $\quad\mid\quad$ **for** *group-sets G of size n* **do**
8 $\quad\mid\quad\quad\mid\quad p_G \leftarrow construct\_plan(g_G)$
9 $\quad\mid\quad\quad\mid\quad$ **if** p_G *is not α-dominated by any other plan in $\mathcal{P}_\alpha$* **then** $\mathcal{P}_\alpha.add(p_G)$
10 $\quad\mid\quad$ **end**
11 **end**
12 **return** $\mathcal{P}_\alpha$

---

p_2 from p_1 by replacing p_{11} by p_{21} and p_{12} by p_{22} where p_{21} and p_{22} are sub-plans of p_2. Then $f(G_{21}) \geq f(G_{11}) \times \alpha$ and $f(G_{22}) \geq f(G_{12}) \times \alpha$ together imply $f(G_2) \geq f(G_1) \times \alpha$. The extension for a minimization objective is straightforward.

We have formally proved that all our objectives satisfy PONO. Proofs are provided in our technical report [12]. Note that among different definitions in the literature for coverage, diversity and rating distribution, we picked the ones that are most intuitive to our problem and that satisfy PONO. For instance, the rating distribution function in [4] does not satisfy PONO.

PONO overrides POO. Thus a new notion of dominance is introduced in Definition 7 to be in line with PONO.

Definition 7 (Approximated Dominance). *Let $\alpha \geq 1$ be the precision value, a plan p_1 α-dominates p_2 if for every maximization objective f (e.g., diversity, coverage, polarization), $f(G_1) \geq f(G_2) \times \alpha$. The extension for a minimization objective is straightforward.*

Definition 8 (Approximated Pareto Plan). *For a precision value α, plan p is an α-approximated Pareto plan if no other plan α-dominates p.*

Generating fewer plans makes a Multi-Objective optimization algorithm run faster [15]. This is because the execution time heavily depends on the number of generated plans. Thus a pruning strategy dictated by PONO is at the core of the α-MOMRI algorithm illustrated in Algorithm 1. In the special case of $\alpha = 1$, the algorithm operates exhaustively. If $\alpha > 1$, the algorithm prunes more and hence is faster. In the latter case, a new plan is only compared with all plans that generate the same result. But a new plan are only inserted into the buffer if no other plan approximately dominates it. This means that α-MOMRI tends to insert fewer plans than the exhaustive algorithm. Note that α-MOMRI

Algorithm 2. Heuristic MOMRI (h-MOMRI)

   Input: σ, k, α, R
   Output: Result set $\mathcal{P}_h$
1 $\mathcal{P}_h \leftarrow \emptyset$
2 $\mathcal{N} \leftarrow$ Set of intervals on *diversity* values
3 **for** n *times* **do**
4      $G_s \leftarrow$ *random_groupset*(k, σ)
5      $G_s^* \leftarrow SHC(G_s)$
6   *interval* $\leftarrow$ *get_interval*(G_s^*)
7      $\mathcal{N}[interval].add(G_s^*)$
8 **end**
9 **for** *interval* $\in \mathcal{N}$ **do**
10   Keep non-dominated plans in *interval* and add them to $\mathcal{P}_h$
11 **end**
12 $\mathcal{P}_h \leftarrow$ *optimize_diameter*$(\mathcal{P}_h)$
13 **return** $\mathcal{P}_h$

---

is objective-independent. In the future, we plan to extend the scope of group discovery to other objectives (as listed in [7]).

4.2 Heuristic Algorithm

A heuristic algorithm has obviously its own advantages and disadvantages. Of course a heuristic algorithm does not provide any approximation guarantee. Eventually, it returns a subset of Pareto set. Nevertheless, the fact that it generates a subset of Pareto makes it faster.

Algorithm 2 illustrates our heuristic algorithm. The algorithm starts by making n different iterations on finding optimal points to avoid local optima (lines 3 to 8). At each iteration, the algorithm begins with a random group-set G_s with k groups whose size is at least σ (line 4). Then a *Shotgun Hill Climbing* [14] local search approach (SHC) is executed (Algorithm 3) to find the group-set with optimal value starting from G_s (line 5). SHC maximizes coverage. Diversity is already divided into intervals $\mathcal{N}$ for each of which a buffer is associated. The resulting group-set of SHC is placed in the buffer whose interval matches the diversity value of the group-set (line 7). Finally, n different solutions are distributed in different interval buffers. The algorithm then iterates over interval buffers to prune dominated plans (lines 9 to 11). Based on Definition 4, a plan is pruned and removed from its buffer if it is dominated by other plans. Finally, for each interval, we report one unique solution that has the maximum/minimum value for diameter based on the requested distribution (line 12).

SHC operates on a generalization/specialization lattice of groups (as in Fig. 2). Navigation of this lattice in a downward fashion satisfies monotonicity property for coverage: given any two groups g_1 and g_2 where g_1 is the parent of g_2, the coverage of g_1 is no smaller than the coverage of g_2. Note that in a bi-objective context, SHC can optimize each one of coverage and diversity.

Algorithm 3. Shotgun Hill Climbing (*SHC*) Algorithm

   Input: Group-set G, R
   Output: Optimized group-set G^*
1 $G^* \leftarrow \emptyset$
2 **while** *true* **do**
3 $\quad$ $C \leftarrow \emptyset$
4 $\quad$ **for** $g \in G$ *and each lattice-based parent* g' *of* g **do**
5 $\quad$ $\quad$ $G' \leftarrow G - \{g\} + \{g\}'$
6 $\quad$ $\quad$ $C.add(G', coverage(G', R))$
7 $\quad$ **end**
8 $\quad$ let $(G'_m, coverage(G'_m, R))$ be the pair with maximum *coverage*
9 $\quad$ **if** $coverage(G'_m, R) \leq coverage(G, R)$ **then**
10 $\quad$ $\quad$ $G^* \leftarrow G$
11 $\quad$ $\quad$ **return** G^*
12 $\quad$ **end**
13 $\quad$ $G \leftarrow G'_m$
14 **end**

However, to benefit from the monotonicity property, we use *SHC* to optimize coverage. *SHC* verifies all local neighbors of a group for an improvement of coverage. If no improvement is achieved, it stops and returns the current group-set. Nevertheless, if we optimize diversity using *SHC*, navigation in the generalization/specialization lattice is nothing but a random walk over the space of groups.

For instance, consider the input group-set $G_s = \{g_1, g_2\}$ where $g_1 = \{\langle\text{gender, male}\rangle, \langle\text{occupation, student}\rangle\}$ and $g_2 = \{\langle\text{location, CA}\rangle, \langle\text{occupation, student}\rangle\}$. These two groups are marked in bold boxes in Fig. 2. We obtain a coverage of 0.79 for G_s. Keeping g_2 fixed, the resulting combinations by swapping g_1 with its parents are either $g_3 = \{\langle\text{gender, male}\rangle\}$ or $g_4 = \{\langle\text{occupation, student}\rangle\}$. For instance, the coverage of $G'_s = \{g_2, g_3\}$ is 0.81. As we observe an improvement, we iterate on this new group-set G'_s to improve coverage.

A detailed discussion on complexity analysis of our proposed algorithms is provided in our technical report [12].

5   Experiments

In this section, we first validate the need for Multi-Objective optimization. Then we compare α-MOMRI and h-MOMRI on the quality of returned groups and the scalability of those algorithms.

We consider two different rating datasets for our study: MOVIELENS and BOOKCROSSING. Due to lack of space, we only show results on MOVIELENS. An exhaustive set of results is presented in our technical report [12]. Both datasets have approximately the same number of ratings. BOOKCROSSING has one order of magnitude more users and items. We consider a 5-star rating system for both datasets.

Table 1. Input sets of rating records

Profile	Movie in MOVIELENS
Highest number of ratings	American Beauty
Lowest number of ratings	Celtic Pride
Highest average rating	Sanjuro
Lowest average rating	Kazaam

MOVIELENS contains four user attributes: gender, age, occupation and zipcode. We convert the numeric age into four categorical attribute values, namely teenager (under 18), young (18 to 35), middle-age (35 to 55) and old (over 55). There are 21 different occupations listed in MOVIELENS e.g., student, artist, doctor, lawyer, etc. We convert zipcodes to states in the USA (or to foreign, if not in USA) by using the USPS zip code lookup.[3] We also enriched MOVIELENS by crawling IMDb[4] using the OMDb API[5] to obtain following item attributes: director, writer and release year and genre.

We implement our prototype system using JDK 1.8.0. All scalability experiments are conducted on an 2.4 GHz Intel Core i5 with 8 GB of memory on OS X 10.9.5 operating system.

For our experiments, we consider four different sets of input rating records described in Table 1. Each item contains at least 50 ratings. We assume that it is straightforward to analyze less than 50 ratings, manually. We also fix $\sigma = 10$ as this value is a border line between frequent ratings and the long tail [12].

5.1 Need for Multi-Objective Optimization

What is the added value of Multi-Objective optimization? We compare first MOMRI with MRI [4], a single-objective approach for group discovery which some authors of this work have already proposed. MRI minimizes *diameter* and considers a lower bound on coverage min_c. Given a set of rating records R for the movie *American Beauty* in MOVIELENS, $k = 3$, $min\_c = 0.7$, one of the returned group-sets by MRI is $G_{MRI} = \{g_1, g_2, g_3\}$ where $g_1 = \{\langle \text{gender, female}\rangle, \langle \text{age, young}\rangle\}$, $g_2 = \{\langle \text{occupation, student}\rangle, \langle \text{age, young}\rangle\}$ and $g_3 = \{\langle \text{gender, male}\rangle, \langle \text{occupation, student}\rangle\}$. The objective values for G_{MRI} are as follows: $coverage(G_{MRI}, R) = 0.81$, $diversity(G_{MRI}, R) = 0.03$ and $diameter(G_{MRI}, R) = 0.13$. However, as diversity is not optimized, there exists huge overlap in groups: many young reviewers are also students.

In the same context, one returned group-set by MOMRI is the one we already discussed in Example 1: $G_{MOMRI} = \{g_4, g_5, g_6\}$ where $g_4 = \{\langle \text{gender, female}\rangle, \langle \text{age, young}\rangle\}$, $g_5 = \{\langle \text{age, young}\rangle, \langle \text{location, DC}\rangle\}$ and $g_6 = \{\langle \text{gender, male}\rangle, \langle \text{age, teen} - \text{ager}\rangle\}$. The objective values for G_{MOMRI}

---

[3] http://zip4.usps.com.
[4] http://www.imdb.com.
[5] http://www.omdbapi.com.

are as follows: $coverage(G_{MOMRI}, R)$=0.79, $diversity(G_{MOMRI}, R)$=0.33 and $diameter(G_{MOMRI}, R) = 0.11$. This group-set has optimized values on all objectives. Specifically, it has a high diversity as only 2 female reviewers for *American Beauty* are both young and residents of DC. It also shows that min_c in MRI is a hard constraint and can easily miss a promising result which has a very high coverage but does not meet the threshold.

We already discussed that consistency of objectives transforms the multi-objective problem into a single-objective one that is trivial to solve (Sect. 4). In this experiment, we verify if our objectives (defined in Sect. 2.1) are consistent. We maximize coverage and observe how values of diversity and diameter evolve. To maximize coverage, we use Algorithm 3. Figure 4 illustrates the results for different sets of input rating records in Table 1. Each point illustrates the objective values for each of 20 runs. Note that this experiment is independent of the heuristic and the approximation algorithms.

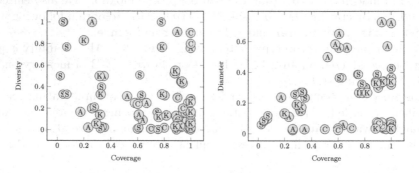

Fig. 4. Conflicting objectives on MovieLens. Movie title initials are illustrated on points.

We observe that in general, no correlation exists between the optimized value of coverage and other objectives. Thus each objective should be optimized independently. The same result was obtained for BookCrossing [12].

5.2 Comparison of Algorithms

In this section, we compare h-MOMRI and α-MOMRI. Our hypothesis is that h-MOMRI has a manageable solution space size compared to α-MOMRI which leads to a reduced execution time.

First we compare the quality of algorithms regarding the dominance of solutions. In Multi-Objective optimization, if for two algorithms X and Y, the majority of X's solutions dominate Y's, it means that X is able to produce solutions with higher quality than Y. In this experiment, we make the same comparison between α-MOMRI and h-MOMRI. For this experiment, we need to compare each pair of α-MOMRI and h-MOMRI solutions. We count the number of times each algorithm dominates the other in pairwise comparison of their results. We consider $\alpha = 1.15$ for α-MOMRI and *nbintervals* = 40 for h-MOMRI. We denote the set of α-MOMRI solutions as $\mathcal{P}_\alpha$ and the set of h-MOMRI solutions as $\mathcal{P}_h$. We observe that for all sets of input rating records in Table 1, at least 62 % of solutions in $\mathcal{P}_h$ are dominated by solutions in $\mathcal{P}_\alpha$. This is because α-MOMRI generates the complete set of α-approximated Pareto plans, while h-MOMRI produces a subset. For instance, for the movie *American Beauty*, α-MOMRI produces 16 times more solutions than the heuristic algorithm. Evidently the solutions in $\mathcal{P}_h$ are either as good as $\mathcal{P}_\alpha$'s or worse. Our results show that although α-MOMRI presents a huge set of all Pareto plans, h-MOMRI can return an acceptable representative subset where almost half of solutions are as good as the set $\mathcal{P}_\alpha$.

Now we compare α-MOMRI and h-MOMRI concerning their performance and the number of solutions they produce. We consider 3 different instances for each algorithm: for α-MOMRI, we consider instances with $\alpha = 2$ (A), $\alpha = 1.5$ (B) and $\alpha = 1.15$ (C), and for h-MOMRI, we consider instances with 5 (D), 10 (E) and 40 (F) intervals. We run this experiment with 4 items having the highest amount of rating records as items with fewer records exhibit similar behavior.

Figure 5 illustrates the results. As expected, in general the number of solutions produced by h-MOMRI is one order of magnitude less than α-MOMRI in both datasets. In both algorithms, the number of ratings records play an important role and increases the number of solutions. In [12], it is shown that the time performance of both algorithms is a function of the group space size. A data-centric observation in Fig. 5 reveals that more rating records lead more groups, hence worse performance (which is the case for *American Beauty*).

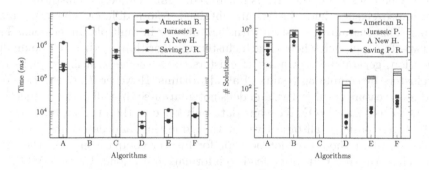

Fig. 5. Comparison of α-MOMRI and h-MOMRI algorithms in execution time (left) and # solutions (right) on MOVIELENS

Choosing between α-MOMRI and h-MOMRI. Both α-MOMRI and h-MOMRI are useful for analysts in different scenarios. α-MOMRI can be used in an *offline* context to produce an exhaustive set of user groups with a precision defined by α for further analysis. For instance, a movie rating website (like IMDb) can index user groups generated offline and execute various user queries like '*what are interesting groups of female teenagers who have rated romantic* *movies'*. On the other hand, in an *online* or *streaming* context, *h*-MOMRI is beneficial because it can immediately produce a representative subset of results. For instance, in a movie rating website an analyst can quickly observe interesting user groups of comedy and romantic movies.

6   Related Work

To the best of our knowledge, no approach has proposed and formalized the problem of discovering user groups for collaborative rating datasets by considering multiple *independent* and *conflicting* quality dimensions. Recent studies[6] have shown an interest in reporting statistics about pre-defined groups, as opposed to our work where we look to discover high-quality user groups on the fly. However our work does relate to a number of others in its aim and optimization mechanism.

Multi-Objective Optimization. There exist different approaches to solve a multi-objective problem [15,16]. We already discussed that Scalarization does not work in our case (Sect. 5.1). Another popular method is ϵ-constraints [13] where one objective is optimized and others are considered as constraints. The approach in [4] can be seen as a relaxed ϵ-constraints version of our problem. Another approach is Multi-Level Optimization [11] which needs a meaningful hierarchy between objectives. In our case, all objectives are independent and conflicting, hence using this mechanism is not feasible.

User Group Discovery. User groups can be discovered by clustering methods [1–3,9] where a single objective is optimized. Multi-Objective clustering [8,10] is an improvement where clusters are obtained from n different clustering algorithms. This guarantees clusters with high quality in multiple dimensions. This is a two-step approach where *i.* each clustering algorithm, applied to one quality dimension, generates its own set of clusters, *ii.* a *goodness* measure picks target clusters by combining results of all algorithms. However, the definition of a goodness measure is subjective and does not guarantee that all desired objectives are optimized. Also MOMRI scans data only once as the pruning technique in α-MOMRI considers all objectives at the same time and determines if a candidate group-set should or not be kept for further comparisons. On the other hand, clustering methods often lead to information overload. Using *h*-MOMRI, the analyst receives a manageable subset of high quality results in a reasonable time. More (precise) results are returned by reducing α for α-MOMRI or increasing *nbintervals* for *h*-MOMRI.

7   Conclusion and Future Work

In this paper, we investigated the question of finding the best group-sets that characterize a database of rating records of the form $\langle i, u, s \rangle$, where $i \in \mathcal{I}$, $u \in \mathcal{U}$,

---

[6] http://blog.testmunk.com/how-teens-really-use-apps/.

and s is the integer rating that user u has assigned to item i. We showed that the problem of finding high-quality group-sets is NP-Complete and proposed a constrained Multi-Objective formulation. Our formulation incorporates local and global group desiderata. We proposed two algorithms that find group-sets as instances of Pareto plans. The first one α-MOMRI, is an α-approximation algorithm and the second, h-MOMRI, is a heuristic-based algorithm. Our extensive experiments on MOVIELENS and BOOKCROSSING datasets show that our approximation finds high quality groups and that our heuristic is very fast without compromising quality.

Our work can be improved in many ways. In particular, we plan to perform an extensive user study to be able to evaluate the quality of returned group-sets. An online poll (about movies or books) could be used to build a ground-truth and will be used to evaluate the usefulness of our group-sets. Also, we plan to investigate an extensive analysis of rating distributions for our algorithms using some dispersion measures.

References

1. Agrawal, R., Gehrke, J., Gunopulos, D., Raghavan, P.: Automatic subspace clustering of high dimensional data for data mining applications. ACM (1998)
2. Agrawal, R., Imielinski, T., Swami, A.N.: Mining association rules between sets of items in large databases. In: SIGMOD (1993)
3. Amiri, B., Hossain, L., Crowford, J.: A multiobjective hybrid evolutionary algorithm for clustering in social networks. In: Proceedings of the 14th Annual Conference Companion on Genetic and Evolutionary Computation. ACM (2012)
4. Das, M., Amer-Yahia, S., Das, G., Yu, C.: Mri: meaningful interpretations of collaborative ratings. In: VLDB (2011)
5. Dutot, P.F., Rzadca, K., Saule, E., Trystram, D.: Multi-objective Scheduling, chap. 9. Chapman and Hall/CRC Press (2009)
6. Ganguly, S., Hasan, W., Krishnamurthy, R.: Query optimization for parallel execution, vol. 21. ACM (1992)
7. Geng, L., Hamilton, H.J.: Interestingness measures for data mining: a survey. ACM Comput. Surv. (CSUR) **38**(3), 9 (2006)
8. Jiamthapthaksin, R., Eick, C.F., Vilalta, R.: A framework for multi-objective clustering and its application to co-location mining. In: Huang, R., Yang, Q., Pei, J., Gama, J., Meng, X., Li, X. (eds.) ADMA 2009. LNCS (LNAI), pp. 188–199. Springer, Heidelberg (2009). doi:10.1007/978-3-642-03348-3_20
9. Kargar, M., An, A., Zihayat, M.: Efficient bi-objective team formation in social networks. In: Flach, P.A., Bie, T., Cristianini, N. (eds.) ECML PKDD 2012. LNCS (LNAI), pp. 483–498. Springer, Heidelberg (2012). doi:10.1007/978-3-642-33486-3_31
10. Law, M.H., Topchy, A.P., Jain, A.K.: Multiobjective data clustering. In: Proceedings of the 2004 IEEE Computer Society Conference on Computer Vision and Pattern Recognition, CVPR 2004, vol. 2, pp. II–424. IEEE (2004)
11. Migdalas, A., Pardalos, P.M., Värbrand, P.: Multilevel optimization: algorithms and applications, vol. 20. Springer Science & Business Media (1997)
12. Omidvar-Tehrani, B., Amer-Yahia, S., Dutot, P.F., Trystram, D.: Multi-objective group discovery on the social web. Research Report RR-LIG-052, LIG, Grenoble, France (2016)

13. Papadimitriou, C.H., Yannakakis, M.: On the approximability of trade-offs and optimal access of web sources. In: FOCS (2000)
14. Russell, S.J., Norvig, P.: Probabilistic reasoning. Artificial intelligence: a modern approach (2003)
15. Trummer, I., Koch, C.: Approximation schemes for many-objectivequery optimization. In: Proceedings of the 2014 ACM SIGMOD International Conference on Management of Data. ACM (2014)
16. Tsaggouris, G., Zaroliagis, C.: Multiobjective optimization: Improved fptas for shortest paths and non-linear objectives with applications. Theory Comput. Syst. **45**(1), 162–186 (2009)

Sequential Data Classification in the Space of Liquid State Machines

Yang Li, Junyuan Hong, and Huanhuan Chen[(✉)]

UBRI, School of Computer Science and Technology,
University of Science and Technology of China, Hefei 230027, Anhui, China
{csly,jyhong}@mail.ustc.edu.cn, hchen@ustc.edu.cn

Abstract. This paper proposes a novel classification approach to carrying out sequential data classification. In this approach, each sequence in a data stream is approximated and represented by one state space model – liquid state machine. Each sequence is mapped into the state space of the approximating model. Instead of carrying out classification on the sequences directly, we discuss measuring the dissimilarity between models under different hypotheses. The classification experiment on binary synthetic data demonstrates robustness using appropriate measurement. The classifications on benchmark univariate and multivariate data confirm the advantages of the proposed approach compared with several common algorithms. The software related to this paper is available at https://github.com/jyhong836/LSMModelSpace.

Keywords: Sequential learning · Classification · Learning in the model space

1 Introduction

Sequential data classification is a fundamental problem in the machine learning community. In the classification, the degree of dissimilarity between sequences needs to be quantified. If the sequential data are of equal length, it is sufficient to use conventional machine learning methods by treating sequences as numerical vectors. Kernel methods could be efficient and might achieve satisfying performances [18], provided that the length of sequence is not long. However, in reality, large amount of sequential data are variable-length.

To deal with sequential data that are variable-length and possibly long, plenty of algorithms, e.g. dynamic time warping [1], autoregressive kernel [8], spectral analysis [11], are proposed.

Searching for a global alignment between variable-length sequences is a way to handle variable-length data. This methodology of non-linear warping and matching segments of two sequences is exemplified by *dynamic time warping* (DTW) [21]. However, due to non-linear warping, the triangular inequality, one of the requisites for the validity of a metric, is not satisfied. The measurement in DTW is not a metric actually, lacking geometric interpretation to the experimental result [9].

© Springer International Publishing AG 2016
P. Frasconi et al. (Eds.): ECML PKDD 2016, Part I, LNAI 9851, pp. 313–328, 2016.
DOI: 10.1007/978-3-319-46128-1_20

Fisher Kernel [12] fits one single generative model (Hidden Markov Model) to sequences and compares how much new incoming sequence "stretches" the average model trained with past sequences. *Fisher Kernel* defines *Fisher Score* as gradients of log-likelihood, $\log p(\boldsymbol{x}|\theta)$, with regard to hidden parameters. As *Fisher Kernel* train the generative model under maximum likelihood principle, it may lead to sub-optimal results. Since a generative model that fits data well may easily get stuck in the local minimum of its log-likelihood, where the gradient representation of data is (nearly) zero [17].

The computation of *Fisher Kernel* of sequences $\boldsymbol{s}_i$ and $\boldsymbol{s}_j$ is defined as:

$$\nabla_{\boldsymbol{\theta}}^T p(\boldsymbol{x}|\theta) \mathcal{I}^{-1} \nabla_{\boldsymbol{\theta}} p(\boldsymbol{x}|\theta) \tag{1}$$

where $\mathcal{I}$ is the *Fisher information matrix*. Computation of *Fisher Kernel* involves the inverse of *Fisher information matrix*. This procedure could be time-consuming. A routinely adopted way to bypass this difficulty is to replace the *Fisher information matrix* with identity matrix, at the cost of losing some precision in the approximation [22].

Fisher Kernel learning [17] leverages the label information so that the objective functions in the same class have similar gradients. It applies idea from metric learning to improve its performance. Both methods show effectiveness but low efficiency in obtaining the representations to data, as more computation is involved in computing gradients, even when the *Fisher information matrix* is assumed to be identity matrix.

Autoregressive Kernel (AR) [8] employs a likelihood profile as features for sequences. The likelihood profile is generated by a *Vector Autoregressive Model* under different parametric settings. The dissimilarity between sequences is computed with Bayesian method. It can be verified that this measurement is a valid Hilbertian metric [8]. AR relaxes the constraint of using a single generative model to explain the whole data as did in *Fisher Kernel* and *Fisher Kernel learning*. However, AR does not use the timestamps in a sequence to improve the prediction [20].

Chen *et al.* approximated time series via echo state networks (ESN) [4,5], and demonstrated that readout weights in ESNs could offer discriminant features for sequences. Under the representation provided by topologically fixed reservoir for the whole data, the readout weights, the only trained part, covers the uniqueness of a specific sequence, bringing in more versatility and flexibility. It was demonstrated that ESN is able to handle continuous sequences in complicated scenario [6]. In addition, a co-learning strategy was devised to strengthen its representation capability on continuous sequences [3]. In this paper, we further extend this methodology to process binary data, and demonstrate the improvement on performance by using liquid state machine (LSM). In LSM, individual node (neuron) has its own "state memory", and responds from its own history and current input signal, while nodes in ESN give responses based on merely their current state. The replacement brings enhancement "memory" to the reservoir, and demonstrates to be beneficial by experiments.

In this paper, we propose a novel approach to representing sequences, which might be of different lengths and of different characteristics, in a higher dimensional space. In this approach, each sequence is represented by a LSM, which gives approximation to the conditional probability of likelihood of the sequence. After obtaining models, the classification is conducted on the models, rather than on the sequences directly. In this paper, we discuss measurements under different assumptions on the "model distributions". The model set, along with the defined measurements, offers a novel space for classification and other possible learning tasks. This space is referred as a model space for a certain data set in this paper.

2   Discriminant Learning in the Model Space

LSM incorporates time into the model of neural network to enhance the level of realism in the simulation, emerging as a new computational model [16]. A LSM consists of two parts (apart from input layer) in its framework. A large collection of nodes that are randomly connected to each other make up the reservoir part. Each node receives inputs from input layer as well as from other nodes. The spatio-temporal pattern of the activations in nodes is read out by the final layer as linear combinations in performing certain tasks. The final layer is the only part that needs training.

We illustrate the scheme diagram of model space and LSM in Fig. 1. In the figure, LSMs are used to give approximations to sequences and in turn the set of LSMs is considered in the learning algorithms.

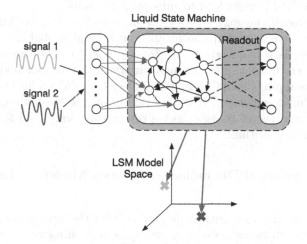

Fig. 1. The schematic diagram for LSM and model space. LSMs provide representations for two sequences. The model space is seen as a high dimensional space, in which the readout weights of LSMs are assembled.

The form of the LSM [14] is generalized as follows:

$$\begin{cases} \boldsymbol{x}(t) = Q(R\boldsymbol{x}(t-1) + V\boldsymbol{s}(t)) \\ \boldsymbol{y}(t) = f(\boldsymbol{x}(t)) = W\boldsymbol{x} \end{cases} \tag{2}$$

where $\boldsymbol{x}(t) \in \Re^n$ is the state vector defined in the real domain. Subscript n is the number of reservoir nodes. Input $\boldsymbol{s}(t) \in \Re^{d+1}$ is input which has been augmented by adding bias as one of its components. R and V are the appropriately defined coefficient matrices. $\boldsymbol{y}(t) \in \Re^{n'}$ and W denote output and readout weights respectively. Superscript n' is the dimensionality of output. $Q(\cdot)$ is the response function defined on the internal nodes.

A LSM is trained by making use of past values and predicting the present value. Readout weights $W \in \Re^{n' \times n}$ are trained by adjusting W in order for $W\boldsymbol{x}(t) = \boldsymbol{s}(t+1)$. The dimensionality n' satisfies $n' = d$ in this scenario.

We consider an arbitrary sequence $\boldsymbol{s} = \{s_0, s_1, \cdots, s_n\} \in \Re^d$, where d is the dimensionality of the sequence. We also use $\boldsymbol{s}(t)$ to denote a sequence which is indexed by t. We assume that the index starts from 0 unless otherwise stated.

The likelihood of a sequence $\boldsymbol{s}$ is expressed as:

$$\ell(\boldsymbol{s}) = \ell(\{s_0, s_1, \cdots, s_n\})$$

which can be further factorized into

$$\ell(\boldsymbol{s}) = \mathcal{P}_0(s_0)\mathcal{P}_1(s_1|s_0)\mathcal{P}_2(s_2|s_1, s_0)\cdots\cdots\mathcal{P}_n(s_n|s_{n-1}, \cdots, s_0)$$

where $\mathcal{P}_i(s_i|s_{i-1}\cdots s_0), i = 0 \cdots n$ is the conditional probability.

In most cases, the assumption is too strong that the conditional probability $\mathcal{P}_i(\cdot|\cdot)$ of a sequence can be generalized and formulated explicitly. Assumptions on the form of $\mathcal{P}_i(\cdot|\cdot)$ might lead to sub-optimal results.

In our approach, we make use of the universal approximating ability [16] of LSM under a weak assumption on the conditional probability distribution, assuming $\mathcal{P}_i(\cdot|\cdot)$ is time-invariant, i.e. $\mathcal{P}_i(\cdot|\cdot) = \mathcal{P}(\cdot|\cdot)$. The universal approximating ability states that, given enough variety in the interior nodes, nonlinear input-output mappings could be approximated by LSM under training of sufficiently long input sequences. Our approach bases the approximation to $\mathcal{P}(\cdot|\cdot)$ on this ability and therefore uses models rather than simplified formulations in the classification algorithm.

2.1   Measurement of Dissimilarity Between Models in the Model Space

The dissimilarity of two sequences is judged from the divergence between two fitting LSMs. Given two sequences $\boldsymbol{s}_i$ and $\boldsymbol{s}_j$, a general measurement of dissimilarity is formulated as follows:

$$\mathcal{D}(\boldsymbol{s}_i, \boldsymbol{s}_j) = (\int_{\boldsymbol{x} \in \mathcal{I}} ||f_i - f_j||^2 d\mu(\boldsymbol{x}))^{1/2} \tag{3}$$

$$= \left(\int_{\boldsymbol{x} \in \mathcal{I}} (W_i\boldsymbol{x} - W_j\boldsymbol{x})^T (W_i\boldsymbol{x} - W_j\boldsymbol{x})d\mu(\boldsymbol{x})\right)^{1/2}$$

$||\cdot||$ is the norm which calculates the disagreement between two model outputs. $\mathcal{I}$ is the change interval for model vector $\boldsymbol{x}$. $\mu(\boldsymbol{x})$ is the probability distribution for $\boldsymbol{x}$.

Uniform distribution over $\boldsymbol{x}$ considers the simplest case, in which the probability distribution $\mu(\boldsymbol{x})$ is assumed to be only dependent on the interval $\mathcal{I}$. Later, this assumption will be relaxed and more general cases will be discussed.

Under the assumption of the uniform distribution, the dissimilarity between sequences $\boldsymbol{s}_i$ and $\boldsymbol{s}_j$ is simplified into

$$\mathcal{D}(\boldsymbol{s}_i, \boldsymbol{s}_j) = \left(\int (W_i\boldsymbol{x} - W_j\boldsymbol{x})^T (W_i\boldsymbol{x} - W_j\boldsymbol{x}) d\mu(\boldsymbol{x}) \right)^{1/2} \tag{4}$$
$$= \mathcal{C}||W_i\boldsymbol{x} - W_j\boldsymbol{x}||.$$

where the irrelevant terms in last formula of Eq. (4) are generalized into constant $\mathcal{C}$.

In more general cases where $\boldsymbol{x}$ is not evenly distributed, but not changes dramatically, we use Gaussian mixture model to approximate the probability distribution $\mu(\boldsymbol{x})$. It fits the probability distribution $\mu(\boldsymbol{x})$ with a mixture of finite Gaussian distributions.

$$\mu(\boldsymbol{x}) = \sum \alpha_i N(\theta_i, \Sigma_i)$$

where α_i are the mixture coefficients for i-th Gaussian distribution. All α_i sum up to 1, $\sum \alpha_i = 1$. Parameters θ_i and Σ_i are mean and variance in i-th Gaussian distribution.

Substitute $\mu(\boldsymbol{x})$ with Gaussian mixture model, the dissimilarity between two sequences is formulated as:

$$\mathcal{D}(\boldsymbol{s}_i, \boldsymbol{s}_j) = \sum_k \alpha_k trace(W_i^T W_j \Sigma_k) + \theta_k^T W_i^T W_j \theta_k \tag{5}$$

Sampling, as a natural alternative to the above approximation method, makes no assumptions on the form of $\mu(\cdot)$. An asymptotic optimal estimation for a probability distribution $\mu(\cdot)$ is guaranteed from the law of large numbers. This estimation may lead to more robust result, if no prior information on $\mu(\cdot)$ exists. Applying sampling to Eq. (3) is straightforward.

$$\mathcal{D}(\boldsymbol{s}_i, \boldsymbol{s}_j) \approx \frac{1}{m} \sum_k ||W_i\boldsymbol{x}_k - W_j\boldsymbol{x}_k|| \tag{6}$$

where m denotes the amount of sampling points.

Assume the deviation $\varepsilon(t)$ between the output of a LSM $y(t) = W\boldsymbol{x}$ and the desired output $\boldsymbol{s}(t+1)$ follows a zero-mean Gaussian distribution $\varepsilon(t) = \mathcal{N}(0, \delta^2 I)$. When the methodology of *Fisher Kernel* is applied, the conditional probability of observing $\boldsymbol{s}(t+1)$ given past values is formulated as:

$$P((\boldsymbol{s}(t+1)|\boldsymbol{s}(1\cdots t)) = (2\pi\delta^2)^{-d/2} exp\left(- \frac{||\boldsymbol{s}(t+1) - W\boldsymbol{x}(t)||}{2\delta^2} \right)$$

The *Fisher score* U between $\boldsymbol{s}_i$ and $\boldsymbol{s}_j$ takes the form of inner product of two derivatives with regard to the hidden parameters. The derivative quantifies how the model adjusts its current parametric setting in order to fit a new sequence. The derivative of probability $\mathcal{P}(\cdot|\cdot)$ in terms of W gives rise to:

$$
\begin{aligned}
U &= \frac{\partial \log \mathcal{P}(\boldsymbol{s}(1 \cdots l))}{\partial W} \\
&= \sum_{t=1}^{l} \frac{\boldsymbol{s}(t)\boldsymbol{x}(t-1)^T - W\boldsymbol{x}(t-1)\boldsymbol{x}(t-1)}{\delta^2}
\end{aligned}
$$

The dissimilarity between $\boldsymbol{s}_i$ and $\boldsymbol{s}_j$ is expressed as:

$$
\mathcal{D}(\boldsymbol{s}_i, \boldsymbol{s}_j) = \mathbf{1}U_i. * U_j\mathbf{1}^T \tag{7}
$$

where .$*$ denotes element-wise multiplication and $\mathbf{1}$ is the all-one vector.

Extending to Binary Data. The sequential data recorded in binary digits $\{0, 1\}$ are more encountered in clinical research, e.g. heart beating signal, signals from neurons. In terms of binary or discrete data, the traditional ways that minimize mean square error (MSE) as did on numerical sequences are infeasible. The traditional ways rely on the gradient of objective function for inference of parameters, while MSE from binary data is non-smooth and thus no gradients exist. LSM is extended to process binary data by replacing MSE with exponential van Rossum metric [23].

A general exponential van Rossum metric $\psi(t, t_0)$ can be formulated as:

$$
\psi(t, t_0) = \begin{cases} -(t - t_0)\dfrac{e^{-(t-t_0)/\tau}}{\tau} & 0 \leq t < \Delta t + t_0 \\ +\infty & \text{otherwise} \end{cases} \tag{8}
$$

where index t_0 is the expected index. Δt is a threshold, restricting the comparison to the affinity of t_0. Argument τ is a penalty on the deviation.

3   Experimental Study

This section presents experiments conducted on synthetic binary data and classifications on benchmark univariate and multivariate data. For a given task, the topology (200 interior nodes) and interior weights between nodes were initialized and kept fixed. In this way, the randomness in LSM was controlled as an invariant factor for comparison purpose. The strategy of restart was adopted in experiments[1].

The implementation of LSM made use of a software simulating the microcircuits of neural network–CSIM [19]. The parameters were set referring the attached examples.

---

[1] The source code is available from https://github.com/jyhong836/LSMModelSpace.

Table 1. The parameters and search ranges

Name	Parameters	Search range
DTW	γ	$\gamma \in \{10^{-6}, 10^{-5}, \cdots, 10^1\}$
AR	γ, ξ, p	$\gamma \in \{10^{-6}, 10^{-5}, \cdots, 10^1\}, \xi \in \{0.1, 0.2, \cdots, 0.9\}, p \in \{1, 2, \cdots, 10\}$
FK	$state$	$state \in \{1, 2, 3, 10\}$
	λ, γ	$\lambda \in \{10^{-6}, 10^{-5}, \cdots, 10^1\}, \lambda \in \{10^{-5}, 10^4, \cdots, 10^1\}$

γ is the parameter in the kernel function.
ξ and p are the weight and the order of the negative definite kernel.
$state$ is the number of states of hidden Markov model defined in Fisher kernel.
λ is the parameter used in ridge regression.

In the implementation of the Gaussian mixture model, the number of Gaussian distribution was auto-determined by the method proposed in [10]. In the sampling, since there existed training sequences that were not sufficiently long, circular block bootstrap was applied. The block length was auto-determined by the method proposed in [15].

LIBSVM [2] was adopted in the classification algorithm. Multi-class data were classified via its default strategy, one-against-one.

The proposed methods were compared with common methods, including *Dynamic Time Warping* (DTW), *Autoregressive Kernel* (AR), *Fast Fisher Kernel* (Fisher), and Reservoir model (RV) proposed in [4,5].

The parameters in the proposed algorithms (regression parameter λ), support vector machine (bandwidth θ and cost $\mathcal{C}$), and the comparison algorithms were tuned with 5-fold-cross-validation[2]. The search ranges for the parameters are detailed in Table 1.

Three classification methods defined with Eqs. (4)–(7) are named as LSM with L^2 norm (L^2-LSM), LSM with Gaussian mixture model (Gaussian-LSM), LSM with sampling method (Sampling-LSM), and LSM with Fisher methodology (Fisher-LSM).

3.1 Synthetic Data

Synthetic binary data were generated following Poisson distribution $p(t) = \frac{\lambda^t e^{-\lambda}}{t!}$. The merit of using Poisson distribution is that it makes the events (bars in Fig. 2) evenly distributed and ensures that no events happen at the same time. The synthetic data were labeled into three classes. Different classes were generated under a slightly changed parameter setting.

For each parametric setting, the simulation lasted 2 s with time unit 10^{-3} s, generating a 2000-length sequence. We generated 55 sequences for each class. In addition, all the sequences were corrupted with Gaussian white noise ($mean = 0, \Sigma = 0.02I$). The Eq. (8) was adopted as cost function in the training algorithm. Figure 2 demonstrates parts of the binary sequences of three classes. From this figure, it is not easy to distinguish class labels.

---

[2] The procedure of cross-validation keeps identical to [4] for comparison.

Fig. 2. The parts of synthetic binary sequences. The data were generated following Poisson distribution and were corrupted with additive Gaussian white noise. Horizontal axis denotes the index. Different classes are drew in different colors, and are separated by a dash line.

The model space in this experiment, which is populated by readout weights of fitting models, is depicted in Fig. 3. In order to visualize the model space, *multidimensional scaling* (MDS) was used to reduce its dimensionality. MDS keeps the original between-objective distance faithfully in a lower dimensional space. Although it was hard to distinguish class labels in the binary data as depicted in Fig. 2, after representing the sequences in the model space, they became separable in Fig. 3.

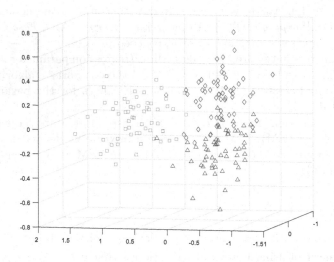

Fig. 3. The model space of synthetic binary data in a 3-dimensional coordinate. Parts of the data are depicted in Fig. 2. The model space was constructed by fitting LSMs to the binary data and extracting data-specific features, i.e. the readout weights, from LSMs. Each point offers representation to an individual binary sequence. Different classes are denoted with different markers.

The sensitivity of proposed method to the additive Gaussian noise was also investigated, in comparison with AR and *Fast Fisher Kernel* (Fisher)[3]. The

---

[3] The methodology of searching a global alignment is unsuitable for binary data, so the experiment of DTW was not reported.

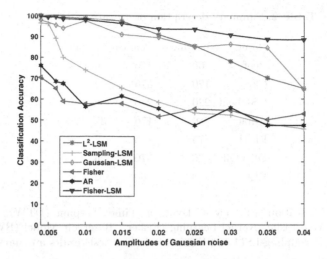

Fig. 4. The classification accuracy versus different amplitudes of Gaussian noise. The horizontal axis denotes the amplitude of noise, and the vertical axis is the classification accuracy on the synthetic data. Overall, the proposed methods demonstrate clear advantages in handling binary sequences in this experimental setting.

classifications were conducted on data with various amplitudes of Gaussian noise. The experimental results are depicted in Fig. 4.

An overall advantage can be observed from Fig. 4. Not surprisingly, Fisher-LSM has the best performance in terms of classification accuracy and robustness to the noise among all the methods. Fisher-LSM assumes that the deviation between observation and true value follows zero-mean Gaussian distribution, which coincides with the noise used in this experiment. Sampling-LSM shows to be less robust to the added noise. Its classification accuracy drops after corrupting data with noise. But as the amplitude of noise grows, its influence on the performance of Sampling-LSM decreases.

3.2 Benchmark Data

The benchmark data sets were obtained from UCR time series classification archive [7] and UCI machine learning repository[4]. Table 2 gives a summary of all data sets. In order to eliminate the influence of different units, all data sets were rescaled into interval $[-1, 1]$.

The experimental results of 5 runs are listed in Table 3. From this table, A general advantage of classifications carried out in the model space of LSM over comparison algorithms can be observed. Among all the proposed methods, learning based on sampling achieved the best performance. The better performance

---

[4] EEG data was obtained from UCI machine learning repository, https://archive.ics. uci.edu/ml/datasets/EEG+Database. And it was preprocessed via *Principle Component Analysis* to reduce its dimensionality.

Table 2. Summary description of univariate data sets.

Name	Instances	Length	Classes
Beef	60	470	6
Car	120	576	4
OSULeaf	220	240	6
Adiac	780	176	37
FISH	175	175	7
OliveOil	60	570	4
EEG	60	60	2

Table 3. Classification accuracy of Dynamic Time Warping (DTW), Autoregressive Kernel (AR), Fisher Kernel Learning (Fisher), Reservoir Model (RV), L^2-LSM, Gaussian-LSM, sampling-LSM and Fisher-LSM. The best results are marked in bold.

Name	DTW	AR	Fisher	RV [4]	L^2-LSM	Gaussian-LSM	Fisher-LSM	Sampling-LSM
Beef	66.67	56.67	58.00	**86.67**	60.00	46.67	53.3	76.67
Car	73.3	60.0	65.00	86.67	78.33	61.67	70.00	**90**
OSULeaf	62.15	73.33	54.96	64.59	72.31	69.00	69.01	**75.20**
Adiac	65.47	64.54	68.03	76.73	76.63	**78.10**	57.54	76.98
FISH	69.86	51.43	57.14	85.71	**89.71**	85.71	68.57	87.43
OliveOil	83.33	42.15	56.67	**90.00**	76.67	80.00	73.33	86.67
EEG	38.0	50.0	50.0	-	48.33	51.67	61.67	**63.33**

of sampling-LSM is largely contributed from the weak hypothesis it imposed on the probability distribution $\mu(\cdot)$. However, Fisher-LSM underperformed on all sequences. A possible reason for its deficiency lies in the pre-assumption over the deviation. When strong autocorrelation exists, the assumption of zero-mean Gaussian noise is unlikely to be true. Compared with good performance achieved on binary sequences, it is more encouraged to be used on binary or discrete sequences.

As the number of Gaussian distributions was auto-determined in Gaussian-LSM and bootstrap was adopted in Sampling-LSM, the computational complexities of proposed approaches are difficult to analyze. We adopted experiments to illustrate the actual time consumption on benchmark data. Experiments were conducted on a sequential data set PEMS[5]. By truncating the sequences and recording the time consumed in obtaining dissimilarities between pairwise sequences, we obtained tuples of time consumption[6] versus length of sequence. And the results are plotted in Fig. 7.

---

[5] PEMS was obtained from UCI machine learning repository. The sequences in PEMS were vectorized to be sufficiently long.

[6] The computational environment is Windows 7 with Intel Core i5 Duo 3.2 GHz CPU and 8 G RAM.

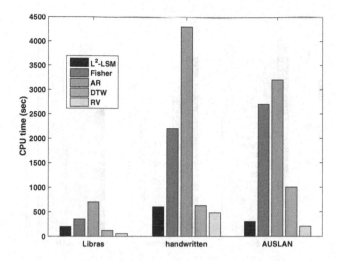

Fig. 5. The time consumptions on three multivariate data sets. The vertical axis denotes the time consumption. It is measured in the unit of CPU time (sec).

In the Fig. 7, the time consumptions of all proposed approaches grow slowly after the sequential length becomes large (beyond 1800). The lines of Gaussian-LSM and L^2-LSM grow in a similar pattern. However, Sampling-LSM maintains a (roughly) consistent time usage, even when the training sequences are short. The reason is, in order to compensate the approximation loss when the training sequences were not sufficiently long, more sampling had to been done. The computation of Fisher-LSM involves matrix multiplication, which makes it grow (roughly) linearly with the sequential length in our experiments (not shown).

In contrast, the time complexity of DTW is $O(m_i m_j)$, where m_i is the length of i-th sequence. An improved variation [13] speeds up DTW by using piece-wise line of length c to approximate the time series. It is reported to have time complexity $O(\frac{m_i m_j}{c^2})$. Autoregressive kernel [8] have time complexity $(m_i + m_j - 2p)^3$, where p is the order of employed model, far less than $min(m_i, m_j)$. So compared with the above algorithms, Gaussian-LSM and L^2-LSM show computational advantage.

Multivariate Sequences. The experiments of classifications on three multivariate data sets, *Brazilian sign language* (Libras), *handwritten* characters and Australian language of signs (AUSLAN) were conducted. Notably, *handwritten* and AUSLAN are also variable-length. The summary of three data sets is listed in Table 4.

In this experiment, we compared L^2-LSM against comparison algorithms in multivariate data. And the experimental results of 5 runs are plotted in Fig. 6.

From Fig. 6, L^2-LSM outperforms all comparison algorithms on *handwritten*. It also gains a slight advantage on data set *Libras*. On high dimensional data set AUSLAN, L^2-LSM only surpasses AR. The hypothesis of uniform distribu-

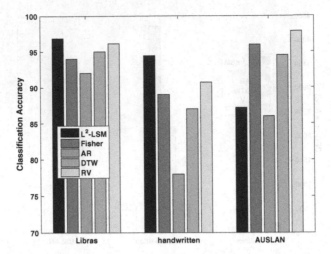

Fig. 6. The classification accuracies carried out on three multivariate data sets.

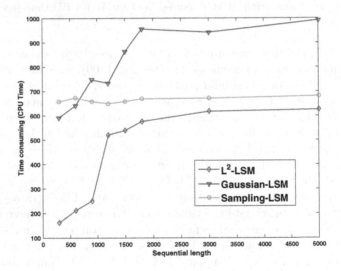

Fig. 7. The time consumptions under different length of sequential data. The vertical axis denotes the time consumption. It is measured in the unit of CPU time (sec). The horizontal axis denotes the sequential length.

tion fails to hold in the high dimensional data set AUSLAN, which leads to a suboptimal result for L^2-LSM.

The time consumption of L^2-LSM and comparison algorithms are plotted in Fig. 5. Generally, L^2-LSM demonstrates an advantage on its efficiency. It has comparative time usage with RV. On high dimensional data set AUSLAN, L^2-LSM and RV build a classifier using less time over other algorithms, and the difference within these two algorithms is not obvious.

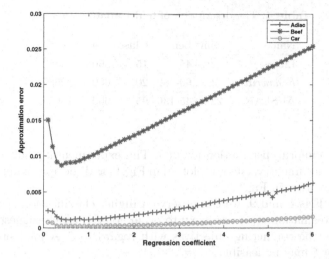

Fig. 8. The approximating errors under different regression coefficients ξ.

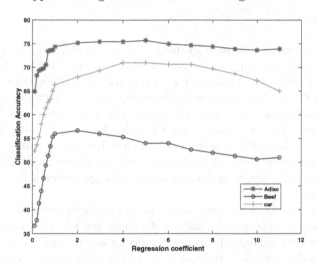

Fig. 9. The classification accuracies under different regression coefficients ξ.

Parametric Sensitivity Analysis. The performance achieved in the model space of LSM are jointly determined by two factors, i.e. the representations offered by LSMs to the sequences and the separation of LSMs in the corresponding space. An unsettled issue is the relationship between these two goals. In the approach, *regression coefficient* ξ is the parameter which needs careful tuning for a better trade-off between the approximation to the sequences and separation of LSMs in the model space.

In this experiment, three data sets were used as benchmark data sets. And experiments were conducted with different settings of parameter ξ. For simplicity,

Table 4. The description of multivariate data sets.

Name	Dim	Len	Class	Train	Test
Libras	2	45	15	360	585
handwritten	3	60-182	20	600	2268
AUSLAN	22	45-136	95	600	1865

we assume a uniform distribution for $\mu(\cdot)$. The experimental result in terms of classification accuracies versus ξ is plotted in Fig. 9, and the approximation errors versus ξ are plotted in Fig. 8.

Compare Figs. 8 and 9, we can observe a higher classification accuracy and a lower approximation error are likely to occur jointly, which suggests that two goals may not be conflicting objectives with regard to ξ. A joint optimization procedure for ξ may be feasible.

4   Conclusion

This paper proposes model space learning for the sequential data on the basis of LSM. LSM is used as a universal approximating tool to fit the conditional probability of a sequence. The models offer representations for sequences of training data. As a result, the learning strategy is carried out in the model space instead of on the original data. From the experiments, the benefits brought by replacing the "memoryless" response function with node that has its own "history" are clear. Fisher-LSM is shown to be robust and effective on processing binary data. An overall improvement of classification accuracy on benchmark data has been observed via experiments. Sampling-LSM is encouraged when the dimensionality of training data is not high.

This paper also discusses measuring the dissimilarity between two LSMs in the model space. A set of models, instead of a single model, is used to give approximations to the training data. Learning in model space relaxes the requirement to use a single model to explain the whole data. The relationship between approximating capability to sequences and separation of LSMs is studied. The result shows the feasibility to implement joint optimization on two seemingly conflict targets.

In general, this paper proposes an approach to constructing data representation without need of assuming a parametric formulation. Its applications on lower dimensional data have been demonstrated to be effective. Promising future work includes improving the model space learning on high dimensional data without sacrificing its efficiency.

Acknowledgements. This work is supported by the National Ket Research and Development plan under Grant 2016YFB1000905, and the National Natural Science Foundation of China under Grants 91546116, 61511130083, 61673363. The authors would like to thank Dr. Hongfei Xing for her valuable comments.

References

1. Berndt, D.J., Clifford, J.: Using dynamic time warping to find patterns in time series. In: KDD Workshop, vol. 10, pp. 359–370 (1994)
2. Chang, C.C., Lin, C.J.: Libsvm: a library for support vector machines. ACM Trans. Intell. Syst. Technol. **2**(3), 27 (2011)
3. Chen, H., Tang, F., Tino, P., Cohn, A.G., Yao, X.: Model metric co-learning for time series classification. In: Proceedings of the Twenty-Fourth International Joint Conference on Artificial Intelligence, pp. 3387–3394. AAAI Press (2015)
4. Chen, H., Tang, F., Tino, P., Yao, X.: Model-based kernel for efficient time series analysis. In: Proceedings of the 19th ACM SIGKDD International Conference on Knowledge Discovery and Data Mining, pp. 392–400. ACM (2013)
5. Chen, H., Tino, P., Rodan, A., Yao, X.: Learning in the model space for cognitive fault diagnosis. IEEE Trans. Neural Netw. Learn. Syst. **25**(1), 124–136 (2014)
6. Chen, H., Tiňo, P., Yao, X.: Cognitive fault diagnosis in tennessee eastman process using learning in the model space. Comput. Chem. Eng. **67**, 33–42 (2014)
7. Chen, Y., Keogh, E., Hu, B., Begum, N., Bagnall, A., Mueen, A., Batista, G.: The UCR time series classification archive, July 2015. www.cs.ucr.edu/~eamonn/time_series_data/
8. Cuturi, M., Doucet, A.: Autoregressive kernels for time series. arXiv preprint arXiv:1101.0673 (2011)
9. Cuturi, M., Vert, J.P., Birkenes, O., Matsui, T.: A kernel for time series based on global alignments. In: IEEE International Conference on Acoustics, Speech and Signal Processing, vol. 2, pp. 413–416 (2007)
10. Figueiredo, M.A.T., Jain, A.K.: Unsupervised learning of finite mixture models. IEEE Trans. Pattern Anal. Mach. Intell. **24**(3), 381–396 (2002)
11. Granger, C.W.J., Hatanaka, M., et al.: Spectral Analysis of Economic Time Series. Princeton University Press, Princeton (1964)
12. Jebara, T., Kondor, R., Howard, A.: Probability product kernels. J. Mach. Learn. Res. **5**, 819–844 (2004)
13. Keogh, E.J., Pazzani, M.J.: Scaling up dynamic time warping for datamining applications. In: Proceedings of the Sixth ACM SIGKDD International Conference on Knowledge Discovery and Data Mining, pp. 285–289. ACM (2000)
14. Kitagawa, G.: A self-organizing state-space model. J. Am. Stat. Assoc. **93**, 1203–1215 (1998)
15. Lahiri, S.N.: Theoretical comparisons of block bootstrap methods. Ann. Stat. **27**, 386–404 (1999)
16. Maass, W., Natschläger, T., Markram, H.: Real-time computing without stable states: a new framework for neural computation based on perturbations. Neural Comput. **14**(11), 2531–2560 (2002)
17. Maaten, L.: Learning discriminative fisher kernels. In: Proceedings of the 28th International Conference on Machine Learning, pp. 217–224 (2011)
18. Müller, K.-R., Smola, A.J., Rätsch, G., Schölkopf, B., Kohlmorgen, J., Vapnik, V.: Predicting time series with support vector machines. In: Gerstner, W., Germond, A., Hasler, M., Nicoud, J.-D. (eds.) ICANN 1997. LNCS, vol. 1327, pp. 999–1004. Springer, Heidelberg (1997). doi:10.1007/BFb0020283
19. Natschläger, T., Markram, H., Maass, W.: Computer models and analysis tools for neural microcircuits. In: Kötter, R. (ed.) Neuroscience Databases, pp. 123–138. Springer, New York (2003)

20. Sahoo, D., Sharma, A., Hoi, S.C., Zhao, P.: Temporal kernel descriptors for learning with time-sensitive patterns. In: Proceedings of the First SIAM Conference on Data Mining (2016)
21. Sakoe, H., Chiba, S.: Dynamic programming algorithm optimization for spoken word recognition. IEEE Trans. Acoust. Speech Sig. Process. **26**(1), 43–49 (1978)
22. Shawe-Taylor, J., Cristianini, N.: Kernel Methods for Pattern Analysis. Cambridge University Press, New York (2004)
23. Van Rossum, M.C.: A novel spike distance. Neural Comput. **13**(4), 751–763 (2001)

Semigeometric Tiling of Event Sequences

Andreas Henelius[1]([✉]), Isak Karlsson[2], Panagiotis Papapetrou[2],
Antti Ukkonen[1], and Kai Puolamäki[1]

[1] Finnish Institute of Occupational Health, PO Box 40, 00251 Helsinki, Finland
{andreas.henelius,antti.ukkonen,kai.puolamaki}@ttl.fi
[2] Department of Computer and Systems Sciences, Stockholm University,
Forum 100, 164 40 Kista, Sweden
{isak-kar,panagiotis}@dsv.su.se

Abstract. Event sequences are ubiquitous, e.g., in finance, medicine, and social media. Often the same underlying phenomenon, such as television advertisements during Superbowl, is reflected in independent event sequences, like different Twitter users. It is hence of interest to find combinations of temporal segments and subsets of sequences where an event of interest, like a particular hashtag, has an increased occurrence probability. Such patterns allow exploration of the event sequences in terms of their evolving temporal dynamics, and provide more fine-grained insights to the data than what for example straightforward clustering can reveal. We formulate the task of finding such patterns as a novel matrix tiling problem, and propose two algorithms for solving it. Our first algorithm is a greedy set-cover heuristic, while in the second approach we view the problem as time-series segmentation. We apply the algorithms on real and artificial datasets and obtain promising results. The software related to this paper is available at https://github.com/bwrc/semigeom-r.

Keywords: Event sequences · Tiling · Covering · Binary matrices

1 Introduction

Phenomena that evolve over time appear in a wide range of application domains including finance (e.g., stock markets [16]), process monitoring (e.g., telecommunications systems [19]), medicine (e.g., biosignals or electronic patient records, [2]), geoscience (e.g., weather or geological measurements [26]), and mobile sensors [15]. Data from such domains can often be represented as *event sequences*, i.e., sequences of labels that correspond to various events associated with a timestamp of the occurrence of the event. Many processes generate such sequences naturally, or a low level signal can be discretised into an event sequence by applying some suitable method, such as SAX [25].

Given multiple time-aligned event sequences, an important problem is to find similarities between them, allowing the detection of underlying higher-level patterns in the data. This problem has been approached using several different techniques, such as segmentation [3], motif detection [29], and clustering [12,14,31].

© Springer International Publishing AG 2016
P. Frasconi et al. (Eds.): ECML PKDD 2016, Part I, LNAI 9851, pp. 329–344, 2016.
DOI: 10.1007/978-3-319-46128-1_21

Finding similarities becomes more challenging when the event sequences are non-stationary, which is often the case in real application domains, such as volatile stock markets or rapidly changing social media streams. In a collection of non-stationary event sequences, interesting local patterns emerge as subsets of event sequences synchronise and desynchronise over short periods of time. Hence, different event sequences are related to each other during different time periods, forming groupings of intra-related sequences. More importantly, these groupings are not static, but they can also evolve over time.

In this paper, we study the following problem: *given multiple event sequences, identify continuous time segments where subgroups of these sequences exhibit similar behaviour.* This formulation is generic and goes beyond state-of-the-art sequence clustering and segmentation problems, since the objective is to identify subgroups of sequence segments that share dominant local trends. Such subgroups can reveal local temporal similarities and dependencies between the sequences belonging to the same subgroup, which would otherwise be hidden by the global trends and structure of the sequences. Our problem is applicable to several domains. For example, in stock market analysis, we may want to identify subgroups of stocks that exhibit similar trends within different time periods. Identifying such groupings of trends and dependencies can provide insights and reveal potential underlying socio-economic events partly affecting the market.

We approach our problem as a *matrix tiling problem*, where event sequences are compactly represented as a matrix, where each row holds an event sequence and each column corresponds to a time point. Hence, our task now becomes equivalent to finding *tiles* in the matrix. A tile consists of a consecutive range of columns (time points) and an arbitrary set of rows (event sequences) of the input matrix. Unlike tiles that are fully geometric (both rows and columns must be consecutive), or combinatorial (both rows and columns can be chosen arbitrarily), we call our tiles *semigeometric*, since only one dimension (time) must form a contiguous segment. A semigeometric tile, thus, represents a group formed by a subset of event sequences sharing the same dominant feature for the duration of the tile. We illustrate our approach with an example using stock index data.

Example. Figure 1 shows daily closing values of ten stock market indices during 1995–2000. Segments representing patterns of economic decline are shown in (a) while segments of economic growth are shown in (b). The discretisation process used for this dataset is described in detail in Sect. 4. Each of the four panels shows a *tiling* of the stock indices using the `MaxTile` and `GlobalTile` algorithms presented in this paper. Vertically aligned segments with the same colour belong to the same *tile* and represent a region where the event series share the dominant feature; here economic growth or decline. More precisely, the coloured tiles in (a) represent segments of economic decline, whereas in (b) the tiles represent segments with economic growth. Applying the proposed tiling method allows us to discover interesting temporal patterns in the data that can be explained by econo-political events. For instance, all algorithms here detect the concurrent economic decline due to the Russian economic crisis in the autumn of 1998 (and the concurrent rebound later the same year).

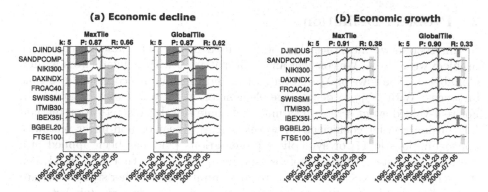

Fig. 1. (a) Periods of economic decline (`stock indices (decline)`) and (b) growth (`stock indices (growth)`) for 10 different stock indices. The coloured regions are identified using the algorithms presented in this paper. The horizontal axis shows days between the years 1995–2000. Vertically aligned segments with the same colour are part of the same tile. Note that the tiles during decline in Figure (a) do not overlap with the tiles during growth in Figure (b). (Color figure online)

Related Work. The problem of finding regions in data matrices with characteristic properties has been extensively studied in multiple contexts; e.g., biclustering (e.g., [5,14]), segmentation (e.g., [12]), tiling (e.g., [10,11,21,33]) and data streams (e.g., [22]). The problem studied in this paper differs from the traditional biclustering and tiling problems in the sense that while we are interested in simultaneously clustering dimensions (rows) over time (columns) based on a given similarity feature of the data, we require the columns in a tile to be consecutive. As discussed below, the temporal ordering has significant impact on the computational efficiency and implementation of the algorithms. In previous work on column-coherent biclusters (see, e.g., [23,27,34]) a specific structure is enforced on the structure column structure of the clusters. The problem of local correlation patterns discussed in [32] is related, but relates to local correlations in time and not precision of tiles. In contrast to these, we define the quality of our tiles in terms of precision and recall, and frame the task of finding a tiling as a covering problem that allows us to build upon existing efficient algorithms.

Contributions. In this paper, we introduce and formulate the novel problem of semigeometric tiling of event sequences and present two algorithms for solving it. The first algorithm, called `MaxTile` is a greedy approach based on the set-cover problem, while the second algorithm, called `GlobalTile`, employs dynamic programming. In addition, we introduce three metrics to assess the quality of a tiling. We also discuss the complexity of the problem and show its connection to two well-known **NP**-hard problems. Finally, we demonstrate the utility of the proposed methods through an extensive empirical evaluation on both real and synthetic datasets.

2 Problem Definition

In this section we introduce the notation used in the paper and formalise the problem we address. We consider an $n \times m$ matrix X with elements from a finite alphabet Σ. The rows of X are event series and the columns correspond to time instances. With X_{ij} we denote the element of X on row i and in column j. For real-valued event series we assume that some suitable discretisation method (e.g., using Fourier coefficients or wavelets [1,4], linear or non-linear piecewise approximations [6,17] or symbolic representations [25]) has been applied in a pre-processing step. Unless otherwise specified, we assume that $\Sigma = \{0, 1\}$, i.e., X is a binary matrix. Generalisation to larger alphabets is rather straightforward and is discussed below, though for simplicity we focus on binary alphabets.

Given a matrix X, our objective is to identify *tiles*. These are consecutive segments in which a subset of the rows of X contain mostly 1s. Formally, a tile t is a three-element tuple (R, a, b) where R is a set of rows of X, and a and b are endpoints of the tile, i.e., column indices corresponding to the beginning and end of the tile, respectively. Below, we use R_t, a_t, and b_t to refer to the tuple elements for a tile t. The *coverage* of the tile t, denoted $C(t)$, is a set of matrix elements belonging to the tile, i.e.,

$$C(t) = \{(i, j) \mid i \in R_t \text{ and } a_t \leq j \leq b_t\}.$$

Tiles t and t' *overlap* unless their intersection $C(t) \cap C(t')$ is empty. Finally, a *tiling* T is a set of possibly overlapping tiles. The coverage of T, denoted $C(T)$, is the union of the covers of each tile in T, and the *weight* $W(T)$ of T is the sum of the elements in $C(T)$, i.e., $C(T) = \bigcup_{t \in T} C(t)$ and $W(T) = \sum_{(i,j) \in C(T)} X_{ij}$. Intuitively, a tiling is good if it (a) has a large weight (covers mainly ones and only a few zeros), (b) covers as many of the ones in X as possible, and (c) uses as few tiles as possible. Requirement (c) is easily achieved by constraining the cardinality of the tiling, while (a) and (b) can be naturally formalised in terms of *precision* and *recall*. Precision is the fraction of ones in the coverage $C(T)$ and recall is the fraction of all ones in X belonging to $C(T)$, i.e.,

$$\text{precision}(T) = \frac{W(T)}{|C(T)|}, \text{ and } \text{recall}(T) = \frac{W(T)}{\sum_i \sum_j X_{ij}}.$$

Optimising both precision and recall simultaneously requires formulating a bicriteria optimisation problem, or using some aggregate objective function, such as the F1-measure. Here we opt for maximising recall with a lower bound constraint on precision and an upper bound constraint on the cardinality of the tiling T. This is a natural definition, as it aims to find a tiling that explains as much of the 1s in X as possible without using tiles that are too noisy (precision constraint), nor is too complex (cardinality constraint). Importantly, in most practical situations the constraint on precision is easy for the user to set given that the density of X (i.e., fraction of 1s) is known. It should be noted that the parameters could also be chosen using an approach based on, e.g., the minimum length description (MDL) principle.

We now formulate the main problem studied in this paper:

Problem 1. SEMIGEOMETRIC TILING Given an $n \times m$ binary matrix X, an integer k, and a real number $\alpha \in [0, 1]$, find the tiling T maximising recall(T) subject to the constraints precision($\{t\}$) $\geq 1 - \alpha$ for each tile $t \in T$ and $|T| \leq k$.

Problem Complexity. Upon first inspection SEMIGEOMETRIC TILING seems very similar to two well-known problems: the nongeometric tiling problem (i.e., the combinatorial tiling problem) [10,11], where *both rows and columns are unordered*, and to the rectilinear picture compression problem [9, problem SR25], where *both columns and rows have a fixed order*. In the semigeometric case, the smallest tiling covering all 1s in the input matrix can be found in polynomial time.

Theorem 1. *A minimum-cardinality semigeometric tiling for an $n \times m$ binary matrix X with perfect precision and recall can be found in polynomial time.*

The proof follows from known results for the problem of covering a vertically convex polygon with rectangles [8,13,20].

Hence, the problem of semigeometric tiling is computationally different from the nongeometric and fully geometric tiling problems. Notice, however, that Theorem 1 does not imply that SEMIGEOMETRIC TILING is tractable; it merely suggests that the complexity of this problem is not trivially **NP**-hard. There are also other reasons to argue why this might be the case. Namely, for the nongeometric tiling problem it is easy to establish that merely finding the largest tile is **NP**-hard. In our case this sub-problem is also easy, as we discuss below. However, the actual complexity of Problem 1 is an open question.

Generalisation to Multi-letter Alphabets. It is important to note that generalisation to multi-letter alphabets follows directly from the above given definitions. Multi-letter tilings are constructed by first finding the individual tiling for each letter, after which recall is found by selecting k tiles from the individual tilings, such that recall is maximised. If different precisions are used for different alphabet letters, the minimum precision of the multi-letter tiling corresponds to the smallest single-letter precision.

3 Algorithms

In this section we describe in detail two algorithms (`MaxTile` and `GlobalTile`) for solving the SEMIGEOMETRIC TILING problem (Problem 1).

3.1 Overview

We approach Problem 1 using two slightly different, yet standard strategies that make use of the property that the columns of the input matrix X are ordered. An example of how our algorithms work on a small example dataset containing three overlapping tiles is shown in Fig. 2.

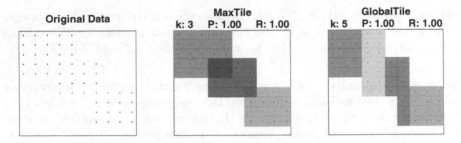

Fig. 2. Example comparing the output between the two algorithms. **k** stands for cardinality, **P** for precision and **R** for recall. The maximum cardinality was set to 5 and the lower bound for precision to 0.95. The solution found by GlobalTile (rightmost figure) is also equal to the vertical decomposition (see Sect. 3.3 and Proposition 2) of the original tiling.

Both MaxTile and GlobalTile algorithms are based on the simple observation that given X and two endpoints a and b, we can easily find the *maximum recall tile* (R, a, b) satisfying the precision constraint:

Definition 1. *Given the $n \times m$ binary matrix X and the precision constraint α, the* maximum recall tile *with endpoints a and b is defined as (R^*, a, b), where R^* is found by maximising recall subject to the precision constraint α.*

We can find the maximum recall tile for every choice of a and b in $O(n \log n)$ time by using cumulative sums of rows to find the number of ones within a row for a given interval and by sorting the rows by the number of ones. Some choices of a and b may result in an empty maximum recall tile, e.g., if there are no rows with enough 1s to satisfy the precision constraint. We omit such tiles, and obtain the set T of at most $\binom{m}{2}$ candidate tiles.

MaxTile and GlobalTile differ in the way the tiling is constructed from candidate tiles. MaxTile greedily solves a cardinality-constrained set-cover problem, while GlobalTile treats the problem as a time-series segmentation task and finds the tiling using dynamic-programming, a side effect of which is that GlobalTile cannot produce overlapping tilings. MaxTile has no such constraint. This is seen in Fig. 2, where the MaxTile tiling is overlapping and the GlobalTile tiling consists of vertical non-overlapping stripes.

3.2 MaxTile

The MaxTile algorithm is based on a straightforward mapping of the tiling problem to a set cover problem. MaxTile first finds all maximum recall tiles (Definition 1) in the data matrix X with a precision of at least $1 - \alpha$. A subset of k, possibly overlapping, tiles maximising recall is then chosen from these tiles. The MaxTile algorithm is presented in Algorithm 1 and has a complexity of $\mathcal{O}\left(m^2 n \log n\right)$. The algorithm maintains a tiling T, and always adds a tile $t \notin T$ maximising the marginal gain, i.e., $\text{recall}(T \cup t) - \text{recall}(T)$.

input : $n \times m$ binary matrix X, precision threshold α, max no of tiles k.
1 Let $\mathcal{T} \leftarrow \cup_{1 \le a \le b \le m} t_{ab}$, where $t_{ab} = (R, a, b)$ is a tile with $R \subseteq [n]$ with the highest recall and precision $\ge \alpha$
2 Greedily find the tiling $T \subseteq \mathcal{T}$, $\mid T \mid = k$ maximising recall of X.
3 **return** The tiling T.

Algorithm 1. MaxTile

input : $n \times m$ binary matrix X, precision threshold α, max no of tiles k.
1 Use dynamic programming to compute all l-segmentations S_l up to $l = 2k + 1$.
2 $\mathcal{T} \leftarrow \emptyset$
3 **for** $l = k$ to $2k + 1$ **do**
4    $T_l \leftarrow$ all tiles from S_l
5   **if** T_l *has more than* k *tiles* **then**
6 | $T_l \leftarrow$ the k tiles $t \in T_l$ having highest $W(t)$
7   **end**
8   insert T_l into $\mathcal{T}$
9 **end**
10 **return** *The tiling* $T \in \mathcal{T}$ *that has the highest* $W(T)$

Algorithm 2. GlobalTile

Proposition 1. *The function* recall(T) *is submodular, i.e., when* $T' \subset T$, *and t is some tile not in either T or T',*

$$\text{recall}(T' \cup t) - \text{recall}(T') \ge \text{recall}(T \cup t) - \text{recall}(T).$$

The proof follows the usual argument for covering functions.

Since recall(T) is submodular, we can employ the same optimisation as in, e.g., [24]: we maintain a priority queue of known marginal gains for every tile and only recompute the marginal gain for a given tile if it is larger than the best marginal gain observed for the current T.

Also, note that recall($\emptyset$) = 0, and that recall(T) increases monotonically as T grows. This, together with the submodularity of recall(T) and well-known results from [28], yields that the greedy algorithm has a constant approximation factor of $1 - 1/e$, i.e., the recall of the MaxTile tiling is at least $1 - 1/e$ times the optimal recall with k tiles.

3.3   GlobalTile

To design the GlobalTile algorithm we view the SEMIGEOMETRIC TILING problem as a time-series segmentation task, the objective of which is to partition a given time-series into l non-overlapping, contiguous intervals, called *segments*. The partition is found by optimising the sum of segment-wise scores. We call this partition an *l-segmentation*. The input matrix X can be viewed as a multidimensional time-series, where a contiguous range of columns from column a to column b forms a segment. The segment-specific score is defined as the recall of the maximum recall tile (R^*, a, b). The mapping between an l-segmentation and a tiling is thus straightforward. Every segment in an l-segmentation has two endpoints, a and b, given which we can compute the maximum recall tile (R^*, a, b) as specified in Definition 1. Since (R^*, a, b) may be empty for some choices of a and b, some segments may correspond to empty tiles. Therefore, every l-segmentation maps to a tiling of cardinality *at most* l. In summary, given the matrix X and the integer k, the idea of GlobalTile is to (a) compute an optimal $l \geq k$-segmentation of X, and (b) turn this l-segmentation into a tiling of size $k \leq l$.

Importantly, an optimal l-segmentation can be computed in polynomial time with dynamic programming [3]. Unfortunately this does not give a polynomial time algorithm for Problem 1, because the parameters l and k do not match one-to-one, and the resulting segmentation must be non-overlapping. This constraint is not present in Problem 1. However, we can still use an l-segmentation as a building block in our algorithm as follows. Given the input X and the integer k, the GlobalTile algorithm first finds all l-segmentations of X from $l = k$ up to $l = 2k + 1$. Let T_l denote the tiling corresponding to a given l-segmentation. Since T_l may contain up to $l \leq 2k + 1$ tiles, the algorithm keeps only the k largest tiles of every T_l to maximise recall. This pruning step is repeated for every tiling T_l with $l \in \{k, \ldots, 2k + 1\}$, and the algorithm returns the tiling T_l having the highest recall. Pseudocode of GlobalTile is shown in Algorithm 2.

Next we discuss some properties of the GlobalTile algorithm. In the following we call a tiling T of cardinality k a *k-tiling*. In any k-tiling, tiles (R_1, a_1, b_1) and (R_2, a_2, b_2) are *vertically non-overlapping* if $b_1 < a_2$ and $b_2 < a_1$. A tiling is vertically non-overlapping if all of its tiles are non-overlapping. The *vertical decomposition* of T is the vertically non-overlapping tiling T' with smallest cardinality that has *exactly the same cover* as T. See the rightmost panel of Fig. 2 for an example of a vertical decomposition. We make the following observation:

Proposition 2. *Any tiling T of cardinality k has a vertical decomposition of cardinality at most $2k - 1$.*

The aim of computing the l-segmentation is to find the vertical decomposition of the underlying tiling. First, observe that if the underlying optimal k-tiling is vertically non-overlapping, the dynamic programming algorithm finds it directly (provided l is set appropriately). For other k-tilings, we know from Proposition 2 that their vertical decompositions correspond to segmentations with at most $2k+1$ segments (including possible empty segments before and after the leftmost and rightmost tile, respectively). For inputs with optimal noise-free tilings (i.e., tilings with precision equal to 1), we obtain the following:

Proposition 3. *Given an input matrix X containing a noise-free k-tiling T^* and the integer k, the dynamic programming algorithm finds a k'-tiling T', with $\mathrm{recall}(T') = \mathrm{recall}(T^*)$, and $k \leq k' \leq 2k - 1$.*

Since the GlobalTile algorithm is given the cardinality constraint k, it may not return the entire vertical decomposition of size $k' \geq k$. In the worst case all $2k - 1$ tiles in the vertical decomposition T' have exactly the same weight, and GlobalTile drops $k - 1$ of these, i.e., $\mathrm{recall}(T) \geq \frac{k}{2k-1} \mathrm{recall}(T^*)$. Hence, GlobalTile finds a tiling T with $\mathrm{recall}(T) \geq \frac{1}{2} \mathrm{recall}(T^*)$. Notice that in practice GlobalTile can perform even better, as it looks for the best k-tiling from all segmentations *up to* size $2k + 1$, and the vertical decomposition may in practice contain fewer than $2k - 1$ tiles.

Propositions 2 and 3 above imply in practice that the `GlobalTile` algorithm can find tilings with recall very close to optimum, at the cost of using at most twice the amount of tiles used in an optimal solution.

4   Experiments

In this section we empirically investigate the relationship between (i) precision and recall and (ii) between recall and cardinality. We test the performance and behaviour of both tiling algorithms on four real and two synthetic datasets. We also compare the tilings we find for real data with tilings of a randomised version of the data having the same row marginals. The purpose of this comparison is to establish that the tilings we have found indeed reflect meaningful structure instead of noise.

The algorithms are implemented in R and are released as the `semigeom` R-package. This package and all source code used for the experiments is available for download[1].

4.1   Datasets

We consider four real and two synthetic datasets, described in detail below. Properties of the datasets are summarised in Table 1.

1. **Stock indices** The `stock indices` data contains daily stock exchange closing index data from 1995 to 2000[2] for 10 stock exchanges. We discretised the data into a two-letter alphabet by first calculating the gradient of the linear trend of each stock index in a 30-day windows and assigning 1 if the trend was rising (gradient positive) and −1 if the trend was decreasing (gradient negative). Since both labels are of interest, `stock indices (growth)` will denote the dataset when the label of interest is 1, and `stock indices (decline)` will denote the dataset when the label of interest is −1.
2. **Stock prices** The `stock prices` dataset contains daily closing prices between 03.01.2011–31.12.2015 (1257 trading days) for 400 randomly selected stocks. The data was discretised by calculating the change (in %) in closing price from the previous day. Instances for which the closing price dropped more than −6% were set to 1 and all other instances were set to zero.
3. **Paleo data** The `paleo` dataset is a binary 139×124 matrix describing the presence or absence of species (139) of Cenozoic mammals in different sites (124) [7,30]. The sites in the dataset have a temporal order determined by specialists.

---

[1] https://github.com/bwrc/semigeom-r.
[2] Data from http://www.economicswebinstitute.org/data/stockindexes.zip.

4. **EEG dataset** The EEG dataset is part of the ISRUC-Sleep dataset [18], consisting of EEG polysomnography recordings from 10 healthy people. There are six EEG channels located on the frontal (channels F3, F4), central (channels C3, C4), and occipital (channels O1, O2) parts of the scalp, all sampled at 200 Hz. We chose to use data from the first 6.5 hours of sleep, where data from all subjects was available, giving a total of 4680000 samples per channel. The data was discretised by calculating the EEG frequency spectrum in four-second windows, and for each window determining which of the following 5 frequency bands was dominant (delta: 1–4 Hz, theta: 4–7 Hz, alpha: 8–15 Hz, beta: 16–31 Hz, and gamma: 16–100 Hz). Each window was then assigned a letter corresponding to the dominant band. After the discretisation the data contains 5850 samples for each of the six channels.

5. **Synthetic data** The synthetic datasets are randomly generated binary matrices containing a given number of randomly inserted segments with a fraction p of ones. We used two types of synthetic datasets; (i) k randomly inserted tiles of size $l \times l$ and (ii) tiles formed from k randomly inserted varying-width segments spanning l rows.

Table 1. Properties of the datasets used in the experiments. Size is the number of rows and columns in the original data and the discrete size are the dimensions of the discretised data used as input to the algorithms. Density is the fraction of 1 s in the input matrix.

Dataset	Size	Size discrete	Density
Stock indices (growth)	10×1353	10×45	0.62
Stock indices (decline)	10×1353	10×45	0.37
Stock prices	400×1258	400×1257	0.02
Paleo	139×124	139×124	0.11
EEG subject 4	6×4680000	6×5850	0.14
EEG subject 7	6×4680000	6×5850	0.21
Synthetic 1	50×100	50×100	0.28
Synthetic 2	50×100	50×100	0.16

4.2 Basic Results

Precision-Recall Trade-Off. We first study the precision-recall trade-off of the algorithms. Precision-recall curves for two different cardinalities (3 and 7) are shown in Fig. 3 for the synthetic1, paleo, stock prices, and stock indices (decline). The curves were computed by modifying the precision constraint parameter. The results suggest that both MaxTile and GlobalTile perform similarly in terms of balance between precision and recall.

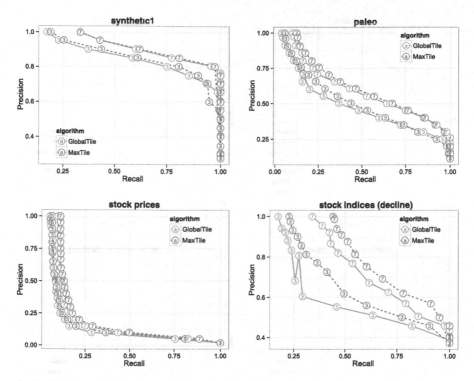

Fig. 3. Precision-recall curve for the algorithms for different datasets using cardinalities of 3 and 7.

Recall in Original vs. Randomised Data. To study if the found tilings reflect meaningful structure, we compare the recall of the tiling of the original data with the recall of a tiling for randomised data, using otherwise the same parameters. The randomised data are generated by shuffling the original values of every data row uniformly at random. This simple randomisation scheme breaks any temporal connections the event sequences may have, and thus any tiling found after randomisation only contains meaningless structure. Tilings found in the original unshuffled data should have a higher recall.

The cardinality-recall curves for the algorithms on the `paleo`, `stock prices`, `stock indices (growth)`, and `stock indices (decline)` datasets are shown in Fig. 4. The precision constraint was set to the value $P_{\min}$ in Table 2. The CR-curves show that for every dataset, both algorithms find a tiling with higher recall than what can be found in the randomised data. This indicates that tilings of the original data indeed contain meaningful structure. Moreover, although the algorithms perform quite similarly, `MaxTile` in general achieves a higher recall for a given cardinality, as expected, since the algorithm allows overlapping tiles. The performance of the algorithms on the `paleo` dataset is very similar. The structure of the dataset fits well with the assumption of vertical segments of

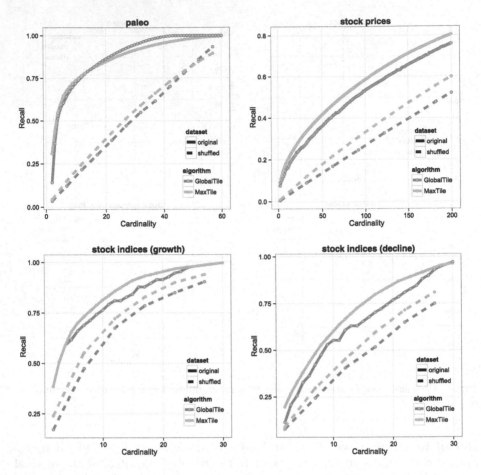

Fig. 4. Cardinality-recall curve for the algorithms for different datasets.

GlobalTile and hence GlobalTile achieves a higher recall than MaxTile for the same cardinality.

Running Times. Typical wall-clock running times (using unoptimised R-code running on a 1.8 GHz Intel Core i7 CPU) as well as examples of recall and cardinality for a given precision constraint for all datasets and both algorithms are shown in Table 2.

4.3   Example Tilings

Paleo dataset. The analysis results of the synthetic1 and paleo datasets are shown in Fig. 5. Both algorithms manage to successfully capture the structure in the synthetic1 dataset. Notice the difference in terms of selected tiles, due to allowance of overlapping tiles by MaxTile. The paleo dataset is very sparse,

Table 2. Numerical results from the experiments. P, R, and k stand for precision, recall, and cardinality, respectively. The time-column gives the calculation time in seconds.

Dataset	Algorithm	P_{min}	k_{max}	P	R	k	time [s]
stock indices	GlobalTile	0.85	5	0.87	0.62	5	0.12
(growth)	MaxTile	0.85	5	0.87	0.66	5	0.03
stock indices	GlobalTile	0.85	5	0.90	0.33	5	0.11
(decline)	MaxTile	0.85	5	0.91	0.38	5	0.03
stock prices	GlobalTile	0.30	10	0.32	0.16	10	112.90
	MaxTile	0.30	10	0.30	0.19	10	104.41
paleo	GlobalTile	0.30	3	0.30	0.85	3	1.32
	MaxTile	0.30	3	0.30	0.82	3	0.60
EEG	GlobalTile	0.60	10	0.60	0.79	10	1451.61
(alpha, subject 4)	MaxTile	0.60	10	0.60	0.79	10	461.38
EEG	GlobalTile	0.60	10	0.60	0.94	10	1456.06
(alpha, subject 7)	MaxTile	0.60	10	0.60	0.94	10	473.34
synthetic1	GlobalTile	0.75	5	0.76	0.99	5	0.47
	MaxTile	0.75	5	0.75	1.00	4	0.21
synthetic2	GlobalTile	0.75	5	0.80	1.00	5	0.48
	MaxTile	0.75	5	0.76	1.00	5	0.22

so a precision threshold of 0.3 was used. The structure in the three geological epochs in the data are clearly visible in the tilings and both algorithms achieve a high recall already with three tiles. The performance of the two algorithms is similar in terms of precision, and the discovered structure is also equivalent. The middle tile (in red) is wider for `MaxTile`, since the algorithm allows overlapping tiles.

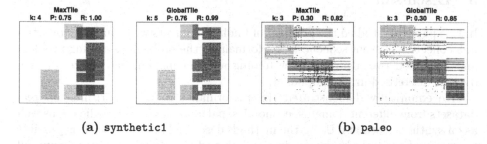

(a) synthetic1   (b) paleo

Fig. 5. Tiling of the synthetic1 and paleo datasets. (Color figure online)

EEG dataset. We chose to consider only one letter corresponding to the alpha band from each channel and calculated the tilings with a maximal cardinality

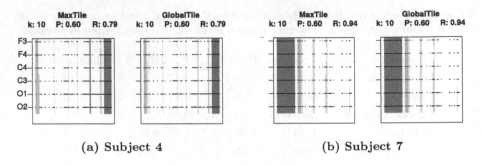

(a) Subject 4   (b) Subject 7

Fig. 6. Tiling of the EEG activity in the alpha band for subjects 4 and 7. (Color figure online)

of 10 and a precision threshold of 0.60. This analysis, hence, reveals whether there are segments with similarly dense alpha activity on the different channels. It is expected that neighbouring channels in the same front-back location (frontal, central, and occipital) should be similar and channels in different front-back locations should be somewhat different. We omitted singleton tiles, i.e., tiles with a cardinality less than two, since we are interested in the relationships between different channels. Figure 6 shows the tiling obtained for the two subjects with prominent patterns. The MaxTile and GlobalTile tilings are very similar, with some small differences. In MaxTile tiling of subject 4 there is a relatively wide segment (green tile), where the neighbouring occipital channels and the C3 channel share a similar pattern of alpha activity. Both algorithms find a similar pattern of alpha activity in all channels of subject 4 (red tile at right end) and subject 7 (red tile at left end). Information regarding the relationships and spatiotemporal distributions of different EEG rhythms could be useful, e.g., when investigating the neural correlates of sleep patterns.

5   Discussion

In this paper we studied the problem of finding segments with similar properties in a collection of event sequences. We formalised the problem as a tiling problem and presented two algorithms for finding a tiling maximising the recall of a dataset discretised into an alphabet.

We empirically demonstrated the performance of the algorithms on real datasets from different domains; economics, palaeontology, and medicine, as well as on synthetic datasets. Using the methods described in this paper it is possible to uncover temporal phenomena in the data and the results provide meaningful insight into the interpretation of the underlying processes. The two algorithms emphasise slightly different properties of the data. For instance, the tilings found by the MaxTile algorithm consist of large, overlapping, submatrices satisfying the precision requirement, whereas the tilings produced by the GlobalTile algorithm are always non-overlapping.

Different datasets require different choices of parameters. The cardinality and precision threshold must be set depending on factors, such as the data density. As shown in the experimental evaluation, the proportion of the dataset explained by the tiles (i.e., the recall) varies depending on both the cardinality and precision. A reasonable heuristic for setting the precision constraint is to consider the overall data density, and set α to a smaller value so that the tiles indeed reflect denser regions in the input matrix.

Interestingly, we here showed that the special case of semigeometric tiling of a dataset, where we find a minimum full cover with perfect precision is tractable in polynomial time despite the related problems of combinatorial and fully geometric tiling being NP-hard. However, our main problem of covering a data matrix with a precision of $1 - \alpha$ with k tiles appears to be challenging, and we suspect it is **NP**-hard. The proof of this, however, remains an open question. Another important direction for future work is to consider the significance of the tilings found. In this paper we use a simple sanity check where the recall of the resulting tiling is compared against the recall of a tiling of the randomised data matrix. Methods to evaluate the significance of individual tiles are also of interest.

Acknowledgements. AH, AU, and KP were supported by Tekes (Revolution of Knowledge Work project) and Academy of Finland (decision 288814) and IK and PP by Swedish Foundation for Strategic Research (grant IIS11-0053).

References

1. Agrawal, R., Faloutsos, C., Swami, A.: Efficient similarity search in sequence databases. In: Lomet, D.B. (ed.) FODO 1993. LNCS, vol. 730, pp. 69–84. Springer, Heidelberg (1993). doi:10.1007/3-540-57301-1_5
2. Batal, I., Fradkin, D., Harrison, J., Moerchen, F., Hauskrecht, M.: Mining recent temporal patterns for event detection in multivariate time series data. KDD **2012**, 280–288 (2012)
3. Bellman, R.: On the approximation of curves by line segments using dynamic programming. Commun. ACM **4**(6), 284 (1961)
4. Chan, K.P., Fu, A.W.C.: Efficient time series matching by wavelets. ICDE **1999**, 126–133 (1999)
5. Cheng, Y., Church, G.M.: Biclustering of expression data. ISMB **8**, 93–103 (2000)
6. Faloutsos, C., Jagadish, H., Mendelzon, A.O., Milo, T.: A signature technique for similarity-based queries. SEQUENCES **1997**, 2–20 (1997)
7. Fortelius, M.: Coordinator: New and old worlds database of fossil mammals (NOW) (2016). University of Helsinki. http://www.helsinki.fi/science/now/
8. Franzblau, D.S., Kleitman, D.J.: An algorithm for covering polygons with rectangles. Inf. Control **63**(3), 164–189 (1984)
9. Garey, M.R., Johnson, D.S.: Computers and Intractability: A Guide to the Theory of NP-completeness. W.H. Freeman & Co, New York (1979)
10. Geerts, F., Goethals, B., Mielikäinen, T.: Tiling databases. In: Suzuki, E., Arikawa, S. (eds.) DS 2004. LNCS (LNAI), vol. 3245, pp. 278–289. Springer, Heidelberg (2004). doi:10.1007/978-3-540-30214-8_22
11. Gionis, A., Mannila, H., Seppänen, J.K.: Geometric and combinatorial tiles in 0–1 data. In: Boulicaut, J.-F., Esposito, F., Giannotti, F., Pedreschi, D. (eds.) PKDD 2004. LNCS, vol. 3202, pp. 173–184. Springer, Heidelberg (2004). doi:10.1007/978-3-540-30116-5_18

12. Gionis, A., Mannila, H., Terzi, E.: Clustered segmentations. In: 3rd Workshop on Mining Temporal and Sequential Data, KDD 2004 (2004)
13. Györi, E.: A minimax theorem on intervals. J. Comb. Theor. Ser. B **37**(1), 1–9 (1984)
14. Hartigan, J.A.: Direct clustering of a data matrix. J. Am. Stat. Assoc. **67**(337), 123–129 (1972)
15. Hemminki, S., Nurmi, P., Tarkoma, S.: Accelerometer-based transportation mode detection on smartphones. In: SenSys 2013, p. 13 (2013)
16. Huang, C.F.: A hybrid stock selection model using genetic algorithms and support vector regression. Appl. Soft Comput. **12**(2), 807–818 (2012)
17. Keogh, E., Chakrabarti, K., Pazzani, M., Mehrotra, S.: Dimensionality reduction for fast similarity search in large time series databases. Knowl. Inf. Syst. **3**(3), 263–286 (2001)
18. Khalighi, S., Sousa, T., Santos, J.M., Nunes, U.: Isruc-sleep: a comprehensive public dataset for sleep researchers. Comput. Methods Program. Biomed. **124**, 180–192 (2015)
19. Klemettinen, M., Mannila, H., Toivonen, H.: Rule discovery in telecommunication alarm data. J. Network Syst. Manage. **7**(4), 395–423 (1999)
20. Knuth, D.E.: Irredundant intervals. J. Exp. Algorithmics (JEA) **1** (1996)
21. Kontonasios, K.N., De Bie, T.: An information-theoretic approach to finding informative noisy tiles in binary databases. In: SDM 2010, p. 153 (2010)
22. Lam, H.T., Pei, W., Prado, A., Jeudy, B., Fromont, É.: Mining top-k largest tiles in a data stream. In: ECML PKDD 2014, pp. 82–97. Springer, Heidelberg (2014)
23. Lee, J., Lee, Y., Jun, C.H.: A biclustering method for time series analysis. Ind. Eng. Manage. Syst. **9**, 129–138 (2010)
24. Leskovec, J., Krause, A., Guestrin, C., Faloutsos, C., VanBriesen, J.M., Glance, N.S.: Cost-effective outbreak detection in networks. KDD **2007**, 420–429 (2007)
25. Lin, J., Keogh, E., Lonardi, S., Chiu, B.: A symbolic representation of time series, with implications for streaming algorithms. SIGMOD **2003**, 2–11 (2003)
26. Lionello, P.: The climate of the venetian and north adriatic region: variability, trends and future change. Phys. Chem. Earth Parts A/B/C **40**, 1–8 (2012)
27. Madeira, S.C., Oliveira, A.L.: A linear time biclustering algorithm for time series gene expression data. In: Casadio, R., Myers, G. (eds.) WABI 2005. LNCS, vol. 3692, pp. 39–52. Springer, Heidelberg (2005). doi:10.1007/11557067_4
28. Nemhauser, G.L., Wolsey, L.A., Fisher, M.L.: An analysis of approximations for maximizing submodular set functionsi. Math. Program. **14**(1), 265–294 (1978)
29. Patel, P., Keogh, E., Lin, J., Lonardi, S.: Mining motifs in massive time series databases. ICDM **2002**, 370–377 (2002)
30. Puolamäki, K., Fortelius, M., Mannila, H.: Seriation in paleontological data using markov chain monte carlo methods. PLoS Comput. Biol. **2**(2), e6 (2006)
31. Rakthanmanon, T., Keogh, E.J., Lonardi, S., Evans, S.: MDL-based time series clustering. Knowl. Inf. Syst. **33**(2), 371–399 (2012)
32. Ukkonen, A.: Mining local correlation patterns in sets of sequences. In: Gama, J., Costa, V.S., Jorge, A.M., Brazdil, P.B. (eds.) DS 2009. LNCS (LNAI), vol. 5808, pp. 347–361. Springer, Heidelberg (2009). doi:10.1007/978-3-642-04747-3_27
33. Xiang, Y., Jin, R., Fuhry, D., Dragan, F.F.: Succinct summarization of transactional databases: An overlapped hyperrectangle scheme. Knowl. Discov. Data Min. **23**, 758–766 (2008)
34. Zhang, Y., Zha, H., Chu, C.H.: A time-series biclustering algorithm forrevealing co-regulated genes. Int. Conf. Informationtechnology: Coding Comput. **1**, 32–37 (2005)

Attribute Conjunction Learning with Recurrent Neural Network

Kongming Liang[1,2], Hong Chang[1(✉)], Shiguang Shan[1], and Xilin Chen[1,2]

[1] Key Lab of Intelligent Information Processing
of Chinese Academy of Sciences (CAS), Institute of Computing Technology, CAS,
Beijing 100190, China
kongming.liang@vipl.ict.ac.cn, {changhong,sgshan,xlchen}@ict.ac.cn
[2] University of Chinese Academy of Sciences, Beijing 100049, China

Abstract. Searching images with multi-attribute queries shows practical significance in various real world applications. The key problem in this task is how to effectively and efficiently learn from the conjunction of query attributes. In this paper, we propose Attribute Conjunction Recurrent Neural Network (AC-RNN) to tackle this problem. Attributes involved in a query are mapped into the hidden units and combined in a recurrent way to generate the representation of the attribute conjunction, which is then used to compute a multi-attribute classifier as the output. To mitigate the data imbalance problem of multi-attribute queries, we propose a data weighting procedure in attribute conjunction learning with small positive samples. We also discuss on the influence of attribute order in a query and present two methods based on attention mechanism and ensemble learning respectively to further boost the performance. Experimental results on aPASCAL, ImageNet Attributes and LFWA datasets show that our method consistently and significantly outperforms the other comparison methods on all types of queries. The software related to this paper is available at https://github.com/GriffinLiang/AC-RNN.

Keywords: Attribute learning · Multi-label learning · Image retrieval

1 Introduction

Attribute learning provides a promising way for computer to understand image content in a fine-grained manner. Beyond traditional object categories, attribute contains abundant information from holistic perception (e.g., color, shape, etc.) to the presence or absence of local parts for images. By bridging the gap between low-level features and high-level categorization, attributes benefit many object recognition and classification problems (e.g., object recognition [26], face verification [15] and zero-shot classification [16]).

Compared with single attribute learning, learning attribute conjunctions shows more practical significance. An attractive application is to retrieve relevant images based on multi-attribute query. For example, it can be used to discover objects with

© Springer International Publishing AG 2016
P. Frasconi et al. (Eds.): ECML PKDD 2016, Part I, LNAI 9851, pp. 345–360, 2016.
DOI: 10.1007/978-3-319-46128-1_22

specified characteristics [5, 22], search people of certain facial descriptions [14] and match products according to users' requirements [13]. In this scenario, a user may describe the visual content of interest by specifying a few attributes. Then the image search engine will calculate the similarity between the input attribute conjunction and the images in some datasets. The most similar images will be returned as the search results.

A common approach [5, 16] to tackle multi-attribute query is transforming the problem into multiple single-attribute learning tasks. Specifically, a binary classifier is built for each single attribute, then the result of multi-attribute prediction is generated by summing up the scores of all single attribute classifiers. Though this kind of combination is simple and shows good scalability, it has two main drawbacks. Firstly, the correlation between attributes is ignored because of the separate training of each attribute classifier. Secondly, attribute classification results are sometimes unreliable since abstract linguistic properties can have very diverse visual manifestations especially when they come across different categories. This situation may get worse with larger number of attributes appearing in a query. For example, when the query length is three, an unreliable attribute classifier may affect $\binom{A-1}{2}$ query results (A is the total number of attributes).

Instead of training a classifier for each attribute separately, a more promising approach is to learn from the attribute conjunctions. Since conjunctions of multiple attributes may lead to very characteristic appearances, training a classifier that detects the conjunctions as a whole may produce more accurate results. For example, training a classifier to predict whether the animal is (black & white & stripe) leads to a specific concept "Zebra". However, straightforward training classifiers from attribute conjunctions is not a good choice. Firstly, the length of multi-attribute query is not fixed and the number of attribute conjunctions grows exponentially w.r.t. the query length. For a three-attribute query, we need to build $\binom{A}{3}$ classifiers for all possible attribute conjunctions. Secondly, there are only a small number of positive examples for each multi-attribute query (a positive sample must have multiple query attributes simultaneously), which brings the learning process more difficulties. With data bias problem, some attribute conjunction classifiers may perform even worse than simply adding the scores of disjoint single-attribute classifiers. Thirdly, the correlation between attribute conjunctions is not well explored, since the queries which share common attributes are considered to be independent from each other.

In this paper, we propose a novel attribute conjunction recurrent neural network (AC-RNN) to tackle multi-attribute based image retrieval problem. As shown in Fig. 1, the input sequence of AC-RNN are the attributes appearing in a query with a predefined order. Each of the input attributes is then embedded into the hidden units and combined in a recurrent way to generate the representation for the attribute conjunction. The conjunction representation is further used to compute the classifier for the input multi-attribute query. As the multiple attributes in each conjunction are processed by the network recurrently, the number of parameters of our model do not increase with the length

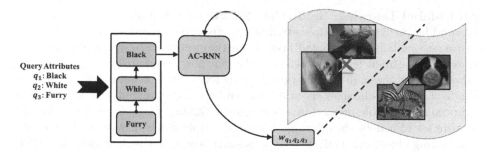

Fig. 1. An Illustration of Attribute Conjunction Recurrent Neural Network (AC-RNN).

of query. Compared with straightforward multi-attribute learning methods, our AC-RNN model is more appropriate to model the complex relationship among different attribute conjunctions. We also introduce a data weighting procedure to address the data bias problem in attribute conjunction learning. Finally, we discuss on the influence of attribute order in our learning framework and propose two methods based on attention mechanism and ensemble learning respectively to improve the performance of AC-RNN.

The rest of this paper is organized as follows. We first introduce some related works in the following section. In Sect. 3, we present the attribute conjunction recurrent neural network in detail. Experimental results are then shown in Sect. 4. Finally, Sect. 5 concludes the paper.

2   Related Work

Multi-Attribute Query[1]: Guillaumin et al. [8] propose to use a weighted nearest-neighbour model to predict the tags of a test image which can directly support multi-word query based image retrieval. Petterson et al. [19] present a reverse formulation to retrieve sets of images by considering labels as input. In this way, they can directly optimize the convex relaxations of many popular performance measures. By leveraging the dependencies between multiple attributes, Siddiquie et al. [23] explicitly model the correlation between query-attributes and non-query attributes. For example, for a query such as "man with sunglasses", the correlated attributes like beard and mustache can also be used to retrieve releant images. Since training a classifier for a combination of query attributes may not always perform well, Rastegari et al. [20] propose an approach to determine whether to merge or split attributes based on the geometric quantities. Different from the above methods, we propose to explicitly learn all the single attribute embeddings and combine them in a recurrent way to generate the representation of attribute conjunction.

---

[1] The multi-attribute we denote here may refer to other statements such as keywords or multi-label in other literature.

Multi-label Learning: Read et al. [21] extend traditional binary relevance method by predicting multiple attribute progressively in an arbitrary order. For instance, one label is predicted first. Then the prediction result is appended at the end of the input feature which is used as the new feature to predict the second label. Finally, the multiple label predictions form into a classifier chain. Since a single standalone classifier chain model can be poorly ordered, the authors also propose an ensemble method in a vote scheme. Zhang et al. [29] exploit different feature set to benefit the discrimination of multiple labels. This method exploits conducting clustering analysis on the positive and negative instances and then performs training and testing referring to the clustering results. Different from multi-label learning problems, the task we deal with here is to model attribute conjunctions instead of multiple separate labels.

Label Embedding: Though deep learning provides a powerful way to learn data representations, how to represent labels is also a key issue for machine learning methods. A common way is Canonical Correlation Analysis (CCA) [9] which maximizes the correlation between data and labels by projecting them into a common space. Another promising way is to learn label embedding by leveraging other possible sources as prior information. Akata et al. [1] propose to embed category labels into attribute space under the assumption that attributes are shared across categories. Frome et al. [6] represent category labels with the embedding learned from textual data. Hwang et al. [11] jointly embed all semantic entities including attributes and super-categories into the same space by exploiting taxonomy information. But so far, there is no work on learning the conjunction representation of multiple labels to the best of our knowledge.

3   Our Method

The problem we aim to address here is to retrieve relevant images according to the user's query. Intuitively, multi-attribute queries are conjunctions of single attributes and the correlation between them is usually strong. Therefore it is critical to learn from all attribute conjunctions jointly. Firstly, we propose to use the recurrent neural network to model complex attribute conjunctions. The model can not only reveal the representation of attribute conjunctions but also output the multi-attribute classifiers. Secondly, we integrate the generated classifiers into traditional logistic regression model. The parameters of recurrent neural network and logistic regression are optimized simultaneously using back propagation. We also propose a weighted version of our model to tackle data imbalance problem. Thirdly, we study the influence of attribute order in each query and propose two methods to further improve the original formulation.

3.1   Attribute Conjunction Learning

Let $\mathcal{Q} = \{\mathbf{Q}^1, \mathbf{Q}^2, ..., \mathbf{Q}^M\}$ be a set of M multi-attribute queries. The mth query is represented as a matrix $\mathbf{Q}^m = (\mathbf{q}_1^m, \mathbf{q}_2^m, ..., \mathbf{q}_{T_m}^m) \in \{0,1\}^{A \times T_m}$, where A is the number of predefined attributes and T_m is the number of attributes appearing in the mth query. $\mathbf{q}_t^m = [\mathbf{q}_{1t}^m, \ldots, \mathbf{q}_{At}^m]^T$ is a one-hot query vector, where $q_{at}^m = 1$ if the tth attribute in the current query is attribute a and $q_{at}^m = 0$ otherwise.

Our model takes multi-attribute query as input and outputs the representation of the query as well as the multi-attribute classifiers. More specifically, for the mth multi-attribute query, the one-hot query vectors $\mathbf{q}_t^m$ $(t = 1, \ldots, T_m)$ corresponding to the attributes involved in the query are input sequentially to our model. The subscript t decides the input order. We learn the multi-attribute conjunction in a recurrent way, as illustrated in Fig. 2. In this model, the first t attributes of the mth query can be represented as:

$$\begin{aligned}
\mathbf{h}_t^m &= f_h(\mathbf{q}_t^m, \mathbf{h}_{t-1}^m) \\
&= \sigma(\mathbf{W}_v \mathbf{q}_t^m + \mathbf{W}_h \mathbf{h}_{t-1}^m + \mathbf{b}_h),
\end{aligned} \tag{1}$$

where f_h is a *conjunction function* to model the relationship of all the attributes belonging to the mth query. $\mathbf{W}_v \in \mathbb{R}^{H \times A}$ and $\mathbf{W}_h \in \mathbb{R}^{H \times H}$ are *embedding* and *conjunction matrix* respectively, where H is the number of hidden units of the recurrent network. $\mathbf{h}_0^m \equiv \mathbf{h}_0$ represents the initial hidden state. $\mathbf{b}_h$ is the bias and $\sigma(\cdot)$ is an element-wise non-linear function which is chosen to be sigmoid in this paper.

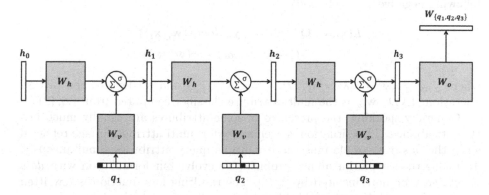

Fig. 2. An Illustration of Attribute Conjunction Recurrent Neural Network (AC-RNN) with triple attribute query.

From Eq. (1), we can see that each column of parameter matrix $\mathbf{W}_v$ can be considered as single attribute representation, noting that the query vector is in one-hot form. Therefore, all input queries, including long queries with many user specified attributes, share the same attribute-level representations. In this way, the parameter growth problem for long queries is addressed.

After computing $\mathbf{h}_{T_m}^m$ of the last query vector with the recurrent network, we actually obtain the hidden representation of the whole query. Then, we stack one layer on top of the recurrent units and compute the *multi-attribute classifier* $\mathbf{w}_m$ as the output of the neural network:

$$\mathbf{w}_m = f_o(\mathbf{h}_{T_m}^m) = \mathbf{W}_o \mathbf{h}_{T_m}^m + \mathbf{b}_o. \tag{2}$$

Here, the regression function f_o is chosen to be in a linear form, though more complex form can be considered. The parameter $\mathbf{W}_o$ and $\mathbf{b}_o$ are the output matrix and bias respectively. In this way, the attribute embeddings of the current query are combined in a recurrent way to learn the complex relationship between the attributes. After that we use the output conjunction as the mth multi-attribute classifier to retrieve relevant images. The model parameters of the conjunction and output functions are denoted as $\Theta = \{\mathbf{W}_v, \mathbf{W}_h, \mathbf{W}_o, \mathbf{b}_h, \mathbf{b}_o, \mathbf{h}_0\}$.

3.2 Multi-attribute Classification

Suppose there are N labeled images, $\{\mathbf{x}_i, \mathbf{y}_i\}_{i=1}^N$, where $\mathbf{x}_i \in \mathbb{R}^D$ denotes the D-dimensional image feature vector, and $\mathbf{y}_i \in \{0,1\}^A$ indicates the absence and presence of all attributes. The attribute label can be expressed in matrix form as $\mathbf{Y} = [\mathbf{y}_1, \mathbf{y}_2, ..., \mathbf{y}_N] \in \{0,1\}^{A \times N}$. In order to retrieve relevant images given a multi-attribute query $\mathbf{Q}^m$, we resort to multi-attribute classification to estimate the labels $\mathbf{Y}$.

Since attribute learning is a binary classification problem, we make use of logistic regression to predict the absence or presence of multiple attributes. The loss function with respect to the mth multi-attribute query is expressed as the following negative log likelihood:

$$L(\mathbf{x}_i, \mathbf{y}_i, \mathbf{Q}^m; \Theta) = -\tilde{\mathbf{y}}_{im} log(\sigma(\mathbf{w}_m^T \mathbf{x}_i)) \\ -(1 - \tilde{\mathbf{y}}_{im}) log(1 - \sigma(\mathbf{w}_m^T \mathbf{x}_i)), \tag{3}$$

where $\tilde{\mathbf{y}}_{im} = (\mathbf{y}_i^T \mathbf{q}_1^m \ \& \ \mathbf{y}_i^T \mathbf{q}_2^m \ \& \ \cdots \ \& \ \mathbf{y}_i^T \mathbf{q}_{T_m}^m)$ and $\&$ denotes the bitwise operation AND. $\mathbf{w}_m$ is the multi-attribute classifier computed from Eq. (2).

Generally speaking, the presences of some attributes are usually much less than its absence. This situation is even worse for multi-attribute image retrieval since the positive sample must have multiple query attributes simultaneously. To tackle the sample imbalance problem, we evolve our formulation with *data weighting* procedure inspired by [8,12]. The resulting loss function is rewritten as the following weighted log likelihood:

$$L_w(\mathbf{x}_i, \mathbf{y}_i, \mathbf{Q}^m; \Theta) = -c_m^+ \tilde{\mathbf{y}}_{im} log(\sigma(\mathbf{w}_m^T \mathbf{x}_i)) \\ -c_m^-(1 - \tilde{\mathbf{y}}_{im}) log(1 - \sigma(\mathbf{w}_m^T \mathbf{x}_i)), \tag{4}$$

where $c_m^+ = N/(2 \times N_m^+)$ and $c_m^- = N/(2 \times N_m^-)$ which make the loss weights of all the data sum up to N. N_m^+ (N_m^-) is the number of positive (negative) images for the mth multi-attribute query. The experimental results show that the weighted loss function performs better than the original logistic regression.

By combining the attribute conjunction and multi-attribute classification into a unified framework, the final objective function is formulated in the following form:

$$\underset{\Theta}{\arg\min} \ \frac{1}{N} \sum_{i=1}^N \sum_{m=1}^M L_*(x_i, y_i, Q^m; \Theta) + \lambda \Omega(\Theta), \tag{5}$$

where $\Omega(\cdot)$ is the weight decay term used to increase the model generalization ability. The parameters λ is used to balance the relative influence of the regularization terms. The loss function can also be replaced with the weighted form as defined in Eq. (4).

We solve the above optimization problem by using L-BFGS. The derivatives of the logistic regression parameters are calculated and back propagated into the output units of AC-RNN. Then the derivatives of Θ can be easily computed with the backpropagation through time algorithm [27]. In this way, our model can be trained in an end-to-end manner.

3.3   Attribute Order in AC-RNN

Recurrent neural networks are well suited to model sequential data. However, the input query attributes are not naturally organized as a sequence since the underlying conditional dependency between attributes are not known. The performance of our model is somewhat sensitive to the input order of attributes. To tackle this problem, we propose two methods by using recurrent attention mechanism and ensemble learning respectively.

Attention Mechanism Based. Attention mechanism has been successfully applied in generating image caption [28], handwriting recognition [7] and machine translation [2]. And it have been used to model the input and output structure of a sequence to sequence framework in a recent paper [25]. Inspired by the previous works, we propose to integrate the attention mechanism into our model to tackle the ordering problem. The pipeline is shown in Fig. 3. The proposed network reads all the attributes according to an attention vector, instead of processing query attributes one by one at each step. The attention vector is a probability vector indicating the relevance of all pre-defined attributes to the current query. And it is automatically modified at each processing step and recurrently contributes to the representation of the attribute conjunction. Intuitively, we first initialize the input attention vector for the mth query as:

$$\mathbf{p}_1^m = \frac{\sum_{i=1}^{T_m} \mathbf{q}_i^m}{T_m}. \tag{6}$$

In this way, the network will first take the query attributes into attention. Then we refined the attention vector and learn the attribute conjunction step by step using the recurrent neural network with attention mechanism. In the tth step, the attribute conjunction and attention vector are generated as follows:

$$\mathbf{h}_t^m = f_h(\mathbf{p}_t^m, \mathbf{h}_{t-1}^m) = \sigma(\mathbf{W}_v\mathbf{p}_t^m + \mathbf{W}_h\mathbf{h}_{t-1}^m + \mathbf{b}_h), \tag{7}$$

$$\mathbf{p}_{t+1}^m = Softmax(\mathbf{U}\mathbf{h}_t^m) = \frac{1}{\sum_{j=1}^{A} e^{\mathbf{U}_j^T\mathbf{h}_t^m}} \begin{bmatrix} e^{\mathbf{U}_1^T\mathbf{h}_t^m} \\ e^{\mathbf{U}_2^T\mathbf{h}_t^m} \\ \vdots \\ e^{\mathbf{U}_A^T\mathbf{h}_t^m} \end{bmatrix}, \tag{8}$$

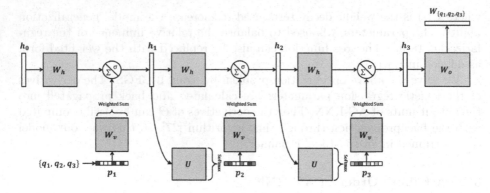

Fig. 3. AC-RNN with attention mechanism (AC-RNN-ATN).

where the *attention matrix* $\mathbf{U} \in \mathbb{R}^{H \times A}$ transforms the hidden units into the attention vector of the next processing step. The other parameters are consistent with the definition in Sect. 3.1.

By using a recurrent attention model, the output attribute conjunction is invariant to the input order. In addition, by taking the non-query attributes into consideration, this model can leverage the co-occurrence information to enhance the query attribute conjunction. Therefore, an unreliable query attribute might piggyback on an co-occurring attribute that has abundant training data and easier to predict.

Ensemble Based. Another method to alleviate the influence of attribute order is directly using the ensemble of original models. Therefore, we present Ensembles of AC-RNN to reduce the negative effect of poorly ordered input. The ensemble framework can be trained in parallel without increasing the overall time cost. Instead of using the pre-defined attribute order, a random order of all the attributes is generated for each model. Then the input attributes are rearranged according to the generated order for each multi-attribute query (Fig. 4).

The model parameters of each AC-RNN are learned to obtain a multi-attribute classifier. At the last stage, the outputs of all the independent models are averaged to obtain the final multi-attribute classifier:

$$\mathbf{w}_m = \frac{1}{C} \sum_{c=1}^{C} \mathbf{w}_m^c, \tag{9}$$

where C is the number of models in the ensemble and $\mathbf{w}_m^c$ represents the weight of multi-attribute classifier generated by the cth model. The ensemble of multi-attribute classifier is further used to retrieve the relevant images.

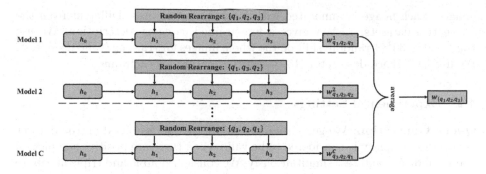

Fig. 4. AC-RNN based on ensemble learning.

4 Experiments

We evaluate our method[2] on three widely used datasets: aPascal [5], ImageNet Attributes [22] and LFWA [17]. Then we verify the effectiveness of the weighted version and visualize the ground truth correlation matrix and the learned similarity matrix for comparison. Finally, we present the experimental results of the proposed two methods to evaluate the influence of attribute order.

4.1 Datasets

aPASCAL. This dataset contains 6430 training images and 6355 testing images from Pascal VOC 2008 challenge. Each image comes from twenty object categories and annotated with 64 binary attribute labels. We use the pre-defined test images for testing and randomly split ten percent images from training set for validation. The feature we used for all the comparison methods are called DeCAF [4] which are extracted by the Convolutional Neural Networks (CNN). Since attributes are only defined for objects instead of the entire image, we use the object bounding box as the input of CNN.

ImageNet Attributes (INA). ImageNet Attribute dataset contains 9,600 images from 384 categories. Each image is annotated with 25 attributes describing color, patterns, shape and texture. 3–4 workers are asked to provide a binary label indicating whether the object in the image contains the attribute or not. When there is no consensus among the workers, the attribute will be labeled as ambiguous for this image. The data with ambiguous attribute are not used for training and evaluating for the queries which contains the corresponding attribute. We use $\{60\,\%, 10\,\%, 30\,\%\}$ random split for training/validation/test. And we also use DeCAF to do feature extraction.

LFWA. The labelled images on this dataset are selected from the widely used face dataset LFW [10]. It contains 5,749 identities with totally 13,233

---

[2] The code of our method is available at https://github.com/GriffinLiang/AC-RNN.

images. Each image is annotated with forty face attributes. Different from the above two datasets, LFWA gives a fine-grained category description. We use $\{60\%, 10\%, 30\%\}$ random split on the whole images for training/validation/test. We use VGG-Face descriptor [18] to extract feature for each image.

4.2 Experimental Settings

Query Generation: We generate multi-attribute queries based on the dataset annotation. A query is considered to be valid when there are positive samples on train/validation and test simultaneously. We consider double and triple attribute queries for comparison. The detail information is shown in Table 1.

Table 1. Valid multi-attribute queries

Dataset	# of attributes	Double queries	Triple queries
aPascal	64	546	2224
INA	25	186	262
LFWA	40	771	9126

Evaluation Metric: We use the AUC (Area Under ROC) and AP (Average Precision) as the evaluation metric for each query. Since the number of attribute conjunctions is large, the resulting performance is hard to visualize for comparison. Therefore, we choose to use the mean AUC and mean AP to reflect the average performance of all the methods.

Comparison Methods: We compare our approach with four baseline methods: TagProp [8], RMLL [19], MARR [23] and LIFT [29]. For TagProp, we use the logistic discriminant model with distance-based weights and the number of K nearest neighbours is chosen on the validation set. For RMLL and MARR, the loss function to be optimized is Hamming loss which is also used in the original papers. Since TagProp, RMLL and LIFT do not support for multi-attribute query directly, we sum up the single attribute prediction scores as the confidence of multi-attribute query following the suggestion in [23]. The ratio parameter r of LIFT is set to be 0.1 as suggested in the paper. For our methods, the original version and the variant based on attention mechanism are presented. The optimal value of λ and the number of hidden units are chosen based on validation set by grid search.

4.3 Comparison on Image Retrieval

We calculate the mean AUC and mean AP of double and triple attribute queries for all the comparison methods. The rank for all the methods is also given for each dataset in terms of a single evaluation metric. Then we average the ranks on all the three datasets to demonstrate the overall performance. The experimental

Table 2. Experimental results (mAUC/mAP rank) for double attribute query.

Data set	Evaluate metric	TapProp		RMLL		MARR		LIFT		AC-RNN		AC-RNN-ATN	
aPascal	mAUC	0.8807	5	0.8876	4	0.9040	3	0.8797	6	0.9356	2	**0.9371**	**1**
	mAP	0.3361	4	0.3274	6	0.3336	5	0.3383	3	0.3758	2	**0.3869**	**1**
INA	mAUC	0.8832	6	0.9166	3	0.8945	4	0.8902	5	0.9436	2	**0.9450**	**1**
	mAP	0.2269	3	0.2126	4	0.1780	6	0.1953	5	**0.2794**	**1**	0.2605	2
LFWA	mAUC	0.8113	6	0.8293	3	0.8210	4	0.8205	5	0.8482	2	**0.8549**	**1**
	mAP	0.4075	6	0.4097	5	0.4209	4	**0.4372**	**1**	0.4223	3	0.4370	2
Avg.Rank		5.00	6	4.17	3	4.33	5	4.17	3	2.00	2	1.33	1
Total	**AC-RNN-ATN $\succ$ AC-RNN $\succ$ LIFT = RMLL $\succ$ MARR $\succ$ TagProp**												

Table 3. Experimental results (mAUC/mAP rank) for triple attribute query.

Data set	Evaluate metric	TapProp		RMLL		MARR		LIFT		AC-RNN		AC-RNN-ATN	
aPascal	mAUC	0.8921	5	0.8988	4	0.9139	3	0.8910	6	0.9336	2	**0.9360**	**1**
	mAP	0.2723	3	0.2497	6	0.2582	5	0.2640	4	0.2828	2	**0.3034**	**1**
INA	mAUC	0.8927	6	0.9539	3	0.9375	4	0.9163	5	0.9627	2	**0.9677**	**1**
	mAP	0.2001	3	0.1829	4	0.1375	6	0.1521	5	**0.2743**	**1**	0.2726	2
LFWA	mAUC	0.8177	6	0.8367	3	0.8284	4	0.8247	5	0.8594	2	**0.8665**	**1**
	mAP	0.2273	5	0.2218	6	0.2355	4	0.2473	2	0.2372	3	**0.2499**	**1**
Avg.Rank		4.67	6	4.33	3	4.33	3	4.50	5	2.00	2	1.17	1
Total	**AC-RNN-ATN $\succ$ AC-RNN $\succ$ RMLL = MARR $\succ$ LIFT $\succ$ TagProp**												

results are shown in Tables 2 and 3 for double and triple attribute queries respectively. Comparing the results of RMLL and MARR, we can see that MARR surpasses RMLL on different types of queries on aPascal and LFWA but fails on INA dataset. This is because the number of attribute on INA is too small and the correlation between them is not as strong as aPascal and LFWA. So the performance of MARR decreases since this method relies on strong attribute correlation. On the three datasets, we can see that our methods AC-RNN and AC-RNN-ATN achieve better performance on all types of multi-attribute queries. Therefore recurrent neural network is beneficial for modelling the complex relationship of multiple attributes. Compared to the original method, AC-RNN-ATN can leverage non-query attributes to enhance the retrieval performance and avoid the order problem by using a recurrent attention vector. It surpasses AC-RNN on both double and triple attribute queries according to the final rank and its optimal processing step is two.

Data Weighting Procedure. Images possessing all the query attributes are considered to be positive for training. Therefore, the positive samples are usually scarce when a query contains multiple attributes. To explicitly show this phenomenon, we calculate the positive sample ratio (N_m^+/N) for the queries in which the number of attributes ranges from one to three. For each type of the queries, we partition them into five parts according to the positive sample ratio and calculate the corresponding proportion for each part. The results are shown

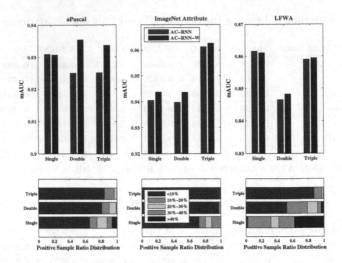

Fig. 5. Comparison on the influence of data weighting on AC-RNN. The second row shows the positive sample ratio distribution on the three datasets.

in the second row of Fig. 5. On all the three datasets, most of double and triple queries contain less than 10 % of the total data as positive samples. Therefore, how to solve data imbalance problem is essential for multi-attribute query based image retrieval. Then we train the proposed models by using the loss functions defined in Eqs. (3) and (4) respectively. The performance of the two versions with and without using data weighting for AC-RNN are shown in the first row of Fig. 5. From the results, we can see the method using data weighting procedure consistently performs better than the original version when the positive data is imbalance.

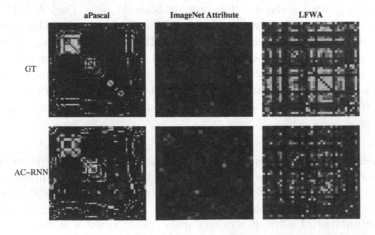

Fig. 6. Attribute similarity matrix on the embedding space.

Attribute Embedding. In this section, we validate the quality of the learned embedding matrix. We first calculate the ground truth correlation matrix which can reflect the correlation information between attributes. Let $\mathbf{R} \in \mathbb{R}^{A \times A}$ be the correlation matrix, where the correlation score between attribute i and j is computed following [24]:

$$R_{i,j} = \frac{\mathbf{Y}_{i,:}^T \mathbf{Y}_{j,:}}{\mathbf{Y}_{i,:}^T \mathbf{1} + \mathbf{Y}_{j,:}^T \mathbf{1} - \mathbf{Y}_{i,:}^T \mathbf{Y}_{j,:}}. \tag{10}$$

From the definition, we can see that two attributes are strongly correlated if they have a large number of images in common. Intuitively, the attribute embedding learned by AC-RNN is expected to reflect the correlation between attributes. So we visualize the similarity matrix of the learned attribute embeddings on all the three datasets in Fig. 6. The similarity score for a pair of attributes is calculated by using their cosine distance. Comparing the ground truth correlation matrix and the learned similarity matrix, we can see most of the correlated attributes are close on the embedding space.

4.4 Influence on Attribute Order

Attention Mechanism Based. We validate the effectiveness of learning attribute conjunction with a recurrent attention vector in this section. A promising perspective of AC-RNN-ATN is the learned conjunction is order invariant. So the dependence of attributes is no longer needed but learned in an automatic way. Moreover, the non-query attributes which are correlated with the current query are also used to generate the final representation of attribute conjunction.

The performance on the three datasets are shown in Tables 2 and 3 for double and triple attribute queries respectively. AC-RNN-ATN achieves superior performance on the both evaluation metrics. The performance gap between AC-RNN and AC-RNN-ATN is larger on aPascal and LFWA than on INA. This is because the number of attributes on INA is smaller than the other two datasets. Therefore, the correlation between query and non-query attributes is hard to exploit. Then we visualize some of the attention vectors learned by our method for both double and triple attribute queries on LFWA dataset. As shown in Fig. 7, the query contains "Chubby" will also take "Double Chin" into attention to generate the representation of attribute conjunction and an "Attractive" person with "Bushy Eyebrows" is probably "Male".

Ensemble Based. As discussion in the Sect. 3.3, attribute order for generating attribute conjunction influences the performance of AC-RNN while the optimal query order is hard to explore. Here we use an ensemble method to tackle the above problem. In detail, we repeatedly train AC-RNN for ten times on aPascal dataset. At the start of each training stage, we randomly change the attribute order in the queries instead of using the attribute order predefined by the dataset. The output matrix of each random model are combined to generate the final result for the multi-attribute query.

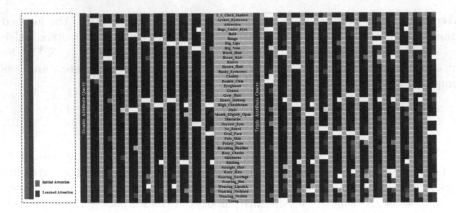

Fig. 7. Attention vector visualization on LFWA dataset.

From the results in Fig. 8, we can see the ensemble method consistently outperforms the best single model on two types of attribute queries. And the performance of ensemble based method increases rapidly at first and then becomes flat. Combining weak models may decrease the overall performance. Comparing the two methods for tackling attribute order problem, we find the ensemble based method performs better in terms of mAUC but inferior to AC-RNN-ATN according to mAP. Since our problem suffering from data imbalance, using an evaluation metric of Precision-Recall is better than Receiver-Operating-Characteristic AUC as mention in [3]. Therefore, the attention based method seems more promising.

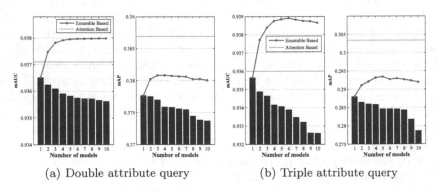

(a) Double attribute query   (b) Triple attribute query

Fig. 8. Performance of ensemble based method in terms of mAUC and mAP on aPascal dataset. The blue bar indicates the performance of single model. (Color figure online)

5   Conclusion

We propose the attribute conjunction recurrent neural network for multi-attribute based image retrieval. Different from previous methods, our model explicitly learns the attribute embedding and generates the representation of attribute conjunction by recurrently combining the learned attribute embeddings. In addition, we propose a variant of our method using data weighting to mitigate the data imbalance problem. Finally, we have a discussion about the influence of attribute order on our method and present two methods to boost the performance based on attention mechanism and ensemble learning respectively. Experimental results on three widely used datasets show the significant improvement over the other comparison methods on all types of queries.

Acknowledgments. This work is partially supported by 973 Program under contract No. 2015CB351802, and Natural Science Foundation of China under contracts Nos. 61390515, 61272319 and 61272321.

References

1. Akata, Z., Perronnin, F., Harchaoui, Z., Schmid, C.: Label-embedding for attribute-based classification. In: 2013 IEEE Conference on Computer Vision and Pattern Recognition (CVPR), pp. 819–826. IEEE (2013)
2. Bahdanau, D., Cho, K., Bengio, Y.: Neural machine translation by jointly learning to align and translate. In: International Conference on Learning Representations (2015)
3. Davis, J., Goadrich, M.: The relationship between precision-recall and roc curves. In: Proceedings of the 23rd International Conference on Machine Learning, pp. 233–240. ACM (2006)
4. Donahue, J., Jia, Y., Vinyals, O., Hoffman, J., Zhang, N., Tzeng, E., Darrell, T.: Decaf: a deep convolutional activation feature for generic visual recognition. In: Proceedings of the 31st International Conference on Machine Learning, pp. 647–655 (2014)
5. Farhadi, A., Endres, I., Hoiem, D., Forsyth, D.: Describing objects by their attributes. In: IEEE Conference on Computer Vision and Pattern Recognition, CVPR 2009, pp. 1778–1785. IEEE (2009)
6. Frome, A., Corrado, G.S., Shlens, J., Bengio, S., Dean, J., Mikolov, T., et al.: Devise: A deep visual-semantic embedding model. In: Advances in Neural Information Processing Systems, pp. 2121–2129 (2013)
7. Graves, A.: Supervised Sequence Labelling. Springer (2012)
8. Guillaumin, M., Mensink, T., Verbeek, J., Schmid, C.: Tagprop: discriminative metric learning in nearest neighbor models for image auto-annotation. In: 2009 IEEE 12th International Conference on Computer Vision, pp. 309–316. IEEE (2009)
9. Hardoon, D.R., Szedmak, S., Shawe-Taylor, J.: Canonical correlation analysis: an overview with application to learning methods. Neural Comput. 16(12), 2639–2664 (2004)
10. Huang, G.B., Ramesh, M., Berg, T., Learned-Miller, E.: Labeled faces in the wild: A database for studying face recognition in unconstrained environments. Technical report
11. Hwang, S.J., Sigal, L.: A unified semantic embedding: relating taxonomies and attributes. In: Advances in Neural Information Processing Systems, pp. 271–279 (2014)

12. King, G., Zeng, L.: Logistic regression in rare events data. Polit. Anal. **9**(2), 137–163 (2001)
13. Kovashka, A., Parikh, D., Grauman, K.: Whittlesearch: Image search with relative attribute feedback. In: 2012 IEEE Conference on Computer Vision and Pattern Recognition (CVPR), pp. 2973–2980. IEEE (2012)
14. Kumar, N., Belhumeur, P., Nayar, S.: FaceTracer: a search engine for large collections of images with faces. In: Forsyth, D., Torr, P., Zisserman, A. (eds.) ECCV 2008. LNCS, vol. 5305, pp. 340–353. Springer, Heidelberg (2008). doi:10.1007/978-3-540-88693-8_25
15. Kumar, N., Berg, A.C., Belhumeur, P.N., Nayar, S.K.: Attribute and simile classifiers for face verification. In: 2009 IEEE 12th International Conference on Computer Vision, pp. 365–372. IEEE (2009)
16. Lampert, C.H., Nickisch, H., Harmeling, S.: Attribute-based classification for zero-shot visual object categorization. IEEE Trans. Pattern Anal. Mach. Intell. **36**(3), 453–465 (2014)
17. Liu, Z., Luo, P., Wang, X., Tang, X.: Deep learning face attributes in the wild. In: Proceedings of the IEEE International Conference on Computer Vision, pp. 3730–3738 (2015)
18. Parkhi, O.M., Vedaldi, A., Zisserman, A.: Deep face recognition. In: British Machine Vision Conference (2015)
19. Petterson, J., Caetano, T.S.: Reverse multi-label learning. In: Advances in Neural Information Processing Systems, pp. 1912–1920 (2010)
20. Rastegari, M., Diba, A., Parikh, D., Farhadi, A.: Multi-attribute queries: to merge or not to merge? In: 2013 IEEE Conference on Computer Vision and Pattern Recognition (CVPR), pp. 3310–3317. IEEE (2013)
21. Read, J., Pfahringer, B., Holmes, G., Frank, E.: Classifier chains for multi-label classification. Mach. Learn. **85**(3), 333–359 (2011)
22. Russakovsky, O., Fei-Fei, L.: Attribute learning in large-scale datasets. In: Kutulakos, K.N. (ed.) ECCV 2010. LNCS, vol. 6553, pp. 1–14. Springer, Heidelberg (2012). doi:10.1007/978-3-642-35749-7_1
23. Siddiquie, B., Feris, R.S., Davis, L.S.: Image ranking and retrieval based on multi-attribute queries. In: 2011 IEEE Conference on Computer Vision and Pattern Recognition (CVPR), pp. 801–808. IEEE (2011)
24. Sigurbjörnsson, B., Van Zwol, R.: Flickr tag recommendation based on collective knowledge. In: Proceedings of the 17th International Conference on World Wide Web, pp. 327–336. ACM (2008)
25. Vinyals, O., Bengio, S., Kudlur, M.: Order matters: Sequence to sequence for sets. arXiv preprint arXiv:1511.06391 (2015)
26. Wang, Y., Mori, G.: A discriminative latent model of object classes and attributes. In: Daniilidis, K., Maragos, P., Paragios, N. (eds.) ECCV 2010. LNCS, vol. 6315, pp. 155–168. Springer, Heidelberg (2010). doi:10.1007/978-3-642-15555-0_12
27. Werbos, P.: Backpropagation through time: what it does and how to do it. Proc. IEEE **78**(10), 1550–1560 (1990)
28. Xu, K., Ba, J., Kiros, R., Cho, K., Courville, A., Salakhudinov, R., Zemel, R., Bengio, Y.: Show, attend and tell: neural image caption generation with visual attention. In: Proceedings of the 32nd International Conference on Machine Learning, pp. 2048–2057 (2015)
29. Zhang, M.L., Wu, L.: Lift: Multi-label learning with label-specific features. IEEE Trans. Pattern Anal. Mach. Intell. **37**(1), 107–120 (2015)

Efficient Discovery of Sets of Co-occurring Items in Event Sequences

Boris Cule[1(✉)], Len Feremans[1], and Bart Goethals[1,2]

[1] University of Antwerp, Antwerp, Belgium
{boris.cule,len.feremans}@uantwerpen.be
[2] Monash University, Melbourne, Australia

Abstract. Discovering patterns in long event sequences is an important data mining task. Most existing work focuses on frequency-based quality measures that allow algorithms to use the anti-monotonicity property to prune the search space and efficiently discover the most frequent patterns. In this work, we step away from such measures, and evaluate patterns using cohesion—a measure of how close to each other the items making up the pattern appear in the sequence on average. We tackle the fact that cohesion is not an anti-monotonic measure by developing a novel pruning technique in order to reduce the search space. By doing so, we are able to efficiently unearth rare, but strongly cohesive, patterns that existing methods often fail to discover. The data and software related to this paper are available at https://bitbucket.org/len_feremans/sequencepatternmining_public.

1 Introduction

Pattern discovery in sequential data is a well-established field in data mining. The earliest attempts focused on the setting where data consisted of many (typically short) sequences, where a pattern was defined as a (sub)sequence that re-occurred in a high enough number of such input sequences [2].

The first attempt to identify patterns in a single long sequence of data was proposed by Mannila et al. [8]. The presented WINEPI method uses a sliding window of a fixed length to traverse the sequence, and a pattern is then considered frequent if it occurs in a high enough number of these sliding windows. An often-encountered critique of this method is that the obtained frequency is not an intuitive measure, since it does not correspond to the actual number of occurrences of the pattern in the sequence. For example, given sequence $axbcdayb$, and a sliding window length of 3, the frequency of itemset $\{a, b\}$ will be equal to 2, as will the frequency of itemset $\{c, d\}$. However, pattern $\{a, b\}$ occurs twice in the sequence, and pattern $\{c, d\}$ just once, and while the method is motivated by the need to reward c and d for occurring right next to each other, the reported frequency values remain difficult to interpret.

Laxman et al. [7] attempted to tackle this issue by defining the frequency as the maximal number of non-intersecting minimal windows of the pattern in the sequence. In this context, a minimal window of the pattern in the sequence

© Springer International Publishing AG 2016
P. Frasconi et al. (Eds.): ECML PKDD 2016, Part I, LNAI 9851, pp. 361–377, 2016.
DOI: 10.1007/978-3-319-46128-1_23

is defined as a subsequence of the input sequence that contains the pattern, such that no smaller subsequence also contains the pattern. However, while the method uses a relevance window of a fixed length, and disregards all minimal windows that are longer than the relevance window, the length of the minimal windows that do fit into the relevance window is not taken into account at all. For example, given sequence $axyzbcd$, with a relevance window larger than 4, the frequency of both itemset $\{a, b\}$ and itemset $\{c, d\}$ would be equal to 1.

Cule et al. [4] propose an amalgam of the two approaches, defining the frequency of a pattern as the maximal sum of weights of a set of non-overlapping minimal windows of the pattern, where the weight of a window is defined as the inverse of its length. However, this method, too, struggles with the interpretability of the proposed measure. For example, given sequence $axbcdayb$ and a relevance window larger than 3, frequency of $\{a, b\}$ would be 2/3, while frequency of $\{c, d\}$ would be 1/2. On top of this, as the input sequence grows longer, the sum of these weights will grow, and the defined frequency can take any real positive value, giving the user no idea how to set a sensible frequency threshold.

All of the techniques mentioned above use a frequency measure that satisfy the so-called APRIORI property [1]. This property implies that the frequency of a pattern is never smaller than the frequency of any of its superpatterns (in other words, frequency is an *anti-monotonic* quality measure). While this property is computationally very desirable, since large candidate patterns can be generated from smaller patterns, and generating unnecessary candidates can be avoided, the undesirable side-effect is that larger patterns, which are often more useful to the end users, will never be ranked higher than all their subpatterns. On top of this, all these methods focus solely on how often certain items occur near each other, and do not take occurrences of these items far away from each other into account. Consequently, if two items occur frequently, and through pure randomness often occur near each other, they will form a frequent itemset, even though they are, in fact, in no way correlated.

In another work, Cule et al. [3] propose a method that steps away from anti-monotonic quality measures, and introduce a new interestingness measure that combines the coverage of the pattern with its cohesion. Cohesion is defined as a measure of how near each other the items making up an interesting itemset occur on average. However, the authors define the coverage of an itemset as the sum of frequencies of all items making up the itemset, which results in a massive bias towards larger patterns instead. Furthermore, this allows for a very infrequent item making its way into an interesting itemset, as long as all other items in the itemset are very frequent and often occur near the infrequent item. As a result, the method is not scalable for any sequence with a large alphabet of items, which makes it unusable in most realistic data sets.

Hendrickx et al. [6] tackle a related problem in an entirely different setting. Given a graph consisting of labelled nodes, they attempt to discover which labels often co-occur. In this context, they aim to discover cohesive itemsets (sets of labels), by computing average distances between the labels, where the distance between two nodes is defined as the length of the shortest path between them (expressed as the number of edges on this path). While the authors also experimented with sequential data, after first converting an input sequence into a graph, by converting each event into a node labelled by the event type, and connecting neighbouring events by an edge, this approach is not entirely suitable for sequential data. More precisely, in a graph setting, an itemset can only be considered fully cohesive if all its occurrences form a clique in the graph. Clearly, in a sequence, for any itemset of size larger than 2, it would be impossible to form a clique, since each node (apart from the first one and the last one) has exactly two edges – one connecting the node to the event that occurred last before the event itself, and the other connecting it to the event that occurred first after the event itself. For example, given sequence $abcd$ (converted into a graph), the cohesion of itemset $\{a, b\}$ would be equal to 1, the cohesion of $\{a, b, c\}$ would be $3/4$, and the cohesion of $\{a, b, c, d\}$ would be $3/5$ (we omit the computational details here), which is clearly not intuitive.

In this work, we use the cohesion introduced by Cule et al. [3] as a single measure to evaluate cohesive itemsets. We consider itemsets as potential candidates only if each individual item contained in the itemset is frequent in the dataset. This allows us to filter out the infrequent items at the very start of our algorithm, without missing out on any cohesive itemsets. However, using cohesion as a single measure brings its own computational problems. First of all, cohesion is not an anti-monotonic measure, which means that a superset of a non-cohesive itemset could still prove to be cohesive. However, since the size of the search space is exponential in the number of frequent items, it is impossible to evaluate all possible itemsets. We solve this by developing a tight upper bound on the maximal possible cohesion of all itemsets that can still be generated in a particular branch of the depth-first-search tree. This bound allows us to prune large numbers of potential candidate itemsets, without having to evaluate them at all. Furthermore, we present a very efficient method to identify minimal windows that contain a particular itemset, necessary to evaluate its cohesion. Our experiments show that our method discovers patterns that existing methods struggle to rank highly, while dismissing obvious patterns consisting of items that occur frequently, but are not at all correlated. We further show that we achieve these results quickly, thus demonstrating the efficiency of our algorithm.

The rest of the paper is organised as follows. In Sect. 2 we formally describe the problem setting and define the patterns we aim to discover. Section 3 provides a detailed description of our algorithm, while in Sect. 4 we present a thorough experimental evaluation of our method, in comparison with a number of existing methods. We present an overview of the most relevant related work in Sect. 5, before summarising our main conclusions in Sect. 6.

2   Problem Setting

The dataset consists of a single event sequence $s = (e_1, \ldots, e_n)$. Each event e_k is represented by a pair (i_k, t_k), with i_k an event type (coming from the domain of all possible event types) and t_k an integer time stamp. For any $1 < k \leq n$, it holds that $t_k > t_{k-1}$. For simplicity, we omit the time stamps from our examples, and write sequence $(e_1, \ldots, e_n)$ as $i_1 \ldots i_n$, implicitly assuming the time stamps are consecutive integers starting with 1. In further text, we refer to event types as *items*, and sets of event types as *itemsets*.

For an itemset $X = \{i_1, \ldots, i_m\}$, we denote the set of occurrences of items making up X in a sequence s with $N(X) = \{t | (i,t) \in s, i \in X\}$. For an item i, we define the *support* of i in an input sequence s as the number of occurrences of i in s, $sup(i) = |N(\{i\})|$. Given a user-defined support threshold min_sup, we say that an itemset X is *frequent* in a sequence s if for each $i \in X$ it holds that $sup(i) \geq min\_sup$.

To evaluate the cohesiveness of an itemset X in a sequence s, we must first identify minimal occurrences of the itemset in the sequence. For each occurrence of an item in X, we will look for the minimal window within s that contains that occurrence and the entire itemset X. Formally, given a time stamp t, such that $(i,t) \in s$ and $i \in X$, we define the *size of the minimal occurrence of X around t* as $W_t(X) = \min\{t_e - t_s + 1 | t_s \leq t \leq t_e \text{ and } \forall i \in X \exists (i,t') \in s, t_s \leq t' \leq t_e\}$.

We further define the *size of the average minimal occurrence of X in s* as

$$\overline{W}(X) = \frac{\sum_{t \in N(X)} W_t(X)}{|N(X)|}.$$

Finally, we define the *cohesion* of itemset X, with $|X| > 0$, in a sequence s as $C(X) = \frac{|X|}{\overline{W}(X)}$. If $|X| = 0$, we define $C(X) = 1$.

Given a user-defined cohesion threshold min_coh, we say that an itemset X is *cohesive* in a sequence s if it holds that $C(X) \geq min\_coh$.

Note that the cohesion is higher if the minimal occurrences are smaller. Furthermore, a minimal occurrence of itemset X can never be smaller than the size of X, so it holds that $C(X) \leq 1$. If $C(X) = 1$, then every single minimal occurrence of X in s is of length $|X|$.

A single item is always cohesive, so to avoid outputting all frequent items, we will from now on consider only itemsets consisting of 2 or more items. An optional parameter, *max_size*, can be used to limit the size of the discovered patterns. Formally, we say that an itemset X is a *frequent cohesive itemset* if $1 < |X| \leq max\_size$, $\forall i \in X : sup(i) \geq min\_sup$ and $C(X) \geq min\_coh$.

Cohesion is not an anti-monotonic measure. A superset of a non-cohesive itemset could turn out to be cohesive. For example, given sequence $abcxacbybac$, we can see that $C(\{a,b\}) = C(\{a,c\}) = C(\{b,c\}) = 6/7$, while $C(\{a,b,c\}) = 1$. While this allows us to eliminate bias towards smaller patterns, it also brings computational challenges which will be addressed in the following section.

3   Algorithm

In this section we present a detailed description of our algorithm. We first show how we generate candidates in a depth-first manner, before explaining how we can prune large numbers of potential candidates by computing an upper bound

---

Algorithm 1. FCI_{SEQ} finds frequent cohesive itemsets in a sequence

---

1 $FI = $ all frequent items;
2 sort FI on support in ascending order;
3 $FC = \emptyset$;
4 $DFS(\langle \emptyset, FI \rangle)$;
5 **return** FC

Algorithm 2. $DFS(\langle X, Y \rangle)$ depth-first search

---

1 **if** $C_{max}(X, Y) \geq min\_coh$ **then**
2   **if** $Y = \emptyset$ **then**
3   **if** $|X| > 1$ **then**
4             $FC = FC \cup \{X\}$;
5   **else**
6          $a = first(Y)$;
7   **if** $|X \cup \{a\}| \leq max\_size$ **then**
8             $DFS(\langle X \cup \{a\}, Y \setminus \{a\} \rangle)$;
9          $DFS(\langle X, Y \setminus \{a\} \rangle)$;

---

of the cohesion of all itemsets that can be generated within a branch of the search tree, and we end the section by providing an efficient method to compute the sum of minimal windows of a particular itemset in the input sequence.

3.1 Depth-First-Search

The main routine of our FCI_{SEQ} algorithm is given in Algorithm 1. We begin by scanning the input sequence, identifying the frequent items, and storing their occurrence lists for later use. We then sort the set of frequent items on support in ascending order (line 2), initialise the set of frequent cohesive itemsets FC as an empty set (line 3), and start the depth-first-search process (line 4). Once the search is finished, we output the set of frequent cohesive itemsets FC (line 5).

The recursive DFS procedure is shown in Algorithm 2. In each call, X contains the candidate itemset, while Y contains items that are yet to be enumerated. In line 1, we evaluate the pruning function $C_{max}(X, Y)$ to decide whether to search deeper in the tree or not. This function will be described in detail in Sect. 3.2. If the branch is not pruned, there are two possibilities. If we have reached a leaf node (line 2), we add the discovered cohesive itemset to the output (provided its size is greater than 1). Alternatively, if there are more items to be enumerated, we pick the first such item a (line 6) and make two recursive calls to the DFS function—the first with a added to X (this is only executed if $X \cup \{a\}$ satisfies the max_size constraint), and the second with a discarded.

3.2 Pruning

At any node in the search tree, X denotes all items currently making up the candidate itemset, while Y denotes all items that are yet to be enumerated. Starting from such a node, we can still generate any itemset Z, such that $X \subseteq Z \subseteq X \cup Y$ and $|Z| \leq max\_size$. In order to be able to prune the entire branch of the search tree, we must therefore be certain that for every such Z, the cohesion of Z cannot satisfy the minimum cohesion constraint.

In the remainder of this section, we first define an upper bound for the cohesion of all itemsets that can be generated in a particular branch of the search tree, before providing a detailed proof of its soundness. Given itemsets X and Y, with $|X| > 0$ and $X \cap Y = \emptyset$, the $C_{max}(X, Y)$ pruning function used in line 1 of Algorithm 2 is defined as

$$C_{max}(X,Y) = \frac{\min(max\_size, |X \cup Y|)(|N(X)| + N_{max}(X,Y))}{\sum_{t \in N(X)} W_t(X) + \min(max\_size, |X \cup Y|)N_{max}(X,Y)},$$

where

$$N_{max}(X,Y) = \max_{\substack{Y_i \subseteq Y, \\ |Y_i| \leq max\_size - |X|}} |N(Y_i)|.$$

For $|X| = 0$, we define $C_{max}(X, Y) = 1$.

Note that if $Y = \emptyset$, $C_{max}(X, Y) = C(X)$, which is why we do not need to evaluate $C(X)$ before outputting X in line 4 of Algorithm 2.

Before proving that the above upper bound holds, we will first explain the intuition behind it. When we find ourselves at node $\langle X, Y \rangle$ of the search tree, we will first evaluate the cohesion of itemset X. If X is cohesive, we need to search deeper in the tree, as supersets of X could also be cohesive. However, if X is not cohesive, we need to evaluate how much the cohesion can still grow if we go deeper into this branch of the search tree. Logically, starting from $C(X) = \frac{|X|}{\overline{W}(X)} = \frac{|X||N(X)|}{\sum_{t \in N(X)} W_t(X)}$, the value of this fraction will grow maximally if the nominator is maximised, and the denominator minimised. Clearly, as we add items to X, the nominator will grow, and it will grow maximally if we add as many items to X as possible. However, as we add items to X, the denominator must grow, too, so the question is how it can grow minimally. In the worst case, each new window added to the sum in the denominator will be minimal (i.e., its length will be equal to the size of the new itemset), and the more such windows we add to the sum, the higher the overall cohesion will grow.

For example, given sequence acb and a cohesion threshold of 0.8, assume we find ourselves in node $\langle \{a, b\}, \{c\} \rangle$ of the search tree. We will then first find the smallest windows containing $\{a, b\}$ for each occurrence of a and b, i.e., $W_1(\{a, b\}) = W_3(\{a, b\}) = 3$. It turns out that $C(\{a, b\}) = \frac{2 \times 2}{3 + 3} = \frac{2}{3}$, which is not cohesive enough. However, if we add c to itemset $\{a, b\}$, we know that the size of the new itemset will be 3, we know the number of occurrences of items from the new itemset will be 3, and the nominator will therefore be equal to 9.

For the denominator, we have no such certainties, but we know that, in the worst case, the windows for the occurrences of a and b will not grow (i.e., each smallest window of $\{a,b\}$ will already contain an occurrence of c), and the windows for all occurrences of c will be minimal (i.e., of size 3). Indeed, when we evaluate the above upper bound, we obtain $C_{max}(\{a,b\},\{c\}) = \frac{3 \times (2+1)}{6+3 \times 1} = \frac{9}{9} = 1$. We see that even though the cohesion of $\{a,b\}$ is $\frac{2}{3}$, the cohesion of $\{a,b,c\}$ could, in the worst case, be as high as 1. And in our sequence acb, that is indeed the case. The above example also demonstrates the tightness of our upper bound, as the computed value can, in fact, turn out to be equal to the actual cohesion of a superset yet to be generated.

We now present a full formal proof of the soundness of the proposed upper bound. In order to do this, we will need the following lemma.

Lemma 1. *For any six positive numbers a, b, c, d, e, f, with $a \leq b$, $c \leq d$ and $e \leq f$, it holds that*

1. *if $\frac{a+c+e}{b+e} < 1$ then $\frac{a+c+e}{b+e} \leq \frac{a+d+f}{b+f}$.*
2. *if $\frac{a+c+e}{b+e} \geq 1$ then $\frac{a+d+f}{b+f} \geq 1$.*

Proof. We begin by proving the first claim. To start with, note that if $\frac{a+c+e}{b+e} < 1$, then $\frac{a+c}{b} < 1$. For any positive number f with $e \leq f$, it therefore follows that $\frac{a+c+e}{b+e} \leq \frac{a+c+f}{b+f}$. Finally, for any positive number d, with $c \leq d$, it holds that $\frac{a+c+f}{b+f} \leq \frac{a+d+f}{b+f}$, and therefore $\frac{a+c+e}{b+e} \leq \frac{a+d+f}{b+f}$. For the second claim, it directly follows that if $\frac{a+c+e}{b+e} \geq 1$, then $\frac{a+c}{b} \geq 1$, $\frac{a+d}{b} \geq 1$, and $\frac{a+d+f}{b+f} \geq 1$. □

Theorem 1. *Given itemsets X and Y, with $X \cap Y = \emptyset$, for any itemset Z, with $X \subseteq Z \subseteq X \cup Y$ and $|Z| \leq max\_size$, it holds that $C(Z) \leq C_{max}(X,Y)$.*

Proof. We know that $C(Z) \leq 1$, so the theorem holds if $|X| = 0$. Assume now that $|X| > 0$. First recall that $C(Z) = \frac{|Z|}{\overline{W}(Z)} = \frac{|Z||N(Z)|}{\sum_{t \in N(Z)} W_t(Z)}$. We can rewrite this expression as

$$C(Z) = \frac{(|X| + |Z \setminus X|)(|N(X)| + |N(Z \setminus X)|)}{\sum_{t \in N(X)} W_t(Z) + \sum_{t \in N(Z \setminus X)} W_t(Z)}.$$

Further note that for a given time stamp in $N(X)$, the minimal window containing Z must be at least as large as the minimal window containing only X, and for a given time stamp in $N(Z \setminus X)$, the minimal window containing Z must be at least as large as the size of Z. It therefore follows that

$$\sum_{t \in N(X)} W_t(Z) \geq \sum_{t \in N(X)} W_t(X) \text{ and } \sum_{t \in N(Z \setminus X)} W_t(Z) \geq |Z||N(Z \setminus X)|,$$

and, as a result,

$$C(Z) \leq \frac{|X||N(X)| + |Z\backslash X||N(X)| + |Z||N(Z\backslash X)|}{\sum_{t\in N(X)} W_t(X) + |Z||N(Z\backslash X)|}.$$

Finally, we note that, per definition, $|Z\backslash X| \leq \min(max\_size, |X\cup Y|) - |X|$, and, since Z is generated by adding items from Y to X, until either $|Z| = max\_size$ or there are no more items left in Y, $|N(Z\backslash X)| \leq N_{max}(X,Y)$.

At this point we will use Lemma 1 to take the proof further. Note that, per definition, $C(X) = \frac{|X||N(X)|}{\sum_{t\in N(X)} W_t(X)} \leq 1$. We now denote

$$a = |X||N(X)| \text{ and } b = \sum_{t\in N(X)} W_t(X).$$

Furthermore, we denote

$$c = |Z\backslash X||N(X)|, \ d = (\min(max\_size, |X \cup Y|) - |X|)|N(X)|,$$

$$e = |Z||N(Z\backslash X)|, \text{ and } f = \min(max\_size, |X \cup Y|)N_{max}(X,Y).$$

Since a, b, c, d, e and f satisfy the conditions of Lemma 1, we know that it holds that

1. if $\frac{|X||N(X)|+|Z\backslash X||N(X)|+|Z||N(Z\backslash X)|}{\sum_{t\in N(X)} W_t(X)+|Z||N(Z\backslash X)|} < 1$ then
$\frac{|X||N(X)|+|Z\backslash X||N(X)|+|Z||N(Z\backslash X)|}{\sum_{t\in N(X)} W_t(X)+|Z||N(Z\backslash X)|} \leq$
$\frac{|X||N(X)|+(\min(max\_size,|X\cup Y|)-|X|)|N(X)|+\min(max\_size,|X\cup Y|)N_{max}(X,Y)}{\sum_{t\in N(X)} W_t(X)+\min(max\_size,|X\cup Y|)N_{max}(X,Y)}.$

2. if $\frac{|X||N(X)|+|Z\backslash X||N(X)|+|Z||N(Z\backslash X)|}{\sum_{t\in N(X)} W_t(X)+|Z||N(Z\backslash X)|} \geq 1$ then
$\frac{|X||N(X)|+(\min(max\_size,|X\cup Y|)-|X|)|N(X)|+\min(max\_size,|X\cup Y|)N_{max}(X,Y)}{\sum_{t\in N(X)} W_t(X)+\min(max\_size,|X\cup Y|)N_{max}(X,Y)} \geq 1.$

Finally, note that
$\frac{|X||N(X)|+(\min(max\_size,|X\cup Y|)-|X|)|N(X)|+\min(max\_size,|X\cup Y|)N_{max}(X,Y)}{\sum_{t\in N(X)} W_t(X)+\min(max\_size,|X\cup Y|)N_{max}(X,Y)} =$
$\frac{\min(max\_size,|X\cup Y|)(|N(X)|+N_{max}(X,Y))}{\sum_{t\in N(X)} W_t(X)+\min(max\_size-|X|,|Y|)N_{max}(X,Y)} = C_{max}(X,Y).$

From the first claim above, it follows that if $\frac{|X||N(X)|+|Z\backslash X||N(X)|+|Z||N(Z\backslash X)|}{\sum_{t\in N(X)} W_t(X)+|Z||N(Z\backslash X)|}$ < 1, then $C(Z) \leq C_{max}(X,Y)$. From the second claim, it immediately follows that if $\frac{|X||N(X)|+|Z\backslash X||N(X)|+|Z||N(Z\backslash X)|}{\sum_{t\in N(X)} W_t(X)+|Z||N(Z\backslash X)|} \geq 1$, then $C_{max}(X,Y) \geq 1$, and since, per definition $C(Z) \leq 1$, $C(Z) \leq C_{max}(X,Y)$. This completes the proof. $\square$

Since an important feature of computing an upper bound for the cohesion is to establish how much cohesion could grow *in the worst case*, we need to figure out which items from Y should be added to X to reach this worst case. As has been discussed above, the worst case is actually materialised by adding as many as possible items from Y, and by first adding those that have the most occurrences. However, if the *max_size* parameter is used, it is not always possible to add all items in Y to X. In this case, we can only add $max\_size - |X|$ items to X, which is why we defined $N_{max}(X,Y)$ as

$$N_{max}(X, Y) = \max_{\substack{Y_i \subseteq Y, \\ |Y_i| \leq max\_size - |X|}} |N(Y_i)|.$$

Clearly, if $|X \cup Y| \leq max\_size$, $N_{max}(X, Y) = N(Y)$. If not, at first glance it may seem computationally very expensive to determine $|N(Y_i)|$ for every possible Y_i. However, we solve this problem by sorting the items in Y on support in ascending order. In other words, if $Y = \{y_1, \ldots, y_n\}$, with $sup(y_i) \leq sup(y_{i+1})$ for $i = 1, \ldots, n - 1$, then we can compute $N_{max}(X, Y)$ as

$$N_{max}(X, Y) = \sum_{i \in \{1, \ldots, max\_size - |X|\}} |N(\{y_{n-i+1}\})|.$$

As a result, the only major step in computing $C_{max}(X, Y)$ is that of computing $\sum_{t \in N(X)} W_t(X)$, as the rest can be computed in constant time. The procedure for computing $\sum_{t \in N(X)} W_t(X)$ is explained in detail in Sect. 3.3.

3.3   Computing the Sum of Minimal Windows

The algorithm for computing the sum of minimal windows is shown in Algorithm 3. For a given itemset X, the algorithm keeps a list of all time stamps at which items of X occur in the *positions* variable. The *nextpos* variable keeps a list of next time stamps for each item, while *lastpos* keeps a list of the last occurrences for each item. Since we need to compute the minimal window for each occurrence, we keep on doing this until we have either computed them all, or until the running sum has become large enough to safely stop, knowing that the branch can be pruned (line 7). Concretely, by rewriting the definition of $C_{max}(X, Y)$, we know we can stop if we are certain the sum will be larger than

$$W_{max}(X, Y) = \frac{\min(max\_size, |X \cup Y|)(|N(X)| + (N_{max}(X, Y)(1 - min\_coh)))}{min\_coh}.$$

When a new item comes in (line 14), we update the working variables, and compute the first and last position of the current window (line 17). If the smallest time stamp of the current window has changed, we go through the list of active windows and check whether a new shortest length has been found. If so, we update it (line 21). We then remove all windows for which we are certain that they cannot be improved from the list of active windows (line 23), and update the overall sum (line 24). Finally, we add the new window for the current time stamp to the list of active windows (line 25).

Note that the sum of minimal windows is independent of Y, the items yet to be enumerated. Therefore, if the branch is not pruned, the recursive DFS procedure shown in Algorithm 2 will be called twice, but X will remain unchanged in the second of those calls (line 9), so we will not need to recompute the sum of windows, allowing us to immediately evaluate the upper bound in the new node of the search tree.

Algorithm 3. SUM_MIN_WINS($\langle X, Y \rangle$) sums minimal windows of X

---

1 $smw \leftarrow 0$; $index \leftarrow 0$;
2 $positions \leftarrow$ position for every item in X;
3 $nextpos \leftarrow \{positions[i_1][0], positions[i_2][0], positions[i_3][0], ...\}$;
4 $lastpos \leftarrow \{-\infty, -\infty, -\infty, ...\}$;
5 $prev\_min \leftarrow -\infty$; $active\_windows \leftarrow \emptyset$;
6 **while** $index < |N(X)|$ **do**
7   **if** $smw + (|N(X)| + |active\_windows| - index) \times |X| > W_{max}(X, Y)$ **then**
8   **return** ∞;

9      $current\_pos \leftarrow \infty$;
10     $current\_item \leftarrow \emptyset$;
11   **for** i **in** X **do**
12   **if** $current\_pos > nextpos[i]$ **then**
13             $current\_pos \leftarrow nextpos[i]$;
14             $current\_item \leftarrow i$;

15     $lastpos[current\_item] \leftarrow current\_pos$;
16     $nextpos[current\_item] \leftarrow next(positions[current\_item], current\_pos)$;
17     $minpos \leftarrow \min(lastpos)$; $maxpos \leftarrow \max(lastpos)$;
18   **if** $minpos \neq -\infty$ **and** $minpos > prev\_min$ **then**
19   **for** $window \in active\_windows$ **do**
20             $newwidth \leftarrow maxpos - \min(minpos, window.pos) + 1$;
21             $window.width \leftarrow \min(window.width, newwidth)$;
22   **if** $window.pos < minpos$ **or** $window.width == |X|$ **or**
                $window.width < (maxpos - window.pos + 1)$ **then**
23                 $active\_windows \leftarrow active\_windows \backslash \{window\}$;
24                 $smw \leftarrow smw + window.width$;

25     $active\_windows \leftarrow$
        $active\_windows \cup \{window(current\_pos, maxpos - minpos + 1)\}$;
26     $prev\_min \leftarrow minpos$; $index \leftarrow index + 1$;
27 $smw \leftarrow smw + \text{sum}(window.width | window \in active\_windows)$;
28 **return** smw;

We illustrate how the algorithm works on the following example. Assume we are given the input sequence *aabccccacb*, and we are evaluating itemset $\{a, b, c\}$. Table 1 shows the values of the main variables as the algorithm progresses. As each item comes in, we update the values of *nextpos* and *lastpos* (other variables are not shown in the table). In each iteration, we compute the current best minimal window for the given time stamp as $\max(lastpos) - \min(lastpos) + 1$. We also update the values of any previous windows that might have changed for the better (this can only happen if $\min(lastpos)$ has changed), using either the current window above if it contains the time stamp of the window's event, or the window stretching from the relevant time stamp to $\max(lastpos)$. Finally, before proceeding with the next iteration, we remove all windows for which we are certain that they cannot get any smaller from the list of active windows.

Table 1. Computation of minimal windows.

Time	Item	nextpos	lastpos	w_1	w_2	w_3	w_4	w_5	w_6	w_7	w_8	w_9	w_{10}
0	-	$(1,3,4)$	$(-\infty,-\infty,-\infty)$	-	-	-	-	-	-	-	-	-	-
1	a	$(2,3,4)$	$(1,-\infty,-\infty)$	∞	-	-	-	-	-	-	-	-	-
2	a	$(8,3,4)$	$(2,-\infty,-\infty)$	∞	∞	-	-	-	-	-	-	-	-
3	b	$(8,10,4)$	$(2,3,-\infty)$	∞	∞	∞	-	-	-	-	-	-	-
4	c	$(8,10,5)$	$(2,3,4)$	**4**	**3**	**3**	**3**	-	-	-	-	-	-
5	c	$(8,10,6)$	$(2,3,5)$	-	-	-	-	4	-	-	-	-	-
6	c	$(8,10,7)$	$(2,3,6)$	-	-	-	-	4	5	-	-	-	-
7	c	$(8,10,9)$	$(2,3,7)$	-	-	-	-	4	5	6	-	-	-
8	a	$(\infty,10,9)$	$(8,3,7)$	-	-	-	-	**4**	5	6	6	-	-
9	c	$(\infty,10,\infty)$	$(8,3,9)$	-	-	-	-	-	5	6	6	7	-
10	b	(∞,∞,∞)	$(8,10,9)$	-	-	-	-	-	**5**	**4**	**3**	**3**	**3**

In the table, windows that are not active are marked with '-', while definitively determined windows are shown in bold. We can see that, for example, at time stamp 4, we have determined the value of the first four windows. Window w_1 cannot be improved on, since time stamp 1 has already dropped out of *lastpos*, while the other three windows cannot be improved since 3 is the absolute minimum for a window containing three items. At time stamp 8, we know that the length of w_5 must be equal to 4, since any new window to come must stretch at least from time stamp 5 to a time stamp in the future, i.e., at least 9. Finally, once we have reached the end of the sequence, we mark all current values of still active windows as determined.

4   Experiments

In order to demonstrate the usefulness of our method, we chose datasets in which the discovered patterns could be easily discussed and explained. We used two text datasets, the *Species* dataset containing the complete text of *On the Origin of Species by Means of Natural Selection* by Charles Darwin[1], and the *Moby* dataset containing *Moby Dick* by Herman Melville[2]. We processed both sequences using the Porter Stemmer[3] and removed the stop words. After preprocessing, the length of the *Species* dataset was 85 450 items and the number of distinct items was 5 547, and the length of *Moby* was 88 945 and the number of distinct items was 10 221.

We performed two types of experiments. In Sect. 4.1, we qualitatively compare our output to that of three existing methods, while in Sect. 4.2, we provide

---

[1] Taken from http://www.gutenberg.org/etext/22764.

[2] Taken from http://www.gutenberg.org/etext/15.

[3] http://tartarus.org/~martin/PorterStemmer/.

a performance analysis of our FCI_{SEQ} algorithm. To ensure reproducibility, we have made our implementations and datasets publicly available[4].

4.1 Quality of Output

In the first set of experiments, we compared the patterns we discovered to those found by three existing pattern mining algorithms—WINEPI, LAXMAN[5] and MARBLES$_w$[6]. As discussed in Sect. 1, these algorithms use a variety of frequency-based quality measures to evaluate the patterns. Since the available implementations were made with the goal of discovering partially ordered episodes, we had to post-process the output in order to filter out only itemsets. Therefore, making any kind of runtime comparisons would be unfair on these methods, since they generate many more candidates. Consequently, in this section we limit ourselves to a qualitative analysis of the output.

For all methods, we set the relevant thresholds low enough in order to generate tens of thousands of patterns. We then sorted the output on the respective quality measures—the sliding window frequency for WINEPI, the non-overlapping minimal window frequency for LAXMAN, the weighted window frequency for MARBLES$_w$, and cohesion for FCI_{SEQ}. We used pattern size to break any ties in all four methods, and the sum of support of individual items making up an itemset as the third criteria for FCI_{SEQ}. Patterns that were still tied were ordered alphabetically. The frequency threshold was set to 30 for WINEPI in both datasets, 5 for LAXMAN in *Origin* and 4 in *Moby*, and 1 for MARBLES$_w$ in both datasets, with the sliding window size set to 15. We ran FCI_{SEQ} with the cohesion threshold set to 0.01, and the support threshold to 5 for *Origin* and 4 for *Moby*. Since none of the existing methods produced any itemsets consisting of more than 5 items, we limited the *max_size* parameter to 5.

The top 5 patterns discovered by the different methods are shown in Table 2. We can see that there are clear differences between the patterns we discovered and those discovered by the existing methods, which all produced very similar results. First of all, the patterns ranked first and second in our output for the *Origin* dataset are of size 3, which would be theoretically impossible for WINEPI and MARBLES$_w$, and highly unlikely for LAXMAN, since all three use anti-monotonic quality measures. Second, we observe that the patterns we discover are in fact quite rare in the dataset, but they are very strong, since all occurrences of these patterns are highly cohesive. Concretely, the phrase *tierra del fuego* occurs seven times in the book, and none of these words occur anywhere else in the book. The value of this pattern is therefore quite clear—if we encounter any one of these three words, we can be certain that the other two can be found nearby. However, the only other method that ranked this pattern in the top 10 000 was MARBLES$_w$, which ranked it 8 357th. On the other hand, pattern *mobi dick* is

---

[4] https://bitbucket.org/len_feremans/sequencepatternmining_public.

[5] The algorithm was given no name by its authors.

[6] The implementations of all three methods were downloaded from http://users.ics.aalto.fi/ntatti/software/closedepisodeminer.zip.

Table 2. Top 5 patterns discovered by the different methods.

Dataset	FCI$_{\text{SEQ}}$	WINEPI	LAXMAN	MARBLES$_{\text{W}}$
Species	tierra del fuego	natur select	natur select	natur select
	natura facit saltum	speci varieti	speci form	speci varieti
	del tierra	speci form	speci varieti	speci distinct
	del fuego	speci natur	speci natur	speci form
	natura facit	speci distinct	speci distinct	life condit
Moby	mobi dick	whale sperm	whale boat	whale sperm
	vinegar cruet	whale boat	ship whale	whale white
	deuteronomi deacon	ship whale	whale sperm	ship whale
	defend plaintiff	whale white	head whale	whale boat
	erskin defend plaintiff	head whale	sea whale	head mast

both cohesive and frequent, and was ranked 14th by WINEPI, 22nd by LAXMAN and 7th by MARBLES$_{\text{W}}$. None of the other patterns in our top 5 in either dataset were ranked in the top 3 000 patterns by any of the other algorithms. We conclude that in order to find the very strong, but rare, patterns, such as *tierra del fuego* or *vinegar cruet*, with the existing methods the user would need to wait a long time before a huge output was generated, and would then need to trawl through tens of thousands of itemsets in the hope of finding them. Our algorithm, on the other hand, ranks them at the very top.

The patterns discovered by other methods typically consist of words that occur very frequently in the book, regardless of whether the occurrences of the words making up the itemset are correlated or not. For example, words *speci* and *varieti* occur very often, and, therefore, also often co-occur. In fact, this pattern was ranked 82 261st by FCI$_{\text{SEQ}}$, with a cohesion of just over 0.01, indicating that the average distance between nearest occurrences of *speci* and *varieti* was close to 100, which clearly demonstrates that this pattern is spurious. Clearly, while the very top patterns seem very different, there was still some overlap between the output generated by the various methods. For example, pattern *natur select*, ranked first in the *Origin* dataset by the existing methods, was ranked 17th by FCI$_{\text{SEQ}}$, which shows that our method is also capable of discovering very frequent patterns, as long as they are also cohesive. Similarly, pattern *life condit* ranked 66th in our output, and pattern *speci distinct* 129th.

Table 3 shows the size of the overlap between the patterns discovered by FCI$_{\text{SEQ}}$ and those discovered by the other three methods. We compute the size of the overlap within the top k patterns for each method, for varying values of k. We note that, in relative terms, the overlap actually drops as k grows, since our method ranks many large patterns highly, which is not the case for the other three methods. For example, the top 500 patterns discovered by FCI$_{\text{SEQ}}$ on the *Species* dataset contain 388 patterns of size larger than 3, while the top 500 patterns produced by the other three methods do not contain a single pattern of

Table 3. Overlap in the top k patterns discovered by FCI$_{\text{SEQ}}$ and other methods.

Dataset	k	WINEPI	LAXMAN	MARBLES$_W$
Species	100	12	13	11
	500	28	35	25
	2 500	80	102	68
Moby	100	11	13	11
	500	27	28	26
	2 500	56	66	49

size larger than 3. Even in the top 100 patterns we discovered in both datasets less than half were of size 2, while the top 100 produced by all other methods on either dataset always contained at least 96 patterns of size 2. This demonstrates the benefit of using a non-anti-monotonic quality measure, which allows us to rank the best patterns on top regardless of size, while frequency-based methods will, per definition, rank all subpatterns of a large pattern higher than (or at least as high as) the pattern itself.

4.2 Performance Analysis

We tested the behaviour of our algorithm when varying the three thresholds. The results are shown in Fig. 1. As expected, we see that the number of patterns increases as the cohesion and support thresholds are lowered. In particular, when the cohesion threshold is set too low, the size of the output explodes, as even random combinations of frequent items become cohesive enough. However, as the support threshold decreases, the number of patterns stabilises, since rarer items typically only make up cohesive itemsets with each other, so only a few new patterns are added to the output (when we lower the support threshold to 2, we see another explosion as nearly the entire alphabet is considered frequent). In all settings, it took no more than a few minutes to find tens of thousands of patterns. With reasonable support and cohesion thresholds, we could even set the *max_size* parameter to ∞ without encountering prohibitive runtimes, allowing us to discover patterns of arbitrary size (in practice, the size of the largest pattern is limited due to the characteristics of the data, so output size stops growing at a certain point). Since existing methods use a relevance window, defining how far apart two items may be in order to still be considered part of a pattern, the existing methods can never achieve this. For example, using a window of size 15 implies that no pattern consisting of more than 15 items can be discovered. Finally, while we kept both *min_sup* and *min_coh* relatively high in the presented experiments with *max_size* set to ∞, it should be noted that a much lower *min_sup* could be used in combination with a higher *min_coh* in order to quickly find only the most cohesive patterns, including the rare ones.

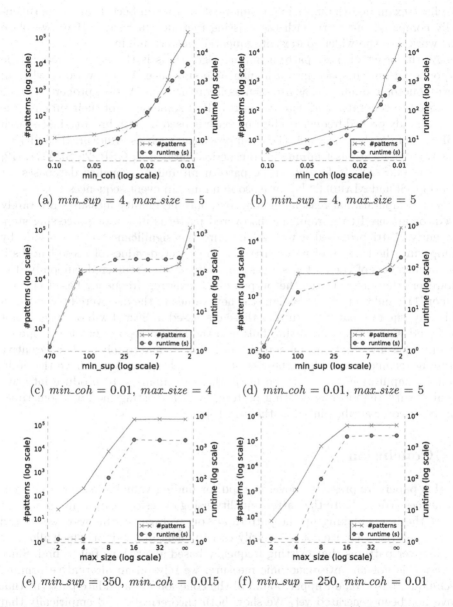

Fig. 1. Impact of various thresholds on output size and runtime. (a) Varying min_coh on *Species*.(b) Varying min_coh on *Moby*. (c) Varying min_sup on *Species*. (d) Varying min_sup on *Moby*. (e) Varying max_size on *Species*. (f) Varying max_size on *Moby*.

5 Related Work

We have examined the most important related work in Sect. 1, and experimentally compared our work with the existing methods in Sect. 4. Here, we place our work into the wider context of sequential pattern mining.

At the heart of most pattern mining algorithms is the need to reduce the exponential search space into a manageable subspace. When working with an anti-monotonic quality measure, such as frequency, the Apriori property can be deployed to generate candidate patterns only if some or all of their subpatterns have already proved frequent. This approach is used in both breadth-first-search (BFS) and depth-first-search (DFS) approaches, such as APRIORI [1] and FP-GROWTH [5] for itemset mining in transaction databases, GSP [2], SPADE [12] and PREFIXSPAN [9] for sequential pattern mining in sequence databases, or WINEPI [8] and MARBLES [4] for episode mining in event sequences.

For computational reasons, non-anti-monotonic quality measures are rarely used, or are used to re-rank the discovered patterns in a post-processing step. Recently, Tatti proposed a way to measure the significance of an episode by comparing the lengths of its occurrences to expected values of these lengths if the occurrences of the patterns' constituent items were scattered randomly [10]. However, the method uses the output of an existing frequency-based episode miner [11], and then simply assigns the new values to the discovered patterns. In this way, the rare patterns, such as those discussed in Sect. 4 will once again not be found. Our FCI_{SEQ} algorithm falls into the DFS category, but the proposed quality measure is not anti-monotonic, and we have had to rely on an alternative pruning technique to reduce the size of the search space. We believe the additional computational effort to be justified, as we manage to produce intuitive results, with the most interesting patterns, which existing methods sometimes fail to discover at all, ranked at the very top.

6 Conclusion

In this paper, we present a novel method for finding valuable patterns in event sequences. We evaluate the pattern quality using cohesion, a measure of how far apart the items making up the pattern are on average. In this way, we reward strong patterns that are not necessarily very frequent in the data, which allows us to discover patterns that existing frequency-based algorithms fail to find. Since cohesion is not an anti-monotonic measure, we rely on an alternative pruning technique, based on an upper bound of the cohesion of candidate patterns that have not been generated yet. We show both theoretically and empirically that the method is sound, the upper bound tight, and the algorithm efficient, allowing us to discover large numbers of patterns reasonably quickly. While the proposed approach concerns itemset mining, most of the presented work can be applied to mining other pattern types, such as sequential patterns or episodes.

References

1. Agrawal, R., Srikant, R.: Fast algorithms for mining association rules. In: International Conference on Very Large Data Bases, pp. 487–499 (1994)
2. Agrawal, R., Srikant, R.: Mining sequential patterns. In: International Conference on Data Engineering, pp. 3–14 (1995)
3. Cule, B., Goethals, B., Robardet, C.: A new constraint for mining sets in sequences. In: SIAM International Conference on Data Mining, pp. 317–328 (2009)
4. Cule, B., Tatti, N., Goethals, B.: Marbles: mining association rules buried in long event sequences. Stat. Anal. Data Min. **7**(2), 93–110 (2014)
5. Han, J., Pei, J., Yin, Y., Mao, R.: Mining frequent patterns without candidate generation: a frequent-pattern tree approach. Data Min. Knowl. Disc. **8**(1), 53–87 (2004)
6. Hendrickx, T., Cule, B., Goethals, B.: Mining cohesive itemsets in graphs. In: International Conference on Discovery Science, pp. 111–122 (2014)
7. Laxman, S., Sastry, P.S., Unnikrishnan, K.: A fast algorithm for finding frequent episodes in event streams. In: ACM SIGKDD Conference on Knowledge Discovery and Data Mining, pp. 410–419 (2007)
8. Mannila, H., Toivonen, H., Verkamo, A.I.: Discovery of frequent episodes in event sequences. Data Min. Knowl. Disc. **1**(3), 259–289 (1997)
9. Pei, J., Han, J., Mortazavi-Asl, B., Wang, J., Pinto, H., Chen, Q., Dayal, U., Hsu, M.C.: Mining sequential patterns by pattern-growth: the prefixspan approach. IEEE Trans. Knowl. Data Eng. **16**(11), 1424–1440 (2004)
10. Tatti, N.: Discovering episodes with compact minimal windows. Data Min. Knowl. Disc. **28**(4), 1046–1077 (2014)
11. Tatti, N., Cule, B.: Mining closed strict episodes. Data Min. Knowl. Disc. **25**(1), 34–66 (2012)
12. Zaki, M.J.: SPADE: an efficient algorithm for mining frequent sequences. Mach. Learn. **42**(1–2), 31–60 (2001)

Collaborative Expert Recommendation for Community-Based Question Answering

Congfu Xu, Xin Wang[(✉)], and Yunhui Guo

College of Computer Science and Technology,
Zhejiang University, Hangzhou 310027, Zhejiang, China
{xucongfu,cswangxinm,gyhui}@zju.edu.cn

Abstract. With the development of Internet, users can share knowledge by asking and answering questions on community question answering (CQA) websites. How to find related experts to contribute their answers is hence worthy of studying. In this paper, we propose a recommendation algorithm called collaborative expert recommendation (CER) for this purpose. We take full advantage of the heterogeneous information including question tags, content, answer's votes, which are considered important for identifying experts. Moreover, we combine such information by a causal assumption of questions and answers, and inner connection exploitation among different types of information such as (questioner, question), (answer, question) and (answerer, question, answer) correlations, which are more explicable and reasonable comparing with the existing methods. Experiments carried out on six real-world datasets prove that CER has a better performance.

1 Introduction

Nowadays, with the development of Internet techniques, people can easily share a wealth of knowledge on the community question answering (CQA) websites. Because the accumulation of many useful domain-specific questions and answers, CQA based websites such as Stack *Exchange* and *Quora* have attracted more and more users. For instance, the *Stack Overflow* website, which features questions and answers on a wide range of topics in computer programming, has over 2 million registered users and 7 million questions in just a few years.

Users expect to receive high quality answers after submitting their questions, prompting some recommendation techniques to be applied to attract more relevant users to provide their ideas and solutions. One way to speed up the delivery is to filter key words in the question titles and texts, and remind questioners to add more relevant tags, expecting the additional information to help classify the questions. Currently, some websites (e.g., *Zhihu.com* and *Quora.com*) have adopted an optimized solution by incorporating an expert invitation system, through which the registered users can invite their friends to contribute answers. Although those systems have improved CQA recommendation performance, they overly relay on users, and lack recommendation stability, because the websites cannot guarantee to deliver each question to the related experts, and whether

© Springer International Publishing AG 2016
P. Frasconi et al. (Eds.): ECML PKDD 2016, Part I, LNAI 9851, pp. 378–393, 2016.
DOI: 10.1007/978-3-319-46128-1_24

they can be answered in time. In fact, there are still many unanswered questions posted months and years ago. Therefore, recently, some related researches focusing on CQA systems prefer to introduce more active and intelligent solutions to help recommend experts automatically by involving recommendation algorithms and some related data mining techniques. With those proposals, each question has more chance to be reviewed by relevant users in time and can receive more high quality answers.

Understanding the CQA work flow is helpful for designing expert recommendation systems. To simplify it, we treat each question and its answers as a unit which contains questions, answers, users, tags and votes.

Question and Questioner. Each question consists of a title, a body, and one or more tags. Tags are usually the most brief and concise aspects for describing question topics; the title is more related to the content and can be regarded as a summary of it, while the body is more concrete and gives more details. Questions can also reflect questioners' expertise. A user who often posts questions about a particular topic may be interested in that topic, but may not have considerable expertise in solving those questions.

Answer and Answerer. Users are encouraged to provide their answers according to their own experience and knowledge. Some of them may contribute satisfying answers with systematic and detailed analysis, while others may not. Most CQA websites try to evaluate the quality of answers with the help of vote buttons, by which users or visitors can vote up and vote down each answer. The interactions result in a ranked answer list, in which the most useful answer will receive the highest voting score and be placed at the top, while the most valueless answer will be placed at the end of the page. We generally estimate the answerers' expertise according to the votes. If their answers receive higher voting scores than other users', we believe they have a high degree of expertise in the similar questions. This empirical conclusion plays an important role on designing our expert recommendation algorithms.

Till now, a few researches (e.g., [11,17–19]) have directly or indirectly leveraged some aforementioned information for CQA expert recommendation, however, they suffer from some drawbacks. First, the valuable information such as tags, questions, answer content and votes are not fully considered. Correctly revealing their inner connections and taking all of them into account will undoubtedly improve recommendation accuracy. Second, Questions and answers are usually modeled on the same level, which is not correct because usually the answer's content is determined by the related question's content and the answerer's expertise. Third, the (answerer, question) correlation is commonly represented as and modeled by (answerer, answer) and (answer, question) correlations separately, hence the expertise estimation is inefficient comparing with directly modeling (answerer, question, answer) triplets.

To overcome the aforementioned drawbacks, in this paper, we propose a novel collaborative expert recommendation (CER) model. We directly correlate users and questions by latent factor techniques, and model (question, answer, answerer) triplet by a reasonable causal assumption based on topic models.

The details of CER is discussed in Sect. 2. In Sect. 3, we analyze the performance of our models with six real world datasets. In Sect. 4, we talk about some related works and finally we conclude our study in Sect. 5.

2   Our Model

In this section, we start with the notation used in our paper. Then we present our assumptions, collaborative expert recommendation (CER) method and the learning algorithm.

2.1   Notation

The notation we use throughout the paper is defined in Table 1. Note that for some symbols such as z_q and $w_{q,i}$, we will omit their subscripts if they are clear in the context.

Table 1. Notation

Symbol	Description
Q	Set of questions $q \in Q$
A	Set of answers $a \in A$
U	Set of users $u \in U$
W, T	Set of words $w \in W$ and tags $t \in T$
$\mathcal{U}_u, \mathcal{Q}_q$	Latent factor for user u and question q
Q_u	Set of questions answered by $u \in U$
Q'_u	Set of questions posted by $u \in U$
A_u	Set of answers contributed by $u \in U$
z, θ_q, θ_a	Latent topic, and topic distribution of q and a
$\phi_{w,z}, \phi_{t,z}$	Word and tag distribution given topic z
$n(d, w)$	Number of occurrences of w in document d
$n(d, t)$	Number of occurrences of t in document d
$s_{u,q}$	Voting score for the answer given by answerer u to question q

2.2   Construction

We assume each user can ask questions, answer questions and give a vote to the existing answers; each question contains a title, a body and some tags; each answer contains a body and has a voting score. We make four assumptions about their correlations in the following three sub-sections, and finally integrate them into CER model in the final sub-section.

(User, Question) Correlation. To find the potential relevant experts, we directly model (user, question) correlations. Here, we use voting scores to quantify answers' quality. Generally, if most users believe an answer can solve a question perfectly, this answer and the answerer will receive a high voting score. We note that this scenario is very similar to item recommendation tasks if we treat answerers as users, questions as items and votes as rating scores. Therefore, we here apply collaborative filtering methods to modeling (user, question) correlation. Let $\mathcal{U}_u \in \mathbb{R}^{D \times |U|}$ and $\mathcal{Q}_q \in \mathbb{R}^{D \times |Q|}$ be the latent factors of answerer u and question q, and $s_{u,q}$ be the voting score for the answer given by answerer u to question q. We assume that $\mathcal{U}_u^T \mathcal{Q}_q$ is positive correlated with $s_{u,q}$, and adopt the latent factor method to model them. Because $s_{u,q}$ is discrete, we here adopt binomial distribution to model it as follows:

$$\mathcal{B}(s_{u,q}|\mathcal{U}_u, \mathcal{Q}_q) = \binom{S}{s_{u,q}} \sigma(u,q)^{s_{u,q}} (1 - \sigma(u,q))^{S - s_{u,q}}, \tag{1}$$

where S is the largest voting score in the data, and $\sigma(u, q)$ is the sigmoid function,

$$\sigma(u,q) = \frac{1}{1 + \exp(-\mathcal{U}_u^T \mathcal{Q}_q)}. \tag{2}$$

By this way, $\mathcal{U}_u^T \mathcal{Q}_q$ can be used to approximate u's expertise in q. For questioner u' who proposes q, we assume (s)he is not good at the similar topics or questions. We also model this assumption into our model by setting $s_{u',q} = 0$ for u'. Note that because the voting scores have a large span, in practice we need to rescale all s to a small range.

Text and Tags. We apply probabilistic latent semantic analysis (pLSA) [5] to modeling the text, including titles and bodies, of each question. Because tags are usually consisted with single word or phrases, we combine them together into a single structure which likes Link-LDA model [3]. Question q's proportion is defined as θ_q(i.e., $p(z|q)$). We use $\phi_{w,z}$ (i.e., $p(w|z)$) and $\phi_{t,z}$ (i.e., $p(t|z)$) to represent word and tag distributions for topic z. Then the likelihood of question corpus for words can be written as

$$\prod_{q \in Q} \prod_{w \in W} \{\sum_z \theta_{q,z}, \phi_{w,z}^q\}^{n(q,w)}, \tag{3}$$

where we multiply all words and questions in the corpus. The likelihood for tags is similar to Eq. (3), which we do not discuss here. Note that the number of tags is relatively less than words for each question, hence they are up-sampled to improve their priority.

Since q can be represented by both latent factor $\mathcal{Q}_q$ and topic distribution θ_q, we combine them into a more compact form for a unified expression. Note that we cannot combine them directly because $\mathcal{Q}_q$ is a random vector in $\mathbb{R}^{D \times |Q|}$ and θ_q is stochastic. Simply let $\mathcal{Q}_q$ be stochastic will reduce the power of the latent factor assumption, while let θ_q be a random vector without restriction will lose probabilistic interpretation of the topics. Inspired by [12], we design a transform function which allows $\mathcal{Q}_q \in \mathbb{R}^{D \times |Q|}$ and enforces $\sum_z \theta_{q,z} = 1$. The function is defined as follows:

$$\theta_{q,z} = \frac{\exp(\beta \mathcal{Q}_{q,z})}{\sum_{z'} \exp(\beta \mathcal{Q}_{q,z'})}, \tag{4}$$

where β controls the influence of each dimension in $\mathcal{Q}_q$. If β is large, q will be mainly about the topic with the highest value in $\mathcal{Q}_q$. If β is small, each topic tends to be evenly discussed.

Similarly, for each answer, we use θ_a (i.e., $p(z|a)$) to denote its topic. We also note that although the details of a question and the related answers may be quite different, the discussed topics revealed by tags are very similar. To take advantage of this information, we also consider tags of related question for each answer.

(Answerer, Question, Answer) Correlation. Beyond words and tags, we assume answer's topic proportion θ_a is determined by the related answerer u and question q, which we will give a more detailed explanation. (1) An answer's content and quality are undoubtedly affected by the related writer's expertise and skill. For instance, an answerer who is familiar with a certain type of questions is more likely to provide reasonable answers. (2) Because an answer is given for solving a particular question, the content and topics are also determined by the question's topics and features. Hence, we assume each answer is correlated with the answerer and the question. Specifically, we model them by the following two assumptions. First, each question and its answers share the similar topics. Second, if an answerer is good at a particular topic, (s)he will emphasize on that topic. In order to capture the correlation, we propose the following transformation function,

$$\theta_{a,z} = \frac{\exp(\gamma \mathcal{U}_{u,z} \mathcal{Q}_{q,z})}{\sum_{z'} \exp(\gamma \mathcal{U}_{u,z'} \mathcal{Q}_{q,z'})}, \tag{5}$$

where γ is a parameter which controls topic influence. Intuitively, large γ means that answerer u tends to focus on a particular topic of question q, while small γ means that answerer u discusses all topics of question q evenly. When γ is fixed, if u is familiar with topic z, and q is mainly about topic z, the value of $\mathcal{U}_{u,z} \mathcal{Q}_{q,z}$ will be larger, which means the answer is mainly about this topic. Therefore, Eq. (5) can correctly reflect the causal relationships among answerers, questions and answers.

2.3   CER Model

We construct our model by linking latent factors $\mathcal{Q}_q$ and $\mathcal{U}_u$ to topic proportions θ_q and θ_a. The graphical representation of CER is shown in Fig. 1.

We do not fit $\mathcal{U}_u$, $\mathcal{Q}_q$, θ_a and θ_q simultaneously since some parameters uniquely define the others. In practice, θ_a and θ_q are updated through $\mathcal{U}_u$ and $\mathcal{Q}_q$.

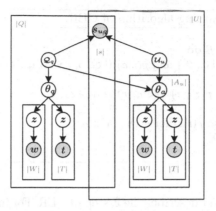

Fig. 1. The plate notation of CER. Note that the (question, questioner) correlation is not shown in the model. In the learning process, we model it together with the (question, answerer) coorrelation.

According to the generative process, our model finally combines users, questions, answers, tags and words altogether. The log likelihood of CER is as follows:

$$
\begin{aligned}
\mathcal{L} = &\sum_{u \in U} \sum_{q \in \{Q_u, Q'_u\}} \log \mathcal{B}(s_{u,q} | \mathcal{U}_u, \mathcal{Q}_q) \\
&+ \sum_{q \in Q} \sum_{w \in W} n(q, w) \log \sum_{z \in Z} p^q(w|z) p(z|q) \\
&+ \sum_{a \in A} \sum_{w \in W} n(a, w) \log \sum_{z \in Z} p^a(w|z) p(z|a) \\
&+ \sum_{q \in Q} \sum_{t \in T} n(q, t) \log \sum_{z \in Z} p^q(t|z) p(z|q) - \frac{\lambda_u}{2} \sum_{u \in U} \|\mathcal{U}_u\|^2 \\
&+ \sum_{a \in A} \sum_{t \in T} n(a, t) \log \sum_{z \in Z} p^a(t|z) p(z|a) - \frac{\lambda_q}{2} \sum_{q \in Q} \|\mathcal{Q}_q\|^2,
\end{aligned}
\tag{6}
$$

s.t Eq. (4) and Eq. (5),

where $p(w|z), p(t|z), p(z|q), p(z|a)$ are probabilistic expressions of $\phi_{w,z}, \phi_{t,z}, \theta_{q,z}$ and $\theta_{a,z}$.[1] λ_u and λ_q are regularization parameters for u and q.

We use EM (expectation-maximization) algorithm to train CER. Our goal is to optimize the parameters associated with voting scores $\Theta = \{\mathcal{U}_u, \mathcal{Q}_q\}$ and the parameters about topics $\Phi = \{p(z|q, w), p(z|q, t), p(z|a, w), p(z|a, t)\}$, $\Omega = \{p^q(w|z), p^a(w|z), p^q(t|z), p^a(t|z)\}$. In the E step, we compute the parameters in Φ for q and a given the other parameters. In the M step, we optimize the parameters in Θ and Ω. E and M steps are updated alternatively, and the optimization procedure is summarized in Algorithm 1.

---

[1] To make the learning process more readable, here we use two symbols to indicate one variable(e.g., $p(z|q)$ equals to $\theta_{q,z}$)).

Algorithm 1. The learning algorithm of CER.

1: Initialize $\mathcal{U}_u$ and $\mathcal{Q}_q$
2: **for** $\ell = 1, 2, ..., ITER$ **do**
3:   Compute $p(z|q)$ (i.e., θ_q) and $p(z|a)$ (i.e., θ_a) via Eq. (4) and Eq. (5)
4:   **E step:**
5:   Update $p(z|q, w)$, $p(z|q, t)$, $p(z|a, w)$ and $p(z|a, t)$
6:   **M step:**
7:   Update $p(w|z)$ and $p(t|z)$ for each q and a
8:   Update $\mathcal{U}_u$ and $\mathcal{Q}_q$
9: **end for**

Line 5 of Algorithm 1 describes the E step of CER. For instance, the updating details for w and k-th topic in the n-th iteration are shown below:

$$p_{n+1}(z^{(k)}|q^{(i)}, w^{(j)}) = \frac{p_n(z^{(k)}|q^{(i)})p_n(w^{(j)}|z^{(k)})}{\sum_{z' \in Z} p_n(z'|q^{(i)})p_n(w^{(j)}|z')}, \tag{7}$$

$$p_{n+1}(z^{(k)}|a^{(i)}, w^{(j)}) = \frac{p_n(z^{(k)}|a^{(i)})p_n(w^{(j)}|z^{(k)})}{\sum_{z' \in Z} p_n(z'|a^{(i)})p_n(w^{(j)}|z')}, \tag{8}$$

In line 7 of Algorithm 1, we update word and tag distributions for each topic. As an example, $p(w^{(j)}|z^{(k)})$ for q and a in the n-th iteration are updated as follows:

$$p_{n+1}^q(w^{(j)}|z^{(k)}) = \frac{\sum_{q \in Q} n(q, w^{(j)})p_{n+1}(z^{(k)}|q, w^{(j)})}{\sum_{w' \in W} \sum_{q \in Q} n(q, w')p_{n+1}(z^{(k)}|q, w')}, \tag{9}$$

$$p_{n+1}^a(w^{(j)}|z^{(k)}) = \frac{\sum_{a \in A} n(a, w^{(j)})p_{n+1}(z^{(k)}|a, w^{(j)})}{\sum_{w' \in W} \sum_{a \in A} n(a, w')p_{n+1}(z^{(k)}|a, w')}, \tag{10}$$

In line 8 of Algorithm 1, we update $\mathcal{U}_u$ and $\mathcal{Q}_q$ by gradient descent. We repeat the process multiple times in each iteration to approximate the local optimal value. The details of gradient descent are as follows:

$$\frac{\partial \mathcal{L}}{\partial \mathcal{U}_{u,k}} = \sum_{q \in \{Q_u, Q'_u\}} s_{u,q}\sigma(-\mathcal{U}_u^T \mathcal{Q}_q)\mathcal{Q}_{q,k} - (S - s_{u,q})\sigma(\mathcal{U}_u^T \mathcal{Q}_q)\mathcal{Q}_{q,k}$$

$$+ \gamma \sum_{a \in A_u} \sum_{w \in W} n(a, w)\frac{\phi_{w,k}^a \theta_{a,k}(1 - \theta_{a,k})\mathcal{Q}_{q,k}}{\sum_{k'} \phi_{w,k'}^a \theta_{a,k'}} \tag{11}$$

$$+ \gamma \sum_{a \in A_u} \sum_{t \in T} n(a, t)\frac{\phi_{t,k}^a \theta_{a,k}(1 - \theta_{a,k})\mathcal{Q}_{q,k}}{\sum_{k'} \phi_{t,k'}^a \theta_{a,k'}} - \lambda_u \mathcal{U}_{u,k},$$

$$\frac{\partial \mathcal{L}}{\partial \mathcal{Q}_{q,k}} = \sum_{u \in U_q} s_{u,q}\sigma(-\mathcal{U}_u^T \mathcal{Q}_q)\mathcal{U}_{u,k} - (S - s_{u,q})\sigma(\mathcal{U}_u^T \mathcal{Q}_q)\mathcal{U}_{u,k}$$

$$+ \beta \sum_{w \in W} n(q,w)\frac{\phi_{w,k}^q \theta_{q,k}(1 - \theta_{q,k})}{\sum_{k'} \phi_{w,k'}^q \theta_{q,k'}}$$

$$+ \beta \sum_{t \in T} n(q,t)\frac{\phi_{t,k}^q \theta_{q,k}(1 - \theta_{q,k})}{\sum_{k'} \phi_{t,k'}^q \theta_{q,k'}} - \lambda_q \mathcal{Q}_{q,k} \qquad (12)$$

$$+ \gamma \sum_{a \in A_q} \sum_{w \in W} n(a,w)\frac{\phi_{w,k}^a \theta_{a,k}(1 - \theta_{a,k})\mathcal{U}_{u,k}}{\sum_{k'} \phi_{w,k'}^a \theta_{a,k'}}$$

$$+ \gamma \sum_{a \in A_q} \sum_{t \in T} n(a,t)\frac{\phi_{t,k}^a \theta_{a,k}(1 - \theta_{a,k})\mathcal{U}_{u,k}}{\sum_{k'} \phi_{t,k'}^a \theta_{a,k'}},$$

where U_q are the users who post or answer question q; A_q are all answers to q, and u is the related answerer. For CER, we use $\mathcal{U}_u^T \mathcal{Q}_q$ to predict u's expertise in q.

Comparing with the existing methods, CER combines ideas from content analysis and collaborative filtering to model the correlations. It has the following advantages.

Table 2. Basic statistics of the six real-world datasets, including number of users, number of questions, number of answers, number of words, number of tags, average number of questions per user, average number of answers per user.

dataset	#users	#questions	#answers	#words	#tags	#avg.user_q	#avg.user_a
physics	1,709	854	4,138	5,797	403	0.4997	2.4213
unix	4,867	5,782	17,097	11,460	1,109	1.1880	3.5128
tex	3,824	6,301	18,675	20,141	757	1.6478	4.8836
super user	4,450	1,074	7,365	5,361	775	0.2413	1.6551
math	19,047	101,002	210,413	65,768	1,023	5.3028	11.0470
english	1,995	11,130	27,453	17,692	768	5.5789	13.7609

1. CER considers both word and tag information. In applications, the text in different domains may contain different types of noisy words. For instance, many questions and answers on *Stack Overflow* contain a large number of code snippets, which have negative impacts on topic learning. On the contrary, tags are always selected carefully by users and more effective for uncovering question's topics.
2. For each user, we model questioners' latent factors and answerers' latent factors separately based on the fact that each user plays two roles in CQA: a questioner and an answerer. Our model can not only discover what types of questions a user is familiar with, but also avoid providing their unfamiliar questions to them.

3. We assume each answer's topics are directly determined by the related answerer's and question's features. This causal assumption is more reasonable than existing models which learn latent topics of answers and questions separately.

4. CER directly correlates answerers with questions rather than construct (answerer, answer) and (answer, question) correlations separately. With those advantages, we can optimize (questioner, question), (answerer, question), (answerer, question, answer) correlations simultaneously, which results to better expert recommendations.

3 Experiments

3.1 Datasets

In our experiments, we sample six real-world datasets from *Stack Exchange* website to test our models. All the datasets are publicly available through official Data Dump Service[2]. Each sampled data contains plenty of community-contributed information in a particular domain on *Stack Exchange*. We preprocess the data by removing inactive users and questions. Specifically, we treat a user as an inactive user if (s)he has less than $n_u \in \{1, 3, 5, 7\}$ questions and answers and treat a question as an inactive question if it has less than $n_q \in \{1, 3, 5, 7\}$ answers. To test algorithm stability, we randomly choose n_u and n_q for each dataset. The statistic information of questions, users, tags and answers for each dataset is listed in Table 2. For each experiment, we use 80 % of the data for training, 10 % for validation and the left 10 % for test.

3.2 Baseline Methods and Evaluation Metrics

We evaluate the effectiveness of our models by comparing them to the state-of-the-art recommendation methods: Topic Expertise Model (TEM) [19], Tri-Role Topic Model (TRTM) [11], Dependent Dual Role Model (DDRM) [17] and Tag-Based Expert Recommendation (TBER) [18].

TEM jointly models users' topic interests and expertise by combining textual content model and link structure analysis, in which the answers' votes are modeled by Gaussian mixture hybrid and questions' topics are modeled based on LDA [1].

TRTM is a very recent question recommendation method which models different rules of users and tries to find their fine-grained topics by separating questions and answers into different topic models. It adopts *exponential KL − divergence* to correlate users, questions and answers.

DDRM is also a topic model based method which considers {questioners, answerers, question content}. It assumes questioners, questions' text, and answers are dependent on each other, and all of them determine question topics.

---

[2] https://archive.org/details/stackexchange.

Table 3. Top words of 5 topics learned by CER (on the *unix* dataset).

t1	http, network, firefox, internet, run
t2	driver, launch, desktop, set, system
t3	i386, linux, install, bash, ubuntu
t4	ubuntu, window, pack, help, linux
t5	bash, run, apt-get, write, copy

Table 4. Top tags of 5 topics learned by CER (on the *unix* dataset).

t1	google, dnsmasq, installed-programs, web, webcam
t2	cpu-load, mount, canon, monitor, pci
t3	time, text-mode, cpu-load, pidgin, unity
t4	thumbnails, ntfs, system-tray, customization, doc
t5	amazon-ec2, coreutils, derivatives, python, source

TBER is a recent proposed expert recommendation model which is mainly based on {answerers, question tags} information. It applies probabilistic matrix factorization to modeling answerers and tags. Given a question, the (user, question) correlation is estimated by the related (user, tag) correlations.

For the methods based on topic models, we use *gensim*[3] package to preprocess the text. For our models, we set $\lambda = 0.01$, $\eta = 0.005$, $\lambda_u = 0.01$, $\lambda_q = 0.01$ and rescale s to $[0, 5]$. β and γ are selected from $\{0.1, 1, 10\}$.

We use some popular ranking evaluation metrics to evaluate the recommendation efficiency of compared models, including mean reciprocal rank (MRR) [14], discounted cumulative gain (nDCG) [6], precision and recall [15].

3.3 Experimental Results

Topic Analysis. We first study word and tag distributions in each topic for CER. As an example, the results on *unix* dataset are listed in Tables 3 and 4. The former shows the top 5 words and the latter shows the top 5 tags for each topic. We list some observations below.

1. The top words under the same topic are strong correlated. For instance, the words in topic 1 are mainly about *network*, and the top words in topic 5 are closely related to *terminals* and *scripts*.
2. The correlations of the top 5 tags in each topic are still obvious. Furthermore, we find that by involving tags, CER can learn phrase level information such as "installed-programs", "cpu-load", which convey more useful semantic information than the bag-of-words.

---

[3] https://radimrehurek.com/gensim/.

Table 5. NDCG, precision, recall and MRR results of the compared methods.

NDCG@10

models	physics	unix	tex	super user	math	english
TEM	0.0602	0.0017	0.0042	0.0327	0.0021	0.0360
TRTM	0.0858	0.0012	0.0164	0.0734	0.0014	0.0342
DDRM	0.1009	0.0014	0.0170	0.0571	0.0025	0.0362
TBER	0.0802	0.0020	0.0177	0.0840	0.0034	0.0358
CER	**0.1185**	**0.0132**	**0.0190**	**0.0940**	**0.0043**	**0.0397**

Pre@10

models	physics	unix	tex	super user	math	english
TEM	0.0292	0.0010	0.0039	0.0128	0.0017	0.0225
TRTM	0.0402	0.0008	0.0129	0.0245	0.0011	0.0207
DDRM	0.0450	0.0008	0.0135	0.0198	0.0017	0.0225
TBER	0.0366	0.0012	0.0143	0.0278	0.0020	0.0230
CER	**0.0499**	**0.0073**	**0.0150**	**0.0302**	**0.0024**	**0.0237**

Rec@10

models	physics	unix	tex	super user	math	english
TEM	0.0910	0.0019	0.0063	0.0486	0.0020	0.0448
TRTM	0.1398	0.0014	0.0176	0.1311	0.0005	0.0434
DDRM	0.1599	0.0020	**0.0190**	0.0948	0.0029	0.0459
TBER	0.1207	0.0029	0.0180	0.1511	0.0047	0.0453
CER	**0.1838**	**0.0248**	0.0178	**0.1668**	**0.0063**	**0.0519**

MRR

models	physics	unix	tex	super user	math	english
TEM	0.0854	0.0260	0.0387	0.0509	0.0080	0.0702
TRTM	0.1126	0.0361	0.0367	0.0884	0.0049	0.0687
DDRM	0.1305	0.0386	0.0365	0.0749	0.0084	0.0690
TBER	0.1082	0.0237	0.0416	0.0975	0.0095	0.0711
CER	**0.1523**	**0.0392**	**0.0426**	**0.1080**	**0.0102**	**0.0757**

3. The top words in each topic are related to the top tags. For instance, the first topic in Table 3 is mainly about the *network* which is correlated to the tags such as "google", "dnsmasq" and "web" in the first line in Table 4. Hence, CER can exploit words' and tags' consistency and take advantage of them, which shows a reason why our method is better than the compared baselines.

Recommendation Efficiency. We investigate expert recommendation performance on some ranking metrics. For all compared models, the number of topics and the dimension of latent factors are set to 5. We set the recommendation list

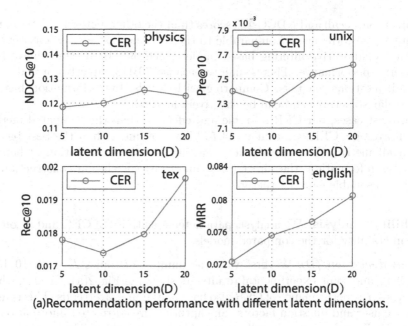

(a)Recommendation performance with different latent dimensions.

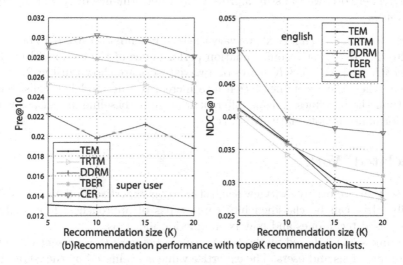

(b)Recommendation performance with top@K recommendation lists.

Fig. 2. Performance comparison with different latent dimensions and recommendation list size.

size to 10 for nDCG, precision and recall. The results on all metrics are shown in Table 5. We summarize our observations as follows.

1. The results on different datasets vary greatly due to data characteristics and model stability. We notice that some topic model based methods are sensitive to the data in different domains. For instance, TEM achieves very low

precision, recall and NDCG@10 scores than the other methods on *tex* dataset, but its performance is comparable to the baselines' on *english* dataset.

2. The collaborative filtering based models such as TBER can achieve better overall performance. This is mainly because TBER directly correlates users and questions via tags. Comparing with words, tag information has more regular structures and can be effectively learned.

3. In most cases, our CER is stable and effective than the compared methods. On average, CER can achieve 14.76 % improvement than the best baselines on all metrics. The effectiveness may due to the exploitation of heterogeneous information, the direct (user, question) correlation construction, and the reasonable causal structure.

Stability Analysis. We analysis parameter sensitivity of CER and recommendation stability for the compared models.

Latent dimension (D). We first vary the number of topics $D \in \{5, 10, 15, 20\}$ for CER, and some experimental results are shown in Fig. 2(a). Clearly, when D increases, the performance of CER also has general upward trends. This is mainly because user and question factors can capture more correlated information with larger D. Therefore, we can still improve CER's performance by increasing topic dimensions.

Recommendation size (K). Next, we vary the recommendation size K to be values in $\{5, 10, 15, 20\}$ to test recommendation performance of the compared models. The $Pre@K$ and $NDCG@K$ scores on two datasets are shown in Fig. 2(b). We find that when the recommendation size increases, our model is consistently better than the baselines. This is different from some baseline models, of which the performance is not relatively stable.

4   Related Work

In this section, we briefly overview some studies on expertise modeling for CQA. Broadly, they can be classified into three types of approaches: link analysis, content-based techniques and matrix completion methods.

The link analysis based methods construct a network to represent the interaction of questions and users. The expertise value is evaluated by the connection degree of each line in the graph. For instance, to recommend experts in Sun Java Forum, Zhang et al. [20] construct a directed bipartite graph for users and posts. They apply some network-based ranking algorithms, including PageRank and HITS, to identify the users with high expertise. The models based on link analysis are dataset specific and not general. In addition, they are not effective to find the latent correlations and ignore question and answer text which carry abundant useful information.

The content-based methods are proposed to utilize topic-level information [4,9,17], which are mainly based on pLSA [5] and LDA [1]. To leverage more useful information, Yang Liu et al.([19]) also try to integrate tags and voting scores into PLSA, assuming votes are sampled from Gaussian mixture models. Some similar assumptions can also be found in [7,8,10,13]. The most recent related work for CQA recommendation is TRTM proposed in [11]. TRTM explicitly separates askers, answerers, questions and answers into different parts and use *exponential KL−divergence* to link users to questions and answers. Beyond those methods, some approaches try to combine the network and topic models such as [2,16,21,23].

Matrix completion assumption is another method for expert recommendation. For instance, TBER [18] applies matrix factorization to modeling users and tags. The users' expertise in questions is translated to (user, tag) scores. Zhao et al. [22] try to leverage users' social networks to help find more relevant experts with graph regularized matrix completion. Such methods have some limitations. For example, TBER is not effective if the number of tags for each question is small. The approach proposed in [22] is not a generalized method for all CQA websites.

5   Conclusion and Future Work

In this paper, we propose a novel expert recommendation model(CER) for CQA. Specifically, we design two transformation functions and a causal construction to model (questioner, question), (answerer, question) and (answerer, question, answer) correlations. Comparing with the state-of-the-art models, CER can capture more useful information and reflect their inner connections, which leads to better experimental results. We apply our models to six real-world datasets to test the performance of expert recommendation, which is measured from different perspectives. The experimental results show that our methods can achieve better overall performance than the state-of-the-art models.

In the future, we will consider more comprehensive information, including (1) the followers of each question, and (2) the comments of the followers to each answer.

Acknowledgements. This research is supported by the National Natural Science Foundation of China (NSFC) No. 61272303.

References

1. Blei, D.M., Ng, A.Y., Jordan, M.I.: Latent dirichlet allocation. J. Mach. Learn. Res. (JMLR) **3**, 993–1022 (2003)
2. Chen, B.C., Guo, J., Tseng, B., Yang, J.: User reputation in a comment rating environment. In: Proceedings of the 17th ACM SIGKDD International Conference on Knowledge Discovery and Data Mining (SIGKDD), pp. 159–167. ACM (2011)
3. Erosheva, E., Fienberg, S., Lafferty, J.: Mixed-membership models of scientific publications. J. National Acad. Sci. (PNAS) **101**(suppl 1), 5220–5227 (2004)
4. Guo, J., Xu, S., Bao, S., Yu, Y.: Tapping on the potential of q&a community by recommending answer providers. In: Proceedings of the 17th ACM Conference on Information and Knowledge Management (CIKM), pp. 921–930. ACM (2008)

5. Hofmann, T.: Probabilistic latent semantic indexing. In: Proceedings of the 22nd Annual International ACM SIGIR Conference on Research and Development in Information Retrieval (SIGIR), pp. 50–57. ACM (1999)
6. Järvelin, K., Kekäläinen, J.: Cumulated gain-based evaluation of ir techniques. ACM Trans. Inf. Syst. (TOIS) **20**(4), 422–446 (2002)
7. Ji, Z., Wang, B.: Learning to rank for question routing in community question answering. In: Proceedings of the 22nd ACM International Conference on Information and Knowledge Management (CIKM), pp. 2363–2368. ACM (2013)
8. Jurczyk, P., Agichtein, E.: Discovering authorities in question answer communities by using link analysis. In: Proceedings of the 16th ACM International Conference on Information and Knowledge Management (CIKM), pp. 919–922. ACM (2007)
9. Liu, J., Song, Y.I., Lin, C.Y.: Competition-based user expertise score estimation. In: Proceedings of the 34th International ACM SIGIR Conference on Research and Development in Information Retrieval (SIGIR), pp. 425–434. ACM (2011)
10. Liu, X., Croft, W.B., Koll, M.: Finding experts in community-based question-answering services. In: Proceedings of the 14th ACM International Conference on Information and Knowledge Management (CIKM), pp. 315–316. ACM (2005)
11. Ma, Z., Sun, A., Yuan, Q., Cong, G.: A tri-role topic model for domain-specific question answering. In: Proceedings of the 29th AAAI Conference on Artificial Intelligence (AAAI) (2015)
12. McAuley, J., Leskovec, J.: Hidden factors and hidden topics: understanding rating dimensions with review text. In: Proceedings of the 7th ACM Conference on Recommender Systems, pp. 165–172. ACM (2013)
13. Qu, M., Qiu, G., He, X., Zhang, C., Wu, H., Bu, J., Chen, C.: Probabilistic question recommendation for question answering communities. In: Proceedings of the 18th International Conference on World Wide Web (WWW), pp. 1229–1230. ACM (2009)
14. Shi, Y., Karatzoglou, A., Baltrunas, L., Larson, M., Oliver, N., Hanjalic, A.: Climf: learning to maximize reciprocal rank with collaborative less-is-more filtering. In: Proceedings of the 6th ACM Conference on Recommender Systems, pp. 139–146. ACM (2012)
15. Wang, X., Pan, W., Xu, C.: Hgmf: Hierarchical group matrix factorization for collaborative recommendation. In: Proceedings of the 23rd ACM International Conference on Conference on Information and Knowledge Management, pp. 769–778. ACM (2014)
16. Weng, J., Lim, E.P., Jiang, J., He, Q.: Twitterrank: finding topic-sensitive influential twitterers. In: Proceedings of the 3rd ACM International Conference on Web Search and Data Mining (WSDM), pp. 261–270. ACM (2010)
17. Xu, F., Ji, Z., Wang, B.: Dual role model for question recommendation in community question answering. In: Proceedings of the 35th International ACM SIGIR Conference on Research and Development in Information Retrieval (SIGIR), pp. 771–780. ACM (2012)
18. Yang, B., Manandhar, S.: Tag-based expert recommendation in community question answering. In: Social Networks Analysis and Mining (ASONAM), pp. 960–963. IEEE (2014)
19. Yang, L., Qiu, M., Gottipati, S., Zhu, F., Jiang, J., Sun, H., Chen, Z.: Cqarank: jointly model topics and expertise in community question answering. In: Proceedings of the 22nd ACM International Conference on Information and Knowledge Management (CIKM), pp. 99–108. ACM (2013)

20. Zhang, J., Ackerman, M.S., Adamic, L.: Expertise networks in online communities: structure and algorithms. In: Proceedings of the 16th International Conference on World Wide Web (WWW), pp. 221–230. ACM (2007)
21. Zhao, T., Bian, N., Li, C., Li, M.: Topic-level expert modeling in community question answering. In: Proceedings of the 13th SIAM International Conference on Data Mining (SDM) 13, pp. 776–784 (2013)
22. Zhao, Z., Zhang, L., He, X., Ng, W.: Expert finding for question answering via graph regularized matrix completion. IEEE Trans. Knowl. Data Eng. (TKDE) **27**(4), 993–1004 (2015)
23. Zhou, G., Lai, S., Liu, K., Zhao, J.: Topic-sensitive probabilistic model for expert finding in question answer communities. In: Proceedings of the 21st ACM International Conference on Information and Knowledge Management (CIKM), pp. 1662–1666. ACM (2012)

Link Prediction in Dynamic Networks Using Graphlet

Mahmudur Rahman and Mohammad Al Hasan[✉]

Indiana University Purdue University Indianapolis, Indianapolis, USA
{mmrahman,alhasan}@iupui.edu

Abstract. Predicting the link state of a network at a future time given a collection of link states at earlier time is an important task with many real-life applications. In existing literature this task is known as link prediction in dynamic networks. Solving this task is more difficult than its counterpart in static networks because an effective feature representation of node-pair instances for the case of dynamic network is hard to obtain. In this work we solve this problem by designing a novel graphlet transition based feature representation of the node-pair instances of a dynamic network. We propose a method GRATFEL which uses unsupervised feature learning methodologies on graphlet transition based features to give a low-dimensional feature representation of the node-pair instances. GRATFEL models the feature learning task as an optimal coding task where the objective is to minimize the reconstruction error, and it solves this optimization task by using a gradient descent method. We validate the effectiveness of the learned feature representations by utilizing it for link prediction in real-life dynamic networks. Specifically, we show that GRATFEL, which use the extracted feature representation of graphlet transition events, outperforms existing methods that use well-known link prediction features. The data and software related to this paper are available at https://github.com/DMGroup-IUPUI/GraTFEL-Source.

1 Introduction

Understanding the dynamics of an evolving network is an important research problem with numerous applications in various fields, including social network analysis, information retrieval, recommendation systems, epidemiology, security, and bioinformatics. A key task towards this understanding is to predict the likelihood of a future association between a pair of nodes, given the existing state of the network. This task is commonly known as the *link prediction* problem. Since, its formal introduction to the data mining community by Liben-Nowell et al. [9] about a decade ago, this problem has been studied extensively by many researchers from a diverse set of disciplines. Comprehensive surveys on link prediction methods are available for interested readers [7].

This research is supported by Mohammad Hasan's NSF CAREER Award (IIS-1149851).

© Springer International Publishing AG 2016
P. Frasconi et al. (Eds.): ECML PKDD 2016, Part I, LNAI 9851, pp. 394–409, 2016.
DOI: 10.1007/978-3-319-46128-1_25

The majority of the existing works on link prediction consider a static snapshot of the given network, which is the state of the network at a given time [6,9–11]. Nevertheless, for many networks, additional temporal information, such as the time of link creation and deletion, is available over a time interval. For example, in an online social or a professional network, we may know the time when two persons have become friends; for collaboration events, such as, a group performance or a collaborative academic work, we can extract the time of the event from an event calendar. The networks built from such data can be represented by a *dynamic network*, which is a collection of temporal snapshots of the network. The link prediction task on such a network is defined as follows: *for a given pair of nodes, predict the link probability between the pair at time $t + 1$ by training the model on the link information at times* $1, 2, \cdots, t$. We will refer this task as dynamic link prediction[1].

A key challenge of dynamic link prediction is finding a suitable feature representation of the node-pair instances which are used for training the prediction model. For the static setting, various topological metrics (common neighbors, Adamic-Adar, Jaccard's coefficient) are used as features, but they cannot be extended easily for the dynamic setting having multiple snapshots of the network. In fact, when multiple (say t) temporal snapshots of a network are provided, each of these scalar features becomes a t-sized sequence. Flattening the sequence into a t-size vector distorts the inherent temporal order of the features. Authors of [5] overcome this issue by modeling a collection of time series, each for one of the topological features; but such a model fails to capture signals from the neighborhood topology of the edges. There exist a few other works on dynamic link prediction, which use probabilistic (nonparametric) and matrix factorization based models. These works consider a feature representation of the nodes and assume that having a link from one node to another is determined by the combined effect of all pairwise node feature interactions [4,18,22]. While this is a reasonable assumption, the accuracy of such models are highly dependent on the quality of the node features, as well as the validity of the above assumption.

Graphlets, which are collection of small induced subgraphs, are increasingly being used for large-scale graph analysis. For example, frequencies of various graphlets are used for classifying networks from various domains [17]. They are also used for designing effective graph kernels [19]. In biological domain, graphlet frequencies are used for comparing structures of different biological networks [15]. In all these works, graphlets are used as a topological building block of a static network. Nevertheless, as new edges are added or existing edges are removed from the given dynamic network, the graphlets which are aligned with the affected edge transition to different graphlets. For illustration, let us consider a dynamic network with two temporal snapshots G_1 and G_2 (Fig. 1). In this example, G_2 has one more edge $(2, 3)$ than G_1. We observe three different types of transition events, where a type of graphlet is changed into another type in the subsequent snapshot (see the table in Fig. 1). Here, all the events are triggered by the edge

---

[1] Strictly speaking, this task should be called link forecasting as the learning model is not trained on partial observation of link instances at time $t + 1$; however, we refer it as link prediction due to the popular usages of this phrase in the data mining literature.

$(2, 3)$. In this work, we use the frequency of graphlet transition events associated with a node-pair for predicting link between the node-pairs in a future snapshot of the dynamic network.

A key challenge of using graphlet transition event for dynamic link prediction is to obtain a good feature representation for this task. This is necessary because graphlet transition event matrix is sparse, and on such dataset, low dimen-

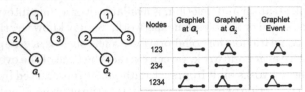

Fig. 1. A Toy Dynamic Network. G_1 and G_2 are two snapshots of the Network. Three different types of graphlet events are observed.

sional feature representation effectively captures the latent dependency among different dimensions of the data. There exist a growing list of recent works which use unsupervised methodologies for finding features from raw data representations of various complex objects, including images [13] and audio [8]. For graph data, we are aware of only one such work, namely DeepWalk [14], which obtains the feature representation of nodes for solving a node classification task. However, no such work exists for finding feature representation of node-pair instances for the purpose of link prediction.

In this work, we propose a novel learning method GRATFEL (**Gra**phlet **T**ransition and **F**eature **E**xtraction for **L**ink Prediction) for obtaining feature representation of node-pair instances from graphlet transition events in the observed snapshots of the given network. GRATFEL considers the feature learning task as an optimal coding problem such that the optimal code of a node-pair is the desired feature representation. The learning can be considered as a two-step process (compression and reconstruction), where the first step compresses the input representation of a node-pair into a code by a non-linear transformation, and the second step reconstructs the input representation from the code by a reverse process and the optimal code is the one which yields the least amount of reconstruction error. The input representation of a node-pair is given as a vector of graphlet transition events (GTEs) associated with the corresponding node-pair. After obtaining an appropriate feature representation of the node-pairs, a traditional supervised learning technique is used (we use SVM and AdaBoost) for predicting link states at future times in the given dynamic network. Below we summarize our contributions in this work:

- We use graphlet transition events (GTEs) for preforming link prediction in a dynamic network. To the best of our knowledge we are the first to use GTEs for solving a prediction task over a dynamic network.
- We propose a learning model (GRATFEL) for unsupervised feature extraction of node-pairs for the purpose of link prediction over a dynamic network.
- We compare the performance of GRATFEL with multiple state-of-the-art methods on three real-life dynamic networks. This comparison results show that our method is superior than each of the competing methods.

2   Related Works

Graphlets have been used successfully in static network setup for a multitude of applications. For a given network, Pržulj et al. [16] count the frequencies of various graphlets in the network for designing a fingerprint of a biological graph. In [12], the authors define *signature similarity* function to characterize the similarity of two vertices of a network, thus allowing a user to cluster vertices based on their structural similarity. Sampling based methods are used for computing graphlet degree distribution efficiently [1,17]. While the above methods address graphlet based analysis of static networks, works, exploring dynamic network characterization using graphlet, are yet to come.

A multitude of methodologies have been developed for link prediction on dynamic networks. The methods proposed by Güneş et al. [5] captures temporal patterns in a dynamic network using a collection of time series on topological features. But this approach fails to capture signals from neighborhood topology, as each time series model is trained on a separate t-size feature sequence of a node-pair. The method proposed by Bliss et al. [2] applies covariance matrix adaptation evolution strategy (CMA-ES) to optimize weights which are used in a linear combination of sixteen neighborhood and node similarity indices. Matrix and tensor factorization based solutions are presented in [4], considering a tensor representation of a dynamic network. The nonparametric link prediction method presented in [18] uses features of the node-pairs, as well as the local neighborhood of node-pairs. This method works by choosing a probabilistic model based on features (common neighbor and last time of linkage) of node-pairs. Stochastic block model based approaches divide nodes in a network into several groups and generates edges with probabilities dependent on the group membership of participant nodes [22]. While probabilistic model based link prediction performs well on small graphs, they become computationally prohibitive for large networks.

3   Problem Definition

Assume $G(V, E)$ is an undirected network, where V is the set of nodes and E is the set of edges $e(u, v)$ such that $u, v \in V$. A dynamic network is represented as a sequence of snapshots $\mathbb{G} = \{G_1, G_2, \ldots, G_t\}$, where t is the number of time stamps for which we have graph snapshots and $G_i(V_i, E_i)$ is a graph snapshot at time stamp $i : 1 \le i \le t$. In this work, we assume that the vertex set remains the same across different snapshots, i.e., $V_1 = V_2 = \cdots = V_t = V$. However, the edges appear and disappear over different time stamps. We also assume that, beside the link information, no other attribute data for the nodes or edges are available. We use n to denote the number of nodes ($|V|$), and m to denote all node-pairs $\binom{|V|}{2}$ in the network.

Problem Statement: Given a sequence of snapshots $\mathbb{G} = \{G_1, G_2, \ldots, G_t\}$ of a network, and a pair of nodes u and v, the link prediction task on a dynamic network predicts whether u and v will have a link in G_{t+1}. Note that, we assume that no link information regarding the snapshot G_{t+1} is available, except the fact that G_{t+1} contains the identical set of vertices.

4   Methods

A key challenge for dynamic link prediction is choosing an effective feature set for this task. Earlier works choose features by adapting topological features for static link prediction or by considering the feature values of different snapshots as a time series. GRATFEL uses graphlet transition events (GTEs) as features for link prediction. For a given node-pair, the value of a specific GTE feature is a normalized count of the observed GTE involving those node-pairs over the training data. The strength of GTEs as feature for dynamic link prediction comes from the fact that for a given node-pair, GTEs involving those nodes capture both the local topology and their transition over the temporal snapshots.

We consider GTEs involving graphlets up to size five (total 30 graphlets), of which graphlets of size four are shown in Fig. 2². The graphlet size upper bound of five is inspired by the fact that more than 95% new links in a dynamic network happen between vertices that are at most 3 distances apart in all three real-life dynamic networks that we use in this work. So, for a given node, GTE of a five vertex graphlet in the neighborhood of that node covers a prospective link formation event as a graphlet transition event. Another reason for limiting the graphlet size is the consideration of computation burden, which increases exponentially with the size of graphlets. There are 30 different graphlets of size up to 5 and the number of possible transition event (GTE) is $O(30^2)$. Increasing the size of graphlets to 6 increases the number of GTE to $O(142^2)$.

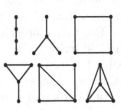

Fig. 2. Graphlets of size 4.

Feature representation for a node-pair in a dynamic network is constructed by concatenating GTE features from a continuous set of graph snapshots. Concatenation, being the simplest form of feature aggregation across a set of graph snapshots is not essentially the best feature representation to capture temporal characteristics of a node-pair. So, GRATFEL uses unsupervised feature extraction (UFE) to get optimal feature representation from GTE features. UFE provides a better feature representation by discovering dependency among different data dimensions, which cannot be achieved by simple aggregation. It also reduces the data dimension and overcomes the sparsity issue in GTE features. Once the optimal feature representation of a node-pair is known, GRATFEL uses that for solving the link prediction task using a supervised classification model.

The discussion of the proposed method GRATFEL can be divided into three steps: (1) graphlet transition event based feature extraction (Sect. 4.1), (2) unsupervised (optimal) feature learning (Sect. 4.2), and (3) supervised learning for obtaining the link prediction model (Sect. 4.3).

---

² There are only one graphlet of size 2, two graphlets of size 3 and twenty-one graphlets of size 5. These graphlets are not shown due to space constraint.

4.1   Graphlet Transition Based Feature Extraction

Say, we are given a dynamic network $\mathbb{G} = \{G_1, G_2, \ldots, G_t\}$, and we are computing the feature vector for a node-pair (u, v), which constitute a row in our training data matrix. We use each of $G_i : 1 \le i \le t-1$ (time window $[1, t-1]$) for computing the feature vector and G_t for computing the target value (1 if edge exist between u and v, 0 otherwise). We use $e^{uv}_{[1,t-1]}$ to represent this vector. It has two components: graphlet transition event (GTE) and link history (LH).

The first component, Graphlet Transition Event (GTE), $g^{uv}_{[1,t-1]}$ is constructed by concatenating GTE feature-set of (u, v) for each time stamp. i.e., $g^{uv}_{[1,t-1]} = g^{uv}_1 \parallel g^{uv}_2 \parallel \ldots \parallel g^{uv}_{t-1}$. Here, the symbol $\parallel$ represents concatenation of two horizontal vectors (e.g., $0\ 1\ 0 \parallel 0.5\ 0\ 1 = 0\ 1\ 0\ 0.5\ 0\ 1$) and g^{uv}_i represents (u, v)'s GTE feature-set for time stamp i, and it captures the impact of edge (u, v) at its neighborhood structure at time stamp i. We construct g^{uv}_i by enumerating all graphlet based dynamic events, that are triggered when edge (u, v) is added with G_i.

For example, consider the toy dynamic network in Fig. 3. We want to construct the GTE feature vector g^{36}_1, which is the GTE feature representation of node-pair $(3, 6)$ at G_1. We illustrate the construction process in Fig. 4. In this figure, we show all the graphlet transitions triggered by edge $(3, 6)$ when it is added in G_1. These transition events are listed in center table of Fig. 4. Column titled *Nodes*

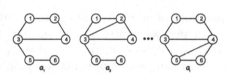

Fig. 3. A Toy Dynamic Network with t snapshots. First two and last snapshots are shown in this figure.

lists the sets of nodes where the graphlet transitions are observed and Column *Current Graphlet* shows the current graphlet structure induced by these nodes. Column *Transformed Graphlet* shows the graphlet structure after $(3, 6)$ is added. The last column *Graphlet Event* is a visual representation of the transition events, where the transition is reflected by the red edges. Once all the transition events are enumerated, we count the frequencies of these events (Table on the right side of Fig. 4). Graphlet transition frequencies can be substantially different for different edges, so the GTE vector is normalized by the largest value of graphlet transition frequencies associated with this edge. Also note that, all possible graphlet transition events are not observed for a given edge. So, among all the possible types of GTE, those that are observed in at least one node-pair in the training dataset are considered in GTE feature-set.

The second component of node-pair feature vector is Link History (LH) of node-pair, which is not captured by GTE feature-set, $g^{uv}_{[1,t-1]}$. Link History, $\mathbf{lh}^{uv}_{[1,t-1]}$ of a node-pair (u, v) is a vector of size $t - 1$, denoting the edge occurrences between the participating nodes over the time window $[1, t - 1]$. It is defined as, $\mathbf{lh}^{uv}_{[1,t-1]} = \mathbf{G}_1(u, v) \parallel \mathbf{G}_2(u, v) \parallel \ldots \parallel \mathbf{G}_{t-1}(u, v)$. Here, $\mathbf{G}_i(u, v)$ is 1 if edge (u, v) is present in snapshot G_i and 0 otherwise.

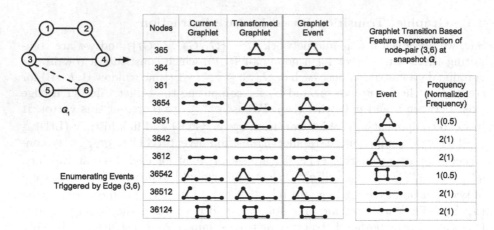

Fig. 4. Construction of graphlet transition based feature representation g_1^{36} of node-pair $(3,6)$ at 1st snapshot of the toy network. (Color figure online)

An appearance of an edge in recent time indicates a higher chance of the edge to reappear in near future. So, we consider weighted link history, $\mathbf{wlh}_{[1,t-1]}^{uv} = w_1 \cdot \mathbf{G}_1(u,v) \parallel w_2 \cdot \mathbf{G}_2(u,v) \parallel \dots \parallel w_{t-1} \cdot \mathbf{G}_{t-1}(u,v)$. here, $w_i = i/(t-1)$ (a time decay function) represents the weight of link history for time stamp i. Finally, a frequent appearance of an edge over time indicates a strong tendency of the edge to reincarnate in the future. This motivates us to reward such events by considering cumulative sum. We define Weighted Cumulative Link History, $\mathbf{wclh}_{[1,t-1]}^{uv} = CumSum(\mathbf{wlh}_{[1,t-1]}^{uv})$.

Finally, the feature vector of a node-pair (u,v), $\mathbf{e}_{[1,t-1]}^{uv}$, is the concatenation of GTE feature-set ($g_{[1,t-1]}^{uv}$) and LH feature-set ($\mathbf{wclh}_{[1,t-1]}^{uv}$); i.e., $\mathbf{e}_{[1,t-1]}^{uv} = g_{[1,t-1]}^{uv} \parallel \mathbf{wclh}_{[1,t-1]}^{uv}$. For predicting dynamic links in time stamp $t+1$, we right-shift the time window by one. In other words, we construct graphlet feature representation $\mathbf{e}_{[2,t]}^{uv}$ by using snapshots from time window $[2,t]$. Final feature representation for all node-pairs,

$$\hat{\mathbf{E}} = \{\mathbf{e}_{[1,t-1]}^{uv}\}_{u,v \in V}$$
$$\bar{\mathbf{E}} = \{\mathbf{e}_{[2,t]}^{uv}\}_{u,v \in V} \tag{1}$$

Here, $\hat{\mathbf{E}}$ is the training dataset and $\bar{\mathbf{E}}$ is the prediction dataset. Both $\hat{\mathbf{E}}$ and $\bar{\mathbf{E}}$ can be represented as matrices of dimensions (m,k). The size of the feature vector is $k = |\mathbf{e}_{[1,t-1]}^{uv}| = c*(t-1)+t-1$, where c is the total number of distinct GTEs that we consider.

GTE enumeration. We compute GTEs by using a local growth algorithm. For computing g_i^{uv}, we first enumerate all graphlets of G_i having both u and v. Starting from the edge graphlet $gl = \{u,v\}$, in each iteration of growth we add a new vertex w from the immediate neighborhood of gl to obtain a larger graphlet

$gl = gl \cup \{w\}$. Growth is terminated when $|gl| = 5$. The enumeration process is identical to our earlier work [17]. After enumeration, GTE is easily obtained from graphlet embedding by marking the edge (u, v) as the transition trigger (see Fig. 4). The computation of GTEs of different node-pairs are not dependent on each other, this makes GTE enumeration task embarrassingly parallel.

4.2 Unsupervised Feature Extraction

GRATFEL performs the task of unsupervised feature extraction as learning an optimal coding function h. Lets consider, e is a feature vector from either $\hat{\mathbf{E}}$ or $\bar{\mathbf{E}}$ ($e \in \hat{\mathbf{E}} \cup \bar{\mathbf{E}}$). Now, the coding function h compresses e to a code vector α of dimension l, such that $l < k$. Here l is a user-defined parameter which represents the code length and k is the size of feature vector. Many different coding functions exist in the dimensionality reduction literature, but GRATFEL chooses the coding function which incurs the minimum loss in the sense that from the code α we can reconstruct e with the minimum error over all possible $e \in \hat{\mathbf{E}} \cup \bar{\mathbf{E}}$. We frame the learning of h as an optimization problem, which we discuss below through two operations: Compression and Reconstruction.

Compression: It obtains α from e. This transformation can be expressed as a nonlinear function of linear weighted sum of the graphlet transition features.

$$\alpha = f(\mathbf{W}^{(c)}e + \mathbf{b}^{(c)}) \tag{2}$$

$\mathbf{W}^{(c)}$ is a (k, l) dimensional matrix. It represents the weight matrix for compression and $\mathbf{b}^{(c)}$ represents biases. $f(\cdot)$ is the Sigmoid function, $f(x) = \frac{1}{1+e^{-x}}$.

Reconstruction: It performs the reverse operation of compression, i.e., it obtains the graphlet transition features $\mathbf{e}$ from α which was constructed during the compression operation.

$$\beta = f(\mathbf{W}^{(r)}\alpha + \mathbf{b}^{(r)}) \tag{3}$$

$\mathbf{W}^{(r)}$ is a matrix of dimensions (l, k) representing the weight matrix for reconstruction, and $\mathbf{b}^{(r)}$ represents biases.

The optimal coding function h constituted by the compression and reconstruction operations is defined by the parameters $(\mathbf{W}, \mathbf{b}) = (\mathbf{W}^{(c)}, \mathbf{b}^{(c)}, \mathbf{W}^{(r)}, \mathbf{b}^{(r)})$. The objective is to minimize the reconstruction error. Reconstruction error for a graphlet transition feature vector (e) is defined as, $J(\mathbf{W}, \mathbf{b}, e) = \frac{1}{2} \| \beta - e \|^2$. Over all possible feature vectors, the average reconstruction error augmented with a regularization term yields the final objective function $J(\mathbf{W}, \mathbf{b})$ which we want to minimize:

$$J(\mathbf{W}, \mathbf{b}) = \frac{1}{2m} \sum_{e \in \hat{\mathbf{E}} \cup \bar{\mathbf{E}}} (\frac{1}{2} \| \beta - e \|^2) + \frac{\lambda}{2}(\| \mathbf{W}^{(c)} \|_F^2 + \| \mathbf{W}^{(r)} \|_F^2) \tag{4}$$

Here, λ is a user assigned regularization parameter, responsible for preventing over-fitting. $\| \cdot \|_F$ represents the Frobenius norm of a matrix.

Training: The training of optimal coding defined by parameters (W, b) begins with random initialization of the parameters. Since the cost function $J(W, b)$ defined in Eq. (4) is non-convex in nature, we obtain a local optimal solution using the gradient descent approach. The parameter update of the gradient descent is similar to the parameter update in Auto-encoder in machine learning. Unsupervised feature representation of any node-pair (u, v) can be obtained by taking the outputs of compression stage (Eq. (2)) of the trained optimal coding (W, b).

$$\alpha^{uv}_{[1,t-1]} = f(W^{(c)} e^{uv}_{[1,t-1]} + b^{(c)}) = h(e^{uv}_{[1,t-1]})$$

$$\alpha^{uv}_{[2,t]} = f(W^{(c)} e^{uv}_{[2,t]} + b^{(c)}) = h(e^{uv}_{[2,t]})$$

Computational Cost: We use Matlab implementation of optimization algorithm L-BFGS (Limited-memory Broyden-Fletcher-Goldfarb-Shanno) for learning optimal coding. Non-convex nature of cost function allows us to converge to local optima. We execute the algorithm for limited number of iterations to obtain unsupervised features within a reasonable period of time. Each iteration of L-BFGS executes two tasks for each edge: back-propagation to compute partial differentiation of cost function, change the parameters (W, b). For each edge the time complexity is $O(kl)$; here, k is the length of basic feature representation, l is length of unsupervised feature representation. Therefore, the time complexity of one iteration is $O(mkl)$, where m is the number of possible edges.

4.3 Supervised Link Prediction Model

Training dataset, $\hat{E}$ is feature representation for time snapshots $[1, t - 1]$, The ground truth $(\hat{y})$ is constructed from G_t. After training the supervised classification model using $\hat{\alpha} = h(\hat{E})$ and $\hat{y}$, prediction dataset $\bar{E}$ is used to predict links at G_{t+1}. For this supervised prediction task, we experiment with several classification algorithms. Among them SVM (support vector machine) and AdaBoost perform the best.

4.4 Algorithm

The pseudo-code of GRATFEL is given in Algorithm 1. For training link prediction model, we split the available network snapshots into two overlapping time windows, $[1, t - 1]$ and $[2, t]$. GTE features $\hat{E}$ and $\bar{E}$ are constructed in Lines 2 and 4, respectively. Then we learn optimal coding for node-pairs using graphlet transition features $(\hat{E} \cup \bar{E})$ in Line 5. Unsupervised feature representations are constructed using learned optimal coding (Lines 6 and 7). Finally, a classification model C is learned (Line 8), which is used for predicting link in G_{t+1}(Line 9).

Algorithm 1. GRATFEL

1: **procedure** GRATFEL($\mathbb{G}, t$)

      Input : $\mathbb{G}$: Dynamic Graph, t: Time steps

      Output: $\bar{\mathbf{y}}$: Forecasted links at time step $t+1$

2:    $\hat{\mathbf{E}}$=GraphletTransitionFeature($\mathbb{G}$,1,$t-1$)

3:    $\hat{\mathbf{y}}$=Connectivity(G_t)

4:    $\bar{\mathbf{E}}$=GraphletTransitionFeature($\mathbb{G}$,2,t)

5:    h=LearningOptimalCoding($\hat{\mathbf{E}} \cup \bar{\mathbf{E}}$)

6:    $\hat{\boldsymbol{\alpha}}$=$h(\hat{\mathbf{E}})$

7:    $\bar{\boldsymbol{\alpha}}$=$h(\bar{\mathbf{E}})$

8:    C=TrainClassifier($\hat{\boldsymbol{\alpha}}, \hat{\mathbf{y}}$)

9:    $\bar{\mathbf{y}}$=LinkForecasting($C, \bar{\boldsymbol{\alpha}}$)

10:    **return** $\bar{\mathbf{y}}$

11: **end procedure**

---

5 Experimental Results

5.1 Dataset

We use three real world dynamic networks for evaluating the performance of GRATFEL. These datasets are Enron, Collaboration and Facebook. The Enron email corpus consists of email exchanges between 184 Enron employees; the vertices of the network are employees, and the edges are email events between a pair of employees. The dataset has 11 temporal snapshots which are prepared identically as in [22]. The Collaboration dataset is constructed by using citation data from ArnetMiner (arnetminer.org). Each vertex in this dataset is an author and the edges represent co-authorship. We consider the data from years 2000-2009—total 10 snapshots considering each year as a time stamp. Many authors in the original dataset have published in only one or two years out of all ten years, so we select only those who participate in a minimum of 2 collaboration edges in at least 7 time stamps. The final dataset has 315 authors.

Lastly, the dynamic social network of Facebook wall posts [20] has 9 time stamps. We follow the same time-stamp setting and node filtering as [21], resulting in 9 time stamps of data, and 663 nodes. The dynamic link prediction task of all three datasets is to predict links on last snapshot, given the link information of all the previous snapshots. Statistics of the dynamic networks are given in Table 1. Although the number of vertices in these networks are in hundreds,

Table 1. Basic statistics of the datasets used.

	Enron	Collaboration	Facebook
Time Snaps	11	10	9
Nodes	184	315	663
Avg. Edge	217	255	1299
Node−Pairs	16, 836	49, 455	219, 453
Avg. Density	.013	.005	.006

these are large datasets considering the possible node-pairs and multiple temporal snapshots.

5.2   Competing Methods

To compare the performance of GRATFEL, we choose link prediction methods from two categories: (1) topological feature based methods and (2) feature time series based methods [5].

For topological feature based method, we consider three leading topological features: common Neighbors (CN), Adamic-Adar (AA), and Jaccard's Coefficient (J). However, in existing works these features are defined for static network only; so we adapt these features for the dynamic network setting by taking the weighted average of the feature values at different time stamps, where the weight of the t'th time stamp is 1 and the weight of the i'th time stamp is $\frac{1}{t-i+1}$. The justification of such a weighting is due to the common belief that recent interaction is a good indicator of a repeat interaction. We also tried different feature weighting, but the above weighting gives the best performance for the topological feature based dynamic link prediction. We also combine the above three features to construct a combined feature vector of length 3 $(CNAAJ)$ and use it with a classifier to build a supervised link prediction method, and include this model in our comparison.

We also compare GRATFEL with time series based neighborhood similarity scores proposed in [5]. In this work, the authors consider several neighborhood-based node similarity scores combined with connectivity information (historical edge information). Authors use time series of similarities to model the change of node similarities over time. Among 16 proposed methods, we consider 4 which are relevant to the link prediction task on unweighted networks, and also have the best performance. TS-CN-Adj represents time series on normalized score of Common Neighbors and connectivity values at time stamps $[1, t]$. Similarly, we get time series based scores for Adamic-Adar $(TS$-AA-$Adj)$, Jaccard's Coefficient $(TS$-J-$Adj)$ and Preferential Attachment $(TS$-PA-$Adj)$.

5.3   Implementation

For implementation of GRATFEL we use a combination of Python and Matlab. Graphlet transition is enumerated using a python implementation. Feature vector construction and unsupervised feature extraction are done using Matlab. The unsupervised feature extraction method runs for a maximum of 50 iterations or until it converges to a local optima. We use coding size $l = 200$ for all three datasets[3]. For link prediction we used several Matlab provided classification algorithms, namely AdaBoostM1, RobustBoost, and Support Vector Machine (SVM). We use Matlab for computing the feature values (CN, AA, J) that we

---

[3] We experiment with different coding sizes ranging from 100 to 800. The change in link prediction performance is not very sensitive to the coding size. At most 2.9 % change in PRAUC was observed for different coding sizes.

use in other competing methods. Time series methods are implemented using Python. We use the ARIMA (autoregressive integrated moving average) time series model implemented in Python module **statsmodels**. Datasets and source code are available at https://github.com/DMGroup-IUPUI/GraTFEL-Source.

5.4 Evaluation Metrics

For evaluating the proposed methods we use three metrics; namely, area under Receiver Operator Characteristics curve (AUC), area under Precision-Recall curve (PRAUC), and Normalized Discounted Cumulative Gain (NDCG). The AUC value for a link prediction problem quantifies the probability that a randomly chosen edge is ranked higher than a randomly chosen node pairs without edge. However, real world datasets for link prediction are generally skewed; the number of edges ($|E|$) is very small compared to the number of possible node-pairs ($\binom{|V|}{2}$) in the graph. In such scenarios, PRAUC gives a more informative picture of the algorithm's performance. The reason why PRAUC is more suitable for the skewed problem is that it does not factor in the effect of true negatives. In skewed data where the number of negative examples is huge compared to the number of positive examples, true negatives are not that meaningful. The last metric, NDCG, is an information retrieval metric which is widely used by the recommender systems community. NDCG measures the performance of link prediction system based on the graded relevance of the recommended links. $NDCG_p$ varies from 0.0 to 1.0, with 1.0 representing the ideal ranking of edges; p is a user-defined parameter, which represents the number of links ranked by the method. We use $p = 50$ (unless stated otherwise). NDCG is suitable when it is important to return the ranked list of top p predicted links.

5.5 Performance Comparison with Competing Methods

In this section, we present performance comparison of GRATFEL with several competing methods (see Fig. 5). The bar charts in the top, middle, and bottom rows of Fig. 5 display the results for Enron, Collaboration, and Facebook datasets, respectively. The bar charts in a row show comparison results using AUC, PRAUC, and $NDCG_{50}$ (from left to right). In total, there are 9 bars in a chart, each representing a link prediction method, where the height of a bar is indicative of the performance metric value of the corresponding method. From left to right, the first four bars (blue) correspond to the topological feature based methods, the next four (green) represent time series based methods, and the final bar (brown) represents GRATFEL.

We observe that GRATFEL (the last bar) outperforms the remaining eight methods in all the nine charts in this figure. The performance difference using PRAUC (Fig. 5(b, e, f), the charts in the middle column) is more pronounced than the performance difference using the other two metrics. Since, PRAUC is the most informative metric for classification performance on a skewed dataset, the performance difference on this metric is a strong endorsement of the superiority of GRATFEL over other methods. We first analyze the performance

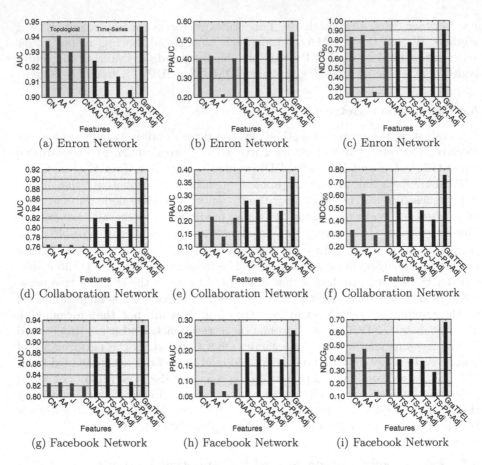

(a) Enron Network   (b) Enron Network   (c) Enron Network

(d) Collaboration Network (e) Collaboration Network (f) Collaboration Network

(g) Facebook Network   (h) Facebook Network   (i) Facebook Network

Fig. 5. Comparison with competing link prediction methods. Each bar represents a method and the height of the bar represents the value of the performance metrics. Results for Enron are presented in charts(a)–(c), results of Collaboration are in charts (d)–(f), and results of Facebook data are presented in charts (g)–(i). The group of bars in a chart are distinguished by color, so the figure is best viewed on a computer screen or color print. (Color figure online)

comparison between GRATFEL and topological feature based methods (first four bars). The best of the topological feature based methods have a PRAUC value of 0.41, 0.21, and 0.095 in Enron, Collaboration, and Facebook dataset, whereas the respective PRAUC values for GRATFEL are 0.546, 0.36, and 0.26, which translates to 33.2 %, 71.4 %, and 173.7 % improvement of PRAUC by GRATFEL for these datasets. Superiority of GRATFEL over all the topological feature based baseline methods can be attributed to the capability of graphlet transition based feature representation to capture temporal characteristics of local neighborhood. Finally, the performance of time series based method (four green bars) is generally better than the topological feature based methods. The best of the time series based method has a PRAUC value of 0.503, 0.28, and 0.19 on these datasets, and GRATFEL's PRAUC values are better than these values by 8.5%, 28.6%, and 36.8%, respectively. Time series based methods, though model the temporal behavior well, probably fail to capture signals from the neighborhood topology of the node-pairs.

Now we focus on the performance comparison using the information retrieval metric $NDCG_{50}$ (Fig. 5(c, f, i)). This metric puts higher weight on the top-ranked predicted links than the lower ranked predicted links. GRATFEL shows substantial improvement over the other methods using this metric also. The performance improvement of GRATFEL in $NDCG_{50}$ over the best among the remaining 8 methods are 9.8%, 24%, and 46.8% on the three datasets. An interesting observation using this metric is that for all the datasets, the best among the topological feature based methods is better than the best among the time series based methods. It indicates that the top ranked predicted links are explained better by the neighborhood topology than the time series of inter-action history. Finally, we discuss the comparison results for the AUC metric (Fig. 5(a, d, g)). GRATFEL is the best performer among all the methods for all three datasets with a percentage improvement over the second best method between 2% to 17%. Note that, the performance gap among all the methods are relatively small using AUC. For instance, the AUC values for all the methods in Enron dataset (Fig. 5(a)) are localized around 0.93. In general, AUC has a poor discrimination ability among classifiers in a highly skewed datasets and due to this reason PRAUC should be the preferred metric for such datasets [3].

When we compare the performance of all the algorithms across different datasets, we observe varying performance. For example, across all the metrics, the performance of dynamic link prediction on Facebook graph is lower than the performance on Collaboration graph, which, subsequently, is lower than the performance on Enron graph, indicating that link prediction in Facebook data is a harder problem to solve. The performance improvement of GRATFEL over the second best method for Facebook is the largest among the three datasets across all three metrics.

5.6 Contribution of Unsupervised Feature Extraction

GRATFEL has two novel aspects: first, utilization of graphlet transition events (GTEs) as features, and the second is unsupervised feature learning by optimal coding. In this section, we compare the relative contribution of these two aspects in the performance of GRATFEL. For this comparison, we build a version of GRATFEL which we call GTLiP. GTLiP uses GTEs just like GRATFEL, but it does not use optimal coding, rather it uses the GTEs directly as features. In Fig. 6, we show the comparison between GRATFEL, and GTLiP using $NDCG_p$ for different p values for all the datasets. The superiority of GRATFEL over GTLiP for all the datasets over a range of p values is clearly evident from the three charts. GTLiP also outperforms all the competing methods. Comparison of GTLiP and other competing methods is not shown in this figure, but the $NDCG_{50}$ from this figure can be compared with $NDCG_{50}$ charts in Fig. 5

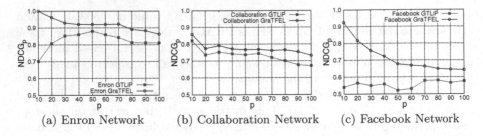

(a) Enron Network   (b) Collaboration Network   (c) Facebook Network

Fig. 6. Performance comparison between link prediction methods with (GRATFEL) and without (GTLIP) unsupervised feature extraction. Y-axis represents the $NDCG_p$ score and X-axis represents the value of p.

for confirming this claim. This shows that GTE based features, irrespective of unsupervised coding, improve the dynamic link prediction performance over the existing state of-art dynamic link prediction methods.

6 Conclusion

In this paper, we present a dynamic link prediction method, which uses unsupervised coding of graphlet transition events. We demonstrate that the proposed method performs better than several topological feature based methods and several time series based methods on three real-life dynamic graph datasets.

References

1. Bhuiyan, M.A., Rahman, M., Rahman, M., Hasan, M.A.: GUISE: uniform sampling of graphlets for large graph analysis. In: Proceedings of 12th International Conference on Data Mining, pp. 91–100 (2012)
2. Bliss, C.A., Frank, M.R., Danforth, C.M., Dodds, P.S.: An evolutionary algorithm approach to link prediction in dynamic social networks. J. Comput. Sci. 5(5), 750–764 (2014)
3. Davis, J., Goadrich, M.: The relationship between precision-recall and ROC curves. In: Proceedings of the 23rd International Conference on Machine Learning, pp. 233–240 (2006)
4. Dunlavy, D.M., Kolda, T.G., Acar, E.: Temporal link prediction using matrix and tensor factorizations. ACM Trans. Knowl. Disc. Data 5(2), 10:1–10:27 (2011)
5. Güneş, İ., Gündüz-Öğüdücü, Ş., Çataltepe, Z.: Link prediction using time series of neighborhood-based node similarity scores. Data Min. Knowl. Disc. 30(1), 147–180 (2015)
6. Hasan, M.A., Chaoji, V., Salem, S., Zaki, M.: Link prediction using supervised learning. In: Proceeding of SDM 2006 workshop on Link Analysis, Counterterrorism and Security (2006)
7. Hasan, M., Zaki, M.: A survey of link prediction in social networks. In: Social Network Data Analytics, pp. 243–275 (2011)

8. Lee, H., Pham, P., Largman, Y., Ng, A.Y.: Unsupervised feature learning for audio classification using convolutional deep belief networks. In: Advances in neural information processing systems. pp. 1096–1104 (2009)
9. Liben-Nowell, D., Kleinberg, J.: The link-prediction problem for social networks. J. Am. Soc. Inform. Sci. Technol. **58**(7), 1019–1031 (2007)
10. Lichtenwalter, R.N., Lussier, J.T., Chawla, N.V.: New perspectives and methods in link prediction. In: Proceedings of the 16th ACM SIGKDD International Conference on Knowledge Discovery and Data Mining, pp. 243–252 (2010)
11. Menon, A.K., Elkan, C.: Link prediction via matrix factorization. In: Proceedings of the 2011 European Conference on Machine Learning and Knowledge Discovery in Databases - volume Part II, pp. 437–452 (2011)
12. Milenković, T., Pržulj, N.: Uncovering biological network function via graphlet degree signatures. Cancer Inform. **6**, 257–273 (2008)
13. Ng, J., Yang, F., Davis, L.: Exploiting local features from deep networks for image retrieval. In: Proceedings of the IEEE Conference on Computer Vision and Pattern Recognition Workshops, pp. 53–61 (2015)
14. Perozzi, B., Al-Rfou, R., Skiena, S.: Deepwalk: Online learning of social representations. In: Proceedings of the 20th ACM SIGKDD Conference on Knowledge Discovery and Data Mining, pp. 701–710 (2014)
15. Pržulj, N.: Biological network comparison using graphlet degree distribution. Bioinformatics **23**(2), e177–e183 (2007)
16. Pržulj, N., Corneil, D.G., Jurisica, I.: Modeling interactome: scale-free or geometric? Bioinformatics **20**(18), 3508–3515 (2004)
17. Rahman, M., Bhuiyan, M.A., Hasan, M.A.: Graft: An efficient graphlet counting method for large graph analysis. IEEE Trans. Knowl. Data Eng. **26**(10), 2466–2478 (2014)
18. Sarkar, P., Chakrabarti, D., Jordan, M.I.: Nonparametric link prediction in dynamic networks. In: Proceedings of International Conference on Machine Learning, pp. 1687–1694 (2012)
19. Shervashidze, N., Petri, T., Mehlhorn, K., Borgwardt, K.M., Vishwanathan, S.: Efficient graphlet kernels for large graph comparison. In: Proceeding of the 12th International Conference on Artificial Intelligence and Statistics, pp. 488–495 (2009)
20. Viswanath, B., Mislove, A., Cha, M., Gummadi, K.P.: On the evolution of user interaction in facebook. In: Proceedings of the 2nd ACM workshop on Online social networks, pp. 37–42 (2009)
21. Xu, K.S.: Stochastic block transition models for dynamic networks. In: Proceedings of the 18th International Conference on Artificial Intelligence and Statistics, pp. 1079–1087 (2015)
22. Xu, K.S., Hero III, A.O.: Dynamic stochastic blockmodels for time-evolving social networks. IEEE J. Sel. Top. Sign. Proces. **8**(4), 552–562 (2014)

CHADE: Metalearning with Classifier Chains for Dynamic Combination of Classifiers

Fábio Pinto[✉], Carlos Soares, and João Mendes-Moreira

INESC TEC/Faculdade de Engenharia, Universidade do Porto,
Rua Dr. Roberto Frias s/n, 4200-465 Porto, Portugal
fhpinto@inesctec.pt, {csoares,jmoreira}@fe.up.pt

Abstract. Dynamic selection or combination (DSC) methods allow to select one or more classifiers from an ensemble according to the characteristics of a given test instance x. Most methods proposed for this purpose are based on the nearest neighbours algorithm: it is assumed that if a classifier performed well on a set of instances similar to x, it will also perform well on x. We address the problem of dynamically combining a pool of classifiers by combining two approaches: metalearning and multi-label classification. Taking into account that diversity is a fundamental concept in ensemble learning and the interdependencies between the classifiers cannot be ignored, we solve the multi-label classification problem by using a widely known technique: Classifier Chains (CC). Additionally, we extend a typical metalearning approach by combining metafeatures characterizing the interdependencies between the classifiers with the base-level features. We executed experiments on 42 classification datasets and compared our method with several state-of-the-art DSC techniques, including another metalearning approach. Results show that our method allows an improvement over the other metalearning approach and is very competitive with the other four DSC methods.

Keywords: Ensembles · Classifier Chains · Dynamic combination · Metalearning

1 Introduction

Ensemble methods are still one of the favorite techniques used by researchers and practitioners to deal with classification problems. The high predictive performance reported together with the increased computational power that we have available has led to a widespread of these techniques. Our proposal focuses on how to aggregate the output of the classifiers in order to achieve a final prediction. We propose a dynamic combination method to combine a subset of classifiers for each test instance. The method is based on a widely known multi-label classification technique, Classifier Chains (CC) [16].

In a typical classification problem, an instance x, represented by a vector of d attributes values, belongs to only one class, y. However, in a multi-label classification problem, an instance x can belong to a subset of L labels, therefore

© Springer International Publishing AG 2016
P. Frasconi et al. (Eds.): ECML PKDD 2016, Part I, LNAI 9851, pp. 410–425, 2016.
DOI: 10.1007/978-3-319-46128-1_26

$Y = \{y_1, ..., y_L\}$. Each label $l, 1 \ leql \leq L$, is associated with x if $y_l = 1$ or not if $y_l = 0$.

A common technique to deal with multi-label classification problems in the so called *problem transformation*. It consists in decomposing the multi-label problem into several binary problems. Therefore, instead of using a classifier that has the ability to deal with multiple outputs, one binary classifier can be trained for each label. However, this technique has a major drawback: it does not take into account the intrinsic interdependencies that can exist between the labels. This is very important for our problem since it is well known that diversity is a fundamental concept of ensembles [10] and can only be managed by a meta-model that has information about the classifiers interdependencies.

Dynamic classifier selection is the problem of deciding which subset of models from an ensemble should be used to generate the prediction for an instance x. It can be addressed as a multi-label classification problem. An ensemble F consists of $K, 1 \leq k \leq K$ individual classifiers, $F = \{f_1, ..., f_K\}$. Each classifier is represented by a label $Y = \{y_{1,x}, ..., y_{k,x}\}, y_{k,x} \in \{0, 1\}$. If $y_{k,x} = 1$, it means that the classifier k correctly classified the instance x; otherwise, $y_{k,x} = 0$. Therefore, using the CC method, we can train a meta-model that relates the attributes of a dataset with the output of each classifier from the ensemble. Given a new test instance x, the meta-model is able to predict which classifiers should be used for the final prediction.

Since our method includes a step in which a model is learned at a higher level (meta), we consider it as a metalearning approach [1]. Therefore, we compare it with other dynamic selection or combination methods (DSC) methods, including an alternative metalearning approach [4].

We executed experiments on 42 classification datasets from the UCI repository [11]. To the best of our knowledge this is largest set of experiments comparing DSC methods. For each dataset, we generated a bagging ensemble of 100 decision stumps. Then, we tested several state-of-the-art DSC methods. Results show the competitiveness of our method, not only at the base-level but also at the meta-level. We also explored the performance of our method in the widely known XOR problem in order to achieve a better understanding of its behaviour.

The main contributions of this paper are:

1. new methods for dynamic combination of classifiers, CHADE and E-CHADE
2. an extensive experimental evaluation to demonstrate the competitiveness of CHADE and E-CHADE
3. a novel way of using landmarkers as metafeatures for metalearning

The paper is organized as follows. In Sect. 2 we present a summary of the state-of-the-art for DSC methods. Section 3 presents our method for dynamic combination of classifiers. In Sect. 4 we describe the experiments that were carried out. Section 5 includes a discussion about the characteristics of our method in the light of the results that were obtained in the XOR problem. Finally, Sect. 6 concludes the paper and sets directions of future work.

2 Related Work

We organized the state-of-the-art on dynamic approaches for ensembles of classifiers into two groups: dynamic selection, for the methods that only select one classifier for each test instance x; and dynamic combination, for the methods that can select more than one classifier for each test instance.

2.1 Dynamic Selection

The first paper concerning dynamic selection of classifiers is due to Ho et al. [7]. In that paper, the authors proposed a selection based on a partition of training examples. The individual classifiers are evaluated on each partition to find the best one for each. Then, the test instance to be predicted is categorized into a partition and classified by the corresponding best classifier.

The DS-LA LCA based method and the DS-LA OLA-based method are often used as benchmark in comparative studies [3,23]. For abbreviation purposes we refer to these as OLA and LCA, respectively. Both methods calculate an estimation of accuracy of the base classifiers in the local region of the feature space close to the test instance in the training dataset. In OLA, it is the percentage of correct classifications within the local region; in LCA, it is the percentage of correct classifications within the local region but considering only those examples where the classifier has given the same class as the one it gives for the test instance. In both methods, only one classifier is selected for the final prediction.

Concerning metalearning approaches, Todorovski and Džeroski [18] proposed the meta decision trees, a method to select the best predictor of an ensemble of decision trees for a given test instance. The decision is made by a meta-model that learns the prediction patterns of the ensemble summarized in a set of three metafeatures.

Yankov et al. [24] proposed a method to select from an ensemble of two k-NN models the one best suited for a given instance. The selection is done by a Support Vector Machine model using metafeatures extracted from the instances.

2.2 Dynamic Combination

The first dynamic combination approach was introduced by Merz [13]. However, the results showed that a simple majority combination was superior to their dynamic approach.

Kuncheva and Rodríguez [9] proposed a method based on the oracle concept. Essentially, each classifier of the ensemble consists of two sub-classifiers and an oracle that decides which of the two sub-classifiers is going to be used to predict the test instance. The oracle is a random linear function. They claimed that the random oracle idea works because it adds diversity to the ensemble. Later, Ko et al. [8] improved this idea by adding a k-nearest neighbours approach and proposed the KNORA-E and KNORA-U methods. In the former, only the classifiers that correctly classify all the K-nearest patterns are used; in the latter, the classifiers that correctly classify any of the k-nearest neighbours are used - a single classifier can be selected more than once.

Tsymbal [19] and Tsymbal and Puuronen [21] combined dynamic integration with classifier ensembles using bagging and boosting algorithms. Results suggest that dynamic integration improves significantly the performance of the ensembles instead of the more typical majority voting integration. Later, Tsymbal et al. [20] also presented experiments in which a dynamic integration approach was added to Random Forest, instead of the simple majority voting.

Santana et al. [17] proposed a method that explicitly uses accuracy and diversity to select a subset of classifiers. The method sorts the classifiers in decreasing order of accuracy and in increasing order of diversity. They presented two versions: DS-KNN, which is very similar to LCA and OLA but it takes into account diversity; and DS-Cluster, that uses a clustering process to divide the validation set into clusters and, for each cluster. The most promising classifiers are selected.

Liyanage et al. [12] proposed a dynamically weighted ensemble classification framework whereby an ensemble of multiple classifiers are trained on clustered features. The decisions from these multiple classifiers are dynamically combined based on the distances of the cluster centers to each test data sample being classified. Results showed that their method is significantly better than a Support Vector Machine baseline classifier.

Recently, Cruz et al. [4] proposed a method that uses metalearning for dynamic combination of classifiers. This method uses a meta-model to decide if a base classifier is competent to classify a given test instance. The meta-model is learned using metafeatures that capture the prediction patterns of the base classifiers in its regions of competence and in the overall decision space.

3   CHADE

We propose CHADE (CHAined Dynamic Ensemble) for dynamic combination of ensembles, combining metalearning and multi-label learning. The CHADE method, as presented in Fig. 1, is composed by three stages: (1) generation, (2) meta-training and (3) generalization.

The generation Stage simply consists in training an ensemble of K classifiers. We used bagging [2] for experimental purposes, but this is not a requirement of CHADE. For future work, we plan to explore other ensemble learning algorithms, both for homogeneous and heterogeneous ensembles.

Stage 2 is the meta-training phase. The ensemble of classifiers is used to make predictions on a validation set. These predictions are then used by a 0–1 loss function, I, that compares them with the true target from the validation set. This generates k binary variables (metafeatures), that represent the performance of each classifier for each example in the validation set. Therefore, the meta-training data D', is composed by the independent variables of the base-level data, X_{val}; and the K binary variables generated, $W = \{W_1, W_2, ..., W_K\}$.

Note that the extension of the data, W, can be regarded as landmarkers. However, in traditional metalearning, landmarkers are aggregated measures of

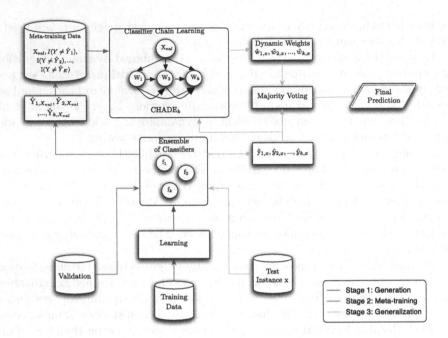

Fig. 1. CHADE framework.

performance of a simple algorithm. In CHADE, since we are interested in capturing patterns at the instance level, it is more useful to not aggregate the performance of the models over all the instances in the validation set. We can also relate this approach with stacking [22], since the dataset is extended with information obtained from the predictions made by the base-level classifiers, although in stacking the predictions are used directly and here we use an indicator of the accuracy of the prediction. The meta-training dataset D' shows the morphology of a typical multi-label classification problem, as shown in the example presented in Table 1.

Table 1. Example of a meta-training dataset D'.

X_1	X_2	X_3	W_1	W_2	...	W_K
10	yes	1	0	1	...	1
8	no	2	0	1	...	1
12	no	5	1	1	...	0
...	...	...	...	...	...	...

After generating the meta-training data, the CC algorithm [16] for multi-label classification is used for training the meta-model *CHADE*. The pseudo code for Stage 2 is described in Algorithm 1. The training can be done with any sort of learning algorithm for classification.

Input: $D = (X_{val}, Y_{val}), F_k = \{f_1, ..., f_K\}$
Output: $CHADE$
for $k \in 1 ... |W|$ **do**
 meta-label computation and training
 $D' \leftarrow \{\}$
 for $(x_{val}, y_{val}) \in D$ **do**
  $D' \leftarrow D' \cup ((x, w_1, ..., w_{k-1}), w_k)$
  Train MC_k to predict w_k
  $MC_k : D' \rightarrow w_k \in \{0, 1\}$
 end
end
Algorithm 1. CHADE training pseudocode.

Finally, in Stage 3, the meta-model trained in Stage 2 is used together with the ensemble generated in Stage 1 for a dynamic combination of the classifiers. The pseudocode for this Stage is presented in Algorithm 2. The weight of each classifier for a test instance x is assigned by the meta-model $CHADE$. The weights $\hat{W}$ are then combined with the base-level predictions $\hat{Y}$ for x by majority voting. This results in the final prediction $\hat{y}$.

Input: $x_{test}, F_k = \{f_1, ..., f_K\}, CHADE$
Output: $\hat{y}$
$\hat{Y} \leftarrow \{\}$
for $k \in 1 ... |W|$ **do**
 $\hat{w}_k \leftarrow MC_k : (x_{test}, \hat{w}_1, ..., \hat{w}_{k-1})$
 $\hat{Y} \leftarrow \hat{Y} \cup (\hat{w}_k \times \hat{f}_k \leftarrow f_k : (x_{test}))$
end
$MajorityVoting(\hat{Y})$
Algorithm 2. CHADE generalization pseudocode.

CHADE does not require parameter tuning. In comparison with the other DSC techniques that we mentioned in the previous section, CHADE presents a major advantage difference: it does not rely on the nearest neighbours algorithm. This can make CHADE particularly useful in datasets with a large number of training examples, since distance computation can be quite costly in those cases.

The paper that introduced the CC method for multi-label classification also proposed the Ensemble of Classifier Chains (ECC) [16]. In ECC, several CC models are trained. However, the order in which the classifiers are *chained* is different and each CC model is trained in a bootstrap sample of the training data. In comparison with CC, ECC allows to reduce the variance component of the error. Therefore, we also did experiments with an ensemble version of CHADE that we named E-CHADE.

4   Experiments

The experiments that we carried out aimed to answer the following research questions:

1. Can CHADE/E-CHADE improve the performance of Bagging?
2. How does the performance of CHADE/E-CHADE relates with other Metalearning approaches for dynamic combination of classifiers?
3. How does the performance of CHADE/E-CHADE relates with other state-of-the-art DSC techniques?

4.1   Setup

A total of 42 datasets were used in the experiments, all of them obtained from the UCI machine learning repository [11]. The selection was done randomly although we took into consideration the variety of the learning problems in order to obtain a diverse set of datasets. A brief description of each dataset is given in Table 2. To the best of our knowledge, these are the experiments for dynamic combination of classifiers with the largest number of datasets.

The datasets were split into training (50 %), validation (25 %) and test set (25 %). The split was done using stratified sampling. Each experiment was repeated 10 times and the results were averaged. We used accuracy as evaluation metric. The methodology proposed by Demšar [5] was used for statistical validation of the results.

For each learning problem, we generated a bagging ensemble of 100 decision stumps. We chose to use weak learners since it has been reported that this approach enhances the detection of differences between dynamic approaches [3].

We selected a few DSC techniques for comparison with our approach. The first is META-DES [4], a metalearning method that uses a set of five different meta-features to learn a meta-model that predicts if a base classifier is competent to classify a given test instance. To the best of our knowledge this is the only metalearning method for dynamic combination of classifiers proposed in the literature. Our experimental setup is slightly different from the one used by Cruz et al. [4]. Therefore, we made two changes to adapt it to our experimental setup: (1) the meta-model is learned with a decision tree algorithm instead of the Levenberg-Marquadt algorithm. Preliminary results showed that the choice of the learning algorithm for the meta-training did not have a significant impact on the method. Therefore, we decided to use decision trees for the meta-training stage of META-DES and CHADE; (2) one metafeature (perpendicular distance between the input sample and the decision boundary of the base classifier) could not be used since its dependent of the base classifier used in the experiments, which, originally, was a perceptron classifier and, here, we use decision stumps. The remaining features of the algorithm are implemented as detailed in the original paper. Concerning the parameters of META-DES, we used the ones that achieved the best results in the original experiments.

Table 2. Datasets summary. In all the datasets marked with * we used a sample of 5000 instances in order to speed up the process.

Dataset	No. of Instances	Dimensionality	No. of Classes
abalone	4177	9	28
allbp	3772	28	3
allhyper	3772	28	5
allhypo	3772	28	4
allrep	3772	28	4
ann	7200*	22	3
c_class_flares	1389	11	8
car	1728	7	4
contraceptive	1473	10	3
crx	690	16	2
diabetes	768	9	2
dis	3772	28	2
fluid	537	31	9
german_numb	10010	25	2
german_symb	1000	21	2
glass	214	10	6
ibm_stock_val	8087*	18	3
ionosphere	351	35	2
krvskp	3196	37	2
led7	3200	8	10
led24	3200	25	10
m_class_flares	1389	11	6
mushrooms	8124*	22	2
optical	5620*	65	10
page	5473*	11	5
parity	1024	11	2
pyrimidines	6996*	55	2
quisclas	5891*	19	3
recljan2jun97	33170*	20	2
sat	6435*	37	6
segment	2310	20	7
shuttle	58000*	10	7
sick	3772	28	2
sick-euthyroid	3163	25	2
tic-tac-toe	958	10	2
titanic	2201	4	2
vehicle	846	19	4
vowel	990	11	11
waveform21	5000	22	3
waveform40	5000	41	3
x_class_flares	1389	11	3
yeast	1484	9	10

As for other state-of-the-art DSC techniques, we compare CHADE with OLA, LCA, KNORA-E and KNORA-U. These techniques were selected since they have shown good results in several experimental studies [3]. Regarding parameters, k was set to 10 in all experiments. Also, in the base-level experiments, we compare the DSC methods with an abstract model, the Oracle. This model selects the classifier that correctly predicts the label for any given test instance, if such classifier exists. This comparison assesses whether the DSC techniques have room for improvement or not.

Finally, we tested two variations of CHADE: the single meta-model version (*CHADE*) and the ensemble version (*E-CHADE*). The number of meta-models in E-CHADE was set to 10 since it was reported to be a good value for ensembles of classifiers chains [16]. The details of these methods can be found in Sect. 3 of this paper.

4.2   Comparison with Another Metalearning Approach

With the aim of a more complete comparison with the alternative approach META-DES, we compared the performance of the metalearning methods both at the base-level and meta-level. The purpose of the meta-level evaluation is to assess if the methods are combining the correct models and leaving out the ones that fail the predictions. We also compared the metalearning methods with a baseline, that is the majority class in the meta-training data. This is important to evaluate the quality of the combinations that are being made by the metalearning approaches.

It is important to notice that a better performance at the meta-level does not necessarily imply a better performance at the base-level. For instance, consider the classifiers c_1, c_2 and c_3, and the meta-models $meta\text{-}model_1$ and $meta\text{-}model_2$. Given a test instance x, $meta\text{-}model_1$ recommends combining c_1 and c_2; on the other hand, $meta\text{-}model_2$ recommends combining c_1, c_2 and c_3. We then verify that the predictions made by the classifiers are the same for the three of them and they are all correct. This scenario implies that $meta\text{-}model_1$ has an accuracy of 66.6 % and $meta\text{-}model_2$ has an accuracy of 100 %. However, the base-level prediction is the same for both cases. Therefore, $meta\text{-}model_1$ and $meta\text{-}model_2$ have the same base-level accuracy.

Figure 2 shows the Critical Difference diagram for the comparison of the metalearning approaches at the meta-level. E-CHADE clearly presents the best performance. On one hand, the difference between E-CHADE and the other methods evaluated is statistically significant. On the other hand, although CHADE achieves a better mean rank than META-DES and the baseline, there is no evidence in these experiments that the difference is statistically significant.

Considering now the base-level performance, Fig. 3 shows that E-CHADE and CHADE have a better performance than META-DES and the baseline, Bagging. The difference between E-CHADE and CHADE in comparison with Bagging is statistically significant.

These results indicate that the answer to our first question is positive: CHADE and E-CHADE do improve the performance of Bagging.

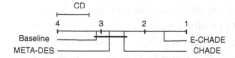

Fig. 2. Critical Difference diagrams (with $\alpha = 0.05$) for the comparison with META-DES at the meta-level. The null hypothesis of the Friedman's test is rejected for $\alpha = 0.01, 0.05$ and 0.1.

However, the difference between E-CHADE, CHADE and META-DES is not statistically significant. The results suggest that CHADE and E-CHADE allow an improvement over META-DES but this statement is not statistically validated. The same conclusion can be made for the difference between META-DES and Bagging. This answers the second research hypothesis that we stated previously.

Interestingly, the results obtained at the base-level are in accordance with the ones obtained at the meta-level. In fact, E-CHADE presents the best overall performance, although the difference is more clear for the comparison made at the meta-level. We must also state the Oracle model has indisputably the best performance of all the methods compared, by far difference. This indicates that there is room for improvement by the dynamic approaches.

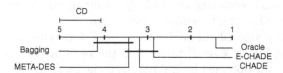

Fig. 3. Critical Difference diagrams (with $\alpha = 0.05$) for the comparison with META-DES at the base-level. The null hypothesis of the Friedman's test is rejected for $\alpha = 0.01, 0.05$ and 0.1.

4.3 Comparison with Other Dynamic Selection/combination Methods

In this section, we extend the comparison made in the previous section to other state-of-the-art DSC methods. The comparison is only made at the base-level since the majority of the methods do not follow a metalearning approach.

Figure 4 presents the Critical Difference diagram of the experiments. The first result that stands out is that the majority of methods obtain a similar performance. At the top of the ranking, the Oracle model appears isolated; at the bottom, LCA and Bagging appear very close to each other.

A more detailed analysis shows that OLA and E-CHADE present the best performance, followed closely by CHADE and KNORA-E. The difference between these four methods to META-DES and KNORA-U is not statistically significant; however, if we compare them with LCA and Bagging we see that the difference is now statistically significant. Given these results, we find it difficult to extract conclusions. However, the fact that E-CHADE and CHADE are two of the three techniques (excluding the Oracle model) with the best mean ranking is a very promising result.

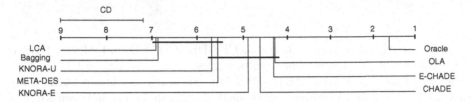

Fig. 4. Critical Difference diagrams (with $\alpha = 0.05$) for the comparison with several dynamic selection/combination methods at base-level. The null hypothesis of the Friedman's test is rejected for $\alpha = 0.01, 0.05$ and 0.1.

We must also state that these results are consistent with conclusions in a recent survey about DSC [3]. In that survey the authors concluded that there is no evidence that one specific technique may win over all the others for any classification problem. They also suggest that it should be put effort into developing a method that recommends which dynamic approach should be used for a specific dataset. Our results reinforce this claim.

We further investigated the performance of E-CHADE in comparison with CHADE. Specifically, we wanted to verify if the performance of CHADE could be as good as the one obtained by E-CHADE if the meta-model was trained in a specific order. The top plot of Fig. 5 shows the mean rank as more meta-models are added to E-CHADE; at the bottom, the same figure shows the individual mean rank of each meta-model that is added to E-CHADE. Figure 6 presents the same graphs but for the meta-level.

We can see in Fig. 5 that none of the individual CHADE meta-models achieves a mean rank as good as the one obtained by E-CHADE. This shows that the ensemble framework for classifier chains is effective in improving the performance of CHADE. This result is consistent with experiments made in multi-label classification datasets at the base-level [16].

Still regarding Fig. 5, its possible to verify that the performance of E-CHADE stabilizes after the 4th/5th meta-model is added to the ensemble. This indicates that E-CHADE does not requires the 10 meta-models that we trained in our experiments and therefore the computational cost of the method can be reduced.

Finally, Fig. 6 shows the same graphs presented in Fig. 5 but for the meta-level. The results are quite similar. However, we can see a more pronounced improvement by adding more meta-models to E-CHADE at the meta-level than in the base-level.

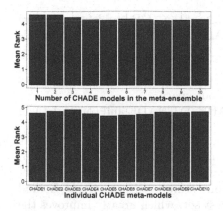

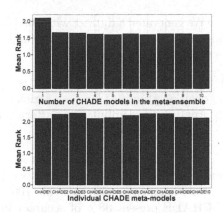

Fig. 5. Evolution of the base-level mean rank as more meta-models are added to E-CHADE.

Fig. 6. Evolution of the meta-level mean rank as more meta-models are added to E-CHADE.

5   Further Analysis

In this Section we discuss and analyse the behaviour of CHADE and E-CHADE in comparison with the Oracle model. For this, we carried out experiments with the XOR problem [14]. The XOR problem is a classic example of a dataset in which a linear model will not perform well. As we can see in Fig. 7, there does not exist a linear model that can separate the blue and red points. Therefore, a decision stump cannot solve this problem. However, a bagging ensemble of decision stumps together with a DSC technique can be used successfully. For that, the dynamic component needs to be able to select the appropriate decision stump(s) for each test instance.

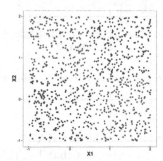

Fig. 7. XOR problem. (Color figure online)

We generated a dataset with 1000 data points of the XOR problem to improve our understanding of CHADE's (and E-CHADE's) behaviour. The aim of this experiment is twofold:

1. Assess if CHADE/E-CHADE is able to correctly identify the appropriate decision stump(s) for each region of the input space.
2. Analyze the combination patterns of CHADE/E-CHADE and compare them with the Oracle.

The XOR data was split into training (50%), validation (25%) and test set (25%) using stratified sampling. Once more, we generated bagging ensembles of 100 decision stumps. Both CHADE and E-CHADE present 88% of accuracy in the test set, which greatly improves the 49.6% achieved by Bagging. By definition, as the data is not noisy, the Oracle model achieves 100%.

Figure 8 shows three heat maps that represent the combination of classifiers made by CHADE, E-CHADE and Oracle, for each test instance. A red square means that the classifier (column) was selected for the corresponding example (row). We can see that it is possible to identify four regions in the heat maps, each region corresponding to the four regions of the input space visible in Fig. 7. This result is more clearly seen for CHADE and the Oracle than for E-CHADE. Also, we see that the heat map for CHADE is more similar to the one for the Oracle than the heat map of E-CHADE.

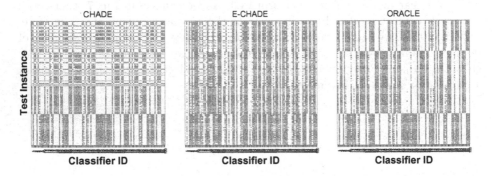

Fig. 8. Heat maps showing the combination of classifiers made by each technique. (Color figure online)

Still regarding Fig. 8, it seems that E-CHADE's heat map has more red squares than the other two. This suggests that E-CHADE combines larger sets of classifiers than CHADE or Oracle. We confirmed this result by analysing the distribution of the number of classifiers selected per instance by each method, presented in Fig. 9; and the distribution of the number of times each classifier was selected, presented in Fig. 10. E-CHADE selects larger sets of classifiers than the other two; on the other hand, there seems to be no difference between

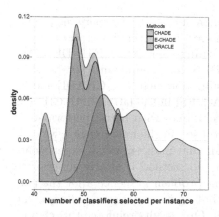

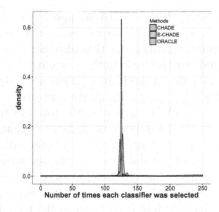

Fig. 9. Distribution of the number of classifiers selected per instance by each method.

Fig. 10. Distribution of the number of times each classifier was selected by each method.

CHADE and Oracle regarding this statistic. Figure 10 also shows an interesting result: CHADE and Oracle select the classifiers in a more even manner than E-CHADE. All the classifiers are used at least 120 times by the first two, while E-CHADE often discards some of the classifiers from the generalization phase.

Finally, we also computed the percentage of duplicated sets of classifiers combined by each method, shown in Table 3. We can see that E-CHADE presents the lower percentage of duplicated classifiers, followed closely by the Oracle model. This result is consistent with the previous one, since we also verified that E-CHADE combines more classifiers per instance than the other methods. Furthermore, this might also indicate that E-CHADE is somehow more *dynamic* than the other techniques, which can be useful in some datasets. We plan to further investigate this characteristic in future work.

Table 3. Percentage of duplicated sets of classifiers combined by each method.

CHADE	E-CHADE	Oracle
90 %	82.8 %	84.4 %

6 Conclusions and Future Work

This paper proposes a method for dynamic combination of classifiers that uses the widely known multi-label classification technique, Classifier Chains. In order to do so, we transform the problem of dynamically combining a set of classifiers into a multi-label classification problem. We propose two versions of the method: CHADE and E-CHADE, based on an ensemble variant of CC, ECC.

We evaluated CHADE and E-CHADE in a large set of experiments with 42 classification datasets. Several state-of-the-art DSC techniques were implemented, including one that also uses a metalearning approach (META-DES). We tested our methods initially against META-DES and then against several DSC techniques. In the former experiment, the results obtained with CHADE and E-CHADE suggest an improvement over META-DES; in the latter, both CHADE and E-CHADE appear in the top 3 of the techniques with better performance. Given the large number of datasets that were used in the experiments, we consider these results as very promising. We recall that most empirical experiments comparing several DSC techniques show that no method is clearly superior to the others. The characteristics of the learning problem influence a lot the performance of the methods [3]. It is expected that as more datasets are used, the smaller the difference between the techniques. Our results reinforce this claim.

For a better understanding of the behaviour of CHADE and E-CHADE, we tested it in the XOR problem. We showed visually that both CHADE and E-CHADE are able to identify the classifiers that are more suitable for each one of the four regions of the input space. We also showed that the combination patterns obtained by CHADE are more similar to the ones made by the Oracle than the ones created by E-CHADE. This can be justified by the fact that E-CHADE combines, on average, more classifiers than CHADE.

As for future work, we plan to study strategies for defining in a non-random way the order in which CHADE and E-CHADE are trained. This could not only improve the performance of the methods but also reduce its computational cost (if less CC's need to be trained). We will start by exploring some strategies already proposed for ECC [15].

The experiments that we carried out in this paper were conducted with homogeneous ensembles. We are interested in verifying the performance of CHADE and E-CHADE in a heterogeneous ensembles scenario. Since heterogeneous ensembles are usually more diverse, they should make the task of the dynamic method more difficult. This could also be an opportunity to study the relation between the diversity of an ensemble and the performance of the DSC method.

Finally, DSC methods are one of the techniques that can be used for dealing with concept drift in data streams [6]. We plan to test CHADE and E-CHADE for that purpose.

Acknowledgments. This research has received funding from the ECSEL Joint Undertaking, the framework programme for research and innovation horizon 2020 (2014–2020) under grant agreement number 662189-MANTIS-2014-1.

References

1. Brazdil, P., Carrier, C.G., Soares, C., Vilalta, R.: Metalearning: Applications to Data Mining. Springer, Heidelberg (2009)
2. Breiman, L.: Bagging predictors. Mach. Learn. **24**(2), 123–140 (1996)
3. Britto, A.S., Sabourin, R., Oliveira, L.E.: Dynamic selection of classifiersa comprehensive review. Pattern Recogn. **47**(11), 3665–3680 (2014)

4. Cruz, R.M., Sabourin, R., Cavalcanti, G.D., Ren, T.I.: META-DES: a dynamic ensemble selection framework using meta-learning. Pattern Recogn. **48**(5), 1925–1935 (2015)
5. Demšar, J.: Statistical comparisons of classifiers over multiple data sets. J. Mach. Learn. Res. **7**, 1–30 (2006)
6. Gama, J., Žliobaitė, I., Bifet, A., Pechenizkiy, M., Bouchachia, A.: A survey on concept drift adaptation. ACM Comput. Surv. (CSUR) **46**(4), 44 (2014)
7. Ho, T.K., Hull, J.J., Srihari, S.N.: Decision combination in multiple classifier systems. Trans. Pattern Anal. Mach. Intell. **16**(1), 66–75 (1994)
8. Ko, A.H., Sabourin, R., Britto Jr., A.S.: From dynamic classifier selection to dynamic ensemble selection. Pattern Recogn. **41**(5), 1718–1731 (2008)
9. Kuncheva, L., Rodriguez, J.J., et al.: Classifier ensembles with a random linear oracle. Trans. Knowl. Data Eng. **19**(4), 500–508 (2007)
10. Kuncheva, L.I., Whitaker, C.J.: Measures of diversity in classifier ensembles and their relationship with the ensemble accuracy. Mach. Learn. **51**(2), 181–207 (2003)
11. Lichman, M.: UCI machine learning repository (2013). http://archive.ics.uci.edu/ml
12. Liyanage, S.R., Guan, C., Zhang, H., Ang, K.K., Xu, J., Lee, T.H.: Dynamically weighted ensemble classification for non-stationary EEG processing. J. Neural Eng. **10**(3), 036007 (2013)
13. Merz, C.J.: Dynamical selection of learning algorithms. In: Fisher, D., Lenz, H.-J. (eds.) Learning from Data, vol. 112, pp. 281–290. Springer, New York (1996)
14. Minsky, M., Papert, S.: Perceptrons. MIT press, Cambridge (1969)
15. Read, J., Martino, L., Luengo, D.: Efficient monte carlo methods for multi-dimensional learning with classifier chains. Pattern Recogn. **47**(3), 1535–1546 (2014)
16. Read, J., Pfahringer, B., Holmes, G., Frank, E.: Classifier chains for multi-label classification. Mach. Learn. **85**(3), 333–359 (2011)
17. Santana, A., Soares, R.G., Canuto, A.M., de Souto, M.C.: A dynamic classifier selection method to build ensembles using accuracy and diversity. In: Brazilian Symposium on Neural Networks, pp. 36–41. IEEE (2006)
18. Todorovski, L., Džeroski, S.: Combining classifiers with meta decision trees. Mach. Learn. **50**(3), 223–249 (2003)
19. Tsymbal, A.: Decision committee learning with dynamic integration of classifiers. In: Štuller, J., Pokorný, J., Thalheim, B., Masunaga, Y. (eds.) ADBIS/DASFAA -2000. LNCS, vol. 1884, pp. 265–278. Springer, Heidelberg (2000). doi:10.1007/3-540-44472-6_21
20. Tsymbal, A., Pechenizkiy, M., Cunningham, P.: Dynamic integration with random forests. In: Fürnkranz, J., Scheffer, T., Spiliopoulou, M. (eds.) ECML 2006. LNCS (LNAI), vol. 4212, pp. 801–808. Springer, Heidelberg (2006). doi:10.1007/11871842_82
21. Tsymbal, A., Puuronen, S.: Bagging and boosting with dynamic integration of classifiers. In: Zighed, D.A., Komorowski, J., Żytkow, J. (eds.) PKDD 2000. LNCS (LNAI), vol. 1910, pp. 116–125. Springer, Heidelberg (2000). doi:10.1007/3-540-45372-5_12
22. Wolpert, D.H.: Stacked generalization. Neural Netw. **5**(2), 241–259 (1992)
23. Woods, K., Bowyer, K., Kegelmeyer Jr., W.P.: Combination of multiple classifiers using local accuracy estimates. In: Computer Society Conference on Computer Vision and Pattern Recognition, pp. 391–396. IEEE (1996)
24. Yankov, D., DeCoste, D., Keogh, E.: Ensembles of nearest neighbor forecasts. In: Fürnkranz, J., Scheffer, T., Spiliopoulou, M. (eds.) ECML 2006. LNCS (LNAI), vol. 4212, pp. 545–556. Springer, Heidelberg (2006). doi:10.1007/11871842_51

Aggregating Crowdsourced Ordinal Labels via Bayesian Clustering

Xiawei Guo[(✉)] and James T. Kwok

Department of Computer Science and Engineering,
Hong Kong University of Science and Technology, Clear Water Bay, Hong Kong
{xguoae,jamesk}@cse.ust.hk

Abstract. Crowdsourcing allows the collection of labels from a crowd of workers at low cost. In this paper, we focus on ordinal labels, whose underlying order is important. Crowdsourced labels can be noisy as there may be amateur workers, spammers and/or even malicious workers. Moreover, some workers/items may have very few labels, making the estimation of their behavior difficult. To alleviate these problems, we propose a novel Bayesian model that clusters workers and items together using the nonparametric Dirichlet process priors. This allows workers/items in the same cluster to borrow strength from each other. Instead of directly computing the posterior of this complex model, which is infeasible, we propose a new variational inference procedure. Experimental results on a number of real-world data sets show that the proposed algorithm is more accurate than the state-of-the-art, and is more robust to sparser labels.

1 Introduction

In many real-world classification applications, acquisition of labels is difficult and expensive. Recently, crowdsourcing provides an attractive alternative. With platforms like the Amazon Mechanical Turk, cheap labels can be efficiently obtained from non-expert workers. However, the collected labels are often noisy because of the presence of inexperienced workers, spammers and/or even malicious workers.

To clean these labels, a simple approach is majority voting [14]. By assuming that most workers are reliable, labels on a particular item are aggregated by selecting the most common label. However, this ignores relationships among labels provided by the same worker. To alleviate this problem, one can assume that labels of each worker are generated according to an underlying confusion matrix, which represents the probability that the worker assigns a particular label conditioned on the true label [7,19]. Others have also modeled the difficulties in labeling various items [2,13,23–25] and workers' dedications to the labeling task [2].

On the other hand, besides the commonly encountered binary and multiclass labels, labels can also be ordinal. For example, in web search, the relevance of a query-URL pair can be labeled as "irrelevant", "relevant" and "highly-relevant". Unlike nominal labels, it is important to exploit the underlying order of ordinal

© Springer International Publishing AG 2016
P. Frasconi et al. (Eds.): ECML PKDD 2016, Part I, LNAI 9851, pp. 426–442, 2016.
DOI: 10.1007/978-3-319-46128-1_27

labels. In particular, adjacent labels are often more difficult to differentiate than those that are further apart.

To solve the aforementioned problem, Lakshminarayanan and Teh [16] assumed that the ordinal labels are generated by the discretization of some continuous-valued latent labels. The latent label for each worker-item pair is drawn from a normal distribution, with its mean equal to the true label and its variance related to the worker's reliability and the item's difficulty. While this model is useful for "good" workers, it is not appropriate for malicious workers whose labels can be very different from the true label. Moreover, it can be too simplistic to use only one reliability (resp. difficulty) parameter to model each worker (resp. item).

A more recent model is the minimax entropy framework [26], which is extended from the minimax conditional entropy approach for multiclass label aggregation [25]. To encode ordinal information, they compare the worker and item labels with a reference label that can take all possible label values. The confusion for each worker-item pair as obtained from the model is then constrained to be close to its empirical counterpart. Finally, the true labels and probabilities are obtained by solving an optimization problem derived from the minimax entropy principle. In comparison with [16], ordering of the ordinal labels is now explicitly considered.

In crowdsourcing applications, some workers may only provide very few labels. Similarly, some items may receive very few labels. Parameter estimation for these workers and items can thus be unreliable. To alleviate this problem, one can consider the latent connections among workers and items. Intuitively, workers with similar characteristics (e.g., gender, age, and nationality) tend to have similar behaviors, and similarly for items. By clustering them together, one can borrow strength from one worker/item to another. Kajino et al. [12] formulated label aggregation as a multitask learning problem [8]. Each worker is modeled as a classifier, and the classifiers of similar workers are encouraged to be similar. However, the ground-truth classifier, which generates the true labels, is required to lie in one of the worker clusters. Moreover, this algorithm requires access to item features and cannot be used with ordinal labels. Venanzi et al. [21] proposed to cluster multiclass labels by using the Dirichlet distribution. However, the number of clusters needs to be pre-specified, which may not be practical. Moreover, item grouping is not considered. Lakkaraju et al. [15] modeled both item and worker groups. However, they again require the use of both worker and item features. Moreover, clustering and label inference are performed as separate tasks.

In this paper, motivated by the conditional probability derived in the dual of [26], we propose a novel algorithm to aggregate ordinal labels. Different from [26], a full Bayesian model is constructed, in which clustering of both workers and items are encouraged via the Dirichlet process (DP) [9] priors. DP is a nonparametric model which is advantageous in that the number of clusters does not need to be pre-specified. The resultant Bayesian model allows detection of clustering structure, learning of worker/item characteristics and label aggregation be performed simultaneously. Empirically, it also significantly outperforms the state-of-the-art. However, as we use DP priors with non-conjugate base distributions, exact inference is infeasible. To address this problem, we extend the techniques in [11], and derive a mean field variational inference algorithm for parameter estimation.

2 Ordinal Label Aggregation by Minimax Conditional Entropy

Let there be N workers, M items, and m ordinal label classes. We use i, j, m to index the workers, items, and labels, respectively. The true label of item j is denoted Y_j, with probability distribution Q. The label assigned by worker i to item j is X_{ij}, and Ξ is the set of (i, j) tuples with X_{ij}'s observed. We assume that there is at least one observed X_{ij} for each worker i, and at least one observed X_{ij} for each item j.

Zhou *et al.* [26] formulated label aggregation as a constrained minimax optimization problem, in which $H(X|Y) - H(Y) - \frac{1}{\alpha}\Omega(\xi) - \frac{1}{\beta}\Psi(\zeta)$ is maximized w.r.t. $P(X_{ij} = k \mid Y_j = c)$ and minimized w.r.t. $Q(Y_j = c)$. Here, $H(X|Y)$ is the conditional entropy of X given Y, $H(Y)$ is the entropy of Y, $\Omega(\xi) = \sum_{i,s}(\xi_{is}^{\Delta,\nabla})^2, \Psi(\zeta) = \sum_{j,s}(\zeta_{js}^{\Delta,\nabla})^2$ are ℓ_2-regularizers on the slack variables $\xi_{is}^{\Delta,\nabla}, \zeta_{js}^{\Delta,\nabla}$ (in (1) and (2)), and α, β are regularization parameters. Let $\phi_{ij}(c, k) = Q(Y_j = c)P(X_{ij} = k|Y_j = c)$ be the expected confusion from label c to label k by worker i on item j, and $\hat{\phi}_{ij}(c, k) = Q(Y_j = c)\mathbb{I}(X_{ij} = k)$ be its empirical counterpart. Besides the standard normalization constraints on probability distributions P and Q, Zhou *et al.* [26] requires $\phi_{ij}(c, k)$ be close to the empirical $\hat{\phi}_{ij}(c, k)$:

$$\sum_{c\Delta s}\sum_{k\nabla s}\sum_j [\phi_{ij}(c, k) - \hat{\phi}_{ij}(c, k)] = \xi_{is}^{\Delta,\nabla}, \forall i, \forall 2 \le s \le m, \tag{1}$$

$$\sum_{c\Delta s}\sum_{k\nabla s}\sum_i [\phi_{ij}(c, k) - \hat{\phi}_{ij}(c, k)] = \zeta_{js}^{\Delta,\nabla}, \forall j, \forall 2 \le s \le m. \tag{2}$$

Here, s is a reference label for comparing the true label c with worker label k, and Δ, ∇ is a binary relation operator (either $\ge$ or $<$). Together, they allow consideration of the four cases: (i) $c < s, k < s$; (ii) $c < s, k \ge s$; (iii) $c \ge s, k < s$; and (iv) $c \ge s, k \ge s$.

At optimality, it can be shown that

$$P(X_{ij} = k|Y_j = c) = \exp[\sigma_i(c, k) + \tau_j(c, k)]/Z_{ijc}, \tag{3}$$

where Z_{ijc} is a normalization factor,

$$\sigma_i(c, k) = \sum_{1\le s\le m}\sum_{\Delta,\nabla}\sigma_{is}^{\Delta,\nabla}\mathbb{I}(c \Delta s, k\nabla s), \tag{4}$$

$$\tau_j(c, k) = \sum_{1\le s\le m}\sum_{\Delta,\nabla}\tau_{js}^{\Delta,\nabla}\mathbb{I}(c \Delta s, k\nabla s), \tag{5}$$

$\sigma_{is}^{\Delta,\nabla}$, $\tau_{js}^{\Delta,\nabla}$ are Lagrange multipliers for the constraints (1) and (2), respectively, and $\mathbb{I}(\cdot)$ is the indicator function. Note that $\sigma_i(c,k)$ controls how likely worker i assigns label k when the true label is c, and $\tau_j(c,k)$ controls how likely item j is assigned label k when the true label is c. Equations (4) and (5) can be written more compactly as $\sigma_i(c,k) = \mathbf{t}_{ck}^T \boldsymbol{\sigma}_i$ and $\tau_j(c,k) = \mathbf{t}_{ck}^T \boldsymbol{\tau}_j$, where $\boldsymbol{\sigma}_i = [\sigma_{is}^{\Delta,\nabla}]$, $\boldsymbol{\tau}_j = [\tau_{js}^{\Delta,\nabla}]$, and $\mathbf{t}_{ck} = [\mathbb{I}(c\Delta s, k\nabla s)]$. Moreover, let $\mathbf{X} = [X_{ij}]_{(i,j)\in\Xi}$, and $\mathbf{Y} = [Y_j]$. Equation (3) can be rewritten as

$$P(\mathbf{X}|\mathbf{Y}) = \prod_{(i,j)\in\Xi} \prod_{c,k} P(X_{ij} = k|Y_j = c)^{\mathbb{I}(X_{ij}=k,Y_j=c)}$$

$$= \prod_{(i,j)\in\Xi} \prod_{c,k} \left[\frac{1}{Z_{ijc}} \exp[\mathbf{t}_{ck}^T(\boldsymbol{\sigma}_i + \boldsymbol{\tau}_j)] \right]^{\mathbb{I}(X_{ij}=k,Y_j=c)}. \tag{6}$$

3 Bayesian Clustering of Workers and Items

Note that each worker i (resp. item j) has its own set of variables $\{\sigma_i(c,k)\}$ (resp. $\{\tau_j(c,k)\}$). When the data are sparse, i.e., the set Ξ of observed labels is small, an accurate estimation of these variables can be difficult. In this section, we alleviate this data sparsity problem by clustering workers and items. While the minimax optimization framework in [26] can utilize ordering information in the ordinal labels, it is non-Bayesian and clustering cannot be easily encouraged. In this paper, we propose a full Bayesian model, and encourage clustering of workers and items using the Dirichlet process (DP) [9]. The DP prior is advantageous in that the number of clusters does not need to be specified in advance. However, with the non-conjugate priors and DPs involved, inference of the proposed model becomes more difficult. By extending the work in [11], we derive a variational Bayesian inference algorithm to infer the parameters and aggregate labels.

3.1 Model

Recall that $\sigma_i(c,k) = \mathbf{t}_{ck}^T \boldsymbol{\sigma}_i$ controls how likely worker i assigns label k when the true label is c. To encourage worker clustering, we define a prior G_a on $\{\boldsymbol{\sigma}_i\}_{i=1}^N$. G_a is drawn from the Dirichlet process $\mathrm{DP}(\beta_a, G_{a0})$, where β_a is the concentration parameter, and G_{a0} is the base distribution (here, we use the normal distribution $\mathcal{N}(\boldsymbol{\mu}_0, \boldsymbol{\Sigma}_0)$). Similarly, as $\tau_j(c,k) = \mathbf{t}_{ck}^T \boldsymbol{\tau}_j$ controls how likely item j assigns label k when the true label is c, we define a prior $G_b \sim \mathrm{DP}(\beta_b, G_{b0})$ on $\{\boldsymbol{\tau}_j\}_{j=1}^M$ to encourage item clustering (with $G_{b0} = \mathcal{N}(\boldsymbol{\nu}_0, \boldsymbol{\Omega}_0)$).

To make variational inference possible, we use the stick-breaking representation [20] and rewrite G_a as $\sum_{l=1}^{\infty} \phi_l \delta_{\boldsymbol{\sigma}_l^*}$, where $\phi_l = V_l \prod_{d=1}^{l-1}(1 - V_d), V_l \sim \mathrm{Beta}(1, \beta_a)$, and $\boldsymbol{\sigma}_l^* \sim G_{a0}$. When we draw $\boldsymbol{\sigma}$ from G_a, $\delta_{\boldsymbol{\sigma}_l^*} = 1$ if $\boldsymbol{\sigma} = \boldsymbol{\sigma}_l^*$, and 0 otherwise. Similarly, $G_b = \sum_{i=1}^{\infty} \varphi_h \delta_{\boldsymbol{\tau}_h^*}$, where $\varphi_h = H_h \prod_{d=1}^{h-1}(1 - H_d), H_h \sim \mathrm{Beta}(1, \beta_b)$ and $\boldsymbol{\tau}_h^* \sim G_{b0}$. Similar to $\sigma_i(c,k), \tau_j(c,k)$ in [26], $\sigma_l^*(c,k) = \mathbf{t}_{ck}^T \boldsymbol{\sigma}_l^*$

controls how likely workers in cluster l assign label k when the true label is c, and $\tau_h^*(c, k) = \mathbf{t}_{ck}^T \boldsymbol{\sigma}_h^*$ controls how likely items in cluster h are assigned label k when the true label is c Let z_i (resp. u_j) indicate the cluster that worker i (resp. item j) belongs to. We then have $\sigma_i(c, k) = \mathbf{t}_{ck}^T \boldsymbol{\sigma}_{z_i}^*$ and $\tau_j(c, k) = \mathbf{t}_{ck}^T \boldsymbol{\tau}_{u_j}^*$ for all i, j, c, k. Putting these into (6), we obtain the conditional probability as

$$P(\mathbf{X}|\mathbf{Y}, \mathbf{z}, \mathbf{u}, \boldsymbol{\sigma}^*, \boldsymbol{\tau}^*) = \prod_{(i,j)\in\Xi} \prod_{c,k} \left[\frac{1}{Z_{ijc}} \exp\left[\mathbf{t}_{ck}^T (\boldsymbol{\sigma}_{z_i}^* + \boldsymbol{\tau}_{u_j}^*) \right] \right]^{\mathbb{I}(X_{ij}=k, Y_j=c)},$$

(7)

where $Z_{ijc} = \sum_k \exp[\mathbf{t}_{ck}^T(\boldsymbol{\sigma}_{z_i}^* + \boldsymbol{\tau}_{u_j}^*)]$, $\boldsymbol{\sigma}^* = [\boldsymbol{\sigma}_l^*], \boldsymbol{\tau}^* = [\boldsymbol{\tau}_h^*], \mathbf{z} = [z_i]$, and $\mathbf{u} = [u_j]$. In other words, rating X_{ij} is generated from a softmax function [3] conditioned on $Y_j, z_i, u_j, \boldsymbol{\sigma}^*, \boldsymbol{\tau}^*$. Finally, the true label Y_j of item j is drawn from the multinomial distribution $\text{Mult}(\pi_1, \pi_2, \ldots, \pi_m)$, where $\pi_1, \ldots, \pi_m$ are drawn from a Dirichlet prior with hyperparameter α. The whole label generation process is shown in Algorithm 1. A graphical representation of the Bayesian model, which will be called Cluster-based Ordinal Label Aggregation (COLA) in the sequel, is shown in Fig. 1.

Algorithm 1. The proposed generation process.

1: **for** $j = 1, 2, \ldots, M$ **do**   ▷ Generate true labels
2:     draw $\boldsymbol{\pi} = [\pi_1, \ldots, \pi_m] \sim \text{Dir}(\alpha/m, \alpha/m, \ldots, \alpha/m)$; // Dirichlet distribution
3:     draw $Y_j \sim \text{Mult}(\boldsymbol{\pi})$;
4: **end for**
5: **for** $l = 1, 2, \ldots$ **do**   ▷ Generate worker clusters
6:     draw $V_l \sim \text{Beta}(1, \beta_a)$;
7:     $\phi_l = V_l \prod_{d=1}^{l-1}(1 - V_d)$;
8:     draw $\boldsymbol{\sigma}_l^* \sim G_{a0} = \mathcal{N}(\boldsymbol{\mu}_0, \boldsymbol{\Sigma}_0)$;
9: **end for**
10: **for** $i = 1, 2, \ldots, N$ **do**   ▷ Generate workers from worker clusters
11:     draw $z_i \sim \text{Mult}(\boldsymbol{\phi})$;
12: **end for**
13: **for** $h = 1, 2, \ldots$ **do**   ▷ Generate item clusters
14:     draw $H_h \sim \text{Beta}(1, \beta_b)$;
15:     $\varphi_h = H_h \prod_{d=1}^{h-1}(1 - H_d)$;
16:     draw $\boldsymbol{\tau}_h^* \sim G_{b0} = \mathcal{N}(\boldsymbol{\nu}_0, \boldsymbol{\Omega}_0)$;
17: **end for**
18: **for** $j = 1, 2, \ldots, M$ **do**   ▷ Generate items from item clusters
19:     draw $u_j \sim \text{Mult}(\boldsymbol{\varphi})$;
20: **end for**
21: **for** $i = 1, 2, \ldots, N; j = 1, 2, \ldots, M$ **do**   ▷ Generate worker labels
22:     draw $X_{ij} \sim P(X_i|Y_j, z_i, u_j, \boldsymbol{\sigma}^*, \boldsymbol{\tau}^*)$;
23: **end for**

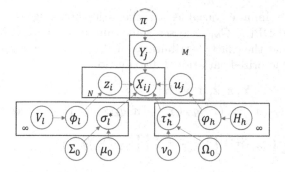

Fig. 1. Graphical representation of the proposed model.

3.2   Inference Procedure

The joint distribution can be written as

$$P(\mathbf{X}, \mathbf{Y}, \boldsymbol{\sigma}^*, \boldsymbol{\tau}^*, \mathbf{z}, \mathbf{u}, \mathbf{V}, \mathbf{H} | \boldsymbol{\mu}_0, \boldsymbol{\Sigma}_0, \boldsymbol{\nu}_0, \boldsymbol{\Omega}_0, \alpha, \beta_a, \beta_b)$$
$$= P(\mathbf{X} | \mathbf{Y}, \boldsymbol{\sigma}^*, \boldsymbol{\tau}^*, \mathbf{z}, \mathbf{u}) P(\boldsymbol{\sigma}^* | \boldsymbol{\mu}_0, \boldsymbol{\Sigma}_0) P(\boldsymbol{\tau}^* | \boldsymbol{\nu}_0, \boldsymbol{\Omega}_0)$$
$$P(\mathbf{Y} | \boldsymbol{\pi}) P(\boldsymbol{\pi} | \alpha) P(\mathbf{z} | \mathbf{V}) P(\mathbf{V} | \beta_1) P(\mathbf{u} | \mathbf{H}) P(\mathbf{H} | \beta_2),$$

where $\mathbf{V} = [V_l]$, and $\mathbf{H} = [H_h]$. Monte Carlo Markov Chain (MCMC) sampling
[1] can be used to approximate the posterior distribution. However, it can be slow
and its convergence is difficult to diagnose [4]. Another approach is variational
inference [11], which approximates the posterior distribution by maximizes a
lower bound of the marginal likelihood. However, due to the infinite number of
variables in the DPs and our use of non-conjugate priors, standard variational
inference cannot be used. To solve this problem, we propose an integration of the
techniques in [4,5] with variational inference. Specifically, we infer the variational
parameters of the DPs based on an extension of [4], and handle the non-conjugate
priors by a technique similar to [5].

Let $\boldsymbol{\theta} = \{\boldsymbol{\sigma}^*, \boldsymbol{\tau}^*, \mathbf{Y}, \boldsymbol{\pi}, \mathbf{z}, \mathbf{u}, \mathbf{V}, \mathbf{H}\}$. In variational inference, the posterior
$P(\boldsymbol{\theta}|\mathbf{X})$ is approximated by a distribution $q(\boldsymbol{\theta})$. The log likelihood of the mar-
ginal distribution of $\mathbf{X}$ is $\log P(\mathbf{X}) = \mathcal{L}(q) + \text{KL}(q\|P_{\boldsymbol{\theta}|\mathbf{X}})$, where $\mathcal{L}(q) =$
$\int q(\boldsymbol{\theta}) \log \frac{P(\mathbf{X}, \boldsymbol{\theta})}{q(\boldsymbol{\theta})} d\boldsymbol{\theta}$, and $\text{KL}(q\|P_{\boldsymbol{\theta}|\mathbf{X}}) = \int q(\boldsymbol{\theta}) \log \frac{q(\boldsymbol{\theta})}{P(\boldsymbol{\theta}|\mathbf{X})} d\boldsymbol{\theta}$ is the KL divergence
between q and $P_{\boldsymbol{\theta}|\mathbf{X}}$. As $\text{KL}(q\|P_{\boldsymbol{\theta}|\mathbf{X}}) \geq 0$, we simply maximize the lower bound
$\mathcal{L}(q)$ of $\log P(\mathbf{X})$. Using the variational mean field approach, $q(\boldsymbol{\theta})$ is assumed to
be factorized as $\prod_{n=1}^{S} q_n(\boldsymbol{\theta}_n)$, where S is the number of factors, $\{\boldsymbol{\theta}_1, \dots, \boldsymbol{\theta}_S\}$ is
a partition of $\boldsymbol{\theta}$, and q_n is the variational distribution of $\boldsymbol{\theta}_n$ [22]. We perform
alternating maximization of $\mathcal{L}\left(\prod_{n=1}^{S} q_n(\boldsymbol{\theta}_n)\right)$ w.r.t. q_n's. It can be shown that
the optimal q_n is given by

$$q_n^*(\boldsymbol{\theta}_n) = \exp\left[\mathbb{E}_{q(\boldsymbol{\theta}_{\neg n})} \log P(\mathbf{X}, \boldsymbol{\theta})\right] + \text{constant}, \tag{8}$$

where $\boldsymbol{\theta}_{\neg n}$ is the subset of variables in $\boldsymbol{\theta}$ excluding $\boldsymbol{\theta}_n$, and $\mathbb{E}_{q(\boldsymbol{\theta}_{\neg n})}$ is the corresponding expectation operator.

As there are infinite variables in the stick-breaking representations of $DP(\beta_a, G_{a0})$ and $DP(\beta_b, G_{b0})$, we set a maximum on the numbers of clusters as in [4]. Note that the exact distributions of the stick-breaking process are not truncated. The factorized variational distribution is

$$q(\boldsymbol{\sigma}^*, \boldsymbol{\tau}^*, \mathbf{Y}, \boldsymbol{\pi}, \mathbf{z}, \mathbf{u}, \mathbf{V}, \mathbf{H})$$
$$= q_{\boldsymbol{\sigma}^*}(\boldsymbol{\sigma}^*) q_{\boldsymbol{\tau}^*}(\boldsymbol{\tau}^*) q_{\mathbf{Y}}(\mathbf{Y}) q_{\boldsymbol{\pi}}(\boldsymbol{\pi}) q_{\mathbf{z}}(\mathbf{z}) q_{\mathbf{u}}(\mathbf{u}) q_{\mathbf{V}}(\mathbf{V}) q_{\mathbf{H}}(\mathbf{H})$$
$$= \prod_{l=1}^{K_1} q_{\sigma_l^*}(\sigma_l^*) \prod_{h=1}^{K_2} q_{\tau_h^*}(\tau_h^*) \prod_{j=1}^{M} q_{Y_j}(Y_j) q_{\boldsymbol{\pi}}(\boldsymbol{\pi})$$
$$\prod_{i=1}^{N} q_{z_i}(z_i) \prod_{j=1}^{M} q_{u_j}(u_j) \prod_{l=1}^{K_1} q_{V_l}(V_l) \prod_{h=1}^{K_2} q_{H_h}(H_h),$$

where K_1, K_2 are the truncated numbers of clusters for workers and items, respectively. Using (8), it can be shown that the variational distributions of $\{\mathbf{Y}, \boldsymbol{\pi}, \mathbf{z}, \mathbf{u}, \mathbf{H}, \mathbf{V}\}$ can be easily obtained as:

$$q_{Y_j}^*(Y_j) = \text{Mult}(\mathbf{r}_j^Y), \quad q_{\boldsymbol{\pi}}^*(\boldsymbol{\pi}) = \text{Dir}(\alpha_1, \alpha_2, \ldots \alpha_m),$$
$$q_{z_i}^*(z_i) = \text{Mult}(\mathbf{r}_i^z), \quad q_{u_j}^*(u_j) = \text{Mult}(\mathbf{r}_j^u),$$
$$q_{V_l}^*(V_l) = \text{Beta}(\gamma_{l,1}, \gamma_{l,2}), \quad q_{H_h}^*(H_h) = \text{Beta}(\eta_{h,1}, \eta_{h,2}),$$

where $\{\mathbf{r}_j^Y\}_{j=1}^M, \{\mathbf{r}_i^z\}_{i=1}^N, \{\mathbf{r}_j^u\}_{j=1}^M, \{\alpha_c\}_{c=1}^m, \{\gamma_{l,1}, \gamma_{l,2}\}_{l=1}^{K_1}$, and $\{\eta_{h,1}, \eta_{h,2}\}_{h=1}^{K_2}$ are variational parameters. All these have closed-form updates as

$$r_{jc}^Y \leftarrow \frac{1}{Z_j^Y} \exp\left[\mathbb{E}_{q(\boldsymbol{\theta}_{\neg\mathbf{Y}})}\log\pi_c + \sum_{i:(i,j)\in\Xi}\sum_{l=1}^{K_1}\sum_{h=1}^{K_2} r_{il}^z r_{jh}^u U_{ijlhc}\right],$$

$$r_{il}^z \leftarrow \frac{1}{Z_i^z} \exp\left[\mathbb{E}_{q(\boldsymbol{\theta}_{\neg\mathbf{z}})}\log\phi_l + \sum_{j:(i,j)\in\Xi}\sum_{c=1}^{m}\sum_{h=1}^{K_2} r_{jh}^u r_{jc}^Y U_{ijlhc}\right],$$

$$r_{jh}^u \leftarrow \frac{1}{Z_j^u} \exp\left[\mathbb{E}_{q(\boldsymbol{\theta}_{\neg\mathbf{u}})}\log\varphi_h + \sum_{i:(i,j)\in\Xi}\sum_{c=1}^{m}\sum_{l=1}^{K_1} r_{il}^z r_{jc}^Y U_{ijlhc}\right],$$

$$\alpha_c \leftarrow \frac{\alpha}{m} + \sum_{j=1}^{M} r_{jc}^Y, \quad \gamma_{l,1} \leftarrow 1 + \sum_{i=1}^{N} r_{il}^z, \quad \gamma_{l,2} \leftarrow \beta_a + \sum_{i=1}^{N}\sum_{d=l+1}^{K_1} r_{id}^z,$$

$$\eta_{h,1} \leftarrow 1 + \sum_{j=1}^{M} r_{jh}^u, \quad \eta_{h,2} \leftarrow \beta_b + \sum_{j=1}^{M}\sum_{d=h+1}^{K_2} r_{jd}^u,$$

where Z_j^Y, Z_i^z, Z_j^u are normalization constants,

$$U_{ijlhc} = \sum_{k=1}^{m} I(X_{ij} = k) \mathbf{t}_{ck}^T [\mathbb{E}_{q(\sigma_l^*)}(\sigma_l^*) + \mathbb{E}_{q(\tau_h^*)}(\tau_h^*)] - \mathbb{E}_{q(\sigma_l^*, \tau_h^*)} Z_{lhc}^* \quad (9)$$

and $Z_{lhc}^* = \sum_{k=1}^m \exp[t_{ck}^T(\sigma_l^* + \tau_h^*)]$. Computing U_{ijlhc} requires knowing the variational distributions of σ_l^* and τ_h^*, and will be derived in the following.

Recall that $P(\sigma_l^*), P(\tau_h^*)$ are normal distributions. These are not the conjugate prior of $P(\mathbf{X}|\mathbf{Y}, \mathbf{z}, \mathbf{u}, \sigma^*, \tau^*)$ in (7). Thus, on maximizing $\mathcal{L}$, $q(\sigma^*)$ and $q(\tau^*)$ do not have closed-form solutions. Note that the $\frac{1}{Z_{ijc}}\exp\left[t_{ck}^T(\sigma_{z_i}^* + \tau_{u_j}^*)\right]$ term in (7) is a softmax function similar to that in [5], which uses variational inference to learn discrete choice models. However, while the parameters of different sets of choices in [5] are conditionally independent, here in (7) they are coupled together. Thus, the inference procedure in [5] cannot be directly applied and has to be extended.

First, (7) can be rewritten as

$$P(\mathbf{X}|\mathbf{Y}, \mathbf{z}, \mathbf{u}, \sigma^*, \tau^*) = \prod_{(i,j)\in\Xi} \prod_{c,k,l,h} \left[\frac{\exp\left[t_{ck}^T(\sigma_l^* + \tau_h^*)\right]}{Z_{lhc}^*} \right]^{\mathbb{I}(z_i=l, u_j=h, X_{ij}=k, Y_j=c)}.$$

Since $P(\sigma_l^*), P(\tau_h^*)$ are normal distributions, we constrain the variational distributions of σ_l^* and τ_h^* to be also normal, i.e., $q_{\sigma_l^*}(\sigma_l^*) = \mathcal{N}(\mu_l, \Sigma_l)$, and $q_{\tau_h^*}(\tau_h^*) = \mathcal{N}(\nu_h, \Omega_h)$. Let $\mu = [\mu_l]$, $\nu = [\nu_h]$, $\Sigma = [\Sigma_l]$, and $\Omega = [\Omega_h]$. On maximizing $\mathcal{L}(q)$, it can be shown that the variational parameters $\{\mu, \Sigma, \nu, \Omega\}$ can be obtained as

$$\min_{\mu,\Sigma,\nu,\Omega} \mathbb{E}_{q(\theta)}\left[\log q_{\sigma^*}(\sigma^*) q_{\tau^*}(\tau^*)\right] - \mathbb{E}_{q(\theta)}\left[\log P(\sigma^*) P(\tau^*)\right]$$
$$-\mathbb{E}_{q(\theta)}\left[\log P(\mathbf{X}|\mathbf{Y}, \sigma^*, \tau^*, \mathbf{z}, \mathbf{u})\right]. \quad (10)$$

The first term is the entropy of the normal distribution, and the second term is the cross-entropy of two normal distributions. Both are easy to compute. The last term can be rewritten as $\sum_{(i,j)\in\Xi}\sum_{c,k,l,h} r_{il}^z r_{jh}^u r_{jc}^Y \mathbb{I}(X_{ij} = k)\left(t_{ck}^T(\mu_l + \nu_h) - \mathbb{E}_{q(\sigma^*,\tau^*)}\log Z_{lhc}^*\right)$. The term $\mathbb{E}_{q(\sigma^*,\tau^*)}\log Z_{lhc}^*$, which also appears in (9), can be approximated as in [5]:

$$\log \sum_{k=1}^m \exp(t_{ck}^T \mu_l + \frac{1}{2}t_{ck}^T \Sigma_l t_{ck}) \exp(t_{ck}^T \nu_h + \frac{1}{2}t_{ck}^T \Omega_h t_{ck}).$$

Problem (10) can then be solved via gradient-based methods, such as L-BFGS [17]. Denote the objective by f. It can be shown that

$$\frac{\partial f}{\partial \mu_l} = -\Sigma_0^{-1}(\mu_l - \mu_0) + \sum_{(i,j)\in\Xi}\sum_{c=1}^m\sum_{h=1}^{K_2} r_{il}^z r_{jh}^u r_{jc}^Y \sum_{k=1}^m [\mathbb{I}(X_{ij} = k) - w_{klh}]t_{ck},$$

where

$$w_{klh} = \frac{\exp(t_{ck}^T \nu_h + \frac{1}{2}t_{ck}^T \Omega_h t_{ck}) \exp(t_{ck}^T \mu_l + \frac{1}{2}t_{ck}^T \Sigma_l t_{ck})}{\sum_{k=1}^m \exp(t_{ck}^T \nu_h + \frac{1}{2}t_{ck}^T \Omega_h t_{ck}) \exp(t_{ck}^T \mu_l + \frac{1}{2}t_{ck}^T \Sigma_l t_{ck})}.$$

Moreover, as $\Sigma_l \succeq 0$, we assume that $\Sigma_l = L_l^T L_l$, where L_l is lower-triangular. It can then be shown that

$$\frac{\partial f}{\partial L_l} = L_l^{-T} - \left[\Sigma_0^{-1} - \sum_{(i,j)\in\Xi} \sum_{c=1}^m \sum_{h=1}^{K_2} r_{il}^z r_{jh}^u r_{jc}^Y \left(\sum_{k=1}^m w_{klh} t_{ck}^T t_{ck}\right)\right] L_l.$$

Recall that L_l is lower-triangular, so in updating L_l, we only need the diagonal elements $(L_l^{-T})_{ii} = 1/(L_l)_{ii}$. of the upper-triangular L_l^{-T} [5]. The gradients of f w.r.t. ν and Ω can be obtained in a similar manner.

4 Experiments

4.1 Synthetic Data Set

In this section, we perform experiments on synthetic data. Workers are generated from three clusters $(w1, w2, w3)$, items from three clusters $(i1, i2, i3)$, and ordinal labels in $\{1, 2, 3, 4, 5\}$. The cluster parameters σ^*, τ^* are sampled independently from the normal distribution with means in Table 1 and standard deviation 0.1. Confusion matrices[1] of the clusters are shown in Fig. 2. As can be seen, workers in cluster $w1$ are the least confused in label assignment. This is followed by cluster $w2$, and workers in cluster $w3$ are most confused (spammers). Similarly, items in cluster $i1$ are the least confused, while those in $i3$ are the most confused.

Table 1. Parameter means of the worker clusters $(w1, w2, w3)$ and item clusters $(i1, i2, i3)$.

	$(\sigma_{ls}^*)^{<,<}$	$(\sigma_{ls}^*)^{<,\geq}$	$(\sigma_{ls}^*)^{\geq,<}$	$(\sigma_{ls}^*)^{\geq,\geq}$		$(\tau_{hs}^*)^{<,<}$	$(\tau_{hs}^*)^{<,\geq}$	$(\tau_{hs}^*)^{\geq,<}$	$(\tau_{hs}^*)^{\geq,\geq}$
$w1$	1	0	0	1	$i1$	1	0	0	1
$w2$	1	0.8	0.8	1	$i2$	1	0.8	0.8	1
$w3$	0.3	1	0.5	1	$i3$	1	0.5	1	0.5

We generate two data sets. Both have 300 workers from the 3 clusters $(w1, w2, w3)$, with sizes 200, 50, and 50, respectively. The first data set (D1) has 1200 items coming from the 3 clusters $(i1, i2, i3)$, with sizes 800, 200, and 200, respectively. Each item is labeled by 6 randomly selected workers. The second data set (D2) has 300 items coming from the 3 clusters with sizes 200, 50, and 50, respectively. Each item is labeled by 30 workers.

---

[1] To obtain the confusion matrices of items, we remove the effects of workers by assuming that workers assign labels randomly. Using (7), it can be shown that the (k, c)th entry of the confusion matrix of item cluster h is $\exp\left(t_{ck}^T \tau_h^*\right) / \sum_k \exp\left(t_{ck}^T \sigma_i^*\right)$. Similarly, for worker cluster l, the (k, c)th entry of its confusion matrix is $\exp\left(t_{ck}^T \sigma_i^*\right) / \sum_k \exp\left(t_{ck}^T (\sigma_i^*)\right)$.

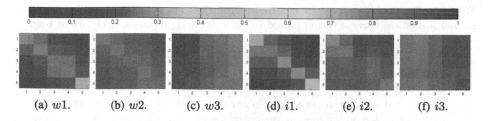

| (a) $w1$. | (b) $w2$. | (c) $w3$. | (d) $i1$. | (e) $i2$. | (f) $i3$. |

Fig. 2. True confusion matrices of the worker and item clusters.

We set the truncated numbers of clusters K_1, K_2 in COLA to 8, and $\boldsymbol{\mu}_0 = \mathbf{0}, \boldsymbol{\nu}_0 = \mathbf{0}, \boldsymbol{\Sigma}_0 = \frac{1}{\lambda_a}\mathbf{I}, \boldsymbol{\Omega}_0 = \frac{1}{\lambda_b}\mathbf{I}$, where $\lambda_b = \lambda_a M/N$, as in [26]. Parameters β_a, β_b, α and λ_a are tuned by maximizing the log-likelihood as in [26]. Latent variables are initialized in an non-informative manner: $\boldsymbol{\mu}_l = \mathbf{0}, \boldsymbol{\nu}_h = \mathbf{0}, \mathbf{r}_i^z = [1/K_1, \ldots, 1/K_1]^T, \mathbf{r}_j^u = [1/K_2, \ldots, 1/K_2]^T, \eta_{h,1} = 1, \eta_{h,2} = \beta_a \gamma_{l,1} = 1, \gamma_{l,2} = \beta_b$, and $\mathbf{r}_j^Y$ is from the empirical probabilities of the observed labels. We compare the proposed algorithm with the following state-of-the-art:

1. Ordinal minimax entropy (OME) [26], with the hyperparameters tuned by the cross-validation method suggested in [26].
2. Ordinal mixture (ORDMIX) [16]: The predicted labels are obtained by discretizing (normally distributed) continuous-valued latent labels.
3. Dawid-Skene model (DS) [7]: A well-known approach for label aggregation, which estimates a confusion matrix for each worker.
4. Majority voting (MV) [14], which has been commonly used as a simple baseline.

To allow statistical significance testing, we learn the model using 90 % of the items and run each experiment for 10 repetitions. As in [16,26], the following measures are used: (i) mean squared error: MSE $= \frac{1}{|S|} \sum_{j \in S}((\mathbb{E}_Q[Y_j] - Y_j^*)^2)$, where S is the set of items with ground-truth labels Y_j^*'s; (ii) ℓ_0 error $= \frac{1}{|S|} \sum_{j \in S} \mathbb{E}_Q[\mathbb{I}(Y_j \neq Y_j^*)]$; (iii) ℓ_1 error $= \frac{1}{|S|} \sum_{j \in S} \mathbb{E}_Q[|Y_j - Y_j^*|]$; and (iv) ℓ_2 error $= \sqrt{\frac{1}{|S|} \sum_{j \in S} \mathbb{E}_Q[(Y_j - Y_j^*)^2]}$.

Results are shown in Table 2. As can be seen, on data set D1, the proposed COLA significantly outperforms the other methods. On D2, with more labels per item, inference becomes easier and all methods have improved performance. COLA is still better in terms of MSE and ℓ_2 error, but can be slightly outperformed by the simpler methods of DS and MV in terms of ℓ_0 and ℓ_1 errors.

Figure 3 shows the normalized sizes of the worker and item clusters obtained by COLA. Recall that $\mathbf{z}$ indicates the cluster memberships of workers. The normalized size of worker cluster l is defined as $\mathbb{E}(S_l^z)/\sum_{l=1}^{K_1} \mathbb{E}(S_l^z)$, where S_l^z is the size of worker cluster l, and $\mathbb{E}(S_l^z) = \mathbb{E}\left[\sum_{i=1}^{N} I(z_i = l)\right] = \sum_{i=1}^{N} P(z_i = l) = \sum_{i=1}^{N} r_{il}^z$. Similarly, the normalized size of item cluster h is $\mathbb{E}(S_h^u)/\sum_{h=1}^{K_2} \mathbb{E}(S_h^u)$, where S_h^u is the size of item cluster h, and $\mathbb{E}(S_h^u) = \sum_{j=1}^{M} r_{jh}^u$. As can be seen,

Table 2. Errors obtained on the synthetic data sets. The best results and those that are not statistically worse (using paired t-test at 95% significance level) are in bold.

		COLA	OME	ORDMIX	DS	MV
D1	MSE	**0.249 ± 0.006**	0.314 ± 0.011	0.341 ± 0.025	0.446 ± 0.009	0.401 ± 0.006
	ℓ_0	**0.180 ± 0.003**	0.228 ± 0.001	0.229 ± 0.006	0.225 ± 0.003	0.225 ± 0.003
	ℓ_1	**0.209 ± 0.002**	0.284 ± 0.004	0.273 ± 0.012	0.289 ± 0.004	0.304 ± 0.005
	ℓ_2	**0.522 ± 0.005**	0.642 ± 0.009	0.616 ± 0.021	0.668 ± 0.007	0.717 ± 0.008
D2	MSE	**0.073 ± 0.007**	0.101 ± 0.005	0.112 ± 0.009	0.089 ± 0.013	0.268 ± 0.012
	ℓ_0	0.080 ± 0.010	0.81 ± 0.005	0.83 ± 0.013	0.074 ± 0.000	**0.073 ± 0.004**
	ℓ_1	0.081 ± 0.010	0.82 ± 0.004	0.84 ± 0.013	**0.079 ± 0.004**	0.083 ± 0.006
	ℓ_2	**0.282 ± 0.017**	0.310 ± 0.013	0.315 ± 0.018	0.298 ± 0.022	0.319 ± 0.020

the sizes of the three dominant worker clusters are close to the ground truth on both data sets. However, the item cluster (normalized) sizes on D1 are less accurate than those on D2. This is due to that each item in D1 only has 6 labels, while each item in D2 has 30 (in comparison, each worker on average has 24 labels for D1 and 30 labels for D2).

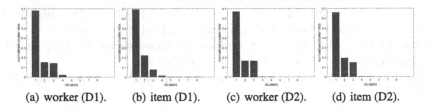

(a) worker (D1).   (b) item (D1).   (c) worker (D2).   (d) item (D2).

Fig. 3. Normalized sizes of the worker and item clusters on D1 (left) and D2 (right). The true normalized sizes of the worker and item clusters are 0.667, 0.167, 0.167.

Next, we show the confusion matrices of the obtained clusters. Since we only have the distributions of σ^* and τ^*, we will use their expectations. Note that $\mathbb{E}[\sigma_l^*] = \mu_l$. From (7), the (c, k)th entry of the confusion matrix of worker cluster l can be represented as $\exp[\mathbf{t}_{ck}^\top \mu_l] / \sum_k \exp[\mathbf{t}_{ck}^\top \mu_l]$, and similarly, that of the item cluster h is $\exp[\mathbf{t}_{ck}^\top \nu_h] / \sum_k \exp[\mathbf{t}_{ck}^\top \nu_h]$. The obtained confusion matrices for worker and item clusters are shown in Fig. 4. Here, we focus on the three largest worker/item clusters, which can be seen to dominate in Fig. 3. Comparing with the ground-truth in Fig. 2, the 3 worker and item clusters can be well detected. Note again that the item clusters obtained on D2 are more accurate than those on D1, as each item in D2 has more labels for inference.

4.2   Real Data Sets

In this section, experiments are performed on three commonly-used data sets (Table 3).

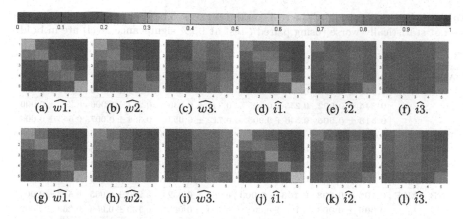

Fig. 4. Confusion matrices of the obtained worker/item clusters on D1 (top) and D2 (bottom).

Table 3. Summary of the data sets used.

	#items		#workers	#classes	#observed labels	#labels/worker			#labels/item		
	total	w/ground truth				min	mean	max	min	mean	max
AC2	11,040	333	825	4	89,799	1	108.8	7551	1	8.1	27
TREC	19,721	3,250	762	3	90,244	1	118.4	7467	1	4.6	34
WEB	2,665	2,665	177	5	15,567	1	87.9	1225	1	5.8	12

1. AC2 [10]: This contains AMT judgments for website ratings, with the 4 levels: "G", "PG", "R", and "X";
2. TREC [6]: This is a web search data set, with the 3 levels: "NR" (non-relevant), "R" (relevant) and "HR" (highly relevant);
3. WEB [26]: This is another web search relevance data set, with the 5 levels: "P" (perfect), "E" (excellent), "G" (good), "F" (fair) and "B" (bad).

The ordinal labels are converted to numbers (e.g., on the AC2 data set, "G", "PG", "R", and "X" are converted to 1, 2, 3, 4, respectively). As can be seen from Table 3, the number of labels provided by the workers can vary significantly.

For the proposed algorithm, we set the truncated numbers of clusters K_1, K_2 to 10 (on WEB, $K_1 = 15$). Larger values do not improve performance. The other parameters of all the algorithms are set as in Sect. 4.1. To allow statistical significance testing, again we learn the model using 90% of the items[2], and repeat this process 10 times.

Results are shown in Table 4. As can be seen, COLA consistently outperforms all the other methods on AC2 and WEB. Moreover, ORDMIX is competitive with COLA on TREC, but much inferior on AC2 and WEB. As AC2 and

---

[2] For AC2 and TREC, since performance can only be evaluated on items with ground-truth labels and these two data sets have fewer such items, all these items (with ground-truth labels) are always selected into the 90% subset.

Table 4. Errors obtained on the real-world data sets. The best results and those that are not statistically worse (using paired t-test at 95 % significance level) are in bold.

		COLA	OME	ORDMIX	DS	MV
AC2	MSE	**0.262 ± 0.003**	0.317 ± 0.001	0.364 ± 0.028	0.302 ± 0.007	0.292 ± 0.000
	ℓ_0	**0.228 ± 0.002**	**0.230 ± 0.004**	0.279 ± 0.015	0.271 ± 0.006	0.241 ± 0.000
	ℓ_1	**0.245 ± 0.002**	0.255 ± 0.005	0.348 ± 0.030	0.283 ± 0.006	0.297 ± 0.000
	ℓ_2	**0.513 ± 0.005**	0.546 ± 0.005	0.712 ± 0.053	0.564 ± 0.007	0.643 ± 0.000
TREC	MSE	0.641 ± 0.006	0.679 ± 0.001	**0.603 ± 0.019**	0.750 ± 0.004	0.649 ± 0.000
	ℓ_0	**0.492 ± 0.003**	**0.495 ± 0.001**	0.557 ± 0.006	0.513 ± 0.003	0.543 ± 0.000
	ℓ_1	**0.602 ± 0.004**	0.615 ± 0.002	**0.606 ± 0.011**	0.635 ± 0.003	0.661 ± 0.000
	ℓ_2	0.886 ± 0.005	0.924 ± 0.003	**0.838 ± 0.013**	0.938 ± 0.004	0.947 ± 0.000
WEB	MSE	**0.105 ± 0.003**	**0.106 ± 0.003**	0.360 ± 0.032	0.230 ± 0.005	0.517 ± 0.004
	ℓ_0	**0.096 ± 0.003**	0.103 ± 0.004	0.194 ± 0.003	0.169 ± 0.006	0.269 ± 0.002
	ℓ_1	**0.108 ± 0.004**	0.117 ± 0.004	0.242 ± 0.010	0.204 ± 0.006	0.425 ± 0.004
	ℓ_2	**0.369 ± 0.007**	0.381 ± 0.008	0.633 ± 0.024	0.534 ± 0.008	0.923 ± 0.006

WEB have more label classes than TREC (Table 3), ORDMIX, which has fewer parameters and is less flexible than COLA, is unable to sufficiently model the confusion matrices of workers and items. The performance of DS is also poor, as ordinal information of the labels is not utilized. Finally, as expected, the simple MV performs the worst overall.

Figure 5(a)–(j) show the confusion matrices of worker clusters obtained on AC2. For most of them ($\widehat{w1} - \widehat{w8}$), the diagonal values for "G" and "X" are high, indicating that most clusters can identify these two types of websites easily. For the largest worker cluster ($\widehat{w1}$), the highest value on each row lies on the diagonal, and so the labels assigned by this cluster are mostly consistent with the ground truth. As for cluster $\widehat{w5}$, the diagonal entries are much larger than the non-diagonal ones. Hence, the worker labels are often the same as the ground truth, suggesting that these workers are experts. This is also confirmed in Table 5, which shows the ℓ_2 error for each cluster. On the other hand, workers in cluster $\widehat{w9}$ almost always predict "G". They are likely to be spammers as observed in many crowdsourcing platforms [18]. In cluster $\widehat{w10}$, the off-diagonal values are larger than the diagonal ones, indicating that workers in this cluster may not understand this website rating task or may even be malicious.

Table 5. ℓ_2 errors for worker clusters obtained on the AC2 data set.

Cluster	$\widehat{w1}$	$\widehat{w2}$	$\widehat{w3}$	$\widehat{w4}$	$\widehat{w5}$	$\widehat{w6}$	$\widehat{w7}$	$\widehat{w8}$	$\widehat{w9}$	$\widehat{w10}$
ℓ_2 error	0.714	0.627	0.789	0.938	0.618	0.922	0.732	0.950	1.133	1.445

Figure 5(k)–(o) show the confusion matrices of the obtained item clusters. In general, as each item has fewer labels than each worker, the clustering structure here is less obvious (as discussed in Sect. 4.1). For the item cluster $\widehat{i1}$, the diagnal

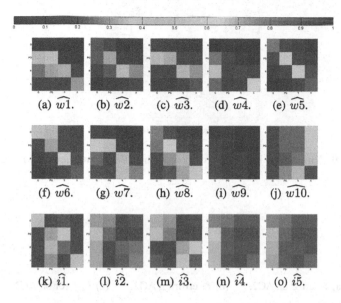

(a) $\widehat{w1}$.   (b) $\widehat{w2}$.   (c) $\widehat{w3}$.   (d) $\widehat{w4}$.   (e) $\widehat{w5}$.

(f) $\widehat{w6}$.   (g) $\widehat{w7}$.   (h) $\widehat{w8}$.   (i) $\widehat{w9}$.   (j) $\widehat{w10}$.

(k) $\widehat{i1}$.   (l) $\widehat{i2}$.   (m) $\widehat{i3}$.   (n) $\widehat{i4}$.   (o) $\widehat{i5}$.

Fig. 5. Confusion matrices for worker clusters (top two rows) and item clusters (bottom row) obtained by COLA on the AC2 data set (clusters are ordered by decreasing size). In each cluster, columns are the cluster-assigned labels (left-to-right: "G", "PG", "R", "X"), and rows are the true labels (top-to-down: "G", "PG", "R", "X"). The five smallest item clusters occupy less than 2 % of the total size, and so are not shown.

elements have high values, indicating that items belonging to this cluster are relatively easy to distinguish. Item cluster $\widehat{i2}$ tends to assign label "G" more often. In $\widehat{i3}$, "G" and "PG" are sometimes confused, and so are "R" and "X".

Varying the Number of Items. In this experiment, we use item subsets of different sizes to learn the model. With a smaller number of items, the number of labels per worker is also reduced (Fig. 6), and estimating the workers' behavior become more difficult. Here, we focus on the two top performers in Table 4, namely, COLA and OME. Figure 7(a)–(i) show the errors averaged over 10 repetitions. As can be seen, as OME does not consider any structure among workers and items, its performance deteriorates significantly with fewer worker labels. On the other hand, COLA clusters workers and items. Thus, information within a cluster can be shared, and the performance is less affected.

Varying the Concentration Parameters. In this experiment, we study the effect of the DP's concentration parameters on the performance of COLA. In general, a smaller concentration parameter encourages fewer clusters, and vice verse. We first fix β_b to the value obtained in the previous experiment, and vary β_a from 0.1 to 3.5.

Figure 8(a) shows how the ℓ_2 error varies with β_a. Because of the lack of space, we only show results on the AC2 data set. COLA has stable performance over a

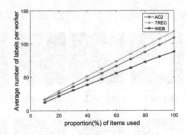

Fig. 6. Number of labels per worker with different numbers of items.

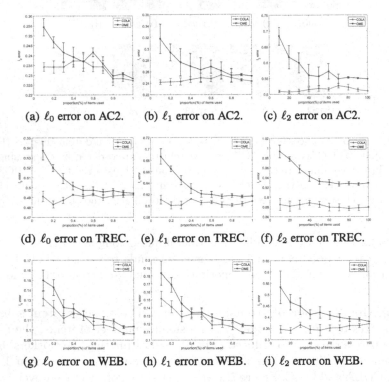

(a) ℓ_0 error on AC2.   (b) ℓ_1 error on AC2.   (c) ℓ_2 error on AC2.

(d) ℓ_0 error on TREC.   (e) ℓ_1 error on TREC.   (f) ℓ_2 error on TREC.

(g) ℓ_0 error on WEB.   (h) ℓ_1 error on WEB.   (i) ℓ_2 error on WEB.

Fig. 7. ℓ_0, ℓ_1 and ℓ_2 errors for COLA and OME with different proportions(%) of items used.

wide range of β_a. When β_a becomes too small, workers can only form very few clusters, and each cluster may not be coherent. When β_a is too large, clusters are split, and each cluster may not have enough data for accurate parameter estimation. Figure 8(b) shows the results on varying β_b. As can be seen, β_b has little influence on the performance. Again, this is consistent with the observation in the previous section that items' cluster structure is more difficult to identify.

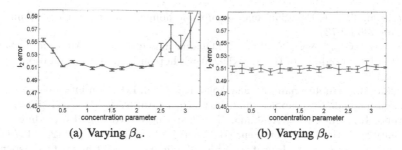

(a) Varying β_a.

(b) Varying β_b.

Fig. 8. ℓ_2 errors of COLA on the AC2 data set, with different values for β_a, β_b.

5 Conclusion

In this paper, we proposed a Bayesian clustering model to aggregate crowd-sourced ordinal labels. Using the Dirichlet process, we encourage the formation of worker and item clusters in the label generating process, which leads to more accurate label estimation. While the probability model is complex and uses non-conjugate DPs, we derive an efficient variational inference procedure to infer the posterior distributions. Experimental results show that the proposed method yields significantly better accuracy than the state-of-the-art, and is more robust to sparser labels. Moreover, it detects meaningful clusters, which can help the user to study the group's behavior.

Acknowledgements. This research was partially supported by the Research Grants Council of the Hong Kong Special Administrative Region (614513).

References

1. Andrieu, C., De Freitas, N., Doucet, A., Jordan, M.I.: An introduction to MCMC for machine learning. Mach. Learn. **50**(1–2), 5–43 (2003)
2. Bi, W., Wang, L., Kwok, J.T., Tu, Z.: Learning to predict from crowdsourced data. In: Proceedings of the Conference on Uncertainty in Artificial Intelligence, pp. 72–82 (2014)
3. Bishop, C.M.: Pattern Recognition and Machine Learning. Springer, New York (2006)
4. Blei, D.M., Jordan, M.I.: Variational inference for Dirichlet process mixtures. Bayesian Anal. **1**(1), 121–143 (2006)
5. Braun, M., McAuliffe, J.: Variational inference for large-scale models of discrete choice. J. Am. Stat. Assoc. **105**(489), 324–335 (2010)
6. Buckley, C., Lease, M., Smucker, M.D.: Overview of the TREC 2010 relevance feedback track (Notebook). In: Proceedings of the Nineteenth Text Retrieval Conference Notebook (2010)
7. Dawid, A.P., Skene, A.M.: Maximum likelihood estimation of observer error-rates using the EM algorithm. Appl. Stat. **28**(1), 20–28 (1979)
8. Evgeniou, T., Pontil, M.: Regularized multi-task learning. In: Proceedings of the 10th International Conference on Knowledge Discovery and Data Mining, pp. 109–117 (2004)

9. Ferguson, T.S.: A Bayesian analysis of some nonparametric problems. Ann. Stat. **1**, 209–230 (1973)
10. Ipeirotis, P.G., Provost, F., Wang, J.: Quality management on Amazon Mechanical Turk. In: Proceedings of the SIGKDD Workshop on Human Computation, pp. 64–67 (2010)
11. Jordan, M.I., Ghahramani, Z., Jaakkola, T.S., Saul, L.K.: An introduction to variational methods for graphical models. Mach. Learn. **37**(2), 183–233 (1999)
12. Kajino, H., Tsuboi, Y., Kashima, H.: Clustering crowds. In: Proceedings of the Twenty-Seventh AAAI Conference on Artificial Intelligence, pp. 1120–1127 (2013)
13. Kamar, E., Kapoor, A., Horvitz, E.: Identifying and accounting for task-dependent bias in crowdsourcing. In: Proceedings of the Third AAAI Conference on Human Computation and Crowdsourcing, pp. 92–101 (2015)
14. Kazai, G., Kamps, J., Koolen, M., Milic-Frayling, N.: Crowdsourcing for book search evaluation: impact of hit design on comparative system ranking. In: Proceedings of the 34th International Conference on Research and Development in Information Retrieval, pp. 205–224 (2011)
15. Lakkaraju, H., Leskovec, J., Kleinberg, J., Mullainathan, S.: A Bayesian framework for modeling human evaluations. In: Proceedings of SIAM International Conference on Data Mining, pp. 181–189 (2015)
16. Lakshminarayanan, B., Teh, Y.: Inferring ground truth from multi-annotator ordinal data: a probabilistic approach. Technical report arXiv:1305.0015 (2013)
17. Nocedal, J.: Updating quasi-Newton matrices with limited storage. Math. Comput. **35**(151), 773–782 (1980)
18. Raykar, V.C., Yu, S.: Eliminating spammers and ranking annotators for crowdsourced labeling tasks. J. Mach. Learn. Res. **13**, 491–518 (2012)
19. Raykar, V.C., Yu, S., Zhao, L.H., Valadez, G.H., Florin, C., Bogoni, L., Moy, L.: Learning from crowds. J. Mach. Learn. Res. **11**, 1297–1322 (2010)
20. Sethuraman, J.: A constructive definition of Dirichlet priors. Stat. Sin. **4**, 639–650 (1994)
21. Venanzi, M., Guiver, J., Kazai, G., Kohli, P., Shokouhi, M.: Community-based Bayesian aggregation models for crowdsourcing. In: Proceedings of the 23rd International Conference on World Wide Web, pp. 155–164 (2014)
22. Wainwright, M.J., Jordan, M.I.: Graphical models, exponential families, and variational inference. Found. Trends Mach. Learn. **1**(1–2), 1–305 (2008)
23. Welinder, P., Branson, S., Belongie, S.J., Perona, P.: The multidimensional wisdom of crowds. In: Advances in Neural Information Processing Systems, pp. 2424–2432 (2010)
24. Yan, Y., Rosales, R., Fung, G., Schmidt, M.W., Valadez, G.H., Bogoni, L., Moy, L., Dy, J.G.: Modeling annotator expertise: learning when everybody knows a bit of something. In: Proceedings of the 13th International Conference on Artificial Intelligence and Statistics, pp. 932–939 (2010)
25. Zhou, D., Basu, S., Mao, Y., Platt, J.C.: Learning from the wisdom of crowds by minimax entropy. In: Advances in Neural Information Processing Systems, pp. 2204–2212 (2012)
26. Zhou, D., Liu, Q., Platt, J., Meek, C.: Aggregating ordinal labels from crowds by minimax conditional entropy. In: Proceedings of the 31st International Conference on Machine Learning, pp. 262–270 (2014)

Joint Learning of Entity Semantics and Relation Pattern for Relation Extraction

Suncong Zheng, Jiaming Xu[✉], Hongyun Bao, Zhenyu Qi, Jie Zhang, Hongwei Hao, and Bo Xu

Institute of Automation, Chinese Academy of Sciences,
Beijing 100190, People's Republic of China
{suncong.zheng,jiaming.xu,hongyun.bao,zhenyu.qi,
jie.zhang,hongwei.hao,bo.xu}@ia.ac.cn

Abstract. Relation extraction is identifying the relationship of two given entities in the text. It is an important step in the task of knowledge extraction, which plays a vital role in automatic construction of knowledge base. When extracting entities' relations from sentences, some keywords can reflect the relation pattern, besides, the semantic properties of given entities can also help to distinguish some confusing relations. Based on the above observations, we propose a mixture convolutional neural network for the task of relation extraction, which can simultaneously learn the semantic properties of entities and the keyword information related to the relation. We conduct experiments on the SemEval-2010 Task 8 dataset. The method we propose achieves the state-of-the-art result without using any external information. Additionally, the experimental results also show that our approach can learn the semantic relationship of the given entities effectively.

Keywords: Relation extraction · Convolutional neural network · Entity embedding · Keywords extraction

1 Introduction

Relation extraction is identifying semantic relation of the entity pairs in a sentence, which is also called relation classification. It serves as an intermediate step in knowledge extraction from unstructured texts, which plays an important role in automatic knowledge base construction

Classical methods for the task of relation extraction focus on designing effective handcrafted features to obtain better classification performance [1,2,9]. These handcrafted features are extracted by analyzing the text and using different natural language processing (NLP) tools. However, these methods need complicated feature engineering and heavily rely on the supervised NLP toolkits, which might lead to the error propagation. In order to reduce the manual work in feature extraction, recently, deep neural networks [3,5–7] have been applied to obtain effective relation features from sentences directly. Although these models

© Springer International Publishing AG 2016
P. Frasconi et al. (Eds.): ECML PKDD 2016, Part I, LNAI 9851, pp. 443–458, 2016.
DOI: 10.1007/978-3-319-46128-1_28

Table 1. Instances in the SemEval-2010 Task 8 dataset.

Entity-Destination(e1,e2):
(1) Mayans charted venuss motion across the sky poured $[chocolate]_{e1}$ **into** $[jars]_{e2}$ and interred them with the dead
(2) Both his $[feet]_{e1}$ have been **moving into** the $[ball]_{e2}$ union members
Cause-Effect(e2,e1):
(3) Plantar $[warts]_{e1}$ are **caused by** a $[virus]_{e2}$ that infects layer of skin
(4) A wind speed associated with the $[devastation]_{e1}$ **caused by** the $[tornado]_{e2}$
Other:
(5) Frequent agitations throw academic $[life]_{e1}$ **into** $[disarray]_{e2}$
(6) Painting shows a historical view of the $[damage]_{e1}$ **caused by** the 1693 catania earthquake and the $[reconstruction]_{e2}$

can learn related features from given sentences without complicated feature engineering work, most of them focus on learning the semantic representation of the whole sentence, and they pay a little attention to keyword information related to the relation. Besides, they also fail to take full advantage of the entities' semantic properties.

Based on our observations, we find that most relation pattern can be reflected by some keywords in a sentence, especially the words between the given entities. We randomly select some instances from the SemEval-2010 Task 8 dataset [9] as Table 1 shows. If entity $e1$ and entity $e2$ satisfy the relation of *"Entity-Destination(e1,e2)"*, the words between $e1$ and $e2$ may be direction words such as: "into". If given entities satisfy the relation of *"Cause-Effect(e2,e1)"*, the words between $e1$ and $e2$ are more inclined to past participles such as: "caused by". Therefore, when compared with learning a semantic embedding of the whole sentence, extracting keyword information between given entities can better reflect the relation pattern in the sentence. Convolutional neural networks (CNN) [11–13] have achieved great success in sentence's semantic representation. It is able to preserve sequence information and extract the keyword information in a sentence. Therefore, in order to extract the keyword information which can reflect the relation pattern, we adopt a CNN architecture to model the sub-sentence between given entities instead of modeling the whole sentence.

However, it is hard to distinguish the confusing relation category by only using the keyword information. As Table 1 shows, the sub-sentence between $e1$ and $e2$ in sentence 5 seems to describe the relation of *"Entity-Destination(e1,e2)"*, but the semantic information of given entities show that they do not have the relationship. Hence, making good use of given entities' semantic properties can help to distinguish the confusing relationship. The semantic properties of given entities can be reflected by their contextual words. Some entities' properties may be reflected by the former (next) one word, some may be reflected by the former (next) two words or more. We set the entity word as the center, and select different sub-sentences around the entity as the entity's contexts. Extracting the entity's semantic properties is mining the semantic information of these contextual sub-sentences. To achieve this, we apply the operation of mixture convolution to extract the entities' different contextual features, then use max-pooling operation to select the most suitable contextual features as the entity's semantic properties. In this manner, we can solve the problem of unknown entity words, and represent the semantic relationship of entities effectively.

Therefore, the model we propose, in this paper, is a kind of mixture convolutional neural network, which can simultaneously learn the semantic properties of entities and the keyword information related to the relation. Different components of the mixture model focus on extracting different information, and all information is merged in the output layer to fix the task of relation classification.

The main work and contributions of this paper can be summarized as follows: (1) We propose a mixture convolutional neural network for the task of relation classification, which can simultaneously learn the semantic properties of entities and the keyword information related to the relation. (2) The method we proposed achieves the state-of-the-art results on the SemEval-2010 Task 8 dataset without using any external information such as: Word-Net or NLP tools. (3) We also conduct experiments to analyze the entity embedding produced by our method. The experimental results also show that our approach can represent semantic relationship of given entities effectively, when compared with word2vec [14].

2   Related Work

Over the years, relation classification is a widely studied task in the NLP community. To accomplish the task, various approaches have been proposed. Existing methods for relation classification can be divided into handcrafted feature based methods [1,2], neural network based methods [3–7] and the other valuable methods [6,10].

The handcrafted feature based methods focus on using different natural language processing (NLP) tools and knowledge resources to obtain effective handcrafted features. Then, they use some statistical classifier such as Support Vector Machines (SVM) [23] or Maximum Entropy (MaxEnt) [24] to get the right relation class based on the handcrafted features. The early work [2] employs Maximum Entropy model to combine diverse lexical, syntactic and semantic features derived from the text. Rink et al. [1] further designs 16 kinds of features that are extracted by using many supervised NLP toolkits and resources. It can get the best result at SemEval-2010 Task 8 when compared with other handcrafted features based methods.

In recent years, deep neural models have made significant progress in the task of relation classification. These models learn effective relation features from the given sentence without complicated feature engineering. The most common neural-network based models applied in this task are Convolutional Neural Networks (CNN) [3,4,8] and sequential neural networks such as Recursive Neural Networks (RecNN) [7] and Long Short Term Memory Networks (LSTM) [5].

Zeng [3] early explores convolutional neural network to represent the sentence level features. But the method still need to use features derived from lexical resources such as Word-Net to achieve the state-of-the-art results. Santos [4] and Xu [8] also apply convolutional neural network to classify relation classes. Santos [4] uses a pair-wise ranking method instead of softmax function on the top of CNN to reduce the effect of the confusing relation "Other" and Xu [8] proposes a negative sampling strategy to improve the assignment of subjects and objects. Some other deep learning approaches [5,7] focus on learning the whole sentences' semantic representation. Their differences mainly concentrate in the model architectures they used. There also exists other valuable methods such as the kernel-based methods [10] and the compositional embedding model [6].

In this paper, we find that keyword information between given entities and the semantic properties of given entities are important factors to reveal relationship. Therefore, we propose a kind of mixture convolutional model by joint learning the entity semantic properties and relation keywords for the task of relation classification.

3 Our Method

In order to extract the relation pattern and reduce the effect of confusing relations for the task of relation classification, we propose the mixture convolutional neural network (MixCNN), an unified model by joint learning of entities' semantic properties and relation pattern. In the following sections, we firstly present the architecture of our method shown in Fig. 1 and then detail each component of the model. After that, we introduce the objective function and training details.

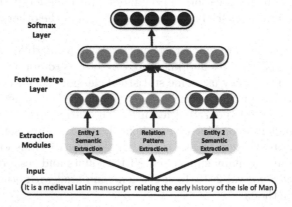

Fig. 1. The architecture of mixture convolutional neural network.

3.1   The Architecture of MixCNN

The framework of MixCNN is shown in Fig. 1, which mainly contains the entity semantic extraction module (ESE) and the relation pattern extraction module (RPE). When given a sentence, the entity semantic extraction module focuses on extracting the semantic properties of the given entities based on their surrounding words. The relation pattern extraction module focuses on extracting the keyword information between the two given entities which can reflect the relation pattern [4]. We merge the information of entities and relation pattern, obtained from the ESE and RPE modules, then fed the merged information into a softmax layer to fix the task of relation classification. In what follows, we describe these modules in detail.

3.2   The Module of Relation Pattern Extraction

Based on our observations and Santos's analysis [4], we find that most relation patterns can be reflected by a few keywords between the given two entities. Hence, the module of RPE aims to extract the keyword information which is related to the target relation. Convolutional neural network (CNN) [11–13] is able to preserve the sequence information and extract the keyword information in a sentence. [3,4,8] also validate the effectiveness of CNN to extract the related keyword information. Therefore, in order to extract the keyword information which can reflect relation pattern, we adopt the CNN architecture [12] to model the sub-sentence between the given entities instead of representing the whole sentence as Fig. 2 shows.

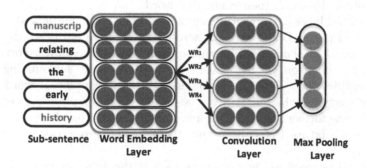

Fig. 2. The module of relation pattern extraction.

Firstly, each word is represented by a word embedding. In our experiments, we employ the word2vec[1] [14] to produce the word embeddings based on Wikipedia corpus. Out-of-vocabulary words are initialized randomly. The dimension of word embeddings is denoted as d. We define $X \in \mathbb{R}^{|V| \times d}$ as the set of word embeddings and the size of vocabulary is $|V|$.

---

[1] https://code.google.com/p/word2vec/.

When given a sentence s, we let $x_i \in \mathbb{R}^d$ be the d-dimensional word vector corresponding to the i-th word in the sentence. Hence, a sentence with the length of L is represented as a matrix: $s = (x_1; x_2; ...; x_L)$. In convolution layer, we use $WR^{(i)} \in \mathbb{R}^{k \times d}$ to represent the i-th convolution filter and $br^{(i)} \in \mathbb{R}$ to represent the bias term accordingly, where k is the context window size of the filter. Filter $WR^{(i)}$ will slide through the sentence s to get the latent features of sentence s. The sliding process can be represented as:

$$z_l^{(i)} = \sigma(WR^{(i)} * s_{l:l+k-1} + br^{(i)}), \tag{1}$$

where $z_l^{(i)}$ is the feature extracted by filter $WR^{(i)}$ from word x_l to word x_{l+k-1}. Hence, the latent features of the given sentence s are denoted as: $z^{(i)} = [z_1^{(i)}, ..., z_{L-k+1}^{(i)}]$. In order to extract keyword information of the sub-sentence, we apply the max-pooling operation to reserve the most prominent feature of filter $WR^{(i)}$ and denote it as:

$$z_{max}^{(i)} = max\{z^{(i)}\} = max\{z_1^{(i)}, ..., z_{L-k+1}^{(i)}\}. \tag{2}$$

We use multiple filters to extract multiple features. Therefore, the relation pattern of the given sub-sentence is represented as: $R_s = [z_{max}^{(1)}, ..., z_{max}^{(nr)}]$, where nr is the number of filters on RPE module.

3.3   The Module of Entity Semantic Extraction

The semantic properties of given entities contribute to reduce the impact of confusing relations. In this module, we focus on extracting the semantic properties of given entities based on their contextual words.

Word embeddings have been shown to preserve the semantic and syntactic information of words. But if we come across the unknown entity words, we still cannot obtain their semantic information from word embeddings. Fortunately, the properties of given entities can be reflected by their surrounding words. Different entities have different dependency on their contextual words. Some entities' property may be reflected by the former (next) one word, some may be reflected by the former (next) two words or more. Based on these motivations, we propose a mixture CNN to capture the semantic properties of entities as Fig. 3 shows.

We set entity word as the center, and select the sub-sentences with different scales around the entity as the entity's contexts. Extracting the entity's semantic properties is mining the semantics of these contexts. We still use CNN to extract the entities' contextual features. As Fig. 3 shows that $CNN \pm 1$ focuses on extracting the contextual semantic which is from word "early" to "of". $CNN \pm j$ mines the semantic information of context, which contains $2*j$ surrounding words of entity "history". The architectures of CNNs we used here are the same as Sect. 3.2 described. We use $WE1_j^{(i)}$ to represent the i-th filter of $CNN \pm j$ on the ESE module for entity $e1$ and $WE2_j^{(i)}$ to represent the i-th filter of $CNN \pm j$ on the ESE module for entity $e2$. Entity $e1$'s feature extracted by

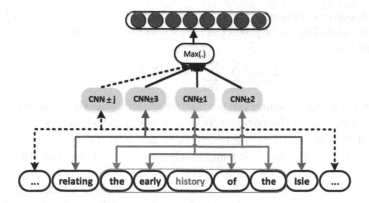

Fig. 3. The module of entity semantic extraction.

$WE1_j^{(i)}$ are denoted as $ze_j^{(i)}$. Hence, the j-th contextual information of entity $e1$ can be represented as $E1_j = [ze_j^{(1)}, ..., ze_j^{(ne)}]$, where ne is the number of filters on ESE module. Considering that different entities have different dependency on the contextual words, we apply a kind of max-pooling operation to merge the features extracted by $CNN \pm (1, 2...j)$. Namely,

$$E1_s = \begin{pmatrix} max(ze_1^{(1)} \ ... \ ze_j^{(1)}) \\ ... \quad ... \quad ... \\ max(ze_1^{(n)} \ ... \ ze_j^{(n)}) \end{pmatrix}. \tag{3}$$

3.4 Output Layer and Objective Function

After obtaining the semantic properties of given entities and relation pattern based on modules described in Sects. 3.2 and 3.3, we then merge these features by a concatenate manner which can be denoted as $f = [E1_s, R_s, E2_s]$. The output layer is the softmax classifier [15] with dropout:

$$y = W \cdot (f \circ r) + b, \tag{4}$$

$$p_i = \frac{exp(y_i)}{\sum\limits_{j=1}^{m} exp(y_j)}, \tag{5}$$

where $W \in \mathbb{R}^{m \times (nr+2 \cdot ne)}$ is the weights between the merge layer and the layer of labels. m is the total number of relation classes. Symbol $\circ$ denotes the element-wise multiplication operator and $r \in \mathbb{R}^{(nr+2 \cdot ne)}$ is a binary mask vector drawn from bernoulli with probability ρ. Dropout guards against overfitting and makes the model more robust. In Formula 5, p_i means the probability that the merge features reflect the relation i.

The objective function of the method is to minimise the cross entropy errors between the distribution of predicted labels and the distribution of actual labels. It is defined as:

$$L = -\sum_{s \in S} \sum_{i=1}^{m} -log(P(y_i|s, \Theta)),$$

(6)

where S represents the sentences in training set and y_i is the correct class of the given sentence s. Θ is the parameters of the model, which can be concluded as: $\Theta = \{X, WR^{(i)}, br^{(i)}, WE1_j^{(i)}, be1_j^{(i)}, WE2_j^{(i)}, be2_j^{(i)}, W, b\}$.

The model is optimized by using stochastic gradient descent [16]. The gradients are obtained via backpropagation. Gradients are backpropagated only through the unmasked units in the layer with dropout. Besides, the learned parameters of weight, in the dropout layer, need to be scaled by ρ such that $W = \rho \cdot W$.

4   Experimental Setup

Dataset. To evaluate the performance of our method, we use SemEval-2010 Task 8 dataset [9] that is the widely used for relation classification. The dataset contains 8,000 sentences for training, and 2,717 sentences for testing. There are 9 directional relations and one additional "other" relation, which is used to represent the relation that does not belong to any of the nine main relations. The directional relations are "Cause-Effect(C-E)", "Component-Whole(C-W)", "Content-Container(C-C)", "Entity-Destination(E-D)", "Entity-Origin(E-O)", "Instrument-Agency(I-A)", "Member-Collection(M-C)", "Message-Topic(M-T)" and "Product-Producer(P-P)". Especially, "Cause-Effect(e1,e2)" and "Cause-Effect(e2,e1)" are different relations. "Cause-Effect(e1,e2)" means that e1 causes e2 and "Cause-Effect(e2,e1)" means e1 is caused by e2. Hence, there are 19 relation classes in total.

Metric. To compare the performance of different methods, we adopt the official metric, the macro-averaged F1 score defined by Hendrickx [9]. The metric computes the macro-averaged F1-scores for the nine actual relations (excluding other) and takes the directionality into consideration [4].

Baselines. The baselines we used are recent methods for the SemEval-2010 Task 8 and they can be mainly cast into two main categories: the handcrafted feature based methods and the neural network based methods.

The handcrafted feature based methods are proposed by Rink [1]. All of these methods use a considerable amount of resources (WordNet, and FrameNet, for example) then employ SVM [23] or MaxEnt [24] as the classifier. The results of handcrafted feature based methods are shown in the first five rows of Table 3.

Recently, neural network models have made significant progress in the task of relation classification. The neural network models are Convolutional Neural Network (CNN) based methods [3,4,8], Recursive Neural Network (RecNN) based methods [7] and Long Short Term Memory Network (LSTM) based methods [5].

- **CNN** [3] is the early work that exploits a convolutional deep neural network to extract lexical and sentence level features for the task of relation classification.
- **CR-CNN** [4] also applies CNN to classify relation classes. Instead of using softmax function on the top layer of CNN, it employs a pair-wise ranking strategy to reduce the effect of the confusing relation "Other".
- **depLCNN** [8] learns relation representations from shortest dependency paths through a convolution neural network. Besides, it also proposes a negative sampling strategy to improve the assignment of subjects and objects and can achieve the state-of-the-art results by using the external resources such as WordNet.
- **RNN** [7] introduces a recursive neural network model that learns compositional vector representations for sentences. Then it uses the sentences representations for the task of relation classification.
- **MV-RNN** [7] finds the path between the two entities in the constituent parse tree and learns the distributed representation of its highest node. It uses that node's vector as feature to classify the relationship.
- **SDP-LSTM** [5] leverages the shortest dependency path (SDP) between two entities; multichannel recurrent neural networks, with long short term memory (LSTM) units, pick up heterogeneous information along the SDP. It is the first to use LSTM-based recurrent neural networks for the relation classification task.
- **FCM** [6] decomposes the sentence into substructures and extracts features for each of them, forming substructure embeddings. These embeddings are combined by sum-pooling and inputed into a softmax classifier.

Hyper Parameter Settings. The hyper parameters used in these experiments are summarized in Table 2. On the ESE module, we set a series of CNNs to model entities' contextual information. The context window size of each CNN on ESE module is set to 5. If the length of input contextual sentence is less than 5, the context window size of this CNN is set to the length of the input.

Table 2. Hyper parameters of the mixture convolutional neural network (MixCNN)

Parameter symbol	Parameter description	Parameter value
d	Dimension of word embedding	300
nr	The filter number of CNN on RPE module	300
ne	The filter number of CNN on ESE module	1000
k	Context window size of RPE module	20
j	The number of CNNs on ESE module	4
ρ	The ratio of dropout in merged layer	0.3

5   Results

5.1   Comparison with the Baselines on Relation Classification

We compare our method with the baselines which are recently published for the SemEval-2010 Task 8. In order to achieve state-of-the-art results, some approaches need to add external information such as: Word-Net, FrameNet or other NLP resources, which is actually an unfair comparison. Because different external resources have different effect on improving the predicted results. Besides, methods using external information have limitations. For example, if a method uses WordNet, it only suits for the task in English. To better illustrate the effectiveness of our method, we do not use any external information except word embedding in this experiment. We report the results of different methods as Table 3 shows.

In Table 3, only using word embedding as input features, our method achieves F1 of 84.8 %, which is the best results comparing with other methods. It shows that joint learning of entities' semantic properties and relation keywords is good for the task of relation classification. [3] is the early work that using CNN to classify relation. Although CNN [3] can extract sentence level features, it cannot achieve good results when only using word embedding features. Santos [4] also employs a kind of CNN method, called CR-CNN, to do the task by proposing a new pairwise ranking loss function. It can achieve the result of 84.1 %. The pairwise ranking loss function can reduce the impact of "Other" class. If it uses log-loss instead of the task-specific pairwise ranking loss function, the *F1* value is only 82.5 % which also has two percentage points worse than our method. Although our method uses the softmax, it can be also superior to CR-CNN with the pairwise ranking loss function. depLCNN [8] combines the dependency path and CNN to represent the sentence and can achieve the results of 81.3 %. Apart from the CNN methods, there are many sequential neural networks [5,7], which achieve results from 74.8 % to 82.4 %. [6] is a factor based compositional embedding model that only achieves the *F1* of 80.4 %.

We also compare our method with these baselines by adding external resources as Table 3 shows. Although we do not use any external information except word embeddings, our method still defeats most baselines which use lexical resources or NLP tools. If depLCNN only uses a negative sampling strategy to increase the number of training samples, our method can still has *+0.8 %* improvement. Besides, if depLCNN uses WordNet and negative sampling strategy simultaneously, we can also get comparable results to theirs under the circumstance that our training set is the half of theirs and without using WordNet.

5.2   The Effectiveness for Extracting Entity Semantic

In this paper we are not only focusing on achieving the state-of-the-art results on relation classification without using any external information, but also providing an effective manner to extract the semantic properties of given entities. In order

Table 3. Comparison of methods with adding different external resources. The external resources can be WordNet or other information obtained by NLP tools. Different resources have different effect on improving the predicted results. To better illustrate the effectiveness of our method, we do not use any external information except word embedding in this experiment.

Method	External resources	F1(%)
SVM [1]	POS, stemming, syntactic patterns	60.1
SVM [1]	word pair, words in between	72.5
SVM [1]	POS, stemming, syntactic patterns, WordNet	74.8
MaxEnt [1]	WordNet, FrameNet, Google n-grams, morphological	77.6
SVM [1]	WordNet, FrameNet, Google n-grams, morphological	82.2
RNN [7]	—	74.8
RNN [7]	POS, NER, WordNet	77.6
MVRNN [7]	—	79.1
MVRNN [7]	POS, NER, WordNet	82.4
FCM [6]	—	80.6
FCM [6]	Dependency parse, NER	83.0
SDP-LSTM [5]	—	82.4
SDP-LSTM [5]	POS, WordNet, Grammar relation	83.7
CNN [3]	—	69.7
CNN [3]	WordNet	82.7
depLCNN [8]	—	81.3
depLCNN [8]	Negative sampling	84.0
depLCNN [8]	WordNet	83.7
depLCNN [8]	WordNet, Negative sampling	**85.6**
CNN + softmax [4]	—	82.5
CNN + CR [4]	—	84.1
MixCNN + CNN	—	**84.8**

to further illustrate the effectiveness of ESE module on representing the semantic properties of given entities, we also conduct cluster experiments.

We use $ESE(e)$ to represent the semantic embedding of entity e that is extracted by module ESE. Hence the semantic relation between an entity pair $(e1, e2)$ can be denoted as $R_{ese}(e1, e2) = ESE(e1) - ESE(e2)$.

Word embeddings have been empirically shown to preserve semantic relation between words [14]. For example, $v(king) - v(queen) \approx v(man) - v(woman)$. $v(w)$ is the word embedding of word w. We use $R_v(e1, e2) = v(e1) - v(e2)$ to represent semantic relation between $e1$ and $e2$, which is initialized by word2vec. Here, we use $R_v(e1, e2)$ as our baseline.

Given the datasets, we obtain the relation embeddings of each entity pair: $R^i_{ese}(e1, e2)$ and $R^i_v(e1, e2)$. Then we employ the K-means algorithm [18] to cluster relation embeddings produced by the above manners. The clustering performance is evaluated by comparing the clustering results of texts with the relation labels provided by the datasets. Two metrics, the accuracy (ACC) [20] and the normalized mutual information (NMI) metrics [19], are used to measure the clustering performance [22]. Given a text x_i, let c_i be the predicted cluster label and y_i be the true label provided by corpus. Then the accuracy is defined as:

$$ACC = \frac{\sum\limits_{i=1}^{n} \delta(y_i, c_i)}{n}, \tag{7}$$

where n is the size of dataset and $\delta(x, y)$ is the indicator function that equals one if $x = y$ and equals zero otherwise. Normalized mutual information is a popular metric used for evaluating clustering tasks. It is defined as:

$$NMI(\mathbf{Y}, \mathbf{C}) = \frac{MI(\mathbf{Y}, \mathbf{C})}{\sqrt{H(\mathbf{Y})H(\mathbf{C})}}, \tag{8}$$

where $MI(\mathbf{Y}, \mathbf{C})$ is the mutual information between the predicted label set $\mathbf{Y}$ and the target label set $\mathbf{C}$. $H(.)$ is the entropy and $\sqrt{H(\mathbf{Y})H(\mathbf{C})}$ is used for normalizing the mutual information [22].

We run 100 times for each experiment and obtain the final results as Table 4 shows. The experimental results show that R_{ese} significantly better than R_v on both the accuracy (ACC) and the normalized mutual information (NMI) metrics. Although word embeddings can preserve the semantic and syntactic information of words, when we come across the unknown entity words, word embedings can do nothing. Besides, word embeddings contain much complex semantic information, so the semantic relation of word embedding is not obvious. ESE extracts the semantic properties of given entities by using their contextual information, which can solve the problem of unknown entity words. Furthermore, ESE focuses on mining relation properties of entities instead of modeling the complex semantic and syntactic information. Therefore, the R_{ese} significantly better than R_v.

Table 4. Comparison of ACC and NMI of K-means cluster algorithm based on different relation representations.

Dataset	Train		Test	
Relation Embedding	R_v	R_{ese}	R_v	R_{ese}
ACC(%)	26.98 ± 1.11	76.17 ± 5.13	23.99 ± 0.99	60.12 ± 2.95
NMI(%)	22.19 ± 0.81	84.72 ± 1.96	20.58 ± 0.76	61.16 ± 0.91

We also visualize the clustering results by using t-SNE [21] as Fig. 4 shows. In the embedding space produced by ESE, the entity pairs with same relation are more close to each other and the entity pairs with different relation are far from each other. On the contrary, there is no such obvious rule in the word embedding space produced by word2vec. The results further illustrate the effectiveness of ESE module on representing the semantic properties of given entities.

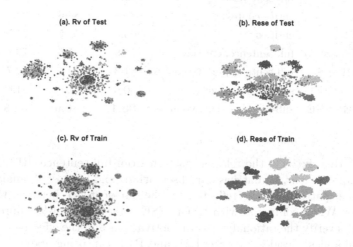

Fig. 4. The t-SNE visualization of the relation embeddings. Figure (a) and (c) are the relation embeddings produced by word2vec on training set and testing set. Figure (b) and (d) are produced by ESE module.

6   Analysis and Discussion

6.1   Module Analysis

In order to extract the relation pattern and obtain the semantic properties of given entities, we set two modules: RSE described in Sect. 3.2 and ESE in Sect. 3.3. In this section, we focus on analyzing the properties of these two modules.

At first, we allow each module with different configurations to perform the task of relation classification. We adopt a CNN architecture [12] called RPE to model the sub-sentence, which is the words between the given entities, instead of the whole sentence. We also compare the results of full sentence configuration which is marked as RPE1. Besides, we propose ESE module to extract the semantic properties of given entities based on their contextual words. In order to better verify the effectiveness of ESE module, we directly use entity word embedding to represent entity information. The embedding of unknown entity word is initialized randomly. We mark this configuration as ESE1. In addition to testing the effects of each module alone, we also test their various combinations. In this paper, our method is the combination of RPE and ESE.

From Table 5, we know that RPE and ESE can also achieve comparable results of $F1$ when compared with most of the baselines. Besides, when compared

Table 5. Comparison of the modules on the task of relation classification.

Methods	Input Text	Prec.(%)	Rec.(%)	F1(%)
RPE	sub-sentence	80.9	84.6	82.6
RPE1	full-sentence	72.4	74.5	73.3
ESE	entity-context	81.8	84.6	83.1
ESE1	entities	65.4	58.5	61.5
RPE1+ESE1	full-sentence, entities	70.9	75.6	73.1
RPE1+ESE	full-sentence, entity-context	81.3	84.2	82.7
RPE+ESE1	sub-sentence, entities	80.6	82.8	81.6
RPE+ESE	sub-sentence, entity-context	**83.1**	**86.6**	**84.8**

with RPE1 that extracts the relation pattern from full sentence, RPE achieves a +10 % improvement. It matches our observations and Santos' [4] analysis that most relation pattern can be reflected by the sub-sentence between the given two entities. When compared with ESE1, ESE achieves a +26 % improvement. These results verify the rationality of our motivations and the effectiveness of the proposed modules. Besides, merging ESE and RPE can bring about 2 points of improvement in *F1* value, which shows the complementarity of the two modules as well as the necessity of module integration.

6.2   Error Analysis

We conduct extensive qualitative and quantitative analysis of errors to better understand our method in terms of learning and predicting quality. We visualized the model's predicted results as Fig. 5 shows.

The diagonal region indicates the correct prediction results and the other regions reflect the distribution of error samples. The highlighted diagonal region means that our method can perform well on each relation class. However, from Fig. 5, we also can see that the distribution of predicted relation is relatively dispersed on the last column of "Other". Besides, most of the specific relation classes can be predicted as the "Other", which reflected from the last row shows in Fig. 5. The class of "Other" is a kind of confusing and heterogeneous class. It contains many different kinds of relation classes. Although our method can reduce the impact of confusing classes, it still need further improvement for the class of "Other". Apart from the class "Other", the class "I-A(e1,e2)" perform worse than the other 17 classes. Based on our observations, we find there are many samples in the class "I-A(e1,e2)" have the property that the given two entities are usually close to each other at the beginning of a sentence. For examples: "Elevator(e1) operator(e2) is a meditation on the..." and "Camera(e1) operator(e2) is that person ...". Because, there is no indicative words between two entities and there are few contextual words around entities. Our method is inadequate to deal with this case.

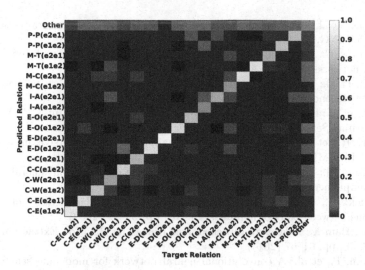

Fig. 5. The distribution of the predicted results for each relation class. The horizontal axis is the target relation and each target relation corresponds to a column of predicted relations. Point (X, Y) means the ratio that the target relation is X and the predicted relation is Y. The sum of each column value equal to 1.

7   Conclusion

In this paper, we propose a mixture convolutional neural network, which is an unified model by jointly learning entities' semantic properties and relation pattern, to fix the task of relation classification. It can achieve the state-of-the-art results on the SemEval-2010 Task 8 dataset without using any external information. Besides, we also conduct experiments to show that the entity embedding generated by our approach can reflect the relation properties of given entities.

Although our method can help to reduce the impact of the confusing relation, it still need further improvement for the class of "Other". In the future, we will focus on solving the problem of the special class "Other" and test our method on more related datasets.

Acknowledgements. This work is also supported by the National High Technology Research and Development Program of China (863 Program) (Grant No. 2015AA015402), the Hundred Talents Program of Chinese Academy of Sciences (No. Y3S4011D31), the NSFC project (No. 61501463) and National Natural Science Foundation (Grant No. 71402178).

References

1. Rink, B., et al.: UTD: Classifying semantic relations by combining lexical and semantic resources. In: 5th SE, pp. 256–259 (2010)
2. Kambhatla, N.: Combining lexical, syntactic, and semantic features with maximum entropy models for extracting relations. In: 43th ACL, pp. 22–26 (2004)

3. Zeng, D., et al.: Relation classification via convolutional deep neural network. In: 25th COLING, pp. 2335–2344 (2014)
4. dos Santos, C., Nogueira, et al.: Classifying relations by ranking with convolutional neural networks. In: 53th ACL, pp. 626–634 (2015)
5. Xu, Y., et al.: Classifying relations via long short term memory networks along shortest dependency paths. In: EMNLP (2015)
6. Yu, M., et al.: Factor-based compositional embedding models. In: NIPS Workshop on Learning Semantics (2014)
7. Socher, R., et al.: Semantic compositionality through recursive matrix-vector spaces. In: EMNLP, pp. 1201–1211 (2012)
8. Xu, K., et al.: Semantic relation classification via convolutional neural networks with simple negative sampling. In: EMNLP (2015)
9. Hendrickx, I., et al.: Semeval-2010 task 8: multi-way classification of semantic relations between pairs of nominals. In: SE, pp. 94–99 (2009)
10. Sun, L., Han, X.: A Feature-Enriched Tree Kernel for Relation Extraction. In: the 52th ACL, pp. 61–67 (2014)
11. Blunsom, P., et al.: A convolutional neural network for modelling sentences. In: 52th ACL, (2014)
12. Kim, Y.: Convolutional neural networks for sentence classification. In: EMNLP, pp. 1746–1751 (2014)
13. Collobert, R., et al.: Natural language processing (almost) from scratch. In: JMLR, pp. 2493–2537 (2011)
14. Mikolov, T., et al.: Distributed representations of words and phrases and their compositionality. In: NIPS, pp. 3111–3119 (2013)
15. Duan, K., et al.: Multi-category classification by soft-max combination of binary classifiers. J. Multiple Classifier Syst., 125–134 (2003)
16. Bottou, L.: Stochastic gradient learning in neural networks. J. Neuro-Nımes **9** (1991)
17. Fu, R., et al.: Learning semantic hierarchies via word embeddings. In: 52th ACL, pp. 1199–1209 (2014)
18. Wagstaff, K., et al.: Constrained k-means clustering with background knowledge. In: 18th ICML, pp. 577–584 (2001)
19. Chen, W.-Y., et al.: Parallel spectral clustering in distributed systems. J. TPAMI, 568–586 (2011)
20. Cai, D., et al.: Document clustering using locality preserving indexing. IEEE Trans. J. Knowl. Data Eng., 1624–1637 (2005)
21. Van der Maaten, L., et al.: Visualizing data using t-SNE. J. Mach. Learn. Res. **9**, 2579–2605 (2008)
22. Jiaming, X., Peng, W., et al.: Short text clustering via convolutional neural networks. In: The NAACL, pp. 62–69 (2015)
23. Hearst, M.A., Dumais, et al.: Support vector machines. In: IEEE Intelligent Systems and their Applications, pp. 18–28 (1998)
24. Phillips, S.J., et al.: Maximum entropy modeling of species geographic distributions. In: Ecological modelling, pp. 231–259 (2006)

BASS: A Bootstrapping Approach for Aligning Heterogenous Social Networks

Xuezhi Cao[(⊠)] and Yong Yu

Apex Data and Knowledge Management Lab,
Shanghai Jiao Tong University, Shanghai, China
{cxz,yyu}@apex.sjtu.edu.cn

Abstract. Most people now participate in more than one online social network (OSN). However, the alignment indicating which accounts belong to same natural person is not revealed. Aligning these isolated networks can provide united environment for users and help to improve online personalization services. In this paper, we propose a bootstrapping approach BASS to recover the alignment. It is an unsupervised general-purposed approach with minimum limitation on target networks and users, and is scalable for real OSNs. Specifically, we jointly model user consistencies of usernames, social ties, and user generated contents, and then employ EM algorithm for the parameter learning. For analysis and evaluation, We collect and publish large-scale data sets covering various types of OSNs and multi-lingual scenarios. We conduct extensive experiments to demonstrate the performance of BASS, concluding that our approach significantly outperform state-of-the-art approaches.

Keywords: Network alignment · Heterogenous networks · User modeling

1 Introduction

Online social network (OSN) is playing an important role in multiple aspects of our lives. We have different OSNs for various needs, e.g. Facebook for friendship, LinkedIn for professional relations, Instagram and Pinterest for content discovery. To fully keep in touch with friends or to explore various kinds of contents, most people participate in multiple OSNs. However, the alignment indicating which accounts belong to the same natural person remains unrevealed.

Benefits of aligning OSNs include but not limited to the followings. (a) Providing an united environment for users to easily keep up-to-date with friends' online activities [22]. (b) Achieving better user modeling by aggregating action histories [25]. (c) Alleviating cold-start problem in recommender system by borrowing data from aligned networks [1,16]. (d) Providing prerequisite information for cross-network behavior analysis [12].

There are platforms trying to recover the alignment by having users manually associate their accounts, e.g. About.Me[1]. However, not all users understand the

---

[1] http://about.me.

© Springer International Publishing AG 2016
P. Frasconi et al. (Eds.): ECML PKDD 2016, Part I, LNAI 9851, pp. 459–475, 2016.
DOI: 10.1007/978-3-319-46128-1_29

Table 1. Summary of existing approaches

Property	Name-Based [14,23]	Profile-Based [4,16,19,22]	Network-Based [10,21]	Specific Sites [9,10]	Our Solution BASS
Target Site	general	general	social relation	tag/geo-based	general
Target User	similar name	with profile	general	general	general
Full Mapping	no	no/yes	yes	yes	no
Leverage UGC	no	no, but possible	no	homogeneous	heterogeneous
Learning Method	statistics	rule/supervised	supervised	rule/supervised	unsupervised
Scalability	excellent, $O(N)$	worst at $O(N^2 K)$	poor, $O(N^2 K)$	poor, $O(N^2)$	good, $O(ND^2)$

benefits and do it willingly. It is preferred if we recover the alignment automatically by mining accessible information. Attempts are made by employing information such as username [14], user profile (email, location, education) [16], [19] and social tie [21]. Due to the task's recency, limitations still exist (summarized in Table 1):

Generality: Limitations on target users implicitly exist in username-based and profile-based approaches. They target only at users with same/similar usernames and users with complete profile respectively. There are also works target at specific types of networks, e.g. tagging system [9] and location-based networks [10]. Besides, most works assume all users participate in both networks (full mapping assumption), i.e. the set of common users is known as prior knowledge. However, mining this information itself is not a trival task.

Learning Method: Beside rule-based methods, supervised learning is widely used in existing works. However, acquiring enough training data for real OSNs (10 %-30 % according to existing experiments) is impractical.

Scalability[2]**:** For real OSN applications, scalability must be achieved. However, only few existing works discuss this issue. By detailed analysis, several works have theoretical time complexity of over $O(N^2)$ thus not scalable for OSN scale.

In this paper, we propose BASS, a bootstrapping approach that is freed from aforementioned limitations. It captures user consistencies of usernames, social ties and preferences jointly. To model the consistencies, partial alignment is required as pre-knowledge. Instead of using training data as in traditional approaches, we model the alignment as unobserved latent variables and employ EM-fashioned algorithm for the learning, leading to an unsupervised approach. For scalability, we achieve time complexity of $O(ND^2)$, which can be considered as linear to the size of the network. Detailed comparisons are listed in Table 1.

Note that aligning social networks will not result in privacy leak. The alignment is recovered using only public available information user revealed in OSNs. In other word, such alignment already exists online just not explicitly revealed yet. For users who don't want to be aligned, understanding how their identities got revealed can be the guidance for future actions to prevent it.

The paper is organized as follows. In Sect. 2, we discuss the related works. Then we define the task in Sect. 3. We introduce the data sets and preliminary

---

[2] Notations: N-number of nodes, D-network degree, K-feature space.

analysis in Sect. 4. BASS is proposed and discussed in Sect. 5. We present the experiment results in Sect. 6. Finally we draw conclusions in Sect. 7.

2   Related Work

2.1   Social Network Alignment

Due to the flexibility of the task and the variety of information available, researchers tackle this task from different angles:

Username. As the identification in OSNs, it is highly valuable for this task. Zafarani et al. make several assumptions upon usernames [23]. However, they are later claimed to be false in 75.47 % cases by analysis in [14]. Liu et al. further divide the task into alias-disambiguation (differentiating accounts with same username) and alias-conflation (linking accounts with different usernames) [14]. They model alias-disambiguation as binary classification task and leave alias-conflation unsolved. However, the coverage of alias-disambiguation is limited. By our analyze only 21.52 % users have same username across networks.

User Profile. Vosecky et al. represent profiles as features and propose feature selection and similarity calculation accordingly [22]. Nunes et al. tackle it with classification models (SVM and Random Forest) [19]. How to handle missing data is discussed in [16]. These approaches depend highly on the information availability. However, the availability may be limited due to privacy setting or incomplete profile. As reported in [13], there is a growing trend of users' awareness of privacy. The accessibility might be further restricted. Besides, user profiles may be heterogeneous, partly missing or with false information [13], making the profile modeling harder and require heavy manually work [14].

Social Relationship. Friend relations and group relations are also considered [10]. Tan et al. use latent vector to capture the graph information, and then combine it with username using rule-based method [21]. The benefit of social relationship is its accessibility. As reported in [11], friends lists can be easily accessed in ten of the twelve analyzed OSNs.

There are also works focus on certain types of OSNs. Iofciu et al. aim at aligning across tagging systems [9]. Geo-location and writing style are considered in [7]. Liu. et al. take advantage of user behavior and topic modeling [15]. Despite the performance, they do not directly lead to general solutions.

Most existing works employ supervised learning technics [19,21,22]. Large amount of training data is required, mostly proportional to the network size. We need heavy manual work to apply such approaches for real OSNs.

2.2   Author Identification

Although the task is to some extent similar with author identification, there are still differences. Because authors mostly use real name or same pseudonym in all articles, author identification focuses more on author-disambiguation. On the other hand, in this task we also need to align accounts with different usernames. For techniques, author identification focuses more on linguistic and writing style analysis [8,26], while we need to leveraging various heterogeneous user generated contents. Further, missing information and untruthful information do not emerge in author identification for most cases. Therefore, author identification approaches can not be directly borrowed for aligning OSNs.

2.3 Security and Privacy

This task is also considered as a security and privacy issue [2,6,13,18]. They focus on answering whether the current OSNs are safe in the sense of anonymity protection. Thus they aim at re-identifying only a part of the users and focus on precision instead of recall. However, our goal is to recover the whole alignment. The focus also shifts toward recall and large scale.

3 Problem Definition

We first formulate online social networks and then define the alignment task.

Definition 1. *An online social network is:* $S = (U, E, O, P)$, *where* U *is the set of user accounts;* E *is the set of social relations;* $O(u)$ *is the ownership oracle indicting who owns the account;* $P(u)$ *is the profile and user generated contents.*

Definition 2. *Social network alignment task is: Given two social network* S_A, S_B, *generate alignment* $\hat{R} \subset U_A \times U_B$ *where* $(u, v) \in \hat{R}$ *indicates that accounts* u, v *belong to same natural person according to the algorithm. The ground truth is:*

$$R = \{(u, v) | u \in U_A, v \in U_B, O_A(u) = O_B(v)\} \tag{1}$$

Following this definition, we remove two constraints that widely exist in previous works. The first is one-to-one constraint [10], forcing each account to align with at most one account in the other network. The other is full mapping assumption, assuming all users participate in both networks (or the set of common users is already known).

Preferred Properties. Recall the existing limitations we summarized in introduction (Table 1). We prefer the solution to have the following three properties. **Generality:** Minimize assumptions on target sites and target users. **Unsupervised:** Minimize human effort needed for training. **Scalability:** Scalable to real social network scale (billion-scale).

4 Data Set

We collect and publish two data sets for comprehensive evaluation[3]. The data sets cover both English and Chinese sites, and include general OSNs, microblogging and movie rating sites.

---

[3] http://dataset.apexlab.org/bass/.

Facebook-Twitter: About.Me is a third party platform for associating one's accounts from different OSNs including Facebook and Twitter. We collect 1,107,695 About.Me accounts as well as the corresponding social links. For this data set we have 328,224 aligned pairs.

Weibo-Douban: Weibo and Douban[4] are one of China's largest microblogging and movie rating sites respectively. Alignment between them is revealed explicitly in Douban's user profile (self descriptions). In total we have 141,614 aligned users. Besides the network, we also collect movie rating histories (123.49 per user) and microblogs (343.78 per user) as user generated contents (UGCs).

4.1 Consistency Analysis

For further insight, we conduct analysis on user consistencies across networks.

Username Consistency. We employ edit distance to measure similarity between usernames. Define $P_{ed}(d)$ to be the pairs of accounts with edit distance less than or equal to d, precision and recall as follow:

$$Prec@d = \frac{|R \cap P_{ed}(d)|}{|P_{ed}(d)|}, Rec@d = \frac{|R \cap P_{ed}(d)|}{|R|} \qquad (2)$$

where R is the ground truth alignment (Eq. 1). Figure 1(a) depicts the result, indicating strong relation between username and network alignment. However, the recall is limited. Only 30 % can be achieved for an acceptable precision.

Social Tie Consistency. Social links in OSNs reflect user's social ties or interests to some extent. We demonstrate social tie consistency by analyzing whether one's social relations in different OSNs tend to be overlapping. We use Jaccard Similarity Coefficient to capture the overlapping level. As only the relative value is required, we normalize the coefficient according to each user.

$$J(u, v) = \frac{O_A(E_A(u)) \cap O_B(E_B(v))}{O_A(E_A(u)) \cup O_B(E_B(v))}, J_n(u, v) = \frac{J(u, v)}{\max_{v' \in U_B} J(u, v')} \qquad (3)$$

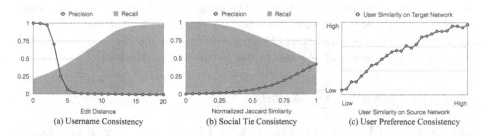

(a) Username Consistency   (b) Social Tie Consistency   (c) User Preference Consistency

Fig. 1. Consistency across OSNs (Username, Social Tie and User Preference)

---

[4] http://www.weibo.com/, http://www.douban.com/.

where $E_A(u), E_B(u)$ are the neighbors of u in network A, B respectively and O_A, O_B are the ownership oracles. Result in Fig. 1(b) indicates the existence of social tie consistency. However, it also indicates that even the alignment is given except the target pair, precision based only on social tie is not ideal ($\sim 40\%$).

User Preference Consistency. We demonstrate user preference consistency by showing that users with similar UGCs in one network tend to be similar in the other. Specifically, we employ topic model (LDA [3]) for modeling text-based UGCs and Jaccard Similarity Coefficient for item-based UGCs (rating/purchasing logs). Result in Fig. 1(c) supports the assumption that user preference consistency exists.

5   Bootstrapping Approach

In this section we propose our approach BASS. We first present the work flow, then discuss the algorithm details, and finally tackle the scalability issues.

5.1   Work Flow

We aim at recovering the alignment by mining consistencies of usernames, social ties and user preferences across the networks. However, we need partial alignment as pre-knowledge to model such consistencies. Specifically, to model social ties we need the alignment over target user's friends, and to model user preferences we need large scale aligned pairs to learn the preference transfer between heterogeneous networks. Traditionally, researchers employ labeled training data to solve this, leading to supervised approach. Instead, we model the alignment as unobserved latent data along with the consistency model, and then employ Expectation-Maximization algorithm for the parameter learning. Following this, we achieve an unsupervised bootstrapping approach.

We have two sub-models in our framework. One is the consistency model $\mathcal{C}(X, R)$ that captures the aforementioned consistencies based on observed data X as well as the given alignment R, where X contains usernames, social relations and user generated contents in both networks. The other is the classification model $\mathcal{Y}$ that takes in the features generated by $\mathcal{C}(X, R)$ and estimates the probability that two accounts belong to the same natural person. For notations, we use $\Theta = \{\Theta_{\mathcal{C}}, \Theta_{\mathcal{Y}}\}$ for the set of unknown parameters for the two parts. Our goal is to estimate the parameters Θ by maximizing the likelihood $L(\Theta; X)$, and then recover the alignment $\hat{R}$ based on the estimated parameters $\hat{\Theta}$.

$$L(\Theta; X) = p(X|\Theta) = \sum_R p(X, R|\Theta)$$

$$p(X, R|\Theta) = \prod_{(u,v) \in R} \mathcal{Y}(\mathcal{C}_{u,v}(X, R, \Theta_{\mathcal{C}}), \Theta_{\mathcal{Y}}) \prod_{(u,v) \notin R} (1 - \mathcal{Y}(\mathcal{C}_{u,v}(X, R, \Theta_{\mathcal{C}}), \Theta_{\mathcal{Y}}))$$

$$\hat{\Theta} = \arg\max_{\Theta} L(\Theta; X), \ \hat{R} = \arg\max_R p(X, R|\hat{\Theta})$$

$$(4)$$

By viewing the alignment R as the unobserved latent data, we can employ Expectation-Maximization algorithm for the learning process. For E-step, we update the alignment R based on current parameters. And for M-step, we update the parameters for both consistency model and classification model based on the just-computed R. The overall work flow is depicted in Fig. 2.

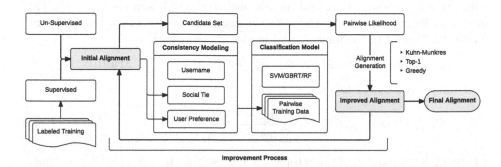

Fig. 2. Work Flow of **BASS**: **B**ootstrapping approach for **A**ligning **S**ocial networ**kS**

The benefit of bootstrapping strategy is multi-folded. Firstly, we achieve unsupervised approach thus large-scale labeled training data is no longer required. Secondly, we employ the whole network for training instead of only the labeled pairs in traditional approaches. Thirdly, we can directly adopt the bootstrapping approach for incremental online learning by considering new users as unaligned, therefore is suitable for real OSN application.

5.2 Algorithm Details

The key components are consistency model and classification model for M-step, and alignment generation for E-step. We first discuss the components and then extend the approach for unsupervised learning in the following subsections.

Consistency Model. Based on previous analysis, we target at consistencies of username, social tie and user generated content.

Username is the easiest as it doesn't depend on the alignment. Consistency features include exact/substring match, edit distance and naming patterns [24].

Social tie comes next. It depends on the alignment but fortunately is of homogeneous format (consider all relationships as undirected). We extend Jaccard Similarity Coefficient to capture the social tie consistency. Given current alignment $\hat{R}$, it can be defined by:

$$com(u, v, \hat{R}) = (E_A(u) \times E_B(v)) \cap \hat{R}$$

$$J(u, v, \hat{R}) = \frac{|com(u, v, \hat{R})|}{|E_A(u)| + |E_B(v)| - |com(u, v, \hat{R})|} \tag{5}$$

The most challenging one is modeling the user generated contents consistency (or user preferences consistency) because it depends on current alignment and is heterogeneous across networks. Most UGCs can be categorized by text-based and item-based ones. Text-based UGCs include microblogging, forum and etc., and item-based ones include rating log, purchase log and etc. Therefore we target at these two types of UGCs. We model them using multi-modal Latent Dirichlet Allocation [3], where each modal corresponds to the UGCs in one network. The model is depicted in Fig. 3. The detailed distributions for upper modal (text-based site Weibo) of the multi-modal model are as below:

$$\theta_i \sim Dir(\alpha), \quad z_{ij}^w \sim Multi(\theta_i), \quad \phi_k^w \sim Dir(\beta^w), \quad w_{ij} \sim Multi(\phi_{z_{ij}}^w) \quad (6)$$

And the inference is as follows:

$$P(z_{ij}^w = k | z_{\neg ij}^w, w, \phi^w, \cdot) \propto \frac{n_{ik}^{w,\neg ij} + \alpha_k}{\sum_q (n_{iq}^{w,\neg ij} + \alpha_q)} \cdot \frac{m_{kw_{ij}}^{w,\neg ij} + \beta_{w_{ij}}}{\sum_{w'} (m_{kw'}^{w,\neg ij} + \beta_{w'})} \quad (7)$$

where n_{ik}^w is the number of times topic k being assigned to user i (number of times $z_{i*}^w = k$) and m_{kj}^w is the number of times word j being assigned to topic k. The distribution as well as the inference for lower modal with z^m, m, ϕ^m are similar with above. After sufficient sampling iterations, the preference distribution θ_i can be estimated by:

$$\hat{\theta}_{ij} = \frac{n_{ik}^w + n_{ik}^m + \alpha_k}{\sum_q (n_{iq}^w + n_{iq}^m + \alpha_q)} \quad (8)$$

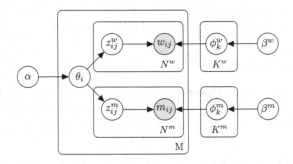

Fig. 3. Multi-modal topic model

The learning process of the multi-modal topic model is as follows. We consider each alignment (u, v) in current alignment $\hat{R}$ as a user instance with actions in both modals. With these as anchor links, we learn the correlation and transfer between heterogeneous modals. We also consider each account $u \in U_A$ and $v \in U_B$ as a user instance with actions only in one modal.

With the multi-modal topic model, we can embed each account in both networks into the universal topic space (θ). Then we can quantify the consistency between accounts u, v using cosine similarity, L1 distance and Kullback-Leibler Divergence over θ_u and θ_v.

Classification Model. We use binary classification model to separate pairs of accounts by whether they belong to same natural person. Specifically, we employ Support Vector Machine [20] here in BASS.

Following the likelihood function in Eq. (4), the current alignment $\hat{R}$ serves as the ground truth when updating the classification model. Specifically, all pairs of accounts that are aligned in the current alignment $((u,v) \in \hat{R})$ are considered as positive instances and the rest as negative instances. Following this, we have $O(N^2)$ training instances, where N is the number of accounts. Such amount is too large for scalability. We will return to this issue in Sect. 5.3.

Note that the current alignment $\hat{R}$ is not the actual ground truth. Because the classification model is considered to be noise-robust, applying it to the noisy alignment $\hat{R}$ can still optimizing the likelihood function. As long as the likelihood is being optimized, the bootstrapping approach can work properly.

Alignment Generation. With the consistency model and classification model trained, we can estimate the pairwise likelihood by $S(u,v) = \mathcal{Y}(\mathcal{C}_{u,v}(X, R, \Theta_{\mathcal{C}}), \Theta_{\mathcal{Y}})$. Therefore, the remaining task is: Given pairwise score S where $S(u,v)$ indicates the likelihood of accounts u, v belonging to the same natural person, generate the required alignment R.

If we align each account to at most one corresponding account, the alignment is actually a partial one-to-one mapping $M : E_A \rightarrow \{E_B \cup \bot\}$, where $M(u) = \bot$ indicates no corresponding account for u. The objective function is:

$$M^* = \arg\max_M \prod_u S(u, M(u))$$
$$R^* = \{(u, M^*(u)) | u \in E_A, M^*(u) \neq \bot\}$$

(9)

where we define $S(u, \bot) = \tau$ as the penalty for mismatch.

By considering each account as a node and $\log S(u,v) - \log \tau$ to be the weight of the edge between node u and v, this task can be deduced to a maximum matching problem on weighted bipartite graph. Such problem can be solved perfectly in $O(N^4)$ by using Kuhn-Munkres (KM) algorithm [17], also known as Munkres assignment algorithm. The algorithm is later improved by Karp et al., achieving a time complexity of $O(N^3)$. However, this is still not scalable for real social network scale. Compromise must be made by using alternative algorithms instead of perfect matching. We discuss it later in Sect. 5.3.

Extend to Unsupervised Version. To minimize human effort needed, it is preferred that the algorithm can run in an unsupervised manner. In other words, the initial alignment need to be automatically generated instead of using labeled training data. In BASS, the initial alignment serves as seeds for alignment propagation. Therefore, precision is strongly required while recall is not. Previous analysis show that alignment generated by username matching fulfill the requirement and can be considered as the initial alignment.

There is another potential issue. As BASS uses current alignment to generate training data for the classification model, there might be a chance that the classification model converge to the rules that we used to generate the initial alignment. To prevent this from happening, we introduce noises into the data intentionally during training process. Specifically, we randomly alter some of the usernames (10 % in experiment) during the consistency modeling process for initial alignment so that username features do not fully suppress other features. Experiments show that the performance is not very sensitive to this parameter if set within reasonable ranges.

5.3   Scalability Issue

Due to tremendous size of OSN users, scalability must be considered. It is normal for social network to be of billion-scale, so even $O(N^2)$ approach is impractical. As stated previously, the size of pairwise training samples and the time complexity of matching algorithm are not scalable under current model settings. Besides, there is a hidden violation that we have $O(N^2)$ candidate pairs.

Training Data Subsampling. Based on learning process proposed previously, a training set of size $O(N^2)$ will be generated, with $O(N)$ positive samples and $O(N^2)$ negative ones. Therefore, we subsample over the negative samples and keep only a constant number (k) of negative samples for each account.

Conducting the subsampling effectively is not trival. Because most negative samples can be easily separated from the positive ones and provide almost no valuable information, purely random sample would highly jeopardized the performance. We follow the idea that boundary samples, the ones that cannot be easily separated by the classification model, are more valuable for the training process. Similar ideas are also used in other scenarios. For example, in Support Vector Machine [20], support vectors are actually boundary samples.

By assuming the models in two consecutive iterations are similar, we consider the negative samples with high but not top scores in previous iteration as boundary samples. The final approach is: for each account u, consider its current alignment $(u, v) \in \hat{R}$ as positive sample, and (u, v') as a negative sample for all v' ranked in top k according to $S(u, v')$ in last iteration. Where k is the parameter to balance between performance and scalability. By setting k to infinity, the process degenerates to original version.

Candidate Pair Generation. As we cannot consider all N^2 pairs due to scalability concern, candidate set must be generated. Analysis in Fig. 1(b) shows that for almost all true alignment ($> 99\%$), their friendships overlap to some extent (with Jaccard coefficient > 0). Therefore, we consider only accounts pairs with common neighbors according to current alignment $\hat{R}$ as candidates. Formally we construct candidate set by $C = \bigcup_{(u,v) \in \hat{R}} E_A(u) \times E_B(v)$. Following this we obtain $O(ND^2)$ candidate pairs, where D is the degree of the network. Note that no matter how the network grows, the degrees of normal users are still limited. So $O(ND^2)$ can be considered as linear to the network's size.

Alternative Alignment Generation. Although perfect alignment can be achieved, it requires large amount of computation. Therefore, we need efficient alternative alignment generation method. Here we propose two candidates:

Top-1 Alignment. Match each user account u to v that maximize $S(u, v)$:

$$\hat{R} = \{(u, \arg\max_v S(u, v))\} \cup \{(\arg\max_u S(u, v), v)\} \tag{10}$$

Similar as previous, we ignore alignment with $S(u, v) < \tau$. For users with no alignment, we consider them as single-site users.

Stable-Marriage Alignment. A matching between two sets of elements is a stable marriage matching if there does not exist a pair of elements that both elements prefer each other than their current alignment. We borrow this definition for social network alignment problem. The algorithm for original stable marriage problem is: first selecting an unaligned element u and its most preferred element v that u has not proposed yet; if v is available then link (u, v); otherwise if v also prefers u over its current alignment then link (u, v) and release v's original alignment. This algorithm runs in time complexity of $O(N^2)$. Fortunately, in this setting we have another property: the preference matrix is symmetric. This property enables us to further speed up the computation. Specifically, if we traverse the candidate pairs (u, v) in descend order of $S(u, v)$ instead of randomly selecting u, we will never need to replace existing alignment as in the traditional algorithm. Thus we can achieve time complexity of $O(|C| \log |C|)$ where C is the candidate set. Similar as previous, we link only when $S(u, v) > \tau$.

6 Experiments

6.1 Experiment Setting

The data sets are discussed in Sect. 4. Recall the data includes general-purposed, microblogging and movie review OSNs, and covers multi-lingual (English and Chinese) scenarios. Now we explain the metrics and comparing algorithms.

Evaluation Metrics: As defined in Sect. 3, the task is a retrieval task that mines the aligned pairs of accounts. F-1 Score is used as the metric, defined by:

$$Prec(\hat{R}) = \frac{|\hat{R} \cap R|}{|\hat{R}|}, \; Recall(\hat{R}) = \frac{|\hat{R} \cap R|}{|R|}, \; F_1(\hat{R}) = 2 \cdot \frac{Prec(\hat{R}) Recall(\hat{R})}{Prec(\hat{R}) + Recall(\hat{R})} \tag{11}$$

where R is the ground truth alignment (Eq. (1)) and $\hat{R}$ is the prediction.

Comparing Algorithms: We compare with the state-of-the-art approaches that based on similar information with BASS (username, social tie and user preference), which are listed and described as follows:

BASS: The bootstarpping approach proposed in this paper. Support Vector Machine (LibSVM [5]) is employed as the classification model. Stable-Marriage alignment is used unless indicated otherwise.

BASS-H: BASS without UGC modeling (user preference consistency).

BASS-U: Unsupervised version of BASS (Sect. 5.2).

MNA: Multi-Network Anchoring proposed by Kong et al. in [10].

MAH: The Manifold Alignment on Hypergraph approach, by Tan et al. [21].

MOBIUS: Aligning by modeling user behaviors, proposed by Zafarani in [24].

NAME: Aligning accounts with same username. Precision for exact name matching is almost 1, the only concern is coverage. So it can be seemed as an upper-bound for works focusing on alias-disambiguation [14].

There are also approaches not compared due to limitations on target sites or target users (tagging system [9], profile-based [4,16,19,22] and etc.).

6.2 Results and Analysis

We first conduct experiments based on the full mapping assumption (the assumption in most existing works): all users participate in both networks thus a perfect alignment exists. 30 % of alignment is given as training data (except for the unsupervised version). Parameters such as α, β, τ are set by cross validation.

We show the results in Table 2. Our approaches, both supervised and unsupervised versions, achieve significantly better performance comparing to state-of-the-art algorithms. Note that UGCs are only available in Weibo-Douban data set, therefore BASS-H and BASS are the same over Facebook-Twitter data set. Precision and recall of MAH and MOBIUS are the same respectively because they always produce full mapping, and in this experiment setting a perfect mapping exists (full mapping assumption).

Note that as OSN varies, experimental results using different data sets are not numerically comparable. For example, In our data set only 21 % users share same username while 52 % in data used in [21]. The differences may due to the collecting methodology and the data source used. As previous works did not

Table 2. Experimental results with 30 % training data

Approach	Facebook-Twitter			Weibo-Douban		
	Precison	Recall	F1-Score	Precison	Recall	F1-Score
BASS	84.40 %	**79.75 %**	**0.8201**	79.49 %	**76.22 %**	**0.7782**
BASS-H	84.40 %	79.75 %	0.8201	76.56 %	73.26 %	0.7487
BASS-U	82.52 %	78.10 %	0.8025	77.64 %	74.88 %	0.7623
MAH	42.82 %	42.82 %	0.4282	41.74 %	41.74 %	0.4174
MNA	67.64 %	62.21 %	0.6481	64.39 %	61.41 %	0.6287
MOBIUS	55.48 %	55.48 %	0.5548	51.37 %	51.37 %	0.5137
NAME	**100.00 %**	21.52 %	0.3541	**100.00 %**	14.70 %	0.2562

publish their data, we can only compare the approaches using our data (we also publish the data sets to the community). The scale of the data set also contribute to the difficulty of aligning. It is rather difficult to find the corresponding accounts from millions of accounts comparing to from hundreds of accounts.

Heterogeneous UGC Modeling: By comparing BASS and BASS-H in Table 2, we show that UGC modeling do improve the alignment quality. To gain further insight, we list part of the resulting word-dictionary along with the movie-dictionary in Table 3. Correlation can be noticed, indicating the multi-modal LDA can capture the heterogeneous UGCs and embed users by their underlying general preferences. For example, the first topic indicates the users tweet about pets are more likely to enjoy comedies, cartoons etc.

Effect of Training Size: We vary the size of training size from 10% to 50%. Results showed in Table 4 indicate that our unsupervised version BASS-U defeats the existing supervised approaches even when 50% data is given. It is also noticeable that the first 20% data does not result in great improvement in BASS comparing to BASS-U, indicating that using unsupervised approach can capture most common knowledge for the aligning task.

Single Network Users: Most existing works assume that all users participate in both networks (full mapping assumption). However, it is not the real-world scenario. Now we break this limitation and expand the data set by adding users that participate in only one OSN. To make it more challenging, we add friends of existing users instead of purely random users. We show the results in Table 4 (right part), where p indicates the ratio of users participate in both networks (traditional approaches still aim at aligning all users). Decreasing p makes the task more challenging. Note that it has more impact on approaches based on social ties or user preferences, while less on ones based on only usernames (performance drops dramatically on MAH and MNA while slightly on MOBIUS). Because of the comprehensiveness of our consistency model (with no heavy dependency on specific aspect), the performance remains mostly the same with only minor drop.

Improvement Process: The effect of the iterative bootstrapping is depicted in Fig. 4(a). A clear trend of improvement, or the snowball effect, can be noticed in both supervised and unsupervised approaches. Note that only few rounds are needed before convergence, so it does not raise a complexity issue.

Social Tie Strength: As the social tie varies from site to site, we are interested in how social tie strength effects the performance of aligning. For comprehensive analysis, we generate synthetic data based on real data. We first union the social relations from both networks and consider them as the potential links. Then we plug each link into the two networks with probability of p, q respectively. Here p, q are parameters controlling the density and coupling strength. User profiles as well as UGCs are kept the same as the real data. By varying p, q from 0 to 1, we can simulate different social tie scenarios. Specifically, we have: (1) *Isomorphic* by setting $p = q = 1$; (2) *Subnetwork* by setting $p = 1, q < 1$ or vice versa; (3) *Partially Overlapping* by setting $p, q < 1$.

Table 3. Dictionaries for Multi-modal topic model over Weibo & Douban

ID	Top Words	Top Movies
1	Food, Cat, Friend, Dog, Home, Like, Life	Hotaru no haka, The Pursuit of Happyness, Jeux d'enfants The Devil Wears Prada, Up, Tonari no Totoro, Ratatouille
2	China, Article, Book, Issue, America, Country, Society	Inception, Social Network, Source Code, Avatar, WALL·E V for Vendetta, The Lord of the Rings, Argo The Shawshank Redemption, The Bourne Identity, Titanic
3	We, Love, Myself, Life, Want, Time, Like, World	Amour, Love Letter, Amlie, Forrest Gump, Before Sunrise Before Sunset, Flipped, Love Actually, The Notebook

Table 4. Varying size of Training Data & Common User Ration (results in Table 4 are according to Facebook-Twitter set.)

F1 Score	Training Size					Common User Ration p				
	10 %	20 %	30 %	40 %	50 %	100 %	90 %	80 %	70 %	60 %
MAH	0.3789	0.4065	0.4282	0.4330	0.4426	0.4282	0.4026	0.3706	0.3444	0.3253
MNA	0.6021	0.6259	0.6481	0.6549	0.6612	0.6481	0.6201	0.5928	0.5503	0.5196
MOBIUS	0.5327	0.5457	0.5548	0.5617	0.5620	0.5548	0.5479	0.5465	0.5431	0.5390
BASS	0.8073	0.8093	0.8201	0.8265	0.8374	0.8201	0.8205	0.8164	0.8158	0.8087
BASS-U	0.8025					0.8025	0.8047	0.7985	0.7990	0.7973

Results for the $p = q$ cases are showed in Fig. 5(a). Our approaches out perform existing approaches except when p, q are too small, which indicates almost no social relations available. In these cases the social ties become total noises and the consistency propagation may be restricted. Hence, username-based approaches have great advantage. We depict results for all p, q in Fig. 5(b). Except for the cases that one network's social relationship is too weak ($\min(p, q) \leq 0.1$), our approach has a satisfying performance.

6.3 Performance vs Scalability

Subsampling. Recall that we conduct subsampling over training data for the classification model. A parameter k controls the number of negative samples for each user. Larger k leads to larger computational complexity but better performance. By varying k, we show the results in Fig. 4(b). We conclude that when k is larger than some small threshold (3–5), keep increasing k does not achieve a significantly better performance. This indicates that our subsampling

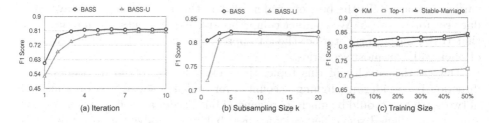

Fig. 4. Detail analysis - Iteration, Alignment Algorithm, Subsampling

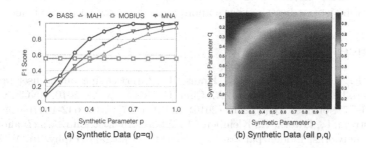

Fig. 5. Experimental results over synthetic data

strategy can shrink the training set of size $O(N^2)$ into of size $O(N)$ with ignorable sacrifice of performance.

Alignment Algorithms. Due to the high time complexity of KM algorithm, two alternative alignment algorithms (Top-1 alignment and Stable-Marriage alignment) are proposed. We show the results using different alignment algorithms in Fig. 4(c). As expected, KM algorithm always gives the best result. Stable-Marriage alignment can always achieve a compatible result with a much lower time complexity. Thus a great balance between the scalability and performance can be achieved using it.

7   Conclusions and Future Works

In this paper, we focus on aligning heterogeneous social networks. To tackle the problem, we propose a bootstrapping approach BASS, which starts with an unperfect alignment and refines it iteratively based on consistency propagation over usernames, social ties and user preferences. The advantage is multi-folded. Firstly, it is a general-purpose approach with minimum limitation on target sites and users, and can be adopted for various heterogeneous scenarios. Secondly, full-mapping constraint is removed. Thirdly, we achieve unsupervised approach. Finally, it is scalable for large-scale social networks without jeopardizing the performance. We also collect and publish large-scale real-world data sets covering various scenarios. To the best of our knowledge, this is the first public available data set for this task. We conduct comprehensive experiments. Results indicate that BASS outperform state-of-the-art approaches with a relative improvement of about 40 % in most scenarios.

Due to the novelty of this topic, there exist plenty of future works. One direction is considering aligning accounts over multiple social networks instead of two, such task would be much more general but harder as well. Further, we can employ the aligned social networks for user behavior analysis across sites. It is also an interesting topic to reveal the underlying real friend relation, i.e. the true friendship among natural persons. Following this work, we can also study what online actions jeopardies the user's anonymity and how to prevent accordingly.

References

1. Abel, F., Henze, N., Herder, E., Krause, D.: Interweaving public user profiles on the web. In: De Bra, P., Kobsa, A., Chin, D. (eds.) UMAP 2010. LNCS, vol. 6075, pp. 16–27. Springer, Heidelberg (2010). doi:10.1007/978-3-642-13470-8_4
2. Backstrom, L., Dwork, C., Kleinberg, J.: Wherefore art thou r3579x?: anonymized social networks, hidden patterns, and structural steganography. In: WWW, pp. 181–190. ACM (2007)
3. Blei, D.M., Ng, A.Y., Jordan, M.I.: Latent Dirichlet allocation. J. Mach. Learn. Res. **3**, 993–1022 (2003)
4. Carmagnola, F., Cena, F.: User identification for cross-system personalisation. Inf. Sci. **179**(1), 16–32 (2009)
5. Chang, C.C., Lin, C.J.: Libsvm: a library for support vector machines. ACM Trans. Intell. Syst. Technol. (TIST) **2**(3), 27 (2011)
6. Frankowski, D., Cosley, D., Sen, S., Terveen, L., Riedl, J.: You are what you say: privacy risks of public mentions. In: SIGIR, pp. 565–572. ACM (2006)
7. Goga, O., Lei, H., Parthasarathi, S.H.K., Friedland, G., Sommer, R., Teixeira, R.: Exploiting innocuous activity for correlating users across sites. In: WWW, pp. 447–458. International World Wide Web Conferences Steering Committee (2013)
8. Houvardas, J., Stamatatos, E.: N-gram feature selection for authorship identification. In: Euzenat, J., Domingue, J. (eds.) AIMSA 2006. LNCS (LNAI), vol. 4183, pp. 77–86. Springer, Heidelberg (2006). doi:10.1007/11861461_10
9. Iofciu, T., Fankhauser, P., Abel, F., Bischoff, K.: Identifying users across social tagging systems. In: ICWSM (2011)
10. Kong, X., Zhang, J., Yu, P.S.: Inferring anchor links across multiple heterogeneous social networks. In: CIKM. ACM (2013)
11. Krishnamurthy, B., Wills, C.E.: On the leakage of personally identifiable information via online social networks. In: Proceedings of the 2nd ACM Workshop on Online Social Networks, pp. 7–12. ACM (2009)
12. Kumar, S., Zafarani, R., Liu, H.: Understanding user migration patterns in social media. In: AAAI (2011)
13. Labitzke, S., Taranu, I., Hartenstein, H.: What your friends tell others about you: low cost linkability of social network profiles. In: Proceeding of 5th International ACM Workshop on Social Network Mining and Analysis, San Diego (2011)
14. Liu, J., Zhang, F., Song, X., Song, Y.I., Lin, C.Y., Hon, H.W.: What's in a name?: an unsupervised approach to link users across communities. In: WSDM, pp. 495–504. ACM (2013)

15. Liu, S., Wang, S., Zhu, F., Zhang, J., Krishnan, R.: Hydra: Large-scale social identity linkage via heterogeneous behavior modeling. In: SIGMOD (2014)
16. Malhotra, A., Totti, L., Meira Jr., W., Kumaraguru, P., Almeida, V.: Studying user footprints in different online social networks. In: Proceedings of the 2012 International Conference on Advances in Social Networks Analysis and Mining (ASONAM 2012), pp. 1065–1070. IEEE Computer Society (2012)
17. Munkres, J.: Algorithms for the assignment and transportation problems. J. Soc. Ind. Appl. Math. **5**(1), 32–38 (1957)
18. Narayanan, A., Shmatikov, V.: De-anonymizing social networks. In: 30th IEEE Symposium on Security and Privacy, pp. 173–187. IEEE (2009)
19. Nunes, A., Calado, P., Martins, B.: Resolving user identities over social networks through supervised learning and rich similarity features. In: Proceedings of the 27th Annual ACM Symposium on Applied Computing, pp. 728–729. ACM (2012)
20. Suykens, J.A., Vandewalle, J.: Least squares support vector machine classifiers. Neural Process. Lett. **9**(3), 293–300 (1999)
21. Tan, S., Guan, Z., Cai, D., Qin, X., Bu, J., Chen, C.: Mapping users across networks by manifold alignment on hypergraph. In: AAAI (2014)
22. Vosecky, J., Hong, D., Shen, V.Y.: User identification across multiple social networks. In: First International Conference on Networked Digital Technologies, NDT 2009, pp. 360–365. IEEE (2009)
23. Zafarani, R., Liu, H.: Connecting corresponding identities across communities. In: ICWSM (2009)
24. Zafarani, R., Liu, H.: Connecting users across social media sites: a behavioral-modeling approach. In: SIGKDD, pp. 41–49. ACM (2013)
25. Zhang, J., Kong, X., Yu, P.S.: Transferring heterogeneous links across location-based social networks. In: WSDM, pp. 303–312. ACM (2014)
26. Zheng, R., Li, J., Chen, H., Huang, Z.: A framework for authorship identification of online messages: writing-style features and classification techniques. JASIST **57**(3), 378–393 (2006)

Multi-graph Clustering Based on Interior-Node Topology with Applications to Brain Networks

Guixiang Ma[1], Lifang He[2(✉)], Bokai Cao[1], Jiawei Zhang[1],
Philip S. Yu[1,3], and Ann B. Ragin[4]

[1] Department of Computer Science, University of Illinois at Chicago,
Chicago, IL, USA
{gma4,caobokai,jzhan9,psyu}@uic.edu
[2] School of Computer Science and Software Engineering,
Shenzhen University, Shenzhen, China
lifanghescut@gmail.com
[3] Institute for Data Science, Tsinghua University, Beijing, China
[4] Department of Radiology, Northwestern University, Chicago, IL, USA
ann-ragin@northwestern.edu

Abstract. Learning from graph data has been attracting much attention recently due to its importance in many scientific applications, where objects are represented as graphs. In this paper, we study the problem of multi-graph clustering (*i.e.*, clustering multiple graphs). We propose a multi-graph clustering approach (MGCT) based on the interior-node topology of graphs. Specifically, we extract the interior-node topological structure of each graph through a sparsity-inducing interior-node clustering. We merge the interior-node clustering stage and the multi-graph clustering stage into a unified iterative framework, where the multi-graph clustering will influence the interior-node clustering and the updated interior-node clustering results will be further exerted on multi-graph clustering. We apply MGCT on two real brain network data sets (*i.e.*, ADHD and HIV). Experimental results demonstrate the superior performance of the proposed model on multi-graph clustering.

Keywords: Multi-graph clustering · Interior-node topology · Brain network

1 Introduction

In recent years, graph mining has been a popular research area because of numerous applications in social network analysis, computational biology and computer networking. In addition, many new kinds of data can be represented as graphs. For example, from common brain images such as the functional magnetic resonance imaging (fMRI) data of multiple subjects, we can construct a brain connectivity network for each of them, where each node represents a brain region, and each link represents the functional/structural connectivity between two brain regions [12]. These multiple brain networks provide us with an unprecedented

© Springer International Publishing AG 2016
P. Frasconi et al. (Eds.): ECML PKDD 2016, Part I, LNAI 9851, pp. 476–492, 2016.
DOI: 10.1007/978-3-319-46128-1_30

opportunity to explore the inner structure and activity of the human brain, serving as valuable supportive information for clinical diagnosis of neurological disorders [18]. Therefore, mining on graphs becomes a crucial task and may benefit various real-world applications.

Among the existing works on graph learning, quite a few of them fall into supervised learning, which usually aim to select frequent substructures such as connected subgraph patterns in a database of graphs and then feed these subgraph features into classifiers [6,11]. These methods typically work well when the graph database is very large or the access to side information is assumed. However, the number of subgraphs is exponential to the size of graphs, thus the subgraph enumeration process is both time and memory consuming which makes it infeasible to explore the complete subgraph space. Moreover, in many real-world cases, only a small number of labeled graphs are available. Therefore, finding discriminative subgraph patterns from a large number of candidate patterns based on such limited instances is not reliable. While supervised methods focus on training classifiers, unsupervised clustering could provide exploratory techniques for finding hidden patterns in multiple graphs. In this paper, we investigate the unsupervised scenarios by exploring the multi-graph clustering based on the interior-node topology of graphs. Topology is the mathematics of neighborhood relationships in space, which is independent of the distance metric, thus the interior-node topology of graphs could provide complementary local structure information for the original linkage, which can only characterize the global structure information of graph. Despite its value and significance, to our best knowledge, the interior-node topology of graphs has not been studied in the problem of multi-graph clustering so far. There are two major challenges in this multi-graph clustering problem:

- How to capture the interior-node topology of each graph? Conventional approaches extract graph-theoretical measures, *e.g.*, clustering coefficients, to quantify the prevalence of clustered connectivity [10,23]. However, assigning a predefined measure to specific nodes in a graph might not fully characterize the subtle local topological structure of the graph.
- How to effectively leverage the extracted topological structure information to facilitate the process of multi-graph clustering? The original linkage metric describes the global connectivity structure in the graph, while the topological structure depicts the local neighborhood relationships. An effective multi-graph clustering model should fuse these two complementary structural information together.

To address the above challenges, we propose a framework of multi-graph clustering with interior-node topology. The contributions of this work are twofold:

- We propose to consider both the global structure and the local topological structure of graphs for the multi-graph clustering task. Specifically, we utilize interior-node clustering to capture local topological structure of graphs.
- Considering the fact that graphs with a high similarity tend to have a similar interior-node topology, we propose to merge the multi-graph clustering stage

and interior-node clustering process into a unified iterative framework called MGCT, where the results of interior-node clustering are exerted on multi-graph clustering and the multi-graph clustering will in turn improve interior-node clustering of each graph, thus achieving a mutual reinforcement.

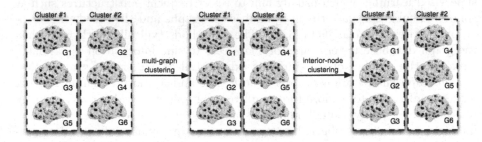

Fig. 1. The framework of the proposed model.

In the scenario of brain network analysis for multiple subjects, the proposed framework of multi-graph clustering can be illustrated with the example shown in Fig. 1. There are two stages in each iteration of the framework: multi-graph clustering and interior-node clustering. In the multi-graph clustering stage, the given six brain networks are clustered into two clusters, and then in the second stage, the interior-node clustering of each graph will be updated with a weighted influence from their neighbor graphs in the same cluster, after which the new interior-node clustering results will be utilized for the multi-graph clustering in the next iteration. After the model converges, we will obtain the final optimal multi-graph clustering results, which can be used for further analysis, for example, the neurological disorder identification.

We evaluate the proposed method on two real brain network data sets (ADHD and HIV). Experimental results illustrate the superior performance of the proposed approach for multi-graph clustering in brain network analysis.

2   Preliminaries

In this section we establish key definitions and notational conventions that simplify the exposition in later sections.

Throughout this paper, matrices are written as boldface capital letters and vectors are denoted as boldface lowercase letters. For a matrix $\mathbf{M} \in \mathbb{R}^{n \times m}$, its elements are denoted by m_{ij}, and its i-th row, j-th column are denoted by $\mathbf{m}^i$, $\mathbf{m}_j$ respectively. The Frobenius norm of $\mathbf{M}$ is defined as $\|\mathbf{M}\|_F = \sqrt{\sum_{i=1}^{n} \|\mathbf{m}^i\|_2^2}$, and the $\ell_{2,1}$ norm of $\mathbf{M}$ is defined as $\|\mathbf{M}\|_{2,1} = \sum_{i=1}^{n} \|\mathbf{m}^i\|_2$. For any vector $\mathbf{u} \in \mathbb{R}^n$, $Diag(\mathbf{u}) \in \mathbb{R}^{n \times n}$ is the diagonal matrix whose diagonal elements are u_i. $\mathbf{I}_n$ denotes an identity matrix with size n. $\|\mathbf{u}\|_0$ is the ℓ_0 norm, which counts the number of nonzero elements in the vector $\mathbf{u}$.

Definition 1 (Multi-graph Clustering). An undirected graph can be formally represented as $G = (V, E, \mathbf{A})$, where V is the set of vertices, $E \subset V \times V$ is the set of edges, and $\mathbf{A}$ is the weighted affinity matrix whose entry denotes the affinity between a pair of nodes. Given a set of such graphs $D = \{G_1, G_2, \cdots, G_n\}$, the goal of multi-graph clustering is to cluster the graphs in D into c subsets.

Definition 2 (Interior-node Clustering). Given an undirected graph $G = (V, E, \mathbf{A})$, the goal of interior-node clustering is to group the nodes of the graph into k clusters $C = \{C_1, \cdots, C_k\}$, with $V = C_1 \cup \cdots \cup C_k$ and $C_i \cap C_j = \varnothing$ for every pair i, j with $i \neq j$.

Definition 3 (Topology). Topology is the mathematics of neighborhood relationships in space independent of metric distance. In the context of graph structures, such neighborhood relationships often correspond to the *connectivity* of nodes, *i.e.,* how nodes are connected to each other.

3 Methodology

In this section, we first introduce the proposed multi-graph clustering framework MGCT, where we formulate the multi-graph clustering stage and the interior-node clustering stage, both of which can be formulated as optimization problems. We then present an iterative algorithm based on half-quadratic optimization to solve this minimization problem.

3.1 An Iterative Framework: MGCT

In the literature of multi-graph clustering, the pairwise distance is mainly measured based on the structure of each graph, and graphs with highly similar structures tend to be clustered into the same group. In other words, the graphs that are clustered into the same group tend to have highly similar topological structure [3]. Following these observations, we propose an iterative framework called MGCT for multiple-graph clustering based on interior-node topology. In each iteration, there are two stages: the interior-node clustering and the multi-graph clustering, where the interior-node clustering results which imply local topological structure are used together with the global structure of graph for clustering multiple graphs, and then the multi-graph clustering results will be utilized in turn to improve the interior-node clustering. Through this iterative mutual reinforcement of interior-node clustering and multi-graph clustering, we can finally achieve a refined multi-graph clustering result.

Multi-graph Clustering. In this part, we focus on the formulation of the multi-graph clustering stage. Since the multi-graph clustering and interior-node clustering depend on each other and are performed alternatively, here we assume we have obtained the interior-node clustering results of the graphs, which can be used for the multi-graph clustering. The formulation of the interior-node clustering problem and the overall iterative process will be discussed later.

Given a set of graphs $D = \{G_1, G_2, \cdots, G_n\}$, with the corresponding set of affinity matrices $A = \{\mathbf{A}_1, \mathbf{A}_2, \cdots, \mathbf{A}_n\}$, where $\mathbf{A}_i \in \mathbb{R}^{m \times m}$ is the weighted affinity matrix of G_i, and its entry denotes the pairwise affinity between nodes in G_i, suppose we have performed interior-node clustering on each of these graphs and obtained a set of clustering indicators $F = \{\mathbf{F}_1, \mathbf{F}_2, \cdots, \mathbf{F}_n\}$, where $\mathbf{F}_i \in \mathbb{R}^{m \times k}$ is the interior-node clustering indicator of G_i, we build a similarity matrix $\mathbf{S} \in \mathbb{R}^{n \times n}$, where s_{ij} denotes the similarity between the two graphs G_i and G_j, and we define it as:

$$s_{ij} = \delta(-\|\mathbf{A}_i - \mathbf{A}_j\|_F^2) + (1 - \delta)(-\|\mathbf{F}_i - \mathbf{F}_j\|_F^2) \tag{1}$$

which is a weighted combination of the similarity based on the original affinity matrix of each graph and the similarity based on interior-node clustering results, where δ is the weight parameter balancing the two parts. In this way, the interior-node topology characterized by the interior-node clustering indicator matrix can be incorporated for multi-graph clustering. With this similarity matrix, we can formulate the clustering of graphs in D as a spectral clustering problem, where graphs with a higher pairwise similarity tend to be grouped into the same cluster. Let $\mathbf{H} \in \mathbb{R}^{n \times c}$ be the multi-graph clustering indicator matrix, then the optimal $\mathbf{H}$ can be obtained by solving the following objective function [22]:

$$\min_{\mathbf{H}} \mathrm{Tr} \left(\mathbf{H}^\mathsf{T} \mathbf{L} \mathbf{H} \right)$$

$$\text{s.t. } \mathbf{H}^\mathsf{T} \mathbf{H} = \mathbf{I}_c \tag{2}$$

where $\mathbf{L} = \mathbf{D}^{-\frac{1}{2}}(\mathbf{D} - \mathbf{S})\mathbf{D}^{-\frac{1}{2}}$ is the symmetric normalized Laplacian matrix, and $\mathbf{D}$ is a diagonal matrix with $d_{ii} = \sum_{j=1}^n s_{ij}$.

Interior-Node Clustering. We now study the problem of interior-node clustering of graph in the context of multi-graph clustering.

In graph theory, a cluster is described as a set of nodes more densely connected with each other than with the rest nodes of the graph. Given a graph G with m nodes and the weighted affinity matrix $\mathbf{A} \in \mathbb{R}^{m \times m}$, the goal of interior-node clustering is to group the m nodes into k clusters, i.e., to find a cluster indicator matrix $\mathbf{F} \in \mathbb{R}^{m \times k}$, whose entry indicates which cluster a node may belong to.

Intuitively, nodes with a higher correlation should have a similar cluster indicator. With this assumption, a graph regularization can be embedded to learn the cluster indicator matrix $\mathbf{F}$, which is formulated as the following minimization problem on the basis of the spectral analysis [22]:

$$\min_{\mathbf{F}} \sum_{i,j=1}^m a_{ij} \left\| \frac{f^i}{\sqrt{d_{ii}}} - \frac{f^j}{\sqrt{d_{jj}}} \right\|_2^2 = \mathrm{Tr} \left(\mathbf{F}^\mathsf{T} \mathbf{L}' \mathbf{F} \right)$$

$$\text{s.t. } \mathbf{F}^\mathsf{T} \mathbf{F} = \mathbf{I}_k \tag{3}$$

where $\mathbf{L}' = \mathbf{D}'^{-\frac{1}{2}}(\mathbf{D}' - \mathbf{A})\mathbf{D}'^{-\frac{1}{2}}$ is the symmetric normalized Laplacian matrix, and $\mathbf{D}'$ is a diagonal matrix with $d_{ii} = \sum_{j=1}^m a_{ij}$.

The above formulation provides a measure of the smoothness of $\mathbf{F}$ over the edges in G. Notice that when a node connects to the nodes in different clusters, it will lead to a relatively large value of $\mathrm{Tr}\left(\mathbf{F}^{\mathrm{T}}\mathbf{L}'\mathbf{F}\right)$ [21]. Therefore, it is expected to identify these boundary-spanning nodes to moderate this influence. In the following, we show how to model and leverage the topology of interior-node to achieve this goal.

From the definition of the topology, we know it is the mathematics of neighborhood relationships in space independent of metric distance. In the context of graph structures, such neighborhood relationships often correspond to the *connectivity* of nodes, *i.e.*, how nodes are connected to each other. In view of the involvement of graph, a naïve approach is that the value of $\mathbf{f}^i$ at every node v_i is the weighted average of $\mathbf{f}^i$ at neighbors of v_i, with the weights being proportional to the edge weights in adjacency matrix $\mathbf{A}$, which can be fitted as

$$\min_{\mathbf{F}} \left\| \mathbf{F} - \mathbf{D}'^{-1}\mathbf{A}\mathbf{F} \right\|_F^2 \tag{4}$$

Since there are some boundary-spanning nodes across clusters, and their neighbors naturally occur in different clusters, to exploit the formulation of (4) on interior-node clustering more effectively, it is crucial for the clustering indicator matrix $\mathbf{F}$ to have discriminative ability for such boundary-spanning nodes, *i.e.*, promoting row-wise sparsity to discriminate relevant boundary-spanning nodes, and thus achieving only characterizing interior nodes. Inspired by [8], we introduce the $\ell_{2,1}$-norm penalty to make it and thus we have the following optimization problem:

$$\min_{\mathbf{F}} \mathrm{Tr}\left(\mathbf{F}^{\mathrm{T}}\mathbf{L}'\mathbf{F}\right) + \alpha\left\|\mathbf{F} - \mathbf{D}'^{-1}\mathbf{A}\mathbf{F}\right\|_{2,1}$$
$$\text{s.t.} \quad \mathbf{F}^{\mathrm{T}}\mathbf{F} = \mathbf{I}_k \tag{5}$$

where α is a parameter balancing two terms (*i.e.*, smoothness and sparsity). It can be seen the sparsity-inducing property of $\ell_{2,1}$ norm pushes the clustering indicator matrix $\mathbf{F}$ to be sparse in rows. More specifically, $\mathbf{f}^i$ shrinks to zero if the neighbors of node v_i belongs to different clusters. In particular, the more nodes having neighbors belonging to different clusters, the larger $\left\|\mathbf{f}_i - \mathbf{D}'^{-1}\mathbf{A}\mathbf{f}_i\right\|_2^2$ tends to be, so the value of $\mathbf{f}^i$ gets penalized more harshly. We can thus obtain a better clustering indicator $\mathbf{F}$ for interior nodes.

As we discussed earlier, the graphs clustered into the same group tend to have more similar topological structure, in each iteration of our framework, we hope to further improve the interior-node clustering of each graph by incorporating the interior-node clustering results of its neighbors, *i.e.*, the graphs clustered into the same group by the multi-graph clustering stage of the previous iteration. For two graphs in the same cluster, the closer they are, the more similar interior-node clustering they tend to have. Based on this assumption, for graph G_i, we consider only the graphs that are in the same cluster with G_i, and we aim to infer the weights of influence they should have on G_i.

Suppose we have a set of feature matrices $X = \{\mathbf{X}_1, \mathbf{X}_2, \cdots, \mathbf{X}_n\}$, where $\mathbf{X}_i$ can represent both the global and local structure of G_i, we aim to infer a weight matrix $\mathbf{W}$ by solving the following minimization problem:

$$\min_{\mathbf{W}} \sum_i \left\| \mathbf{X}_i - \sum_j w_{ij} \mathbf{X}_j \right\|_F^2$$

$$\text{s.t.} \quad \sum_j w_{ij} = 1 \tag{6}$$

where w_{ij} denotes the weight of G_j for G_i, which will be used to control the extent that $\mathbf{F}_j$ will be used to influence $\mathbf{F}_i$ in the next iteration, and G_j can only be a graph from the cluster containing G_i. A larger w_{ij} implies a closer distance between G_i and G_j in the same cluster.

Now we can improve the interior-node clustering of each graph by adding a weighted influence from the neighbour graphs based on the multi-graph clustering. For a graph G_i, the interior-node clustering can be obtained by solving the following objective function extended from Eq. (5):

$$\min_{\mathbf{F}_i} \text{Tr} \left(\mathbf{F}_i^T \mathbf{L}_i \mathbf{F}_i \right) + \alpha \left\| \mathbf{F}_i - \mathbf{D}_i^{-1} \mathbf{A}_i \mathbf{F}_i \right\|_{2,1} + \beta \left\| \mathbf{F}_i - \sum_j w_{ij} \mathbf{F}_j \right\|^2$$

$$\text{s.t.} \quad \mathbf{F}_i^T \mathbf{F}_i = \mathbf{I}_k \tag{7}$$

where $\mathbf{A}_i$ is the weighted affinity matrix of G_i, $\mathbf{D}_i$ is the diagonal matrix, and $\mathbf{L}_i$ is the symmetric normalized Laplacian matrix.

With the two stages illustrated above, we can formulate the overall iterative process. We first obtain an initial multi-graph clustering indicator matrix $\mathbf{H}_0$ by Eq. (2), where $\mathbf{S}$ is computed by Eq. (1) with $\delta = 1$. Then we can infer the weight matrix $\mathbf{W}$ by solving (6), which will be used for optimizing the interior-node clustering of each graph in (7). With the resulted $\mathbf{F}_i$ for each graph G_i, a new similarity matrix can be created by Eq. (1), which leads to another iteration of multi-graph clustering by Eq. (2). The overall iterative algorithm with optimization solutions will be discussed in the following section.

3.2 Optimization

Since the minimization problem in Eq. (2) is a typical spectral clustering problem, we can directly solve it by computing the first c generalized eigenvectors of the eigenproblem as illustrated in [20].

To solve the minimization problem (7), we propose an iterative algorithm based on the half-quadratic minimization [16] and the following lemma [9].

Lemma 1. Let $\phi(.)$ be a function satisfying the conditions: $x \to \phi(x)$ is convex on R; $x \to \phi(\sqrt{x})$ is convex on R_+; $\phi(x) = \phi(-x), \forall x \in R$; $\phi(x)$ is C^1 on R; $\phi''(0^+) \geq 0$, $\lim_{x \to \infty} \phi(x)/x^2 = 0$. Then for a fixed $\|\mathbf{u}^i\|_2$, there exists a dual potential function $\varphi(.)$, such that

$$\phi(\|\mathbf{u}^i\|_2) = \inf_{p \in R} \{p\|\mathbf{u}^i\|_2^2 + \varphi(p)\} \tag{8}$$

where p is determined by the minimizer function $\varphi(.)$ with respect to $\phi(.)$.

Let $\mathbf{P}_i = \mathbf{F}_i - \mathbf{D}_i^{-1}\mathbf{A}_i\mathbf{F}_i$. According to the analysis for the $\ell_{2,1}$ norm in [9], if we define $\phi(x) = \sqrt{x^2 + \epsilon}$, we can replace $\|\mathbf{P}_i\|_{2,1}$ with $\sum_{j=1}^n \phi(\|\mathbf{p}_i^j\|_2)$. Thus, based on Lemma 1, we reformulate the objective function of Eq. (7) as follows:

$$\min_{\mathbf{F}_i} \text{Tr}\left(\mathbf{F}_i^T\mathbf{L}_i\mathbf{F}_i\right) + \alpha\text{Tr}\left(\mathbf{P}_i^T\mathbf{Q}\mathbf{P}_i\right) + \beta\left\|\mathbf{F}_i - \sum_j w_{ij}\mathbf{F}_j\right\|^2$$

$$\text{s.t.} \quad \mathbf{F}_i^T\mathbf{F}_i = \mathbf{I}_k \tag{9}$$

where $\mathbf{Q} = Diag(\mathbf{q})$, and $\mathbf{q}$ is an auxiliary vector of the $\ell_{2,1}$ norm. The elements of $\mathbf{q}$ are computed by $q_j = \frac{1}{2\sqrt{\|\mathbf{p}_i^j\|_2^2 + \epsilon}}$, where ϵ is a smoothing term and is usually set to be a small constant value (we set $\epsilon = 10^{-4}$ in this paper).

The quadratic optimization problem with orthogonal constraint have been well studied, and can be solved by a lot of solvers [1,24]. Here we employ the solver Algorithm 2 in [24] to solve Eq. (9), which is a more efficient optimization algorithm with publicly available code.

Another optimization problem we need to solve is Eq. (6). In [19], such a minimization problem with respect to vectors is solved as a constrained least squares problem for locally linear embedding. Since the Frobenius norm for matrices is a straightforward generalization of the l_2 norm for vectors, we can directly obtain the following equation based on the analysis in [19].

$$\left\|\mathbf{X}_i - \sum_j w_{ij}\mathbf{X}_j\right\|_F^2 = \sum_{jr} w_{ij}w_{ir}\mathbf{C}_{jr} \tag{10}$$

where G_j and G_r denote two neighbors of G_i, *i.e.*, G_j and G_r are in the cluster containing G_i. $\mathbf{C}_{jr}$ is the local covariance matrix, which can be obtained by

$$\mathbf{C}_{jr} = \frac{1}{2}(M_j + M_r - m_{jr} - M_0) \tag{11}$$

where $m_{jr} = -s_{jr}$ denotes the squared distance between the jth and rth neighbors of G_i, thus can be obtained by Eq. (1), $M_j = \sum_z m_{jz}$, $M_r = \sum_z m_{rz}$ and $M_0 = \sum_{jr} m_{jr}$. Then the optimal weights can be obtained by:

$$w_{ij} = \frac{\sum_r \mathbf{C}_{jr}^{-1}}{\sum_{lz} \mathbf{C}_{lz}^{-1}} \tag{12}$$

For details about the derivation of the above solution, readers can refer to [19]. Based on the above analysis, we summarize the overall optimization algorithm of MGCT in Algorithm 1.

Algorithm 1. MGCT

Input: $D = \{G_1, G_2, \cdots, G_n\}$, c, k
Output: Assignments to c clusters
1: Initialize $\mathbf{H}_0$ s.t. $\mathbf{H}_0^T \mathbf{H}_0 = \mathbf{I}_c$;
2: **while** not converge **do**
3:     Compute $\mathbf{W}$ according to Eq. (12);
4:     **for** $i = 1; i <= n; i++$ **do**
5:         Initialize $\mathbf{F}_{i0}$ s.t. $\mathbf{F}_{i0}^T \mathbf{F}_{i0} = \mathbf{I}_k, t \leftarrow 0$;
6:         **while** not converge **do**
7:             Set $\mathbf{Q}_t \leftarrow Diag(\frac{1}{2\sqrt{\|\mathbf{P}_t^i\|_2^2 + \epsilon}})$;
8:             Compute $\mathbf{F}_{it+1}$ by solving Eq. (9);
9:             $t \leftarrow t+1$;
10:         **end while**
11:     **end for**
12:     Update $\mathbf{H}$ by solving Eq. (2);
13:     Cluster $\mathbf{H}$ by k-means;
14: **end while**

---

4   Experiments

In order to empirically evaluate the effectiveness of the proposed multi-graph clustering approach for brain network analysis, we test our model on real fMRI brain network data and compare with several state-of-the-art baselines.

4.1   Data Collection and Preprocessing

In this work, we use two real resting-state fMRI datasets as follows:

- *Human Immunodeficiency Virus Infection (HIV):* This dataset is collected from Chicago Early HIV Infection Study in Northwestern University [18]. The clinical cohort in this study includes 77 subjects, 56 of which are early HIV patients (positive) and the other 21 are seronegative controls (negative). The two groups did not differ in the demographic characteristics including age, gender, racial composition and education level.
- *Attention Deficit Hyperactivity Disorder (ADHD):* This dataset is collected from ADHD-200 global competition dataset[1], which contains the resting-state fMRI images of 768 subjects. Subjects are either ADHD patients or normal controls. In particular, the patient group in ADHD involves three stages of ADHD disease, which can be treated as three different groups, making the total number of groups be 4.

We use DPARSF toolbox[2] for fMRI data preprocessing. A time series of responds is extracted from each of the 116 anatomical volumes of interest (AVOI), which represents the 116 different brain regions. We perform the standard fMRI brain image processing steps, including functional images realignment, slice timing correction and normalization. Afterwards, spatial smoothing is performed on these images with an 8-mm Gaussian kernel for increasing signal-to-noise ratio, followed by the band-pass filtering (0.01–0.08 Hz) and the linear

---

[1] http://neurobureau.projects.nitrc.org/ADHD200/.
[2] http://rfmri.org/DPARSF.

trend removing of the time series. We also apply linear regression to reduce spurious variance coming from hardware reasons or subject factors such as thermal motion of electrons. After all these preprocessing steps, we compute the brain activity correlations among different brain regions based on the obtained time series for each of them, and then we use the positive correlations to form the links among the regions. Finally, we exclude the 26 cerebellar regions, and each brain is represented as a graph with 90 nodes, which correspond to the 90 cerebral regions.

4.2 Baselines and Metrics

We use four clustering methods as baselines.

- **k-means:** a classic clustering method [4]. We convert the adjacency matrix of each subject graph into vectors and then apply the k-means algorithm to cluster all the subject graphs. For the implementation of the k-means algorithm, we adopt the Litekmeans [5], which has been proven to be a fast MATLAB implementation of the k-means algorithm.
- **Spectral Clustering (SC)** [7]: a method for constructing graph partitions based on eigenvectors of the adjacency matrix of graph. In the experiment, we apply the normalized spectral clustering algorithm proposed in [20]. We first construct the similarity matrix for the multiple graphs only based on their adjacency matrices and then use the similarity matrix as the input for normalized spectral clustering of the multiple graphs.
- **Clustering Coefficient (CC):** the k-means clustering with clustering coefficient [17] as the feature representation of each graph.
- **Two-layer Spectral Clustering(TSC):** We adapt the typical spectral clustering into both of the two stages in our framework, where spectral clustering on the multi-graph is based on the spectral clustering on each graph. We call the model TSC.
- **MGCT:** our proposed multi-graph clustering method based on interior-node topology. To evaluate the discriminative ability of the sparsity-inducing term, i.e., the $\ell_{2,1}$-norm penalty term in Eq. (7), we employ MGCT with and without the sparsity-inducing term and denote them as **MGCT** and **MGCT**$_{nonST}$ respectively.

We adopt the following two measures for the evaluation.

- *Accuracy.* Let c_i represent the clustering label result of a multi-graph clustering algorithm and y_i represent the corresponding ground truth label of the graph G_i. Then *Accuracy* is defined as: $Accuracy = \frac{\sum_{i=1}^{n} \delta(y_i, map(c_i))}{n}$, where δ is the Kronecker delta function, and $map(c_i)$ is the best mapping function that permutes clustering labels to match the ground truth labels using the KuhnMunkres algorithm [13]. A larger *Accuracy* indicates better clustering performance.

- *Purity*. *Purity* is a measure used to evaluate the clustering method's ability to recover the groundtruth class labels, and it is defined as: $Purity = \frac{1}{n}\sum_{q=1}^{k}\max_{1\leq j\leq l} n_q^j$, where n is the total number of samples, and n_q^j is the number of samples in cluster q that belongs to original class j. Therefore, the purity is a real number in $[0,1]$. The larger the *Purity*, the better the clustering performance.

The main parameters in our framework include the weight parameters α, β, and δ as well as the number of interior-node clusters k. Note that in the rest part of this paper, we use k specifically to denote the number of interior-node clusters in each graph although it might has been used for denoting other general variables in the equations above. For the convenience of evaluation, we directly use the number of distinct labels in each dataset as the number of clusters in multi-graph clustering. Since there are four possible labels of the samples in ADHD datasets, we set the number of clusters to be 4. For HIV dataset, we have two possible labels (positive, negative), so we set the cluster number to be 2. We apply the grid search to find the optimal values for α, β and δ. We do grid search for α in $\{10^{-2}, 10^{-1}, \cdots 10^2\}$, β in $\{10^{-4}, 10^{-3}, \cdots 10^4\}$, and δ in $\{0.1, 0.2, \cdots 0.9\}$. The optimal k is selected by the grid search from $\{2, 3, \cdots, 12\}$. For fair comparisons of all the methods, we employ Litekmeans [5] for all the k-means clustering step if it is needed in the implementation of the six methods listed above. We repeat clustering for 20 times with random initialization as k-means depends on initialization. For the evaluation, we repeat running the program of each methods for 50 times and report the average accuracy and purity as the results.

4.3   Performance Evaluations

As shown in Tables 1 and 2, our MGCT method performs the best on the two datasets in terms of both *accuracy* and *purity*. Among the six clustering methods, the first two methods (*i.e.*, k-means, Spectral Clustering) directly use the original matrix of each graph in the data set for calculating the distance or similarity between each pair of the graphs, which is utilized for the final multi-graph clustering. From Tables 1 and 2, we can see that the clustering accuracy and purity of these two methods are quite low. This is probably because that they do not consider the interior-node topology of these graphs when doing clustering. The CC achieves a slightly better result compared to k-means and Spectral Clustering. This is mainly due to the fact that CC does consider some local structure information while calculating the clustering coefficient. However, since it only assigns a single predefined measure to each node in the graph, which represents each brain region in the brain networks, the subtle topological structure of each brain network might not be fully characterized.

Comparatively, the last three methods (*i.e.*, TSC, MGCT, MGCT$_{nonST}$) all utilize the topological structure information but at different level. The TSC method first performs spectral clustering on each graph, and the resulted matrix containing the clustering indicator vectors are used in the multi-graph spectral clustering. This process does include the topological structure, but it only has the one-way and one-time influence on the multi-graph clustering task. The result of multi-graph clustering does not have influence on the interior-node clustering. Different from TSC, the two methods we proposed namely the MGCT and MGCT$_{nonST}$ perform the task in an iterative way, and achieves the mutual reinforcement by leveraging the topology structure into multi-graph clustering and inferring a better topology structure for each graph from the multi-graph clustering result alternatively. According to Tables 1 and 2, we can also see that the proposed MGCT method outperforms the MGCT$_{nonST}$ in both accuracy and purity. This indicates the importance of the $\ell_{2,1}$ norm we add in Eq. (7), which has the sparsity-inducing property.

In order to evaluate the effectiveness of MGCT for interior-node topology extraction of brain networks, we investigate the resulted brain networks with interior-node clusters detected by MGCT and show the results of two brain networks in Fig. 2. We can find from the figure that the interior nodes of the normal brain network have been well grouped into several clusters, while the cluster boundaries in the patient's brain network are very blurred and the nodes widely spread out. Usually, the correlated regions of human brain will work together towards a task, and tend to present an approximately synchronized trend in their time series. Thus, the nodes representing these correlated regions would become more possible to be grouped into the same cluster. Therefore, the fuzzy cluster boundaries of the patient's interior nodes indicate that the collaboration activity

Table 1. Clustering *Accuracy*.

Methods	Accuracy	
	ADHD ($k = 6$)	HIV ($k = 9$)
k-means	52.0%	60.3%
Spectral Clustering	55.2%	60.9%
CC	56.8%	63.7%
TSC	57.6%	62.5%
MGCT$_{nonST}$	**59.3%**	**64.9%**
MGCT	**62.8%**	**68.1%**

Table 2. Clustering *Purity*.

Methods	Purity	
	ADHD ($k = 6$)	HIV ($k = 9$)
k-means	0.55	0.63
Spectral Clustering	0.59	0.65
CC	0.57	0.66
TSC	0.57	0.64
MGCT$_{nonST}$	**0.62**	**0.69**
MGCT	**0.67**	**0.72**

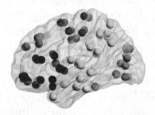

(a) a typical normal control   (b) a stage-2 ADHD patient

Fig. 2. Comparison of two brain networks with interior-node topology captured by MGCT from two subject graphs in ADHD dataset

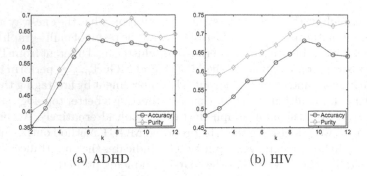

(a) ADHD   (b) HIV

Fig. 3. Accuracy and purity with different k

of different regions might not be very organized. These observations imply that our proposed framework can be further used for distinguishing subjects with neurological disorders from healthy controls.

4.4 Parameter Sensitivity

In this section, we explore the sensitivity and effects of the four main parameters in our proposed method, including α, β, δ and k. We first evaluate the clustering performance of MGCT with different k values, ranging from 2 to 12. Figure 3 shows the clustering performance of MGCT in accuracy and purity with different k on both ADHD and HIV datasets. As we can see from the figure, the multi-graph clustering performance is very sensitive to the value of k, especially when the value for k keeps very small. For example, as shown in Fig. 3(a), the accuracy increases dramatically when the value of k goes from 2 to 6 before it reaches the peak value at 6. The main reason for such high sensitivity is that when k is set to be a small number, the interior-node clusters identified from each brain network tend to have large sizes, which could not capture the interior-node topological structure very well, resulting in a less discriminative measure for distinguishing subjects in different neurological states. A similar changing trend is shown for the purity, while noticeably the peak purity value shows up when $k = 9$ instead of $k = 6$. This can be traced back to the definition of *purity*. Since it counts the number of nodes in the dominated class for each cluster instead of counting the number of nodes only when they match the correct groundtruth labels. Thus, when the number of clusters increases, each cluster becomes easier to be dominated by one class, leading to a higher purity.

Now, we analyze the sensitivity of MGCT to δ, which balances the weights from the original affinity matrix and the interior-node clustering indicator matrix when creating the similarity matrix among multiple graphs. As shown in Fig. 4, MGCT achieves different level of accuracy and purity when the value of δ varies. For ADHD, the highest accuracy is achieved when $\delta = 0.4$, while for HIV, it achieves the highest accuracy when $\delta = 0.7$, and similar situations for the purity. These results indicate that both the global structure and the interior-node topo-

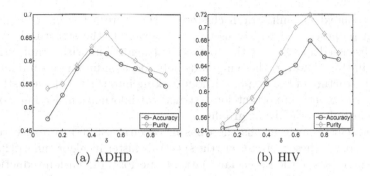

(a) ADHD   (b) HIV

Fig. 4. Accuracy and purity with different δ

logical structure are important for the multi-graph clustering analysis, and their weights need to be determined for specific practical situations. Next, we evaluate the sensitivity of MGCT to α and β. We set k to be 6 and run the MGCT method with different values for α and β on ADHD and HIV data. The clustering accuracy of MGCT is plotted versus the values for α and β in Fig. 5. As shown in the figure, MGCT achieves the best performance when $\alpha = 10^2, \beta = 10^3$ on ADHD dataset, and $\alpha = 10^2, \beta = 10^2$ on HIV dataset. Parameter α controls the sparsity while parameter β controls the influence of iterative multi-graph clustering results on interior-node clustering. If the value for α is very small, then it will not really enforce the sparsity. Similarly, if the value for β is quite small, the iterative process would barely have influence on interior-node clustering optimizing. In these cases, the performance will decline. However, when the values for them are too large, they would enforce too much sparsity or influence, which might make the performance drop as well. Therefore, an optimal combination of the two parameters is crucial for improving the performance of MGCT.

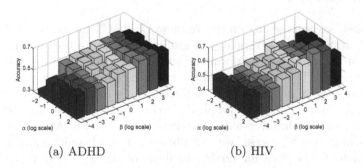

(a) ADHD   (b) HIV

Fig. 5. Accuracy with different α, β

5   Related Work

Our work relates to several bodies of studies, including the multi-graph clustering, node clustering in graphs, and brain network analysis.

In the context of multi-graph clustering, there are a few of strategies that have been proposed and widely used [3], for example the structural summary method discussed in [2], and the hierarchical algorithm with graph structure proposed in [15]. However, these methods only focus on finding a summary from the global structure of the graphs without looking into the topological structure, thus would lose very important local structural information, leading to a less effective clustering of multiple graphs.

For node clustering in graphs, there has also been a vast literature of works. One classic category of these methods are the spectral clustering algorithms [22], which use the eigenvalues of the Laplacian matrix to perform dimension reduction and then cluster the data in fewer dimensions. Recently, new methods of node clustering have been proposed for various applications, such as the works for social network analysis [25, 26], which utilize the heterogeneous information in aligned networks for node clustering. Although these work use information from multiple graphs, they focus on the mutual relationship of graphs at the node level instead of the graph-graph neighbourhood relationship as we consider.

Brain network analysis has become a hot research topic of medical data mining these years. A major task in brain network analysis is to identify the difference of a healthy subject and a neurological demented subject in brain network structure. In the past decade, quite a few of works have been done to solve this problem. In [11], a discriminative subgraph mining method is proposed for classifying brain networks. In [14], they find a unified cut and a contrast cut of multiple graphs for studying brain networks of multiple subjects. This work is the most related one of ours. However, they study the brain networks when the labels of subjects (healthy or demented) are given, while we cluster the unlabeled subjects into groups with their brain network features.

6   Conclusions

In this paper, we propose an iterative framework MGCT for multi-graph clustering based on interior-node topology of graphs. To capture the local topological structure of the graphs, we perform the sparsity-inducing interior-node clustering on each graph. In this framework, the interior-node clustering and the multi-graph clustering are performed alternatively, where the results of interior-node clustering are exerted on multi-graph clustering and the multi-graph clustering in turn improves the interior-node clustering of each graph. After this iterative mutual reinforcement process, we can obtain a refined multi-graph clustering result, which can be used for further analysis of the graphs. Experiments on two real brain network datasets demonstrate the superior performance of the proposed model in multi-graph clustering for brain network analysis.

Acknowledgments. This work is supported in part by NSF through grants III-1526499, NSFC through grants 61503253 and 61472089, and NIH through grant $R01 - MH080636$.

References

1. Absil, P.A., Mahony, R., Sepulchre, R.: Optimization Algorithms on Matrix Manifolds. Princeton University Press, Princeton (2009)
2. Aggarwal, C.C., Ta, N., Wang, J., Feng, J., Zaki, M.: XProj: a framework for projected structural clustering of XML documents. In: KDD, pp. 46–55. ACM (2007)
3. Aggarwal, C.C., Wang, H.: A survey of clustering algorithms for graph data. In: Aggarwal, C.C., Wang, H. (eds.) Managing and Mining Graph Data, pp. 275–301. Springer, New York (2010)
4. Berkhin, P.: A survey of clustering data mining techniques. In: Kogan, J., Nicholas, C., Teboulle, M. (eds.) Grouping Multidimensional Data, pp. 25–71. Springer, Heidelberg (2006)
5. Cai, D.: Litekmeans: the fastest matlab implementation of kmeans. Software (2011). http://www.zjucadcg.cn/dengcai/Data/Clustering.html
6. Cao, B., Kong, X., Zhang, J., Philip, S.Y., Ragin, A.B.: Mining brain networks using multiple side views for neurological disorder identification. In: ICDM, pp. 709–714. IEEE (2015)
7. Donath, W.E., Hoffman, A.J.: Lower bounds for the partitioning of graphs. IBM J. Res. Dev. **17**(5), 420–425 (1973)
8. He, L., Lu, C.T., Ma, J., Cao, J., Shen, L., Philip, S.Y.: Joint community and structural hole spanner detection via harmonic modularity. In: KDD. ACM (2016)
9. He, R., Tan, T., Wang, L., Zheng, W.S.: $\ell_{2,1}$ regularized correntropy for robust feature selection. In: CVPR, pp. 2504–2511 (2012)
10. Jie, B., Zhang, D., Gao, W., Wang, Q., Wee, C., Shen, D.: Integration of network topological and connectivity properties for neuroimaging classification. Biomed. Eng. **61**(2), 576 (2014)
11. Kong, X., Ragin, A.B., Wang, X., Yu, P.S.: Discriminative feature selection for uncertain graph classification. In: SDM, pp. 82–93. SIAM (2013)
12. Kong, X., Yu, P.S.: Brain network analysis: a data mining perspective. ACM SIGKDD Explor. Newsl. **15**(2), 30–38 (2014)
13. Kuhn, H.W.: The hungarian method for the assignment problem. Naval Res. Logistics Q. **2**(1–2), 83–97 (1955)
14. Kuo, C.T., Wang, X., Walker, P., Carmichael, O., Ye, J., Davidson, I.: Unified and contrasting cuts in multiple graphs: application to medical imaging segmentation. In: KDD, pp. 617–626. ACM (2015)
15. Lian, W., Mamoulis, N., Yiu, S.M., et al.: An efficient and scalable algorithm for clustering XML documents by structure. IEEE Trans. Knowl. Data Eng. **16**(1), 82–96 (2004)
16. Nikolova, M., Ng, M.K.: Analysis of half-quadratic minimization methods for signal and image recovery. SIAM J. Sci. Comput. **27**(3), 937–966 (2005)
17. Onnela, J.P., Saramäki, J., Kertész, J., Kaski, K.: Intensity and coherence of motifs in weighted complex networks. Phys. Rev. E **71**(6), 065103 (2005)
18. Ragin, A.B., Du, H., Ochs, R., Wu, Y., Sammet, C.L., Shoukry, A., Epstein, L.G.: Structural brain alterations can be detected early in HIV infection. Neurology **79**(24), 2328–2334 (2012)
19. Saul, L.K., Roweis, S.T.: An introduction to locally linear embedding (2000). http://www.cs.toronto.edu/~roweis/lle/publications.html
20. Shi, J., Malik, J.: Normalized cuts and image segmentation. IEEE Trans. Pattern Anal. Mach. Intell. **22**(8), 888–905 (2000)

21. Spielman, D.A.: Algorithms, graph theory, and linear equations in Laplacian matrices. In: ICM, vol. 4, pp. 2698–2722 (2010)
22. Von Luxburg, U.: A tutorial on spectral clustering. Stat. Comput. **17**(4), 395–416 (2007)
23. Wee, C.Y., Yap, P.T., Zhang, D., Denny, K., Browndyke, J.N., Potter, G.G., Welsh-Bohmer, K.A., Wang, L., Shen, D.: Identification of MCI individuals using structural and functional connectivity networks. Neuroimage **59**(3), 2045–2056 (2012)
24. Wen, Z., Yin, W.: A feasible method for optimization with orthogonality constraints. Math. Program. **142**(1–2), 397–434 (2013)
25. Zhang, J., Yu, P.S.: Community detection for emerging networks. In: SDM. SIAM (2015)
26. Zhang, J., Yu, P.S.: Mutual clustering across multiple heterogeneous networks. In: IEEE BigData Congress (2015)

Interactive Learning from Multiple Noisy Labels

Shankar Vembu[1(✉)] and Sandra Zilles[2]

[1] Donnelly Centre for Cellular and Biomolecular Research,
University of Toronto, Toronto, ON, Canada
shankar.vembu@utoronto.ca
[2] Department of Computer Science, University of Regina, Regina, SK, Canada
zilles@cs.uregina.ca

Abstract. Interactive learning is a process in which a machine learning algorithm is provided with meaningful, well-chosen examples as opposed to randomly chosen examples typical in standard supervised learning. In this paper, we propose a new method for interactive learning from multiple noisy labels where we exploit the disagreement among annotators to quantify the easiness (or meaningfulness) of an example. We demonstrate the usefulness of this method in estimating the parameters of a latent variable classification model, and conduct experimental analyses on a range of synthetic and benchmark datasets. Furthermore, we theoretically analyze the performance of perceptron in this interactive learning framework.

1 Introduction

We consider binary classification problems in the presence of a teacher, who acts as an intermediary to provide a learning algorithm with meaningful, well-chosen examples. This setting is also known as curriculum learning [1,9,21] or self-paced learning [7,8,10] in the literature. Existing practical methods [10,11] that employ such a teacher operate by providing the learning algorithm with easy examples first and then progressively moving on to more difficult examples. Such a strategy is known to improve the generalization ability of the learning algorithm and/or alleviate local minima problems while optimizing non-convex objective functions.

In this work, we propose a new method to quantify the notion of easiness of a training example. Specifically, we consider the setting where examples are labeled by multiple (noisy) annotators [4,14,18,20]. We use the disagreement among these annotators to determine how easy or difficult the example is. If a majority of annotators provide the same label for an example, then it is reasonable to assume that the training example is easy to classify and that these examples are likely to be located far away from the decision boundary (separating hyperplane). If, on the other hand, there is a strong disagreement among annotators in labeling an example, then we can assume that the example is difficult to classify, meaning it is located near the decision boundary. In the paper by Urner et al. [19], a strong annotator always labels an example according to

P. Frasconi et al. (Eds.): ECML PKDD 2016, Part I, LNAI 9851, pp. 493–508, 2016.
DOI: 10.1007/978-3-319-46128-1_31

the true class probability distribution, whereas a weak annotator is likely to err on an example whose neighborhood is comprised of examples from both classes, i.e., whose neighborhood is label heterogeneous. In other words, both strong and weak annotators do not err on examples far away from the decision boundary, but weak annotators are likely to provide incorrect labels near the decision boundary where the neighborhood of an example is heterogeneous in terms of its labels. There are a few other theoretical studies where weak annotators were assumed to err in label heterogeneous regions [6,12]. The notion of annotator disagreement also shows up in the multiple teacher selective sampling algorithm of Dekel *et al.* [3]. This line of research indicates the potential of using annotator disagreement to quantify the easiness of a training example.

To the best of our knowledge, there has not been any work in the literature that investigates the use of annotator disagreement in designing an interactive learning algorithm. We note that a recent paper [17] used annotator disagreement in a different setting, namely as privileged information in the design of classification algorithms. Self-paced learning methods [7,8,10] aim at simultaneously estimating the parameters of a (linear) classifier and a parameter for each training example that quantifies its easiness. This results in a non-convex optimization problem that is solved using alternating minimization. Our setting is different as the training example is comprised of not just a single (binary) label but multiple noisy labels provided by a set of annotators, and we use the disagreement among these annotators (which is fixed) to determine how easy or difficult a training example is. We note that it is possible to parameterize the easiness of an example as described in Kumar *et al.*'s paper [10] in our framework and use it in conjunction with the disagreement among annotators.

Learning from multiple noisy labels [4,14,18,20] has been gaining traction in recent years due to the availability of inexpensive annotators from crowdsourcing websites like Amazon's *Mechanical Turk*. These methods typically aim at learning a classifier from multiple noisy labels and in the process also estimate the annotators' expertise levels. We use one such method [14] as a test bed to demonstrate the usefulness of our interactive learning framework.

1.1   Problem Definition and Notation

Let $\mathcal{X} \subseteq \mathbb{R}^n$ denote the input space. The input to the learning algorithm is a set of m examples with corresponding (noisy) labels from L annotators denoted by $S = \left\{ \left(x_i, y_i^{(1)}, y_i^{(2)}, \ldots, y_i^{(L)} \right) \right\}_{i=1}^m$ where $\left(x_i, y_i^{(\ell)} \right) \in \mathcal{X} \times \{\pm 1\}$, for all $i \in \{1, \ldots, m\}$ and $\ell \in \{1, \ldots, L\}$. Let $z_1, z_2, \ldots, z_L \in [0, 1]$ denote the annotators' expertise scores, which is not known to the learning algorithm. A strong annotator will have a score close to one and a weak annotator close to zero. The goal is to learn a classifier $f : \mathcal{X} \to \{\pm 1\}$ parameterized by a weight vector $w \in \mathbb{R}^n$, and also estimate the annotators' expertise scores $\{z_1, z_2, \ldots, z_L\}$. In this work, we consider linear models $f(x) = \langle w, x \rangle$, where $\langle \cdot, \cdot \rangle$ denotes the dot-product of input vectors.

2   Learning from Multiple Noisy Labels

One of the algorithmic advantages of interactive learning is that it can potentially alleviate local minima problems in latent variable models [10] and also improve the generalization ability of the learning algorithm. A latent variable model that is relevant to our setting of learning from multiple noisy labels is the one proposed by Raykar *et al.* [14] to learn from crowdsourced labels. For squared loss function,[1] i.e., regression problems and a linear model,[2] the weight vector w and the annotators' expertise scores (the latent variable) $\{z_\ell\}$ can be simultaneously estimated using the following iterative updates:

$$\hat{w} = \operatorname*{argmin}_{w \in \mathbb{R}^n} \frac{1}{m} \sum_{i=1}^{m} \left(\langle w, x_i \rangle - \hat{y}_i \right)^2 + \lambda \|w\|^2 , \quad \text{with } \hat{y}_i = \frac{\sum_{\ell=1}^{L} \hat{z}_\ell y_i^{(\ell)}}{\sum_{\ell=1}^{L} \hat{z}_\ell} ;$$

$$\frac{1}{\hat{z}_\ell} = \frac{1}{m} \sum_{i=1}^{m} \left(y_i^{(\ell)} - \langle \hat{w}, x_i \rangle \right)^2 , \quad \text{for all } \ell \in \{1, \dots, L\} , \tag{1}$$

where λ is the regularization parameter. Intuitively, the updates estimate the score z of an annotator based on her performance (measured in terms of squared error) with the current model $\hat{w}$, and the label of an example is adjusted $\{\hat{y}_i\}$ by taking the weighted average of all its noisy labels from the annotators. In practice, the labels $\{\hat{y}_i\}$ are initialized by taking a majority vote of the noisy labels. The above updates are guaranteed to converge only to a locally optimum solution.

We now use the disagreement among annotators in the regularized risk minimization framework. For each example x_i, we compute the disagreement d_i among annotators as follows:

$$d_i = \sum_{\ell=1}^{L} \sum_{\ell'=1}^{L} \left(y_i^{(\ell)} - y_i^{(\ell')} \right)^2 , \tag{2}$$

and solve a weighted least-squares regression problem:

$$\hat{w} = \operatorname*{argmin}_{w \in \mathbb{R}^n} \frac{1}{m} \sum_{i=1}^{m} g(d_i) \left(\langle w, x_i \rangle - \hat{y}_i \right)^2 + \lambda \|w\|^2 , \tag{3}$$

where $g : \mathbb{R} \to [0, 1]$ is a monotonically decreasing function of the disagreement among annotators, and iteratively update $\{z_\ell\}$ using:

$$\frac{1}{\hat{z}_\ell} = \frac{1}{m} \sum_{i=1}^{m} g(d_i) \left(y_i^{(\ell)} - \langle \hat{w}, x_i \rangle \right)^2 , \quad \text{for all } \ell \in \{1, \dots, L\} . \tag{4}$$

---

[1] We consider squared loss function to describe our method and in our experiments for the sake of convenience. The method can be naturally extended to the classification model described in Raykar *et al.*'s paper [14]. Also, we note that it is perfectly valid to minimize squared loss function for classification problems [15].

[2] Although we consider linear models in our exposition, we note that our method can be adapted to accommodate any classification algorithm that can be trained with weighted examples.

In our experiments, we use $g(d) = (1 + e^{\alpha d})^{-1}$. The parameter α controls the reweighting of examples. Large values of α place a lot of weight on examples with low disagreement among labels, and small values of α reweight all the examples (almost) uniformly as shown in Fig. 1. The parameter α is a hyperparameter that the user has to tune akin to tuning the regularization parameter.

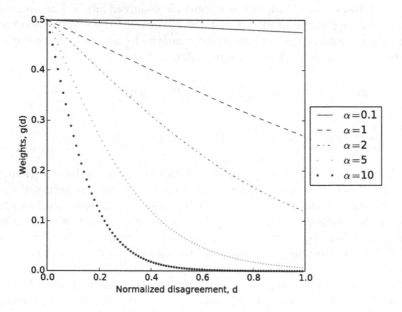

Fig. 1. Example reweighting function.

The optimization problem (3) has a closed-form solution. Let $X \in \mathbb{R}^{m \times n}$ denote the matrix of inputs, $D \in \mathbb{R}^{m \times m}$ denote a diagonal matrix whose diagonal entries are $g(d_i)$, for all $i \in \{1, \ldots, m\}$ and $\hat{y}$ denote the (column) vector of labels. The solution is given by: $\hat{w} = (X^\top DX + \lambda \mathbf{I})^{-1} X^\top D\hat{y}$, where $\mathbf{I}$ is the identity matrix. Hence, optimization solvers used to estimate the parameters in regularized least-squares regression can be adapted to solve this problem by a simple rescaling of inputs via $X \leftarrow \sqrt{D}X$ and $\hat{y} \leftarrow \sqrt{D}\hat{y}$.

In the above description of the algorithm, we fixed the weights $g(\cdot)$ on the examples. Ideally, we would want to reweight the examples uniformly as learning progresses. This can be done in the following way. Let $\mathcal{P}_{\mathcal{X}}$ denote some probability distribution induced on the examples via $g(\cdot)$. In every iteration t of the learning algorithm, we pick one of $\mathcal{P}_{\mathcal{X}}$ or the uniform distribution based on a Bernoulli trial with success probability $1/t^c$ for some fixed positive integer c to ensure that the distribution on examples converges to a uniform distribution as learning progresses. Unfortunately, we did not find this to work well in practice and the parameters of the optimization problem did not converge as smoothly as when fixed weights $g(\cdot)$ were used throughout the learning process. We leave this as an open question and use fixed weights in our experiments.

3   Mistake Bound Analysis

In this section, we analyze the mistake bound of perceptron operating in the interactive learning framework. The algorithm is similar to the classical perceptron, but the training examples are sorted based on their distances from the separating hyperplane and fed to the perceptron starting from the farthest example. The theoretical analysis requires estimates of margins of all examples. We describe a method to estimate the margin of an example and also its ground-truth label (from the multiple noisy labels) in the Appendix. We would like to remark that the margin of examples is needed only to prove the mistake bound. In practice, the perceptron algorithm can directly use the disagreement among annotators (2).

Theorem 1 (Perceptron [13]**).** *Let* $((x_1, y_1), \ldots, (x_T, y_T))$ *be a sequence of training examples with* $\|x_t\| \leq R$ *for all* $t \in \{1, \ldots, T\}$. *Suppose there exists a vector* u *such that* $y_t \langle u, x_t \rangle \geq \gamma$ *for all examples. Then, the number of mistakes made by the perceptron algorithm on this sequence is at most* $(R/\gamma)^2 \|u\|^2$.

The above result is the well-known mistake bound of perceptron and the proof is standard. We now state the main theorem of this paper.

Theorem 2. *Let* $((x_1, \hat{y}_1, \hat{\gamma}_1), \ldots, (x_T, \hat{y}_T, \hat{\gamma}_T))$ *be a sequence of training examples along with their label and margin estimates, sorted in descending order based on the margin estimates, and with* $\|x_t\| \leq R$ *for all* $t \in \{1, \ldots, T\}$. *Let* $\hat{\gamma} = \min(\hat{\gamma}_1, \ldots, \hat{\gamma}_T) = \hat{\gamma}_T$ *and* $K = \lceil R/\hat{\gamma} \rceil - 1$. *Suppose there exists a vector* u *such that* $\hat{y}_t \langle u, x_t \rangle \geq \hat{\gamma}$ *for all examples. Divide the input space into* K *equal regions, so that for any example* x_{t_k} *in a region* k *it holds that* $\hat{y}_{t_k} \langle x_{t_k}, u \rangle \geq k\hat{\gamma}$. *Let* $\{\varepsilon_1, \ldots, \varepsilon_K\}$ *denote the number of mistakes made by the perceptron in each of the* K *regions, and let* $\varepsilon = \sum_k \varepsilon_k$ *denote the total number of mistakes. Define* $\varepsilon_s = \sqrt{1/K \sum_k (\varepsilon_k - \varepsilon/K)^2}$ *to be the standard deviation of* $\{\varepsilon_1, \ldots, \varepsilon_K\}$.
Then, the number of mistakes ε *made by the perceptron on the sequence of training examples is bounded from above via:*

$$\sqrt{\varepsilon} \leq \frac{R\|u\| + \sqrt{R^2\|u\|^2 + \varepsilon_s K(K+1)^2 \sqrt{K-1}\hat{\gamma}^2}}{\hat{\gamma}(K+1)}.$$

We will use the following inequality in proving the above result.

Lemma 1 (Laguerre-Samuelson Inequality [16]**).** *Let* $(r_1, \ldots, r_n)$ *be a sequence of real numbers. Let* $\bar{r} = \sum_i r_i/n$ *and* $s = \sqrt{1/n \sum_i (r_i - \bar{r})^2}$ *denote their mean and standard deviation, respectively. Then, the following inequality holds for all* $i \in \{1, \ldots, n\}$: $\bar{r} - s\sqrt{n-1} \leq r_i \leq \bar{r} + s\sqrt{n-1}$.

Proof. Using the margin estimates $\hat{\gamma}_1, \ldots, \hat{\gamma}_T$, we divide the input space into $K = \lceil R/\hat{\gamma} \rceil - 1$ equal regions, so that for any example x_{t_k} in a region k, $\hat{y}_{t_k} \langle x_{t_k}, u \rangle \geq k\hat{\gamma}$. Let $T_1, \ldots, T_K$ be the number of examples in these regions, respectively. Let τ_t be an indicator variable whose value is 1 if the algorithm makes a prediction mistake on example x_t and 0 otherwise. Let $\varepsilon_k = \sum_{t=i}^{T_k} \tau_{t_i}$ be the number of mistakes made by the algorithm in region k, $\varepsilon = \sum_k \varepsilon_k$ be the total number of mistakes made by the algorithm.

We first bound $\|w_{T+1}\|^2$, the weight vector after seeing T examples, from above. If the algorithm makes a mistake at iteration t, then $\|w_{t+1}\|^2 = \|w_t + \hat{y}_t x_t\|^2 = \|w_t\|^2 + \|x_t\|^2 + 2\hat{y}_t \langle w_t, x_t \rangle \leq \|w_t\|^2 + R^2$, since $\hat{y}_t \langle w_t, x_t \rangle < 0$. Since $w_1 = 0$, we have $\|w_{T+1}\|^2 \leq \varepsilon R^2$.

Next, we bound $\langle w_{T+1}, u \rangle$ from below. Consider the behavior of the algorithm on examples that are located in the farthest region K. When a prediction mistake is made in this region at iteration $t_K + 1$, we have $\langle w_{t_K+1}, u \rangle = \langle w_{t_K} + \hat{y}_{t_K} x_{t_K}, u \rangle = \langle w_{t_K}, u \rangle + \hat{y}_{t_K} \langle x_{t_K}, u \rangle \geq \langle w_{t_K}, u \rangle + K\hat{\gamma}$. The weight vector moves closer to u by at least $K\hat{\gamma}$. After the algorithm sees all examples in the farthest region K, we have $\langle w_{T_K+1}, u \rangle \geq \varepsilon_K K\hat{\gamma}$ (since $w_1 = 0$), and similarly for region $K - 1$, $\langle w_{T_{(K-1)}+1}, u \rangle \geq \varepsilon_K K\hat{\gamma} + \varepsilon_{K-1}(K-1)\hat{\gamma}$, and so on for other regions. Therefore, after the algorithm has seen T examples, we have

$$\langle w_{T+1}, u \rangle \geq \sum_{k=1}^{K} \varepsilon_k k \hat{\gamma} \geq \left(\frac{\varepsilon}{K} - \varepsilon_s \sqrt{K-1} \right) \left(\frac{K(K+1)}{2} \right) \hat{\gamma}.$$

where we used the Laguerre-Samuelson inequality to lower-bound ε_k for all k, using the mean ε/K and standard deviation ε_s of $\{\varepsilon_1, \ldots, \varepsilon_K\}$.

Combining these lower and upper bounds, we get the following quadratic equation in $\sqrt{\varepsilon}$:

$$\left(\frac{\varepsilon}{K} - \varepsilon_s \sqrt{K-1} \right) \left(\frac{K(K+1)}{2} \right) \hat{\gamma} - \sqrt{\varepsilon} R \|u\| \leq 0,$$

whose solution is given by:

$$\sqrt{\varepsilon} \leq \frac{R\|u\| + \sqrt{R^2 \|u\|^2 + \varepsilon_s K(K+1)^2 \sqrt{K-1}\hat{\gamma}^2}}{\hat{\gamma}(K+1)}.$$

□

Note that if $\varepsilon_s = 0$, i.e., when the number of mistakes made by the perceptron in each of the regions is the same, then we get the following mistake bound:

$$\varepsilon \leq \frac{4R^2 \|u\|^2}{\hat{\gamma}^2 (K+1)^2},$$

clearly improving the mistake bound of the standard perceptron algorithm. However, $\varepsilon_s = 0$ is not a realistic assumption. We therefore plot x-fold improvement of the mistake bound as a function of ε_s for a range of margins $\hat{\gamma}$ in Fig. 2. The y-axis is the ratio of mistake bounds of interactive perceptron to standard perceptron with all examples scaled to have unit Euclidean length ($R = 1$) and $\|u\| = 1$. From the figure, it is clear that even when $\varepsilon_s > 0$, it is possible to get non-trivial improvements in the mistake bound.

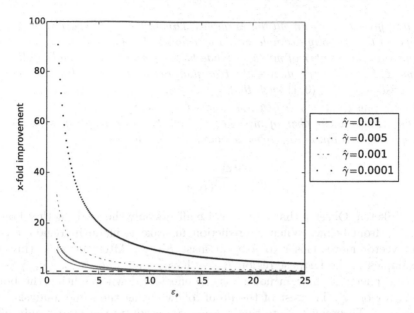

Fig. 2. Illustration of the improvement in the mistake bound of interactive perceptron when compared to standard perceptron. The dashed line is $y = 1$.

The above analysis uses margin and label estimates, $\hat{\gamma}_1, \ldots, \hat{\gamma}_T, \hat{y}_1, \ldots, \hat{y}_T$, from our method described in the Appendix, which may not be exact. We therefore have to generalize the mistake bound to account for noise in these estimates. Let $\{\gamma_1, \ldots, \gamma_T\}$ be the true margins of examples. Let $\epsilon_{\gamma_u}, \epsilon_{\gamma_\ell} \in (0, 1]$ denote margin noise factors such that $\hat{\gamma}_t / \epsilon_{\gamma_\ell} \geq \gamma_t \geq \epsilon_{\gamma_u} \hat{\gamma}_t$, for all $t \in \{1, \ldots, T\}$. These noise factors will be useful to account for overestimation and underestimation in $\hat{\gamma}_t$, respectively.

Label noise essentially makes the classification problem linearly inseparable, and so the mistake bound can be analyzed using the method described in the work of Freund and Schapire [5] (see Theorem 2). Here, we define the deviation of an example x_t as $\delta_t = \max(0, \hat{\gamma} / \epsilon_{\gamma_\ell} - \hat{y}_t \langle u, x_t \rangle)$ and let $\Delta = \sqrt{\sum_t \delta_t^2}$. As will become clear in the analysis, if $\hat{\gamma}$ is overestimated, then it does not affect the worst-case analysis of the mistake bound in the presence of label noise. If the labels were accurate, then $\delta_t = 0$, for all $t \in \{1, \ldots, T\}$. With this notation in place, we are ready to analyze the mistake bound of perceptron in the noisy setting. Below, we state and prove the theorem for $\varepsilon_s = 0$, i.e., when the number of mistakes made by the perceptron is the same in all the K regions. The analysis is similar for $\varepsilon_s > 0$, but involves tedious algebra and so we omit the details in this paper.

Theorem 3. *Let* $((x_1, \hat{y}_1, \hat{\gamma}_1), \ldots, (x_T, \hat{y}_T, \hat{\gamma}_T))$ *be a sequence of training examples along with their label and margin estimates, sorted in descending order based on the margin estimates, and with* $\|x_t\| \leq R$ *for all* $t \in \{1, \ldots, T\}$. *Let* $\hat{\gamma} = \min(\hat{\gamma}_1, \ldots, \hat{\gamma}_T) = \hat{\gamma}_T$ *and* $K = \lceil R/\hat{\gamma} \rceil - 1$. *Suppose there exists a vector* u

such that $\hat{y}_t \langle u, x_t \rangle \geq \hat{\gamma}$ for all the examples. Divide the input space into K equal regions, so that for any example x_{t_k} in a region k it holds that $\hat{y}_{t_k} \langle x_{t_k}, u \rangle \geq k\hat{\gamma}$. Assume that the number of mistakes made by the perceptron is equal in all the K regions. Let $\{\gamma_1, \ldots, \gamma_T\}$ denote the true margins of the examples, and suppose there exists $\epsilon_{\gamma_u}, \epsilon_{\gamma_\ell} \in (0, 1]$ such that $\hat{\gamma}_t/\epsilon_{\gamma_\ell} \geq \gamma_t \geq \epsilon_{\gamma_u} \hat{\gamma}_t$ for all $t \in \{1, \ldots, T\}$. Define $\delta_t = \max(0, \hat{\gamma}/\epsilon_{\gamma_\ell} - \hat{y}_t \langle u, x_t \rangle)$ and let $\Delta = \sqrt{\sum_t \delta_t^2}$.

Then, the total number of mistakes ε made by the perceptron algorithm on the sequence of training examples is bounded from above via:

$$\varepsilon \leq \frac{4(\Delta + R\|u\|)^2}{\epsilon_{\gamma_u}^2 \, \hat{\gamma}^2 (K + 1)^2}.$$

Proof. (Sketch) Observe that margin noise affects only the analysis that bounds $\langle w_{T+1}, u \rangle$ from below. When a prediction mistake is made in region K, the weight vector moves closer to u by at least $K\epsilon_{\gamma_u}\hat{\gamma}$. After the algorithm sees all examples in the farthest region K, we have $\langle w_{TK+1}, u \rangle \geq \varepsilon_K K \epsilon_{\gamma_u} \hat{\gamma}$ (since $w_1 = 0$). Therefore, margin noise has the effect of down-weighting the bound by a factor of ϵ_{γ_u}. The rest of the proof follows using the same analysis as in the proof of Theorem 2. Note that margin noise affects the bound only when $\hat{\gamma}_t$ is overestimated because the margin appears only in the denominator when $\varepsilon_s = 0$.

To account for label noise, we use the proof technique in Theorem 2 of Freund and Schapire's paper [5]. The idea is to project the training examples into a higher dimensional space where the data becomes linearly separable and then invoke the mistake bound for the separable case. Specifically, for any example x_t, we add T dimensions and form a new vector such that the first n coordinates remain the same as the original input, the $(n+t)$'th coordinate gets a value equal to C (a constant to be specified later), and the remaining coordinates are set to zero. Let $x'_t \in \mathbb{R}^{n+T}$ for all $t \in \{1, \ldots, T\}$ denote the examples in the higher dimensional space. Similarly, we add T dimensions to the weight vector u such that the first n coordinates remain the same as the original input, and the $(n+t)$'th coordinate is set to $\hat{y}_t \delta_t/C$, for all $t \in \{1, \ldots, T\}$. Let $u' \in \mathbb{R}^{n+T}$ denote the weight vector in the higher dimensional space.

With the above construction, we have $\hat{y}_t \langle u', x'_t \rangle = \hat{y}_t \langle u, x_t \rangle + \delta_t \geq \hat{\gamma}/\epsilon_{\gamma_\ell}$. In other words, examples in the higher dimensional space are linearly separable by a margin $\hat{\gamma}/\epsilon_{\gamma_\ell}$. Also, note that the predictions made by the perceptron in the original space are the same as those in the higher dimensional space. To invoke Theorem 2, we need to bound the length of the training examples in the higher dimensional space, which is $\|x'_t\|^2 \leq R^2 + C^2$. Therefore the number of mistakes made by the perceptron is at most $4(R^2 + C^2)(\|u\|^2 + \Delta^2/C^2)/(\epsilon_{\gamma_u}^2 \hat{\gamma}^2(K + 1^2))$. It is easy to verify that the bound is minimized when $C = \sqrt{R\Delta/\|u\|}$, and hence the number of mistakes is bounded from above by $4(\Delta + R\|u\|)^2/(\epsilon_{\gamma_u}^2 \hat{\gamma}^2(K + 1)^2)$. $\square$

4   Empirical Analysis

We conducted experiments on synthetic and benchmark datasets.[3] For all datasets, we simulated annotators to generate (noisy) labels in the following way. For a given set of training examples, $\{(x_i, y_i)\}_{i=1}^{m}$, we first trained a linear model $f(x) = \langle w, x \rangle$ with the true binary labels and normalized the scores f_i of all examples to lie in the range $[-1, +1]$. We then transformed the scores via $\tilde{f} = 2 \times (1 - (1/(1 + \exp(p \times -2.5 * |f|))))$, so that examples close to the decision boundary with $f_i \approx 0$ get a score $\tilde{f}_i \approx 1$ and those far away from the decision boundary with $f_i \approx \pm 1$ get a score $\tilde{f}_i \approx 0$ as shown in Fig. 3.

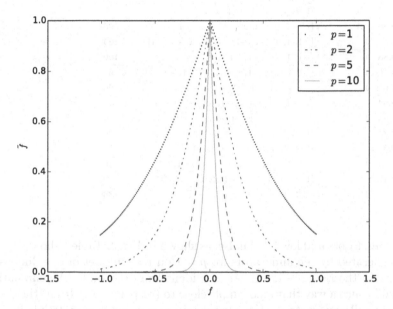

Fig. 3. Illustration of the function used to convert the score from a linear model to a probability that is used to simulate noisy labels from annotators.

For each example x_i, we generated L copies of its true label, and then flipped them based on a Bernoulli trial with success probability $\tilde{f}_i/2$. This has the effect of generating (almost) equal numbers of labels with opposite sign and hence maximum disagreement among labels for examples that are close to the decision boundary. In the other extreme, labels of examples located far away from the decision boundary will not differ much. Furthermore, we flipped the sign of all labels based on a Bernoulli trial with success probability $\tilde{f}_i$ if the majority of labels is equal to the true label. This ensures that the majority of labels are noisy for examples close to the decision boundary. The noise parameter p controls the amount of noise injected into the labels – high values result in weak disagreement

---

[3] Software is available at https://github.com/svembu/ilearn.

Table 1. Labels provided by a set of 10 simulated annotators for a one-dimensional dataset in the range $[-1, +1]$

f	$\tilde{f}$	True label	Noisy labels	Label disagreement, d (Eq. (2))
−1.0	0.15	−1	$[-1, -1, -1, -1, -1, -1, -1, -1, -1, -1]$	0
−0.9	0.19	−1	$[1, -1, -1, -1, -1, -1, -1, -1, -1, -1]$	36
−0.8	0.24	−1	$[-1, -1, -1, -1, 1, -1, -1, -1, -1, -1]$	36
−0.7	0.3	−1	$[-1, 1, 1, -1, -1, -1, -1, 1, -1, -1]$	84
−0.6	0.36	−1	$[-1, -1, -1, -1, -1, -1, -1, -1, -1, -1]$	0
−0.5	0.45	−1	$[1, -1, 1, -1, -1, -1, -1, -1, -1, -1]$	64
−0.4	0.55	−1	$[1, 1, 1, 1, 1, 1, -1, 1, 1, -1]$	64
−0.3	0.64	−1	$[1, 1, 1, -1, 1, 1, -1, -1, 1, 1]$	84
−0.2	0.76	−1	$[1, 1, 1, -1, 1, 1, 1, 1, 1, -1]$	64
−0.1	0.88	−1	$[1, -1, 1, -1, 1, -1, 1, 1. -1, 1]$	96
0	1	1	$[-1, 1, 1, -1, -1, -1, -1, 1, -1, -1]$	84
0.1	0.88	1	$[-1, 1, -1, -1, -1, 1, 1, -1, 1, 1]$	100
0.2	0.76	1	$[-1, -1, 1, -1, 1, -1, -1, -1, -1, -1]$	64
0.3	0.64	1	$[-1, -1, 1, -1, -1, -1, 1, -1, 1, 1]$	96
0.4	0.54	1	$[1, -1, -1, 1, -1, -1, -1, -1, 1, -1]$	84
0.5	0.45	1	$[1, -1, -1, -1, -1, -1, -1, -1, -1, -1]$	36
0.6	0.36	1	$[1, 1, 1, 1, 1, 1, 1, 1, -1, 1]$	36
0.7	0.3	1	$[1, 1, 1, 1, 1, -1, -1, 1, -1, 1]$	84
0.8	0.24	1	$[-1, -1, -1, -1, 1, -1, -1, -1, -1, -1]$	36
0.9	0.19	1	$[1, 1, 1, 1, 1, 1, 1, 1, 1, 1]$	0
1.0	0.15	1	$[1, 1, 1, 1, 1, 1, 1, 1, 1, -1]$	36

among annotators and low label noise, as shown in Fig. 3. Table 1 shows the noisy labels generated by ten annotators for $p = 1$ on a simple set of one-dimensional examples in the range $[-1, +1]$. As is evident from the table, the simulation is designed in such a way that an example close to (resp. far away from) the decision boundary will have a strong (resp. weak) disagreement among its labels.

4.1   Synthetic Datasets

We considered binary classification problems with examples generated from two 10-dimensional Gaussians centered at $\{-0.5\}^{10}$ and $\{+0.5\}^{10}$ with unit variance. We generated noisy labels using the procedure described above. Specifically, we simulated 12 annotators – one of them always generated the true labels, another flipped all the true labels, the remaining 10 flipped labels using the simulation procedure described above. We randomly generated 100 datasets, each of them having 1000 training examples equally divided between the two classes. We used half of the data set for training and the other half for testing. In each experiment, we tuned the regularization parameter (λ in Eq. (3)) by searching over the range $\{2^{-14}, 2^{-12}, \ldots, 2^{12}, 2^{14}\}$ using 10-fold cross validation on the training set, retrained the model on the entire training set with the best-performing parameter, and report the performance of this model on the test set. We experimented with a range of (α, p) values. Recall that the parameter α influences the reweighting of examples with small values placing (almost) equal weights on all the examples and large values placing a lot of weight on examples whose labels have a large disagreement (Fig. 1). The parameter p as mentioned before controls label noise. We compared the performance of the algorithm in interactive and non-interactive modes described in Sect. 2. The non-interactive algorithm is the one described in Raykar *et al.*'s paper [14].

The results are shown in Table 2. We use area under the receiver operating characteristic curve (AU-ROC) and area under the precision-recall curve (AU-PRC) as performance metrics. In the table, we show the number of times the AU-ROC and the AU-PRC of the interactive algorithm is higher than its non-interactive counterpart (#wins out of 100 datasets). We also show the two-sided p-value from the Wilcoxon signed-rank test. From the results, we note that the performance of the interactive algorithm is not significantly better than its non-interactive counterpart for small and large values of α. This is expected because small values of α reweight examples (almost) uniformly and so there is not much to gain when compared to running the algorithm in the non-interactive mode. In the other extreme, large values of α tend to discard a large number of examples close to the decision boundary thereby degrading the overall performance of the algorithm in the interactive mode. $\alpha = 2$ gives the best performance. We also note that for high values of p, i.e., weak disagreement among annotators and hence low label noise, the interactive algorithm offers no statistically significant gains when compared to the non-interactive algorithm. This, again, is as expected.

Table 2. Experimental results on synthetic datasets. Also shown in the table are two-sided p-values of the Wilcoxon signed-rank test.

Parameters	AU-ROC (#wins)	p-value	AU-PRC (#wins)	p-value
$\alpha = 0.1, p = 1$	61	0.0542	61	0.0535
$\alpha = 1, p = 1$	75	4.11×10^{-11}	75	2.97×10^{-11}
$\alpha = 2, p = 1$	88	6.26×10^{-14}	89	3.65×10^{-14}
$\alpha = 5, p = 1$	73	3.56×10^{-5}	75	2.26×10^{-5}
$\alpha = 0.1, p = 2$	50	0.5684	49	0.6199
$\alpha = 1, p = 2$	51	0.1822	50	0.1799
$\alpha = 2, p = 2$	61	0.0007	61	0.0009
$\alpha = 5, p = 2$	49	0.5075	49	0.7463
$\alpha = 0.1, p = 5$	32	0.4784	34	0.8479
$\alpha = 1, p = 5$	44	0.0615	42	0.0817
$\alpha = 2, p = 5$	48	0.2334	49	0.3661
$\alpha = 5, p = 5$	37	0.0562	33	0.028

Table 3. Datasets used in the experiments

Name	No. of training examples	No. of test examples	No. of features
a1a	1,605	30,956	123
a2a	2,265	30,296	123
a3a	3,185	29,376	123
a4a	4,781	27,780	123
a5a	6,414	26,147	123
Australian	690	-	14
Breast-cancer	683	-	10
Diabetes	768	-	8
Fourclass	862	-	8
German.nuner	1000	-	24
Heart	270	-	13
Ionosphere	351	-	34
Liver-disorders	345	-	6
Splice	1,000	2,175	60
Sonar	208	-	60
w1a	2,477	47,272	300
w2a	3,470	46,279	300
w3a	4,912	44,837	300
w4a	7,366	42,383	300
w5a	9,888	39,861	300

4.2 Benchmark Datasets

We used LibSVM benchmark[4] datasets in our experiments. We selected binary classification datasets with at most 10,000 training examples and 300 features (Table 3), so that we could afford to train multiple linear models (100 in our experiments) for every dataset using standard solvers and also afford to tune hyperparameters carefully in a reasonable amount of time. We generated noisy labels with the same procedure used in our experiments on synthetic data. Also, we tuned the regularization parameter in an identical manner. For datasets with no predefined training and test splits, we randomly selected 75 % of the examples for training and used the rest for testing. For each dataset, we randomly generated 100 sets of noisy labels from the 12 annotators resulting in 100 different random versions of the dataset. The results are shown in Table 4. In the table, we again show the number of times the AU-ROC and the AU-PRC of the inter-active algorithm is higher than its non-interactive counterpart (#wins out of 100 datasets). We report results on only a subset of (α, p) values that were found to

---

[4] https://www.csie.ntu.edu.tw/~cjlin/libsvmtools/datasets/.

Table 4. Experimental results on benchmark datasets. Statistically *insignificant* results (p-value > 0.01) are indicated with an asterisk (∗).

Dataset	$\alpha = 1, p = 1$ AU-ROC | AU-PRC	$\alpha = 2, p = 1$ AU-ROC | AU-PRC	$\alpha = 5, p = 1$ AU-ROC | AU-PRC
a1a	59 | 65	79 | 73	66 | 53*
a2a	64 | 68	83 | 79	66 | 57
a3a	74 | 76	88 | 83	71 | 63
a4a	80 | 83	88 | 84	67 | 70
a5a	82 | 83	95 | 92	79 | 74
Australian	44* | 43*	47* | 50*	41* | 43*
Breast-cancer	51 | 54	60 | 63	61 | 62
Diabetes	84 | 80	81 | 76	67 | 65
Fourclass	38* | 37*	41* | 39*	46* | 41
German.numer	79 | 75	73 | 67	49* | 48*
Heart	55* | 52*	63* | 58*	57* | 50*
Ionosphere	56* | 56*	64 | 65	49* | 54*
Liver-disorders	67 | 64	61 | 60*	52* | 54*
Splice	93 | 93	90 | 90	70 | 68
Sonar	62 | 66	66 | 64	56* | 50*
w1a	38* | 37*	48* | 46*	28 | 32
w2a	64 | 57	69 | 61	46* | 44*
w3a	54* | 48*	52* | 48*	34 | 33
w4a	85 | 81	79 | 74	71 | 59*
w5a	89 | 80	89 | 78	75 | 66

give good results based on our experimental analysis with synthetic data. From the table, it is clear that the interactive algorithm performs significantly better than its non-interactive counterpart on the majority of datasets. On datasets where its performance was worse than that of the non-interactive algorithm, the results were not statistically significant across all parameter settings.

As a final remark, we would like to point out that the performance of the interactive algorithm dropped on some of the datasets with class imbalance. We therefore subsampled the training sets (using a different random subset in each of the 100 experiments for the given dataset) to make the classes balanced. We believe the issue of class imabalance is orthogonal to the problem we are addressing, but needs further investigation and so we leave this open for future work.

5   Concluding Remarks

Our experiments clearly demonstrate the benefits of interactive learning and how disagreement among annotators can be utilized to improve the performance of supervised learning algorithms. Furthermore, we presented theoretical evidence by analyzing the mistake bound of perceptron. The question as to whether annotators in real world scenarios behave according to our simulation model, i.e., if they tend to disagree more on difficult examples located close to the decision boundary when compared to easy examples farther away, is an open one. However, if this assumption holds then our experiments and theoretical analysis show that learning can be improved.

In real-world crowdsourcing applications, an example is typically labeled only by a subset of annotators. Although we did not consider this setting, we believe we could still use the disagreement among annotators to reweight examples, but the algorithm would require some modifications to handle missing labels. We leave this setting open for future work.

Appendix: Estimating the Margin of an Example

The margin of an example x with respect to a linear function $f(\cdot)$ is defined as $\gamma(x; f) = |f(x)| = |\langle w, x \rangle|$. Examples close to the decision boundary will have a small margin and those that are farther away will have a large margin. We assume that an annotator labels an example x using the true labels of all neighboring examples in a ball of some radius centered at x. The size of an annotator's ball is inversely proportional to her strength (expertise). This model of annotators is similar to the one used in Urner et $al.$'s analysis [19]. Note that neither the true labels nor the size of the annotator's ball is known to us. Our only input is a set of m examples with corresponding (noisy) labels from L annotators. Given this input, the goal is to estimate the radius of the annotator's ball. This will then allow us to estimate the margin of an example, i.e., its distance from the separating hyperplane. We proceed in two steps: first, we describe a method to estimate the annotators' expertise scores $\{z_1, z_2, \ldots, z_L\}$ and the ground-truth labels; second, we use these estimates to compute the radii of the annotators' balls.

Estimating an Annotator's Expertise, z. We use a variant of kernel target alignment [2] to estimate the expertise score of each annotator. Let K denote the (centered) kernel matrix on the input examples, i.e., $K \in [-1, +1]^{m \times m}$ with $k_{ij} = \langle \phi(x_i), \phi(x_j) \rangle$, where $\phi(\cdot)$ is a feature map. For linear models, the entries of the kernel matrix are pairwise dot-products of training examples. We consider the following optimization problem to estimate an annotator's expertise score:

$$\hat{z} = \operatorname*{argmin}_{z \in [0,1]^L} \sum_{i=1}^{m} \sum_{j=1}^{m} \left(k_{ij} - \frac{1}{L} \sum_{\ell} z_\ell y_i^{(\ell)} y_j^{(\ell)} \right)^2.$$

This is a constrained least-squares regression problem. The complexity of this optimization problem is quadratic in the number of examples. However, we can use stochastic (projected) gradient descent to remove the dependence on the number of examples.

The ground-truth label of an example x_i can be estimated by taking the weighted average of labels provided by the annotators, i.e., for each given tuple $\left(x_i, y_i^{(1)}, y_i^{(2)}, \ldots, y_i^{(L)}\right)$, we form a new training example $(x_i, \hat{y}_i)$ with $\hat{y}_i = \text{sgn}\left(\sum_\ell z_\ell y_i^{(\ell)}\right)$ and let $\hat{S} = \{(x_i, \hat{y}_i), \ldots, (x_i, \hat{y}_m)\}$.

Estimating the Radius of an Annotator's Ball. Let $r_1, r_2, \ldots, r_L \geq 0$ denote the radii of the annotators' balls. Let $B_r(x) = \{z \in \mathcal{X} \mid d(x, z) \leq r\}$ denote the ball of radius r centered at x, with $d(\cdot, \cdot)$ being a distance metric, such as the Euclidean distance for linear models, defined on the input space $\mathcal{X}$. Given the expertise score z_ℓ for an annotator ℓ, we estimate the radius r_ℓ of her ball by solving the following univariate optimization problem:

$$\hat{r}_\ell = \operatorname*{argmin}_{r \in \mathbb{R}_+} \sum_{i=1}^m \left(\frac{\sum_{(z, \hat{y}) \in B_r(x_i) \cap \hat{S}} \hat{y}}{|B_r(x_i) \cap \hat{S}|} - y_i^{(\ell)} \right)^2 .$$

Intuitively, the above optimization problem is trying to estimate the radius of the annotator's ball by minimizing the squared difference between the (noisy) label of the annotator and the average of the estimates of true labels of all neighboring examples in the ball.

Putting it all Together. Given a training example x, its noisy labels $(y^{(1)}, \ldots, y^{(L)})$, an estimate of the ground-truth label $\hat{y}$, and the radius estimates of the annotators' balls, we compute a lower bound on the margin of x, i.e., its distance from the decision boundary, as follows. Centered at x, we draw nested balls of increasing size, one for each annotator using her radius. Starting from the annotator with the smallest ball, we compare her noisy label with the ground-truth label estimate. At some ball/expert, the noisy label and the ground-truth label estimate will differ, and the radius of this ball is a lower bound on the distance of x from the decision boundary.

References

1. Bengio, Y., Louradour, J., Collobert, R., Weston, J.: Curriculum learning. In: Proceedings of the International Conference on Machine Learning (2009)
2. Cristianini, N., Shawe-Taylor, J., Elisseeff, A., Kandola, J.S.: On kernel-target alignment. In: Proceedings of the Annual Conference on Neural Information Processing Systems (2001)
3. Dekel, O., Gentile, C., Sridharan, K.: Selective sampling and active learning from single and multiple teachers. J. Mach. Learn. Res. **13**, 2655–2697 (2012)

4. Dekel, O., Shamir, O.: Good learners for evil teachers. In: Proceedings of the International Conference on Machine Learning (2009)
5. Freund, Y., Schapire, R.E.: Large margin classification using the perceptron algorithm. Mach. Learn. **37**(3), 277–296 (1999)
6. Jabbari, S., Holte, R.C., Zilles, S.: PAC-learning with general class noise models. In: Glimm, B., Krüger, A. (eds.) KI 2012. LNCS (LNAI), vol. 7526, pp. 73–84. Springer, Heidelberg (2012). doi:10.1007/978-3-642-33347-7_7
7. Jiang, L., Meng, D., Yu, S.I., Lan, Z., Shan, S., Hauptmann, A.G.: Self-paced learning with diversity. In: Proceedings of the Annual Conference on Neural Information Processing Systems (2014)
8. Jiang, L., Meng, D., Zhao, Q., Shan, S., Hauptmann, A.G.: Self-paced curriculum learning. In: Proceedings of the AAAI Conference on Artificial Intelligence (2015)
9. Khan, F., Zhu, X.J., Mutlu, B.: How do humans teach: on curriculum learning and teaching dimension. In: Proceedings of the Annual Conference on Neural Information Processing Systems (2011)
10. Kumar, M.P., Packer, B., Koller, D.: Self-paced learning for latent variable models. In: Proceedings of the Annual Conference on Neural Information Processing Systems (2010)
11. Lee, Y.J., Grauman, K.: Learning the easy things first: self-paced visual category discovery. In: Proceedings of the IEEE Conference on Computer Vision and Pattern Recognition (2011)
12. Malagò, L., Nicolò Cesa-Bianchi, J.M.R.: Online active learning with strong and weak annotators. In: Proceedings of the NIPS 2014 Workshop on Crowdsourcing and Machine Learning (2012)
13. Novikoff, A.B.: On convergence proofs on perceptrons. In: Proceedings of the Symposium on the Mathematical Theory of Automata (1962)
14. Raykar, V.C., Yu, S., Zhao, L.H., Valadez, G.H., Florin, C., Bogoni, L., Moy, L.: Learning from crowds. J. Mach. Learn. Res. **11**, 1297–1322 (2010)
15. Rifkin, R., Yeo, G., Poggio, T.: Regularized least-squares classification. In: Advances in Learning Theory: Methods, Model and Applications. NATO Science Series III:Computer and Systems Sciences, vol. 190, pp. 131–153 (2003)
16. Samuelson, P.: How deviant can you be? J. Am. Stat. Assoc. **63**(324), 1522–1525 (1968)
17. Sharmanska, V., Hernández-Lobato, D., Hernández-Lobato, M., Quadrianto, N.: Ambiguity helps: classification with disagreements in crowdsourced annotations. In: Proceedings of the IEEE Conference on Computer Vision and Pattern Recognition (2016)
18. Snow, R., O'Connor, B., Jurafsky, D., Ng, A.Y.: Cheap and fast - but is it good? Evaluating non-expert annotations for natural language tasks. In: Proceedings of the Conference on Empirical Methods in Natural Language Processing (2008)
19. Urner, R., Ben-David, S., Shamir, O.: Learning from weak teachers. In: Proceedings of the International Conference on Artificial Intelligence and Statistics (2012)
20. Yan, Y., Rosales, R., Fung, G., Ramanathan, S., Dy, J.G.: Learning from multiple annotators with varying expertise. Mach. Learn. **95**(3), 291–327 (2014)
21. Zhu, X.J.: Machine teaching for Bayesian learners in the exponential family. In: Proceedings of the Annual Conference on Neural Information Processing Systems (2013)

An Online Gibbs Sampler Algorithm for Hierarchical Dirichlet Processes Prior

Yongdai Kim[1(✉)], Minwoo Chae[2], Kuhwan Jeong[1], Byungyup Kang[3],
and Hyoju Chung[3]

[1] Department of Statistics, Seoul National University, Seoul, South Korea
ydkim0903@gmail.com, kuhemian@gmail.com
[2] Department of Mathematics, The University of Texas at Austin, Austin, USA
minwoo.chae@gmail.com
[3] NAVER Corp., Seongnam, South Korea
{byungyup.kang,hyoju.chung}@navercorp.com

Abstract. The hierarchical Dirichlet processes (HDP) is a Bayesian nonparametric model that provides a flexible mixed-membership to documents. In this paper, we develop a novel mini-batch online Gibbs sampler algorithm for the HDP which can be easily applied to massive and streaming data. For this purpose, a new prior process so called the generalized hierarchical Dirichlet processes (gHDP) is proposed. The gHDP is an extension of the standard HDP where some prespecified topics can be included in the top-level Dirichlet process. By analyzing various datasets, we show that the proposed mini-batch online Gibbs sampler algorithm performs significantly better than the online variational algorithm for the HDP.

Keywords: Topic model · hierarchical Dirichlet processes · Mini-batch online algorithm · generalized hierarchical Dirichlet processes

1 Introduction

Hierarchical Bayesian modeling has received much attention in machine learning and applied statistics. Bayesian models provide a natural way to model complicated observed data efficiently by the use of latent variables. Probabilistic topic model is a popularly used hierarchical Bayesian model for sparse vectors of count data such as bags of words for documents and bags of features for images, where the posterior distribution reveals a latent semantic structure that can be used for many applications.

There are two popularly used Bayesian topic models - latent Dirichlet allocation (LDA, [2]) and its nonparametric counterpart hierarchical Dirichlet processes (HDP, [11]). Both the LDA and HDP are derived from a generative model which provides a flexible mixed-membership (called a topic) to documents. While it is easy to understand, the LDA requires a prior choice of the number of topics that is not easy in practice. In contrast, the HDP can infer the number

© Springer International Publishing AG 2016
P. Frasconi et al. (Eds.): ECML PKDD 2016, Part I, LNAI 9851, pp. 509–523, 2016.
DOI: 10.1007/978-3-319-46128-1_32

of topics from the data though it is conceptually and technically much more involved.

Posterior inference for the HDP is analytically intractable, and various computational algorithms to approximate the posterior distribution have been proposed. Among these, the Gibbs sampler [5,11,14] and variational Bayesian methods [12,16] are widely used. Both methods attempt to approximate the posterior distribution of parameters as well as latent topics. As is well-known, the Gibbs sampler algorithm generates samples from a Markov chain whose stationary distribution is equal to the posterior distribution, while the variational Bayesian method finds the nearest distribution to the posterior with respect to the Kullback-Leibler divergence among a class of variational distributions. Since algorithms based on the variational Bayesian method usually have closed form updates, they can be implemented faster than the Gibbs sampler algorithms. Moreover, using the stochastic gradient descent method, the online algorithm can be easily devised where the posterior distribution can be sequentially approximated as data arrive [6,16].

While being computationally efficient, the performance of variational Bayesian algorithms is not fully satisfied because the variational distribution could be significantly different from the true posterior distribution. In particular, the discrepancy between two distributions becomes large when the variational distribution assumes that the parameters are independent while the correlations between the parameters are large in the full posterior distribution [13]. Moreover, it is difficult to infer the number of topics by a variational Bayesian method for the HDP. Therefore, there is still a need to devise computationally efficient Gibbs sampler algorithms to analyze a large collection of documents.

In this paper, we propose a novel mini-batch online Gibbs sampler algorithm for the HDP. For this purpose, we propose a new prior process so called the *generalized hierarchical Dirichlet processes (gHDP)*. The gHDP is an extension of the standard HDP where some prespecified topics can be included. The main idea of the proposed algorithm is to approximate the current posterior distribution with the gHDP and use it as a prior for the next update of the posterior distribution. To be more specific, suppose that the whole dataset is divided into two groups, say $\mathbf{x} = (\mathbf{x}^{\text{old}}, \mathbf{x}^{\text{new}})$. Then, the Bayes rule implies

$$p(\theta|\mathbf{x}) \propto p(\mathbf{x}^{\text{new}}|\theta)p(\theta|\mathbf{x}^{\text{old}}),\tag{1}$$

where θ is the parameters. We approximate $p(\theta|\mathbf{x})$ by replacing $p(\theta|\mathbf{x}^{\text{old}})$ in (1) with the gHDP $p^{\text{gHDP}}(\theta|\eta)$ with parameter η as its proxy, where the parameter η is estimated based on the posterior samples generated from $p(\theta|\mathbf{x}^{\text{old}})$. By analyzing various datasets, we demonstrate that the proposed method outperforms the variational Bayesian method.

The rest of this paper is organized as follows. We briefly review the HDP in Sect. 2. In the next section, the gHDP is proposed and a Gibbs sampler algorithm is provided to infer parameters in the gHDP. Based on the proposed gHDP, a mini-batch online algorithm for HDP is developed in Sect. 4. Theoretical and experimental results are presented in Sects. 5 and 6, respectively. Concluding remarks follow in Sect. 7.

2   Review of Hierarchical Dirichlet Processes

In this section, we briefly review the HDP proposed by [11]. Since the main application of the HDP is document analysis, we focus on the HDP with multinomial distributions. For a given positive integer W, that is the number of words in the dictionary, let $\mathcal{S} = \{(p_1, \ldots, p_W) : p_w \geq 0, \sum_{w=1}^{W} p_w = 1\}$ be the $(W-1)$-dimensional simplex. Denote $\mathcal{DP}(\alpha, H)$ the Dirichlet process with a precision parameter $\alpha > 0$ and a mean probability measure H. See [4] for the definition of Dirichlet processes.

Let $(\mathbf{x}_j)_{j=1}^{n}$ be the vector of observed words in n documents, where $\mathbf{x}_j = (x_{ji})_{i=1}^{n_j}$ and x_{ji} is the word at the ith position in the jth document that is an element of the dictionary $\{1, \ldots, W\}$. Here, n_j is the number of words in the jth document. For given probability distributions G_j's on $\mathcal{S}$, we assume that $\mathbf{x}_1, \ldots, \mathbf{x}_n$ are conditionally independent with

$$p(\mathbf{x}_j | G_j) = \prod_{i=1}^{n_j} \int p(x_{ji} | \boldsymbol{\phi}) dG_j(\boldsymbol{\phi}),$$

where $p(x_{ji} | \boldsymbol{\phi})$ is a multinomial distribution with the cell probabilities $\boldsymbol{\phi} = (\phi_1, \ldots, \phi_W)$. As a prior for G_j's, the HDP is composed of

$$G_0 \sim \mathcal{DP}(\alpha_0, \mathcal{D}(\beta, \ldots, \beta)), \quad G_j | G_0 \overset{\text{indep}}{\sim} \mathcal{DP}(\alpha_1, G_0), \quad j = 1, \ldots, n. \quad (2)$$

Here α_0, α_1 and β are hyperparameters and assumed to be fixed constants throughout this paper.[1] We denote this prior as $\text{HDP}(\alpha_0, \alpha_1, \beta)$.

Since a Dirichlet process has a discrete sample path almost surely, we can write

$$G_0 = \sum_{k=1}^{\infty} \pi_k \delta_{\phi_k}, \quad G_j = \sum_{k=1}^{\infty} \theta_{jk} \delta_{\phi_k}, \quad j = 1, \ldots, n,$$

for some nonnegative π_k's and θ_{jk}'s with $\sum_{k=1}^{\infty} \pi_k = \sum_{k=1}^{\infty} \theta_{jk} = 1$ and $\phi_k \in \mathcal{S}$, where $\delta_{\phi}(\cdot)$ denotes the Dirac measure which assigns mass 1 at the point ϕ. Let $\boldsymbol{\pi} = (\pi_k)_{k=1}^{\infty}$, $\boldsymbol{\phi} = (\phi_k)_{k=1}^{\infty}$, $\phi_k = (\phi_{kw})_{w=1}^{W}$, $\boldsymbol{\theta} = (\theta_j)_{j=1}^{n}$ and $\boldsymbol{\theta}_j = (\theta_{jk})_{k=1}^{\infty}$. By using the stick breaking representation of Dirichlet processes [10], we can rewrite the HDP model as the following Bayesian hierarchical model:

$$\boldsymbol{\pi} \sim \text{GEM}(\alpha_0), \quad \boldsymbol{\theta}_j | \boldsymbol{\pi} \overset{\text{indep}}{\sim} \mathcal{DP}(\alpha_1, H_{\boldsymbol{\pi}}), \quad \phi_k \overset{\text{indep}}{\sim} \mathcal{D}(\beta, \ldots, \beta),$$

$$p(z_{ji} = k | \boldsymbol{\theta}_j) = \theta_{jk}, \quad p(\mathbf{x} | \mathbf{z}, \boldsymbol{\phi}) = \prod_{j=1}^{n} \prod_{i=1}^{n_j} p(x_{ji} | z_{ji}, \phi_{z_{ji}}),$$

where $\mathbf{z} = (\mathbf{z}_j)_{j=1}^{n}$, $\mathbf{z}_j = (z_{ji})_{i=1}^{n_j}$ and $H_{\boldsymbol{\pi}}$ is the distribution on the set of all natural numbers whose probabilities are given by $H_{\boldsymbol{\pi}}(\{k\}) = \pi_k$. Here $\boldsymbol{\pi} \sim \text{GEM}(\alpha)$, where

---

[1] These hyperparameters also can be inferred with additional hyperpriors. See for example [11].

the letters stand for Griffiths-Engen-McCloskey [8], means that $\pi_1 = \pi_1'$ and $\pi_k = \pi_k' \prod_{l=1}^{k-1}(1 - \pi_l')$ for $k \geq 2$ where $\pi_k' \overset{\text{iid}}{\sim} \text{Beta}(1, \alpha)$.

There are various Gibbs sampler algorithms to infer the posterior distribution of the HDP model [5,11,14], among which we use the partially collapsed Gibbs sampler algorithm [3,11], which integrates out θ and generates π, ϕ, z and auxiliary variables $\mathbf{m} = (m_{jk} : 1 \leq j \leq n, 1 \leq k < \infty)$ iteratively as follows. For given $\mathbf{z}$, let N_{jkw} be the cardinality of the set $\{i : x_{ji} = w, z_{ji} = k\}$ and let $\cdot$ in the subscripts represent the summation over the corresponding indices. For example, $N_{jk\cdot} = \sum_{w=1}^{W} N_{jkw}$ and $N_{j\cdot\cdot} = \sum_{k=1}^{\infty} \sum_{w=1}^{W} N_{jkw}$. When $\mathbf{z}$ is given, we relabel z_{ji}'s so that $z_{ji} \in \{1, \ldots, K\}$ and rearrange π, ϕ and N_{jkw} accordingly. Here K is the integer such that $N_{\cdot k\cdot} > 0$ for $k \leq K$ and $N_{\cdot k\cdot} = 0$ for $k > K$. Let $\pi_u = \sum_{k>K} \pi_k$. Then, we sample $\pi, \phi, \mathbf{z}$ and auxiliary variables $\mathbf{m} = (m_{jk} : 1 \leq j \leq n, 1 \leq k < \infty)$ iteratively from the following conditional distributions:

$$p(m_{jk} = m | \mathbf{x}, \mathbf{z}) \propto s(N_{jk\cdot}, m)(\alpha_1 \pi_k)^m \text{ for } m = 1, \ldots, N_{jk\cdot}, \quad (3)$$

$$\phi_k | \mathbf{z}, \mathbf{x} \sim \mathcal{D}(\beta + N_{\cdot k1}, \ldots, \beta + N_{\cdot kW}), \quad \text{for } k \geq 1, \quad (4)$$

$$(\pi_1, \ldots, \pi_K, \pi_u) | \mathbf{m} \sim \mathcal{D}(m_{\cdot 1}, \ldots, m_{\cdot K}, \alpha_0), \quad (5)$$

$$(\pi_k)_{k>K} | \pi_u \sim \pi_u \text{GEM}(\alpha_0), \quad (6)$$

$$p(z_{ji} = k | \mathbf{x}, \mathbf{z}^{-ji}, \pi, \phi) \propto (N_{jk\cdot}^{-ji} + \alpha_1 \pi_k)\phi_{kx_{ji}}. \quad (7)$$

Here $s(n, m)$ is the unsigned Stirling's number of the first kind and the superscript $-ji$ means that the corresponding word is excluded in the counts. For the interpretation of m_{jk} in the Chinese restaurant franchise metaphor, see [11].

Remark 1. While the collapsed Gibbs sampler algorithm, which generates $\pi, \mathbf{z}$ and $\mathbf{m}$ after integrating out ϕ as well as θ, is popularly used in many applications [7,11], we use the partially collapsed Gibbs sampler algorithm since samples of ϕ are needed in the proposed online algorithm.

3   Generalized Hierarchical Dirichlet Processes

For a given nonnegative integer K_0, let μ_k and $\beta_k = (\beta_{kw})_{w=1}^{W}$ be given positive numbers for $k = 1, \ldots, K_0$. For a given $\beta > 0$, let $\beta = (\beta_k)_{k=1}^{\infty}$, where $\beta_k = (\beta, \ldots, \beta)$ for $k > K_0$. The gHDP prior with parameters $\alpha_0, \alpha_1, \beta$ and $\mu = (\mu_1, \ldots, \mu_{K_0})$, denoted by gHDP$(\alpha_0, \alpha_1, \beta, \mu, K_0)$, is defined as

$$\phi_k \overset{\text{indep}}{\sim} \mathcal{D}(\beta_{k1}, \ldots, \beta_{kW}), \quad k = 1, \ldots, K_0,$$

$$G_0 | \phi_1, \ldots, \phi_{K_0} \sim \mathcal{DP}\left(\alpha_0 + \mu_\cdot, w_0 \mathcal{D}(\beta, \ldots, \beta) + \sum_{k=1}^{K_0} w_k \delta_{\phi_k}\right),$$

$$G_j | G_0 \overset{\text{indep}}{\sim} \mathcal{DP}(\alpha_1, G_0), \quad j = 1, \ldots, n,$$

where $\mu_\cdot = \mu_1 + \cdots + \mu_{K_0}, w_0 = \alpha_0/(\mu_\cdot + \alpha_0)$ and $w_k = \mu_k/(\mu_\cdot + \alpha_0)$ for $k = 1, \ldots, K_0$. The gHDP is an extension of the HDP$(\alpha_0, \alpha_1, \beta)$ in (2) where the prior information about topics, represented by $\boldsymbol{\mu}$ and $\boldsymbol{\beta}$, is incorporated into G_0. For example, the w_k's reflect the prior belief of the proportions of the topics ϕ_k in the whole corpus.

As in the HDP, we can rewrite the gHDP model by using the stick breaking representation as follows. We write the sample path of G_0 by $\sum_{k=1}^{\infty} \pi_k \delta_{\phi_k}$ where

$$\phi_k \overset{\text{indep}}{\sim} \mathcal{D}(\beta_{k1}, \ldots, \beta_{kW}) \text{ for } k \geq 1,$$
$$(\pi_1, \ldots, \pi_{K_0}, \pi_u) \sim \mathcal{D}(\mu_1, \ldots, \mu_{K_0}, \alpha_0),$$
$$(\pi_k)_{k > K_0} | \pi_u \sim \pi_u \text{GEM}(\alpha_0).$$

Note that these are quite similar to the conditional posterior distributions of the HDP in (4)–(6). Therefore, we expect that we can closely approximate the posterior distribution of $\boldsymbol{\pi}$ and $\boldsymbol{\phi}$ with the HDP$(\alpha_0, \alpha_1, \beta)$ prior to the gHDP$(\alpha_0, \alpha_1, \beta, \mu, K_0)$ prior distribution. Here β, μ and K_0 are estimated using samples from the posterior distribution. This is the main motivation of considering the gHDP.

The novelty of the gHPD is that the posterior distribution can be inferred by a Gibbs sampler algorithm similar to that for the HDP which generates $\boldsymbol{\pi}, \boldsymbol{\phi}, \mathbf{z}$ and auxiliary variables $\mathbf{m} = (m_{jk} : 1 \leq j \leq n, 0 \leq k < \infty)$ iteratively from their conditional posteriors as follows. For given $\mathbf{z}$, let $\mathcal{Z} = \{z_{ji} : z_{ji} \notin \{1, \ldots, K_0\}\}$. Similarly to the HDP, we relabel z_{ji} in $\mathcal{Z}$ such that $\mathcal{Z} = \{K_0 + 1, \ldots, K\}$ for $K \geq K_0$, and rearrange the corresponding quantities accordingly. Then, the conditional distributions are given as

$$p(m_{jk} = m | \mathbf{z}, \boldsymbol{\pi}) \propto s(N_{jk\cdot}, m)(\alpha_1 \pi_k)^m \text{ for } m = 1, \ldots, N_{jk\cdot},$$
$$\phi_k | \mathbf{z}, \mathbf{x} \sim \mathcal{D}(\beta_{k1} + N_{\cdot k1}, \ldots, \beta_{kW} + N_{\cdot kW}), \text{ for } k \geq 1,$$
$$(\pi_1, \ldots, \pi_K, \pi_u) | \mathbf{m} \sim \mathcal{D}(\mu_1 + m_{\cdot 1}, \ldots, \mu_{K_0} + m_{\cdot K_0}, m_{\cdot (K_0+1)}, \ldots, m_{\cdot K}, \alpha_0),$$
$$(\pi_k)_{k > K} | \pi_u \sim \pi_u \text{GEM}(\alpha_0),$$
$$p(z_{ji} = k | \mathbf{x}, \mathbf{z}^{-ji}, \boldsymbol{\pi}, \boldsymbol{\phi}) \propto (N_{jk\cdot}^{-ji} + \alpha_1 \pi_k) \phi_{kx_{ji}}.$$

Similarly to the case of HDP, $\boldsymbol{\phi}$ can be integrated out from the Gibbs sampler algorithm to have the collapsed Gibbs sampler algorithm.

4 Mini-batch Online Algorithm for HDP

We develop a mini-batch online algorithm for HDP$(\alpha_0, \alpha_1, \beta)$ in this section. In the HDP, parameters can be distinguished by corpus-level parameters $\boldsymbol{\pi}$ and $\boldsymbol{\phi}$ (or G_0) and document-level parameter $\boldsymbol{\theta}$. Once the corpus-level parameters are given, it is not difficult to infer $\boldsymbol{\theta}$, because every documents are conditionally independent given $\boldsymbol{\pi}$ and $\boldsymbol{\phi}$. Therefore, we focus on the mini-batch online algorithm for the posterior distribution of $\boldsymbol{\pi}$ and $\boldsymbol{\phi}$.

Suppose that the whole data $\mathbf{x}$ can be divided into two groups as $\mathbf{x} = (\mathbf{x}^{\text{old}}, \mathbf{x}^{\text{new}})$. The Bayes rule implies

$$p(\boldsymbol{\pi}, \boldsymbol{\phi} | \mathbf{x}^{\text{old}}, \mathbf{x}^{\text{new}}) \propto p(\boldsymbol{\pi}, \boldsymbol{\phi} | \mathbf{x}^{\text{old}}) p(\mathbf{x}^{\text{new}} | \boldsymbol{\pi}, \boldsymbol{\phi}).$$

Algorithm 1. The mini-batch online algorithm for the HDP

1: Divide $\mathbf{x}$ into mini-batches $\mathbf{x} = (\mathbf{x}^{[1]}, \mathbf{x}^{[2]}, \ldots, \mathbf{x}^{[S]})$.
2: Initialize $K_0^{[0]} = 0$.
3: **for** $s = 1, \ldots, S$ **do**
4:     Run the Gibbs sampler with $\mathrm{gHDP}(\alpha_0, \alpha_1, \beta^{[s-1]}, \mu^{[s-1]}, K_0^{[s-1]})$ as prior and $\mathbf{x}^{[s]}$ as data to collect samples $(\pi^{(m)}, \phi^{(m)})_{m=1}^M$.
5:     Estimate $\mu^{[s]}, \beta^{[s]}$ and $K_0^{[s]}$ based on the MCMC samples $(\pi^{(m)}, \phi^{(m)})_{m=1}^M$.
6: **end for**

That is, the full posterior distribution $p(\pi, \phi | \mathbf{x}^{\mathrm{old}}, \mathbf{x}^{\mathrm{new}})$ can be understood as the posterior distribution of π and ϕ given the new data $\mathbf{x}^{\mathrm{new}}$ with the prior $p(\pi, \phi | \mathbf{x}^{\mathrm{old}})$. The main idea of the proposed mini-batch Gibbs sampler algorithm is to approximate $p(\pi, \phi | \mathbf{x}^{\mathrm{old}})$ by the $\mathrm{gHDP}(\alpha_0, \alpha_1, \beta, \mu, K_0)$, where β, μ and K_0 are estimated based on MCMC samples $(\pi^{(m)}, \phi^{(m)}, \mathbf{z}^{(m)})_{m=1}^M$ generated from $p(\pi, \phi, \mathbf{z} | \mathbf{x}^{\mathrm{old}})$.

To estimate β, μ and K_0, let $\mathcal{K} = \{k : N_{\cdot k \cdot}^{(M)} > 0\}$. Initially, we estimate K_0 by the cardinality of $\mathcal{K}$. Once we relabel the topics so that $\mathcal{K} = \{1, \ldots, K_0\}$, we estimate $w_k = \sum_{m=1}^M \pi_k^{(m)}/M$ for $k = 1, \ldots, K_0$ and $w_0 = 1 - \sum_{k=1}^{K_0} w_k$, which are the method of moments estimators. Then, we estimate $\mu_k = (\mu_\cdot + \alpha_0) w_k$ for $k = 1, \ldots, K_0$ with $\mu_\cdot = \alpha_0/w_0 - \alpha_0$. Since topics with a small proportions are likely to cause overfitting, we delete topics from $\mathcal{K}$ when $w_k < \epsilon$, where ϵ is a prespecified small positive constant (so called the topic threshold). Finally, we let K_0 be the cardinality of $\mathcal{K}$, relabel the topics so that $\mathcal{K} = \{1, \ldots, K_0\}$ and modify μ accordingly.

Once $\mathcal{K}$ is given, we estimate $\beta_k, k = 1, \ldots, K_0$ as follows. We assume that $\phi_k^{(m)} \sim \mathcal{D}(\beta_{k1}, \ldots, \beta_{kW})$. First, we estimate $\gamma_{kw} = \beta_{kw}/\beta_{k\cdot}$ by the corresponding method of moments estimator $\sum_{m=1}^M \phi_{kw}^{(m)}/M$ for $w = 1, \ldots, W$. To estimate $\beta_{k\cdot}$, we choose a set $I \subset \{1, \ldots, W\}$ and consider $\phi_{kI}^{(m)} = \sum_{w \in I} \phi_{kw}$. Since $\phi_{kI}^{(m)} \sim \mathrm{Beta}(\beta_{k\cdot} \sum_{w \in I} \gamma_{kw}, \beta_{k\cdot} \sum_{w \notin I} \gamma_{kw})$, we estimate $\beta_{k\cdot}$ by the maximum likelihood estimator based on $\phi_{kI}^{(m)}, m = 1, \ldots, M$. For the choice of I, we recommend an I such that $\sum_{w \in I} \gamma_{kw}$ is close to $1/2$.

This two step procedure can be generalized for streaming data $\mathbf{x}$ which are divided into batches as $\mathbf{x} = (\mathbf{x}^{[1]}, \ldots, \mathbf{x}^{[S]})$, where $\mathbf{x}^{[s]}$'s are disjoint collections of documents in the corpus $\mathbf{x}$. We first set $K_0^{[0]} = 0$ and generate MCMC samples of (π, ϕ) from the posterior distribution $p(\pi, \phi | \mathbf{x}^{[1]})$. Then, we estimate $K_0^{[1]}, \mu^{[1]}$ and $\beta^{[1]}$ using the aforementioned method. For given $K_0^{[s-1]}, \mu^{[s-1]}$ and $\beta_k^{[s-1]}$, we obtain MCMC samples of (π, ϕ) from the approximated posterior distribution that is proportional to $p(\mathbf{x}^{[s]} | \pi, \phi) p^{\mathrm{gHDP}}(\pi, \phi | \alpha_0, \alpha_1, \beta^{[s-1]}, \mu^{[s-1]}, K_0^{[s-1]})$. Then, we estimate $K_0^{[s]}, \mu^{[s]}$ and $\beta_k^{[s]}$ accordingly. The proposed mini-batch online Gibbs sampler algorithm is summarized in Algorithm 1.

A novel aspect of the proposed algorithm is that we do not need to store MCMC samples $(\pi^{(m)}, \phi^{(m)})_{m=1}^M$ which requires a large memory space. This is because to estimate parameters for the next batch, it suffices to have $\sum_m \pi_k^{(m)}$, $\sum_m \sum_{w \in I} \phi_{kw}^{(m)}$, $\sum_m \log(\sum_{w \in I} \phi_{kw}^{(m)})$ and $\sum_m \log(1 - \sum_{w \in I} \phi_{kw}^{(m)})$.

Remark 2. The proposed algorithm approximates $p(\boldsymbol{\pi}, \boldsymbol{\phi} | \mathbf{x}^{\text{old}})$ by gHDP(γ) with $\gamma = (\alpha_0, \alpha_1, \boldsymbol{\beta}, \boldsymbol{\mu}, K_0)$, and estimates γ based on the MCMC samples. Suppose that γ is estimated by the maximum likelihood estimator. Then the proposed algorithm can be considered as an algorithm to search a distribution which minimizes the Kullback-Leibler divergence between $p(\boldsymbol{\pi}, \boldsymbol{\phi} | \mathbf{x}^{\text{old}})$ and gHDP(γ) with respect to $p(\boldsymbol{\pi}, \boldsymbol{\phi} | \mathbf{x}^{\text{old}})$. That is, the algorithm finds γ which minimizes

$$\mathbb{E}_{p(\boldsymbol{\pi}, \boldsymbol{\phi} | \mathbf{x}^{\text{old}})} \left\{ \log p(\boldsymbol{\pi}, \boldsymbol{\phi} | \mathbf{x}^{\text{old}}) - \log \text{gHDP}(\gamma) \right\} . \tag{8}$$

The variational Bayes method, on the other hand, would find γ which minimizes

$$\mathbb{E}_{\text{gHDP}(\gamma)} \left\{ \log \text{gHDP}(\gamma) - \log p(\boldsymbol{\pi}, \boldsymbol{\phi} | \mathbf{x}^{\text{old}}) \right\} . \tag{9}$$

The main difference of the proposed algorithm and variational Bayes method is that the expectation operators in the calculation of the K-L divergence are different. Analytically, the K-L divergence (8) is easier to be minimized compared to (9) since the parameter γ affects only $\log \text{gHDP}(\gamma)$ in (8) while it affects $\mathbb{E}_{\text{gHDP}(\gamma)}$ as well as $\log \text{gHDP}(\gamma)$. It is almost impossible to find γ minimizing (9) and in general a distribution with much simpler form (e.g. all parameters are assumed to be independent) is used instead of gHDP(γ) in the variational Bayes method, which explains why the proposed algorithm is superior to the variational Bayes method in approximating the posterior distribution. In the algorithm, we used the mixture of the method of moments and maximum likelihood estimators instead of the maximum likelihood estimators since the former is more stable.

5   Theoretical Results

In this section, we study theoretical properties of the proposed mini-batch online algorithm for the HDP. In particular, we provide sufficient conditions for measuring the degree of approximation of the posterior distribution.

The proposed online Gibbs sampler algorithm approximates the posterior distribution $p(\theta | \mathbf{x}^{\text{old}})$ by a certain distribution $g(\theta)$. Then, the full posterior distribution

$$p(\theta | \mathbf{x}^{\text{new}}, \mathbf{x}^{\text{old}}) \propto p(\mathbf{x}^{\text{new}} | \theta) p(\theta | \mathbf{x}^{\text{old}})$$

is approximated by

$$p^a(\theta | \mathbf{x}^{\text{new}}) \propto p(\mathbf{x}^{\text{new}} | \theta) g(\theta) .$$

Therefore, the accuracy of the approximated posterior distribution depends on the degree of approximation of $p(\theta | \mathbf{x}^{\text{old}})$ by $g(\theta)$. The following theorem provides a method of measuring the degree of approximation of the full posterior distribution. We use the capital letter P to denote the corresponding probability measure of the density p.

Theorem 1. *Suppose that there exists an $\epsilon > 0$ such that*

$$P^a \left(\{\theta : |\log p(\theta|\mathbf{x}^{\mathrm{old}}) - \log g(\theta)| < \epsilon \} | \mathbf{x}^{\mathrm{new}} \right) > 1 - \epsilon. \tag{10}$$

Assume also that there exists a constant $c > 0$ such that

$$\frac{p(\theta|\mathbf{x}^{\mathrm{old}})}{g(\theta)} \leq c \tag{11}$$

for all $\theta \in \Theta$. Then, for any constant $M > 0$, there exists a constant $C > 0$, depending only on M and c, such that

$$\left| \mathbb{E}^f (h(\theta)|\mathbf{x}^{\mathrm{new}}, \mathbf{x}^{\mathrm{old}}) - \mathbb{E}^a (h(\theta)|\mathbf{x}^{\mathrm{new}}) \right| \leq C\epsilon$$

for every bounded function $h : \Theta \to [-M, M]$, where $\mathbb{E}^f$ and $\mathbb{E}^a$ denote the expectations with respect to the full and approximated posterior distributions.

Proof. It is easy to see that

$$\mathbb{E}^f (h(\theta)|\mathbf{x}^{\mathrm{new}}, \mathbf{x}^{\mathrm{old}}) = \frac{\mathbb{E}^a \{h(\theta)p(\theta|\mathbf{x}^{\mathrm{old}})/g(\theta)|\mathbf{x}^{\mathrm{new}}\}}{\mathbb{E}^a \{p(\theta|\mathbf{x}^{\mathrm{old}})/g(\theta)|\mathbf{x}^{\mathrm{new}}\}}.$$

Hence, it suffices to show that there exists a constant $C_1 > 0$, depending only on M and c, such that

$$\left| \mathbb{E}^a \{h(\theta)p(\theta|\mathbf{x}^{\mathrm{old}})/g(\theta)|\mathbf{x}^{\mathrm{new}}\} - \mathbb{E}^a (h(\theta)|\mathbf{x}^{\mathrm{new}}) \right| \leq C_1 \epsilon.$$

We may assume that $h(\theta) \geq 0$ for all θ because $h(\theta) = h(\theta)I(h(\theta) \geq 0) + h(\theta)I(h(\theta) < 0)$, where I denotes the indicator function. Let

$$A = \{\theta : |\log p(\theta|\mathbf{x}^{\mathrm{old}}) - \log g(\theta)| < \epsilon\}.$$

Then, we can write

$$\mathbb{E}^a \{h(\theta)p(\theta|\mathbf{x}^{\mathrm{old}})/g(\theta)|\mathbf{x}^{\mathrm{new}}\}$$
$$= \int_A \frac{h(\theta)p(\theta|\mathbf{x}^{\mathrm{old}})}{g(\theta)} p^a(\theta|\mathbf{x}^{\mathrm{new}})d\theta + \int_{A^C} \frac{h(\theta)p(\theta|\mathbf{x}^{\mathrm{old}})}{g(\theta)} p^a(\theta|\mathbf{x}^{\mathrm{new}})d\theta$$
$$\leq e^\epsilon \int_A h(\theta)p^a(\theta|\mathbf{x}^{\mathrm{new}})d\theta + \int_{A^C} \frac{h(\theta)p(\theta|\mathbf{x}^{\mathrm{old}})}{g(\theta)} p^a(\theta|\mathbf{x}^{\mathrm{new}})d\theta$$
$$\leq \mathbb{E}^a \{h(\theta)|\mathbf{x}^{\mathrm{new}}\} + R_u,$$

where

$$R_u \leq (e^\epsilon - 1) \int_A h(\theta)p^a(\theta|\mathbf{x}^{\mathrm{new}})d\theta + McP^a(A^c|\mathbf{x}^{\mathrm{new}}).$$

Hence, we can choose $c_u > 0$ such that

$$\mathbb{E}^a \{h(\theta)p(\theta|\mathbf{x}^{\mathrm{old}})/g(\theta)|\mathbf{x}^{\mathrm{new}}\} - \mathbb{E}^a (h(\theta)|\mathbf{x}^{\mathrm{new}}) \leq c_u \epsilon$$

because $P^a(A^c|\mathbf{x}^{\mathrm{new}}) < \epsilon$.

For the lower bound, write

$$\mathbb{E}^a\{h(\theta)p(\theta|\mathbf{x}^{\text{old}})/g(\theta)|\mathbf{x}^{\text{new}}\} \geq \int_A \frac{h(\theta)p(\theta|\mathbf{x}^{\text{old}})}{g(\theta)}p^a(\theta|\mathbf{x}^{\text{new}})d\theta$$
$$= \mathbb{E}^a\{h(\theta)|\mathbf{x}^{\text{new}}\} - R_l,$$

where

$$R_l \leq M\left[(1 - e^{-\epsilon}) + P^a(A^c|\mathbf{x}^{\text{new}})\right].$$

Thus, there exists $c_l > 0$ such that

$$\mathbb{E}^a\{h(\theta)p(\theta|\mathbf{x}^{\text{old}})/g(\theta)|\mathbf{x}^{\text{new}}\} - \mathbb{E}^a(h(\theta)|\mathbf{x}^{\text{new}}) \geq -c_l\epsilon.$$

By letting $C_1 = \max\{c_l, c_u\}$, the proof is done.   □

Condition (10) requires that $g(\theta)$ approximates $p(\theta|\mathbf{x}^{\text{old}})$ well on the area where the mass of the approximated posterior is large. In Sect. 6.3, we propose a way of measuring the degree of approximation ϵ in practice.

In general, condition (11) would hold when $g(\theta)$ has heavier tails than $p(\theta|\mathbf{x}^{\text{old}})$, which is a standard requirement for approximating a density. In the gHDP approximation, every parameter is supported on a compact set, so the condition (11) is easily satisfied.

Typically, the centers of $p(\theta|\mathbf{x}^{\text{old}})$ and $p(\theta|\mathbf{x}^{\text{new}}, \mathbf{x}^{\text{old}})$ are close, while the dispersion of $p(\theta|\mathbf{x}^{\text{old}})$ is larger than that of $p(\theta|\mathbf{x}^{\text{new}}, \mathbf{x}^{\text{old}})$. Also, $p(\theta|\mathbf{x}^{\text{old}})$ usually has most of its mass at a small neighborhood of its center. Hence, if $g(\theta)$ approximates $p(\theta|\mathbf{x}^{\text{old}})$ well around the center of $p(\theta|\mathbf{x}^{\text{old}})$, the approximated posterior works well and has most of its mass around the center of $p(\theta|\mathbf{x}^{\text{old}})$ too. Hence, condition (10) would be satisfied when $g(\theta)$ approximates $p(\theta|\mathbf{x}^{\text{old}})$ well around the center of $p(\theta|\mathbf{x}^{\text{old}})$.

When we apply the mini-batch online algorithm with the $(s+1)$-th batch, we can apply Theorem 1 repeatedly to conclude that there exists a constant $C > 0$, depending only on M and c, such that

$$\left|\mathbb{E}^f(h(\theta)|\mathbf{x}^{[0]}, \ldots, \mathbf{x}^{[s+1]}) - \mathbb{E}^{a\otimes s}(h(\theta)|\mathbf{x}^{[s+1]})\right| \leq sC\epsilon$$

provided conditions (10) and (11) are satisfied at each batch, where $\mathbb{E}^{a\otimes s}$ denotes the expectation with respect to the approximated posterior distribution after applying the $(s+1)$-th batch.

6   Experimental Results

In this section, we evaluate the performance of the proposed mini-batch algorithm compared with the full Gibbs sampler algorithm (3)–(7) and the online variational algorithm of [16] by analyzing various datasets.

Table 1. Basic information of the datasets used in the experiments

	KOS	NIPS	ENRON	NYTIMES	PUBMED
Number of documents	3,430	1,500	39,961	300,000	8,200,000
Dictionary size	6,909	12,419	28,102	102,660	141,043

6.1 Data

Our experiments are based on five datasets which can be found in UCI machine learning repository https://archive.ics.uci.edu/ml/datasets/ Bag+of+Words. Table 1 shows the summary of the datasets.

We eliminate words whose popularity (the number of documents containing the word) is less than a threshold. After elimination, empty documents are also removed. Table 2 shows thresholds and results of words elimination. For each dataset, we divide the dataset randomly into training and test datasets, whose sizes are given in Table 3.

Table 2. Basic information after elimination

	KOS	NIPS	ENRON	NYTIMES	PUBMED
Threshold	50	50	100	1,000	5,000
Number of documents	3,430	1,500	39,859	299,645	8,199,998
Dictionary size	1,663	3,002	5,918	9,542	11,140

Table 3. The numbers of documents in training and test datasets

	KOS	NIPS	ENRON	NYTIMES	PUBMED
Train	3,000	1,300	36,000	290,000	8,190,000
Test	430	200	3,859	9,645	9,998

In our experiments, we want to investigate how well the proposed algorithm approximates the full posterior. For this purpose, it should be able to run the full Gibbs sampler algorithm within a reasonable amount of time. To run the full Gibbs sampler algorithm for PUBMED dataset, having more than 8,000,000 documents, we used a parallel computation of the partially collapsed Gibbs sampler algorithm, which was proposed implicitly in [11], with several dozens of CPUs. This is why we didn't analyze really huge and streaming data. We believe that applying the proposed algorithm for such big data is not that difficult.

6.2   Results

We use the test log-likelihood as an evaluation metric. First, we fix π and ϕ obtained from analyzing the training dataset. Then, we generate $\mathbf{z}^{\text{test}}$ from (7) for 1,000 times. We estimate $\boldsymbol{\theta}^{\text{test}}$ from last 500 samples of $\mathbf{z}^{\text{test}}$. Finally, we calculate the test log-likelihood by

$$\frac{1}{N^{\text{test}}} \sum_{j,i} \log \left(\sum_k \theta_{jk}^{\text{test}} \phi_{kx_{ji}^{\text{test}}} \right),$$

where N^{test} is the total number of words in $\mathbf{x}^{\text{test}}$.

Wang *et al.* (2011, [16]) used hold-out words instead of hold-out documents. We thought that using hold-out documents is easier than using hold-out words since we don't have to care about how much words we should hold-out from each document in the test dataset. A similar metric to ours is also used in [15].

For prior parameters, we set $\alpha_0 = 3, \alpha_1 = 0.5, \beta = 0.01$ for the HDP. These values are chosen so that the posterior distribution of the number of topics converges after reasonable amount of iterations of the full Gibbs sampler algorithm. For the mini-batch online Gibbs sampler algorithm, we divided the training dataset into 10 batches. For each batch $\mathbf{x}^{[s]}$, we first ran 500 gHDP Gibbs sampler iterations as a burn-in period and the next 500 samples are collected to estimate $\boldsymbol{\mu}^{[s]}, \boldsymbol{\beta}^{[s]}$ and $K_0^{[s]}$. For the topic threshold, we set $\epsilon = 10^{-4}$. For the full Gibbs sampler algorithm, we ran 1000, 2000, 3000, 4000 and 5000 HDP Gibbs sampler iterations with the whole training dataset. For the online variational algorithm of [16], we also divided the training dataset into 10 batches and set $\kappa = 0.9, \tau = 1024, K = 150, T = 15$ as [16] and the running time of each dataset as Table 4.

Table 4. The running times of the online variational HDP algorithm

	KOS	NIPS	ENRON	NYTIMES	PUBMED
Running time (seconds)	5,000	5,000	10,000	20,000	30,000

We plot the test log-likelihoods of different algorithms for five datasets in Fig. 1(a). Solid black line represents the test log-likelihood obtained from the online Gibbs sampler algorithm. The bottom solid red line is the test log-likelihood of the online variational algorithm after processing the whole batches, and five horizontal lines with various color are the test log-likelihoods of the full Gibbs sampler algorithm after 1000, 2000, 3000, 4000 and 5000 iterations.

The test log-likelihood of the online variational algorithm is much lower than the others. Surprisingly, this is less than the test log-likelihood of the online Gibbs sampler algorithm after processing only one batch. This may be due to that the correlations between the parameters are large so that the assumption of independent parameters used in the variational algorithm does not work well.

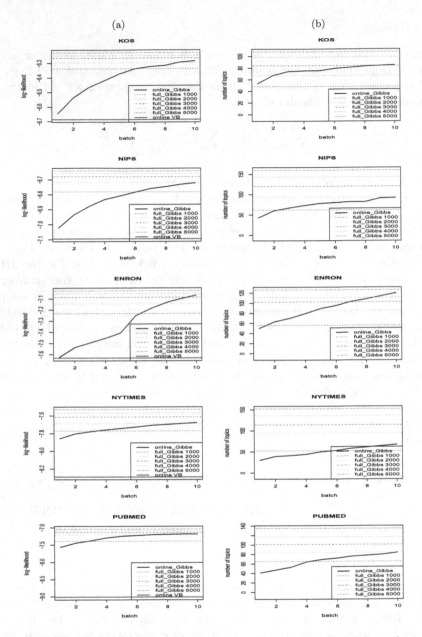

Fig. 1. (a) Solid black line represents the test log-likelihood obtained from the online Gibbs sampler algorithm with a topic threshold $\epsilon = 10^{-4}$. The bottom solid red line is the test log-likelihood of the online variational algorithm, and five horizontal lines with various color in the top of the graph are the test log-likelihoods of the full Gibbs sampler algorithm after 1000, 2000, 3000, 4000 and 5000 iterations. (b) The number of topics of the online Gibbs sampler algorithm as well as those of the full Gibbs sampler algorithm with different iterations. The number of topics in online variational algorithm is 150. Since it is a predetermined value, we omit this from the graph. (Color figure online)

In contrast, the test log-likelihood of the online Gibbs sampler algorithm keeps increasing as it processes more batches, and the final test log-likelihood after processing all batches is not much smaller than those of the full Gibbs sampler algorithm.

Figure 1(b) shows the number of topics of the online Gibbs sampler algorithm as well as those of the full Gibbs sampler algorithm with different iterations. The number of topics of the online Gibbs sampler algorithm increases as more batches are processed, which indicates that the online Gibbs sampler algorithm has a capability of discovering new topics, which lacks in the online variational algorithm.

6.3   A Diagnostic Tool for Approximation

A novel feature of Theorem 1 is that there is a natural way to check the condition (10) in practice. Suppose $\theta_1, \ldots, \theta_N$ are posterior samples generated from $p(\theta|\mathbf{x}^{\mathrm{old}})$. Then, we can evaluate the degree of approximation by

$$\sup_{j \in J} | \log p(\theta_j|\mathbf{x}^{\mathrm{old}}) - \log g(\theta_j) |,$$

where J is an index set whose cardinality $|J|$ satisfies $|J|/N \geq 1 - \epsilon$ for given $\epsilon > 0$. This method would be applicable only when the form of $p(\theta|\mathbf{x}^{\mathrm{old}})$ and $g(\theta)$ are explicitly given. For the case of the HDP, however, the closed forms of $p(\theta|\mathbf{x}^{\mathrm{old}})$ and $g(\theta)$ are not available. Nonetheless, condition (10) gives a guide to check indirectly whether the approximation of $g(\theta)$ is sufficient as follows. We first choose a parameter of interest γ, which is a functional of θ and whose marginal density derived from the density $g(\theta)$ can be easily calculated. With a slight abuse of notation, we also denote the marginal density of γ derived from $p(\theta|\mathbf{x}^{\mathrm{old}})$ and $g(\theta)$ by $p(\gamma|\mathbf{x}^{\mathrm{old}})$ and $g(\gamma)$, respectively. Then, we can evaluate the degree of approximation by $\sup_{j \in J} | \log p(\gamma_j|\mathbf{x}^{\mathrm{old}}) - \log g(\gamma_j) |$, where $p(\gamma|\mathbf{x}^{\mathrm{old}})$ can be estimated by the use of any method of density estimation based on the MCMC samples of γ.

In the HDP topic model, the number of topics tends to increase as more documents are involved, so parameters that are directly affected by the number of topics are not appropriate for γ. Hence, we propose to use $\sum_{w \in I} \phi_{kw}$ as the parameter of interests, where k is a fixed integer and I is an index set with which $\sum_{w \in I} \phi_{kw} \approx 1/2$. Note that $g(\gamma)$ is a density of a beta distribution.

Figure 2 draws the plots of $\sup_{j \in J} | \log p(\gamma_j|\mathbf{x}^{\mathrm{old}}) - \log g(\gamma_j) |$ with various values of ϵ for the KOS and NIPS datasets. We set $k = 1$, $s = 1$ and $I = \{1, 3, 5, \ldots\}$. It shows that $\sup_{j \in J} | \log p(\gamma_j|\mathbf{x}^{\mathrm{old}}) - \log g(\gamma_j) |$ is small and stable until $1 - \epsilon < 0.9$, which implies that the loss of information in the gHDP approximation is around 10 % of the full posterior distribution.

Figure 3 compares $p(\gamma|\mathbf{x}^{\mathrm{old}})$ and $g(\gamma)$ for the KOS and NIPS datasets. The shapes of the two densities are similar. However, some discrepancies around the mode of the posterior distributions are observed, which suggests that considering a larger class of densities for estimating $g(\gamma)$ other than the Dirichlet distribution would be helpful to improve the degree of approximation.

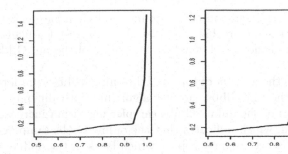

Fig. 2. Degree of approximation $\sup_{j \in J} |\log p(\gamma_j|\mathbf{x}^{\text{old}}) - \log g(\gamma_j)|$ (y-axis) with various values of $1 - \epsilon$ (x-axis) for the the KOS (left) and NIPS (right) datasets.

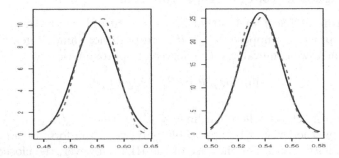

Fig. 3. Plots of $p(\gamma|\mathbf{x}^{\text{old}})$ (red dashed line) and $g(\gamma)$ (black solid line) for the KOS (left) and NIPS (right) datasets. (Color figure online)

7   Discussion

The numerical results confirmed that the proposed online Gibbs sampler algorithm is a very promising alternative to the online variational algorithm. Moreover, certain theoretical results about the degree of approximation are available, with which a diagnostic tool has been developed.

In the gHDP, the hyperparameters α_0, α_1 and β also can be learned with additional priors on them as is done for the HDP [11]. The corresponding mini-batch online Gibbs sampler algorithm can be implemented easily, where the posterior distribution of the hyperparameters after processing each batch is used as a prior for the next batch. The posterior distribution of the hyperparameters based on MCMC samples can be estimated with standard parametric or nonparametric approaches.

In this paper, we assume that the underlying probability model is the same for all the batches. When there is a structural change of the underlying model as time goes [1,9], we can reflect the structural change into the online Gibbs algorithm by modifying the gHDP accordingly while it is not possible for the full Gibbs sampler algorithm. We leave this topic as a future work.

Acknowledgement. This research was supported in part by Basic Science Research Program through the National Research Foundation of Korea (NRF) grant funded by the Korea government (MSIP) (No. 2014R1A2A2A01004496).

References

1. Ahmed, A., Xing, E.P.: Timeline: A dynamic hierarchical Dirichlet process model for recovering birth/death and evolution of topics in text stream. arXiv preprint arXiv:1203.3463 (2012)
2. Blei, D.M., Ng, A.Y., Jordan, M.I.: Latent Dirichlet allocation. J. Mach. Learn. Res. **3**, 993–1022 (2003)
3. Chae, M., Chung, H., Kang, B., Jeun, W.C., Kim, Y.: Distributed algorithms for hierarchical Dirichlet process via partially collapsed Gibbs sampler, pp. 1–8 (2015)
4. Ferguson, T.S.: A Bayesian analysis of some nonparametric problems. Ann. Stat. **1**, 209–230 (1973)
5. Heinrich, G.: Infinite LDA implementing the HDP with minimum code complexity. Technical note, Feb 170 (2011)
6. Hoffman, M.D., Blei, D.M., Bach, F.R.: Online learning for latent Dirichlet allocation. In: NIPS, vol. 2, p. 5 (2010)
7. Newman, D., Asuncion, A., Smyth, P., Welling, M.: Distributed algorithms for topic models. J. Mach. Learn. Res. **10**, 1801–1828 (2009)
8. Pitman, J.: Combinatorial Stochastic Processes: Ecole D'Eté de Probabilités de Saint-Flour XXXII-2002. Springer (2006)
9. Ren, L., Dunson, D.B., Carin, L.: The dynamic hierarchical Dirichlet process. In: Proceedings of the 25th International Conference on Machine Learning, pp. 824–831. ACM (2008)
10. Sethuraman, J.: A constructive definition of Dirichlet priors. Stat. Sinica **4**(2), 639–650 (1994)
11. Teh, Y.W., Jordan, M.I., Beal, M.J., Blei, D.M.: Hierarchical Dirichlet processes. J. Am. Stat. Assoc. **101**, 1566–1581 (2006)
12. Teh, Y.W., Kurihara, K., Welling, M.: Collapsed variational inference for HDP. In: NIPS (2007)
13. Teh, Y.W., Newman, D., Welling, M.: A collapsed variational Bayesian inference algorithm for latent Dirichlet allocation. NIPS **6**, 1378–1385 (2006)
14. Wang, C., Blei, D.M.: A split-merge MCMC algorithm for the hierarchical Dirichlet process. arXiv preprint arXiv:1201.1657 (2012)
15. Wang, C., Blei, D.M.: Truncation-free online variational inference for bayesian nonparametric models. In: Advances in Neural Information Processing Systems, pp. 413–421 (2012)
16. Wang, C., Paisley, J.W., Blei, D.M.: Online variational inference for the hierarchical Dirichlet process. In: International Conference on Artificial Intelligence and Statistics. pp. 752–760 (2011)

Structure Pattern Analysis and Cascade Prediction in Social Networks

Bolei Zhang, Zhuzhong Qian$^{(\boxtimes)}$, and Sanglu Lu

State Key Laboratory for Novel Software Technology,
Nanjing University, Nanjing, China
zhangbolei@dislab.nju.edu.cn, {qzz,sanglu}@nju.edu.cn

Abstract. As information spreads across social links, it may reach different people and become cascades in social networks. However, the elusive micro-foundations of social behaviors and the complex underlying social networks make it very difficult to model and predict the information diffusion process precisely. From a different perspective, we can often observe the interplay between information diffusion and the cascade structures. On one hand, information driven by different mechanics may evolve into diverse structures; On the other hand, different cascade structures will reach different groups people and thus affect the diffusion process.

In this paper, we explore the relationships between information diffusion and the cascade structures in social networks. By embedding the cascades in a lower dimensional space and employing spectral clustering algorithm, we find that the cascades generally evolve into five typical structure patterns with distinguishable characteristics. In addition, these patterns can be identified by observing the initial footprints of the cascades. Based on this observation, we propose to predict cascade growth with the structure patterns. The experiment results show that the accuracy of predicting both the structure and virality of cascades can be improved significantly.

1 Introduction

How does information spread in social networks? When users observe information from their neighbors in the social network, they may make decisions and share the information to their friends. Starting from some root node, the information could then spread out and become cascades with tree structures. This phenomenon of information cascades has been observed ubiquitously in social networks across various domains. With the proliferation and emergence of online social networks, understanding and predicting how will the cascades evolve have attracted enormous attentions in opinion monitoring, advertising prediction and rumor control.

In a micro scope, we can model each user's behavior to predict the process of information diffusion [2,11,13]. The diffusion can often be described as a stochastic process where the nodes spread the information according to some

P. Frasconi et al. (Eds.): ECML PKDD 2016, Part I, LNAI 9851, pp. 524–539, 2016.
DOI: 10.1007/978-3-319-46128-1_33

predefined probability. According to the social sciences, the information diffusion probability is mainly dependent on two factors: homophily and social influence [3]. Homophily is the phenomenon that people tend to build social relationship and spread information with users in similar interests or background [18], while social influence occurs when users emotions or opinions are influenced by others. Despite extensive researches in studying the process of information diffusion, the complex micro-foundations of social diffusion and unpredictability of user decisions make it very difficult to extract the diffusion models precisely.

Instead, we can often observe the interplay between information diffusion and the cascade structures. On one hand, different mechanics of information diffusion can cause different cascade structures [6,8]. For example, influence-driven cascades usually evolve as rapid, complex structures, whereas homophily-driven cascades may become simple and star-like structures [3]. On the other hand, according to the widely used information diffusion models [13], different structures of the cascades may reach different people or communities, and thus affect the spread of information. Such phenomenon motivates us to study the relationships between information diffusion and cascade structures, despite that the intricate and diverse structures of the information cascades are often very difficult to analyze [15,20].

In this paper, we dive into the structures of information cascade in social networks and explore the relationships with information diffusion empirically. We propose that the structure patterns can be predictive of the cascade growth. In our first experiment, by dimension reduction and defining similarity measure between the cascades, we find that the information cascades in social networks generally evolve into five typical structure patterns with distinguishable statistics. In addition, these patterns can be detected from the early footprints of the cascades. Based on this observation, we predict the growth of the cascades by incorporating the structure patterns. The results show that the accuracy of predicting both the structure and virality of the cascades can be improved significantly when considering the structure patterns.

Contributions. The main contributions of this paper are:

- We propose a novel method for embedding the cascades in a lower dimensional space by incorporating social influence and homophily at the same time.
- By dimension reduction and spectral clustering, we find that the information cascades generally evolve into five typical structure patterns with distinguishable characteristics.
- We propose that the structure patterns can be predictive of the growth of information cascades. The experiment results show that the accuracy of predicting the growth of the cascades can be improved significantly by using the structure patterns as new features.

Organization. The rest of this paper is organized as follows: In Sect. 2, we review the research works related to this paper. Then we will introduce some preliminaries including the data set, the theories of information diffusion and cascade structures in Sect. 3. Section 4 formally presents the method for finding the structure patterns of the cascades. In Sect. 5, we present the experiments of predicting the cascade growth. Finally, the paper is discussed and concluded in Sect. 6.

2   Related Works

There are three threads of researches related to our work: the mechanics of information diffusion, the prediction of cascade virality and analysis of cascade structures. We now introduce each of them respectively.

Modeling information diffusion. Modeling information diffusion is a central problem in studying social networks. Earlier research works proposed that information spreads like epidemics. The epidemic model generally assumes homogeneous networks [2,4] where the network is full clique and each person has the same probability for spreading the information. Typical diffusion models include the SIS (Susceptible, Infected), SIR (Susceptible, Infected, Recovered) etc. In [17], the authors proposed to fit the temporal curves of the cascade spikes using SI-like model. Compared to the biological viruses, extensive researches have been proposed to model the rumors, ideas, memes using social influence models, such as the independent cascade model [13], threshold models [11,23] and coverage models [21] etc. Following works have studied how to select the most influential nodes to maximize the spread of information diffusion [7,14]. Despite the algorithmic progress on selecting the most influential nodes, how to infer the diffusion models accurately remains a challenge.

Cascade prediction. From a macro scope, we can omit the diffusion process and predict the statistics of a cascade from its early footprints directly. Previous works usually considered the task as a regression problem [5,22] or a binary classification problem [12,24]. The growth of cascades may originate from multiple factors. In [20,22], the authors proposed to dive into the content of the information diffusion and analyze the spread of the information. The temporal features are also often used to predict the evolution of cascades [17,25]. The temporal dynamics of online usually falls into six different patterns [25]. In [17], the authors fit the model with one unified model. Recently, the structures of the social network are also taken into account to predict the evolution of cascades [1,6,19,24]. Generally, cascades that spread across multiple communities are considered to be more viral than those trapped in a single community. In comparison, we try to analyze the structure pattern of the cascade to observe how it evolves from time to time.

Cascade structures. When information starts to spread in the social network, it usually generates a tree structure. The properties of the structures have been studied over years. In [15], the authors find that the cascades in email network are usually very narrow and continually reaching people several hundred levels away. There are also implications that cascades with different topics spread in different structures [20]. For example, the political topics are usually more persistent than the conversational idioms. To quantitatively differ the structures of the cascades, the authors in [9] proposed to use wiener index to characterize the structure of information cascades. The wiener index is the average distance between each pair of nodes in the cascade, which is often used to describe the complexity of the structure of a graph. Due to the intricate and diverse structures of social cascades, it is necessary to fully understand the cascade structures and explore the relationships with information diffusion.

3   Data Set, Information Diffusion, and Structure Patterns

In this section, we first present the collected data set for analysis. Then we will briefly introduce the mechanics of information diffusion in social networks. In the last part, we show how the information could shape the structure of the cascades.

3.1   Data Collection

We choose Weibo (http://weibo.com) as the basis social network platform for analysis. Weibo is a Twitter-like micro-blog platform in China. The network can be modeled as a directed graph $G = (V, E)$ where the nodes set V represents users and the edges E represent following/follower relationships between users. For a user i, we denote the followers of i as $N(i)$. In Weibo, each user can post a message with at most 140 (Chinese) characters publicly. Once the message is posted, the neighbor users may observe the occurrence and share the message.

In Weibo, we can trace the information diffusion path by analyzing the spreading contents. For a message with content "A: xxx//@B: yyy//@C: zzz", it means that, user B first shares the origin message with content "yyy" from user C; then, user A shares B's message with content "xxx". A diffusion path as $C \rightarrow B \rightarrow A$ can be constructed from the above message. To get the full trace of each cascade, we start from the root message and get all the shared messages. Each cascade can then be modeled as a tree structure with the origin message from the root node. Figure 1 presents an example of the tree structure of a cascade.

In total, we crawled 16, 439, 997 messages from Weibo during April 1st 2015 to April 30th 2015, and extracted 33, 214 cascades with at least 10 shares. The social network has 6,738, 199 users and 11,271,789 following relationships. By adopting text clipping on the text and training a multinomial Naive Bayes classifier with a labeled data set, the messages are classified into 8 topics, including: Economy, Education, Technology, Culture, Sports, Health, Politics and Travel.

3.2   The Mechanics of Information Diffusion

Information diffusion is ubiquitous in online social networks. There are complex factors that drive the diffusion of information between people. Among them,

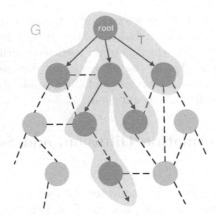

Fig. 1. Cascade example. A cascade example in the social network G, where the lines are social links between users. The cascade starts from the root node to spread. The cascade include the nodes with orange color, denoted as T. The green nodes are those in the social network but not engaged in the cascade. The information spreads only on the solid line. (Color figure online)

homophily and social influence are usually considered to be the most important ones [3]. *Homophily* refers to the tendency of individuals to associate with similar peers, such as age, gender, religion etc. In addition, researches show that homophily also contributes a lot to the information diffusion [26], since the behaviors of similar users are often correlated. Accordingly, information driven by homophily is more likely to spread to the close neighbors. In comparison, *Social influence* occurs when one's opinions are affected by others. When the information is driven by the social influence, it is more likely to spread across communities and in long distances. Thus, the information diffusion is more unpredictable in this case.

Due to complex micro-foundations of user decisions, it is often difficult to distinguish the two factors from information diffusion processes. Moreover, in most cases, both the factors may play a role in driving the spread of information. From a different perspective, we can often observe the interplay between information diffusion and the cascade structures. Thus, if we could identify the different structure patterns of information cascade, it may help us understand the mechanics of information diffusion and accordingly predict the cascade growth.

3.3   Cascade Structures

We propose that the cascade structures can often reflect the mechanics of the information diffusion. For instance, the deep and complex structure of a cascade tree means that the information is spreading in different communities and long distances, which may be the result of the social influence. Such observations motivate us to explore the structure patterns of information cascades.

In Fig. 2 we show two illustrative examples of the evolution of cascade structures in social networks. The topic of the first message belongs to Culture, which is possibly driven by homophily. As shown in the figure, it has a star-like structure, since the information mainly spreads to users not too far away. While in the second example, the message of the information is about Technology, which is more likely to be driven by social influence. There may be professional discussion and complex diffusion in this topic. But the information may not be interested to a wide range of the near neighbors, since the social links are probably built on homophily.

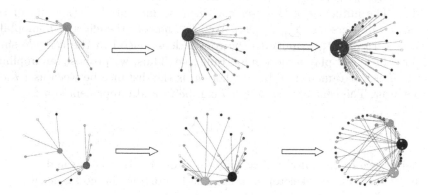

Fig. 2. Evolution of the cascades. Two illustrative examples of the cascade structures. The size of the nodes reflects the degree of the nodes in the tree.

4 Finding the Structural Patterns

In this section, we formally present the method for finding the structural patterns. We first embed the cascades into a lower dimensional space, and then apply spectral clustering to find structure patterns. Finally we will show the properties of the patterns.

4.1 Cascade Embedding

Due to the intricate structures of the information cascades, we first try to embed the cascades in a lower dimensional space to find the structure patterns. To preserve the structure characteristics, the following prerequisites should be considered: 1. The embedded cascades can distinguish cascades with different structures, such as star-like trees and the trees with deep complex structure; 2. It can reflect the virality of the cascades; 3. In the embedded cascade, users with close proximity are less important since information is more likely to spread between these users as a result of homophily.

Considering the above aspects, we propose to extract a centrality value for each of the nodes in the cascades. The basic idea is that the centrality increases with the influence of a user, but is offset by the proximity to the root of the cascade. First, we assume that the information diffusion on each edge depends on the difference of the posting time: $\Delta_{ij} = |t_i - t_j|$, which has been extensively studied in papers such as [10,16]. The weight of influence strength from i to j can be formulated as:

$$w_{ij} = \alpha_{ij} e^{-\alpha_{ij}\Delta_{ij}} = \alpha_{ij} e^{-\alpha_{ij}(t_j - t_i)}$$

Thus, an edge with larger weight indicates the information is more "viral" on the edge. The influence of the user can then be measured as the sum of the influence to all users, i.e. $\sum_{j \in N(i)} w_{ij}$. We also consider the effect of homophily: the users with close proximity to the root is less central in the cascade since the homophily may play a more important role. Thus, we propose an amplified factor of $d_i + \gamma$ parameterized by γ, where d_i is the distance between user i and the root user. The centrality of a user i can be formally represented as:

$$w_i = (d_i + \gamma) \sum_{j \in N(i)} w_{ij}$$

In practice, we empirically choose uniform value for α_{ij} as 1.0 and γ as 5.0. Finally, we extract the skeleton of a cascade by sorting the nodes according to their centralities in decreasing order as a vector $\mathbf{w}$.

4.2   Spectral Clustering

After embedding the cascade in a vector space, the distance between two embedded trees $\mathbf{w}^i$ and $\mathbf{w}^j$ can be computed with the Euclidean distance, which is defined as:

$$d(\mathbf{w}^{(i)}, \mathbf{w}^{(j)}) = \sum_t ||\mathbf{w}_t^{(i)} - \mathbf{w}_t^{(j)}||$$

where $|| \cdot ||$ is the l_2 norm. The Euclidean distance is then converted to similarity as:

$$S_{ij} = e^{-\frac{d(\mathbf{w}^{(i)}, \mathbf{w}^{(j)})}{\sigma}}$$

where σ is the standard variance of the derived distance matrix. Given the similarity measure, we use spectral clustering to find the structure patterns of the cascades, so that trees with similar structures are clustered into the same group. The spectral clustering techniques make use of the spectrum (eigenvalues) of the similarity matrix of the data to perform dimensionality reduction. The first step of the spectral clustering algorithm is to get the graph Laplacian matrix L.

$$L = D - S$$

where D is the diagonal matrix $D_{ii} = \sum_j S_{ij}$. Then, we get the first k eigenvectors of L, denote as $u_1, u_2, ..., u_k$. Let y_i be a row vector of the matrix $U = [u_1, u_2, ..., u_k]$, i.e., $y_i = [u_{1i}, u_{2i}, ..., u_{ki}]$. The results can be derived by applying k-means clustering on **y** to get the cluster results.

4.3   Structure Patterns of Cascades

In this experiment, we will apply spectral clustering algorithm on the diffusion trees to find the structure patterns. Specifically, we explore the relationship between the early structure patterns and the eventual growth of the cascades. The clustering algorithm is employed on the first $k = 30$ nodes of each cascade. The value of k is chosen empirically and the reason will be explained in Sect. 5.

To apply the clustering algorithm, the first step is to choose the number of clusters. Here we use the average silhouette score to evaluate the performance of the results of clustering algorithm. The higher silhouette score indicates better cluster results. We find that the scores are almost the same ranging from 0.1 to 0.2 with cluster number from 3 to 8. We choose 5 as the number of the clusters when the silhouette score is 0.14. Despite that the silhouette score is higher with fewer clusters, it reveals less information about the structures.

After clustering the cascades, we observe the statistics of the cascades in each cluster in Table 1. We show the average size, average depth, largest degree and average wiener index of the cascades in each cluster. Generally, higher wiener index indicates more complex structures. According to the statistics, even though the clustering algorithm is employed on the initial structures (first 30 nodes) of the cascades, the eventual statistics in the clusters are quite distinguishable from each other. For example, The cascades in C2 have almost the same size (243.14) as the cascades in C3 (292.34). However, the cascades depth (4.07) is much higher than that in C3 (2.62). In C4, the cascades have more nodes (308.30) than both C2 and C3, but the average depth (3.56) is between them. In C1 and C5, the cascades have significantly more nodes (1265.57 and 2448.77) than other clusters. And the depth in C1 (4.70) is almost the same as that in C5 (4.93). The cascades in C1 and C5 usually have a dominated node since the largest degree is very close to the size of the cascades. Moreover, the cascades in C2 have more complex structures, since the wiener index (2.28) is higher than other cascades.

Table 1. Table of statistics.

	C1	C2	C3	C4	C5
Cascade number	4537	12482	9409	3762	3024
Average size	1265.57	243.14	292.34	308.30	2448.77
Average depth	4.70	4.07	2.62	3.56	4.93
Largest degree	999.25	128.71	56.69	218.27	2054.28
Average wiener	2.14	2.28	2.01	2.09	2.13

In addition to the statistics, we also plot the cumulative distribution of the cascades size and depth in each cluster in Fig. 3. It can be observed that cascades in the same clusters tend to have the similar statistics, since the cumulative distribution curves increase steeply around the average size or depth. This also strongly implies that the early structure of a cascade may be predictive of the cascade growth.

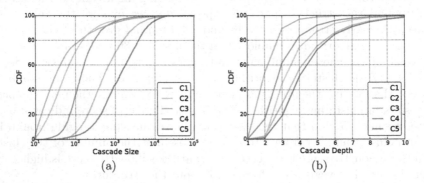

(a)   (b)

Fig. 3. CDF of cascade depth and size. In the first figure, the x-axis is the log of the cascade size. The cascades in different clusters are distinguished in different colors. (Color figure online)

Now we plot the representatives of from C1 to C5 to observe their structures in Fig. 4. The representatives are selected as the cascade which has the closest distance to all other cascades in the same cluster. As observed from the figure, the structures of cascades from different clusters vary significantly from each other. The representatives of C3 and C4 have fewer nodes and star-like structure; and the representatives of C1 and C5 have more nodes and complex structures. In particular, in C2, there are relevant as many nodes as that in C3 and C4, while the structure of C2 seems to be much more complex.

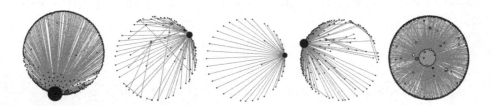

Fig. 4. Representative cascade in each cluster.

Finally, we explore the topic distributions of different clusters of the cascades. Figure 5 shows the proportion of the cascades of each cluster in the 8 topics. As observed, the cascades with topics such as Culture, Health, Travel are more likely to have wide and simple structures in C3, C4 and C5. These topics may be related to user interests or public opinions that may be driven by homophily. Meanwhile, the cascades with topics such as Economics, Education, Politics, Sports, Technology are more likely to grow as complex structures in C1 and C2. The topics are more likely to be controversial or professional that may cause social influence between users. This result is consistent with the "persistence" of topics as introduced in [20].

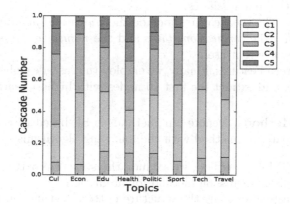

Fig. 5. The proportion of cascades in each topics.

5 Predicting Cascade Growth

By spectral clustering, we identify five patterns of cascades with distinguishable characteristics. The observation implies that the early structures can be predictive of the future growth of the cascades. In this section, we use the patterns to predict the structure and virality of the cascades growth respectively.

5.1 Experiment Setup

In predicting the cascade growth, our general method is to use the machine learning techniques to predict the labels of the target data with a set of features. Here, we use the method of *logistic regression* algorithm. Other classification methods such as random forest, SVM etc. were also tried. The results show that the performances of different classification methods do not vary a lot from each other. We use 5-fold cross validation for training and testing.

Features. We select a set of features that might be correlated with the growth of the cascade. The features are extracted from the first k nodes of the cascades on 5 dimensions: content, temporal, structural, root and structure pattern features. In total, we extract 50 features for each of the cascade.

- Content Features. The content features include some origin content related statistics: whether the content has a *link*, a *hashtag* or a *mention*; and whether the content belongs to one of the 8 topics.
- Temporal Features. In the temporal features, we first use the *average time* between the first and the last $k/2$ shares; and the temporal features also include the time elapsed between the original and the first 10 shares.
- Structural Features. In the structural features, we have the total number of *friends* of the root node in the first k shares, the total number of *uninfected friends* of the first k sharers, the *average depth* of the first k users, and the out-degree of the first 10 shares.
- Root Features. The root features include the *number of followers* of the root user, the *gender*, the *verification status* and the *number of messages* that has been posted by the root user.
- Structure Pattern Features. Finally, we employ the cascade embedding on each of the cascade, and extract the first 10 nodes with highest centralities.

Comparison Methods. Denote our method as SP-based (Structure Pattern based). We compare our method with the following algorithms:

- SP-blind: In SP-blind method, we exclude the structure pattern features and employ the logistic regression method for prediction.
- PCA-based: Instead of using the structure pattern features, in PCA method, we use PCA (principle component analysis) for dimension reduction of the cascade matrix T and get the eigenvector of the covariance matrix with largest eigenvalue as features.
- Wiener-based: Similar to PCA-based method, we compute the wiener index of the first k nodes as a feature for prediction.
- Random: We randomly guess the result to be positive or negative.

5.2 Predicting Cascade Structure

We begin by predicting the structures of the cascades. The intuition is that: a cascade that is initially wide is more likely to evolve as star-like structure, while a cascade with complex structure initially would also grow to be complex in the future.

First, we observe the average Pearson correlation between the node centralities and the wiener index of the eventual structure of the cascades. The higher absolute value of the correlation implies higher importance of the node centrality in the feature space. Figure 6 shows the changes of the feature importance of the first 6 nodes with respect to different values of k. As expected, the node centralities are positive correlated with the wiener index (or complexity) of the cascade. This is because a node with high centrality has high influence and low homophily, which is more likely to cause rapid and complex structures. The correlation grows with the number of observed nodes k in the early cascade. But when k reaches 30, the correlation has diminishing returns and stablizes around a certain value.

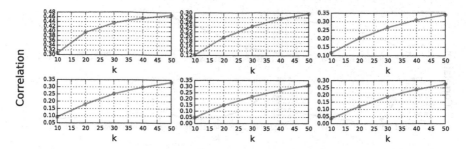

Fig. 6. Average Pearson correlation between node centralities and the cascade wiener index.

According to the above observation, we now predict whether a cascade will have a wide structure or a deep structure by observing the first k nodes of the cascade. We use wiener index to measure the structure of a cascade. Generally, a tree with wiener index approximately 2.0 is wide shallow, while a tree with higher index indicates it has complex structure. In Table 2, we show our results for predicting whether the wiener index of the cascade will evolve to be below or above a chosen value of wiener index. The results include the precision, recall, F1-score and accuracy. To avoid the imbalance of data set, we set the value of wiener index for prediction as 2.05. As presented in Table 2, our SP-based method can reach the best result in almost all cases, showing the effectiveness of the structure pattern features. In comparison to our method, the SP-blind, PCA-based and Wiener-based based method have worse results, since they did not consider the mechanics of information diffusion. According to the selection of wiener index, the random method reaches almost 50 % in every result.

Table 2. Predicting cascade structures.

Algorithm	Prec.	Rec.	F1	Accu.
SP-based	**0.775**	0.641	**0.697**	**0.722**
SP-blind	0.591	**0.643**	0.613	0.596
PCA-based	0.597	0.628	0.607	0.598
Wiener-based	0.596	0.642	0.615	0.601
Random	0.504	0.505	0.500	0.499

5.3   Predicting Cascade Virality

Another important application of the structure pattern is to predict the virality of a cascade by observing its early footprints. As shown in Fig. 3, the cascades in the same cluster tend to have the same size. Based on this observation, in

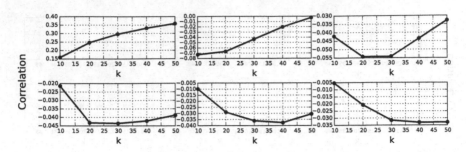

Fig. 7. Average Pearson correlation between the cascade size and node centralities.

this section, we aim to predict the virality of the cascades using the structural patterns.

We start by observing the Pearson correlation between the node centralities and the size of the eventual cascades in Fig. 7. Surprisingly, we find that from the second node, the centralities are negative correlated with the size of the cascades. This implies that the cascades are more likely to be viral if the structures of cascade are simple and wide. The centrality of the second node becomes less important when observing more nodes. This may be the reason that a node with high centrality is also common in homophily-driven cascades. And in most other cases, the absolute value of the correlation would increase with the size initial cascade.

Table 3. Predicting cascade virality with respect to different thresholds.

Threshold	Algorithm	Prec.	Rec.	F1	Accu.
100	SP-based	**0.790**	**0.783**	**0.775**	**0.809**
	SP-blind	0.742	0.732	0.727	0.753
	PCA-based	0.768	0.776	0.755	0.788
	Wiener-based	0.742	0.737	0.730	0.755
	Random	0.463	0.498	0.477	0.499
200	SP-based	**0.769**	**0.719**	**0.727**	**0.836**
	SP-blind	0.690	0.493	0.534	0.738
	PCA-based	0.698	0.465	0.517	0.738
	Wiener-based	0.674	0.485	0.529	0.742
	Random	0.327	0.495	0.395	0.499
400	SP-based	0.755	**0.545**	**0.589**	**0.847**
	SP-blind	0.731	0.368	0.426	0.817
	PCA-based	**0.771**	0.455	0.516	0.835
	Wiener-based	0.733	0.368	0.425	0.813
	Random	0.229	0.493	0.310	0.499

Next, we predict whether the cascade will reach a certain number of nodes. We try different values of the threshold as $100, 200$ and 400. Table 3 shows the results of the predictions. Obviously, our SP-based method performs the best in almost all cases. And according to the setting of the experiment, the accuracy of the Random method is almost around 50% when the threshold is 100. When the threshold increases, generally, the precision and the recall will decrease as the cascades become more and more unpredictable. However, our SP-based algorithm could still reach a high F1-measure. The significant improvement in the results validates the effectiveness of the importance of the structure pattern features. In predicting the virality, we should try to identify as many viral cascades as possible. Thus, recall is often a critical measure. And in all cases, our SP-based method can has the highest recall of all.

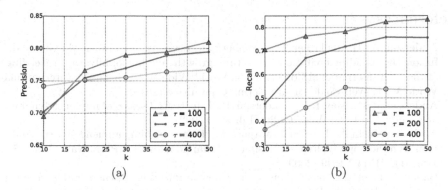

Fig. 8. Predicting virality with respect to different values of k.

Finally, we examine the effect of the prediction with respect to different values of k from 10 to 50. The results are shown in Fig. 8. As presented, the precision and recall are almost monotone with the increase of k. The values increases quickly at early steps. But when k exceeds 30, the values are not likely to grow too much. This also demonstrates that $k = 30$ is reasonable for predicting the growth of the cascades.

6   Conclusions

In this paper, we studied the structures of cascades in online social networks and explore the relationships with information diffusion. By embedding the cascades in a lower dimensional space and employing the spectral clustering algorithm, we can identify five typical patterns of the cascade structures with distinguishable characteristics. In addition, since the structure patterns of the cascades can be identified based on the early footprints, we can incorporate the structure patterns to predict the growth of information cascades.

The analysis and experiments of the results are based on the Weibo platform. We believe that the Weibo data set is comprehensive for the information cascades since it has large scale of data and includes topics across different disciplines. For the future work, first, we would empirically analyze the effect of our algorithm on other data sets; On the other hand, we would use the structure patterns to guide the micro analysis of user behaviors in social networks, so that we can predict even the individual behaviors more accurately.

Acknowledgements. This work was partially supported by the National Natural Science Foundation of China under Grant Nos. 61321491, 61472181, 91218302, the Natural Science Foundation of Jiangsu Province of China under Grant No. BK20151392, Jiangsu Key Technique Project (industry) under Grant No. BE2013116, EU FP7 IRSES MobileCloud Project under Grant No. 612212, the Program B for Outstanding PhD candidate of Nanjing University, and the Collaborative Innovation Center of Novel Software Technology and Industrialization.

References

1. Anderson, A., Huttenlocher, D., Kleinberg, J., Leskovec, J., Tiwari, M.: Global diffusion via cascading invitations: structure, growth, and homophily. In: Proceedings of the 24th International Conference on World Wide Web, pp. 66–76. International World Wide Web Conferences Steering Committee (2015)
2. Anderson, R.M., May, R.M., Anderson, B.: Infectious diseases of humans: dynamics and control, vol. 28. Wiley Online Library (1992)
3. Aral, S., Muchnik, L., Sundararajan, A.: Distinguishing influence-based contagion from homophily-driven diffusion in dynamic networks. Proc. Natl. Acad. Sci. **106**(51), 21544–21549 (2009)
4. Bailey, N.T., et al.: The mathematical theory of infectious diseases and its applications. Charles Griffin & Company Ltd., 5a Crendon Street, High Wycombe, Bucks HP13 6LE (1975)
5. Bakshy, E., Karrer, B., Adamic, L.A.: Social influence and the diffusion of user-created content. In: Proceedings of the 10th ACM Conference on Electronic Commerce, pp. 325–334. ACM (2009)
6. Budak, C., Agrawal, D., El Abbadi, A.: Structural trend analysis for online social networks. Proc. VLDB Endowment **4**(10), 646–656 (2011)
7. Chen, W., Wang, Y., Yang, S.: Efficient influence maximization in social networks. In: Proceedings of the 15th ACM SIGKDD International Conference on Knowledge Discovery and Data Mining, pp. 199–208. ACM (2009)
8. Cheng, J., Adamic, L., Dow, P.A., Kleinberg, J.M., Leskovec, J.: Can cascades be predicted? In: Proceedings of the 23rd International Conference on World Wide Web, pp. 925–936. ACM (2014)
9. Goel, S., Watts, D.J., Goldstein, D.G.: The structure of online diffusion networks. In: Proceedings of the 13th ACM Conference on Electronic Commerce, pp. 623–638. ACM (2012)
10. Gomez Rodriguez, M., Leskovec, J., Krause, A.: Inferring networks of diffusion and influence. In: Proceedings of the 16th ACM SIGKDD International Conference on Knowledge Discovery and Data Mining, pp. 1019–1028. ACM (2010)
11. Granovetter, M.: Threshold models of collective behavior. Am. J. Sociol. **83**, 1420–1443 (1978)
12. Hong, L., Dan, O., Davison, B.D.: Predicting popular messages in twitter. In: Proceedings of the 20th International Conference Companion on World Wide Web, pp. 57–58. ACM (2011)

13. Kempe, D., Kleinberg, J., Tardos, É.: Maximizing the spread of influence through a social network. In: Proceedings of the Ninth ACM SIGKDD International Conference on Knowledge Discovery and Data Mining, pp. 137–146. ACM (2003)
14. Leskovec, J., Krause, A., Guestrin, C., Faloutsos, C., VanBriesen, J., Glance, N.: Cost-effective outbreak detection in networks. In: Proceedings of the 13th ACM SIGKDD International Conference on Knowledge Discovery and Data Mining, pp. 420–429. ACM (2007)
15. Liben-Nowell, D., Kleinberg, J.: Tracing information flow on a global scale using internet chain-letter data. Proc. Natl. Acad. Sci. **105**(12), 4633–4638 (2008)
16. Malmgren, R.D., Stouffer, D.B., Motter, A.E., Amaral, L.A.: A poissonian explanation for heavy tails in e-mail communication. Proc. Natl. Acad. Sci. **105**(47), 18153–18158 (2008)
17. Matsubara, Y., Sakurai, Y., Prakash, B.A., Li, L., Faloutsos, C.: Rise and fall patterns of information diffusion: model and implications. In: Proceedings of the 18th ACM SIGKDD International Conference on Knowledge Discovery and Data Mining, pp. 6–14. ACM (2012)
18. McPherson, M., Smith-Lovin, L., Cook, J.M.: Birds of a feather: Homophily in social networks. Ann. Rev. Sociol. **27**, 415–444 (2001)
19. Nematzadeh, A., Ferrara, E., Flammini, A., Ahn, Y.Y.: Optimal network modularity for information diffusion. Phys. Rev. Lett. **113**(8), 088701 (2014)
20. Romero, D.M., Meeder, B., Kleinberg, J.: Differences in the mechanics of information diffusion across topics: idioms, political hashtags, and complex contagion on twitter. In: Proceedings of the 20th International Conference on World Wide Web, pp. 695–704. ACM (2011)
21. Singer, Y.: How to win friends and influence people, truthfully: influence maximization mechanisms for social networks. In: Proceedings of the Fifth ACM International Conference on Web Search and Data Mining, pp. 733–742. ACM (2012)
22. Tsur, O., Rappoport, A.: What's in a hashtag?: content based prediction of the spread of ideas in microblogging communities. In: Proceedings of the Fifth ACM International Conference on Web Search and Data Mining, pp. 643–652. ACM (2012)
23. Ugander, J., Backstrom, L., Marlow, C., Kleinberg, J.: Structural diversity in social contagion. Proc. Natl. Acad. Sci. **109**(16), 5962–5966 (2012)
24. Weng, L., Menczer, F., Ahn, Y.Y.: Virality prediction and community structure in social networks. Sci. Rep. **3**, 2522 (2013)
25. Yang, J., Leskovec, J.: Patterns of temporal variation in online media. In: Proceedings of the Fourth ACM International Conference on Web Search and Data Mining, pp. 177–186. ACM (2011)
26. Yavaş, M., Yücel, G.: Impact of homophily on diffusion dynamics over social networks. Soc. Sci. Comput. Rev. **32**, 0894439313512464 (2014)

Graph-Margin Based Multi-label Feature Selection

Peng Yan and Yun Li[✉]

School of Computer Science and Technology,
Nanjing University of Posts and Telecommunications,
Wenyuanlu 9, Nanjing 210023, China
yanpeng9008@hotmail.com, liyun@njupt.edu.cn

Abstract. Since instances in multi-label problems are associated with several labels simultaneously, most traditional feature selection algorithms for single label problems are inapplicable. Therefore, new criteria to evaluate features and new methods to model label correlations are needed. In this paper, we adopt the graph model to capture the label correlation, and propose a feature selection algorithm for multi-label problems according to the graph combining with the large margin theory. The proposed multi-label feature selection algorithm GMBA can efficiently utilize the high order label correlation. Experiments on real world data sets demonstrate the effectiveness of the proposed method. The codes of the experiment of this paper are available at https://github.com/Faustus-/ECML2016-GMBA.

Keywords: Feature selection · Multi-label learning · Graph · Margin

1 Introduction

Multi-label learning studies the problem in which each instance is associated with a set of labels simultaneously. It usually occurs in text categorization, automatic annotation and bioinformatics, etc. [24]. For example, each music in emotions [15] data set can be associated with at most six different emotion tags simultaneously. A straightforward method to solve the multi-label problem is to decompose the problem into a series of single label binary classification problems, such as Binary Relevance [2] and ML-kNN [23]. However, this strategy neglects the label correlation which is usually helpful for improving the performance of a multi-label learning algorithm. To complement this, various multi-label learning algorithms with the consideration of label correlation have been proposed, such as [4,7,8,11,13,19,22]. According to the utilization of label correlation, these algorithms can be divided into three orders [24]: (a) the first order algorithms predict labels for an unseen instance one by one. They are very simple while neglecting label correlation [2,23]. (b) the second order algorithms consider pairwise relation between labels, which usually leads to a label ranking problem [4,8]. (c) the high order algorithms capture more complex correlation between labels, but they are computationally expensive [7,11,13,19,22].

© Springer International Publishing AG 2016
P. Frasconi et al. (Eds.): ECML PKDD 2016, Part I, LNAI 9851, pp. 540–555, 2016.
DOI: 10.1007/978-3-319-46128-1_34

Similar to other machine learning tasks, multi-label learning also suffers from the curse of dimensionality. Redundant and irrelevant features make data intractable, resulting in unreliable model and degraded learning performance. Feature selection is an efficient and popular technique to reduce dimensionality. Several feature selection algorithms for the multi-label problem have been presented. For example, feature selection algorithms for multi-label naive bayes classifier and Rank-SVM classifier are introduced in [5,21] respectively. These multi-label feature selection algorithms belong to the wrapper model [14], which evaluates features according to predictive results of the specified learning algorithm, thus they share bias of the learning algorithm and it is prohibitively expensive to run for data with a large number of features. In [6], two classic single-label feature selection algorithms, F-Statistic and ReliefF, are extended to handle multi-label problems. These algorithms belong to the filter model [14], which evaluates features by measuring the statistics of a multi-label data set. Algorithms belonging to the filter model are independent of specified classifiers and more flexible than those belonging to the wrapper model.

In this paper, a graph-margin based multi-label feature selection algorithm (GMBA) is proposed. GMBA firstly describes multi-label data with a graph, which has good discrimination capability and shares similar expression capability to the hypergraph applied in [13,19]. Then, it measures features based on the graph combining with the large margin theory. Since GMBA evaluates features according to the graph derived from the training data, it is independent of a specified learning algorithm and belongs to the filter model. We will introduce GMBA in the following order. In Sect. 2, we define a similarity measure for multi-label instances and describe multi-label data by a graph. The discrimination capability and expression capability of the graph are also discussed in this section. In Sect. 3, we define a margin for multi-label data and derive GMBA depending on the graph combining with the margin. In addition, experimental results on real world data sets are reported in Sect. 4 and paper concludes in Sect. 5.

Notations. Before introducing the algorithm, we will give the notations in this paper. n, D and Q denote the number of training instances, the data dimensionality, and the number of labels, respectively. F^d denotes the dth feature and l^q denotes the qth label, where $1 \leq d \leq D$ and $1 \leq q \leq Q$. n_q denotes the number of training samples associated with l^q. (x_i, y_i) denotes the ith instance in the training data.

$x_i = \left(x_i^1, ..., x_i^d, ..., x_i^D \right)$ denotes the features of the ith instance in the training data, where x_i^d denotes the dth component of the ith instance, or the ith instance has value x_i^d for F^d.

$x^d = \left(x_1^d, ..., x_i^d, ..., x_n^d \right)^T$ denotes a feature vector of the dth feature. The superscript T means the transpose of a vector or matrix.

$y_i = \left(y_i^1, ..., y_i^q, ..., y_i^Q \right)$ denotes the relationship between labels and the ith instance in the training data. If the ith instance is associated with l^q then $y_i^q = 1$, or $y_i^q = 0$. For single-label problems, there is a constraint that $|y_i| = 1$, $1 \leq i \leq n$, where $|\cdot|$ denotes the 1-norm of the vector.

$s_{(i,i')}$ denotes the similarity between the ith and i'th instances.

$\mathbf{G} = (\mathbf{V}, \mathbf{E})$ denotes a graph, where $\mathbf{V}$ and $\mathbf{E}$ denote the vertex set and the edge set of the graph, respectively. $\mathbf{A}_G$ denotes the adjacent matrix of $\mathbf{G}$, $\mathbf{D}_G$ denotes the degree matrix of $\mathbf{G}$. $\mathbf{L}_G = \mathbf{D}_G - \mathbf{A}_G$ is the corresponding Laplacian matrix.

$[\pi]$ returns 1 if predicate π holds, and 0 otherwise.

$|\cdot|$ and $\|\cdot\|$ returns 1-norm and 2-norm respectively.

ω is a weight vector of features.

2   Graph Model for Multi-label Data

2.1   Graph Definition

Graph is a widely used model for its powerful expression capability. For example, well-known page rank and image segmentation algorithm in [3] are based on the graph model. In this paper, we adopt the graph to capture the correlation between labels and instances for multi-label data.

Suppose, in a graph, each vertex $v_i \in \mathbf{V}$ represents an instance and an edge $e_{(i,i')} \in \mathbf{E}$ connecting two vertexes denotes the similarity of the corresponding instances, then a simple undirected graph $\mathbf{G} = (\mathbf{V}, \mathbf{E})$ can be built to model the correlation between instances. The key of building the graph depends on how one measures the instance similarity. For a single-label problem, the similarity between two instances x_i and $x_{i'}$ are usually defined as Eq. 1 [25].

$$s_{single}(i, i') = \begin{cases} \frac{1}{n_q}, & y_i^q = y_{i'}^q = 1 \\ 0, & otherwise \end{cases} \tag{1}$$

which means that instances in the same class share the same similarity, while similarity between instances from different classes is 0. However, when it comes to multi-label problems, an instance is associated with several labels (classes) simultaneously and it is ambiguous to compare the belongingness of two different instances. Therefore, Eq. (1) is not suitable when solving multi-label problems and we define Eq. 2 to measure the similarity between two multi-label instances.

$$s_{multi}(i, i') = \begin{cases} \frac{\sum_{q=1}^{Q} n_q \cdot [y_i^q = 1 \wedge y_{i'}^q = 1]}{\sum_{q=1}^{Q} n_q \cdot [y_i^q = 1 \vee y_{i'}^q = 1]}, & \sum_{q=1}^{Q} n_q \cdot [y_i^q = 1 \vee y_{i'}^q = 1] \neq 0 \\ 0, & otherwise \end{cases} \tag{2}$$

In Eq. 2, the numerator counts the labels two instances shared, and the denominator counts the labels at least one of the two instances associated with. n_q is the number of training samples associated with l_q and it is applied as a weight to tune the importance of different labels. Equation 2 is a variation of the Jaccard similarity, which measures the ratio of the size of intersection and the size of union for two sets. Then, the multi-label data can be represented as a graph using an adjacent matrix $\mathbf{A}_G$ definded in Eq. 3, where $A_G(i, i')$ is the element in the ith row, i'th column of $\mathbf{A}_G$.

$$A_G(i, i') = \begin{cases} s_{multi}(i, i'), & i \neq i' \\ 0, & otherwise \end{cases} \tag{3}$$

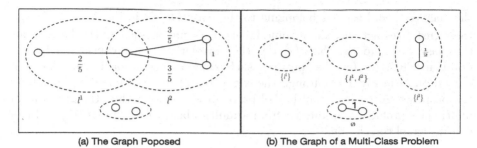

(a) The Graph Poposed   (b) The Graph of a Multi-Class Problem

Fig. 1. (a) Our proposed graph for multi-label data, (b) The graph for multi-class data transformed from the multi-label data. The edges in graphs are denoted with solid lines and circles are vertexes. The fractions on the edges represent the similarity weight. Circles fallen in the same ellipse (dash line) represent instances associated with the same label/class. The circle fallen in the intersection of two ellipses means the instance is associated with two labels simultaneously. And the two instances associated with no labels are put in the ellipse below.

2.2 Discrimination Capability

To explain the discrimination capability of the proposed graph, an example is presented below. Assuming that there are Q different labels, we have $y_i \in \{0, 1\}^Q$. Without loss of generality, we set $Q = 2$ and two labels are named l^1 and l^2. We also assume that a multi-label training data set consists of one instance associated with l^1, two instances associated with l^2, one instance associated with l^1 and l^2 simultaneously and two instances associated with no labels. The proposed graph to describe these instances is given in Fig. 1(a). Then, if one can split the graph into different parts (such as the ellipses of dash line), instances associated with different labels will be discriminated. Hence multi-label instances are discriminable in the proposed graph. Moreover, some off the shelf algorithms can be applied to finish this task, such as normalized cut [12], ration cut [18], etc.

In addition, the discrimination capability of the proposed graph is similar to the one derived from label power set algorithms as in [16,17], while the proposed graph is smoother and can capture label correlation. More specifically, a label power set algorithm usually transforms a multi-label problem into a multi-class problem in which each class corresponds to a label power set. For the multi-label problem mentioned above, a label power set algorithm will transform it into a multi-class problem with 4 different classes: $\emptyset$, $\{l^1\}$, $\{l^2\}$ and $\{l^1, l^2\}$, and each instance is associated with one class. Since a multi-class problem belongs to the single-label learning problem, the similarity between instances can be measured by Eq. 1. The resulting graph is shown in Fig. 1(b), which includes 4 unconnected subgraphs. The partitions of the graph (ellipses of dash line) are similar to the ones in Fig. 1(a), hence they have similar discrimination capability. However, in multi-class problems, the similarity between instances from different classes is 0, and there are no edges connecting them, such as the instance belonging to the class $\{l^1\}$ and the one belonging to the class $\{l^1, l^2\}$ in Fig. 1(b). Although these instances actually share some labels in common, such as the label l^1 for the class $\{l^1\}$ and the class $\{l^1, l^2\}$, the correlation is not considered by the graph in Fig. 1(b). On the contrary, such kind of correlation is considered in our graph as in Fig. 1(a) through the edges weight between 0 and 1. Therefore, the proposed graph for a multi-label problem is smoother than the graph for a multi-class problem transformed from a multi-label problem in [16,17] and can capture label correlation.

2.3 Expression Capability

Though the proposed graph in Sect. 2.1 is a simple-graph, it has similar expression capability to a hypergraph, which has been successfully applied to capture high order label correlation in [13,19].

Different from edges in a simple-graph, an edge, which is called hyperedge, in a hypergraph connects more than two vertexes simultaneously. Hence multi-label data can be described by a hypergraph as follows: in a hypergraph $\mathbf{G}_H = (\mathbf{V}_H, \mathbf{E}_H)$, each vertex $v_i \in \mathbf{V}_H$ corresponds to an instance in the multi-label data set, each hyperedge $e_q \in \mathbf{E}_H$ is a subset of $\mathbf{V}_H$, where $e_q = \{v_i \mid y_i^q = 1, 1 \leq i \leq n\}$. The degree of each hyperedge $d(e_q)$ is defined as the number of vertexes on that hyperedge, namely n_q, and we may set the weight of a edge, $w(e_q)$, equals to its degree.

If we apply Clique Expansion [1,13,19] to expand the hypergraph above, we obtain a simple-graph $\mathbf{G}_C = (\mathbf{V}_C, \mathbf{E}_C)$, where $\mathbf{V}_C = \mathbf{V}_H$ and $\mathbf{E}_C = \{e_{(i,i')} \mid v_i \in e_q \wedge v_{i'} \in e_q, e_q \in \mathbf{E}_H\}$. The weight of $e_{(i,i')}$ is defined as Eq. 4.

$$w\left(e_{(i,i')}\right) = \sum_{v_i \in e_q \wedge v_{i'} \in e_q, e_q \in \mathbf{E}_H} w(e_q) = \sum_{q=1}^{Q} n_q \cdot [y_i^q = 1 \wedge y_{i'}^q = 1] \qquad (4)$$

Normalizing it to obtain Eq. 5, we find that Eq. 5 is the same to the similarity defined in Eq. 2

$$\hat{w}\left(e_{(i,i')}\right) = \frac{\sum_{v_i \in e_q \wedge v_{i'} \in e_q, e_q \in \mathbf{E}_H} w(e_q)}{\sum_{v_i \in e_q \vee v_{i'} \in e_q, e_q \in \mathbf{E}_H} w(e_q)} = \frac{\sum_{q=1}^{Q} n_q \cdot [y_i^q = 1 \wedge y_{i'}^q = 1]}{\sum_{q=1}^{Q} n_q \cdot [y_i^q = 1 \vee y_{i'}^q = 1]} \qquad (5)$$

Thus our proposed graph is the same to the simple-graph expanded from a hypergraph by Clique Expansion. According to [1,13], both the hypergraph and the expanded simple-graph, as well as the proposed graph, can capture similar high order correlation and therefore they share similar expression capability for multi-label data.

3   Graph-Margin Based Multi-label Feature Selection (GMBA)

In Sect. 2, we propose a discriminative graph to describe multi-label data. According to the similarity measure defined in Eq. 2, the graph reflects the relations of data in label space. However, these relations in label space are usually different from the one in feature space. We will illustrate this case in Fig. 2(a) and (b). For an instance denoted by star in Fig. 2(a), its several nearest neighbors in label space are represented by squares. That is to say, the similarity measured by Eq. 2 between a square and the star is greater than a threshold s_{min}, and these squares are the closest instances to the star in the proposed graph as in Fig. 2(a). However, if we estimate similarities among instances in feature space, such as using a radial basis function, an instance represented by triangle could be more similar (closer) to the star than squares. This means that the graph built in feature space as in Fig. 2(b) is inconsistent with the graph in label space as in Fig. 2(a).

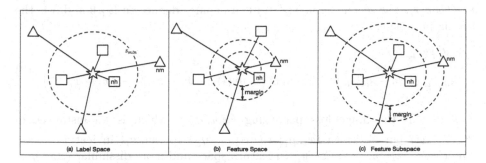

Fig. 2. A comparison of graphs built in different spaces. Each star, square or triangle represents an instance. The edges connect two different shapes denote the similarity between them. The shorter an edge is, the more similar two instances are. We omit the edges that do not connect with the star.

Furthermore, for classification problems, the target of a classifier is using features to divide instances into different classes, which means that we have to use features to predict the partitions of the graph in label space. Although the graph built in label space is discriminative as analyzed in Sect. 2.2, an inconsistent counterpart in feature space does not maintain its discrimination power and may lead to wrong partition. Thus we propose a multi-label feature selection algorithm GMBA, which will choose a subset of features that the graph built in this feature subspace, as in Fig. 2(c), is similar to the one in label space, as Fig. 2(a). In addition, a margin, as depicted in Fig. 2(c), is applied in GMBA to guarantee the generalization capability of the selected features.

3.1 Loss Function

To evaluate the inconsistency described above, we design a loss function based on margin. Firstly, we apply $sim\,(i)$ and $dissim\,(i)$ to represent the instance subsets similar and dissimilar to $(\boldsymbol{x}_i, \boldsymbol{y}_i)$ in label space respectively. They are described in Eqs. 6 and 7.

$$sim\,(i) = \{(\boldsymbol{x}_{i'}, \boldsymbol{y}_{i'}) \mid s_{multi}\,(i, i') \geq s_{min}, 1 \leq i' \leq n \text{ and } i \neq i'\} \qquad (6)$$

$$dissim\,(i) = \{(\boldsymbol{x}_{i'}, \boldsymbol{y}_{i'}) \mid s_{multi}\,(i, i') < s_{min}, 1 \leq i' \leq n \text{ and } i \neq i'\} \quad (7)$$

where s_{min} is a given threshold and $s_{multi}\,(i, i')$ is the similarity defined in Eq. 2. Then, the loss function is designed in Eq. 8 to evaluate the inconsistency between the graph in label space and the one in feature space for $(\boldsymbol{x}_i, \boldsymbol{y}_i)$.

$$Loss\,(i) = \sum_{i' \in neighbor(i)} s_{multi}\,(i, i') \|\boldsymbol{x}_i - \boldsymbol{x}_{i'}\|^2 + \lambda \sum_{i'' \in dissim(i)} \delta\,(i', i'') \quad (8)$$

where $neighbor\,(i)$ denotes a instance subset with k instances those are both nearest to $(\boldsymbol{x}_i, \boldsymbol{y}_i)$ in feature space and belong to $sim\,(i)$. The first term of Eq. 8 penalizes large distance between $(\boldsymbol{x}_i, \boldsymbol{y}_i)$ and its neighbors $(\boldsymbol{x}_{i'}, \boldsymbol{y}_{i'})$ in $neighbor\,(i)$. The second term $\delta\,(i', i'')$ is a penalty defined in Eq. 9 and λ is the tuning parameter.

$$\delta\,(i', i'') =$$
$$(s_{multi}\,(i, i') - s_{multi}\,(i, i'')) \cdot \max\left(0, m\,(i) + \|\boldsymbol{x}_i - \boldsymbol{x}_{i'}\|^2 - \|\boldsymbol{x}_i - \boldsymbol{x}_{i''}\|^2\right) \quad (9)$$

$\delta\left(i', i''\right)$ is the hinge loss penalizing $(\boldsymbol{x}_{i''}, \boldsymbol{y}_{i''})$, which is an instancec in $dissmiss\,(i)$ but closer to $(\boldsymbol{x}_i, \boldsymbol{y}_i)$ than $(\boldsymbol{x}_{i'}, \boldsymbol{y}_{i'})$ to $(\boldsymbol{x}_i, \boldsymbol{y}_i)$ in feature space. The closer $(\boldsymbol{x}_{i''}, \boldsymbol{y}_{i''})$ to $(\boldsymbol{x}_i, \boldsymbol{y}_i)$ in feature space and more dissimilar $(\boldsymbol{x}_{i''}, \boldsymbol{y}_{i''})$ to $(\boldsymbol{x}_i, \boldsymbol{y}_i)$ in label space, the larger the penalty. $m\,(i)$ is the margin defined in Eq. 10, where nh (i) and nm (i) are the nearest instances from $sim\,(i)$ and $dissim\,(i)$ respectively to the $(\boldsymbol{x}_i, \boldsymbol{y}_i)$ in feature spaces.

$$m\,(i) = \left| \|\boldsymbol{x}_i - \boldsymbol{x}_{\text{nh}(i)}\|^2 - \|\boldsymbol{x}_i - \boldsymbol{x}_{\text{nm}(i)}\|^2 \right| \quad (10)$$

We will illustrate the penalty defined Eq. 9 for the case depicted in Fig. 2(b). Assuming that the star represents $(\boldsymbol{x}_i, \boldsymbol{y}_i)$, the margin $m\,(i)$ is the absolute value of the square Euclidean distance between the square marked nh and the star minus the square Euclidean distance between the triangle marked nm and the star. If the square Euclidean distance between any triangle and the star is smaller than the square Euclidean distance between a square and the star plus this margin, it will be penalized by Eq. 9.

3.2   Feature Ranking

Based on the loss function in Eq. 8, one can evaluate the inconsistency between the graph in label space and the one in feature space by summing up the loss of all training data as depicted in Eq. 11. The smaller Eq. 11 is, the more consistent two graphs are. In addition, for feature selection, it is key to find a feature subspace that minimize Eq. 11.

$$Loss\,(\mathbf{G}) = \sum_{i=1}^{n} Loss\,(i) \quad (11)$$

However, it suffers from the complexity of $O\left(2^D\right)$ to find the best subspace for Eq. 11. As a result, according to [9,10], we evaluate the fitness of features by a weight vector $\boldsymbol{\omega}$ and find the best $\boldsymbol{\omega}$ by gradient descent method. Specifically, searching for the best $\boldsymbol{\omega}$ can be formulated as Eq. 12

$$\min_{\boldsymbol{\omega}} Loss\left(\boldsymbol{\omega}, \mathbf{G}\right) = \min_{\boldsymbol{\omega}} \sum_{i=1}^{n} Loss\left(\boldsymbol{\omega}.i\right) \tag{12}$$

where

$$Loss\left(\boldsymbol{\omega}.i\right) = \sum_{i'\in neighbor(i)} s_{multi}\left(i, i'\right) \left\|\boldsymbol{x}_i - \boldsymbol{x}_{i'}\right\|_{\boldsymbol{\omega}}^2 + \lambda \sum_{i''\in dissim(i)} \delta\left(\boldsymbol{\omega}, i', i''\right) \tag{13}$$

$$\delta\left(\boldsymbol{\omega}, i', i''\right)$$
$$= \left(s_{multi}\left(i, i'\right) - s_{multi}\left(i, i''\right)\right) \cdot \max\left(0, m\left(i\right) + \left\|\boldsymbol{x}_i - \boldsymbol{x}_{i'}\right\|_{\boldsymbol{\omega}}^2 - \left\|\boldsymbol{x}_i - \boldsymbol{x}_{i''}\right\|_{\boldsymbol{\omega}}^2\right) \tag{14}$$

and $\left\|\boldsymbol{z}\right\|_{\boldsymbol{\omega}} = \sqrt{\sum_{d=1}^{D}\left(\omega^d z^d\right)^2}$.

Then Eq. 12 can be solved by the gradient descent and the algorithm is summarized as follows.

Step 1: Initialize $\boldsymbol{\omega} = \left(1, 1, 1, ..., 1\right)$, and set the number of iterations I.
Step 2: For i =1, 2, ... , I.
(a) Pick up an instance $\left(\boldsymbol{x}_i, \boldsymbol{y}_i\right)$, and find $sim\left(i\right)$ and $dissim\left(i\right)$ according to Eqs. 2, 6 and 7.
(b) Find k nearest instances to $\left(\boldsymbol{x}_i, \boldsymbol{y}_i\right)$ in feature space from $sim\left(i\right)$ as $neighbor\left(i\right)$.
(c) Find nh(i) and nm(i) from $sim\left(i\right)$ and $dissim\left(i\right)$ respectively.
(d) Calculate $m\left(i\right)$ according to Eq. 10
(e) For d =1, 2, ... , D

$$\nabla^d = 2\omega^d \sum_{i'\in neighbor(i)} s_{multi}\left(i, i'\right) \left\|x_i^d - x_{i'}^d\right\|^2 + \lambda \sum_{i'\in dissim(i)} \frac{\partial\delta\left(\boldsymbol{\omega}, i', i''\right)}{\partial\omega^d},$$

where $\frac{\partial\delta\left(\boldsymbol{\omega}, i', i''\right)}{\partial\omega^d}$ is the partial derivative of $\delta\left(\boldsymbol{\omega}, i', i''\right)$ given in Eqs. 15 and 16

$$\frac{\partial\delta\left(\boldsymbol{\omega}, i', i''\right)}{\partial\omega^d} = \begin{cases} 0, & m\left(i\right) + \left\|\boldsymbol{x}_i - \boldsymbol{x}_{i'}\right\|^2 < \left\|\boldsymbol{x}_i - \boldsymbol{x}_{i''}\right\|^2 \\ diff\left(d\right), & otherwise \end{cases} \tag{15}$$

$$diff\left(d\right) = 2\omega^d \left(s_{multi}\left(i, i'\right) - s_{multi}\left(i, i''\right)\right) \left(\left\|x_i^d - x_{i'}^d\right\|^2 - \left\|x_i^d - x_{i''}^d\right\|^2\right) \tag{16}$$

(f)$\boldsymbol{\omega} = \boldsymbol{\omega} - \beta\boldsymbol{\nabla}/\left\|\boldsymbol{\nabla}\right\|$, where β is a decay factor.
Step 3: Ranking features based on $\boldsymbol{\omega}$. The greater the ω^d, the better the F^d.

4   Experiments

To demonstrate the effectiveness of the proposed GMBA, we empirically compare the GMBA with the multi-label F-Statistic (MLFS)[6] and the multi-label ReliefF (MLRF) [6].

In addition, spectral feature selection framework (SPEC) [25] is an algorithm which selects features based on the graph structure for single label problems. It measures features according to Eq. 17.

$$\phi\left(F^d\right) = \left(\hat{x}^d\right)^T \mathcal{L}_G \hat{x}^d = \sum_{1 \le i,i' \le n} \frac{s_{single}\left(i, i'\right)}{\sqrt{d\left(i\right) d\left(i'\right)}} \left\| \hat{x}_i^d - \hat{x}_{i'}^d \right\|^2 \tag{17}$$

where $d(i)$ is the degree of vertex v_i, $\hat{x}^d = \frac{D^{\frac{1}{2}} x^d}{\left\| D^{\frac{1}{2}} x^d \right\|}$ is the normalized feature vector and $\mathcal{L}_G = D_G^{-\frac{1}{2}} L_G D_G^{-\frac{1}{2}}$ is the normalized Laplacian matrix. The smaller the Eq. 17, the better the F^d. We adapt it to multi-label problems by replacing the $s_{single}(i, i')$ with the proposed similarity $s_{multi}(i, i')$, so that it will select features consistent with the proposed graph structure for multi-label data.

4.1 Data Sets

Eight benchmark multi-label data sets from different domains are used for experiments, which are downloaded from MULAN[1]. Details about data sets are listed in Table 1. All numerical features are normalized with zero mean and unit variance in experiments. Features with variance 0 are eliminated.

Table 1. Summary of 8 benchmark data sets

Name	Instance	Features	Labels	Domain	Name	Instance	Features	Labels	Domain
bibtex	7395	1836	159	text	mediamill	43907	120	101	video
emotions	593	72	6	music	medical	978	1449	45	text
enron	1702	1001	53	text	scene	2407	294	6	image
genebase	662	1186	27	biology	yeast	2417	103	14	biology

4.2 Classifiers and Parameters

Binary Relevance [2] (1_{st} order algorithm) and Classifier Chain [11] (high order algorithm) are used as multi-label learning strategy respectively, 3-Nearest Neighbor (3-NN) classifier in scikit-learn[2] is applied as the base classifier. Number of neighbors for MLRF and $neighbor(i)$ in GMBA are set 3. The threshold s_{min} and tuning parameter λ are 1. The number of iterations I equals to the number of training data n. The decay factor β is 0.9. Experiments[3] are carried on under the environment of Python 2.7.

---

[1] http://mulan.sourceforge.net/.
[2] http://scikit-learn.org/stable/.
[3] Codes can be acquired at https://github.com/Faustus-/ECML2016-GMBA.

4.3 Evaluations

Three different measurements [24], i.e., Hamming loss ($\downarrow$), micro ($\uparrow$) and macro ($\uparrow$) F1-Measure, are applied to validate the performance of the selected features for multi-label learning. ($\downarrow$) denotes the smaller the better, while ($\uparrow$) denotes the larger the better. Except for mediamill and bibtex, all results reported in this paper are the average of 10-cross validation. Since the big size of mediamill and bibtex, we randomly select 1800 instances and other 10 percent of total instances for training and testing respectively. The results reported are the average of 10 trials of experiments.

4.4 Results

Experimental results are shown in Figs. 3, 4, 5 and 6. For space limitation, we display the Hamming Loss for the bibtex, emotions, enron and genebase, macro F1-Measure metrics for mediamill, medical, scene and yeast when the multi-label learning strategy is Binary Relevance. We also display the Hamming Loss for mediamill, medical, scene and yeast, macro F1-Measure metrics for bibtex, emotions, enron and genebase when the multi-label learning strategy is Classifier Chain. Complete results of micro F1-Measure metrics are displayed in Figs. 5 and 6.

Experimental results show that features selected by proposed GMBA obtain better classifying performance than others in most cases. For emotions and scene data sets, all algorithms achieve similar performance, which might result from the fact that there are only 6 labels, causing a weak discrimination power of graphs built in the label space. In addition, GMBA and the adapted SPEC are suitable for more data sets than MLFS and MLRF, since the performance of MLFS and MLRF vary from different data sets.

5 Discussions and Conclusions

According to experimental results, GMBA performs better than other algorithms, and both GMBA and SPEC are suitable for more data sets than MLFS and MLRF. In addition, while GMBA and SPEC all aim to find a feature subset that the graph built in this subspace is consistent with the graph built in label space, GMBA is superior to the SPEC in most cases. This results from the margin we applied in GMBA, since a margin usually leads to better discrimination and generalization, such as LMNN in [20] and the classic SVM. More specifically, as illustrated in Fig. 2(c), similar instances are *pushed* close to each other and dissimilar instances are *pulled* away from them according to the margin. In this way, the margin makes features in this subspace become more discriminative.

In conclusion, based on the graph and the large margin theory, the proposed GMBA can capture high order label correlation and guarantee generalization capability. Experimental results on different real world data sets indicate the effectiveness and good performance of the proposed algorithm.

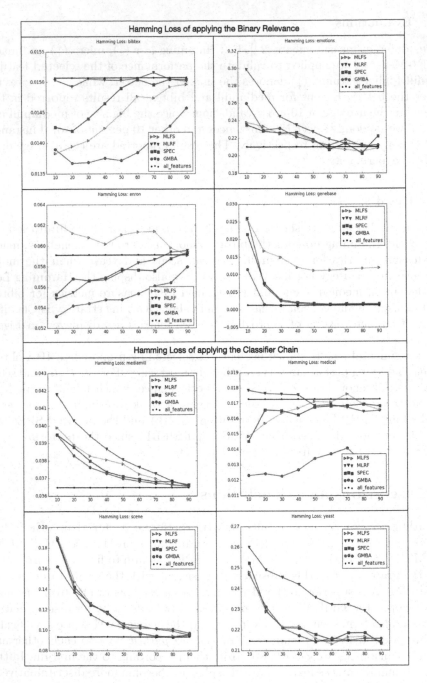

Fig. 3. Hamming loss (↓). The first 4 diagrams show the hamming loss of applying the Binary Relevance while the rest show the results from the Classifier Chain. Y-axis corresponds to different metrics and X-axis denotes the percentage of features selected. The horizontal lines are the results of classifying with all features

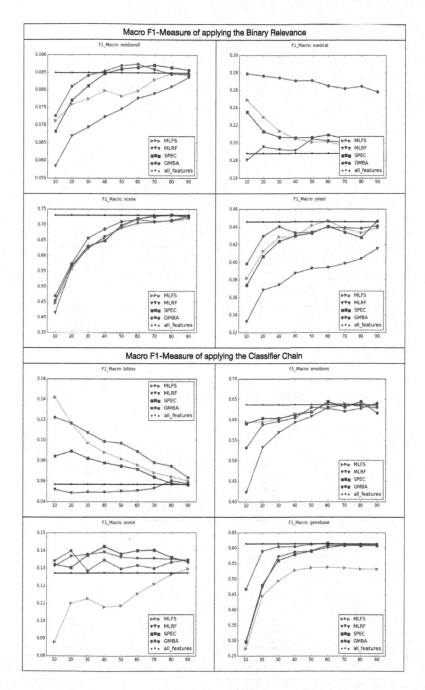

Fig. 4. macro F1-Measure (↑). The first 4 diagrams show the macro F1-Measure of applying the Binary Relevance while the rest show the results from the Classifier Chain. Y-axis corresponds to different metrics and X-axis denotes the percentage of features selected. The horizontal lines are the results of classifying with all features

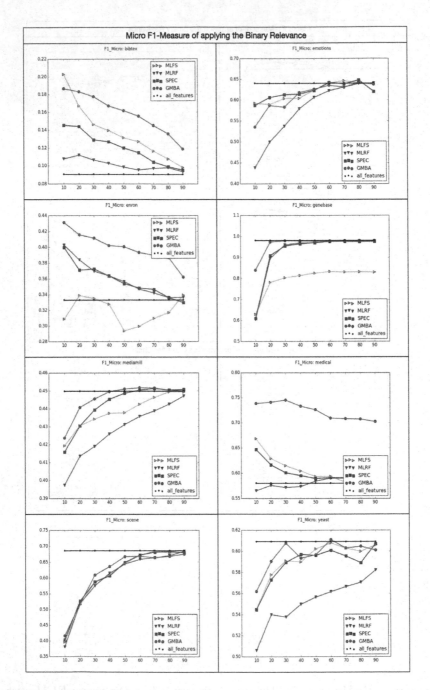

Fig. 5. The micro F1-Measure (↑) of applying the Binary Relevance. Y-axis corresponds to the metrics and X-axis denotes the percentage of features selected. The horizontal lines are the results of classifying with all features

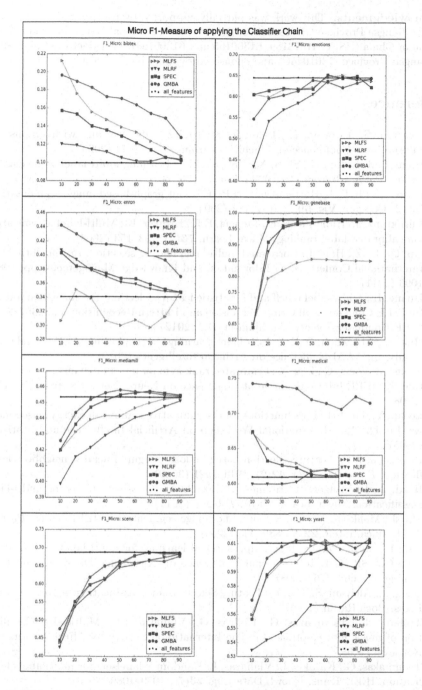

Fig. 6. The micro F1-Measure ($\uparrow$) of applying the Classifier Chain. Y-axis corresponds to the metrics and X-axis denotes the percentage of features selected. The horizontal lines are the results of classifying with all features

Acknowledgments. This work was partially supported by Natural Science Foundation of Jiangsu Province (BK20131378, BK20140885), National Natural Science Foundation of China (NSFC 41573189, 61300165 and 61300164), Post-doctoral Foundation of Jiangsu Province (1401045C) and sponsored by Qing Lan Project.

References

1. Agarwal, S., Branson, K., Belongie, S.: Higher order learning with graphs. In: International Conference on Machine Learning, pp. 17–24 (2006)
2. Boutell, M.R., Luo, J., Shen, X., Brown, C.M.: Learning multi-label scene classification. Pattern Recogn. **37**(9), 1757–1771 (2004)
3. Felzenszwalb, P.F., Huttenlocher, D.P.: Efficient graph-based image segmentation. Int. J. Comput. Vis. **59**(2), 167–181 (2004)
4. Fürnkranz, J., Hüllermeier, E., Mencía, E.L., Brinker, K.: Multi-label classification via calibrated label ranking. Mach. Learn. **73**(2), 133–153 (2008)
5. Gu, Q., Li, Z., Han, J.: Correlated multi-label feature selection. In: the 20th ACM International Conference on Information and Knowledge Management, pp. 1087–1096 (2011)
6. Huang, H.: Multi-label relieff and f-statistic feature selections for image annotation. In: IEEE Conference on Computer Vision and Pattern Recognition, pp. 2352–2359. IEEE Computer Society, Washington, DC (2012)
7. Huang, S.J., Zhou, Z.H.: Multi-label learning by exploiting label correlations locally. In: AAAI Conference on Artificial Intelligence (2012)
8. Jiang, A., Wang, C., Zhu, Y.: Calibrated rank-svm for multi-label image categorization. In: IEEE International Joint Conference on Neural Networks, pp. 1450–1455 (2008)
9. Lecun, Y., Fu, J.H.: Loss functions for discriminative training of energy-based models. In: The 10th International Workshop on Artificial Intelligence and Statistics (2005)
10. Li, Y., Lu, B.L.: Feature selection based on loss-margin of nearest neighbor classification. Pattern Recogn. **42**(9), 1914–1921 (2009)
11. Read, J., Pfahringer, B., Holmes, G., Frank, E.: Classifier chains for multi-label classification. Mach. Learn. **85**(3), 254–269 (2011)
12. Shi, J., Malik, J.: Normalized cuts and image segmentation. IEEE Trans. Pattern Anal. Mach. Intell. **22**(8), 888–905 (2000)
13. Sun, L., Ji, S., Ye, J.: Hypergraph spectral learning for multi-label classification. In: ACM SIGKDD International Conference on Knowledge Discovery and Data Mining, pp. 668–676 (2008)
14. Tang, J., Alelyani, S., Liu, H.: Feature Selection for Classification: A Review. CRC Press, Boca Raton (2012)
15. Trohidis, K., Tsoumakas, G., Kalliris, G., Vlahavas, I.P.: Multi-label classification of music into emotions. In: The International Society for Music Information Retrieval (2008)
16. Tsoumakas, G., Katakis, I., Vlahavas, I.: Random k-labelsets for multilabel classification. IEEE Trans. Knowl. Data Eng. **23**(7), 1079–1089 (2010)
17. Tsoumakas, G., Vlahavas, I.: Random k-labelsets: an ensemble method for multilabel classification. In: Kok, J.N., Koronacki, J., Mantaras, R.L., Matwin, S., Mladenič, D., Skowron, A. (eds.) ECML 2007. LNCS (LNAI), vol. 4701, pp. 406–417. Springer, Heidelberg (2007). doi:10.1007/978-3-540-74958-5_38

18. Wang, S., Siskind, J.M.: Image segmentation with ratio cut. IEEE Trans. Pattern Anal. Mach. Intell. **25**(6), 675–690 (2003)
19. Wang, Y., Li, P., Yao, C.: Hypergraph canonical correlation analysis for multi-label classification. Signal Process. **105**(12), 258–267 (2014)
20. Weinberger, K.Q., Blitzer, J., Saul, L.K.: Distance metric learning for large margin nearest neighbor classification. In: Weiss, Y., Schölkopf, B., Platt, J.C. (eds.) Advances in Neural Information Processing Systems 18, pp. 1473–1480. MIT Press (2006)
21. Zhang, M.L., Peña, J.M., Robles, V.: Feature selection for multi-label naive bayes classification. Inf. Sci. **179**(19), 3218–3229 (2009)
22. Zhang, M.L., Zhang, K.: Multi-label learning by exploiting label dependency. In: Acm SIGKDD International Conference on Knowledge Discovery and Data Mining, pp. 999–1008 (2010)
23. Zhang, M.L., Zhou, Z.H.: Ml-knn: A lazy learning approach to multi-label learning. Pattern Recogn. **40**(7), 2038–2048 (2007)
24. Zhang, M.L., Zhou, Z.H.: A review on multi-label learning algorithms. IEEE Trans. Knowl. Data Eng. **26**(8), 1819–1837 (2014)
25. Zhao, Z., Liu, H.: Spectral feature selection for supervised and unsupervised learning. In: the 24th International Conference on Machine Learning, pp. 1151–1157 (2007)

Pure Exploration for Max-Quantile Bandits

Yahel David[(✉)] and Nahum Shimkin

Department of Electrical Engineering,
Technion – Israel Institute of Technology, 32000 Haifa, Israel
yahel83@gmail.com

Abstract. We consider a variant of the pure exploration problem in Multi-Armed Bandits, where the goal is to find the arm for which the λ-quantile is maximal. Within the PAC framework, we provide a lower bound on the sample complexity of any (ϵ, δ)-correct algorithm, and propose algorithms with matching upper bounds. Our bounds sharpen existing ones by explicitly incorporating the quantile factor λ. We further provide experiments that compare the sample complexity of our algorithms with that of previous works.

1 Introduction

In the classical multi-armed bandit (MAB) problem, the learning agent faces a set K of stochastic arms, from which it chooses arms sequentially. In each round, the agent observes a random reward that depends on the selected arm. The goal of the agent is to maximize the cumulative reward (in the regret formulation), or to identify the arm with the highest expected reward (in the pure exploration problem). The MAB model has been studies extensively in the statistical and learning literature, see [2] for a comprehensive survey.

In this paper, we consider a quantile-based variant of the pure exploration MAB problem (quantile-MAB). In this variant, for a given $0 < \lambda < 1$, the goal is to identify the arm for which the λ-quantile is the largest among all arms (here, as usual the λ-quantile is such that the probability of observing a larger reward is at least λ). More precisely, considering the PAC framework, the goal is to identify an (ϵ, δ)-*correct* arm, namely an arm for which the $(\lambda - \epsilon)$-quantile is not smaller than the largest λ-quantile among all arms, with a probability larger than $1 - \delta$. In addition, we wish to minimize the sample complexity, i.e., the expected number of samples observed until the learning algorithm terminates.

For the standard MAB problem, algorithms that find the best arm (in terms of its expected reward) in the PAC sense were presented in [1,5–8,10], and lower bounds on the sample complexity were presented in [1,9,11].

Similar to the present quantile-MAB problem is the variant of the MAB problem in which the goal is to find the arm from which the largest possible sample can be obtained. This is known as the max k-armed bandit problem, and was first introduced in [3]. For this variant, algorithms that find the best arm in the PAC sense were provided in [4,13], and a lower bound was presented in [4]. In contrast to the current quantile-MAB problem, in the max k-armed

© Springer International Publishing AG 2016
P. Frasconi et al. (Eds.): ECML PKDD 2016, Part I, LNAI 9851, pp. 556–571, 2016.
DOI: 10.1007/978-3-319-46128-1_35

setting, it is necessary to assume a lower bound on the tail probabilities of the arms. When the tail functions of the arms are known, and $\epsilon = \lambda$, the algorithms for the max k-armed bandit setting can be applied in the present quantile-MAB problem. However, their sample complexity upper bounds are larger than those of the algorithms presented in this paper.

More related to the present quantile-MAB problem is the work [15] which consider a measure of risk called value-at-risk (see [12]). The value-at-risk of a given random variable (R.V.) X is actually the same as the quantile of the R.V. $-X$. An algorithm with an upper bound on the sample complexity that increases as $\frac{\lambda |K|}{\epsilon^2 \delta D}$, (where D is the upper bound on the density functions) was provided in [15], that algorithm is computationally demanding since at each iteration it solves a non-linear constrained and integer-valued optimization problem. Recently, the quantile-MAB problem was studied in [14]. They provided a lower bound for the case in which $\lambda = 3/4$ and an algorithm with an upper bound on the sample complexity of the order of $\sum_{k \in K} \frac{1}{(\max(\epsilon, \Delta_{k,\lambda}))^2} \ln(\frac{|K|}{\delta \max(\epsilon, \Delta_{k,\lambda})})$, where $\Delta_{k,\lambda}$ is the difference between the λ-quantile of arm k and that of the best arm.

In this paper, for certain arm distributions, we provide a lower bound of the order of $\sum_{k \in K} \frac{\lambda(1-\lambda) \ln(\frac{1}{\delta})}{(\max(\epsilon, \Delta_{k,\lambda}))^2}$ on the sample complexity of every (ϵ, δ)-correct algorithm. That lower bound improves the bound in [14] in the sense of considering the quantile factor $\lambda(1 - \lambda)$. This is significant when λ is close to 1 or to 0. Furthermore, for general distribution functions, we provide two algorithms that attain the lower bound up to the logarithmic terms $\ln(|K|/\epsilon)$ and $\ln(|K| \log_2(\epsilon))$ respectively. The upper bounds of these algorithms are smaller than that in [14] by a factor of λ and a logarithmic factor in ϵ for the second algorithm.

The paper proceeds as follows. In the next section we present our model. In Sect. 3, a lower bound on the sample complexity of every (ϵ, δ)-correct algorithm is presented. Then in Sect. 4 we present our (ϵ, δ)-correct algorithms, and provide upper bounds on their sample complexity. The second algorithm is bases on applying the doubling trick on the first one. Then, in Sect. 5 we provide experiments that illustrate the improved sample complexity of our algorithms compared with the results presented in [14]. In Sect. 6 we close the paper with some concluding remarks.

2   Model Definition

We consider a finite set of arms, denoted by K. At each stage $t = 1, 2, \ldots$ the learning agent chooses an arm $k \in K$, and a real valued reward is obtained from that arm. The rewards obtained from each arm k are independent and identically distributed, with a distribution function (CDF) $F_k(x)$, $x \in \mathbb{R}$. We denote the quantile function of arm $k \in K$ by $Q_k : [0, 1] \to \mathbb{R}$, and define it as follows.

Definition 1. *For every arm $k \in K$, the quantile function $Q_k(\lambda)$ is defined by*

$$Q_k(\lambda) \triangleq \inf\{x \in \mathbb{R} | 1 - \lambda < F_k(x)\}.$$

Note that $P\left(x_k \geq Q_k(\lambda)\right) \geq 1 - \lambda$ where x_k stands for a random variable with distribution F_k. Clearly, if F_k is continuous at the point $Q_k(\lambda)$, we have equality, namely, $P\left(x_k \geq Q_k(\lambda)\right) = 1 - \lambda$.

An algorithm for the quantile-MAB problem samples an arm at each time step, based on the observed history so far (i.e., the previously selected arms and observed rewards). We require the algorithm to terminate after a random number T of samples, which is finite with probability 1, and return an arm k'. An algorithm is said to be (ϵ, δ)-*correct* if the returned arm is ϵ-optimal with a probability larger than $1 - \delta$, (see a precise definition later in this section). The expected number of samples $E[T]$ taken by the algorithm is the *sample complexity*, which we wish to minimize.

We next provide some definitions and notations which we use later in this paper. A λ-quantile optimal arm is defined as follows.

Definition 2. *Arm* $k \in K$ *is* λ-*quantile optimal if*

$$Q_k(\lambda) = x_\lambda^* \triangleq \max_{k' \in K} Q_{k'}(\lambda).$$

We use the following quantity which represents the distance of an arm from being optimal,

$$\Delta_{k,\lambda} = \sup\{F_k(x) | x < x_\lambda^*\} - (1 - \lambda). \tag{1}$$

If F_k is continuous, then $\Delta_{k,\lambda} = F_k(x_\lambda^*) - (1 - \lambda)$. Furthermore, note that for every suboptimal arm k, namely, an arm for which $Q_k(\lambda) < x_\lambda^*$, it follows by the monotonicity of CDF functions that $\Delta_{k,\lambda} > 0$.

Now we are ready to precisely define an (ϵ, δ)-*correct* algorithm.

Definition 3. *For* λ *and* ϵ *such that* $0 < \epsilon < \lambda < 1$ *and* $\delta > 0$, *an algorithm is* (ϵ, δ)-correct *if*

$$P\left(Q_{k'}(\lambda - \epsilon) \geq x_\lambda^*\right) \geq 1 - \delta$$

where k' *stands for the arm returned by the algorithm.*

3   Lower Bound

Before presenting our algorithms, we provide a lower bound on the sample complexity of any (ϵ, δ)-*correct* algorithm for certain arm distributions. The lower bound is provided in the following Theorem.

Theorem 1. *Assume that* F_k *is continuous for every* $k \in K$. *Fix some* ϵ_0 *such that* $0 < \epsilon_0 < \frac{1}{4}$. *For every* $\lambda \in [2\epsilon_0, 1 - 2\epsilon_0]$, $\epsilon \in (0, \epsilon_0]$ *and* $\delta \leq 0.15$, *there exist some set of arm distributions* $\{F_k\}_{k \in K}$, *such that for every* (ϵ, δ)-correct *algorithm,*

$$E[T] \geq \sum_{k \in K \setminus k^*} \frac{\lambda(1 - \lambda)}{2\left(\max\left(\epsilon, \Delta_{k,\lambda}\right)\right)^2} \ln\left(\frac{1}{2.4\delta}\right) \tag{2}$$

where k^* *denote some optimal arm, with* $Q_{k^*}(\lambda) = x_\lambda^*$.

The above lower bound refines the one presented in [14] in the sense that here the size of the quantile λ is considered in the bound. To illustrate the lower bound, we provide an example.

Example 1. Let $\{\mu_k\}_{k \in K}$ be a set of constants, and let $\mu^* = \max_{k \in K} \mu_k$. Suppose that the rewards of each arm $k \in K$ are uniformly distributed on the interval $(\mu_k - 1, \mu_k)$. Since, $\mu_k - x_\lambda^* \leq 1$, for every arm $k \in K$ it follows that

$$\sup\{F_k(x)|x < x_\lambda^*\} = F_k(x_\lambda^*) = \begin{cases} 1 - (\mu_k - x_\lambda^*), & \mu_k \geq x_\lambda^* \\ 1, & \mu_k < x_\lambda^* \end{cases}.$$

As $x_\lambda^* = \mu^* - \lambda$, Eq. (1), implies that

$$\Delta_{k,\lambda} = \min\left(\lambda, \mu^* - \mu_k\right).$$

Since $\epsilon < \lambda$, the denominator term in Eq. (2) can be seen to be

$$\max\left(\epsilon, \Delta_{k,\lambda}\right) = \begin{cases} \epsilon, & \mu^* - \mu_k < \epsilon \\ \mu^* - \mu_k, & \epsilon \leq \mu^* - \mu_k < \lambda \\ \lambda, & \lambda \leq \mu^* - \mu_k \end{cases}.$$

Proof. (Theorem 1). First we assume that the quantile value of the optimal arm, namely, x_λ^* is known. Moreover, we assume that for every arm $k \in K$, the conditional probabilities $P\left(\boldsymbol{x}_k | \boldsymbol{x}_k \geq x_\lambda^*\right)$ and $P\left(\boldsymbol{x}_k | \boldsymbol{x}_k < x_\lambda^*\right)$ are also known. Therefore, the learning algorithm needs only to estimate the parameters

$$p_k \triangleq P\left(\boldsymbol{x}_k \geq x_\lambda^*\right), \quad \forall k \in K.$$

Now, by the continuity of the distribution functions it follows that $\max_{k \in K} p_k = \lambda$. Also, by Eq. (1) it follows that

$$\max_{k \in K} p_k - p_{k'} = \Delta_{k',\lambda}.$$

Therefore, finding an arm k' such that $\Delta_{k',\lambda} \leq \epsilon$ is the same as finding a Bernoulli arm k', such that its expected value is ϵ-optimal, namely, $\max_{k \in K} p_k - p_{k'} \leq \epsilon$. So, our problem is the same as the standard Bernoulli bandit problem with $\{p_k\}_{k \in K}$ as the Bernoulli parameters.

Then, by Remark 5 in [9], in which a lower bound for the standard MAB problem with Bernoulli arms is provided for $\delta \leq 0.15$, we have

$$E[T] \geq \left(\frac{|S_\epsilon| - 1}{KL\left(\lambda, \lambda - \epsilon\right)} + \sum_{k \in \{K \backslash S_\epsilon\}} \frac{1}{KL\left(p_k, \lambda + \epsilon\right)}\right) \ln \frac{1}{2.4\delta}, \tag{3}$$

where $S_\epsilon \triangleq \{k | k \in K, p_k \geq \lambda - \epsilon\}$ and $KL\left(p, q\right)$ stands for the Kullback-Leibler divergence between two Bernoulli distributions with parameters p and q respectively.

We note that $\ln(1 + x) \le x$. Hence,

$$KL\left(p, q\right) = p\ln\left(\frac{p}{q}\right) + (1 - p)\ln\left(\frac{1 - p}{1 - q}\right) \le p\frac{p - q}{q} + (1 - p)\frac{q - p}{1 - q} = \frac{(p - q)^2}{q(1 - q)}. \quad (4)$$

Therefore, by Eqs. (3) and (4) it follows that

$$E[T] \ge \left(\frac{(\lambda - \epsilon)(1 - \lambda + \epsilon)(|S_\epsilon| - 1)}{\epsilon^2} + \sum_{k \in \{K \backslash S_\epsilon\}} \frac{(\lambda + \epsilon)(1 - \lambda - \epsilon)}{(\lambda + \epsilon - p_k)^2}\right) \ln\frac{1}{2.4\delta}.$$

Hence, by the facts that $2\epsilon \le \lambda$ and $2\epsilon \le (1 - \lambda)$ and since $\Delta_{k,\lambda} = \lambda - p_k$, Eq. (2) is obtained. $\qquad\square$

4 Algorithms

In this section we provide two related algorithms. The first one is simpler and attains the lower bound in Theorem 1 up to a logarithmic term. The second algorithm is based on applying the doubling trick on the first one and hence its upper bound attains Theorem 1 up to a double logarithmic term.

4.1 The Max-Q Algorithm

Here we present our Max-Q algorithm. The algorithm is (ϵ, δ)-correct and based on sampling the arm which has the highest potential λ-quantile value.

The Max-Q algorithm starts by sampling a fixed number of times from each arm. Then, for each arm, the algorithm associates a value that has been sampled from its quantile in a large probability and choses the arm for which the value is maximal. If the number of times that arm has been sampled is larger than a certain threshold, the algorithm stops returns that arm, else it samples one more time from the chosen arm.

The fundamental difference between the Max-Q algorithm and the algorithm presented in [14] is the fact that in the latter the entire CDF is estimated, while in this paper, just the value of the quantile is estimated. That difference leads to a bound on the sample complexity of the Max-Q algorithm which is smaller by a factor of λ, compared to that in [14].

Theorem 2. *For every $\lambda \in (0, 1)$, $\epsilon \in (0, \lambda)$ and $\delta \in (0, 1)$, Algorithm 1 is (ϵ, δ)-correct with a sample complexity bound of*

$$E[T] \le \sum_{k \in K} \frac{10\lambda L}{(\max(\epsilon, \Delta_{k,\lambda}))^2} + |K| + 1, \quad (5)$$

where $L = 6\ln\left(|K|\left(1 + \frac{-10\lambda \ln(\delta)}{\epsilon^2}\right)\right) - \ln(\delta)$ as defined in the algorithm.

Algorithm 1. Maximal Quantile (Max-Q) Algorithm

1: **Input:** Quantile $\lambda \in (0,1)$, constants $\delta > 0$ and $\epsilon > 0$.
   Define $L = 6 \ln \left(|K| \left(1 + \frac{-10\lambda \ln(\delta)}{\epsilon^2} \right) \right) - \ln(\delta)$.
2: **Initialization:** Counters $C(k) = N_0$, $k \in K$,
   where $N_0 = \lfloor \frac{3L}{\lambda} \rfloor + 1$.
3: Sample N_0 times from each arm.
4: Set $k^* \in \arg\max_{k \in K} V^k$ (with ties broken arbitrary), where V^k is the m_k-th largest
   reward observed so far from arm k and

$$m_k = \lfloor \lambda C(k) - \sqrt{3\lambda C(k)L} \rfloor + 1.$$

5: **if** $C(k^*) > \frac{10\lambda L}{\epsilon^2}$ **then**
6:     Stop and return arm k^*.
7: **else**
8:     Sample once from arm k^*, set $C(k^*) = C(k^*) + 1$ and return to step 4.
9: **end if**

It may be observed that for $\lambda \le \frac{1}{2}$, the upper bound provided in Theorem 2 is of the same order as the lower bound in Theorem 1, up to a logarithmic factor.

To establish Theorem 2, we first bound the probability of the event under which the m-th largest sample of one of the optimal arm is below the λ-quantile. Then, we bound the number of samples needed to be observed from each suboptimal arm such that the m-th largest value (obtained from that arm) is below the $(\lambda - \epsilon)$-quantile. For establishing these bounds in a way that the multiplicative factor of λ remains in the bounds, we use Bernstein's inequality for bounding the difference between the empirical mean and the mean value of a Bernoulli R.V. which is one if the sampled value is above the quantile and zero otherwise.

Proof. (Theorem 2). We denote the time step of the algorithm by t, the value of the counter $C(k)$ at time step t by $C^t(k)$ and we use the notations $L' = L + \ln(\delta)$ and x^* as a short for x^*_λ. Recall that T stands for the random final time step. By the condition in step 5 of the algorithm, for every arm $k \in K$, it follows that,

$$C^{T-1}(k) \le \lfloor \frac{10\lambda (L' - \ln(\delta))}{\epsilon^2} \rfloor + 1. \tag{6}$$

Note that by the facts that for $x \ge 6$ it follows that $\frac{d6 \ln(x)}{dx} \le 1$, and that for $x_0 = 20$ it follows that $x_0 > 6 \ln(x_0) + 1$, it is obtained that

$$L'' \triangleq |K| \left(\frac{-10\lambda \ln(\delta)}{\epsilon^2} + 1 \right) > 6 \ln \left(|K| \left(\frac{-10\lambda \ln(\delta)}{\epsilon^2} + 1 \right) \right) + 1 = L' + 1,$$

for $L'' \ge 20$. So, by the fact that $T = \sum_{k \in K} C^{T-1}(k) + 1$, for $L'' \ge 20$ it follows that

$$T \le |K| \left(\frac{10\lambda (L' - \ln(\delta))}{\epsilon^2} + 1 \right) + 1 < |K| \left(\frac{10\lambda (L'' - \ln(\delta))}{\epsilon^2} + 1 \right) \tag{7}$$

$$\le L''^2 = e^{\frac{L'}{3}}.$$

We proceed to establish the (ϵ, δ)-correctness of the algorithm. Let $V_N^k(m)$ stand for the m-th largest value obtained from arm k after sampling it for N times and assume w.l.o.g. that $Q_1(\lambda) = x_\lambda^*$. Then, for $N \geq N_0$ and $m = \lfloor \lambda N - \sqrt{3\lambda N L} \rfloor + 1$, as stated in the algorithm, by Lemma 1 below it follows that

$$P\left(V_N^1(m) < x^*\right) \leq \delta e^{-L'}. \tag{8}$$

Hence, at every time step t, by Eqs. (7) and (8), applying the union bound obtains

$$P\left(V^{t,1} < x^*\right) \leq \sum_{N=N_0}^{\exp\left(\frac{L'}{3}\right)} P\left(V_N^1(m) < x^*\right) = \delta e^{-\frac{2L'}{3}}. \tag{9}$$

where $V^{t,k}$ stands for the value of V^k at time step t.

Let k_T^* stand for the arm returned by the algorithm. Also, by Lemma 1, for $N > \frac{10\lambda\left(L' - \ln(\delta)\right)}{\epsilon^2}$, it follows that

$$P\left(V_N^k(m) > Q_k(\lambda - \epsilon)\right) \leq \delta e^{-L'}. \tag{10}$$

So, since by the condition in step 5, it is obtained that $C(k_T^*) > \frac{10\lambda\left(L' - \ln(\delta)\right)}{\epsilon^2}$, it follows by Eq. (10) and the union bound that

$$P\left(V^{T,k_T^*} > Q_{k_T^*}(\lambda - \epsilon)\right) \leq \sum_{k \in K} \sum_{t=1}^{\exp\left(\frac{L'}{3}\right)} \sum_{N=1}^{\exp\left(\frac{L'}{3}\right)} \delta e^{-L'} = |K|\delta e^{-\frac{L'}{3}}. \tag{11}$$

Also, by Eq. (9) and the union bound it follows that

$$P\left(V^{T,1} < x^*\right) \leq \sum_{t=1}^{\exp\left(\frac{L'}{3}\right)} P\left(V^{t,1} < x^*\right) \leq \delta e^{-\frac{L'}{3}}. \tag{12}$$

So, since by step 4 of the algorithm, $V^{T,k_T^*} \geq V^{T,1}$, it follows by Eqs. (11) and (12) that

$$P\left(Q_{k_T^*}(\lambda - \epsilon) < x^*\right) \leq P\left(V^{T,k_T^*} > Q_{k_T^*}(\lambda - \epsilon)\right) + P\left(V^{T,1} < x^*\right) < \delta.$$

It follows that the algorithm returns an ϵ-optimal arm with a probability larger than $1 - \delta$. Hence, it is (ϵ, δ)-correct.

To prove the bound on the expected sample complexity of the algorithm, we define the following sets:

$$M(\epsilon) = \{l \in K | \Delta_{k,\lambda} \leq \epsilon\} \quad \text{and} \quad N(\epsilon) = \{l \in K | \Delta_{k,\lambda} > \epsilon\}.$$

As before, we assume w.l.o.g. that $Q_1(\lambda) = x^*$. Then, for the case in which

$$E_1 \triangleq \bigcap_{1 \leq t \leq T} \{V^{t,1} \geq x^*\}$$

occurs, for every arm $k \in K$, a necessary condition for $C^T(k) > N'_k$, where $N'_k = \lfloor \frac{10\lambda(L'-\ln(\delta))}{\Delta^2_{k,\lambda}} \rfloor + 1$ is

$$E_k \triangleq \left\{ V^k_{N'_k}(m'_k) \geq x^* \right\},$$

where $m'_k = \lfloor \lambda N'_k - \sqrt{3\lambda N'_k (L' - \ln(\delta))} \rfloor + 1$.

Now, by using the bound in Eq. (6) and the fact that $\sum_{k \in K} C^T(k) = \sum_{k \in K} C^{T-1}(k) + 1$ for the arms in the set $M(\epsilon)$, N'_k as a bound for the arms in the set $N(\epsilon)$, and the bound in Eq. (7), it is obtained that

$$E[T] \leq (1 - P(E_1)) e^{\frac{L'}{3}} + P(E_1) \sum_{k \in N(\epsilon)} \left((1 - P(E_k|E_1)) \Phi_k(\epsilon) + e^{\frac{L'}{3}} P(E_k|E_1) \right)$$

$$+ \sum_{k \in M(\epsilon)} \Phi_k(\epsilon) + 1,$$

$$(13)$$

where $\Phi_k(\epsilon) = \lfloor \frac{10\lambda(L'-\ln(\delta))}{(\max(\epsilon, \Delta_{k,\lambda}))^2} \rfloor + 1$. But, by Eq. (9) it follows that

$$P(E_1) \geq 1 - \sum_{t=1}^{\exp\left(\frac{L'}{3}\right)} P\left(V^{t,1} < x^*\right) \geq 1 - \delta e^{\frac{-2L'}{3}} e^{\frac{L'}{3}} = 1 - \delta e^{\frac{-L'}{3}}. \quad (14)$$

Also, since $Q'_k \triangleq Q_k \left(\lambda - \sqrt{\frac{10\lambda(L'-\ln(\delta))}{N'_k}} \right) < x^*$ for $k \in N(\epsilon)$, it follows by Lemma 1 that

$$P(E_k|E_1) P(E_1) \leq P(E_k) \leq P\left(V^k_{N'_k}(m'_k) > Q'_k \right) \leq \delta e^{-L'}, \quad \forall k \in N(\epsilon) \quad (15)$$

Therefore, by Eqs. (13), (14) and (15) and the definition of $\Phi_k(\epsilon)$, the bound on the sample complexity is obtained. □

Lemma 1. *For every arm $k \in K$, let $V^k_N(m)$ stand for the m-th largest value obtained from arm k after sampling it for N times. Then, for any positive integers m and N such that $m < N$, and every $\lambda \in [0,1]$, it follows that,*

1. *If $\frac{m}{N} > \lambda$, then*
$$P\left(V^k_N(m) > Q_k(\lambda) \right) \leq f_0(m, N, \lambda).$$

2. *If $\frac{m}{N} < \lambda$, then*
$$P\left(V^k_N(m) < Q_k(\lambda) \right) \leq f_0(m, N, \lambda),$$

where $f_0(m, N, \lambda) = \exp\left(-\frac{|m-N\lambda|^2}{2(N\lambda + |m-N\lambda|/3)} \right)$.

The proof is based on Bernstein's inequality.

Proof. In this proof, we omit the arm index k for short. We start with claim (1). Let x_i stand for the i-th sampled value from the arm, and let $\{X_i(\lambda)\}$ and $\{Y_i(\lambda)\}$ be random variables for which

$$X_i(\lambda) = \begin{cases} 1 & w.p \ \lambda \\ 0 & w.p \ 1 - \lambda \end{cases} \quad \text{and} \quad Y_i(\lambda) = \begin{cases} 1 & x_i > Q(\lambda) \\ 0 & x_i \le Q(\lambda) \end{cases} . \tag{16}$$

Note that the variables $\{Y_i(\lambda)\}$ are i.i.d. The variables $\{X_i(\lambda)\}$ are i.i.d as well. Then, since $P\left(Y_i(\lambda) = 1\right) \le P\left(X_i(\lambda) = 1\right)$, after sampling N times,

$$P\left(V_N(m) > Q(\lambda)\right) = P\left(\frac{1}{N}\sum_{i=1}^{N} Y_i(\lambda) \ge \frac{m}{N}\right) \le P\left(\frac{1}{N}\sum_{i=1}^{N} X_i(\lambda) \ge \frac{m}{N}\right)$$

$$= P\left(\frac{1}{N}\sum_{i=1}^{N} \widetilde{X}_i(\lambda) \ge \frac{m}{N} - E[X_1(\lambda)]\right) \triangleq \Upsilon(\lambda, m, N),$$

$$\tag{17}$$

where $\widetilde{X}_i(\lambda) = X_i(\lambda) - E[X_1(\lambda)]$. So, $\left\{\widetilde{X}_i(\lambda)\right\}$ satisfies the conditions of Bernstein's inequality with $\sigma^2 = \lambda\left(1 - \lambda\right)$, and $E[X_1(\lambda)] = \lambda$. Therefore

$$\Upsilon(\lambda, m, N) \le \exp\left(-\frac{(m - N\lambda)^2}{2\left(N\lambda\left(1 - \lambda\right) + (m - N\lambda)/3\right)}\right)$$

$$\le \exp\left(-\frac{(m - N\lambda)^2}{2\left(N\lambda + (m - N\lambda)/3\right)}\right) . \tag{18}$$

Proceeding to claim (2), let $\{Z_i(\lambda)\}$ be random variables for which

$$Z_i(\lambda) = \begin{cases} 1 & x_i \ge Q(\lambda) \\ 0 & x_i < Q(\lambda) \end{cases} . \tag{19}$$

Note that $\{Z_i\}$ are i.i.d. Then, since $P\left(Z_i(\lambda) = 1\right) \ge P\left(X_i(\lambda) = 1\right)$,

$$P\left(V_N(m) < Q(\lambda)\right) = P\left(\frac{1}{N}\sum_{i=1}^{N} Z_i(\lambda) < \frac{m}{N}\right) \le P\left(\frac{1}{N}\sum_{i=1}^{N} X_i(\lambda) \le \frac{m}{N}\right)$$

$$= P\left(\frac{1}{N}\sum_{i=1}^{N} \widetilde{X}_i(\lambda) \le \frac{m}{N} - E[X_1(\lambda)]\right) \triangleq \hat{\Upsilon}(\lambda, m, N) \tag{20}$$

and by symmetry

$$\hat{\Upsilon}(\lambda, m, N) \le \exp\left(-\frac{(N\lambda - m)^2}{2\left(N\lambda\left(1 - \lambda\right) + (N\lambda - m)/3\right)}\right)$$

$$\le \exp\left(-\frac{(N\lambda - m)^2}{2\left(N\lambda + (N\lambda - m)/3\right)}\right) . \tag{21}$$

□

Algorithm 2. Doubled Maximal Quantile (Max-Q) Algorithm

1: **Input:** Quantile $\lambda \in (0,1)$, constants $\delta > 0$ and $\epsilon > 0$.

Define $L_D = 6 \ln \left(|K| \log_2 \left(\frac{-20\lambda \ln(\delta)}{\epsilon^2} \right) \right) - \ln(\delta)$.

2: **Initialization:** Counters $C(k) = N_0$, $k \in K$,

where $N_0 = \lfloor \frac{3L_D}{\lambda} \rfloor + 1$.

3: Sample N_0 times from each arm.

4: Set $k^* \in V^k$ (with ties broken arbitrary), where V^k is the m_k-th largest reward observed so far from arm k and

$$m_k = \lfloor \lambda C(k) - \sqrt{3\lambda C(k) L_D} \rfloor + 1.$$

5: **if** $C(k^*) > \frac{10\lambda L_D}{\epsilon^2}$ **then**

6:     Stop and return arm k^*.

7: **else**

8:     Sample $C(k^*)$ times from arm k^*, set $C(k^*) = 2C(k^*)$ and return to step 4.

9: **end if**

4.2   The Doubled Max-Q Algorithm

Here we improve on the previous algorithm by resorting to the doubling trick. The Doubled Max-Q Algorithm is based on the same principle as the Max-Q Algorithm. However, instead of observing one sample at each time step, here the algorithm doubles the number of samples of the chosen arm. Consequently, the number of times at which the algorithm needs to choose an arm is roughly logarithmic compared to that under the previous algorithm, leading to a tighter bound. Algorithm 2 presents the proposed Doubled Max-Q algorithm.

Theorem 3. *For every* $\lambda \in (0,1)$, $\epsilon \in (0,\lambda)$ *and* $\delta \in (0,1)$, *Algorithm 2 is* (ϵ, δ)-*correct with a sample complexity bound of*

$$E[T] \leq \sum_{k \in K} \frac{20\lambda L_D}{\left(\max \left(\epsilon, \Delta_{k,\lambda} \right) \right)^2} + |K| + 1, \tag{22}$$

where $L_D = 6 \ln \left(|K| \log_2 \left(\frac{-20\lambda \ln(\delta)}{\epsilon^2} \right) \right) - \ln(\delta)$ *as defined in the algorithm.*

Here, the upper bound is of the same order as the lower bound in Theorem 1, up to a double-logarithmic order.

The proof of Theorem 3 is established by some adjustments of the proof of Theorem 2.

Proof. As before, we denote the time step of the algorithm by t, the value of the counter $C(k)$ at time step t by $C^t(k)$ and we use the notations $L'_D = L_D + \ln(\delta)$ and x^* as a short for x^*_λ. We note that here, at each time step, there may be more than a single sample, so T, the sample complexity, may be different than the final time step. Hence, here we denote the (random) final time step by T_D. By the condition in step 5 of the algorithm, for every arm $k \in K$, it follows that,

$$C^{T_D-1}(k) \leq \frac{10\lambda \left(L'_D - \ln(\delta) \right)}{\epsilon^2}. \tag{23}$$

Note that by the facts that for $x \geq 6$ it follows that $\frac{d6\ln(x)}{dx} \leq 1$, and that for $x_0 = 20$ it follows that $x_0 > 6\ln(x_0) + 1$ it is obtained that

$$L_D'' \triangleq |K| \log_2 \left(\frac{-20\lambda \ln(\delta)}{\epsilon^2} \right) > 6\ln \left(|K| \log_2 \left(\frac{-20\lambda \ln(\delta)}{\epsilon^2} \right) \right) + 1 = L_D' + 1,$$

for $L_D'' \geq 20$. So, by the fact that $T = \sum_{k \in K} \log_2 \left(2C^{T_D - 1}(k) \right)$, for $L_D'' \geq 20$ it follows that

$$T \leq |K| \log_2 \left(\frac{20\lambda \left(L_D' - \ln(\delta) \right)}{\epsilon^2} \right) < |K| \log_2 \left(\frac{20\lambda \left(L_D'' - \ln(\delta) \right)}{\epsilon^2} \right) \tag{24}$$

$$\leq |K| \log_2 \left(L_D'' \right) + L_D'' \leq L_D'' \left(\log_2 \left(L_D'' \right) + 1 \right) \leq \frac{(L_D'')^2}{2} = \frac{1}{2} e^{\frac{L_D'}{3}}.$$

Recall that x^* is used as a short for x_λ^*. Now, we begin with proving the (ϵ, δ)-correctness property of the algorithm. We let $V_N^k(m)$ stands for the m-th largest value obtained from arm k after sampling it for N times and we assume w.l.o.g. that $Q_1(\lambda) = x^*$. Then, for $N \geq N_0$ and $m = \lfloor \lambda N - \sqrt{3\lambda N L_D} \rfloor + 1$, as stated in the algorithm, by Lemma 1 it follows that

$$P\left(V_N^1(m) < x^* \right) \leq \delta e^{-L_D'} \tag{25}$$

Hence, at every time step t, by Eqs. (24) and (25), by applying the union bound, for $N_i = 2^i N_0$ it follows that

$$P\left(V^{t,1} < x^* \right) \leq \sum_{i=0}^{\frac{1}{2} \exp\left(\frac{L_D'}{3} \right)} P\left(V_{N_i}^1(m) < x^* \right) = \delta e^{-\frac{2L_D'}{3}}. \tag{26}$$

where $V^{t,k}$ stands for the value of V^k at time step t.

Now, we let $k_{T_D}^*$ stands for the arm returned by the algorithm. Also, by Lemma 1, for $N > \frac{10\lambda \left(L_D' - \ln(\delta) \right)}{\epsilon^2}$, it follows that

$$P\left(V_N^k(m) > Q_k(\lambda - \epsilon) \right) \leq \delta e^{-L_D'}. \tag{27}$$

So, since by the condition in step 5, it is obtained that $C(k_{T_D}^*) > \frac{10\lambda \left(L_D' - \ln(\delta) \right)}{\epsilon^2}$, it follows by Eq. (27) and the union bound that

$$P\left(V^{T_D, k_{T_D}^*} > Q_{k_{T_D}^*}(\lambda - \epsilon) \right) \leq \sum_{k \in K} \sum_{t=1}^{\frac{1}{2} \exp\left(\frac{L_D'}{3} \right)} \sum_{i=0}^{\frac{1}{2} \exp\left(\frac{L_D'}{3} \right)} \delta e^{-L_D'} = |K| \delta e^{-\frac{L_D'}{3}}. \tag{28}$$

Also, by Eq. (26) and applying the union bound it follows that

$$P\left(V^{T_D, 1} < x^* \right) \leq \sum_{t=1}^{\frac{1}{2} \exp\left(\frac{L_D'}{3} \right)} P\left(V^{t,1} < x^* \right) \leq \delta e^{-\frac{L_D'}{3}} \tag{29}$$

So, since by step 4 of the algorithm, $V^{T_D, k_{T_D}^*} \geq V^{T_D, 1}$, it follows by Eqs. (28) and (29) that

$$P\left(Q_{k_{T_D}^*}(\lambda - \epsilon) < x^*\right) \leq P\left(V^{T_D, k_{T_D}^*} > Q_{k_{T_D}^*}(\lambda - \epsilon)\right) + P\left(V^{T_D, 1} < x^*\right) < \delta$$

Therefore, it follows that the algorithm returns an ϵ-optimal arm with a probability larger than $1 - \delta$. So, it is (ϵ, δ)-correct.

For proving the bound on the expected sample complexity of the algorithm we define the following sets:

$$M(\epsilon) = \{l \in K | \Delta_{k,\lambda} \leq \epsilon\} \quad \text{and} \quad N(\epsilon) = \{l \in K | \Delta_{k,\lambda} > \epsilon\}.$$

As before, we assume w.l.o.g. that $Q_1(\lambda) = x^*$. For the case in which

$$E_1 \triangleq \bigcap_{1 \leq t \leq T} \{V^{t,1} \geq x^*\},$$

occurs, for every arm $k \in K$, a necessary condition for $C^{T_D}(k) > N'_{k,D}$, where

$$N'_{k,D} \triangleq \min\left\{N_i | N_i > \frac{10\lambda (L'_D - \ln(\delta))}{\Delta^2_{k,\lambda}}, i \in \mathbb{N}\right\}$$

is

$$E_{k,D} \triangleq \left\{V^k_{N'_{k,D}}(m'_{k,D}) \geq x^*\right\},$$

where $m'_{k,D} = \lfloor \lambda N'_{k,D} - \sqrt{3\lambda N'_{k,D}(L'_D - \ln(\delta))}\rfloor + 1$.

Then for

$$\Phi_{k,D}(\epsilon) = \frac{20\lambda (L'_D - \ln(\delta))}{(\max(\epsilon, \Delta_{k,\lambda}))^2}$$

for $k \in N(\epsilon)$ it follows that

$$N'_{k,D} \leq \Phi_{k,D}(\epsilon)$$

So, by using the bound in Eq. (23) and the fact that $\sum_{k \in K} C^{T_D}(k) = 2\sum_{k \in K} C^{T_D - 1}(k)$ for the arms in the set $M(\epsilon)$, $N'_{k,D}$ as a bound for the arms in the set $N(\epsilon)$ and the bound in Eq. (24), it is obtained that

$$E[T] \leq (1 - P(E_1)) e^{\frac{L'_D}{3}}$$

$$+ P(E_1) \sum_{k \in N(\epsilon)} \left((1 - P(E_{k,D}|E_1)) \Phi_{k,D}(\epsilon) + e^{\frac{L'_D}{3}} P(E_{k,D}|E_1)\right) \quad (30)$$

$$+ \sum_{k \in M(\epsilon)} \Phi_{k,D}(\epsilon) + 1,$$

But, by Eq. (26) it follows that

$$P\left(E_1\right) \geq 1 - \sum_{t=1}^{\exp\left(\frac{L'_D}{3}\right)} P\left(V^{t,1} < x^*\right) \geq 1 - \delta e^{-\frac{2L'_D}{3}} e^{\frac{L'_D}{3}} = 1 - \delta e^{-\frac{L'_D}{3}} \quad (31)$$

Also, since $Q_k\left(\lambda - \sqrt{\frac{10\lambda\left(L'_D - \ln(\delta)\right)}{N'_{k,D}}}\right) < x^*$ for $k \in N(\epsilon)$, it follows by Lemma 1 that

$$P\left(E_{k,D}|E_1\right)P\left(E_1\right) < \delta e^{-L'_D}, \quad \forall k \in N(\epsilon) \quad (32)$$

Therefore, by Eqs. (30), (31) and (32) and the definition of $\Phi_{k,D}(\epsilon)$, the bound on the sample complexity is obtained. $\qquad\square$

5   Experiments

In this section we investigate numerically the Max-Q and the Double-Max-Q algorithms presented in this paper and compare them with the QPAC algorithm presented in [14].

In Fig. 1, we present the average sample complexity of 10 runs vs. the quantile λ for $\delta = 0.01$ and various values of ϵ. As shown in Fig. 1, and detailed in Tables 1, 2, 3 and 4, the Max-Q and the Double-Max-Q algorithms significantly outperform the QPAC algorithm. The arms distribution functions used here were uniform with an interval of length 1.

Table 1. $\log_{10}$ of the average sample complexity of the Max-Q, the Double-Max-Q and the QPAC algorithms. The number of arms was 10 and the averages were computed over 10 runs.

$\log_{10}(E[T])$	Algorithm		
	QPAC	Double-Max-Q	Max-Q
	$\epsilon = 0.005$	$\epsilon = 0.005$	$\epsilon = 0.005$
$1 - \lambda = 0.8$	7.26	7.43	7.6
$1 - \lambda = 0.85$	7.26	6.47	6.75
$1 - \lambda = 0.9$	7.26	6.35	6.57
$1 - \lambda = 0.95$	7.26	6.05	6.25
$1 - \lambda = 0.96$	7.26	5.85	6.16
$1 - \lambda = 0.97$	7.26	5.68	6.04
$1 - \lambda = 0.98$	7.26	5.58	5.88
$1 - \lambda = 0.99$	7.28	5.42	5.72

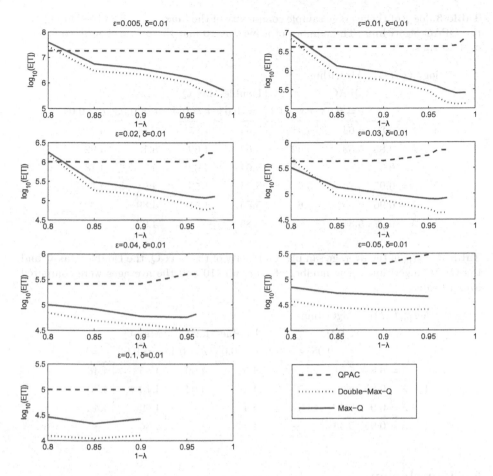

Fig. 1. The average sample complexity of the Max-Q, the Double-Max-Q and the QPAC algorithms for various of parameters settings. The number of arms was 10 and the averages were computed over 10 runs.

Table 2. $\log_{10}$ of the average sample complexity of the Max-Q, the Double-Max-Q and the QPAC algorithms. The number of arms was 10 and the averages were computed over 10 runs.

$\log_{10}(E[T])$	Algorithm					
	QPAC		Double-Max-Q		Max-Q	
	$\epsilon = 0.01$	$\epsilon = 0.02$	$\epsilon = 0.01$	$\epsilon = 0.02$	$\epsilon = 0.01$	$\epsilon = 0.02$
$1 - \lambda = 0.8$	6.22	6.01	6.83	6.2	6.96	6.26
$1 - \lambda = 0.85$	6.63	6.01	5.86	5.26	6.11	5.48
$1 - \lambda = 0.9$	6.63	6.01	5.74	5.15	5.93	5.33
$1 - \lambda = 0.95$	6.63	6.01	5.46	4.92	5.64	5.12
$1 - \lambda = 0.96$	6.63	6.03	5.28	4.8	5.57	5.09
$1 - \lambda = 0.97$	6.63	6.23	5.15	4.76	5.47	5.07
$1 - \lambda = 0.98$	6.66	6.23	5.11	4.81	5.4	5.1
$1 - \lambda = 0.99$	6.85	—	5.12	—	5.41	—

Table 3. $\log_{10}$ of the average sample complexity of the Max-Q, the Double-Max-Q and the QPAC algorithms. The number of arms was 10 and the averages were computed over 10 runs.

| $\log_{10}(E[T])$ | Algorithm | | | | | |
| | QPAC | | Double-Max-Q | | Max-Q | |
	$\epsilon = 0.03$	$\epsilon = 0.04$	$\epsilon = 0.03$	$\epsilon = 0.04$	$\epsilon = 0.03$	$\epsilon = 0.04$
$1 - \lambda = 0.8$	5.64	5.41	5.64	4.85	5.51	5
$1 - \lambda = 0.85$	5.63	5.42	5.63	4.69	5.13	4.92
$1 - \lambda = 0.9$	5.64	5.41	5.64	4.62	5	4.78
$1 - \lambda = 0.95$	5.74	5.6	5.74	4.53	4.9	4.76
$1 - \lambda = 0.96$	5.85	5.6	5.85	4.51	4.89	4.82
$1 - \lambda = 0.97$	5.85	—	5.85	—	4.92	—

Table 4. $\log_{10}$ of the average sample complexity of the Max-Q, the Double-Max-Q and the QPAC algorithms. The number of arms was 10 and the averages were computed over 10 runs.

| $\log_{10}(E[T])$ | Algorithm | | | | | |
| | QPAC | | Double-Max-Q | | Max-Q | |
	$\epsilon = 0.05$	$\epsilon = 0.1$	$\epsilon = 0.05$	$\epsilon = 0.1$	$\epsilon = 0.05$	$\epsilon = 0.1$
$1 - \lambda = 0.8$	5.31	5	4.57	4.09	4.85	4.46
$1 - \lambda = 0.85$	5.3	5	4.44	4.04	4.73	4.33
$1 - \lambda = 0.9$	5.3	5	4.4	4.1	4.69	4.42
$1 - \lambda = 0.95$	5.48	—	4.41	—	4.66	—

6   Conclusion

In this paper we studied the pure exploration problem where the goal is to find the arm with the maximal λ-quantile. Under the PAC framework, we provided a lower bound and algorithms that attain it up to a logarithmic term (for the first algorithm) and a double-logarithmic term (for the second algorithm).

A challenge for future work is closing the logarithmic gap between the lower and upper bounds.

References

1. Audibert, J.Y., Bubeck, S.: Best arm identification in multi-armed bandits. In: Proceeding of the 23rd Conference on Learning Theory (COLT), pp. 41–53 (2010)
2. Bubeck, S., Cesa-Bianchi, N.: Regret analysis of stochastic and nonstochastic multi-armed bandit problems. Found. Trends Mach. Learn. **5**(1), 1–122 (2012)
3. Cicirello, V.A., Smith, S.F.: The max k-armed bandit: a new model of exploration applied to search heuristic selection. Proc. Ntl. Conf. Artif. Intell. **20**, 1355–1361 (2005)

4. David, Y., Shimkin, N.: PAC lower bounds and efficient algorithms for the max k-armed bandit problem. In: Proceedings of the 33rd International Conference on Machine Learning (ICML), pp. 878–887 (2016)
5. Even-Dar, E., Mannor, S., Mansour, Y.: PAC bounds for multi-armed bandit and markov decision processes. In: Ben-David, S. (ed.) EuroCOLT 1997. LNCS, vol. 1208, pp. 255–270. Springer, Heidelberg (2002). doi:10.1007/3-540-45435-7_18
6. Gabillon, V., Ghavamzadeh, M., Lazaric, A.: Best arm identification: a unified approach to fixed budget and fixed confidence. Adv. Neural Inf. Process. Syst. **25**, 3212–3220 (2012). Curran Associates, Inc
7. Kalyanakrishnan, S., Tewari, A., Auer, P., Stone, P.: PAC subset selection in stochastic multi-armed bandits. In: Proceedings of the 29th International Conference on Machine Learning (ICML), pp. 655–662 (2012)
8. Karnin, Z.S., Koren, T., Somekh, O.: Almost optimal exploration in multi-armed bandits. In: Proceedings of the 30th International Conference on Machine Learning (ICML), pp. 1238–1246 (2013)
9. Kaufmann, E., Cappé, O., Garivier, A.: On the complexity of best-arm identification in multi-armed bandit models. J. Mach. Learn. Res. **17**(1), 1–42 (2016)
10. Kaufmann, E., Kalyanakrishnan, S.: Information complexity in bandit subset selection. In: Proceeding of the 26th Conference on Learning Theory (COLT), pp. 228–251 (2013)
11. Mannor, S., Tsitsiklis, J.N.: The sample complexity of exploration in the multi-armed bandit problem. J. Mach. Learn. Res. **5**, 623–648 (2004)
12. Schachter, B.: An irreverent guide to value at risk. Finan. Eng. News **1**(1), 17–18 (1997)
13. Streeter, M.J., Smith, S.F.: An asymptotically optimal algorithm for the max k-armed bandit problem. Proc. Ntl. Conf. Artif. Intell. **21**, 135–142 (2006)
14. Szörényi, B., Busa-Fekete, R., Weng, P., Hüllermeier, E.: Qualitative multi-armed bandits: A quantile-based approach. In: Proceedings of the 32nd International Conference on Machine Learning (ICML), pp. 1660–1668 (2015)
15. Yu, J.Y., Nikolova, E.: Sample complexity of risk-averse bandit-arm selection. In: Proceedings of the 23rd International Joint Conference on Artificial Intelligence (IJCAI) (2013)

A Hybrid Knowledge Discovery Approach for Mining Predictive Biomarkers in Metabolomic Data

Dhouha Grissa[1,3], Blandine Comte[1], Estelle Pujos-Guillot[2],
and Amedeo Napoli[3(✉)]

[1] INRA, UMR1019, UNH-MAPPING, 63000 Clermont-Ferrand, France
[2] INRA, UMR1019, Plateforme D'Exploration du Métabolisme,
63000 Clermont-Ferrand, France
[3] LORIA, B.P. 239, 54506 Vandoeuvre-lès-Nancy, France
Amedeo.Napoli@loria.fr

Abstract. The analysis of complex and massive biological data issued from metabolomic analytical platforms is a challenge of high importance. The analyzed datasets are constituted of a limited set of individuals and a large set of features where predictive biomarkers of clinical outcomes should be mined. Accordingly, in this paper, we propose a new hybrid knowledge discovery approach for discovering meaningful predictive biological patterns. This hybrid approach combines numerical classifiers such as SVM, Random Forests (RF) and ANOVA, with a symbolic method, namely Formal Concept Analysis (FCA). The related experiments show how we can discover among the best potential predictive biomarkers of metabolic diseases thanks to specific combinations of classifiers mainly involving RF and ANOVA. The visualization of predictive biomarkers is based on heatmaps while FCA is mainly used for visualization and interpretation purposes, complementing the computational power of numerical methods.

1 Introduction

The analysis of metabolomic data using data mining methods is one main challenge addressed in this paper. This analysis can be considered as a hard knowledge discovery task since data generated by analytical metabolomic platforms, e.g., mass spectrometry (MS), are massive, complex and noisy. In such data, one of the major objectives is to identify, among thousands of features, predictive biomarkers of disease development [11]. More precisely, in the current study, we aim at identifying early predictive biomarkers of T2D, i.e. type 2 diabetes, a few years before occurrence of the disease, in homogeneous populations considered as healthy at the time of analysis. In general, the considered datasets have a limited set of individuals and very large sets of features (or variables) and thus require a specific data processing. In addition, it is desirable that the data analysis methods differentiate a two state clinical feature, i.e. healthy vs. not healthy, and contribute to explanations about this difference.

© Springer International Publishing AG 2016
P. Frasconi et al. (Eds.): ECML PKDD 2016, Part I, LNAI 9851, pp. 572–587, 2016.
DOI: 10.1007/978-3-319-46128-1_36

For carrying out the knowledge discovery task, we have to pay attention to the reduction of dimensionality, i.e. feature selection, and to avoid overfitting. Accordingly, we defined a "knowledge discovery workflow" (KD workflow) based on various data mining methods for the discovery of predictive biomarkers from metabolomic data. This KD workflow is based on an original combination of numerical data mining methods to analyze a data table with a large number of numerical features, e.g. molecules or fragments of molecules, and a limited number of individuals (samples), and one binary target variable (having or not the disease at the follow-up, 5 years after the time of analysis). The resulting reduced dataset is then transformed as binary data table that can be used as a context for applying Formal Concept Analysis (FCA [5]) and discovering candidate biomarkers.

This hybrid knowledge discovery process involve several numerical classifiers including Random Forests (RF) [2], Support vector Machine (SVM) [15], and ANOVA [3]. RF, SVM and ANOVA are used to discover relevant biological patterns which are then organized within a binary table. The numerical classifiers are used for feature selection with the help of filtering methods based on the correlation coefficient and mutual information, for eliminating redundant/dependent features, reducing the size of the data table and preparing the application of RF, SVM and ANOVA.

Among the numerous combinations of RF, SVM and ANOVA with the filtering methods, we defined ten reference combinations of classifiers (CC) in agreement with the wishes of biologists (especially w.r.t. metabolomic data and usage). Then a comparative study was run to identify the top-k features in computing the so-called "top-ranking degree" a feature, i.e. the number of times that a feature is classified among the first features for each CC. Actually, we retained the best ranked features having a top-ranking degree greater or equal to 6. A binary table can then be built, where features are lying in rows and combination of classifiers (CC) are lying in columns. A cell (i, j) in the table is marked with 1 when the feature i has a sufficient top-ranking degree for the CC j. Such a binary table can be in turn considered as a formal context and a starting point for Formal Concept Analysis, and used for the identification of the best features. These last features, which have the best ranking w.r.t. the ten combinations of classifiers, are considered as candidates to be "potential predictive biomarkers".

For experts in metabolomics, it is crucial to compute the ability of the potential biomarkers in predicting the disease. This is usually done thanks to ROC analysis [18] which returns a short list of the best features retained as a core set of predictive biomarkers. Based on ROC analysis and FCA, we are able to identify a list of the best combinations of classifiers that provide the best ranking of potential biomarkers. One final objective of this study is to provide a short list of at most 10 biomarkers that can be used in clinical assays, where the simplest combination of metabolites producing an effective predictive outcome must be found. In this way, we can measure the actuality of our knowledge discovery results. This whole process defines an original and hybrid knowledge discovery approach where numerical and symbolic classifiers are combined.

The remainder of this paper is organized as follows. Section 2 provides a description of related works. Section 3 presents the proposed hybrid knowledge discovery approach and explains the analysis of biomarker identification. Section 4 describes the experiments performed on a real-world metabolomic data set and discusses the results, while Sect. 5 concludes the paper.

2   State of the Art

In [13], authors provide an overview on fundamental aspects of univariate and multivariate analysis related to the analysis of metabolomic data. They make precise the main differences between possible approaches and explain several experiments on real and simulated metabolomic data. In this case, the analysis of such data is performed by supervised learning techniques, such as PLS-DA (partial least squares-discriminant analysis), PC-DFA (Principal component-discriminant function analysis), LDA (Linear discriminant analysis), RF and SVM.

In [7], authors show that PLS-DA outperforms other approaches in terms of feature selection and classification. In a more detailed study [8], authors compare different variable selection approaches such as LDA, PLS-DA, SVM-Recursive Feature Elimination (RFE), RF (with accuracy and gini), for identifying the best suited method for analyzing metabolomic data and classifying the Gram-positive bacteria Bacillus. They conclude that RF with accuracy and gini and SVM with RFE [9] provide the best results. However, these studies also show that the choice of appropriate algorithms is highly dependent of the dataset characteristics and on the objective of the data mining process.

In the field of biomarker discovery, SVM and RF algorithms proved to be robust for extracting relevant chemical and biological knowledge from complex data, especially in metabolomics [8]. RF is a highly accurate classifier, based on a robust model to outlier detection. One main advantage is its power to deal with overfitting and missing data [1], as well as its ability to handle large datasets.

Finally, in [12], authors discuss recent papers on applying a symbolic classification method such as Formal Concept Analysis in biology and medicine. For example, in [6], authors use a classifier based on FCA to identify combinatorial biomarkers of breast cancer from genes expression values. However, according to literature, no working approach combining supervised and unsupervised data mining techniques was proposed so far for processing metabolomic data. This is precisely the objective of the present paper to fill this gap and to propose an original combination of numerical and symbolic classifiers for mining metabolomic data.

3   The Design of a Hybrid Knowledge Discovery Approach for Metabolomic Data

In this section, we explain how to design a hybrid knowledge discovery process in agreement with experts in biology and in combining various numerical classifiers, e.g. RF, SVM, and ANOVA, with a symbolic knowledge discovery method such as FCA, to discover the top-k biological features having a high predictive ability.

3.1   The Reduction of Dimensionality

Here, the reduction of dimensionality is mainly based on feature selection. This is one of the most important operations that can be carried out, especially considering the data table at hand, with small sets of individuals but very large sets of features. Such an operation requires a careful choice of appropriate feature selection methods [14]. Two main types of approaches can be considered:

- "Filtering" (or filter methods) consists in selecting features using statistical test. Metabolomic data usually contain highly correlated features, leading to some problems when using RF for example [7]. Filter methods allow to select "good features", such as the "coefficient of correlation" (Cor) or the "mutual information" (MI) measures. Cor and MI can be used to discard highly correlated features for keeping a reasonable number of features to be analyzed.
- The so-called "embedded methods" are searching for an optimal subset of features based on a reference classifier such as RF or SVM [10]. Embedded methods are dependent on the classifier and try to optimize the results of this classifier.

Based on that, we will consider two kinds of classifiers, the first kind using one of the two filters, i.e. correlation coefficient "Cor" and mutual information "MI", and the other not using any filtering, as this is illustrated by the KD workflow in Fig. 1. Filtering based on "Cor" and "MI" eliminates redundant/dependent features, i.e. highly correlated are filtered out and features with MI average values smaller than a given threshold are selected [16]. The result of the filtering is used an an input for the application of the RF and SVM classifiers. Regarding embedded methods, "Recursive Feature Elimination" (RFE), i.e. a backward elimination method proposed in [9] for improving the classification process, is most of the time used for lowering correlation between features when it is still high, either with RF or SVM. Accordingly, we can build three different classifiers, namely RF, RF-RFE and SVM-RFE.

The selection of the top-ranked features can be completed by the use of accuracy measures, including "MdGini[1]", "MdAcc[2]", and "Kappa[3]". One general idea supporting these measures is to permute the values of each feature and then to measure the decrease in accuracy of the classifier.

In parallel, even if filter methods are generally robust against overfitting, they may still fail to select the best features. For being able to consider this problem, we can apply the classifiers without any filtering, directly working on data. We decided to consider the two classifiers RF and ANOVA alone, and a classifier with an embedded method, namely SVM-RFE. In these last cases, we

---

[1] Mean decrease in Gini index (MdGini) provides a measure of the internal structure of the data.

[2] Mean decrease in accuracy (MdAcc) measures the importance/performance of each feature to the classification.

[3] Cohens Kappa (Kappa) is a statistical measure which compares an "observed accuracy" with an "expected accuracy".

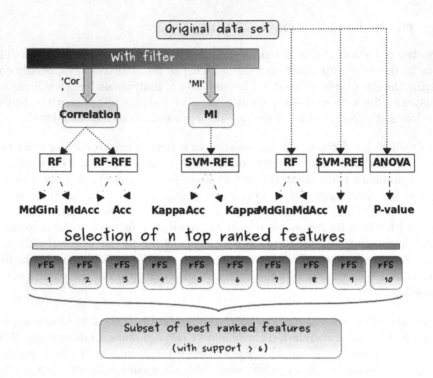

Fig. 1. The KD workflow and the ten combinations of classifiers.

still can choose accuracy measures such as "MdGini" and "MdAcc". In addition, we also decided to try the "features weight W", i.e. the weight magnitude of features, with SVM-RFE, and and the "p-value" with ANOVA, for improving the classification of features and the identification of the features with the highest discriminant ability.

Finally, we design ten different combinations of classifiers (CC) as illustrated in Fig. 1. Applying each CC to the original dataset produces a set of "best ranked features" corresponding to the ten datasets called rFS_i (for "reduced Feature Sets"). The ten rFS_i include the best ranked features w.r.t. the corresponding CC_i.

3.2 Identification of the "top-k Best Features"

Now, we have at our disposal ten reduced features sets denoted by rFS_i of best ranked features. We will compare these feature datasets and try to discover which are the "top-k features", i.e. the features which have the best ranking considering all rFS_i. This can be likened to a problem of preferences and decision-making, and this is precisely here that symbolic methods can play a major role, and they do... A binary table relating features and combinations of classifiers, i.e. $features \times CC$ can be designed (see Table 1), where the rows are related to best ranked features and the columns are related to combinations of classifiers (10 CC_i).

Each feature has a "top-ranking degree" w.r.t. a given CC. Then, we retain all features which have a top-ranking degree superior or equal to 6, i.e., features belonging to an rFS_i $(i = 1 \ldots 10)$ and which are top-ranked by at least 6 CC out of 10.

3.3   Selecting Predictive Biomarkers (prediction)

For evaluating the predictive power of the best ranked features selected as explained just above, we used the RF classifier again with several configurations, taking the set of the best ranked features as a training set and considering the whole set of available features. In this way, we are able to obtain several different sets of ranked features that can be evaluated thanks to specific evaluation measures, namely "sensitivity[4]", "specificity[5]", "accuracy[6]", "precision[7]", "OOB error[8]" and "misclassification rate[9]". Since the number of features to propose as potential predictive biomarkers should be low and of high biological relevance, we should find the best classifier w.r.t. these evaluation measures.

A second feature selection algorithm based on RF, namely "VarSelRF" [4], can be applied for prediction purpose. VarSelRF is based on a "backward variable elimination" for selecting small sets of non-redundant features and provides a reduced set of predictive features. Several trials can be carried out, each producing a different reduced set of relevant features, until obtaining the lowest OOB error rate.

Then, the results of RF and VarSelRF can be combined, as well as computing the p-values of the selected predictive features using T-tests[10]. The core set of best features with the smallest p-values and the highest accuracy values is selected to finally obtain a short list of potential predictive biomarkers.

---

[4] Sensitivity evaluates the efficiency of the classifier in identifying the true positive instances.

[5] Specificity also called true negative rate, measures the proportion of correctly identified negative instances relative to all real relative ones.

[6] Accuracy evaluates the overall performance of the feature selection method, since it measures the ability of the predictive model to correctly classify both positive and negative instances.

[7] Precision rates the predictive power of a method, by measuring the proportion of the true positive instances relative to all the predicted positive ones.

[8] A Random Forest classifier returns a measure of error rate based on the out-of-bag (OOB) cases for each fitted tree.

[9] This rate refers to the misclassification rate of the learning model, by estimating the proportion of wrongly classified negative and positive features.

[10] T-test or Student T-test is a statistical hypothesis test which can be used to determine if two sets of data are significantly different from each other. If the p-value is below the threshold chosen for statistical significance (usually the 0.10, the 0.05, or 0.01 level), then the null hypothesis is rejected in favor of the alternative hypothesis.

3.4 Visualization and Interpretation

Now we can consider the core set of best ranked features identified in the previous prediction step for visualization and interpretation purposes. Visualization can be carried out using heatmaps, which are currently used in metabolomics and which provide useful insights about the understanding of the metabolomic changes w.r.t. experimental settings and sample groupings. Heatmaps are very useful for patterns recognition in mass spectrometry-based metabolomic domain. They can be used to visualize the results of "biclustering", e.g. classification w.r.t. the both sets of features and of individuals. Heatmaps represent with different colors the features (or molecules) which predict the set of individuals that are affected or not by the disease. Hotter areas indicate a more intense presence of the feature(s) among individuals. Cooler areas show a lower level of importance.

For completing interpretation of the resulting sets of features and the identification of predictive features, we should find the best classifiers to apply on the metabolomic data at hand. We would like to help the expert to rank the classifiers w.r.t. their ability to detect the best ranked predictive features among a large set of features. A short list of potential predictive biomarkers can be noticed in the binary Table 1 where they are denoted by bold 1. Then, the related combinations of classifiers can be recommended for reducing dimensionality of metabolomics data and for identifying the best predictive features. Moreover, FCA can be also applied to such a binary table as discussed farther.

4 Experiments

In this section, we discuss the experiments related to our hybrid knowledge discovery approach. Practically, we used a Dell machine running Ubuntu 14.04 LTS, a 3.60 GHZ $\times 8$ CPU and $15, 6$ GB RAM. All experiments were performed in the Rstudio software environment (Version 0.98.1103, R 3.1.1).

4.1 The Dataset and Its Preparation

The reference dataset is composed of homogeneous individuals considered healthy at the beginning of the study. The binary variable describing the two target classes, i.e. healthy and not healthy, is based on the health status of the same individuals at another time, actually five years after the initial analysis. Meanwhile, some individuals developed the disease. In particular, discriminant features which enable a good separation between target data classes (healthy vs. not healthy) are not necessarily the best features predicting the disease development five years later.

More precisely, the dataset to be analyzed is based on a case-control study from the GAZEL French population-based cohort (20000 subjects). This set includes numeric and symbolic data about 111 male subjects (54-64 years old) free of T2D at baseline. Fifty five subjects who developed T2D at the follow-up

belong to class "1" (non healthy or diabetic subjects) while 56 subjects belong
to class "−1" (controls or healthy subjects). 3000 features are generated for
each individuals after carrying out mass spectrometry (MS) analysis, resulting
in a dataset containing peak intensities (continuous numerical values of these
measurements).

The metabolomic database contains thousands of features with a wide intensity value range. A data preprocessing step is mandatory for adjusting the importance weights allocated to the features. Thus, before applying any classifier,
data are transformed using zero mean normalization and Unit-Variance scaling method. This method removes the average and divides each feature value
by its standard deviation. This enables all features to have the same chance to
contribute in the classification model when they have an equal unit variance.
Finally, the transformed dataset including 1195 features is used as input for all
combinations of classifiers.

4.2   The Combination of Classifiers

Following the KD workflow as introduced in Sect. 3.1 and as depicted in Fig. 1,
we defined ten different combinations of classifiers ($CC_i, i = 1\ldots10$) for feature
selections purposes. These ten CC_i are detailed hereafter. For example, "Cor-RF-
MdAcc" denotes the sequence of three operations, i.e. the correlation coefficient
"Cor" is used on the original dataset for retaining features whose correlation
value is less than a given threshold, then the classifier Random Forests (RF) is
applied, and the final ranking is provided according to the "MdAcc" accuracy
measure.

The nine other CC_i are named accordingly as (2) "Cor-RF-MdGini", (3)
"Cor-RF-RFE-Acc", (4) "Cor-RF-RFE-Kap", (5) "MI-SVM-RFE-Acc", (6)
"MI-SVM-RFE-Kap", (7) "RF-MdAcc", (8) "RF-MdGini", (9) "SVM-RFE-W"
and (10) "ANOVA-pValue".

To work only with important features, we retain the 200 first ranked features
from each of the ten CC_i, except for the CC "ANOVA-pValue" from which we
only retained 107 features having a "reasonable" p-value (i.e. lower than 0.1).

Then, to analyze and interpret the relative importance of each feature, the
reduced feature sets rFS_i related to each CC_i ($i = 1\ldots10$) are compared.
Since we are looking for the best ranked features according to the different CC_i,
features which are among the best ranked features in at least 6 CC_i are selected.
This leads to the identification of 48 features and the generation of the binary
Table 1 whose dimension is 48 × 10, where features are lying in rows and CC_i in
columns.

In this binary table, we can identify four features, namely "m/z 383", "m/z
227", "m/z 114" and "m/z 165" as the best ranked features for all CC_i ($i =
1\ldots10$) because they generate a "maximum rectangle full of 1" (the four first
rows in Table 1), i.e., they are best ranked in all the 10 CC_i. Furthermore, we
can see that some other features are also best ranked by a high number of CC_i
such as "m/z 284", "m/z 204", "m/z 132", "m/z 187", "m/z 219", "m/z 203",

"m/z 109", "m/z 97" and "m/z 145". Moreover, among these 48 best ranked features, 39 are significant w.r.t ANOVA, i.e. the p-value is less that 0.05.

4.3   The Search for Predictive Biomarkers

Here we intend to use two feature selection algorithms, namely "VarSelRF" and "Random Forests", for prediction purposes. The first algorithm is based on a subset selection method and the second one is based on a feature ranking method as introduced previously.

During the application of "VarSelRF", it was decided to train the algorithm 100 times and to retain the stable features identified w.r.t. the different replications results. Experiments were performed on the subset of 48 best ranked features and revealed 5 features common to all repeated tests, i.e. "m/z 145", "m/z 162", "m/z 263", "m/z 268" and "m/z 97", as potential predictive biomarkers.

When using the RF classifier, we are highly interested in measuring the impact of each feature on the accuracy of classification. Thus, we first split the data into a training set and a test set. Then, we apply the RF classifier on the set of best ranked features including the 48 features using the "MdAcc" measure for ranking. 100 replications of the procedure are performed and the classification with the lowest error is retained. A confusion matrix is generated where a new set of 48 ranked features denoted by "48-RF-MdAcc" is obtained. From "48-RF-MdAcc", 5 additional sets of features, namely "40-RF", "30-RF", "20-RF", "10-RF" and "5-RF" are built, containing respectively 40, 30, 20, 10 and 5 best ranked features.

Table 2 summarizes the scores obtained from the six common evaluation metrics, starting from the set of 1195 features, through the set of 200 features (the 200 best ranked features w.r.t. RF with "MdAcc") until the reduced set of the 5 best ranked features according to RF-MdAcc on the set of 48 best ranked features. The table shows that training RF on the whole data set gives the lowest values. However, reducing data dimensionality to 48 features, better values are obtained. As there is not a set of features which outperforms all the others, the smallest set of the 5 top ranked features, i.e. "m/z 219", "m/z 268", "m/z 145", "m/z 97", and "m/z 325", is retained.

In addition, Table 2 shows that only a small fraction of features is discriminant, highlighting the importance of feature selection methods for obtaining the best performances of predictive classifiers. Actually, the RF classifier is able to handle thousands of features, but when applied to the original dataset (1195-RF), it does not achieve a good accuracy (26.1% of OOB error). The set of 48 features (48-RF) gives the best value for "Recall" only. The highest values for "precision" and "specificity" are obtained with the set of 30 features i.e. 30-RF. The measures "Accuracy" (which gives an overall estimate of the performance of a classifier) and "F-measure" are better for the set of 20 features. The worst values are measured for the whole data set of 1195 features. This underlines another time the fact that reducing the dimension of the original data table to identify relevant features is essential.

Table 1. The 48×10 binary table relating the 48 features which are the best ranked w.r.t. the 10 combinations of classifiers.

Features	Cor-RF-MdGini	Cor-RF-MdAcc	Cor-RF-RFE-Acc	Cor-RF-RFE-Kap	RF-MdGini	RF-MdAcc	MI-SVM-RFE-Acc	MI-SVM-RFE-Kap	SVM-RFE-W	ANOVA-pValue
m/z 383	1	1	1	1	1	1	1	1	1	1
m/z 227	1	1	1	1	1	1	1	1	1	1
m/z 114	1	1	1	1	1	1	1	1	1	1
m/z 165	1	1	1	1	1	1	1	1	1	1
m/z 145	1	1	1	1	1	1	1	1		1
m/z 97	1	1	1	1	1	1	1	1		1
m/z 441	1	1	1	1	1	1	1		1	1
m/z 109	1	1	1	1	1	1	1	1		1
m/z 203	1	1	1	1	1	1	1	1		1
m/z 219	1	1	1	1	1	1	1	1		1
m/z 198	1		1	1	1	1	1	1	1	1
m/z 263	1	1	1		1	1	1	1	1	1
m/z 187	1	1	1	1	1	1	1	1		1
m/z 132	1	1	1	1	1	1	1	1		1
m/z 204	1	1	1	1	1	1	1	1		1
m/z 261	1	1	1	1	1	1	1	1	1	
m/z 162	1	1			1	1	1	1	1	1
m/z 284	1	1	1	1	1	1	1	1		1
m/z 603	1	1	1	1	1	1			1	1
m/z 148	1	1	1	1	1	1			1	
m/z 575	1	1	1	1	1	1	1	1		
m/z 69	1	1		1	1	1	1		1	
m/z 325	1	1			1	1	1	1		1
m/z 405	1	1			1	1	1		1	1
m/z 929	1	1	1	1	1	1	1	1		
m/z 58	1	1	1	1	1	1	1	1		
m/z 336	1	1	1	1	1	1	1			1
m/z 146	1		1	1	1		1	1	1	
m/z 104	1		1	1	1		1		1	
m/z 120		1	1	1	1	1	1	1	1	
m/z 558	1	1			1	1	1	1	1	
m/z 231					1	1	1	1	1	1
m/z 132*	1	1	1	1	1	1	1			1
m/z 93		1	1	1	1	1	1	1		1
m/z 907		1	1	1		1	1	1	1	
m/z 279		1	1	1	1	1	1	1		
m/z 104*	1	1			1	1	1	1		1
m/z 90	1	1			1	1	1	1		1
m/z 268					1	1	1	1	1	1
m/z 288*	1	1	1	1	1	1				1
m/z 287	1	1	1	1	1	1				1
m/z 167	1	1			1	1	1	1		1
m/z 288	1	1	1	1	1	1				1
m/z 252	1	1			1	1	1	1		1
m/z 141	1		1	1	1		1	1		1
m/z 275			1	1	1	1		1	1	1
m/z 148*			1	1	1			1	1	1
m/z 92	1	1				1	1	1	1		1

Table 2. The values of measures for several sets of features computed with RF and accuracy.

Metrics	Recall	Specificity	F-measure	Accuracy	Precision	OOB error
1195-RF	0.81	0.65	0.75	0.73	0.71	0.261
200-RF	0.86	0.82	0.85	0.84	0.84	0.154
48-RF	0.93	0.80	0.88	0.87	0.83	0.131
40-RF	0.85	0.88	0.86	0.87	0.87	0.131
30-RF	0.83	0.90	0.86	0.87	0.90	0.131
20-RF	0.90	0.85	0.88	0.88	0.86	0.119
10-RF	0.85	0.86	0.85	0.85	0.85	0.142
5-RF	0.86	0.85	0.86	0.85	0.86	0.142

In parallel, using ANOVA, we also retained the 5 best ranked features w.r.t. an ascending order of their p-value, i.e. "m/z 383", "m/z 145", "m/z 97", "m/z 268", "m/z 263". In metabolomics, it is usually interesting to consider features with a small p-value for prediction.

Finally, considering the three features, i.e. "m/z 145", "m/z 268" and "m/z 97", which are common to RF and VarSelRF classifiers, plus the top five ANOVA features, we obtain 8 potential predictive features.

4.4 Interpretation of the Potential Biomarkers

Now we want to show that the 8 selected features are potential predictive biomarkers. This validation is based on the value of AUC and T-tests. Table 3 shows how the 8 selected features are ranked w.r.t. AUC (univariate ROC curves). This analysis was performed thanks to the ROCCET[11] tool. If we only keep features with an AUC higher than or equal to 0.75, and with significantly small T-test values (i.e. smaller than 10E−5), we should exclude two features, namely "m/z 219" and "m/z 162", leading to a short list of 6 features as potential predictive biomarkers.

In multifactorial diseases such as T2D, a combination of a multiple "weak" multivariate biomarkers instead of a single "strong" individual biomarker often provides the required high levels of discrimination and confidence. Therefore, the performances of the top ranked features (top 8 and top 6) previously obtained are evaluated and compared (see Table 4) using the ROCCET RF tool. The results show that the multivariate top features (top 8 and top 6) are very accurate w.r.t. the single features (RF-top5, VarSelRF-top5, ANOVA-top5), with an AUC higher than 0.81. For comparison, we select the six first features having an AUC higher than 0.75, and a significant small T-test value for building a multivariate ROC curve. The combination of these single features did not show any improvement in prediction accuracy compared to the multivariate features. Finally, prediction based on the six top ranked features (top 6) shows a misclassification rate of 20.7% which is close to the rate of 19.8% obtained by RF-top5.

---

[11] http://www.roccet.ca.

Table 3. The 8 best AUC ranked features.

Name	AUC	T-tests	95 % CI
m/z 145	0.79	1.4483E-6	0.657 - 0.896
m/z 383	0.79	5.0394E-7	0.703 - 0.876
m/z 97	0.78	1.5972E-6	0.657 - 0.898
m/z 325	0.77	2.2332E-5	0.627 - 0.896
m/z 268	0.75	4.564E-6	0.614 - 0.866
m/z 263	0.75	5.996E-6	0.642 - 0.874
m/z 219	0.71	1.177E-4	0.162 - 0.798
m/z 162	0.65	0.00195	0.225 - 0.710

Table 4. The 5 predictive classifiers.

Name	AUC	95 % CI	Misclassification (%)
RF-top5	0.83	0.749 - 0.923	19.8
VarSelRF-top5	0.841	0.765 - 0.924	22.5
ANOVA-top5	0.826	0.755 - 0.906	20.7
top 8	0.827	0.694 - 0.918	21.6
top 6	0.812	0.714 - 0.903	20.7

4.5 Visualization

For visualization purposes, we also used "heatmaps" as an easy-to-use inter-active tool for exploring data and results, as heatmaps are commonly used in metabolomics [17]. The rows of the heatmap table represent the features while the columns correspond to the samples or individuals. The color gradient denotes the normalized abundance of each feature among the samples. Heatmaps can be used to visualize feature classification vs. individual classification. Hotter areas indicate a higher "intensity" of the feature(s) among the individuals.

Figure 2 presents the heatmap corresponding to the 6×111 data matrix relating the 6 predictive features and the 111 individuals (healthy and not healthy individuals), as a mean to visualize the classification of individuals w.r.t. predictive biomarkers. The relationships that can be discovered are very useful for the experts in biology and allow to identify subgroups of individuals who share same metabolite (linked to a feature) levels. For example, from the set of 6 predictive features (input), we identify that 4 of them, namely "m/z 383", "m/z 325", "m/z 97" and "m/z 145", which are highly correlated according to Fig. 3, characterize a group of individuals with intensity values ranging between 2 and 6 (this group of individuals is described by the highlighted rectangle in Fig. 2). Meanwhile both remaining features have very low intensity values (between −4 and 0) for the same group of individuals, but, by contrast, are more present among healthy individuals. These results show the need of combining markers to be able to predict the disease within a whole heterogeneous population.

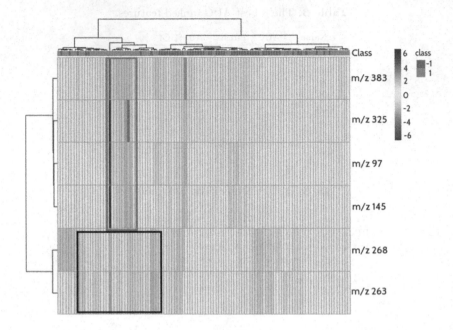

Fig. 2. A heatmap matrix displaying the predictive power of the 6 best features w.r.t. the 111 individuals. Colors represent the distribution of the data ranging from −6 to 6. From −6 to 0, the features are not representative, while from 0 to 6, the features are more and more representative. The idea is to show the functional relationships among the 6 features and the 111 samples by means of a color-coded matrix elements and adjacent dendrograms. The class −1 represents healthy individuals while class 1 represents not healthy individuals (or patients). (Color figure online)

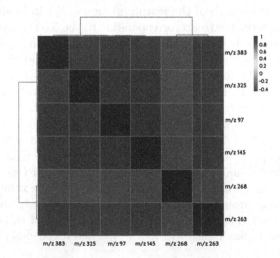

Fig. 3. The correlation network based on the 6 best predictive features.

4.6   The Role of Symbolic KD Methods

A close examination of the relationships between the best predictive features can contribute to a better understanding of the results Actually, quite strong correlations and associations can be found within the 6 best features. This can help the experts, firstly, in the identification of the structure of metabolites related to predictive features and secondly, in the biological interpretation, as metabolites from a same metabolic pathway should be linked.

Here, it is also time to go back to the role that can be played by symbolic methods in such an hybrid knowledge discovery process. In this way, FCA can be used for information retrieval and visualization purposes, and also to identify the combination of classifiers (CC_i) that shows the best behavior. Considering the final set of six potential predictive features, we built a binary table corresponding to the part of Table 1 with bold 1. The associated concept lattice can be seen in Fig. 4, involving 5 combinations of classifiers, namely "ANOVA-pValue", "MI-SVM-RFE-Kappa", "MI-SVM-RFE-Acc", "RF-MdAcc" and "RF-MdGini" and the 6 features. The interpretation of this concept lattice is still under discussion before validation.

In a second time, we consider the 6 best predictive features and their rankings w.r.t the 5 CC_i above. Table 5 shows that RF-based techniques and ANOVA usually give a good ranking to the 6 features by contrast with "MI-SVM-RFE-MdAcc" and "MI-SVM-RFE-Kappa". For example, "m/z 145" is ranked first according to "RF-MdAcc", "RF-MdGini", second according to "ANOVA-pValue", 100th for "MI-SVM-RFE-Acc" and 125th for "MI-SVM-RFE-Kappa". The feature "m/z 268" is ranked 9th according to "RF-MdAcc", 6th for "RF-MdGini", 168th for "MI-SVM-RFE-Acc", 181th for "MI-SVM-RFE-Kappa", and 4th for "ANOVA-pValue". Consequently, the top list combination of classifiers for predictive biomarker identification from metabolomic data is based on RF and ANOVA. However, the choice of appropriate feature selection methods is highly dependent of the dataset characteristics. Moreover, it is also clear that, so far, there is no universal combination of classifiers [8].

Based on these results, one recommendation could be to explore the combination of ANOVA and RF-MdGini methods for reducing the dimensionality of datasets in metabolomic data, especially when predictive biomarkers are searched.

Table 5. Rankings of the 6 predictive features.

Feature	RF-MdAcc	RF-MdGini	MI-SVM-RFE-Acc	MI-SVM-RFE-Kappa	ANOVA-pValue
m/z 145	1	1	100	125	2
m/z 383	3	3	40	39	1
m/z 97	2	2	63	67	3
m/z 325	5	5	38	37	8
m/z 268	9	6	168	181	4
m/z 263	8	7	28	27	5

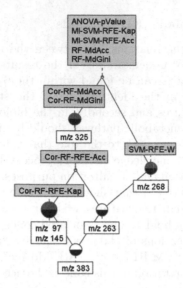

Fig. 4. The concept lattice of the 6 best predictive features.

5   Conclusion

In this paper, we presented a new hybrid knowledge discovery process for the identification of relevant predictive biomarkers in metabolomic data. Such data are usually highly correlated and noisy. Accordingly, the reduction of dimensionality and feature selection are two tasks of the higher importance. We used classifiers such as Random Forests, SVM and ANOVA, completed by the use of measures for minimizing noise and feature correlations.

This study shows that different combinations of classifiers and measures can be designed and that some of them are better applied to specific datasets, such as metabolomic datasets. Several experiments were performed to assess the predictive power of the best ranked features and visualization tools such as heatmaps allowed a deeper interpretation of the results.

In addition, a symbolic knowledge discovery method such as Formal Concept Analysis was used for visualization and interpretation purposes. Such an association of numerical and symbolic classifiers is original and should be further studied. The present paper is a first step in this direction and more extended theoretical studies and experiments remain to be done.

References

1. Biau, G.: Analysis of a random forests model. J. Mach. Learn. Res. **13**(1), 1063–1095 (2012)
2. Breiman, L.: Random forests. Mach. Learn. **45**(1), 5–32 (2001)

3. Cho, H., Kim, S., Jeong, M., Park, Y., Miller, N., Ziegler, T., Jones, D.: Discovery of metabolite features for the modelling and analysis of high-resolution nmr spectra. Int. J. Data Min. Bioinform. **2**(2), 176–192 (2008)
4. Díaz-Uriarte, R., de Andrés, S.A.: Gene selection and classification of microarray data using random forest. BMC Bioinform. **7**(1), 1–13 (2006)
5. Ganter, B., Wille, R.: Formal Concept Analysis - Mathematical Foundations. Springer, Heidelberg (1999)
6. Gebert, J., Motameny, S., Faigle, U., Forst, C., Schrader, R.: Identifying genes of gene regulatory networks using formal concept analysis. J. Comput. Biol. **2**, 185–194 (2008)
7. Gromski, P., Muhamadali, H., Ellis, D., Xu, Y., Correa, E., Turner, M., Goodacre, R.: A tutorial review: metabolomics and partial least squares-discriminant analysis-a marriage of convenience or a shotgun wedding. Anal. Chim. Acta **879**, 10–23 (2015)
8. Gromski, P., Xu, Y., Correa, E., Ellis, D., Turner, M., Goodacre, R.: A comparative investigation of modern feature selection and classification approaches for the analysis of mass spectrometry data. Ana. Chim. Acta **829**, 1–8 (2014)
9. Guyon, I., Weston, J., Barnhill, S., Vapnik, V.: Gene selection for cancer classification using support vector machines. Mach. Learn. **46**, 389–422 (2002)
10. Lal, T., Chapelle, O., Weston, J., Elisseeff, A.: Feature Extraction: Foundations and Applications. In: Guyon, I., Nikravesh, M., Gunn, S., Zadeh, L. (eds.) Embedded Methods, pp. 137–165. Springer, Heidelberg (2006)
11. Mamas, M., Dunn, W., Neyses, L., Goodacre, R.: The role of metabolites and metabolomics in clinically applicable biomarkers of disease. Arch. Toxicol. **85**(1), 5–17 (2011)
12. Poelmans, J., Kuznetsov, S., Ignatov, D., Dedene, G.: Formal concept analysis in knowledge processing: a survey on models and techniques. Expert Syst. Appl. **40**(16), 6601–6623 (2013)
13. Saccenti, E., Hoefsloot, H., Smilde, A., Westerhuis, J., Hendriks, M.: Reflections on univariate and multivariate analysis of metabolomics data. Metabolomics **10**(3), 361–374 (2014)
14. Saeys, Y., Inza, I., Larraaga, P.: A review of feature selection techniques in bioinformatics. Bioinformatics **23**(19), 2507–2517 (2007)
15. Vapnik, V.: Statistical Learning Theory. Wiley, New York (1998)
16. Wang, H., Khoshgoftaar, T., Wald, R.: Measuring stability of feature selection techniques on real-world software datasets. In: Information Reuse and Integration in Academia and Industry, pp. 113–132. Springer (2013)
17. Wilkinson, L., Friendly, M.: The history of the cluster heat map. Am. Statist. **62**, 179–184 (2009)
18. Xia, J., Broadhurst, D., Wilson, M., Wishart, D.: Translational biomarker discovery in clinical metabolomics: an introductory tutorial. Metabolomics **9**(2), 280–299 (2013)

Native Advertisement Selection and Allocation in Social Media Post Feeds

Iordanis Koutsopoulos[(✉)] and Panagiotis Spentzouris

Department of Informatics, Athens University of Economics and Business,
Athens, Greece
jordan@aueb.gr

Abstract. We study native advertisement selection and placement in social media post feeds. In the prevalent pay-per-click model, each ad click leads to certain amount of revenue for the platform. The probability of click for an ad depends on attributes that are either inherent to the ad (*e.g.*, ad quality) or related to user profile and activity or related to the post feed. While the first two types of attributes are also encountered in web-search advertising, the third one fundamentally differentiates native from web-search advertising, and it is the one we model and study in this paper. Evidence from online platforms suggests that the main attributes of the third type that affect ad clicks are the relevance of ads to preceding posts, and the distance between consecutively projected ads; *e.g.*, the fewer the intervening posts between ads, the smaller the click probability is, due to user saturation.

We model the events of ad clicks as Bernoulli random variables. We seek the ad selection and allocation policy that optimizes a metric which is a combination of (*i*) the platform expected revenue, and (*ii*) uncertainty in revenue, captured by the variance of provisionally consumed budget of selected ads. Uncertainty in revenue should be minimum, since this translates into reduced profit or wasted advertising opportunities for the platform. On the other hand, the expected revenue from ad clicking should be maximum. The constraint is that the expected revenue attained for each selected ad should not exceed its apriori set budget. We show that the optimization problem above reduces to an instance of a resource-constrained minimum-cost path problem on a weighted directed acyclic graph. Through numerical evaluation, we assess the impact of various parameters on the objective, and the way they shape the tradeoff between revenue and uncertainty.

Keywords: Native advertising · Advertisement allocation · Social media feeds · Mathematical modeling · Optimization · Shortest-path problem

1 Introduction

Internet advertising in its early form more than 15 years ago consisted in sponsored search advertising, whereby personalized targeted advertisements were dis-

© Springer International Publishing AG 2016
P. Frasconi et al. (Eds.): ECML PKDD 2016, Part I, LNAI 9851, pp. 588–603, 2016.
DOI: 10.1007/978-3-319-46128-1_37

played in certain order next to web-search results after a search query. The proliferation of social media, micro-blogging and social-networking platforms has created novel opportunities for advertising. Facebook, Twitter, Tumblr, Pinterest and other online platforms aim at user engagement by providing services or showing content of interest, and they leverage the user base for marketing campaigns run by advertisers who pay to have their advertisements displayed. Advertising is performed by inserting sponsored posts in certain positions in the post feed on the user screen as the user scrolls (Fig. 1). A *feed* or *timeline* is a set of posts displayed on a user's screen, such as news, posts, updates, photos or videos. A sponsored post can be an ad adhering to the pay-per-click or the pay-per-impression model, a sponsored news item, or a promoted item such as a video or image.

The term coined for this type of advertising is *native* or *in-stream* advertising, as the ad format is assimilated into that of other content shown and is least intrusive for the user. Native advertising is becoming a multi-billion business with projected spend at $6.4 billion by 2017 only in USA. Selection, ranking and pricing of ads are realized with the Generalized Second-Price (GSP) auction as in web-search advertising. Advertisers bid an amount to pay per ad click, and ads are selected and displayed in prespecified positions in the post feed. Since the user scrolls on the screen and presumably views ads at a rate of one ad every some tens of seconds, a plausible scenario is that advertisers do not change their bids in between consecutive ad positions. Hence, the auction is run *once* before user scroll, and ads end up in their positions according to their rank as a result of the auction.

Fig. 1. Native advertisements in Facebook and Twitter online post feeds.

Click probability of an ad depends on three types of attributes: those that are inherent to the ad (*e.g.*, ad quality), those related to the user profile and activity, and those related to the post feed itself. The main inherent feature of an ad is its quality, which is reflected into ad design format, nature of the advertised product, accompanying text, and quality of the landing page. On the other hand, the relevance of the ad to the user profile captures the similarity of an ad to user preferences, search activity, posted items, and so on. For example an ad about a restaurant seems more appealing to a user that usually places food-related posts than it is to a user that posts text about books.

The third type of attributes concern the placement of ads in the user post feed. Based on evidence from online platforms [1–3], in native advertising, the ad click probability may depend on: (i) Relevance (i.e. context similarity) of the ad to preceding posts. For example, an ad about a hotel may be more likely to be clicked if shown directly after a post discussing vacation than after one on politics. (ii) Distance between consecutively shown ads. It is plausible that, the fewer the intervening posts between two ads, the smaller the click probability for the latter ad is, due to user saturation or fatigue effects. (iii) Position of the ad in the stream. If an ad is shown earlier in a feed, it will be clicked with higher probability than if shown later, since the user may quit scrolling.

In web search, the ranked list of ads is created through the GSP auction and is displayed next to search results. The rank of an ad is determined by the product of bid and click probability. On the other hand, in native advertising, the notion of rank is less clear, since native ads are placed in between diverse user posts. Further, in web search, the expected revenue of an ad decreases with its rank in the list. In native advertising, this is not the case, since the ad click probability depends on the precise *placement* of an ad in the post feed. As explained above, ads that appear earlier in the feed are not necessarily more likely to be clicked at, unless they are relevant to preceding posts or they are projected sparsely enough. In addition, in web search, the click probability of an ad depends only on its own rank and not on other ads. In native advertising the situation is more complex. Placing an ad at a certain position in the feed affects the click probability of this ad but also the click probabilities of subsequent ads. If these are placed close to the first ad and close to each other in general, user saturation due to frequent ad projection may lead to reduced click probability.

While the first two types of attributes above are also encountered in web-search advertising, the third one fundamentally distinguishes native from web-search advertising, and it is the one we model and study in this paper.

1.1   Our Contribution

We study optimal native ad selection and placement in social media post feeds, and the way it impacts the platform revenue. A set of ads emerge out of a GSP auction. Each ad comes with its apriori budget. In the prevalent pay-per-click model, each ad click entails a given amount of revenue for the platform and a corresponding amount of reduction for ad budget. The budget of each ad is renewed after a certain time interval. The product of click probability and revenue per ad click is the expected revenue from the ad. In our model, the ad click events are represented by Bernoulli random variables, and the ad click probability depends on the *relevance* of the ad to the preceding post, and on the *distance* between consecutive ads in the feed.

We seek the ad selection and allocation policy that optimizes a metric which combines (*i*) platform expected revenue, and (*ii*) uncertainty in revenue, captured by the variance of provisionally consumed budget of selected ads. The constraint is that the expected revenue attained for each selected ad should not exceed its apriori set budget. Uncertainty in revenue should be minimum, since this translates into reduced profit or wasted advertising opportunities for the platform. An ad selection and allocation policy that would lead to few clicks and small expected consumed budget is not preferable, since the revenue of the platform is smaller that it could potentially be. On the other hand, an ad allocation policy that would involve too many ad clicks while the ad budget is exhausted is also not desirable, since it leads to wasted advertising opportunities for the platform that are provided for free. That is, ad clicks that correspond to ad budget beyond the apriori one do not incur additional revenue for the platform. Furthermore, the expected revenue from ad clicking should be maximum.

To the best of our knowledge, both the problem and the model are novel in the literature. For clarity purposes, we study the problem for the post feed of one user. The model is amenable to multiple users as well as to extensions that include the other two types of features *i.e.*, those inherent to the ad or related to user profile. The contributions of our work are as follows.

- We provide a model and mathematical formulation for the problem of minimizing a combined metric of (*i*) uncertainty in ad-click generated revenue, which is quantified as the variance of provisionally consumed budget, and (*ii*) total expected revenue. The constraint is that the expected consumed budget for each selected ad should be no more than its apriori budget.
- We showcase how the relative positioning of ads in the post feed alters the ad click probabilities and therefore the revenue and uncertainty about it, and we specify the way in which the joint positioning of ads needs to be engineered so as to achieve the optimization objective above.
- We show that the problem above is an instance of a resource-constrained minimum-cost path one on an appropriately defined directed acyclic graph. The solution path reflects the policy of selecting which ads to show in the feed (out of a given set of ads), and in which positions to place them.

Through numerical evaluation results, we verify the tradeoff between revenue and uncertainty. The work in [27] is the most relevant to ours. Compared to that work, we consider the relevance of ads to posts as a factor that shapes ad click probability. Further, rather than adhering to the cumulative effect of projected ads on click probability as in [27], we assume that click probability of an ad depends on the distance from the previously shown ad. This essentially translates to a dependence of click probability on the average ad projection rate *i.e.*, the average number of posts elapsed until an ad is shown, which is a more plausible scenario that captures user annoyance. The work [21] is also relevant, in the sense that the measure of "regret" could be seen as similar to variance. In that work, the emphasis is on controlling social-network diffusion through user targeting without considering post feed aspects, while our work considers ad placement in the post feed.

The paper is organized as follows. In Sects. 2 and 3 we present the model, problem formulation and solution. Numerical results are provided in Sect. 4, literature overview is provided in Sect. 5, and the paper is concluded in Sect. 6. In the sequel, we use the words "ad" and "advertisement" interchangeably.

2   Model

We consider a set $\mathcal{T}$ of T posts displayed on a user screen in a social-media platform. Posts are displayed as a stream in a certain order e.g. most recently occurred, and the user scrolls through the posts. There is also a set $\mathcal{A}$ of N native ads with $N < T$. Typically, $N = \beta T$, $0 < \beta < 1$, with β in the range 10^{-2} to 10^{-1}. The set of ads $\mathcal{A}$ is the outcome of a bidding process among competing advertisers, and the social media platform selects the ones to display and their positions. Non-selected ads may be placed in subsequent feeds.

We assume the pay-per-click payment model; each time the user clicks on a displayed ad a, the advertiser is committed to pay the platform an amount b_a which emerges from the auction. Each ad comes with an apriori budget B_a, and each time an ad is clicked by a user, its budget is reduced by b_a. The budget of each ad is renewed after a certain time interval, and we focus our attention in studying the ad allocation policy in such a time interval.

For each ad $a \in \mathcal{A}$ and post $t \in \mathcal{T}$, let $r_{at} \in [0,1]$ be the *relevance* of ad a and post t. Relevance quantifies context similarity between the ad and the post. For example, an ad about a hotel is more relevant to a post on vacation than to one on politics. Relevance may be computed through cosine similarity [4, Chap. 9] or other metrics on vectors of words that are representative of the post and the ad. These may be defined *e.g.*, with the Term-Frequency-Inverse Document Frequency (TF-IDF) metric from information retrieval [4, Chap. 1].

We consider two main determinants of probability of click for an ad: ad relevance to the preceding post (after which the ad is placed), and distance (in elapsed posts) from the previously displayed ad. We assume that we learn from historical data the probability $p(r, d)$ that an ad is clicked if placed at distance d posts from the previous ad, $d = 1, \ldots, T$, and if it is displayed after a post of relevance r to it. Logistic regression or other machine-learning tools can be used to train a model and learn $p(r, d)$. Recall that the ad click probability may depend on other attributes such as user profile or ad quality, or its position (early/late) on the stream. However, we choose not to include them here for clarity of presentation, and because we wish to focus on attributes that are peculiar to native advertising.

The nature of native advertising implies that, besides the style and layout of the ad, the content of the ad should be assimilated to that of other content shown on the platform *e.g.*, the post feed. Hence, we consider the relevance of the ad to the preceding post as an attribute that affects the ad click probability. Further, the distance from the previously shown ad is essentially mapped to ad projection rate *i.e.*, average number of posts elapsed until an ad is shown. Our rationale for selecting these attributes to map to ad click probability is spurred by realistic marketing principles in online social networks and media. Platforms aim at high user experience while ensuring that ads obtain substantial attention. For instance, Facebook does not show too many ads and too frequent so as to prevent negative impact on user experience and engagement. It shows adequate number of ads so as to consume their budget and have advertisers satisfied.

Let us index post positions by $t = 1, \ldots, T$, and ads by $a = 1, \ldots, N$. Define binary variables x_{at}, for $a = 1, \ldots, N$, and $t = 1, \ldots, T$, with $x_{at} = 1$, if ad a is placed after post t, and $x_{at} = 0$ otherwise. An *ad selection and allocation policy* is a $NT \times 1$ *ad allocation vector* $\mathbf{x} = (x_{at} : a \in \mathcal{A}, t \in \mathcal{T})$. For ad a, let $t_{\mathbf{x}}(a)$ be the post after which ad a is placed according to policy $\mathbf{x}$. For ad a, define the distance from the previous ad, $d_a(\mathbf{x})$, for allocation policy $\mathbf{x}$, as

$$
d_a(\mathbf{x}) = \begin{cases} t_{\mathbf{x}}(a), & \text{if } t_{\mathbf{x}}(a) = \min_{a' \in \mathcal{A}} t_{\mathbf{x}}(a'), \\ t_{\mathbf{x}}(a) - \max_{\substack{a' \in \mathcal{A}: \\ t_{\mathbf{x}}(a') < t_{\mathbf{x}}(a)}} t_{\mathbf{x}}(a'), & \text{otherwise}. \end{cases} \tag{1}
$$

If an ad is the first one placed in the feed, its distance is just the index of the preceding post (after which the ad is placed). Otherwise, the distance is the difference between the index of the preceding post of that ad, and that of the preceding post of the immediately previously placed ad. We define $d_a(\mathbf{x}) = 0$ if ad a is not allocated in the feed, i.e. if $x_{at} = 0$ for all t.

Given an ad allocation policy $\mathbf{x}$, each allocated ad a has a click probability $p(r_{at}, d_a(\mathbf{x}))$. For notational simplicity, and with a little abuse of notation, let us in the sequel use notation $p_{at}(\mathbf{x})$ to denote $p(r_{at}, d_a(\mathbf{x}))$. The event of click of each allocated ad a when placed after post t according to allocation policy $\mathbf{x}$ may be represented by a Bernoulli random variable $X_{at}(\mathbf{x})$ having as parameter the probability $p_{at}(\mathbf{x})$. Thus,

$$
X_{at}(\mathbf{x}) = \begin{cases} b_a, & \text{w.p. } p_{at}(\mathbf{x}), \\ 0, & \text{w.p. } 1 - p_{at}(\mathbf{x}), \end{cases} \tag{2}
$$

with expectation $\mathbb{E}[X_{at}(\mathbf{x})] = b_a p_{at}(\mathbf{x})$ and variance

$$
\text{var}[X_{at}(\mathbf{x})] = b_a^2 \, p_{at}(\mathbf{x}) \cdot (1 - p_{at}(\mathbf{x})). \tag{3}
$$

Denote by $R(\mathbf{x})$ the random variable that shows platform revenue as function of the allocation policy $\mathbf{x}$, with

$$
R(\mathbf{x}) = \sum_{a=1}^{N} \sum_{t=1}^{T} X_{at}(\mathbf{x}) x_{at}. \tag{4}
$$

The total expected revenue for the platform for an ad allocation policy $\mathbf{x}$ is

$$
\mathbb{E}[R(\mathbf{x})] = \sum_{a=1}^{N} \sum_{t=1}^{T} b_a p_{at}(\mathbf{x}) x_{at}. \tag{5}
$$

The underlying assumption in (5) is that the Bernoulli random variables that show click events of different ads in the feed are independent from each other. This is a plausible assumption, since the probability that an ad is clicked does not seem to depend on whether a previous ad was clicked or not. On the other hand, the model recognizes that the probability of ad click depends on *how many posts elapse* since the appearance of the most recent ad in the feed, and thus it captures in a sense the annoyance caused to the user; this is reflected through the dependence of click probability on distance. The variance of revenue is

$$\mathrm{var}[R(\mathbf{x})] = \sum_{a=1}^{N} \sum_{t=1}^{T} \mathrm{var}[X_{at}(\mathbf{x})] x_{at} = \sum_{a=1}^{N} \sum_{t=1}^{T} b_a^2 p_{at}(\mathbf{x})(1 - p_{at}(\mathbf{x})) x_{at}. \quad (6)$$

3   Problem Formulation and Solution

We are interested in the ad selection and allocation policy $\mathbf{x}^*$ that minimizes a metric of the form, $(\mathrm{var}[R(\mathbf{x})] - \lambda \mathbb{E}[R(\mathbf{x})])$, where $\lambda \geq 0$ is a calibration parameter that determines the relative emphasis on total expected revenue and its variance. The rationale for selecting this metric is that it arises in the constrained optimization problems of maximizing expected revenue subject to keeping variance of revenue less than a given value, and that of minimizing variance of revenue subject to keeping expected revenue larger than a value. The optimal policy selects a number of ads to place in the feed and may place an ad several times in the feed. The number of selected ads may be small or large, depending on what is better for the objective above. The optimal policy may place an ad more times after posts that induce small click probability or fewer times after posts that induce larger click probability. An allocation policy $\mathbf{x}$ changes the expected values and the variances of individual random variables that correspond to placed ads in the feed, and hence it affects $\mathbb{E}[R(\mathbf{x})]$ and $\mathrm{var}[R(\mathbf{x})]$. There exists a non-trivial coupling in the problem. The probability associated with a certain ad placed depends on the specific position (post) through the ad-post relevance, but it also depends on the post distance from the previously placed ad.

If $\lambda = 0$ or it is very small, the objective is to minimize deviation of revenue from the expected one. As the total expected revenue has no or little weight in the objective, the platform is selective in choosing a subset of ads to allocate in the feed such that the uncertainty in revenue is minimized. A large uncertainty translates to potentially reduced revenue or to wasted advertising opportunities for the platform. Specifically, an ad allocation policy that would lead to an ad clicking profile with few clicks would result in a smaller expected consumed budget than the one that could potentially be consumed. On the other hand, an ad allocation policy with an ad clicking profile with too many clicks is also not desirable, since it would translate into wasted advertising opportunities and advertising service that the platform would provide for free. That is, ad clicks that correspond to ad budget beyond the apriori defined one do not incur revenue until the budget is renewed. If $\lambda = \infty$ or it is very large, the aim is to

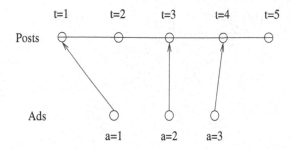

Fig. 2. Example of a feasible ad placement in $T = 5$ posts. Ad 1 is placed after post 1, ad 2 is placed after post 3, and ad 3 is placed after post 4. The expected revenue for this assignment is $b_1p(r_{11}, 1) + b_2p(r_{23}, 2) + b_3p(r_{34}, 1)$. For example, for ad 2, click probability is $p(r_{23}, 2)$ since the ad is placed after post 3 (hence the relevance factor r_{23}) and at distance $d_2 = 2$ posts from the previous ad, ad 1. Here, each ad is placed in exactly one position in the feed; in our case, an ad may be repeated.

maximize total expected profit. In that case, the platform does not place emphasis on revenue uncertainty, and hence it is more tolerant to wasted advertising opportunities.

The problem of ad selection and allocation so as to minimize the combined metric above is formulated as follows:

$$\min_{\mathbf{x}} \left(\text{var}[R(\mathbf{x})] - \lambda \mathbb{E}[R(\mathbf{x})]\right) = \min_{\mathbf{x}} \sum_{a=1}^{N} \sum_{t=1}^{T} \left(b_a^2 p_{at}(\mathbf{x})(1 - p_{at}(\mathbf{x})) - \lambda b_a p_{at}(\mathbf{x})\right) x_{at} \tag{7}$$

subject to:

$$\sum_{t=1}^{T} b_a p_{at}(\mathbf{x}) x_{at} \le B_a, \ \forall \text{ ad } a \in \mathcal{A}, \tag{8}$$

and

$$\sum_{a=1}^{N} x_{at} \le 1, \ \forall \text{ post } t \in \mathcal{T}, \tag{9}$$

with $x_{at} \in \{0, 1\}$. Constraint (8) says that for each ad, the expected revenue from policy $\mathbf{x}$ should be no more than its apriori budget, B_a. Further, constraint (9) says that at most one ad is placed after a post. Figure 2 depicts an example allocation policy of ads to posts, where each ad is displayed exactly once.

Problem (7)–(9) is a non-standard one. If ad positions were known, distances $d_a(\cdot)$ between ads would also be known, and the problem would be to select the ads to place in these positions. Even in that case, the problem would be a generalized assignment (GAP) one on the bipartite graph of nodes $\mathcal{A} \cup \mathcal{T}$ with link weights $b_a^2 p_{at}(1 - p_{at}) - \lambda b_a p_{at}$, for each link connecting ad a and post t. The GAP problem is already NP-Hard [5]. The need to determine distances $d_a(\cdot)$ further complicates the problem, since the decision on distance $d_a(\cdot)$ of an ad a

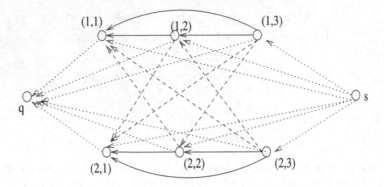

Fig. 3. A graph G corresponding to $N = 2$ ads and $T = 3$ posts that illustrates the mapping of the ad selection and placement problem to a shortest-path one. For example, if the shortest path is $s \to (1,3) \to (2,1) \to q$, this means that placing ad 1 after post 3 and ad 2 after post 1 minimizes the value of objective (7).

would affect the weights of links emanating from ad a but also the weights of links for the ad to be placed after ad a.

3.1   Graph Model and Solution

We construct the following directed graph G. For each pair of ad $a \in \mathcal{A}$ and post $t \in \mathcal{T}$, we define a node (a,t). The easiest way to visualize it is if we place nodes (a,t) in rows and columns; for each ad a, nodes (a,t), $t = 1,\ldots,T$ are in one row, and nodes corresponding to different ads are in different rows. Node (a,t) represents the tentative placement of ad a after post t. There also exist two other nodes s and q.

Next, we add links as follows. We add a link from each node (a,t) to nodes (a',t') with $t' < t$. That is, for each ad a, we add links between nodes (a,t) and (a,t') for $t' < t$. We also add links between (a,t) and (a',t'), for $a' \neq a$ and $t' < t$. The weight of each link that points from node (a,t) to (a',t') is

$$w_{(a,t),(a',t')} = b_a^2 p(r_{at}, t - t')[1 - p(r_{at}, t - t')] - \lambda b_a p(r_{at}, t - t'). \tag{10}$$

We also add a link from node s to each node (a,t) in the graph with weight 0 and a link from each node (a,t) to q with weight

$$w_{(a,t),q} = b_a^2 p(r_{at}, t)[1 - p(r_{at}, t)] - \lambda b_a p(r_{at}, t). \tag{11}$$

For each node (a,t), let $\mathcal{O}_{(a,t)}$ be the set of outgoing links of node (a,t), i.e. the set of links that originate from (a,t). The resulting graph is a weighted directed acyclic graph (DAG).

Main observation. A path from s to q in the weighted graph G corresponds to an ad selection and placement policy. A minimum-cost path from s to q corresponds to a policy that leads to minimum value in the objective (7). Nodes

(a, t) that are part of the minimum-cost path correspond to ads a that are assigned in positions t. An example graph G for $N = 2$ ads and $T = 3$ posts is shown in Fig. 3.

First, consider the problem with the optimization objective (7) subject only to constraint (9), while the ad budget constraint (8) is relaxed. From the discussion above, we deduce that finding a policy that minimizes the objective (7) subject to constraint (9) is equivalent to finding a minimum-cost path from s to q in the graph above. The minimum-cost path from s to q can be found by running the Bellman-Ford (BF) algorithm, which also applies to graphs with negative weights, as long as there are no negative cycles. In our case, the graph is a DAG with possibly negative link weights, but with no cycles. In fact, a variant of the BF algorithm can find the minimum-cost path for a DAG in $\Theta(|V| + |E|)$ time, where $|V|$ is the number of nodes and $|E|$ is the number of links of the graph [6, Sect. 24.2]. In G, there exist $(NT + 2)$ nodes and $O(NT^2)$ number of links, thus the algorithm runs in $O(NT^2)$ time.

Now, consider our problem with the optimization objective (7) subject to constraints (8) and (9). In our formulation, the expected consumed budget for each ad should not exceed B_a. We associate each ad a with a "resource type" a. For link $e \in \mathcal{O}_{(a,t)}$ from node (a, t) to node (a', t') let ad a have budget consumption $d_a^e = b_a p(r_{at}, t - t')$ while other ads $a' \neq a$ have consumption $d_{a'}^e = 0$ for that link. Furthermore, link e from node (a, t) to node q has consumption level $d_a^e = b_a p(r_{at}, t)$ for ad a, and 0 for all other ads. Links from node s to nodes in the graph have consumption level equal to 0 for all ads. The total consumption level for ad a in a path p is $d_a(p) = \sum_{\ell \in p} d_a^\ell$.

A resource-constrained path p (where "constrained" refers to the total consumed budget of a resource type, $i.e.$, ad in it) from s to q is $feasible$, if and only if $d_a(p) \leq B_a$ for all ads $a \in \mathcal{A}$, i.e. if it comprises links such that the total consumed budget for each ad included in the path is more than B_a. For ads that are not included in the path, the inequality trivially holds since these ads do not consume budget. The problem in this case is equivalent to a resource-constrained shortest-path one, which is NP-Hard [5]. There exist several heuristics proposed in the literature for solving the problem, see $e.g.$, [7,8].

4   Numerical Evaluation

4.1   Setup and Data

We approach user ad click behavior as an instance of the two-class probabilistic classification problem, where the two classes C_0 and C_1 correspond to the alternatives of not clicking and clicking an ad respectively. The way the user weighs the attributes associated with an ad a (namely the relevance r_a to the post, and the distance d_a from the previous ad) so as to reach a decision is modeled through a logistic regression model. Given a vector of values $\mathbf{x}$ for the two ad attributes, the ad is clicked with probability

$$\Pr(C_1|\mathbf{x}) = \frac{1}{1 + e^{-\mathbf{w} \cdot \mathbf{x}}} = \sigma(\mathbf{w} \cdot \mathbf{x}), \tag{12}$$

where $\sigma(y) = (1 + e^{-y})^{-1}$ denotes the logistic sigmoid function, while $\mathbf{w} \cdot \mathbf{x}$ denotes vector dot product, and $\mathbf{w}$ is the vector of attribute weights. These weights are learned from historical data and capture the significance that the user places on the two different attributes and their values in reaching a decision. Similarly, ad a is not clicked with probability $\Pr(C_0|\mathbf{x}) = 1 - \Pr(C_1|\mathbf{x})$. An important property of logistic regression is that the objective for learning weights $\mathbf{w}$ is convex, so there are no local optima involved.

Real datasets for native advertising are scarce to find, and studies on native ads are either purely theoretical with no data experiments, e.g., [23,27], or they use company proprietary data, e.g., [24,25] [26, Chap. 7]. Hence in this work, we employ synthetic datasets to justify our claims and test our model. The training dataset consists of triads of the form (r_a, d_a, c_a), where each triad represents an ad a. In each triad, $r_a \in [0,1]$ is the relevance to the post after which the ad was placed, while $d_a \in [0,1]$ is the distance from the previous ad, normalized with a defined maximum possible distance between two ads, and $c_a \in \{0,1\}$ denotes whether the ad a was clicked or not. To set the value of c_a to 0 or 1, we calculate metric $0.5 \times r_a + 0.5 \times d_a$, and if this is greater than a configurable threshold, which we take here to be equal to 0.75, then we set $c_a = 1$, else we set $c_a = 0$.

The classical loss function minimization approach with regularization was used to train our algorithm [9]. For training, we generate 50 triads and for testing we generated $1,000$ (r,d)-pairs. Given the trained logistic-regression model, we estimate the ad click probabilities through the sigmoid function. We create a pool of 50 ads and in each experiment, we select $T = 20$ posts and we draw from the pool a certain maximum number of ads, $K_{\max}$ to include in the feed. For simplicity we take the bid $b_a = 1$ and the apriori budget $B_a = 5$ for each ad.

The resource-constrained minimum-cost problem (7)–(9) was solved with the Lagrangian relaxation heuristic from [7], and the parameters were selected so that problem feasibility was not an issue. Based on [7], we can show that the

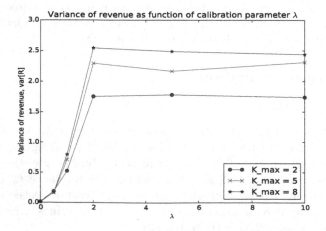

Fig. 4. Variance of revenue as a function of λ for different values of ad pool size, $K_{\max} = 2, 5, 8$.

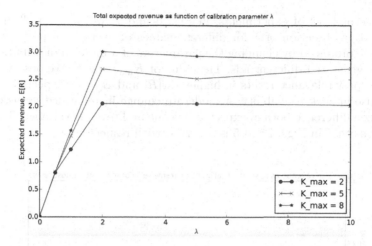

Fig. 5. Expected revenue as function of λ for different ad-post relevance.

algorithm runs in $O(N^2T^4\log^4(NT^2))$ time. Note that the algorithm does not need to solve the problem in an online fashion, but rather it pre-computes the position of ads in a post feed, hence the algorithm requirements in execution time (and thus, complexity) are not so stringent. Although the scale of the problem will be larger in practice, our conjecture is that the trends will remain the same, as only the associated parameters of the optimization problem will change.

4.2   Numerical Results

In the first set of numerical experiments, we assess the impact of calibrating parameter λ. In Figs. 4 and 5 we depict $\text{var}[R]$ and $\mathbb{E}[R]$ respectively as a function of λ for a maximum number of ads to be placed in the feed $K_{\max} = 2, 5$ and 8. For each value of λ, we solve the resource-constrained shortest-path problem, and for the solution path we measure $\text{var}[R]$ and $\mathbb{E}[R]$ by summing the corresponding link costs over path links. Each value in the plot corresponds to the average value over ten experiments, where for each experiment the pool of ads and the set of $T = 20$ posts are varied. Both the variance of revenue and the expected revenue are seen to increase as λ increases up to a certain value $i.e.$, $\lambda = 2$, while for values $\lambda > 2$, the respective values of variance and expected revenue are almost stabilized, or they change slightly. This fact demonstrates the tradeoff that, if the platform wishes to increase revenue, it would have to tolerate higher variance, $i.e.$, higher uncertainty in revenue. As λ increases, link costs decrease, and therefore more ads tend to be placed in principle in the feed. This results both in higher $\text{var}[R]$ and $\mathbb{E}[R]$.

The second observation from Figs. 4 and 5 is that as the value of $K_{\max}$ increases, both the expected revenue and the revenue variance increase, albeit the difference in the increase of these metrics decreases, as $K_{\max}$ increases.

In a second set of experiments, in Figs. 6 and 7, we plot var[R] and $\mathbb{E}[R]$ respectively as function of λ for different values of average ad-post relevance, which were produced by changing the parameters of a uniform distribution with which we generated different ads. The value of K_{max} was 5. We observe that a higher ad-post relevance resuts in higher var[R] and $\mathbb{E}[R]$ as expected, because of the raise in click probability. A moderate change in relevance seems capable of making a difference both in expected revenue and revenue variance. The same behavior as that in Figs. 4 and 5 is observed with respect to λ.

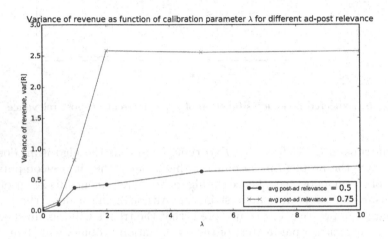

Fig. 6. Variance of revenue as function of λ for different ad-post relevance.

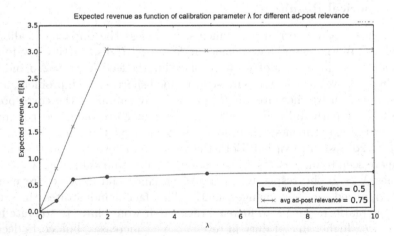

Fig. 7. Expected revenue as function of λ for different ad-post relevance.

5   Related Work

In sponsored-search auctions that are used in web search advertising, ads are ranked based on expected profit, which is the product of bid and click-through-ratio (CTR), *i.e.*, the probability that the ad will be clicked. When the user clicks on an ad at the k-th position with bid b_k and CTR_k, the advertiser pays $b_{k+1} \times \text{CTR}_{k+1}$ according to the Generalized Second-Price (GSP) auction [10,11]. In [12], a variation of GSP is presented, where each ad bid undergoes a fine, equal to a metric of negative impact of the ad on user experience. Advertisers that are charged with large fines are less willing to enter the competition, and thus the platform becomes more attractive to other advertisers. Under certain conditions, the winners' gains in this new less crowded setting may supercede payments due to fines, and thus overall winners are benefited.

A recent thread concerns ad allocation through stochastic optimization. In [13], the optimal-auction framework is used for the single-slot revenue maximization. The optimal policy is to allocate the slot to users in decreasing order of $q_i \nu_i$ where q_i, ν_i are the selling probability and valuation of user i. In [14], the authors use Lyapunov optimization for the problem of maximizing long-term average revenue for a web-search service provider by dynamically allocating ads to webpage slots in the presence of dynamic keyword query arrivals, subject to a long-term average budget constraint. The work [15] studies the problem of allocating budget-constrained advertisers in each keyword auction round so as maximize the likelihood of ad click or to reduce advertiser cost per click. Dynamic actions under limited budget over the entire horizon are studied in [16] through the lens of multi-armed bandit theory.

Another thread that relates to advertising is *social influence*, which involves positive externalities, namely that the benefit from influencing a user comes also from indirect influence of that user onto others. The seminal paper [17] formulates the problem of influence maximization as one of selecting a subset of users (seeds) to advertise to so that the cascading effect in the graph reaches the maximum number of users. They show that the problem is NP-Hard and propose a greedy algorithm with constant-factor approximation guarantees based on sub-modularity properties of the set function of anticipated influence. Various extensions have been considered, *e.g.*, on optimal marketing strategies that include pricing and the sequence of offers to social-network users that may or may not respond strategically [18,19], and on user targeting for global opinion maximization under a game-theoretic model for opinion formation [20]. The work [21] studies ad allocation to social-network users under a certain diffusion model and topic-based influence. The host aims on leveraging virality to improve advertising efficacy, while avoiding giving away free service due to uncontrolled virality. The problem is to allocate ads to minimize regret, defined as the absolute difference between the expected revenue and the budget of each ad. Social diffusion in ads is also considered in [22], where the joint problem of targeting ad impressions to users and of scheduling them in time is studied with the aim to maximize expected number of clicks. The work [23] considers the problem of selecting a set of ads and the number of times to display each ad so as to minimize the error in estimating the true CTR of ads.

Native ads have spurred interest of the research community in the last two years or so, aiming at improved used experience, see *e.g.*, [24,25] [26, Chap. 7]. The works [24,25] aim to predict ad quality by focusing respectively on post-click user behavior on the ad landing page, and on user feedback about offensive ads. In a related work [27], the problem of ad placement in a stream is addressed. The model involves a probability that the user will reach to the ad, which is decreasing function of the number of ads shown previously. Given a set of ads, a reward and a set of candidate positions to place each ad, the objective is to find an ad placement that maximizes total reward. An approximation algorithm is proposed, albeit the computational complexity of the problem is not characterized. The authors also use the optimal-auction framework to design a mechanism that is truthful and approximately optimal in terms of revenue.

6   Conclusion

We study native advertisement selection and placement in the post feed of a user, and we optimize a metric that combines expected platform revenue and revenue uncertainty. Ad click probabilities are derived with a machine-learning model that maps the key attributes of post-ad relevance and distance from the previous ad to click probability. Next, these ad click probabilities are engineered and adapted through ad selection and placement in the feed to achieve the objective. We showed that the problem becomes a resource-constrained minimum-cost path one.

To the best of our knowledge, both the problem and the model are novel. The model is amenable to various extensions such as the one for multiple user feeds, thus encapsulating the relevance of an ad to personalized user profile as a factor that shapes click probability. In that case, personalized user models for ad click behavior would be needed. Other attributes that shape ad click probability could be included in the model as well, such as ad quality. We are currently in the process of designing a larger-scale real-life experiment with training and test data that come through some hundreds of real users on tentative Facebook feeds and ads presented to them through a mobile app.

References

1. https://www.bigfin.com/blog/facebook-to-provide-relevant-news-feed-ads/
2. http://www.buzzfeed.com/mattlynley/this-is-how-an-ad-gets-placed-in-your-facebook-news-feed#.gq48arOBY
3. http://marketingland.com/facebook-raises-limits-daily-frequency-news-feed-ads-96517
4. Leskovec, J., Rajaraman, A., Ullman, J.: Mining Massive Datasets. Cambridge University Press, New York (2014)
5. Garey, M.R., Johnson, D.S.: Computers and Intractability: A guide to the Theory of NP-Completeness. Freeman, New York (1979)
6. Cormen, T., Leiserson, C.E., Rivest, R.L., Stein, C.: Introduction to Algorithms, 3rd edn. MIT Press, Cambridge (2009)

7. Juttner, A., Szviatovszki, B., Mecs, I., Rajko, Z.: Lagrange relaxation based method for the QoS routing problem. In: Proceedings of IEEE INFOCOM (2001)
8. Boland, N., Dethridge, J., Dumitrescu, I.: Accelerated label setting algorithms for the elementary resource constrained shortest path problem. Oper. Res. Lett. **34**(1), 58–68 (2006)
9. Flach, P.: Machine Learning: The Art and Science of Algorithms that Make Sense Out of Data. MIT Press, Cambridge (2012)
10. Varian, H.: Online ad auctions. Am. Econ. Rev. **99**(2), 430–434 (2009)
11. Narahari, Y., Garg, D., Narayanam, R., Prakash, H.: Game Theoretic Problems in Network Economics and Mechanism Design Solutions. Advanced Information and Knowledge Processing. Springer, London (2009)
12. Stourm, V., Bax, E.: Pigovian taxes can increase platform competitiveness: the case of online display advertising. arXiv preprint arXiv:1411.0710
13. Menache, I., Ozdaglar, A., Srikant, R., Acemoglu, D.: Dynamic online-advertising auctions as stochastic scheduling. In: Proceedings of NetEcon (2009)
14. Tan, B., Srikant, R.: Online advertising, optimization and stochastic networks. IEEE Trans. Autom. Control **57**(11), 2854–2868 (2012)
15. Karande, C., Mehta, A., Srikant, R.: Optimizing budget constrained spend in search advertising. In: Proceedings of ACM Web Search and Data Mining (WSDM) Conference (2013)
16. Wu, H., Srikant, R., Liu, X., Jiang, C.: Algorithms with logarithmic or sublinear regret for constrained contextual bandits. In: Proceedings of the Neural Information Processing Systems Conference (2015)
17. Kempe, D., Kleinberg, J.M., Tardos, E.: Maximizing the spread of influence through a social network. In: Proceedings of ACM Knowledge Discovery Data Mining (KDD) (2003)
18. Hartline, J.D., Mirrokni, V.S., Sundararajan, M.: Optimal marketing strategies over social network. In: Proceedings of ACM World Wide Web (WWW) Conference (2008)
19. Candogan, O., Bimpikis, K., Ozdaglar, A.: Optimal pricing in the presence of local network effect. In: Proceedings of Conference on Web and Internet Economics (WINE) (2010)
20. Gionis, A., Terzi, E., Tsaparas, P.: Opinion maximization in social networks. In: Proceedings of SIAM International Conference on Data Mining (SDM) (2013)
21. Aslay, C., Lu, W., Bonchi, F., Goyal, A., Lakshman, L.V.S.: Viral marketing meets social advertising: ad allocation with minimum regret. In: Proceedings of Very Large DataBases (VLDB) Conference (2015)
22. Abbassi, Z., Bhaskara, A., Misra, V.: Optimizing display advertising in online social networks. In: Proceedings of WWW Conference (2015)
23. Gopalakrishnan, R., Bax, E., Chitrapura, K.P., Garg, S.: Portfolio allocation for sellers in online advertising. arXiv preprint arXiv:1506.02020
24. Lalmas, M., Lehmann, J., Shaked, G., Silvestri, F., Tolomei, G.: Promoting positive post-click experience for In-stream Yahoo Gemini users. In: Proceedings of ACM Knowledge Discovery Data Mining (KDD) (2015)
25. Zhou, K., Redi, M., Haines, A., Lalmas, M.: Predicting pre-click quality for native advertisements. In: Proceedings of ACM World Wide Web (WWW) Conference (2016, to appear)
26. Lehmann, J.: From site to inter-site user engagement: fundamentals and applications. Ph.D. thesis, Universitat Pompeu Fabra (2014)
27. Ieong, S., Mahdian, M., Vassilvitskii, S.: Advertising in a stream. In: Proceedings of ACM WWW Conference (2014)

Asynchronous Feature Extraction for Large-Scale Linear Predictors

Shin Matsushima[(✉)]

The University of Tokyo, Tokyo, Japan
shin_matsushima@mist.i.u-tokyo.ac.jp

Abstract. Learning from datasets with a massive number of possible features to obtain more accurate predictors is being intensively studied. In this paper, we aim to perform effective learning by using the L1 regularized risk minimization problems regarding both time and space computational resources. This is accomplished by concentrating on the effective features from among a large number of unnecessary features. To achieve this, we propose a multithreaded scheme that simultaneously runs processes for developing seemingly important features in the main memory and updating parameters regarding only the important features. We verified our method through computational experiments, showing that our proposed scheme can handle terabyte-scale optimization problems with one machine.

1 Introduction

In this paper, we consider the regularized risk minimization problems using L1 regularization, which is formulated as minimization of the following convex function [12]:

$$P(\mathbf{w}) \triangleq \|\mathbf{w}\|_1 + C \sum_{i=1}^{n} \ell(\langle \mathbf{w}, \phi(\mathbf{x}_i) \rangle, y_i). \tag{1}$$

Here, $\mathbf{x}_i \in \mathcal{X}$ and $y_i \in \mathcal{Y}$ represent the input and output of a datapoint, $\mathbf{w} \in \mathbb{R}^p$ is a vector of parameters, $\ell : \mathbb{R} \times \mathcal{Y} \to \mathbb{R}$ defines a convex function for any given $y \in \mathcal{Y}$, and $\phi : \mathcal{X} \to \mathbb{R}^p$ explicitly denotes a certain feature function. This formulation includes many important problems in machine learning. For example, $\ell(\langle \mathbf{w}, \phi(\mathbf{x}_i) \rangle, y_i) = (\langle \mathbf{w}, \phi(\mathbf{x}_i) \rangle - y_i)^2$ corresponds to an equivalent problem in the LASSO introduced in [16]. The setting of $\ell(\langle \mathbf{w}, \phi(\mathbf{x}_i) \rangle, y_i) = \log(1 + \exp(-y_i \langle \mathbf{w}, \phi(\mathbf{x}_i) \rangle))$ corresponds to L1 logistic regression in binary classification and $\ell(\mathbf{w}, \mathbf{x}_i, y_i) = -\langle \mathbf{w}, \phi(\mathbf{x}_i) \otimes y_i \rangle + \log\left(\sum_y \exp(\langle \mathbf{w}, \phi(\mathbf{x}_i) \otimes y \rangle)\right)$ corresponds to multiclass classification.

With the availability of a larger number of datapoints because of recent developments in information technology, it has become important to consider larger number of features to avoid underfitting, fully utilize data, and enhance the performance of a predictor. Recent intensive research has revealed that complex

© Springer International Publishing AG 2016
P. Frasconi et al. (Eds.): ECML PKDD 2016, Part I, LNAI 9851, pp. 604–618, 2016.
DOI: 10.1007/978-3-319-46128-1_38

multilayer nonlinear models perform better than simple linear models without overfitting despite a much larger hypothesis space [7]. This suggests that preparing a large class of feature functions and adaptively choosing informative feature functions can enhance the performance of the predictor for linear models as well.

Conventionally, when we solve (1) in practice, the process of computing $\Phi \triangleq \{\phi_j(\mathbf{x}_i)\}_{1 \leq i \leq n, 1 \leq j \leq p}$, and finding a solution for the formulation are separated. We develop the features $\phi_j(\mathbf{x}_i)$ for all i and j before optimization (typically into the lower level of memory such as hard disk), and run an optimization algorithm to find the solution. However, when the number of features is extremely large, their extraction leads to huge memory costs, associated with storing all the values of Φ, and substantial computational time costs. Especially, when Φ cannot fit into the main memory, running an optimization algorithm with frequent accesses to lower levels of memory is impractical. Moreover, finding the solution requires considerable computational time because of the increasing size of the optimization problem. Furthermore, designing feature spaces containing an exponentially large number of features is straightforward in most cases, as discussed in Sect. 3.1. Therefore, the entire process (or scheme), including the optimization algorithm and methods of developing the feature Φ, needs to be efficient.

The block minimization scheme presented by Yu et al. was the first scheme to propose solving of the regularized risk minimization problem when the data does not fit in the memory [20]. In this scheme, it was proposed that the data would be split into several blocks, each containing a relatively small number of datapoints to fit each block into a higher level of memory. By considering both primal and dual variables, they obtained a subproblem in which only datapoints in a single block can be accessed at a time. They showed that the global solution can be achieved by solving subproblems successively and repeatedly. Matsushima et al. proposed the dual cached loop method, which uses multithreading to run a reading thread that accesses a disk to read each datapoint and a training thread that updates parameters simultaneously [8]. The aforementioned schemes are all focused on L2 regularized risk minimization problems, in which it is preferable to solve the dual problem rather than the primal problem. For the L1 regularized problem, which has rarely been focused upon, it is preferable to solve the primal problem than the dual problem.

From an algorithmic perspective, a key insight that has been used to scale up this risk minimization problem is that there are several optimization methods that do not require all the information within Φ at once. Stochastic gradient methods [14] are widely used in large-scale optimization problems because they require only one instance at a time. Similarly, coordinate descent methods [9] can update parts of parameters, j-th component of $\mathbf{w}$ in this case, with only information from the j-th column of Φ, as explained in Sect. 2. Moreover, the L1 regularization problem (1) implies that most features are redundant and contribute no information to the estimated predictor. In particular, the following statement is said to hold true [14].

Property 1. Let w_j^* be the j-th component of the solution of (1) based on $(\varPhi, \mathbf{y})$. Further, let J^* represent columns $\{j | w_j^* \neq 0\}$ and $\varPhi_{J^*}$ be a matrix containing only the j-th column such that $j \in J^*$. Then, $\hat{w}_j^*$, the component of the solution of (1) based on $(\varPhi_{J^*}, \mathbf{y})$ corresponding to j, coincides with w_j^*.

This implies that the intrinsic size of the optimization problem (1) is much smaller than it appears; the optimization problem can be reduced, given the information in which components of the parameter vector will be annihilated. This suggests that computational efforts to not only optimize by considering such parts of the parameter but also to develop such features and load them into memory are inefficient.

The dual cached loop scheme, similar to the block minimization scheme, can be easily integrated into the stochastic gradient descent (SGD) method, although this is not explicitly indicated in prior research [8]. Therefore, the L1 regularized problem when the dataset cannot fit into memory can be solved using the SGD method on the basis of the aforementioned schemes. However, as SGD requires access to one row of the data matrix, we cannot exploit the fact that the intrinsic size of the optimization problem is much smaller than it initially appears.

In this study, we develop a scheme to efficiently compute the solution of (1) with a large value of $\varPhi$ that cannot fit in the main memory by utilizing structures of coordinate descent algorithms and L1 regularization problems. We call our scheme *feature cached loops* (FCL). In this scheme, two threads run asynchronously and simultaneously: one for extracting features and another for updating the solutions to the optimization problem. We thus aim to efficiently extract effective features for the temporal values of parameters. As discussed in Sect. 3.2, this algorithm can be said to operate on a principle similar to that used in boosting algorithms [13].

The remainder of this paper is organized as follows. Section 2 reviews the coordinate descent method, which is the most important building block of our scheme. In Sect. 3, we explain our scheme in detail and discuss similarities with the boosting algorithm. Section 4 presents the experimental evaluation of our method. Finally, Sect. 5 concludes this paper.

2   Coordinate Descent Method

Coordinate descent methods are well-known optimization methods used for minimizing convex functions. Recently, several studies have highlighted the methods computational efficiency, fast theoretical convergence rate, and suitability to large scale learning [1,9]. Coordinate descent methods aim to find the solution by selecting one component of the parameters, w_j in case of (1), and then solving the one-variable optimization problem to update w_j. This procedure is repeated for each choice of a parameter component. The coordinate choice can be random, cyclic, or sampled from an arbitrary distribution. In the LASSO formulation, that is, minimizing (1) when $\ell(\langle \mathbf{w}, \phi(\mathbf{x}_i) \rangle, y_i) = (\langle \mathbf{w}, \phi(\mathbf{x}_i) \rangle - y_i)^2$, the one-variable optimization can be solved analytically as follows:

$$w_j^{t+1} = \underset{w_j}{\operatorname{argmin}} \, P(\mathbf{w}^t + (w_j - w_j^t)\mathbf{e}_j)$$

$$= \begin{cases} w_j^t - \dfrac{\sum_i(\langle \mathbf{w}^t, \phi(\mathbf{x}_i)\rangle - y_i)\phi_j(\mathbf{x}_i) + \frac{1}{2C}}{\sum_i \phi_j^2(\mathbf{x}_i)} & w_j^t > \dfrac{\sum_i(\langle \mathbf{w}^t, \phi(\mathbf{x}_i)\rangle - y_i)\phi_j(\mathbf{x}_i) + \frac{1}{2C}}{\sum_i \phi_j^2(\mathbf{x}_i)} \\[2ex] w_j^t - \dfrac{\sum_i(\langle \mathbf{w}^t, \phi(\mathbf{x}_i)\rangle - y_i)\phi_j(\mathbf{x}_i) - \frac{1}{2C}}{\sum_i \phi_j^2(\mathbf{x}_i)} & w_j^t < \dfrac{\sum_i(\langle \mathbf{w}^t, \phi(\mathbf{x}_i)\rangle - y_i)\phi_j(\mathbf{x}_i) - \frac{1}{2C}}{\sum_i \phi_j^2(\mathbf{x}_i)} \\[2ex] 0 & \text{o.w.} \end{cases}$$

In binary logistic regression, $\ell(\langle \mathbf{w}, \mathbf{x}_i\rangle, y_i) = \log(1 + \exp(-y_i \langle \mathbf{w}, \phi(\mathbf{x}_i)\rangle))$, implying that we cannot solve the subproblem analytically. Therefore, it is suggested that the quadratic approximation of the function $P(\mathbf{w}^t + \delta \mathbf{e}_j) \sim P_j^t(w_j^t + \delta)$ must be utilized as follows [21]:

$$P_j^t(w_j^t + \delta) \triangleq |w_j^t + \delta| + \nabla_j L(\mathbf{w}^t)\delta + \frac{1}{2}\nabla_{jj}L(\mathbf{w}^t)\delta^2,$$

where

$$\nabla_j L(\mathbf{w}) \triangleq P_j^{t\prime}(w_j^t) = C \sum_{i=1}^n \frac{y_i \phi_j(\mathbf{x}_i)}{1 + \exp(y_i \langle \mathbf{w}, \phi(\mathbf{x}_i)\rangle)},$$

$$\nabla_{jj} L(\mathbf{w}) \triangleq P_j^{t\prime\prime}(w_j^t) = C \sum_{i=1}^n \frac{\phi_j^2(\mathbf{x}_i)\exp(y_i \langle \mathbf{w}, \phi(\mathbf{x}_i)\rangle)}{(1 + \exp(y_i \langle \mathbf{w}, \phi(\mathbf{x}_i)\rangle))^2}.$$

To stabilize the algorithm, a sufficient decrease condition is examined while the stepsize βd is geometrically discounted [17]. This can be said to be the modified version of Armijo's rule and is denoted as

$$P(\mathbf{w}) - P(\mathbf{w} + \beta\delta \mathbf{e}_j) \geq \sigma\beta\left(\nabla_j L(\mathbf{w}^t)\delta + |w_j + \delta| - |w_j|\right), \tag{2}$$

where $0 < \beta \leq 1$ and $\sigma > 0$ is a fixed value throughout the optimization. First, condition (2) is verified by setting $\beta = 1$ and $\delta = \operatorname{argmin} P_j^t(w_j^t + \delta)$. Next β is decreased geometrically until (2) is satisfied. The resulting update is written as

$$w_j^{t+1} = w_j^t + \beta\delta. \tag{3}$$

A remarkable property of coordinate descent methods regarding the solving of (1) is that we only need to look at ϕ_j, one column of Φ, while updating. In LASSO, by monitoring $u_i = \langle \mathbf{w}, \phi(\mathbf{x}_i)\rangle$, the update can be given as

$$w_j^{t+1} = \begin{cases} w_j^t - \dfrac{\sum_i(u_i - y_i)\phi_j(\mathbf{x}_i) + \frac{1}{2C}}{\sum_i \phi_j^2(\mathbf{x}_i)} & w_j^t > \dfrac{\sum_i(u_i - y_i)\phi_j(\mathbf{x}_i) + \frac{1}{2C}}{\sum_i \phi_j^2(\mathbf{x}_i)} \\[2ex] w_j^t - \dfrac{\sum_i(u_i - y_i)\phi_j(\mathbf{x}_i) - \frac{1}{2C}}{\sum_i \phi_j^2(\mathbf{x}_i)} & w_j^t < \dfrac{\sum_i(u_i - y_i)\phi_j(\mathbf{x}_i) - \frac{1}{2C}}{\sum_i \phi_j^2(\mathbf{x}_i)} \\[2ex] 0 & \text{o.w.} \end{cases},$$

where $\Omega_j \triangleq \{j | \phi_j(\mathbf{x}_i) \neq 0\}$ and u_i is updated as

$$u_i^{t+1} = u_i^t + (w_j^{t+1} - w_j^t)y_i x_{ij}$$

in $O(|\Omega_j|)$ time. Similarly, in case of logistic regression, we can compute $\nabla_j L(\mathbf{w}^t)$ and $\nabla_{jj} L(\mathbf{w}^t)$ by monitoring $u_i^t = \exp(y_i \langle \mathbf{w}^t, \phi(\mathbf{x}_i) \rangle)$,

$$\nabla_j L(\mathbf{w}^t) = C \sum_{i \in \Omega^j} \frac{y_i \phi_j(\mathbf{x}_i)}{1 + u_i^t}, \tag{4}$$

$$\nabla_{jj} L(\mathbf{w}^t) = C \sum_{i \in \Omega^j} \frac{u_i^t \phi_j^2(\mathbf{x}_i)}{(1 + u_i^t)^2}, \tag{5}$$

each time we update u_i as

$$u_i^{t+1} = u_i^t \exp(\beta \delta y_i x_{ij})$$

This holds for any function of form ℓ that depends only on $\langle \mathbf{w}, \phi(\mathbf{x}) \rangle$. We maintain $u_i^t = \exp(y_i \langle \mathbf{w}^t, \phi(\mathbf{x}_i) \rangle)$ in logistic regression problems to reduce the number of exponential and log computations.

Another remarkable property of coordinate descent methods is that it is possible to solve the optimization problem efficiently by concentrating on updating parameters that are not zero at the optimal solution, i.e., the value of a parameter remains at 0 after a certain update point for j such that $w_j^* = 0$ [21]. In other words, for sufficiently large t,

$$-1 < C\nabla_j L(\mathbf{w}^t) < 1 \tag{6}$$

will hold for all j, such that $w_j^* = 0$. This suggests that it is unlikely to observe a j such that

$$-1 + M^t < C\nabla_j L(\mathbf{w}^t) < 1 - M^t \tag{7}$$

and $w_j^* \neq 0$ holds simultaneously for a given large t. Here, M^t is an amount that expresses a suboptimality level or a closeness to the optimal solution. The value of M^t that was used in the implementation of the L1 problem solver in [5] (`liblinear`) is formally written as

$$M^t \triangleq \frac{\max_{\tau = \lceil t/n \rceil n - n + 1, \dots, \lceil t/n \rceil n} v_j^\tau}{n},$$

where

$$v_j^t \triangleq \begin{cases} |\nabla_j L(\mathbf{w}^t) - 1| & w_j^t < 0 \\ |\nabla_j L(\mathbf{w}^t) + 1| & w_j^t > 0 \\ \max\{\nabla_j L(\mathbf{w}^t) - 1, -\nabla_j L(\mathbf{w}^t) - 1, 0\} & w_j^t = 0. \end{cases}$$

3   Proposed Scheme

In this section, we explain our proposed FCL scheme, which can handle datasets with large feature spaces. This scheme is flexible because the class of basis functions (feature space) can be arbitrary. Therefore, after introducing our FCL

Algorithm 1. Pseudo-algorithm of `writer` thread in binary L1 logistic regression

---

1: $J \leftarrow \varnothing, \Phi \leftarrow \varnothing$
2: share J, Φ, $\mathbf{w}$, $\mathbf{u}$ with Algorithm 2
3: **while** Algorithm 2 running **do**
4:     randomly choose $j \in \{1, \ldots, p\}$
5:     extract $\phi_j = \{\phi_j(\mathbf{x}_i)\}$
6:     compute $\nabla_j L(\mathbf{w})$ using (4)
7:     **if** $C|\nabla_j L(\mathbf{w})| > 1$ **then**
8:         $\Phi \leftarrow \Phi \cup \{\phi_j\}$
9:         $J \leftarrow J \cup \{j\}$
10:    **end if**
11:    **while** $|J| >$ capacity of feature cache **do**
12:        randomly select $j' \in J$
13:        $J \leftarrow J \setminus \{j'\}$
14:        $\Phi \leftarrow \Phi \setminus \{\phi_{j'}\}$
15:    **end while**
16: **end while**
17: // no output

---

scheme, we show two cases in which combinatorial features and random Fourier features are applied. In addition, we discuss the relationship between our algorithm and the boosting algorithm in Sect. 3.2.

In the FCL scheme, two types of threads are prepared asynchronously: the `writer` and `trainer` threads. The `writer` thread sequentially reads datapoints and writes a column of the matrix Φ into the main memory repeatedly. If the extracted column does not improve the current solution, it is discarded without being shared with the `trainer` thread to save space in the limited memory. The condition for discarding the column can be written as

$$-1 < C\nabla_j L(\mathbf{w}^t) < 1. \tag{8}$$

Note that this amount can be easily computed by allowing the value of $\mathbf{u}$ to be shared. If the amount of memory used by the columns of Φ exceeds a prespecified amount, the thread discards columns randomly and places a new column in the freed location. The pseudo-code of this algorithm is shown in Algorithm 1.

In contrast, the `trainer` thread selects one random column uniformly performs the standard coordinate descent method, explained in Sect. 2. If the coordinate is not effective for learning, that is, if (8) holds, the column is discarded from the memory. Note that this condition (8) is stricter than that used in [5] and other studies as discussed in the previous section, and thus may not correctly discriminate against columns that correspond to 0 in the optimal solution. This enables the coordinate descent to update in the `trainer` thread and become more efficient, while the entire scheme is still guaranteed to reach the optimal solution because the reader thread repeatedly checks the condition in (8) for all j. The pseudo-code of this algorithm is shown in Algorithm 2.

Algorithm 2. Pseudo-algorithm of trainer thread in binary L1 logistic regression

---

1: $\mathbf{w} \leftarrow \mathbf{0}, \mathbf{u} \leftarrow \mathbf{1}$
2: share $J, \Phi, \mathbf{w}, \mathbf{u}$ with Algorithm 1
3: **while** not converged **do**
4:     randomly choose $j \in J$
5:     compute $\nabla_j L(\mathbf{w}), \nabla_{jj} L(\mathbf{w})$ using (4) and (5)
6:     **if** $C|\nabla_j L(\mathbf{w})| < 1$ **then**
7:         $J \leftarrow J \setminus \{j\}$
8:         $\Phi \leftarrow \Phi \setminus \{\phi_j\}$
9:     **end if**
10:     $\delta \leftarrow \operatorname{argmin} |w_j + \delta| + \nabla_j L(\mathbf{w})\delta + \frac{1}{2}\nabla_{jj} L(\mathbf{w})\delta^2$
11:     **while** (2) does not hold **do**
12:         $\delta \leftarrow \beta\delta$
13:     **end while**
14:     $w_j \leftarrow w_j + \delta$
15:     $u_i \leftarrow u_i \exp\left(\delta y_i \phi_j(\mathbf{x}_i)\right)$ for $i \in \Omega_j$
16: **end while**
17: output $\mathbf{w}$

3.1   Examples of Large Feature Spaces and Their Relation to Learning with Kernels

In this section, we show examples of feature spaces containing large number of features.

Use of Combinatorial Features. When $\mathbf{x}_i$ is already embedded in a Euclid space, that is, $\mathcal{X} = \mathbb{R}^l$ for some natural number l, we can create a new feature by combining their components.

$$\phi_j(\mathbf{x}_i) = c_j \prod_k \left\langle \mathbf{x}_i, \mathbf{e}_{j_k(j)} \right\rangle,$$

where $c_j \in \mathbb{R}$ and $j_k(j) \in \{1, \ldots, l\}$. For example, if the datapoint and its component correspond to a document and the number of occurrences of a word, ϕ_j expresses the co-occurrence of a certain combination of words. When c_j and j_k are chosen appropriately, the set of features is equivalent to that induced by a polynomial kernel and is expressed as

$$k(\mathbf{x}, \mathbf{x}') = (\langle \mathbf{x}, \mathbf{x}' \rangle + c_1)^{c_2}.$$

with $c_1 \in \mathbb{R}$ and $c_2 \in \mathbb{N}$ given.

Use of Random Fourier Features. Again, we consider the case of $\mathcal{X} = \mathbb{R}^l$. By generating a random vector $\omega_j \in \mathbb{R}^l$, we can produce a feature function as

$$\phi_j(\mathbf{x}) = c_j \cos(\langle \omega_j, \mathbf{x} \rangle).$$

Algorithm 3. General boosting method template

1: set an initial distribution on examples $d^{(0)}$
2: **while** $t = 0, 1, 2, \ldots, T$ **do**
3:     The oracle gives a hypothesis $h^{(t)}$ under the current distribution $d^{(t)}$
4:     **if** The oracle gives no hypothesis **then**
5:         break
6:     **end if**
7:     find a distribution $d^{(t+1)}$ under hypotheses $\{h^{(s)}\}_{s=1\ldots,t}$
8: **end while**
9: output the final strong hypothesis using $d^{(T)}$ and $\{h^{(s)}\}_{s=1,\ldots,T}$

This class of feature functions can be said to be induced by the Gaussian kernel,

$$k(\mathbf{x}, \mathbf{x}) = \exp(-\mu \left\| \mathbf{x} - \mathbf{x}' \right\|^2),$$

by appropriately setting c_j and sampling ω_j from the specific distributions. Furthermore, an arbitrary shift-invariant kernel can be approximated as in [10, 11].

These examples imply that our scheme can handle several types of kernelized versions of L1 regularized risk minimization problems. Note that a representer theorem cannot be applied with L1 regularization; therefore, applying a kernel function to L1 regularization is usually difficult.

3.2 Relation to Boosting

A boosting algorithm consists of two processes: first, an oracle hypothesizes that $h^{(t)} \in \mathcal{H}$ under the current distribution of each datapoint $d^{(t)} \in \mathbb{R}^n_{\geq 0}$, then the next distribution $d^{(t+1)}$ is determined, for which the past hypothesis performs poorly. Each boosting algorithm differs in choosing a new hypothesis and updating the distribution over the datapoints. The abstraction of the algorithm is summarized in Algorithm 3. It is well known that a greedy coordinate descent method with respect to empirical risk minimization problems can be described as a boosting method [13]. Furthermore, a similarity to the problem in the form of (1) is reported in [3, 4]. Therefore, we explain in this section that the `writer` thread of our scheme continuously generates new Hypotheses, whereas the `trainer` thread of our scheme continuously updates the adversarial distributions.

For simplicity, we focus on binary classification problems. In the context of boosting, the parameter $\mathbf{w}$ in our scheme can be interpreted as the parameter defining unnormalized distribution d such that

$$d_i(\mathbf{w}) = -y_i \nabla \ell(y_i \langle \mathbf{w}, \phi(\mathbf{x}_i) \rangle), \tag{9}$$

if we aim to minimize (1), where $\ell(\langle \mathbf{w}, \phi(\mathbf{x}) \rangle, y_i)$ is a form of $\ell(y_i \langle \mathbf{w}, \phi(\mathbf{x}) \rangle)$. Furthermore, a hypothesis h corresponds to a feature expression ϕ_j if the range of our feature functions is restricted to $\{+1, -1\}$.

Generating Hypothesis h. In conventional boosting algorithms, such as Adaboost introduced in [6], at the t-th iteration, the oracle formulates the following hypothesis h that maximizes the edge $\gamma(h) \triangleq \sum_i d_i^{(t)} y_i h(\mathbf{x}_i)$ among all possible hypothesis $h \in \mathcal{H}$. That is,

$$h^{(t)} = \underset{h \in \mathcal{H}}{\operatorname{argmax}} \, \gamma(h). \qquad (10)$$

In contrast, the `writer` in our scheme repeatedly searches for ϕ_j such that the derivative is sufficiently large to satisfy the following condition:

$$\left| C \sum_i \nabla \ell(y_i \langle \mathbf{w}, \phi(\mathbf{x}_i) \rangle) y_i \phi_j(\mathbf{x}_i) \right| > 1, \qquad (11)$$

depending on the currently available parameter $\mathbf{w}$. By defining and substituting

$$h(\cdot) \triangleq \operatorname{sign} \left(\sum_i -\nabla \ell(y_i \langle \mathbf{w}, \phi(\mathbf{x}_i) \rangle) y_i \phi_j(\mathbf{x}_i) \right) \phi_j(\cdot),$$

into (11), the condition can be rewritten as

$$\sum_i d_i^{(t)} y_i h(\mathbf{x}_i) > C^{-1}.$$

Therefore, the strategy of `writer` is to accept all hypotheses that show larger edge than a certain threshold. Note that this property is inherited from random coordinate descent methods, whereas the greedy coordinate descent method corresponds to the strategy of (10).

The Distribution Update. As in [13], the updates of the distribution over datapoints are conventionally formulated as follows:

$$d^{(t+1)} = \underset{d}{\operatorname{argmin}} \, \operatorname{RE}(d|d^{(t)}) \qquad (12)$$

$$\text{subject to } \sum_i d_i y_i h^{(t)}(\mathbf{x}_i) = 0, \qquad (13)$$

where $\operatorname{RE}(d|d^{(t)})$ denotes a certain type of relative entropy between d and $d^{(t)}$ in case of (nonregularized) empirical risk minimization. Duchi and Singer provided an alternative formulation of distribution update corresponding to the L1 regularized risk minimization, in which (13) is replaced by

$$\sum_i d_i^{(t+1)} y_i h^{(t)}(\mathbf{x}_i) \leq \nu, \qquad (14)$$

while minimizing a relative entropy. Those strategies of the distribution updates have a unified consistent viewpoint as coordinate descent methods for minimizing

$$\sum_{i=1}^n \ell(y_i \langle \mathbf{w}, \phi(\mathbf{x}_i) \rangle) + \sum_{j=1}^p r(w_j),$$

where $r(w_j) = 0$ or $r(w_j) = C|w_j|$. From this point of view, the exact form of relative entropy varies depending on the underlying function ℓ. The one-variable subproblem that the coordinate descent method defines can be reformulated as follows:

$$\operatorname*{argmin}_{w_j} \sum_{i=1}^{n} \ell(y_i \langle \mathbf{w}, \phi(\mathbf{x}_i) \rangle) + r(w_j)$$

$$= \operatorname*{argmin}_{w_j} \sum_{i=1}^{n} \max_{d_i} d_i y_i \langle \mathbf{w}, \phi(\mathbf{x}_i) \rangle - \ell_i^*(d_i) + r(w_j),$$

for a fixed j, while any other components of $\mathbf{w}$ are fixed. Here, ℓ^* denotes the Fenchel dual of ℓ. Therefore, the dual problem can be formulated as

$$d = \operatorname*{argmax}_{d} \sum_{i=1}^{n} -\ell^*(d_i) + \sum_{j' \neq j} d_i w_{j'} y_i \phi_{j'}(\mathbf{x}_i) - r^* \left(\sum_{i=1}^{n} d_i y_i \phi_j(\mathbf{x}_i) \right).$$

The primal-dual relationship can be written as in (9). Moreover, when $r(w_j) = 0$, r^* is 0 if (13) holds and ∞ otherwise. Furthermore, when $r(w_j) = C|w_j|$, r^* is 0 if (14) is true for $\nu = C^{-1}$ and ∞ otherwise. Therefore, relative entropy terms correspond to $\sum_{i=1}^{n} -\ell^*(d_i) + \sum_{j' \neq j} d_i w_{j'} y_i \phi_{j'}(\mathbf{x}_i)$, which coincides with various relative entropies except the difference of constant when ℓ is appropriately set.

A similar consistency holds for the totally corrective boosting algorithm [18]. The updates to the distribution are provided by selecting d^{t+1} that satisfies

$$\sum_i d_i^{(t+1)} y_i h^{(s)}(\mathbf{x}_i) \leq \nu,$$

for all $s = 1, \dots, t$, while some relative entropy is minimized. With a similar argument, we can see that this corresponds to the dual problem of (1), in which we restrict to hold $w_j = 0$ if $w_j^t = 0$ for all j and set $\nu = C^{-1}$. Therefore, the procedure of the `trainer` thread in our scheme, which aims to solve a subproblem restricted to a feature cache, is similar to the strategy of updating distributions over datapoints defined by totally corrective boosting algorithms.

4 Experimental Results

In this section, we verify the effectiveness of our scheme for large-scale optimization of L1 logistic regression that uses $\ell(\langle \mathbf{w}, \phi(\mathbf{x}_i) \rangle, y_i) = \log(1 + \exp(y_i \langle \mathbf{w}, \mathbf{x} \rangle))$ in (1). We first demonstrate that optimization can be performed efficiently by the asynchronous feature extraction of the FCL scheme, and secondly show that the method can be used effectively on problems with a large number of features that is overly difficult for any other known scheme. We consider the binary classification of spam e-mail recognition in our first set of experiments, and splice site recognition by using over one million DNA sequences for the second set of experiments. Further, we implemented the proposed scheme and the DCL scheme using C++. All experiments were performed on Intel Xeon CPU X5690 3.47 GHz processor with 96 GB memory.

Table 1. Dataset configuration

Dataset	# of instances	# of features	# of non-zero elements	Sparsity (%)	Data size (GB in text file)
Webspam	280,000	16,609,163	1.231×10^9	0.022	20.03
Splice($d = 8$)	50,000,000	41,875,000	875.6×10^9	0.410	$< 9,000$
(sampled)	50,000,000	200,000	4.179×10^9	0.418	44.44
Splice($d = 9$)	50,000,000	207,812,500	1.702×10^{12}	0.164	$< 18,000$
Splice($d = 10$)	50,000,000	1,031,250,000	3.38×10^{12}	0.066	$< 36,000$

4.1 Spam Recognition

In the initial experiments, we verified that our scheme can find the optimal solution efficiently even when the feature cache cannot fit all the data matrix using of the largest public dataset for binary classification. We used webspam dataset [19], in which trigram features are already developed. We limited the memory capacity for storing the columns of the data matrix to 20 GB, whereas the dataset utilizes more than 20 GB in the text format. We plotted the relative value of the objective function as a function of elapsed time and compared it to the dual cached loops with SGD. The relative value of the objective function refers to $\frac{P(\mathbf{w}^t)}{P(\mathbf{w}^t) - P^*}$ for a given time t. For comparison, we also implemented the modified version of the dual cached loop scheme with SGD, as discussed in Sect. 1. We omitted a comparison with the block minimization scheme because the dual cached loops are reported to be consistently superior [8]. Following [2], we used scheduling of the step size denoted as $\eta_t \triangleq \eta_0 \left(1 + \frac{1}{Cn}\eta_0 t\right)^{-1}$. For the value of the hyper-parameter, we used $C = 0.1, 1, 10$, and 100.

The results are shown in Figs. 1 and 2. FCL achieves the optimal solution more quickly than the dual cached loops scheme by using any step size for any value of C. This is not only because the coordinate descent method enjoys a faster convergence rate for this specific form of optimization problem but also because the selective deletion of columns facilitates to focus updates on parameters corresponding to important features.

4.2 Splice Site Recognition

In the second set of experiments, we examined the performance of the FCL scheme on the binary splice site classification problem for a DNA sequence. Each $\mathbf{x}_i$ and y_i represent a DNA sequence of length 141 and a label of whether the corresponding sequence has a splice site respectively [15]. The dataset can be obtained from http://sonnenburgs.de/soeren/item/coffin/. We used 50,000,000 sequences as the training data, and used the first 100,000 sequences from the test data to compute the testing indicators. Owing to the dataset being extremely imbalanced, only 143,688 datapoints out of 50,000,000 have positive labels, we evaluated the learning performance by using the area under precision-recall curve (AUPRC).

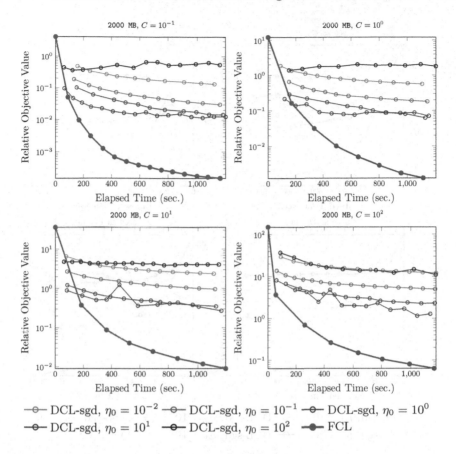

Fig. 1. Relative objective function value versus elapsed time (sec)

We consider simple combinatorial features using the following methods. We denoted a sequence of $\{A,T,C,G,?\}$, with length d that does not begin with "?," with b and a natural number less than $141 - (d-1)$ with k, and consider a one-to-one correspondence between (b, k) and $j = 1, \ldots, p$. The value of $\phi_j(\mathbf{x}_i)$ is defined as 1 if the subsequence of $\mathbf{x}_i$ from k to $k + d$ matches the regular expression represented by b, and 0 otherwise. Therefore, the total possible number of features p is $(141 - (d-1)) \times 4 \times 5^{d-1}$ and a single datapoint consists of $(141 - (d-1)) \times 2^{d-1}$ nonzero elements. The entire data matrix requires approximately 3,000 GB of memory even when formulated as a sparse matrix, assuming that one nonzero element requires 4 bytes when $d = 8$. We examined the cases for $d = 8, 9, 10$ and $C = 0.1, 0.01$, by setting the capacity of feature cache as $70 \times 2^{d-1}$. To accelerate feature extraction, we asynchronously ran 5 `writer` threads for this experiment. Furthermore, to evaluate the performance of those results, we examined the performance obtained by `liblinear`. We randomly sampled a part of the

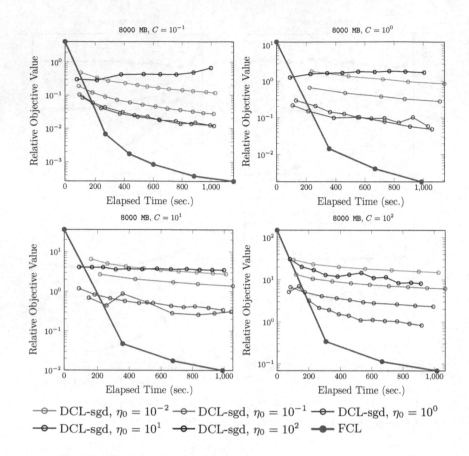

Fig. 2. Relative objective function value versus elapsed time (sec)

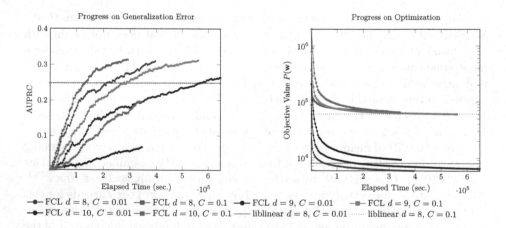

Fig. 3. Progress of feature cached loops (FCL) in splice site recognition

features so that the required amount of memory becomes 60 GB, as described in Table 1[1].

Figure 3 plots the objective function value versus elapsed time and AUPRC versus elapsed time for different values of C and d. As shown, $d = 8$ achieves the highest value of AUPRC and the lowest value of the objective function. $d = 10$ achieves lower AUPRC and higher objective values. This result contradicts the function value at the optimal solutions as the feature space of $d = 8$ is strictly included by that of $d = 9$, which is similarly included by $d = 10$. This indicates that when $d = 8$, the optimization progresses more efficiently by focusing on important features. For $d = 10$, a more efficient extraction of features is required to accelerate the entire scheme.

Notably, even for developing sampled features used by `liblinear`, more than two days were required for each dataset. Therefore, our scheme could obtain higher values of AUPRC than those obtained by randomly sampling features in shorter time when the same hyper-parameters are set. In addition, our scheme keeps decreasing the objective values and increasing AUPRC for more than a week. This implies that the optimization is not completed yet and AUPRC could be further improved. This observation also indicates that a more efficient scheme is desired for better performance for solving even larger feature spaces.

5   Conclusion

In this paper, we proposed a scheme for solving the L1 regularization problem with a large number of datapoints and an even larger number of features. By simultaneously developing the data matrix Φ and optimizing the parameters, the proposed scheme can efficiently learn the parameters, while only the seemingly important features are developed. The experiments show that the proposed scheme can efficiently learn parameters with richer feature spaces than could be used in the past. In our research, we demonstrated the performance of our scheme by using combinatorial features. In the future, we will consider an extensive application of various types of feature spaces and distributed optimization schemes.

Acknowledgments. The author thanks S.V.N. Vishwanathan for the illuminating discussion to initiate the basic idea and Hiroshi Iida and Hiroshi Nakagawa for their suggestive trials and arguments on preceding schemes. This work is partially supported by MEXT KAKENHI Grant Number 23240019 and JST-CREST.

References

1. Bertsekas, D.P.: Convex Optimization Algorithms. Athena Scientific, Belmont (2015)
2. Bottou, L.: Stochastic gradient tricks. In: Montavon, G., Orr, G.B., Müller, K.-R. (eds.) Neural Networks: Tricks of the Trade, Reloaded. LNCS, vol. 7700, pp. 430–445. Springer, Heidelberg (2012)

---

[1] Liblinear required more memory than expected; therefore, the experiments are conducted in machines other than those described earlier.

3. Demiriz, A., Bennett, K.P., Shawe-Taylor, J.: Linear programming boosting via column generation. Mach. Learn. **46**(1), 225–254 (2002)
4. Duchi, J., Singer, Y.: Boosting with structural sparsity. In: Proceedings of International Conference on Machine Learning, pp. 297–304 (2009)
5. Fan, R.E., Chang, K.W., Hsieh, C.J., Wang, X.R., Lin, C.J.: Liblinear: a library for large linear classification. J. Mach. Learn. Res. **9**, 1871–1874 (2008)
6. Freund, Y., Schapire, R.E.: A decision-theoretic generalization of on-line learning and an application to boosting. J. Comput. Syst. Sci. **55**(1), 119–139 (1997)
7. LeCun, Y., Bengio, Y., Hinton, G.: Deep learning. Nature **521**, 436–444 (2015)
8. Matsushima, S., Vishwanathan, S., Smola, A.J.: Linear support vector machines via dual cached loops. In: Proceedings of Knowledge Discovery and Data Mining, pp. 177–185 (2012)
9. Nesterov, Y.: Efficiency of coordinate descent methods on huge-scale optimization problems. SIAM J. Optim. **22**(2), 341–362 (2012)
10. Rahimi, A., Recht, B.: Random features for large-scale kernel machines. In: Proceedings of Conference on Neural Information Processing Systems, pp. 1177–1184 (2007)
11. Rahimi, A., Recht, B.: Weighted sums of random kitchen sinks: replacing minimization with randomization in learning. In: Proceedings of Conference on Neural Information Processing Systems, pp. 1313–1320 (2008)
12. Rish, I., Grabarnik, G.: Sparse Modeling: Theory, Algorithms, and Applications. CRC Press Inc., Boca Raton (2014)
13. Schapire, R.E., Freund, Y.: Boosting: Foundations and Algorithms. The MIT Press, Cambridge (2012)
14. Shalev-Shwartz, S., Ben-David, S.: Understanding Machine Learning: From Theory to Algorithms. Cambridge University Press, New York (2014)
15. Sonnenburg, S., Franc, V.: COFFIN: a computational framework for linear SVMs. In: Proceedings of International Conference on Machine Learning, pp. 999–1006 (2010)
16. Tibshirani, R.: The lasso method for variable selection in the cox model. In: Statistics in Medicine, pp. 385–395 (1997)
17. Tseng, P., Yun, S.: A coordinate gradient descent method for nonsmooth separable minimization. Math. Program. **117**(1–2), 387–423 (2009)
18. Warmuth, M.K., Liao, J.: Totally corrective boosting algorithms that maximize the margin. In: Proceedings of International Conference on Machine Learning, pp. 1001–1008 (2006)
19. Webb, S., Caverlee, J., Pu, C.: Introducing the webb spam corpus: using email spam to identify web spam automatically. In: Proceedings of the Third Conference on Email and Anti-Spam (2006)
20. Yu, H.F., Hsieh, C.J., Chang, K.W., Lin, C.J.: Large linear classification when data cannot fit in memory. In: Proceedings of Knowledge Discovery and Data Mining, pp. 833–842 (2010)
21. Yuan, G.X., Chang, K.W., Hsieh, C.J., Lin, C.J.: A comparison of optimization methods and software for large-scale l1-regularized linear classification. J. Mach. Learn. Res. **11**, 3183–3234 (2010)

F-Measure Maximization
in Multi-Label Classification
with Conditionally Independent Label Subsets

Maxime Gasse[(⊠)] and Alex Aussem

LIRIS, UMR 5205, University of Lyon 1, 69622 Lyon, France
{maxime.gasse,alexandre.aussem}@liris.cnrs.fr

Abstract. We discuss a method to improve the exact F-measure maximization algorithm called GFM, proposed in [2] for multi-label classification, assuming the label set can be partitioned into conditionally
independent subsets given the input features. If the labels were all independent, the estimation of only m parameters (m denoting the number
of labels) would suffice to derive Bayes-optimal predictions in $O(m^2)$
operations [10]. In the general case, $m^2 + 1$ parameters are required by
GFM, to solve the problem in $O(m^3)$ operations. In this work, we show
that the number of parameters can be reduced further to m^2/n, in the
best case, assuming the label set can be partitioned into n conditionally
independent subsets. As this label partition needs to be estimated from
the data beforehand, we use first the procedure proposed in [4] that finds
such partition and then infer the required parameters locally in each label
subset. The latter are aggregated and serve as input to GFM to form the
Bayes-optimal prediction. We show on a synthetic experiment that the
reduction in the number of parameters brings about significant benefits
in terms of performance. The data and software related to this paper are
available at https://github.com/gasse/fgfm-toy.

Keywords: Multi-label classification · F-measure · Bayes optimal
prediction · Label dependence

1 Introduction

Multi-label classification (MLC) has received increasing attention in the last
years from the machine learning community. Unlike in the case of multi-class
learning, in MLC each instance can be assigned simultaneously to multiple binary
labels. Formally, learning from multi-label examples amounts to finding a mapping from a space of features to a space of labels. Given a multi-label training set
$\mathcal{D}$, the goal of multi-label learning is to find a function which is able to map any
unseen example to its proper set of labels. From a Bayesian point of view, this
problem amounts to modeling the conditional joint distribution $p(\mathbf{y}|\mathbf{x})$, where
$\mathbf{x}$ is a random vector in $\mathbb{R}^d$ associated with the input space, $\mathbf{y}$ a random vector
in $\{0,1\}^m$ associated with the labels, and p the probability distribution defined

© Springer International Publishing AG 2016
P. Frasconi et al. (Eds.): ECML PKDD 2016, Part I, LNAI 9851, pp. 619–631, 2016.
DOI: 10.1007/978-3-319-46128-1_39

over $(\mathbf{x}, \mathbf{y})$. Knowing the label conditional distribution $p(\mathbf{y}|\mathbf{x})$ still leaves us with the question of deciding what prediction $\mathbf{y}$ should be made given $\mathbf{x}$ in order to minimize the loss. Dembczynski et al. [3] show that the expected benefit of exploiting label dependence depends on the type of loss to be minimized and, most importantly, one cannot expect the same MLC method to be optimal for different types of losses at the same time. In particular, optimizing the subset 0/1 loss, the F-measure loss or the *Jaccard* index requires some knowledge of the dependence structure among the labels that cannot be inferred from the marginals $p(y_i|\mathbf{x})$ alone.

The F-measure is a standard performance metric in information retrieval that was used in a variety of prediction problems including binary classification, multi-label classification and structured output prediction. Let $\mathbf{y} = (y_1, \ldots, y_m)$ denote the label vector associated with a single instance $\mathbf{x}$ in MLC, and $\mathbf{h} = (h_1, \ldots, h_m) \in \{0,1\}^m$ denote the prediction for $\mathbf{x}$, the F-measure is defined as follows:

$$F(\mathbf{y}, \mathbf{h}) = \frac{2(\mathbf{y} \cdot \mathbf{h})}{\mathbf{y} \cdot \mathbf{y} + \mathbf{h} \cdot \mathbf{h}}, \tag{1}$$

where $\cdot$ denotes the dot product operator[1] and $0/0 = 1$ by definition. Optimizing the F-measure is a statistically and computationally challenging problem, since no closed-form solution exists and few theoretical studies of the F-measure were carried out. Very recently, Waegeman et al. [9] presented a new Bayes-optimal algorithm regardless of the underlying distribution that is statistically consistent. Assuming the underlying probability distribution p is known, the optimal prediction $\mathbf{h}^*$ that maximizes the expected F-measure is given by

$$\mathbf{h}^* = \arg\max_{\mathbf{h} \in \{0,1\}^m} \mathbb{E}_{\mathbf{y}}[F(\mathbf{y}, \mathbf{h})] = \arg\max_{\mathbf{h} \in \{0,1\}^m} \sum_{\mathbf{y} \in \{0,1\}^m} p(\mathbf{y})F(\mathbf{y}, \mathbf{h}). \tag{2}$$

The corresponding optimization problem is non-trivial and cannot be solved in closed form. Moreover, a brute-force search is intractable, as it would require checking all 2^m combinations of prediction vector $\mathbf{h}$ and summing over an exponential number of terms in each combination. As a result, many works reporting the F-measure in experimental studies rely on optimizing a surrogate loss like the Hamming loss and the subset zero-one loss as an approximation of (2). However, Waegeman et al. [9] have shown that these surrogate loss functions yield a high *worst-case regret*.

Apart from optimizing surrogates, a few other approaches for finding the F-measure maximizer have been presented but they explicitly rely on the restrictive assumption of independence of the Y_i [5,10]. This assumption is not tenable in domains like MLC and structured output prediction. Algorithms based on independence assumptions or marginal probabilities are not statistically consistent when arbitrary probability distributions p are considered.

---

[1] In a binary setting the dot product $\mathbf{h} \cdot \mathbf{y}$ offers a convenient notation to count the number of positives values common to both $\mathbf{h}$ and $\mathbf{y}$.

In [4], we established several results to characterize and compute disjoint label subsets called *irreducible label factors* (ILFs) that appear in the factorization of $p(\mathbf{y}|\mathbf{x})$ (i.e., minimal subsets $\mathbf{Y}_{LF} \subseteq \mathbf{Y}$ such that $\mathbf{Y}_{LF} \perp\!\!\!\perp \mathbf{Y} \setminus \mathbf{Y}_{LF} \mid \mathbf{X}$) under various assumption underlying the probability distribution. In that paper, the emphasis was placed on the *subset zero-one* loss minimization. In the present work, we show that ILF decomposition can also benefit to the F-measure maximization problem in the MLC context.

Section 2 introduces the General F-measure Maximizer method (GFM) from [2]. Section 3 discusses some key concepts about irreducible label factors, and addresses the problem of exploiting a label factor decomposition within GFM, with an exact procedure called Factorized GFM (F-GFM). Section 4 presents a practical calibrated parametrization method for GFM and F-GFM, and finally Sect. 5 presents a synthetic experiment to corroborate our theoretical findings.

2 The General F-Measure Maximizer Method

We start by reviewing the General F-measure Maximizer method presented in Dembczynski et al. [2]. Jansche [5] noticed that (2) can be solved via outer and inner maximization. The inner maximization step is

$$\mathbf{h}^{(k)} = \arg\max_{\mathbf{h} \in \mathcal{H}_k} \mathbb{E}_{\mathbf{y}}[F(\mathbf{y}, \mathbf{h})], \tag{3}$$

where $\mathcal{H}_k = \{\mathbf{h} \in \{0,1\}^m | \mathbf{h} \cdot \mathbf{h} = k\}$, followed by an outer maximization

$$\mathbf{h}^* = \arg\max_{\mathbf{h} \in \{\mathbf{h}^{(0)}, \dots, \mathbf{h}^{(m)}\}} \mathbb{E}_{\mathbf{y}}[F(\mathbf{y}, \mathbf{h})]. \tag{4}$$

The outer maximization (4) can be done in linear time by simply checking all $m + 1$ possibilities. The main effort is then devoted to solving the inner maximization (3). For convenience, Waegeman et al. [9] introduce the following quantities:

$$s_{\mathbf{y}} = \mathbf{y} \cdot \mathbf{y}, \quad \Delta_{ik} = \sum_{\mathbf{y} \in \mathcal{Y}_i} \frac{2p(\mathbf{y})}{s_{\mathbf{y}} + k},$$

with $\mathcal{Y}_i = \{\mathbf{y} \in \{0,1\}^m | y_i = 1\}$. The first quantity is the number of ones in the label vector $\mathbf{y}$, while Δ_{ik} is a specific marginal value for the i-th label. Using these quantities, the maximizer in (3) becomes

$$\mathbf{h}^{(k)} = \arg\max_{\mathbf{h} \in \mathcal{H}_k} \sum_{i=1}^{m} h_i \Delta_{ik},$$

which boils down to selecting the k labels with the highest Δ_{ik} value. In the special case of $k = 0$, we have $\mathbf{h}^{(0)} = 0$ and $\mathbb{E}_{\mathbf{y}}[F(\mathbf{y}, \mathbf{h}^{(0)})] = p(\mathbf{y} = \mathbf{0})$. As a result, it is not required to estimate the 2^m parameters of the whole distribution $p(\mathbf{y})$ to find the F-measure maximizer $\mathbf{h}^*$, but only $m^2 + 1$ parameters: the values of Δ_{ik} which take the form of an $m \times m$ matrix Δ, plus the value of $p(\mathbf{y} = \mathbf{0})$.

The resulting algorithm is referred to as General F-measure Maximizer (GFM), and yields the optimal F-measure prediction in $O(m^2)$ (see [9] for details). In order to combine GFM with a training algorithm, the authors decompose the $\mathbf{\Delta}$ matrix as follows. Consider the probabilities

$$p_{is} = p(y_i = 1, s_{\mathbf{y}} = s), \quad i, s \in \{1, \ldots, m\}$$

that constitute an $m \times m$ matrix $\mathbf{P}$, along an $m \times m$ matrix $\mathbf{W}$ with elements

$$w_{sk} = \frac{2}{s + k},$$

then it can be easily shown that

$$\mathbf{\Delta} = \mathbf{PW}. \tag{5}$$

If the matrix $\mathbf{P}$ is taken as an input by the algorithm, then its complexity is dominated by the matrix multiplication (5), which is solved naively in $O(m^3)$.

In view of this result, Dembczynski et al. [2] establish that modeling pairwise or higher degree dependences between labels is not necessary to obtain an optimal solution, only a proper estimation of marginal quantities p_{is} is required to take the number of co-occurring labels into account. In this work, we will show that modeling high degree dependences between labels can help to obtain a better estimation of p_{is}, and thereby better predictions within the GFM framework.

3 Factorized GFM

In the following we will show that, assuming a factorization of the conditional distribution of the labels, the p_{is} parameters can be reconstructed from a smaller number of parameters that are estimated locally in each label factor, at a computational cost of $O(m^3)$.

3.1 Label Factor Decomposition

We now introduce the concept of *label factor* that will play a pivotal role in the factorization of $p(\mathbf{y}|\mathbf{x})$ [4].

Definition 1. *A label factor is a label subset $\mathbf{Y}_F \subseteq \mathbf{Y}$ such that $\mathbf{Y}_F \perp\!\!\!\perp \mathbf{Y} \backslash \mathbf{Y}_F \mid \mathbf{X}$. Additionally, a label factor is said irreducible when it is non-empty and has no other non-empty label factor as proper subset.*

The key idea behind irreducible label factors (ILFs as a shorthand) is the decomposition of the conditional distribution of the labels into a product of factors,

$$p(\mathbf{y} \mid \mathbf{x}) = \prod_{k=1}^{n} p(\mathbf{y}_{F_k} \mid \mathbf{x}),$$

where $\{\mathbf{Y}_{F_k}\}_{k=1}^n$ is a partition of $\mathbf{Y} = \{Y_1, Y_2, \ldots, Y_m\}$. From the above definition, we have that $\mathbf{Y}_{F_i} \perp\!\!\!\perp \mathbf{Y}_{F_j} \mid \mathbf{X}, \forall i \neq j$. To illustrate the concept of label factor decomposition, consider the following example.

Example 1. Suppose p is faithful to one of the DAGs displayed in Fig. 1. In DAG 1a, it is easily shown using the d-separation criterion that $\{Y_1\} \perp\!\!\!\perp \{Y_2, Y_3\} \mid \mathbf{X}$, so both $\{Y_1\}$ and $\{Y_2, Y_3\}$ are label factors. However, we have $\{Y_2\} \not\!\perp\!\!\!\perp \{Y_1, Y_3\} \mid \mathbf{X}$ and $\{Y_3\} \not\!\perp\!\!\!\perp \{Y_1, Y_2\} \mid \mathbf{X}$, so $\{Y_2\}$ and $\{Y_3\}$ are not label factors. Therefore $\{Y_1\}$ and $\{Y_2, Y_3\}$ are the only irreducible label factors. Likewise, in DAG 1b the only irreducible label factor is $\{Y_1, Y_2, Y_3\}$. Finally, in DAG 1c we have that $\{Y_1\} \not\!\perp\!\!\!\perp \{Y_2, Y_3\} \mid \mathbf{X}$, $\{Y_2\} \perp\!\!\!\perp \{Y_1, Y_3\} \mid \mathbf{X}$ and $\{Y_3\} \not\!\perp\!\!\!\perp \{Y_1, Y_2\} \mid \mathbf{X}$, so $\{Y_2\}$ and $\{Y_1, Y_3\}$ are the irreducible label factors.

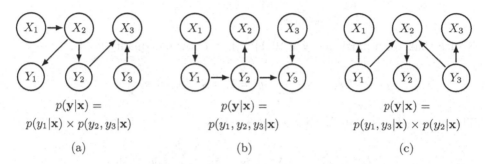

Fig. 1. Three Bayesian networks for illustration purposes, along with the induced factorization of $p(\mathbf{y}|\mathbf{x})$.

For convenience, the conditioning on $\mathbf{X}$ will be made implicit in the remainder of this work. Let m_k denote the number of labels in a particular label factor, we introduce for every label factor $\mathbf{Y}_{F_k} = \{Y_1, \ldots, Y_{m_k}\}$ the following terms,

$$p_{is}^k = p(y_i = 1, s_{\mathbf{y}_{F_k}} = s), \quad i, s \in \{1, \ldots, m_k\},$$

which constitute an $m_k \times m_k$ matrix $\mathbf{P}^k$.

Given a factorization of the label set into label factors, our proposed method called F-GFM requires to estimate, for each label factor, a local matrix $\mathbf{P}^k$ of size m_k^2, and then combine them to reconstruct the global matrix $\mathbf{P}$ of size m^2. The total number of parameters is therefore reduced from m^2 to $\sum_{k=1}^n m_k^2$. It is easily shown that, in the best case, the total number of parameters is m^2/n when $m_k = m/n$ for every label factor, and the worst case is $(n-1)+(m-n+1)^2$ when all the label factors, but one, are singletons. In both cases the number of parameters is reduced, which results in a better estimation of these parameters and a better robustness of the model. In the following, we describe a procedure to recover $\mathbf{P}$ and $p(\mathbf{y} = \mathbf{0})$ from the individual $\mathbf{P}^k$ matrices in $O(m^3)$.

3.2 Recovering $\mathbf{d}^k$

Consider, for every label factor $\mathbf{Y}_{F_k}$, the following probabilities,

$$d_s^k = p(s_{\mathbf{y}_{F_k}} = s), \quad s \in \{0, \ldots, m_k\},$$

which form a vector $\mathbf{d}^k$ of size $m_k + 1$. Instead of estimating these additional terms, they are extracted directly from $\mathbf{P}^k$ in $m_k{}^2$ operations. Extracting these parameters is done prior to recovering $\mathbf{P}$. We now describe how to recover a particular $\mathbf{d}^k$ vector from a $\mathbf{P}^k$ matrix. Note that the same method holds to recover $\mathbf{d}$ from $\mathbf{P}$, therefore in the following we will drop the superscript k to keep our notations uncluttered. Consider the following expression for p_{is} and d_s,

$$p_{is} = \sum_{\mathbf{y} \in \{0,1\}^m} p(\mathbf{y}) \cdot \mathbb{I}[s_{\mathbf{y}} = s] \cdot \mathbb{I}[y_i = 1],$$

$$d_s = \sum_{\mathbf{y} \in \{0,1\}^m} p(\mathbf{y}) \cdot \mathbb{I}[s_{\mathbf{y}} = s].$$

Notice that, for a particular $\mathbf{y} \in \{0,1\}^m$, the following equality holds,

$$\mathbb{I}[s_{\mathbf{y}} = s] \cdot \sum_{i=1}^{m} \mathbb{I}[y_i = 1] = s \cdot \mathbb{I}[s_{\mathbf{y}} = s].$$

Therefore, when $s > 0$, d_s can be expressed as

$$d_s = \sum_{\mathbf{y} \in \{0,1\}^m} p(\mathbf{y}) \cdot \mathbb{I}[s_{\mathbf{y}} = s] \cdot \frac{1}{s} \sum_{i=1}^{m} \mathbb{I}[y_i = 1].$$

This expression can be further simplified in order to express d_s as a composition of p_{is} terms,

$$d_s = \frac{1}{s} \sum_{i=1}^{m} p_{is}, \quad \forall s \in \{1, \dots, m\}.$$

We may recover d_0 from

$$d_0 = 1 - \sum_{s=1}^{m} d_s.$$

As a result, each vector $\mathbf{d}^k$ can be obtained from $\mathbf{P}^k$ in $m_k{}^2$ operations. Interestingly, because $p(\mathbf{y} = \mathbf{0}) = d_0$, this additional parameter can actually be inferred from $\mathbf{P}$ at the expense of m^2 operations, thereby reducing the number of parameters required by GFM to m^2 instead of $m^2 + 1$.

3.3 Recovering P

We will now show how the whole $\mathbf{P}$ matrix can be recovered from the $\mathbf{P}^k$ matrices in $O(m^3)$.

When $n = 2$. Let us first assume that there are only two label factors $\mathbf{Y}_{F_1}$ and $\mathbf{Y}_{F_2}$. Consider a label Y_i that belongs to $\mathbf{Y}_{F_1}$, from the marginalization rule p_{is} may be decomposed as follows,

$$p_{is} = \sum_{s'} p(y_i = 1, s_{\mathbf{y}} = s, s_{\mathbf{y}_{F_1}} = s'). \tag{6}$$

The inner term of this sum factorizes because of the label factor assumption. First, recall that $s_{\mathbf{y}} = s_{\mathbf{y}_{F_1}} + s_{\mathbf{y}_{F_2}}$, which allows us to write

$$p(y_i = 1, s_{\mathbf{y}} = s, s_{\mathbf{y}_{F_1}} = s') = p(y_i = 1, s_{\mathbf{y}_{F_1}} = s', s_{\mathbf{y}_{F_2}} = s - s').$$

Second, due to the label factor assumption, i.e. $\mathbf{Y}_{F_1} \perp\!\!\!\perp \mathbf{Y}_{F_2}$, we have

$$p(y_i, s_{\mathbf{y}}, s_{\mathbf{y}_{F_1}}) = p(y_i, s_{\mathbf{y}_{F_1}}) \cdot p(s_{\mathbf{y}_{F_2}}). \tag{7}$$

We may combine (7) and (6) to obtain

$$p_{is} = \sum_{s'} p(y_i = 1, s_{\mathbf{y}_{F_1}} = s') \cdot p(s_{\mathbf{y}_{F_2}} = s - s'). \tag{8}$$

Finally, we have necessarily $s' \leq s$ and $s' \leq m_1$, which implies $s' \leq min(s, m_1)$. Also, $s - s' \leq m_2$ and $s' \geq 1$ because $y_i = 1$, which implies $s' \geq max(1, s - m_2)$. So we can re-write (8) as follows,

$$p_{is} = \sum_{s'=max(1,s-m_2)}^{min(s,m_1)} p_{is'}^1 \cdot d_{s-s'}^2. \tag{9}$$

In the case where $Y_i \in \mathbf{Y}_{F_2}$, we obtain a similar result. In the end, given that both $\mathbf{P}^k$ and $\mathbf{d}^k$ are known for $\mathbf{Y}_{F_1}$ and $\mathbf{Y}_{F_2}$, (9) allows us to recover all term in $\mathbf{P}$ in $(m_2+1)m_1{}^2+(m_1+1)m_2{}^2$ operations. Assuming that only the $\mathbf{P}^k$ matrices are known, we must add up the additional cost for recovering the $\mathbf{d}^k$ vectors, which brings the total computational burden to $(m_2 + 2)m_1{}^2 + (m_1 + 2)m_2{}^2$.

For any n. The same procedure can be used iteratively to merge $\mathbf{P}^1$ and $\mathbf{P}^2$ into a matrix $\mathbf{P}'$ of size $(m_1 + m_2)^2$, then combine this matrix with $\mathbf{P}^3$ to form a new matrix of size $(m_1 + m_2 + m_3)^2$, and so on until every label factor is merged into a matrix of size m^2. In the end we obtain $\mathbf{P}$ in a total number of operations equal to

$$\sum_{i=2}^{n}(m_i + 2)(\sum_{j=1}^{i-1} m_j)^2 + m_i^2(2 + \sum_{j=1}^{i-1} m_j).$$

To avoid tedious calculations, we can easily compute a tight upper bound of the number of computations, i.e.

$$\max_{m_1,\ldots,m_n} \sum_{i=2}^{n}(m_i + 2)\left(\sum_{j=1}^{i-1}(m_j + 2)\right)\left(\sum_{j=1}^{i}(m_j + 2)\right) \quad \text{s.t.} \sum_{i=1}^{n} m_i = m.$$

Solving $\nabla \mathcal{L}(m_1, \ldots, m_n, \lambda) = 0$ yields

$$m_i = \left((m + 2n)^2 - \lambda\right)^{1/2} + 2n, \quad \forall i \in \{1, \ldots, n\},$$

which implies that all the label factors have equal size. As a result, with $m_i = m/n$ for every label factor we obtain an upper bound on the worst case number of operations equal to $(\frac{m}{n} + 2)^3(n^2 - 1)$. Thus, the overall complexity to recover $\mathbf{P}$ is bounded by $O(m^3)$.

3.4   The F-GFM Algorithm

Given that the label factors are known and that every $\mathbf{P}^k$ matrix has been estimated, the whole procedure for recovering $\mathbf{P}$ and then $\mathbf{h}^*$ is presented in Algorithm 1. As shown in the previous section, the overall complexity of F-GFM is $O(m^3)$, just as GFM.

4   Parameter Estimation

Our proposed method F-GFM requires to estimate for each label factor $\mathbf{Y}_{F_k}$ the $m_k \times m_k$ matrix $\mathbf{P}^k$, instead of the whole $m \times m$ matrix $\mathbf{P}$ in GFM. Still, the

---

Algorithm 1. Factorized-GFM

Require: $\mathbf{Y}$ the label set, $\mathbf{Y}_{F_1}, \ldots, \mathbf{Y}_{F_n}$ the label factors, $m_1, \ldots, m_n$ their size and
    $\mathbf{P}^1, \ldots, \mathbf{P}^n$ their matrix of $p_{i,s}^k$ parameters.
Ensure: $\mathbf{h}^*$ the F-measure maximizing prediction.
1: Initialize $m \leftarrow 0$, $\mathbf{P} \leftarrow \emptyset$, $\mathbf{d} \leftarrow \{1\}$
2: **for all** $k \in \{1, \ldots, n\}$ **do**
3:     $m' \leftarrow m$, $\mathbf{P}' \leftarrow \mathbf{P}$, $\mathbf{d}' \leftarrow \mathbf{d}$, $m \leftarrow m' + m_k$
4:     Initialize $\mathbf{d}^k = \{d_0, \ldots, d_{m_k}\}$ a vector of size $m_k + 1$
5:     **for all** $s \in \{1, \ldots, m_k\}$ **do**   ▷ 1) recover $\mathbf{d}^k$ from $\mathbf{P}^k$
6:         $d_s^k \leftarrow s^{-1} \sum_{i=1}^{m_k} p_{i,s}^k$
7:     **end for**
8:     $d_0^k \leftarrow 1 - \sum_{s=1}^{m_k} d_s^k$
9:     Initialize $\mathbf{P}$ a zero matrix of size $m \times m$
10:    **for all** $i \in \{1, \ldots, m_k\}$ **do**   ▷ 2) merge $\mathbf{P}^k$ and $\mathbf{d}'$ into $\mathbf{P}$
11:        **for all** $s1 \in \{1, \ldots, m_k\}$ **do**
12:            **for all** $s2 \in \{0, \ldots, m'\}$ **do**
13:                $p_{i,s1+s2} \leftarrow p_{i,s1+s2} + p_{i,s1}^k \cdot d_{s2}'$
14:            **end for**
15:        **end for**
16:    **end for**
17:    **for all** $i \in \{1, \ldots, m'\}$ **do**   ▷ 3) merge $\mathbf{P}'$ and $\mathbf{d}^k$ into $\mathbf{P}$
18:        **for all** $s1 \in \{1, \ldots, m'\}$ **do**
19:            **for all** $s2 \in \{0, \ldots, m_k\}$ **do**
20:                $p_{i+m_k,s1+s2} \leftarrow p_{i+m_k,s1+s2} + p_{i,s1}' \cdot d_{s2}^k$
21:            **end for**
22:        **end for**
23:    **end for**
24:    Initialize $\mathbf{d}$ a zero vector of size $m + 1$
25:    **for all** $s \in \{1, \ldots, m\}$ **do**   ▷ 4) recover $\mathbf{d}$ from $\mathbf{P}$
26:        $d_s \leftarrow s^{-1} \sum_{i=1}^{m} p_{i,s}$
27:    **end for**
28:    $d_0 \leftarrow 1 - \sum_{s=1}^{m} d_s$
29: **end for**
30: $\mathbf{h}^* \leftarrow GFM(\mathbf{P}, d_0)$ ▷ 5) obtain $\mathbf{h}^*$ from $\mathbf{P}$ and d_0
31: Rearrange $\mathbf{h}^*$ to match the order of the labels in $\mathbf{Y}$.

---

problem of parameter estimation in GFM and F-GFM is essentially the same, that is, estimating the matrix $\mathbf{P}$ (resp. $\mathbf{P}^k$) for a particular input $\mathbf{x}$, given a set of training samples $(\mathbf{x}, \mathbf{y})$ (resp. $(\mathbf{x}, \mathbf{y}_{F_k})$).

Dembczynski et al. [1] propose a solution to estimate the p_{is} terms directly, by solving m multinomial logistic regression problems with $m + 1$ classes. For each label Y_i the scheme of the reduction is the following:

$$(\mathbf{x}, \mathbf{y}) \to (\mathbf{x}, y = y_i \cdot s_{\mathbf{y}}).$$

However, we observed that the parameters estimated with this approach are inconsistent, that is, they often result in a negative probability for d_0 when trying to recover $\mathbf{d}$ from $\mathbf{P}$. To overcome this numerical problem, we found a straightforward and effective approach. Instead of estimating the p_{is} terms directly, we can proceed in two steps. From the chain rule of probabilities, we have that

$$p(y_i, s_{\mathbf{y}}|\mathbf{x}) = p(s_{\mathbf{y}}|\mathbf{x}) \cdot p(y_i|s_{\mathbf{y}}, \mathbf{x}). \tag{10}$$

The idea is to estimate each of these two terms independently. First, the $p(s_{\mathbf{y}} = s|\mathbf{x})$ terms are obtained by performing multinomial logistic regression with $m + 1$ classes, using the following mapping:

$$(\mathbf{x}, \mathbf{y}) \to (\mathbf{x}, y = s_{\mathbf{y}}).$$

Second, for each label Y_i we estimate the $p(y_i = 1|s_{\mathbf{y}} = s, \mathbf{x})$ terms with a binary logistic regression model, using the following mapping:

$$(\mathbf{x}, \mathbf{y}) \to ((\mathbf{x}, s_{\mathbf{y}}), y = y_i).$$

To summarize, for each label factor, one multinomial logistic regression model with $m_k + 1$ classes, and m_k binary logistic regression models are trained. In order to estimate the p_{is}^k terms, we combine the outputs of the multinomial and the binary models according to (10). This approach has the desirable advantage of producing calibrated $\mathbf{P}^k$ matrices and $\mathbf{d}^k$ vectors, which appears to be crucial for the success of F-GFM. Notice that in our experiments this approach was also very beneficial to GFM in terms of MLC performance.

5   Experiments

In this section, we compare GFM and F-GFM on a synthetic toy problem to assess the effective improvement in classification performance due to the label factorization. The code to reproduce this experiment was made available online[2].

---

[2] https://github.com/gasse/fgfm-toy.

5.1   Setup Details

Consider $\mathbf{Y} = \{Y_1, \ldots, Y_8\}$ 8 labels and $\mathbf{X} = \{X_1, \ldots, X_6\}$ 6 binary random variables. The true joint distribution $p(\mathbf{x}, \mathbf{y})$ is encoded in a Bayesian network (one example is displayed in Fig. 2) which imposes different label factor decompositions and serves as a data-generative model. In this BN structure (a directed acyclic graph, DAG for short), each of the features X_1, X_2, X_3, X_4 is a parent node to every label, to enable a relationship between $\mathbf{X}$ and $\mathbf{Y}$. The remaining features X_5 and X_6 are totally disconnected in the graph, and thus serve as irrelevant features. Each label factor $\mathbf{Y}_{F_k}$ is made fully connected by placing an edge $Y_i \rightarrow Y_j$ for every $Y_i, Y_j \in \mathbf{Y}_{F_k}$, $i < j$. As a result each label factor is conditionally independent of the other labels given $\mathbf{X}$, yet it exhibits conditional dependencies between its own labels. We consider 4 distinct structures encoding the following label factor decompositions:

- DAG 1: $\{Y_1, Y_2\}, \{Y_3, Y_4\}, \{Y_5, Y_6\}, \{Y_7, Y_8\}$;
- DAG 2: $\{Y_1, Y_2, Y_3, Y_4\}, \{Y_5, Y_6, Y_7, Y_8\}$;
- DAG 3: $\{Y_1, Y_2, Y_3, Y_4, Y_5, Y_6\}, \{Y_7, Y_8\}$;
- DAG 4: $\{Y_1, Y_2, Y_3, Y_4, Y_5, Y_6, Y_7, Y_8\}$.

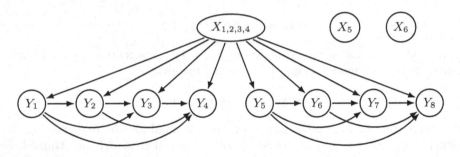

Fig. 2. BN structure of our toy problem with DAG 2, i.e. two label factors $\{Y_1, Y_2, Y_3, Y_4\}$ and $\{Y_5, Y_6, Y_7, Y_8\}$. Note that nodes X_1, X_2, X_3 and X_4 are grouped up for readability.

Once these BN structures are fixed, the next step is to generate random distributions $p(\mathbf{x}, \mathbf{y})$ to sample from. For each BN structure we generate a probability distribution by sampling uniformly the conditional probability table of each node given its parents, $p(x|\mathbf{pa}_x)$, from a unit simplex as discussed in [7]. The process is repeated 100 times randomly, and each time we generate 7 data sets with 50, 100, 200, 500, 1000, 2000 and 5000 training samples, and 5000 test samples. We report the comparative performance of GFM and F-GFM on the test samples with respect to each scenario (DAG structure) and each training size, averaged over the 100 repetitions.

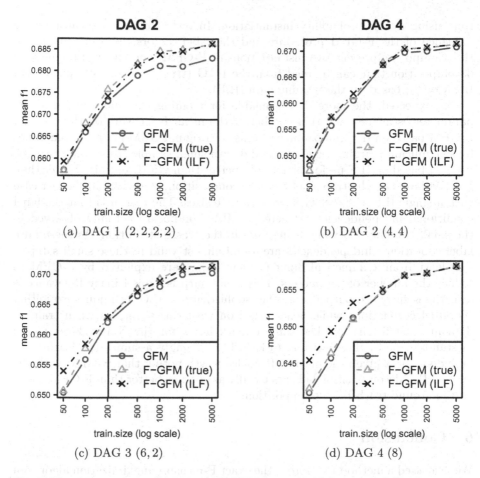

Fig. 3. Mean F-measure of GFM and F-GFM on each DAG, with 7 different training sizes (50, 100, 200, 500, 1000, 2000, 5000) displayed on a logarithmic scale, averaged over 100 repetitions with random distributions. F-GFM (true) uses the true decomposition, while F-GFM (ILF) uses the decomposition learned with ILF-Compo from the training data.

5.2   Implementation Details

To extract the irreducible label factors, we employ the ILF-Compo algorithm proposed in [4], with $\alpha = 0.01$. To estimate the parameters, we use the standard multinomial logistic regression model from the *nnet*[8] R package, with weight decay regularization and λ chosen over a 3-fold cross validation.

5.3   Results

The comparative performance results for GFM and F-GFM are displayed in Fig. 3 in terms of mean F-measure on the test set averaged over 100 runs, each time using a new probability instantiation. In order to assess separately the influence of the F-GFM procedure and the label factors discovery procedure ILF-Compo, we present two instantiations of F-GFM: one which uses the true decomposition that can be read from the DAG (true), and one obtained from ILF-Compo based on the training data (ILF).

As expected, the more date available for training, the more accurate the parameter estimates, and thus the better the mean F-measure on the test set. F-GFM based on ILF-compo outperforms the original GFM method, sometimes by a significant margin (see Fig. 3c and d with small sample sizes). Interestingly, F-GFM based on ILF performs not only better than GFM, but also better than F-GFM based on the true label factor decomposition, especially in the last case with a single ILF of size 8 with small sample sizes. The reason is that the label conditional independencies extracted by ILF-Compo are actually observed in the small training sets while being false in the true distribution. As these false label conditional independencies are found almost valid in these small samples - at least from a numerical point of view - they are exploited by F-GFM to reduce the number of parameters. This is not surprising as Binary Relevance is sometimes shown to outperform other sophisticated MLC techniques exploiting the label correlations while being based on wrong assumptions when training data are insufficient [6]. The same remark holds for the Naive Bayes model in standard multi-class learning tasks, which wrongly assumes the features to be independent given the output. It is also worth noting that F-GFM with the learned ILF decomposition behaves usually as good or better than F-GFM based on the ground truth ILF decomposition.

6   Conclusion

We discussed a method to improve the exact F-measure maximization algorithm (GFM), for multi-label classification, assuming the label set can be partitioned into conditionally independent subsets given the input features. In the general case, $m^2 + 1$ parameters are required by GFM, to solve the problem in $O(m^3)$ operations. In this work, we show that the number of parameters can be reduced further to m^2/n, in the best case, assuming the label set can be partitioned into n conditionally independent subsets. As the label partition needs to be estimated from the data beforehand, we use first the procedure proposed in [4] that finds such partition and then infer the required parameters locally in each label subset. The latter are aggregated and serve as input to GFM to form the Bayes-optimal prediction. Our experimental results on a synthetic problem exhibiting various forms of label inpendencies demonstrate noticeable improvements in terms of F-measure over the standard GFM approach. Interestingly, F-GFM was shown to take advantage of purely fortuitous label independencies in small training sets, despite being false in the underlying distribution, to reduce further the number of parameters, while performing better than F-GFM based on the true decomposition. This is not surprising as Binary Relevance is sometimes shown to outperform other sophisticated MLC techniques exploiting the label correlations while being based on wrong assumptions when training data are insufficient [6]. Future work will be aimed at reducing further the number of parameters and the overall complexity of the inference algorithm. Large real-world MLC problems will also be considered in the future.

Acknowledgements. This work was funded by both the French state trough the Nano 2017 investment program and the European Community through the European Nanoelectronics Initiative Advisory Council (ENIAC Joint Undertaking), under grant agreement no 324271 (ENI.237.1.B2013).

References

1. Dembczynski, K., Jachnik, A., Kotlowski, W., Waegeman, W., Hüllermeier, E.: Optimizing the f-measure in multi-label classification: plug-in rule approach versus structured loss minimization. In: ICML (3). JMLR Proceedings vol. (28), pp. 1130–1138. JMLR.org (2013)
2. Dembczynski, K., Waegeman, W., Cheng, W., Hüllermeier, E.: An exact algorithm for f-measure maximization. In: Shawe-Taylor, J., Zemel, R.S., Bartlett, P.L., Pereira, F.C.N., Weinberger, K.Q. (eds.) NIPS, pp. 1404–1412 (2011)
3. Dembczynski, K., Waegeman, W., Cheng, W., Hüllermeier, E.: On label dependence and loss minimization in multi-label classification. Mach. Learn. **88**(1–2), 5–45 (2012)
4. Gasse, M., Aussem, A., Elghazel, H.: On the optimality of multi-label classification under subset zero-one loss for distributions satisfying the composition property. In: Bach, F.R., Blei, D.M. (eds.) ICML. JMLR Proceedings, vol. 37, pp. 2531–2539. JMLR.org (2015)
5. Jansche, M.: A maximum expected utility framework for binary sequence labeling. In: Carroll, J.A., van den Bosch, A., Zaenen, A. (eds.) ACL. The Association for Computational Linguistics (2007)
6. Luaces, O., Díez, J., Barranquero, J., del Coz, J.J., Bahamonde, A.: Binary relevance efficacy for multilabel classification. Prog. AI **1**(4), 303–313 (2012)
7. Smith, N., Tromble, R.: Sampling Uniformly from the Unit Simplex. Johns Hopkins University, Technical report (2004)
8. Venables, W.N., Ripley, B.D.: Modern Applied Statistics with S, 4th edn. Springer, New York (2002). ISBN 0-387-95457-0
9. Waegeman, W., Dembczynski, K., Jachnik, A., Cheng, W., Hüllermeier, E.: On the bayes-optimality of f-measure maximizers. J. Mach. Learn. Res. **15**(1), 3333–3388 (2014)
10. Ye, N., Chai, K.M., Lee, W.S., Chieu, H.L.: Optimizing f-measure: a tale of two approaches. In: Langford, J., Pineau, J. (eds.) ICML. pp. 289–296. ICML 2012, Omnipress, New York, NY, USA, July 2012

Cost-Aware Early Classification of Time Series

Romain Tavenard[1]([✉]) and Simon Malinowski[2]

[1] LETG-Rennes COSTEL / IRISA, University of Rennes 2, Rennes, France
romain.tavenard@irisa.fr
[2] IRISA, University of Rennes 1, Rennes, France
simon.malinowski@irisa.fr

Abstract. In time series classification, two antagonist notions are at stake. On the one hand, in most cases, the sooner the time series is classified, the more rewarding. On the other hand, an early classification is more likely to be erroneous. Most of the early classification methods have been designed to take a decision as soon as sufficient level of reliability is reached. However, in many applications, delaying the decision with no guarantee that the reliability threshold will be met in the future can be costly. Recently, a framework dedicated to optimizing a trade-off between classification accuracy and the cost of delaying the decision was proposed, together with an algorithm that decides online the optimal time instant to classify an incoming time series. On top of this framework, we build in this paper two different early classification algorithms that optimize a trade-off between decision accuracy and the cost of delaying the decision. These algorithms are non-myopic in the sense that, even when classification is delayed, they can provide an estimate of when the optimal classification time is likely to occur. Our experiments on real datasets demonstrate that the proposed approaches are more robust than existing methods. The data and software related to this paper are available at https://github.com/rtavenar/CostAware_ECTS.

Keywords: Time-series classification · Early classification

1 Introduction

Time series classification has received a lot of attention from researchers in the last years due to the increasing amount of data available and its applicability in many domains like medicine, environment, business, *etc.* Classical methods for time series classification have been designed to make a decision based on complete time series. However, in many applications where observations arrive one at a time, it is rewarding to be able to classify a time series even without knowing it entirely. In this case, each time a new sample arrives, one can decide either to perform classification or to wait for more data. It is more valuable to

Electronic supplementary material The online version of this chapter (doi:10.1007/978-3-319-46128-1_40) contains supplementary material, which is available to authorized users.

© Springer International Publishing AG 2016
P. Frasconi et al. (Eds.): ECML PKDD 2016, Part I, LNAI 9851, pp. 632–647, 2016.
DOI: 10.1007/978-3-319-46128-1_40

make a decision as soon as possible, but premature decisions are more likely to be erroneous. Consequently, two antagonist notions are at stake. For example, in applications related to business, higher profits can be made thanks to early classification of customer behaviours, provided that the classification is accurate. Hence, two different costs clearly appear from this point of view: the cost of delaying the decision and the cost of misclassification.

Most of the approaches for early classification in the literature focus on the design of algorithms that make a reliable decision from incomplete time series, without explicitly accounting for the cost of delaying the decision [1, 4–7, 10, 13, 14]. In these works, sufficient guarantee of accuracy triggers the decision.

In this paper, we focus on cost-aware early classification of time series. Let us illustrate the advantages of cost-aware early classification of time series on a concrete example. In a surveillance scenario, one would want the system to warn the police as soon as a crime scene is detected. The optimal time to make a decision depends on two antagonist notions: accuracy and earliness. For this scenario, warning the police too early would induce a cost if the scene is actually not a crime scene, but waiting too long would prevent the police from stepping in the crime scene. It hence becomes desirable to allow end-users to set a trade-off between earliness and accuracy, which is the goal of cost-aware early classification of time series.

Recently, Dachraoui et al. [3] introduced a framework for cost-aware early classification that is dedicated to optimizing a trade-off between the accuracy of the prediction and the time at which it is performed. This framework hence involves both costs defined above: a misclassification cost and a cost of delaying the decision. The authors of [3] derive an algorithm that decides, at test time, whether the decision should be made at a given time instant t or if more data should be waited for (based on both costs defined above). In addition, the method of [3] has two other interesting properties: it is adaptive and non-myopic. It is adaptive in the sense that the optimal time of decision depends on the time series to be classified. It is non-myopic because at every time instant t, the algorithm not only decides if t is the optimal time instant to classify, but it also estimates when this optimal time is likely to happen, if not at time t. Such a property is essential since it can help end users of such systems to get prepared for action when the decision should soon be made (e.g. for medical applications).

In this paper, we propose two new early-classification schemes built upon the framework defined in [3]. This framework is based on approximating an expected cost through clustering, which tends to bring vagueness in the process. Our schemes improve on this existing work by removing the clustering step. They optimize a trade-off between a misclassification cost and a cost of delaying the decision, they are adaptive and non-myopic.

The rest of this paper is organized as follows. Section 2 explains the problem statement and reviews the related work on early classification of time series. Section 3 presents both proposed approaches and how they differ from [3]. Experiments on real time series data sets are presented in Sect. 4. Notations used throughout this paper are summarized in Table 1.

Table 1. Mathematical notations used throughout the paper.

Notation	Description
$\mathbf{x} = (x_1, \dots, x_T)$	The time series to be classified
$\mathbf{x_t} = (x_1, \dots, x_t)$	Time series $\mathbf{x}$ truncated after t observations, $t \leq T$
$\tau(\mathbf{x})$	Time index at which early classification of $\mathbf{x}$ is performed, also referred to as classification time
$\tau^*(\mathbf{x})$	Optimal time index for early classification of $\mathbf{x}$
$\hat{\tau}(\mathbf{x_t})$	Time index (greater than current time t) at which classification is most likely to occur (for non-myopic methods only)
$y \in \mathcal{Y}$	Class label of time series $\mathbf{x}$
$\mathcal{T} = \{(\mathbf{x}^i, y^i)\}_{1 \leq i \leq N}$	A training set of N time series and their labels
$\mathcal{T}_t = \{(\mathbf{x}_t^i, y^i)\}_{1 \leq i \leq N}$	A training set of N truncated time series and their labels
$C_m(\hat{y}, y)$	Cost of misclassifying a time series of class y into class $\hat{y}$
$C_d(t)$	Cost of delaying classification after t time instants
$C(\mathbf{x}, y)$	Overall early classification cost
$E_t(\mathbf{x})$	Expected cost of performing classification of time series $\mathbf{x}$ at time t
$\mathcal{H} = \{h_t\}_{1 \leq t \leq T}$	A set of classifiers for predicting the class of time series of length t
$\mathcal{D} = \{d_t\}_{1 \leq t \leq T}$	A set of classifiers for predicting whether early classification should be performed or not at time t (for *2Step* method only)
$\mathcal{M} = \{m_t\}_{1 \leq t \leq T}$	A set of regressors predicting when early classification is likely to happen (for *2Step* method only)

2   Related Work

A wide range of methods tackle the early classification problem without explicitly accounting for the cost of delaying the decision [1,4–7,10,13,14]. These methods differ in (i) the design of the early classifiers and (ii) the estimation of the optimal time to make a decision. They mostly wait for sufficient confidence to make their decision, using varied procedures to define this confidence. The authors of [7] design an early classification scheme based on the boosting procedure for classification, where one weak classifier is built at each time stamp. This method is able to predict the class of a test time series at any time, but the issue of estimating the optimal time to make the decision is not adressed. Hatami *et al.* [5] propose an ensemble classifier with a reject option. A decision is made as soon as the agreement between all classifiers is above a certain threshold. Otherwise, the reject option is activated and more data is waited for. In [14], the authors rely on feature extraction for early classification. They first extract, from a training set of time series, a set of features called shapelets with a high discriminating power. Shapelets having high utility are then selected, where utility is an extension of the F-measure which also takes earliness into account. At test time, classification is performed as soon as one of the selected shapelets is found in the testing time series. Ghalwash *et al.* [4] extend this work so that it also provides an estimation of uncertainty associated with the decision. An alternative of these works designed for multivariate time series is proposed in [6]. Antonucci *et al.* [1] use Hidden Markov Model Classifiers with set valued parameters to determine whether the model is confident enough about its prediction (if not, several possible outputs are returned, which illustrates the uncertainty of the model). Early classification happens when there exists no ambiguity anymore about the prediction.

These methods do not attempt to infer whether future observations could improve classification accuracy. For difficult classification problems, they might tend to delay the decision even if future observations are unlikely to help. To bypass this limitation, Xing *et al.* [13] present a method that relies on neighborhood properties. For a given training time series $\mathbf{x^i} \in \mathcal{T}$, optimal prediction time $\tau(\mathbf{x^i})$ is set to the smallest τ^i such that the set of reverse nearest neighbors for $\mathbf{x^i}$ is stable for all $t \geq \tau^i$. At test time, if a time series has $\mathbf{x_t^i}$ as nearest neighbor after $t \geq \tau(\mathbf{x^i})$ time steps, classification is performed at this point. To improve on this setting, clustering can be performed on training time series so that stability can be robustly evaluated at the cluster level. Parrish *et al.* [10] design a probabilistic approach based on Quadratic Discriminant Analysis which aims at maximizing the *reliabilty*, *i.e.* the probability that the early decision leads to the same classification result as the classification of the complete time series.

One drawback of these methods is that they do not explicitly optimize a trade-off between earliness and accuracy, hence leading to sub-optimal solutions in this regard. In order to overcome this limitation, Dachraoui *et al.* [3] propose a framework for early classification, where the cost of delaying the decision is included in the optimization function. This framework considers a time series classification task for which two cost functions are given:

- $C_m(\hat{y}, y)$ is the misclassification cost function, *i.e.* the cost of predicting class label $\hat{y}$ whereas the effective class label was y;
- $C_d(t)$ is the cost of delaying the decision to time t.

In this framework, the cost of classifying a time series $\mathbf{x}$ of class y using a set $\mathcal{H} = \{h_t\}_{1 \leq t \leq T}$ of classifiers is given by:

$$C(\mathbf{x}, y) = C_m(h_{\tau(\mathbf{x})}(\mathbf{x}), y) + C_d(\tau(\mathbf{x})), \qquad (1)$$

where $\tau(\mathbf{x})$ is the time index at which classification is performed and $h_{\tau(\mathbf{x})}(\mathbf{x})$ is the class predicted at time $\tau(\mathbf{x})$ by the *ad hoc* classifier. To optimize on Eq. (1), the authors compute an expected cost for current and future time instants. If the current expected cost is lower than all future ones, classification is performed. This method is further discussed in Sect. 3.

The authors of [9] also design an early classification scheme that takes the cost of delaying the decision into account. They define a stopping rule that depends on 4 parameters (to be fine tuned), on the time of the decision and on the posterior probabilities output by the classifiers. The optimal parameters for the stopping rule are learned through cross-validation so that they minimize a cost function that is similar in spirit to Eq. (1):

$$C'(\mathbf{x}, y) = \alpha \times C_m(h_{\tau(\mathbf{x})}(\mathbf{x}), y) + (1 - \alpha) \times C_d(\tau(\mathbf{x})), \tag{2}$$

where α is a parameter that designs a trade-off between C_m and C_d. Unlike the method proposed in [3], the one from [9] is myopic in the sense that at a given time t, if the algorithm decides that it is better to wait to perform classification, it cannot provide an estimate $\hat{\tau}(\mathbf{x_t}) > t$ of the optimal classification time.

In this paper, we propose two non-myopic early classification methods taking into account the cost of delaying the decision. These methods are improvements over the framework defined in [3]. In the next section, we detail this framework and position the methods we propose with respect to it.

3 Proposed Early Classification Schemes

The problem tackled in this paper is to estimate, for an incoming time series $\mathbf{x}$, the optimal time $\tau^*(\mathbf{x})$ at which the classification of $\mathbf{x}$ should be done in order to minimize the cost given by Eq. (1). In Sect. 3.1, we first give a detailed review of the method proposed in [3] and its limitations. We then introduce an improved version of this approach from which we derive two early classification methods presented in Sects. 3.2 and 3.3.

3.1 Computing Expected Costs for Training Time Series

As an attempt to optimize on Eq. (1), Dachraoui *et al.* [3] intend to perform classification at a time t such that the expected cost:

$$f(\mathbf{x_t}) = \sum_{y \in \mathcal{Y}} P(y|\mathbf{x_t}) \sum_{\hat{y} \in \mathcal{Y}} P(\hat{y}|y, \mathbf{x_t}) \, C_m(\hat{y}, y) + C_d(t) \tag{3}$$

is minimized. The term $P(y|\mathbf{x_t})$ is unknown by definition of the learning problem, which prevents us from computing this expected cost directly.

In order to approximate Eq. (3), authors rely on a clustering $\mathcal{C} = \{\mathbf{c^1}, \ldots, \mathbf{c^K}\}$ of the set $\mathcal{T}$ of complete training time series. At any given instant t, the expected cost of classifying $\mathbf{x}$ at future time instants $t + \delta$ (for $\delta \geq 0$) is computed as:

$$f_\delta(\mathbf{x_t}) = \sum_{k=1}^{K} P(\mathbf{c^k}|\mathbf{x_t}) \sum_{y \in \mathcal{Y}} P(y|\mathbf{c^k}) \sum_{\hat{y} \in \mathcal{Y}} P_{t+\delta}(\hat{y}|y, \mathbf{c^k}) \, C_m(\hat{y}, y) + C_d(t + \delta)$$

$$= \sum_{k=1}^{K} P(\mathbf{c^k}|\mathbf{x_t}) \, E_{t+\delta}(\mathbf{c^k}), \tag{4}$$

where $E_{t+\delta}(\mathbf{c^k})$ is the expected cost of performing classification at time $t + \delta$ for series in cluster $\mathbf{c^k}$, that is computed as follows. At every time t, probabilities $P_t(\hat{y} \mid y, \mathbf{c^k})$ are estimated using cross-validation on the training set. Each of these

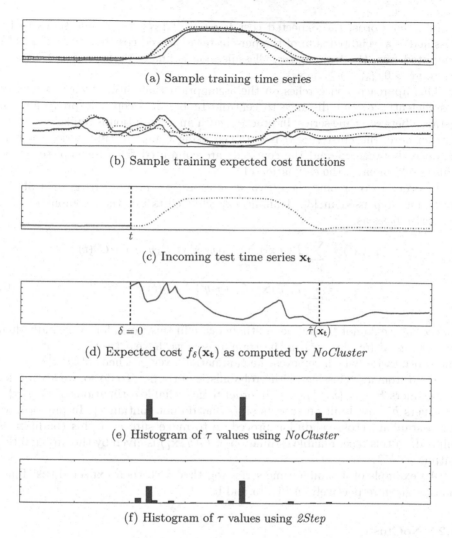

(a) Sample training time series

(b) Sample training expected cost functions

(c) Incoming test time series $\mathbf{x_t}$

(d) Expected cost $f_\delta(\mathbf{x_t})$ as computed by *NoCluster*

(e) Histogram of τ values using *NoCluster*

(f) Histogram of τ values using *2Step*

Fig. 1. Illustration of proposed methods on *Gun_Point* time series.

probabilities correspond to a bin of the confusion matrix associated to cluster $\mathbf{c^k}$ at time t. The computation of the expected cost function $f_\delta(\mathbf{x_t})$ also depends on the probabilities $P(\mathbf{c^k} \mid \mathbf{x_t})$. In [3], these probabilities are computed as:

$$P(\mathbf{c^k}|\mathbf{x_t}) = \frac{s^k}{\sum_{i=1}^{K} s^i}, \quad \text{where } s^k = \frac{1}{1 + \exp^{-\lambda \Delta_t^k}}, \tag{5}$$

λ is some positive constant and $\Delta_t^k = (\bar{D}_t - d_t^k)/\bar{D}_t$ is the normalized difference between the average $\bar{D}_t$ of the distances between $\mathbf{x_t}$ and all the clusters, and the distance d_t^k between $\mathbf{x_t}$ and the cluster $\mathbf{c^k}$. We refer the interested reader to [3] for more details on these equations.

In other words, the expected cost function $f_\delta(\mathbf{x_t})$ for a time series to be classified is a weighted sum of the per-cluster expected cost functions $E_{t+\delta}(\mathbf{c^k})$ computed from the training set. Classification is finally performed at time t if, for any $\delta > 0$, $f_\delta(\mathbf{x_t}) \geq f_0(\mathbf{x_t})$.

This approach hence relies on the assumption that intra-cluster variability is sufficiently low for distance to centroid to be an acceptable proxy for the distance between time series. In practice, such an assumption is unlikely to hold. Furthermore, the clustering $\mathcal{C}$ is obtained using complete time series whereas distances between $\mathbf{x_t}$ and the centroids are computed from incomplete series. This surely impacts the estimation of $\tau^*(\mathbf{x})$.

In order to overcome these two weaknesses, we propose to get rid of the clustering step used in [3]. Training expected costs are then computed on a per-series basis as:

$$E_t(\mathbf{x^i}) = \sum_{y \in \mathcal{Y}} P(y|\mathbf{x^i}) \sum_{\hat{y} \in \mathcal{Y}} P_t(\hat{y}\,|\,\mathbf{x_t^i})\, C_m(\hat{y}, y^i) + C_d(t)$$

$$= \sum_{\hat{y} \in \mathcal{Y}} P_t(\hat{y}\,|\,\mathbf{x_t^i})\, C_m(\hat{y}, y^i) + C_d(t). \qquad (6)$$

Indeed, as we do not tackle the multilabel case in this work, for every individual time series $\mathbf{x^i}$ (of class y^i) in the training set, we have $P(y\,|\,\mathbf{x^i}) = 1$ if $y = y^i$ and 0 otherwise, which removes the summation over y. Then, $P_t(\hat{y}\,|\,\mathbf{x_t^i})$ can be obtained from any classifier with probabilistic outputs.[1] To do so, we learn a set of classifiers $\mathcal{H}^{-i} = \{h_t^{-i}\}_{1 \leq t \leq T}$. In order to get reliable estimations of $P_t(\hat{y}\,|\,\mathbf{x_t^i})$, classifiers h_t^{-i} are built on subsets of $\mathcal{T}_t$ that do not contain $\mathbf{x_t^i}$. In practice, we use a standard cross-validation procedure to make sure that this condition is fulfilled. In this case, reliable estimation of $P_t(\hat{y}\,|\,\mathbf{x_t^i})$ is given by the probabilities output by h_t^{-i}.

An example of 4 sample time series together with their expected cost functions is given respectively in Fig. 1a and b.

3.2   NoCluster

Our first proposed method, which we denote *NoCluster*, consists in computing the expected cost for a test time series as a weighted sum of training expected costs calculated using Eq. (6). In other words, present and future expected costs for a time series to be classified are computed, at time t, as:

$$\forall \delta \geq 0, \quad f_\delta(\mathbf{x_t}) = \sum_{i=1}^{N} \frac{1}{K_0} \frac{1}{1 + \exp^{-\lambda \Delta_t^i}}\, E_{t+\delta}(\mathbf{x^i}), \qquad (7)$$

where $K_0 = \sum_{i=1}^{N} \frac{1}{1 + \exp^{-\lambda \Delta_t^i}}$ and Δ_t^i is computed the way Δ_t^k is in Eq. (5), where distance to centroids are replaced by distance to training time series. As in [3],

---

[1] Note that even when using classifiers that do not inherently compute probability estimates, cross validation can be used to get such estimates, as done in [2].

classification is performed at the smallest time t such that, for any $\delta > 0$, we get $f_\delta(\mathbf{x_t}) \geq f_0(\mathbf{x_t})$. This time instant is denoted $\tau(\mathbf{x_t})$. At test time, the Δ_t^i are computed for all times t up to $\tau(\mathbf{x})$, while when clustering was used, only K such normalized distance computations were required per time t. As this method is derived from [3], it preserves its non-myopia and adaptiveness properties.

Figure 1c shows a sample test time series of length t, and Fig. 1d shows its associated expected cost from time t, and the estimation of the optimal decision time $\hat{\tau}(\mathbf{x_t})$. Figure 1e depicts the histogram of τ values obtained by *NoCluster* for all the test time series of the *Gun_Point* dataset. Note that peaks of this histogram (which also correspond to valleys of the expected cost presented in Fig. 1d) are related to regions of high interest in the time series for this dataset: a first increasing part of the time series, followed by a plateau and a final decrease. Algorithmic description of this method is provided in Algorithms 1 and 2.

Algorithm 1. Offline training of *NoCluster* method

Input: $\mathcal{T} = \{(\mathbf{x}^i, y^i)\}_{i \in \{1..N\}}, \{\mathcal{H}^{-i}\}_{i \in \{1..N\}}$
Output: $\{E_t(\mathbf{x}^i)\}_{i \in \{1..N\}, t \in \{1..T\}}$
    for $i \in \{1..N\}$ do
        for $t \in \{1..T\}$ do
            Compute $P_t(\hat{y} \mid \mathbf{x_t^i})$ for all $\hat{y} \in \mathcal{Y}$ using h_t^{-i}
            Compute $E_t(\mathbf{x}^i)$ using $P_t(\hat{y} \mid \mathbf{x_t^i})$ according to Eq. (6)
        end for
    end for
    return $\{E_t(\mathbf{x}^i)\}_{i \in \{1..N\}, t \in \{1..T\}}$

Algorithm 2. Classification using *NoCluster* method

Input: $\{E_t(\mathbf{x}^i)\}_{i \in \{1..N\}, t \in \{1..T\}}, \mathbf{x}, \mathcal{H} = \{h_t\}_{t \in \{1..T\}}, t_{\min}$
Output: $\hat{y}, \tau(\mathbf{x})$
    for $t \in \{t_{\min}..T\}$ do
        for $i \in \{1..N\}$ do
            $\Delta_t^i \leftarrow \|\mathbf{x_t} - \mathbf{x_t^i}\|_2$
        end for
        for $\delta \in \{0..T - t + 1\}$ do
            Compute $f_\delta(\mathbf{x_t})$ using Δ_t^i and $E_t(\mathbf{x}^i)$, according to Eq. (7)
        end for
        if $\forall \delta \geq 0, f_0(\mathbf{x_t}) \leq f_\delta(\mathbf{x_t})$ then
            break
        end if
    end for
    return $h_t(\mathbf{x_t}), t$

3.3   2Step

As explained above, the computational complexity of *NoCluster* at test time is higher than that of the baseline, which can be a limitation for some applications. In this section, we present another early classification method that we call *2Step*. This method has a lower complexity at test time, at the cost of possibly higher training complexity. *2Step* also relies on the computation of the expected costs for every training time series (Eq. (6)), but uses them in a different manner. A set of classifiers $\mathcal{D} = \{d_t\}_{1 \leq t \leq T}$ is built as follows. The classifier d_t is built on the training set $\mathcal{T}_t$. Its target variable is not the class of the time series but a binary variable γ_t indicating if t is the expected optimal time to classify the time series. In other words,

---

Algorithm 3. Offline training of *2Step* method

Input: $\mathcal{T} = \{(\mathbf{x}^i, y^i)\}_{i \in \{1..N\}}$, $\{\mathcal{H}^{-i}\}_{i \in \{1..N\}}$
Output: $\mathcal{D} = \{d_t\}_{t \in \{1..T\}}$
  for $i \in \{1..N\}$ **do**
    for $t \in \{1..T\}$ **do**
      Compute $P_t(\hat{y} \mid \mathbf{x}_t^i)$ for all $\hat{y} \in \mathcal{Y}$ using h_t^{-i}
      Compute $E_t(\mathbf{x}^i)$ using $P_t(\hat{y} \mid \mathbf{x}_t^i)$ according to Eq. (6)
    end for
    for $t \in \{1..T\}$ **do**
      Compute $\gamma_t(\mathbf{x}_t^i)$ according to Eq. (8).
    end for
  end for
  for $t \in \{1..T\}$ **do**
    Learn classifier d_t with training set $\mathcal{T}_t$ and target variables $\{\gamma_t(\mathbf{x}_t^i)\}_{i \in \{1..N\}}$
  end for
  return $\{d_t\}_{t \in \{1..T\}}$

$$\forall \mathbf{x}_\mathbf{t}^\mathbf{i} \in \mathcal{T}_t, \ \gamma_t(\mathbf{x}_\mathbf{t}^\mathbf{i}) = \begin{cases} 1 \text{ if } E_t(\mathbf{x^i}) = \min_{\delta \geq 0} E_{t+\delta}(\mathbf{x^i}) \\ \\ 0 \text{ otherwise.} \end{cases} \qquad (8)$$

A graphical example showing how γ_t can be computed from the expected cost of classification of a time series is given in Fig. 2. In particular, this figure illustrates that γ_t is not monotonically increasing with t.

At test time, time series $\mathbf{x}_t$ is given to classifier d_t. If the output of d_t is 1, then $\tau(\mathbf{x}) = t$ and h_t is used to classify $\mathbf{x}_t$. Otherwise, the classification of $\mathbf{x}_t$ is delayed. Figure 1f depicts the histogram of τ values obtained by *2Step* for all the test time series of the *Gun_Point* dataset. One can note from Fig. 1e and f that both proposed methods are adaptive, in the sense that predicted timings τ vary across test time series. Algorithmic description of this method is provided in Algorithms 3 and 4.

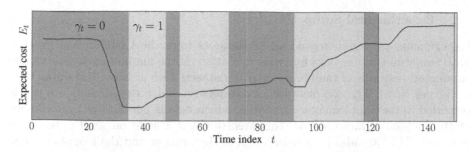

Fig. 2. Expected early classification cost for a training time series from the *Gun_Point* dataset, together with the corresponding γ_t values. For all time instants that are depicted in pink, $\gamma_t = 0$, and for time instants depicted in green, $\gamma_t = 1$. Best viewed in color

---

Algorithm 4. Classification using *2Step* method

Input: $\mathcal{D} = \{d_t\}_{t \in \{1..T\}}, \mathcal{H} = \{h_t\}_{t \in \{1..T\}}, \mathbf{x}, t_{\min}$
Output: $\hat{y}, \tau(\mathbf{x})$
    for $t \in \{t_{\min}..T\}$ **do**
        if $d_t(\mathbf{x_t}) = 1$ **then**
            break
        end if
    end for
    return $h_t(\mathbf{x_t}), t$

The early classification method as presented in this section is myopic. In order to keep the non-myopic property, a set of regressors $\mathcal{M} = \{m_t\}_{1 \leq t \leq T}$ should be learned in addition to the set $\mathcal{D}$ of classifiers. The target variable of regressor m_t should represent the time difference between t and the expected time of minimal cost, that is $\arg\min_{\delta \geq 0} (E_{t+\delta}(\mathbf{x_t}))$.

4   Experimental Results

In this section, we design extensive experiments in order to evaluate the performance of both *NoCluster* and *2Step* early classification methods. The method presented in [3] is used as our main baseline, as (i) our proposed methods are built on top of it and (ii) all three methods share the desirable properties of non-myopia and adaptiveness. However, comparison with other state-of-the-art methods is also provided in Sects. 4.4 and 4.5. The cost function defined in Eq. (1) (which defines a trade-off between accuracy and earliness of the decision) is used as our main performance indicator. Other criteria such as classification accuracy, earliness (expressed as the average prediction time $\bar{\tau}$) and timings are also considered.

4.1   Experimental Setup

Experiments are conducted on all 76 datasets from the UCR archive [8] that have sufficient training data for cross-validation on the method parameters to be conducted (as a rule of thumb, we keep all datasets with at least 10 training time series per class). The complete list of datasets on which experiments are run is provided in the Supplementary material, which details per-dataset performance of the methods. Datasets from this archive cover a wide range of application domains. They are also diverse in terms of the number and the lengths of time series in each dataset. The datasets are split into a training and a test set in the archive and we keep this separation in our experiments. For this set of experiments, the cost of delaying the decision is chosen linear in t:

$$C_d(t) = \beta \times t. \tag{9}$$

So as not to focus on a single trade-off between accuracy and earliness, we vary β in the set $\{0.0005, 0.001, 0.005, 0.01\}$ in all experiments. In real-world applications, β should be set according to expert knowledge and effective cost of waiting for more observations. For instance, in remote sensing applications where buying new images can be costly, high β values shall be used, unlike applications such as ECG monitoring for which new observations are made at almost no cost.

Results are reported for all setups (i.e., all datasets and β values) for which any of the compared methods finishes in less than 7 days. Presented results (for all criteria) are medians over 4 runs for each method.

Classifiers from sets $\mathcal{D}$ and $\mathcal{H}$ are linear Support Vector Machines trained using scikit-learn and LIBSVM Python binding [2,11]. Parameters C from the SVM and λ from the expected cost functions are learned through cross-validation on the training set. Five values for parameter C are sampled regularly from a logarithmic scale between 10^{-1} and 10^1. Five values for λ are sampled in a similar way between 10^1 and 10^3. As done in [3], we do not perform early classification before time $t_{min} = 4$. For the baseline method, the number of clusters is chosen so as to maximize the silhouette factor [12], as suggested in [3].

Following the principle of reproducible research, the Python code used in these experiments (including both proposed method and our implementation of the baseline) is made publicly available for download[2].

4.2   Sensitivity to β

Figure 3 presents results obtained for dataset *FISH* in terms of accuracy, earliness and overall early classification cost function ($C(\mathbf{x}, y)$) for different values of β. As expected, when parameter β increases, accuracy and earliness both drop, for all methods. Indeed, high values of β indicate that it is more costly to wait for more data. Decision is hence made earlier, leading to worse accuracy. On the contrary, overall early classification costs increases with β, whatever the method. For this dataset (see Sect. 4.5 for more results) and for the whole range

---

[2] https://github.com/rtavenar/CostAware_ECTS.

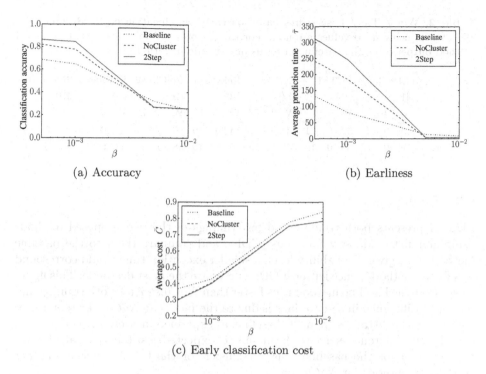

(a) Accuracy

(b) Earliness

(c) Early classification cost

Fig. 3. Early classification performance as a function of β for dataset *FISH*.

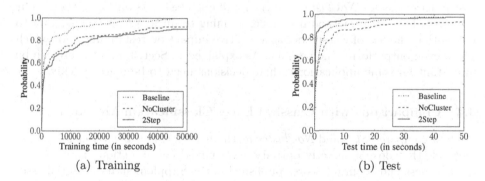

(a) Training

(b) Test

Fig. 4. Cumulative density functions of training (a) and test (b) times over all datasets and β values.

of β values considered here, *2Step* and *NoCluster* methods both outperform the baseline in terms of early classification cost, which is the indicator all three methods optimize on (remind that the lower cost, the better).

Table 2. Win / Tie / Lose scores based on early classification costs. Results from all datasets using all β values are used. For these scores, "Win" means that proposed method outperforms the baseline in terms of cost minimization.

	EDSC [14]			ECTS1 [13]			ECTS2 [13]			RelClass1 [10]			RelClass2 [10]			RelClass3 [10]			RelClass4 [10]		
	W	T	L	W	T	L	W	T	L	W	T	L	W	T	L	W	T	L	W	T	L
2Step	49	0	21	52	0	21	51	0	22	54	0	19	56	0	17	57	0	16	54	0	19
NoCluster	54	0	17	56	0	19	56	0	19	56	0	19	58	0	17	58	0	17	57	0	18

4.3   Timings

Figure 4 presents both training and test timings of the two proposed methods using cumulative density functions (CDF), and compare them to the baseline method. At a given probability level, a smaller execution time would correspond to a faster method, hence leftmost CDF curves are the most desirable. This figure shows that the baseline method runs faster than *NoCluster* for both training and test. The difference in test timings is due to the fact that *NoCluster* requires to compute more distances than the baseline, as explained in Sect. 3.2. For training timings, the difference is due to the number of expected cost function calculations that is lower for the baseline (one per cluster for the baseline versus one per training time series for *NoCluster*).

Concerning *2Step* method, training consists in computing the same expected cost functions as for *NoCluster* and then building classifiers on top of them, so it is expected that this method has higher training timings. Yet, it is important to notice that the use of classifiers has a positive impact on test timings, for which *2Step* even outperforms the baseline. As explained in Sect. 3, this feature can be important for some applications where decisions need to be made quickly.

4.4   Comparisons with Classical Early Classification Methods

We first compare *2Step* and *NoCluster* methods with classical early classification methods that do not directly optimize on a mixed cost function. To do so, we use earliness and accuracy scores published in the Supplementary material associated to [9]. Such scores are available for EDSC [14], strict and loose variants of ECTS [14] (denoted ECTS1 and ECTS2 respectively) and RelClass [10] with four different reliability thresholds (0.001, 0.1, 0.5, 0.9 for methods RelClass1 to RelClass4 respectively). Based on these scores, we can compute early classification costs for all methods with varying β values. Win/Tie/Lose scores computed from these costs are presented in Table 2. These results show that both proposed methods outperform all these baselines. To assess statistical significance, we perform one-sided Wilcoxon signed rank tests between each proposed method and each baseline. All tests show significance with p-values lower than 10^{-6}.

This performance improvement was expected, as baseline methods considered in this Section do not optimize on the cost function that mixes earliness

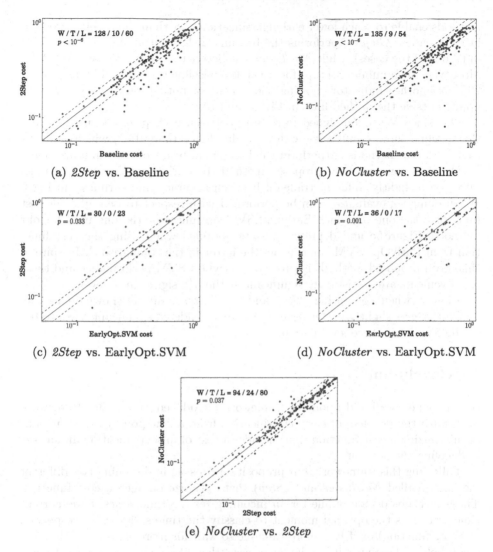

(a) *2Step* vs. Baseline

(b) *NoCluster* vs. Baseline

(c) *2Step* vs. EarlyOpt.SVM

(d) *NoCluster* vs. EarlyOpt.SVM

(e) *NoCluster* vs. *2Step*

Fig. 5. Comparison of early classification costs between proposed methods and the baselines from [3] and [9]. Results from all datasets using all β values are gathered here. Solid line indicates equal costs and dashed lines correspond to the +/-20 % cost interval.

and accuracy. To further evaluate performance of our proposed methods, we compare them to other cost-aware methods in the following Section.

4.5   Comparisons with Cost-Aware Methods

Figure 5 presents comparisons of early classification costs between both our proposed methods and cost-aware baselines. Presented results show that proposed methods enable to reach lower early classification cost than the baseline from [3] in most cases: *2Step* outperforms the baseline in 128 out of the 198 cases (i.e., in 64.6 % of the cases), while *NoCluster* has better results in 135 cases (68.2 %). Moreover, the number of setups for which the baseline is improved by more than 20 % is another indicator showing the benefit of both proposed methods (cf. points outside the dashed lines in Fig. 5a and b).

One-sided Wilcoxon signed rank tests between each proposed method and the baseline show significance with a p-value lower than 10^{-6}, which confirms that both our methods outperform the baseline in terms of cost minimization.

We then compare both proposed methods to EarlyOpt.SVM [9], that optimizes on a slightly different trade-off between accuracy and earliness, in Fig. 5 When doing so, evaluation can be performed with respect to two different cost functions: the one on which EarlyOpt.SVM optimizes or the one used for our methods. Figure 5c and d present results obtained when using the cost function from EarlyOpt.SVM, that favors the latter in the comparison[3]. In spite of this, both proposed methods improve on EarlyOpt.SVM performance and these improvements are statistically significant at the 5 % significance level.

Finally, when comparing *2Step* and *NoCluster* costs as shown in Fig. 5e, *NoCluster* gets slightly better results that are considered significant when tested at the 5 % significance level ($p = 0.037$).

5   Conclusion

In this paper, we build upon the framework introduced in [3]. This framework deals with the problem of early classification from a new point of view: it aims at minimizing a cost function that includes a cost of misclassification and a cost of delaying the decision.

Following this framework, our proposition consists in designing two different methods (called *NoCluster* and *2Step*) that optimize on such a cost function. These methods decide online for an incoming testing time series if the current time instant is the optimal moment to classify the time series (with respect to the cost function) or if it is more valuable to wait for more samples of the time series before classifying it. In the latter case, they also give an estimate of when the optimal moment is more likely to occur. Both methods are hence adaptive and non myopic.

We have performed extensive experiments on a large and widely used set of datasets in order to evaluate the appropriateness of these methods, in terms of performance and timings. It turns out that (i) *2Step* outperforms the state-of-the art in terms of both cost minimization and classification time and (ii) *NoCluster* is an even better option in terms of early classification cost at the expense of slower classification. As future work, we will consider the use of time series specific classifiers, which should improve the overall performance of these approaches.

---

[3] See Supplementary material for more details on this comparison.

Acknowledgements. This work has been partly funded by ANR project ASTERIX (ANR-13-JS02- 0005-01) and CNES-TOSCA project VEGIDAR. Authors would like to thank Antoine Cornuéjols for his insight on the state-of-the-art, as well as data donors.

References

1. Antonucci, A., Scanagatta, M., Mauá, D.D., de Campos, C.P.: Early classification of time series by hidden markov models with set-valued parameters. In: Proceedings of the NIPS Time Series Workshop (2015)
2. Chang, C.C., Lin, C.J.: LIBSVM: a library for support vector machines. ACM Trans. Intell. Syst. Technol. **2**, 1–27 (2011). software. http://www.csie.ntu.edu.tw/~cjlin/libsvm
3. Dachraoui, A., Bondu, A., Cornuéjols, A.: Early classification of time series as a non myopic sequential decision making problem. In: Appice, A., Rodrigues, P.P., Santos Costa, V., Soares, C., Gama, J., Jorge, A. (eds.) ECML PKDD 2015. LNCS (LNAI), vol. 9284, pp. 433–447. Springer, Heidelberg (2015). doi:10.1007/978-3-319-23528-8_27
4. Ghalwash, M.F., Radosavljevic, V., Obradovic, Z.: Utilizing temporal patterns for estimating uncertainty in interpretable early decision making. In: Proceedings of the ACM SIGKDD International Conference on Knowledge Discovery and Data Mining, pp. 402–411 (2014)
5. Hatami, N., Chira, C.: Classifiers with a reject option for early time-series classification. In: Proceedings of the IEEE Symposium on Computational Intelligence and Ensemble Learning (2013)
6. He, G., Duan, Y., Peng, R., Jing, X., Qian, T., Wang, L.: Early classification on multivariate time series. Neurocomputing **149**, 777–787 (2015). Part B
7. Ishiguro, K., Sawada, H., Sakano, H.: Multi-class boosting for early classification of sequences. In: Proceedings of the British Machine Vision Conference, pp. 24.1–24.10 (2010)
8. Keogh, E., Zhu, Q., Hu, B., Hao, Y., Xi, X., Wei, L., Ratanamahatana, C.A.: The UCR Time Series Classification/Clustering Homepage (2011). http://www.cs.ucr.edu/~eamonn/time_series_data/
9. Mori, U., Mendiburu, A., Dasgupta, S., Lozano, J.A.: Early classification of time series from a cost minimization point of view. In: Proceedings of the NIPS Time Series Workshop (2015)
10. Parrish, N., Anderson, H.S., Gupta, M.R., Hsiao, D.Y.: Classifying with confidence from incomplete information. J. Mach. Learn. Res. **14**, 3561–3589 (2013)
11. Pedregosa, F., Varoquaux, G., Gramfort, A., Michel, V., Thirion, B., Grisel, O., Blondel, M., Prettenhofer, P., Weiss, R., Dubourg, V., Vanderplas, J., Passos, A., Cournapeau, D., Brucher, M., Perrot, M., Duchesnay, E.: Scikit-learn: machine learning in python. J. Mach. Learn. Res. **12**, 2825–2830 (2011)
12. Rousseeuw, P.J.: Silhouettes: a graphical aid to the interpretation and validation of cluster analysis. J. Computat. Appl. Math. **20**, 53–65 (1987)
13. Xing, Z., Pei, J., Yu, P.S.: Early classification on time series. Knowl. Inf. Syst. **31**(1), 105–127 (2012)
14. Xing, Z., Pei, J., Yu, P.S., Wang, K.: Extracting interpretable features for early classification on time series. In: Proceedings of the SIAM International Conference on Data Mining, pp. 247–258 (2011)

The Matrix Generalized Inverse Gaussian Distribution: Properties and Applications

Farideh Fazayeli[✉] and Arindam Banerjee

Department of Computer Science and Engineering,
University of Minnesota,Twin Cities, USA
{farideh,banerjee}@cs.umn.edu

Abstract. While the Matrix Generalized Inverse Gaussian ($\mathcal{MGIG}$) distribution arises naturally in some settings as a distribution over symmetric positive semi-definite matrices, certain key properties of the distribution and effective ways of sampling from the distribution have not been carefully studied. In this paper, we show that the $\mathcal{MGIG}$ is unimodal, and the mode can be obtained by solving an Algebraic Riccati Equation (ARE) equation [7]. Based on the property, we propose an importance sampling method for the $\mathcal{MGIG}$ where the mode of the proposal distribution matches that of the target. The proposed sampling method is more efficient than existing approaches [32,33], which use proposal distributions that may have the mode far from the $\mathcal{MGIG}$'s mode. Further, we illustrate that the the posterior distribution in latent factor models, such as probabilistic matrix factorization (PMF) [24], when marginalized over one latent factor has the $\mathcal{MGIG}$ distribution. The characterization leads to a novel Collapsed Monte Carlo (CMC) inference algorithm for such latent factor models. We illustrate that CMC has a lower log loss or perplexity than MCMC, and needs fewer samples.

1 Introduction

Matrix Generalized Inverse Gaussian ($\mathcal{MGIG}$) distributions [3,10] are a family of distributions over the space of symmetric positive definite matrices and has been recently applied as the prior for covariance matrix [20,32,33]. $\mathcal{MGIG}$ is a flexible prior since it contains Wishart, and Inverse Wishart distributions as special cases. We anticipate the usage of $\mathcal{MGIG}$ as prior for statistical machine learning models to grow with potential applications in Bayesian dimensionality reduction and Bayesian matrix completion (Sect. 4).

Some properties of the $\mathcal{MGIG}$ distribution and its connection with Wishart distribution has been studied in [10,26,27]. However, to best of our knowledge, it is not yet known if the distribution is unimodal and, if it is unimodal, how to obtain the mode of $\mathcal{MGIG}$. Besides, it is difficult to analytically calculate mean of the distribution and sample from the $\mathcal{MGIG}$ distribution. Monte Carlo methods like the importance sampling can in principle be applied to infer the mean of $\mathcal{MGIG}$ but one needs to design a suitable proposal distribution [21,23].

There is only one important sampling procedure for estimating the mean of $\mathcal{MGIG}$ [32,33]. In this approach, $\mathcal{MGIG}$ is viewed as a product of the Wishart

© Springer International Publishing AG 2016
P. Frasconi et al. (Eds.): ECML PKDD 2016, Part I, LNAI 9851, pp. 648–664, 2016.
DOI: 10.1007/978-3-319-46128-1_41

and Inverse Wishart distributions and one of them is used as the proposal distribution. However, we illustrate that the mode of the proposal distribution in [32,33] may be far away from the $\mathcal{MGIG}$'s mode. As a result, the proposal density is small in a region where the $\mathcal{MGIG}$ density is large yielding to an ineffective sampler and drastically wrong estimate of the mean (Figs. 1 and 2).

In this paper, we first illustrate that the $\mathcal{MGIG}$ distribution is unimodal where the mode can be obtained by solving an *Algebraic Riccati Equation (ARE)* [7]. This characterization leads to an effective importance sampler for the $\mathcal{MGIG}$ distribution. More specifically, for estimating the expectation $\mathbb{E}_{X \sim \mathcal{MGIG}}[g(X)]$, we select a proposal distribution over space of symmetric positive definite matrices like Wishart or Inverse Wishart distribution such that the mode of the proposal matches the mode of the $\mathcal{MGIG}$. As a result, unlike the current sampler [32,33], by aligning the shape of the proposal and the $\mathcal{MGIG}$, the density of the proposal gets higher values in the high density regions of the target, yielding to a good approximation of $\mathbb{E}_{X \sim \mathcal{MGIG}}[g(X)]$.

Further, we discuss a new application of the $\mathcal{MGIG}$ distribution in latent factor models such as probabilistic matrix factorization (PMF) [24] or Bayesian PCA (BPCA) [4]. In these settings, the given matrix $X \in \mathbb{R}^{N \times M}$ is approximated by a low-rank matrix $\hat{X} = UV^T$ where $U \in \mathbb{R}^{N \times D}$ and $V \in \mathbb{R}^{M \times D}$ with Gaussian priors over the latent matrices U and V. We show that after analytically marginalizing one of the latent matrices in PMF (or BPCA), the posterior over the other matrix has the $\mathcal{MGIG}$ distribution. This illustration yields to a novel Collapsed Monte Carlo (CMC) inference algorithm for PMF. In particular, we marginalize one of the latent matrices, say V, and propose a direct Monte Carlo sampling from the posterior of the other matrix, say U. Through extensive experimental analysis on synthetic, SNP, gene expression, and MovieLens datasets, we show that CMC has lower log loss or perplexity with fewer samples than Markov Chain Monte Carlo (MCMC) inference approach for PMF [25].

The rest of the paper is organized as follows. In Sect. 2, we cover background materials. In Sect. 3, we show that $\mathcal{MGIG}$ is unimodal and give a novel importance sampler for $\mathcal{MGIG}$. We provide the connection of $\mathcal{MGIG}$ with PMF in Sect. 4, present the results in Sect. 5, and conclude in Sect. 6.

2 Background and Preliminary

In this section, we provide some background on the relevant topics and tools that will be used in our analysis. We start by an introduction to importance sampling, $\mathcal{MGIG}$ distribution, followed by a brief overview of the ARE.

Notations: Let $\mathbb{S}_{++}^N$ and $\mathbb{S}_+^N$ denote the space of symmetric $(N \times N)$ positive definite and positive semi-definite matrix, respectively. Let $|.|$ denote the determinant of matrix, $\text{Tr}(.)$ be the matrix trace. A matrix $\Lambda \in \mathbb{S}_{++}^N$ has a Wishart distribution denoted as $\mathcal{W}_N(\Lambda | \Phi, \tau)$ where $\tau > N - 1$ and $\Phi \in \mathbb{S}_{++}^N$ [31]. A matrix $\Lambda \in \mathbb{S}_{++}^N$ has an Inverse Wishart distribution denoted as $\mathcal{IW}_N(\Lambda | \Psi, \alpha)$ where $\alpha > N - 1$ and $\Psi \in \mathbb{S}_{++}^N$ is the scale matrix. We denote $\mathbf{x}_{:m}$ as the mth column of matrix $X \in \mathbb{R}^{N \times M}$ and $\mathbf{x}_n$ as the nth row of X.

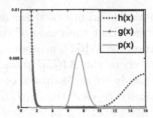

Fig. 1. An illustration of bad proposal distribution in importance sampling. Let $p(x) = h^*(x)g^*(x)/Z_p \propto h(x)g(x)$. Neither $h(x) = h^*(x)/Z_h$ nor $g(x) = g^*(x)/Z_g$ are a good candidate proposal distribution since their modes are far away from the one of $p(x)$.

2.1 Importance Sampling

Consider distribution $p(x) = \frac{1}{Z_p}p^*(x)$ where Z_p is the partition function which plays the role of a normalizing constant. Importance sampling is a general technique for estimating $\mathbb{E}_{x \sim p(x)}[g(x)]$ where sampling from $p(x)$ (the target distribution) is difficult but we can evaluate the value of $p^*(x)$ at any given x [21]. The idea is to draw S samples $\{x_i\}_{i=1}^S$ from a similar but easier distribution denoted by proposal distribution $q(x) = \frac{1}{Z_q}q^*(x)$. Define $w(x_i) = \frac{p^*(x_i)}{q^*(x_i)}$ as the weight of each sample i. Then, we calculate the expected value as follows

$$\mathbb{E}_{x \sim p}[g(x)] = \mathbb{E}_{x \sim q}\left[\frac{g(x)p(x)}{q(x)}\right] \approx \frac{\sum_{i=1}^S g(x_i)w(x_i)}{\sum_{i=1}^S w(x_i)},$$

The efficiency of importance sampling depends on how closely the proposal approximates the target in the shape. One way for monitoring the efficiency of importance sampling is the effective sample size $ESS = \frac{(\sum_{i=1}^S w(x_i))^2}{\sum_{i=1}^S w^2(x_i)}$ [15]. Very small value of ESS indicates a big discrepancy between the proposal and target (for example when the mode of the proposal distribution is far away from the target's mode) leading to a drastically wrong estimate of $\mathbb{E}_{x \sim p}[g(x)]$ [21].

2.2 $\mathcal{MGIG}$ Distribution

$\mathcal{MGIG}$ distribution was first introduced in [3] as a distribution over the space of symmetric $(N \times N)$ positive definite matrices defined as follows.

Definition 21. *A matrix-variate random variable $\Lambda \in \mathbb{S}_{++}^N$ is $\mathcal{MGIG}$ distributed [3,10] and is denoted as $\Lambda \sim \mathcal{MGIG}_N(\Psi, \Phi, \nu)$ if the density of Λ is*

$$f(\Lambda) = \frac{|\Lambda|^{\nu - (N+1)/2}}{|\frac{\Psi}{2}|^\nu \ B_\nu(\frac{\Phi}{2}\frac{\Psi}{2})} \exp\{\mathrm{Tr}(-\frac{1}{2}\Psi\Lambda^{-1} - \frac{1}{2}\Phi\Lambda)\},$$

where $B_\nu(.)$ is the matrix Bessel function [13] defined as

$$B_\nu\left(\frac{\Phi}{2}\frac{\Psi}{2}\right) = |\frac{\Phi}{2}|^{-\nu} \int_{\mathbb{S}_{++}^N} |S|^{-\nu - \frac{N+1}{2}} \exp\{\mathrm{Tr}(-\frac{1}{2}\Psi S^{-1} - \frac{1}{2}\Phi S)\}dS. \qquad (1)$$

When $N = 1$, the $\mathcal{MGIG}$ is the generalized inverse Gaussian distribution $\mathcal{GIG}$ [14] which is often used as the prior in several domains [6,12]. If $\Psi = 0$, the $\mathcal{MGIG}$ distribution reduces to the Wishart, and if $\Phi = 0$, it becomes the Inverse Wishart distribution.

Proposition 1. *[32, Proposition 2] If matrix $\Lambda \sim \mathcal{MGIG}_N(\Psi, \Phi, \nu)$, then $\Lambda^{-1} \sim \mathcal{MGIG}_N(\Phi, \Psi, -\nu)$.*

Sampling Mean of $\mathcal{MGIG}$: The sufficient statistics of $\mathcal{MGIG}$ are $\log|\Lambda|$, Λ, and Λ^{-1}. It is, however, difficult to analytically calculate the expectations $\mathbb{E}_{\Lambda \sim \mathcal{MGIG}}[\Lambda]$ and $\mathbb{E}_{\Lambda \sim \mathcal{MGIG}}[\Lambda^{-1}]$. Importance sampling can be applied to approximate those quantities. Note that based on the result of Proposition 1, the importance sampling procedure for estimating mean of $\mathcal{MGIG}$, i.e., $\mathbb{E}_{\Lambda \sim \mathcal{MGIG}}[\Lambda]$, can also be applied to infer the reciprocal mean i.e. $\mathbb{E}_{\Lambda \sim \mathcal{MGIG}}[\Lambda^{-1}]$.

An importance sampling procedure proposed in [32,33], where the $\mathcal{MGIG}$ is viewed as a product of Inverse Wishart and Wishart distributions and one of the multiplicands is used as the natural choice of the proposal distribution. In particular, in [32,33], the $\mathcal{MGIG}$ is viewed as

$$\mathcal{MGIG}_N(\Lambda|\Psi, \Psi, \nu) \propto \underbrace{e^{\mathrm{Tr}(-\frac{1}{2}\Phi\Lambda)}}_{T_1} \underbrace{\mathcal{IW}_N(\Lambda\,|\,\Psi, -2\nu_u)}_{T_2} \propto \underbrace{e^{\mathrm{Tr}(-\frac{1}{2}\Psi\Lambda^{-1})}}_{T_3} \underbrace{\mathcal{W}_N(\Lambda\,|\,\Phi, 2\nu_u)}_{T_4}.$$

In [32,33], authors advocate using T_2 (or T_4) as the proposal distribution which simplify the weight calculation to the evaluation of T_1 (or T_3). However, it is not studied how close T_2 (or T_4) are to the $\mathcal{MGIG}$ distribution in shape. For example, consider the $1-$dimensional $\mathcal{MGIG}$ distribution

$$\mathcal{MGIG}_1(\Lambda\,|\,35, 10, 10) \propto \underbrace{e^{\mathrm{Tr}(-\frac{35}{2}\Lambda^{-1})}}_{T_3} \underbrace{\mathcal{W}_1(\Lambda\,|\,10, 20)}_{T_4}. \tag{2}$$

In [32,33], $T_4 : \mathcal{W}_1(\Lambda\,|\,10, 20)$ is considered as the proposal distribution, but the mode of T_4 is far away from the mode of $\mathcal{MGIG}_1(\Lambda\,|\,35, 10, 10)$ (Fig. 2(a)). As a result, samples drawn from T_4 will be on the tail of the $\mathcal{MGIG}_1(\Lambda\,|\,10, 20)$ distribution, and will end up getting low weights from the $\mathcal{MGIG}_1(\Lambda\,|\,10, 20)$ distribution. Such a sampling procedure will be wasteful, i.e., drawing samples from the tails of the target $\mathcal{MGIG}_1$ distribution, leading to a very low ESS. Similar behavior is observed with several different choices of parameters for the $\mathcal{MGIG}$, here we only show three of them in Fig. 2 due to the lack of space.

2.3   Algebraic Riccati Equation

An algebraic Riccati equation (ARE) is

$$A^T X + XA + XRX + Q = 0, \tag{3}$$

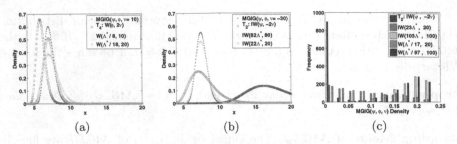

(a)   (b)   (c)

Fig. 2. (a,b) Comparison of different proposal distribution (a) Wishart ($\mathcal{W}$) and (b) Inverse Wishart ($\mathcal{IW}$) for sampling mean of $\mathcal{MGIG}_1(\Psi, \Phi, \nu)$ where Λ^* is the mode of $\mathcal{MGIG}$. The blue curves are the proposal distribution defined in [32,33] which can not recover the mode of the $\mathcal{MGIG}$ distribution. (c) Density of $\mathcal{MGIG}_2(\Psi, \Phi, \nu)$ for 1000 samples generated by each proposal distribution is calculated. More than 90 % of samples generated by the previous proposal distribution in [32,33] ($\mathcal{IW}(\psi, -2\nu)$) have zero $\mathcal{MGIG}$ density leading to $ESS = 40$. Whereas, the new proposal distribution $IW(23\Lambda^*, 20)$ has the $ESS = 550$ which has a very similar shape to the target $\mathcal{MGIG}$. (Color figure online)

where $A \in \mathbb{R}^{N \times N}, Q \in \mathbb{S}_+^N$, and $R \in \mathbb{S}_+^N$. We associate a $2N \times 2N$ matrix called the Hamiltonian matrix H with the ARE (3) as $H = \begin{bmatrix} A & R \\ -Q & -A^T \end{bmatrix}$. The ARE (3) has a unique positive definite solution if and only if the associated Hamiltonian matrix H has no imaginary eigenvalues (Section 5.6.3 of [7]).

There have been offered various numerical methods to solve the ARE which can be reviewed in [1]. The key of numerical technique to solve ARE (3) is to convert the problem to a stable invariant subspace problem of the Hamiltonian matrix i.e., finding the invariant subspace corresponding to the eigenvalues of H with negative real parts. The usual ARE solvers such as the Schur vector method [16], SR methods [9], the matrix sign function [2,11] require in general $O(n^3)$ flops [19]. For special cases, faster algorithms such as [19] can be applied which solves such an ARE with 20k dimensions in seconds. In this paper, we use Matlab ARE solver (*care*) to find the solution of ARE.

3 $\mathcal{MGIG}$ Properties

Some properties of the $\mathcal{MGIG}$ and its connection with Wishart distribution has been studied in [10,26,27]. However, to best of our knowledge, it is not yet known if the distribution is unimodal and how to obtain the mode of $\mathcal{MGIG}$. In the following Lemma we show that the $\mathcal{MGIG}$ distribution is unimodal.

Lemma 1. *Consider the $\mathcal{MGIG}$ distribution $\mathcal{MGIG}_N(\Lambda | \Psi, \Phi, \nu)$. The mode of $\mathcal{MGIG}$ is the solution of the following Algebraic Riccati Equation (ARE)*

$$-2\alpha\Lambda + \Lambda\Phi\Lambda - \Psi = 0, \tag{4}$$

where $\alpha = (\nu - \frac{N+1}{2})$. ARE in (4) has a unique positive definite solution, thus the $\mathcal{MGIG}$ distribution is a unimodal distribution.

Proof. The log-density of $\mathcal{MGIG}_N(\Lambda|\Psi,\Phi,\nu)$ is

$$\log f(\Lambda) = \alpha \log|\Lambda| - \frac{1}{2}\mathrm{Tr}(\Psi\Lambda^{-1} + \Phi\Lambda) + C, \tag{5}$$

where $\alpha = (\nu - \frac{N+1}{2})$, and C is a constant which does not depend on Λ. The mode of $\mathcal{MGIG}_N$ is obtained by setting derivative of (5) to zero as follows

$$\nabla f(\Lambda) = -2\alpha\Lambda + \Lambda\Phi\Lambda - \Psi = 0, \tag{6}$$

which is a special case of ARE (3). The associated Hamiltonian matrix for (6) is $H = \begin{bmatrix} -\alpha\mathbb{I}_N & \Phi \\ \Psi & \alpha\mathbb{I}_N \end{bmatrix}$. Recall that ARE has a unique positive definite solution if and only if the associated Hamiltonian matrix H has no imaginary eigenvalues (Section 5.6.3 of [7]). Thus, to show the unimodality of $\mathcal{MGIG}$, it is enough to show that the corresponding characteristic polynomial $|H - \lambda\mathbb{I}_{2N}| = 0$ has no imaginary solution.

$$|H - \lambda\mathbb{I}_{2N}| = |-(\alpha+\lambda)\mathbb{I}_N| \, |(\alpha-\lambda)\mathbb{I}_N + (\alpha+\lambda)^{-1}\Psi\Phi|$$

$$= |(\alpha-\lambda)\mathbb{I}_N| \, |-(\alpha+\lambda)\mathbb{I}_N - (\alpha-\lambda)^{-1}\Phi\Psi| = \prod_{i=1}^{N}\{-(\alpha^2-\lambda^2) - \tilde{\lambda}_i\} = 0,$$

which yields to $\lambda^2 = \tilde{\lambda}_i + \alpha^2$ where $\tilde{\lambda}_i$ is the ith eigenvalue of $\Phi\Psi$. Note $\tilde{\lambda}_i > 0$ since Φ and Ψ are positive definite and product of two positive definite matrix has positive eigenvalue. As a result, (7) has no imaginary solution and H does not have any imaginary eigenvalue. As a result, ARE in (6) has a unique positive definite solution. This completes the proof. □

Importance Sampling for $\mathcal{MGIG}$: Since $\mathcal{MGIG}$ is a unimodal distribution, we propose an efficient importance sampling procedure for $\mathcal{MGIG}$ by mode matching. We select a proposal distribution over space of positive definite matrices by matching the proposal's mode to the mode of $\mathcal{MGIG}$ (mode matching) which aligns the proposal and $\mathcal{MGIG}$ shapes. As a result, the proposal $q(x)$ is large in a region where the target distribution $\mathcal{MGIG}$ is large leading to a good approximation of $\mathbb{E}_{\Lambda\sim\mathcal{MGIG}}[g(\Lambda)]$. An example of such proposal distribution is Inverse Wishart or Wishart distribution.

Let Λ^* be the mode of $\mathcal{MGIG}_N(\Lambda|\Psi,\Phi,\nu)$ which can be found by solving the ARE (6). The mode of $\mathcal{W}_N(\Lambda|\Sigma,\rho)$ distribution is $\Sigma^* = (\rho - N - 1)\Sigma$. To match the mode of $\mathcal{W}_N(\Lambda|\Sigma,\rho)$ with that of $\mathcal{MGIG}_N(\Lambda|\Psi,\Phi,\nu)$, we choose the scale parameter Σ of the Wishart distribution by setting $\Sigma^* = \Lambda^*$. In particular,

$$\Sigma^* = \Lambda^* = (\rho - N - 1)\Sigma \quad \Rightarrow \quad \Sigma = \frac{\Lambda^*}{\rho - N - 1}. \tag{7}$$

Thus, we suggest using $\mathcal{W}_N(\frac{\Lambda^*}{\rho-N-1}, \rho)$ as the proposal distribution. At each iteration, we draw a sample $\Lambda_i \sim \mathcal{W}_N(\frac{\Lambda^*}{\rho-N-1}, \rho)$, and calculate $w(\Lambda_i)$. More specifically, the density of Wishart distribution is

---

Algorithm 1. Random Generator of $\mathcal{W}_N(\Lambda|\Sigma, \rho, L)$ [28]

1: Note $L^T L = \Sigma$ is the Cholesky factorization of Σ
2: $P_{ii} \sim \sqrt{\chi^2(\rho - (i-1))}$ for all $i = 1 \cdots N$.
3: $P_{ij} \sim \mathcal{N}(0,1)$ for $i < j$.   ▷ P: upper triangular
4: $R = PL$.
5: Return $\Lambda = R^T R$, $\Lambda^{-1} = R^{-1} R^{-1^T}$, and $|\Lambda| = [\prod_{i=1}^N R_{ii}]^2$.

$$q(\Lambda) = \frac{q^*(\Lambda)}{Z_q}, \qquad \text{where} \quad q^*(\Lambda) = |\Lambda|^{\frac{\rho - N - 1}{2}} \exp\{-\frac{1}{2}\operatorname{Tr}(\Sigma^{-1}\Lambda)\}. \qquad (8)$$

Then, the importance weight can be calculated as

$$w(\Lambda_i) = |\Lambda_i|^{\nu - \frac{\rho}{2}} \exp\left\{-\frac{1}{2}\operatorname{Tr}\left(\Psi\Lambda_i^{-1} + [\Phi - \Sigma^{-1}]\Lambda_i\right)\right\}. \qquad (9)$$

As a result, we approximate $\mathbb{E}_{\Lambda \sim \mathcal{MGIG}}[g(\Lambda)] \approx \frac{\sum_{i=1}^S w(\Lambda_i)g(\Lambda_i)}{\sum_{j=1}^S w(\Lambda_j)}$. A similar argument holds when the proposal distribution is an Inverse Wishart distribution.

Note that the weight calculation requires to calculate the inverse and determinant of sampled matrix Λ_i. However, as illustrated in Algorithm 1, the random samples generator from $\mathcal{W}$ [28] returns the upper triangular matrix R where $\Lambda = R^T R$. Hence the inverse and determinant of Λ can be calculated efficiently from the inverse and diagonal of the triangular matrix R, respectively. Therefore, the cost of weight calculation is reduced to the cost of solving a linear system and upper triangular matrix production at each iteration.

Figure 2 illustrates that the proposed importance sampling outperforms the one in [32,33] for three examples of $\mathcal{MGIG}$. In particular, more than 90 % of samples drawn from the proposal distribution T_2 in [32,33] have zero weights leading to $ESS = 40$ (Fig. 2 (c)). Whereas, our proposal distribution achieved $ESS = 550$ leading to a better approximation of the mean of $\mathcal{MGIG}$. Similar behavior is observed with several different choices of parameters for the $\mathcal{MGIG}$.

4   Connection of $\mathcal{MGIG}$ and Bayesian PCA

In this section, we illustrate that the mapping matrix V in Bayesian PCA can be marginalized or 'collapsed' yielding a Matrix Generalized Inverse Gaussian ($\mathcal{MGIG}$) [3,10] posterior distribution over the latent matrix U denoting as the marginalized posterior distribution. Then, we explain the derivation of the marginalized posterior for data with missing values, followed by a collapsed Monte Carlo Inference for PMF.

4.1 PMF, PPCA, and Bayesian PCA

First, we give a review of PMF [24], Probabilistic PCA (PPCA) [29], and Bayesian PCA (BPCA) [4], to illustrate the similarity and differences between the existing ideas and our approach. A related discussion appears in [18]. All these models focus on an (partially) observed data matrix $X \in \mathbb{R}^{N \times M}$. Given latent factors $U \in \mathbb{R}^{N \times D}$ and $V \in \mathbb{R}^{M \times D}$, the rows of X are assumed to be generated according to $\mathbf{x}_{:m} = U\mathbf{v}_m^T + \epsilon$, where $\epsilon \in \mathbb{R}^N$. The different models vary depending on how they handle distributions or estimates of the latent factors U, V. Without loss of generality, in this paper, we are considering a fat matrix X where $M > N$.

PMF and BPMF: In PMF [24], one assumes independent Gaussian priors for all latent vectors $\mathbf{u}_n$ and $\mathbf{v}_m$, i.e., $\mathbf{u}_n \sim \mathcal{N}(0, \sigma_u^2 \mathbb{I})$, $[n]_1^N$ and $\mathbf{v}_m \sim \mathcal{N}(0, \sigma_v^2 \mathbb{I})$, $[m]_1^M$. Then, one obtains the following posterior over (U, V)

$$p(U, V | X, \theta) = \prod_{n,m} [\mathcal{N}(x_{nm} | \langle \mathbf{u}_n, \mathbf{v}_m \rangle, \sigma^2)]^{\delta_{nm}} \prod_n \mathcal{N}(\mathbf{u}_n | 0, \sigma_u^2 \mathbb{I}) \prod_m \mathcal{N}(\mathbf{v}_{:m} | 0, \sigma_v^2 \mathbb{I}) ,$$

where $\delta_{nm} = 0$ if x_{nm} is missing and $\delta = \{\sigma^2, \sigma_u^2, \sigma_v^2\}$. PMF obtains point estimates $(\hat{U}, \hat{V})$ by maximizing the posterior (MAP), based on alternating optimization over U and V [24].

Bayesian PMF (BPMF) [25] considers independent Gaussian priors over latent factors with full covariance matrices, i.e., $\mathbf{u}_n \sim \mathcal{N}(0, \Sigma_u)$, $[n]_1^N$ and $\mathbf{v}_m \sim \mathcal{N}(0, \Sigma_v)$, $[m]_1^M$. Inference is done using Gibbs sampling to approximate the posterior $P(U, V | X)$. At each iteration t, $U^{(t)}$ is sampled from the conditional probability of $p(U | V^{(t-1)}, X)$, followed by sampling V from $p(V | U^{(t)}, X)$.

Probabilistic PCA: In PPCA [29], one assumes independent Gaussian prior over $\mathbf{u}_n$, i.e., $\mathbf{u}_n \sim \mathcal{N}(0, \sigma_u^2 \mathbb{I})$, but V is treated as a parameter to be estimated. In particular, V is chosen so as to maximize the marginalized likelihood of X

$$p(X | V) = \int_U p(X | U, V) p(U) dU = \prod_{n=1}^N \mathcal{N}(\mathbf{x}_n | 0, \sigma_u^2 V V^T + \sigma^2 \mathbb{I}). \qquad (10)$$

Interestingly, as shown in [29], the estimate $\hat{V}$ can be obtained in closed form. For such a fixed $\hat{V}$, the posterior distribution over $U | X, \hat{V}$ can be obtained as

$$p(U | X, \hat{V}) = \frac{p(X | U, \hat{V}) p(U)}{p(X | \hat{V})} = \prod_{n=1}^N \mathcal{N}\left(\mathbf{u}_n | \Gamma^{-1} \hat{V}^T \mathbf{x}_n, \sigma^{-2} \Gamma\right), \qquad (11)$$

where $\Gamma = \hat{V}^T \hat{V} + \sigma_u^{-2} \sigma^{-2} \mathbb{I}$. Note that the posterior of the latent factor U in (11) depends on both X and $\hat{V}$. For applications of PPCA in visualization, embedding, and data compression, any point x_n in the data space can be summarized by its posterior mean $E[\mathbf{u}_n | \mathbf{x}_n, \hat{V}]$ and covariance $Cov(\mathbf{u}_n | \hat{V})$ in the latent space.

Bayesian PCA: In Bayesian PCA [4], one assumes independent Gaussian priors for all $\mathbf{u}_n$ and $\mathbf{v}_m$, i.e., $\mathbf{u}_n \sim \mathcal{N}(0, \sigma_u^2 \mathbb{I})$ and $\mathbf{v}_m \sim \mathcal{N}(0, \sigma_v^2 \mathbb{I})$, $[m]_1^M$. Bayesian posterior inference by Bayes rule considers $p(U, V | X) = p(X | U, V) p(U) p(V) / p(X)$, which includes the intractable partition function

$$p(X) = \int_U \int_V p(X|U,V)p(U)p(V)dUdV \ . \tag{12}$$

The literature has considered approximate inference methods, such as variational inference [5], gradient descent optimization [18], MCMC [25], or Laplace approximation [4,22].

While PPCA and Bayesian PCA were originally considered in the context of embedding and dimensionality reduction, PMF and BPMF have been widely used in the context of matrix completion where the observed matrix X has many missing entries. Nevertheless, as seen from the above exposition, the structure of the models are closely related (also see [17,18]).

4.2   Closed Form Posterior Distribution in Bayesian PCA

The key challenge in models such as Bayesian PCA or BPMF is that joint marginalization over both latent factors U, V is intractable. PPCA gets around the problem by considering one of the variables, say V, to be a constant. In this section, we show that one can marginalize or 'collapse' one of the latent factors, say V, and obtain the marginalized posterior $P(U|X)$ over the other variable denoted. In fact, we obtain the posterior with respect to the covariance structure $\Lambda_u = \beta_u \mathbb{I} + UU^T$, for a suitable constant β_u, which is sufficient to do Bayesian inference on new test points x_{test}. Note that

$$p(U|X) \propto p(U)P(X|U) = p(U)\int_V P(X|U,V)p(V)dV \ , \tag{13}$$

and, based on the posterior over U, one can obtain the probability on a new point as $p(x_{\text{test}}|X) = \int_U p(x_{\text{test}}|U)p(U|X)dU$.

Next, we show that the posterior over U as in (13), rather the distribution over $\Lambda_u = \beta_u \mathbb{I} + UU^T$, can be derived analytically in *closed form*. The distribution is the Matrix Generalized Inverse Gaussian ($\mathcal{MGIG}$) distribution.

Similar to (10), marginalizing V gives

$$p(X|U) = \int_V p(X|U,V)p(V)dV = \prod_{m=1}^{M} \mathcal{N}\left(\mathbf{x}_{:m}\,|\,0, \sigma_v^2 \Lambda_u\right), \tag{14}$$

where $\Lambda_u = \beta_v \mathbb{I} + UU^T$ and $\beta_v = \frac{\sigma^2}{\sigma_v^2}$. Then, the marginalized posterior of U is

$$p(U|X) \propto p(X|U)\, p(U) = |\,\Lambda_u\,|^{-M/2} \exp\left\{ \frac{-\operatorname{Tr}\left(\Lambda_u^{-1} \sum_{m=1}^{M} \mathbf{x}_{:m}\mathbf{x}_{:m}^T\right)}{2\sigma_v^2} \right\}$$

$$\times \exp\left\{ \frac{-\operatorname{Tr}(\Lambda_u)}{2\sigma_u^2} \right\} \times \exp\left\{ \frac{\operatorname{Tr}(\beta_u \mathbb{I})}{2\sigma_u^2} \right\} \tag{15}$$

$$= |\,\Lambda_u\,|^{-M/2} \exp\left\{ \operatorname{Tr}(-\frac{1}{2}\Lambda_u^{-1}\Psi_u - \frac{1}{2}\Lambda_u \Phi_u) \right\}$$

$$\sim \mathcal{MGIG}(\Lambda_u\,|\,\Psi_u, \Phi_u, \nu_u), \tag{16}$$

where $\Psi_u = \frac{1}{\sigma_v^2}XX^T$, $\Phi_u = \frac{1}{\sigma_u^2}\mathbb{I}$, and $\nu_u = \frac{N-M+1}{2}$.

Therefore, by marginalizing or collapsing V, the posterior over $\Lambda_u = \beta_v \mathbb{I} + UU^T$ corresponding to the latent matrix U can be characterized exactly with a $\mathcal{MGIG}$ distribution with parameters depending only on X. Note that this is in sharp contrast with (11) for PPCA, where the posterior covariance of $\mathbf{u}_n$ is $\sigma^{-2}\Gamma$ which in turn depends on the point estimate for $\hat{V}$.

4.3 Posterior Distribution with Missing Data

In this section, we consider the matrix completion setting, when the observed matrix X has missing values. In presence of missing values, the likelihood of the observed sub-vector in any column of X is given as

$$p\left(\mathbf{x}_{n_m,m} \mid U, V\right) = \mathcal{N}\left(\mathbf{x}_{n_m,m} \mid \tilde{U}_m \mathbf{v}_m^T, \sigma^2 \mathbb{I}\right). \tag{17}$$

where n_m is a vector of size $\tilde{N}_m$ containing indices of non-missing entries in column m of X, and $\tilde{U}_m$ is a sub-matrix of U with size of $\tilde{N}_m \times D$ where each row correspond to a non-missing entry in the mth column of X. The marginalized likelihood (14) is $p\left(X \mid U\right) = \prod_{m=1}^M \mathcal{N}\left(\mathbf{x}_{n_m,m} \mid 0, \sigma_v^2 \Lambda_{un}\right)$, where $\Lambda_{un} = \beta_v \mathbb{I} + \tilde{U}_n \tilde{U}_n^T$ and $\beta_v = \frac{\sigma^2}{\sigma_v^2}$. The marginalized posterior is given by

$$p(U \mid X) \propto \exp\left\{-\frac{\mathrm{Tr}(UU^T)}{2\sigma_u^2}\right\} \prod_{m=1}^M \mid \Lambda_{un} \mid^{-M/2} \exp\left\{-\frac{1}{2}\mathbf{x}_{n_m,m}^T \Lambda_{un}^{-1}\mathbf{x}_{n_m,m}\right\}.$$

Thus, in presence of missing values, the posterior cannot be factorized as in (15) because each column $\mathbf{x}_{:m}$ contributes to different blocks Λ_{un} of Λ.

We propose to address the missing value issue by gap-filling. In particular, if one can obtain a good estimate of the covariance structure in X, so that $\Psi_u = \frac{1}{\sigma_v^2}XX^T$ in (16) can be approximated well, one can use the $\mathcal{MGIG}$ posterior to do approximate inference. We consider two simple approaches to approximate the covariance structure of X: (i) by zero-padding the missing value matrix X (assuming $E[X] = 0$ or centering the data in practice), and estimating the covariance structure based on the zero-padded matrix [30], and (ii) by using a suitable matrix completion method, such as PMF, to get point estimates of the missing entries in X, and estimating the covariance structure based on the completed matrix. We experiment with both approaches in Sect. 5, and the zero-padded version seems to work quite well.

4.4 Collapsed Monte Carlo Inference for PMF

Given that $\Lambda_u \sim \mathcal{MGIG}_N$, we predict the missing values as follows. Let $\mathbf{x} = [\mathbf{x}^o, \mathbf{x}^*] \sim \mathcal{N}(0, \Lambda)$, where $\mathbf{x}^o \in \mathbb{R}^p$ is the observed partition of $\mathbf{x} \in \mathbb{R}^N$ and $\mathbf{x}^* \in \mathbb{R}^{N-p}$ is missing. Accordingly, partition Λ as $\Lambda_u = \begin{pmatrix} \Lambda_{oo} & \Lambda_{o*} \\ \Lambda_{*o} & \Lambda_{**} \end{pmatrix} \begin{matrix} p \\ N-p \end{matrix}$.

Algorithm 2. CMC Inference for PMF

1: Let $\Psi_u = \frac{ZZ^T}{\sigma_v^2}$, $\Phi_u = \frac{1}{\sigma_u^2}$, and $\nu_u = \frac{N-M+1}{2}$.
2: Solve (6) to find mode Λ^* of $\mathcal{MGIG}(\Psi_u, \Phi_u, \nu_u)$.
3: Let $L^T L = \Lambda^*$ be the Cholesky factorization of Λ^*. Let $\tilde{L} = \frac{L}{\sqrt{\rho - M - 1}}$.
4: **for** $t = 1 \cdots T$ **do**
5:     Let $\Lambda^{(t)} \sim \mathcal{W}_N(\frac{\Lambda^*}{\rho - M - 1}, \rho, \tilde{L})$ ▷ Algorithm 1
6:     Let $w^t = \frac{\mathcal{MGIG}_N(\Lambda^{(t)}|\Psi_u,\Phi_u,\nu_u)}{\mathcal{W}_N(\Lambda^{(t)}|\frac{\Lambda^*}{\rho-M-1},\rho,\tilde{L})}$.
7:     Let $\mu^t = \Lambda_{*o}^{(t)} \Lambda_{oo}^{(t)-1} \mathbf{x}^o$. Let $\Sigma^t = \Lambda_{**}^{(t)} - \Lambda_{*o}^{(t)} \Lambda_{oo}^{(t)-1} \Lambda_{o*}^{(t)}$.
8:     Let $\bar{\mu} = \bar{\mu} + w^t \mu^t$.
        Let $\bar{\Sigma} = \bar{\Sigma} + w^t \Sigma^t$.
9: Report the distribution of $\mathbf{x}^* \sim \mathcal{N}(\tilde{\mu}^*, \tilde{\Sigma}^*)$ where $\tilde{\mu}^* = \frac{\bar{\mu}}{\sum_{t=1}^T w^t}$ and $\tilde{\Sigma}^* = \frac{\bar{\Sigma}}{\sum_{t=1}^T w^t}$.

Then the conditional probability of $\mathbf{x}^*$ given $\mathbf{x}^o$ and Λ is

$$p(\mathbf{x}^* \mid \mathbf{x}^o, \Lambda) \sim \mathcal{N}(\underbrace{\Lambda_{*o}\Lambda_{oo}^{-1}\mathbf{x}^o}_{\mu^*}, \underbrace{\Lambda_{**} - \Lambda_{*o}\Lambda_{oo}^{-1}\Lambda_{*o}}_{\Sigma^*}) \tag{18}$$

where $\mathbf{y} = \Lambda_{*o}\Lambda_{oo}^{-1}$ is the solution of the linear system $\Lambda_{oo}\mathbf{y} = \Lambda_{*o}^T$ and can be calculated efficiently. Since sampling from $\mathcal{MGIG}$ is difficult, we propose to use importance sampling to infer the missing values as

$$p(x_n^*|x_n^o) = \mathbb{E}_{\Lambda \sim \mathcal{MGIG}}\left[p(x_n^*|x_n^o, \Lambda)\right] = \mathbb{E}_{\Lambda \sim q}\left[\frac{p(x_n^*|x_n^o, \Lambda)\mathcal{MGIG}_N(\Lambda|\Psi_u, \Phi_u, \nu_u)}{q(\Lambda)}\right],$$

where q is the proposal distribution as discussed above and sampling $\Lambda^{(t)}$ from q yields to the estimate of

$$\tilde{\mu}^* = \frac{\sum_{t=1}^T \Lambda_{*o}^{(t)}\Lambda_{oo}^{(t)-1}\mathbf{x}^o w(\Lambda^{(t)})}{\sum_{t=1}^t w(\Lambda^{(t)})}, \quad \tilde{\Sigma}^* = \frac{\sum_{t=1}^T [\Lambda_{**}^{(t)} - \Lambda_{*o}^{(t)}\Lambda_{oo}^{(t)-1}\Lambda_{*o}^{(t)}]w(\Lambda^{(t)})}{\sum_{t=1}^t w(\Lambda^{(t)})}.$$

Algorithm 2 illustrates the summary of the collapsed Monte Carlo (CMC) inference for predicting the missing values. A practical approximation to avoid the calculations in Lines 7–8 of Algorithm 2 at each iteration is to simply estimate the mean of the posterior $\bar{\Lambda} = \frac{\sum_{t=1}^T \Lambda^{(t)} w^t}{\sum_{t=1}^T w^t}$ with samples drawn from the proposal distribution (line 6), then do the inference based on $\bar{\Lambda}$.

5   Experimental Results

We compared the performance of MCMC and CMC on both log loss and running times. We evaluated the models on 4 datasets: (1) **SNP**: single nucleotide polymorphism (SNP) is important for identifying gene-disease associations where the data usually has 5 to 20 % of genotypes missing [8]. We used phased SNP dataset for chromosome 13 of the CEU population[1]. We randomly dropped 20 %

---

[1] http://hapmap.ncbi.nlm.nih.gov/downloads/phasing/.

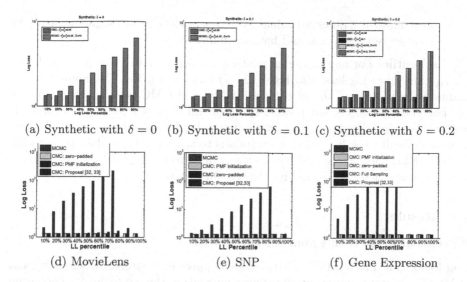

(a) Synthetic with $\delta = 0$ (b) Synthetic with $\delta = 0.1$ (c) Synthetic with $\delta = 0.2$

(d) MovieLens   (e) SNP   (f) Gene Expression

Fig. 3. Log loss (LL) of CMC and MCMC for different log loss percentile on different datasets presented in the log scale (δ denotes the missing proportion). CMC consistently achieves lower LL compared to MCMC. LL of MCMC increases exponentially (linearly in log scale) by adding data points with higher log loss. Proposal in [30,31] achieved infinity LL for MovieLens. Empty bar represents infinity LL (e.g. 90 % and 100 % percentile in (d)

of the entries. (2) **Gene Expression:** DNA microarrays provides measurement of thousands of genes under a certain experimental condition where suspicious values are usually regarded as missing values. Here we used gene expression dataset for Breast Cancer (BRCA)[2]. We randomly dropped 20 % of the entries. (3) **MovieLens:** we used MovieLens[3] dataset with 1M rating represented as a fat matrix $X \in \mathbb{R}^{N \times M}$ where $M = 3900$ movies and $N = 6040$ users. (4) **Synthetic:** first the latent matrices U and V are generated by randomly choosing each $\{\mathbf{u}_n\}_{n=1}^N$ and $\{\mathbf{v}_m\}_{m=1}^M$ from $\mathcal{N}(0, \sigma_u^2 \mathbb{I})$ and $\mathcal{N}(0, \sigma_v^2 \mathbb{I})$, respectively. Then, matrix X is built by sampling each x_{nm} from $\mathcal{N}(\langle \mathbf{u}_n, \mathbf{v}_m \rangle, \sigma^2)$. The parameters are set to $N = 100$, $M = 6000$, $\sigma_u^2 = \sigma_v^2 = 0.05$, and $\sigma^2 = 0.01$. We dropped random entries using Bernoulli distributions with $\delta = 0.1, 0.2$.

5.1 Methodology

We compared CMC with MCMC inference for PMF. Gibbs sampling with diagonal covariance prior over the latent matrices is used for MCMC. For the model evaluation, average of log loss (LL) is reported over 5-fold cross-validation. LL measures how well a probabilistic model q predicts the test sample defined as $LL = -\frac{1}{T} \sum_{i=1}^N \sum_{j=1}^M \delta_{ij} \log q(x_{ij})$ where $q(x_{ij})$ is the inferred probability and T

---

[2] http://cancergenome.nih.gov/.
[3] www.movielens.umn.edu.

is the total number of observed values. A better model q assign higher probability $q(x_{ij})$ to observed test data, and have a smaller value of LL.

LL Percentile: For any posterior model $q(x)$, a test data point x_{test} with low $q(x_{\text{test}})$ has large log loss, and high $q(x_{\text{test}})$ has low log loss. To comparatively evaluate the posteriors obtained from MCMC and CMC, we consider their log loss percentile plots. For any posterior, we sort all the test data points in ascending order of their log loss, and plot the mean log loss in 10 percentile batches. More specifically, the first batch corresponds to the top 10 % of data points with the lowest log loss, the second batch corresponds to the top 20 % of data points with the lowest log loss (including the first 10 % percentile), and so on.

5.2   Results

We summarize the results from different aspects:

Log loss: CMC has a small log loss across all percentile batches, whereas log loss of MCMC increases exponentially (linear increase in the log scale) for percentile batches with higher log loss i.e., smaller predicting probability, (Fig. 3). Thus, MCMC assigned extremely low probability to several test points as compared to CMC. Figure 4(a) illustrates that log loss of MCMC continues to decrease with growing sample size up to 2000 samples, implying that MCMC has not yet converged to the equilibrium distribution. Note that log loss of CMC with 200 samples (Fig. 4(b)) is 10 times less than log loss of MCMC with 2000 samples. We also compared the results with the previous proposal [32,33], and observed that for MovieLens the results are worse than our proposed result as they achieved Inf LL on the last batch.

Effective number of samples: For the synthetic, SNP, and gene expression datasets, we generated 10,000 samples using MCMC. The burn-in period is set to 500 with a lag of 10 yielding to 1000 effective samples. For the MovieLens, we generated 5,000 samples using MCMC with the burn-in period of 1000 and a lag of 2 yielding to 2000 effective samples. We initialized the latent matrices

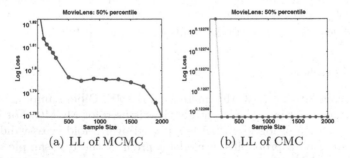

(a) LL of MCMC   (b) LL of CMC

Fig. 4. LL of CMC and MCMC for different sample size of MovieLens data in the log scale. LL of both CMC and MCMC is decreasing by adding more samples. LL of MCMC is in magnitude 10 times more than CMC.

U and V with the factors estimated by PMF, to help the convergence of MCMC. Sample size in CMC procedure is set to 1,000 for all datasets. Note that MCMC alternately sample both latent matrices U and V from a Markov chain and the quality of the posterior improves with increasing number of samples. For the proposed CMC procedure, the bigger matrix V is marginalized and only samples from the smaller U matrix is drawn directly from the true posterior distribution. Hence, CMC has considerably improved sample utilization.

Initialization: As discussed in Sect. 4.3, in order to use the $\mathcal{MGIG}$ posterior for inference, the covariance structure of matrix X should be estimated. Here we evaluate two approaches to approximate the covariance structure of X: (i) by zero-padding the missing value matrix X, and (ii) by computing the point estimates of the missing entries in X with PMF. CMC with zero-padded initialization has a similar log loss behavior as point estimate initialization with PMF (Figs. 3 (d-f)).

Full sampler vs Mean sampler: Figure 3(f) shows the result of the full sampler (Algorithm 2), and the mean sampler (approximating the inference by estimating $\bar{\Lambda} = \mathbb{E}_{\Lambda \sim \mathcal{MGIG}}[\Lambda]$ as discussed in Sect. 4.4) on gene expression data. Since the log losses are similar with both samplers, and the behavior is typical, we presented log loss results on the other datasets only based on the mean sampler, which is around 100 times faster.

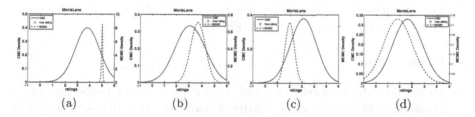

(a)   (b)   (c)   (d)

Fig. 5. Density of CMC and MCMC for several data input on MovieLens data with LL of (a) CMC:-1.78, MCMC:-Inf, (b) CMC:-3, MCMC:-17, (c) CMC:-4.2, MCMC:-6.4, (d) CMC:-1.4, MCMC:-2.04. CMC achieves lower LLs compared to MCMC. (Color figure online)

Comparison of inferred posterior distributions: To emphasize the importance of choosing the right measure for comparison, e.g., log loss vs RMSE, we illustrate the inferred posterior distributions over several missing entries/ratings in MovieLens obtained from MCMC and CMC in Fig. 5. Note that the scales for CMC (red) and MCMC (blue) are different. Overall, the posterior from CMC tends to be more conservative (not highly peaked), and obtains lower log loss across a range of test points. Interestingly, as shown in Fig. 5(a), MCMC can make mistakes with high confidence, i.e., predicts 5 stars with a peaked posterior whereas the true rating is 3 stars. Such troublesome behavior is correctly assessed with log loss, but not by RMSE since it does not consider the confidence in the prediction. As shown in Fig. 5(d), for some test points, both MCMC and

Table 1. Time Comparison of CMC and MCMC on different datasets. At each step of MCMC, rows of U and V can be sampled in parallel denoted by MCMC parallel. The running time is reported over 1000 steps for both methods. Note that the effective number of samples of MCMC is less than 1000 and more steps is required to obtain enough samples. The number of iterations for convergence of CMC is less than 1000.

Dataset	Size	MCMC(200 samples)		CMC(1000 samples)	
		Serial	Parallel	Serial	Parallel
Synthetic	100 × 6,000	728s	404s	6s	4s
SNP	120 × 104,868	12,862s	5,859s	75s	22s
Gene Expression	591 × 17,814	3,478s	2,278s	140s	90s
MovieLens	3,233 × 6,040	2,350s	2,100s	5,387s	2,058s

CMC inferred similar posterior distributions with a bias difference where the mean of CMC is closer to the true value.

Time Comparison: We have compared running time in both serial and in parallel over 1000 steps yielding to 200 and 1000 samples for MCMC and CMC, respectively. We implement the algorithms in Matlab. The computation time is estimated on a PC with a 3.40 GHz Quad core CPU and 16.0 G memory. The average run time results are reported in Table 1. For Synthetic, SNP, and gene expression datasets, MCMC converges very slowly. For MovieLens dataset, the running time of both are very close but note that MCMC requires more number of samples for convergence than CMC (Fig. 4).

6   Conclusion

We studied the $\mathcal{MGIG}$ distribution and provided certain key properties with a novel sampling technique from the distribution and its connection with the latent factor models such as PMF or BPCA. With showing that the $\mathcal{MGIG}$ distribution is unimodal and the mode can be obtained by solving an ARE, we proposed a new importance sampling approach to infer the mean of $\mathcal{MGIG}$. The new sampler, unlike the existing sampler [32,33], chooses the proposal distribution to have the same mode as the $\mathcal{MGIG}$. This characterization leads to a far more effective sampler than [32,33] since the new sampler align the shape of the proposal to the target distribution. Although, the $\mathcal{MGIG}$ distribution has been recently applied to Bayesian models as the prior for the covariance matrix, here, we introduced a novel application of the $\mathcal{MGIG}$ in PMF or BPCA. We showed that the posterior distribution in PMF or BPCA has the $\mathcal{MGIG}$ distribution. This illustration, yields to a new CMC inference algorithm for PMF.

Acknowledgements. The research was supported by NSF grants IIS-1447566, IIS-1447574, IIS-1422557, CCF-1451986, CNS-1314560, IIS-0953274, IIS-1029711, NASA grant NNX12AQ39A, and gifts from Adobe, IBM, and Yahoo. F.F. acknowledges the support of DDF (2015–2016) from the University of Minnesota.

References

1. Anderson, B., Moore, J.: Optimal control: linear quadratic methods (2007)
2. Bai, Z., Demmel, J.: Using the matrix sign function to compute invariant subspaces. SIAM J. Matrix Anal. Appl. **19**(1), 205–225 (1998)
3. Barndorff-Nielsen, O., Blæsild, P., et al.: Exponential transformation models. Proc. Roy. Soc. London Ser. A **379**(1776), 41–65 (1982)
4. Bishop, C.M.: Bayesian PCA. NIPS **11**, 382–388 (1999)
5. Bishop, C.M.: Variational principal components. In: ICANN (1999)
6. Blei, D., Cook, P., Hoffman, M.: Bayesian nonparametric matrix factorization for recorded music. In: ICML, pp. 439–446 (2010)
7. Boyd, S., Barratt, C.: Linear controller design: limits of performance (1991)
8. Brevern, A.D., Hazout, S., Malpertuy, A.: Influence of microarrays experiments missing values on the stability of gene groups by hierarchical clustering. BMC Bioinform. **5**(1), 114 (2004)
9. Bunse-Gerstner, A., Mehrmann, V.: A symplectic QR like algorithm for the solution of the real algebraic Riccati equation. IEEE Trans. Autom. Control **31**, 1104–1113 (1986)
10. Butler, R.W.: Generalized inverse Gaussian distributions and their Wishart connections. Scand. J. Statist. **25**(1), 69–75 (1998)
11. Byers, R.: Solving the algebraic Riccati equation with the matrix sign function. Linear Algebra Appl. **85**, 267–279 (1987)
12. Eberlein, E., Keller, U.: Hyperbolic distributions in finance. Bernoulli **1**, 281–299 (1995)
13. Herz, C.: Bessel functions of matrix argument. Ann. Math. **23**, 77–87 (1955)
14. Jørgensen, B.: Statistical properties of the generalized inverse Gaussian distribution
15. Kong, A., Liu, J., Wong, W.: Sequential imputations and bayesian missing data problems. JASA **89**(425), 278–288 (1994)
16. Laub, A.: A Schur method for solving algebraic Riccati equations. IEEE Trans. Autom. Control **24**(6), 913–921 (1979)
17. Lawrence, N.: Probabilistic non-linear principal component analysis with Gaussian process latent variable models. JMLR **6**, 1783–1816 (2005)
18. Lawrence, N., Urtasun, R.: Non-linear matrix factorization with gaussian processes. In: ICML (2009)
19. Li, T., Chu, E., et al.: Solving large-scale continuous-time algebraic Riccati equations by doubling. J. Comput. Appl. Math. **237**(1), 373–383 (2013)
20. Li, Y., Yang, M., Qi, Z., Zhang, Z.: Bayesian multi-task relationship learning with link structure. In: ICDM (2013)
21. MacKay, D.: Information theory, inference, and learning algorithms (2003)
22. Minka, T.P.: Automatic choice of dimensionality for PCA. In: NIPS (2000)
23. Owen, A.: Monte Carlo theory, methods and examples (2013)
24. Salakhutdinov, R., Mnih, A.: Probabilistic matrix factorization. In: NIPS (2007)
25. Salakhutdinov, R., Mnih, A.: Bayesian probabilistic matrix factorization using markov chain monte carlo. In: ICML (2008)
26. Seshadri, V.: Some properties of the matrix generalized inverse Gaussian distribution. Stat. Methods Pract. **69**, 47–56 (2003)
27. Seshadri, V., Wesołowski, J.: More on connections between Wishart and matrix GIG distributions. Metrika **68**(2), 219–232 (2008)
28. Smith, W., Hocking, R.: Algorithm as 53: Wishart variate generator. Appl. Statist. **21**, 341–345 (1972)

29. Tipping, M.E., Bishop, C.M.: Probabilistic principal component analysis. J. Royal Statist. Soc. Seri. B **61**(3), 611–622 (1999)
30. Wang, H., Fazayeli, F., et al.: Gaussian copula precision estimation with missing values. In: AISTATS (2014)
31. Wishart, J.: The generalised product moment distribution in samples from a normal multivariate population. Biometrika **20A**, 32–52 (1928)
32. Yang, M., Li, Y., Zhang, Z.: Multi-task learning with Gaussian matrix generalized inverse Gaussian model. In: ICML (2013)
33. Yoshii, K., Tomioka, R.: Infinite positive semidefinite tensor factorization for source separation of mixture signals. In: ICML (2013)

On the Convergence of a Family of Robust Losses for Stochastic Gradient Descent

Bo Han, Ivor W. Tsang$^{(\boxtimes)}$, and Ling Chen

Centre for Quantum Computation and Intelligent Systems,
University of Technology Sydney, Sydney, Australia
Bo.Han@student.uts.edu.au.com, {Ivor.Tsang,Ling.Chen}@uts.edu.au.com

Abstract. The convergence of Stochastic Gradient Descent (SGD) using convex loss functions has been widely studied. However, vanilla SGD methods using convex losses cannot perform well with noisy labels, which adversely affect the update of the primal variable in SGD methods. Unfortunately, noisy labels are ubiquitous in real world applications such as crowdsourcing. To handle noisy labels, in this paper, we present a family of robust losses for SGD methods. By employing our robust losses, SGD methods successfully reduce negative effects caused by noisy labels on each update of the primal variable. We not only reveal the convergence rate of SGD methods using robust losses, but also provide the robustness analysis on two representative robust losses. Comprehensive experimental results on six real-world datasets show that SGD methods using robust losses are obviously more robust than other baseline methods in most situations with fast convergence.

1 Introduction

To handle large-scale optimization problems, a popular strategy is to employ Stochastic Gradient Descent (SGD) methods because of two advantages. First, they do not need to compute all gradients over the whole dataset in each iteration, which lowers computational cost per iteration. Secondly, they only process a mini-batch of data points [1] or even one data point [2] in each iteration, which vastly reduces the memory storage. Therefore, many researchers have extensively studied and applied various SGD methods [3,4]. For instance, Large-Scale SGD [5] has been substantially applied to the optimization of deep learning models [6]. Primal Estimated Sub-Gradient Solver (Pegasos) [7] is employed to speed up the Support Vector Machines (SVM) methods, which is suitable for large-scale text classification problems. However, vanilla SGD methods suffer from the label noise problem since the noisy labels adversely affect the update of the primal variable in SGD methods. Unfortunately, the label noise problems are

---

I.W. Tsang—An Australian Future Fellow and Associate Professor with the Centre for Quantum Computation and Intelligent Systems, at the University of Technology Sydney.

© Springer International Publishing AG 2016
P. Frasconi et al. (Eds.): ECML PKDD 2016, Part I, LNAI 9851, pp. 665–680, 2016.
DOI: 10.1007/978-3-319-46128-1_42

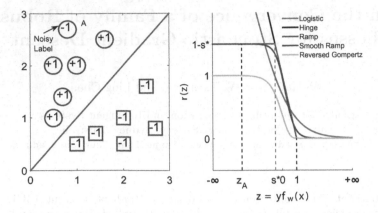

$$z = yf_w(x)$$

Fig. 1. Left Panel: Squares represent real negative instances. Circles denote real positive instances, however one circle instance "A" is erroneously annotated as negative class, which creates a noisy label. **Right Panel:** Red curve and blue curve respectively denote Ramp Loss and Smooth Ramp Loss parameterized by s^*. Magenta curve, black curve and green curve correspond to Logistic Loss, Hinge Loss and Reversed Gompertz Loss accordingly. It can be observed that the incorrectly labeled instance "A" in the left panel can be regarded as the outlier of negative class, and its loss value $r(z_A)$ is upper bounded by Ramp Loss, Smooth Ramp Loss, and Reversed Gompertz Loss (see "z_A" in the right panel). (Color figure online)

very common in real-world applications. For instance, Amazon Mechanical Turk (MTurk) is a crowdsourcing Internet platform that takes advantage of human intelligence to provide supervision, such as labeling different kinds of bird pictures and annotating keywords according to geoscience records. However, the quality of annotations is not always satisfactory because many workers are not sufficiently trained to label or annotate such specific data [8]. Another situation is where the data labels are automatically inferred from user online behaviors or implicit feedback. For example, the existing recommendation algorithms usually consider a user clicking on an online item (e.g., advertisements on Youtube or eBay) as a positive label indicating user preference, whereas users may click the item for different reasons, such as curiosity or clicking by mistake. Therefore, the labels inferred from online behaviors are often noisy.

The aforementioned issues lead to a challenging question- if the majority of data labels are incorrectly annotated, can we reduce the negative effects on SGD methods caused by these noisy labels? Our high-level idea is to design a robust loss function with a threshold for SGD methods. We illustrate our idea by using a binary classification example. In the left panel of Fig. 1, we notice that the instance $\mathbf{x}_A$ (i.e., data point "A") is incorrectly annotated with the label $y_A = -1$, which is opposite to its predicted label value $(+1)$ according to the hyperplane. Moreover, this instance is far away from the distribution of negative class. Therefore, this instance $\mathbf{x}_A$ with the noisy label y_A can be regarded as the outlier of negative class.

Let the output of the classifier $f_\mathbf{w}$ for a given $\mathbf{x}$ be $f_\mathbf{w}(\mathbf{x})$. Let z be the product of the real label and the predicted label of an instance $\mathbf{x}$ (i.e., $z = yf_\mathbf{w}(\mathbf{x})$). Then, given the outlier $\{\mathbf{x}_A, y_A\}$ in the left panel of Fig. 1, we have $z_A = y_A f_\mathbf{w}(\mathbf{x}_A) < 0$. As illustrated in the right panel of Fig. 1, with z on the x-axis, the gradient of Hinge Loss is non-zero on the z_A, which will mislead the update of the primal variable $\mathbf{w}$ in SGD methods. However, if the loss function has a threshold, for example Ramp Loss [9] in Fig. 1 with a threshold $1 - s^*$, the gradient of Ramp Loss on the z_A is zero, which minimizes the negative effects caused by this outlier on the update. Therefore, it is reasonable to employ the loss with a threshold for SGD methods in the label noise problem.

Although the Ramp Loss is robust to outliers, it is computationally hard to optimize due to its nonsmoothness and nonconvexity [10]. Therefore, we consider to relax the Ramp Loss into smooth and locally strongly-convex loss. With random initialization, SGD methods can converge into a qualified local minima with a fast speed. Our main contributions are summarized as follows.

1. We present a family of robust losses, which specifically benefit SGD methods to reduce the negative effects introduced by noisy labels, even under a high percentage of noisy labels.
2. We reveal the convergence rate of SGD methods using the proposed robust losses. Moreover, we provide the robustness analysis on two representative robust losses.
3. Comprehensive experimental results on varying scale datasets with noisy labels show that SGD methods using robust losses are obviously more robust than other baseline methods in most situations with fast convergence.

2 Related Works

First, our work is closely related to SGD methods. For example, Xu proposes the Averaged Stochastic Gradient Descent (ASGD) method [11] to lower the testing error rate of the SGD [5]. However, their work is based on the assumption that the data is clean, which significantly limits their applicability to the label noise problem. Ghahdimi & Lan introduce a randomized stochastic algorithm to solve nonconvex problems [12], and then generalize the accelerated gradient method to improve the convergence rate if the problem is nonconvex [13]. However they do not focus on learning with noisy labels specifically, and do not consider strongly convex regularizer.

Second, our work is also related to bounded nonconvex losses for robust classification. For example, Collobert et al. propose the bounded Ramp Loss for support vector machine (SVM) classification problems. Wang et al. further propose a robust SVM based on a smooth version of Ramp Loss for suppressing the outliers [14]. Their models are commonly inferred by Concave-Convex Procedure (CCCP) [9]. However, both of them do not consider that SGD methods suffer from the label noise problem. In other words, our robust losses are tailor-made for SGD methods to alleviate the effect of noisy labels while their loss is designed only for robust SVM.

Finally, our work is highly related to noisy labels. For instance, Reed & Sukhbaatar focus on training deep neural networks using noisy labels [15]. Natarajan et al. propose a probabilistic model for handling label noise problems [16]. However, all these works are unrelated to SGD methods. Moreover, they cannot be used in real-time or large-scale applications due to their high computational cost. It is also demonstrated that the 0-1 loss function is robust for outliers. However, the 0-1 loss is neither convex nor differentiable, and it is intractable for real learning algorithms in practice. Even though the surrogates of 0-1 loss is convex [17], they are very sensitive to outliers. To the best of our knowledge, the problem of SGD methods for noisy labels has not yet been successfully addressed. This paper therefore studies this problem and provides an answer with theoretical analysis and empirical verification.

3   A Family of Robust Losses for Stochastic Gradient Descent

In this section, we begin with the definition of a family of robust losses for SGD methods. Under this definition, we introduce two representative robust losses: Smooth Ramp Loss and Reversed Gompertz Loss. Then, we reveal the convergence rate of SGD methods using robust losses, and provide the robustness analysis on two representative robust losses.

3.1   Notations and Definitions

Let $\mathcal{D} = \{\mathbf{x}_i, y_i\}_{i=1}^n$ be the training data, where $\mathbf{x}_i \in \mathbb{R}^d$ denotes the ith instance and $y_i \in \{-1, +1\}$ denotes its binary label. The basic support vector machine model for classification is represented as

$$\min_{\mathbf{w}} G(\mathbf{w}) = \min_{\mathbf{w}} \frac{1}{n} \sum_{i=1}^n g_i(\mathbf{w}) \tag{1}$$

where $\mathbf{w} \in \mathbb{R}^d$ is the primal variable. Specifically, $g_i(\mathbf{w}) = \rho_\lambda(\mathbf{w}) + r(\mathbf{w}; \{\mathbf{x}_i, y_i\})$ where λ is the regularization parameter, $\rho_\lambda(\mathbf{w})$ is the regularizer and $r(\mathbf{w}; \{\mathbf{x}_i, y_i\})$ is a loss function.

Based on Restricted Strong Convexity (RSC) and Restricted Smoothness (RSM) [18,19], we propose two extended definitions. We use $\|\cdot\|$ to denote the Euclidean norm, and $B_d(\mathbf{w}^*, \gamma)$ to denote the d dimensional Euclidean ball of radius γ centered at local minima $\mathbf{w}^*$. And we assume that function G and g_i are continuously differentiable.

Definition 1 (Augmented Restricted Strong Convexity (ARSC)). *If there exists a constant $\alpha > 0$ such that for any $\mathbf{w}, \tilde{\mathbf{w}} \in B_d(\mathbf{w}^*, \gamma)$, we have*

$$G(\mathbf{w}) - G(\tilde{\mathbf{w}}) - \langle \nabla G(\tilde{\mathbf{w}}), \mathbf{w} - \tilde{\mathbf{w}} \rangle \geq \frac{\alpha}{2} \|\mathbf{w} - \tilde{\mathbf{w}}\|^2 \tag{2}$$

then G satisfies Augmented Restricted Strong Convexity.

Definition 2 (Augmented Restricted Smoothness (ARSM)). *If there exists a constant $\beta > 0$ such that for any $i \in \{1, \cdots, n\}$ and $\mathbf{w}, \tilde{\mathbf{w}} \in B_d(\mathbf{w}^*, \gamma)$, we have*

$$g_i(\mathbf{w}) - g_i(\tilde{\mathbf{w}}) - \langle \nabla g_i(\tilde{\mathbf{w}}), \mathbf{w} - \tilde{\mathbf{w}} \rangle \leq \frac{\beta}{2} \|\mathbf{w} - \tilde{\mathbf{w}}\|^2 \tag{3}$$

then g_i satisfies Augmented Restricted Smoothness.

3.2   A Family of Robust Losses

We first present the motivation and definition of a family of robust losses. Take Support Vector Machines (SVM) with convex hinge loss as an example. SGD methods are commonly used to optimize the SVM model for large-scale learning. However, if data points with noisy labels deviate significantly from the hyperplane, these mislabeled data points can be equally viewed as outliers. These outliers will severely mislead the update of the primal variable in SGD methods. Therefore, it is intuitive to design a loss function with a threshold, which truncates the value that exceeds the threshold. Inspired by Ramp Loss [9], we consider whether we can design a family of bounded, locally strongly-convex and smooth losses. If we combine this new loss with strongly-convex regularizer, the objective then satisfies the ARSC (i.e., Definition 1) and ARSM (i.e., Definition 2) simultaneously. Here, we define a family of robust losses $r(z)$ for SGD methods, where z is the variable of loss function in the x-axis of Fig. 1.

Definition 3. *A loss function $r(z)$ is robust for SGD methods if it simultaneously meets the following conditions:*

1. *Upper bound condition - it should be bounded such that $\lim_{z \to -\infty} r'(z) = 0$.*
2. *Locally λ-strongly convex condition - it should be locally λ-strongly convex if there exists a constant $\lambda > 0$ such that $r(z) - \frac{\lambda}{2}\|z\|^2$ is convex when $z \in B_1(z^*, \gamma)$, where $B_1(z^*, \gamma)$ denotes the 1 dimensional Euclidean ball of radius $\gamma > 0$ centered at local minima z^*.*
3. *Smoothly decreasing condition - it should be monotonically decreasing and continuously differentiable.*

Remark 1. We explain three conditions on Definition 3. (1) Since the upper bound can be equally viewed as the threshold, it is natural that the negative effects introduced by outliers are removed by the upper bound. (2) The loss function should be locally λ-strongly convex. If the loss function is locally λ-strongly convex and the regularizer is globally λ-strongly convex (e.g., $\frac{\lambda}{2}\|\mathbf{w}\|^2$), the objective $G(\mathbf{w})$ is locally strongly-convex. Then, objective $G(\mathbf{w})$ satisfies the ARSC. (3) If the loss function is monotonically decreasing, we reasonably assume that the objective is non-increasing around some local minima, which is convenient to prove the convergence rate. If the loss function is differentiable at every point, $g_i(\mathbf{w})$ satisfies the ARSM when $\frac{\lambda}{2}\|\mathbf{w}\|^2$ is used.

Then a family of robust losses for SGD methods can be acquired under these conditions. Here, we propose two representative robust losses that perfectly satisfy the above three conditions. Both of them are presented in Fig. 1 and employed through the whole paper.

The first one is the Smooth Ramp Loss (4), which is the smooth version of Ramp Loss[1]. If we smooth the Ramp Loss around s^* and around 1, it is much easier to optimize and satisfy the ARSM. Therefore, we employ reversed sigmoid function to represent the Smooth Ramp Loss.

$$r(s^*, z) = \frac{1 - s^*}{1 + e^{\alpha_{s^*} \cdot (z + \beta_{s^*})}} \tag{4}$$

where we set the s^* of Ramp Loss, then the parameters α_{s^*} and β_{s^*} of Smooth Ramp Loss are determined by minimizing the difference between Smooth Ramp Loss and Ramp Loss.

The second one is the Reversed Gompertz Loss, which is a special case of the Gompertz function and we reverse the Gompertz function by the y-axis.

$$r(c^*, z) = e^{-e^{c^* \cdot z}} \tag{5}$$

where the curve of this loss is controlled by parameter c^*. The aforementioned losses are integrated into the SVM model and SGD methods are employed to update the primal variable $\mathbf{w}$.

By employing two above robust losses, we finally summarize the robust SGD algorithm - Stochastic Gradient Descent with Robust Losses in Algorithm 1. Specifically, the generalized algorithm consists of two special cases. For Stochastic Gradient Descent with Smooth Ramp Loss, the algorithm employs "Set I and Update I". For Stochastic Gradient Descent with Reversed Gompertz Loss, the algorithm employs "Set II and Update II". In practical implementations, we often choose option A and also provide averaging option B.

3.3 Convergence Analysis

When we apply SGD methods to SVM model with proposed robust losses, it converges into the qualified local minima. According to the detailed explanation about the three conditions in Sect. 3.2, the objective $G(\mathbf{w})$ satisfies the ARSC and $g_i(\mathbf{w})$ satisfies the ARSM. Based on the ARSC and ARSM, we can analyze the convergence rate of SGD methods using robust losses. We use $\mathbb{E}[\cdot]$ to denote the expectation.

Theorem 1. *Consider that $G(\mathbf{w})$ satisfies Augmented Restricted Strong Convexity and $g_i(\mathbf{w})$ satisfies Augmented Restricted Smoothness. Define $\mathbf{w}^*$ as a local minima and β as the parameter of Augmented Restricted Smoothness. Assume that learning rate η is sufficient to let $G(\mathbf{w}^{(t)})$ be a non-increasing update. After T iterations, we have*

$$G(\mathbf{w}^{(T)}) - G(\mathbf{w}^*) \leq \frac{\mathbb{E}[\|\mathbf{w}^{(0)} - \mathbf{w}^*\|^2]}{(2\eta - 12\eta^2\beta) \cdot T}$$

---

[1] The common optimization method for Ramp Loss is using Concave-Convex Procedure (CCCP). However, CCCP is time-consuming compared to SGD methods.

Algorithm 1. Stochastic Gradient Descent with Robust Losses (**SGDRL**)

Input: $\lambda \geq 0$, s^*, c^*, the learning rate η, the max number of epochs T_{max}, and the
  training set $\mathcal{D} = \{\mathbf{x}_i, y_i\}_{i=1}^n$

Initialize: $\tilde{\mathbf{w}}^{(0)} = \mathbf{0}$

Set: $\begin{cases} I : f(\alpha_{s^*}, \beta_{s^*}, g) = e^{\alpha_{s^*}(g + \beta_{s^*})} \\ II : f(c^*, g) = c^* g - e^{c^* g} \end{cases}$

for $epoch = 1, 2, \ldots, T_{max}$ **do**

  Preprocess: $\mathbf{w}^{(0)} = \tilde{\mathbf{w}}^{(epoch-1)}$ and randomly shuffle n training instances in $\mathcal{D}$

  for $t = 1, \ldots, n$ **do**

    Sequentially pick: $\{\mathbf{x}_{it}, y_{it}\}$ from $\mathcal{D}$, $it \in \{1, \ldots, n\}$

    Compute: $g(\mathbf{w}^{(t-1)}) = (\langle \mathbf{w}^{(t-1)}, \mathbf{x}_{it} \rangle + b) y_{it}$

    $\mathbf{w}^{(t)} = \begin{cases} I : \mathbf{w}^{(t-1)} - \eta[\lambda \mathbf{w}^{(t-1)} - (1 - s^*)\alpha_{s^*} \mathbf{x}_{it} y_{it} \dfrac{f(\alpha_{s^*}, \beta_{s^*}, g(\mathbf{w}^{(t-1)}))}{(1 + f(\alpha_{s^*}, \beta_{s^*}, g(\mathbf{w}^{(t-1)})))^2}] \\ II : \mathbf{w}^{(t-1)} - \eta[\lambda \mathbf{w}^{(t-1)} - c^* \mathbf{x}_{it} y_{it} e^{f(c^*, g(\mathbf{w}^{(t-1)}))}] \end{cases}$

  end

  option A: $\tilde{\mathbf{w}}^{(epoch)} = \mathbf{w}^{(n)}$ or **option B:** $\tilde{\mathbf{w}}^{(epoch)} = \frac{1}{n} \sum_{t=1}^n \mathbf{w}^{(t)}$

end

Output: $\tilde{\mathbf{w}}^{(T_{max})}$

Proof Sketch for Theorem 1

Proof. Due to space constraints, here we focus on key steps, and the detailed proof is in the arXiv version[2]. According to stochastic gradient descent update rule $\mathbf{w}^{(t)} = \mathbf{w}^{(t-1)} - \eta \nabla g_{it}(\mathbf{w}^{(t-1)})$ where random number $it \in \{1, \ldots, n\}$, and $\mathbb{E}[\nabla g_{it}(\mathbf{w}^{(t-1)})] = \nabla G(\mathbf{w}^{(t-1)})$ by (1), we construct the following inequality

$$\begin{aligned} \mathbb{E}&[\|\mathbf{w}^{(t)} - \mathbf{w}^*\|^2] \\ &= \mathbb{E}[\|\mathbf{w}^{(t-1)} - \mathbf{w}^*\|^2] + \eta^2 \mathbb{E}[\|\nabla g_{it}(\mathbf{w}^{(t-1)})\|^2] \\ &\quad - 2\eta \langle \nabla G(\mathbf{w}^{(t-1)}), \mathbf{w}^{(t-1)} - \mathbf{w}^* \rangle \\ &\leq \mathbb{E}[\|\mathbf{w}^{(t-1)} - \mathbf{w}^*\|^2] + \eta^2 \mathbb{E}[\|\nabla g_{it}(\mathbf{w}^{(t-1)})\|^2] \\ &\quad - 2\eta[G(\mathbf{w}^{(t-1)}) - G(\mathbf{w}^*)] \end{aligned} \tag{6}$$

where the inequality employs the ARSC. Then we construct an auxiliary function $\varphi_i(\mathbf{w})$

$$\varphi_i(\mathbf{w}) = g_i(\mathbf{w}) - g_i(\mathbf{w}^*) - \langle \nabla g_i(\mathbf{w}^*), \mathbf{w} - \mathbf{w}^* \rangle \tag{7}$$

And it is obvious that

$$\varphi_i(\mathbf{w}^*) = g_i(\mathbf{w}^*) - g_i(\mathbf{w}^*) = 0 \tag{8a}$$

$$\nabla \varphi_i(\mathbf{w}) = \nabla g_i(\mathbf{w}) - \nabla g_i(\mathbf{w}^*) \tag{8b}$$

$$\nabla \varphi_i(\mathbf{w}^*) = \nabla g_i(\mathbf{w}^*) - \nabla g_i(\mathbf{w}^*) = 0 \tag{8c}$$

---

[2] Please search the arXiv version in https://arxiv.org/abs/1605.01623.

Thus, $\mathbf{w}^*$ is local minima of $\varphi_i(\mathbf{w})$ by (8c) and we construct the following inequality from (7)

$$
\begin{aligned}
0 = \varphi_i(\mathbf{w}^*) &\leq \min \varphi_i(\mathbf{w} - \gamma \nabla \varphi_i(\mathbf{w})) \\
&\leq \min \varphi_i(\mathbf{w}) + \frac{\beta \gamma^2}{2} \|\nabla \varphi_i(\mathbf{w}))\|^2 - \gamma \|\nabla \varphi_i(\mathbf{w}))\|^2 \\
&= \varphi_i(\mathbf{w}) - \frac{1}{2\beta} \|\nabla \varphi_i(\mathbf{w})\|^2
\end{aligned}
\tag{9}
$$

where the last inequality satisfies the ARSM and the function is minimized at the parameter $\gamma = \frac{1}{\beta}$. We construct the following inequality based on (7), (8b) and (9)

$$
\begin{aligned}
&\|\nabla g_i(\mathbf{w}) - \nabla g_i(\mathbf{w}^*)\|^2 \\
&\leq 2\beta \big[g_i(\mathbf{w}) - g_i(\mathbf{w}^*) - \langle \nabla g_i(\mathbf{w}^*), \mathbf{w} - \mathbf{w}^* \rangle \big]
\end{aligned}
\tag{10}
$$

Therefore, we have

$$
\begin{aligned}
&\mathbb{E} \big[\|\nabla g_i(\mathbf{w}) - \nabla g_i(\mathbf{w}^*)\|^2 \big] \\
&\leq 2\beta \big[G(\mathbf{w}) - G(\mathbf{w}^*) - \langle \nabla G(\mathbf{w}^*), \mathbf{w} - \mathbf{w}^* \rangle \big] \\
&\leq 4\beta \big[G(\mathbf{w}) - G(\mathbf{w}^*) \big]
\end{aligned}
\tag{11}
$$

where the second last inequality satisfies the ARSC. Because $\|A + B + C\|^2 \leq 3\|A\|^2 + 3\|B\|^2 + 3\|C\|^2$ and $\mathbf{w}^*$ is a local minima, we have the following inequality with $\nabla G(\mathbf{w}^*) = 0$ and (11)

$$
\begin{aligned}
&\mathbb{E} \big[\|\nabla g_{it}(\mathbf{w}^{(t-1)})\|^2 \big] \\
&\leq 3\mathbb{E} \big[\|\nabla g_{it}(\mathbf{w}^{(t-1)}) - \nabla g_{it}(\mathbf{w}^*)\|^2 \big] \\
&\quad + 3\mathbb{E} \big[\|\nabla g_{it}(\mathbf{w}^*) - \nabla G(\mathbf{w}^*)\|^2 \big] + 3\mathbb{E} \big[\|\nabla G(\mathbf{w}^*)\|^2 \big] \\
&\leq 12\beta \big[G(\mathbf{w}^{(t-1)}) - G(\mathbf{w}^*) \big]
\end{aligned}
\tag{12}
$$

Therefore, (6) equals to the following inequality

$$
\begin{aligned}
&\mathbb{E} \big[\|\mathbf{w}^{(t)} - \mathbf{w}^*\|^2 \big] \\
&\leq \mathbb{E} \big[\|\mathbf{w}^{(t-1)} - \mathbf{w}^*\|^2 \big] + \eta^2 \mathbb{E} \big[\|\nabla g_{it}(\mathbf{w}^{(t-1)})\|^2 \big] \\
&\quad - 2\eta \big[G(\mathbf{w}^{(t-1)}) - G(\mathbf{w}^*) \big] \\
&\leq \mathbb{E} \big[\|\mathbf{w}^{(t-1)} - \mathbf{w}^*\|^2 \big] \\
&\quad + (12\eta^2 \beta - 2\eta) \big[G(\mathbf{w}^{(t-1)}) - G(\mathbf{w}^*) \big]
\end{aligned}
\tag{13}
$$

Based on (13), when t varies from $1 \cdots T$, we get T inequalities respectively, and then simultaneously add the left hand side and right hand side of T inequalities to get

$$
\begin{aligned}
\mathbb{E} \big[\|\mathbf{w}^{(T)} - \mathbf{w}^*\|^2 \big] &\leq \mathbb{E} \big[\|\mathbf{w}^{(0)} - \mathbf{w}^*\|^2 \big] \\
&\quad + (12\eta^2 \beta - 2\eta) \Big[\sum_{t=1}^{T} G(\mathbf{w}^{(t-1)}) - T \cdot G(\mathbf{w}^*) \Big]
\end{aligned}
\tag{14}
$$

Under the assumption of a non-increasing update, we have the following inequality

$$(2\eta - 12\eta^2\beta)\left[T \cdot G(\mathbf{w}^{(T)}) - T \cdot G(\mathbf{w}^*)\right]$$

$$\leq (2\eta - 12\eta^2\beta)\left[\sum_{t=1}^{T} G(\mathbf{w}^{(t-1)}) - T \cdot G(\mathbf{w}^*)\right] \qquad (15)$$

$$\leq \mathbb{E}\left[\|\mathbf{w}^{(0)} - \mathbf{w}^*\|^2\right] - \mathbb{E}\left[\|\mathbf{w}^{(T)} - \mathbf{w}^*\|^2\right]$$

We thus obtain

$$G(\mathbf{w}^{(T)}) - G(\mathbf{w}^*) \leq \frac{\mathbb{E}\left[\|\mathbf{w}^{(0)} - \mathbf{w}^*\|^2\right] - \mathbb{E}\left[\|\mathbf{w}^{(T)} - \mathbf{w}^*\|^2\right]}{(2\eta - 12\eta^2\beta) \cdot T}$$

$$\leq \frac{\mathbb{E}\left[\|\mathbf{w}^{(0)} - \mathbf{w}^*\|^2\right]}{(2\eta - 12\eta^2\beta) \cdot T} = \frac{d}{\eta \cdot T} = \epsilon \qquad (16)$$

where $d = \frac{\mathbb{E}\left[\|\mathbf{w}^{(0)} - \mathbf{w}^*\|^2\right]}{(2 - 12\eta\beta)}$. Therefore we conclude that when $T = \frac{d}{\eta \cdot \epsilon}$, SGD methods using robust losses have ϵ-solution and the convergence rate is $\mathcal{O}(1/T)$. Therefore, to achieve a ϵ-solution, the complexity of Algorithm 1 is $\mathcal{O}(\frac{n \cdot d}{\eta \cdot \epsilon})$.

4   Robustness Analysis

Theorem 2. *Assume that an instance* $\mathbf{x}_i$ *is annotated with noisy label* y_i, *which means* $y_i(\mathcal{K}_i^T \alpha + b) < 0$. *Its corresponding weighted coefficient* ϕ_i *for Smooth Ramp Loss with* $(s^*, \alpha_{s^*}, \beta_{s^*})$ *is*

$$\phi_i = \frac{(1 - s^*)\alpha_{s^*}\delta e^{\alpha_{s^*}(y_i\mathcal{K}_i^T\alpha + y_ib)}}{(1 - (y_i\mathcal{K}_i^T\alpha + y_ib))(1 + \delta e^{\alpha_{s^*}(y_i\mathcal{K}_i^T\alpha + y_ib)})^2}$$

for Reversed Gompertz Loss with c^* *is*

$$\phi_i = \frac{c^* e^{c^*(y_i\mathcal{K}_i^T\alpha + y_ib)} - e^{c^*(y_i\mathcal{K}_i^T\alpha + y_ib)}}{1 - (y_i\mathcal{K}_i^T\alpha + y_ib)}$$

if $|f_w(\mathbf{x}_i)| = |(\mathcal{K}_i^T\alpha + b)|$ *increases, which means* $\mathbf{x}_i$ *with noisy label* y_i *becomes an outlier, then both* ϕ_i *will definitely decrease. It indicates that the proposed Robust Losses do reduce the negative effects introduced by noisy labels.*

Proof Sketch for Theorem 2

Proof. Firstly, we assume that $\{\mathbf{x}_i, y_i\}_{i=1}^k$ is a random subset of training data $\mathcal{D}$ and $f_\mathbf{w}$ is the decision function, according to the representer theorem, $z_i = y_i f_\mathbf{w}(\mathbf{x}_i) = y_i(\sum_{j=1}^k \mathcal{K}(\mathbf{x}_j, \mathbf{x}_i)\alpha_j + b) = y_i\mathcal{K}_i^T\alpha + y_ib$, where $\alpha = (\alpha_1, \alpha_2, ..., \alpha_k)'$, $\mathcal{K} = (\mathcal{K}_1, \mathcal{K}_2, ..., \mathcal{K}_k)'$ and $\mathcal{K}_i = (\mathcal{K}(\mathbf{x}_1, \mathbf{x}_i), \mathcal{K}(\mathbf{x}_2, \mathbf{x}_i), ..., \mathcal{K}(\mathbf{x}_k, \mathbf{x}_i))'$. $\lambda > 0$ is a regularizer parameter, $\mathcal{K}$ is a mercer kernel and $\mathcal{H}_\mathcal{K}$ is a Reproducing Kernel Hilbert Space (RKHS). For a family of robust losses $r(z)$, we define two functions

$\rho(z)$ and $\varrho(z)$ such that $r(z) = \rho(1 - z)$ and $\varrho(z) = \frac{\rho'(z)}{z}$. Therefore, our robust model can be presented as

$$
\begin{aligned}
f_{\mathbf{w}}^* &= \arg\min_{f_{\mathbf{w}}} \frac{1}{k} \sum_{i=1}^{k} r(z_i) + \frac{\lambda}{2} \|f_{\mathbf{w}}\|^2 \\
&= \arg\min_{f_{\mathbf{w}}} \frac{1}{k} \sum_{i=1}^{k} r(f_{\mathbf{w}}(\mathbf{x}_i) \cdot y_i) + \frac{\lambda}{2} f_{\mathbf{w}}^T f_{\mathbf{w}} \\
&= \arg\min_{\alpha,b} \frac{1}{k} \sum_{i=1}^{k} \rho(1 - y_i \mathcal{K}_i^T \alpha - y_i b) + \frac{\lambda}{2} \alpha^T \mathcal{K} \alpha
\end{aligned}
\tag{17}
$$

The last equation satisfies the second condition of robust losses $r(z) = \rho(1-z)$. Due to $\varrho(z) = \frac{\rho'(z)}{z}$, we define coefficient $\phi_i = \varrho(1 - y_i \mathcal{K}_i^T \alpha - y_i b)$, then

$$
\rho'(1 - y_i \mathcal{K}_i^T \alpha - y_i b) = (1 - y_i \mathcal{K}_i^T \alpha - y_i b)\phi_i
\tag{18}
$$

Because our proposed loss is nonconvex, we assume that $(\hat{\alpha}, \hat{b})$ is one of the critical points for above minimization problem (17). Let's set $Q(\alpha, b) = \frac{1}{k} \sum_{i=1}^{k} \rho(1 - y_i \mathcal{K}_i^T \alpha - y_i b) + \frac{\lambda}{2} \alpha^T \mathcal{K} \alpha$, therefore: $\frac{\partial Q(\hat{\alpha}, \hat{b})}{\partial \alpha} = 0$ and $\frac{\partial Q(\hat{\alpha}, \hat{b})}{\partial b} = 0$. Then, we have two equations below

$$
\frac{1}{k} \sum_{i=1}^{k} (1 - y_i \mathcal{K}_i^T \hat{\alpha} - y_i \hat{b})(y_i \mathcal{K}_i)\phi_i - \lambda \mathcal{K}^T \hat{\alpha} = 0
\tag{19}
$$

$$
\frac{1}{k} \sum_{i=1}^{k} (1 - y_i \mathcal{K}_i^T \hat{\alpha} - y_i \hat{b}) y_i \phi_i = 0
\tag{20}
$$

The solution $(\hat{\alpha}, \hat{b})$ of Eqs. (19) and (20) can be achieved by solving the following L2-SVM

$$
\min_{\alpha,b} \frac{1}{k} \sum_{i=1}^{k} (y_i - \mathcal{K}_i^T \alpha - b)^2 \phi_i + \frac{\lambda}{2} \alpha^T \mathcal{K} \alpha
\tag{21}
$$

When $k = 1$, we solve it by streaming stochastic gradient descent. If $k > 1$, we solve it by mini-batch stochastic gradient descent. Currently, we consider ϕ_i as an important coefficient that affects the update of stochastic dual variable α, and therefore, we analyze robust statistics briefly from coefficient ϕ_i view.

If an instance $\mathbf{x}_i$ is annotated with noisy label y_i, it means that $y_i f_{\mathbf{w}}(\mathbf{x}_i) < 0$. By the representer theorem, we can easily find $y_i(\mathcal{K}_i^T \alpha + b) < 0$ for this instance. We consider $|(\mathcal{K}_i^T \alpha + b)|$ as the degree where this instance is far away from the hyperplane. So we define $\phi_i = \varrho(1 - y_i \mathcal{K}_i^T \alpha - y_i b)$. To analyze the robustness of $r(z)$, we only take Smooth Ramp Loss as an example here due to space constraints. And the robustness analysis of Reversed Gompertz loss can be found in the arXiv version. We define $\delta = e^{\alpha_{s^*} \beta_{s^*}}$ and according to our inference

$$\phi_i = \varrho(1 - y_i\mathcal{K}_i^T\alpha - y_ib)$$

$$= \frac{(1 - s^*)\alpha_{s^*}\delta e^{\alpha_{s^*}(y_i\mathcal{K}_i^T\alpha + y_ib)}}{(1 - (y_i\mathcal{K}_i^T\alpha + y_ib))(1 + \delta e^{\alpha_{s^*}(y_i\mathcal{K}_i^T\alpha + y_ib)})^2} \tag{22}$$

Remark 2. If $\{\mathbf{x}_i, y_i\}$ is an instance with a noisy label $(y_i(\mathcal{K}_i^T\alpha + b) < 0)$, then the mislabeled instance becomes an outlier when $|f_{\mathbf{w}}(\mathbf{x}_i)| = |(\mathcal{K}_i^T\alpha + b)|$ increases. It means this mislabeled instance is far away from the hyperplane. The coefficient ϕ_i will then decrease because $1 - (y_i\mathcal{K}_i^T\alpha + y_ib)$ will increase while $\frac{e^{\alpha_{s^*}(y_i\mathcal{K}_i^T\alpha + y_ib)}}{(1 + \delta e^{\alpha_{s^*}(y_i\mathcal{K}_i^T\alpha + y_ib)})^2}$ will decrease. This indicates that the coefficient ϕ_i will decrease with the increase of $|f_{\mathbf{w}}(\mathbf{x}_i)|$ for outlier instance $\mathbf{x}_i$ and does not play a significant role in the update of the dual variable. Therefore, Smooth Ramp Loss can reduce the negative effects introduced by noisy labels.

5 Experiments

In this section, we mainly perform experiments on noisy datasets to verify the convergence and robustness of SGD methods with two representative robust losses. The datasets range from small to large scale. For convenience, we abbreviate SGD with Smooth Ramp Loss as SGD(SRamp) and SGD with Reversed Gompertz Loss as SGD(RGomp) respectively.

5.1 Experimental Settings

All experimental datasets come from the LIBSVM datasets webpage[3]. The statistics of the datasets are summarized in the Table of the arXiv version. Among them, REAL-SIM, COVTYPE, MNIST38 and IJCNN1 are manually split into the training set and testing set by about 4 : 1. We normalize the data by scaling each feature to [0,1]. To generate the datasets with noisy labels, we follow the settings in [16]. Specifically, we proportionally flip the class label of training data. For example, we randomly flip 20 % of data labels from −1 to 1 or 1 to −1, and assume that the data has 20 % of noisy labels. We then repeat the same process to produce 40 % and 60 % of noisy labels on all datasets.

In the experiments, the baseline methods are classified into two categories. The first category consists of SGD methods with different losses ranging from convex losses to robust nonconvex losses, which can verify the convergence and robustness of SGD methods with two representative losses for noisy labels. For example, we choose SGD with Logistic Loss (SGD(Log)), Hinge Loss (SGD(Hinge)) and Ramp Loss (SGD(Ramp)). We also choose ASGD [11] with Logistic Loss (ASGD(Log)) and PEGASOS [7] as baseline methods. For the second category, we compare proposed methods with LIBLINEAR (We abbreviate L2-regularized L2-loss SVM Primal solution as LIBPrimal and Dual solution as LIBDual) due to its wide popularity in large-scale machine learning. All the

---

[3] http://www.csie.ntu.edu.tw/~cjlin/libsvmtools/datasets/.

Table 1. Testing error rate (in %) with standard deviation on datasets without noisy labels. Methods are indicated by "-"due to running out of memory.

Methods	A7A	IJCNN1	REAL-SIM	COVTYPE	MNIST38	SUSY
LIBPRIMAL	**14.99**	8.25	2.57	24.35	5.71	21.34
LIBDUAL	15.02	8.20	2.67	24.25	6.09	35.32
PEGASOS	17.62 ± 1.56	8.50 ± 0.19	3.32 ± 0.06	26.36 ± 1.99	-	-
SGD(Log)	15.16 ± 0.06	9.08 ± 0.48	2.62 ± 0.03	25.07 ± 0.28	5.73 ± 0.09	20.93 ± 0.01
ASGD(Log)	14.99 ± 0.14	8.04 ± 0.04	2.54 ± 0.01	24.38 ± 0.01	**5.54 ± 0.01**	20.83 ± 0.09
SGD(Hinge)	15.45 ± 0.09	8.40 ± 0.22	2.69 ± 0.13	24.62 ± 0.54	5.77 ± 0.16	20.89 ± 0.08
SGD(Ramp)	15.54 ± 0.54	8.50 ± 0.03	4.02 ± 0.02	24.22 ± 0.10	6.04 ± 0.08	21.36 ± 0.05
SGD(SRamp)	15.11 ± 0.06	6.49 ± 0.12	2.55 ± 0.03	23.69 ± 0.04	5.76 ± 0.06	**20.81 ± 0.03**
SGD(RGomp)	15.10 ± 0.01	**6.45 ± 0.02**	**2.45 ± 0.03**	**23.29 ± 0.03**	5.56 ± 0.01	20.94 ± 0.01

methods are implemented in C++. Experiments are performed on a computer with a 3.20 GHz Inter CPU and 8 GB main memory running on a Windows 7.

The regularization parameter λ is chosen by 10-fold cross validation for all methods in the range of $\{10^{-6}, 10^{-5}, 10^{-4}, 10^{-3}, 10^{-2}, 10^{-1}, 1, 10\}$. For SGD methods with different losses, the number of epochs is normally set to 15 for convergence comparison and the primal variable $\mathbf{w}$ is initialized to $\mathbf{0}$. For LIB-LINEAR, we set the bias b to 1 and the stopping tolerance ϵ to 10^{-2} for primal solution and 10^{-1} for dual solution by default. For PEGASOS, the number of epochs for convergence is set to $\frac{10}{\lambda}$ by default and the block size k is set to 1 for training efficiency. For SGD(SRamp), the parameter s^* is chosen by 10-fold cross validation in the range of $[-2, 0]$ according to real-world datasets. Therefore, the parameter $(s^*, \alpha_{s^*}, \beta_{s^*})$ is optimized to $(-0.7, 3, -0.15)$, $(-1, 2, -0.03)$ or $(-2, 1.5, 0.5)$. For SGD(RGomp), the parameter c^* is randomly fixed to 2. All the experiments are repeated ten times and the results are averaged over the 10 trials. Methods are indicated by "-"in Table 1 due to running out of memory. Methods are not reported in Figs. 3 and 4 due to running out of memory or too long training time.[4]

5.2   The Performance of Convergence

First, we verify the convergence of SGD methods with two representative losses for noisy labels. Due to the limit of space, we provide the primal objective value of SGD(SRamp) with the number of epochs on representative small-scale IJCNN1 and large-scale SUSY datasets in the arXiv version. We observe that SGD(SRamp) converges within 15 epochs. This observation is consistent with our convergence analysis in Sect. 3.3. Since SGD(SRamp) and SGD(RGomp) are very similar, the convergence curve of SGD(RGomp) is also similar to that of SGD(SRamp). Thus, we do not report the results of SGD(RGomp).

Then, we further observe the convergence comparison of SGD methods with different losses for noisy labels in Fig. 2 where, with the increase of number

---

[4] On MNIST38 and SUSY datasets, PEGASOS run out of memory, and the training time of LIBDual is several orders of magnitude more than that of other baselines.

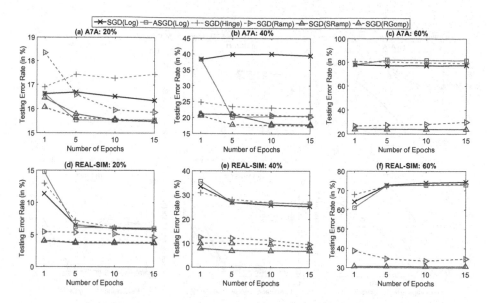

Fig. 2. Testing error rate (in %) with the number of epochs on A7A and REAL-SIM. Datasets have varying percentages (in %) of noisy labels (20 %, 40 % and 60 %). For PEGASOS, the number of epochs for convergence is set to $\frac{10}{\lambda}$ by default. Therefore, we do not report its result

of epochs, the testing error rate of SGD(SRamp) and SGD(RGomp) not only decrease faster than that of other baseline methods but also keep relative stable in the most cases. In other words, our method takes 1–5 epochs to converge while SGD(Hinge) takes more than 15 epochs to converge. Even worse, SGD(Hinge) diverges in presence of 60 % of noisy labels.

5.3   The Performance of Robustness

Finally, we verify the robustness of SGD methods with two representative losses for noisy labels. Figures 3 and 4 respectively report testing error rate and variance with varying percentages of noisy labels. From Figs. 3 and 4, we have the following observations. (a) On all datasets, SGD(SRamp) and SGD(RGomp) obviously outperform the other baseline methods in testing error rate beyond 40 % of noisy labels. Between 0 % to 40 %, SGD(SRamp) and SGD(RGomp) still have comparative advantages. In particular, for a high-dimensional dataset REAL-SIM, the advantage of SGD(SRamp) and SGD(RGomp) is extremely obvious in the whole range of the x-axis. (b) Meanwhile, we notice that the variance of testing error rate for baseline methods (e.g., PEGASOS) gradually increases with the growing percentage of noisy labels, but the variance of testing error rate for SGD(SRamp) and SGD(RGomp) remains at the lowest level in the most cases. Therefore, the robustness of SGD(SRamp) and SGD(RGomp) have been validated by their testing error rate and variance. Although two losses are comparable in the performance of robustness, the parameter of SGD(RGomp) is easier to tune.

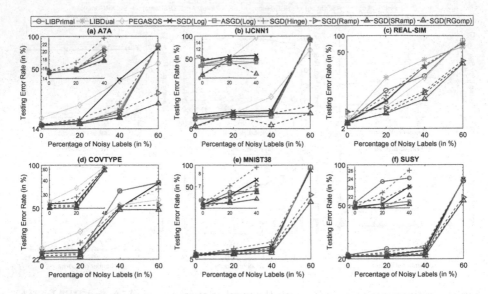

Fig. 3. Testing error rate (in %) on datasets with varying percentages (in %) of noisy labels. We provide the subfigures to compare the testing error rate with 0 % to 40 % of noisy labels on all datasets except for REAL-SIM. The y-axis is in log-scale.

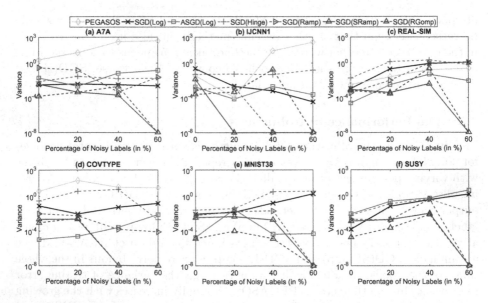

Fig. 4. Variance on datasets with varying percentages (in %) of noisy labels. The y-axis is in log-scale. Note that there is no variance for LIBPrimal and LIBDual because in each update of the primal variable, they compute full gradients instead of stochastic gradients.

In the most cases, the proposed SGD(SRamp) and SGD(RGomp) outperform other baseline methods not only on datasets with varying percentage of noisy labels but also on clean datasets. For example, Table 1 demonstrates that in terms of the testing error rate with the standard deviation, SGD(SRamp) and SGD(RGomp) outperform other baseline methods on IJCNN1, REAL-SIM, COVTYPE and SUSY datasets without noisy labels.

6   Conclusions

This paper studies SGD methods with a family of robust losses for the label noise problem. For convenience, we mainly introduce two representative robust losses including Smooth Ramp Loss and Reversed Gompertz Loss. Our theoretical analysis not only reveals the convergence rate of SGD methods using robust losses, but also proves the robustness of two representative robust losses. Comprehensive experimental results show that, on real-world datasets with varying percentages of noisy labels, SGD methods using our proposed losses are robust enough to reduce negative effects caused by noisy labels with fast convergence. In the future, we will extend our proposed robust losses to improve the performance of SGD methods for regression problems with noisy labels.

Acknowledgments. This work was supported in part by the Australia Research Council (ARC) Discovery Project under Grant No. DP140100545. Dr. Ivor W. Tsang is grateful for the support from the ARC Future Fellowship FT130100746 and ARC Linkage Project under Grant No. LP150100671. Bo Han would like to thank Dr. Chen Gong for many helpful discussions.

References

1. Cotter, A., Shamir, O., Srebro, N., Sridharan, K.: Better mini-batch algorithms via accelerated gradient methods. In: Advances in Neural Information Processing Systems (NIPS), pp. 1647–1655 (2011)
2. Mitliagkas, I., Caramanis, C., Jain, P.: Memory limited, streaming PCA. In: Advances in Neural Information Processing Systems (NIPS), pp. 2886–2894 (2013)
3. Nesterov, Y.: Efficiency of coordinate descent methods on huge-scale optimization problems. SIAM J. Optim. (SIAM) **22**(2), 341–362 (2012)
4. Agarwal, A., Foster, D.P., Hsu, D., Kakade, S.M., Rakhlin, A.: Stochastic convex optimization with bandit feedback. SIAM J. Optim. (SIAM) **23**(1), 213–240 (2013)
5. Bottou, L.: Large-scale machine learning with stochastic gradient descent. In: Proceedings of the 19th International Conference on Computational Statistics (COMPSTAT), pp. 177–187 (2010)
6. Le, Q.V., Ngiam, J., Coates, A., Lahiri, A., Prochnow, B., Ng, A.Y.: On optimization methods for deep learning. In: Proceedings of the 28th International Conference on Machine Learning (ICML), pp. 265–272 (2011)
7. Shalev-shwartz, S., Singer, Y., Srebro, N., Cotter, A.: Pegasos: primal estimated sub-gradient solver for SVM. Math. Program. **127**(1), 3–30 (2011)

8. Yan, Y., Rosales, R., Fung, G., Schmidt, M., Hermosillo, G., Bogoni, L., Moy, L., Dy, J.-G.: Modeling annotator expertise: learning when everybody knows a bit of something. In: Proceedings of the 13th International Conference on Artificial Intelligence and Statistics (AISTATS), pp. 932–939 (2010)

9. Collobert, R., Sinz, F., Weston, J., Bottou, L.: Trading convexity for scalability. In: Proceedings of the 23rd International Conference on Machine Learning (ICML), pp. 201–208 (2006)

10. Yu, Y.-L., Yang, M., Xu, L.-L., White, M., Schuurmans, D.: Relaxed clipping: a global training method for robust regression and classification. In: Advances in Neural Information Processing Systems (NIPS), pp. 2532–2540 (2010)

11. Xu, W.: Towards optimal one pass large scale learning with averaged stochastic gradient descent. arXiv preprint arXiv:1107.2490 (2011)

12. Ghadimi, S., Lan, G.-H.: Stochastic first-and zeroth-order methods for nonconvex stochastic programming. SIAM J. Optim. (SIAM) **23**(4), 2341–2368 (2013)

13. Ghadimi, S., Lan, G.-H.: Accelerated gradient methods for nonconvex nonlinear and stochastic programming. Math. Program. **156**, 59–99 (2015)

14. Wang, L., Jia, H.-D., Li, J.: Training robust support vector machine with smooth ramp loss in the primal space. Neurocomputing **71**(13), 3020–3025 (2008)

15. Sukhbaatar, S., Bruna, J., Paluri, M., Bourdev, L., Fergus, R.: Training convolution networks with noisy labels. In: Proceedings of the International Conference on Learning Representations (ICLR) (2015)

16. Natarajan, N., Dhillon, I.S., Ravikumar, P.K., Tewari, A.: Learning with noisy labels. In: Advances in Neural Information Processing Systems (NIPS), pp. 1196–1204 (2013)

17. Bartlett, P.L., Jordan, M.I., McAuliffe, J.D.: Convexity, classification, and risk bounds. J. Am. Stat. Assoc. **101**(473), 138–156 (2006)

18. Agarwal, A., Negahban, S., Wainwright, M.J.: Fast global convergence of gradient methods for high-dimensional statistical recovery. Ann. Stat. **40**(5), 2452–2482 (2012)

19. Loh, P.-L., Wainwright, M.J.: Regularized M-estimators with nonconvexity: statistical and algorithmic theory for local optima. J. Mach. Learn. Res. (JMLR) **16**, 559–616 (2015)

Composite Denoising Autoencoders

Krzysztof J. Geras[(✉)] and Charles Sutton

Institute of Adaptive Neural Computation, School of Informatics,
The University of Edinburgh, Edinburgh EH8 9AB, UK
k.j.geras@sms.ed.ac.uk, csutton@inf.ed.ac.uk

Abstract. In representation learning, it is often desirable to learn features at different levels of scale. For example, in image data, some edges will span only a few pixels, whereas others will span a large portion of the image. We introduce an unsupervised representation learning method called a composite denoising autoencoder (CDA) to address this. We exploit the observation from previous work that in a denoising autoencoder, training with lower levels of noise results in more specific, fine-grained features. In a CDA, different parts of the network are trained with different versions of the same input, corrupted at different noise levels. We introduce a novel cascaded training procedure which is designed to avoid types of bad solutions that are specific to CDAs. We show that CDAs learn effective representations on two different image data sets.

Keywords: Denoising autoencoders · Unsupervised learning · Neural networks

1 Introduction

In most applications of representation learning, we wish to learn features at different levels of scale. For example, in image data, some edges will span only a few pixels, whereas others, such as a boundary between foreground and background, will span a large portion of the image. Similarly, in speech data, different phonemes and different words vary a lot in their duration. In text data, some features in the representation might model specialized topics that use only a few words. For example a topic about electronics would often use words such as "big", "screen" and "tv". Other features model more general topics that use many different words. Good representations should model both of these phenomena, containing features at different levels of granularity.

Denoising autoencoders [12, 28, 29] provide a particularly natural framework to formalise this intuition. In a denoising autoencoder, the network is trained to be able to reconstruct each data point from a corrupted version. The noise process used to perform the corruption is chosen by the modeller, and is an important aspect of the learning process that affects the final representation. On a digit recognition task, Vincent et al. [29] noticed that using a low level of noise leads to learning blob detectors, while increasing it results in obtaining detectors of strokes or parts of digits. They also recognise that either too low or

© Springer International Publishing AG 2016
P. Frasconi et al. (Eds.): ECML PKDD 2016, Part I, LNAI 9851, pp. 681–696, 2016.
DOI: 10.1007/978-3-319-46128-1_43

too high level of noise harms the representation learnt. The relationship between the level of noise and spatial extent of the filters was also noticed by Karklin and Simoncelli [18] for a different feature learning model. Despite impressive practical results with denoising autoencoders (e.g. [13,23]), how to choose the noise distribution is not fully understood.

In this paper, we introduce *composite denoising autoencoders* (CDA), in which different parts of the network receive versions of the input that are corrupted with different levels of noise. This encourages different hidden units of the network to learn features at different scales. A key challenge is that finding good parameters in a CDA requires some care, because naive training methods will cause the network to rely mostly on the low-noise corruptions, without fully training the features for the high-noise corruptions, because after all the low noise corruptions provide more information about the original input. We introduce a training method specifically for CDA that sidesteps this problem.

On two different data sets of images, we show that CDAs learn significantly better representations that standard DAs. In particular, we achieve to our knowledge the best accuracy on the CIFAR-10 data set with a permutation invariant model, outperforming scheduled denoising autoencoders [10].

2   Background

The core idea of learning a representation by learning to reconstruct artificially corrupted training data dates back at least to the work of Seung [24], who suggested using a recurrent neural network for this purpose. Using unsupervised layer-wise learning of representations for classification purposes appeared later in the work of Bengio et al. [3] and Hinton et al. [16].

The denoising autoencoder (DA) [28] is based on the same intuition as the work of Seung [24] that a good representation should contain enough information to reconstruct corrupted versions of an original input. In its simplest form, it is a single-layer feed-forward neural network. Let $\mathbf{x} \in \mathbb{R}^d$ be the input to the network. The output of the network is a hidden representation $\mathbf{y} \in \mathbb{R}^{d'}$, which is simply computed as $f_\theta(\mathbf{x}) = h(\mathbf{Wx} + \mathbf{b})$, where the matrix $\mathbf{W} \in \mathbb{R}^{d' \times d}$ and the vector $\mathbf{b} \in \mathbb{R}^{d'}$ are the parameters of the network, and h is a typically nonlinear transfer function, such as a sigmoid. We write $\theta = (\mathbf{W}, \mathbf{b})$. The function f is called an *encoder* because it maps the input to a hidden representation. In an autoencoder, we have also a *decoder* that "reconstructs" the input vector from the hidden representation. The decoder has a similar form to the encoder, namely, $g_{\theta'}(\mathbf{y}) = h'(\mathbf{W'y} + \mathbf{b'})$, except that here $\mathbf{W'} \in \mathbb{R}^{d \times d'}$ and $\mathbf{b'} \in \mathbb{R}^d$. It can be useful to allow the transfer function h' for the decoder to be different from that for the encoder. Typically, $\mathbf{W}$ and $\mathbf{W'}$ are constrained by $\mathbf{W'} = \mathbf{W}^T$, which has been justified theoretically by Vincent [27].

During training, our objective is to learn the encoder parameters $\mathbf{W}$ and $\mathbf{b}$. As a byproduct, we will need to learn the decoder parameters $\mathbf{b'}$ as well. We do this by defining a *noise distribution* $p(\tilde{\mathbf{x}}|\mathbf{x}, \nu)$. The amount of corruption is controlled by a parameter ν. We train the autoencoder weights to be able to reconstruct a random input from the training distribution $\mathbf{x}$ from its corrupted version $\tilde{\mathbf{x}}$ by running the encoder and the decoder in sequence. Formally, this process is described by minimising the autoencoder reconstruction error with respect to the parameters θ^* and θ'^*, i.e.,

$$\theta^*, \theta'^* = \arg\min_{\theta, \theta'} \mathbb{E}_{(X, \tilde{X})} \left[L \left(X, g_{\theta'}(f_\theta(\tilde{X})) \right) \right], \tag{1}$$

where L is a loss function over the input space, such as squared error. Typically we minimize this objective function using SGD with mini-batches, where at each iteration we sample new values for both the uncorrupted and corrupted inputs.

In the absence of noise, this model is known simply as an autoencoder or autoassociator. A classic result [2] states that when $d' < d$, then under certain conditions, an autoencoder learns the same subspace as PCA. If the dimensionality of the hidden representation is too large, i.e., if $d' > d$, then the autoencoder can obtain zero reconstruction error simply by learning the identity map. In a denoising autoencoder, in contrast, the noise forces the model to learn interesting structure even when there are a large number of hidden units. Indeed, in practical denoising autoencoders, the best results are found with *overcomplete* representations for which $d' > d$.

There are several choices to be made here, including the noise distribution, the transformations h and h' and the loss function L. For the loss function L, for continuous $\mathbf{x}$, squared error can be used. For binary $\mathbf{x}$ or $\mathbf{x} \in [0, 1]$, as we consider in this paper, it is common to use the *cross entropy* loss,

$$L(\mathbf{x}, \mathbf{z}) = - \sum_{i=1}^{D} \left(x_i \log z_i + (1 - x_i) \log (1 - z_i) \right).$$

For the transfer functions, common choices include the sigmoid $h(v) = \frac{1}{1+e^{-v}}$ for both the encoder and decoder, or to use a rectifier $h(v) = \max(0, v)$ in the encoder paired with sigmoid decoder.

One of the most important parameters in a denoising autoencoder is the noise distribution p. For continuous $\mathbf{x}$, Gaussian noise $p(\tilde{\mathbf{x}}|\mathbf{x}, \nu) = N(\tilde{\mathbf{x}}; \mathbf{x}, \nu)$ can be used. For binary $\mathbf{x}$ or $\mathbf{x} \in [0, 1]$, it is most common to use *masking noise*, that is, for each $i \in 1, 2, \ldots d$, we sample $\tilde{x}_i$ independently as

$$p(\tilde{x}_i | x_i, \nu) = \begin{cases} 0 & \text{with probability } \nu, \\ x_i & \text{otherwise.} \end{cases} \tag{2}$$

In either case, the level of noise ν affects the degree of corruption of the input. If ν is high, the inputs are more heavily corrupted during training. The noise level has a significant effect on the representations learnt. For example, if the input data are images, masking only a few pixels will bias the process of learning the representation to deal well with local corruptions. On the other hand, masking many pixels will push the algorithm to use information from more distant regions.

It is possible to train multiple layers of representations with denoising autoencoders by training a denoising autoencoder with data mapped to a representation learnt by the encoder of another denoising autoencoder. This model is known as the stacked denoising autoencoder [28,29]. As an alternative to stacking, constructing deep autoencoders with denoising autoencoders was explored by Xie et al. [30].

Although the standard denoising autoencoders are not, by construction, generative models, Bengio et al. [5] proved that, under mild regularity conditions, denoising autoencoders can be used to sample from a distribution which consistently estimates the data generating distribution. This method, which consists of alternately adding noise to a sample and denoising it, yields competitive performance in terms of estimated log-likelihood of the samples. An important connection was also made by Vincent [27], who showed that optimising the training objective of a denoising autoencoder is equivalent to performing score matching [17] between the Parzen density estimator of the training data and a particular energy-based model.

3   Composite Denoising Autoencoders

Composite denoising autoencoders learn a diverse representation by leveraging the observation that the types of features learnt by the standard denoising autoencoders differ depending on the level of noise. Instead of training all of the hidden units to extract features from data corrupted with the same level of noise, we can partition the hidden units, training each subset of model's parameters with a different noise level.

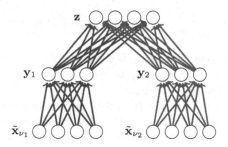

Fig. 1. A composite denoising autoencoder using two levels of noise. (Color figure online)

More formally, let $\boldsymbol{\nu} = (\nu_1, \nu_2, \ldots, \nu_S)$ denote the set of noise levels that is to be used in the model. For each noise level ν_s the network includes a vector $\mathbf{y}_s \in \mathbb{R}^{D_s}$ of hidden units and a weight matrix $\mathbf{W}_s \in \mathbb{R}^{D_s \times d}$. Note that different noise levels may have different numbers of hidden units. We use $\mathbf{D} = (D_1, D_2, \ldots D_S)$ to denote a vector containing the number of hidden units for each noise level.

When assigning a representation to a new input $\mathbf{x}$, the CDA is very similar to the DA. In particular, the hidden representation is computed as

$$\mathbf{y}_s = h(\mathbf{W}_s \mathbf{x} + \mathbf{b}_s) \qquad \forall s \in 1, \ldots, S, \tag{3}$$

where as before h is a nonlinear transfer function such as the sigmoid. The full representation $\mathbf{y}$ for $\mathbf{x}$ is constructed by concatenating the individual representations as $\mathbf{y} = (\mathbf{y}_1, \ldots, \mathbf{y}_S)$.

Where the CDA differs from the DA is in the training procedure. Given a training input $\mathbf{x}$, we corrupt it S times, once for each level of noise, yielding corrupted vectors

$$\tilde{\mathbf{x}}_s \sim p(\tilde{\mathbf{x}}_s | \mathbf{x}, \nu_s) \qquad \forall s. \tag{4}$$

Then each of the corrupted vectors are fed into the corresponding encoders, yielding the representation

$$\mathbf{y}_s = h(\mathbf{W}_s \tilde{\mathbf{x}}_s + \mathbf{b}_s) \qquad \forall s \in 1, \ldots, S. \tag{5}$$

The reconstruction $\mathbf{z}$ is computed by taking all of the hidden layers as input

$$\mathbf{z} = h' \left(\sum_{s=1}^{S} \mathbf{W}_s^\top \mathbf{y}_s + \mathbf{b}' \right), \tag{6}$$

where as before h' is a nonlinear transfer function, potentially different from h. Finally given a loss function L, such as squared error, an update to the parameters can be made by taking a gradient step on $L(\mathbf{z}, \mathbf{x})$.

This procedure can be seen as a stochastic gradient on an objective function that takes the expectation over the corruptions:

$$\mathbb{E}_{(X, \tilde{X}_{\nu_1}, \ldots, \tilde{X}_{\nu_S})} \left[L \left(X, h' \left(\sum_{s=1}^{S} \mathbf{W}_s^\top h \left(\mathbf{W}_s \tilde{\mathbf{x}}_{\nu_s} + \mathbf{b}_s \right) + \mathbf{b}' \right) \right) \right], \tag{7}$$

This architecture is illustrated in Fig. 1 for two levels of noise, where we use the different colours to indicate the weights in the network that are specific to a single noise level.

3.1 Learning

A CDA could be trained by standard optimization methods, such as stochastic gradient descent on the objective (7). As we will show, however, it is difficult to achieve good performance with these methods (Sect. 4.1). Instead, we propose a new cascaded training procedure for CDAs, which we describe in this section.

Cascaded training is based on two ideas. First, previous work [10] found that pretraining at high noise levels helps learning the parameters for the low noise levels. Second, and more interesting, the problem with taking a joint gradient step on (7) is that low noise levels provide more information about the original input $\mathbf{x}$ than high noise levels, which can cause a training procedure to get stuck in a local optimum in which it relies on the low noise features without using the high noise features. Cascaded training first trains the weights that correspond to high noise levels, and then freezes them before moving on to low noise levels. This way the hidden units trained with lower levels of noise are trained to correct what the hidden units associated with higher noise levels missed.

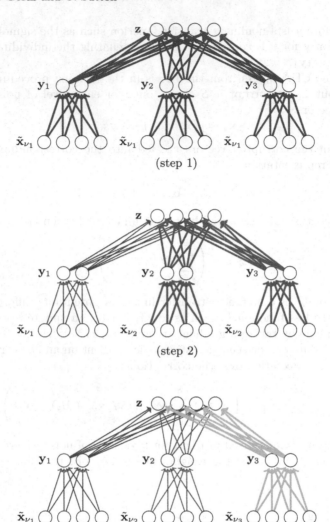

(step 1)

(step 2)

(step 3)

Fig. 2. The cascaded training procedure for a composite denoising autoencoder with three noise levels. We use the notation $\mathbf{y}_{1:3} = (\mathbf{y}_1, \mathbf{y}_2, \mathbf{y}_3)$. First all parameters are trained using the level of noise ν_1. In the second step, the blue parameters remain frozen and the red parameters are trained using the noise ν_2. Finally, in the third step, only the green parameters are trained, using the noise ν_3. This is more formally described in Algorithm 1. (Color figur online)

Putting these ideas together, cascaded training works as follows. We assume that the noise levels are ordered so that $\nu_1 > \nu_2 > \cdots > \nu_S$. Then the first step is that we train *all* of the parameters $\mathbf{W}_1 \ldots \mathbf{W}_S, \mathbf{b}_1, \ldots \mathbf{b}_S, \mathbf{b}'$, but using only the noise level ν_1 to corrupt all S copies $\tilde{\mathbf{x}}_1 \ldots \tilde{\mathbf{x}}_S$ of the input. Once this is done,

---

Algorithm 1. Training the composite denoising autoencoder

---

  for R in $1, \ldots, S$ **do**
    for K_R steps **do**
      Randomly choose a training input $\mathbf{x}$
      Sample $\tilde{\mathbf{x}}_s \sim p(\cdot|\mathbf{x}, \nu_s)$ for $s \in \{1, 2, \ldots, R-1\}$
      Sample $\tilde{\mathbf{x}}_s \sim p(\cdot|\mathbf{x}, \nu_R)$ for $s \in \{R, R+1, \ldots, S\}$
      Compute $\mathbf{y}_s$ for all s as in (5)
      Compute reconstruction $\mathbf{z}$ as in (6)
      Take a gradient step

$$\mathbf{W}_s \leftarrow \mathbf{W}_s - \alpha \nabla_{\mathbf{W}_s} L(\mathbf{z}, \mathbf{x})$$
$$\mathbf{b}_s \leftarrow \mathbf{b}_s - \alpha \nabla_{\mathbf{b}_s} L(\mathbf{z}, \mathbf{x})$$
$$\mathbf{b}' \leftarrow \mathbf{b}' - \alpha \nabla_{\mathbf{b}'} L(\mathbf{z}, \mathbf{x})$$

      for $s \in \{R, R+1, \ldots S\}$
    end for
  end for

---

we freeze the weights $\mathbf{W}_1, \mathbf{b}_1$ and we do not alter them again during training. Then we train the weights $\mathbf{W}_2 \ldots \mathbf{W}_S, \mathbf{b}_2 \ldots \mathbf{b}_S, \mathbf{b}'$, where the corrupted input $\tilde{\mathbf{x}}_1$ is as before corrupted with noise ν_1, and the $S-1$ corrupted copies $\tilde{\mathbf{x}}_2 \ldots \tilde{\mathbf{x}}_S$ are all corrupted with noise ν_2. We repeat this process until at the end we are training the weights $\mathbf{W}_S, \mathbf{b}_S, \mathbf{b}'$ using the noise level ν_S. This process is illustrated graphically in Fig. 2 and in pseudocode in Algorithm 1. To keep the exposition simple, this algorithm assumes that we employ SGD with only one training example per update, although in practice we use mini-batches.

The composite denoising autoencoder builds on several ideas and intuitions. Firstly, our training procedure can be considered an application of the idea of curriculum learning [4, 14]. That is, we start by training all units with high noise level, which serves as a form of *unsupervised pretraining* for the units that will be trained later with lower levels of noise, giving them a good starting point for further optimisation. We experimentally show that the training of a denoising autoencoder learning with data corrupted with high noise levels needs less training epochs to converge, therefore, it can be considered an *easier* problem. This is shown in Fig. 3. Secondly, we are inspired by multi-column neural networks (e.g. Ciresan et al. [7]), which achieve excellent performance for supervised problems. Finally, our work is similar in motivation to scheduled denoising autoencoders [10], which learn a diverse set of features thanks to the training procedure which involves using a sequence of levels of noise. Composite denoising autoencoders achieve this goal more explicitly thanks to their training objective.

3.2   Recovering the Standard Denoising Autoencoder

If, for every training example, the corrupted inputs $\tilde{\mathbf{x}}_{\nu_i}$ were always identical, $[\mathbf{W}_1, \ldots, \mathbf{W}_S]$ were initialised randomly from the same distribution as $\mathbf{W}$ in the standard denoising autoencoder, $\mathbf{b}_i$ and $\mathbf{b}'$ were initalised to $\mathbf{0}$ and $\mathbf{V}_i$

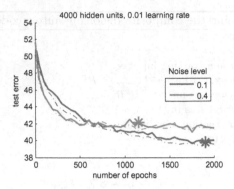

Fig. 3. Classification results with the CIFAR-10 data set yielded by representations learnt with standard denoising autoencoders and data corrupted with two different noise levels. Dashed lines indicate the errors on the validation set. The stars indicate the test errors for the epochs at which the validation errors had its lowest value. The DA trained with high noise level learns faster at the beginning but stops to improve earlier. See Sect. 4 for the details of the experimental setup.

were constrained to be $\mathbf{V}_i = \mathbf{W}_i^T$, then this model is exactly equivalent to the standard denoising autoencoder described in Sect. 2. Therefore, it is natural to incrementally corrupt the training examples shown to the composite denoising autoencoders in such a way that when all the noise levels are the same, this equivalency holds. For example, when working with masking noise, consider two noise levels ν_i and ν_j such that $\nu_i > \nu_j$. Denote the random variables indicating the presence of corruption of a pixel in a training datum by C_{ν_i} and C_{ν_j}. Assuming $C_{\nu_j} \sim \text{Bernoulli}(\nu_j)$, we want $C_{\nu_i} \sim \text{Bernoulli}(\nu_i)$, such that when $C_{\nu_j} = 1$ then also $C_{\nu_i} = 1$. It can be easily shown that this is satisfied when $C_{\nu_i} = \max(C_{\nu_j} + C_{\nu_j \to \nu_i}, 1)$, where $C_{\nu_j \to \nu_i} \sim \text{Bernoulli}(\frac{\nu_i - \nu_j}{1 - \nu_j})$. We use this incremental noising procedure in all our experiments.

4   Experiments

We used two image recognition data sets to evaluate the CDA, the CIFAR-10 data set [19] and a variant of the NORB data set [21]. To evaluate the quality of the learnt representations, we employ a procedure similar to that used by Coates et al. [8] and by many other works[1]. That is, we first learn the representation in an unsupervised fashion and then use the learnt representation within a linear classifier as a measure of its quality. For both data sets, in the unsupervised feature learning stage, we use masking noise as the corruption process, a sigmoid encoder and decoder and cross entropy loss (Eq. 2) following Vincent et al. [28, 29]. To do optimisation, we use stochastic gradient descent with mini-batches.

---

[1] We do not use any form of pooling, keeping our setup invariant to the permutation of the features.

For the classification step, we use $L2$-regularised logistic regression with the regularisation parameter chosen to minimise the validation error. Additionally, with the CIFAR-10 data set, we also trained a single-layer supervised neural network using the parameters of the encoder we learnt in the unsupervised stage as an initialisation. When conducting our experiments, we first find the best hyperparameters using the validation set, then merge it with the training set, retrain the model with the hyperparameters found in the previous step and report the error achieved with this model.

We implemented all neural network models using Theano [6] and we used logistic regression implemented by Fan et al. [9]. We followed the advice of Glorot and Bengio [11] on random initialisation of the parameters of our networks.

4.1   CIFAR-10

This data set consists of 60000 colour images spread evenly between ten classes. There are 50000 training and validation images and 10000 test images. Each image has a size of 32×32 pixels and each pixel has three colour channels, which are represented with a number in $\{0, \ldots, 255\}$. We divide the training and validation set into 40000 training instances and 10000 validation instances. The only preprocessing step we use is dividing the intensity of every pixel by 255 to get numbers in $[0, 1]$.

In our experiments with this data set we trained autoencoders with the total number of 2000 hidden units (undercomplete representation) and 4000 hidden units (overcomplete representation).

Training the Baselines. The simplest possible baseline, logistic regression trained with raw pixel values, achieved 59.4 % test error. To get the best possible baseline denoising autoencoder we explored combinations of different learning rates, noise levels and numbers of training epochs. For 2000 hidden units we considered $\nu \in \{0.05, 0.1, 0.2, 0.3, 0.4, 0.5\}$ and for 4000 hidden units we also additionally considered $\nu = 0.15$. For both sizes of the hidden layers we tried learning rates $\in \{0.01, 0.02, 0.04\}$. Each model was trained for up to 2000 training epochs and we measured the validation error every 50 epochs. The best baselines we got achieved the test errors of 40.71 % (2000 hidden units) and 38.35 % (4000 hidden units).

Concatenating Representations Learnt Independently. To demonstrate that diversity in noise levels improves the representation, we evaluate representations yielded by concatenating the representations from two different DAs, trained independently. We will combine DAs trained with noise levels $\nu \in \{0.1, 0.2, \ldots, 0.5\}$ for each noise level training three DAs with different random seeds. Denote parameters learnt by a DA with the noise level ν and using the random seed R by $\left(\mathbf{W}^{(R,\nu)}, \mathbf{b}^{(R,\nu)}, \mathbf{b}'^{(R,\nu)} \right)$ and denote by E_{ij}^{kl} the classification error on the test set yielded by the concatenating the representations of

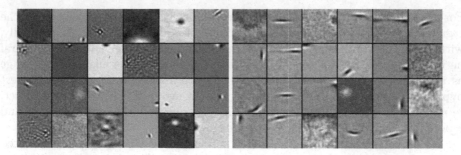

Fig. 4. Example filters (columns of the matrix **W**) learnt by standard denoising autoencoders with $\nu = 0.1$ (left) and $\nu = 0.5$ (right).

two independently trained DAs, the first trained with random seed R_k and noise level ν_i, and the second trained by random seed R_l and noise level ν_j. For each pair of noise levels (ν_i, ν_j), we measure the average error across random seeds, that is, $\bar{E}_{ij} = \frac{1}{2\binom{K}{2}} \left(\sum_{k \neq l} E_{ij}^{kl} + E_{ji}^{kl} \right)$. The results of this experiment are shown in Fig. 5. For every ν we used, it was optimal to concatenate the representation learnt with ν with a representation learnt with a different noise level. To understand this intuition, we visually examine features from DAs with different noise levels (Fig. 4). From this figure it can be seen that features at higher noise levels depend on larger regions of the image. This demonstrates the benefit of using a more diverse representation for classification.

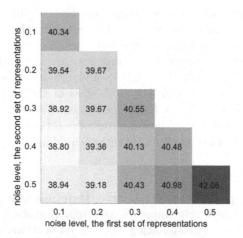

Fig. 5. Classification errors for representations constructed by concatenating representations learnt independently.

Comparison of CDA to DA. The CDA offers freedom to choose the number of noise levels, the value ν_s for each noise level, and the number D_s of hidden units at each noise level.

For computational reasons, we limit the space of possible combinations of hyperparameters in the following manner (of course, expanding the search space would only make our results better). We considered models containing up to four different noise levels. We first consider only the models with two noise levels and hidden units divided equally between them. For 2000 total hidden units, we consider all possible pairs of noise levels drawn from the set $\{0.5, 0.4, 0.3, 0.2, 0.1, 0.05\}$. Once we have found the value of ν that minimizes that validation error for $D_1 = D_2$, we try splitting hidden units such that the ratio $D_1 : D_2 = 1 : 3$ or $D_1 : D_2 = 3 : 1$. Similarly, for four noise levels, we consider the following sets of noise levels $\nu \in \{(0.5, 0.4, 0.3, 0.2), (0.4, 0.3, 0.2, 0.1), (0.3, 0.2, 0.1, 0.05)\}$. We select the value of ν that has lowest validation error for an equal split $D_1 = \cdots = D_4$, and then try splitting the hidden units with different ratios: $D_1 : D_2 : D_3 : D_4 = 3 : 1 : 1 : 1$, $D_1 : D_2 : D_3 : D_4 = 9 : 1 : 1 : 1$ and the permutations of these ratios. As for the learning rate, we train each of the cascaded DAs with the learning rate that had the best validation error for the first noise level ν_1. The models were trained for up to 500 epochs at each consecutive noise level and we computed the validation error every 50 training epochs. Note that when training with four noise levels, it is possible that the lowest validation error occurs before the training procedure has moved on to the final noise level. In this circumstance, it is possible that the final model will have only two or three noise levels instead of four.

We trained the models with 4000 hidden units the same way, except that we used different sets of noise levels for this higher number of hidden units. This is because our experience with the baseline DAs was that units with 4000 hidden units do better with lower noise levels. For the CDAs with four noise levels, we compared three difference choices for ν: $(0.4, 0.3, 0.2, 0.1)$, $(0.3, 0.2, 0.1, 0.05)$, and $(0.2, 0.15, 0.1, 0.05)$. For the models with two noise levels the values were drawn from $\{0.4, 0.3, 0.2, 0.15, 0.1, 0.05\}$.

For either number of hidden units, we find that CDAs perform better than simple DAs. The best models with 2000 hidden units and 4000 hidden units we found achieved the test errors of 38.86 % and 37.53 % respectively, thus yielding a significant improvement over the representations trained with a standard DA. These results are compared to the baselines in Table 1. It is also noteworthy that a CDA performs better than concatenating two indepedently trained DAs with different noise levels (cf. Fig. 5).

Table 1. Classification errors of standard denoising autoencoders and composite denoising autoencoders.

Hidden units	Best DA	Test error	Best CDA	Test error
2000	$\nu = 0.2$	40.71 %	$\nu = (0.3, 0.2, 0.1)$, $\mathbf{D} = (500, 500, 1000)$	38.86 %
4000	$\nu = 0.1$	38.35 %	$\nu = (0.3, 0.05)$, $\mathbf{D} = (1000, 3000)$	37.53 %

Comparison of Optimization Methods. One could consider several simpler alternatives to the cascaded training procedure from Sect. 3.1. The simplest alternative, which we call *joint SGD*, is to train all of the model parameters jointly, at every iteration sampling each corrupted input $\tilde{\mathbf{x}}_s$ using its corresponding noise level ν_s. This is simply SGD on the objective (7). A second alternative, which we call *alternating SGD*, is block coordinate descent on (7), where we assign each weight matrix $\mathbf{W}_s$ to a separate block. In other words, at each iteration we choose a different parameter block $\mathbf{W}_s$, and take a gradient update only on $\mathbf{W}_s$ (note that this requires computing a corrupted input $\tilde{\mathbf{x}}_s$ for all noise levels ν_s). Neither of these simpler methods try to prevent undertraining of the parameters for the high noise levels in the way that cascaded training does.

Figure 6 shows a comparison of joint SGD, alternating SGD, and our cascaded SGD methods on a CDA with four noise levels $\nu = (0.4, 0.3, 0.2, 0.1)$ and $\mathbf{D} = (500, 500, 500, 500)$. We ran both joint SGD and cascaded SGD until they converged in validation error, and then we ran alternating SGD until it had made the same number of parameter updates as joint SGD. This means that alternating SGD was run for four times as many iterations as joint SGD, because alternating SGD only updates one-quarter of the parameters at each iteration. Cascaded SGD was stopped early when it converged according to validation error. The vertical dashed lines in the figure indicate the epochs at which alternating SGD switched between parameter blocks.

From these results, it is clear that the cascaded training procedure is significantly more effective than either joint or alternating SGD. Joint SGD seems to have converged to much worse parameters than cascaded SGD. We hypothesize that this is because the parameters corresponding to the high noise levels are undertrained. To verify this, in Fig. 7 we show the features learned by a composite CDA with joint training at two different noise levels. Note that at the higher noise level (at right) there are many filters that are mostly noise; this is not observed at the lower noise or to the same extent in an standard DA. Alternating SGD seems to converge fairly slowly. It is possible that its error would continue to decrease, but even after 8000 iterations its solution is still much worse than that found by cascading SGD after only 3500 iterations.

We have made similar comparisons for other choices of ν and found a similar difference in performance between joint, alternating, and cascaded SGD. One exception to this is that alternating SGD seems to work much better on models with only two noise levels ($S = 2$) than those with four noise levels. In those situations, the performance of alternating SGD often equals, but usually does not exceed, that of cascaded SGD.

Fine-Tuning. We also trained a supervised single-layer neural network using parameters of the encoder as the initialisation of the parameters of the hidden layer of the network. This procedure is known as fine-tuning. We did that for the best standard DAs and CDAs with 4000 hidden units. The learning rate, the same for all parameters, was chosen from the set $\{0.00125, 0.00125 \cdot 2^{-1}, \ldots, 0.00125 \cdot 2^{-4}\}$ and the maximum number of training epochs was 2000

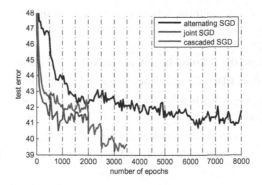

Fig. 6. Classification errors achieved by three different methods of optimising the objective in (7).

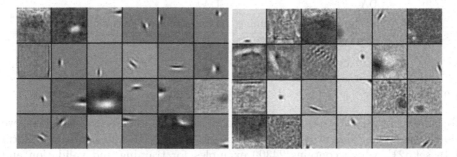

Fig. 7. Example filters (columns of the matrix **W**) learnt by composite denoising autoencoders with $\nu = 0.1$ (left) and $\nu = 0.4$ (right) when all the parameters were optimise using joint SGD. While the filters associated with $\nu_2 = 0.1$ have managed to learn interesting features, many of these associated with $\nu_1 = 0.4$ remained undertrained. These hard to interpret filters are much more rare with cascaded SGD.

(we computed the validation error after each epoch). We report the test error for the combination of the learning rate and the number of epochs yielding the lowest validation error. The results are shown in Table 3. Fine-tuning makes the performance of DA and CDA much more similar, which is to be expected since the fine-tuning procedure is identical for both models. However, note that the result achieved with a standard denoising autoencoder and supervised fine-tuning we present here is an extremely well tuned one. In fact, its error is lower than any previous result achieved by a permutation-invariant method on the CIFAR-10 data set. Our best model, yielding the error of 35.06 % is, by a considerable margin, more accurate than any previously considered permutation-invariant model for this task, outperforming a variety of methods. A summary of the best results reported in the literature is shown in Table 2.

Table 2. Summary of the results on CIFAR-10 among permutation-invariant methods.

Model	Test error
Composite Denoising Autoencoder	**35.06 %**
Scheduled Denoising Autoencoder [10]	35.7 %
Zero-bias Autoencoder [22]	35.9 %
Fastfood FFT [20]	36.9 %
Nonparametrically Guided Autoencoder [25]	43.25 %
Deep Sparse Rectifier Neural Network [12]	49.52 %

Table 3. Test errors on CIFAR-10 data set for the best DA and CDA models trained without supervised fine-tuning and their fine-tuned versions.

DA		CDA	
no fine-tuning	fine-tuning	no fine-tuning	fine-tuning
38.35 %	35.30 %	37.53 %	35.06 %

4.2 NORB

To show that the advantage of our model is consistent across data sets, we did the same experiment use a variant of the small NORB normalized-uniform data set [21], which contains 24300 examples for training and validation and 24300 test examples. It contains images of 50 toys belonging to five generic categories: animals, human figures, airplanes, trucks, and cars. The 50 toys are evenly divided between the training and validation set and the test set. The objects were photographed by two cameras under different lighting conditions, elevations and azimuths. Every example consists of a stereo pair of grayscale images, each of size 96×96 pixels whose intensities are represented as a number $\in \{0, \ldots, 255\}$. We transform the data set by taking the middle 64×64 pixels from both images in a pair and dividing the intensity of every pixel by 255 to get numbers in $[0, 1]$. The simplest baseline, logistic regression using raw pixels, achieved the test error of 42.32 %.

In the experiments with learning the representations with this data set we used the hidden layer with 1000 hidden units and adapted the set up we used for CIFAR-10. To find the best possible standard DA we considered all combinations of the noise levels $\in \{0.1, 0.2, 0.3, 0.4\}$ and the learning rates $\in \{0.005, 0.01, 0.02\}$. The representation learnt by the best denoising autoencoder yielded 18.75 % test error when used with logistic regression. By contrast, a composite denoising autoencoder with $\nu = (0.4, 0.3, 0.2, 0.1)$ and $\mathbf{D} = (250, 250, 250, 250)$ results in a representation that yields a test error of 17.03 %.

5   Discussion

We introduced a new unsupervised representation learning method, called a composite denoising autoencoder, by modifying the standard DA so that different parts of the network were exposed to corruptions of the input at different noise levels. Naive training procedures for the CDA can get stuck in bad local optima, so we designed a cascaded training procedure to avoid this. We showed that CDAs learned more effective representations than DAs on two different image data sets.

A few pieces of prior work have considered related techniques. In the context of RBMs, the benefits of learning a diverse representation was also noticed by Tang and Mohamed [26], achieving diversity by manipulating the resolution of the image. Also, ensembles of denoising autoencoders, where each member of the ensemble is trained with a different level or different type of noise, have been considered by Agostinelli et al. [1]. This work differs from ours because in their method all DAs in the ensemble are trained independently, whereas we show that training the different representations together is better than independent training. The cascaded training procedure has some similarities in spirit to the incremental training procedure of Zhou et al. [31], but that work considered only DAs with one level of noise. Usefulness of varying the level of noise during training of neural nets was also noticed by Gulcehre et al. [15], who add noise to the activation functions. Our training procedure also resembles the walkback training suggested by Bengio et al. [5], however, we do not require our training loss to be interpretable as negative log-likelihood. Understanding the relative merits of walkback training, scheduled denoising autoencoders and composite denoising autoencoders would be an interesting future challenge.

References

1. Agostinelli, F., Anderson, M.R., Lee, H.: Robust image denoising with multi-column deep neural networks. In: NIPS (2013)
2. Baldi, P., Hornik, K.: Neural networks and principal component analysis: learning from examples without local minima. Neural Netw. **2**, 53–58 (1989)
3. Bengio, Y., Lamblin, P., Popovici, D., Larochelle, H.: Greedy layer-wise training of deep networks. In: NIPS (2007)
4. Bengio, Y., Louradour, J., Collobert, R., Weston, J.: Curriculum learning. In: ICML (2009)
5. Bengio, Y., Yao, L., Alain, G., Vincent, P.: Generalized denoising auto-encoders as generative models. In: NIPS (2013)
6. Bergstra, J., Breuleux, O., Bastien, F., Lamblin, P., Pascanu, R., Desjardins, G., Turian, J., Warde-Farley, D., Bengio, Y.: Theano: a CPU and GPU math expression compiler. In: SciPy (2010)
7. Ciresan, D., Meier, U., Schmidhuber, J.: Multi-column deep neural networks for image classification. In: CVPR (2012)
8. Coates, A., Ng, A.Y., Lee, H.: An analysis of single-layer networks in unsupervised feature learning. In: AISTATS (2011)

9. Fan, R.E., Chang, K.W., Hsieh, C.J., Wang, X.R., Lin, C.J.: LIBLINEAR: a library for large linear classification. JMLR **9**, 1871–1874 (2008)
10. Geras, K.J., Sutton, C.: Scheduled denoising autoencoder. In: ICLR (2015)
11. Glorot, X., Bengio, Y.: Understanding the difficulty of training deep feedforward neural networks. In: AISTATS (2010)
12. Glorot, X., Bordes, A., Bengio, Y.: Deep sparse rectifier networks. In: AISTATS (2011)
13. Glorot, X., Bordes, A., Bengio, Y.: Domain adaptation for large-scale sentiment classification: a deep learning approach. In: ICML (2011)
14. Gulcehre, C., Bengio, Y.: Knowledge matters: Importance of prior information for optimization. JMLR **17**, 1–32 (2016)
15. Gulcehre, C., Moczulski, M., Denil, M., Bengio, Y.: Noisy activation functions (2016). arXiv:1603.00391
16. Hinton, G.E., Osindero, S., Teh, Y.W.: A fast learning algorithm for deep belief nets. Neural Comput. **18**(7), 1527–1554 (2006)
17. Hyvärinen, A.: Estimation of non-normalized statistical models by score matching. J. Mach. Learn. Res. **6**, 695–708 (2005)
18. Karklin, Y., Simoncelli, E.P.: Efficient coding of natural images with a population of noisy linear-nonlinear neurons. In: NIPS (2011)
19. Krizhevsky, A.: Learning multiple layers of features from tiny images. University of Toronto, Technical report (2009)
20. Le, Q., Sarlós, T., Smola, A.: Fastfood-computing Hilbert space expansions in loglinear time. In: ICML (2013)
21. LeCun, Y., Huang, F.J., Bottou, L.: Learning methods for generic object recognition with invariance to pose and lighting. In: CVPR (2004)
22. Memisevic, R., Konda, K., Krueger, D.: Zero-bias autoencoders and the benefits of co-adapting features. In: ICLR (2015)
23. Mesnil, G., Dauphin, Y., Glorot, X., Rifai, S., Bengio, Y., Goodfellow, I.J., Lavoie, E., Muller, X., Desjardins, G., Warde-Farley, D., Vincent, P., Courville, A.C., Bergstra, J.: Unsupervised and transfer learning challenge: a deep learning approach. In: ICML Unsupervised and Transfer Learning Workshop (2012)
24. Seung, H.S.: Learning continuous attractors in recurrent networks. In: NIPS (1998)
25. Snoek, J., Adams, R.P., Larochelle, H.: Nonparametric guidance of autoencoder representations using label information. JMLR **13**, 2567–2588 (2012)
26. Tang, Y., Mohamed, A.r.: Multiresolution deep belief networks. In: AISTATS (2012)
27. Vincent, P.: A connection between score matching and denoising autoencoders. Neural Comput. **23**, 1661–1674 (2011)
28. Vincent, P., Larochelle, H., Bengio, Y., Manzagol, P.A.: Extracting and composing robust features with denoising autoencoders. In: ICML (2008)
29. Vincent, P., Larochelle, H., Lajoie, I., Bengio, Y., Manzagol, P.A.: Stacked denoising autoencoders: learning useful representations in a deep network with a local denoising criterion. JMLR **11**, 3371–3408 (2010)
30. Xie, J., Xu, L., Chen, E.: Image denoising and inpainting with deep neural networks. In: NIPS (2012)
31. Zhou, G., Sohn, K., Lee, H.: Online incremental feature learning with denoising autoencoders. In: AISTATS (2012)

Trust Hole Identification in Signed Networks

Jiawei Zhang[1(✉)], Qianyi Zhan[2], Lifang He[3],
Charu C. Aggarwal[4], and Philip S. Yu[1,5]

[1] University of Illinois at Chicago, Chicago, IL, USA
jzhan9@uic.edu, psyu@cs.uic.edu
[2] Nanjing University, Nanjing, China
zhanqianyi@gmail.com
[3] Shenzhen Univeristy, Guangdong, China
lifanghescut@gmail.com
[4] IBM T. J. Watson Research Center, Yorktown Heights, NY, USA
charu@us.ibm.com
[5] Institute for Data Science, Tsinghua University, Beijing, China

Abstract. In the trust-centric context of signed networks, the social links among users are associated with specific polarities to denote the attitudes (trust vs distrust) among the users. Different from traditional unsigned social networks, the diffusion of information in signed networks can be affected by the link polarities and users' positions significantly. In this paper, a new concept called *"trust hole"* is introduced to characterize the advantages of specific users' positions in signed networks. To uncover the trust holes, a novel trust hole detection framework named "Social Community based tRust hOLe expLoration" (SCROLL) is proposed in this paper. Framework SCROLL is based on the signed community detection technique. By removing the potential trust hole candidates, SCROLL aims at maximizing the community detection cost drop to identify the optimal set of trust holes. Extensive experiments have been done on real-world signed network datasets to show the effectiveness of SCROLL.

Keywords: Trust hole detection · Signed networks · Data mining

1 Introduction

In traditional works on sociology and social networks, the concept *structural hole* refers to individuals who act as intermediaries or bridges between others who are not directly connected [9]. Via these *structural holes*, information can propagate to separated individuals in different communities, or those who are otherwise not interacting with each other. As a result, the *structural holes* who take these bridging positions in society or social networks will accrue significant advantages than other users [9]. In traditional social networks with regular friendship connections among users, structural holes related problems have been studied for years, and dozens of papers on it have already been published [1,3,5,9]

© Springer International Publishing AG 2016
P. Frasconi et al. (Eds.): ECML PKDD 2016, Part I, LNAI 9851, pp. 697–713, 2016.
DOI: 10.1007/978-3-319-46128-1_44

Meanwhile, in some online social networks like Epinions[1], the connections connected to users are associated with specific polarities (e.g., *positive* vs *negative*) to denote different attitudes among users (e.g., *trust* vs *distrust*). Such a kind of online social networks are formally represented as the *signed networks* [11]. Different from traditional regular unsigned social networks, in the trust-centric context of signed networks, diffusion of information can be affected by the link polarities significantly. For instance, in signed networks, information tends to propagate via the trust links between users who trust each other instead of those distrusted ones. Viewed in this way, users who bridge different cliques via distrust links actually cannot transmit information across these cliques. Therefore, the traditional *structural holes* (i.e., the inter-community users in unsigned networks) concept [9] can no longer work for the signed networks.

To characterize the advantages of specific users' positions in signed networks, a new concept named *trust holes* is introduced in this paper. Depending on the polarities of links attached to them, the *trust hole* concept has two variants: (1) *positive trust holes* who connect multiple isolated social communities via positive links, and (2) *negative trust holes* who connect users within communities via negative links instead. Via the positive trust holes, information can propagate between different social communities, as people will trust information propagated from these hole users. Meanwhile, via the negative trust holes, the intra-community information dissemination will be blocked instead, as few of the neighbors will believe the information from the people they distrust. Therefore, the positions of both *positive trust holes* and *negative trust holes* will have great advantages in passing information among users in signed networks. The formal definition of the *trust hole* concept is available in Sect. 2. Specifically, the inter-community nodes attached with negative links and the intra-community nodes attached with positive links are not *trust holes*, as their position own no advantages in propagating information in signed networks. We will clarify that in detail in Sect. 2.

Problem Studied: In this paper, we aim at identifying the *trust holes* from the signed networks, and the problem is referred to as the "Signed network trust HolE iDentification" (SHED) problem formally.

The SHED problem is an interesting research problem, and it is also very important for many concrete applications, e.g., community structure [6,10], and information diffusion [13,14] (existence of the holes can help disseminate the information more broadly) in signed networks. In addition, the SHED problem is a novel problem and we are the first to study it in signed networks. Different from existing works about structural holes in unsigned networks [3,9], the networks studied in this paper are *signed networks*, and the target to be identified are the *trust holes* instead. For more information about related works, please refer to Sect. 5.

---

[1] http://www.epinions.com.

The SHED problem is very challenging to solve due to the following reasons:

- *definition of trust hole*: The *trust hole* proposed in this paper is a new concept. A formal definition of the *trust hole* concept is needed before studying the SHED problem.
- *formulation of the* SHED *problem*: In the signed network setting, how to formulate the SHED problem with clear motivations and objectives is still an open problem.
- *solution to the* SHED *problem*: The SHED itself is a difficult problem. Some trivial methods, like isolated trust hole identification in the positive sub-graph (and negative sub-graph) along, will face great challenges in both obtaining the positive and negative trust holes independently and fusing the trust hole results from these two sub-graphs to get the final consistent results. An integrated *trust hole* detection framework based on the whole signed network is desired.

To address the above challenges, an integrated trust hole detection framework named SCROLL (Social Community based tRust hOLe expLoration) is proposed in this paper. Before introducing SCROLL, we will first define the *trust hole* concept with full considerations about the link polarities in the signed networks. SCROLL formulates the SHED problem from the community detection perspective. By removing the potential trust hole users, SCROLL aims at maximizing the community detection cost drop to identify the optimal set of trust hole candidates, which maps the SHED to a max-min optimization problem. A new concept named "*signed normalized cut decrease*" is proposed in SCROLL to quantify the cost drop formally based on the *signed normalized cut* measure introduced in this paper. The SHED problem is shown to be NP-hard, but based on such a formulation, SCROLL can solve the SHED approximately with an alternative updating schema based schema.

The following parts of this paper are organized as follows. Terminology definition and problem formulation are given in Sect. 2. The method is introduced in Sect. 3, which is evaluated in Sect. 4. Finally, Sect. 5 is about the related works and Sect. 6 concludes this paper.

2 Problem Formulation

2.1 Terminology Definition

The networks studied in this paper are *signed networks*, where links are associated with different polarities.

Definition 1 (Signed Network): A signed network can be represented as $G = (\mathcal{V}, \mathcal{E}, s)$, where $\mathcal{V}$ ($|\mathcal{V}| = n$) and $\mathcal{E}$ ($|\mathcal{E}| = m$) are the sets of users and links respectively. Sign mapping $s : \mathcal{E} \rightarrow \{+1, -1\}$ projects links to their different polarities, where polarities $+1$ and -1 denote that the links are the trust and distrust links respectively.

Users in regular social networks will form social communities based on the connections among them, where intra-community connections are more dense compared with those between different communities [17]. The *social communities* formed in signed networks can be different due to the polarities attached to links.

Definition 2 (Signed Social Community): Given a signed network G, we can represent the communities formed by users in G as $C = \{C_1, C_2, \cdots, C_k\}$, where k is the community number, $C_i \subseteq V, \forall i \in \{1, 2, \cdots, k\}$ and $\bigcup_{i=1}^{k} C_i = V$. Generally speaking, in the trust-centric context, users connected by positive links tend to trust each other and will be grouped in the same community. Meanwhile, for those connected by distrust links, they will have very few social interactions and will be partitioned into different communities.

However, in the real scenario, the signed social communities formed by people cannot fit the definitions exactly, and there may still exist a large number of inter-community positive links and intra-community negative links. In such a case, the positions of individuals connecting different communities via positive links, as well as those connecting individuals within the same community via negative links will have significant advantages in information dissemination (as introduced in Sect. 1).

Definition 3 (Signed Trust Hole): Given a signed network G (with signed social community C), literally, the *signed trust holes* in G denote a subset of users in G (i.e., $\mathcal{H} \subset V$) occupying positions of the largest advantages. More specifically, the signed trust holes in $\mathcal{H}$ are the users either (1) connecting different communities via positive links (connected with users in them), who are referred to as the *positive trust holes*; or (2) connecting users within the same community with negative links, who are called the *negative trust holes* respectively.

To help illustrate this concept more clearly, we also give an example in Fig. 1, where the colored regions denote different communities in the network. In plots A and B, we show the signed networks, where the links are associated with different polarities (i.e., positive vs negative). In the signed networks, nodes bridging different groups are not necessarily the trust holes. For instance, in plot A, the "*Green*" nodes which connecting different groups via positive links is defined as the positive trust hole, while the "*Red*" node bridging groups with negative links is not, as information will not propagate between groups via him/her. Besides the inter-community nodes, in signed networks, the intra-community nodes can also be the *trust holes*. For instance, in plot B, we observe that, in the "*Blue*" group, the central node connects to other nodes via both positive and negative links, which will partially block the dissemination of information within the group as some of his neighbors distrust information from him/her (he/she is still in the group as some others tend to trust him). The remaining intra-community nodes are not *trust holes* on the other hand. As a result, the positions of both the "*Green*" node in plot A and the central "*Blue*" node in plot B have significant advantages, which are called the *positive* and *negative trust holes* respectively.

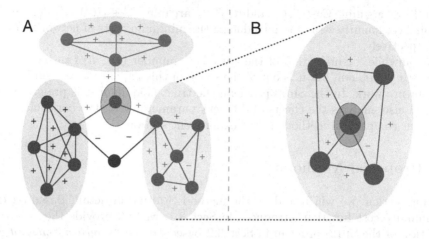

Fig. 1. Trust Holes vs Structural Hole. (A: positive trust hole, B: negative trust hole) (Color figure online)

2.2 Problem Statement

In this paper, we aim at identifying the set of *trust holes* from signed networks. Let $G = (\mathcal{V}, \mathcal{E}, s)$ be the signed network and $\mathcal{C}$ be the community structures detected from G. Various cost functions can be utilized to measure the quality of the detected community structure $\mathcal{C}$, which can be denoted as $\text{cost}(\mathcal{C}, G)$. Meanwhile, let $G' = (\mathcal{V}', \mathcal{E}')$ be the network obtained after removing the *positive* and *negative trust holes*, where $\mathcal{V}' = \mathcal{V} \setminus \mathcal{H}$ and $\mathcal{E}' = \{(u, v)|(u, v) \in \mathcal{E}, u \notin \mathcal{H}, v \notin \mathcal{H}\}$, and $\mathcal{C}'$ be the new community structures of G', which will lead to cost $\text{cost}(\mathcal{C}', G')$.

According to the definition of *trust holes*, the existence of *positive/negative trust hole* will not only influence the dissemination of information, but also blurring the network community structure. Removal of the *trust holes* from network G will also delete the inter-community positive links and intra-community negative links attached to them, and better community structures can be identified from G.

Therefore, in this paper, we propose to formulate the SHED problem from the community detection perspective. The optimal *trust holes* set $\mathcal{H}$ of size h can be identified by removing potential trust hole candidates from the network. The users removal of whom introducing the maximum community detection cost drop will be the optimal result. Formally, the objective function of the SHED problem can be represented as

$$\max_{\mathcal{H}} \text{Cost}(\mathcal{C}^*, G) - \text{Cost}((\mathcal{C}')^*, G')$$
$$s.t. \quad |\mathcal{H}| = h,$$

where $\text{Cost}(\cdot)$ denote the costs introduced by the community structure in the network, and its concrete representations will be introduced in the following sections.

And $\mathcal{C}^* = \arg\min_\mathcal{C} \mathrm{Cost}(\mathcal{C}, G)$ and $(\mathcal{C}')^* = \arg\min_{\mathcal{C}'} \mathrm{Cost}((\mathcal{C}'), G')$ denote the optimal community structure introducing the minimum costs in networks G and G' respectively.

Meanwhile, identification of the trust hole number (i.e., h) can be another interesting problem, but it is out of the scope of this paper, and we will leave it as a future work. In the SHED problem, the trust hole number is pre-given but we will also analysis the effects of different parameter h on the performance of different comparison methods in the experiment section.

3   Proposed Methods

In this section, we will introduce the method SCROLL in detail. Based on the community cost function introduced in Sect. 3.1, we will provide the objective function of the SHED problem in Sect. 3.2 based on the "*signed normalized cut decrease*" concept. With some simple analysis, the SHED problem is shown to be NP-hard. An alternative updating schema based solution will be applied to address the objective function in Sect. 3.3.

3.1   Signed Normalized Cut Cost Function

Any community quality measures, e.g., *entropy, normalized dbi*, can be applied to define the community cost function. In this paper, we propose to use the *normalized cut* [20]. In this part, we will first do a quick review about the *normalized cut* measure for unsigned networks. Next, we propose to extend it to the *signed network* setting with full considerations of the constraints introduced by link polarities.

Traditional Normalized Cut Measure for Unsigned Networks. Given a traditional unsigned social network G^u, based on the connection among users in which, we can define its adjacency matrix as $\mathbf{A}^u$. Its corresponding Laplace matrix can be represented as $\mathbf{L}^u = \mathrm{Diag}(\mathbf{A}^u) - \mathbf{A}^u$, where $\mathrm{Diag}(\mathbf{A}^u)$ denotes the corresponding diagonal matrix of $\mathbf{A}$ and $\mathrm{Diag}(A^u)(i,i) = \sum_j A(i,j)$. Meanwhile, given the social structure $\mathcal{C}^u = \{C_1^u, C_2^u, \cdots, C_k^u\}$, we can define the corresponding indicator matrix as $\mathbf{X} = (\mathbf{x}_1, \mathbf{x}_2, \cdots, \mathbf{x}_k)$, where $\mathbf{x}_j = (x_{1,j}, x_{2,j}, \cdots, x_{n,j})^\top$, and entry $\mathbf{x}_j(i) = x_{i,j}$ denotes whether user u_i is in cluster C_j^u or not. Traditional unsigned *normalized cut* cost function [16] is defined to be

$$\mathrm{Ncut}(\mathcal{C}^u, G^u) = \sum_{i=1}^{i=k} \mathbf{x}_i^\top \mathbf{L}^u \mathbf{x}_i = \mathrm{Tr}(\mathbf{X}^\top \mathbf{L}^u \mathbf{X}),$$

where $\mathrm{Tr}(\cdot)$ denotes the trace of a matrix, and $\mathbf{X}$ is subject to constraint $\mathbf{X}^\top \mathbf{X} = \mathbf{I}$.

Extended Signed Normalized Cut Measure for Signed Networks. However, in signed networks studied in this paper, the polarities associated to links can post extra constraints [22] on the community structures:

- *constraint of positive links*: From the trust-centric point of view, trust links (i.e., positive links) are stronger indicators of the closeness among users in signed networks. Generally, users who trust each other are more likely to share information and can be in the same community.
- *constraint of negative links*: Meanwhile, on the other hand, distrust links (i.e., negative links) can show the negative attitudes among users in signed networks. Users who distrust each other tend to have less social interactions, and will stay in different communities.

To handle the constraints introduced by the polarities of these signed links, in this paper, we propose to extend the traditional *normalized cut* concept to the trust-centric signed networks. Based on the positive links in network signed network G, we propose to construct the positive Laplace matrix $\mathbf{L}^+$. The cost introduced by detected communities $\mathcal{C}$ in cutting positive links can be represented as the *positive normalized cut* cost function:

$$\mathrm{Ncut}(\mathcal{C}, G)^+ = \sum_{i=1}^{i=k} \mathbf{x}_i^\top \mathbf{L}^+ \mathbf{x}_i = \mathrm{Tr}(\mathbf{X}^\top \mathbf{L}^+ \mathbf{X}).$$

Meanwhile based on the negative links in signed network G, we can construct the negative Laplace matrix $\mathbf{L}^-$. The cost introduced by detected communities $\mathcal{C}$ in cutting the negative links can be represented as the following *negative normalized cut* cost function:

$$\mathrm{Ncut}(\mathcal{C}, G)^- = \sum_{i=1}^{i=k} \mathbf{x}_i^\top \mathbf{L}^- \mathbf{x}_i = \mathrm{Tr}(\mathbf{X}^\top \mathbf{L}^- \mathbf{X}).$$

By considering the polarities of links in signed networks, in the trust-centric context, the optimal community structure should cut the minimum positive links but the maximum negative links. Viewed in this way, we introduce the *signed normalized cut* cost function for network G to be

$$\begin{aligned}
\mathrm{Ncut}(\mathcal{C}, G) &= \alpha \cdot \mathrm{Ncut}(\mathcal{C}, G)^+ - (1 - \alpha) \cdot \mathrm{Ncut}(\mathcal{C}, G)^- \\
&= \alpha \cdot \mathrm{Tr}(\mathbf{X}^\top \mathbf{L}^+ \mathbf{X}) - (1 - \alpha) \cdot \mathrm{Tr}(\mathbf{X}^\top \mathbf{L}^- \mathbf{X}) \\
&= \mathrm{Tr}\left(\mathbf{X}^\top (\alpha \cdot \mathbf{L}^+ - (1 - \alpha) \cdot \mathbf{L}^-) \mathbf{X}\right) \\
&= \mathrm{Tr}\left(\mathbf{X}^\top \mathbf{L} \mathbf{X}\right),
\end{aligned}$$

where matrix $\mathbf{L} = (\alpha \cdot \mathbf{L}^+ - (1 - \alpha) \cdot \mathbf{L}^-)$ and α is the weight of the positive normalized cut cost term.

Moreover, the optimal community structure $\mathcal{C}^*$ (i.e., the optimal indicator matrix $\mathbf{X}^*$) which can minimize the *signed normalized cut* cost function can be represented as:

$$X^* = \arg\min_{X} \operatorname{Tr}\left(X^{\top} L X\right),$$

$$s.t. \quad X^{\top} X = I.$$

Constraint $X^{\top} X = I$ ensures the obtained indicator matrix X is orthogonal. The discrete binary value constraint on X is usually relaxed, which can actually take any real values in range $[0, 1]$.

3.2 Objective Function of the SHED Problem

As introduced in Sect. 2, the *trust holes* either connecting different communities via positive links or linking individuals within communities via negative links will make the social community structure of the network hard to distinguish. Therefore, we propose to remove potential *trust hole* candidates together with their attached links from the network. The users removal of whom from the network can lead to the maximum community detection cost drop will be the optimal *trust holes* to be identified in the SHED problem. Based on the *signed normalized cut* cost function introduced in the previous section, we will define the concrete objective function of SHED in this section.

Let G, C^* and G', $(C')^*$ be the networks and their optimal community structures before and after removing the *signed trust holes* $\mathcal{H}$ respectively. Considering that, in the *signed normalized cut* cost function, network information is actually stored in the Laplace matrix, next we will first study how to represent the Laplace matrix of network G' (i.e., L') after removing the *signed trust holes* from the original Laplace matrix of network G (i.e., L).

Let matrix $I \in \{0, 1\}^{|\mathcal{V}| \times |\mathcal{V}|}$ be the identity matrix with 1s on its diagonal only. Given the user set $\mathcal{V}$ and structure hole set $\mathcal{H} = \{u_i, u_j, \cdots, u_m\}$, we define the corresponding transformation matrix $T \in \{0, 1\}^{(|\mathcal{V}| - |\mathcal{H}|) \times |\mathcal{V}|}$ based on I, where the rows corresponding structure holes in $\mathcal{H}$ are all removed. For instance, if u_i is identified as a trust hole, after removing u_i, we can define the corresponding transformation to be $T \in \{0, 1\}^{(|\mathcal{V}| - 1) \times |\mathcal{V}|}$, where rows $T(l, :) = I(l, :), \forall l \in \{1, 2, \cdots, i - 1\}$ and $T(l, :) = I(l + 1, :), \forall l \in \{i, i + 1, \cdots, |\mathcal{V}| - 1\}$. Therefore, given a set of structure holes $\mathcal{H}$, we can define a unique transformation matrix T for it. In this paper, we will misuse matrix T to denote the signed trust hole for simplicity. With transformation matrix T, we can represent the Laplace matrix to be $L' = \operatorname{Diag}(TAT^{\top}) - TAT^{\top}$, where A is the signed adjacency matrix of G weighted by parameter α.

Definition 4 (Signed NCut Decrease): Based on the Laplace matrices L and L' as well as transformation matrix T, we can define the *signed ncut decrease* introduced by matrix T to be

$$\text{NCut-Decrease}(T) = \text{Ncut}(C^*, G) - \text{Ncut}((C')^*, G')$$

$$= \min_{X} \operatorname{Tr}\left(X^{\top} L X\right) - \min_{X'} \operatorname{Tr}\left((X')^{\top} L'(X')\right).$$

Furthermore, the objective function for detecting the optimal *signed trust hole* $\mathcal{H}^*$ can be represented as

$$\mathcal{H}^* = \arg\max_{\mathcal{H}} \text{NCut-Decrease}(\mathbf{T})$$

$$= \arg\max_{\mathbf{T}} \left(\min_{\mathbf{X}} \text{Tr}\left(\mathbf{X}^{\top}\mathbf{L}\mathbf{X}\right) - \min_{\mathbf{X}'} \text{Tr}\left((\mathbf{X}')^{\top}\mathbf{L}'(\mathbf{X}')\right) \right),$$

$$s.t.\ \mathbf{X}^{\top}\mathbf{X} = \mathbf{I}, (\mathbf{X}')^{\top}(\mathbf{X}') = \mathbf{I}, \mathbf{T}\mathbf{T}^{\top} = \mathbf{I},$$

$$|\mathcal{H}| = h, \mathbf{T} \in \{0,1\}^{(|\mathcal{V}|-|\mathcal{H}|) \times |\mathcal{V}|}.$$

Considering that matrix $\mathbf{T}$ is obtained from the identity matrix by removing rows corresponding to the structure hole users, the last constraint is added to ensure each row of $\mathbf{T}$ should contain only one entry with value 1, while the remaining entries are all 0s.

3.3   Solution to the SHED Problem

In this section, we will first analyze the objective function of the SHED problem first, and after that we will introduce an approximated method to address it.

Objective Function Analysis. By studying the objective equation, we observe that the first constrained minimization equation is actually a constant, removal of which has no effects on the solutions. Therefore, we can simplify the objective function as follows

$$\mathcal{H}^* = \arg\min_{\mathbf{T}} \min_{\mathbf{X}'} \text{Tr}\left((\mathbf{X}')^{\top}\mathbf{L}'(\mathbf{X}')\right),$$

$$s.t.\ (\mathbf{X}')^{\top}(\mathbf{X}') = \mathbf{I}, \mathbf{T}\mathbf{T}^{\top} = \mathbf{I}, |\mathcal{H}| = h, \mathbf{T} \in \{0,1\}^{(|\mathcal{V}|-|\mathcal{H}|) \times |\mathcal{V}|}.$$

As we can see, the objective function is actually a joint min-min non-linear integer programming problem involving multiple variables simultaneously, joint optimization of which is shown to be NP-hard [7]. Therefore, a new approximated method SCROLL is proposed in this paper to address the objective function based on an alternative updating schema. Constraint $|\mathcal{H}| = h$ will be removed from the objective function since the number of detected trust holes has been denoted by the dimension of matrix $\mathbf{T}$ already.

Solution SCROLL with Alternative Updating. As introduced in the previous section, Laplace matrix $\mathbf{L}'$ can be represented as $\mathbf{L}' = \text{Diag}(\mathbf{T}\mathbf{A}\mathbf{T}^{\top}) - \mathbf{T}\mathbf{A}\mathbf{T}^{\top}$. The transformation matrix $\mathbf{T}$ is also involved in the $\text{Diag}(\cdot)$, which will make the partial derivatives calculation of the objective function about variable $\mathbf{T}$ infeasible. To address this problem, in this paper, we propose approximate the representation of $\mathbf{L}'$ as $\mathbf{L}' \approx \mathbf{T}\mathbf{L}\mathbf{T}^{\top}$ instead. The introduced deviation by such a approximation will be $\mathbf{T}\mathbf{L}\mathbf{T}^{\top} - (\text{Diag}(\mathbf{T}\mathbf{A}\mathbf{T}^{\top}) - \mathbf{T}\mathbf{A}\mathbf{T}^{\top}) = \mathbf{T} \cdot \text{Diag}(\mathbf{A}) \cdot \mathbf{T}^{\top} - \text{Diag}(\mathbf{T}\mathbf{A}\mathbf{T}^{\top})$, which are mainly about the values (about the out-degrees of the trust holes) on the diagonal of $\mathbf{L}'$. Based on such an approximation, the new objective function can be represented as

$$\mathcal{H}^* = \arg \min_{\mathbf{T}} \min_{\mathbf{X}'} \mathrm{Tr}\left((\mathbf{X}')^\top \mathbf{TLT}^\top (\mathbf{X}')\right),$$

$$s.t. \ (\mathbf{X}')^\top (\mathbf{X}') = \mathbf{I}, \mathbf{TT}^\top = \mathbf{I}.$$

Here the integer constraint matrix $\mathbf{T}$ is relaxed and entries in $\mathbf{T}$ can take any real values in range $[0, 1]$.

We propose to address the objective function with an alternative updating schema: (1) fix $\mathbf{T}$ and update $\mathbf{X}'$; and (2) fix $\mathbf{X}'$ and update $\mathbf{T}$.

Step 1: By fixing variable $\mathbf{T}$ and adding the constraint term $(\mathbf{X}')^\top (\mathbf{X}') = \mathbf{I}$ as a regularizer term, we can represent the objective function to be

$$\min_{\mathbf{X}'} \mathrm{Tr}\left((\mathbf{X}')^\top \mathbf{TLT}^\top (\mathbf{X}')\right) + \rho \left\|(\mathbf{X}')^\top (\mathbf{X}') - \mathbf{I}\right\|_F^2,$$

where parameter ρ denotes the weight of the regularizer term, and it is assigned with very large value (e.g., 10) in the experiment to ensure the constraint can be maintained.

The above objective function is a convex function can be addressed with gradient descent method, and the updating equation of variable $\mathbf{X}'$ can be represented as

$$\mathbf{X}' = \mathbf{X}' - \eta_1 \frac{\partial \left(\mathrm{Tr}\left((\mathbf{X}')^\top \mathbf{TLT}^\top \mathbf{X}'\right) + \rho \left\|(\mathbf{X}')^\top \mathbf{X}' - \mathbf{I}\right\|_F^2\right)}{\partial \mathbf{X}'}$$

$$= \mathbf{X}' - 2\eta_1 \left(\mathbf{TLT}^\top \mathbf{X}' + 2\rho(\mathbf{X}'(\mathbf{X}')^\top \mathbf{X}' - \mathbf{X}')\right),$$

where parameter η_1 denotes the learning step and it is assigned with a small constant value (0.0001) in the experiments.

Step 2: Meanwhile, in a similar way, by fixing parameter $\mathbf{X}'$ and adding the constraint term $\mathbf{TT}^\top = \mathbf{I}$ as a regularizer term, we can find that the resulting objective function is also a convex function. We can further represent the updating equation of variable $\mathbf{T}$ to be

$$\mathbf{T} = \mathbf{T} - 2\eta_2 \left(\mathbf{X}'(\mathbf{X}')^\top \mathbf{TL} + 2\rho(\mathbf{T}(\mathbf{T})^\top \mathbf{T} - \mathbf{T})\right),$$

where parameter η_2 denotes the learning step of updating $\mathbf{T}$.

Therefore, the alternative updating equation of variables $\mathbf{X}'$ and $\mathbf{T}$ at step τ can be represented as

$$\begin{cases} \mathbf{X}'^{(\tau)} & = (\mathbf{X}')^{(\tau-1)} - 2\eta_1 \left(\mathbf{T}^{(\tau-1)} \mathbf{L}(\mathbf{T}^{(\tau-1)})^\top (\mathbf{X}')^{(\tau-1)} \right. \\ & \quad \left. + 2\rho((\mathbf{X}')^{(\tau-1)}((\mathbf{X}')^{(\tau-1)})^\top (\mathbf{X}')^{(\tau-1)} - (\mathbf{X}')^{(\tau-1)})\right), \\ \mathbf{T}^{(\tau)} & = \mathbf{T}^{(\tau-1)} - 2\eta_2 \left((\mathbf{X}')^{(\tau)}((\mathbf{X}')^{(\tau)})^\top (\mathbf{T}')^{(\tau-1)} \mathbf{L} \right. \\ & \quad \left. + 2\rho(\mathbf{T}^{(\tau-1)}(\mathbf{T}^{(\tau-1)})^\top \mathbf{T}^{(\tau-1)} - \mathbf{T}^{(\tau-1)})\right). \end{cases}$$

Such a alternative updating process will continue until both $\mathbf{X}'$ and $\mathbf{T}$ converge. From the result of $\mathbf{T}$, we can recover the rows that are removed from the identified results, which corresponding to the *signed trust holes* of the signed network. Under the constraint that each row and each column can constrain at most one entry being filled with value 1, for entries in $\mathbf{T}^{(\tau)}$, we sort their values in decreasing order to select the entries with the largest values to preserve, which will be assigned with 1. The rest will all be assigned with value 0. In addition, based on matrix $\mathbf{X}'$, we can obtain the community structures formed by users in the signed networks. Depending on the positions and the connections attached to the identified *trust holes* (i.e., intra or inter community, and positive or negative links), we can differentiate the specific categories of *trust holes* (i.e., *positive* and *negative trust holes* respectively) from the results.

4   Experiments

To test the effectiveness of SCROLL in addressing the SHED problem. We conduct extensive experiments on real-world signed network datasets, and compare them with both state-of-art and traditional baseline methods.

4.1   Dataset Description

The real-world signed network dataset used in the experiments include the Epinions network and the Slashdot network. Some basic statistical information about these two datasets is available in Table 1.

Table 1. Properties of different networks

network	# nodes	# links	link type
Epinions	131,828	841,372	directed
Slashdot	77,350	516,575	directed

Reproducible Research?: The dataset used in this paper is public accessible, which can be downloaded from the SNAP site[2].

4.2   Experiment Setting

Based on the positive and negative links, the positive and negative adjacency matrices are constructed respectively. With the positive/negative adjacency matrices, we can define the integrated Laplace matrix and the weight of positive Laplace matrix α is set as 0.5. Framework SCROLL infers the transformation matrix $\mathbf{T}$ and confidence matrix $\mathbf{X}$ simultaneously with the alternative updating schema. From the transformation matrix $\mathbf{T}$ we can recover the trust holes.

---

[2] http://snap.stanford.edu/data/index.html#signnets.

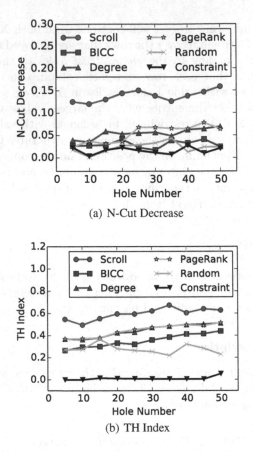

(a) N-Cut Decrease

(b) TH Index

Fig. 2. Experiment results on the Epinions network.

Comparison Methods. The networks studied in this paper are signed, and no existing works have studied the trust hole detection problem based on signed networks before yet. We propose to apply the existing works on traditional structural hole detection problem to the signed networks by discarding the polarity information. In SHED, no community structure information is available about the signed networks, thus the unsigned structural hole detection method proposed in [15] taking the community structure as the input cannot work for the SHED problem. The list of comparison methods used in the experiments are provided as follows:

- SCROLL: Method SCROLL is the trust hold detection method proposed in this paper, which can consider both the links as well as the polarities attached to the links.
- BICC: Method BICC is the state-of-art structural hole detection method for unsigned networks [18]. To accommodate the setting of BICC, we transform the signed networks to a unsigned one by discarding the link polarities.

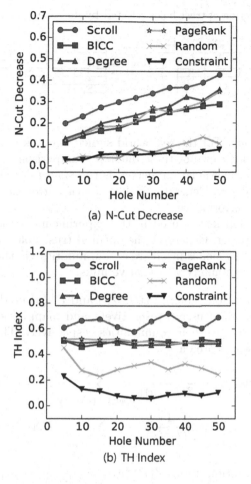

(a) N-Cut Decrease

(b) TH Index

Fig. 3. Experiment results on the Slashdot network.

- CONSTRAINT: Method CONSTRAINT proposed in [2] uses constraint to estimate the importance of each node and selects the nodes with the lowest K constraint scores as the hole candidates.
- PAGERANK: Traditional node ranking algorithm PAGERANK can also be used as a method for structural hole detection in [15], which returns the nodes with the top K page rank scores as the result.
- DEGREE: Method DEGREE selects the users with the top K degree as the hole candidates. Considering that the links in the network datasets are directed, the social degree of user u is identical to the number of neighbors of u in the network (i.e., $|\Gamma(u)| = |\{v | v \in \mathcal{U}, (u, v) \in \mathcal{E} \vee (v, u) \in \mathcal{E}\}|$).
- RANDOM: Method RANDOM randomly selects K users as the hole candidates.

Table 2. Intersection of trust holes selected by different comparison methods.

Epinions:

	SCROLL	BICC	CONSTRAINT	PAGERANK	DEGREE	RANDOM
SCROLL	50	26	1	32	25	4
BICC		50	2	25	26	3
CONSTRAINT			50	1	1	2
PAGERANK				50	22	4
DEGREE					50	4
RANDOM						50

Slashdot:

	RANDOM	DEGREE	PAGERANK	CONSTRAINT	BICC	SCROLL
SCROLL	4	35	27	4	23	50
BICC	5	23	23	4	50	
CONSTRAINT	0	4	5	50		
PAGERANK	6	27	50			
DEGREE	4	50				
RANDOM	50					

Evaluation Metrics. For the real-world signed network datasets, we propose to measure the performance of these different comparison methods two different metrics. One of the evaluation metrics is the *signed normalized cut decrease* introduced in this paper. Generally, higher *signed normalized cut decrease* corresponds to better performance.

Another evaluation metric used in the experiments is the *trust hole index* introduced in this paper. Generally, the optimal trust holes are the user nodes which connect to other nodes in different communities via the positive links or nodes in the same community via negative links.

Definition 5 (Trust Hole Index): Let $\Gamma^-(u)$ and $\Gamma^+(u)$ be the sets of negative and positive neighbors of user u respectively and mapping $c : \mathcal{U} \to \mathcal{C}$ be the function projects users to their communities respectively. The *trust hole index* for user u can be represented as

$$\text{TH-Index}(u) = \frac{1}{Z(\Gamma^+(u))} \sum_{v,w \in \Gamma^+(u), v \neq w} I(c(v) \neq c(w))$$

$$+ \frac{1}{Z(\Gamma^-(u))} \sum_{v,w \in \Gamma^-(u), v \neq w} I(c(v) = c(w)).$$

where the normalization term $Z(\Gamma^+(u)) = \frac{1}{|\Gamma^+(u)| \times (|\Gamma^+(u)|-1)}$, and term $Z(\Gamma^-(u)) = \frac{1}{|\Gamma^-(u)| \times (|\Gamma^-(u)|-1)}$. Function $I(c(v) \neq c(w))$ is 1 if $c(v) \neq c(w)$ and similar for $I(c(v) = c(w))$.

Generally, users with higher *trust hole index* are more likely to be the trust holes, and methods achieving higher *trust hole index* will have better performance.

4.3   Experiment Result

The experiment results on the real-world signed networks, Epinions and Slashdot, are shown in Figs. 2 and 3 respectively.

In Fig. 2(a), we show the performance of different comparison methods evaluated by the *normalized cut decrease* metric at different hole numbers. From the plot, we can observe that SCROLL can outperform other comparison methods for different hole numbers. For instance, when the hole number is 25, the *normalized cut decrease* achieved by SCROLL is 0.150, which is more than two times higher than that obtained by DEGREE and PAGERANK, about 7 times larger than that achieved by BICC, RANDOM and CONSTRAINT. In Fig. 2(b), we show the performance of the comparison methods when the evaluation metric is *trust hole index*. According to the plot, we can observe that DEGREE and PAGERANK has comparable performance, both of which are below the *trust hole index* curve of SCROLL. The constraint method cannot work well when dealing with the signed networks at all, which is the baseline of all the comparison methods.

Similar results of these comparison methods can be observed for the Slashdot network in Fig. 3, and SCROLL can outperform other methods for different hole numbers.

In addition, in Table 2, we show the shared trust holes detected by the different comparison methods in the Epinions and Slashdot networks respectively. From the results, we can observe that the trust holes selected by SCROLL have some overlapping with BICC, DEGREE and PAGERANK. In the Epinions, the number of detected trust holes by SCROLL and BICC is 26, the number of shared trust holes by SCROLL and PAGERANK is 32, while those shared by SCROLL and DEGREE is 25 respectively. Meanwhile, the trust holes detected by these three methods are quite different from those selected by RANDOM and CONSTRAINT. For instance, in Epinions, the trust holes shared by SCROLL and CONSTRAINT is merely 1, and those shared by SCROLL and RANDOM is only 4. Similarly results can be observed for the Slashdot network, and SCROLL can choose some common holes with BICC, PAGERANK and DEGREE, but have quite different results with RANDOM and CONSTRAINT.

5   Related Works

Traditional structural holes are usually correlated to a wide range of indicators about social success, which have been studied in various papers already [1,3,4]. Ahuja [1] proposes to study the effects of a firm's network of relations on innovation from three different perspectives: direct ties, indirect ties and structural holes, where structural holes are discovered to have both positive and negative influences on subsequent innovations. Burt [3] introduces the relation between structural holes and good ideas. Burt discovers that structural holes connecting different groups are more likely to express ideas, less likely to have ideas dismissed, and more likely to have ideas evaluated as valuable.

Later, some works propose to study the formation of structural holes in social networks from the game theory perspective [5,8,9]. Goyal et al. [8] propose that in social networks, individuals form links with others to create surplus, to gain intermediation rents, and to circumvent others, which are the forces in the process of strategic network formation. Kleinberg et al. [9] propose to apply a game-theoretic approach to study the structural holes, and notice that individuals will differentiate themselves in equilibrium of the game, occupying different social strata and receiving different payoffs.

Recently, some works have been done on finding the structure holes from social networks [15,23]. Lou et al. [15] formulate the structure hold mining problem from the information diffusion and community detection perspectives, and discover that the problem is NP-hard based on these two modeling. Vilhena et al. [23] extend the structural hole concept to "culture holes" and propose to find the "culture holes" from the citation networks. For more background knowledge about online social networks and the research works studied based on them, please refer to a recent survey paper written by Shi et al. in [19].

Signed networks since introduced by Leskovec et al. [12] have become a hot research topic, as links in signed networks can denote different attitudes among users, which provide new opportunities for researchers to study the connections among users. Leskovec et al. [11] propose to predict the positive and negative links in signed networks based on the balance theory. Doreian et al. [6] study the partition problem in signed networks. A recent survey about related works in signed networks is given by Tang et al. in [21].

6   Conclusion

In this paper, we have studied the *trust hole* detection problem in signed networks. A formal definition about the *trust hole* concept as well as its two different variants are clearly illustrated in this paper. To identify the set of trust holes from the signed network, a community detection based trust hole detection framework, SCROLL, is introduced in this paper. By identifying the set of users, removal of whole from the network can lead to the maximum *signed normalized cut decrease*, SCROLL can detect the optimal set of trust holes from the signed network. Extensive experiments have been done on real-world signed networks, and the results demonstrate the effectiveness of SCROLL in addressing the SHED problem.

Acknowledgement. This work is supported in part by NSF through grants III-1526499.

References

1. Ahuja, G.: Collaboration networks, structural holes, and innovation: a longitudinal study. Adm. Sci. Q. **45**, 425–455 (2000)
2. Burt, R.: Structural Holes. Cambridge University Press, Cambridge (1992)
3. Burt, R.: Structural holes and good ideas. Am. J. Sociol. **110**(2), 349–399 (2004)
4. Burt, R.: Second-Hand brokerage: evidence on the importance of local structure for managers, bankers, and analysts. Acad. Manage. J. **50**, 119–148 (2006)
5. Buskens, V., van de Rijt, A.: Dynamics of networks if everyone strives for structural holes. Am. J. Soci. **92**(6), 1287–1335 (2008)
6. Doreian, P., Mrvar, A.: Partitioning signed social networks. Soc. Netw. **33**, 196–221 (2009)
7. Gao, D., Sherali, H. (eds.): Advances in Applied Mathematics and Global Optimization. Springer, Heidelberg (2009)
8. Goyal, S., Vega-Redondo, F.: Structural holes in social networks. J. Econ. Theor. **137**, 460–492 (2007)

9. Kleinberg, J., Suri, S., Tardos, E., Wexler, T.: Strategic network formation with structural holes. ACM SIGecom Exchanges **7**, 1–4 (2008)
10. Kunegis, J., Schmidt, S., Lommatzsch, A., Lerner, J.: Spectral analysis of signed graphs for clustering, prediction and visualization. In: SDM (2010)
11. Leskovec, J., Huttenlocher, D., Kleinberg, J.: Predicting positive and negative links in online social networks. In: WWW (2010)
12. Leskovec, J., Huttenlocher, D., Kleinberg, J.: Signed networks in social media. In: CHI (2010)
13. Li, D., Xu, Z., Chakraborty, N., Gupta, A., Sycara, K., Li, S.: Polarity related influence maximization in signed social networks. PLOS **9**(7), e102199 (2014)
14. Li, Y., Chen, W., Wang, Y., Zhang, Z.: Influence diffusion dynamics and influence maximization in social networks with friend and foe relationships. In: WSDM (2013)
15. Lou, T., Tang, J.: Mining structural hole spanners through information diffusion in social networks. In: WWW (2013)
16. Luxburg, U.: A tutorial on spectral clustering. Statist. Comput. **17**, 395–416 (2007)
17. Newman, M.E.J.: Detecting community structure in networks. Europ. Phys. J. B - Condens. Matter Complex Syst. **38**, 321–330 (2004)
18. Rezvani, M., Liang, W., Xu, W., Liu, C.: Identifying top-k structural hole spanners in large-scale social networks. In: CIKM (2015)
19. Shi, C., Li, Y., Zhang, J., Sun, Y., Yu, P.: A survey of heterogeneous information network analysis. CoRR abs/1511.04854 (2015)
20. Shi, J., Malik, J.: Normalized cuts and image segmentation. IEEE Trans. Pattern Anal. Mach. Intell. **22**(8), 888–905 (2000)
21. Tang, J., Chang, Y., Aggarwal, C., Liu, H.: A survey of signed network mining in social media. ACM Computing Surveys, to appear CoRR abs/1511.07569 (2015) http://arxiv.org/abs/1511.07569
22. Traag, V., Bruggeman, J.: Community detection in networks with positive and negative links. Phys. Rev. E **80**(3), 36115 (2009)
23. Vilhena, D., Foster, J., Rosvall, M., West, J., Evans, J., Bergstrom, C.: Finding cultural holes: How structure and culture diverge in networks of scholarly communication. Sociol. Sci. **1**, 221–238 (2014)

Huber-Norm Regularization
for Linear Prediction Models

Oleksandr Zadorozhnyi[1]([✉]), Gunthard Benecke[1], Stephan Mandt[2],
Tobias Scheffer[1], and Marius Kloft[3]

[1] Department of Computer Science, University of Potsdam, Potsdam, Germany
{zadorozh,gunthard.benecke,tobias.scheffer}@uni-potsdam.de
[2] Department of Computer Science, Data Science Institute,
Columbia University, New York City, USA
sm3976@columbia.edu
[3] Department of Computer Science,
Humboldt-Universität zu Berlin, Berlin, Germany
kloft@hu-berlin.de

Abstract. In order to avoid overfitting, it is common practice to regularize linear prediction models using squared or absolute-value norms of the model parameters. In our article we consider a new method of regularization: Huber-norm regularization imposes a combination of ℓ_1 and ℓ_2-norm regularization on the model parameters. We derive the dual optimization problem, prove an upper bound on the statistical risk of the model class by means of the Rademacher complexity and establish a simple type of oracle inequality on the optimality of the decision rule. Empirically, we observe that logistic regression with Huber-norm regularizer outperforms ℓ_1-norm, ℓ_2-norm, and elastic-net regularization for a wide range of benchmark data sets.

1 Introduction

Linear classification and regression models—such as the support vector machine (SVM) and logistic and linear regression—are widely used in machine learning, and regularized empirical-risk minimization is a standard approach to optimizing their parameters. To avoid overfitting, linear models are typically either densely or sparsely regularized. With an ℓ_2 regularizer, one obtains a dense weight vector in which all features contribute to the prediction task. For interpretability, one is often interested in a sparse solution in which many entries of the weight vector are zero. To this end, one may employ an ℓ_1 absolute value norm regularizer [18,25]. While this type of regularization may lead to lower predictive accuracies than ℓ_2 regularization [10], the result focuses only on the most relevant features.

This paper promotes the idea of using a combination of both types of regularization, thus combining the best of both worlds. Instead of using just a single weight vector $\mathbf{w}$ that is either dense *or* sparse, we employ a sum of two weight vectors $\mathbf{w} + \mathbf{v}$. While $\mathbf{w}$ is ℓ_2 regularized and therefore dense, $\mathbf{v}$ is ℓ_1 regularized

© Springer International Publishing AG 2016
P. Frasconi et al. (Eds.): ECML PKDD 2016, Part I, LNAI 9851, pp. 714–730, 2016.
DOI: 10.1007/978-3-319-46128-1_45

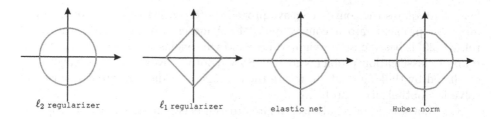

ℓ_2 regularizer ℓ_1 regularizer   elastic net   Huber norm

Fig. 1. Geometrical illustration of the proposed Huber-norm regularizer and comparison to common regularizers.

and therefore sparse. Having two different weight vectors with different regularizations allows linear models to more flexibly fit the data. It comes at a moderate computational cost, since the number of parameters is doubled.

We first show that the proposed combination of two weight vectors is mathematically equivalent to imposing Huber-norm [7] regularization on the empirical risk of a linear model. This approach is known to be statistically more robust [8] in the sense that individual sparse weights do not necessarily involve a huge cost in the loss. This Huber norm involves quadratic costs near the origin and linear costs far away from the origin, this way penalizing outliers less severely. Because of this analogy, we call our method *Huber-norm regularization*. We derive uniform and data-dependent upper bounds on the statistical risk of the model class by means of the Rademacher complexity. We deduce a simple type of oracle inequality on the inference efficiency of the decision rule which measures the deviation of the model's risk from the lowest risk of any model in the class.

Our empirical studies show that *Huber-norm regularized logistic regression* outperforms ℓ_1- and ℓ_2-regularized as well as elastic-net-regularized logistic regression [26] in the majority of cases over a wide range of benchmark problems. To support this claim we provide evidence based on empirical studies on the UCI machine learning repository, where our method performs best among the compared methods on 23 out of 31 data sets. On particular data set—the well-known Iris data set—Huber-norm regularization leads to a prediction accuracy of 0.96 while the next-best method merely achieves 0.84.

Our paper is organized as follows. Section 2 reviews related work. In Sect. 3, we describe our model and its basic properties. We also prove the equivalence of the two weight vectors to Huber-norm regularization in the conventional setting. In Sect. 4 we then present the underlying theoretical foundations of our approach, where we prove an upper bound on the statistical risk. We present our experimental results in Sect. 5 and conclude in Sect. 6.

2 Related Work

Comparisons between ℓ_1-norm and ℓ_2-norm SVMs are ubiquitous in the literature [13,14,25]. A robust alternative to the SVM based on the smooth ramp loss [23] requires the convex-concave procedure to convert this non-convex optimization problem into a convex one [24]. Another way of making the SVM robust [20] is based on the weighted LS-SVM that yields sparse results. Different type of classification problems for the SVM (both convex and non-convex) are discussed by Hailong et al. [6] where the conjugate gradient approach is used to solve the optimization problem.

Our novel type of regularizer relates to the elastic-net regularizer [26] that simply amounts to taking the sum of an ℓ_1 and ℓ_2 regularizer. Our proposed regularizer is very different, as is evident from Fig. 1. The plot shows contours of different regularizers in comparison. As a major difference between the elastic net and our approach, our regularizer grows asymptotically linearly for large weight vectors whereas the elastic net grows asymptotically quadratically. Lastly, our theoretical contributions are based on fundamental work by Vapnik [22].

The Huber norm [7] is frequently used as a loss function; it penalizes outliers asymptotically linearly which makes it more robust than the squared loss. The Huber norm is used as a regularization term of optimization problems in image super resolution [21] and other computer-graphics problems. The *inverse Huber function* [17] has been studied as a regularizer for regression problems. While the Huber norm penalizes large weights asymptotically linearly, the inverse Huber function imposes an asymptotically squared penalty on large weights.

3   Huber-Norm-Regularized Linear Models

In this section, after formally introducing the problem setting and optimization criterion, we show that this optimization criterion has an equivalent formulation in which the Huber norm becomes explicit. We derive the dual form and show how *Huber-norm regularization* for linear models can be implemented.

3.1   Problem Setting and Preliminaries

We consider the standard supervised prediction setup, where we are given a training sample $S = \{\mathbf{x}_i, y_i\}_{i=1}^n$ from a space $\mathcal{X} \times \mathcal{Y}$ with $\mathcal{X} = \mathbb{R}^d$. We aim at finding a linear function f that predicts well. A common way to achieve this is to first define a loss function $\ell : \mathbb{R} \times \mathcal{Y} \to \mathbb{R}_+ \cup \{0\}$ that measures the deviation of the prediction $f(\mathbf{x})$ from the correct value y, such as the logistic loss $\ell(f(\mathbf{x}), y) := \log(1 + \exp(-yf(\mathbf{x})))$ or hinge loss $\ell(f(\mathbf{x}), y) := \max(0, 1 - yf(\mathbf{x}))$. The empirical risk is then the averaged loss over the training sample, $\hat{L}_n(f) = \frac{1}{n}\sum_{i=1}^n \ell(f(\mathbf{x}_i), y_i)$ of f.

In this paper we consider methods that employ linear prediction functions $f(\mathbf{x}) = \mathbf{w}^\top \mathbf{x}$. To avoid overfitting, one usually uses a regularizer such as the ℓ_1 regularizer $R_1(\mathbf{w}) = ||\mathbf{w}||_1$, the ℓ_2 regularizer $R_2(\mathbf{w}) = ||\mathbf{w}||_2^2$, or the elastic-net regularizer $R_{en}(\mathbf{w}) = ||\mathbf{w}||_1 + ||\mathbf{w}||_2^2$. This results in the *regularized empirical risk minimization* or short *reg-ERM* problem:

$$\min_{\mathbf{w}} \lambda R(\mathbf{w}) + \frac{1}{n}\sum_{i=1}^n \ell(y_i, \mathbf{w}^\top \mathbf{x}_i).$$

The ℓ_1 and elastic-net regularizers produce sparse, ℓ_2-norm regularizer dense weight vectors. Hence, depending on the problem, the regularizer can be chosen to match the underlying sparsity of the problem.

3.2 Linear Models with Sums of Dense and Sparse Weights

Using ℓ_1-, ℓ_2-, or elastic-net-regularized ERM either produces dense or sparse solutions. In this paper, we argue it can be beneficial to produce dense solutions with pronounced feature weights as in ℓ_1-norm regularized methods. We propose to consider linear models of the form $f(\mathbf{x}) := (\mathbf{v} + \mathbf{w})^\top \mathbf{x}$ (for notational convenience, we disregard constant offsets and assume that the first element of each $\mathbf{x}$ is a constant 1) and the regularizer $R_H(\mathbf{v}, \mathbf{w}) = \lambda ||\mathbf{v}||_1 + \mu ||\mathbf{w}||_2^2$, hence resulting in the following optimization problem.

Optimization Problem 1 (Sums of dense and sparse weights). *Given* $\lambda, \mu > 0$ *and loss function* $\ell(t, y)$, *solve:*

$$(\hat{\mathbf{w}}, \hat{\mathbf{v}}) = \arg\min_{\mathbf{v}, \mathbf{w}} G(\mathbf{w}, \mathbf{v}, S)$$

$$\text{with } G(\mathbf{w}, \mathbf{v}, S) = \lambda ||\mathbf{v}||_1 + \mu ||\mathbf{w}||_2^2 + \frac{1}{n} \sum_{i=1}^{n} \ell(y_i, (\mathbf{w} + \mathbf{v})^\top \mathbf{x}_i), \qquad (1)$$

where $|| \cdot ||_2$ *and* $|| \cdot ||_1$ *denote standard* ℓ_2-*norm and* ℓ_1-*norm correspondingly.*

For reasons that will become clear in the section below we call the method *Huber-regularized empirical risk minimization* or short *Huber-regERM*. Note that by letting $\lambda \to \infty$, we obtain the classic ℓ_2-norm regularization, while letting $\mu \to \infty$ leads to ℓ_1-norm regularization. Thus these methods are obtained as limit cases of our method. *Elastic-net-regularization* is not a special case of this framework, but it could be obtained by enforcing an additional constraint $\mathbf{v} = \mathbf{w}$.

3.3 Geometry of the Huber Norm

The following geometrical interpretation lets us compare linear models with sums of dense and sparse weights to the ℓ_1, ℓ_2, and elastic-net regularizers. We prove that Problem 1 is equivalent to the following problem.

Optimization Problem 2 (Equivalent Huber-Norm Problem). *Optimization Problem 1 can equivalently be formulated as:*

$$\hat{\mathbf{z}} = \arg\min_{\mathbf{z}} R_H(\mathbf{z}) + \frac{1}{n} \sum_{i=1}^{n} \ell(y_i, \mathbf{z}^\top \mathbf{x}_i) \qquad (2)$$

where $R_H(\mathbf{z}) = \sum_{i=1}^{d} r_H(z_i)$, *and* $r_H(z_i) = \begin{cases} \lambda \left(|z_i| - \frac{\lambda}{4\mu} \right) & \text{if } |z_i| \geq \frac{\lambda}{2\mu} \\ \mu z_i^2, & \text{otherwise} \end{cases}$.

Note that $R_H(\mathbf{z})$ is the Huber norm of $\mathbf{z}$. While the Huber norm is often used as a robust loss function that is less sensitive to outliers, Optimization Problem 2 employs the Huber norm as regularizer. Intuitively, this results in a regularization scheme that is less sensitive to individual features which have a stong impact on f than ℓ_2 regularization. Figure 1 illustrates isotropic lines for the Huber-norm regularizer and known regularizers for $\lambda = \mu = 1$. The Huber norm is composed of linear and squared segments. While it does not encourage sparsity as the ℓ_1 regularizer does, it encourages that most attributes only have a small impact on the decision function.

Proof (Equivalence of Optimization Problems 1 and 2). Let $\mathbf{z} = \mathbf{w} + \mathbf{v}$. Problem 1 can then be formulated as

$$\min_{\mathbf{w},\mathbf{v}} G(\mathbf{w}, \mathbf{v}, S) = \min_{\mathbf{z},\mathbf{v}} \lambda ||\mathbf{v}||_1 + \mu ||\mathbf{z} - \mathbf{v}||_2^2 + \frac{1}{n} \sum_{i=1}^{n} \ell(y_i, \mathbf{z}^\top \mathbf{x}_i)$$

$$= \min_{\mathbf{z}} \left(\mu \min_{\mathbf{v}} \left(\frac{\lambda}{\mu} ||\mathbf{v}||_1 + ||\mathbf{z} - \mathbf{v}||_2^2 \right) + \frac{1}{n} \sum_{i=1}^{n} \ell(y_i, \mathbf{z}^\top \mathbf{x}_i) \right). \quad (3)$$

Let us define $R(\mathbf{v}, \mathbf{z}) := \bar{c} ||\mathbf{v}||_1 + ||\mathbf{z} - \mathbf{v}||_2^2$ where $\bar{c} := \frac{\lambda}{\mu}$. It remains to be shown that $\min_{\mathbf{v}} R(\mathbf{v}, \mathbf{z})$ is a Huber-norm regularizer.

Simplifying $R = \mathbf{v}^\top \mathbf{v} - 2\mathbf{v}^\top \left(\mathbf{z} - \frac{\bar{c}}{2} \mathrm{sgn}(\mathbf{v}) \right) + \mathbf{z}^\top \mathbf{z}$, we find

$$\min_{\mathbf{v}} R = \min_{\mathbf{v}=(v_1,\dots,v_d)} \left(\sum_{i=1}^{d} v_i^2 - 2v_i(z_i - \frac{\bar{c}}{2} \mathrm{sgn}(v_i)) \right) + \sum_{i=1}^{d} z_i^2. \quad (4)$$

For each $i \in \{1, ..., d\}$ we minimize $R_i := v_i^2 - 2v_i(z_i - \frac{\bar{c}}{2} \mathrm{sgn}(v_i))$ with respect to v_i. This is equivalent to:

$$\begin{bmatrix} \min_{v_i} v_i^2 - 2(z_i - \frac{\bar{c}}{2})v_i \ \text{if}\, v_i > 0 \\ \min_{v_i} v_i^2 - 2(z_i + \frac{\bar{c}}{2})v_i \ \text{if}\, v_i \le 0. \end{bmatrix}$$

We can minimize each of these two quadratic terms analytically:

$$\begin{bmatrix} -(z_i - \frac{\bar{c}}{2})^2 \ \text{if}\, z_i \in \mathcal{A} := \{z \in \mathbb{R} : |z| \ge \frac{\bar{c}}{2}\} \\ 0 \qquad\qquad \text{if}\, z_i \in \mathcal{A}_c := \{z \in \mathbb{R} : |z| < \frac{\bar{c}}{2}\}. \end{bmatrix}$$

This means, that for Eq. 4 we have explicitly:

$$\min_{\mathbf{v}} R = \sum_{i=1}^{d} \left(z_i^2 - \left(z_i - \frac{\bar{c}}{2} \right)^2 \mathbb{I}_{z_i \in \mathcal{A}} \right) = \sum_{i=1}^{d} \left(z_i^2 \mathbb{I}_{z_i \in \mathcal{A}_c} + \bar{c} \left(|z_i| - \frac{\bar{c}}{4} \right) \mathbb{I}_{z_i \in \mathcal{A}} \right).$$

This is exactly the Huber-norm regularizer $R_H(\mathbf{z})$ of Optimization Problem 2.

$\square$

3.4   Dual Problem

In order to classify a training point, we need to compute the scalar product $(\mathbf{w} + \mathbf{v})^\top \mathbf{x}$ which may be expensive when the dimension of vectors $\mathbf{w}, \mathbf{v}$ is large. One possible solution to overcome this consists in considering a weighted sum of constraints together with an objective function computed on the training sample. This leads to a dual approach. Steinwart [19] gives a general overview of dual optimization problems for SVMs using ℓ_2- and ℓ_1-norm regularizers. The dual form of the optimization problem depends on the loss function. We complete Steinwart's overview by deriving the dual form of the *Huber-norm regularized SVM* in the following.

Optimization Problem 3 (Dual Huber-Norm SVM Problem). *Optimization Problem 1 with hinge-loss loss function (Huber-Norm SVM) has an equivalent dual form which can be formulated as follows:*

$$\max_{\alpha \in \mathbb{R}^n} \sum_{i=1}^n \alpha_i - \frac{1}{2} \sum_{i,j=1}^n \alpha_i \alpha_j y_i y_j \mathbf{x}_i^\top \mathbf{x_j}$$

$$s.t. \ \alpha \in [0, C]^n \wedge ||\mathbf{X}^\top \alpha||_\infty \leq \frac{\lambda}{2\mu}, \tag{5}$$

where $C := \frac{1}{2n\mu}$ and $\mathbf{X} = (\mathbf{x}_1, ..., \mathbf{x}_n) \in \mathbb{R}^{n \times d}$.

Proof. The Lagrangian $L(\mathbf{w}, \mathbf{v}, \xi, \alpha, \eta)$ that corresponds to Eq. 1 is given as follows:

$$L(\mathbf{w}, \mathbf{v}, \xi, \alpha, \eta) := C \sum_{i=1}^n \xi_i + \frac{\lambda}{2\mu} ||\mathbf{v}||_1 + \frac{1}{2} ||\mathbf{w}||_2^2 +$$

$$\sum_{i=1}^n \alpha_i \left(1 - y_i (\mathbf{w}^\top + \mathbf{v}^\top)\mathbf{x}_i - \xi_i\right) - \sum_{i=1}^n \eta_i \xi_i, \tag{6}$$

where $\alpha = (\alpha_1, ..., \alpha_n) \in [0, \infty)^n$ and $\eta = (\eta_1, ..., \eta_n) \in [0, \infty)^n$. So the dual problem [3] can be written as:

$$\max_{\alpha, \eta} \inf_{\mathbf{w}, \mathbf{v}, \xi} L(\mathbf{w}, \mathbf{v}, \xi, \alpha, \eta). \tag{7}$$

Grouping the terms in the Lagrangian gives us:

$$L(\mathbf{w}, \mathbf{v}, \xi, \alpha, \eta) = \sum_{i=1}^n (C - \alpha_i - \eta_i)\xi_i + \frac{\lambda}{2\mu} ||\mathbf{v}||_1$$

$$- \sum_{i=1}^n \alpha_i y_i \mathbf{v}^\top \mathbf{x}_i + \frac{1}{2} ||\mathbf{w}||_2^2 - \sum_{i=1}^n \alpha_i y_i \mathbf{w}^\top \mathbf{x}_i + \sum_{i=1}^n \alpha_i.$$

Now, considering the infimum with respect to $\mathbf{v}$ and $\mathbf{w}$ separately, and using the definition of a conjugate function [3,19] we obtain:

$$\inf_{\mathbf{v}} \frac{\lambda}{2\mu}||\mathbf{v}||_1 - \sum_{i=1}^{n} \alpha_i y_i \mathbf{v}^\top \mathbf{x}_i = -\sup_{\mathbf{v}} \frac{\lambda}{2\mu}||\mathbf{v}||_1 + \mathbf{v}^\top \sum_{i=1}^{n} \alpha_i y_i \mathbf{x}_i$$

$$= \begin{cases} 0, & \text{when } |\mathbf{X}^\top \alpha||_\infty \leq \frac{\lambda}{2\mu} \\ -\infty, & \text{otherwise,} \end{cases} \tag{8}$$

where $\mathbf{X} = (\mathbf{x}_1, ..., \mathbf{x}_n) \in \mathbb{R}^{n \times d}$ is the data matrix whose rows $\mathbf{x}_i^\top$ are the instances and $||\cdot||_\infty$ -supremum norm in $\mathbb{R}^d$. Analogously, for $\mathbf{w}$ we have:

$$\inf_{\mathbf{w}} \frac{1}{2}||\mathbf{w}||_2^2 - \sum_{i=1}^{n} \alpha_i y_i \mathbf{w}^\top \mathbf{x}_i = -\sup_{\mathbf{w}} -\frac{1}{2}||\mathbf{w}||_2^2 + \mathbf{w}^\top \sum_{i=1}^{n} \alpha_i y_i \mathbf{x}_i$$

$$= \frac{1}{2} \left(\sum_{i=1}^{n} \alpha_i y_i \mathbf{x}_i \right)^\top \left(\sum_{i=1}^{n} \alpha_i y_i \mathbf{x}_i \right). \tag{9}$$

Finally, computing the gradient with respect to ξ gives that for each $i \in \{1, ...n\}$:

$$C - \eta_i - \alpha_i = 0 \Leftrightarrow \alpha_i = C - \eta_i. \tag{10}$$

Now, for fixed λ, μ, and $\mathbf{X}$, define $P = \{\alpha | \alpha \in [0, C]^n \wedge ||\mathbf{X}^\top \alpha||_\infty \leq \frac{\lambda}{2\mu}\}$,where $\mathbf{y} = (y_1, ..., y_n) \in \mathbb{R}^n$. Substituting Eqs. 8, 9, and 10 into Eq. 7 gives the following dual problem:

$$\max_{\alpha \in P} \sum_{i=1}^{n} \alpha_i - \frac{1}{2} \sum_{i,j=1}^{n} \alpha_i \alpha_j y_i y_j \mathbf{x}_i^\top \mathbf{x}_j, \tag{11}$$

which is a quadratic optimization problem within set P and can be solved with known methods.   □

By close inspection of Eq. 11, we observe that our dual optimization problem closely resembles the one for *SVM* using ℓ_2 regularization, but with a difference in the form of the domain P of the optimization problem.

3.5   Algorithm and Implementation

Algorithm 1 implements *Huber-regularized empirical risk minimization for linear models*. The algorithm works by alternatingly minimizing the occurring ℓ_1-norm and ℓ_2-norm regularized minimization problems, respectively. For each step of optimization procedure we use gradient descent, assuming that the other vector is constant. The gradient of the ℓ_1 norm of $\mathbf{v}$ is not defined for $\mathbf{v} = 0$; here, we use subgradients [3].

4   Theoretical Analysis

In this section we present a theoretical analysis of the proposed *Huber-norm regularizer for linear models*. We obtain bounds on the statistical risk based on the established framework of Rademacher complexities [2,16] and, consequently, on the norms of the vectors $\mathbf{v}, \mathbf{w}$ and number of training samples n [2].

Algorithm 1. Optimization Procedure

1: **Input:** $S = \{\mathbf{x}_i, y_i\}_{i=1}^n$
2: $\mathbf{w} = \mathbf{0}, \mathbf{v} = \mathbf{0}$.
3: **repeat**
4:      solve $\hat{\mathbf{w}} := \arg\min_{\mathbf{w}} G(\mathbf{w}, \mathbf{v}, S)$ by gradient descent,
5:      solve $\hat{\mathbf{v}} := \arg\min_{\mathbf{v}} G(\hat{\mathbf{w}}, \mathbf{v}, S)$ by gradient descent,
6:      let $\mathbf{w}, \mathbf{v} = (\hat{\mathbf{w}}, \hat{\mathbf{v}})$.
7: **until** convergence.
8: **Output:** $\mathbf{w}, \mathbf{v}$

4.1   Preliminaries and Aim

Let $S = \{\mathbf{x}_i, y_i\}_{i=1}^n$ be a sample of n training points that are independently drawn from one and the same distribution $P_{X,Y}$ over $\mathcal{X} \times \mathcal{Y}$, where $\mathcal{X} = \mathbb{R}^d$; let the output space $\mathcal{Y}$ be discrete for classification and continuous for regression. In this theoretical analysis, we study the *Huber-regERM* model class

$$\mathcal{F} := \{f : \mathbf{x} \mapsto (\mathbf{w} + \mathbf{v})^\top \mathbf{x} : \mathbb{R}^d \to \mathbb{R} | \|\mathbf{w}\|_2 \leq W, \|\mathbf{v}\|_2 \leq V\}, \qquad (12)$$

where W and V are initially unknown constants. Loss function $\ell : \mathbb{R} \times \mathcal{Y} \to \mathbb{R}_+ \cup \{0\}$ may be any convex loss function that is $\mathcal{L}$-Lipschitz continuous and absolutely bounded by constant $B \in \mathbb{R}$. The aim of our theoretical analysis is to obtain bounds on the deviation of the *risk* $L(f) = E_{P_{X,Y}}[\ell(f(\mathbf{x}), y)]$ of the model $f \in \mathcal{F}$ from *empirical risk* $\hat{L}_n(f) = \frac{1}{n}\sum_{i=1}^n \ell(f(\mathbf{x}_i), y_i)$.

Let $\{\sigma_i\}_{i=1}^n$ be independent Rademacher random variables, meaning that each of them is uniformly distributed over $\{-1, +1\}$. Denote by Σ the joint uniform distribution of $\sigma_1, \ldots, \sigma_n$. Then the *empirical Rademacher complexity* is defined as

$$\hat{\mathfrak{R}}_S(\ell \circ \mathcal{F}) := E_\Sigma \left[\sup_{f \in \mathcal{F}} \frac{1}{n} \sum_{i=1}^n \sigma_i \ell(f(\mathbf{x}_i), y_i)) \right], \qquad (13)$$

and the (theoretical) *Rademacher complexity* [2,16] is defined as $\mathfrak{R}_n(\ell \circ \mathcal{F}) := E_S[\hat{\mathfrak{R}}_S(\ell \circ \mathcal{F})]$. Here, the expectation is taken under the distribution of the sample S. It has been shown [2,16] that when ℓ is $\mathcal{L}$-Lipschitz continuous in the second argument, then with probability at least $1 - \delta$, for all $f \in \mathcal{F}$:

$$L(f) \leq \hat{L}_n(f) + 2\mathcal{L}E_S\left[\hat{\mathfrak{R}}_S(\mathcal{F})\right] + B\sqrt{\frac{\log \delta^{-1}}{2n}}. \qquad (14)$$

4.2   Bounds on the Risk of Huber-Regularized Linear Models

Our main theoretical contributions are bounds on statistical risk based on data-dependent and uniform upper bounds on the Rademacher complexity of the model class $\mathcal{F}$ defined by Eq. 12.

Theorem 1 (Uniform risk bound for Huber regularization). *Let $\mathcal{F}$ be defined by Eq. 12, let ℓ be a $\mathcal{L}$-Lipschitz continuous loss function, and let R be a constant such that $|\ell(t, y)| \leq R$ for all $t \in \mathbb{R}$ and $y \in \mathcal{Y}$. Let the ℓ_2 norm of all instances is bounded by $\|\mathbf{x}\|_2 \leq R_{\mathbf{x}}$ with probability 1 by some $R_{\mathbf{x}}$. Then, for every $\delta \in (0, 1)$, with probability at least $1 - \delta$ the following holds for all $f \in \mathcal{F}$:*

$$L(f) \leq \hat{L}_n(f) + 2\mathcal{L}\sqrt{\frac{2(W^2 + V^2)}{n}} R_{\mathbf{x}} + R\sqrt{\frac{\log \delta^{-1}}{2n}} \tag{15}$$

where $W = \sqrt{\frac{R}{\mu}}, V = \frac{R}{\lambda}$.

Instead of relying on a uniform bound $R_{\mathbf{x}}$ on the data $\mathbf{x}_i$, we can give the following data-dependent bound on the risk.

Proposition 1 (Data-dependent risk bound for Huber regularization). *Let $\mathcal{F}$ be defined by Eq. 12, and let ℓ be a $\mathcal{L}$-Lipschitz continuous loss function. Then, for every $\delta \in (0, 1)$, with probability at least $1 - \delta$ the following holds for all $f \in \mathcal{F}$, where W, V, and R as defined as in Theorem 1:*

$$L(f) \leq \hat{L}_n(f) + 2\mathcal{L}\frac{\sqrt{2(W^2 + V^2)\sum_{i=1}^{n} \|\mathbf{x}_i\|^2}}{n} + (2\mathcal{L} + 1)R\sqrt{\frac{\log(\frac{2}{\delta})}{2n}}. \tag{16}$$

4.3   Lemmata and Auxilary Results

The risk bounds are based on the following three lemmas.

Lemma 1. *For the functional class $\mathcal{F}$ of Eq. 12, the following data-dependent bound on the empirical Rademacher complexity holds:*

$$\hat{\mathfrak{R}}_S(\mathcal{F}) \leq \frac{\sqrt{2(W^2 + V^2)\sum_{i=1}^{n} \|\mathbf{x}_i\|^2}}{n}. \tag{17}$$

Lemma 2. *For the functional class $\mathcal{F}$ of Eq. 12, the (theoretical) Rademacher complexity is bounded as follows:*

$$\mathfrak{R}_n(\mathcal{F}) = E_S[\hat{\mathfrak{R}}_S(\mathcal{F})] \leq \sqrt{\frac{2(W^2 + V^2)}{n}} R_{\mathbf{x}}. \tag{18}$$

where $R_{\mathbf{x}}$ is a constant such that $\|\mathbf{x}\|_2 \leq R_{\mathbf{x}}$ almost surely under P_X.

Lemma 3. *Let $(\hat{\mathbf{w}}, \hat{\mathbf{v}}) = \arg\min_{\mathbf{v}, \mathbf{w}} G(\mathbf{w}, \mathbf{v}, S)$. Then $\|\hat{\mathbf{w}}\|_2 \leq \sqrt{\frac{R}{\mu}}$, $\|\hat{\mathbf{v}}\|_2 \leq \frac{R}{\lambda}$, where R as in Theorem 1.*

Proof (Lemma 1). Following the ideas presented by Mohri [16], we rewrite the empirical Rademacher complexity using the Cauchy-Schwartz inequality:

$$
\begin{aligned}
\hat{\mathfrak{R}}_S(\mathcal{F}) &= \frac{1}{n} E_\sigma \left[\sup_{\|\mathbf{w}\|_2 \leq W, \|\mathbf{v}\|_2 \leq V} \sum_{i=1}^{n} (\sigma_i (\mathbf{w} + \mathbf{v})^\top \mathbf{x}_i) \right] \\
&= \frac{1}{n} E_\sigma \left[\sup_{\|\mathbf{w}\|_2 \leq W, \|\mathbf{v}\|_2 \leq V} (\mathbf{w} + \mathbf{v})^\top \sum_{i=1}^{n} \sigma_i \mathbf{x}_i \right] \\
&\leq \frac{1}{n} E_\sigma \left[\sup_{\|\mathbf{w}\|_2 \leq W, \|\mathbf{v}\|_2 \leq V} \|\mathbf{w} + \mathbf{v}\|_2 \left\| \sum_{i=1}^{n} \sigma_i \mathbf{x}_i \right\|_2 \right].
\end{aligned}
\tag{19}
$$

Using the inequality $\sum_{i=1}^{d}(w_i + v_i)^2 \leq 2 \sum_{i=1}^{d}(w_i^2 + v_i^2)$ for the right-hand side of Eq. 19, according to the restrictions on the norms of $\mathbf{w}, \mathbf{v}$ we get:

$$
E_\sigma \left[\sup_{\|\mathbf{w}\|_2 \leq W, \|\mathbf{v}\|_2 \leq V} \|\mathbf{w} + \mathbf{v}\|_2 \left\| \sum_{i=1}^{n} \sigma_i \mathbf{x}_i \right\|_2 \right] \leq \sqrt{2(W^2 + V^2)} E_\sigma \left[\left\| \sum_{i=1}^{n} \sigma_i \mathbf{x}_i \right\|_2 \right]
\tag{20}
$$

and because of Jensen's inequality for $E_\sigma [\|| \cdot |\|]$, linearity of expectation and independence of σ_i, σ_j for $j \neq i$ we obtain:

$$
\begin{aligned}
E_\sigma \left[\left\| \sum_{i=1}^{n} \sigma_i \mathbf{x}_i \right\|_2 \right] &\leq \sqrt{ E_\sigma \left[\sum_{i,j=1}^{n} \sigma_i \sigma_j \mathbf{x}_i{}^t \mathbf{x}_j \right] } \\
&= \sqrt{ \sum_{i=1}^{n} E_\sigma [\|\mathbf{x}_i\|_2^2] } = \sqrt{ \sum_{i-1}^{n} \|\mathbf{x}_i\|_2^2 }.
\end{aligned}
\tag{21}
$$

Uniting the results of Inequality 20 and Eq. 21 in Eq. 19 we get the statement of Lemma 1. □

Proof (Lemma 2). Using Lemma 1 and the assumption that the $\mathbf{x}_i$ are uniformly bounded by constant $R_\mathbf{x}$ we obtain:

$$
\hat{\mathfrak{R}}_S(\mathcal{F}) \leq \sqrt{ \frac{2(W^2 + V^2)}{n} } R_\mathbf{x}.
\tag{22}
$$

Equation 22 no longer depends on the sample, and therefore Lemma 2 follows. □

Naturally, one may not have any *a-priori* knowledge about the constants W and V that restrict the possible values of $\mathbf{w}$ and $\mathbf{v}$ in Inequality 18. Despite that, for a given optimization problem that includes the current class of models, one can apply certain arguments from which one can infer bounds for W and V. Lemma 3 gives us such bounds for Optimization Problem 1.

Proof (Lemma 3). When $(\hat{\mathbf{w}}, \hat{\mathbf{v}})$ is a solution of optimization problem (1), then

$$G(\hat{\mathbf{w}}, \hat{\mathbf{v}}, S) \leq G(\mathbf{0}, \mathbf{0}, S) \leq R$$

This implies that the optimal solution necessarily satisfies the following condition: $\lambda||\mathbf{v}||_1 + \mu||\mathbf{w}||_2 \leq R$. As far as $||\mathbf{v}||_1 \geq ||\mathbf{v}||_2$ we have that in order to be an optimal solution $\hat{\mathbf{v}}$ should satisfy following constraint: $||\mathbf{v}||_2 \leq \frac{R}{\lambda}$. For $\hat{\mathbf{w}}$ we obtain straightforward necessary condition, that $||\mathbf{w}||_2^2 \leq \frac{R}{\mu}$ which implies the claim of Lemma 3. □

Lemma 3 implies that the norms of the vectors $\mathbf{v}$ and $\mathbf{w}$ of a solution of Optimization Problem 1 necessary have to lie within balls with radius $W := \sqrt{\frac{R}{\mu}}$ for $\mathbf{w}$ and of radius $V := \frac{R}{\lambda}$ for $\mathbf{v}$, centered in the origin.

4.4 Proof of the Huber-Norm Risk Bounds

We are now equipped to prove Theorem 1.

Proof (Theorem 1). Lemma 2 gives us a bound on the Rademacher complexity of the functional class of Eq. 12, and Lemma 3 gives us necessary constraints on the norms W and V. Inserting both into Inequality 14, we obtain Theorem 1. □

Proof (Proposition 1). Lemma 1 gives us a data-dependent bound on the empirical Rademacher complexity of the functional class of Eq. 12. Adapting Inequality (3.14) from theorem 3.1 in Mohri et al. [16] for our needs, we have with probability at least $1 - \frac{\delta}{2}$:

$$\mathfrak{R}_n(\mathcal{F}) \leq \hat{\mathfrak{R}}_S(\mathcal{F}) + R\sqrt{\frac{\log(\frac{2}{\delta})}{2n}}. \tag{23}$$

Using the union bound for Inequality 14 (with $\frac{\delta}{2}$ instead of δ and constant R from Theorem 1) and Inequality 23, we get with probability $1 - \delta$:

$$L(f) \leq \hat{L}_n(f) + 2\mathcal{L}\hat{\mathfrak{R}}_S(\mathcal{F}) + 2\mathcal{L}R\sqrt{\frac{\log(\frac{2}{\delta})}{2n}} + R\sqrt{\frac{\log(\frac{2}{\delta})}{2n}}. \tag{24}$$

Together with Lemma 1 this yields the claim of Proposition 1. □

4.5 Corollaries

In practice, we will be interested in obtaining upper bounds for concrete loss functions such as the hinge loss $\ell(t, y) = \max(0, 1 - yt)$ or logistic loss $\ell(y, t) = \log(1 + \exp(-yt))$ in case of two-class classification problems. Since these loss functions are 1-Lipschitz [19], Theorem 1 produces therefore following corollaries.

Corollary 1. *For Optimization Problem 1 under the assumptions of Theorem 1 with loss-function $\ell(y,t) = \max(0, 1 - yt), t \in \mathbb{R}, y \in \{-1,1\}$ one obtains that, with probability at least $1 - \delta$ for all $f \in \mathcal{F}$:*

$$L(f) \le \hat{L}_n(f) + 2\sqrt{\frac{2(W^2 + V^2)}{n}} R_\mathbf{x} + B\sqrt{\frac{\log \delta^{-1}}{2n}} \tag{25}$$

where $W = \sqrt{\frac{1}{\mu}}$, $V = \frac{1}{\lambda}$, $B = 1 + \sqrt{2(W^2 + V^2)}R_\mathbf{x}$.

Corollary 2. *For Optimization Problem 1 under the assumptions of Theorem 1 with loss-function $\ell(y,t) = \log(1 + \exp(-yt)), t \in \mathbb{R}, y \in \{-1,1\}$ one obtains that with probability at least $1 - \delta$ for all $f \in \mathcal{F}$:*

$$L(f) \le \hat{L}_n(f) + 2\sqrt{\frac{2(W^2 + V^2)}{n}} R_\mathbf{x} + B^l\sqrt{\frac{\log \delta^{-1}}{2n}} \tag{26}$$

where $W = \sqrt{\frac{\log 2}{\mu}}$, $V = \frac{\log 2}{\lambda}$, $B^l := \frac{\exp(R_\mathbf{x}\sqrt{2(W^2+V^2)})}{\sqrt{\exp(R_\mathbf{x}\sqrt{2(W^2+V^2)})+1}}$.

Proof. For the hinge loss, under the conditions of Theorem 1, we have that for any $\mathbf{x} \in \mathbb{R}^d$, s.t, $||\mathbf{x}||_2 \le R_\mathbf{x}$ the loss is bounded by $1 + |(\mathbf{w}+\mathbf{v})^\top \mathbf{x}|$, which is upper-bounded by $1 + \sqrt{2(W^2 + V^2)}R_\mathbf{x}$ as the combination of bounds on $||\mathbf{w} + \mathbf{v}||_2$ and $||\mathbf{x}||_2$. So, $|\ell(t,y)| \le B := 1 + \sqrt{2(W^2 + V^2)}R_\mathbf{x}$. Then the conclusion follows by applying Theorem 1. The proof for the logistic loss is analogous.   □

4.6   Discussion of Results

We will now compare the generalization performance of the developed *Huber-norm regularizer* with the performance of known regularizers.

Comparison to ℓ_1 and ℓ_2-Norm Regularization. The optimization problems of the ℓ_2-norm and ℓ_1-norm empirical risk minimization are

$$\hat{\mathbf{w}} = \arg\min_\mathbf{w} \mu||\mathbf{w}||_2^2 + \frac{1}{n}\sum_{i=1}^n \ell(y_i, \mathbf{w}^T \mathbf{x}_i) \text{ and} \tag{27}$$

$$\hat{\mathbf{v}} = \arg\min_\mathbf{v} \lambda||\mathbf{v}||_1 + \frac{1}{n}\sum_{i=1}^n \ell(y_i, \mathbf{v}^T \mathbf{x}_i), \tag{28}$$

respectively. Theoretical upper bounds on the statistical risk for both Eqs. 27 and 28 result from Mohri [16] for the Rademacher complexity of linear models. In these cases, the upper bound on the Rademacher complexity is also of the order of $\sqrt{\frac{1}{n}}$ and depends as well on the bounds on norms of the vectors W, V (for each case separately) and on the bounds on the data.

Comparison to Elastic Net. The optimization problem of the *empirical risk minimization with elastic-net regularizer* is

$$\hat{\mathbf{w}} = \arg\min_{\mathbf{w}} \lambda||\mathbf{w}||_1 + \mu||\mathbf{w}||_2^2 + \frac{1}{n}\sum_{i=1}^n \ell(y_i, \mathbf{w}^\top \mathbf{x}_i) \qquad (29)$$

with $\ell(y, 0) = 1$ [26]. From a similar argumentation as in Theorem 1 [11,16] one can infer that upper bounds on the Rademacher complexity for this procedure will also be of order $\mathcal{O}(\sqrt{\frac{W^2 R_\mathbf{x}^2}{n}})$, where now $W = \sqrt{\frac{1}{\lambda+\mu}}$ and $R_\mathbf{x}$ as before.

Oracle Inequality. We will relate the generalization performance of the model to the performance of the best possible model in that class—which is unknown in practice—using an oracle-type inequality [4,12]. As a corollary of Theorem 1, we can obtain an oracle-type inequality in high probability for $\mathcal{F}$:

$$G(\hat{\mathbf{w}}, \hat{\mathbf{v}}, S) \leq \arg\min_{(\mathbf{w},\mathbf{v})} G(\mathbf{w}, \mathbf{v}, S) + 2\Delta,$$

where Δ is the parameter that defines the complexity of $(\hat{\mathbf{w}}, \hat{\mathbf{v}}) \in \mathcal{F}$ and is given explicitly in the following Proposition 2 that follows from Theorem 1.

Proposition 2. *Let all conditions of Theorem 1 hold, let $(\hat{\mathbf{w}}, \hat{\mathbf{v}}) = \arg\min_{\mathbf{w},\mathbf{v}} G(\mathbf{w}, \mathbf{v}, S)$, and let $W, V, R_\mathbf{x}$, and R be defined in Theorem 1. Then with probability at least $1 - \delta$:*

$$G(\hat{\mathbf{w}}, \hat{\mathbf{v}}, S) - \arg\min_{\mathbf{w},\mathbf{v}} G(\mathbf{w}, \mathbf{v}, S) \leq 2\mathcal{L}\sqrt{\frac{2(W^2 + V^2)}{n}} R_\mathbf{x} + R\sqrt{\frac{\log \delta^{-1}}{2n}}. \qquad (30)$$

Tightness Comparison. Comparing the order of our upper risk bound with classical results for empirical risk minimization problems [1], [5] one can see that our bound is tight, and of order $\sqrt{\frac{1}{n}}$.

5   Experiments

This section compares *logistic regression with Huber-norm regularization* to *logistic regression with ℓ_1, with ℓ_2, and with elastic-net regularization*.

5.1   Experimental Setting

We conduct experiments on benchmark problems from the UCI repository [15]. In order to avoid a possible selection bias, we select the 31 first (in alphabetical order) classification problems that use matrix data format. We skip trivial problems for which all models achieve perfect accuracy. We transform categorical features into binary values using one-hot coding. For multi-class problems, we removed classes that have fewer instances than the number of cross-validation folds. All features are centered and scaled to unit variance. Missing values are

Table 1. Accuracies and standard errors for UCI data sets

Data Set	ℓ_1 regularization	Elastic-net reg.	ℓ_2 reg.	Huber reg.
Abalone	0.236 ± 0.008	0.236 ± 0.008	0.238 ± 0.015	**0.262 ± 0.016**∗
Arrhythmia	0.687 ± 0.044	0.683 ± 0.049	0.634 ± 0.053	**0.722 ± 0.033**∗
Audiology	0.576 ± 0.071	0.688 ± 0.045	0.738 ± 0.044	**0.748 ± 0.055**
Balance-scale	0.907 ± 0.016	0.910 ± 0.015	0.910 ± 0.015	**0.957 ± 0.019**∗
Bank	0.899 ± 0.001	0.899 ± 0.000	0.899 ± 0.000	**0.901 ± 0.001**∗
Banknote	0.977 ± 0.011	0.976 ± 0.011	0.977 ± 0.011	**0.991 ± 0.004**
Blood	0.770 ± 0.010	0.769 ± 0.010	0.771 ± 0.013	**0.774 ± 0.012**
Breast-canc	0.689 ± 0.029	0.692 ± 0.039	0.696 ± 0.052	**0.710 ± 0.065**
Breast-canc-wisc	0.963 ± 0.017	0.970 ± 0.009	0.953 ± 0.013	**0.973 ± 0.012**
Breast-canc-wisc-dia	0.952 ± 0.032	0.952 ± 0.031	0.959 ± 0.019	**0.977 ± 0.017**
Breast-tissue	0.879 ± 0.083	0.878 ± 0.060	0.878 ± 0.060	**0.907 ± 0.044**
Car	0.841 ± 0.012	0.842 ± 0.011	0.841 ± 0.010	**0.896 ± 0.005**∗
Climate-model	0.915 ± 0.004	0.915 ± 0.004	0.915 ± 0.004	**0.955 ± 0.018**∗
Congress-voting	**0.956 ± 0.029**	0.956 ± 0.029	0.954 ± 0.025	0.954 ± 0.032
Conn-sonar	0.746 ± 0.050	0.760 ± 0.071	0.736 ± 0.036	**0.770 ± 0.036**
Contraceptive	0.506 ± 0.041	0.505 ± 0.041	0.508 ± 0.042	**0.512 ± 0.035**
Credit-approval	0.851 ± 0.012	0.855 ± 0.018	0.859 ± 0.015	**0.862 ± 0.009**
Cylinder-bands	0.746 ± 0.014	0.780 ± 0.020	**0.802 ± 0.025**	0.798 ± 0.016
Dermatology	**0.975 ± 0.027**	0.965 ± 0.031	0.970 ± 0.025	0.970 ± 0.026
Echocardiogram	0.757 ± 0.058	0.770 ± 0.075	0.784 ± 0.119	**0.797 ± 0.106**
Ecloi	0.840 ± 0.034	0.840 ± 0.034	0.837 ± 0.032	**0.871 ± 0.070**
First-order	0.822 ± 0.001	**0.822 ± 0.001**	0.822 ± 0.002	0.821 ± 0.001
Flags	0.675 ± 0.046	**0.691 ± 0.029**	0.659 ± 0.032	0.670 ± 0.022
Glass	0.588 ± 0.042	0.583 ± 0.033	0.592 ± 0.056	**0.603 ± 0.052**
Haberman-survival	**0.735 ± 0.005**	0.726 ± 0.019	0.684 ± 0.114	0.709 ± 0.038
Hepatitis	0.800 ± 0.075	0.806 ± 0.067	0.806 ± 0.077	**0.815 ± 0.111**
Horse-colic	0.831 ± 0.025	**0.848 ± 0.041**	0.826 ± 0.025	0.845 ± 0.024
Image-segmentation	0.829 ± 0.010	0.833 ± 0.011	0.846 ± 0.007	**0.865 ± 0.127**
Ionosphere	**0.880 ± 0.034**	0.866 ± 0.012	0.878 ± 0.038	0.878 ± 0.042
Iris	0.840 ± 0.060	0.833 ± 0.047	0.833 ± 0.071	**0.960 ± 0.015**∗
Leaf	0.644 ± 0.048	0.665 ± 0.065	0.675 ± 0.036	**0.834 ± 0.036**∗

filled in using mean imputation for continuous values and are represented as a separate one-hot coded attribute for categorical values.

We run nested stratified cross validation with an outer loop of five folds. Regularization parameters $[\lambda, \mu]$ are tuned by an inner loop of three-fold cross validation on the training portion over the grid of $[10^{-5}, ..., 10^3] \times [10^{-3}, ..., 10^4]$.

5.2 Results

Table 1 shows the accuracies of different regularizers. For each problem the highest empirical accuracy is typeset in bold face; asterisks mark models that are significantly better than the best of the other three models, based on a paired t test with $p < 0.05$. *logistic regression with Huber-norm regularization* achieves the highest empirical accuracy for 23 out of 31 problems; its accuracy is significantly higher than the accuracy of any other model for 8 problems. No reference methods outperform *Huber-norm regularization* significantly.

The UCI repository reflects a certain distribution $P(S)$ over data sets. We state the null hypothesis A that the probability of *Huber-norm regularization* outperforming all three reference methods on a randomly drawn problem under $P(S)$ does not exceed 0.5, and the null hypothesis B that the probability of *Huber-norm regularization* outperforming all three reference methods on a randomly drawn problem under $P(S)$ is below 0.5. We count each cross-validation fold of each UCI data set as a single observation of a binary random variable and determine the binomial likelihood of observing the outcomes which are reflected in Table 1. *Logistic regression with Huber-norm regularization* achieves a higher empirical accuracy than all three baselines in 86 out of 155 cross-validation folds, and an equally high accuracy as the best baseline in an additional 24 cases. We can therefore reject the null hypothesis A at $p = 0.09$ and null hypothesis B even at $p < 0.001$. We conclude that for the distribution of UCI problems, the *Huber-norm regularization* is the best-performing regularizer among the ℓ_1, ℓ_2, *elastic-net* and *Huber regularization*.

6   Conclusions

We proposed a new way of regularizing linear prediction models based on a combination of dense and sparse weight vectors. In more detail, we employ a linear weight vector that is the sum of two terms, $\mathbf{w} + \mathbf{v}$, where $\mathbf{w}$ is ℓ_2 regularized and $\mathbf{v}$ is ℓ_1 regularized. This results in an effective *Huber-norm regularizer* for $\mathbf{w} + \mathbf{v}$, which is very different from an elastic net. Starting with theoretical considerations, we first derived bounds on the statistical risk based on the framework of Rademacher complexities. In our subsequent experimental study, our algorithm showed higher predictive accuracies on a majority of UCI data sets, where we compared against ℓ_1, ℓ_2, and elastic-net regularization. In future work, we would like to study extensions to non-linear kernel functions and multiple kernels [9].

Acknowledgments. MK acknowledges support from the German Research Foundation (DFG) award KL 2698/2-1 and from the Federal Ministry of Science and Education (BMBF) award 031L0023A. The authors would like to thank Christoph Lippert, Gilles Blanchard, Florian Wenzel, and Shinichi Nakajima for helpful discussions.

References

1. Bartlett, P.L., Bousquet, O., Mendelson, S.: Local rademacher complexities. Ann. Stat. **33**, 1497–1537 (2005)
2. Bartlett, P.L., Mendelson, S.: Rademacher and gaussian complexities: risk bounds and structural results. In: Helmbold, D.P., Williamson, B. (eds.) COLT 2001 and EuroCOLT 2001. LNCS (LNAI), vol. 2111, pp. 224–240. Springer, Heidelberg (2001)
3. Boyd, S., Vandenberghe, L.: Convex Optimization. Cambridge University Press, New York (2004)
4. Clarke, B., Fokoué, E., Zhang, H.: Principles and Theory for Data Mining and Machine Learning. Springer, New York (2009)
5. Devroye, L., Lugosi, G.: Lower bounds in pattern recognition and learning. Pattern Recogn. **28**(7), 1011–1018 (1995)
6. Hailong, H., Haien, L., Jianwei, L.: P-norm regularized SVM classifier by non-convex conjugate gradient algorithm. In: Proceedings of the Chinese Control and Decision Conference, vol. 3, pp. 2685–2690. IEEE (2013)
7. Huber, P.: Robust estimation of a location parameter. Ann. Math. Stat. **53**, 73–101 (1964)
8. Huber, P.J.: Robust Statistics. Springer, Heidelberg (2011)
9. Kloft, M., Brefeld, U., Sonnenburg, S., Zien, A.: lp-norm multiple kernel learning. J. Mach. Learn. Res. **12**, 953–997 (2011)
10. Kloft, M., Brefeld, U., Laskov, P., Müller, K.R., Zien, A., Sonnenburg, S.: Efficient and accurate lp-norm multiple kernel learning. In: Advances in Neural Information Processing Systems, vol. 22, pp. 997–1005. Curran Associates, Inc. (2009)
11. Kloft, M., Rückert, U., Bartlett, P.L.: A unifying view of multiple kernel learning. In: Balcázar, J.L., Bonchi, F., Gionis, A., Sebag, M. (eds.) ECML PKDD 2010. LNCS (LNAI), vol. 6322, pp. 66–81. Springer, Heidelberg (2010). doi:10.1007/978-3-642-15883-4_5
12. Koltchinskii, V.: Oracle Inequalities in Empirical Risk Minimization and Sparse Recovery Problems, vol. 2033. Springer, Heidelberg (2011)
13. Koshiba, Y., Abe, S.: Comparison of l1 and l2 support vector machines. In: Proceedings of the International Joint Conference on Neural Networks, vol. 3, pp. 2054–2059. IEEE (2003)
14. Kujala, J., Aho, T., Elomaa, T.: A walk from 2-norm SVM to 1-norm SVM. In: Proceedings of the IEEE Conference on Data Mining, pp. 836–841. IEEE (2009)
15. Lichman, M.: UCI machine learning repository (2013). http://archive.ics.uci.edu/ml
16. Mohri, M., Rostamizadeh, A., Talwalkar, A.: Foundations of Machine Learning. MIT Press, Cambridge (2012)
17. Owen, A.: A robust hybrid of lasso and ridge regression. Contemp. Math. **443**, 59–72 (2007)
18. Robert, T.: The Lasso method for variable selection in the Cox model. Stat. Med. **16**, 385–395 (1997)
19. Steinwart, I., Christmann, A.: Support Vector Machines. Springer, New York (2008)
20. Suykens, J.A., De Brabanter, J., Lukas, L., Vandewalle, J.: Weighted least squares support vector machines: robustness and sparse approximation. Neurocomputing **48**(1), 85–105 (2002)

21. Unger, M., Pock, T., Werlberger, M., Bischof, H.: A convex approach for variational super-resolution. In: Goesele, M., Roth, S., Kuijper, A., Schiele, B., Schindler, K. (eds.) Pattern Recognition. LNCS, vol. 6376, pp. 313–322. Springer, Heidelberg (2010)
22. Vapnik, V.: The Nature of Statistical Learning Theory. Springer, New York (1995)
23. Wang, L., Jia, H., Li, J.: Training robust support vector machine with smooth ramp loss in the primal space. Neurocomputing **71**(13), 3020–3025 (2008)
24. Yuille, A.L., Rangarajan, A.: The concave-convex procedure. Neural Comput. **15**(4), 915–936 (2003)
25. Zhu Ji, S.R., Trevor, H., Rob, T.: 1-norm support vector machines. Advances of Neural Information Processing Systems (2004)
26. Zou, H., Hastie, T.: Regularization and variable selection via the elastic net. J. Roy. Stat. Soc. B **67**(2), 301–320 (2005)

Robust Dictionary Learning
on the Hilbert Sphere in Kernel Feature Space

Suyash P. Awate$^{(\boxtimes)}$ and Nishanth N. Koushik

Computer Science and Engineering Department,
Indian Institute of Technology (IIT) Bombay, Mumbai, India
suyash@cse.iitb.ac.in

Abstract. This paper presents a novel dictionary learning method in
kernel feature space that is part of a reproducing kernel Hilbert space
(RKHS). Our method focuses on several popular kernels, e.g., radial basis
function kernels like the Gaussian, that implicitly map data to a Hilbert
sphere, a Riemannian manifold, in RKHS. Our method exploits this manifold structure of the mapped data in RKHS, unlike typical methods for
kernel dictionary learning that use linear methods in RKHS. We show that
dictionary learning on a Hilbert sphere in RKHS is possible without the
need of the explicit lifting map underlying the kernel, but using solely the
Gram matrix. Unlike the typical L^1 norm sparsity prior, we incorporate
the non-convex L^p quasi-norm based penalty, with $p < 1$, on coefficients to
enforce a stronger sparsity prior and achieve more robust dictionary learning in the presence of corrupted training data. We evaluate our method for
image classification on two large publicly available datasets and demonstrate the improved performance of our method over the state of the art
dictionary learning methods.

1 Introduction

Methods for modeling data as sparse linear combinations of a set of basis elements, often referred to as a dictionary, have found a wide spectrum of applications in machine learning, signal and image processing, and statistical analyses [20,30–32,48,54]. Each element of the dictionary is often referred to as an
atom. Dictionary-based modeling leads to optimization problems that can be
thought of as posterior mode estimation problems [45], where the dictionary fit
relates to the likelihood term and the sparsity-based regularization relates to a
prior term on the coefficients in the linear combination. The problem of fitting
a given dictionary to the data is a sparse-regression problem for which different
sparsity penalties lead to different forms of regression such as the Lasso [54] and
subset selection [30], where the efficacy of the Lasso formulation exploits the convexity of the underlying optimization problem [18,31]. The principles of sparse

We thank funding through IIT Bombay Seed Grant 14IRCCSG010.

© Springer International Publishing AG 2016
P. Frasconi et al. (Eds.): ECML PKDD 2016, Part I, LNAI 9851, pp. 731–748, 2016.
DOI: 10.1007/978-3-319-46128-1_46

modeling also play a key role in compressed sensing [20] during reconstruction of signals from corrupted and missing data.

When the data exhibits a nonlinear manifold structure within Euclidean space, the representation of the data using a linear combination of atoms can become inefficient because the linearity in the dictionary representation can fail to adapt to the nonlinearity of the data distribution. Better fits may be obtained by increasing the number of atoms in the dictionary, but doing that increases the complexity of the model and makes the problems of dictionary learning and fitting more difficult, often leading to higher variance [30]. A standard way of adapting to the nonlinearity in the data in input space is to use a nonlinear *kernel* [49,50] to (implicitly) map the data to a high-dimensional kernel feature space, where the nonlinearity in the distribution of the mapped data in kernel feature space is significantly reduced. This mapping to the kernel feature space is typically denoted by $\Phi(\cdot)$. Subsequently, dictionary learning methods can employ standard Euclidean/linear learning in kernel feature space, utilizing the kernel trick to avoid the need to explicitly map the data to the high-dimensional kernel feature space. Our method also exploits kernels and the kernel trick for learning a dictionary model, but, furthermore, exploits the additional spherical structure of the mapped data in the kernel feature space associated with several popular kernels.

In some special cases, the data, in input space, resides on a *Riemannian manifold* that is analytically known, e.g., the space of symmetric positive definite matrices [6,7,14,21,47,51], the Grassmann manifold [15,55], the hypersphere [40,53], or shape space [27,33,36]. In such cases, the dictionary model and the learning [14,29,52,61,65] can exploit the known structure of the manifold to provide atoms that also, desirably, reside on the manifold and efficiently capture the variability in the data. Our method shares the spirit of these approaches in exploiting any known geometrical structure of the data for better modeling and improving performance in practice.

Our method relies on standard kernels that (implicitly) map the data in input space to a known manifold in kernel feature space, specifically, the unit Hilbert sphere in a reproducing kernel Hilbert space (RKHS). Such a mapping occurs for (i) several common kernels [23] including the radial basis function (RBF) kernels like the Gaussian, (ii) kernel normalization, which is common, e.g., in pyramid match kernel [26], and (iii) polynomial and sigmoid kernels when the input points have constant L^2 norm, which is common in certain image analyses [49]. This special structure arises because for these kernels $\kappa(\cdot, \cdot)$, the self similarity of any data point x equals unity, i.e., $\kappa(x, x) = 1$. The kernel defines the inner product in the RKHS $\mathcal{H}$, and thus, $\langle \Phi(x), \Phi(x) \rangle_{\mathcal{H}} = 1$, which, in turn, equals the distance of the mapped point $\Phi(x)$ from the origin in $\mathcal{H}$. Thus, all of the mapped points $\Phi(x)$ lie on a Hilbert sphere in RKHS. Figure 1(a) illustrates this behavior. Subsequently, we exploit the Riemannian structure of the Hilbert sphere in RKHS, on which the mapped data resides, to perform dictionary learning in RKHS.

An ideal notion of sparsity relates to the subset-selection problem [30] or, equivalently, constraints or penalties related to the L^0 pseudo-norm [17] of the coefficients for atoms in the dictionary fit. However, because this ideal notion of sparsity leads to combinatorial optimization problems that are NP hard, typical approaches constrain or penalize the L^1 norm of the coefficients. We know that the L^0 pseudo-norm penalty is actually the limiting case of penalizing the p-th power of the L^p quasi-norm as p approaches zero. In fact, there are close theoretical relationships [22,64] between L^0 regularization/combinatorial optimization and L^p-based regularization for sufficiently small $p > 0$. In practice, compared to the squared-L^2-norm penalty, the use of the L^1 norm leads to increased robustness to corruption and outliers in the data [56,58] via a stronger sparsity-promoting penalty. Extending the argument, for our dictionary learning framework, we incorporate the non-convex L^p-to-the-power-p penalties with $p < 1$ on coefficients to enforce a stronger sparsity-promoting penalty and achieve more robust dictionary learning, compared to the L^1-norm penalty.

This paper makes the following contributions. First, we propose a novel dictionary learning framework adapted to the Riemannian manifold of the Hilbert sphere in RKHS. In this way, we combine the advantages of kernel-based frameworks together with the benefits of adapting the analysis to the Riemannian manifold of the (mapped) data. We show that such a learning algorithm does *not* require the knowledge of the explicit mapping function $\Phi(\cdot)$ implied by the kernel $\kappa(\cdot, \cdot)$, but only requires the Gram matrix. Second, our method relies on a stronger notion of sparsity, i.e., the L^p quasi-norm for $p < 1$, which is similar to the desired L^0 sparsity for sufficiently small p, thereby increasing the robustness of the learning algorithm to the noise and outliers in the data. Third, it uses empirical evaluations on two publicly available large real-world datasets to demonstrate the advantages of (i) Riemannian modeling in RKHS as well as (ii) sparsity priors stronger than the typical L^1 norm.

2   Related Work

This section describes the relationships of our method to several related works in the literature, including the state of the art.

Some recent works have presented dictionary learning on the manifolds of symmetric positive definite matrices [14,52,61] and the Grassmann manifold [29,65], which rely on adapting the analyses to the Riemannian metric underlying the manifold. However, the data can have significantly nonlinear distributions even within such manifolds, where typical dictionary models, relying on linear combinations of atoms, can be challenged. Unlike these methods, we rely on (i) kernel-based mapping to reduce the nonlinearity in the data (even when it may lie on a manifold) and (ii) a stronger L^p quasi-norm based sparsity model for robustness to large amounts of corruptions in the training data.

The recent interesting work in [41] exploits kernels for learning dictionaries in RKHS, but does *not* exploit the manifold structure of the mapped data in the RKHS. The method in [41] relies on a greedy approximation algorithm (orthogonal matching pursuit) to handle the sparsity constraint. Another very recent work [28] adapts dictionary learning for non-Euclidean data by designing the kernel based on the Riemannian metric of the manifold (in input space) on which the data lies. Thus, while [28] exploits the manifold structure in the input space, it ignores the manifold structure in the RKHS. Our framework allows exploitation of the manifold structure of the data in both the input space (via kernel design) and the RKHS (via spherical statistics). We also demonstrate robustness to large levels of noise and outliers.

Several previous works [1,9,16,19,25,49] perform statistical analyses on the Hilbert sphere in RKHS, but are outside the realm of dictionary learning and sparse modeling. Nevertheless, the spirit of these works, in adapting the analyses to the Hilbert sphere in RKHS to improve performance, is akin to the spirit of our approach. This prior knowledge of the mapped data (and atoms) lying on a sphere leads to added regularization that can prevent overfitting in high-dimension low-sample-size scenarios where Euclidean analysis is known to be unstable and error prone as it interacts with the curvature of the sphere on which the data resides [2]. For directional data, although distributions modeling the covariance structure exist in the literature, the underlying parameter estimation is intractable in high-dimensional spaces [10,40,46]; furthermore, these distributions don't exploit sparsity-based regularization. Indeed Gaussian and Factor models have exploited sparsity as a regularizer [37,59].

Several applications require learning that is robust to large levels of corruption in the (training) data. Some approaches [39,44] achieve robustness to non-Gaussianity of the data distributions, resulting from non-Gaussianity of the noise/likelihood model, by replacing the squared-L^2-norm penalty on the residual arising from the data fitting term by the L^1-norm penalty. Other approaches [34] further propose limiting such a L^1-norm penalty by capping the penalty at a fixed level. Alternate strategies [13] continue to penalize the squared L^2 norm of the residual, but modify the residual to explicitly include outlier variables such that residual distribution remains close to a Gaussian. Instead of modifying the data-fit (likelihood) term, robustness can result from using a stronger sparsity prior that prevents the overfitting of the dictionary to the (highly) corrupted data. In this spirit, some approaches achieve robustness by changing the squared-L^2-norm penalty on the coefficients to the L^1-norm penalty on the coefficients [56,58]; this is akin to replacing ridge regression by Lasso. Other approaches [42] use the L^1-norm penalty for the data fit along with a mixed $l_{2,1}$-norm penalty on the sparse matrix of coefficients to obtain robustness. In contrast to these approaches, our method employs L^p quasi-norm penalties for $p \in (0,1)$ for the sparsity prior to demonstrate improved robustness to corruption in data in the form of noise and outliers.

Alternate approaches to sparsity include the local coordinate coding or, rather, locally constrained sparse coding [62,63] that relies on the empirical observation that the sparse code for a datum is likely to exhibit non-zero coefficients for atoms that are in the locality of the datum. Subsequently, [62,63] redesign the standard L^1 penalty on coefficients as a weighted L^1 penalty, where, for each datum, the weight increases the penalty for atoms far from that datum. A fast approximate extension of local coordinate coding for image classification appears in [57] that replaces the dictionary with a local dictionary (nearest few atoms) for coding each datum; such schemes have also been used in RKHS [29]. Complementary to these approaches, our approach also adds to the conventional notion of L^1 sparsity, but it does so by using the L^p quasi-norm and tuning the parameter p to get solutions closes to the ones found with the L^0 pseudo-norm.

3   Riemannian Geometry of the Hilbert Sphere in RKHS

Many popular kernels are associated with an infinite-dimensional RKHS. So, the analysis in this paper focuses on such spaces. Nevertheless, analogous theory holds for other important kernels (e.g., normalized polynomial) where the RKHS is finite dimensional.

Let X be a random variable taking values x in *input space* $\mathcal{X}$. Let $\{x_n\}_{n=1}^N$ be a set of observations in input space. Let $\kappa(\cdot, \cdot)$ be a real-valued Mercer kernel with an associated map $\Phi(\cdot)$ that maps x to $\Phi(x) := \kappa(\cdot, x)$ in a RKHS $\mathcal{H}$ [5,49]. Consider two points in RKHS, within the span of the mapped data, represented as $f := \sum_{i=1}^I \alpha_i \Phi(x_i)$ and $f' := \sum_{j=1}^J \beta_j \Phi(x_j)$ where the weights $\alpha_i \in \mathbb{R}$ and $\beta_j \in \mathbb{R}$. The inner product $\langle f, f' \rangle_{\mathcal{H}} := \sum_{i=1}^I \sum_{j=1}^J \alpha_i \beta_j \kappa(x_i, x_j)$. The norm $\|f\|_{\mathcal{H}} := \sqrt{\langle f, f \rangle_{\mathcal{H}}}$.

Let $Y := \Phi(X)$ be the random variable taking values y in RKHS. Previous methods for kernel-based dictionary learning [28,41] model each y as a sparse linear combination of atoms in RKHS. The analysis in this paper applies to kernels that map points in input space to a Hilbert sphere in RKHS, i.e., $\forall x : \kappa(x, x) = \theta$, a constant (without loss of generality, we assume $\theta = 1$). Methods for non-Euclidean dictionary learning in input space [29,52,61,65] exploit the Riemannian structure of the manifold on which the data lie to propose dictionary learning on the manifold. The proposed dictionary-learning method exploits the property of y lying on the Riemannian manifold of the unit Hilbert sphere [4,11] in RKHS. Thus, we now introduce differential geometric constructs specific to this Hilbert sphere in RKHS [9].

Consider a and b on the unit Hilbert sphere in RKHS represented, in general, as $a := \sum_n \gamma_n \Phi(x_n)$ and $b := \sum_n \delta_n \Phi(x_n)$. The logarithmic map, or Log map, of a with respect to b is the vector

$$\text{Log}_b(a) = \frac{a - \langle a, b \rangle_{\mathcal{H}} b}{\|a - \langle a, b \rangle_{\mathcal{H}} b\|_{\mathcal{H}}} \arccos(\langle a, b \rangle_{\mathcal{H}}) \tag{1}$$

$$= \sum_n \zeta_n \Phi(x_n), \text{ where } \forall n : \zeta_n \in \mathbb{R}. \tag{2}$$

Clearly, $\text{Log}_b(a)$ can always be written as a linear combination of points $\{\Phi(x_n)\}_{n=1}^N$. The tangent vector $\text{Log}_b(a)$ lies in the tangent space, at b, of the unit Hilbert sphere. The tangent space to the Hilbert sphere in RKHS inherits the

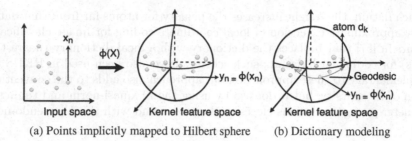

(a) Points implicitly mapped to Hilbert sphere   (b) Dictionary modeling

Fig. 1. Dictionary Modeling on a Hilbert Sphere in RKHS. (a) Points x_n in input space get mapped *implicitly*, via several popular Mercer kernels, to $y_n := \Phi(x_n)$ on a Hilbert sphere in RKHS. **(b)** Dictionary atoms d_k, on the Hilbert sphere in RKHS, being used to fit to a point y_n.

same structure (inner product) as the ambient space and, thus, is also a RKHS. The geodesic distance between a and b is $d_g(a, b) = \|\mathrm{Log}_b(a)\|_{\mathcal{H}} = \|\mathrm{Log}_a(b)\|_{\mathcal{H}}$.

Now, consider a tangent vector $t := \sum_n \beta_n \Phi(x_n)$ lying in the tangent space at b. The exponential map, or Exp map, of t with respect to b is

$$\mathrm{Exp}_b(t) = \cos(\|t\|_{\mathcal{H}})b + \sin(\|t\|_{\mathcal{H}})\frac{t}{\|t\|_{\mathcal{H}}} \tag{3}$$

$$= \sum_n \omega_n \Phi(x_n), \text{ where } \forall n : \omega_n \in \mathbb{R}. \tag{4}$$

Clearly, $\mathrm{Exp}_b(t)$ can always be written as a linear combination of points $\{\Phi(x_n)\}_{n=1}^N$. $\mathrm{Exp}_b(t)$ maps a tangent vector t to the unit Hilbert sphere, i.e., $\|\mathrm{Exp}_b(t)\|_{\mathcal{H}} = 1$.

4   Robust Dictionary Learning on a Hilbert Sphere in RKHS

Motivated by principles underlying sparse modeling, we model each $y = \Phi(x)$ in RKHS, where $\|y\|_{\mathcal{H}} = 1$, to have been generated as a sparse, but nonlinear, combination of atoms that also lie on the unit Hilbert sphere in RKHS. We consider the dictionary to have K atoms $\{d_k \in \mathcal{H}\}_{k=1}^K$, each with $\|d_k\|_{\mathcal{H}} = 1$. Given data $\{x_n\}_{n=1}^N$ and the number of atoms K, we learn the atoms $\{d_k\}_{k=1}^K$. Figure 1(b) gives an illustration.

In Euclidean spaces, dictionary learning typically penalizes the squared norm of the residual between the datum and its representation as a sparse linear combination of atoms. The natural generalization of this residual to a Riemannian manifold is the logarithmic map. The notion of sparsity in Euclidean spaces is generalized to the notion of affine sparsity in non-Euclidean spaces [61–63], which makes the dictionary representation independent of a coordinate frame shift or a notion of origin in the Riemannian space. Affine sparsity constraints the sum of weights, in the dictionary fit, to 1 and is important in the nonlinear sparse-coding setting because Riemannian analyses treats y as a point, unlike linear sparse coding in Euclidean space that treats y as a vector. Another motivation is as follows. Consider K atoms on the Hilbert sphere in RKHS in general position, i.e., these points are *not* contained in any great $(K - 2)$-sphere [43]. Given K points in general position on a Hilbert sphere, there is a unique great $(K - 1)$-sphere containing them [43]. Given K atoms in general position and a sparsity level $M \leq K$, principles underlying sparse modeling motivate us to model each y as lying within the unique great $(M - 1)$-sphere containing some M atoms. Thus, ideally we would desire to fit y by, say, $\widehat{y}$, where $\widehat{y}$ lies within such a great $(M-1)$-sphere. When $\widehat{y}$ lies within this great $(M-1)$-sphere, there exists a set of weights $w_m \in \mathbb{R}$ where

$$\sum_{m=1}^{M} w_m = 1 \text{ and } \sum_{m=1}^{M} w_m \text{Log}_{\widehat{y}}(d_m) = 0, \tag{5}$$

i.e., in the tangent space at $\widehat{y}$, there exist weights w_m that satisfy the affine constraint and make the weighted combination of vectors $\text{Log}_{\widehat{y}}(d_m)$ coincide with the origin. Thus, our fitting problem for y, given the dictionary, reduces to finding weights w_m that, under the sparsity prior and the affine constraints, minimize

$$\| \sum_{m=1}^{M} w_m \text{Log}_y(d_m) \|_{\mathcal{H}}. \tag{6}$$

This fitting term is in the same spirit as the fitting terms in [24,61].

However, because the ideal sparsity prior on weights (L^0 pseudo-norm [17] equivalent to subset selection [30]) leads to a combinatorial optimization problem that is NP hard, the sparsity prior is typically relaxed by (i) using all K atoms, instead of some subset of size $M < K$, for the fitting and (ii) penalizing the p-th power of the L^p norm of the weight vector w associated with the fit, where $p \in \mathbb{R}_{>0}$ is a free parameter. In the Euclidean case, the typical relaxed penalty is the L^1 norm of the weight vector, i.e., $p = 1$, that is motivated by the desire to retain the convexity of the optimization problem when the dictionary-fitting term is the squared residual between the datum and the linear combination of atoms. As with other manifolds, in the case of the Hilbert sphere, the dictionary-fitting term is itself nonconvex. Furthermore, choosing $p \in (0,1)$ leads to the penalty $\|w\|_p^p = \sum_m |w_m|^p$ that tends to the L^0 pseudo-norm as $p \to 0$ and improves the robustness of the dictionary learning to outliers.

We now formulate the robust dictionary learning problem. Let $\{x_n\}_{n=1}^{N}$ be the data points in input space. Let $y_n := \Phi(x_n)$ be the (implicitly) mapped points in RKHS where the mapping $\Phi(\cdot)$ ensures that $\|y_n\|_{\mathcal{H}} = 1$. We do *not* need the mapping $\Phi(\cdot)$ explicitly, but use the kernel trick for all the analysis in RKHS. Let the dictionary D comprise K atoms d_k, as described before. Let W be the weight matrix comprising columns w_n that comprise the weights w_{nk} for the contribution of the k-th atom in the fit for the n-th point y_n. Then, we propose to formulate robust dictionary learning as

$$\arg\min_D \left[\min_W \sum_{n=1}^{N} \left(\|\sum_{k=1}^{K} w_{nk} \mathrm{Log}_{y_n}(d_k)\|_{\mathcal{H}}^2 + \lambda \|w_n\|_{p,\epsilon}^p \right) \right]$$

$$\text{under the constraints: } \forall n, \sum_{k=1}^{K} w_{nk} = 1 \text{ and } \forall k, \|d_k\|_{\mathcal{H}} = 1, \tag{7}$$

where $\lambda > 0$ is the regularization parameter balancing the data-fitting/fidelity term and the sparsity prior term, $p \in (0,1)$ is a free parameter, and $\|w_n\|_{p,\epsilon}^p$ is the p-th power of the ϵ-regularized L^p quasi-norm

$$\|w_n\|_{p,\epsilon}^p := \sum_{k=1}^{K} (|w_{nk}|^2 + \epsilon)^{p/2} \tag{8}$$

that is smooth and amenable to gradient-based optimization. The affine constraint is useful here, without which the formulation leads to the trivial solution: $w_{nk} = 0, \forall n, k$.

Because the dictionary fit attempts to minimize the norm of a linear combination of tangent vectors $\mathrm{Log}_{y_n}(d_k)$, it is natural to have atoms d_k within a subsphere $\mathcal{S}$ that corresponds to the intersection of (i) the linear subspace in RKHS representing the span of the set of points y_n with (ii) the unit Hilbert sphere in RKHS. For instance, if all but one atom d_l lies outside the span of the data, then either (i) that atom remains unused, i.e., all weights w_{nl} for atom d_l are zero, or (ii) the use of atom d_l leads to an increase in the norm of the linear combination, as compared to using an atom that is the projection of d_l on the subsphere $\mathcal{S}$, because it leads to an additional vector component in the linear combination along a direction orthogonal to the other log-mapped atoms $\mathrm{Log}_{y_n}(d_l)$ in the tangent space at y_n. Thus, if d_l is to be used, replacing d_l with its projection on the subsphere $\mathcal{S}$ will reduce the objective function and be a better solution. Take another instance where multiple atoms d_k lie outside the span of the data, or equivalently outside the lowest-dimensional subsphere $\mathcal{S}$ that contains all the mapped points y_n. In this case, if there is a sparse combination of atoms d_k that exactly fits the data, i.e., $\|\sum_{k=1}^{K} w_{nk} \mathrm{Log}_{y_n}(d_k)\|_{\mathcal{H}} = 0$, then we can use the same weights w_{nk} and use the projections of atoms d_k on the subsphere $\mathcal{S}$ to again get the exact fit. Thus, it is unnecessary to use atoms d_k outside the subsphere $\mathcal{S}$. So, we represent each atom as

$$d_k := \sum_{n=1}^{N} \gamma_{kn} \Phi(x_n), \text{ where } \forall n : \gamma_{kn} \in \mathbb{R} \text{ and } \|d_k\|_{\mathcal{H}} = 1. \tag{9}$$

This representation for atoms ensures that each logarithmic map $\mathrm{Log}_{y_n}(d_k)$ can be represented as a linear combination of the $\Phi(y_n)$.

Within the first term in the objective function, $\forall y_n$, the norm $\|\sum_{k=1}^{K} w_{nk} \mathrm{Log}_{y_n}(d_k)\|_{\mathcal{H}}^2$ can be represented purely in terms of the Gram matrix and the coefficients γ_{kn} underlying the atoms' representation, as follows. Each tangent vector $\mathrm{Log}_{y_n}(d_k)$ can be represented as a weighted combination of $\Phi(x_n)$.

Thus, the linear combination $\sum_{k=1}^{K} w_{nk}\mathrm{Log}_{y_n}(d_k)$ is also weighted combination of $\Phi(x_n)$. Then, the norm

$$\|\sum_{k=1}^{K} w_{nk}\mathrm{Log}_{y_n}(d_k)\|_{\mathcal{H}}^2 = \|\sum_{p=1}^{N} \xi_p \Phi(x_p)\|_{\mathcal{H}}^2, \text{ for some weights } \xi_p \in \mathbb{R} \qquad (10)$$

$$= \sum_{n'=1}^{N}\sum_{n''=1}^{N} \xi_{n'}\xi_{n''}\langle \Phi(x_{n'}), \Phi(x_{n''})\rangle_{\mathcal{H}} = \sum_{n'=1}^{N}\sum_{n''=1}^{N} \xi_{n'}\xi_{n''}G_{n'n''}, \qquad (11)$$

where G is the Gram matrix such that $G_{n'n''} := \langle \Phi(x_{n'}), \Phi(x_{n''})\rangle_{\mathcal{H}}$ with all diagonal elements $G_{nn} := \langle \Phi(x_n), \Phi(x_n)\rangle_{\mathcal{H}} = 1$. The scalar weights ξ_p are

$$\xi_p = \sum_{k=1}^{K} w_{nk}\frac{\arccos(\langle d_k, y_n\rangle_{\mathcal{H}})}{\|d_k - \langle d_k, y_n\rangle_{\mathcal{H}}y_n\|_{\mathcal{H}}}\left(\gamma_{kp} - \mathbf{I}_{n,p}\langle d_k, y_n\rangle_{\mathcal{H}}\right), \qquad (12)$$

where the indicator function $\mathbf{I}_{n,p} = 1$ if $n = p$ and $\mathbf{I}_{n,p} = 0$ otherwise.

The constraint on the atoms, $\|d_k\|_{\mathcal{H}} := \|\sum_{n=1}^{N} \gamma_{kn}\Phi(x_n)\|_{\mathcal{H}} = 1$, can also be specified in terms of the Gram matrix G as

$$\sum_{n'=1}^{N}\sum_{n''=1}^{N} \gamma_{kn'}\gamma_{kn''}G_{n'n''} = 1. \qquad (13)$$

This confirms that the optimization problem, both the objective function and the constraints, can be specified purely in terms of the variables $\{\{w_{nk} \in \mathbb{R}\}_{k=1}^{K}\}_{n=1}^{N}$ and $\{\{\gamma_{kn} \in \mathbb{R}\}_{n=1}^{N}\}_{k=1}^{K}$, given the Gram matrix G. Hence, we do *not* need the explicit mapping $\Phi(\cdot)$ for our dictionary learning method.

We optimize the atoms d_k, represented through the parameters $\{\gamma_{kn} \in \mathbb{R}\}$, and weights $w_n \in \mathbb{R}^K$ alternatingly using projected gradient descent that guarantees convergence to a stationary point. For each atom d_k, the projection is on the unit Hilbert sphere, which involves a simple rescaling of the parameters γ_{kn}. For each weight vector w_n, the projection is on the hyperplane through the origin $\sum_{k=1}^{K} w_{nk} = 1$, which is also straightforward. We adjust the step sizes, using a line search, to ensure that the objective function decreases. To alleviate the difficulty of the non-convexity resulting from the ϵ-regularized L^p quasi-norm, we use continuation-based optimization [3,60] (a general strategy for optimizing non-convex functions; similar to annealing) that starts with a relatively large value of $\epsilon = 1$ and gradually reduces ϵ to 0.001. In practice, such continuation strategies help find better local minima for non-convex objective functions.

Given the number of atoms K, we initialize atoms d_k using kernel k-means using geodesic distances on the Hilbert sphere in RKHS. We initialize the k-means using the k-means++ algorithm [8]. After initializing the atoms d_k, we initialize the weight w_{nk} in inverse proportion to the distance of y_n from d_k $\mathrm{Log}_{y_n}(d_k)$, normalized such that the weights of any input sum to 1 as required by the affine constraint, i.e.,

$$w_{nk}^{\mathrm{init}} := \frac{\|\mathrm{Log}_{y_n}(d_k)\|_{\mathcal{H}}^{-1}}{\sum_{k'=1}^{K}\|\mathrm{Log}_{y_n}(d_{k'})\|_{\mathcal{H}}^{-1}}. \qquad (14)$$

For each y_n, this gives larger weights w_{nk} for those atoms d_k that are closer to y_n.

We summarize the proposed *algorithm for dictionary learning* below:

1. **Inputs:** A set of points $\{x_n\}_{n=1}^N$ in input space. Gram matrix G underlying a kernel such that all diagonal elements $G_{nn} = 1$. Number of atoms K. Parameters $\lambda > 0$ and $p \in (0,1)$. Set iteration number $i = 0$.
2. Initialize the dictionary D^i comprising atoms $\{d_k^i\}_{k=1}^N$ using kernel k-means adapted to the Hilbert sphere in RKHS, as described previously. Each atom d_k is represented using parameters $\{\gamma_{kn}^i\}$, as described in Eq. 9.
3. Initialize the weights matrix W^i, as described before.
4. Fix $\epsilon = 1$.
5. Fixing the weights W^i, use projected gradient descent to optimize for $\{\gamma_{kn}\}$ based on the formulation in (7) to get the updated parameters $\{\gamma_{kn}^{i+1}\}$ used to represent atoms $\{d_k^{i+1}\}_{k=1}^N$ in the dictionary D^{i+1}.
6. Fixing the parameters $\{\gamma_{kn}^{i+1}\}$, use projected gradient descent to optimize for W based on the formulation in 7 to get the updated weight matrix W^{i+1}.
7. If the relative change in the values of the objective function, in 7, evaluated at (W^i, D^i) and (W^{i+1}, D^{i+1}), is less than a small threshold, then terminate, otherwise increment i by 1 and repeat the last 2 steps.
8. Reduce $\epsilon \leftarrow \epsilon/10$. If $\epsilon < 0.001$, then terminate, otherwise repeat the last 3 steps (projected gradient descent optimization) with the initial solution as that obtained for the previous value of ϵ.
9. **Outputs:** A set of parameters $\{\gamma_{kn}^*\}$ representing the optimal atoms d_k^* in the optimal dictionary D^* and an optimal weight matrix W^* representing the coefficients w_n, for all atoms, in the fit to each y_n.

5   Application: Image Classification

We apply the proposed dictionary learning framework for image classification. The framework is general and considers image feature vectors x_n extracted from each image as the training data. We assume a kernel, e.g., Gaussian, such that the self similarity $\kappa(x, x) = 1$ for all feature vectors x. We first learn a dictionary from training data and then use it to code each training datum, where each datum's code is the vector of coefficients obtained from the dictionary fit. We then train a linear support vector machine (SVM) to learn a classifier on these codes.

We summarize the proposed *algorithm for learning a dictionary-based classifier* for the purpose of image classification.

1. **Inputs:** For the Q classes (denoted by $q = 1, 2, \cdots, Q$), N_q feature vectors $\{x_{qn}\}_{n=1}^{N_q}$ for class q. Gram matrix G, for the pooled dataset, underlying a kernel such that all diagonal elements equal 1. The number of atoms K_q in the dictionary D_q for each class q. Parameters $\lambda > 0$ and $p \in (0,1)$.
2. For all Q classes, use the dictionary learning algorithm in Sect. 4 to learn dictionary D_q of K_q atoms each for class q.

3. Pool all the dictionaries $\{D_q\}_{q=1}^Q$ to create a dictionary D having $K :=$ $\sum_{q=1}^Q K_q$ atoms.

4. Use the pooled dictionary D to fit to the mapped data $\{y_{qn} := \Phi(x_{qn})\}_{n=1}^{N_q}$ for all classes q, using the algorithm in Sect. 4, but keeping the dictionary fixed to D. This gives weight matrices W_q for all feature vectors in each class q, where each column w_{qn} has length K.

5. Learn a classifier $\mathcal{C}$ based on feature vectors $\{w_{qn} \in \mathbb{R}^K\}_{n=1}^{N_q}$ for each class q, using a linear SVM to classify any w into one of the Q classes. We do so by training Q one-versus-all SVM classifiers [12].

6. **Outputs:** Pooled dictionary D and the classifier $\mathcal{C}$.

We summarize the proposed *algorithm for dictionary-based image classification*.

1. **Inputs:** Pooled dictionary D and the Gram matrix G for which it is learned. The classifier $\mathcal{C}$. Test image x to be classified along with the extension of the Gram matrix (one row/column) for this test image's feature vector x, giving kernel similarity of the test image's feature vector x with all training image feature vectors.

2. Use the pooled dictionary D to fit to the mapped datum $y := \Phi(x)$ using the algorithm in Sect. 4, but keeping the dictionary fixed to D. This gives weight vector w of length K.

3. Use classifier $\mathcal{C}$ to classify the weight vector w into one of the Q classes, say q'.

4. **Output:** Class q'.

6   Results and Discussion

This section shows the results of empirical evaluation of our dictionary-learning based method, compared with the state of the art, on two large publicly available real-world image datasets of handwritten digits: (i) the MNIST dataset [38] available at http://yann.lecun.com/exdb/mnist/ and (ii) the USPS dataset [30] available at www-stat.stanford.edu/~tibs/ElemStatLearn/data.html. For evaluation on each dataset, we consider each raw image, after vectorization, as an input feature vector x_n. We use the popular Gaussian kernel $\kappa(x_i, x_j) :=$ $\exp(-0.5\|x_i - x_j\|_2^2/\sigma^2)$ and set σ^2, as per convention, to the average squared Euclidean distance between all pairs (x_i, x_j). We measure performance by the classification accuracy, i.e., the percentage of correctly classified images out of the total number of image classified.

We compare our method against state-of-the-art approaches involving kernel-based dictionary learning (all methods use the same dictionary size) and varying sparsity priors. First, we compare our Riemannian dictionary learning on the unit Hilbert sphere in RKHS with the alternate strategy that assumes the mapped data to lie in a linear space and performs standard linear dictionary learning in RKHS, e.g., in [28,41]. This compares two different data-fitting strategies in kernel feature space, i.e., our method using Hilbert-sphere modeling in RKHS and

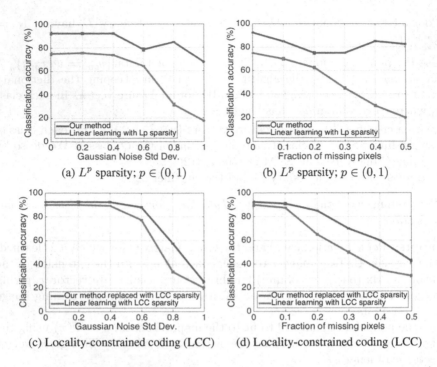

Fig. 2. USPS Handwritten Digit Image Recognition. Recognition rates (averages and standard deviations obtained by bootstrap sampling of the training dataset) for images with varying levels of corruptions (i.i.d. additive zero-mean Gaussian noise or missing pixels), when the sparsity prior is the **(a)-(b)** L^p quasi norm, with optimally tuned value of $p \leq 1$ for each noise level, and **(c)-(d)** locality-constrained coding, with optimally tuned value of $\rho > 0$ for each noise level.

the state of the art involving Euclidean modeling in RKHS. Second, we evaluate both aforementioned methods for two different kinds of sparsity priors, i.e., our L^p quasi-norm for $p < 1$ and the alternate strategy based on locality constrained coding, e.g., in [29,57,62,63]. However, unlike the faster, but approximate, versions [29,57] that use the few nearest atoms for coding each datum, we use the version with the weighted L^1 penalty over all atoms, i.e.,

$$\lambda \sum_{n=1}^{N} \sum_{k=1}^{K} \left(w_{nk} \exp(\rho \|y_n - d_k\|) \right)^2 \tag{15}$$

where $\rho \in \mathbb{R}_{>0}$ is a free parameter. Unlike the heuristic of choosing the nearest atoms for coding, this penalty lends itself to optimization via gradient descent that guarantees the reduction in the objective function at each step. When we use Hilbert-sphere modeling in RKHS, we choose the distance $\|y_n - d_k\|$ as the geodesic distance $\|\mathrm{Log}_{y_n}(d_k)\|_{\mathcal{H}}$; otherwise, we choose the norm $\|y_n - d_k\|_{\mathcal{H}}$ in RKHS.

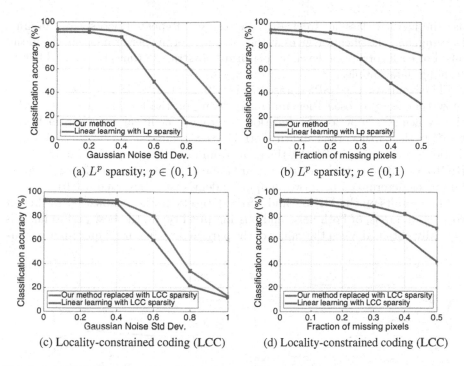

(a) L^p sparsity; $p \in (0,1)$

(b) L^p sparsity; $p \in (0,1)$

(c) Locality-constrained coding (LCC)

(d) Locality-constrained coding (LCC)

Fig. 3. MNIST Handwritten Digit Image Recognition. Recognition rates (averages and standard deviations obtained by bootstrap sampling of the training dataset) for images with varying levels of corruptions (i.i.d. additive zero-mean Gaussian noise or missing pixels), when the sparsity prior is the **(a)-(b)** L^p quasi norm, with optimally tuned value of $p \le 1$ for each noise level, and **(c)-(d)** locality-constrained coding, with optimally tuned value of $\rho > 0$ for each noise level.

We evaluate the robustness of the dictionary learning methods under two kinds of corruptions in the *training* data: (i) independent and identically distributed (i.i.d.) additive zero-mean Gaussian noise and (ii) missing data at randomly chosen pixels in the image. In case of images with missing intensities at specific pixels, we replace the missing intensity values with the average intensity over the pixels where the intensities are observed. We found this strategy of filling in missing information to be easy, parameter free, and leading to stable results. In the real-world, missing-pixel scenarios arise naturally in solving recognition problems for partially visible/occluded objects [35]. For both these kinds of corruptions, we evaluate performance over a range of corruption levels. At each corruption level, we tune the parameters underlying each method, i.e., the regularization parameter λ and the parameter underlying the sparsity prior (i.e., either p or ρ) using 5-fold cross validation [12] on the chosen training data.

To measure the variability of the classification performance with respect to the variability in the training data sample, we use bootstrap sampling to randomly select 90 % of the available training set to learn the classifier and, then, use the learned classifier to classify the test set. Repeated bootstrap sampling, learning, and classification gives us the reliability of the performance of the methods. For each corruption level, we perform bootstrap 25 times to get 25 different training data samples.

The results on the USPS dataset (in Fig. 2) and the MNIST dataset (in Fig. 3) show that the proposed Riemannian modeling on the Hilbert-sphere in RKHS clearly achieves superior classification accuracy over linear modeling in RKHS for (i) both sparsity priors and (ii) both kinds of corruptions. Compared to linear dictionary modeling in RKHS, the gains from our Hilbert-spherical modeling in RKHS are often more than 10 % and are almost always statistically significantly better, as determined by a two-sample Student's t-test (p-value < 0.01).

The results on the MNIST and USPS datasets in Fig. 4 show that for higher corruption levels, of both noise and missing-pixel types, the best performance is typically obtained when the values of the parameter p in our L^p quasi-norm spar-

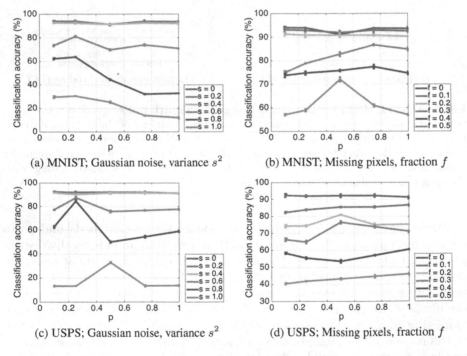

(a) MNIST; Gaussian noise, variance s^2 (b) MNIST; Missing pixels, fraction f

(c) USPS; Gaussian noise, variance s^2 (d) USPS; Missing pixels, fraction f

Fig. 4. Robustness of Dictionary Learning to Corrupted Training Data. Recognition rates (averages and standard deviations obtained by bootstrap sampling of the training dataset) for images with varying levels of corruptions (i.i.d. additive zero-mean Gaussian noise or missing pixels), with various values of the parameter $p < 1$ in the L^p quasi-norm sparsity penalty for **(a)** MNIST dataset with i.i.d. additive zero-mean Gaussian noise, **(b)** MNIST dataset with a fraction of pixels with missing intensity values, **(c)** USPS dataset with i.i.d. additive zero-mean Gaussian noise, and **(d)** USPS dataset with a fraction of pixels with missing intensity values.

sity penalty is less than 1. For very small values of p, the performance expectedly degrades possibly because of the increasing non-convexity of the penalty and the greater tendency to get stuck in a local minimum. This can also happen in some rare cases, where better continuation strategies can help. Nevertheless, for large corruption levels, the optimal performance almost always occurs for values of p significantly less than 1; for these datasets, the optimal p is typically in the range 0.2 and 0.8. This confirms the benefits of our proposed L^p quasi-norm sparsity penalty for tuned values of $p < 1$.

7   Conclusion

This paper presents a new method for kernel-based dictionary learning that addresses the hyperspherical geometry of the (implicitly) mapped points in RKHS, which naturally arises from many popular kernels and kernel normalization. In the same spirit that motivates manifold-based dictionary learning in input space, we perform manifold based dictionary learning in RKHS. We also propose stronger sparsity priors in the form of the non-convex L^p quasi-norm penalties that we deal with practically using a continuation-based optimization algorithm. We utilize the new dictionary learning algorithm for recognizing handwritten digits on large standard datasets and clearly demonstrate improved performances resulting from the modeling the Hilbert-sphere geometry in RKHS. We also demonstrate the gain in the robustness of the dictionary learning, to corruptions in the training data, arising from the stronger sparsity constraint.

References

1. Ah-Pine, J.: Normalized kernels as similarity indices. In: Proceeding of Pacific-Asia Conference Advances in Knowledge Discovery and Data Mining, vol. 2, pp. 362–373 (2010)
2. Ahn, J., Marron, J.S., Muller, K., Chi, Y.Y.: The high-dimension, low-sample-size geometric representation holds under mild conditions. Biometrika **94**(3), 760–766 (2007)
3. Allgower, E., Georg, K.: Introduction to Numerical Continuation Methods. SIAM (2003)
4. Amari, S., Nagaoka, H.: Methods of Information Geometry. Oxford Univ. Press, New York (2000)
5. Aronszajn, N.: Theory of reproducing kernels. Trans. Amer. Math. Soc. **68**(3), 337–404 (1950)
6. Arsigny, V., Fillard, P., Pennec, X., Ayache, N.: Log-Euclidean metrics for fast and simple calculus on diffusion tensors. Mgn. Reson. Med. **56**(2), 411–421 (2006)
7. Arsigny, V., Fillard, P., Pennec, X., Ayache, N.: Geometric means in a novel vector space structure on symmetric positive-definite matrices. SIAM J. Mat. Anal. Appl. **29**(1), 328–347 (2007)
8. Arthur, D., Vassilvitskii, S.: k-means++: the advantages of careful seeding. In: Proceeding Annual ACM-SIAM Symposium on Discrete Algorithms, pp. 1027–1035 (2007)

9. Awate, S.P., Yu, Y.-Y., Whitaker, R.T.: Kernel principal geodesic analysis. In: Calders, T., Esposito, F., Hüllermeier, E., Meo, R. (eds.) ECML PKDD 2014, Part I. LNCS, vol. 8724, pp. 82–98. Springer, Heidelberg (2014)
10. Banerjee, A., Dhillon, I., Ghosh, J., Sra, S.: Clustering on the unit hypersphere using von Mises-Fisher distributions. J. Mach. Learn. Res. **6**, 1345–1382 (2005)
11. Berger, M.: A Panoramic View of Riemannian Geometry. Springer, Heidelberg (2007)
12. Bishop, C.: Pattern Recognition and Machine Learning. Springer, New York (2006)
13. Chen, Z., Wu, Y.: Robust dictionary learning by error source decomposition. In: Proceeding International Conference on Computer Vision, pp. 2216–2223 (2013)
14. Cherian, A., Sra, S.: Riemannian sparse coding for positive definite matrices. In: Proceeding European Conference on Computer Vision, pp. 299–314 (2014)
15. Common, P., Golub, G.: Tracking a few extreme singular values and vectors in signal processing. Proc. IEEE **78**(8), 1327–1343 (1990)
16. Courty, N., Burger, T., Marteau, P.: Geodesic analysis on the Gaussian RKHS hypersphere. In: European conference Machine Learning Practice of Knowledge Discovery Data, vol. 1, 299–313 (2012)
17. Donoho, D., Elad, M.: Optimally sparse representation in general (nonorthogonal) dictionaries via $l1$ minimization. Proc. Nat. Acad. Sci. **100**(5), 2197–2202 (2003)
18. Efron, B., Hastie, T., Johnstone, I., Tibshirani, R.: Least angle regression. Ann. Stat. **32**(2), 407–451 (2004)
19. Eigensatz, M.: Insights into the geometry of the Gaussian kernel and an application in geometric modeling. Master thesis. Swiss Federal Institute of Technology (2006)
20. Elad, M.: Sparse and Redundant Representations: From Theory to Applications in Signal and Image Processing. Springer, New York (2010)
21. Fletcher, P.T., Joshi, S.: Riemannian geometry for the statistical analysis of diffusion tensor data. Signal Process. **87**(2), 250–262 (2007)
22. Fung, G., Mangasarian, O.: Equivalence of minimal l_0- and l_p-norm solutions of linear equalities, inequalities and linear programs for sufficiently small p. J. Optim. Theory Appl. **151**(1), 1–10 (2011)
23. Genton, M.: Classes of kernels for machine learning: a statistics perspective. J. Mach. Learn. Res. **2**, 299–312 (2001)
24. Goh, A., Vidal, R.: Clustering and dimensionality reduction on Riemannian manifolds. In: Proceeding of Computer Vision and Pattern Recognition, pp. 1–7 (2008)
25. Graf, A., Smola, A., Borer, S.: Classification in a normalized feature space using support vector machines. IEEE Trans. Neural Netw. **14**(3), 597–605 (2003)
26. Grauman, K., Darrell, T.: The pyramid match kernel: efficient learning with sets of features. J. Mach. Learn. Res. **8**, 725–760 (2007)
27. Hamsici, O., Martinez, A.: Rotation invariant kernels and their application to shape analysis. IEEE Trans. Pattern Anal. Mach. Intell. **31**(11), 1985–1999 (2009)
28. Harandi, M., Salzmann, M.: Riemannian coding and dictionary learning: Kernels to the rescue. In: Proceeding of Computer Vision and Pattern Recognition, pp. 3926–3935 (2015)
29. Harandi, M., Sanderson, C., Shen, C., Lovell, B.: Dictionary learning and sparse coding on Grassmann manifolds: An extrinsic solution. In: International Conference on Computer Vision, pp. 3120–3127 (2013)
30. Hastie, T., Tibshirani, R., Friedman, J.: The Elements of Statistical Learning: Data Mining, Inference, and Prediction. Springer, New York (2009)
31. Hoyer, P.: Non-negative sparse coding. In: Neural Networks for Signal Processing, pp. 557–565 (2002)

32. Hoyer, P.: Non-negative matrix factorization with sparseness constraints. J. Mach. Learn. Res. **5**, 1457–1469 (2004)
33. Jayasumana, S., Salzmann, M., Li, H., Harandi, M.: A framework for shape analysis via Hilbert space embedding. In: International Conference on Computer Vision, pp. 1249–1256 (2013)
34. Jiang, W., Nie, F., Huang, H.: Robust dictionary learning with capped l_1-norm. In: Proceeding of International Conference on Artificial Intelligence, pp. 3590–3596 (2015)
35. Johnson, J., Olshausen, B.: The recognition of partially visible natural objects in the presence and absence of their occluders. Vision Res. **45**, 3262–3276 (2005)
36. Kendall, D.: A survey of the statistical theory of shape. Statist. Sci. **4**(2), 87–99 (1989)
37. Lan, A., Waters, A., Studer, C., Baraniuk, R.: Sparse factor analysis for learning and content analytics. J. Mach. Learn. Res. **15**(1), 1959–2008 (2014)
38. LeCun, Y., Bottou, L., Bengio, Y., Haffner, P.: Gradient-based learning applied to document recognition. Proc. IEEE **86**(11), 2278–2324 (1998)
39. Lu, C., Shi, J., Jia, J.: Online robust dictionary learning. In: Proceeding Computer Vision and Pattern Recognition, pp. 415–422 (2013)
40. Mardia, K., Jupp, P.: Directional Statistics. Wiley, Chichester (2000)
41. Nguyen, H., Patel, V., Nasrabadi, N., Chellappa, R.: Design of non-linear kernel dictionaries for object recognition. IEEE Trans. Imag. Proc. **22**(12), 5123–5135 (2013)
42. Nie, F., Huang, H., Cai, X., Ding, C.: Efficient and robust feature selection via joint $l2,1$-norms minimization. In: Advances in Neural Information Processing Systems, pp. 1813–1821 (2010)
43. Onishchik, A., Sulanke, R.: Projective and Cayley-Klein Geometries. Springer, Heidelberg (2006)
44. Pan, Q., Kong, D., Ding, C., Luo, B.: Robust non-negative dictionary learning. In: Proceedings AAAI Conference on Artificial Intelligence, pp. 2027–2033 (2014)
45. Park, T., Casella, G.: The Bayesian lasso. Am. Stats. **103**(482), 681–686 (2008)
46. Peel, D., Whiten, W., McLachlan, G.: Fitting mixtures of Kent distributions to aid in joint set identification. J. Amer. Stat. Assoc. **96**, 56–63 (2001)
47. Pennec, X., Fillard, P., Ayache, N.: A Riemannian framework for tensor computing. Int. J. Comp. Vis. **66**(1), 41–66 (2006)
48. Rubinstein, R., Bruckstein, A., Elad, M.: Dictionaries for sparse representation modeling. Proc. IEEE **98**(6), 1045–1057 (2010)
49. Scholkopf, B., Smola, A.: Learning with Kernels. MIT Press, Cambridge (2002)
50. Scholkopf, B., Smola, A., Muller, K.R.: Nonlinear component analysis as a kernel eigenvalue problem. Neural Comput. **10**, 1299–1319 (1998)
51. Sra, S.: A new metric on the manifold of kernel matrices with application to matrix geometric means. In: Advances in Neural Information Processing Systems, pp. 144–152 (2012)
52. Sra, S., Cherian, A.: Generalized dictionary learning for symmetric positive definite matrices with application to nearest neighbor retrieval. In: Proceeding of European Conference on Machine Learning and Principles and Practice of Knowledge Discovery in Databases, pp. 318–332 (2012)
53. Srivastava, A., Jermyn, I., Joshi, S.: Riemannian analysis of probability density functions with applications in vision. In: Proceeding of International Conference Computer Vision and Pattern Recognition, pp. 1–8 (2007)
54. Tibshirani, R.: Regression shrinkage and selection via the lasso. J. Royal Stat. Soc. Ser. B **58**(1), 267–288 (1996)

55. Turaga, P., Veeraraghavan, A., Srivastava, A., Chellappa, R.: Statistical computations on Grassmann and Stiefel manifolds for image and video-based recognition. IEEE Trans. Pattern Anal. Mach. Intell. **33**(11), 2273–2286 (2011)
56. Wagner, A., Wright, J., Ganesh, A., Zhou, Z., Mobahi, H., Ma, Y.: Towards a practical face recognition system: Robust alignment and illumination by sparse representation. IEEE Trans. Pattern Anal. Mach. Intell. **34**(2), 372–386 (2012)
57. Wang, J., Yang, J., Yu, K., Lv, F., Huang, T., Gong, Y.: Locality-constrained linear coding for image classification. In: Proceeding Computer Vision Pattern Recognition, pp. 3360–3367 (2010)
58. Wang, N., Wang, J., Yeung, D.Y.: Online robust non-negative dictionary learning for visual tracking. In: Proceeding International Conference on Computer Vision, pp. 657–664 (2013)
59. Wong, E., Awate, S.P., Fletcher, P.T.: Adaptive sparsity in Gaussian graphical models. In: International Conference Machine Learning, vol. 1, pp. 311–319 (2013)
60. Wu, Z.: The effective energy transformation scheme as a special continuation approach to global optimization with application to molecular conformation. SIAM J. Opt. **6**, 748–768 (2006)
61. Xie, Y., Ho, J., Vemuri, B.: On a nonlinear generalization of sparse coding and dictionary learning. J. Mach. Learn. Res. **28**, 1480–1488 (2013)
62. Yu, K., Zhang, T.: Improved local coordinate coding using local tangents. In: Proceeding International Conference Machine learning, pp. 1215–1222 (2010)
63. Yu, K., Zhang, T., Gong, Y.: Nonlinear learning using local coordinate coding. In: Advances in neural information processing systems, pp. 2223–2231 (2009)
64. Yukawa, M., Amari, S.I.: l_p-regularized least squares $(0 < p < 1)$ and critical path. IEEE Trans. Info. Th. **62**(1), 488–502 (2016)
65. Zeng, X., Bian, W., Liu, W., Shen, J., Tao, D.: Dictionary pair learning on Grassmann manifolds for image denoising. IEEE Trans. Imag. Proc. **24**(11), 4556–4569 (2015)

Is Attribute-Based Zero-Shot Learning an Ill-Posed Strategy?

Ibrahim Alabdulmohsin[1], Moustapha Cisse[2], and Xiangliang Zhang[1(✉)]

[1] Computer, Electrical and Mathematical Sciences and Engineering Division,
King Abdullah University of Science and Technology (KAUST),
Thuwal 23955-6900, Saudi Arabia
{ibrahim.alabdulmohsin,xiangliang.zhang}@kaust.edu.sa
[2] Facebook Artificial Intelligence Research (FAIR), Menlo Park, USA
moustaphacisse@fb.com
http://mine.kaust.edu.sa

Abstract. One transfer learning approach that has gained a wide popularity lately is attribute-based *zero-shot* learning. Its goal is to learn novel classes that were never seen during the training stage. The classical route towards realizing this goal is to incorporate a prior knowledge, in the form of a semantic embedding of classes, and to learn to predict classes indirectly via their semantic attributes. Despite the amount of research devoted to this subject lately, no known algorithm has yet reported a predictive accuracy that could exceed the accuracy of supervised learning with *very few* training examples. For instance, the *direct attribute prediction* (DAP) algorithm, which forms a standard baseline for the task, is known to be as accurate as supervised learning when as few as two examples from each hidden class are used for training on some popular benchmark datasets! In this paper, we argue that this lack of significant results in the literature is not a coincidence; attribute-based zero-shot learning is fundamentally an *ill-posed* strategy. The key insight is the observation that the mechanical task of *predicting* an attribute is, in fact, quite different from the epistemological task of learning the "correct meaning" of the attribute itself. This renders attribute-based zero-shot learning fundamentally ill-posed. In more precise mathematical terms, attribute-based zero-shot learning is equivalent to the mirage goal of learning with respect to one distribution of instances, with the hope of being able to predict with respect to any *arbitrary* distribution. We demonstrate this overlooked fact on some synthetic and real datasets. The data and software related to this paper are available at https://mine.kaust.edu.sa/Pages/zero-shot-learning.aspx.

Keywords: Zero-shot learning · Attribute-based classification · Multilabel classification

1 Introduction

Humans are capable of learning new concepts using a few empirical observations. This remarkable ability is arguably accomplished via *transfer learning*

© Springer International Publishing AG 2016
P. Frasconi et al. (Eds.): ECML PKDD 2016, Part I, LNAI 9851, pp. 749–760, 2016.
DOI: 10.1007/978-3-319-46128-1_47

techniques, such as the *bootstrapping learning strategy*, where agents learn simple tasks first before tackling more complex activities [22]. For instance, humans begin to cruise and crawl before they learn how to walk. Learning to cruise and crawl allows infants to improve their locomotion skills, body balance, control of limbs, and the perception of depth, all of which are crucial pre-requisites for learning the more complex activity of walking [11,16].

In many machine learning applications, a similar transfer learning strategy is desired when labeled examples are difficult to obtain that can faithfully represent the entire target set $\mathcal{Y}$. This is often the case, for example, in image classification and in neural image decoding [14,17]. The transfer learning strategy typically employed in this setting is either called *few-shot*, *one-shot*, or *zero-shot* learning, depending on how many labeled examples are available during the training stage [10,14]. Here, a desired target set $\mathcal{Y}$ (classes) is learned *indirectly* by learning semantic attributes instead. These attributes are, then, used to predict the classes in $\mathcal{Y}$.

The motivation behind the attribute-based learning approach with scarce data is close in spirit to the rationale of the bootstrapping learning strategy. In brief terms, it helps to learn simple tasks first before attempting to learn more complex activities. In the context of classification, semantic attributes are (chosen to be) abundant, where a single attribute spans multiple classes. Hence, labeled examples for the semantic attributes are more plentiful, which makes the task of predicting attributes relatively easy. Moreover, the target set $\mathcal{Y}$ is embedded in the space of semantic attributes, a.k.a. the *semantic space*, which makes it possible, perhaps, to predict classes that were rarely seen, if ever, during the training stage.

In this paper, we focus on the attribute-based zero-shot learning setting, where a finite number of semantic attributes is used to predict *novel* classes that were never seen during the training stage. More formally [14]:

Definition 1 (Attribute-Based Zero-Shot Setting). *In the attribute-based zero-shot setting, we have an instance space $\mathcal{X}$, a semantic space $\mathcal{A}$, and a target set $\mathcal{Y}$, where $|\mathcal{A}| < \infty$ and $|\mathcal{Y}| < \infty$. A sample S comprises of m examples $\{(X_i, Y_i, A_i)\}_{i=1,\dots,m}$, with $X_i \in \mathcal{X}$, $Y_i \in \mathcal{Y}$, and $A_i \in \mathcal{A}$. Moreover, $\mathcal{Y}$ is partitioned into two non-empty subsets: the set of visible classes $\mathcal{Y}_V = \bigcup_{(X_i,Y_i,A_i) \in S}\{Y_i\}$ and the set of hidden classes $\mathcal{Y}_H = \mathcal{Y} \backslash \mathcal{Y}_V$. The goal is to use S to learn a hypothesis $f : \mathcal{X} \to \mathcal{Y}_H$ that can correctly predict the hidden classes $\mathcal{Y}_H$.*

The key part of Definition 1 is the final goal. Unlike the traditional setting of learning, we no longer assume that the sample size m is large enough for all classes in $\mathcal{Y}$ to be seen during the training stage. In general, we allow $\mathcal{Y}$ to be partitioned into two non-empty subsets $\mathcal{Y}_V$ and $\mathcal{Y}_H$, which, respectively, correspond to the *visible* and the *hidden* classes in the given sample S. The classical argument for why the goal of learning to predict the hidden classes is possible in this setting is that the hidden classes $\mathcal{Y}_H$ are *coupled* with the instances $\mathcal{X}$ and the visible classes $\mathcal{Y}_V$ via the semantic space $\mathcal{A}$ [14].

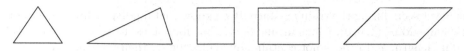

Fig. 1. In the polygon shape recognition problem, the instances are images of polygons and we have five disjoint classes: equilateral triangles, non-equilateral triangles, squares, non-square rectangles, and non-rectangular parallelograms.

To illustrate the traditional argument for attribute-based zero-shot learning, let us consider the simple *polygon shape recognition* problem shown in Fig. 1. In this problem, the instance space $\mathcal{X}$ is the set of images of *polygons*, i.e. two-dimensional shapes bounded by a closed chain of a finite number of line segments, while the target set $\mathcal{Y}$ is the set of the five disjoint classes shown in Fig. 1.

In the traditional setting of learning, a large sample of instances S would be collected and a classifier would be trained on the sample (e.g. using one-vs-all or one-vs-one). One of the fundamental assumptions in the traditional setting of learning for guaranteeing generalization is the *stationarity* assumption; examples in the sample S are assumed to be drawn i.i.d. from the same distribution as the future examples. Along with a few additional assumptions, learning in the traditional setting can be rigorously shown to be feasible [1,5,19,23].

In the zero-shot learning setting, by contrast, it is assumed that the target set $\mathcal{Y}$ is partitioned into two non-empty subsets $\mathcal{Y} = \mathcal{Y}_V \cup \mathcal{Y}_H$. During the training stage, only instances from the visible classes $\mathcal{Y}_V$ are seen. The goal is to be able to predict the hidden classes correctly. This goal is arguably achieved by introducing a coupling between $\mathcal{Y}_V$ and $\mathcal{Y}_H$ via the semantic space. For example, we recognize that the five classes in Fig. 1 can be completely determined by the values of the following three binary attributes:

– a_1: *Does the polygon contain 4 sides?*
– a_2: *Are all sides in the polygon of equal length?*
– a_3: *Does the polygon contain, at least, one acute angle?*

The set of all possible answers to these three binary questions forms a semantic space $\mathcal{A}$ for the polygon shape recognition problem. Given an instance X with the semantic embedding $A = (a_1, a_2, a_3) \in \{0,1\}^3$, its class can be uniquely determined. For example, any *equilateral triangle* has the semantic embedding $(0, 1, 1)$, which means that the latter polygon (1) does not contain four sides, (2) its sides are all of equal length, and (3) it contains some acute angles. Among the five classes in $\mathcal{Y}$, only the class of equilateral triangles satisfy this semantic embedding. Similarly, the four remaining classes have unique semantic embeddings as well.

Because the classes can be inferred from the values of the three binary attributes mentioned above, it is often argued that hidden classes can be predicted by, first, learning to *predict* the values of the semantic attributes based on a sample (training set) S, and, second, by using those predicted attributes to predict the hidden classes in $\mathcal{Y}_H$ via some hand-crafted mappings [7,9,13–15,17,20,21]. In our example in Fig. 1, suppose the class of *non-square rectangles* is never seen during the training stage. If we know that a polygon has the semantic embedding $(1, 0, 0)$, which means that it has four sides, its sides are not all of equal length, and it does not contain any acute angles, then it seems reasonable to conclude that it is a non-square rectangle even if we have not seen any non-square rectangles in the sample S. Does this imply that zero-shot learning is a well-posed approach? We will show that the answer is, in fact, negative. The key ingredient in our argument is the fact that the mechanical task of *predicting* an attribute is quite different from the epistemological task of learning the *correct meaning* of the attribute.

The rest of the paper is structured as follows. We first explain why the two tasks of "predicting" an attribute and "learning" an attribute are quite different from each other. We will illustrate this overlooked fact on the simple shape recognition problem of Fig. 1 and demonstrate it in a greater depth on some synthetic and real datasets afterward. Next, we use such a distinction between "predicting" and "learning" to argue that the attribute-based zero-shot learning approach is fundamentally ill-posed, which, we believe, explains why the previous zero-shot learning algorithms proposed in the literature have not performed significantly better than supervised learning with very few training examples.

2   Why Learning and Predicting Are Two Different Tasks

2.1   The Polygon Shape Recognition Problem

Let us return to the original polygon shape recognition example of Fig. 1. Suppose that the two classes of *non-square rectangles* and *non-rectangular parallelograms* are hidden from the sample S. That is:

$$\mathcal{Y}_V = \{\text{equilateral triangles,}\quad \text{non-equilateral triangles,}\quad \text{squares}\}$$

$$\mathcal{Y}_H = \{\text{non-square rectangles,}\quad \text{non-rectangular parallelograms}\}$$

In the attribute-based zero-shot learning setting, we learn to predict the three semantic attributes (a_1, a_2, a_3) mentioned earlier based on the sample S that only contains examples from the visible three classes. Once we learn to predict them correctly based on the sample S, we are supposed to be able to recognize the two hidden classes via their semantic embeddings. The semantic embedding for non-square rectangles is, in this example, $(1, 0, 0)$, while the semantic embeddings for non-rectangular parallelograms is the set $\{(1, 0, 1), (1, 1, 1)\}$.

To see why this is, in fact, an incorrect approach, we note that the task of *predicting* an attribute aims, by definition, at utilizing *all the relevant information* in the sample S that aid the prediction task. In our example, since only the three visible classes $\mathcal{Y}_V$ are seen in the sample S, a *good predictor* should infer from S the following logical assertions:

1. If a polygon does not contain four sides, then it contains one acute angle.
   Formally:

$$(a_1 = 0) \rightarrow (a_3 = 1)$$

From this, the contrapositive assertion $(a_3 = 0) \to (a_1 = 1)$ is deduced as well.

2. If the sides of a polygon are not of equal length, then it does not contain four sides. Formally:
$$(a_2 = 0) \to (a_1 = 0)$$

Again, its contrapositive assertion also holds.

3. If the polygon does not contain an acute angle, then all of its sides are of equal length. Formally:
$$(a_3 = 0) \to (a_2 = 1)$$

These logical assertions and others are likely to be used by a good predictor, at least implicitly, since they are always true in the sample S. In addition, such a predictor would have a good generalization ability if the instances continued to be drawn i.i.d. *from the same distribution* as the training sample S, i.e. if the set of visible classes remained unchanged.

If, on the other hand, an instance is now drawn from a hidden class in $\mathcal{Y}_H$, then some of these logical assertions would no longer hold and the original algorithm that was trained to predict the semantic attributes would fail. This follows from the fact that instances drawn from the hidden classes have a *different* distribution. Therefore, the fact that classes can be uniquely determined by the values of the semantic attributes is of little importance here because the semantic attributes are likely to be predicted correctly for the visible classes *only*. Needless to mention, this violates the original goal of the attribute-based zero-shot learning setting.

2.2   Optical Digit Recognition

To show that the previous argument on the polygon shape recognition problem is not a contrived argument, let us look into a real classification problem, in which we can visualize the decision rule used by the predictors. We will use the optical digit recognition problem to illustrate our argument. In order to be able to interpret the decision rule used by the predictor, we will use the linear support vector machine (SVM) algorithm [4,6], trained without the bias term using the LIBLINEAR package [8].

One way of introducing a semantic space for the ten digits is to use the *seven-segment display* shown in Fig. 2. That is, the instance space $\mathcal{X}$ is the set of noisy digits, the classes are the ten digits $\mathcal{Y} = \{0, 1, 2, \ldots, 9\}$, and the semantic space is $\mathcal{A} = \{-1, +1\}^7$ corresponding to the seven segments. For example, using the order of segments (a,b,c,d,e,f,g) shown in Fig. 2, the digit 0 in Fig. 2 has the semantic embedding $(1, 1, 1, 0, 1, 1, 1)$ while the digit 1 has the semantic embedding $(0, 0, 1, 0, 0, 1, 0)$, and so on.

In our implementation, we run the experiment as follows[1]. First, a perfect digit is generated, which is later contaminated with noise. In particular, every

---

[1] The MATLAB implementation codes that generate the images in this section are available at https://mine.kaust.edu.sa/Pages/zero-shot-learning.aspx.

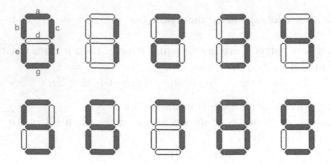

Fig. 2. In optical digit recognition, a semantic space can be introduced using the seven segment display segments shown in this figure.

Fig. 3. The instance space in the optical digit recognition problem is the set of noisy digits, where the value of every pixel is flipped with probability 0.1.

pixel is flipped with probability 0.1. As a result, the instance space is the set of *noisy* digits, as depicted in Fig. 3. Then, five digits are chosen to be in the *visible* set of classes, and the remaining digits are hidden during the training stage. We train classifiers that predict the attributes (i.e. segments) using the visible set of classes, and use those classifiers to predict the hidden classes afterward.

Now, the classical argument for attribute-based zero-shot learning goes as follows:

1. Every digit can be uniquely determined by its seven-segment display. When an exact match is not found, one can carry out a nearest neighbor search [7,9,15,17,20,21] or a maximum a posteriori estimation method [13,14].
2. Every segment in {a,b,c,d,e,f,g} is a concept class by itself that spans multiple digits. Hence, the number of training examples available for each segment is large, which makes it easier to predict.
3. Because each of the seven segments spans multiple classes, we no longer need to see all of the ten digits during the training stage in order to learn to predict the seven segments reliably.

This argument clearly rests on the assumption that "learning" a concept is equivalent to the task of reliably "predicting"it. From the earlier discussion in the polygon shape matching problem, this assumption is, in fact, invalid. Figure 4 shows what happens when a proper subset of the ten digits is seen during the training stage. As shown in the figure, the linear classifier trained using SVM exploits the relevant information available in the training set to maximize its prediction accuracy of the attributes.

For example, when we train a classifier to predict the segment 'a' using the visible classes $\{0, 1, 2, 3, 4\}$, a good predictor would use the fact that the segment 'a', which is the target concept, always co-occurs with the segment 'g'. Therefore, the contrapositive rule implies that the absence of 'g' implies the absence of the segment 'a'. This is clearly seen in Fig. 4 (top left corner). Of course, what the predictor learns is even more complex than this, as shown in Fig. 4. When novel instances from the *hidden* classes are present, these correlations no longer hold and the algorithm fails to predict the semantic attributes correctly. To reiterate, such a failure is fundamentally due to the fact that the hidden classes constitute a *different* distribution of instances from the one seen during the training stage.

The results of applying a linear SVM using *binary relevance* to predict the seven segments is shown in Fig. 4. In this figure, the blue regions correspond to the pixels that contribute positively to the decision rule for predicting the corresponding segment, while the red regions contribute negatively. There are two key takeaways from this figure. First, the prediction rule used by the classifier does not correspond to the "true" meaning of the semantic attribute. After all, the goal of classification is to be able to "predict"the attribute as opposed to learning what it actually means. Second, changing the set of visible classes can change the prediction rule *for the same attribute* quite notably. Both observations challenge the rationale behind the attribute-based zero-shot learning setting.

2.3   Zero-Shot Learning on Popular Datasets

Next, we examine the performance of zero-shot learning on benchmark datasets. Two of the most popular datasets for evaluating zero-shot learning algorithms are the *Animals with Attributes* (AwA) dataset [14] and the *aPascal-aYahoo* dataset [9][2]. We briefly describe each dataset next.

The Animals with Attributes (AwA) Dataset: The AwA dataset was collected by querying search engines, such as Google and Flickr, for images of 50 animals. Afterward, these images were manually handled to remove outliers and duplicates. The final dataset contains 30,475 images, where the minimum number of images per class is 92 and the maximum is 1,168. In addition, 85

---

[2] These datasets are available at:
http://attributes.kyb.tuebingen.mpg.de/
http://vision.cs.uiuc.edu/attributes/.

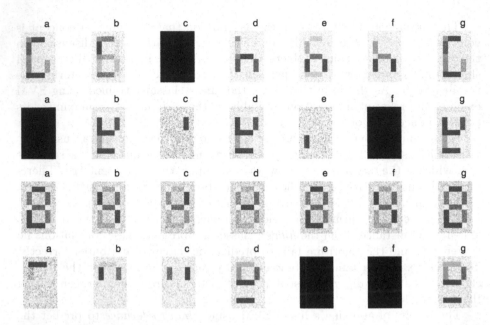

Fig. 4. In this figure, linear SVM was implemented to predict the seven segments of noisy optical images. The columns correspond to the seven segments $\{a, b, c, d, e, f, g\}$ while the rows correspond to different choices of visible classes. From top to bottom, the visible classes are $\{0, 1, 2, 3, 4\}$ for the first row, $\{5, 6, 7, 8, 9\}$ for the second row, $\{0, 2, 4, 6, 8\}$ for the third row, and $\{1, 3, 5, 7, 9\}$ for the fourth row. Every figure depicts the coefficient vector w that is learned by linear SVM, where blue regions correspond to the pixels that contribute positively towards the corresponding segment and red regions contribute negatively. The black figures are not applicable because the training sample S either lacks negative examples for the corresponding attribute or it lacks positive examples. (Color figure online)

attributes are introduced. In the zero-shot learning setting, 40 (visible) classes are used for training and 10 (hidden) classes are used for testing [14].

The aPascal-aYahoo (aP-AY) Dataset: The aP-aY dataset contains 12,695 images, which were chosen from the PASCAL VOC 2008 data set [9]. These images are used during the training stage of the zero-shot learning setting. In addition, a total of 2,644 images were collected from the Yahoo image search engine to be used during the test stage. Both sets of images have disjoint classes. More specifically, the training dataset contains 20 classes while the test dataset contains 12 classes. Moreover, every image has been annotated with 64 binary attributes.

Results: Table 1 presents some fairly-recent reported results on the two datasets AwA and aP-aY. The zero-shot learning algorithms provided in the table are the *direct attribute prediction* (DAP) algorithm proposed in [13,14], which is one of the standard baseline methods for this task, the *indirect attribute prediction* (IAP) algorithm proposed in [13,14], the *embarrassingly simple zero-shot learning* algorithm proposed in [18], and the *zero-shot random forest* algorithm proposed in [12]. The best reported prediction accuracy for the AwA dataset is 49.3 % while the best reported prediction accuracy for the aP-aY dataset is 26.0 %.

Table 1. In this table, some fairly-recent results of zero-shot learning algorithms are presented. The reported figures of the four zero-shot learning algorithms are for the multiclass prediction accuracy on the two datasets AwA and aP-aY, where the figures are taken from the original papers [12–14,18]

Algorithm	AwA	aP-aY
DIRECT ATTRIBUTE PREDICTION	41.4 %	19.1 %
INDIRECT ATTRIBUTE PREDICTION	42.2 %	16.9 %
EMBARRASSINGLY SIMPLE ZERO-SHOT	**49.3 %**	15.1 %
ZERO-SHOT RANDOM FOREST	43.0 %	**26.0 %**
Average # Training Examples/Visible Class	607.38	634.75
Best Reported Accuracy	49.3 %	26.0 %
Equivalent # Training Examples/Hidden Class	≈ 20	≈ 2

In order to properly interpret the reported results, we have also provided in Table 1 the number of training examples from the hidden classes that would suffice, in a traditional supervised learning setting, to obtain the same accuracy reported by the zero-shot learning algorithms in the literature. These latter figures are obtained from the experimental study conducted in [14]. Note that while about 600 examples per visible class are used during the training stage, the best reported zero-shot prediction accuracy on the hidden classes is equivalent to the accuracy of supervised learning using fewer than 20 training examples per hidden class. In fact, the zero-shot learning accuracy reported on the aP-aY is worse than the accuracy of supervised learning when as few as 2 training examples per hidden class are used.

When the area under the curve (AUC) is used as a performance measure, which is known to be more robust to class imbalance than the prediction accuracy, then the apparent merit of zero-shot learning becomes even more questionable. For instance, the popular *direct attribute prediction* (DAP) on the AwA dataset achieves an AUC of 0.81, which is equivalent to the performance of supervised learning using as few as 10 training examples from each hidden class only (c.f. Tables 4 and 7b in [14]). Recall, by contrast, that over 600 examples per visible class are used for training.

3   A Mathematical Formalism

The above argument and empirical evidence on the ill-posedness of attribute-based zero-shot learning can be formalized mathematically. Incidentally, this will allow us to identify paradigms of zero-shot learning for which the above argument no longer holds.

As stated in Sect. 2 and illustrated in Fig. 4, the fundamental problem with attribute-based zero-shot learning is that it aims at learning concept classes (semantic attributes) with respect to one distribution of instances (i.e. when conditioned on the visible set of classes) with the goal of being able to predict those concept classes for an arbitrary distribution of instances (i.e. when conditioned on the unknown hidden set of classes). Clearly, this is an ill-posed strategy that violates the core assumptions of statistical learning theory.

To remedy this problem, we can cast zero-shot learning as a *domain adaptation* problem [18]. In the standard domain adaptation setting, it is assumed that the training examples are drawn i.i.d. from some source distribution $\mathcal{D}_S$ whereas future test examples are drawn i.i.d. from a different target distribution $\mathcal{D}_T$. Let $h : \mathcal{X} \to \mathcal{Y}$ be a predictor. Then, the average misclassification error rate of h with respect to $\mathcal{D}_T$ is bounded by:

$$\mathbb{E}_{(X,Y)\sim\mathcal{D}_T}\{h(X) \neq Y\} \leq \mathbb{E}_{(X,Y)\sim\mathcal{D}_S}\{h(X) \neq Y\} + d_{TV}(\mathcal{D}_S, \mathcal{D}_T), \qquad (1)$$

where $d_{TV}(\mathcal{D}_S, \mathcal{D}_T)$ is the total-variation distance between the two probability distributions $\mathcal{D}_S$ and $\mathcal{D}_T$ [2]. Similar bounds that also hold with a high probability can be found in [3]. Hence, learning a good predictor h with respect to some source distribution $\mathcal{D}_S$ *does not* guarantee a good prediction accuracy with respect to an arbitrary target distribution $\mathcal{D}_T$ unless the two distributions are nearly identical.

Therefore, in order to turn zero-shot learning into a well-posed strategy, it is imperative that a common representation $R(X)$ is used, such that the induced distribution of $R(X)$ remains nearly unchanged when the instances X are conditioned on the visible set of classes or on the hidden set of classes. Then, by learning to predict semantic attributes given $R(X)$, generalization bounds, such as the one provided in Eq. (1), guarantee a good prediction accuracy in the zero-shot setting. One method that can accomplish this goal is to divide the instances X_i into multiple local segments $X_i \to (Z_{i,1}, Z_{i,2}, \ldots) \in \mathcal{Z}^r$ such that a classifier $h : \mathcal{Z} \to \mathcal{A}$ is trained to predict the semantic attributes in every local segment separately. If these local segments have a *stable* distribution across the visible and hidden set of classes, then zero-shot learning is feasible. A prototypical example of this approach is segmenting sounds into phonemes in word recognition systems and using those phonemes to recognize words (classes) [17].

4   Conclusion

Attribute-based zero-shot learning is a transfer learning strategy that has been widely studied in the literature. Its aim is to learn to predict novel classes that are never seen during the training stage by learning to predict semantic attributes instead. In this paper, we argue that attribute-based zero-shot learning is an ill-posed strategy because the two tasks of "predicting" and "learning" an attribute are fundamentally different. We demonstrate our argument on synthetic datasets and use it, finally, to explain the poor performance results that have been reported so far in the literature for various zero-shot learning algorithms on popular benchmark datasets.

Acknowledgment. Research reported in this publication was supported by King Abdullah University of Science and Technology (KAUST) and the Saudi Arabian Oil Company (Saudi Aramco).

References

1. Abu-Mostafa, Y.S., Magdon-Ismail, M., Lin, H.T.: Learning from data (2012)
2. Alabdulmohsin, I.: Algorithmic stability and uniform generalization. In: NIPS, pp. 19–27. Curran Associates, Inc. (2015)
3. Ben-David, S., Blitzer, J., Crammer, K., Pereira, F.: Analysis of representations for domain adaptation. In: Schölkopf, B., Platt, J., Hoffman, T. (eds.) Advances in Neural Information Processing Systems 19, pp. 137–144. MIT Press, Cambridge (2006). http://books.nips.cc/papers/files/nips19/NIPS2006_0838.pdf
4. Boser, B.E., Guyon, I., Vapnik, V.: A training algorithm for optimal margin classifiers. In: Fifth Annual Workshop on Computational Learning Theory, pp. 144–152 (1992)
5. Bousquet, O., Boucheron, S., Lugosi, G.: Introduction to statistical learning theory. In: Bousquet, O., von Luxburg, U., Rätsch, G. (eds.) Machine Learning 2003. LNCS (LNAI), vol. 3176, pp. 169–207. Springer, Heidelberg (2004)
6. Cortes, C., Vapnik, V.: Support-vector networks. Mach. Learn. **20**, 273–297 (1995)
7. Dinu, G., Baroni, M.: Improving zero-shot learning by mitigating the hubness problem. In: ICLR: Workshop Track (2015). arXiv:1412.6568
8. Fan, R.E., Chang, K.W., Hsieh, C.J., Wang, X.R., Lin, C.J.: LIBLINEAR: a library for large linear classification. JMLR **9**, 1871–1874 (2008)
9. Farhadi, A., Endres, I., Hoiem, D., Forsyth, D.: Describing objects by their attributes. In: CVPR, pp. 1778–1785. IEEE (2009)
10. Fei-Fei, L., Fergus, R., Perona, P.: One-shot learning of object categories. IEEE Trans. Pattern Anal. Mach. Intell. **28**(4), 594–611 (2006)
11. Haehl, V., Vardaxis, V., Ulrich, B.: Learning to cruise: Bernstein's theory applied to skill acquisition during infancy. Hum. Mov. Sci. **19**(5), 685–715 (2000)
12. Jayaraman, D., Grauman, K.: Zero-shot recognition with unreliable attributes. In: NIPS, pp. 3464–3472 (2014)
13. Lampert, C.H., Nickisch, H., Harmeling, S.: Learning to detect unseen object classes by between-class attribute transfer. In: IEEE Conference on Computer Vision and Pattern Recognition, CVPR 2009, pp. 951–958. IEEE (2009)
14. Lampert, C.H., Nickisch, H., Harmeling, S.: Attribute-based classification for zero-shot visual object categorization. IEEE Trans. Pattern Anal. Mach. Intell. **36**(3), 453–465 (2014)
15. Liu, J., Kuipers, B., Savarese, S.: Recognizing human actions by attributes. In: CVPR, pp. 3337–3344. IEEE (2011)

16. Rader, N., Bausano, M., Richards, J.E.: On the nature of the visual-cliff-avoidance response in human infants. Child Dev. **51**(1), 61–68 (1980)
17. Palatucci, M., Pomerleau, D., Hinton, G.E., Mitchell, T.M.: Zero-shot learning with semantic output codes. In: NIPS, pp. 1410–1418 (2009)
18. Romera-Paredes, B., Torr, P.: An embarrassingly simple approach to zero-shot learning. In: ICML, pp. 2152–2161 (2015)
19. Shalev-Shwartz, S., Ben-David, S.: Understanding Machine Learning: From Theory to Algorithms. Cambridge University Press, Cambridge (2014)
20. Shigeto, Y., Suzuki, I., Hara, K., Shimbo, M., Matsumoto, Y.: Ridge regression, hubness, and zero-shot learning. In: Appice, A., Rodrigues, P.P., Santos Costa, V., Soares, C., Gama, J., Jorge, A. (eds.) ECML PKDD 2015. LNCS (LNAI), vol. 9284, pp. 135–151. Springer, Heidelberg (2015). doi:10.1007/978-3-319-23528-8_9
21. Socher, R., Ganjoo, M., Manning, C.D., Ng, A.: Zero-shot learning through cross-modal transfer. In: NIPS, pp. 935–943 (2013)
22. Thrun, S., Mitchell, T.M.: Lifelong robot learning. Rob. Auton. Syst. **15**, 25–46 (1995)
23. Vapnik, V.N.: An overview of statistical learning theory. IEEE Trans. Neural Netw. **10**(5), 988–999 (1999)

Stochastic CoSaMP: Randomizing Greedy Pursuit for Sparse Signal Recovery

Dipan K. Pal$^{(\boxtimes)}$ and Ole J. Mengshoel

Carnegie Mellon University, 5000 Forbes Avenue, Pittsburgh, PA 15232, USA
{dipanp,ole.mengshoel}@cmu.edu

Abstract. In this paper, we formulate the K-sparse compressed signal recovery problem with the L_0 norm within a Stochastic Local Search (SLS) framework. Using this randomized framework, we generalize the popular sparse recovery algorithm CoSaMP, creating Stochastic CoSaMP (StoCoSaMP). Interestingly, our deterministic worst case analysis shows that under the Restricted Isometric Property (RIP), even a purely random version of StoCoSaMP is guaranteed to recover a notion of strong components of a sparse signal, thereby leading to support convergence. Empirically, we find that StoCoSaMP outperforms CoSaMP, both in terms of signal recoverability and computational cost, on different problems with up to 1 million dimensions. Further, StoCoSaMP outperforms several other popular recovery algorithms, including StoGradMP and StoIHT, on large real-world gene-expression datasets.

1 Introduction

Sparse Signal Recovery. The fundamental problem of K-sparse signal recovery from compressed samples is to identify the correct support over the measurement matrix atoms or columns. Given an $M \times N$ measurement matrix Φ and a set of M measurements in the form of the measurement vector $\mathbf{y}$, we want to determine an N-dimensional vector $\mathbf{x}$, a K-sparse signal. Moreover, we want to identify which K atoms of Φ (*i.e.*, the support) were used to generate the signal. Once the support is known, it is trivial to recover the signal using least squares estimation. Looking at this problem naively, one sees that the problem is to pick the right support among $\binom{N}{K}$ different ones, which is known to be NP-hard [13].

Early sparse signal recovery algorithms for Compressed Sensing were greedy pursuit algorithms, *e.g.*, OMP [22], ROMP [14], and CoSaMP [15] (among others such as GraDes [7], IHT [2], and AMP [6]). A key assumption that many greedy pursuit recovery algorithms, including OMP and CoSaMP, make is that Φ satisfies the Restricted Isometry Property (RIP) [4]. A Φ matrix that satisfies RIP preserves the norm of K-sparse signals under its transformation, *i.e.*, $\Phi\mathbf{x}$.

Definition 1 (RIP). *A real-valued matrix Φ satisfies the Restricted Isometry Property (RIP) with constant δ_K if for all K-sparse vectors $\mathbf{x}$ we have*

$$(1 - \delta_K)\|\mathbf{x}\|_2^2 \leq \|\Phi\mathbf{x}\|_2^2 \leq (1 + \delta_K)\|\mathbf{x}\|_2^2.$$

© Springer International Publishing AG 2016
P. Frasconi et al. (Eds.): ECML PKDD 2016, Part I, LNAI 9851, pp. 761–776, 2016.
DOI: 10.1007/978-3-319-46128-1_48

OMP, ROMP, and CoSaMP all have a well-defined phase transition boundary beyond which they begin to fail due to RIP breaking down. This is also because of the inherent non-convexity of the search space as the recovery problems become more difficult (higher sparsity K and more columns than rows for Φ). The phase transition diagram characterizes the degree of difficulty of the recovery problem and is parameterized by M, N, and K.

Stochastic Local Search (SLS). Stochastic Local Search (SLS) algorithms seek to obtain approximate solutions for non-convex problems through randomization. SLS has long played a key role in state-of-the-art algorithms for tackling NP-hard and other difficult computational problems [8]. Example problems for which SLS algorithms are competitive include satisfiability of propositional logic formulas [8]; most probable explanation in Bayesian networks [11]; and maximum a posteriori hypothesis in Bayesian networks [18]. SLS algorithms show great variety [8], however they all use pseudo-randomness (often called "noise"in the SLS literature) during initialization, local search, or both. Carefully balancing randomness and greediness typically has a dramatic and positive impact on the run-time of SLS algorithms [10,20], leading us to strive for similar positive results for sparse signal recovery in this paper.

Related Work. A stochastic approach based on Threshold Accepting (a deterministic form of Simulated Annealing) has been developed [1]. This approach uses an objective function as the product between the L_1 norm and the spectral entropy. A more direct method of randomizing atom selection has also emerged [19], based on Matching Pursuit and OMP. Limitations of this work are experimental validation using low-dimensional problems (up to 128 dimensions) and modest theoretical insights. Another randomized approach to Matching Pursuit uses a non-adaptive random sequence of sub-dictionaries in the decomposition process [12]. More recently, StoGradMP [17] has focused on an approach similar to ours, namely randomizing GradMP [16]. GradMP is a generalized version of CoSaMP. However, StoGradMP is based on stochastic gradient descent, randomly picking one support component at every iteration before projecting the gradient onto a $2K$ dimensional subspace before merging. Under RIP, the algorithm is shown to have exponential convergence in error on average.

Contribution. Our StoCoSaMP method is inspired by SLS and generalizes CoSaMP. Thus, we randomly execute a greedy or a stochastic step at every iteration. During a stochastic step, we randomly choose $2K$ atoms to merge into the support (see Sect. 2.1). In contrast to StoGradMP [17], which has only been shown to handle problems with up to 1000 dimensions so far, we show Sto-CoSaMP to be effective even in problems with up to 1 million dimensions. This renders StoCoSaMP immediately available to real-world applications. Further, StoGradMP requires a careful choice of the block size parameter which has a significant impact on performance. StoCoSaMP also has a parameter, namely the probability P_R of a random step. However, as we find in our experiments, StoCoSaMP's performance is robust for a large range of values for P_R.

In this paper, we make the following contributions:

- We formulate the sparse recovery problem within an SLS framework and propose a novel randomized version of CoSaMP: Stochastic CoSaMP (Sto-CoSaMP). Key in StoCoSaMP is to randomize support selection: we randomly choose $2K$ elements of the support, augmenting the greedy selection through a signal proxy as in CoSaMP [15].
- In a worst case deterministic analysis of StoCoSaMP, we find that under RIP, random support selection is sufficient for StoCoSaMP to recover β-strong components of the true sparse signal. We also show that even a purely random version of StoCoSaMP converges to the true support.
- In experiments with random Gaussian and Hadamard measurement matrices, we demonstrate that StoCoSaMP is more efficient than and outperforms CoSaMP for problems with up to 1 million dimensions.
- In experiments focused on the problem of classification of large-scale real-world gene expression cancer datasets, we compare StoCoSaMP with StoGradMP, StoIHT, CoSaMP, OMP, ROMP, IHT, and AMP. We find that StoCoSaMP outperforms all of these algorithms in terms of test error.

2   Stochastic Local Search for Sparse Signal Recovery

2.1   Stochastic CoSaMP (StoCoSaMP)

Signal proxy and CoSaMP. A popular greedy recovery algorithm is CoSaMP [15]. We generalize CoSaMP to Stochastic CoSaMP (StoCoSaMP) as presented in Fig. 1. StoCoSaMP takes as input a measurement matrix $\mathbf{\Phi}$, a measurement vector $\mathbf{y}$, a sparsity level K, and the probability of a random step P_R. It outputs a sparse coefficient vector $\mathbf{a}$.

StoCoSaMP invokes, with probability P_R (line 6 in Fig. 1), a random step (line 7 in Fig. 1). This random step complements the greedy optimization of the most correlated atoms through the signal proxy (lines 9–10 in Fig. 1); see Definition 6. The random step randomly chooses $2K$ atoms to merge into the current support set T (line 12 in Fig. 1). Lines 12 to 16 proceed exactly like CoSaMP. CoSaMP maintains a constant sized support (size K) at every iteration (while loop in Fig. 1) and hence can be represented in an SLS framework (see Sect. 2.2). Theoretically, in Sect. 2.5, we find that a least squares approximation provides guarantees that do not require a greedy selection such as through the signal proxy. Example stopping criteria (line 3 in Fig. 1) are (i) a maximum number of iterations and (ii) a threshold on the difference between reconstruction errors in subsequent iterations.

We now introduce the framework for defining an SLS algorithm and proceed to model StoCoSaMP in that framework.

2.2   Stochastic Local Search Framework

Definition 2 (General SLS model). *An SLS model is a 4-tuple (S, N_b, G, O), where: S is the set of all states in the search space; $G : S \rightarrow \mathbb{R}$ is the objective*

1: StoCoSaMP ($\mathbf{\Phi}, \mathbf{y}, K, P_R$):

2:   $\mathbf{a}^0 \leftarrow 0, \mathbf{v} \leftarrow \mathbf{y}, i \leftarrow 0, T \leftarrow \{\}$ (Initialization)

3: **while** stopping criterion not satisfied **do**

4:     $i \leftarrow i + 1$

5:     $r \leftarrow$ sample uniform $U[0,1]$

6:     **if** $r < P_R$ **then**

7:       $\lambda \leftarrow$ Randomly choose $2K$ atoms to merge (Novel random step)

8:     **else**

9:       $\mathbf{u} \leftarrow \mathbf{\Phi}^T \mathbf{v}$ (Form signal proxy or estimate, *i.e.*, Traditional greedy step from CoSaMP)

10:       $\lambda \leftarrow supp(\mathbf{u_{2K}})$ (Identify $2K$ largest components)

11:     **end if**

12:     $T \leftarrow \lambda \cup supp(\mathbf{a^{k-1}})$ (Merge supports)

13:     $\mathbf{b}|_{\mathbf{T}} \leftarrow \mathbf{\Phi}_T^\dagger \mathbf{y}$ (Estimate signal using least squares)

14:     $\mathbf{b}|_{\mathbf{T^c}} \leftarrow 0$ (Set the complement support to 0)

15:     $\mathbf{a^i} \leftarrow \mathbf{b_K}$ (Next approximation: keep largest K elements)

16:     $\mathbf{v} \leftarrow \mathbf{y} - \mathbf{\Phi}\mathbf{a^i}$

17: **end while**

18: return $\mathbf{a^i}$

Fig. 1. StoCoSaMP algorithm. $supp(\cdot)$ returns the indices of the non-zero atoms. Lines 6 and 7 contain the key randomization step distinguishing StoCoSaMP from CoSaMP.

or evaluation function; O is the set of optimal states, defined as $O = \{s^*|s^* = \arg\max_s G(s)\}$; and N_b denotes the neighborhood relation, i.e., $N_b \subseteq S \times S$.

Sparse signal recovery can be framed as an SLS problem. We consider binary vectors $\mathbf{s} \in \mathbb{B}^N$. If $s_i = 1$, then the i-th atom or column is included in the support estimate while if $s_i = 0$, then it is not. Thus, the cardinality of the search space is $|S| = |\mathbb{B}^N| = 2^N$. Typically, SLS techniques randomly alternate between a greedy step and a random step. The greedy step usually *enumerates* the entire neighborhood search space N_b and chooses the state which produces the lowest error with respect to some objective function. Sometimes, the random step is only invoked when the previous greedy step produced no improvement.

2.3 SLS for Sparse Signal Recovery

We consider a relaxed neighborhood relation definition as in Definition 3 in order to utilize an efficient search method prevalent in sparse signal recovery: the signal proxy [15]. The signal proxy results in a closed form search step of polynomial complexity. CoSaMP and StoCoSaMP on the whole, at every iteration, search for the next best K atoms for approximation. However, the signal proxy step searches for the top $2K$ atoms. Thus, a relaxation in the neighborhood size is required for modelling it in the SLS framework. This also results in StoCoSaMP being modelled as *two* interconnected sub-models in the SLS framework (see Definitions 5 and 6).

Definition 3 (*Relaxed Neighborhood*). *For some* $\eta, K \in \mathbb{N}$, *we define the neighborhood relation* $N_b^{\eta K}(\mathbf{s}) = \{\mathbf{s}' \in \mathbb{B}^N | \ ||\mathbf{s} - \mathbf{s}'||_0 \leq \eta K\}$ *with a neighbor threshold of* ηK.

For binary vectors, the Manhattan distance between two vectors equals the L_1 distance. Also, η is chosen to model an algorithm. For CoSaMP and Sto-CoSaMP, $\eta = 2$. This relaxation on the neighborhood size threshold from 2 to $2K$ enables using the signal proxy as the greedy element of the search. We now model a sparse signal recovery algorithm in the SLS framework that is constrained to maintain a fixed sized support of cardinality K.

Definition 4 (*SLS model for K-sparse signal recovery*). *An SLS model for K-sparse signal recovery is a 4-tuple* $(S, N_b^{\eta K}, G, O)$, *where: S is the set of all states in the search space with each binary vector state* $\mathbf{s}$ *satisfying* $\sum_{i=1}^{N} s_i = K$. *Lastly, $N_b^{\eta K}$ denotes the relaxed neighborhood function, i.e., $N_b^{\eta K} \subseteq S \times S$ as defined in Definition 3.*

In the case of well-conditioned signal recovery problems, $|O| = 1$, *i.e.*, there exists a single unique solution to the problem [3,5]. The framework remains the same even if the solution is not unique.

2.4   StoCoSaMP in the SLS Framework

StoCoSaMP uses a polynomial complexity search step called the signal proxy [22]. We utilize this technique to get around the computational bottleneck of the standard SLS greedy step.

Signal proxy (line 9 in StoCoSaMP): An efficient greedy search. Greedy sparse signal recovery utilizes the top γK components (for some $\gamma \in \mathbb{N}$) of the signal proxy. The signal proxy is defined as $(\mathbf{\Phi}^T \mathbf{v})_{\gamma K}$, where $(\cdot)_{\gamma K}$ chooses the top γK elements, and $\mathbf{v}$ is the current residue. The signal proxy is an efficient way to determine the most likely active components in the residue. This provides an efficient closed form solution to evaluate the greedy step with $\mathbf{s}^* = \arg\max_{\mathbf{s}} H(\mathbf{s})$ where $H(\mathbf{s}) = ||(\mathbf{\Phi}^T \mathbf{v}) \odot \mathbf{s}||_1$, with $\odot$ denoting the Hadamard product. Recall that $\mathbf{s}$ is a binary vector with γK non-zeros. Note that H depends on the residue $\mathbf{v}$ and thus might change with every iteration.

Due to the incorporation of the signal proxy step, a relaxed neighborhood definition was needed. This complicates the SLS model for StoCoSaMP since there are now two different search spaces. One is the overall space of the top K atoms for the current support estimate (line 15 in Fig. 1) and the other is the selection of the top $2K$ atoms through the signal proxy (line 10 in Fig. 1). StoCoSaMP therefore needs a more elaborate SLS model than given in Definition 4. We thus introduce two connected sub-models: Definition 5 models the overall K-sparse support search of StoCoSaMP whereas Definition 6 models the top $2K$ atom selection through the signal proxy.

Definition 5 (*Sub-model 1: SLS model for K-sparse signal recovery with StoCoSaMP*). *This SLS model is a 4-tuple* $(S, N_b^{\eta K}, G, O)$ *and parameterizes Definition 4 with* $S = \{s \in \mathbb{B}^N \mid ||s||_1 = K\}$. *Also,* $\eta = 2$, *therefore* N_b^{2K} *denotes the relaxed neighborhood function for* S *as in Definition 3. Definitions for* G *and* O *remain the same.*

Definition 6 (*Sub-model 2: SLS model for support selection in Sto-CoSaMP*). *The SLS model is a 4-tuple* $(S', N_b^{\eta K}, G', O'_s)$ *and parameterizes Definition 4 with* $S' = \{s' \in \mathbb{B}^N \mid ||s'||_1 = \gamma K\}$. $G' : S \times S' \to \mathbb{R}$ *where* S *belongs to sub-model 1.* $G'_s(s') = ||(\Phi^T(y - \Phi(a^i \odot s))) \odot s')||_1$ *where* a^i *is the current coefficient estimate (line 15 in StoCoSaMP) and* $G'_s(s')$ *is parameterized by* $s' \in S$; $O'_s = \{s'^* \mid s'^* = \arg\max_{s'} G'_s(s')\}$ *denoting the optimal state for* $G'_s(s')$, *which is unique, i.e.,* $|O'_s| = 1$. *Lastly, with* $\eta = 2\gamma$, $N_b^{2\gamma K}$ *denotes the relaxed neighborhood function for* S' *as in Definition 3.*

In StoCoSaMP, the SLS technique is only explicitly used within S' in Sub-model 2 (for support selection) and not in S where the original problem lies. However, as we explain Sect. 2.5, SLS effects in S' allow StoCoSaMP to escape local minima in S as well. In Sub-model 1, G is not evaluated explicitly by the algorithm. G' in Sub-model 2, however, is efficient to evaluate while being parameterized by $s \in S = \{s \in \mathbb{B}^N \mid ||s||_1 = K\}$. Greedily optimizing G' for $s'^* = \arg\max_{s'} G'_s(s')$, such that $||s'||_1 = \gamma K$, offers the exponential recovery guarantees that CoSaMP enjoys. These guarantees also arise due to a least squares approximation in subsequent steps.

2.5 Analysis of StoCoSaMP

Since StoCoSaMP randomly picks a random step or a greedy step, a comprehensive analysis of the phenomenon of escaping local minima is difficult. Experimentally, we observe strong performance of StoCoSaMP as reported in Sect. 3. Analytically, we have some but limited results as reported below.[1]

We first analyze the extreme cases $P_R = 0$ and $P_R = 1$, under RIP, before discussing the general case $0 \leq P_R \leq 1$. Note that our analysis assumes RIP only for $P_R = 0$ and $P_R = 1$. We only hypothesize a condition (when RIP breaks down) under which local minima arise in the general case.

Special Case: $P_R = 0$ (Purely Greedy Pursuit)

Lemma 1. *When* $P_R = 0$, *StoCoSaMP is* **equivalent** *to CoSaMP, i.e.,* $CoSaMP = StoCoSaMP(\Phi, y, K, 0)$.

StoCoSaMP with $P_R = 0$ enjoys the same exponential recovery guarantees as CoSaMP [15]. Unfortunately, it also inherits the propensity of CoSaMP to get trapped in local optima that may not be global.

Special Case: $P_R = 1$ (Purely Random Pursuit). The primary goal here is to show that even if $P_R = 1$, StoCoSaMP will retain strong components per Definition 9.

---

[1] Proofs of all results not found here will be included in the full version of this paper.

Lemma 2. *When $P_R = 1$, StoCoSaMP is equivalent to a random walk or a purely random pursuit.*

We have Φ as a normalized matrix with the RIP constant δ_K for K-sparse signals. Let $\mathbf{x} \in \mathbb{R}^N$ be a K-sparse vector. The following two definitions model lines 13 in Fig. 1 and the true support and the current support estimate $supp(\mathbf{b_K})$ respectively.

Definition 7 (*Least Squares Estimate*). *For any **randomly selected or arbitrary** support set T (see lines 6 and 7 of StoCoSaMP) with $|T| = 2K$, the least squares estimate for the signal at any particular iteration $\mathbf{b}$ is defined as $\mathbf{b}_{|T} = \Phi_T^\dagger \mathbf{y}$ together with $\mathbf{b}_{|T^c} = 0$.*

Definition 8 (*True Support and Current Support*). *We define the true support to be $\lambda^* = supp(\mathbf{x})$ and the current support to be $\lambda = supp(\mathbf{b}_K)$ where $\mathbf{b}_K$ is the best K-sparse approximation of $\mathbf{b}$ at the current StoCoSaMP iteration.*

The following notion of strong components will prove useful in our analysis (see Theorems 1 and 2).

Definition 9 (*β-Strong Component w.r.t. Ψ*). *We define a true signal component $\mathbf{x}_\Omega$ ($|\Omega| = 1$) to be β-strong w.r.t. Ψ, if for a subset $\Psi \subset \lambda^*$ of the indices of the true components, with $|\Psi| \leq K - 1$ and $\Omega \notin \Psi$, we have $\frac{|\mathbf{x}_{|\Omega}|}{||\mathbf{x}_{|\Psi}||_2} \geq \beta$.*

Notation: We now define notations for the rest of Sect. 2.5. We denote, for *one* iteration of StoCoSaMP, a selected support by λ. For the analysis, we also define $\Omega \in F = \{T \cap \lambda^*\}$ with $|\Omega| = 1$; F represents the true components in the current support estimate. $Z = T \backslash \lambda^*$ represents the rest of the false components in the current support set T, which are not active in the actual signal. Lastly $\Psi = F \backslash \Omega$ for every iteration of the random step (lines 6 and 7 of StoCoSaMP).

The following lemma is useful in proving Theorem 1.

Lemma 3. *We have*

$$||((\Phi_F^* \Phi_F)^{-1})^\Omega \mathbf{x}||_2 \lessgtr \left(1 \pm \delta_K^2 \eta\right) |\mathbf{x}_{|\Omega}| \pm \eta ||\mathbf{x}_{|\Psi}||_2$$

where $\eta = \left(\frac{\delta_K}{1 - \delta_K - 1 - \delta_K^2}\right)$ and $((\Phi_F^* \Phi_F)^{-1})^\Omega$ corresponds to the Ωth row of $(\Phi_F^* \Phi_F)^{-1}$. Further, we assume $\Phi_\Psi^* \Phi_\Psi$ is full rank and that Φ has normalized columns.

Lemma 3 involves two inequalities which have been combined in one statement. It is useful since it bounds the projection of $\mathbf{x}$ onto the Ωth row of $(\Phi_F^* \Phi_F)^{-1}$. We now present our main result.

Theorem 1 (*Retaining Strong True Support*). *For StoCoSaMP with $P_R = 1$, if $\delta_Z \leq \delta_{Z+K} \leq 0.03$, $\delta_K \leq \delta_{Z+K}$ and $\mathbf{x}_{|\Omega}$ is β-strong w.r.t. Ψ with $\beta = 0.1$, then $\Omega \in \lambda$.*

Interestingly, under RIP, Theorem 1 shows that even if an algorithm is not greedy and *always* picks a random support, β-strong true components of the signal present in the current support (w.r.t. to the other true components) are guaranteed to be retained. Our proof strategy is to find a condition such that the lower bound for $\|\mathbf{b}_{|\Omega}\|_2$ is greater than the upper bound for $\|\mathbf{b}_{|\Psi}\|_2$. This forces the true component Ω into the top K elements chosen during pruning. We find the lower and upper bounds through RIP.

When the sparse signal is exactly K-sparse, the β-strong constraint might seem restrictive, since it only allows components stronger w.r.t. the rest of the components by a factor of β to be recovered. However, in the case of general signals, where the true sparse vector contains noise, the β-strong constraint is more easily satisfied due the presence of very small noisy components. Thus, the K large components are more likely to be recovered.

Support convergence under β -strong condition: Now let Υ^i be the set of *true* components in the support estimate at the ith StoCoSaMP iteration, *i.e.*, $\Upsilon^i = \lambda^i \cap \lambda^*$. Thus, at every iteration, we would like $|\Upsilon|$ to increase up until the desired cardinality of the support, (*i.e.*, K). We have the following result on support convergence.

Theorem 2 *(Support Convergence for Purely Random StoCoSaMP).* *For StoCoSaMP with* $P_R = 1$, *i.e., a purely random pursuit, if* $\{\delta_K, \delta_Z\} \leq \delta_{Z+K} \leq 0.03$ *and* $\exists \mathbf{x}_{|\Omega}$ *in support* T^i *at iteration i, such that* $\mathbf{x}_{|\Omega}$ *is β-strong w.r.t. some Ψ with $\beta = 0.1$ and $\Omega \notin \lambda^i$, then* $|\Upsilon^{i+1}| \geq |\Upsilon^i|$.

Theorem 2 shows that even for StoCoSaMP's purely random case ($P_R = 1$), the support estimate does not worsen as the algorithm progresses. There will be no improvement when β-strong components are not present: the support estimate does not change. These results are deterministic since they analyze the worst case and are stronger guarantees than the average case analysis typical of randomized algorithms. For $0 < P_R < 1$, the greedy step has already been shown to have exponential reduction in error [15], thus the results presented here ensure that the algorithm does not diverge while executing a random step.

Theorem 2 is a contrast to earlier results suggesting that greed is important for recoverability [21]. Although a greedy algorithm might have stronger guarantees for recoverability, random support selection allows for practical improvements (analogous to those in the SLS literature) in situations where RIP breaks down, (*e.g.*, past the phase transition boundaries of greedy pursuits). Note that in such a case, Theorem 1 will not hold and StoCoSaMP, like CoSaMP, currently has no theoretical guarantees. The SLS properties of StoCoSaMP then assume a larger role, which is difficult to analyze theoretically.[2]

---

[2] Indeed, in our experiments, we find that for $P_R = 1$, when RIP breaks down, Sto-CoSaMP does not perform well. A few greedy steps are needed for convergence (see Fig. 4(a)). Nonetheless, in experiments in Sect. 3 we find that StoCoSaMP converges towards the true solution even in large dimensions for high $P_R = 0.9$ but not for $P_R = 1$ (see Fig. 4(a)). In most cases, we find empirically that StoCoSaMP converges on average faster than CoSaMP.

General Case: $0 \leq P_R \leq 1$ (Randomized Greedy Pursuit). With $0 \leq P_R \leq 1$, under RIP, the individual theoretical guarantees of greedy and random steps still hold. However, when RIP does not hold, the random effects of the SLS model S from Definition 5 become interesting. We now define an active variable for use in the informal discussion about escaping local minima.

Definition 10 (Active Variable). *Let* $\mathbf{y} = \mathbf{\Phi}\mathbf{x}$ *where* $\mathbf{x} \in \mathbb{R}^N$ *is a K-sparse vector. Then the set $A = \{i \mid |x_i| > 0\}$ is called the active set, and the corresponding i-th element of $\mathbf{x}$ is called an active variable.*

Assume that a column or atom τ of $\mathbf{\Phi}$ is approximately linearly dependent on some other set of columns Γ belonging to $\mathbf{\Phi}$, *i.e.*, $\phi_{|\tau} \approx \sum_{i \in \Gamma} \alpha_i \phi_i$ for some α_i. Now, if the signal had each element in Γ as its *active variable*, but not τ, then the signal proxy $\mathbf{\Phi}^T \mathbf{v}$ (line 9 in StoCoSaMP) forces CoSaMP (and StoCoSaMP) to pick τ. The atom τ can be said to be "stronger" than the atoms in the set Γ since τ is more likely to be picked by the signal proxy rather than the true atoms in Γ. This is because picking τ explains much more of the signal.

In this situation, when a "stronger" component τ exists w.r.t. a set Γ, the search falls into a *local minimum*. It would be hard to drop τ from the support estimate, as it approximately explains the components Γ in the signal by itself. This is where StoCoSaMP randomness could help. In randomly choosing $2K$ atoms from $\mathbf{\Phi}$, it is more likely than in a greedy setting that the algorithm might pick a few atoms that are active and "weak" compared to some other atom. Once CoSaMP (and StoCoSaMP) chooses a variable, it explains away that component. Thus, the random step in StoCoSaMP (for $0 < P_R \leq 1$) helps the search to avoid being trapped in a local minimum.

This effective dodging of local minima acts even when RIP might not hold. However, in the case where RIP does hold for $\mathbf{\Phi}$, Theorem 1 shows that greed is not necessary for recovering β-strong components of the signal w.r.t. $\mathbf{\Psi}$. In many of our experiments, such as the real-world gene expression data (Sect. 3.5), we do not check for the RIP condition, but StoCoSaMP still works well.

3   Experimental Results

3.1   Phase Transition Diagrams

Goal. The goal of this experiment is to investigate whether StoCoSaMP can solve a broader and harder range of problems compared to CoSaMP. Specifically, we seek to reconstruct K-sparse i.i.d. Gaussian signals with no noise added.

Method and Data. The inherent dimensionality of the problem was set to $N = 200$. The measurement matrices were i.i.d. sampled from the standard normal distribution, $\mathcal{N}(0, 1)$. For StoCoSaMP, $P_R = 0.3$.[3] As is standard in the

---

[3] For all experiments, we set the maximum number of iterations for both CoSaMP and StoCoSaMP generously to 250.

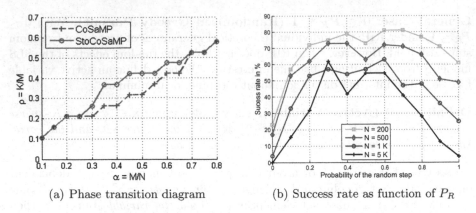

(a) Phase transition diagram   (b) Success rate as function of P_R

Fig. 2. (a) Phase transition boundary diagrams (showing probability of signal recovery) for CoSaMP and StoCoSaMP (with $P_R = 0.3$). (b) Effect of varying the probability of the random step P_R, along the x-axis, on StoCoSaMPs signal recoverability, along the y-axis, for Gaussian measurement matrices. CoSaMP is $P_R = 0$ in (b).

literature [9], intrinsic recovery capability of the algorithm was measured in a noiseless setting with a 90 % threshold in probability of recovery.

Results. We compare the phase transition diagrams of CoSaMP and Sto-CoSaMP in Fig. 2(a). The x-axis is $\alpha = M/N$ and the y-axis is $\rho = K/M$, where K is the sparsity and M, N are the dimensions of Φ. If a point is below the transition boundary, problems of that setting are considered solved given a threshold for the probability of recovery (90 %). The axes denote gradual change in difficulty, with the most difficult setting being the top left corner and the easiest being the bottom right corner. Figure 2(a) shows that StoCoSaMP clearly improves on the phase transition region over CoSaMP, especially for $0.35 \le \alpha \le 0.5$.

3.2 Effect on Recoverability: Random Gaussian Matrices

Goal. The goal of this experiment is to compare the performance of CoSaMP and StoCoSaMP for a broad range of P_R-values.[4] This will (i) shed light on the problem of local optima in sparse signal recovery and (ii) provide an experimental counter-point to Theorem 1.

Method and Data. We constructed 100 normalized synthetic signal recovery problems; the $M \times N$ measurement matrix was sampled i.i.d. from $\mathcal{N}(0,1)$. Then, by varying P_R, we examine the percentage of successful recoveries for Sto-CoSaMP (recovered SNR > 50 dB). We investigate a challenging point on the phase transition, specifically $\alpha = 0.6$ and $\rho = 0.5$ (see Fig. 2(a)). The dimensionality is varied from $N = 200$ to $N = 5000$.

---

[4] For comparative results on real-world data please refer to Sect. 3.5.

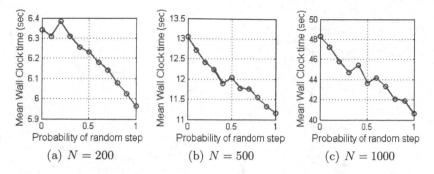

Fig. 3. Effect of varying the random step probability P_R (x-axes) on the mean wall clock run time (y-axes). CoSaMP is $P_R = 0$. The reduction in mean wall clock time is relatively small, but comes in addition to improved accuracy (see Fig. 2(b)).

Results. Figure 2(b) illustrates the serious handicap of purely greedy pursuits in escaping local optima. Specifically, Fig. 2(b) shows that at $\alpha = 0.6$ and $\rho = 0.5$, CoSaMP ($P_R = 0$) performs poorly.[5] For $0.1 \leq P_R \leq 0.9$, StoCoSaMP tends to succeed significantly more often. *For these lower dimensions, even a purely random walk ($P_R = 1$) performs better than CoSaMP ($P_R = 0$), owing to SLS properties and Theorem 1 (random Gaussian matrices are known to satisfy RIP).* The result, though perhaps initially surprising, is consistent with previous studies of the role of randomization in hard combinatorial problems. In problems of high difficulty, the expected time to find a global optimum is minimized when search is close to a random walk [10].

Figure 3(a)-(c) show the mean wall clock run time for the convergence over these 100 problems as P_R in StoCoSaMP was varied. Since the time complexity of a single random step is lower than that of a single greedy step, the overall computational time generally decreases as we increase P_R. Hence, StoCoSaMP can not only outperforms CoSaMP in terms of recoverability, but also in terms of computation time for a significant range of P_R.

3.3   Effect on Recoverability: High-Dimensional Problems

Goal. To handle high-dimensional data such as images or spatio-temporal data, we experiment with Hadamard matrices with up to 1 million dimensions.

Method and Data. We use sets of randomly permuted rows of the Hadamard matrix as the measurement matrix, and set $\alpha = 0.1$ and $\rho = 0.05$, giving us reasonable values for the sparsity K and the number of measurements M. We simulate 100 different problems, and define a strict SNR threshold for a successful recovery at 120 dB.

---

[5] We consider a recovery successful if the SNR of the recovered sparse signal to the ground truth is above a certain threshold (50 dB for this experiment).

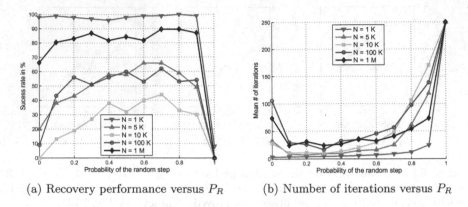

(a) Recovery performance versus P_R (b) Number of iterations versus P_R

Fig. 4. Effect of varying P_R, along x-axis, on: (a) StoCoSaMP's signal recoverability for randomly permuted Hadamard measurement matrices and (b) the mean number of iterations required for convergence by StoCoSaMP. $P_R = 0$ is CoSaMP. (Color figure online)

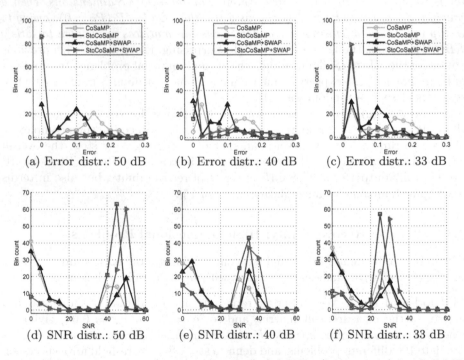

(a) Error distr.: 50 dB   (b) Error distr.: 40 dB   (c) Error distr.: 33 dB

(d) SNR distr.: 50 dB   (e) SNR distr.: 40 dB   (f) SNR distr.: 33 dB

Fig. 5. Histograms of recovery errors (top row) and SNRs (bottom row) for CoSaMP (green), StoCoSaMP (blue), CoSaMP+SWAP (black), and StoCoSaMP+SWAP (red), for different levels of noise (50 dB, 40 dB, or 33 dB) added to the measurement vector. (Color figure online)

Table 1. Mean (μ) and Standard deviations (σ) of recovery errors and SNRs for the comparison of CoSaMP, StoCoSaMP and SWAP initialized by each algorithm (denoted by CoSaMP+SWAP and StoCoSaMP+SWAP), under different noise added to measurement vector (SNR in dB). For a given SNR level, **bold** and *italics* signify the best and the second-best results respectively.

SNR	Algorithm	μ_{Error}	σ_{Error}	μ_{SNR}	σ_{SNR}
50	CoSaMP	0.1135	0.0803	13.934	18.393
	StoCoSaMP	*0.0334*	0.0784	*37.954*	**14.595**
	CoSaMP+SWAP	0.0735	*0.0527*	15.923	20.543
	StoCoSaMP+SWAP	**0.0216**	**0.0516**	**42.157**	*15.942*
40	CoSaMP	0.1031	0.0719	13.222	**13.749**
	StoCoSaMP	*0.0600*	0.0748	*24.157*	*14.002*
	CoSaMP+SWAP	0.0665	**0.0478**	15.134	15.634
	StoCoSaMP+SWAP	**0.0390**	*0.0508*	**27.471**	15.624
33	CoSaMP	0.1189	0.0678	9.633	*10.589*
	StoCoSaMP	*0.0625*	0.0652	*20.974*	**9.718**
	CoSaMP+SWAP	0.0781	*0.0473*	11.331	12.254
	StoCoSaMP+SWAP	**0.0401**	**0.0441**	**24.179**	10.887

Results. Figure 4(a) reports success rates, while Fig. 4(b) reports the number of iterations. The figures suggest that the advantages, both in terms of success rate and computation time (iterations), of StoCoSaMP over CoSaMP are not restricted to the lower-dimensional case. For *success rate* (see Fig. 4(a)), the advantage of StoCoSaMP over CoSaMP is very clear for the high-dimensional cases of $N = 100K$ (blue line) and $N = 1$ million (black line).[6] In image processing applications, for example, $N = 1$ million is typical. As seen in Fig. 4(a), at these high dimensions, a purely random pursuit ($P_R = 1$) fails in many cases, whereas $0.1 \leq P_R \leq 0.9$ does much better.

Using $0.1 \leq P_R \leq 0.6$, StoCoSaMP also gives faster convergence than CoSaMP ($P_R = 0$) in terms of number of iterations, see Fig. 4(b). This and the results of the previous experiment in Sect. 3.2 experimentally validate Theorem 1. From this and the previous experiment, it seems that recoverability is best when StoCoSaMP employs a combination of greedy and random steps. To the practitioner, we suggest to use $0.2 \leq P_R \leq 0.6$, where both high success rates and computational gains are apparent.[7] The theoretical justification is unclear and would be interesting to explore in future work.

---

[6] The variation of success for different dimensions for both StoCoSaMP and CoSaMP is inherent to the performance characteristics of CoSaMP itself.

[7] This does not hold for the smallest $N = 1K$ problems, which on the other hand are the least interesting from a scalability point of view.

3.4   Effect of Noise on Recoverability

Goal. We now focus on noisy measurement signals. We also evaluate the performance of SWAP when it is initialized by CoSaMP and StoCoSaMP. SWAP is a greedy algorithm; it can be used to fine-tune the solutions of other recovery algorithms [2]. Here, we use SWAP to investigate whether a solution obtained by StoCoSaMP on average lies in a better basin than a solution obtained by CoSaMP. Intuitively, SWAP's explicit greediness forces the solution within each basin towards the local optimum of that particular basin.

Method and Data. We experiment with three noise levels in the measurement vector 50 dB (low noise), 40 dB, 33 dB (high noise). We construct 100 synthetic problems with the measurement matrix and the sparse vectors being sampled i.i.d. from $N(0,1)$. We add varying amounts of white Gaussian noise to our measurement vectors, such that the SNR is at one of the three levels of noise. For all problems, we set $P_R = 0.3$ in StoCoSaMP and $N = 200$.

Results. We report the error and SNR statistics and the corresponding histograms for the four algorithmic combinations in Table 1 and Fig. 5. Table 1 suggests that StoCoSaMP achieves higher quality recoveries (SNR) on average compared to CoSaMP over all problems (SNR $\geq$ 33 dB) in the presence of varying levels of noise. In fact, StoCoSaMP in most cases performs better on average than SWAP initialized with CoSaMP. This is powerful, since StoCoSaMP is also computationally less expensive than either CoSaMP (purely greedy) or SWAP (exponential complexity). The histogram plots in Fig. 5 clearly show that both StoCoSaMP and StoCoSaMP+SWAP tend to achieve much lower errors than their CoSaMP counterparts. This experiment suggests that the advantages of StoCoSaMP extend to situations with significant noise.

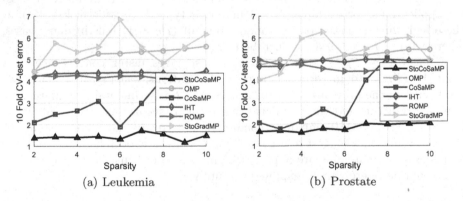

(a) Leukemia   (b) Prostate

Fig. 6. 10-fold cross-validation testing error on the gene-expression datasets for (a) Leukemia data (72 × 5147) (b) and Prostate data (102 × 12533).

3.5   Real-World Data: Classifying Gene Expression

Goal. To evaluate StoCoSaMP's real-world classification performance, we study two large-scale gene expression cancer datasets. We compare StoCoSaMP, CoSaMP, StoGradMP, StoIHT, OMP, ROMP, IHT, and AMP.

Method and Data. The datasets contain the gene expression levels for leukemia ($M = 72$ and $N = 5147$) and prostate cancer ($M = 102$ and $N = 12533$) with a binary label.[8] This is a classification problem where unseen gene expression values are to be classified. We learn a K-sparse linear classifier for this experiment. The labels act as the signal to be expressed as a linear combination of the gene expressions. The linear combination is parameterized by a weight vector that is recovered by a sparse signal recovery algorithm. Following a previous study [23], we explore sparsities ranging from $K = 2$ to $K = 10$.

AMP determines the sparsity level internally through soft thresholding. For AMP, unlike for the other algorithms, we do not enforce K-sparsity. We perform 10-fold cross validation for each level of sparsity (20 trials for StoCoSaMP, StoIHT and StoGradMP for each sparsity) with $P_R = 0.5$ for StoCoSaMP. For StoGradMP and StoIHT, we set the block size to $\min(K, M)$ as in previous work [17]. We then pick the classifier (sparse solution) that minimizes the training error as our model for testing for all stochastic algorithms.

Results. Figure 6(a) and (b) show the experimental results for all algorithms for the two datasets. StoCoSaMP's results (black line) are a consistent improvement over all other algorithms (including StoGradMP) in all cases. AMP achieved an error of 11.10 (with $K = 62$) on the Leukemia dataset and 31.35 (with $K = 89$) on the Prostate dataset, worse than StoCoSaMP. StoIHT performed worse in all cases and reported an error consistently above 10, and is not plotted in the figure.

4   Conclusion

In this paper, we present a Stochastic CoSaMP (StoCoSaMP) method. Under RIP, even a purely random version of StoCoSaMP ($P_R = 1$) will observe support convergence. This provides an interesting addition to previous results, which have suggested that greed is good for recovery [21]. Our experiments show that StoCoSaMP out-performs CoSaMP on a variety of signal recovery problems, and other algorithms (including CoSaMP, StoGradMP and StoIHT) on a real-world large scale classification task.

Acknowledgements. We thank Aswin C. Sankaranarayanan for his helpful comments and for checking the correctness of the proofs.

---

[8] See http://www.biolab.si/supp/bi-cancer/projections/info/leukemia.htm and /prostate.htm.

References

1. Andrecut, M.: Stochastic recovery of sparse signals from random measurements. Eng. Lett. **19**(1), 1–6 (2011)
2. Blumensath, T., Davies, M.E.: Iterative hard thresholding for compressed sensing. Appl. Comput. Harmonic Anal. **27**(3), 265–274 (2009)
3. Candes, E.: Compressive sampling. In: International Congress of Mathematicians, vol. 3, pp. 1433–1452 (2006)
4. Candes, E., Romberg, J., Tao, T.: Robust uncertainty principles: exact signal reconstruction from highly incomplete frequency information. IEEE Trans. Inf. Theory **52**(2), 489–509 (2006)
5. Candes, E., Tao, T.: Near-optimal signal recovery from random projections: universal encoding strategies? IEEE Trans. Inf. Theory **52**(12), 5406–5425 (2006)
6. Donoho, D., Tanner, J.: Precise undersampling theorems. Proc. IEEE **98**(6), 913–924 (2010)
7. Garg, R., Khandekar, R.: Gradient descent with sparsification: an iterative algorithm for sparse recovery with restricted isometry property. In: ICML 2009, New York, NY, USA, pp. 337–344 (2009)
8. Hoos, H.H., Stützle, T.: Stochastic Local Search: Foundations and Applications. Morgan Kaufmann, San Francisco (2005)
9. Maleki, A., Donoho, D.L.: Optimally tuned iterative reconstruction algorithms for compressed sensing. IEEE J. Sel. Topics Signal Process. **4**(2), 330–341 (2010)
10. Mengshoel, O.J.: Understanding the role of noise in stochastic local search: analysis and experiments. Artif. Intell. **172**(8–9), 955–990 (2008)
11. Mengshoel, O.J., Wilkins, D.C., Roth, D.: Initialization and restart in stochastic local search: computing a most probable explanation in Bayesian networks. IEEE Trans. Knowl. Data Eng. **23**(2), 235–247 (2011)
12. Moussallam, M., Daudet, L., Richard, G.: Matching pursuits with random sequential subdictionaries. Signal Process. **92**(10), 2532–2544 (2012)
13. Needell, D.: Topics in Compressed Sensing. ArXiv e-prints, May 2009
14. Needell, D., Vershynin, R.: Signal recovery from incomplete and inaccurate measurements via regularized orthogonal matching pursuit. IEEE J. Sel. Topics Signal Process. **4**(2), 310–316 (2010)
15. Needell, D., Tropp, J.A.: CoSaMp: iterative signal recovery from incomplete and inaccurate samples. Commun. ACM **53**(12), 93–100 (2010)
16. Nguyen, N., Chin, S., Tran, T.: A unified iterative greedy algorithm for sparsity-constrained optimization (2012)
17. Nguyen, N., Needell, D., Woolf, T.: Linear convergence of stochastic iterative greedy algorithms with sparse constraints (2014). arXiv preprint arXiv:1407.0088
18. Park, J.D., Darwiche, A.: Complexity results and approximation strategies for MAP explanations. J. Artif. Intell. Res. **21**, 101–133 (2004)
19. Peel, T., Emiya, V., Ralaivola, L., Anthoine, S.: Matching pursuit with stochastic selection. In: EUSIPCO, pp. 879–883. IEEE (2012)
20. Selman, B., Kautz, H.A., Cohen, B.: Noise strategies for improving local search. In: Proceedings of AAAI 1994, pp. 337–343 (1994)
21. Tropp, J.A.: Greed is good: algorithmic results for sparse approximation. IEEE Trans. Inf. Theory **50**(10), 2231–2242 (2004)
22. Tropp, J.A., Gilbert, A.C.: Signal recovery from random measurements via orthogonal matching pursuit. IEEE Trans. Inf. Theory **53**, 4655–4666 (2007)
23. Vats, D., Baraniuk, R.: When in doubt, swap: high-dimensional sparse recovery from correlated measurements. In: NIPS (2013)

Deep Metric Learning with Data Summarization

Wenlin Wang[1]([⊠]), Changyou Chen[1], Wenlin Chen[2],
Piyush Rai[3], and Lawrence Carin[1]

[1] Department of Electrical and Computer Engineering,
Duke University, Durham, USA
{ww107,cc448,lcarin}@duke.edu
[2] Department of Computer Science and Engineering,
Washington Univeristy in St. Louis, St. Louis, USA
dtccwl@gmail.com
[3] Department of Computer Science and Engineering,
IIT Kanpur, Kanpur, India
piyush.rai@duke.edu

Abstract. We present Deep Stochastic Neighbor Compression (DSNC), a framework to compress training data for instance-based methods (such as k-nearest neighbors). We accomplish this by inferring a smaller set of *pseudo-inputs* in a new feature space learned by a deep neural network. Our framework can equivalently be seen as *jointly* learning a nonlinear distance metric (induced by the deep feature space) and learning a compressed version of the training data. In particular, compressing the data in a deep feature space makes DSNC robust against label noise and issues such as within-class multi-modal distributions. This leads to DSNC yielding better accuracies and faster predictions at test time, as compared to other competing methods. We conduct comprehensive empirical evaluations, on both quantitative and qualitative tasks, and on several benchmark datasets, to show its effectiveness as compared to several baselines.

1 Introduction

In machine learning problems there are situations for which the massive data scale renders learning algorithms infeasible to run in a reasonable amount of time. One solution is to first summarize the data in the form of a small set of *representative* data points that best characterize and represent the original data, and then run the original algorithm on this subset of the data. This may be desirable due to the requirement of making a fast prediction at test time, in problems where the predictions depend on the entire training data, e.g., k-nearest neighbors (kNN) classification [4,8] or kernel methods such as SVMs [21]. For example, in traditional kNN classification, the prediction cost for each test example scales linearly in the number of training examples, which can be expensive if the number of training examples is large. Traditional approaches to speed-up such methods usually rely on cleverly designed data structures or select a compact subset of the original data (e.g., via subsampling [4]). Although such methods may reduce the storage requirements and/or prediction time, the

© Springer International Publishing AG 2016
P. Frasconi et al. (Eds.): ECML PKDD 2016, Part I, LNAI 9851, pp. 777–794, 2016.
DOI: 10.1007/978-3-319-46128-1_49

performance tends to suffer, especially if the original data is high-dimensional and/or noisy.

Recently [24] introduced Stochastic Neighbor Compression (SNC), which learns a set of *pseudo-inputs* for kNN classification by minimizing a stochastic 1-nearest neighbor classification error on the training data. Compared to the data sub-sampling approaches, SNC achieves impressive improvements in test accuracy when using these pseudo-inputs as the new training set. However, since SNC performs data compression in the original data space (or in a linearly transformed lower-dimensional space), it may perform poorly when the data in the original space are highly non-separable and noisy.

Motivated by this, we present Deep Stochastic Neighbor Compression (DSNC), a new framework to jointly perform data summarization akin to the methods like SNC, while also learning a *nonlinear* feature representation of the data via a deep learning architecture. Our framework is based on optimizing an objective function that is designed to learn nonlinear transformations that preserve the neighborhood structure in the data (based on label information), while simultaneously learning a small set of pseudo-inputs that summarize the entire data. Note that, due to the neighborhood preserving property, our framework can also be viewed as performing a nonlinear (deep) distance metric learning [22], while also learning a summarized version of the original data. The data summarization aspect also makes DSNC much faster than other metric learning based approaches which need all the training data. In DSNC, the data summarization and feature learning, both, are performed jointly through backpropagation [31] using stochastic gradient descent, making our framework readily scalable to large data sets. Moreover, our framework is also more general than standard feedforward neural networks which perform simultaneous feature learning and classification but are not designed to learn a summary of the data which may be useful in its own right.

In our comprehensive empirical studies, DSNC achieves superior classification accuracies on the seven datasets we used in the experiments, outperforming SNC by a significant margin. For example, with DSNC, 1-NN is able to achieve 0.67 % test error on MNIST with only ten compressed data samples (one per class) on a 20-dimensional feature space, compared to 7.71 % for SNC. We also report qualitative experiments (via visualization) showing that DSNC is effective in learning a good summary of the data.

2   Background

Throughout this paper, we denote vectors as bold, lower-case letters, and matrices as bold, upper-case letters. $\|\cdot\|$ applied to a vector denotes the standard vector norm, $[\mathbf{X}]_{ij}$ means the (i, j)-th element of matrix $\mathbf{X}$. We denote the training data $\mathbf{X} = \{\mathbf{x}_1, ..., \mathbf{x}_N\}$, where $\mathbf{X} \in \mathbb{R}^{D \times N}$ are N observed data samples of dimensionality D with corresponding labels $\mathbf{Y} = \{y_1, ..., y_N\} \in \mathcal{Y}^N$, with $\mathcal{Y}$ as a discrete set of possible labels.

To motivate our proposed framework DSNC (described in Sect. 3), we first provide an overview of Neighborhood Components Analysis (NCA) [16,32] and

Stochastic Neighbor Compression (SNC) [24], which our proposed framework builds on.

2.1   Neighborhood Components Analysis

Neighborhood Components Analysis (NCA) [16] is a distance metric learning method that learns a mapping $f(\cdot|\mathbf{W})$ with parameters $\mathbf{W}$ to optimize the k-nearest-neighbors classification objective. The optimization is based on preserving the Euclidean distance $d_{ij} = \|f(\mathbf{x}_i) - f(\mathbf{x}_j)\|^2$ in the transformed space for $\mathbf{x}_i$ and $\mathbf{x}_j$, based on their original neighborhood relationship in the original space. Specifically, soft neighbor assignments are used in NCA to directly optimize the mapping f for kNN classification performance. The probability p_{ij} that $\mathbf{x}_i$ is assigned to $\mathbf{x}_j$ as its stochastic nearest-neighbor is modeled with a softmax over distances between $\mathbf{x}_i$ and the other training samples, i.e., $p_{ij} = \frac{\exp(-d_{ij})}{\sum_{k:k\neq i} \exp(-d_{ik})}$. The objective of NCA is to maximize the expected number of correctly classified points, expressed here as a log-minimization problem: $\hat{\mathbf{W}} = \arg\min_{\mathbf{W}} - \sum_{i=1}^{N} \log(p_i)$, where p_i is the probability that the mapped sample $f(\mathbf{x}_i|\mathbf{W})$ is correctly classified with label y_i, i.e., $p_i = \sum_{j:y_i=y_j} p_{ij}$. Although NCA can learn a distance metric adaptively from data, the entire training data still needs to be stored, making it computationally and storage-wise expensive at test time. To extend NCA with nonlinear transformations, [32] defines $f(\cdot|\mathbf{W})$ to be a feedforward neural network parameterized by weights $\mathbf{W}$.

2.2   Stochastic Neighbor Compression (SNC)

Stochastic Neighbor Compression (SNC) is an improvement over NCA by learning a compressed kNN training set by optimizing a soft neighborhood objective [24]. The goal in SNC is to find a subset of $m \ll N$ compressed samples $\mathbf{Z} = [\mathbf{z}_1, ..., \mathbf{z}_m]$ with labels $\hat{\mathbf{Y}} = [\hat{y}_1, ..., \hat{y}_m]$, to best approximate the kNN decision rule on the original set of training samples $\mathbf{X}$ and labels $\mathbf{Y}$. Different from NCA, a compressed set $\mathbf{Z}$ needs to be learned from the whole data. The objective is to maximize the stochastic nearest-neighbor accuracy with respect to $\mathbf{Z}$, i.e., $\hat{\mathbf{Z}} = \arg\min_{\mathbf{Z}} - \sum_{i=1}^{N} \log(p_i)$, where the probability of a correct assignment between a training sample $\mathbf{x}_i$ and the compressed neighbors $\mathbf{z}_i$ is defined as $p_i = \sum_{j:y_i=y_j} \frac{\exp(-\gamma^2 \|\mathbf{x}_i - \mathbf{z}_j\|^2)}{\sum_{k=1}^{m} \exp(-\gamma^2 \|\mathbf{x}_i - \mathbf{z}_k\|^2)}$, where γ is the width of the Gaussian kernel. Given such probabilities, the objective of SNC is constructed as in the case of NCA and is optimized w.r.t. the m pseudo-inputs $\mathbf{Z}$. In [24], a *linear* metric learning extension of this approach was also considered, which defines $p_i = \sum_{j:y_i=y_j} \frac{\exp(\|-\mathbf{A}(\mathbf{x}_i - \mathbf{z}_j)\|^2)}{\sum_{k=1}^{m} \exp(-\gamma^2 \|-\mathbf{A}(\mathbf{x}_i - \mathbf{z}_k)\|^2)}$, in which the pseudo-inputs will be learned in the linearly transformed space. However, in the case of noisy and highly non-separable data sets, the linear transformation may not be able to learn a good set of pseudo-inputs. Our proposed framework, on the other hand, is designed to learn these pseudo-inputs, while simultaneously learning a nonlinear feature representation for these.

3   Deep Stochastic Neighbor Compression

Our proposed framework *Deep Stochastic Neighbor Compression* (DSNC) is
based on the idea of summarizing/compressing data in a *nonlinear* feature space
learned via a deep feedforward neural network. Although methods like SNC
(Sect. 2.2) can achieve a significant data compression, the inferred pseudo-inputs
$\mathbf{Z}$ still belong to the *original* feature space, or a linear subspace of the original
data. In contrast, DSNC learns $\mathbf{Z}$ in a more expressive, nonlinear feature space.
Note that, in our framework, nonlinear feature learning naturally corresponds
to a nonlinear (deep) metric learning.

DSNC consists of a deep feedforward neural network architecture which
jointly learns a compressed set $\mathbf{Z} \in \mathbb{R}^{d \times m}$ with m pseudo-inputs ($m \ll N$),
along with a deep feature mapping $f(\cdot|\mathbf{W})$ from the original feature space $\mathbb{R}^{D}$
to a transformed space $\mathbb{R}^{d}$. The procedure is illustrated in Fig. 1. The set $\mathbf{Z}$ con-
sisting of the inferred pseudo-inputs and the deep feature representation $f(\cdot|\mathbf{W})$
are used as a reference set and feature transformation, respectively, at test-time
of an instance based method such as kNN classification. In the following we
describe the key components of DSNC.

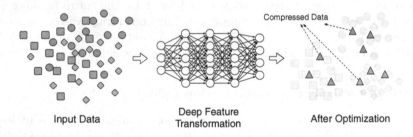

Fig. 1. A conceptual illustration of DSNC, which transforms the data via a deep feed-
forward neural net while simultaneously learning the pseudo-inputs that summarize
the original data.

3.1   Deep Stochastic Reference Set

Let $f(\cdot|\mathbf{W}) : \mathbb{R}^{D} \to \mathbb{R}^{d}$ be a deep neural network mapping function, with $\mathbf{W}$ as
the set of parameters from all layers of the network.[1] Similar to SNC, we aim to
learn a compressed set of pseudo-inputs, $\mathbf{Z} = [\mathbf{z}_1, \cdots, \mathbf{z}_m]$ with $\mathbf{z} \in \mathbb{R}^{d}$, such that
$\mathbf{Z}$ summarizes the original training set in the deep feature space. To this end,
akin to SNC, we define the probability that input $\mathbf{x}_i$ chooses $\mathbf{z}_j$ as its nearest
reference vector as:

$$p_{ij} = \frac{\exp(-\gamma^2 \|f(\mathbf{x}_i) - \mathbf{z}_j\|^2)}{\sum_{k=1}^{m} \exp(-\gamma^2 \|f(\mathbf{x}_i) - \mathbf{z}_k\|^2)} . \tag{1}$$

---

[1] For conciseness, we will typically omit the parameters $\mathbf{W}$ from the notation for the
mapping function, *i.e.*, $f(\cdot) \triangleq f(\cdot|\mathbf{W})$.

In the optimization, in addition to learning the parameters of a deep neural network, the compressed set $\mathbf{Z}$ is also learned from the data. This is done by first initializing $\mathbf{Z}$ with m randomly sampled examples from $\mathbf{X}$, noted as $\mathbf{X}'$, and then computing their deep representation via $\mathbf{Z} = f(\mathbf{X}')$, while recording their original labels. Note that while learning f and $\mathbf{Z}$, these labels are fixed throughout, while $\mathbf{Z}$ and the parameters $\mathbf{W}$ of the deep mapping f are learned jointly with the objective defined below.

3.2   DSNC Objective

To define an objective function for DSNC, we would like to ensure $p_i \triangleq \sum_{j:y_i=y_j} p_{ij} = 1$ for all $\mathbf{x}_i \in \mathbf{X}$, where p_{ij} is defined in (1). This means that the probability p_{ij} corresponding to an input $\mathbf{x}_i$ and a pseudo-input $\mathbf{z}_j$, both having different labels, is zero. We then define the KL-divergence between the "perfect" distribution "1" and p_i as

$$KL(1\|p_i) = -\log(p_i) \tag{2}$$

We wish to find a compressed set $\mathbf{Z}$ such that as many training inputs as possible are classified correctly in the deep feature space. In other words, we would like p_i to be close to 1 for all $\mathbf{x}_i \in \mathbf{X}$. This leads to the following objective:

$$\tilde{\mathcal{L}}(\mathbf{Z}, \mathbf{W}) = -\sum_{i=1}^{n} \log(p_i), \tag{3}$$

where $\mathbf{W}$ denotes the parameters of the deep feedforward neural network.

There are two possible issues that may arise while optimizing the objective (3) for DSNC and need to be properly accounted for. First, since we are jointly learning the deep feature map f and the compressed set $\mathbf{Z}$, without any constraints, it is possible that the mapped samples $f(\mathbf{x}_i)$ are on a different scale than the compressed samples $\mathbf{Z}$ in the deep feature space, while achieving a small value for the objective function (3). To handle this issue, we encourage the distance between $f(\mathbf{x}_i)$ and $\mathbf{z}_j$ to be *small* to avoid an inhomogeneous distribution in the feature space.

Second, it is also possible that all the compressed data samples with the same label collapse into a single point since our objective aims to maximize the classification accuracy. As a result, we also penalize the distribution of the compressed samples to encourage a multi-modal distribution for each label. This is done by *maximizing* the pair-wise distance between two pseudo-inputs $\mathbf{z}_i$ and $\mathbf{z}_j$ with the same label. Consequently, the DSNC objective function combines the KL-divergence term $\tilde{\mathcal{L}}(\mathbf{Z}, \mathbf{W})$ with two additional regularization terms to account for these, and is given by

---

Algorithm 1. DSNC in pseudo-code

1: **Input:** $\{\mathbf{X}, \mathbf{Y}\}$; compressed data set size $\mathbf{m}$
2: Initialize $\mathbf{Z}$ by sampling $\mathbf{m}$ inputs from $\mathbf{X}$, uniformly in each class, and forwarding
  into the initialized deep neural network $f(\cdot)$
3: Learn $\mathbf{Z}$ and the deep networks $f(\cdot)$ with back-propagation using gradients in (5)
  and (6)
4: Return $f(\cdot)$ and $\mathbf{Z}$

$$\mathcal{L}(\mathbf{Z}, \mathbf{W}) = -\sum_{i=1}^{n} \log(p_i) + \lambda_1 \underbrace{\sum_{i=1}^{n}\sum_{j=1}^{m} \|f(\mathbf{x}_i) - \mathbf{z}_j\|^2}_{R_1}$$

$$- \lambda_2 \underbrace{\sum_{i=1}^{m}\sum_{j=1}^{m} \delta(\hat{y}_i, \hat{y}_j)\|\mathbf{z}_i - \mathbf{z}_j\|^2}_{R_2} \tag{4}$$

where λ_1 and λ_2 are regularization coefficients and the delta function $\delta(\hat{y}_i, \hat{y}_j) = 1$ if $\hat{y}_i = \hat{y}_j$, and 0 otherwise. $\{\hat{y}_i\}$ are the labels for the compressed set $\mathbf{Z}$. R_1 regularizes the compressed samples to be close to the training data in the deep feature space, while R_2 encourages compressed samples with the same label to dissociate.

3.3 Learning with Stochastic Gradient Descent

The objective function (4) can be easily optimized via the back-propagation algorithm with stochastic gradient descent [7]. We adopt the RMSProp algorithm [35].

Specifically, there are two components that need to be updated: the parameters $\mathbf{W}$ of the deep neural network, and the compressed set $\mathbf{Z}$. Parameters $\mathbf{W}$ are updated by back-propagation, which requires the gradient of the objective with respect to the output $f(\mathbf{X})$, which is then back-propagated down the neural network. The compressed set $\mathbf{Z}$ can be simply updated with a stochastic gradient descent step. The stochastic gradients for both $\mathbf{Z}$ and $f(\mathbf{X})$ have simple and compact forms. To write down the gradients, we first define the following matrices $\{\mathbf{Q}, \mathbf{P}, \hat{\mathbf{P}}\} \in \mathbb{R}^{n \times m}$, $\mathbf{Q}_1 \in \mathbb{R}^{m \times m}$, $\mathbf{P}_1 \in \mathbb{R}^{d \times n}$, and $\{\mathbf{P}_2, \mathbf{Q}_2\} \in \mathbb{R}^{d \times m}$ as

$$[\mathbf{Q}]_{ij} = (\delta_{y_i, \hat{y}_j} - p_i), \quad [\mathbf{Q}_2]_{ij} = \sum_{i=1}^{m} \mathbf{Q}_{ij}$$

$$[\mathbf{Q}_1]_{ij} = \delta(\hat{y}_i, \hat{y}_j), \quad [\mathbf{P}]_{ij} = \frac{p_{ij}}{p_i}, \quad [\hat{\mathbf{P}}]_{ij} = p_{ij}$$

$$[\mathbf{P}_1]_{ik} = \sum_{j}^{m} \mathbf{z}_{ij}, \quad [\mathbf{P}_2]_{jk} = \sum_{i}^{n} \mathbf{x}_{ji}$$

Here, p_{ij} is defined in (1), x_{ij} and z_{ij} denote the corresponding elements of row/column i/j in X/Z. After some careful algebra, the gradient of $\mathcal{L}$ with respect to the compressed set $\mathbf{Z}$ and $f(\mathbf{X})$ can then be conveniently represented in matrix operations with the above defined symbols, $i.e.$,

$$\frac{\partial \mathcal{L}}{\partial \mathbf{Z}} = -2\gamma^2 \left(\mathbf{X} \left(\mathbf{Q} \circ \mathbf{P} \right) - \mathbf{Z} \operatorname{diag} \left((\mathbf{Q} \circ \mathbf{P})^T \mathbf{1}_n \right) \right) \qquad (5)$$
$$+ 2\lambda_1 \left(n\mathbf{Z} - \mathbf{P}_2 \right) + 2\lambda_2 \left(\mathbf{Z}\mathbf{Q}_1 - \mathbf{Q}_2 \circ \mathbf{Z} \right)$$

$$\frac{\partial \mathcal{L}}{\partial f(\mathbf{X})} = -2\gamma^2 \mathbf{Z} \left(\mathbf{Q} \circ \mathbf{P} - \hat{\mathbf{P}} \right)^T + 2\lambda_1 \left(m\mathbf{X} - \mathbf{P}_1 \right). \qquad (6)$$

where $\circ$ is the Hadamard (element-wise) product, $\mathbf{1}_n$ is the $n \times 1$ vector of all ones, and $\operatorname{diag}(\cdot)$ is the diagonal operator placing a vector along the diagonal of an otherwise 0 matrix. Given the gradients, learning is straightforward by applying the RMSProp algorithm on $\mathbf{Z}$ and the back-propagation for learning W, described in Algorithm 1.

3.4   Relationship with Deep Neural Net with Softmax Output

We now show how DSNC is related to a deep neural network with a softmax output. Note they are comparable only when $m = |Y|$, $i.e.$, the number of pseudo-inputs is equal to the number of classes. Note that, for a deep neural network with a softmax output, the corresponding probability for (1) can be written as

$$p_{ij} = \frac{\exp(f^T(\mathbf{x}_i)\mathbf{z}_j)}{\sum_{k=1}^{|Y|} \exp(f^T(\mathbf{x}_i)\mathbf{z}_k)}.$$

Note that the Euclidean distance in DSNC is replaced by an inner product in softmax function above. When $\gamma^2 = \frac{1}{2}$ and $\|f(\mathbf{x}_i)\|_2^2 = \|\mathbf{z}_j\|_2^2 = 1$, the probability that $\mathbf{x}_i$ belongs to "class" $\mathbf{z}_j$, as given by (1) can be written as

$$p_{ij} = \frac{\exp(-\frac{1}{2}\|f(\mathbf{x}_i)\|^2)\exp(-\frac{1}{2}\|\mathbf{z}_j\|^2)\exp(f^T(\mathbf{x}_i)\mathbf{z}_j)}{\sum_{k=1}^{|Y|} \exp(-\frac{1}{2}\|f(\mathbf{x}_i)\|^2)\exp(-\frac{1}{2}\|\mathbf{z}_k\|^2)\exp(f^T(\mathbf{x}_i)\mathbf{z}_k)} = \frac{\exp(f^T(\mathbf{x}_i)\mathbf{z}_j)}{\sum_{k=1}^{|Y|} \exp(f^T(\mathbf{x}_i)\mathbf{z}_k)}$$

which exactly recovers the softmax output. Therefore, a deep neural network with a softmax output can be viewed as a special case of our DSNC framework.

4   Related Work

Our work is aimed at improving the accuracies of instance based methods, such as kNN, by learning highly discriminative feature representations (equivalently, learning a good distance metric), while also speeding up the test-time predictions. It is therefore related to both feature/distance-metric learning algorithms, as well as data summarization/compression algorithms for instance based methods.

In the specific context of kNN methods, there have been several previous efforts to speed up kNN's test-time predictions. The vast majority of these methods reduce to speeding up the retrieval of k nearest neighbors without modifying the training set. These include space partition algorithms such as ball-trees [6,30] and kd-trees [5], as well as approximate neighbor search like local-sensitive hashing [3,14]. Our paper addresses the problem from the perspective of data compression that reduces the size of the training set. Note that data compression approach is orthogonal to prior efforts on fast retrieval approaches, and thus these two methodologies could be combined.

Perhaps the most straightforward idea for data compression is subsampling the dataset. The seminal work in this area is Condensed Nearest Neighbors (CNN) proposed by [18]. It starts off with two sets, S and T, where S contains an instance of the training set and T contains the rest. CNN repeatedly scans T, looking for an instance in T that is misclassified using the data in S. This instance is then moved from T to S. This process continues until no more data movement can be made. Since this work, there have been several variants of CNN, including MCNN to address the order dependent issue of CNN [11], post-processing method [13], and fast CNN (FCNN) [4]. With these methods, the compressed training set is always a subset of the original training set, which is not necessarily a good representation. Recently, [24] introduce Stochastic Neighbor Compression (SNC), which learns a synthetic set as the compressed set. Assuming the synthetic set is presented as the *design variables*, SNC uses stochastic neighborhood [16,20,26] to model the probability of each training instance being correctly classified by the synthetic set. The synthetic set is obtained through numerical optimization, where the objective is to minimize the KL-divergence between the modeled distribution and the "perfect" distribution in which all training instances are correctly classified.

Among other works on summarizing/compressing massive data sets for machine learning problems includes methods such as coresets [1] for geometric problems (e.g., k-means/k-median clustering, nearest neighbor methods, etc.). Kernel methods are also known to have the problem of having to store the entire training data in the memory and being slow at test time, and several methods such as landmarks based approximations [21,40] have been proposed to address these issues. However all these methods can only perform data compression by learning a set of representatives in the original feature space, and are not suited for data sets that are high-dimensional and exhibit significant nonlinearities.

All of the above methods operate on the original data space, not embracing the superior expressive power of deep learning. With unprecedented generalization performance, deep learning has achieved great success in various important applications, including speech recognition [17,19,29], natural language processing [9,27,28], image labeling [12,23,34,39], and object detection [15,36]. Recently, the kNN classifier has been equipped with deep learning in modern face recognition systems, such as FaceNet [33]. In particular, kNN performs classification on the space mapped by a convolutional net [25]. However, Facenet trains the convolutional net to reflect the actual similarity between images/faces, rather than the accuracy performance of kNN. [32] introduce a method to train a deep neural net for kNN to perform well on the transformed space. Though inspired by this work, our DSNC is fundamentally different in that it not only optimizes the kNN performance but also simultaneously learns a compressed set in a new nonlinear feature space learned by a feedforward deep neural network.

5   Results

We present experimental results on seven benchmark datasets, including four from [24], *i.e.*, MNIST, YALEFACE, ISOLET, ADULT; and three additional, more complex datasets, *i.e.*, 20NEWS, CIFAR10 and CIFAR100. Some statistics are listed in Table 1. Since YALEFACE has no predefined test set, we report the average performance over 10 splits. All other results are reported on predefined test sets. We begin by describing the experimental settings, and then evaluate the test errors, compression ratios, feature representations, sensitivity to hyperparameters and visualization of distributions of the test sets in the deep feature space. Our code is publicly available at http://people.duke.edu/~ww107/.

Table 1. Summary of datasets used in the evaluation.

Dataset	n	$|Y|$	$d\ (d_L)$	N_{max}
MNIST	60000	10	784 (164)	600
YALEFACE	1961	38	8064 (100)	100
ISOLET	3898	26	617 (172)	100
ADULT	32562	2	123 (50)	100
20NEWS	11314	20	2000 (100)	100
CIFAR10	50000	10	3072 (200)	100
CIFAR100	50000	100	3072 (200)	300

5.1   Experimental Setting

To explore the advantages of our deep-learning-based method, we use raw features as the input for DSNC and the corresponding reference deep neural networks[2]. For MNIST, YALEFACE (rescaled to 48×42 pixels [37]), CIFAR10 and CIFAR100, we adopt convolutional neural networks, while ISOLET, ADULT and 20NEWS are fitted with feed-forward neural networks. ReLU is adopted as the activation function after hidden layers for all models. Details of the network structures are shown in Table 2. When comparing the error with varying compressed ratios in Sect. 5.3, we fix d in Hd to be d_L in Table 1, and the time

---

[2] The same network structure as DSNC except with a softmax-output.

Table 2. The feedforward neural network structure used for each dataset. 'Ck' ('Hk') indicates a convolutional (fully-connected) layer with k filters (hidden units). The variable d represents the dimensionality of the output feature representation of the inferred pseudo-inputs **Z**.

Dataset	Network Structure
MNIST	C20-C50-Hd
YALEFACE	C20-C50-Hd
ISOLET	H500-H500-Hd
ADULT	H200-H200-Hd
20NEWS	H800-H800-Hd
CIFAR10	C64-C128-C256-C128-Hd
CIFAR100	C64-C128-C256-C128-Hd

comparing the error with varying dimensions in Sect. 5.4, we keep the compared size m to be N_{max}.

DSNC is implemented using Torch7 [10] and trained on NVIDIA GTX TITAN graphics cards with 2688 cores and 6 GB of global memory. We verify the implementation by numerical gradient checking, and optimize using stochastic gradient descent with RMSprop, using mini-batch in size of 100. For all the datasets, we randomly select 20 % of the training data for cross-validation of hyper-parameters λ_1 and λ_2 and early stopping. In contrast to SNC, our DSNC is not sensitive to γ. Thus we use a constant value 1 for all DSNC experiments set up.

With SNC we follow a similar setup to [24]. For ISOLET and MNIST, the dimensionality is reduced with LMMN as described in [38]. For YALEFACE, we follow [38] and first rescale the images to 48×42 pixels, then reduce the dimensionality with PCA, while omitting the leading five principal components which largely account for lighting variations. Finally we apply large margin nearest neighbor (LMNN) to reduce the dimensionality further to $d = 100$. For CIFAR10 and CIFAR100, we use LMNN to reduce the dimensionality to $d = 200$. In fact, the dimensionality of SNC is determined by LMNN. The parameters used for comparing the test error with varying compression rates and dimensionality are exactly the same as DSNC as we described before. Parameters are listed in Table 1. Notice that LMNN is used as the pre-processing step for all the methods except our DSNC and the corresponding reference networks.

5.2 Baselines

We experiment with two versions of DSNC, one uses the compressed data **Z** as the kNN reference during testing, denoted by *Compression*, the other uses the entire training data, denoted by *ALL*. We compare DSNC against the following related baselines, where the 1-nearest neighbor rule is adopted for all kNN methods.

- kNN without compression, with/without dimensionality reduction with LMNN;
- kNN using *Stochastic Neighbor Compression* (SNC) [24];
- Approximate kNN with *Locality-Sensitive Hashing* (LSH) [2,14];
- kNN using *CNN* [18] and *FCNN* [4] dataset compression;
- Deep neural network classifier with the same network structure as DSNC.

5.3   Errors with Varying Compression Ratios

In this section we experiment with varying compressed ratio of the dataset, defined as the ratio between the compressed data size and the whole data size. The results are plotted in Fig. 2. Several conclusions can be drawn from the results: (1) DSNC outperforms other methods on all data sets. The gap between DSNC and SNC is huge for all the data sets, which indicates the advantage of learning the nonlinear feature space for data compression. (2) DSNC emerges as a stable compression method that is robust to the compression ratio. This is especially true when the compression ratio is small, for example, when the compression data size equals the number of classes ($m = |Y|$), DSNC still performs well, yielding significantly lower errors than LSH, CNN, FCNN and SNC. And, generally, with increasing m, test errors tend to decrease to a certain degree. (3) Compared with reference deep neural networks with softmax outputs, DSNC exhibits better performances on most datasets except ADULT, but with smaller gaps than the other methods. A possible reason could be that the task is binary classification and multi-modality within class distributions may be not that explicit in the dataset. It is notable when $m = |Y|$, DSNC degrades to the reference neural network using Euclidean distance as the metrics in softmax. We can see on 20NEWS, CIFAR10 and CIFAR100, the reference neural networks perform better. However, with an adaptive m, DSNC can always surpass the reference neural networks; while the observation on $YaleFace$ is particular surprising, as there is a big performance gap between DSNC and the corresponding convolutional neural networks. This indicates our motivation of learning a representative feature space for data compression to be effective, as DSNC has more degrees of freedom to adapt the compression data to a weak feature presentation.

5.4   Errors with Varying Feature Dimensions

Next we investigate the impact of feature dimensions on the classification accuracy. To test the adaptive ability of DSNC to extremely low-dimensional feature spaces, we vary the feature space dimensions from 10 to 300 on CIFAR10 and CIFAR100, and from 10 to 100 on the other datasets. The results are plotted in Fig. 3. We can see from the figure that the performance does not deteriorate when learning with a deep nonlinear transformation, *i.e.*, DSNC and DNN/CNN yield almost the same test errors with different feature dimensions on all the datasets, while other methods produce significantly worse performance when the feature dimension is low. Particularly, for MNIST and ISOLET, a 20 dimensional space is found to be powerful enough to express the dataset, while for CIFAR100, a nonlinear transformation into a 100-dimensional space obtains an accuracy that is close to the optimal performance. Interestingly, we also notice that using the compressed data outperforms the one using the entire mapped data. This is because our objective optimizes directly on the compressed data set, which can effectively filter out the noise in the original data set consisting of all the observations.

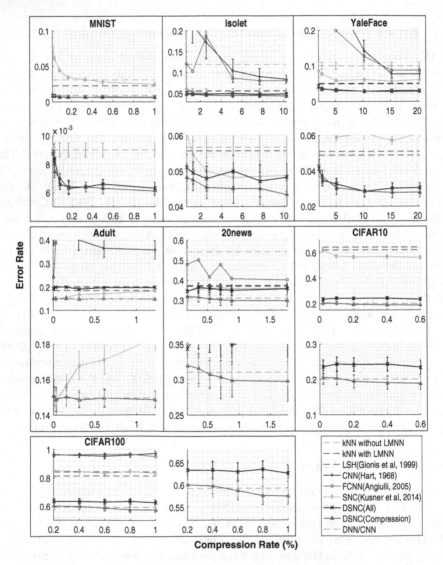

Fig. 2. Test error with varying dataset compression rates. The images below or on the right side in the blue rectangle are the zoom in images (Color figure online)

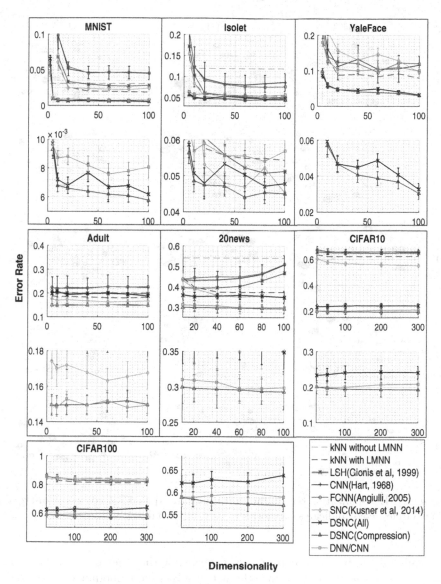

Dimensionality

Fig. 3. Test error rates after mapping into different size of feature space. Zoom in images are organized the same as Fig. 2

5.5   Sensitivity to Hyper-parameters

In contrast to SNC, it is found that our model is not sensitive to the parameter γ in the stochastic neighborhood term. However, the hyper-parameters λ_1 and λ_2 do influence the performance of DSNC, because they control different behaviors of the objective. Specifically, λ_1 tends to pull the compressed data closer to the training sets in the deep feature space, while λ_2 pushes the compressed data with

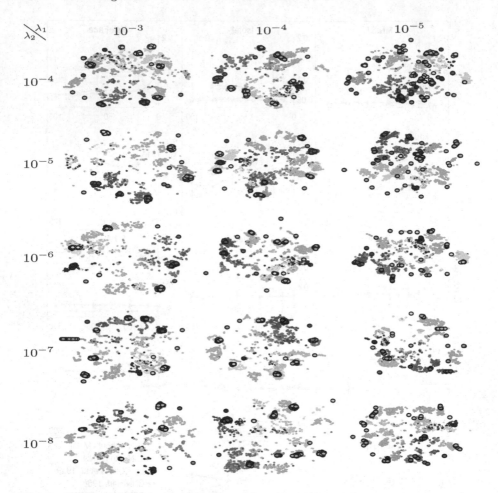

Fig. 4. tSNE visualization on 20NEWS with varying λ_1 and λ_2, compression size 500 (black circle), color indicates categories (Color figure online)

the same label to be far away from each other, such that they do not collapse into a single point and tend to capture the within-class multi-modality. We visualize the effects for these hyper-parameters by embedding the compressed data into 2-dimensional space using tSNE [26]. We use the 20NEWS dataset for visualization in Fig. 4. Consistent with our intuition, we find that with increasing λ_1, the compressed data tends to be condense, and far way from the training data; while increasing λ_2 generally pushes the compressed data in the same class to distribute more scatteringly. This indicates that if we want a larger compressed set of pseudo-inputs (i.e., m is large), a larger value of λ_2 should be set. The accuracies with different values of λ_1 and λ_2 are summarized in Table 3, which indicates suitable choices for λ_1 and λ_2 is essential for good performance.

Table 3. Test errors on 20news with varying the hyper-parameters λ_1 and λ_2 under the networks structure H800-H800-H100. The compressed size m fixes to be 100.

λ_2 \ λ_1	10^{-3}	10^{-4}	10^{-5}	10^{-6}	10^{-7}
10^{-4}	30.36	28.75	30.48	32.80	34.10
10^{-5}	29.96	**28.47**	30.33	32.86	33.50
10^{-6}	30.80	29.27	29.77	33.55	32.70
10^{-7}	29.12	28.74	29.80	32.84	33.20
10^{-8}	29.02	28.54	30.68	33.86	33.27

5.6   Comparison of DSNC with SNC and SOFTMAX

In order to further understand the advantage of DSNC over SNC and softmax-based deep neural networks (SOFTMAX), we visualize them on MNIST. We adopt the same models as the above experiments with a reference set consisting of $m = 100$ pseudo-inputs. This gives us cleaner results in the visualization. The inferred pseudo-inputs in the feature space are plotted in Fig. 5. It can be clearly seen that DSNC is able to learn both separable feature space and representative data, whereas for the SNC, the compressed data does not seem to be separable. In terms of SOFTMAX, even though it can learn centered clusters, its tendency to only learn unimodal within-class distributions lead to poor performance around the decision boundary.

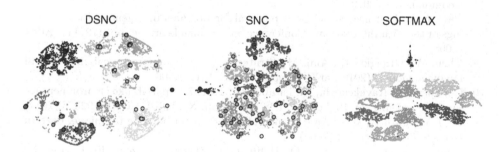

Fig. 5. Comparison of DSNC (left) with SNC (middle), SOFTMAX (right) on MNIST dataset. Circles represent the reference set.

6   Conclusion

We propose DSNC to jointly learn a deep feature space and a subset of compressed data that best represents the whole data. The algorithm consists of a deep neural network component for feature learning, on top of which an objective is proposed to optimize the kNN criteria, leading to a natural extension of the popular softmax-based deep neural networks. We test DSNC on a number of benchmark datasets, obtaining significantly improved performance compared to existing data compression algorithms.

Acknowledgments. This research was supported in part by ARO, DARPA, DOE, NGA and ONR.

References

1. Agarwal, P.K., Har-Peled, S., Varadarajan, K.R.: Geometric approximation via coresets. Comb. Comput. Geom. **52**, 1–30 (2005)
2. Aly, M., Munich, M., Perona, P.: Indexing in large scale image collections: scaling properties and benchmark. In: 2011 IEEE Workshop on Applications of Computer Vision (WACV), pp. 418–425. IEEE (2011)
3. Andoni, A., Indyk, P.: Near-optimal hashing algorithms for approximate nearest neighbor in high dimensions. In: 47th Annual IEEE Symposium on Foundations of Computer Science (FOCS 2006), pp. 459–468. IEEE (2006)
4. Angiulli, F.: Fast condensed nearest neighbor rule. In: Proceedings of the 22nd International Conference on Machine Learning, pp. 25–32. ACM (2005)
5. Bentley, J.L.: Multidimensional binary search trees used for associative searching. Commun. ACM **18**(9), 509–517 (1975)
6. Beygelzimer, A., Kakade, S., Langford, J.: Cover trees for nearest neighbor. In: Proceedings of the 23rd International Conference on Machine Learning, pp. 97–104. ACM (2006)
7. Bottou, L.: Stochastic gradient descent tricks. Tech. rep, Microsoft Research, Redmond, WA (2012)
8. Cayton, L.: Fast nearest neighbor retrieval for bregman divergences. In: Proceedings of the 25th International Conference on Machine Learning, pp. 112–119. ACM (2008)
9. Chen, W., Grangier, D., Auli, M.: Strategies for training large vocabulary neural language models (2015). arXiv preprint arXiv:1512.04906
10. Collobert, R., Kavukcuoglu, K., Farabet, C.: Torch7: a matlab-like environment for machine learning. In: BigLearn, NIPS Workshop. No. EPFL-CONF-192376 (2011)
11. Devi, V.S., Murty, M.N.: An incremental prototype set building technique. Pattern Recogn. **35**(2), 505–513 (2002)
12. Donahue, J., Jia, Y., Vinyals, O., Hoffman, J., Zhang, N., Tzeng, E., Darrell, T.: Decaf: a deep convolutional activation feature for generic visual recognition (2013). arXiv preprint arXiv:1310.1531
13. Gates, G.: The reduced nearest neighbor rule. IEEE Trans. Inf. Theory **18**(3), 431–433 (1972)
14. Gionis, A., Indyk, P., Motwani, R., et al.: Similarity search in high dimensions via hashing. VLDB **99**, 518–529 (1999)
15. Girshick, R., Donahue, J., Darrell, T., Malik, J.: Rich feature hierarchies for accurate object detection and semantic segmentation. In: CVPR (2014)
16. Goldberger, J., Hinton, G.E., Roweis, S.T., Salakhutdinov, R.: Neighbourhood components analysis. In: Advances in neural information processing systems, pp. 513–520 (2004)
17. Graves, A., Mohamed, A.R., Hinton, G.: Speech recognition with deep recurrent neural networks. In: ICASSP (2013)

18. Hart, P.E.: The condensed nearest neighbor rule. IEEE Trans. Inf. Theory **14**, 515–516 (1968)

19. Hinton, G., Deng, L., Yu, D., Dahl, G.E., Mohamed, A.R., Jaitly, N., Senior, A., Vanhoucke, V., Nguyen, P., Sainath, T.N., et al.: Deep neural networks for acoustic modeling in speech recognition: the shared views of four research groups. IEEE Signal Process. Magaz. **29**(6), 82–97 (2012)

20. Hinton, G.E., Roweis, S.T.: Stochastic neighbor embedding. In: Advances in neural information processing systems, pp. 833–840 (2002)

21. Hsieh, C.J., Si, S., Dhillon, I.S.: Fast prediction for large-scale kernel machines. In: Advances in Neural Information Processing Systems, pp. 3689–3697 (2014)

22. Hu, J., Lu, J., Tan, Y.P.: Discriminative deep metric learning for face verification in the wild. In: CVPR (2014)

23. Krizhevsky, A., Sutskever, I., Hinton, G.E.: Imagenet classification with deep convolutional neural networks. In: NIPS (2012)

24. Kusner, M., Tyree, S., Weinberger, K.Q., Agrawal, K.: Stochastic neighbor compression. In: Proceedings of the 31st International Conference on Machine Learning (ICML 2014), pp. 622–630 (2014)

25. LeCun, Y., Boser, B., Denker, J.S., Henderson, D., Howard, R.E., Hubbard, W., Jackel, L.D.: Backpropagation applied to handwritten zip code recognition. Neural Comput. **1**(4), 541–551 (1989)

26. Van der Maaten, L., Hinton, G.: Visualizing data using t-sne. J. Mach. Learn. Res. **9**(2579–2605), 85 (2008)

27. Mikolov, T., Karafiát, M., Burget, L., Cernocký, J., Khudanpur, S.: Recurrent neural network based language model. In: 11th Annual Conference of the International Speech Communication Association (INTERSPEECH 2010), Makuhari, Chiba, Japan, 26–30 September 2010, pp. 1045–1048 (2010)

28. Mikolov, T., Sutskever, I., Chen, K., Corrado, G.S., Dean, J.: Distributed representations of words and phrases and their compositionality. In: NIPS (2013)

29. Mohamed, A.r., Sainath, T.N., Dahl, G., Ramabhadran, B., Hinton, G.E., Picheny, M.A.: Deep belief networks using discriminative features for phone recognition. In: ICASSP (2011)

30. Omohundro, S.M.: Five balltree construction algorithms. International Computer Science Institute Berkeley (1989)

31. Rumelhart, D.E., Hinton, G.E., Williams, R.J.: Learning internal representations by error propagation. Tech. rep, DTIC Document (1985)

32. Salakhutdinov, R., Hinton, G.E.: Learning a nonlinear embedding by preserving class neighbourhood structure. In: International Conference on Artificial Intelligence and Statistics, pp. 412–419 (2007)

33. Schroff, F., Kalenichenko, D., Philbin, J.: Facenet: a unified embedding for face recognition and clustering (2015). arXiv preprint arXiv:1503.03832

34. Sermanet, P., Eigen, D., Zhang, X., Mathieu, M., Fergus, R., LeCun, Y.: Overfeat: Integrated recognition, localization and detection using convolutional networks (2013). arXiv preprint arXiv:1312.6229

35. Tieleman, T., Hinton, G.E.: Lecture 6.5-rmsprop: Divide the gradient by a running average of its recent magnitude. Tech. rep., Coursera: Neural Networks for Machine Learning (2012)

36. Vinyals, O., Toshev, A., Bengio, S., Erhan, D.: Show and tell: a neural image caption generator. arXiv preprint arXiv:1411.4555 (2014)

37. Weinberger, K., Dasgupta, A., Langford, J., Smola, A., Attenberg, J.: Feature hashing for large scale multitask learning. In: ICML (2009)

38. Weinberger, K.Q., Blitzer, J., Saul, L.K.: Distance metric learning for large margin nearest neighbor classification. In: Advances in neural information processing systems, pp. 1473–1480 (2005)
39. Zeiler, M.D., Fergus, R.: Visualizing and understanding convolutional networks. In: Fleet, D., Pajdla, T., Schiele, B., Tuytelaars, T. (eds.) ECCV 2014, Part I. LNCS, vol. 8689, pp. 818–833. Springer, Heidelberg (2014)
40. Zhang, K., Tsang, I.W., Kwok, J.T.: Improved nyström low-rank approximation and error analysis. In: Proceedings of the 25th International Conference on Machine Learning, pp. 1232–1239. ACM (2008)

Linear Convergence of Gradient and Proximal-Gradient Methods Under the Polyak-Łojasiewicz Condition

Hamed Karimi, Julie Nutini, and Mark Schmidt[(✉)]

Department of Computer Science, University of British Columbia,
Vancouver, BC, Canada
hamedkarim@gmail.com, {jnutini,schmidtm}@cs.ubc.ca

Abstract. In 1963, Polyak proposed a simple condition that is sufficient to show a global linear convergence rate for gradient descent. This condition is a special case of the Łojasiewicz inequality proposed in the same year, and it does not require strong convexity (or even convexity). In this work, we show that this much-older Polyak-Łojasiewicz (PL) inequality is actually weaker than the main conditions that have been explored to show linear convergence rates without strong convexity over the last 25 years. We also use the PL inequality to give new analyses of coordinate descent and stochastic gradient for many non-strongly-convex (and some non-convex) functions. We further propose a generalization that applies to proximal-gradient methods for non-smooth optimization, leading to simple proofs of linear convergence for support vector machines and L1-regularized least squares without additional assumptions.

Keywords: Gradient descent · Coordinate descent · Stochastic gradient · Variance-reduction · Boosting · Support vector machines · L1-regularization

1 Introduction

Fitting most machine learning models involves solving some sort of optimization problem. Gradient descent, and variants of it like coordinate descent and stochastic gradient, are the workhorse tools used by the field to solve very large instances of these problems. In this work we consider the basic problem of minimizing a smooth function and the convergence rate of gradient descent methods. It is well-known that if f is strongly-convex, then gradient descent achieves a global linear convergence rate for this problem [28]. However, many of the fundamental models in machine learning like least squares and logistic regression yield objective functions that are convex but not strongly-convex. Further, if f is only convex, then gradient descent only achieves a sub-linear rate.

Electronic supplementary material The online version of this chapter (doi:10.1007/978-3-319-46128-1_50) contains supplementary material, which is available to authorized users.

© Springer International Publishing AG 2016
P. Frasconi et al. (Eds.): ECML PKDD 2016, Part I, LNAI 9851, pp. 795–811, 2016.
DOI: 10.1007/978-3-319-46128-1_50

This situation has motivated a variety of alternatives to strong convexity (SC) in the literature, in order to show that we can obtain linear convergence rates for problems like least squares and logistic regression. One of the oldest of these conditions is the *error bounds* (EB) of Luo and Tseng [22], but four other recently-considered conditions are *essential strong convexity* (ESC) [20], *weak strong convexity* (WSC) [25], the *restricted secant inequality* (RSI) [45], and the *quadratic growth* (QG) condition [2]. Some of these conditions have different names in the special case of convex functions: a convex function satisfying RSI is said to satisfy *restricted strong convexity* (RSC) [45] while a convex function satisfying QG is said to satisfy *optimal strong convexity* (OSC) [19] or (confusingly) WSC [23]. The proofs of linear convergence under all of these relaxations are typically not straightforward, and it is rarely discussed how these conditions relate to each other.

In this work, we consider a much older condition that we refer to as the Polyak-Łojasiewicz (PL) inequality. This inequality was originally introduced by Polyak [31], who showed that it is a sufficient condition for gradient descent to achieve a linear convergence rate. We describe it as the PL inequality because it is also a special case of the inequality introduced in the same year by Łojasiewicz [21]. We review the PL inequality in the next section and how it leads to a trivial proof of the linear convergence rate of gradient descent. Next, in terms of showing a global linear convergence rate to the optimal solution, we show that the PL inequality is *weaker* than all of the more recent conditions discussed in the previous paragraph. This suggests that we can replace the long and complicated proofs under any of the conditions above with simpler proofs based on the PL inequality. Subsequently, we show how this result implies gradient descent achieves linear rates for standard problems in machine learning like least squares and logistic regression that are not necessarily SC, and even for some non-convex problems (Sect. 2.3). In Sect. 3 we use the PL inequality to give new convergence rates for randomized and greedy coordinate descent (implying a new convergence rate for certain variants of boosting), sign-based gradient descent methods, and stochastic gradient methods in either the classical or variance-reduced setting. Next we turn to the closely-related problem of minimizing the sum of a smooth function and a simple non-smooth function. We propose a generalization of the PL inequality that allows us to show linear convergence rates for proximal-gradient methods without SC. This leads to a simple analysis showing linear convergence of methods for training support vector machines. It also implies that we obtain a linear convergence rate for ℓ_1-regularized least squares problems, showing that the extra conditions previously assumed to derive linear converge rates in this setting are in fact not needed.

2   Polyak-Łojasiewicz Inequality

We first focus on the basic unconstrained optimization problem

$$\underset{x \in \mathbb{R}^d}{\mathrm{argmin}} \; f(x), \tag{1}$$

and we assume that the first derivative of f is L-Lipschitz continuous. This means that

$$f(y) \leq f(x) + \langle \nabla f(x), y - x \rangle + \frac{L}{2} \|y - x\|^2, \tag{2}$$

for all x and y. For twice-differentiable objectives this assumption means that the eigenvalues of $\nabla^2 f(x)$ are bounded above by some L, which is typically a reasonable assumption. We also assume that the optimization problem has a non-empty solution set $\mathcal{X}^*$, and we use f^* to denote the corresponding optimal function value. We will say that a function satisfies the PL inequality if the following holds for some $\mu > 0$,

$$\frac{1}{2} \|\nabla f(x)\|^2 \geq \mu(f(x) - f^*), \quad \forall x. \tag{3}$$

This inequality simply requires that the gradient grows faster than a quadratic function as we move away from the optimal function value. Note that this inequality implies that every stationary point is a global minimum. But unlike SC, it does not imply that there is a unique solution. Linear convergence of gradient descent under these assumptions was first proved by Polyak [31]. Below we give a simple proof of this result when using a step-size of $1/L$.

Theorem 1. *Consider problem* (1), *where f has an L-Lipschitz continuous gradient* (2), *a non-empty solution set $\mathcal{X}^*$, and satisfies the PL inequality* (3). *Then the gradient method with a step-size of $1/L$,*

$$x_{k+1} = x_k - \frac{1}{L} \nabla f(x_k), \tag{4}$$

has a global linear convergence rate,

$$f(x_k) - f^* \leq \left(1 - \frac{\mu}{L}\right)^k (f(x_0) - f^*).$$

Proof. By using update rule (4) in the Lipschitz inequality condition (2) we have

$$f(x_{k+1}) - f(x_k) \leq -\frac{1}{2L} \|\nabla f(x_k)\|^2.$$

Now by using the PL inequality (3) we get

$$f(x_{k+1}) - f(x_k) \leq -\frac{\mu}{L}(f(x_k) - f^*).$$

Re-arranging and subtracting f^* from both sides gives us $f(x_{k+1}) - f^* \leq \left(1 - \frac{\mu}{L}\right)(f(x_k) - f^*)$. Applying this inequality recursively gives the result. □

Note that the above result also holds if we use the optimal step-size at each iteration, because of the inequality

$$\min_\alpha f(x_k - \alpha \nabla f(x_k)) \leq f\left(x_k - \frac{1}{L} \nabla f(x_k)\right).$$

A beautiful aspect of this proof is its simplicity; in fact it is *simpler* than the proof of the same fact under the usual SC assumption. It is certainly simpler than typical proofs which rely on the other conditions mentioned in Sect. 1. Further, it is worth noting that the proof does *not* assume convexity of f. Thus, this is one of the few general results we have for global linear convergence on non-convex problems.

2.1   Relationships Between Conditions

As mentioned in the Sect. 1, several other assumptions have been explored over the last 25 years in order to show that gradient descent achieves a linear convergence rate. These typically assume that f is convex, and lead to more complicated proofs than the one above. However, it is rarely discussed how the conditions relate to each other. Indeed, all of the relationships that have been explored have only been in the context of convex functions [19, 25, 44]. In Appendix 2.1, we give the precise definitions of all conditions and also prove the result below giving relationships between the conditions.

Theorem 2. *For a function f with a Lipschitz-continuous gradient, the following implications hold:*

$$(SC) \rightarrow (ESC) \rightarrow (WSC) \rightarrow (RSI) \rightarrow (EB) \equiv (PL) \rightarrow (QG).$$

If we further assume that f is convex then we have

$$(RSI) \equiv (EB) \equiv (PL) \equiv (QG).$$

This result shows that (QG) is the weakest assumption among those considered. However, QG allows non-global local minima so it is not enough to guarantee that gradient descent finds a global minimizer. This means that, among those considered above, *PL and the equivalent EB are the most general conditions* that allow linear convergence to a global minimizer. Note that in the convex case QG is called OSC, but the result above shows that in the convex case it is also equivalent to EB and PL (as well as RSI which is known as RSC in this case).

2.2   Invex and Non-convex Functions

While the PL inequality does not imply convexity of f, it does imply the weaker condition of *invexity*. Invexity was first introduced by Hanson in 1981 [12], and has been used in the context of learning output kernels [8]. Craven and Glover [7] show that a smooth f is invex if and only if every stationary point of f is a global minimum. Since the PL inequality implies that all stationary points are global minimizers, functions satisfying the PL inequality must be invex. Indeed, Theorem 2 shows that all of the previous conditions except (QG) imply invexity. The function $f(x) = x^2 + 3\sin^2(x)$ is an example of an invex but non-convex function satisfying the PL inequality (with $\mu = 1/32$). Thus, Theorem 1 implies gradient descent obtains a global linear convergence rate on this function.

Unfortunately, many complicated models have non-optimal stationary points. For example, typical deep feed-forward neural networks have sub-optimal stationary points and are thus not invex. A classic way to analyze functions like this is to consider a *global convergence phase* and a *local convergence phase*. The global convergence phase is the time spent to get "close" to a local minimum, and then once we are "close" to a local minimum the local convergence phase characterizes the convergence rate of the method. Usually, the local convergence phase starts to apply once we are locally SC around the minimizer. But this means that the local convergence phase may be arbitrarily small: for example, for $f(x) = x^2 + 3\sin^2(x)$ the local convergence rate would not even apply over the interval $x \in [-1, 1]$. If we instead defined the local convergence phase in terms of locally satisfying the PL inequality, then we see that it can be *much* larger ($x \in \mathbb{R}$ for this example).

2.3 Relevant Problems

If f is μ-SC, then it also satisfies the PL inequality with the same μ (see Appendix 2.3). Further, by Theorem 2, f satisfies the PL inequality if it satisfies any of ESC, WSC, RSI, or EB (while for convex f, QG is also sufficient). Although it is hard to precisely characterize the general class of functions for which the PL inequality is satisfied, we note one important special case below.

Strongly-convex composed with linear: This is the case where f has the form $f(x) = g(Ax)$ for some σ-SC function g and some matrix A. In Appendix 2.3, we show that this class of functions satisfies the PL inequality, and we note that this form frequently arises in machine learning. For example, least squares problems have the form

$$f(x) = \|Ax - b\|^2,$$

and by noting that $g(z) \triangleq \|z - b\|^2$ is SC we see that least squares falls into this category. Indeed, this class includes all convex quadratic functions.

In the case of logistic regression we have

$$f(x) = \sum_{i=1}^{n} \log(1 + \exp(b_i a_i^T x)).$$

This can be written in the form $g(Ax)$, where g is strictly convex but not SC. In cases like this where g is only strictly convex, the PL inequality will still be satisfied over any compact set. Thus, if the iterations of gradient descent remain bounded, the linear convergence result still applies. It is reasonable to assume that the iterates remain bounded when the set of solutions is finite, since each step must decrease the objective function. Thus, for practical purposes, we can relax the above condition to "strictly-convex composed with linear" and the PL inequality implies a linear convergence rate for logistic regression.

3 Convergence of Huge-Scale Methods

In this section, we use the PL inequality to analyze several variants of two of the most widely-used techniques for handling large-scale machine learning problems: coordinate descent and stochastic gradient methods. In particular, the PL inequality yields very simple analyses of these methods that apply to more general classes of functions than previously analyzed. We also note that the PL inequality has recently been used by Garber and Hazan [9] to analyze the Frank-Wolfe algorithm. Further, inspired by the resilient backpropagation (RPROP) algorithm of Riedmiller and Braun [32], in Appendix 3 we also give the first convergence rate analysis for sign-based gradient descent methods.

3.1 Randomized Coordinate Descent

Nesterov [29] shows that randomized coordinate descent achieves a faster convergence rate than gradient descent for problems where we have d variables and it is d times cheaper to update one coordinate than it is to compute the entire gradient. The expected linear convergence rates in this previous work rely on SC, but in this section we show that randomized coordinate descent achieves an expected linear convergence rate if we only assume that the PL inequality holds.

To analyze coordinate descent methods, we assume that the gradient is coordinate-wise Lipschitz continuous, meaning that for any x and y we have

$$f(x + \alpha e_i) \leq f(x) + \alpha \nabla_i f(x) + \frac{L}{2}\alpha^2, \quad \forall \alpha \in \mathbb{R}, \quad \forall x \in \mathbb{R}^d, \tag{5}$$

for any coordinate i, and where e_i is the ith unit vector.

Theorem 3. *Consider problem* (1), *where f has a coordinate-wise L-Lipschitz continuous gradient* (5), *a non-empty solution set $\mathcal{X}^*$, and satisfies the PL inequality* (3). *Consider the coordinate descent method with a step-size of $1/L$,*

$$x_{k+1} = x_k - \frac{1}{L}\nabla_{i_k} f(x_k) e_{i_k}. \tag{6}$$

If we choose the variable to update i_k uniformly at random, then the algorithm has an expected linear convergence rate of

$$\mathbb{E}[f(x_k) - f^*] \leq \left(1 - \frac{\mu}{dL}\right)^k [f(x_0) - f^*].$$

Proof. By using the update rule (6) in the Lipschitz condition (5) we have

$$f(x_{k+1}) \leq f(x_k) - \frac{1}{2L}\|\nabla_{i_k} f(x_k)\|^2.$$

By taking the expectation of both sides with respect to i_k we have

$$\mathbb{E}\left[f(x_{k+1})\right] \leq f(x_k) - \frac{1}{2L}\mathbb{E}\left[\|\nabla_{i_k} f(x_k)\|^2\right]$$

$$\leq f(x_k) - \frac{1}{2L}\sum_i \frac{1}{d}\|\nabla_i f(x_k)\|^2$$

$$= f(x_k) - \frac{1}{2dL}\|\nabla f(x_k)\|^2.$$

By using the PL inequality (3) and subtracting f^* from both sides, we get

$$\mathbb{E}[f(x_{k+1}) - f^*] \leq \left(1 - \frac{\mu}{dL}\right)[f(x_k) - f^*].$$

Applying this recursively and using iterated expectations yields the result.   □

As before, instead of using $1/L$ we could perform exact coordinate optimization and the result would still hold. If we have a Lipschitz constant L_i for each coordinate and sample proportional to the L_i as suggested by Nesterov [29], then the above argument (using a step-size of $1/L_{i_k}$) can be used to show that we obtain a faster rate of

$$\mathbb{E}[f(x_k) - f^*] \leq \left(1 - \frac{\mu}{d\bar{L}}\right)^k [f(x_0) - f^*],$$

where $\bar{L} = \frac{1}{d}\sum_{j=1}^{d} L_j$.

3.2   Greedy Coordinate Descent

Nutini et al. [30] have recently analyzed coordinate descent under the greedy Gauss-Southwell (GS) rule, and argued that this rule may be suitable for problems with a large degree of sparsity. The GS rule chooses i_k according to the rule $i_k = \mathrm{argmax}_j |\nabla_j f(x_k)|$. Using the fact that

$$\max_i |\nabla_i f(x_k)| \geq \frac{1}{d}\sum_{i=1}^{d} |\nabla_i f(x_k)|,$$

it is straightforward to show that the GS rule satisfies the rate above for the randomized method.

However, Nutini et al. [30] show that a faster convergence rate can be obtained for the GS rule by measuring SC in the 1-norm. Since the PL inequality is defined on the dual (gradient) space, in order to derive an analogous result we could measure the PL inequality in the ∞-norm,

$$\|\nabla f(x)\|_\infty^2 \geq 2\mu_1(f(x) - f^*).$$

Because of the equivalence between norms, this is not introducing any additional assumptions beyond that the PL inequality is satisfied. Further, if f is μ_1-SC in the 1-norm, then it satisfies the PL inequality in the ∞-norm with the same constant μ_1. By using that $|\nabla_{i_k} f(x_k)| = \|\nabla f(x_k)\|_\infty$ when the GS rule is used, the above argument can be used to show that coordinate descent with the GS rule achieves a convergence rate of

$$f(x_k) - f^* \leq \left(1 - \frac{\mu_1}{L}\right)^k [f(x_0) - f^*],$$

when the function satisfies the PL inequality in the ∞-norm with a constant of μ_1. By the equivalence between norms we have that $\mu/d \leq \mu_1$, so this is faster than the rate with random selection.

Meir and Rätsch [24] show that we can view some variants of boosting algorithms as implementations of coordinate descent with the GS rule. They use the error bound property to argue that these methods achieve a linear convergence rate, but this property does not lead to an explicit rate. Our simple result above thus provides the first explicit convergence rate for these variants of boosting.

3.3 Stochastic Gradient Methods

Stochastic gradient (SG) methods apply to the general stochastic optimization problem

$$\operatorname*{argmin}_{x \in \mathbb{R}^d} f(x) = \mathbb{E}[f_i(x)], \tag{7}$$

where the expectation is taken with respect to i. These methods are typically used to optimize finite sums,

$$f(x) = \frac{1}{n} \sum_i^n f_i(x). \tag{8}$$

Here, each f_i typically represents the fit of a model on an individual training example. SG methods are suitable for cases where the number of training examples n is so large that it is infeasible to compute the gradient of all n examples more than a few times.

Stochastic gradient (SG) methods use the iteration

$$x_{k+1} = x_k - \alpha_k \nabla f_{i_k}(x_k), \tag{9}$$

where α_k is the step size and i_k is a sample from the distribution over i so that $\mathbb{E}[\nabla f_{i_k}(x_k)] = \nabla f(x_k)$. Below, we analyze the convergence rate of stochastic gradient methods under standard assumptions on f, and under both a decreasing and a constant step-size scheme.

Theorem 4. *Consider problem (7). Assume that each f has an L-Lipschitz continuous gradient (2), f has a non-empty solution set $\mathcal{X}^*$, f satisfies the PL inequality (3), and $\mathbb{E}[\|\nabla f_i(x_k)\|^2] \leq C^2$ for all x_k and some C. If we use the SG algorithm (9) with $\alpha_k = \frac{2k+1}{2\mu(k+1)^2}$, then we get a convergence rate of*

$$\mathbb{E}[f(x_k) - f^*] \leq \frac{LC^2}{2k\mu^2}.$$

If instead we use a constant $\alpha_k = \alpha < \frac{1}{2\mu}$, then we obtain a linear convergence rate up to a solution level that is proportional to α,

$$\mathbb{E}[f(x_k) - f^*] \leq (1 - 2\mu\alpha)^k [f(x_0) - f^*] + \frac{LC^2\alpha}{4\mu}.$$

Proof. By using the update rule (9) inside the Lipschitz condition (2), we have

$$f(x_{k+1}) \leq f(x_k) - \alpha_k \langle f'(x_k), \nabla f_{i_k}(x_k) \rangle + \frac{L\alpha_k^2}{2} \|\nabla f_{i_k}(x_k)\|^2.$$

Taking the expectation of both sides with respect to i_k we have

$$\mathbb{E}[f(x_{k+1})] \leq f(x_k) - \alpha_k \langle \nabla f(x_k), \mathbb{E}[\nabla f_{i_k}(x_k)] \rangle + \frac{L\alpha_k^2}{2} \mathbb{E}[\|\nabla f_i(x_k)\|^2]$$

$$\leq f(x_k) - \alpha_k \|f'(x_k)\|^2 + \frac{LC^2 \alpha_k^2}{2}$$

$$\leq f(x_k) - 2\mu\alpha_k (f(x_k) - f^*) + \frac{LC^2 \alpha_k^2}{2},$$

where the second line uses that $\mathbb{E}[\nabla f_{i_k}(x_k)] = f'(x_k)$ and $\mathbb{E}[\|\nabla f_i(x_k)\|^2] \leq C^2$, and the third line uses the PL inequality. Subtracting f^* from both sides yields:

$$\mathbb{E}[f(x_{k+1}) - f^*] \leq (1 - 2\alpha_k\mu)[f(x_k) - f^*] + \frac{LC^2 \alpha_k^2}{2}. \tag{10}$$

Decreasing step size: With $\alpha_k = \frac{2k+1}{2\mu(k+1)^2}$ in (10) we obtain

$$\mathbb{E}[f(x_{k+1}) - f^*] \leq \frac{k^2}{(k+1)^2}[f(x_k) - f^*] + \frac{LC^2(2k+1)^2}{8\mu^2(k+1)^4}.$$

Multiplying both sides by $(k+1)^2$ and letting $\delta_f(k) \equiv k^2 \mathbb{E}[f(x_k) - f^*]$ we get

$$\delta_f(k+1) \leq \delta_f(k) + \frac{LC^2(2k+1)^2}{8\mu^2(k+1)^2}$$

$$\leq \delta_f(k) + \frac{LC^2}{2\mu^2},$$

where the second line follows from $\frac{2k+1}{k+1} < 2$. Summing up this inequality from $k = 0$ to k and using the fact that $\delta_f(0) = 0$ we get

$$\delta_f(k+1) \leq \delta_f(0) + \frac{LC^2}{2\mu^2} \sum_{i=0}^{k} 1 \leq \frac{LC^2(k+1)}{2\mu^2}$$

$$\Rightarrow \qquad (k+1)^2 \mathbb{E}[f(x_{k+1}) - f^*] \leq \frac{LC^2(k+1)}{2\mu^2}$$

which gives the stated rate.

Constant step size: Choosing $\alpha_k = \alpha$ for any $\alpha < 1/2\mu$ and applying (10) recursively yields

$$\mathbb{E}[f(x_{k+1}) - f^*] \leq (1 - 2\alpha\mu)^k [f(x_0) - f^*] + \frac{LC^2\alpha^2}{2} \sum_{i=0}^{k} (1 - 2\alpha\mu)^i$$

$$\leq (1 - 2\alpha\mu)^k [f(x_0) - f^*] + \frac{LC^2\alpha^2}{2} \sum_{i=0}^{\infty} (1 - 2\alpha\mu)^i$$

$$= (1 - 2\alpha\mu)^k [f(x_0) - f^*] + \frac{LC^2\alpha}{4\mu},$$

where the last line uses that $\alpha < 1/2\mu$ and the limit of the geometric series. $\square$

The $O(1/k)$ rate for a decreasing step size matches the convergence rate of stochastic gradient methods under SC [27]. It was recently shown using a non-trivial analysis that a stochastic Newton method could achieve an $O(1/k)$ rate for least squares problems [4], but our result above shows that the basic stochastic gradient method already achieves this property (although the constants are worse than for this Newton-like method). Further, our result does not rely on convexity. Note that if we are happy with a solution of fixed accuracy, then the result with a constant step-size is perhaps the more useful strategy in practice: it supports the often-used empirical strategy of using a constant size for a long time, then halving the step-size if the algorithm appears to have stalled (the above result indicates that halving the step-size will at least halve the sub-optimality).

3.4 Finite Sum Methods

In the setting of minimizing *finite* sums, it has recently been shown that there are methods that have the low iteration cost of stochastic gradient methods but that still have linear convergence rates [33]. While the first methods that achieved this remarkable property required a *memory* of previous gradient values, the stochastic variance-reduced gradient (SVRG) method of Johnson and Zhang [16] does not have this drawback. In Appendix 3.4, we give a new analysis of the SVRG method that shows that it achieves a linear convergence rate under the PL inequality. Similar results for finite-sum methods under the PL inequality recently appeared in the works of Reddi et al. [36,37]. Garber and Hazan [10] have also given a related result in the context of an improved algorithm for principal component analysis (PCA), showing that the f_i do not need to be convex in order to achieve a linear convergence rate. However, their result still assumes that f is SC while our analysis only assumes the PL inequality is satisfied.

4   Proximal-Gradient Generalization

Attouch and Bolte [3] consider a generalization of the PL inequality due to Kurdyak to give conditions under which the classic proximal-point algorithm achieves a linear convergence rate for non-smooth problems (called the KL inequality). However, in practice proximal-*gradient* methods are more relevant to many machine learning problems. While the KL inequality has been used to show local linear convergence of proximal-gradient methods [6,18], in this section we propose a different generalization of the PL inequality that yields a simple global linear convergence analysis.

Proximal-gradient methods apply to problems of the form

$$\operatorname*{argmin}_{x \in \mathbb{R}^d} F(x) = f(x) + g(x), \tag{11}$$

where f is a differentiable function with an L-Lipschitz continuous gradient and g is a simple but potentially non-smooth convex function. Typical examples of simple functions g include a scaled ℓ_1-norm of the parameter vectors, $g(x) = \lambda\|x\|_1$, and indicator functions that are zero if x lies in a simple convex set and are infinity otherwise.

In order to analyze proximal-gradient algorithms, a natural (though not particularly intuitive) generalization of the PL inequality is that there exists a $\mu > 0$ satisfying

$$\frac{1}{2}\mathcal{D}_g(x, L) \geq \mu(F(x) - F^*), \tag{12}$$

where

$$\mathcal{D}_g(x, \alpha) \equiv -2\alpha \min_y [\langle \nabla f(x), y - x \rangle + \frac{\alpha}{2}||y - x||^2 + g(y) - g(x)]. \tag{13}$$

We call this the *proximal-PL* inequality, and we note that if g is constant (or linear) then it reduces to the standard PL inequality. Below we show that this inequality is sufficient for the proximal-gradient method to achieve a global linear convergence rate.

Theorem 5. *Consider problem* (11), *where f has an L-Lipschitz continuous gradient* (2), *F has a non-empty solution set $\mathcal{X}^*$, g is convex, and F satisfies the proximal-PL inequality* (12). *Then the proximal-gradient method with a step-size of $1/L$,*

$$x_{k+1} = \underset{y}{argmin} \ [\langle \nabla f(x_k), y - x_k \rangle + \frac{L}{2}||y - x_k||^2 + g(y) - g(x_k)] \tag{14}$$

converges linearly to the optimal value F^*,

$$F(x_k) - F^* \leq \left(1 - \frac{\mu}{L}\right)^k [F(x_0) - F^*].$$

Proof. By using Lipschitz continuity of the function f we have

$$F(x_{k+1}) = f(x_{k+1}) + g(x_k) + g(x_{k+1}) - g(x_k)$$

$$\leq F(x_k) + \langle \nabla f(x_k), x_{k+1} - x_k \rangle + \frac{L}{2}||x_{k+1} - x_k||^2 + g(x_{k+1}) - g(x_k)$$

$$\leq F(x_k) - \frac{1}{2L}\mathcal{D}_g(x_k, L)$$

$$\leq F(x_k) - \frac{\mu}{L}[F(x_k) - F^*],$$

which uses the definition of x_{k+1} and $\mathcal{D}_g$ followed by the proximal-PL inequality (12). This subsequently implies that

$$F(x_{k+1}) - F^* \leq \left(1 - \frac{\mu}{L}\right)[F(x_k) - F^*], \tag{15}$$

which applied recursively gives the result.   □

We note that the condition $\mu \leq L$ is implicit in the definition of the proximal-PL inequality, but this is not restrictive since we can simply set μ to a smaller value to satisfy this. While other conditions have been proposed to show linear convergence rates of proximal-gradient methods without SC [17, 44], their analyses tend to be much more complicated than the above while, as we discuss in the next section, the proximal-PL inequality includes the standard scenarios where these apply.

4.1   Relevant Problems

As with the PL inequality, we now list several important function classes that satisfy the proximal-PL inequality (12). We give proofs that these classes satisfy the inequality in Appendices 4.1, 4.2, and 4.4.

1. The inequality is satisfied if f satisfies the PL inequality and g is constant. Thus, the above result generalizes Theorem 1.
2. The inequality is satisfied if f is SC. This is the usual assumption used to show a linear convergence rate for the proximal-gradient algorithm [34], although we note that the above analysis is much simpler than standard arguments.
3. The inequality is satisfied if f has the form $f(x) = h(Ax)$ for a SC function h and a matrix A, while g is an indicator function for a polyhedral set.
4. The inequality is satisfied if F is convex and satisfies the QG property. In Appendices 4.2 and 4.4 we show that L1-regularized least squares and the support vector machine dual (respectively) fall into this category, and we discuss these two notable cases further below.

We expect that it is possible to show the proximal-PL inequality holds in other cases where the proximal-gradient achieves a linear convergence rate like the case of group L1-regularization [40] and nuclear-norm regularization [14].

4.2   Least Squares with L1-Regularization

Perhaps the most interesting example of problem (11) is the ℓ_1-regularized least squares problem,

$$\underset{x \in \mathbb{R}^d}{\operatorname{argmin}} \frac{1}{2} \|Ax - b\|^2 + \lambda \|x\|_1,$$

where $\lambda > 0$ is the regularization parameter. This problem has been studied extensively in machine learning, signal processing, and statistics. This problem structure seems well-suited to using proximal-gradient methods, but the first works analyzing proximal-gradient methods for this problem only showed sublinear convergence rates. There subsequently have been a variety of works showing that linear convergence rates can be achieved under additional assumptions. For example, Gu et al. [11] prove that their algorithm achieves a linear convergence rate if A satisfies a *restricted isometry property* (RIP) and the solution is sufficiently sparse. Xiao and Zhang [43] also assume the RIP property and show linear convergence using a homotopy method that slowly decreases the value of λ. Agarwal et al. [1] give a linear convergence rate under a *modified restricted strong convexity* and *modified restricted smoothness* assumption. In Appendix 4.2 we show that *any* L1-regularized least squares problem satisfies the QG property if we use a descent method and thus by convexity also satisfies the proximal-PL inequality. Thus, Theorem 5 implies a global linear convergence rate for these problems without making additional assumptions or making any modifications to the algorithm. A similar result recently appeared in the work of Necoara and Clipici [26] under a generalized EB, but with a much more complicated analysis.

4.3   Proximal Coordinate Descent

It is also possible to adapt our results on coordinate descent and proximal-gradient methods in order to give a linear convergence rate for coordinate-wise proximal-gradient methods for problem (11). To do this, we require the extra assumption that g is a separable function. This means that $g(x) = \sum_i g_i(x_i)$ for a set of univariate functions g_i. The update rule for the coordinate-wise proximal-gradient method is

$$x_{k+1} = \operatorname*{argmin}_{\alpha} \left[\alpha \nabla_{i_k} f(x_k) + \frac{L}{2}\alpha^2 + g_{i_k}(x_{i_k} + \alpha) - g_{i_k}(x_{i_k}) \right], \quad (16)$$

We state the convergence rate result below.

Theorem 6. *Assume the setup of Theorem 5 and that g is a separable function $g(x) = \sum_i g_i(x_i)$, where each g_i is convex. Then the coordinate-wise proximal-gradient update rule (16) achieves a convergence rate*

$$\mathbb{E}[F(x_k) - F^*] \leq \left(1 - \frac{\mu}{dL}\right)^k [F(x_0) - F^*], \quad (17)$$

when i_k is selected uniformly at random.

The proof is given in Appendix 4.3 and although it is more complicated than the proofs of Theorems 4 and 5, it is still simpler than existing proofs for proximal coordinate descent under SC [39]. It is also possible to analyze stochastic proximal-gradient algorithms, and indeed Reddi et al. use the proximal-PL inequality to analyze finite-sum methods in the proximal stochastic case [38].

4.4   Support Vector Machines

Another important model problem that arises in machine learning is support vector machines,

$$\operatorname*{argmin}_{x \in \mathbb{R}^d} \frac{\lambda}{2} x^T x + \sum_{i=1}^n \max(0, 1 - b_i x^T a_i). \quad (18)$$

where (a_i, b_i) are the labelled training set with $a_i \in \mathbb{R}^d$ and $b_i \in \{-1, 1\}$. We often solve this problem by performing coordinate optimization on its dual, which has the form

$$\min_{\bar{w}} f(\bar{w}) = \frac{1}{2}\bar{w}^T M\bar{w} - \sum \bar{w}_i, \quad \bar{w}_i \in [0, U], \quad (19)$$

for a particular matrix M and constant U. This function satisfies the QG property and thus Theorem 6 implies that coordinate optimization achieves a linear convergence rate in terms of optimizing the dual objective. Further, since Hush et al. [15] show that we can obtain an ϵ-accurate solution to the primal problem with an $O(\epsilon^2)$-accurate solution to the dual problem, this also implies a linear convergence rate for stochastic dual coordinate ascent on the primal problem. Global linear convergence rates for SVMs have also been shown by others [23, 41, 42], but again we note that these works lead to much more complicated analyses. Although the constants in these convergence rate may be quite bad (depending on the smallest non-zero singular value of the Gram matrix), we note that the existing sublinear rates still apply in the early iterations while, as the algorithm begins to identify support vectors, the constants improve (depending on the smallest non-zero singular value of the block of the Gram matrix corresponding to the support vectors).

The result of the previous section is not only restricted to SVMs. Indeed, the result of the previous section implies a linear convergence rate for many ℓ_2-regularized linear prediction problems, the framework considered in the stochastic dual coordinate ascent (SDCA) work of Shalev-Shwartz and Zhang [35]. While Shalev-Shwartz and Zhang [35] show that this is true when the primal is smooth, our result gives linear rates in many cases where the primal is non-smooth.

5   Discussion

We believe that this work provides a unifying and simplifying view of a variety of optimization and convergence rate issues in machine learning. Indeed, we have shown that many of the assumptions used to achieve linear convergence rates can be replaced by the PL inequality and its proximal generalization. Throughout the paper, we have also pointed out how our analysis implies new convergence rates for a variety of machine learning models and algorithms. Some of these were previously known, typically under stronger assumptions or with more complicated proofs, but many of these are novel. Note that we have not provided any experimental results in this work, since the main contributions of this work are showing that existing algorithms actually work better on standard problems than we previously thought. We expect that going forward, efficiency will no longer be decided by the issue of whether functions are SC, but rather by whether they satisfy a variant of the PL inequality.

Acknowledgments. We would like to thank Simon LaCoste-Julien and Martin Takáč for valuable discussions. This research was supported by the Natural Sciences and Engineering Research Council of Canada (NSERC RGPIN-06068-2015). Julie Nutini is funded by a UBC Four Year Doctoral Fellowship (4YF) and Hamed Karimi is support by a Mathematics of Information Technology and Complex Systems (MITACS) Elevate Fellowship.

References

1. Agarwal, A., Negahban, S.N., Wainwright, M.J.: Fast global convergence rates of gradient methods for high-dimensional statistical recovery. Ann. Statist. **40**, 2452–2482 (2012)
2. Anitescu, M.: Degenerate nonlinear programming with a quadratic growth condition. SIAM J. Optim. **10**, 1116–1135 (2000)
3. Attouch, H., Bolte, J.: On the convergence of the proximal algorithm for nonsmooth functions involving analytic features. Math. Program. Ser. B **116**, 5–16 (2009)
4. Bach, F., Moulines, E.: Non-strongly-convex smooth stochastic approximation with convergence rate $O(1/n)$. In: Advances in Neural Information Processing Systems (NIPS), pp. 773–791 (2013)
5. Ben-Israel, A., Mond, B.: What is invexity? J. Austral. Math. Soc. **28**, 1–9 (1986)
6. Bolte, J., Nguyen, T.P., Peypouquet, J., Suter, B.W.: From Error Bounds to the Complexity of First-Order Descent Methods for Convex Functions. arXiv:1510.08234 (2015)
7. Craven, B.D., Glover, B.M.: Invex functions and duality. J. Austral. Math. Soc. **39**, 1–20 (1985)
8. Dinuzzo, F., Ong, C.S., Gehler, P., Pillonetto, G.: Learning output kernels with block coordinate descent. In: Proceedings of the 28th ICML, pp. 49–56 (2011)
9. Garber, D., Hazan, E.: Faster rates for the Frank-Wolfe method over strongly-convex sets. In: Proceedings of the 32nd ICML, pp. 541–549 (2015)
10. Garber, D., Hazan, E.: Faster and Simple PCA via Convex Optimization. arXiv:1509.05647v4 (2015)
11. Gu, M., Lim, L.-H., Wu, C.J.: ParNes: a rapidly convergent algorithm for accurate recovery of sparse and approximately sparse signals. Numer. Algor. **64**, 321–347 (2013)
12. Hanson, M.A.: On sufficiency of the Kuhn-Tucker conditions. J. Math. Anal. Appl. **80**, 545–550 (1981)
13. Hoffman, A.J.: On approximate solutions of systems of linear inequalities. J. Res. Nat. Bur. Stand. **49**, 263–265 (1952)
14. Hou, K., Zhou, Z., So, A.M.-C., Luo, Z.-Q.: On the linear convergence of the proximal gradient method for trace norm regularization. In: Advances in Neural Information Processing Systems (NIPS), pp. 710–718 (2013)
15. Hush, D., Kelly, P., Scovel, C., Steinwart, I.: QP algorithms with guaranteed accuracy and run time for support vector machines. J. Mach. Learn. Res. **7**, 733–769 (2006)
16. Johnson, R., Zhang, T.: Accelerating stochastic gradient descent using predictive variance reduction. In: Advances in Neural Information Processing Systems (NIPS), pp. 315–323 (2013)
17. Kadkhodaie, M. Sanjabi, M., Luo, Z.-Q.: On the Linear Convergence of the Approximate Proximal Splitting Method for Non-Smooth Convex Optimization. arXiv:1404.5350v1 (2014)
18. Li, G., Pong, T.K.: Calculus of the Exponent of Kurdyka-Łojasiewicz Inequality and its Applications to Linear Convergence of First-Order Methods. arXiv:1602.02915v1 (2016)
19. Liu, J., Wright, S.J.: Asynchronous stochastic coordinate descent: parallelism and convergence properties. SIAM J. Optim. **25**, 351–376 (2015)
20. Liu, J., Wright, S.J., Ré, C., Bittorf, V., Sridhar, S.: An Asynchronous Parallel Stochastic Coordinate Descent Algorithm. arXiv:1311.1873v3 (2014)

21. Łojasiewicz, S.: A Topological Property of Real Analytic Subsets (in French). Coll. du CNRS, Les équations aux dérivées partielles, vol. 117, pp. 87–89 (1963)
22. Luo, Z.-Q., Tseng, P.: Error bounds and convergence analysis of feasible descent methods: a general approach. Ann. Oper. Res. **46**, 157–178 (1993)
23. Ma, C., Tappenden, T., Takáč, M.: Linear Convergence of the Randomized Feasible Descent Method Under the Weak Strong Convexity Assumption. arXiv:1506.02530 (2015)
24. Meir, R., Rätsch, G.: An introduction to boosting and leveraging. In: Mendelson, S., Smola, A.J. (eds.) Advanced Lectures on Machine Learning. LNCS (LNAI), vol. 2600, pp. 118–183. Springer, Heidelberg (2003). doi:10.1007/3-540-36434-X_4
25. Necoara, I., Nesterov, Y., Glineur, F.: Linear Convergence of First Order Methods for Non-Strongly Convex Optimization. arXiv:1504.06298v3 (2015)
26. Necoara, I., Clipici, D.: Parallel random coordinate descent method for composite minimization: convergence analysis and error bounds. SIAM J. Optim. **26**, 197–226 (2016)
27. Nemirovski, A., Juditsky, A., Lan, G., Shapiro, A.: Robust stochastic approximation approach to stochastic programming. SIAM J. Optim. **19**, 1574–1609 (2009)
28. Nesterov, Y.: Introductory Lectures on Convex Optimization: A Basic Course. Kluwer Academic Publishers, Dordrecht (2004)
29. Nesterov, Y.: Efficiency of coordinate descent methods on huge-scale optimization problems. SIAM J. Optim. **22**, 341–362 (2012)
30. Nutini, J., Schmidt, M., Laradji, I.H., Friedlander, M., Koepke, H.: Coordinate descent converges faster with the Gauss-Southwell rule than random selection. In: Proceedings of the 32nd ICML, pp. 1632–1641 (2015)
31. Polyak, B.T.: Gradient methods for minimizing functionals. Zh. Vychisl. Mat. Mat. Fiz. **3**, 643–653 (1963). (in Russian)
32. Riedmiller, M., Braun, H.: RPROP - a fast adaptive learning algorithm. In: Proceedings of ISCIS VII (1992)
33. Roux, N.L., Schmidt, M., Bach, F.R.: A stochastic gradient method with an exponential convergence rate for finite training sets. In: Advances in Neural Information Processing Systems (NIPS), pp. 2672–2680 (2012)
34. Schmidt, M., Roux, N.L., Bach, F.R.: Convergence rates of inexact proximal-gradient methods for convex optimization. In: Advances in Neural Information Processing Systems (NIPS), pp. 1458–1466 (2011)
35. Shalev-Shwartz, S., Zhang, T.: Stochastic dual coordinate ascent methods for regularized loss minimization. J. Mach. Learn. Res. **14**, 567–599 (2013)
36. Reddi, S.J., Sra, S., Poczos, B., Smola, A.: Fast Incremental Method for Nonconvex Optimization. arXiv:1603.06159 (2016)
37. Reddi, S.J., Hefny, A., Sra, S., Poczos, B., Smola, A.: Stochastic Variance Reduction for Nonconvex Optimization. arXiv:1603.06160 (2016)
38. Reddi, S.J., Sra, S., Poczos, B., Smola, A.,: Fast Stochastic Methods for Nonsmooth Nonconvex Optimization. arXiv:1605.06900 (2016)
39. Richtárik, P., Takáč, M.: Iteration complexity of randomized block-coordinate descent methods for minimizing a composite function. Math. Program. Ser. A **144**, 1–38 (2014)
40. Tseng, P.: Approximation accuracy, gradient methods, and error bound for structured convex optimization. Math. Program. Ser. B **125**, 263–295 (2010)
41. Tseng, P., Yun, S.: Block-coordinate gradient descent method for linearly constrained nonsmooth separable optimization. J. Optim. Theory Appl. **140**, 513–535 (2009)

42. Wang, P.-W., Lin, C.-J.: Iteration complexity of feasible descent methods for convex optimization. J. Mach. Learn. Res. **15**, 1523–1548 (2014)
43. Xiao, L., Zhang, T.: A proximal-gradient homotopy method for the sparse least-squares problem. SIAM J. Optim. **23**, 1062–1091 (2013)
44. Zhang, H.: The Restricted Strong Convexity Revisited: Analysis of Equivalence to Error Bound and Quadratic Growth. arXiv:1511.01635 (2015)
45. Zhang, H., Yin, W.: Gradient Methods for Convex Minimization: Better Rates Under Weaker Conditions. arXiv:1303.4645v2 (2013)

Author Index